D0083979

ELECTRIC MACHINES AND POWER SYSTEMS

Volume I, Electric Machines

McGraw-Hill Series in Electrical and Computer Engineering

Senior Consulting Editor

Stephen W. Director, *Carnegie Mellon University*

Circuits and Systems
Communications and Signal Processing
Computer Engineering
Control Theory
Electromagnetics
Electronics and VLSI Circuits
Introductory
Power and Energy
Radar and Antennas

Previous Consulting Editors

Ronald N. Bracewell, Colin Cherry, James F. Gibbons, Willis W. Harman, Hubert Heffner, Edward W. Herold, John G. Linvill, Simon Ramo, Ronald A. Rohrer, Anthony E. Siegman, Charles Susskind, Frederick E. Terman, John G. Truxal, Ernst Weber, and John R. Whinnery

Power and Energy

Senior Consulting Editor

Stephen W. Director, *Carnegie Mellon University*

Chapman: *Electric Machinery Fundamentals*
Elgerd: *Electric Energy Systems Theory*
Fitzgerald, Kingsley, and Umans: *Electric Machinery*
Gonen: *Electric Power Distribution System Engineering*
Grainger and Stevenson: *Power System Analysis*
Krause and Wasynczuk: *Electromechanical Motion Devices*
Nasar: *Electric Machines and Power Systems: Volume I, Electric Machines*
Stevenson: *Elements of Power System Analysis*
Vithayathil: *Power Electronics: Principles and Applications*

Also Available from McGraw-Hill

Schaum's Outline Series in Electronics & Electrical Engineering

Most outlines include basic theory, definitions and hundreds of example problems solved in step-by-step detail, and supplementary problems with answers.

Related titles on the current list include:

Analog & Digital Communications
Basic Circuit Analysis
Basic Electrical Engineering
Basic Electricity
Basic Mathematics for Electricity & Electronics
Digital Principles
Electric Circuits
Electric Machines & Electromechanics
Electric Power Systems
Electromagnetics
Electronic Circuits
Electronic Communication
Electronic Devices & Circuits
Electronics Technology
Feedback & Control Systems
Introduction to Digital Systems
Microprocessor Fundamentals

Schaum's Solved Problems Books

Each title in this series is a complete and expert source of solved problems with solutions worked out in step-by-step detail.

Related titles on the current list include:

3000 Solved Problems in Calculus
2500 Solved Problems in Differential Equations
3000 Solved Problems in Electric Circuits
2000 Solved Problems in Electromagnetics
2000 Solved Problems in Electronics
3000 Solved Problems in Linear Algebra
2000 Solved Problems in Numerical Analysis
3000 Solved Problems in Physics

Available at most college bookstores, or for a complete list of titles and prices, write to:

Schaum Division
McGraw-Hill, Inc.
1221 Avenue of the Americas
New York, NY 10020

ELECTRIC MACHINES AND POWER SYSTEMS

Volume I, Electric Machines

Syed A. Nasar

University of Kentucky

McGraw-Hill, Inc.

New York St. Louis San Francisco Auckland Bogotá
Caracas Lisbon London Madrid Mexico City Milan
Montreal New Delhi San Juan Singapore Sydney Tokyo Toronto

The editor was George T. Hoffman;
the production supervisor was Friederich W. Schulte.
the cover was designed by John Hite.
R. R. Donnelley & Sons Company was printer and binder.

ELECTRIC MACHINES AND POWER SYSTEMS
Volume I, Electric Machines

Copyright ©1995 by McGraw-Hill, Inc. All rights reserved. Printed in the United States of America. Except as permitted under the United States Copyright Act of 1976, no part of this publication may be reproduced or distributed in any form or by any means, or stored in a data base or retrieval system, without the prior written permission of the publisher.

This book is printed on acid-free paper.

2 3 4 5 6 7 8 9 0 DOC DOC 9 0 9 8 7 6 5

ISBN 0-07-045958-4

Library of Congress Cataloging-in-Publication Data

Nasar, S. A.
Electric machines and power systems / Syed A. Nasar.
p. cm. —(McGraw-Hill series in electrical and computer engineering. Power and energy)
Includes index.
Contents: v. 1. Electric machines
ISBN 0-07-045958-4
1. Electric machinery. 2. Power electronics. I. Title.
II. Series.
TK2182.N37 1995 v.1
621.31—dc20 94-32717

DISCARDED WIDENER UNIVERSITY

INTERNATIONAL EDITION
Copyright 1995. Exclusive rights by McGraw-Hill, Inc. for manufacture and export. This book cannot be re-exported from the country to which it is consigned by McGraw-Hill. The International Edition is not available in North America.

When ordering this title, use ISBN 0-07-113526-X.

ABOUT THE AUTHOR

Syed A. Nasar is Professor and Chairman, Department of Electrical Engineering, University of Kentucky. He earned his Ph.D. degree in Electrical Engineering from the University of California at Berkeley. He is the author or co-author of thirty books and over 100 journal papers in Electrical Engineering. He is a Fellow IEEE and Fellow Institution of Electrical Engineering, United Kingdom. Professor Nasar has been involved in teaching, research, and consulting in the area of electric machines for over 30 years.

CONTENTS

PREFACE

This book is Volume I of a two-volume set on *ELECTRIC MACHINES AND POWER SYSTEMS*. The first volume is dedicated to *electric machines* and deals with magnetic circuits, transformers, electric machines, and power electronics. Volume II is devoted to *electric power systems*.

In the present volume, we give a fairly in-depth and modern coverage of magnetic circuits, transformers, rotating electric machines, and certain aspects of electric motor control via power electronics. More than adequate material is included here for a one-semester course in electric machines. However, topics may be chosen judiciously by the instructor to instill the fundamentals of the subject matter at the junior level in a typical electrical engineering curriculum. For prerequisites, traditional mathematics and electric circuits courses through the sophomore level are considered adequate. An exposure to basic vector calculus would be helpful.

To review the contents of the book, Chapter 1 presents certain basic concepts and a general introduction to electric machine types. Fields, energy, and forces based on a review of electromagnetic field theory are discussed in Chapter 2. This chapter also includes a discussion of magnetic circuits and magnetic materials. Forces of electromagnetic origin are also treated in sufficient detail in this chapter. A fairly detailed treatment of power transformers is given in Chapter 3.

Induction machines (Chapter 4) being similar to transformers (with minor modifications), logically follow Chapter 3, This is a departure from most textbooks on electric machines, where dc machines follow transformers. A traditional treatment of synchronous machines is given in Chapter 5. In other words, we present a sequence of most common ac machines first and then devote a chapter to dc commutator machines (Chapter 6). Control motors and universal motors are included in Chapter 6, just as small induction motors and small synchronous motors are included in Chapters 4 and 5, respectively.

Chapter 7 discusses permanent magnet dc and ac machines. Because of recent significant progress in permanent magnet materials, it was felt important to include this topic in the book. Control of electric motors via power electronics is presented in Chapter 8.

Numerous worked-out examples and chapter problems are given to facilitate the use of the book in the classroom. Finally, appendixes on unit conversion, wire tables, winding diagrams, and certain review material are included for ready reference.

McGraw-Hill and the author would like to thank Dennis P. Carroll, University of Florida; Mariesa L. Crow, University of Missouri–Rolla; and Alvin L. Day, Iowa State University, for their many helpful comments in reviewing the manuscript.

Syed A. Nasar

ELECTRIC MACHINES AND POWER SYSTEMS

Volume I, Electric Machines

CHAPTER 1

INTRODUCTION

Harnessing and utilizing energy has always been a key factor in improving the quality of life. The use of energy has aided our ability to develop socially and live with physical comfort. In this book, our goal is to develop an understanding of electric energy systems. In particular, we will study the various components of an electric power system as well as the overall system. The components of the system include transformers, generators, motors, transmission lines, and protective equipment, and these will constitute the topics of study in the following chapters.

Focusing first on power conversion devices, we notice the importance of these devices in almost every aspect of life. The number of electric motors in the average U.S. residence today is probably a minimum of 10 and can easily exceed 50. There are at least 5 rotating electric machines on even the most spartan compact automobile, and this number is increasing steadily as emission and fuel economy systems are added. In an aircraft there are many more. Electromechanical devices are involved in every industrial and manufacturing process of a technological society. Many rotating machines have been on the moon and play an important role in most aerospace systems. More people travel by means of electrical propulsion each day—in elevators and horizontal people movers—than by any other mode of propulsion. Electrical blackouts in several major U.S. cities are a reminder of the almost total dependence of activity in urban areas on electric power systems.

This book, therefore, deals with a vast and significant topic. An understanding of the principles of electric power devices and systems is important for all who wish to extend the usefulness of electrical technology in order to ameliorate the problems of energy, pollution, and poverty that presently face humanity. Readers should keep sight of the long-range potential usefulness of electromechanical devices and electric machines while using this book.

From the brief earlier listing of the applications of electric machines, it is obvious that many portions of this industry are mature and are meeting the needs of society with relatively little need for research or advanced development. For example, the motor used in the garbage disposal of a modern home is probably designed by a relatively simple computer program and manufactured in a totally automated process. The hundreds of millions of clock motors manufactured each year also are almost totally standardized in design and manufacture. The same can be said of many types of industrial motors.

This is only a part of the picture, however, as one looks at the state of electromechanical development today. Even this apparently placid, static state-of-the-art in the manufacture of conventional motors may be changing drastically in the near future. In an effort to improve the efficiency of small induction motors used in homes, offices, businesses, and industrial plants, it was estimated by the Federal Energy Administration that 1 to 2 million barrels of oil per day could be saved in the United States by improving the efficiency of these motors by 20 percent. This particular effort makes use of changes in capacitor size and winding connections in single-phase induction motors.

Efforts are also continually being made to improve efficiency of such motors by means of improved materials and design, while keeping in mind the availability of materials, the adverse environmental effects in the manufacture and use of materials (especially insulation materials), and the energy cost of manufacturing these materials. For example, aluminum has many desirable electrical characteristics for electromechanical applications and is one of the most abundant metals on the earth, but it is very costly in terms of energy use to process from raw materials.

Besides the changes in the manufacture and operation of conventional machines that are beginning to occur because of the need for energy and for environmental reasons, there are many exciting applications for new machine configurations, unusual operation of old configurations, sophisticated electronic control of all types of machines, and improved understanding of theory and design techniques to achieve more economical and energy-efficient machine designs. Many of the newer applications involve a new look at some old machines, such as redesigning commutator motors or operating a conventional squirrel-cage induction motor supplied from a transistor inverter in order to develop an economically competitive electric car. Others involve the design of totally new motor configurations, such as the brushless dc motors being developed for aerospace, automotive, and industrial applications.

Electronic control of electric machines has been in use almost from the dawn of the electronic era, beginning with the relatively crude mercury-arc, rectifier-controlled motors. However, with the advent and present-day rapid development of solid-state power devices, integrated circuits, and low-cost computer modules, the range, quality, and precision of motor control have become practically unlimited.

The integration of electromechanical devices and electronic circuits has only just begun. The environment has always offered challenges to the design and operation of electric machines. For example, effective and reliable electrical insulation was one of the most difficult problems for early machine designers. Recently, rotating machines and other electromechanical devices have had to be developed for certain environments, including various types of nuclear radiation for nuclear power generating plants and for several space vehicles. Extremes of reliability in these environments have also been required in the space applications. Finally, as new sources of energy come into economic viability, electromechanical energy converters will be needed with characteristics to match energy sources such as solar converters, windmills, various nuclear configurations, coal-to-oil conversion processes, hydrogen systems, and so forth.

In this chapter, besides introducing the exciting possibilities for advanced development of electromechanical devices, we will discuss some of the basic concepts common to most such devices, review the methods of analysis that will be presented in subsequent chapters, and introduce the major classifications of rotating electric machines.

1.1 ELECTROMECHANICAL ENERGY CONVERSION AND ELECTRIC MACHINE TYPES

Fossil fuels are major sources of energy for the generation of electric power. For applications in areas remote from the energy source, this energy must be converted into electrical form and transmitted by transmission lines to the vicinity of utilization of the energy. The bulk of the energy is converted into electrical form by electromechanical energy converters, the most common being the rotary generators in electric power stations.

As the name implies, an electromechanical energy converter converts mechanical energy into electrical energy, and vice versa. A device that converts energy from mechanical to electrical form, and modulates in response to an electrical signal, is a *generator*. When the conversion involved is from electrical to mechanical energy, and the modulating signal is electrical in nature, the device accomplishing such conversion is a *motor*. Incremental-motion electromechanical energy converters, whose main function is to process energy, are called *transducers*. For example, a microphone or an electric strain gauge is a transducer. Focusing on energy converters, there are four principal classes of rotating machines: dc commutator, induction, synchronous, and polyphase commutator machines. There are several other types of machines that do not fit conveniently into any of these classifications; they include stepper motors, which are, in general, synchronous machines operated in a digital manner; torque motors, which are either dc commutator or brushless synchronous machines operated in the torque (zero or low-speed) mode; homopolar machines, which are a variation of the Faraday disc generator in principle and are used to supply low-voltage, high-current for plating loads; and electrostatic machines, which fall into a different category of theory and practice from the electromagnetic machines to be discussed in this book.

1. *DC Commutator Machines.* These are commonly referred to as just "dc machines" and are distinguished by the mechanical switching device known as the commutator. They are widely used in traction and industrial applications and are discussed in Chapter 6.

2. *Induction Machines.* The induction motor is the so-called workhorse of industry, but it is also the principal appliance motor in homes and offices. It is simple, rugged, durable, and long-lived, which accounts for its widespread acceptability in almost all aspects of technology. It can be operated as a generator and is so used in various aerospace and hydroelectric applications. Induction motors, because of their simple rotor structure, can operate at very high speeds. Fig. 1.1 pictures an aerospace induction motor that operates at speeds near 64,000 rpm when driven from a source of 3200 Hz (see Chapter 4).

3. *Synchronous Machines.* The synchronous machine is probably the most diversified machine configuration, and it is often difficult to recognize the many variations that this class of machines can take on. The term synchronous refers to the relationship between the speed and frequency in this class of machines, which is given by (see Chapter 5)

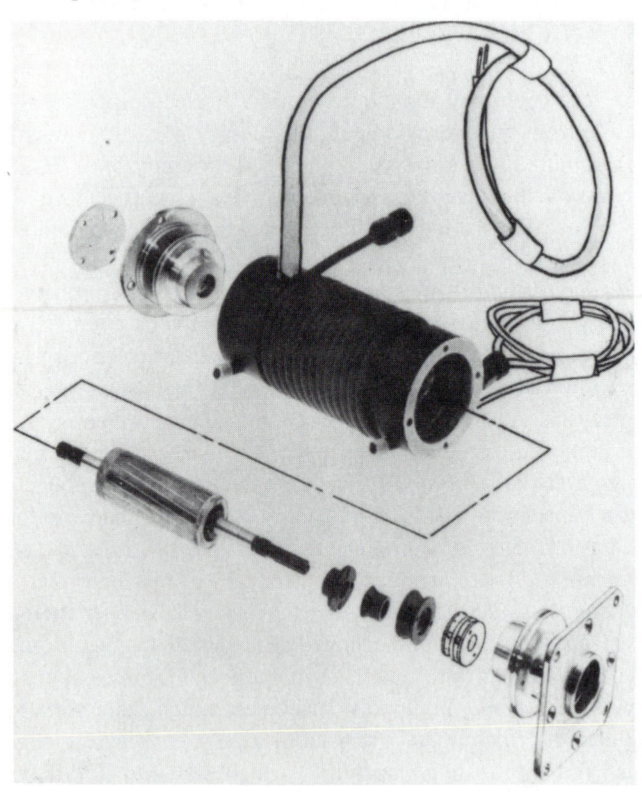

FIGURE 1-1
A 64,000 r/min induction motor for use in the aerospace industry.

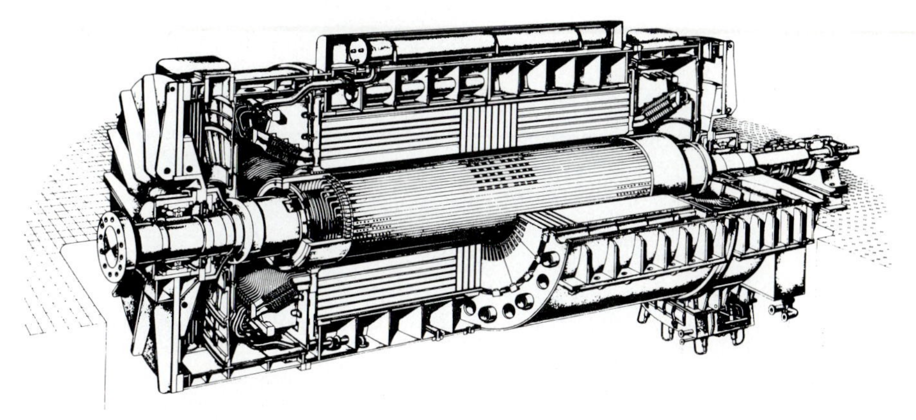

FIGURE 1-2
Cutaway of a water-cooled turbine generator. (Courtesy Brown Boveri Company.)

$$n = 120 \frac{f}{P} \tag{1.1}$$

where

n = machine speed in revolutions per minute
f = frequency of applied source in hertz
P = number of poles on the machine

At the present, we take (1.1) as an end result. Its derivation and further explanation are given in Chapters 4 and 5. See (4.9) and (5.11). A synchronous machine operates only at a synchronous speed; induction machines, often called *asynchronous* machines, operate at speeds somewhat below synchronous speeds. A wide variety of synchronous machines are in use today:

(a) *Conventional.* This is the standard synchronous machine (discussed in Chapter 5) which requires a dc source of excitation in its rotor. It is the machine used in most central-station electrical generating plants (as a generator) and in many motor applications for pumps, compressors, and so forth. A cutaway of a central-station generator is shown in Fig. 1.2.

(b) *Reluctance.* This is a conventional machine without the dc field excitation; it is discussed in Chapters 2 and 5. It is one of the simplest machine configurations and has recently been used in applications conventionally supplied by induction motors. In very small power ratings, it is used for electric clocks, timers, and recording applications.

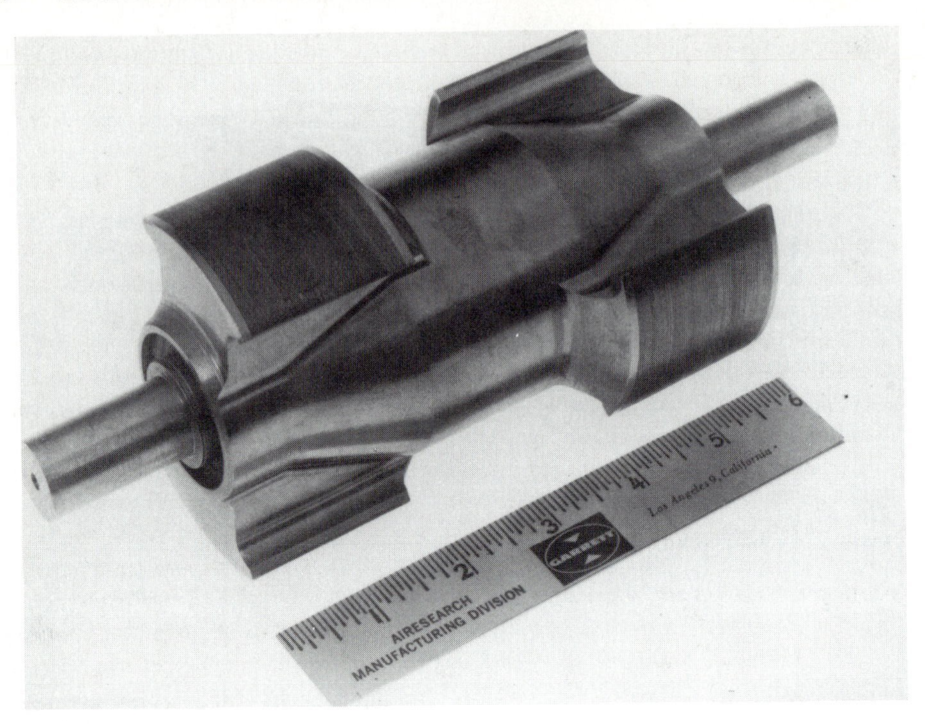

FIGURE 1-3
Rotor of an inductor machine rated at 100 kVA, 120,000 rpm. (Courtesy The Garrett Corporation.)

(c) *Hysteresis.* This configuration, like the reluctance configuration, requires only one electrical input (singly excited). The rotor of a hysteresis motor is a solid cylinder constructed of permanent magnet materials. Hysteresis motors are used in electric clocks, phonograph turntables, and other constant-speed applications. Recently, hysteresis motors have been used in applications requiring larger power output, such as centrifuge drives.

(d) *Rotating Rectifier.* This configuration, in performance, is identical to the conventional synchronous machine except that the field excitation is supplied from an ac auxiliary generator and from rectifiers located on the rotating member.

(e) *Inductor and Flux-Switch.* These are brushless synchronous motors and generators and have been used in many aerospace and traction applications. Like the reluctance configuration, the inductor and flux-switch configurations operate on a variable-reluctance principle. The varying reluctance, as a function of rotor position, is achieved by means of the rotor design. Fig. 1.3 illustrates the rotor construction of an inductor alternator; Fig. 1.4 illustrates a fully assembled inductor machine.

(f) *Lundell.* This configuration is used in automotive alternators and, therefore, is probably the most common type of synchronous machine (above the clock motor size). It is brushless but requires a slip ring for the supply of direct current to its field. The Lundell machine also operates on the principle of varying magnetic reluctance because of the rotor construction.

(g) *Beckey-Robinson and Nadyne Rice.* These are brushless synchronous machines that depend on rotor magnetic structure (varying reluctance) for operation. They have been used in many aerospace applications.

(h) *Permanent Magnet.* This is a conventional synchronous machine in which the field excitation is supplied by a permanent magnet (PM) instead of by a source of electrical energy. It has the potential for very high-energy efficiency, since there are no field losses, and can generally be constructed at a low cost. An example of a rotor of a permanent magnet alternator used in an aerospace application is shown in Fig. 1.5.

High efficiency is a characteristic of PM machines. However, to achieve relatively large power levels, PM machines require the use of permanent magnets of a type that are, at present, relatively costly—such as cobalt-platinum and the cobalt-rare earth alloys. Also, the fixed field excitation of a PM machine eliminates one element of control that is a principal advantage of synchronous machines over induction machines—the field control.

FIGURE 1-4
A fully assembled 520-kVA, 3900-rpm, 12-pole inductor alternator. (Courtesy The Garrett Corporation.)

FIGURE 1-5
Rotor for an eight-pole permanent magnet machine rated 60 kVA, 30,000 rpm, illustrating a machine used in aerospace applications where high efficiency and high specific power (watts per kilogram) are required. (Courtesy The Garrett Corporation.)

1.2 EFFICIENCY, LOSSES, AND DUTY CYCLE

An important factor in the applications of energy conversion devices of all types is the efficiency of the device. Efficiency can have different meanings in different types of physical systems. In fact, it can have a fairly general meaning that is used in everyday conversation, which is "how well a specific job is done." In mechanical systems, use is made of thermal efficiency and mechanical efficiency, which describe the efficiency of two phases of a given process and also "ideal" efficiencies. In the electric systems that will be discussed, efficiency is defined as

$$\eta = \frac{\text{output power or energy}}{\text{input power or energy}} \tag{1.2}$$

This can also be expressed in terms of mechanical and electrical losses in either energy or power terms as

$$\eta = \frac{\text{output}}{\text{output} + \text{losses}} = \frac{\text{input} - \text{losses}}{\text{input}} \tag{1.3}$$

The SI units of power are watts, abbreviated W; SI units of energy are joules, J, or watt-seconds, W-s, or watt-hours, W-h.

Approximate maximum efficiencies of various types of energy converters, discussed previously, in relation to some common energy converters such as the automobile engine, are shown in Fig. 1.6.

The energy use or efficiency of an electric machine is becoming increasingly significant and is one of the more important design criteria today. Therefore it is important to know how to calculate the numerator and denominator of the preceding equations. In electromechanical devices, either the numerator or denominator of these equations is a mechanical power or energy. Mechanical power of a rotating machine can be expressed as

$$P_m = T_{av}\omega_m \tag{1.4}$$

where

T_{av} = shaft torque in newton-meters

ω_m = shaft speed in radians per second

On the electrical side of a machine, power is expressed as

$$P_e = VI \cos \theta \quad \text{(sinusoidal)} \tag{1.5}$$

or

$$P_e = (VI)_{av} \quad \text{(dc or pulse)} \tag{1.6}$$

where

V = terminal voltage in volts

I = terminal current in amperes

θ = power factor angle

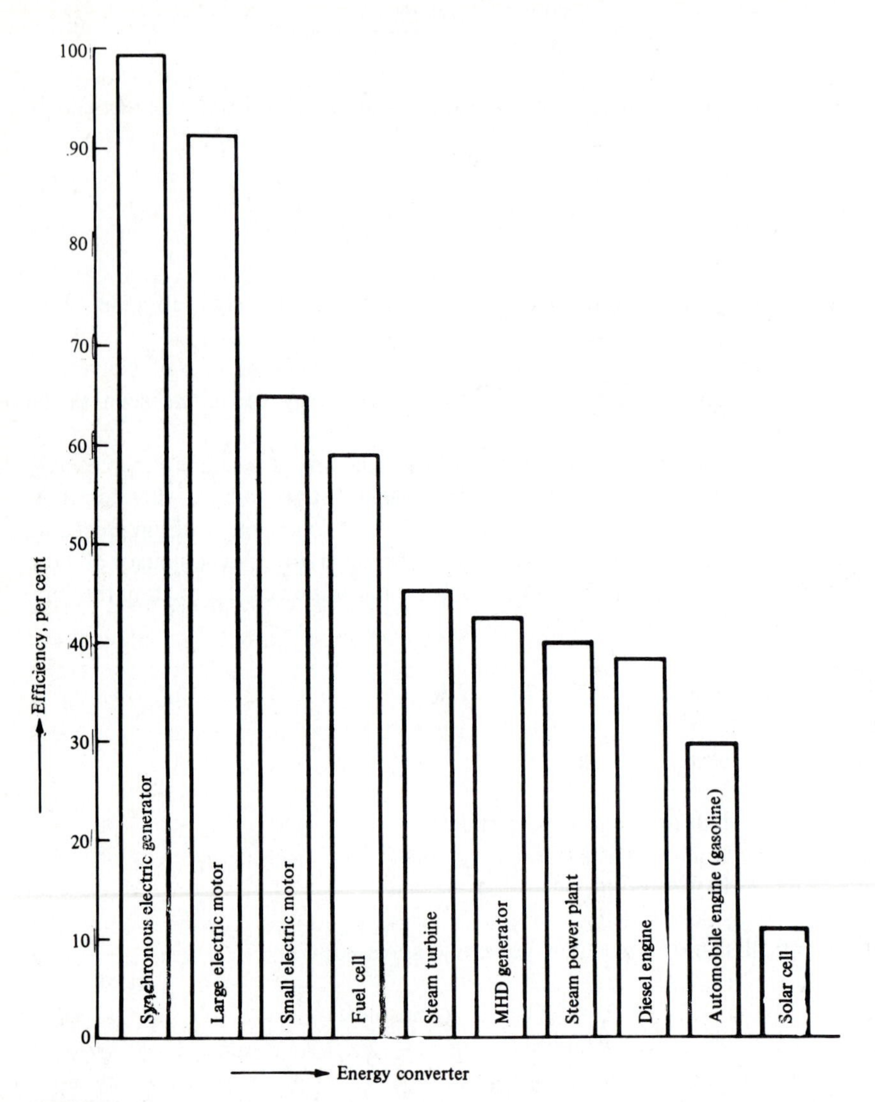

FIGURE 1-6
Approximate maximum efficiencies of energy converters.

In these equations and throughout this book, root-mean-square (rms) parameters are designated by uppercase, unsubscripted symbols; time-averaged parameters are designated by uppercase symbol and the subscript "av." Power calculated by these equations is *average power*. It is also common to have instantaneous quantities on the right side of these equations, in which lowercase symbols would be used and the power on the left would be referred to as *instantaneous power*. The use of both average and instantaneous power is common in the analysis of electromechanical systems.

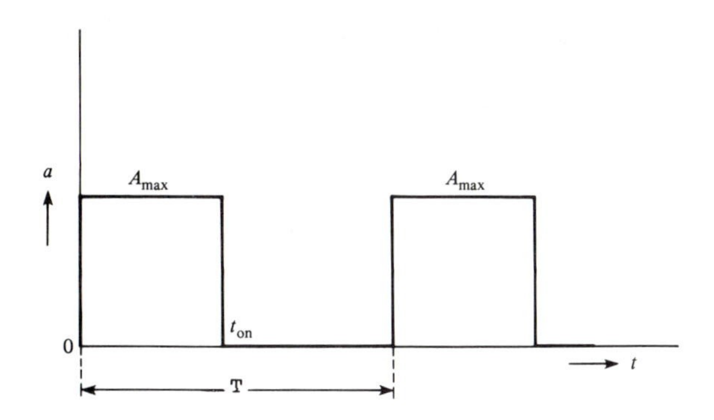

FIGURE 1-7
On/off duty cycle.

Duty Cycle

In many applications electric machines are operated at power levels that change with time instead of being at constant power. When this variation in power level can be described as a periodic function, it is known as a *duty cycle*. Duty-cycle variations are used primarily to describe the variations of loading of a motor, but the concept is equally valid for generators, transformers, and other power devices under such conditions. A duty cycle may be expressed in terms of input or output power, shaft torque, armature current, or other machine parameters that describe the load variation. Examples of applications in which the load characteristics can be so described include automatic washing machines, refrigerator compressors, many types of machine tool processes, and electric vehicles traveling through a prescribed driving cycle.

One simple duty cycle is shown in Fig. 1.7 with a time period *T*; it is typical of many motor loads, such as refrigerator compressors. A generalized symbol is used for the magnitude function in Fig. 1.7, since it may represent a power, torque, current, or other parameter. A more complex duty cycle is shown in Fig. 1.8. This cycle is typical of torque in an electric vehicle, such as a forklift truck, moving through a prescribed set of operations. Note that the function has both positive and negative values. In this application the negative values represent negative or *generator* power, torque, or current and imply a braking operation in which energy is recovered regeneratively. We also discuss duty cycles in connection with the all-day efficiency of transformers, in Chapter 3. DC motor controllers, such as choppers, are also evaluated in terms of duty cycles.

The principal purposes for the use of duty-cycle analysis are (1) to determine the size or the *rating* of a machine to satisfy a particular load requirement, and (2) to determine *energy efficiency*. The latter purpose is important for machines operating at several power levels, such as described by Fig. 1.8, since the *power efficiency* is generally different at each power level in the duty cycle.

Example 1.1 Refer to Fig. 1.7, and let it denote the power absorbed by an electric motor. Let A_{max} correspond to an input power of 900 W during the on-period (t_{on}) of 20 min. The on-period is followed by an off-period of 40 min, and the cycle repeats. Calculate the energy consumed per hour and the average power taken by the motor.

Solution
Energy consumed per hour

$$= (900 \times 20 \times 60 + 0 \times 40 \times 60)$$

$$= 1{,}080{,}000 \text{ W-s (or J)}$$

The average power is

$$P_{av} = \frac{1{,}080{,}000}{60 \times 60} = 300 \text{ W}$$

Example 1.2 In Fig. 1.8, let a be the armature current of the drive motor of an electric vehicle. The currents and their respective durations (in seconds) are:

60 A, $0 \le t \le 10$	-40 A, $40 \le t \le 45$
25 A, $10 \le t \le 30$	-20 A, $45 \le t \le 60$
50 A, $30 \le t \le 40$	0 A, $60 \le t \le 100$

Determine the average and rms values of this current waveform.

Solution
Time period $T = 100$ s

Algebraic sum of the areas under the positive and negative currents

$$= 60 \times 10 + 25 \times 20 + 50 \times 10 - 40 \times 5 - 20 \times 15 + 0 \times 40 = 1100 \text{ A-s}$$

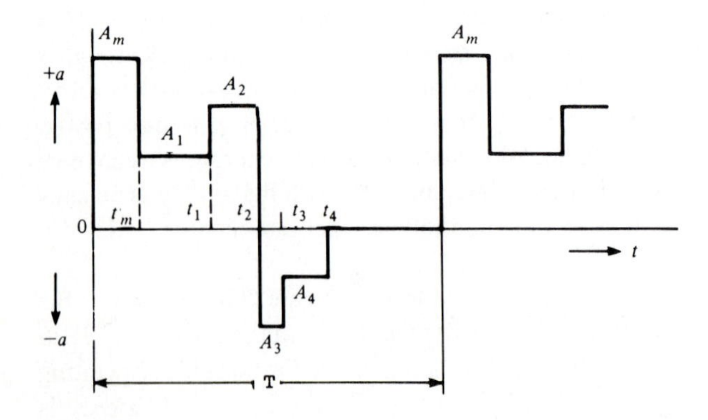

FIGURE 1-8
Duty cycle for electric vehicle with regenerative braking.

$$\text{Average value} = \frac{1100}{100} = 11 \text{ A}$$

The rms value is defined as the square root of the mean of the sum of the squares. Hence

$$\text{rms value} = \left[\frac{1}{100}(60^2 \times 10 + 25^2 \times 20 + 50^2 \times 10 + 40^2 \times 5 + 20^2 \times 15)\right]^{\frac{1}{2}}$$

$$= 29.58 \text{ A}$$

1.3 NAMEPLATE RATINGS

Each commercial rotating machine has a nameplate attached to the machine in some manner. Besides the legal and warranty considerations related to the numerical values listed on the nameplate and the usefulness of these values in applying a machine to a specific load, the nameplate is literally the "name" of the machine. It describes the type of machine; its power capability; its speed, voltage, and current characteristics; and some of its environmental limitations. These nameplate parameters give the set of parameters most often used analytically or experimentally to determine machine performance and are widely used in college laboratory experiments and in machine applications. Therefore, a brief description of the meaning of machine nameplate parameters follows.

1. *Power.* This is expressed in terms of the unit horsepower in motors and watts or kilowatts in generators. It refers to the continuous output power that the machine can deliver. In machines with variable loading, continuous power is equivalent to average load cycle power, as described in Example 1.1. The continuous power rating is primarily a function of the thermal capacity of the machine, which, in turn, is dependent on the frame configuration, sometimes called the machine "package." There are two basic types of packages in commercial machines, with a number of variations within each type determined by the type of environment and external heat transfer equipment in which the machine will be operated: *open* (drip-proof, splashproof, externally ventilated, etc.), and *totally enclosed* (nonventilated, fan-cooled, dustproof, water-cooled, encapsulated, etc.).

2. *Speed.* This is expressed in the unit revolutions per minute (rpm). The speed listed on the nameplate depends on the general type of machine: synchronous—synchronous speed; induction—speed at rated power (synchronous minus slip speed at rated power); dc—usually base speed, the maximum speed at which rated torque can be supplied; universal—usually no-load or light-load speed; dc control—usually no-load speed. Many commercial motors are designed to operate from two different voltage source levels or at two different synchronous speeds, and there will be two sets of speed ratings in such cases.

3. *Voltage.* The *nominal* voltage rating of the source to which the machine windings are to be connected; units are volts. In polyphase machines the rated voltage is always expressed as a *line-to-line* voltage. The rated voltage also gives the *insulation level* at which the machine has been constructed and tested. Insulation levels are standardized in commercial machines.

4. *Current.* The *steady-state current* in the armature or power circuit at rated power output and rated speed, expressed in root mean square in ac machines and average amperes in dc commutator machines. In polyphase machines this current is always a *line current.* In dc commutator machines the *field circuit* current rating is the current "full field," that is, the field required for maximum torque at base speed.

5. *Volt-Amperes.* In ac motors the input volt-ampere rating is derived from the current and voltage ratings. In single-phase motors, this is

   rated volt-amperes = (rated volts) (rated amperes)

   In three-phase motors,

   rated volt-amperes = $\sqrt{3}$ (rated volts) (rated amperes)

   In most ac generators the output volt-ampere rating is stamped on the nameplate in place of the current rating, from which the output power factor can be derived. Both the current and volt-ampere ratings of machines are determined primarily by the thermal characteristics of the windings.

6. *Temperature Rise.* The maximum safe temperature rise (in degrees Celsius) in the "hot spot" of the machine, which is usually the armature or power windings.

7. *Service Factor.* A number that indicates how much over the nameplate power rating the machine can be continuously operated without overheating. It has a value of 1.15 for many commercial motors. In addition, there is a series of *short-time overload ratings* for most commercial motors, which can be obtained from the manufacturer.

8. *Frequency.* The frequency of the supply voltage, in hertz.

9. *Efficiency Index.* A recent addition to nameplates of some manufacturers, which gives an indication of minimum and nominal efficiency.

10. *Torque.* Torque generally is listed on nameplates only for control and torque motors. Units vary considerably, although ounce-inches are most common. Rated steady-state torque in power motors can be derived from nameplate power and speed. Two important torque parameters in ac machines—pullout and starting torques at rated voltage—can be obtained through the manufacturer or by means of relatively simple laboratory tests.

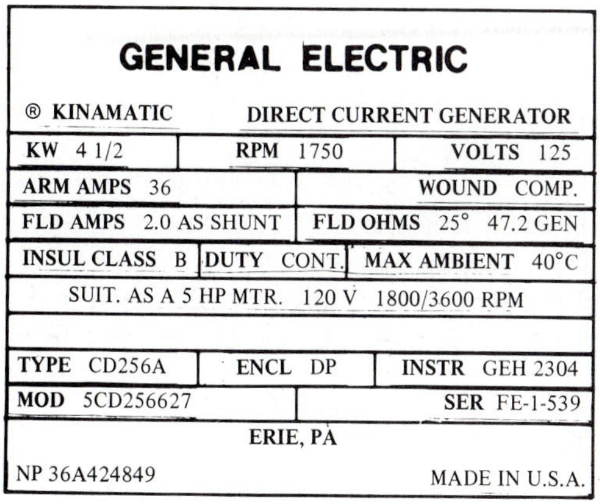

GENERAL ELECTRIC

® KINAMATIC	DIRECT CURRENT GENERATOR	
KW 4 1/2	RPM 1750	VOLTS 125
ARM AMPS 36		WOUND COMP.
FLD AMPS 2.0 AS SHUNT	FLD OHMS 25° 47.2 GEN	
INSUL CLASS B	DUTY CONT.	MAX AMBIENT 40°C
SUIT. AS A 5 HP MTR. 120 V 1800/3600 RPM		
TYPE CD256A	ENCL DP	INSTR GEH 2304
MOD 5CD256627		SER FE-1-539
	ERIE, PA	
NP 36A424849		MADE IN U.S.A.

FIGURE 1-9
A sample nameplate of a dc machine.

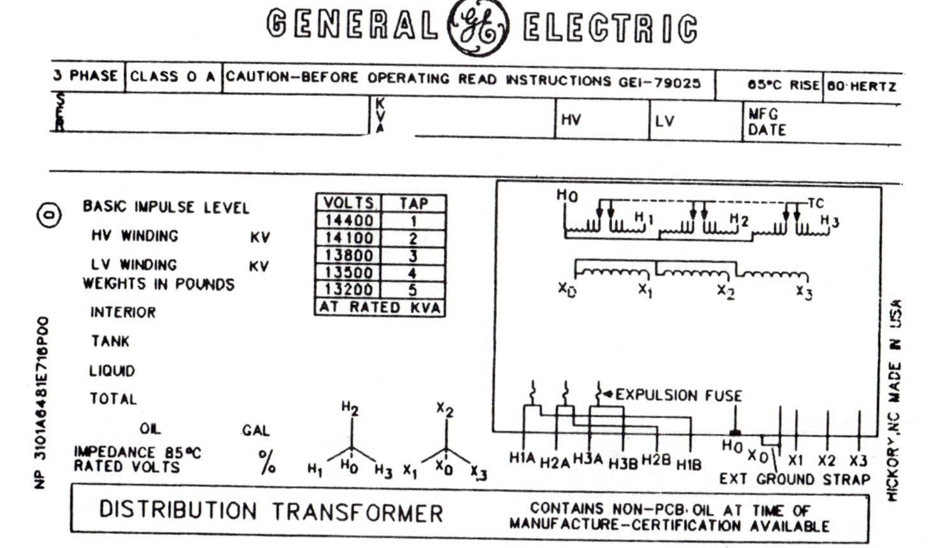

FIGURE 1-10
A sample nameplate of a distribution transformer.

11. *Inertia.* This parameter, the inertia of the rotating member, is a nameplate item only for control motors.

12. *Manufacturer.* The manufacturer's name also appears on the nameplate.

13. *Frame Size.* The frame size is the manufacturer's indication of the physical size of the motor and its mounting dimensions. For motors of ratings greater than 1 hp, a standard system of frame numbering and standardized dimensions of these frame sizes have been developed by the National Electrical Manufacturers Association (NEMA).

A typical nameplate of a dc motor is shown in Fig. 1.9. Some of the other information given on some nameplates includes serial number, style, whether it is open or totally enclosed, and code letter to indicate locked-rotor kVA.

A sample nameplate of a three-phase distribution transformer is shown in Fig. 1.10. Notice that the nameplate contains a wealth of information pertinent to the transformer.

PROBLEMS

1.1. An ac single-phase generator supplies 10 kW power at 220 V. If the generator is 92 percent efficient at this load and runs at 1200 rpm, calculate the torque to drive the generator. Also, calculate the current supplied by the generator if the load power factor is 0.8 lagging.

1.2. A dc motor ideally develops 10 hp at 1160 rpm. What is the developed torque at this speed?

1.3. A dc motor drives an ac generator, which delivers a 50-kVA load at 0.9 power factor at a 90 percent efficiency. The operating efficiency of the motor is 87 percent. If the motor runs at 220 V, calculate the current taken by the motor.

1.4. Refer to the duty cycle of a motor shown in Fig. 1.7. Assume that A_{max} corresponds to 5 kW and the time period T is 60 min. If the on-period (t_{on}) is 15 min, determine the average power taken by the motor.

1.5. The motor of Problem 1.4 is ideal and is driven by a 96-volt source. Calculate the rms and average values of the current taken by the motor for the given duty cycle.

1.6. Refer to the duty cycle of Fig. 1.8, which denotes the current taken by a motor. The durations and the magnitudes of the current are: 30 A for the first 15 sec; 10 A for the second 15 sec; 20 A for the third 15 sec; -25 A for the fourth 15 sec; -15 A for the fifth 15 sec; and 0 A for the last 15 sec. This cycle repeats. Determine the average and rms values of this current waveform.

1.7. A three-phase induction motor is rated 10 hp, 1750 rpm, 220 V, 30 A, 60°C temperature rise, and 1.15 service factor. Determine rated torque, rated input volt-amperes, and maximum continuous output power.

1.8. A major manufacturer of home and office air conditioners uses a 1/2-hp, single-phase induction motor that has an efficiency of 72 percent at its average power output level. Several large-volume customers indicate that, because of a government subsidy of energy-saving installations, they would be willing to pay a larger initial investment in air conditioners if they could recover this increased investment during the warranty period, which happens to be 2 years. The typical office air conditioner in which they are interested runs approximately 8 hours per day during 140 equivalent days of an Atlanta year. The motor supplied to the manufacturer has a wholesale or OEM cost of $45. If the motor supplier could achieve an average efficiency of 85 percent by improving materials and design, how much cost differential could be added to the wholesale motor cost and still satisfy the customer's request?

CHAPTER 2

FIELDS, ENERGY, AND FORCES

In the previous chapter we gave a general introduction to a discussion of electromechanical energy converters. The essential element in all rotating machines and electromechanical devices of the electromagnetic type is an electromagnetic system. The function of this system is to establish and control electromagnetic fields to accomplish the desired process of energy conversion, energy transfer, or energy processing. In the simplest sense, an electromagnetic system consists of electric circuits located in a region of space and with very specific geometry designed to establish the required electromagnetic field relationships. It is possible to describe and analyze many of the functions and performance characteristics of an electromagnetic system in terms of these electric circuits by means of conventional electric circuit theory. However, to understand the basic energy conversion process and to be able to determine the electric circuit parameters of an electric machine, it is necessary to understand the electromagnetic field of the machine and to become familiar with the terms and analytical expressions used to describe this field. Also, much of the process of designing a machine or an electromechanical device is centered around the design of the magnetic system. This chapter will review some basic concepts of magnetic field theory and show how they are of value in the design and analysis of electric machines.

Several important concepts from basic electromagnetic field theory should be recalled throughout this study of electromagnetic fields as applied to electromagnetic energy converters.

1. The term *field* is a concept used to describe a distribution of *forces* throughout a region of space. The electric field describes the force on a unit of charge of electricity, the electron. The magnetic field describes the force on a magnetic dipole. Machines and electromechanical devices of the *electromagnetic type* produce forces or torques resulting from the presence of the magnetic force field. There is a class of electric machines, known as *electrostatic machines*, whose forces result from the presence of electric fields; these will not be treated in this book.

2. Fields are three-dimensional spatial phenomena; their analyses and understanding require some capability to visualize in the abstract. It follows that geometric characteristics are important in the application of force fields to the production of useful forces. Rotary motion results from a rotary arrangement of magnetic fields in this class of machines. In another variety of machines and devices linear motion results from a linear arrangement of magnetic fields.

3. A truly three-dimensional analysis of a field becomes very complex and time consuming and will tie up tremendous blocks of computer storage when used in computer analysis methods. Fortunately, three-dimensional analysis is seldom necessary because of the property of fields known as *symmetry*. Symmetry considerations allow one to resolve the three-dimensional problem into one of two dimensions or even into one dimension within a limited region of space, thus simplifying the analysis and conceptual difficulties. Much of the task of analyzing machines and electromechanical devices rests on identifying the symmetry of its fields. Fortunately, this task has been accomplished for most of the standard configurations by early investigators, but any new configuration presents a very interesting challenge.

Tests for symmetry simplification revolve around answers to two questions:
(a) What dimensional coordinate components of the field do not exist?
(b) With which dimensional coordinate components does the field not vary?
These tests will be applied to examples later in this chapter. Probably the most "symmetrical" of electromagnetic devices is a transformer with a toroidal (doughnut-shaped) core and distributed windings (i.e., windings wound uniformly around the circumference of the toroid). Envision taking a cross-section "slice" perpendicular to the core. No matter where this slice is taken around the circumference, one would expect the magnetic field relationships across the cross section of this slice to be the same, since there is no change in the geometry or in the winding as one moves around the circumference of the toroid. Therefore the magnetic field can be examined on the basis of this two-dimensional cross section of the slice.

4. The *form* of the mathematics used to describe a field depends on the choice of dimensional coordinates. Most of us see, think, and even feel in terms of Cartesian (rectangular) coordinates. However, most rotating machines are best described by means of cylindrical coordinates.

Field equations for systems described by Cartesian coordinates are of the class known as linear homogeneous equations, with which most of us are reasonably familiar. In cylindrical systems the equations result in a form known as Bessel equations. These are less familiar to the average reader, but the numerous tables of Bessel functions that now exist and the availability of computer routines to handle Bessel functions allow this form of mathematics to be handled with almost as much ease as Cartesian coordinates. The third standard set of coordinates, spherical coordinates, is applicable to relatively few configurations of electromechanical devices and will not be treated in this book.

Another set of coordinates is unique to the study of electric machinery and introduces a fourth dimensional concept, motion. It is a means of relating the electrical and magnetic quantities on a moving rotor of an electric machine to the stationary electric circuit connected to the stator of the machine.

2.1 A REVIEW OF ELECTROMAGNETIC FIELD THEORY

The basic laws of electricity are governed by a set of equations called *Maxwell's equations*. Naturally, these equations govern the electromagnetic phenomena in energy-conversion devices too. Because of the presence of moving media, a direct application of Maxwell's equations to energy conversion devices is rather subtle. But certain other aspects of applications, such as parameter determination, are straightforward. In this section, we shall briefly review Maxwell's equations, then derive certain other equations from Maxwell's equations, and finally take up a number of examples which will illustrate some of the applications of Maxwell's equations and of other equations derived therefrom.

We know that charged particles are acted upon by forces when placed in *electric* and *magnetic* fields. In particular, the magnitude and direction of the force $\mathbf{F}$ acting on a charge q moving with a velocity $\mathbf{u}$ in an electric field $\mathbf{E}$ and in a magnetic field $\mathbf{B}$ is given by the Lorentz force equation

$$\mathbf{F} = q(\mathbf{E} + \mathbf{u} \times \mathbf{B}) = \mathbf{F}_E + \mathbf{F}_B \tag{2.1}$$

The electric field intensity $\mathbf{E}$ is thus defined by the following equation:

$$\mathbf{E} = \frac{\mathbf{F}_E}{\Delta q} \tag{2.2}$$

where $\mathbf{F}_E$ is the force on an infinitesimal test charge Δq; the electric field is measured in volts per meter. The magnetic field can also be similarly defined, from (2.1), as the force on a unit charge moving with unit velocity at right angles to the direction of **B**. The quantity B is called the *magnetic flux density* and is measured in webers per square meter, or in tesla (T).

Having defined the electric field, we can now define the *potential difference* dV between two points separated by a distance $d\mathbf{l}$ as

$$dV = -\mathbf{E} \cdot d\mathbf{l} \tag{2.3a}$$

Or, using the vector operator ∇, (2.3*a*) is expressed as

$$\mathbf{E} = -\nabla V \tag{2.3b}$$

We also see that, if **B** is the magnetic flux density, the total *magnetic flux* ϕ can be expressed as

$$\phi = \int_{\mathbf{s}} \mathbf{B} \cdot d\mathbf{s} \tag{2.4}$$

where the integral is over a surface **s**.

With the above definitions in mind, we now recall Faraday's law. It states that an electromotive force (emf) is induced in a closed circuit when the magnetic flux ϕ linking the circuit changes. If the closed circuit consists of an N-turn coil, the induced emf is given by

$$\mathrm{emf} = -N \frac{d\phi}{dt} \tag{2.5}$$

The negative sign is introduced to take into account Lenz's law and to be consistent with the positive sense of circulation about a path with respect to the positive direction of flow through the surface (as shown in Fig. 2.1) when (2.5) is expressed in the integral form as follows.

Because the potential difference has been defined by (2.3*a*,*b*), it is reasonable to define emf as

$$\mathrm{emf} = \oint \mathbf{E} \cdot d\mathbf{l} \tag{2.6}$$

which can be considered as the potential difference about a specific closed path. If we consider "going around the closed path only once," we can put $N = 1$; then from (2.4 to 2.6) it follows that

$$\oint \mathbf{E} \cdot d\mathbf{l} = -\frac{\partial}{\partial t} \int_{\mathbf{s}} \mathbf{B} \cdot d\mathbf{s} \tag{2.7}$$

The partial derivative with respect to time is used to distinguish derivatives with space, since **B** can be a function of both time and space.

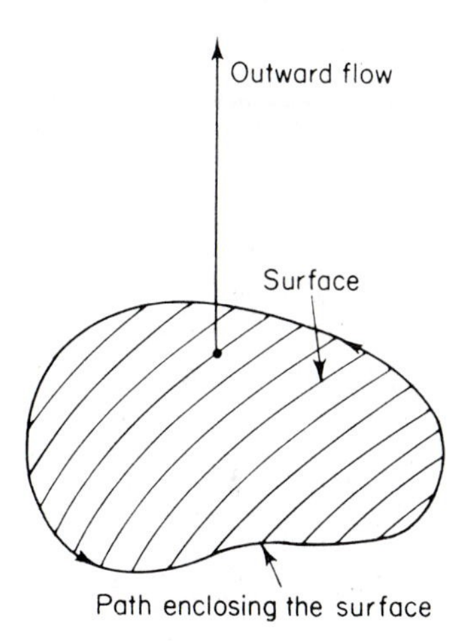

FIGURE 2-1
Positive circulation about a path enclosing a surface and positive flow through the surface.

Equation (2.7) is an expression of Faraday's law in the integral form. We now recall *Stokes' theorem*, which states that

$$\oint \mathbf{E} \cdot d\mathbf{l} = \int_{\mathbf{s}} (\nabla \times \mathbf{E}) \cdot d\mathbf{s} \tag{2.8}$$

where the surface **s** is that enclosed by the closed path, as shown in Fig. 2.1. We notice that the terms on the left-hand sides of (2.7) and (2.8) are identical. Equating the terms on the right-hand sides of these equations we have

$$\nabla \times \mathbf{E} = -\frac{\partial \mathbf{B}}{\partial t} \tag{2.9}$$

which is an expression of Faraday's law in the differential form.

A knowledge of vector algebra shows that the divergence of a curl is zero. Expressed mathematically,

$$\nabla \cdot \nabla \times \mathbf{E} = 0 \tag{2.10}$$

From (2.9) and (2.10), therefore,

$$\nabla \cdot \mathbf{B} = 0 \tag{2.11}$$

In deriving (2.9) we assumed that there was no relative motion between the closed circuit and the magnetic field. If, however, there is relative motion between the circuit and the magnetic field, in addition to the time variation of this field, (2.9) has to be modified. Referring to the Lorentz force equation, (2.1), we notice that the fields **E** and **B** are measured in a stationary reference frame. It is only the charge that is moving with a velocity **u**. If, on the other hand, the charge moves with a velocity **u′** with respect to a moving reference frame having a velocity **u**, the velocity of the charge with respect to the stationary reference frame becomes (**u** + **u′**). The force on the charge is then given by

$$\mathbf{F} = q[\mathbf{E} + (\mathbf{u} + \mathbf{u}') \times \mathbf{B}] = q[(\mathbf{E} + \mathbf{u} \times \mathbf{B}) + \mathbf{u}' \times \mathbf{B}] \tag{2.12}$$

Thus the electric field **E′** in the moving reference frame becomes

$$\mathbf{E}' = \mathbf{E} + \mathbf{u} \times \mathbf{B} \tag{2.13}$$

where **E** is given by (2.9). Here we have not taken into account relativistic effects because in electromagnetic energy conversion devices the velocities involved are considerably small. Using (2.9) and (2.13) we see that the electric field **E** induced in a circuit moving with a velocity **u** in a time-varying field **B** is given by

$$\nabla \times \mathbf{E} = -\frac{\partial \mathbf{B}}{\partial t} + \nabla \times \mathbf{u} \times \mathbf{B} \tag{2.14}$$

In (2.14) we notice that the second term on the right-hand side is introduced to take into account the motion of the circuit. Thus we can identify the first term in (2.14) as due to a *transformer emf* and the second as due to a *motional emf*; that is, the "change in the flux linkage" rule as well as the "flux-cutting" rule of obtaining the emf induced in a circuit are both taken into account in (2.14).

We would like to point out here that (2.5), expressing Faraday's law, is complete. A slightly more tedious analysis than that given above can be shown to lead to (2.14). In practical cases, it is better to identify the transformer and motional voltages separately. A direct application of (2.5), without extra care, may lead to inconsistent results.

So far we have discussed ways of finding the electric field from given magnetic fields. We shall now consider the relationship between given currents and resulting magnetic fields. In this connection, *Ampere's circuital law* expressed as

$$\oint \mathbf{H} \cdot d\mathbf{l} = I \tag{2.15a}$$

gives the relationship between the *magnetic-field intensity* **H** and the current I enclosed by the closed path of integration. If **J** is the surface-current density, (2.15*a*) can be also written as

$$\oint \mathbf{H} \cdot d\mathbf{l} = \int_s \mathbf{J} \cdot d\mathbf{s} \tag{2.15b}$$

Using Stokes' theorem, (2.8), the point form of Ampere's law becomes, from (2.15*b*),

$$\nabla \times \mathbf{H} = \mathbf{J} \tag{2.15c}$$

The surface-current density $\mathbf{J}$ is related to the volume-charge density ρ through the *continuity equation*

$$\nabla \cdot \mathbf{J} = -\frac{\partial \rho}{\partial t} \tag{2.16}$$

Taking the divergence of (2.15*c*) reveals an immediate inconsistency when compared with (2.16), since $\nabla \bullet \nabla \times \mathbf{H} = 0$. If an extra term $\partial \mathbf{D}/\partial t$ is added to the right-hand side of (2.15*c*) such that

$$\nabla \times \mathbf{H} = \mathbf{J} + \frac{\partial \mathbf{D}}{\partial t} \tag{2.17}$$

we find that the continuity equation is satisfied provided that

$$\nabla \cdot \mathbf{D} = \rho \tag{2.18}$$

But (2.18) is perfectly legitimate, since it expresses *Gauss's law* in the differential form. The quantity $\mathbf{D}$ is called the *electric-flux density* and $\partial \mathbf{D}/\partial t$ is known as the *displacement-current* density.

From the preceding considerations we have the following set of equations relating to the various field quantities:

$$\nabla \times \mathbf{E} = -\frac{\partial \mathbf{B}}{\partial t} \tag{2.19a}$$

$$\nabla \times \mathbf{H} = \mathbf{J} + \frac{\partial \mathbf{D}}{\partial t} \tag{2.19b}$$

$$\nabla \cdot \mathbf{B} = 0 \tag{2.19c}$$

$$\nabla \cdot \mathbf{D} = \rho \tag{2.19d}$$

Equations (2.19*a-d*) are generally known as *Maxwell's equations.* We can easily verify that (2.19*c,d*) are actually contained in (2.19*a,b*) respectively.

In order to obtain complete information regarding the various field quantities, in addition to Maxwell's equations certain auxiliary relations are also very useful. These relations are as follows.

Ohm's law. For a conductor of conductivity σ,

$$\mathbf{J} = \sigma \mathbf{E} \tag{2.20}$$

where **J** = surface-current density and **E** = electric-field intensity.

Permittivity. The electric-field intensity and the electric-flux density in a medium are related to each other by

$$D = \varepsilon E \tag{2.21}$$

where ε is called the *permittivity* of the material.

Permeability. The magnetic-field intensity and the magnetic-flux density in a material are related to each other by

$$\mathbf{B} = \mu \mathbf{H} \tag{2.22}$$

where μ is called the *permeability* of the medium.

Finally, we should note that in the majority of energy-conversion devices there are no free charges. Consequently, for these cases the Lorentz force equation takes the form

$$\mathbf{F} = \mathbf{J} \times \mathbf{B} \tag{2.23}$$

which follows from (2.1), since the motion of charges constitutes the flow of current. Because **J** and **B** are the current and flux densities, respectively, (2.23) determines the force density rather than the total force. We will use the simplified form of the force equation, (2.23), more often than the original equation, (2.1).

Because most commonly encountered electric machines have magnetic structures, we now consider some properties of magnetic materials.

2.2 MAGNETIC MATERIALS

Returning to (2.22), in free space **B** and **H** are related by the constant μ_0, known as the *permeability of free space*:

$$\mathbf{B} = \mu_0 \mathbf{H} \tag{2.24}$$

and

$$\mu_0 = 4\pi \times 10^{-7} \text{ H/m}$$

The preceding value for μ_0 is that of the Standard International (SI) system of units; the SI unit of **B** is tesla, and **H** is ampere per meter. Since it is still common for material characteristics to be given in CGS units and sometimes in English units, the units for (2.24) in these two systems of units are given in Appendix I.

Within a material, (2.24) must be modified to describe a magnetic phenomenon different from that occurring in free space:

$$\mathbf{B} = \mu\mathbf{H}, \quad \mu = \mu_R\mu_0 \tag{2.25}$$

where μ is termed *permeability* and μ_R *relative permeability*, a nondimensional constant. Permeability in a material medium as defined by (2.22) must be further qualified as applicable only in regions of homogeneous (uniform quality) and *isotropic* (having the same properties in any direction) materials. In materials not having these characteristics, μ becomes a vector (instead of tensor). Finally, note that for some common materials, (2.22) is *nonlinear*, and μ varies with the magnitude of **B**. This results in several subdefinitions of permeability related to the nonlinear *B-H* characteristic of the material, which will be discussed next.

A material is classified according to the nature of its relative permeability, μ_R, which is actually related to the internal atomic structure of the material and will not be discussed further at this point. Most "nonmagnetic" materials are classified as either *paramagnetic*, for which μ_R is slightly greater than 1.0, or *diamagnetic*, in which μ_R is slightly less than 1.0. However, for all practical purposes, μ_R can be considered as equal to 1.0 for all of these materials.

There is one interesting case of diamagnetism that is becoming of interest in certain types of electromagnetic devices. This is "perfect diamagnetism" (Meissner effect), which occurs in certain types of materials known as *superconductors* at temperatures near absolute zero. In such materials $\mathbf{B} \rightarrow 0$ and μ_R is essentially zero; that is, no magnetic field can be established in the superconducting material. This phenomenon has several potential applications, for instance, in several types of rotating machines and in a switching device.

Magnetic properties in matter are related to the existence of permanent magnetic dipoles within the matter. Such dipoles exist in paramagnetic materials but, as noted previously, the resulting magnetism is so weak as to classify these materials as nonmagnetic. Several other classifications of materials exhibit greater degrees of magnetism, but only two classes are discussed in detail here: *ferromagnetic* and *ferrimagnetic* materials. Ferromagnetic materials are further subgrouped into hard and soft materials, this classification roughly corresponding to the physical hardness of the materials. Soft ferromagnetic materials include the elements iron, nickel, cobalt, and one rare-earth element, most soft steels, and many alloys of the four elements. Hard ferromagnetic materials include the permanent magnet materials such as the alnicos, several alloys of cobalt with the rare-earth elements, chromium steels, certain copper-nickel alloys, and many other metal alloys. Ferrimagnetic materials are the *ferrites* and are composed of iron oxides having the formula MeO • Fe_2O_3, where Me represents a metallic ion. Ferrites are subgrouped into hard and soft ferrites, the former being the permanent magnetic ferrites, usually barium or strontium ferrite. Soft ferrites include the nickel-zinc and manganese-zinc ferrites and are used in microwave devices, delay lines, transformers, and other generally high frequency applications. A third class of magnetic materials of growing importance is made from powdered iron particles or other magnetic materials suspended in a nonferrous matrix such as epoxy or plastic. Sometimes termed *superparamagnetic*, powdered iron parts are formed by compression or injection molding techniques and are widely used in electronics transformers and as cores for inductors. Permalloy (molybdenum-nickel-iron powder) is one of the earliest and best known of the powdered materials.

Several magnetic properties of magnetic materials are important in the study of electromagnetic systems: permeability at various levels of flux density, saturation flux density, H at various levels of flux density, temperature variation of permeability, hysteresis characteristic, electrical conductivity, Curie temperature, and loss coefficients. These parameters vary widely among the different types of materials, so this discussion will be very general. Because of the nonlinear characteristics of most magnetic materials, graphical techniques are generally valuable in describing their magnetic characteristics. The two graphical characteristics of most importance are known as the B-H curve, or magnetization characteristic, and the hysteresis loop. There are many well-known laboratory methods for obtaining these characteristics or for displaying them on an oscilloscope. Fig. 2.2 shows a typical B-H characteristic. This characteristic can be obtained in two ways: the *virgin* B-H curve, obtained from a totally demagnetized sample, or the *normal* B-H curve, obtained as the tips of hysteresis loops of increasing magnitude. There are slight differences between the two methods that are not important for our purposes. The B-H curve is the result of *domain* changes within the magnetic material. In ferromagnetic materials the material is divided into small regions or domains (approximately 10^{-2} to 10^{-5} cm in size); in each region all dipole moments are spontaneously aligned. When the material is completely demagnetized, these domains have random orientation resulting in zero net flux density in any finite sample. As an external magnetizing force, H, is applied to the material, the domains that happen to be in line with the direction of H tend to grow, increasing B (region I in Fig. 2.2). In region II, as H is further increased, the domain walls move rapidly until each crystal of the material is a single domain. In region III the domains rotate in some direction until all domains are aligned with H. This results in *magnetic saturation*, and the flux density within the material cannot increase beyond the *saturation density*, B_s. The small increase that occurs beyond this condition is due to the increase in the space occupied by the material according to the relationship $B = \mu_0 H$. It is often convenient to subtract out this component of "free space" flux density and observe only the flux density variation within the material. Such a curve is known as the *intrinsic magnetization* curve and is of use in the design of permanent magnet devices.

The regions shown in Fig. 2.2 are also of value in describing the nonlinear permeability characteristic. From (2.25) it is seen that permeability is the slope of the B-H curve. In the following discussion relative permeability is assumed; that is, the factor μ_0 is factored out. The slope of the B-H curve is actually properly called *relative differential permeability*, or

$$\mu_d = \frac{1}{\mu_0} \frac{dB}{dH} \tag{2.26}$$

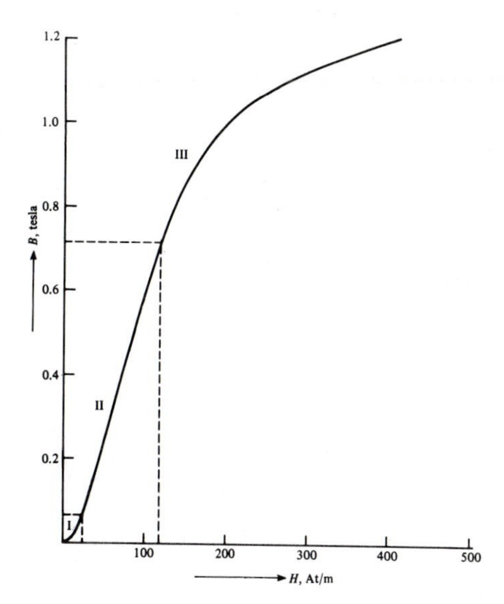

FIGURE 2-2
A typical *B-H* curve.

Relative initial permeability is defined as

$$\mu_i = \lim_{H \to 0} \frac{1}{\mu_0} \left| \frac{B}{H} \right| \tag{2.27}$$

and is seen to be the permeability in region I. This is important in many electronics applications where signal strength is low. It can also mislead one in measuring inductance of a magnetic core device with an inductance bridge because the low signal strength in most bridges will often magnetize the sample only in region I, where permeability is relatively low. In region II the *B-H* curve for many materials is relatively straight, and if a magnetic device is operated only in this region, linear theory can be used. In all regions the most general permeability term is known as *relative amplitude permeability* and is defined as merely the ratio of *B* to *H* at any point on the curve, or

$$\mu_a = \frac{1}{\mu_0} \left| \frac{B}{H} \right| \tag{2.28}$$

In general, permeability has to be defined on the basis of the type of signal exciting the magnetic material. There arc additional definitions for pulsed and sinusoidal excitation that will not be included here. *Maximum permeability* is the maximum value of the amplitude permeability and is important in many electronics applications.

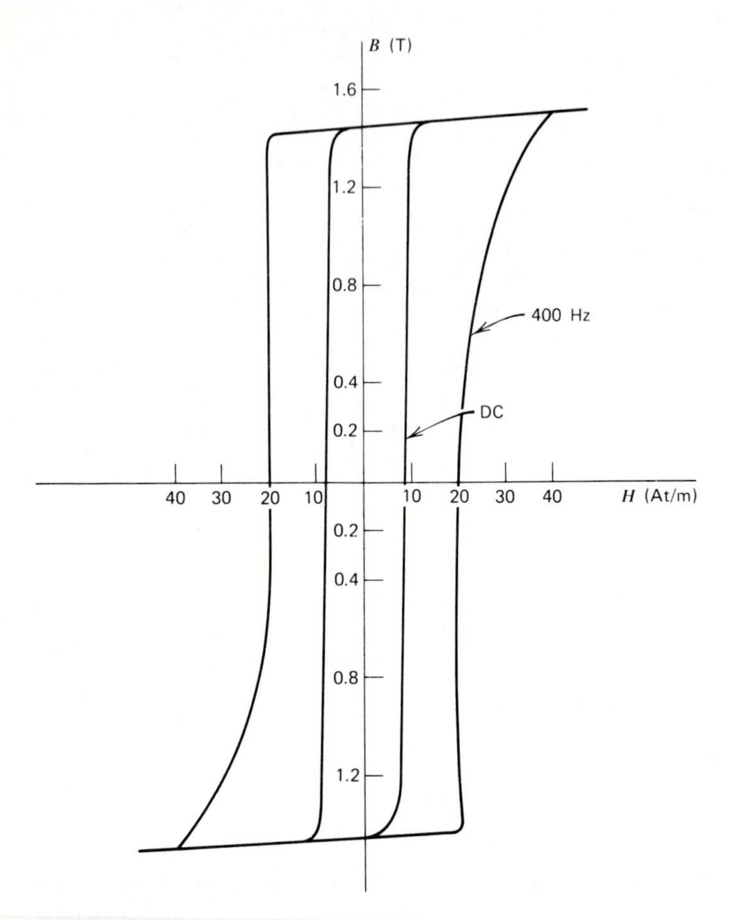

FIGURE 2-3
Deltamax tape-wound core 0.002-in. strip hysteresis loop.

The second graphical characteristic of interest is the hysteresis loop; a typical sample is shown in Fig. 2.3. This is a *symmetrical hysteresis loop*, obtained only after a number of reversals of the magnetizing force between plus and minus H_s. This characteristic illustrates several parameters of most magnetic materials, the most obvious being the property of hysteresis itself. The area within the loop is related to the energy required to reverse the magnetic domain walls as the magnetizing force is reversed. This is a nonreversible energy and results in an energy loss known as the *hysteresis loss.* This area varies with temperature and the frequency of reversal of H in a given material.

The second quadrant of the hysteresis loop is very valuable in the analysis of devices containing permanent magnets.* The intersection of the loop with the hori-

* We will have more to say about permanent magnets in Chapter 7.

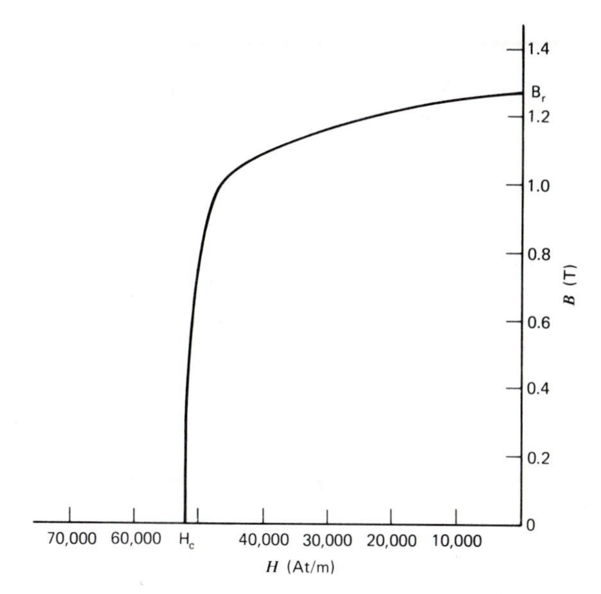

FIGURE 2-4
Demagnetization curve of Alnico V.

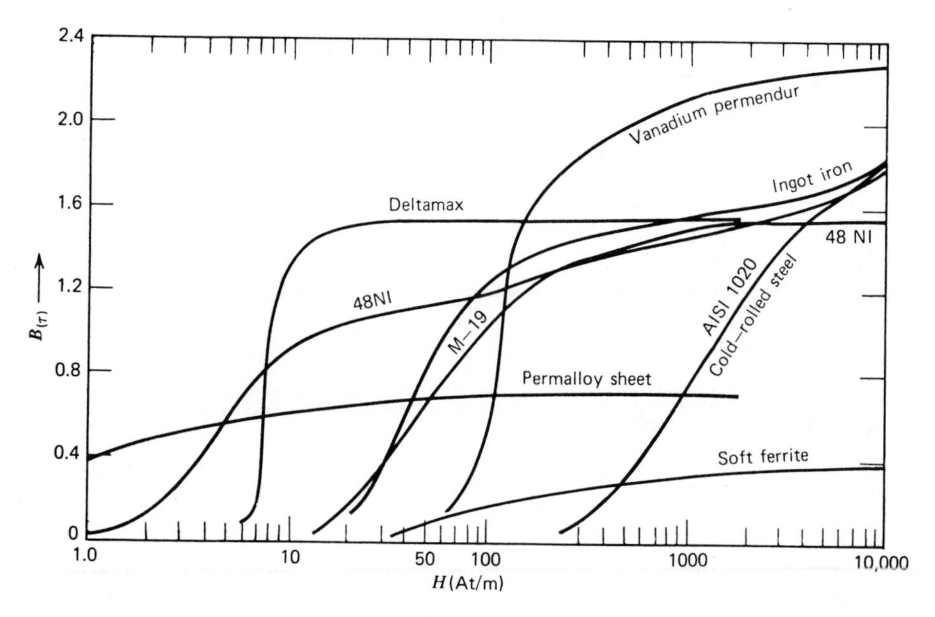

FIGURE 2-5
B-H curves of selected soft magnetic materials.

TABLE 2.1
Characteristics of soft magnetic materials

Trade Name	Principal Alloys	Saturation Flux Density (T)	H at B_{sat} (A/m)	Amplitude Permeability Max μ_m	Coercive Force H_c (A/m)	Electrical Resistivity (μohm-cm)	Curie Temperature (°C)
48 NI	48% Ni	1.25	80	200,000		65	
Monimax	48% Ni	1.35	6,360	100,000	4.0	65	398
High Prm 49	49% Ni	1.1	80			48	
Satmumetal	Ni, Cu	1.5	32	240,000		45	398
Permalloy (sheet)	Ni, Mo	0.8	400	100,000	1.6	55	454
Moly Permalloy (powder)	Ni, Mo	0.7	15,900	125			
Deltamax	50% Ni	1.4	25	200,000	8	45	499
M-19	Si	2.0	40,000	10,000	28	47	
Silectron	Si	1.95	8,000	20,000	40	50	732
Oriented-T	Si	1.6	175	30,000		47	
Oriented M-5	Si	2.0	11,900		26	48	746
Ingot Iron	None	2.15	55,000		80	10.7	
Spermendur	49% Co, V	2.4	15,900	80,000	8	26	
Vanadium Permendur	49% Co, V	2.3	12,700	4,900	92	40	932
Hyprco 27	27% Co	2.36	70,000	2,800	198	19	925
Flake Iron	Carbonal power	≈0.8	5,200	5-130		10^5-10^{15}	
Ferrotron(powder)	Mo, Ni	(Linear)	(Linear)	5-25		10^{16}	
Ferrite	Mg, Zn	0.39	1,115	3,400	13	10^7	135
Ferrite	Mn, Zn	0.453	1,590	10,000	6.3	3×10^7	190
Ferrite	Ni, Zn	0.22	2,000	160	318	10^9	500
Ferrite	Ni, Al	0.28	6,360	400	143		500
Ferrite	Mg, Mn	0.37	2,000	4,000	30	1.8×10^8	210

zontal (H) axis is known as the *coercive force*, H_c, and is a measure of the magnet's ability to withstand demagnetization from external magnetic signals. Often shown on this curve is a second curve, known as the *energy product*, which is the product of B and H plotted as a function of H and is a measure of the energy stored in the permanent magnet. The value of B at the vertical axis is known as the *residual flux density*.

The *Curie temperature*, or Curie point, T_c, is the critical temperature above which a ferromagnetic material becomes paramagnetic.

Up to this point we have not indicated numerical values for these various parameters. The parameter values for several common magnetic materials are given in Table 2.1. Several *B-H* curves are given in Fig. 2.5. It is important to note the typical values of relative permeability for good magnetic materials and to compare them with values of electrical conductivity for good electrical conductors. Some magnetic materials, such as permalloy, supermendur, and other nickel alloys, have a *maximum* relative permeability of over 100,000, giving a ratio to the permeability of a nonmagnetic material, such as air or free space, of 10^5. High permeability of this magnitude can be realized in only a few materials and only over a very limited range of operation. The permeability ratio between good and poor magnetic materials over a typical working range of operation is more like 10^4 at best. However, the ratio of electrical conductivity of a good electrical conductor, such as copper, to a good insulator, such as polystyrene, is of the order of 10^{24}. This implies, correctly, that no material that is a good magnetic insulator exists except for the superconductors mentioned previously. This point will become more apparent as we discuss magnetic circuits.

2.3 MAGNETIC LOSSES

A characteristic of magnetic materials that is very significant in the energy efficiency of an electromagnetic device is the energy loss within the magnetic material itself. The actual physical nature of this loss is still not completely understood, and a theoretical description of the basic mechanism that results in magnetic material losses is beyond the scope of this text. A simple explanation of this complex mechanism is as follows: Energy is used to effect "magnetic domain wall motion" as the domains grow and rotate under the influence of an externally applied magnetic field as described in Section 2.2. When the external field is reduced or reversed from a given value, domain wall motion is irreversible and manifests itself as heat within the magnetic material. The *rate* at which the external field is changed has a strong influence on the magnitude of the loss, and the loss is generally proportional to some function of the frequency of variation of the magnetic field. The metallurgical structure of the magnetic material, including its electrical conductivity, also has a profound effect on the magnitude of the loss. In electric machines and transformers, this loss is generally termed the *core loss*, or sometimes the *magnetizing loss* or *excitation loss.*

Traditionally, core loss has been divided into two components: *hysteresis loss* and *eddy current loss*. The hysteresis loss component has been alluded to previously and is generally held to be equal to the area of the low-frequency hysteresis loop times the frequency of the magnetizing force in sinusoidal systems. The hysteresis loss, P_h, is given by the empirical relationship

$$P_h = k_h f B_m^{1.5 \text{ to } 2.5} \text{ W/kg} \tag{2.29}$$

where k_h is a constant, f is the frequency, and B_m is the maximum flux density.

Eddy current losses are caused by induced electric currents, called eddies, since they tend to flow in closed paths within the magnetic material itself. The eddy current loss in a sinusoidally excited material, neglecting saturation, can be expressed by the relationship

$$P_e = k_e f^2 B_m^2 \text{ W/kg} \tag{2.30}$$

where B_m is the maximum value of flux density, f is the frequency, and k_e is the proportionality constant, depending on the type of material and the lamination thickness.

To reduce the eddy current loss, the magnetic material is *laminated*, that is, divided into thin sheets with a very thin layer of electrical insulation between the sheets. The sheets must be oriented in a direction parallel to the flow of magnetic flux (Fig. 2.6). The eddy current loss is roughly proportional to the square of the lamination thickness and inversely proportional to the electrical resistivity of the material. Lamination thickness varies from about 0.5 to 5 mm in electromagnetic devices used in power applications and from about 0.01 to 0.5 mm for devices used in electronics applications. Many magnetic cores used in electronics transformers and inductors are tape-wound from very thin strips of magnetic material. Laminating a magnetic part usually increases its volume. This increase may be appreciable, depending on the method used to bond the laminations together. The ratio of the volume actually occupied by magnetic material to total volume of a magnetic part is the *stacking factor*. This factor is important in accurately calculating flux densities

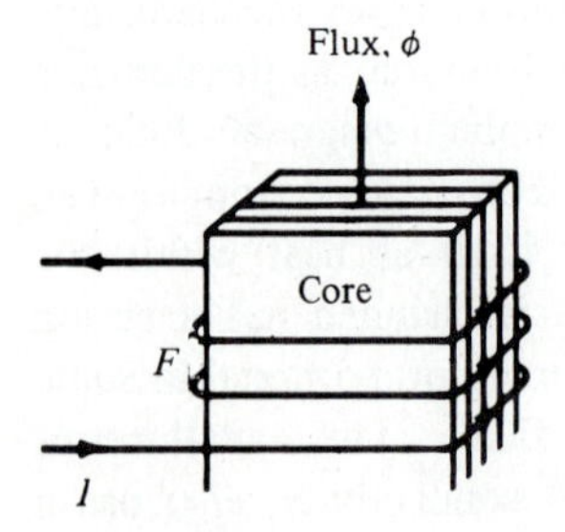

FIGURE 2-6
A portion of a laminated core.

TABLE 2.2
Stacking factor for laminated cores

Lamination Thickness (mm)	Stacking Factor
0.0127	0.50
0.0254	0.75
0.0508	0.85
0.1-0.25	0.90
0.27-0.36	0.95

in magnetic parts. Table 2.2 gives typical stacking factors for the thinner lamination sizes.

Stacking factor approaches 1.0 as the lamination thickness increases. In powdered iron and ferrite magnetic parts, there is an "equivalent stacking factor" that is approximately equal to the ratio of the volume of the magnetic particles to overall volume.

Equations (2.29) and (2.30) and the statements made concerning hysteresis loss are good rules for evaluating variations of these loss components with various field parameters, but they are inadequate for analytical predictions of absolute values of core loss. Therefore core loss should be determined from experimental data. Most manufacturers of magnetic materials have obtained core loss data under the condition of sinusoidal excitation for most of their products. Figs. 2.7 and 2.8 show measured core loss values for two common magnetic materials: M-15, a 3% silicon steel widely used in transformers and small motors, and 48 NI, a nickel alloy used in many electronics applications. These data are obtained by a measurement known as the Epstein frame method on sheet samples of material. Fig. 2.8*(b)* shows measured core loss in a ferrite material.

As a word of caution, many electromagnetic devices are finding increasing application in circuits in which the voltages and currents have waveshapes that cannot be classified by any of the standard waveforms such as sine waves, steady dc, square waves, and so forth. In many such circuits, power levels are relatively large; therefore the measurement of power, losses, and efficiency is a significant factor in their design and application. The source of these nonstandard waveforms is frequently the switching action of semiconductors in systems that include inverters, cycloconverters, controlled rectifiers, and so forth. Core loss data from sinusoidal measurements are generally not adequate for such systems. Generally, core loss measurements should be performed with the device excited from a source whose waveform is as close as possible to that under which the device will be actually operated. Power measurements under conditions of nonsinusoidal excitation generally require the use of Hall-effect, thermal, or electronic-multipler types of wattmeters.

2.4 MAGNETIC CIRCUITS

It is important to emphasize that a magnetic field is a distributed parameter phenomenon; that is, it is distributed over a region of space. As such, rigorous analysis requires the use of the distance variables as contained in the divergence and curl symbols of (2.9) and (2.11), for instance. However, under the proper conditions,

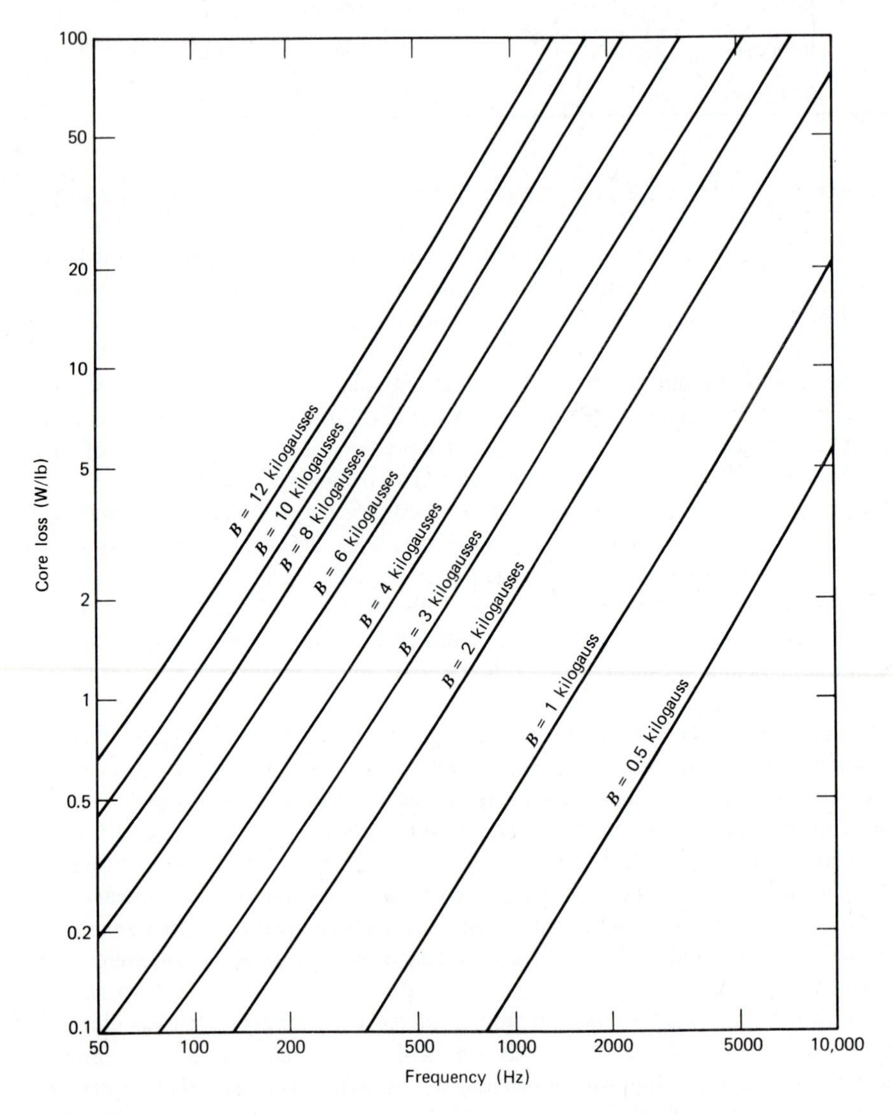

FIGURE 2-7
Core loss for nonoriented silicon steel 0.019-in.-thick lamination. (Courtesy Armco Steel Corporation.)

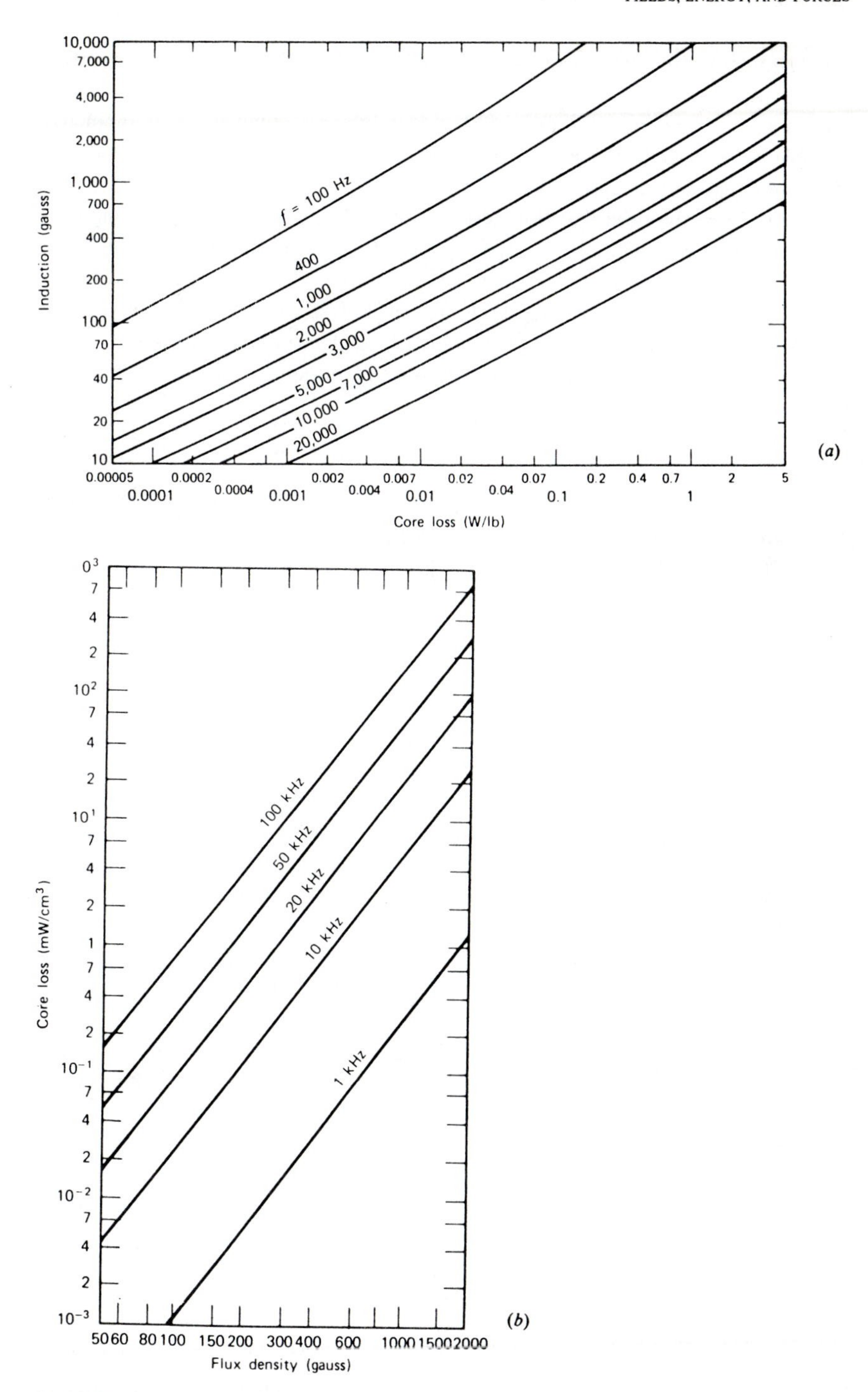

FIGURE 2-8
(a) Core loss for typical 48% nickel alloy 4 mils thick. (Courtesy Armco Steel Corporation.) *(b)* Core loss for Mn-Zn ferrites.

it is possible to apply lumped parameter analysis to certain classes of magnetic field problems just as it is applied in electric circuit analysis. The accuracy and precision of such analysis in the magnetic circuit problem is much less, however, than in electric circuit problems because of the relatively small permeability variation between magnetic conductors and insulators, as previously discussed.

This section briefly describes lumped circuit analysis as applied to magnetic systems, often called *magnetic circuit analysis*. Magnetic circuit analysis follows the approach of simple dc electric circuit analysis and applies to systems excited by dc signals or, by means of an incremental approach, to low-frequency ac excitation. Its usefulness is in sizing the magnetic components of an electromagnetic device during design stages, calculating inductances, and determining airgap flux densities for power and torque calculations.

Let us begin with a few definitions.

1. *Magnetic Potential.* For regions in which no electric current densities exist, which is true for the magnetic circuits we will discuss, the magnetic field intensity, **H**, can be defined in terms of *scalar* magnetic potential, F, as

$$\mathbf{H} = \nabla F; \qquad F = \int \mathbf{H} \cdot d\mathbf{l} \tag{2.31}$$

It is seen that F has the dimension of amperes, although "ampere-turns" is frequently used as a unit for F. For a potential rise or source of magnetic energy, the term *magnetomotive force* (mmf) is frequently used. As a potential drop, the term *reluctance drop* is often used. There are two types of sources of mmf in magnetic circuits: electric current and permanent magnets. The current source usually consists of a coil of a number of turns, N, carrying a current known as the *exciting current.* Note that the number of turns, N, is nondimensional.

2. *Magnetic Flux.* Streamlines or flowlines in a magnetic field are known as lines of magnetic flux, denoted by the symbol ϕ, and having the SI unit weber. Flux is related to **B** by the surface integral [see (2.4)]

$$\phi = \int_s \mathbf{B} \cdot d\mathbf{S} \tag{2.32}$$

3. *Reluctance.* Reluctance is a component of magnetic impedance, somewhat analogous to resistance in electric circuits except that reluctance is not an energy loss component. It is defined by a relationship analogous to Ohm's law:

$$\phi = \frac{F}{\mathfrak{R}} \tag{2.33}$$

The SI unit of magnetic reluctance is henry^{-1}. In regions containing magnetic material that is homogeneous and isotropic and where the magnetic field is uniform, (2.33) gives further insight into the nature of reluctance. If we assume

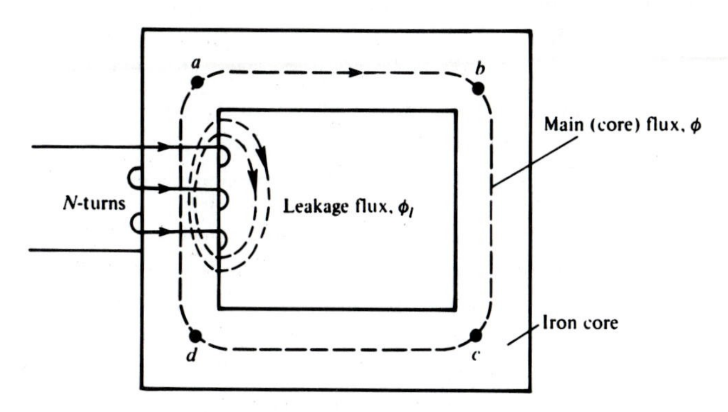

FIGURE 2-9
Leakage flux.

that the flux density has only one directional component, B, and is uniform over a cross section of area A_m, taken perpendicular to the direction of B, (2.32) becomes $\phi = BA_m$. We also assume that H is nonvarying along the length l_m in the direction of B, and (2.33) becomes, with some rearranging,

$$\mathfrak{R} = \frac{F}{\phi} = \frac{Hl_m}{BA_m} = \frac{l_m}{\mu A_m} \tag{2.34}$$

which is similar to the expression for electrical resistance in a region with similarly uniform electrical properties.

4. *Permeance.* The permeance, $\mathcal{P}$, is the reciprocal of reluctance and has the SI unit henry. In electronics transformer analysis the term *induction factor*, A_L, is frequently used and is identical to what is here called permeance. Permeance and reluctance are both used to describe the geometric characteristics of a magnetic field, mainly for purposes of calculating inductances.

5. *Leakage Flux.* Between any two points at different magnetic potentials in space, a magnetic field exists, as shown by (2.31). In any practical magnetic circuit there are many points or, more generally, planes at magnetic potentials different from each other. The magnetic field between these points can be represented by flowlines or lines of magnetic flux. Where these flux lines pass through regions of space—generally air spaces, electrical insulation, or structural members of the system—instead of along the main path of the circuit, they are termed *leakage flux lines*. In coupled circuits with two or more windings the definition of leakage flux is specific: flux links one coil but not the other. Leakage flux is identified, in Fig. 2.9, along the path l whereas the main core flux, also termed mutual flux, is along the path *abcd*.

   Leakage is a characteristic of all magnetic circuits and can never be completely eliminated. At dc or very low ac excitation frequencies, magnetic shielding consisting of thin sheets of high-permeability material can reduce leakage flux. This is done not by eliminating leakage but by establishing new levels of magnetic potential in the leakage paths to better direct the flux lines along the desired path. At higher frequencies of excitation, electrical shielding, such as aluminum foil, can reduce leakage flux by dissipating its energy as induced currents in the shield.

6. *Fringing.* Fringing is somewhat similar to leakage; it describes the spreading of flux lines in an airgap of a magnetic circuit. Fig. 2.10 illustrates fringing at a gap. Fringing results from lines of flux that appear along the sides and edges of the two magnetic members at each side of the gap, which are at different magnetic potentials. Fringing is almost impossible to calculate analytically, except in the simplest of configurations. Fringing has the effect of increasing the effective area of the airgap, which must be considered with the length of the airgap.

Based on (2.33) and (2.34), we may now construct the relationships summarized in Table 2.3. In this table l is the length and A is the cross-sectional area of the path for the flow of current in the electric circuit or for the flux in the magnetic circuit. It may be verified from Table 2.3 that the unit of reluctance is H^{-1}.

Because ϕ is analogous to I and $\Re$ is analogous to R, the laws for series- or parallel-connected resistors also hold for reluctances.

Example 2.1 In analyzing magnetic circuits such as the one shown in Fig. 2.11, we use the concept of mean length. For the circuit of Fig. 2.11*(a)*, we designate the mean lengths by l's and the various areas of cross sections by A's. If $2A_1 = A_2 = 10$ cm^2, $l_1 = 3l_2 = 24$ cm, and the relative permeability of the magnetic material at a certain operating point is 500, calculate the net reluctance as seen by *(a)* the 200-turn coil and *(b)* the 100-turn coil. Also, show electrical analogs for the two cases. Both coils are not assumed to be excited at the same time. Neglect saturation.

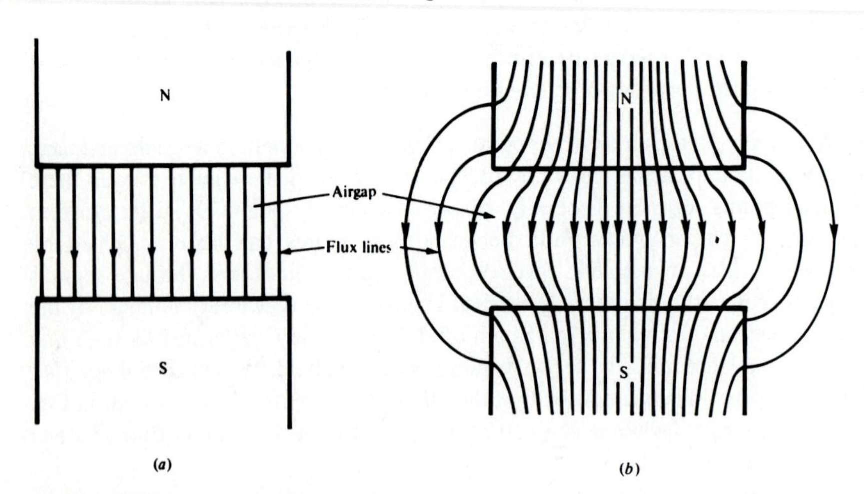

FIGURE 2-10
(a) No fringing of flux; *(b)* fringing of flux.

TABLE 2.3
Analogy between a magnetic circuit and a dc electric circuit

Magnetic Circuit	Electric Circuit
Flux, ϕ	Current, I
MMF, F	Voltage, V
Reluctance, $\mathfrak{R} = l/\mu A$	Resistance, $R = l/\sigma A$
Permeance, $\wp = 1/\mathfrak{R}$	Conductance, $G = 1/R$
Permeability, μ	Conductivity, σ
Ohm's law, $\phi = F/\mathfrak{R}$	Ohm's law, $I = V/R$

Solution

The electrical analogs are shown in Fig. 2.11*(b)* and *(c)*, respectively, for the cases when only the 200-turn coil is excited and when only the 100-turn coil is excited. These analogs, in turn, are respectively reduced to those shown in Fig. 2.11*(d)* and *(e)*. From the given data

$$\mu = \mu_r\mu_0 = 500 \times 4\pi \times 10^{-7} = 6.28 \times 10^{-4} \text{ H/m}$$

$$\mathfrak{R}_1 = \frac{l_1}{\mu A_1} = \frac{24 \times 10^{-2}}{6.28 \times 10^{-4} \times \frac{10}{2} \times 10^{-4}} = 0.764 \times 10^6 \text{ H}^{-1}$$

$$\mathfrak{R}_2 = \frac{l_2}{\mu A_2} = \frac{8 \times 10^{-2}}{6.28 \times 10^{-4} \times 10 \times 10^{-4}} = 0.127 \times 10^6 \text{ H}^{-1}$$

We now use these values of reluctance in Fig. 2.11*(b)-(e)*.

(a) We have from Fig. 2.11*(d)* the reluctances as seen by the 200-turn coil as the sum of the two $\mathfrak{R}_1$'s in parallel, the combination being in series with $\mathfrak{R}_2$. Hence

$$\mathfrak{R}_{(a)} = \frac{1}{2}\mathfrak{R}_1 + \mathfrak{R}_2 = 10^6\left(\frac{1}{2} \times 0.764 + 0.127\right) = 0.509 \times 10^6 \text{ H}^{-1}$$

(b) From Fig. 2.11*(e)* we have the reluctance, as seen by the 100-turn coil, as the sum: $\mathfrak{R}_1$ in parallel with $\mathfrak{R}_2$, the combination being in series with $\mathfrak{R}_1$. Hence

$$\mathfrak{R}_{(b)} = \frac{\mathfrak{R}_1\mathfrak{R}_2}{\mathfrak{R}_1 + \mathfrak{R}_2} + \mathfrak{R}_1 = \frac{10^{12}(0.764 \times 0.127)}{10^6(0.764 + 0.127)} + 10^6 \times 0.764$$

$$= 0.873 \times 10^6 \text{ H}^{-1}$$

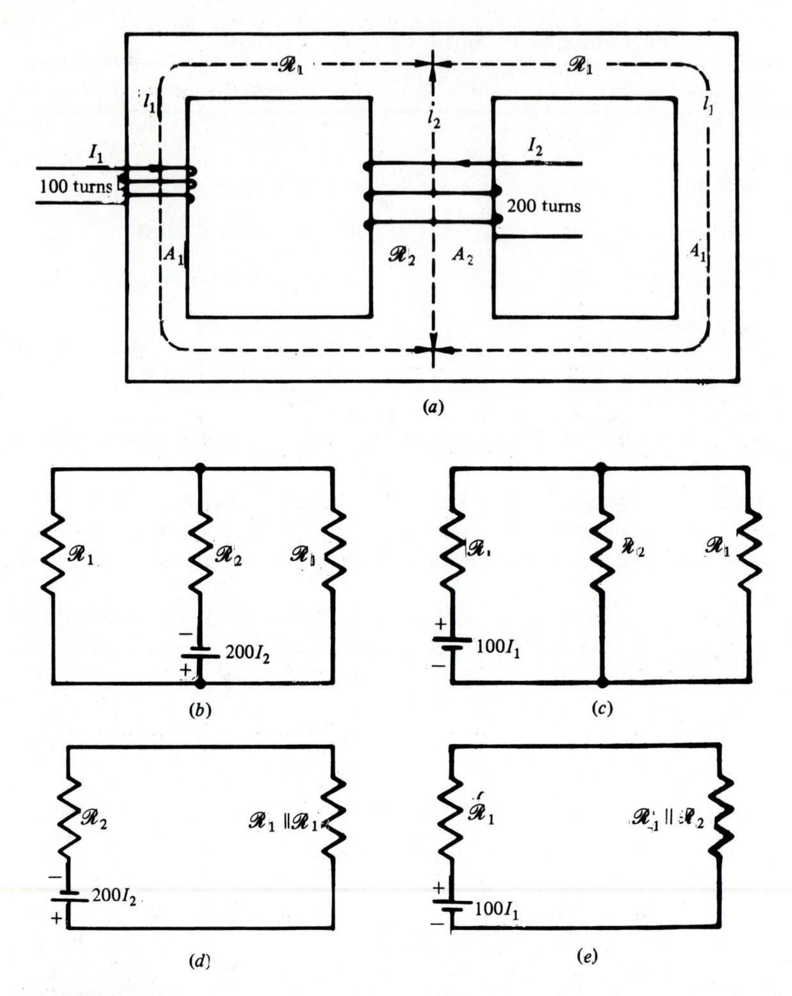

FIGURE 2-11
(a) A magnetic circuit; *(b)* electrical analog when 200-turn coil excited; *(c)* electrical analog when 100-turn coil excited; *(d)* reduction of circuit *(b)*; *(e)* further reduction of circuit *(c)*.

2.5 AMPERE'S LAW APPLIED TO A MAGNETIC CIRCUIT

According to (2.15*a*), the integral around any closed path of the magnetic field intensity, **H**, equals the electric current contained within that path. A word about directions in using this integral expression is in order here. Positive current is defined as flowing in the direction of the advance of a right-handed screw turned in the direction in which the closed path is traversed.

Let us apply Ampere's law to the simple magnetic circuit whose cross section is shown in Fig. 2.12; the circuit consists of a magnetic member of mean length l_m, in series with an airgap of length l_g, around which are wrapped three coils of turns N_1, N_2, and N_3 respectively. The path of magnetic flux ϕ is shown along the mean length of the magnetic member and across the airgap. Let the line integration proceed in a clockwise manner. Current directions are shown in the three coils. Note that for the directions shown, current direction is into the plane of the paper for the conductors enclosed by the integration paths for coils 1 and 3 and out of the plane of the paper for coil 2. From the left side of (2.15*a*), we obtain

$$\oint \mathbf{H} \cdot d\mathbf{l} = \int_0^{l_m} \mathbf{H}_m \cdot d\mathbf{l} + \int_0^{l_g} \mathbf{H}_g \cdot d\mathbf{l} \tag{2.35}$$

If the magnetic material is linear, homogeneous, and isotropic and if leakage flux is neglected, (2.35) becomes

$$\oint \mathbf{H} \cdot d\mathbf{l} = H_m l_m + H_g l_g = \phi(\mathfrak{R}_m + \mathfrak{R}_g) = F_m + F_g \tag{2.36}$$

where $\mathfrak{R}_m$ and $\mathfrak{R}_g$ are the reluctances of the magnetic member and gap, respectively, and F_m and F_g represent the magnetic potential or reluctance drop across these two members of the magnetic circuit. The right side of (2.15*a*) gives

$$I = N_1 I_1 + N_3 I_3 - N_2 I_2 \tag{2.37}$$

Combining (2.36) and (2.37) yields

$$N_1 I_1 + N_3 I_3 - N_2 I_2 - F_m - F_g = 0 \tag{2.38}$$

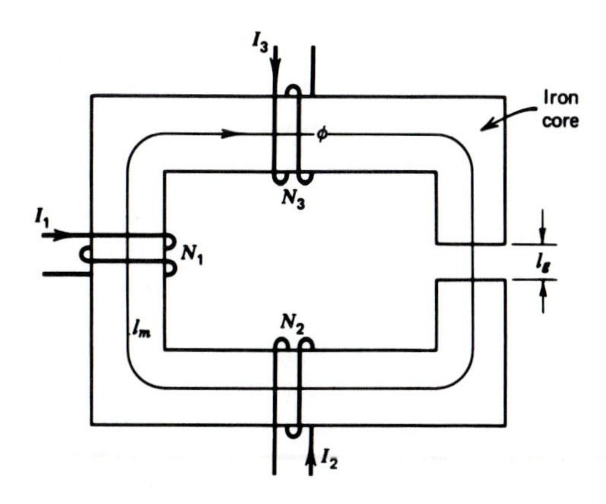

FIGURE 2-12
A composite magnetic circuit, with multiple excitation (mmf's).

We may generalize Ampere's law on the basis of this simple example to state that "the sum of the magnetic potentials around any closed path is equal to zero," which is analogous to Kirchhoff's voltage relationship in electric circuits. Note that this generalization follows even without the simplifying assumptions used to eliminate the integral form of (2.35).

Example 2.2 Let us illustrate the use of (2.38) by assuming some numerical values in the circuit of Fig. 2.12 and solving the following problem. Determine the number of ampere-turns required to establish a flux density of 1 T in the airgap. It will serve no instructive purpose to have three coils, so we will set I_2 and I_3 equal to zero and solve for the product I_1N_1. Assume that the airgap length, l_g, is 0.1 mm; the magnetic member is constructed of laminated M-19 steel with a stacking factor of 0.9 and of length, l_m, equal to 100 mm; and fringing and leakage are neglected.

Solution

The reluctance drops can be calculated by either of the forms in the interior of (2.36); since the airgap flux density has been specified, the form in terms of the magnetic field intensities is simple:

$$H_g = \frac{B_g}{\mu_0} = \frac{1.0}{4\pi \times 10^{-7}} = 7.95 \times 10^5 \text{ A/m}$$

$$F_g = H_g l_g = (7.95 \times 10^5)(10^{-4}) = 79.5 \text{ A}$$

If we neglect fringing and leakage, the flux density in the magnetic member can be assumed to be the airgap density divided by the stacking factor, or

$$B_m = \frac{1.0}{0.9} = 1.11 \text{ T}$$

From the curve for M-19 steel in Fig. 2.5, at this value of flux density,

$$H = 130 \text{ At/m}$$

$$F_m = 130 \times 0.1 = 13 \text{ At}$$

The required ampere-turns in the exciting coil are

$$N_1 I_1 = F_m + F_g = 92.5 \text{ At}$$

Example 2.3 Using the same configuration and numerical values as in Example 2.2, determine the required ampere-turns in the exciting coil for establishing a flux of 0.001 Wb in the airgap. Let us solve this problem using reluctances neglecting leakage but not fringing effects. We will need to know the cross-sectional area of the magnetic member to determine reluctances; assume that A_m = 16 cm^2 (gross).

Solution

Equation (2.34) can be used to determine the airgap reluctance. Assume that fringing effects increase the effective gap area over the area of the steel surface facing it by 10%. The reluctance becomes

$$\Re_g = \frac{10^{-4}}{(4\pi \times 10^{-7})(1.1 \times 0.0016)} = 4.5 \times 10^4$$

If we neglect leakage, the same flux will exist in the magnetic member. The flux density in the magnetic material is

$$B_m = \frac{0.001}{0.9 \times 0.0016} = 0.695 \text{ T}$$

From the M-19 curve in Fig. 2.5, the amplitude permeability is

$$\mu_a = \frac{B_m}{\mu_0 H_m} = \frac{0.695}{(4\pi \times 10^{-7})(54)} = 10{,}240 = \mu_R$$

The reluctance of the magnetic member is

$$\Re_m = \frac{l_m}{\mu_R \mu_0 A_m} = \frac{0.1}{10{,}240(4\pi \times 10^{-7})(0.9 \times 0.0016)} = 0.54 \times 10^4$$

The required exciting ampere-turns are

$$N_1 I_1 = \phi(\Re_g + \Re_m) = 0.001(5.04 \times 10^4) = 50.4 \text{ At}$$

There are several conclusions to observe from these simple examples.

1. The magnetic core with airgap is analogous to a simple series dc circuit, as shown in Fig. 2.13.

2. Because of symmetry about the plane of the paper of the system of Fig. 2.12, two-dimensional representation of the magnetic field is acceptable.

3. The calculation of reluctances is a more cumbersome approach than the use of magnetic field intensities for determining reluctance drops. Therefore, for numerical solutions, the use of $(H_m l_m + H_g l_g)$ is always preferred, as illustrated in Example 2.2. The calculation of magnetic reluctances for magnetic sections

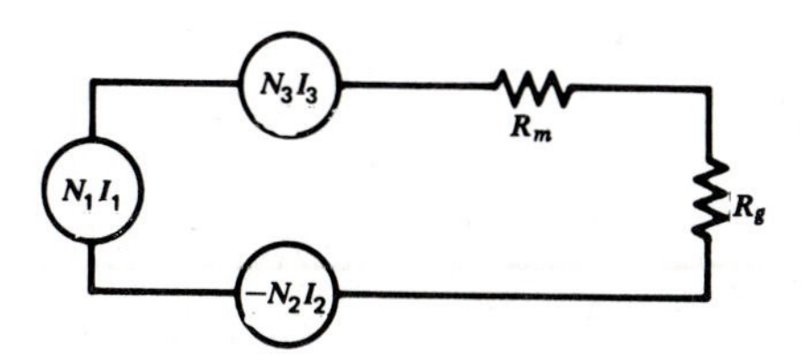

FIGURE 2-13
Approximate equivalent circuit for Fig. 2.12.

requires the determination of amplitude permeability from curves such as Fig. 2.5, the same process required to determine H_m. However, if the reluctance technique is continued, a great deal of additional mathematical manipulation is required that often introduces trivial errors. The reluctance technique is most valuable in qualitative analysis of magnetic circuits, such as in the development of equivalent circuits or in comparing different geometric sections, or in the calculation of inductance.

4. The reader may wonder about solving the inverse problem to Examples 2.2 and 2.3; that is, given the exciting ampere-turns, determine the flux (or flux density) in the airgap. A little thought will show that there is no direct analytical solution to this problem because of the nonlinear B-H characteristic of the magnetic material. The flux density must be known before the field intensity or reluctance of the magnetic member can be determined. There are several graphical techniques for solving this inverse problem. This type of problem is amenable to iterative computer techniques, although considerable computer storage space may be required, since the B-H characteristics of the magnetic material must be stored for repeated access during the iterative process.

Example 2.4 Fig. 2.14 illustrates a parallel magnetic circuit. The magnetic portions of this circuit are constructed from AISI 1020 cold-rolled steel (Fig. 2.5). All dimensions are in millimeters. What exciting current is required in the coil of 1000 turns located on the center leg of the circuit to develop a flux density of 0.775 T in the gap? Neglect fringing and leakage.

Solution

Note that symmetry about the plane of the paper is assumed in this problem, permitting the use of ($H_m l_m = H_g l_g$) instead of the reluctance approach in solving this problem. This implies that the magnetic relationships are invariant throughout the *thickness* of the magnetic circuit (the dimensions perpendicular to the plane of the paper).

In this problem there are two paths that are magnetically in parallel between the points shown as A-A' in Fig. 2.14. Again, note the distributed nature of a magnetic circuit (as contrasted to an electric circuit). Without a much more detailed field analysis, there is no guarantee that at points A-A' the parallel paths divide or separate. However,

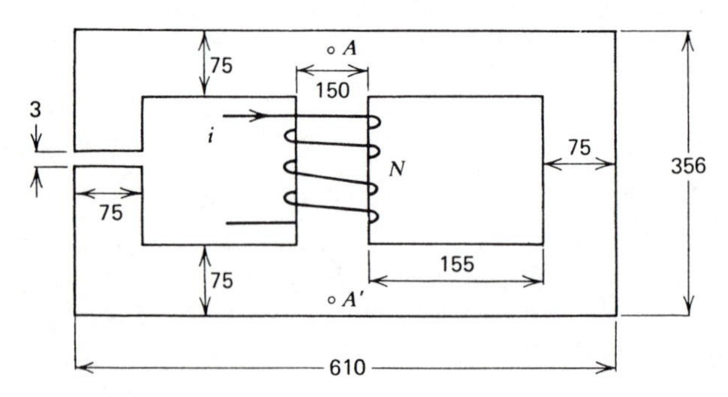

FIGURE 2-14
A parallel magnetic circuit.

as an approximation, we will assume that this is the case and that points A-A' are the midpoints of the cross sections of the upper and lower branches of the circuit. There is no possibility of an a priori determination of the reluctance drop in the right branch, since the permeability of the magnetic material in this branch is not known without knowing the flux density in this branch. We can, however, determine the reluctance drop in the left branch as follows.

The field intensity in the gap is $H_g = 0.775/(4\pi \times 10^{-7}) = 6.17 \times 10^5$. The reluctance drop in the gap is $= 6.17 \times 10^5 \times 0.003 = 1850$ A. Neglecting leakage, the flux density in the magnetic sectors is the same as in the gap, 0.775 T.

From Fig. 2.5, for AISI 1020, $H_m = 1000$ At/m; the length of the left-hand magnetic material circuit is $2 \times (305 - 37.5) + (356 - 75 - 3) = 813$ mm $= 0.813$ m; the reluctance drop in the magnetic portion of the left branch is $(0.813)\ (1000) = 813$ A; the total reluctance drop in the left-hand branch $= 1850 + 813 = 2663$ A. From the concept implied in (2.38), the reluctance drop between A and A' around the right branch of Fig. 2.14 must also equal 2633 A in magnitude. H_m in the right branch is, therefore, $2633/[2 \times (305 - 37.5) + (356 - 75)] = 3.263$ A/mm $= 3263$ A/m. From Fig. 2.5, B_m for AISI 1020 is 1.6 T. The flux density in the center branch (assuming a uniform thickness, d, throughout the magnetic circuit) is $(0.775 \times 0.075 \times d + 1.6 \times 0.075 \times d)/(0.15 \times d) = 1.188$ T; from Fig. 2.5, H_m in the center branch is 2000 A/m. The total mmf required by the coil is $2633 + 2000 \times (356 - 75)/1000 = 2633 + 562 = 3225$ A; required current is 3.225 A.

2.6 LIMITATIONS OF MAGNETIC CIRCUIT APPROACH

The number of problems in practical magnetic circuits that can be solved by the approach outlined in Sections 2.4 and 2.5 is limited, despite the similarity of this approach to simple dc electric circuit theory. The discussion in point 4 that follows Example 2.3 has illustrated only one of the limitations. The purpose of introducing magnetic circuits is to state some very fundamental principles and definitions that are necessary to understand electromagnetic systems, not as a problem-solving technique. The limitations of magnetic circuit theory rest primarily on the nature of magnetic materials as contrasted with conductors, insulators, and dielectric materials. Most of these limitations have already been introduced as "assumptions" in the discussion of magnetic circuits. Let us assess the significance of these assumptions.

1. *Homogeneous Magnetic Material.* Most materials used in practical electromagnetic systems can be considered homogeneous over finite regions of space, allowing the use of the integral forms of Maxwell's equations and calculations of reluctances and permeances.

2. *Isotropic Magnetic Materials.* Many sheet steels and ferrites are oriented by means of the metallurgical process during their production. Oriented materials have a "favored" direction in their grain structure, giving superior magnetic properties when magnetized along this direction.

3. *Nonlinearity.* This is an inherent property of all ferro- and ferrimagnetic materials. However, there are many ways of treating this class of nonlinearity analytically.

(a) As can be seen by observing the *B-H* curve shown in this chapter, a considerable portion of the curve for most materials can be approximated as a straight line, and many electromagnetic devices operate in this region.

(b) Numerous analytical and numerical techniques have been developed for describing the *B-H* and other nonlinear magnetic characteristics. Several of these are discussed in Appendix III.

(c) The nonlinear *B-H* characteristic of magnetic materials manifests itself in their relationship between flux and exciting current in electromagnetic systems; the relationship between flux and induced voltage is a *linear* relationship, as given by Faraday's law, (2.7) or (2.9). It is possible to treat these nonlinear excitation characteristics separately in many systems, such as is done in the equivalent circuit approach to transformers and induction motors.

(d) An inductance whose magnetic circuit is composed of a magnetic material is a nonlinear electric circuit element, such as a coil wound on a magnetic toroid. With an airgap in the magnetic toroid, however, the effect of the nonlinear magnetic material on the inductance is lessened. Rotating machines and many other electromechanical devices have airgaps in their magnetic circuits, permitting the basic theory of these devices to be described by means of linear equations.

4. *Saturation.* All engineering materials and devices exhibit a type of saturation when output fails to increase with input, for instance, in the saturation of an electronic amplifier. Saturation is very useful in many electromagnetic devices, such as magnetic amplifiers and saturable reactors. The magnetic materials frequently used in these and other types of magnetic switching devices are called *square-loop* materials, since their hysteresis loops (Fig. 2.3) can be approximated by a square or rectangle with sides parallel to the *B* and *H* axes.

5. *Leakage and Fringing Flux.* This is a property of all magnetic circuits. It is best treated as a part of the generalized solution of magnetic field distribution in space, often called a boundary value problem. In many rotating machine magnetic circuits, boundaries between regions of space containing different types of magnetic materials (usually, a boundary between a ferromagnetic material and air) are often planes or cylindrical surfaces that, in a two-dimensional cross section, become straight lines or circles. Leakage inductances can frequently be determined in such regions by calculating the reluctance or permeance of the region using fairly simple integral formulations. The spatial or geometric coefficients so obtained are known as permeance coefficients.

2.7 THE IDEAL MAGNETIC CIRCUIT

Using the assumptions discussed in Section 2.6, we can define the ideal magnetic circuit. The ideal magnetic circuit is composed of magnetic materials that are homo-

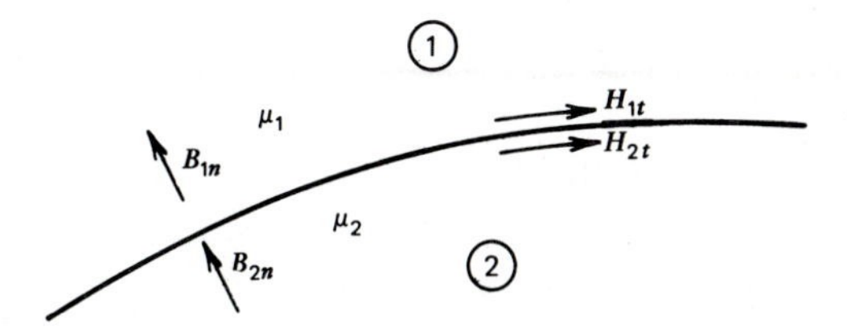

FIGURE 2-15
Magnetic materials interface.

geneous, isotropic, and linear and has *infinite permeability*. Airgap fringing is also usually neglected. A magnetic material with infinite permeability is analogous to a perfect electrical conductor, that is, a conductor with infinite electrical conductivity and the relationships between the **B** and **H** vectors are analogous to the relationships between **J** and **E** in the perfect conductor. These relationships are derived in most texts on field theory and will only be summarized here. Consider Fig. 2.15, which shows the boundary between two regions of different magnetic permeability. Let region 1 be free space with $\mu_1 = \mu_0$; region 2 is the magnetic material with $\mu_2 \rightarrow \infty$. The boundary conditions, in terms of field components normal to the boundary, H_n and B_n, and tangential to the boundary, H_t and B_t, can be shown to be (assuming zero current density on the boundary surface)

$$\begin{aligned} B_{1n} &= B_{2n} \\ H_{1t} &= H_{2t} \end{aligned} \tag{2.39}$$

where the subscripts 1 and 2 refer to the respective regions. Within the magnetic material, region 2,

$$H_{2t} = \frac{B_{2t}}{\mu_2} \rightarrow 0 \quad (\mu_2 \rightarrow \infty) \tag{2.40}$$

showing that the tangential component of field intensity in the magnetic material is zero as the permeability approaches infinity. Therefore, from (2.39), the tangential component of the magnetic field in region 1 at the boundary is also zero. Also, from (2.31), it can be seen that F, the magnetic potential, is zero along a path parallel to the tangential field within the magnetic material. There are two important conclusions from this analysis.

1. Flowlines or lines of magnetic flux are perpendicular to the surface of a perfect magnetic conductor.

2. There is no potential difference or reluctance drop between points or planes at different locations with a perfect magnetic conductor.

These characteristics of an ideal magnetic circuit are frequently used in magnetic circuit analysis.

2.8 ENERGY RELATIONS IN A MAGNETIC FIELD

The energy stored in a magnetic field is defined throughout space by the volume integral

$$W = \frac{1}{2}\int_{vol} \mathbf{B}\cdot\mathbf{H}\,dv = \frac{1}{2}\int_{vol} \mu H^2 dv = \frac{1}{2}\int_{vol} \frac{B^2}{\mu}\,dv \tag{2.41}$$

This equation is valid only in regions of constant permeability. Therefore its usefulness is limited to *static* linear magnetic circuits. Energy relationships in a time-varying magnetic circuit, such as a rotating machine or a relay, including the concept of *coenergy*, will be developed later in the book.

Example 2.5 Determine the magnetic energy stored in the airgap and magnetic material of the magnetic circuit of Example 2.4.

Solution

Under the assumptions used in Example 2.4, the field distribution was uniform in both the gap and the magnetic material, facilitating the use of (2.41). In the gap, B_g = 0.001/(1.1 x 0.0016) = 0.57 T.

$$W = \frac{1}{2}\left(\frac{B_g^2}{\mu_0}\right)(\text{vol}) = \frac{1}{2}\left(\frac{0.57^2}{4\pi \times 10^{-7}}\right)(1.1 \times 0.0016 \times 10^{-4})$$

$$= 0.0227 \text{ J}$$

It is evident that most of the energy is required to establish the flux in the airgap.

2.9 INDUCTANCE

Inductance is one of the three circuit constants in electric circuit theory and is defined as flux linkage per ampere:

$$L = \frac{\lambda}{i} = \frac{N\phi}{i} \tag{2.42}$$

Consider the magnetic toroid around which are wound n distinct coils electrically isolated from each other, as shown in Fig. 2.16. The coils are linked magnetically by the flux, ϕ, some portion of which links each of the coils. A number of inductances can be defined for this system;

$$L_{km} = \frac{\text{flux linking the } k\text{th coil due to the current in the } m\text{th coil}}{\text{current in the } m\text{th coil}}$$

Mathematically, this can be stated as

$$L_{km} = \frac{N_k(K\phi_m)}{i_m} \tag{2.43}$$

where K is the portion of the flux due to coil m that links coil k and is known as the *coupling coefficient.* By definition, its maximum value is 1.0. A value of K less than 1.0 is attributable to leakage flux in the regions between the location of coil k and coil m. When the two subscripts in (2.43) are identical, the inductance is termed *self-inductance*; when different, the inductance is termed *mutual inductance* between coils k and m. Mutual inductances are symmetrical; that is,

$$L_{km} = L_{mk} \tag{2.44}$$

Inductance can be related to the magnetic parameters derived earlier in this chapter. In (2.43), ϕ_m can be replaced, using (2.33), by the magnetic potential of coil m, F_m, divided by reluctance of the magnetic circuit, $\mathfrak{R}$; the magnetic potential of coil m, however, is $N_m I_m$. Making these substitutions in (2.43) gives

$$L_{km} = \frac{KN_kN_m}{\mathfrak{R}} = KN_kN_m\wp \tag{2.45}$$

where $\wp$ is the permeance, the reciprocal of the reluctance. In a simple magnetic circuit such as the toroid of Fig. 2.16, the reluctance, from (2.34), can be substituted into (2.45), giving

$$L_{km} = \frac{KN_kN_mA_t\mu_t}{l_t} \tag{2.46}$$

where μ_t, A_t, and l_t are the permeability, cross-sectional area, and mean length of the toroid, respectively.

Stored energy can be expressed in terms of inductance:

$$W = \frac{1}{2}Li^2 \tag{2.47}$$

By substituting for L from (2.42) and for Ni (magnetic potential) from (2.33), (2.47) can be expressed as

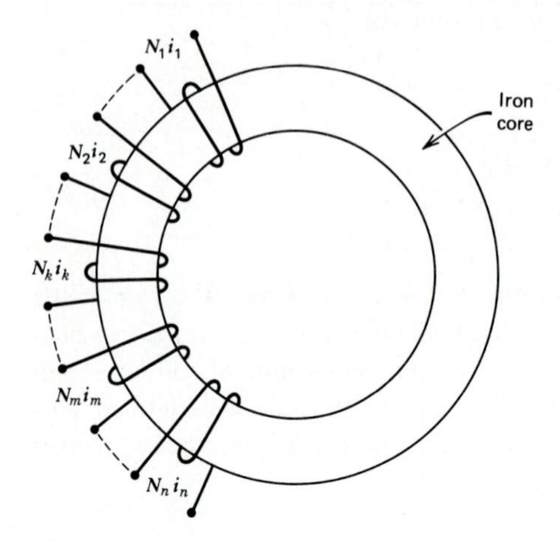

FIGURE 2-16
A toroid with n windings.

$$W = \frac{1}{2} \Re \phi^2 \tag{2.48}$$

At this point, it is interesting to compare these expressions for energy with those in terms of the field quantities as given in (2.41). The two forms are equivalent.

Example 2.6 Calculate the self-inductance of coil 1 in Example 2.3 (Fig. 2.12). To obtain a numerical value for inductance, the number of turns in the coil must be specified. Let us assume 10 turns.

Solution

$$L_{11} = \frac{N_1 \phi_1}{I_1} = \frac{(10)(0.001)}{5.04} = 0.00198 \text{ H}$$

If we should assume 100 turns,

$$L_{11} = \frac{(100)(0.001)}{0.504} = 0.198 \text{ H}$$

Note that the inductance varies with the square of the turns.

Example 2.7 Calculate the mutual inductance between coils 1 and 2 in Example 2.3, assuming 10 turns in coil 1 and 20 turns in coil 2. Since leakage was neglected in Example 2.3, $K = 1.0$; that is, all of the flux produced by coil 1 links coil 2.

$$L_{21} = L_{12} = \frac{(20)(0.001)}{5.04} = 0.00398 \text{ H}$$

Example 2.8 Calculate the inductance of coil 1 of the toroid in Fig. 2.16, assuming material is 48% Ni alloy and that the toroid mean circumference is 0.1 m and the cross-sectional area is 0.0016 m^2, with a stacking factor of 1.0; $N_1 = 10$.

Solution

From the *B-H* curve for 48% Ni alloy in Fig. 2.5, we find that the *absolute* amplitude permeability in the linear range is 0.115. Using (2.46) with $K = 1.0$, we obtain

$$L_{11} = \frac{(10^2)(0.0016)(0.115)}{0.1} = 0.184 \text{ H}$$

It is seen that the inductance of this toroid is much larger than that of the circuit of Fig. 2.12 used in Example 2.3, even though the mean length and area of the two cores and the winding turns are equal. This is attributable to the effect of the gap in the circuit of Fig. 2.12 and also to the higher permeability of the magnetic material in the toroid.

Example 2.9 Determine the inductance of the armature slot whose cross section is shown in Fig. 2.17, assuming that the lower portion of the slot of height y_2 is filled with conductors carrying a current density, J A/m^2, perpendicular to the plane of the paper. Also assume that the steel magnetic material surrounding the slot has infinite permeability. This example calculates an expression for the "slot inductance," an important component of the leakage inductance of many types of rotating machines, and will also illustrate the use of permeance coefficients and the concept of *partial flux linkages*. Furthermore, it will illustrate the use of symmetry because Fig. 2.17 represents the cross section of a portion of an armature of length l_a in the direction perpendicular to the plane of the paper. The two-dimensional analysis is possible, since only one component of the current density, J, is present. The assumption of infinite permeability of the magnetic material on the sides and bottom of the slot means that flux lines leave the sides in a perpendicular direction. Therefore the flux lines can be considered as horizontal components.

Solution

Partial flux linkage is the term used to describe flux lines that link only a portion of an electrical conductor or only a portion of the turns of a coil. In this example, a flux line across the slot at any height below y_2 links only the current below the flux line. (The path of the flux line is closed through the magnetic material, as shown.) The assumption of horizontal flux lines is not exactly correct in a practical configuration, but an analytical solution would be impossible without this assumption.

Consider the differential flux in the "strip" at a distance y from the slot bottom; the strip has height dy, width l_a (into the paper), and length t_2. The magnetic potential enclosed by this strip is

$$F_y = J t_2 y$$

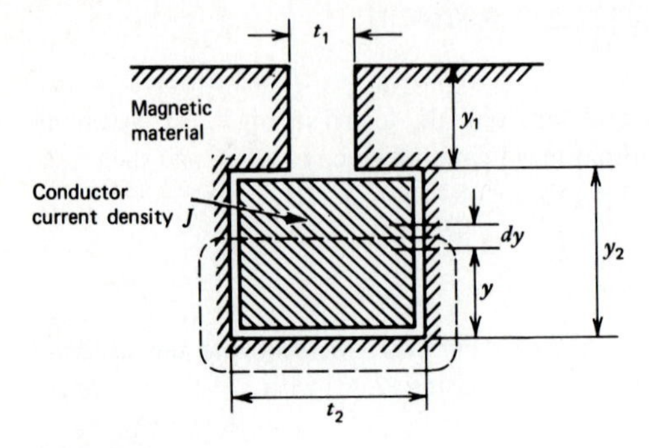

FIGURE 2-17
A slot cross section.

The permeance of the strip is

$$d\wp_y = \frac{\mu_0 l_a dy}{t_2}$$

The flux through the strip is, from (2.33),

$$d\phi = F_y d\wp_y = \mu_0 J l_a y dy$$

The total flux across the gap is

$$\phi_2 = \mu_0 J l_a \int_0^{y_2} y dy = \mu_0 J l_a \frac{y_2^2}{2}$$

The total magnetic potential of the slot is

$$F_s = J y_2 t_2$$

The permeance of the lower portion of the slot is

$$\wp_2 = \frac{\phi_2}{F_s} = \frac{\mu_0 l_a y_2}{2t_2}$$

The permeance of the upper portion of the slot is

$$\wp_1 = \mu_0 y_1 \frac{l_a}{t_1}$$

The slot inductance is

$$L_s = \wp_1 + \wp_2 = \mu_0 \left(\frac{y_1 l_a}{t_1} + \frac{y_2 l_a}{2t_2} \right) \text{H}$$

TABLE 2.4
Characteristics of permanent magnets

Type	Residual Flux Density B_r (G)	Coercive Force H_c (Oe)	Maximum-Energy Product (G-Oe x 10^6)	Average Recoil Permeability
1% Carbon steel	9,000	50	0.18	
3½% Chrome steel	9,500	65	0.29	35
36% Cobalt steel	9,300	230	0.94	10
Alnico I	7,000	440	1.4	6.8
Alnico IV	5,500	730	1.3	4.1
Alnico V	12,500	640	5.25	3.8
Alnico VI	10,500	790	3.8	4.9
Alnico VIII	7,800	1,650	5.0	—
Cunife	5,600	570	1.75	1.4
Cunico	3,400	710	0.85	3.0
Vicalloy 2	9,050	415	2.3	—
Platinum-cobalt	6,200	4,100	8.2	1.1
Barium ferrite-isotropic	2,200	1,825	1.0	1.15
Oriented type A	3,850	2,000	3.5	1.05
Oriented type B	3,300	3,000	2.6	1.06
Strontium ferrite				
Oriented type A	4,000	2,220	3.7	1.05
Oriented type B	3,550	3,150	3.0	1.05
Rare earth-cobalt	8,600	8,000	18.0	1.05
Nd FeB	10,200	9,800	25.0	1.05

2.10 MAGNETIC CIRCUITS CONTAINING PERMANENT MAGNETS

The second type of excitation source commonly used for supplying energy to magnetic circuits used in rotating machines and other types of electromechanical devices is the permanent magnet. The nature of permanent magnets has been briefly described in Section 2.2. There is obviously a great deal of difference in physical appearance between an electrical exciting coil and a permanent magnet source of excitation, so we should expect some differences in methods of analyses used in the two types of magnetic circuits. Actually, these differences are relatively minor and are related to the use of the permanent magnet itself, not to the other portions of the magnetic circuit.

An electrical excitation coil energized from a constant-voltage or constant-current source is relatively unaffected by the magnetic circuit that it excites, except during transient conditions when changes are occurring in the magnetic circuit or in the external electric circuit. Under steady-state conditions with a constant-voltage source, the current in the coil is determined solely by the magnitude of the voltage source and the dc resistance of the coil.

In a circuit excited by a permanent magnet, the operating conditions of the permanent magnet are largely determined by the external magnetic circuit. Also, the operating point and subsequent performance of the permanent magnet are a function of how the magnet is physically installed in the circuit and whether it is magnetized before or after installation. In many applications, the magnet must go through a stabilizing routine before use. These considerations are, of course, meaningless for electrical excitation sources. The details necessary to find the excitation required to establish flux density in an airgap of known dimensions are demonstrated for electrical exciting coils in Examples 2.2 and 2.3. For permanent magnet excitation, the object is to determine the size (length and cross section) of the permanent magnet. The first step in this process is to choose a specific type of permanent magnet, since each type of magnet has a unique characteristic that will partially determine the size of the magnet required. In a practical design this choice will be based on cost factors, availability, mechanical design (hardness and strength requirements), available space in the magnetic circuit, and the magnetic and electrical performance specifications of the circuit. Most permanent magnets are nonmachinable and usually must be used in the circuit as obtained from the manufacturer. Table 2.4 summarizes some of the pertinent characteristics of common permanent magnets.

Permanent magnet excitation is chosen for a specified airgap flux density with the aid of the second-quadrant *B-H* curve, often called the *demagnetization curve*, for a specific type of magnet. This curve has been introduced in Section 2.2 as Fig. 2.4. The *B-H* characteristics of a number of Alnico permanent magnets are shown in Fig. 2.18; Fig. 2.19 shows the characteristics of several ferrite magnets. Also shown on these figures are curves of *energy product*, the product of B in gauss (G) and H in oersteds (Oe), and *permeance ratio*, the ratio of B to H. The energy product is a measure of the magnetic energy that the permanent magnet is capable of supplying to an external circuit as a function of its flux density and field intensity. In general, a permanent magnet is used most efficiently when operated at conditions of B and H that result in the maximum energy product. Permeance coefficients are useful in the design of the external magnetic circuit. This parameter is actually "relative permeability" as defined previously, since μ_0 is 1.0 in the CGS system of units. The symbols B_d for flux density and H_d for field intensity are used to designate the coordinates of the demagnetization curve.

Once the permanent magnet type has been chosen, the design of the magnet's size follows the general approach taken in Section 2.4. From Ampere's law,

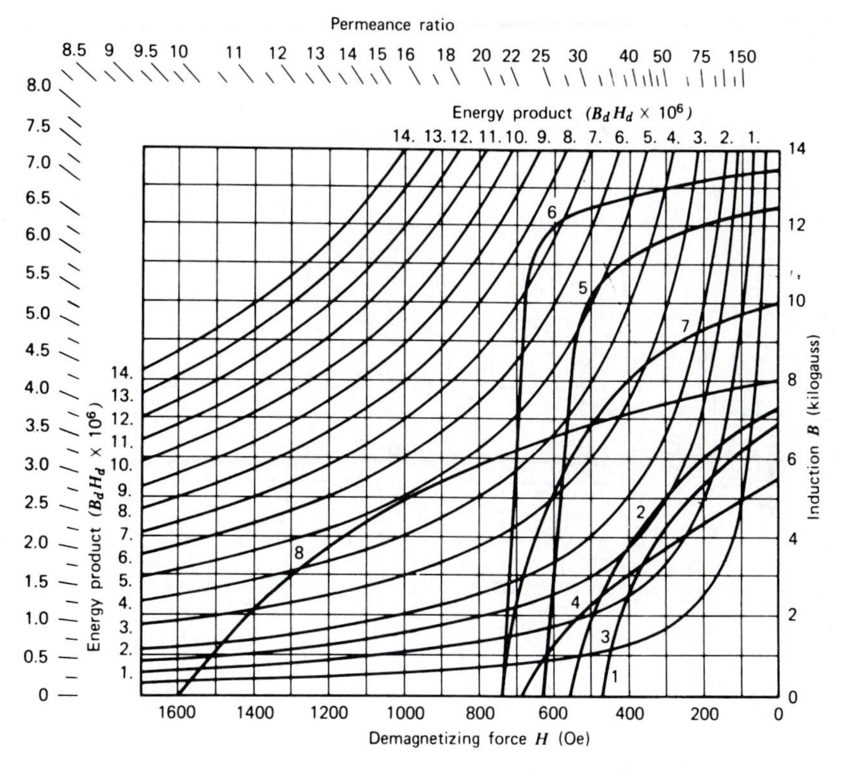

FIGURE 2-18
Demagnetization and energy product curves for Alnicos 1 to 8. Key: 1, Alnico I; 2, Alnico II; 3, Alnico III; 4, Alnico IV; 5, Alnico V; 6, Alnico V-7; 7, Alnico VI; 8, Alnico VIII; 9, rare earth-cobalt.

$$H_d l_m = H_g l_g + V_{mi} \tag{2.49}$$

where

H_d = magnetic field intensity of the magnet in oersteds

l_m = length of magnet in centimeters

H_g = field intensity in the gap = flux density in gap (in CGS units)

l_g = length of gap in centimeters

V_{mi} = reluctance drop in other ferromagnetic portions of the circuit in gilberts

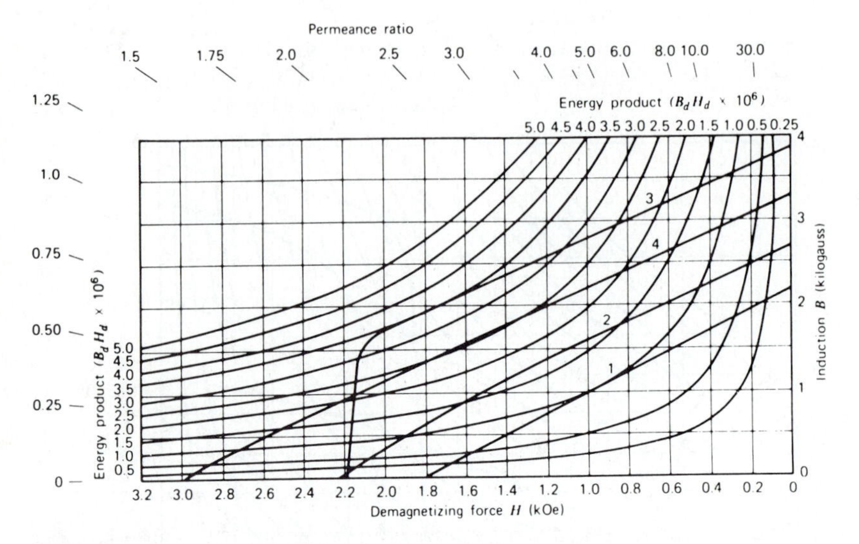

FIGURE 2-19
Demagnetization and energy product curves for Indox ceramic magnets. Key: 1, Indox I; 2, Indox II; 3, Indox V; and 4, Indox VI-A.

The cross-sectional area of the magnet is calculated from the flux required in the airgap as follows:

$$B_d A_m = B_g A_g K_1 \tag{2.50}$$

where

B_d = flux density in the magnet in gauss

A_m = cross-sectional area of magnet in square centimeters

B_g = flux density in the gap in gauss

A_g = cross-sectional area of gap in square centimeters

K_1 = leakage factor

The leakage factor, K_1, is the ratio of flux leaving the magnet to flux in the airgap. The difference between these two fluxes is the leakage flux in the regions of space between the magnet and the airgap. The leakage factor can be determined by the methods described in Sections 2.4 and 2.6 or by other standard and more accurate methods. Standard formulas are available in the literature to find the leakage factor. Two of these formulas will be illustrated in the following examples.

Example 2.10 Determine the length and cross-sectional area of the magnet in Fig. 2.20 to produce a flux density in the airgap of 2500 G. The permanent magnet to be used is Alnico V; the dimensions of Fig. 2.20 are as follows: l_g = 0.4 cm, W = 6.0 cm; and the gap area = 4.0 cm^2 (2 cm x 2 cm). We assume that the reluctance in the soft iron portions of the circuit is negligible, giving a reluctance drop, V_{mi}, of zero; we estimate that the leakage factor is 4.0 and that the magnet is to be operated at its maximum-energy product condition (knee of the demagnetization curve in Fig 2.18).

Solution

From (2.50)

$$A_m = \frac{B_g A_g K_1}{B_d} = \frac{(2500)(4.0)(4.0)}{10{,}500} = 3.8 \text{ cm}^2$$

From (2.49), noting that $H_g = B_g$ in CGS units,

$$l_m = \frac{(2500)(0.4)}{450} = 2.22 \text{ cm} = H \qquad \text{(in Fig. 2.20)}$$

The dimension l_m determines the dimensions H and G in Fig. 2.20. We must now check the assumption of leakage factor. A leakage factor for this configuration, based on experimental measurements and calculations similar to those described earlier in this chapter, is given by

$$K_1 = 1 + \frac{l_g}{A_g}\left(1.7 C_G \frac{G}{G + l_g} + 1.4 W \sqrt{\frac{C_W}{H}} + 0.67 C_H\right) \tag{2.51}$$

where

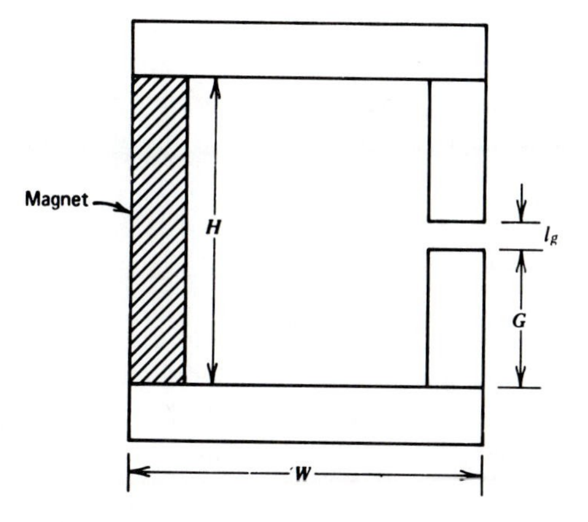

FIGURE 2-20
A magnetic circuit with a permanent magnet.

C_H = the perimeter of the cross section of the circuit of length H

C_W = the perimeter of the cross section of the circuit of length W

C_G = the perimeter of the cross section of the circuit of length G

The factor 0.67 in the third term within the parentheses in (2.51) arises from the fact that permanent magnets have a "neutral zone" that does not contribute to leakage. From the preceding length and area calculations we can determine the length parameters in (2.51) and estimate the perimeters of the various cross sections: $H = 2.22$ cm, $G = 0.91$ cm, $W = 6.0$ cm, $C_G = 8$ cm, $C_W = 8.0$ cm, and $C_H = 7.8$ cm. Substituting these values into (2.51) gives

$$K_1 = 4.062$$

This value could now be put back into (2.50), giving a slight change in magnet area A_m. In turn, this calculation may require a few changes in other dimensions used in (2.51), resulting in a new value of leakage factor. A few iterations of these formulas are usually necessary to obtain a consistent set of dimensions for the total magnetic circuit.

The high value for leakage factor obtained for the configuration of Fig. 2.20 indicates that this is not a very efficient magnetic circuit. Stated another way, the permanent magnet is in the wrong location within the magnetic circuit. The high leakage in this configuration can be readily explained by means of Ampere's law and the simple magnetic circuit theory presented earlier in this chapter. This explanation is left as a problem at the end of this chapter. A much more efficient use of the permanent magnet in this configuration is to locate it adjacent to the airgap, as shown in Fig. 2.21. The leakage factor for Fig. 2.21 is

$$K_1 = 1 + \frac{l_g}{A_g} 0.67 C_G \left(1.7 \frac{0.67G}{0.67G + l_g} + \frac{l_g}{2G} \right) \tag{2.52}$$

Using the same dimensions for all sections of the circuit in Fig. 2.21 as were used for Fig. 2.20 (even though this might result in an oversized permanent magnet) and substituting these into (2.52) gives

$$K_1 = 1.69$$

which is less than half the leakage factor found for the configuration of Fig. 2.20.

It is interesting to observe the volume of permanent magnet material required to establish a given flux in an airgap. Solving for A_m in (2.50) and for l_m in (2.49) (neglecting V_{mi}) and noting that in the CGS system $H_g = B_g$, we obtain

$$\text{vol} = A_m l_m = \frac{B_g^2 A_g l_g K_1}{B_d H_d} \tag{2.53}$$

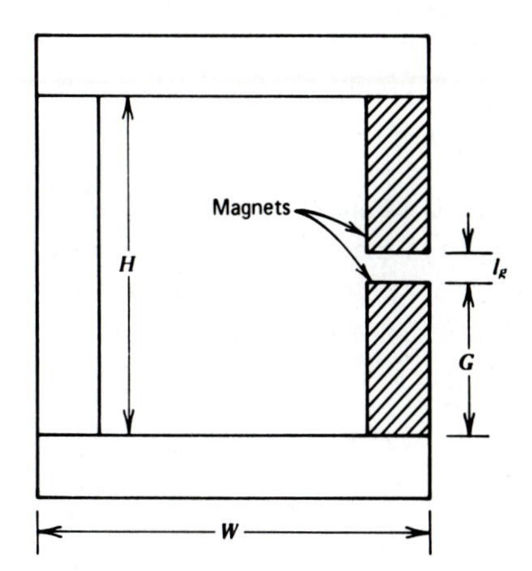

FIGURE 2-21
An efficient use of permanent magnets in a magnetic circuit.

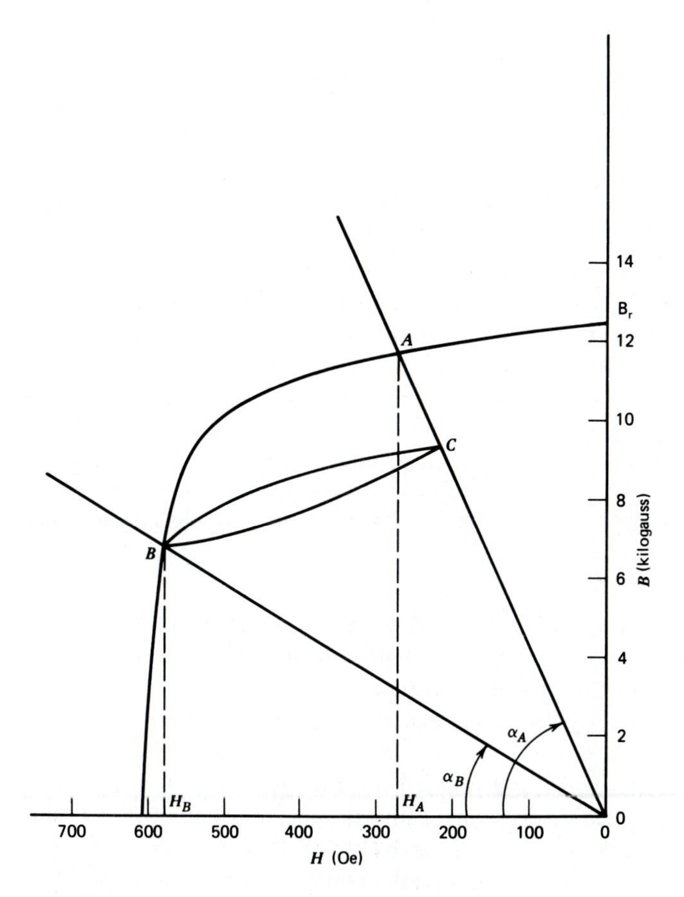

FIGURE 2-22
Second-quadrant B-H characteristic of a permanent magnet.

It is seen that magnet volume is a function of the square of the airgap flux density. The importance of the leakage factor in minimizing the required magnet size is also apparent from this equation. The denominator of (2.53) is the energy product that is a function of the permanent magnet material and the operating point on the demagnetization curve of the magnet.

The parameter, permeance ratio, shown in Figs. 2.18 and 2.19 is the ratio of the equivalent permeance of the external circuit, $A_g K_1/l_g$, to the permeance of the space occupied by the permanent magnet, A_m/l_m, in the CGS system of units. This can be seen by solving for B_d from (2.50) and for H_d from (2.49) (neglecting V_{mi}) and taking the ratio:

$$\frac{B_d}{H_d} = \frac{A_g l_m K_1}{A_m l_g} = \frac{P_{ge}}{P_M} = \tan\alpha \qquad (2.54)$$

Equation (2.54) is deceptively simple in appearance because the task of obtaining analytical expressions for K_1 is very difficult, as has been seen. Also, the reluctance drop in the soft iron portions of the magnetic circuit, V_{mi}, must be included somehow in (2.54). This is even a more difficult task, since the reluctance drop is a function of both the permanent magnet's operating point, $B_d H_d$, and the effects of leakage flux in the iron. The reluctance drop is usually introduced by means of a factor similar to the leakage factor and is based on measurements in practical circuit configurations. The various expressions that make up (2.54) are of value in observing general relationships among the magnetic parameters as the permeance of the external circuit is varied. This leads us to the second type of permanent magnet circuit.

Circuits with a varying airgap will be briefly described with the aid of Fig. 2.22. Keeping (2.54) in mind, let us observe the variation of B and H of a permanent magnet as the external circuit permeance is varied. Fig. 2.22 shows a typical second-quadrant B-H characteristic for a permanent magnet. Theoretically, it is possible to have infinite permeance in the external magnetic circuit that would correspond to $\alpha = 90°$ in (2.54), and the magnet operating point would be at $B_d = B_r$ and $H_d = 0$ in Fig. 2.22. This situation is approximated by a permanent magnet having an external circuit consisting of no airgap and a high-permeability soft iron member, often called a "keeper." In practice, however, there is always a small equivalent airgap and a small reluctance drop in the keeper, and the operating point is to the left of B_r and α is less than 90°.

For a finite airgap, the operating point will be at some point A on the B-H curve, and the permanent magnet will develop the magnetic field intensity H_A to overcome the reluctance drop of the airgap and other portions of the external magnetic circuit. If the airgap is increased, P_{ge} decreases, and the magnet must develop a larger magnetic field intensity, H_d. From (2.54), it is seen that α decreases and the operating point on Fig. 2.22 will move further to the left to some point, say B, at α_B. If the airgap is subsequently returned to its original value, the operating point will not return to A but, instead, to C. If the airgap is successively varied between the two values, the operating point will trace a "minor hysteresis loop" between B and C, as shown in Fig. 2.22. The slope of this loop is known as *recoil permeability*; since it is a slope

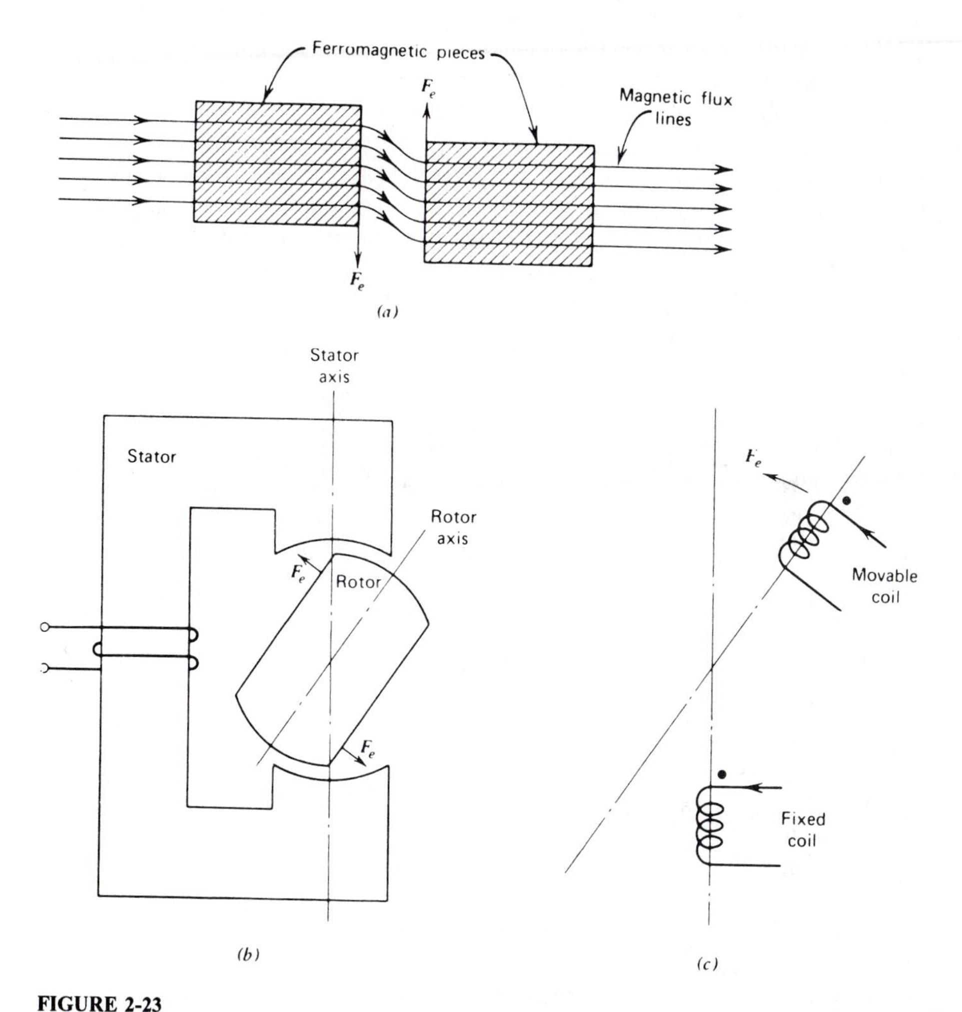

FIGURE 2-23
Magnetic field effects producing electrical force, F_e. *(a)* Alignment of ferromagnetic pieces in a magnetic field. *(b)* A reluctance motor. *(c)* Alignment of two current-carrying coils.

on the *B-H* plane, it is also sometimes called incremental permeability, as defined in (2.26). Recoil permeability is an important parameter of a permanent magnet for applications with varying airgaps; values of this parameter are given in Table 2.4 for the permanent magnets shown.

2.11 FORCES OF ELECTROMAGNETIC ORIGIN

Up to this point we have considered only static magnetic circuits. However, for electromechanical energy conversion we must have mechanical motion and inherent forces to produce this motion (as in an electric motor). The two basic magnetic field effects resulting in the production of mechanical forces are (1) alignment of flux lines, and (2) interaction between magnetic fields and current-carrying conductors. Examples of "alignment" are shown in Fig. 2.23. In Fig. 2.23 the force on the ferromagnetic pieces cause them to align with the flux lines, thus shortening the magnetic flux path and reducing the reluctance. Fig. 2.23 shows a simplified form of a reluctance motor in which electrical force tends to align the rotor axis with that of the stator. Fig. 2.23*(c)* shows the alignment of two current-carrying coils. A few examples of "interaction" are shown in Fig. 2.24, in which current-carrying conductors experience mechanical forces when placed in magnetic fields. For instance, in Fig. 2.24*(b)* a force is produced by the interaction between the flux lines and coil current, resulting in a torque on the moving coil. This mechanism forms the basis of a variety of electrical measuring instruments. Almost all industrial dc motors work on the interaction principle.

Quantitative evaluation of the mechanical force of electromagnetic origin will be considered later. Here, we have simply considered some of the principles of force

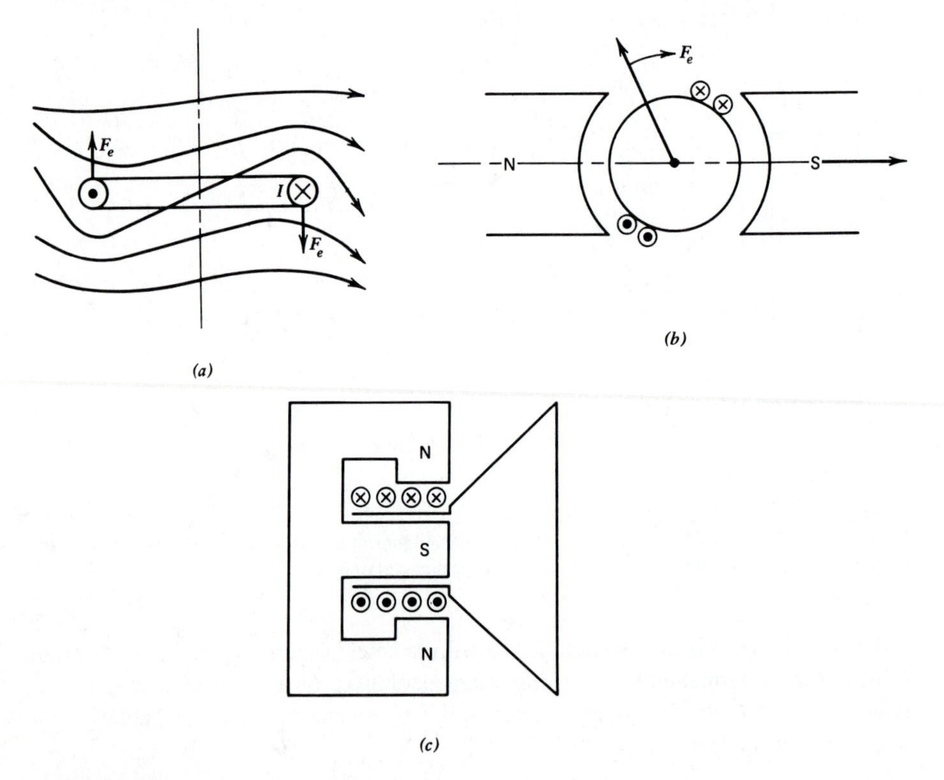

FIGURE 2-24
Electrical force produced by interaction of current-carrying conductors and magnetic fields. *(a)* A one-turn coil in a magnetic field. *(b)* A permanent magnet moving coil ammeter. *(c)* A moving coil loudspeaker.

production. We wish to point out that the force is always in such a direction that the net magnetic reluctance is reduced or the energy stored in the magnetic field isminimized. Forces of relatively small magnitude may also be produced by the deformation of a ferromagnetic material by magnetostriction, which is not included in this book.

2.12 ENERGY CONSERVATION AND ENERGY CONVERSION

We have given a few examples showing how mechanical forces are produced by magnetic fields. Clearly, for energy conversion (i.e., for doing work), mechanical motion is as important as mechanical force. Thus, during mechanical motion, the energy stored in the coupling magnetic field is disturbed. In Fig. 2.24*b,* for instance, most of the magnetic field energy is stored in the airgap separating the rotor from the stator. The airgap field may be termed the coupling field. Electromechanical energy conversion occurs when coupling fields are disturbed in such a way that the energy stored in the fields changes with mechanical motion. A justification of this statement is possible from energy conservation principles, which will enable us to determine the magnitudes of mechanical forces arising from magnetic field effects.

The energy conservation principle in connection with electromechanical systems may be stated in a number of ways. For instance, we may say that

$$\begin{array}{c}\text{input}\\ \text{electrical}\\ \text{energy}\end{array} + \begin{array}{c}\text{input}\\ \text{mechanical}\\ \text{energy}\end{array} = \begin{array}{c}\text{increase}\\ \text{in stored}\\ \text{energy}\end{array} + \begin{array}{c}\text{energy}\\ \text{dissipated}\\ \text{as heat}\end{array} \tag{2.55}$$

or

$$\begin{array}{c}\text{input}\\ \text{electrical}\\ \text{energy}\end{array} = \begin{array}{c}\text{mechanical}\\ \text{work}\\ \text{done}\end{array} + \begin{array}{c}\text{increase}\\ \text{in stored}\\ \text{energy}\end{array} + \begin{array}{c}\text{energy}\\ \text{dissipated}\\ \text{as heat}\end{array} \tag{2.56}$$

Or, if only the conservative (or lossless) portion of the system is considered, we have

$$\text{sum of input energy} = \text{change in stored energy} \tag{2.57}$$

or

$$\begin{array}{c}\text{input}\\ \text{electrical}\\ \text{energy}\end{array} = \begin{array}{c}\text{mechanical}\\ \text{work}\\ \text{done}\end{array} + \begin{array}{c}\text{increase}\\ \text{in stored}\\ \text{energy}\end{array} \tag{2.58}$$

Fig. 2.25 gives a schematic representation of the separation of the conservative part of a system from its dissipative portion. The total stored energy is the sum of the energy stored in the electric and magnetic fields. For practical purposes, however, the energy stored in the electric field is almost negligible.

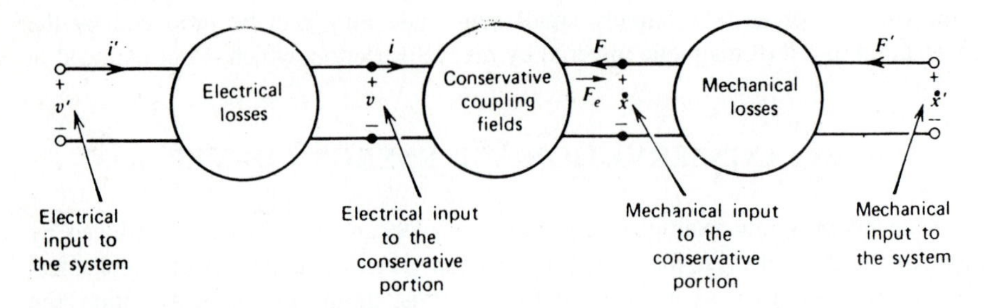

FIGURE 2-25
A representation of an electromechanical system.

2.13 THE FORCE EQUATION

From the preceding remarks we see that energy conversion is possible because of the interchange between electrical and mechanical energy through the coupling fields. This fact leads us to a method of determining mechanical forces that arise from changes in stored energy. Thus, referring to Fig. 2.26, (2.58) implies that

$$F\,dx + vi\,dt = dW \tag{2.59}$$

where

$F\,dx$ = mechanical energy input

$vi\,dt$ = electrical energy input

dW = increase in stored energy.

Now, if F_e is the force of electrical origin and acts against F (Fig. 2.25), that is, $F\,dx = -F_e\,dx$, and if dW_m is the energy stored in the magnetic field (energy stored in the electric field being negligible), (2.59) may be rewritten as

$$F_e\,dx = -dW_m + vi\,dt \tag{2.60}$$

From Faraday's law, voltage v may be expressed in terms of the flux linkage λ as

$$v = \frac{d\lambda}{dt} \tag{2.61}$$

so that (2.60) becomes

$$F_e\,dx = -dW_m + i\,d\lambda \tag{2.62}$$

Before continuing, we should point out the (2.62) is a restatement of (2.58).

In an electromechanical system either (i, x) or (λ, x) may be considered as independent variables. If we consider (i, x) as independent, the flux linkage λ is given by $\lambda = \lambda(i, x)$, which can be expressed in terms of small changes as

$$d\lambda = \frac{\partial \lambda}{\partial i} di + \frac{\partial \lambda}{\partial x} dx \tag{2.63}$$

Also, we have $W_m = W_m(i, x)$, so that

$$dW_m = \frac{\partial W_m}{\partial i} di + \frac{\partial W_m}{\partial x} dx \tag{2.64}$$

Thus (2.63) and (2.64), when substituted into (2.62), yield

$$F_e dx = -\frac{\partial W_m}{\partial x} dx - \frac{\partial W_m}{\partial i} di + i\frac{\partial \lambda}{\partial x} dx + i\frac{\partial \lambda}{\partial i} di$$

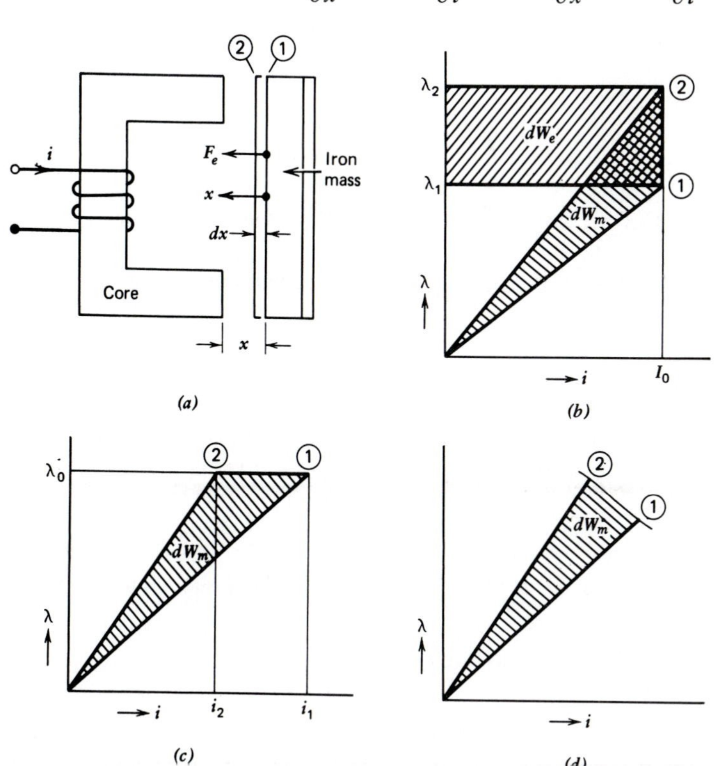

FIGURE 2-26

Energy balance in an electromechanical system. *(a)* A simple system. *(b)* Constant-current operation. *(c)* Constant-voltage (or flux linkage) operation. *(d)* A general case.

or

$$F_e dx = \left(-\frac{\partial W_m}{\partial x} + i\frac{\partial \lambda}{\partial x} \right) dx + \left(-\frac{\partial W_m}{\partial i} + i\frac{\partial \lambda}{\partial i} \right) di \tag{2.65}$$

Because the incremental changes di and dx are arbitrary, F_e must be independent of these changes. Thus, for F_e to be independent of di, its coefficient in (2.65) must be zero. Consequently, (2.65) becomes

$$F_e = -\frac{\partial W_m}{\partial x}(i, x) + i\frac{\partial \lambda}{\partial x}(i, x) \tag{2.66}$$

which is the force equation and holds true if i is the independent variable. To show that the coefficient of di in (2.65) is always zero, that is,

$$-\frac{\partial W_m}{\partial x} + i\frac{\partial \lambda}{\partial x} = 0 \tag{2.67}$$

we recall from (2.61)

$$W_m = \int i d\lambda = i\lambda - \int \lambda di \tag{2.68}$$

Substituting (2.68) into (2.67) shows that (2.67) is always satisfied. Hence (2.66) is valid.

If, on the other hand, λ is taken as the independent variable, that is, if $i = i(\lambda, x)$ and $W_m = W_m(\lambda, x)$, then

$$dW_m = \frac{\partial W_m}{\partial \lambda} d\lambda + \frac{\partial W_m}{\partial x} dx$$

which, when substituted into (2.62) gives

$$F_e dx = -\frac{\partial W_m}{\partial x} dx - \frac{\partial W_m}{\partial \lambda} d\lambda + i d\lambda \tag{2.69}$$

But (2.68) implies that $\partial W_m / \partial \lambda = i$, so (2.69) finally becomes

$$F_e = -\frac{\partial W_m}{\partial x}(\lambda, x) \tag{2.70}$$

Observe that in view of (2.68), the coefficient of $d\lambda$ in (2.69) is zero.

We notice that the preceding derivation assumes either i or λ as an independent variable. The case in which i is the independent variable corresponds to a current-excited system. For an electromechanical system such as that shown in Fig. 2.26*(a)*, the current in the coil is held constant at I_0 during the period that the armature undergoes motion from position 1 to position 2. During this motion, the flux linkage changes from λ_1 to λ_2 and the electrical energy input becomes $dW_e = I_0(\lambda_2 - \lambda_1)$. The electrical energy comes from the current source, as shown in Fig. 2.26(*b*). Also during the process, the increase in field energy is $dW_m = \frac{1}{2}I_0(\lambda_2 - \lambda_1)$.

Thus, from (2.59) we have

$$dW_e + F dx = dW_m$$

or

$$I_0(\lambda_2 - \lambda_1) + F dx = \frac{1}{2}I_0(\lambda_2 - \lambda_1)$$

or

$$F dx = -\frac{1}{2}I_0(\lambda_2 - \lambda_1) \tag{2.71}$$

where F may be considered as an externally applied mechanical force and the negative sign is associated with dx, indicating that motion is against the positive x direction. Clearly, the right side of (2.71) is negative, $\lambda_2 > \lambda_1$; and $F\,dx = -F_e\,dx$. Thus (2.71) becomes

$$-F_e dx = -\frac{1}{2}I_0(\lambda_2 - \lambda_1) \tag{2.72}$$

indicating that for a current-excited system the electrical energy input divides equally between increasing stored energy and doing mechanical work.

Next we consider the case in which the flux linkage is kept constant at λ_0 and the current is allowed to vary from i_1 to i_2 ($i_2 < i_1$), during motion, as shown in Fig. 2.26*(c)*. In this case there is no electrical energy input from the source, as may be seen by comparing Fig. 2.26*(b)* and 2.26*(c)*. The change in stored energy is

$$dW_m = -\frac{1}{2}\lambda_0(i_2 - i_1)$$

which is negative. Thus, from (2.59), we get

$$F dx = dW_m$$

or

$$F dx = -\frac{1}{2}\lambda_0(i_2 - i_1)$$

or

$$-F_e dx = -\frac{1}{2}\lambda_0 (i_2 - i_1) \tag{2.73}$$

indicating that the mechanical work done equals the reduction in stored energy.

In the preceding theoretical discussions we have considered that either i or λ remains constant. In reality, however, neither condition holds, and the change from position 1 to 2 follows a path such as that in Fig. 2.26(*d*). Nevertheless, the principles of energy conservation are still useful in determining the electrical force.

Reconsidering (2.66) and (2.69), neglecting saturation, we have

$$W_m = \frac{1}{2}\lambda i = \frac{1}{2}Li^2$$

so that

$$F_e = \frac{1}{2} i^2 \frac{\partial L}{\partial x} = \frac{1}{2}\lambda \frac{\partial i}{\partial x} \tag{2.74}$$

The Concept of Coenergy and the Force Equation

Consider the conservative system of Fig. 2.26(*a*). Let electrical energy be supplied to the input terminals and let there be no mechanical motion. In this case, the entire energy supplied is stored in the magnetic field, and this energy is

$$W_m = \int_0^{\lambda} i \, d\lambda'$$

This, when integrated by parts, yields

$$W_m = i\lambda - \int_0^{i} \lambda' \, di \tag{2.75}$$

The quantity $\int_0^{i} \lambda' \, di$ in (2.75) is called *magnetic coenergy* W_m'. Thus (2.75) can be written as

$$i\lambda = W_m + W_m' = \int_0^{\lambda} i \, d\lambda' + \int_0^{i} \lambda' \, di \tag{2.76}$$

For a nonlinear magnetic circuit, W_m and W_m' are graphically depicted in Fig. 2.27. Clearly, for a linear magnetic circuit, magnetic energy and coenergy are equal. Following the procedure in deriving the force equations (2.66) and (2.70), it may be shown that in terms of coenergy, the force equations become

$$F_e = \frac{\partial W_m'}{\partial x}(i, x) \tag{2.77}$$

$$F_e = \frac{\partial W_m'}{\partial x}(\lambda, x) - \lambda \frac{\partial i}{\partial x}(\lambda, x) \tag{2.78}$$

A comparison of (2.66) and (2.77) indicates that, with current as the independent variable, (2.77) is simpler than (2.66).

We will illustrate the application of the force equation by the following examples.

Example 2.11 In an electromechanical system (Fig. 2.28) the current, flux linkage, and position are related by

$$i = \lambda^2 + 2\lambda(x - 1)^2, \quad x < 1$$

Find the force on the iron mass at $x = 0.5$.

Solution

Recall that $W_m = \int i\,d\lambda$ and, from the given expression for i, we get

$$W_m = \frac{1}{3}\lambda^3 + \lambda^2(x - 1)^2$$

Consequently, the electrical force calculated from (2.70) becomes

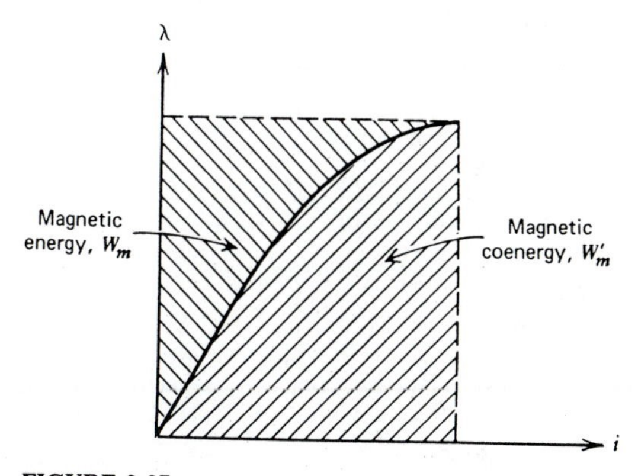

FIGURE 2-27
Magnetic energy and coenergy.

$$F_e = -2\lambda^2(x - 1)$$

The same result can also be achieved by evaluating $\partial i/\partial x$ as

$$\frac{\partial i}{\partial x} = 4\lambda(x - 1)$$

and substituting this into (2.74) we obtain

$$F_e = -2\lambda^2(x - 1)$$

At $x = 0.5$, $F_e = \lambda^2$, which indicates that the force is proportional to the square of the voltage—if leakage is neglected.

2.14 CURRENT VARIATIONS

In a voltage-excited system it is interesting to investigate the variation of the input current as a function of time. The following discussion is meant to be of a qualitative nature only.

Consider the system shown in Fig. 2.28. With no applied voltage the movable iron (armature) is, let us say, at a distance x_0 for the core. The corresponding inductance is L_0—the minimum value of the inductance—and the time constant τ_0 is L_0/R. If the iron is held in the original position and a step voltage is applied, the circuit behaves like an R-L circuit, with time constant τ_0. Final current is V/R, where V is the applied torque. If, however, the iron is allowed to move and its final position is x_f, the inductance of the circuit increases to L_f and the corresponding time constant

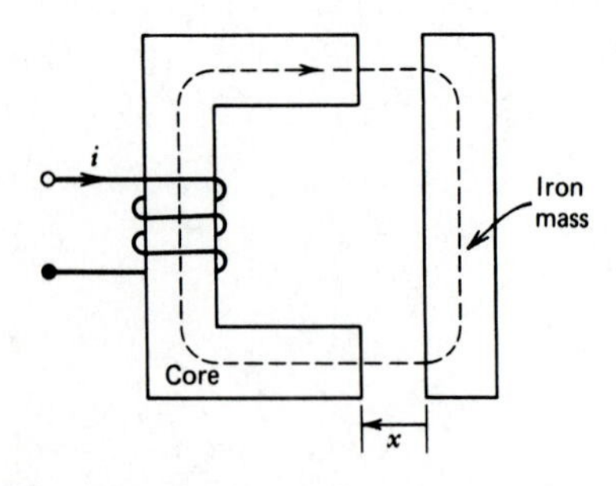

FIGURE 2-28
An electromechanical system.

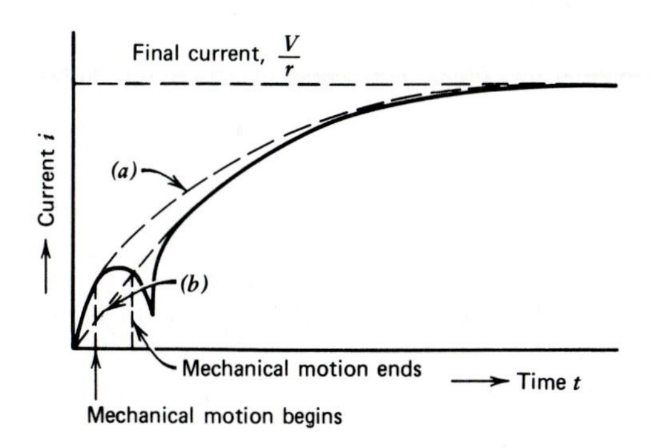

FIGURE 2-29
i(t) for a step input voltage.

is $\tau_f = L_f/R$. It is evident that $\tau_f > \tau_0$. For the initial and final positions of the movable iron, the currents are shown, respectively, by curves *(a)* and *(b)* in Fig. 2.29. However, the transition from curves *(a)* and *(b)* is not smooth because as soon as the iron starts moving, the mechanical time constant τ_m of the system comes into play. It should be recognized that $\tau_m > \tau_f > \tau_0$. This explains the nature of the current variation when the movable iron is allowed to move at a constant voltage.

In case of constant-current excitation, the variations of flux are of interest. For initial position x_0 the reluctance is maximum and the corresponding flux ϕ_0 is a minimum. The flux reaches its maximum value ϕ_f when the motion is complete. The variation of the flux is governed by the mechanical as well as by the electrical time constants, as in the preceding case.

PROBLEMS

2.1. From the magnetic material characteristics shown in Fig. 2.5, determine the relative amplitude permeability at a flux density of 1.0 T for the following materials: *(a)* Deltamax, *(b)* M-19, and *(c)* AISI 1020.

2.2. It has been stated (Section 2.3) that the area enclosed by the dc hysteresis loop of a magnetic material is equal to the hysteresis loss of that material. Referring to Fig. 2.3, derive the SI units of this loss from the coordinates of the hysteresis loop.

2.3. From Fig. 2.5, determine the relative differential permeability for the material 48 NI at a flux density of 0.8 T.

2.4. The 3% silicon steel designated M-19 in Fig. 2.5 is widely used in small power transformers, induction motors, and in other electromagnetic devices in electronic circuits. In this chapter it is stated that there is a portion of the *B-H* characteristic over which the relationship is approximately linear. This does not necessarily show up in Fig. 2.5, since

relationship is approximately linear. This does not necessarily show up in Fig. 2.5, since the curves are plotted on semilog paper in order to describe the characteristic over a wider range. Replot the M-19 curve on rectangular coordinate paper and determine the range of B over which the characteristic can be considered linear. What is the relative differential permeability in this range?

2.5. Derive an equation to describe the straight-line or linear portion of the B-H curve for M-19 obtained in Problem 2.4.

2.6. If you needed to use the B-H characteristics of M-19 in a digital computer design program for an electronics transformer, how would you model this characteristic, including saturation and initial characteristics?

2.7. Discuss the physical significance of the first Maxwell equation on the nature of lines of magnetic flux (magnetic streamlines).

2.8. A certain airgap has an area of 1 in^2 and a length of 1 mm. What are the reluctance and permeance of this gap in SI units?

2.9. A flux of 0.02 Wb exists in the airgap of Problem 2.8. Assuming no fringing effects, what is the magnetic potential (mmf) across the gap?

2.10. Assume a region of space described by conventional cylindrical coordinates r, ϕ, and z. An *infinitely* long solenoid of radius r_1 extends along and is centered about the z-axis. The solenoid has N turns/m of length and carries a current I A. Neglect the small z component of current and assume that the current is purely "circular" at any finite value of z. At any point in space (say at r_a, ϕ_a, z_a), with which dimensional coordinate does the magnetic field intensity H vary? What component of H exists at any point in space? Answer these questions for points both within and outside of the radius r_1.

2.11. For the long solenoid of Problem 2.10, show that $H_z = NI$ for $r < r_1$.

2.12. A toroid of mean radius 0.1 m is constructed of 0.14-mm strips of 48 NI magnetic material. The flux density in the magnetic material is 0.6 T. Including the effects of "stacking factor," determine the current in a 100-turn coil required to maintain this flux density.

2.13. The toroid of Problem 2.12 is cut to have an airgap 1 mm in length. What current is required in the coil to maintain the same flux density as in Problem 2.12? Neglect leakage and fringing.

2.14. For the magnetic circuit shown in Fig. 2.30, determine the mmf of the exciting coil required to produce a flux density of 1.6 T in the airgap. Neglect leakage and fringing. In Fig. 2.30, the mean lengths of portions of the magnetic circuit of constant cross-sectional area are l_{m1} = 45 cm, A_{m1} = 24 cm^2, l_{m2} = 8 cm, A_{m2} = 16 cm^2; l_g = 0.08 cm, and A_g = 16 cm^2. The material is M-19.

2.15. For the magnetic circuit shown in Fig. 2.31, determine the current in a 10-turn coil to establish a flux in the airgap of 10^{-3} Wb. Assume a fringing factor of 1.09 and neglect the leakage flux.

2.16. Fig. 2.32 shows the cross section of a portion of the magnetic circuit found in certain types of reluctance stepper motors. The rectangular cross sections at the bottom of the

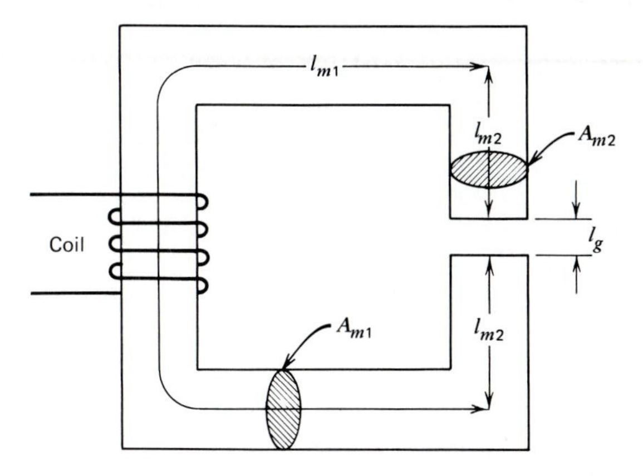

FIGURE 2-30
Problem 2.14.

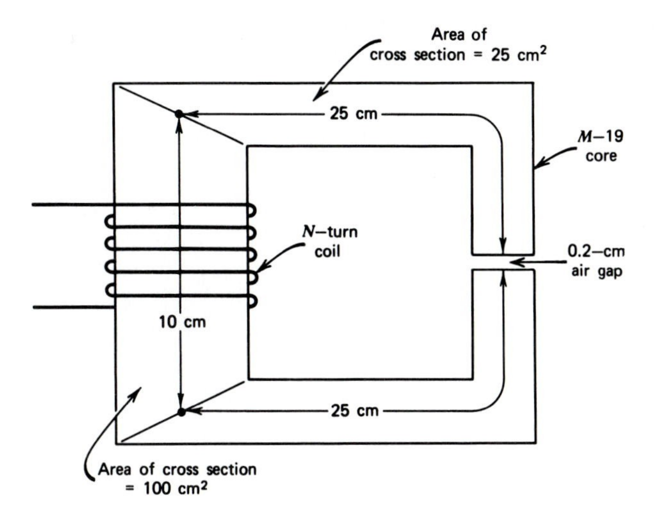

FIGURE 2-31
Problem 2.15.

magnetic circuit are cross sections of magnetic sectors that are located on rotating discs. There are three airgaps, all equal in length, separating these disc sectors from each other and from the U-shaped "return" magnetic circuit. The symbol ϕ_l represents the net equivalent of all leakage flux across the air spaces between the sides of the U magnet and can be assumed to be concentrated at the location in the magnetic circuit shown in Fig. 2.32. Circuit dimensions are as follows:

Gaps: $l_g = 1$ mm, $A_g = 4$ cm^2

Rotor sectors: $l_r = 1$ cm, $A_r = 4$ cm^2, material: 48 NI

U section: l_{m1} = 5 cm, A_{m1} = 4 cm^2, l_{m2} = 12 cm, A_{m2} = 6 cm^2, material: M-19

For the condition when the airgap flux density equals 1.2 T, it was found that ϕ_1 = 0.0003 Wb. For this condition,

(*a*) Find the magnetic potential across the equivalent leakage path.
(*b*) Determine the permeance in SI units of the equivalent leakage path.
(*c*) Determine the flux through the exciting coil.
(*d*) Determine the current required in a 10-turn coil to produce this flux.
(*e*) Calculate the "leakage inductance" of the coil.
(*f*) Draw the equivalent dc circuit (in the manner of Fig. 2.13) for this magnetic circuit.

2.17. Referring to the two permanent magnet configurations shown in Figs. 2.20 and 2.21, explain the large difference in leakage flux between these two configurations. As an aid in this analysis, assume a relative magnetic potential, V_m, across the airgap of, say 1.0, in each circuit, and relate this to the estimated potential between the top and bottom sections of the circuit (of width W).

2.18. As a result of the analysis made in Problem 2.17, where should the exciting coil in the circuit of Fig. 2.12 (assuming only one coil required) be located to minimize leakage flux?

2.19. A nonmagnetic toroid has a square cross section of the dimensions 2 cm x 2 cm. The toroid inner diameter is 8 cm and the outer diameter is 12 cm. A 1200-turn coil is wound uniformly and tightly about the toroid. Assuming zero leakage flux and a uniform flux distribution over the cross section of the toroid:

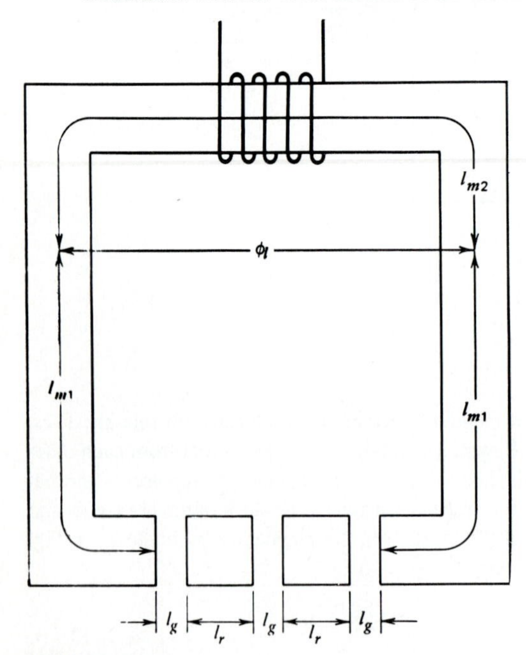

FIGURE 2-32
Problem 2.16.

(a) Calculate the flux density at the mean diameter (center of the toroid's cross section).

(b) Calculate the inductance of the coil.

2.20. Repeat Problem 2.19 without the assumption of uniform flux distribution over the toroid's cross section.

2.21. A 1-turn coil carries a current of 5 A. Determine the inductance of this coil if:

(a) The total energy stored in the magnetic field is 0.01 J.

(b) The total flux enclosed by the coil is 0.005 Wb.

2.22. Fig. 2.33 shows the cross section of a typical solenoid used as a mechanical actuator in many industrial applications, such as a hydraulic valve control. The top section of the plunger is composed of nonmagnetic material. Magnetic force for moving the plunger is created in the airgap region at the bottom of the plunger. The nonmagnetic regions, where the exciting coil is located, is known as the "window" of the solenoid. The "fill factor" of a solenoid or transformer is the ratio of the total cross-sectional area of the copper (or aluminum) in the coil to the window area.

(a) For the dimensions shown, how many turns of No. 24 AWG copper magnet wire can be used in the exciting coil if the fill factor is 75%? (See Appendix II for wire size.)

(b) The airgap shown in Fig. 2.33 is the initial airgap for the solenoid. Determine the current in the coil chosen in part *(a)* to establish a flux density of 1.0 T in the airgap beneath the plunger. Neglect leakage and fringing. The material is M-19.

(c) In the use of solenoids, the speed of response of the plunger is often important. The electrical circuit time delay is related to the time constant, τ, of the *R-L* circuit of the exciting coil. Assuming that the coil is energized from a constant-voltage source, that a fixed "fill factor," airgap flux density, and length are maintained, will changing the number of turns in the exciting coil affect the time constant?

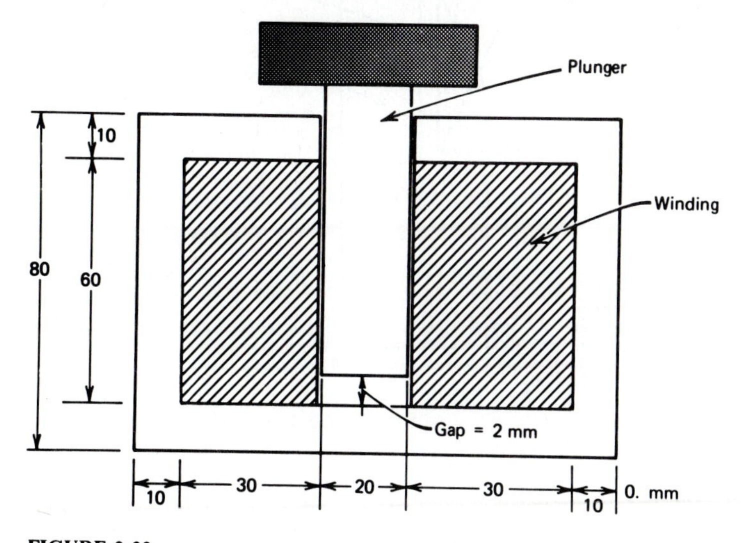

FIGURE 2-33
Problem 2.22. Solenoid cross section. All dimensions in millimeters. Plunger and magnetic housing 40 mm perpendicular to the plane of paper.

(*d*) Assuming the conditions fixed as in part (*c*), will changing the magnetic material from M-19 to Deltamax change the time constant?

2.23. A magnetic circuit of the configuration of Fig. 2.20 is to be constructed of soft iron members and Indox V ferrite permanent magnets. The dimensions are l_g = 1 cm, W = 5 cm, gap area, A_g = 3.5 cm^2. If the flux density in the airgap is to be 0.1 T, determine the length and cross-sectional area of the Indox V magnets. Assume K_1 = 1.5 and neglect fringing.

2.24. Design a permanent magnet excitation for the magnetic circuit of Fig. 2.30 (Problem 2-14) using any of the permanent magnet characteristics described in this chapter. The magnet will replace portions of the soft iron members in Fig. 2.30. Are all types of permanent magnets applicable to this problem?

2.25. For the toroid of Problem 2.12, calculate the potential energy stored at the flux density used in this problem. In Problem 2.13 calculate the potential energy stored in the gap and in the magnetic material at the given flux density. Compare and explain the differences in energy storage between the gapped and nongapped toroids.

2.26. Calculate the inductances of the gapped and nongapped toroids of Problems 2.12 and 2.13 with 100-turn exciting coils. Compare and explain the differences in inductance values between the two configurations.

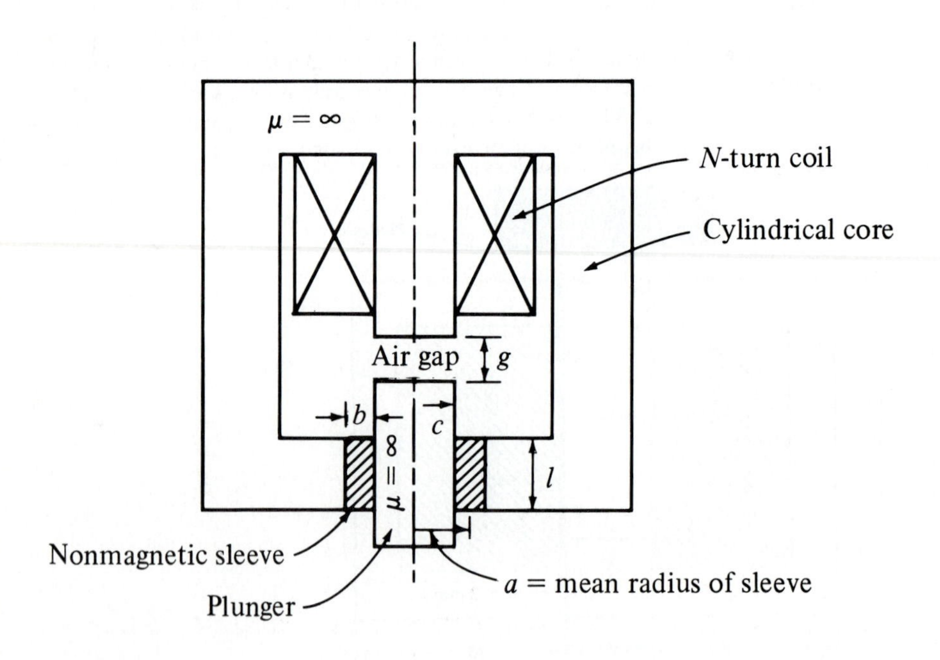

FIGURE 2-34
Problem 2.28.

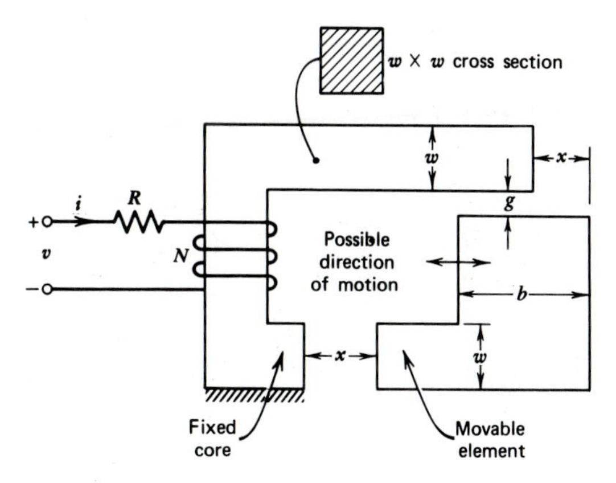

FIGURE 2-35
Problem 2.30.

2.27. Calculate the potential energy stored in the three sections of the magnetic circuit of Problem 2.16 (airgap and two magnetic sectors).

2.28. A solenoid of cylindrical geometry is shown in Fig. 2.34. For the numerical values g = 5 mm, a = 2 cm, b = 2 mm, and l = 4 cm, determine the electromagnetic force on the plunger, if the coil is 10 A dc. Take N = 500 turns.

2.29. For the solenoid of Problem 2.28, calculate the ac coil current at 60 Hz if it is required that the time-average force on the plunger be 600 N.

2.30. The singly excited electromechanical system shown in Fig. 2.35 is constrained to move only horizontally. The pertinent dimensions are shown in the diagram. Determine the electrical force exerted on the movable iron member for:

(a) Current excitation:

$$i = I \cos \omega t$$

(b) Voltage excitation:

$$v = V \cos \omega t$$

For parts *(a)* and *(b)*, neglect the winding resistance, leakage fields, and fringing. Assume that all energy is stored in the airgaps; that is, the permeability of iron is very large compared with that of free space. What modifications have to be made if the winding resistance is not negligible?

2.31. An electromagnetic structure is characterized by the following Θ-dependent inductances.

$$L_{11} = 2 + \cos 2\Theta = L_{22}$$

$$L_{12} = 1 + 0.5\cos\Theta = L_{21}$$

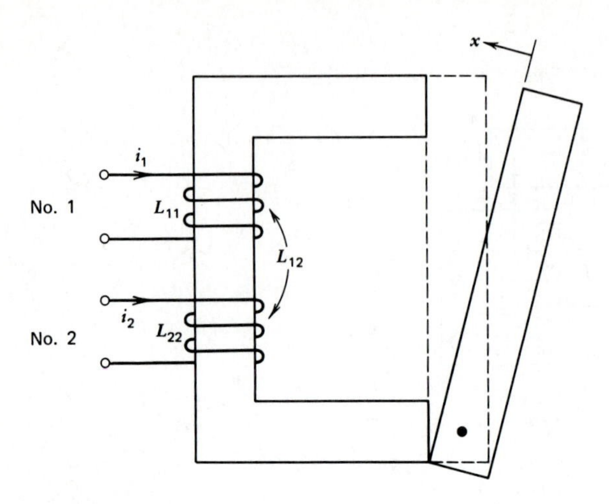

FIGURE 2-36
Problem 2.32.

Assume winding resistance to be zero. Find the torque (as a function of Θ) when both windings are connected to the same source, so that $v_1 = v_2 = 155 \sin 377t$.

2.32. The system shown in Fig. 2.36 carries two coils having self- and mutual inductances L_{11}, L_{12}, and L_{22}. Coil 1 carries a current $i_1 = I_1 \sin \omega_1 t$; coil 2 carries a current $i_2 = I_2 \sin \omega_2 t$. The inductances are $L_{11} = k_1/x$, $L_{22} = k_2/x$, and $L_{12} = k_3/x$; k_1, k_2, and k_3 are constants. Derive an expression (in integral form) for the average force. Find a relation between ω_1 and ω_2 for *(a)* maximum average force, and *(b)* minimum average force. Determine the maximum, minimum, and intermediate values of the average force.

2.33. The (λ,i) relationship for the system of Fig. 2.28 is given by

$$i = 3\lambda^2 + 2\lambda(x - 2)^2$$

where x = airgap clearance. Determine the magnetic energy stored at $x = 1$. What is the value of the electromagnetic force at $x = 1$ if $\lambda = N\phi = 12$?

2.34. For the (λ,i) relationship given in Problem 2.33, evaluate the magnetic coenergy stored at $x = 1$ and, using the concept of coenergy, determine the value of the force developed at $x = 1$ and $\lambda = 12$. Hence show that the result is consistent with that of Problem 2.33.

2.35. The inductances of two mutually coupled coils are: $L_{11} = A = L_{22}$ and $L_{12} = L_{21} = -B \cos \theta$, where A and B are constants. Find the developed electromagnetic torque for the following cases: *(a)* $i_1 = I_0$, $i_2 = 0$; *(b)* $i_1 = i_2 = I_0$; *(c)* $i_1 = I_m \sin \omega t$, $i_2 = I_0$; *(d)* $i_1 = i_2 = I_m \sin \omega t$; *(e)* $i_1 = I_m \sin \omega t$, $i_2 = I_m \cos \omega t$; and *(f)* $i_1 = I_0$ and coil 2 short-circuited. In these cases, I_0 = a constant dc current.

2.36. In an electromagnet the i-λ relationship is

$$i = \lambda^3 + \lambda x$$

where x is some arbitrary airgap. If λ is the independent variable, *(a)* find an expression for the electromagnetic force at $x = 0.5$ and at $x = 1$. *(b)* Does the force change with x? Explain.

2.37. Repeat Problem 2.36 using the concept of coenergy.

2.38. Using the expression for the electromagnetic force given by (2.74) of the text, show that in order for the same time-average force to be produced by an ac excitation $i_{ac} = i(t)$ as that produced by a dc excitation $i_{dc} = I_{dc}$, it is necessary that the root-mean square (rms) value of the alternating current be equal to the magnitude of the direct current.

2.39. Two coils have self- and mutual inductances (in henries) as functions of displacement x (in meters) as follows:

$$L_{11} = L_{22} = 3 + \frac{2}{3x} \qquad \text{and} \qquad L_{12} = L_{21} = \frac{1}{3x}$$

The resistances are negligible. Both coils are excited by the same voltage source $v = v_1 = v_2 = 100 \cos 50t$ volts.

(a) Find an expression for the electrical force (i.e., the force of electrical origin).
(b) Calculate the time-average value of the force at $x = 1$.
(c) Does the force tend to increase or decrease x?

2.40. The system shown in Fig. 2.28 has a reluctance of the form $\Re = a + bx$, where a and b are constants. The coil has N turns and is excited by a voltage source $v = V_m \sin \omega t$. If the coil resistance is R, find the instantaneous and average values of the electrical force.

CHAPTER 3

TRANSFORMERS

In this chapter we will begin to apply the general principles and methods of analysis developed in Chapter 2 to a specific electromagnetic device, the transformer. Since the transformer has a relatively simple electromagnetic structure, it will be useful for illustrating these principles and for developing relationships that will be of value later in the analysis of more complex electromagnetic structures. The transformer is extremely important as a component in many different types of electric circuits, from small-signal electronic circuits to high-voltage power transmission systems. A knowledge of the theory, design relationships, and performance capabilities of transformers is essential for understanding the operation of many electronic, control, and power systems. Therefore, both as a vehicle for understanding some basic electromagnetic principles and as an important component of electric systems, the transformer deserves serious study.

The most common functions of transformers are (1) changing the voltage and current levels in an electric system, (2) impedance matching, and (3) electrical isolation. The first of these functions is probably best known to the reader and is typified by the distribution transformer on the nearby electric pole that steps down the voltage on the distribution lines from, say, 2300 V to the household voltage of 115/230 V. The second function is found in many communication circuits and is used, for example, to match a load to a line for improved power transfer and minimization of standing waves. The third feature is used to eliminate electromagnetic noise in many types of circuits, blocking dc signals, and user safety in electric instruments and appliances.

Transformers are used in circuits of all voltage levels, from the microvolt level of some electronic circuits to the highest voltage used in power systems which, today, is approximately 765 kV. In some pulse applications, even higher voltages may exist. Also, transformers are applied throughout the entire frequency spectrum found in electric circuits, from near dc to hundreds of megahertz, with both continuous sinusoidal and pulse waveforms. The physical size and shape of transformers is also very varied, and transformers come in sizes from not much bigger than a pea up to the size of a small house.

3.1 CONSTRUCTION AND ELECTROMAGNETIC STRUCTURE

The magnetic structure of transformer consists of one or more electrical windings linked or coupled together magnetically by a magnetic circuit or core. The magnetic circuit of most transformers is constructed of a magnetic material, but nonmagnetic materials—often called "air cores"—are found in some applications. Also, in some special applications, the magnetic circuit may consist of magnetic material in series with an air gap. When there are more than two windings on a transformer, two of the windings are usually performing the identical functions. Therefore, in terms of understanding the theory and performance of a multiwinding transformer, only the relationships in two of the windings need to be considered. The autotransformer, with only one winding, will be treated later. The two basic windings are often called primary and secondary. The meaning usually attached to this nomenclature is that the input or source energy is applied to the primary windings and the output energy is taken from the secondary winding. However, since a transformer is a bilateral device and is often operated bilaterally, this meaning is not very significant and these words are used more as a way to distinguish the two windings. It is more common to designate the windings by numbered subscripts or as high-voltage and low-voltage windings. A simple two-winding transformer model is shown in Fig. 3.1.

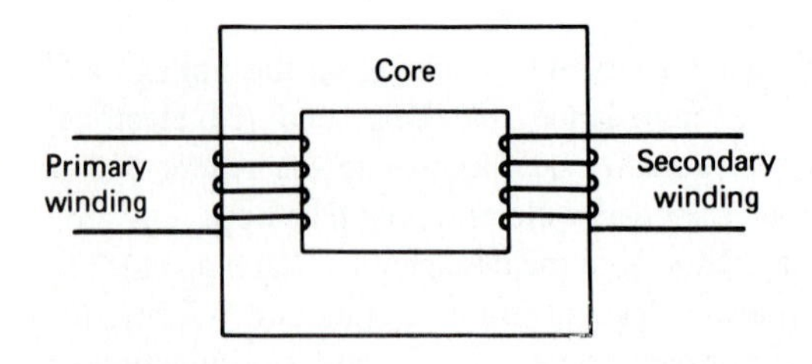

FIGURE 3-1
Elementary model of a transformer.

FIGURE 3-2*(a)*
Elements of an electronics power transformer showing magnetic core with wound strip of laminations and typical windings. (Courtesy GTE Lenkurt.)

The construction of transformers varies greatly, depending on their applications, winding voltage and current ratings, and operating frequencies. Many electronics transformers consist of little more than the electromagnetic structure itself with a suitable means of mounting to a frame. In general, the electromagnetic structure is contained within a housing or case for safety and protection. In several types of transformers the space surrounding the electromagnetic structure is filled with an electrically insulating material to prevent damage to the windings or core and to prevent their movement or facilitate heat transfer between the electromagnetic structure and the case. In electronics transformers a viscous insulating material called "potting compound" is used, and in many power transformers, a nonflammable insulating oil called transformer oil is used.

Transformer oil serves an added function of improving the insulation characteristics of the transformer, since it has a higher dielectric strength than air. In most oil-filled transformers the oil is permitted to circulate through cooling fins or tubes on the outside of the case to improve further the heat transfer characteristics. The fins or tubes are often cooled by forced air; Figs. 3.2 and 3.3 illustrate practical transformers of various types and applications. In larger transformers operating at high voltage and current levels there are other important structural components, some of which can be seen in Fig. 3.3. Such components include porcelain bushings, through which the winding leads are brought for external connection, oil pressure and temperature gauges, and internal structural supports to prevent movement of the leads or windings caused by electromagnetic forces resulting from high current levels.

FIGURE 3-2*(b)*
Constant-voltage ferroresonant transformer. Capacitor shown on left. (Courtesy Sola Electric Division of Sola Basic Industries.)

FIGURE 3-2*(c)*
Pulse transformer, frequently used in thyristor gate trigger circuits. (Courtesy GTE Lenkurt.)

FIGURE 3-3
Power transformer. (Courtesy General Electric Company.)

The magnetic core of a transformer must be constructed in a manner to minimize the magnetic losses. These core constructions are described briefly in Chapter 2. Power transformer cores are generally constructed from soft magnetic materials in the form of punched laminations or wound tapes. Lamination or tape thickness is a function of the transformer frequency. Pulse transformer and high-frequency electronics transformer cores are often constructed of soft ferrites. Laminations used in electronics transformers are often called "alphabet" laminations, since they are in the shape of several letters of the alphabet and designated as *E*, *C*, *I*, *U* cores, and so forth. Ferrite cores are designated by the names cup, pot, sleeve, rod, slug, and so forth, which also describes the general shape of these cores. The most common lamination materials are silicon-iron, nickel-iron, and cobalt-iron alloys. Powdered Permalloy is used for many communication transformer cores.

Transformer windings are constructed of solid or stranded copper or aluminum conductors. The conductor used in electronics transformers—as well as in many small and medium-sized motors and generators—is magnet wire.

In transformers having split cores of the *E*, *C*, *U*, or torodial configurations, the windings are usually wound on an insulating spool or reel known as a bobbin. The wound bobbin can then be conveniently slipped over one leg of the core section; the remaining sections of core can be assembled; and the completed core assembly containing the winding is generally clamped together by means of a metal tape. The purpose of the bobbin is to provide a structural support for the windings, electrically insulate the winding from the core, and prevent abrasion or cutting of the winding at the core edges. Transformer bobbins are constructed of nylon, teflon, and various paper and fiber products. To improve the magnetic coupling between the two windings, a bifiler winding technique is frequently used. This technique consists of laying the conductors to be used in the windings side by side and winding them on the core or bobbin simultaneously. Fig. 3.4 illustrates the various stages of assembling a winding on an electronics transformer core.

The windings of large power transformers generally use conductors with heavier insulation than magnet wire insulation. The windings are assembled with much greater mechanical support, and winding layers are insulated from each other. Larger, high-power windings are often preformed, and the transformer is assembled by stacking the laminations within the preformed coils.

3.2 CLASSIFICATION

Because of the great diversity in size, shape, and application of transformers, there has been some attempt to designate types or classes of transformers. However, the terms used to classify transformers are very loosely defined and there is much overlap among the meanings of these terms. The terms have developed more from the types of circuits in which the transformers are used than from a delineation of transformer characteristics. Since these descriptive terms are widely used in manufacturers' catalogs of transformers and since it will aid in the subsequent description of transformer characteristics and structural features, some of the terms that categorize basic transformer types are listed here.

1. *General Application Classification.*
   (a) **Power Transformers.** Transformers used in power distribution and transmission systems. This class has the highest power or volt-ampere ratings and the highest continuous voltage ratings.

FIGURE 3-4
Various stages of the assembly of an electronics transformer. (Courtesy GTE Lenkurt.)

(*b*) **Electronics Transformers.** Transformers of many different types and applications used in electronic circuits. Sometimes electronics transformers are considered as those transformers with ratings of 300 VA and below. A large class of electronics transformers are called "power transformers" and are used in supplying power to other electronic systems, which confuses this nomenclature considerably.

(*c*) **Instrument Transformers.** Transformers used to sense voltage or current in both electronic circuits and power systems—often called potential transformers and current transformers. The latter are series-connected devices and are operated in a configuration that is, in many respects, the inverse of the conventional voltage or potential transformer.

(*d*) **Specialty Transformers.** This designation covers many styles and operating features and includes devices such as saturating, constant-voltage, constant-current, ferroresonant, and variable-tap transformers.

2. *Classification by Frequency Range.* This is probably the most significant method of classifying transformers for describing electromagnetic design features and includes the following.

(*a*) **Power.** These are generally constant-frequency transformers that operate at the power frequencies (50, 60, 400 Hz, etc.), although other frequency components, including dc, may be present in electronics power transformers and power semiconductor transformers.

(*b*) **Audio.** Used in many communication circuits to operate at audio frequencies.

(*c*) **Ultra High Frequency (UHF).**

(*d*) **Wide-Band.** Electronics transformers operating over a wide range of frequencies.

(*e*) **Narrow-Band.** Electronics transformers designed for a specific frequency range.

(*f*) **Pulse.** Transformers designed for use with pulsed or chopped excitation, both in electronics and power systems applications.

3. *Classification by Number of Windings.* This has already been alluded to and includes one-winding (autotransformers), two-winding (or conventional), and multiwinding transformers, where a winding is defined as a two-terminal electrical circuit with no electrical connection to other electrical circuits.

4. *Classification by Polyphase Connection.* This classification applies mainly to transformers in power systems and refers to the method of connecting individual windings in polyphase applications. In some transformers the entire polyphase set of transformers is contained within a single housing or case, with some resulting reduction of the weight of the magnetic core. The most common connections are the wye (or star) and the delta connections of three-phase systems.

3.3 PRINCIPLE OF OPERATION: THE IDEAL TRANSFORMER

We have so far introduced various terms and descriptions of transformers. Let us now look at the principle of operation of the simple two-winding model of Fig. 3.1. Despite our reservations about the use of the words "primary" and "secondary," as stated in the introduction, we will find it convenient to use these terms and will associate the subscript "1" with the primary and "2" with the secondary. However, keep in mind that these are arbitrary terms and are in no way inherent properties of a transformer. Because a number of these arbitrary designations are required in the analysis of the transformer—and are subsequently transferred to the analysis of many other electromagnetic devices—we will adhere to the conventions used by the international standards organization (IEC) and the Institute of Electrical and Electronics Engineers (IEEE). Also, all formulas are given in terms of the Standard International (SI) units. The simple transformer model of Fig. 3.1 has been repeated as Fig. 3.5, with the addition of pertinent parameters for this analysis.

A fundamental parameter of the transformer is the turns ratio, defined as

$$a = \frac{N_1}{N_2} \tag{3.1}$$

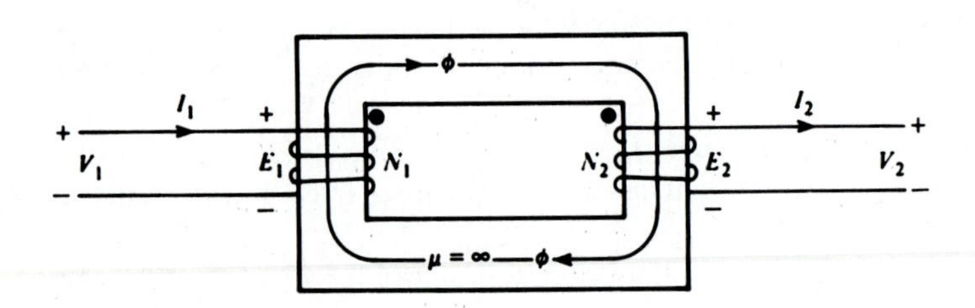

FIGURE 3-5
An ideal transformer model showing polarities and dot convention.

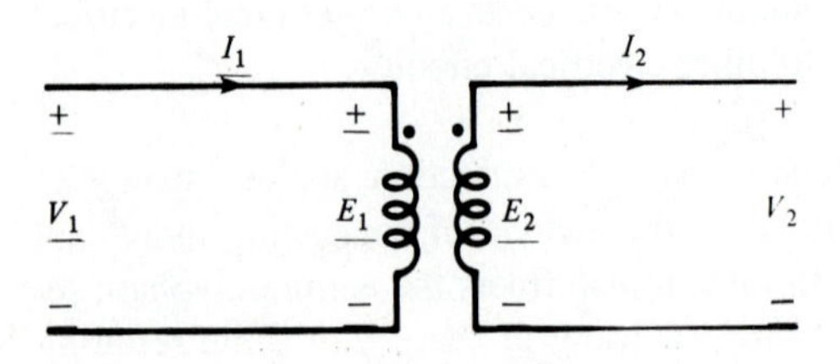

FIGURE 3-6
The dot convention of polarity marking.

The value of the turns ratio is generally known only to the transformer manufacturer or the person winding a laboratory transformer. It can be measured in the laboratory by measuring the induced voltages in the two windings, but always with a certain degree of inaccuracy, which depends on the coefficient of coupling between the two windings. The turns ratio is frequently given as part of the nameplate data by the manufacturer in both electronics and power systems transformers.

In this section we will consider a transformer excited from a single-frequency, sinusoidal voltage source, represented as V_1 in Fig. 3.5. We will deal primarily with the *steady-state* voltages, currents, and magnetic fluxes. We begin with the electromagnetic field relationships in the transformer as an aid in the understanding of the physical phenomena that result in the external transformer characteristics.

The transformer of Fig. 3.5 is ideal in the sense that its core is lossless, is infinitely permeable, and has no leakage fluxes, and the windings have no losses. Absence of leakage flux implies that the entire flux links with both windings completely. In Fig. 3.5 the basic components are the *core*, the *primary winding* having N_1 turns, and the *secondary winding* having N_2 turns. If ϕ is the mutual (or core) flux linking N_1 and N_2, then according to Faraday's law of electromagnetic induction, emf's e_1 and e_2 are induced in N_1 and N_2 due to a time rate of change of ϕ such that

$$e_1 = \pm N_1 \frac{d\phi}{dt} \tag{3.2}$$

and

$$e_2 = \pm N_2 \frac{d\phi}{dt} \tag{3.3}$$

The direction of e_1 is such as to oppose the flux change, according to Lenz's law. The transformer being ideal, $E_1 = V_1$ (Fig. 3.6). The sign, + or - in (3.2) or (3.3), is determined by the assumed polarities at the terminals and must be consistent with Lenz's law. From (3.2) and (3.3), we have

$$\frac{e_1}{e_2} = \frac{N_1}{N_2}$$

which may also be written in terms of rms values as

$$\frac{E_1}{E_2} = \frac{N_1}{N_2} = a \tag{3.4}$$

where a is the *turns ratio*.

If the flux varies sinusoidally, such that

$$\phi = \phi_m \sin\omega t \tag{3.5}$$

then from (3.2) and (3.5) the corresponding induced voltage, e, linking an N-turn winding is given by

$$e = \omega N \phi_m \cos \omega t \tag{3.6}$$

From (3.6) the rms value of the induced voltage is

$$E = \frac{\omega N \phi_m}{\sqrt{2}} = 4.44 f N \phi_m \tag{3.7}$$

In (3.7), $f = \omega/2\pi$ is the frequency in hertz.

Equation (3.7) is known as the *emf equation* of a transformer.

In conjunction with Faraday's law, the direction, or polarity, of the induced voltage is determined by Lenz's law. For our purposes, we mark the polarities by using the *dot convention*, shown in Fig. 3.6. Notice that we have placed a dot at one terminal of each winding. This convention implies that (1) currents entering at the dotted terminals will result in mmf's that will produce fluxes in the same direction and (2) voltages from dotted to undotted terminals have the same sign. The advantage of the dot convention is that we do not have to draw the magnetic circuit, as in Fig. 3.5. Rather, Fig. 3.6, which is simpler to draw, contains the same information as that in Fig. 3.5. In fact, in Fig. 3.5 the dots are redundant and we marked the terminals there simply to introduce the convention.

3.4 VOLTAGE, CURRENT, AND IMPEDANCE TRANSFORMATIONS

As we mentioned earlier, major applications of transformers are in voltage, current, and impedance transformations, and for providing isolation (that is, eliminating direct connections between electric circuits). The voltage transformation property, mentioned in the preceding section, of an ideal transformer is expressed as

$$\frac{V_1}{V_2} = \frac{E_1}{E_2} = a \tag{3.8}$$

where the subscripts 1 and 2 correspond to the primary and secondary sides, respectively.

For an ideal transformer, the net mmf around its magnetic circuit must be zero. In other words, we do not need any mmf (or, we need zero mmf) to establish the flux in the core of an ideal transformer. The relationship between the flux ϕ and the mmf F for a magnetic circuit having a permeance $\wp$ is

$$\phi = F \wp \tag{3.9}$$

Since the core of an ideal transformer is assumed to be infinitely permeable, $\wp = \infty$ in (3.9). Consequently, $F = 0$ for ϕ to be finite. Therefore,

$$F = F_1 + F_2 = N_1 I_1 - N_2 I_2 = 0 \tag{3.10}$$

where I_1 and I_2 are the primary and the secondary currents, respectively. From (3.4) and (3.10) we get

$$\frac{I_2}{I_1} = \frac{N_1}{N_2} = a \tag{3.11}$$

From (3.8) and (3.11), the impedance transformation relationships can be obtained. If Z_1 is the impedance seen at the primary terminals, then

$$\begin{aligned} Z_1 &= \frac{V_1}{I_1} = \frac{E_1}{I_1} = \frac{aE_2}{I_2/a} = a^2 \frac{E_2}{I_2} \\ &= \frac{a^2 V_2}{I_2} = a^2 Z_2 \end{aligned} \tag{3.12}$$

We now illustrate the discussions up to this point by the following examples.

Example 3.1 How many turns must the primary and the secondary windings of a 220/110-V 60-Hz ideal transformer have if the core flux is not allowed to exceed 5 mWb?

Solution

From the emf equation (3.7) we have

$$N = \frac{E}{4.44 f \phi_m}$$

Consequently,

$$N_1 = \frac{220}{4.44 \times 60 \times 5 \times 10^{-3}} \approx 166 \text{ turns}$$

$$N_2 = \frac{1}{2} N_1 = 83 \text{ turns}$$

Example 3.2 A 220/110-V 60-Hz ideal transformer has 166 turns on its primary. What is the instantaneous flux?

Solution

From (3.7) we have

$$\phi_m = \frac{E}{4.44fN} = \frac{220}{4.44 \times 60 \times 166} = 4.975 \text{ mWb} \qquad (a)$$

Since $f = 60$ Hz,

$$\omega = 2\pi f = 2\pi \times 60 = 377 \text{ rad/s}$$

Hence from (3.5),

$$\phi = 4.975 \sin 377t \quad \text{mWb}$$

In general, however, we observe that the flux and voltage are related by (3.2), or

$$\phi = \frac{1}{N} \int e \, dt$$

Example 3.3 The load on an ideal transformer having a turns ratio of 5 is 10 Ω. What is the ohmic value of impedance as seen (or measured) at the primary terminals?

Solution

This problem illustrates a direct application of (3.12), in which we substitute a = 5 and $Z_2 = Z_{\text{load}} = 10\ \Omega$. Hence

$$Z_1 = 5^2 \times 10 = 250\ \Omega$$

The problem shows the impedance transformation property of the transformer. In this particular case, a 10-Ω impedance at the secondary would appear (or measured) as a 250-Ω impedance at the primary.

3.5 THE NONIDEAL TRANSFORMER

A nonideal (or an actual) transformer differs from an ideal transformer in that the former has hysteresis and eddy-current (or core) losses, and has resistive (I^2R) losses in its primary and secondary windings. Furthermore, the core of a nonideal transformer is not perfectly permeable, and the transformer core requires a finite mmf for its magnetization. Also, not all fluxes link with the primary and secondary windings simultaneously because of leakages. Referring to Fig. 3.7, we observe that R_1 and R_2 are the respective resistances of the primary and secondary windings. The flux ϕ_c, which replaces the flux ϕ of Fig. 3.5, is called the *core flux* or *mutual flux*, as it links both the primary and secondary windings. The primary and secondary leakage fluxes are shown as ϕ_{l1} and ϕ_{l2}, respectively. Thus in Fig. 3.7 we have accounted for all the imperfections listed above, except the core losses. We will include the core losses as well as the rest of the imperfections in the equivalent circuit of a nonideal

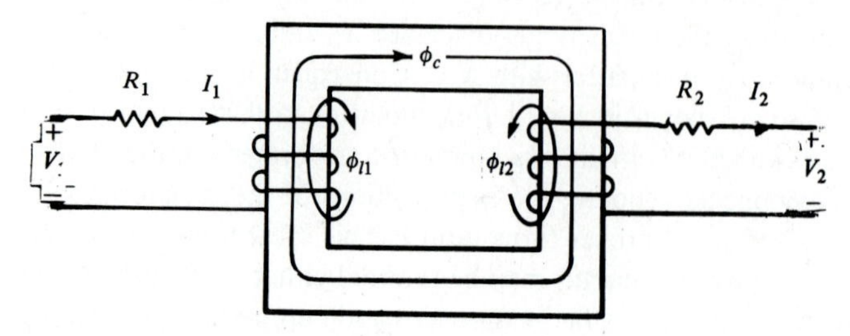

FIGURE 3-7
A nonideal transformer showing the leakage fluxes and the winding resistances.

transformer. This circuit is also known as the *exact equivalent circuit*, as it differs from the idealized equivalent circuit and the various approximate equivalent circuits. We now proceed to derive these circuits.

Considering the primary windings and the nonideal core, first, we assume that the core has a finite and constant permeability, but no losses. We also assume that the coil has no resistance. In such an idealized case the magnetic circuit and the coil can be represented just by an inductance L_m, as given in Fig. 3.8*(a)*, which corresponds to the core flux, ϕ_c. If the coil is excited by a sinusoidal ac voltage $v = V_m \sin \omega t$, it is conventional to express L_m as an inductive reactance $X_m = \omega L_m$. This reactance is known as the *magnetizing reactance*. Thus in Fig. 3.8*(a)* we have X_m across which we show the terminal voltage V_1, equal to the induced voltage E_1. Note that V_1 and E_1 are root-mean-square (rms) values.

Next, we include the hysteresis and eddy-current losses by the resistor R_c in parallel with X_m, such that the voltage V_1 appears across R_c also, the core losses being directly dependent on V_1. We thus obtain Fig. 3.8*(b)*, where again $V_1 = E_1$. Finally, we include the series resistance R_1, the resistance of the coil, and X_1, its *leakage reactance*, which arises from leakage fluxes. Hence we obtain the electrical equivalent shown in Fig. 3.8*(c)*, which represents an iron core excited by an ac mmf. We will use this basic equivalent circuit to obtain the complete equivalent circuit of a transformer.

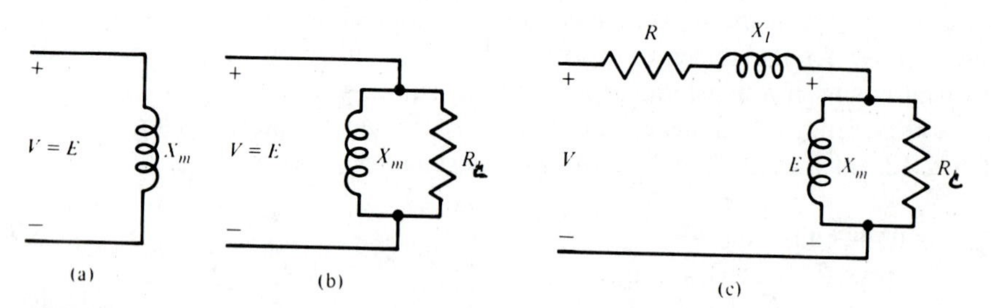

FIGURE 3-8
Equivalent circuits of an iron core excited by an ac mmf.

In summary, in the equivalent circuit of a nonideal transformer, the leakage fluxes ϕ_{l1} and ϕ_{l2} give rise to the leakage reactances, X_1 and X_2 respectively. The mutual flux ϕ_c is represented by the magnetizing reactance X_m and the core losses are represented by a resistance, R_c, in parallel with X_m. Consequently, the equivalent circuit of an ideal transformer, shown in Fig. 3.9*(a)*, modifies to that shown in Fig. 3.9*(b)*, which is the exact equivalent circuit of a nonideal transformer. In Fig. 3.9*(b)*, notice that the circuit components denoting the imperfections of the transformer are coupled by an ideal transformer (of proper turns ratio). This ideal transformer may be removed and the entire equivalent circuit may be referred either to the primary or to the secondary of the transformer by using its transformation properties, as given by (3.8), (3.11), and (3.12).

Referring to the primary side, according to (3.12) an impedance Z_2, on the secondary side, will appear as a^2Z_2 on the primary side. In accordance with (3.11), a current I_2 on the secondary side corresponds to a current I_2/a on the primary side. Finally, to eliminate the ideal transformer from Fig. 3.9*(b)*, and refer the secondary quantities to the primary, we use (3.8) to replace E_2 by $aE_2 = E_1$. The circuit then becomes as shown in Fig. 3.10. Of course, the primary quantities (V_1, I_1, R_1, and X_1) remain unchanged.

At this point, let us pose two questions. First, why is it necessary to obtain an equivalent circuit referred to the primary (or secondary)? Second, what is the physical meaning of referring secondary quantities to the primary (and vice versa)? In answer to the first question, by referring the entire circuit to the primary, we have eliminated the ideal transformer. Thus the transformer can be represented by an *R-L* circuit (Fig. 3.10). Such a representation involves a simpler circuit analysis than that of the circuit of Fig. 3.9. Furthermore, an *R-L* circuit of the type shown in Fig. 3.10 can be approximated by simpler circuits in a straightforward manner, as we shall see later. Turning to the second question, referring the secondary impedance to the primary side implies that the real and reactive powers in an impedance Z_2 through which the secondary current I_2 flows remain the same when the primary current I_1 flows through an equivalent impedance Z_2'. This restriction ensures that the performance of a transformer as calculated from a circuit referred to the primary or from one referred to the secondary remains the same as the results obtained from the circuit of Fig. 3.9. In simple terms, the circuits of Figs. 3.10 and 3.9 must give the same results.

To obtain the equivalent circuit referred to the secondary, we again use (3.8), (3.11), and (3.12). Thus, we get the equivalent circuit shown in Fig. 3.11. It is left as an exercise for the student to verify that the circuit of Fig. 3.11 is indeed an equivalent circuit of a transformer referred to its secondary.

A phasor diagram for the circuit of Fig. 3.10, for lagging power factor, is shown in Fig. 3.12. In Figs. 3.10 to 3.12, the various symbols are as follows:

a = turns ratio
E_1 = primary induced voltage
E_2 = secondary induced voltage
V_1 = primary terminal voltage
V_2 = secondary terminal voltage

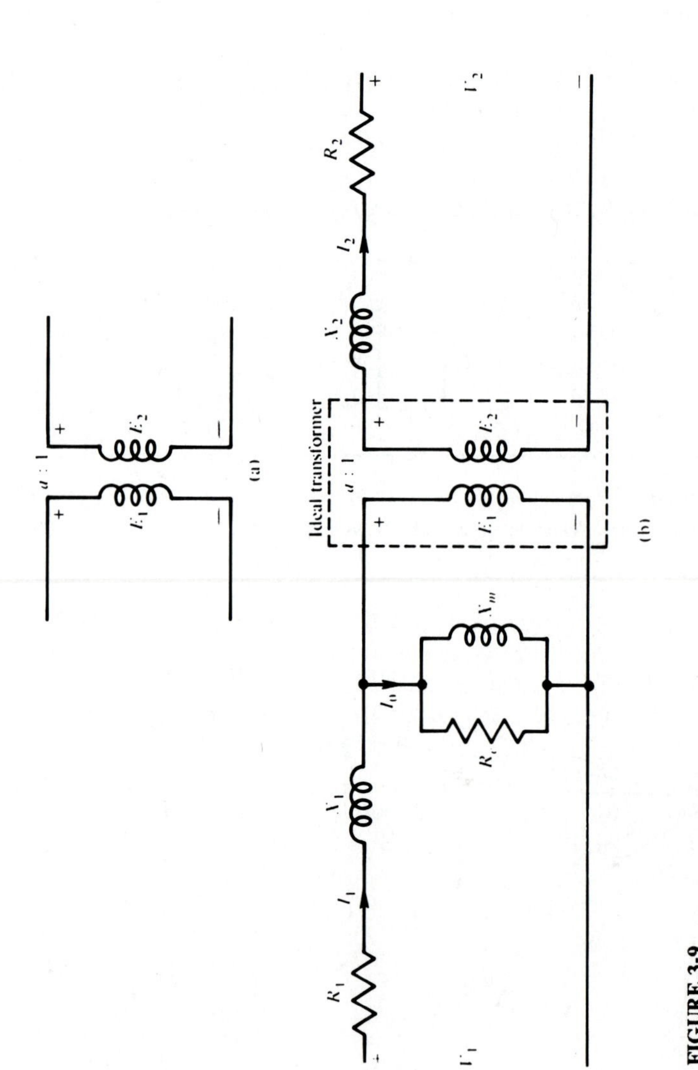

FIGURE 3-9
Equivalent circuits of (*a*) ideal and (*b*) nonideal transformers.

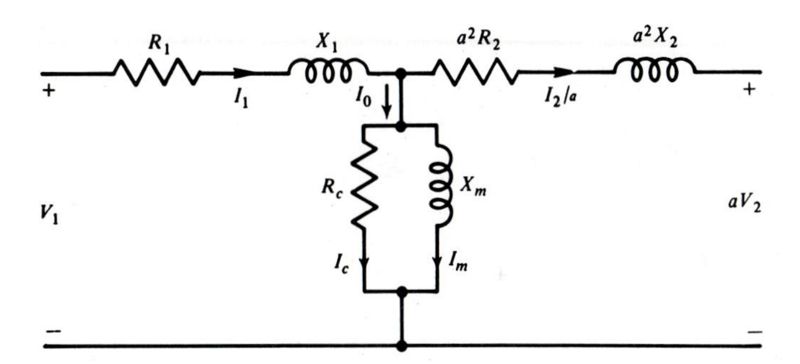

FIGURE 3-10
Equivalent circuit referred to primary (high-voltage side).

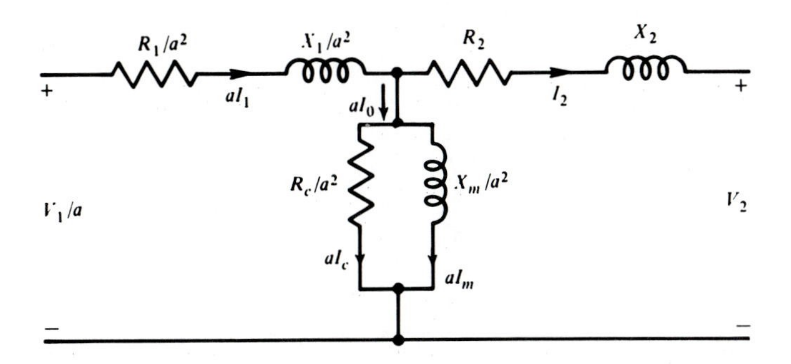

FIGURE 3-11
Equivalent circuit referred to secondary (low-voltage side).

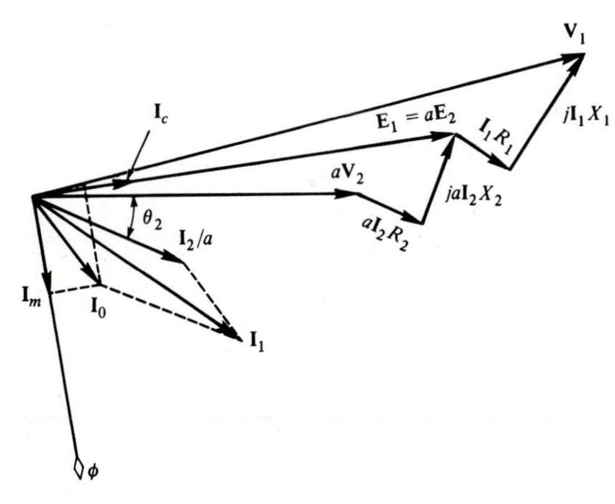

FIGURE 3-12
Phasor diagram corresponding to Fig. 3.10.

I_1 = primary current
I_2 = secondary current
I_0 = no-load (primary) current
R_1 = resistance of the primary winding
R_2 = resistance of the secondary winding
X_1 = primary leakage reactance
X_2 = secondary leakage reactance
I_m, X_m = magnetizing current and reactance
I_c, R_c = current and resistance accounting for the core losses.

Whereas the following governing equations apply to the equivalent circuit of Fig. 3.10, these equations may be better visualized from the phasor diagram shown in Fig. 3.12. Choosing $a\bar{V}_2$* as the reference phasor, we have referred to the high-voltage side:

load (secondary) terminal voltage = $a\bar{V}_2 \angle 0°$ (3.13)

load (secondary) current = $\dfrac{\bar{I}_2}{a} \angle -\theta_2$

(3.14)

secondary induced voltage, $a\bar{E}_2 = a\bar{V}_2 + \dfrac{\bar{I}_2}{a}(a^2R_2 + ja^2X_2)$ (3.15)

magnetizing current, $\bar{I}_m = \dfrac{a\bar{E}_2}{jX_m}$ (3.16)

current corresponding to core losses, $\bar{I}_c = \dfrac{a\bar{E}_2}{R_c}$ (3.17)

no-load current, $\bar{I}_0 = \bar{I}_m + \bar{I}_c$ (3.18)

primary current, $\bar{I}_1 = \bar{I}_0 + \dfrac{\bar{I}_2}{a}$ (3.19)

primary (applied) voltage, $\bar{V}_1 = a\bar{E}_2 + \bar{I}_1(R_1 + jX_1)$ (3.20)

From the above equations, the complete performance characteristics of the transformer may be obtained, as illustrated by the following examples.

---

* In much of the text we have *not* used overbars to denote phasors.

Example 3.4 In a nonideal transformer, we assume that the windings have nonzero resistances. A 220/110-V 10-kVA nonideal transformer has a primary winding resistance of 0.25 Ω and a secondary winding resistance of 0.06 Ω. Determine *(a)* the primary and secondary currents at rated load; and the total resistance *(b)* referred to the primary and *(c)* referred to the secondary.

Solution

(a) The transformation ratio is

$$a = \frac{220}{110} = 2$$

The primary current is

$$I_1 = \frac{10 \times 10^3}{220} = 45.45 \text{ A}$$

The secondary current is

$$I_2 = aI_1 = 2 \times 45.45 = 90.9 \text{ A}$$

(b) The secondary winding resistance referred to the primary is

$$a^2 R_2 = 2^2 \times 0.06 = 0.24 \ \Omega$$

The total resistance referred to the primary is

$$R_e' = R_1 + a^2 R_2 = 0.25 + 0.24 = 0.49 \ \Omega$$

(c) The primary winding resistance referred to the secondary is

$$\frac{R_1}{a^2} = \frac{0.25}{4} = 0.0625 \ \Omega$$

The total resistance referred to the secondary is

$$R_e'' = \frac{R_1}{a^2} + R_2 = 0.0625 + 0.06 = 0.1225 \ \Omega$$

Example 3.5 Determine the I^2R loss in each winding of the transformer of Example 3.4, and thus find the total I^2R loss in the two windings. Verify that the same result can be obtained by using the equivalent resistance referred to the primary winding.

Solution

$$I_1^2 R_1 \text{ loss} = (45.45)^2 \times 0.25 = 516.529 \text{ W}$$

$$I_2^2 R_1 \text{ loss} = (90.9)^2 \times 0.06 = 495.867 \text{ W}$$

$$\text{total } I^2 R \text{ loss} = 1012.39 \text{ W}$$

For the equivalent resistance, R_e',

$$I_1^2 R_e' = (45.45)^2 \times 0.49 = 1012.39 \text{ W}$$

which is consistent with the preceding result.

Example 3.6 A 150-kVA 2400/240-V transformer has the following parameters: $R_1 = 0.2\ \Omega$, $R_2 = 0.002\ \Omega$, $X_1 = 0.45\ \Omega$, $X_2 = 0.0045\ \Omega$, $R_c = 10{,}000\ \Omega$, and $X_m = 1550\ \Omega$, where the symbols are shown in Fig. 3.9(*b*). Refer the circuit to the primary. From this circuit, calculate the primary voltage of the transformer at rated load with a 0.8 lagging power factor.

Solution

The circuit referred to the primary is shown in Fig. 3.10. From the data given, we have $V_2 = 240$ V, $a = 10$, and $\theta_2 = \cos^{-1} 0.8 = 36.8°$.

$$I_2 = \frac{150 \times 10^3}{240} = 625 \text{ A}$$

$$\frac{I_2}{a} = 62.5 \angle{-36.8°} = 50 - j37.5 \text{ A}$$

$$aV_2 = 2400 \angle 0° = 2400 + j0 \text{ V}$$

$$a^2 R_2 = 0.2\ \Omega \quad \text{and} \quad a^2 X_2 = 0.45\ \Omega$$

Hence

$$E_1 = (2400 + j0) + (50 - j37.5)(0.2 + j0.45)$$

$$= 2427 + j15 = 2427 \angle 0.35° \text{ V}$$

$$I_m = \frac{2427 \angle 0.35°}{1550 \angle 90°} = 1.56 \angle{-89.65} = 0.0095 - j1.56 \text{ A}$$

$$I_c = \frac{2427 + j15}{10{,}000} \approx 0.2427 + j0 \text{ A}$$

$$I_0 = I_c + I_m = 0.25 - j1.56 \text{ A}$$

$$I_1 = I_0 + \frac{I_2}{a} = 50.25 - j39.06 = 63.65 \angle{-37.850°} \text{ A}$$

Thus, the primary voltage is

$$V_1 = (2427 + j15) + (50.25 - j39.06)(0.2 + j0.45)$$

$$= 2455 + j30 = 2455 \angle 0.7° \text{ V}$$

3.6 CORE SATURATION AND MAGNETIZING CURRENT

In the preceding section we have considered some of the imperfections in the electric and magnetic circuits of a transformer to obtain the equivalent circuit of a nonideal transformer. We recall that the imperfect magnetic circuit leads to core losses and a finite magnetizing current. However, we paid no attention to the waveform of this current.

A manifestation of the presence of magnetic materials in the core is the nonlinear *B-H* curve. In transformer analysis the use of core flux is more convenient than flux density because of the relationship between flux and voltage as given in (3.2). Likewise, instead of *H*, either magnetizing current (which is related to *H* by means of the mean magnetic core length) or magnetizing ampere-turns is used. The *B-H* characteristic, therefore, becomes a ϕ vs. I_m characteristic in the transformer and most other electromagnetic systems. The magnetizing current I_m has been assumed as a linear, sinusoidal component in (3.18). In *voltage-excited* transformers, it is seldom sinusoidal and usually nonlinear, as illustrated in the following example.

Example 3.7 The flux versus ampere-turn relationship of the magnetic circuit of a certain transformer is given in Fig. 3.13. This transformer has a primary winding of 120 turns, which is excited from a sinusoidal voltage source of 110 V (rms) and 60 Hz. Determine the maximum value of magnetic flux and the general waveform of the magnetizing current. Neglect the winding resistance.

Solution

From *(a)* of Example 3.2 we have

$$\phi_m = \frac{110}{4.44 \times 60 \times 120} = 3.44 \text{ mWb}$$

The waveform of the flux is sinusoidal; phase relationships are not required in this problem. The resulting magnetizing current is found by the graphical construction illustrated in Fig. 3.13. Values of flux from the sine wave of flux with maximum value of 0.00344 Wb are transferred to the magnetic saturation curve, and the resultant magnetic potential or mmf for each value of flux is read from the abscissa of this curve. The magnetizing current is found by dividing the mmf by the turns, that is, by 120.

It is seen from this example that the magnetizing current in a transformer with saturation will be nonsinusoidal in waveform. There are other circuit parameters that will also affect this waveshape in an actual practical transformer. Hysteresis of the magnetic characteristic will alter the waveform somewhat, and there are usually other circuit parameters in series with the winding, such as winding resistance, that will influence the waveform. In some cases, the nature of the excitation source may change the waveform, especially if the source is a voltage- or current-regulated electronic power supply. The relative magnitude of the magnetic core losses may also influence the current waveform. In many transformers the magnetic circuit is designed so that the maximum flux required at rated applied voltage results in operation at approximately the knee of the saturation curve. This will minimize the amount of distortion from a sine wave of magnetizing current. If the transformer is operated at a voltage/frequency ratio in excess of its design value, however, the distortion of the magnetizing current may be greatly increased. For example, if the transformer in Example 3.7 were to be excited from a 110-V, 50-Hz source, the maximum flux would be 0.0041 Wb. Inspection of the graphical construction of Fig. 3.13 shows that this operation would greatly increase the spike at the center of the current half-wave.

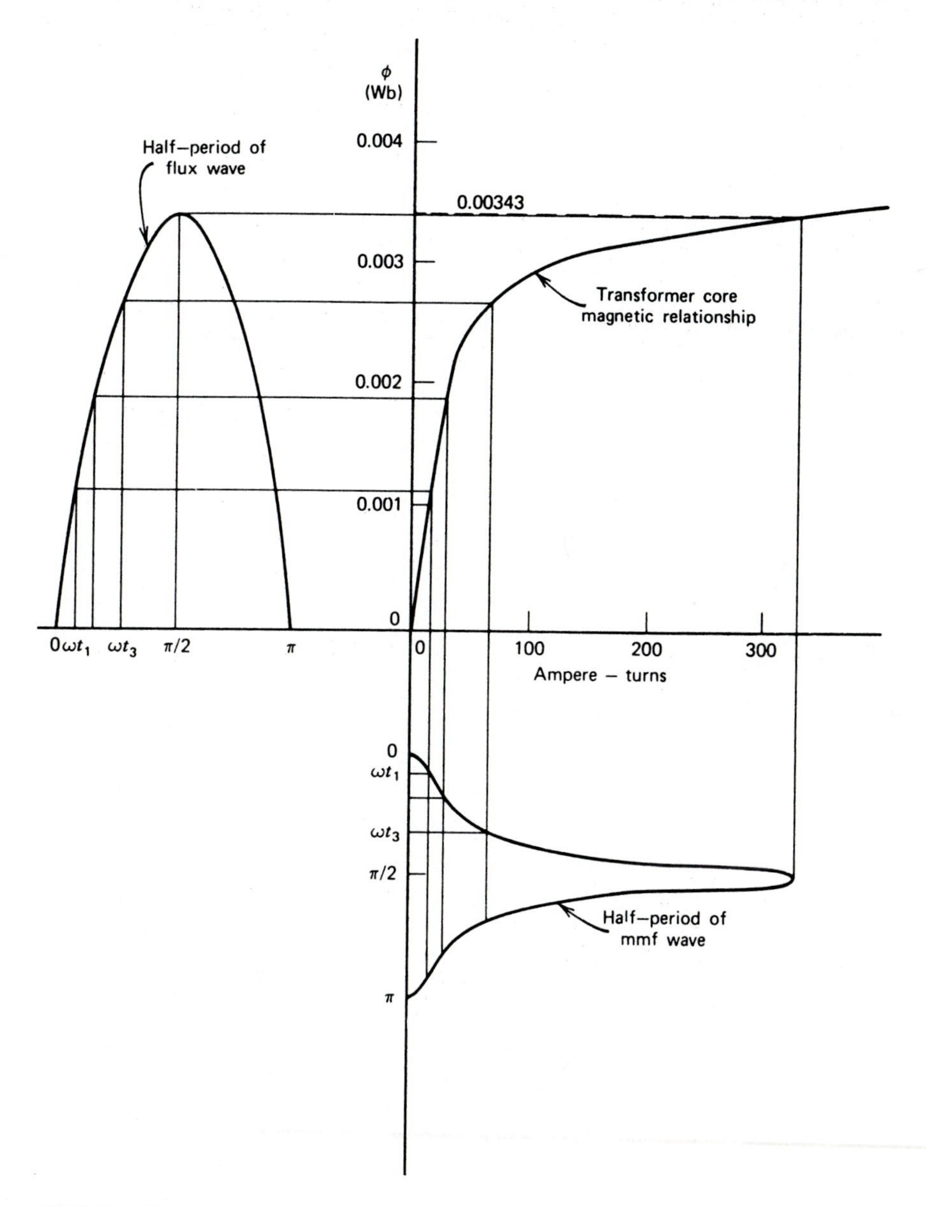

FIGURE 3-13
Construction for obtaining mmf or exciting current waveform.

Our discussion so far has been concerned with the excitation of a transformer from a sinusoidal voltage source. This mode of excitation is frequently termed sine-flux excitation. In electronics transformers, many other forms of excitation are frequently encountered. Excitation from a square-wave voltage source is common, for example. Many electronics transformers are excited from a constant-current source. When it is sinusoidal, this mode of excitation is frequently termed sine-current excitation. By analogy to the graphical procedure demonstrated in Example 3.7, it is seen that sine-current excitation results in a nonsinusoidal flux in the core and, hence, a nonsinusoidal induced voltage. A problem at the end of this chapter is also given to aid in developing the flux-voltage relationships for this mode of excitation (see Problem 3.15).

3.7 NONIDEAL TRANSFORMATION RATIO

We have noted in (3.4) that the turns ratio is equal to the ratio of voltage transformation only in an ideal transformer. We will define terminal voltage transformation ratio as

$$\frac{V_1}{V_2} = a' \tag{3.21}$$

From (3.20), it is seen that the terminal voltage ratio will be affected by the leakage impedance of the primary; it follows that it will also be affected by the leakage impedances of the secondary. It can be shown* that the voltage transformation ratio may be expressed as

$$a' = \frac{Z_{10} + k\sqrt{Z_{10}Z_{20}}}{Z_{20} + k\sqrt{Z_{10}Z_{20}}} \tag{3.22}$$

where

$$Z_{10} = R_1 + j(X_1 + X_m)$$

$$Z_{20} = R_2 + j\left(X_2 + \frac{X_m}{a^2}\right)$$

$$k = \text{coefficient of coupling} = \frac{X_m}{X_1 + X_m}$$

---

*Tests for Electronics Transformers and Inductors, *IEEE* Standard No. 389-1978, New York, NY, 1978.

Equation (3.22) does not include the effects of core loss, that is, the effect of the equivalent resistance, R_c; and Z_{10} and Z_{20} are the *open-circuit impedances*, measured on the primary and secondary sides, respectively, using rated voltage on the respective sides.

Example 3.8 Calculate the voltage transformation ratio, a', for a 2400:480 V transformer having $X_m = 400\ \Omega$, $X_1 = 0.29\ \Omega$, $X_2 = 0.012\ \Omega$, $R_1 = 0.058\ \Omega$, and $R_2 = 0.002\ \Omega$.

Solution
The numerical values are calculated as follows:

$$k = \frac{400}{400 + 0.29} = 0.9993$$

$$Z_{10} = 0.058 + j(400.29)$$

$$Z_{20} = 0.002 + j(16.012)$$

Substituting into (3.22) gives $a' = 5.0023$. It is seen that there is negligible difference between actual turns ratio and transformation ratio in this particular case, which is a high-power, closely coupled transformer. In electronics transformers there is usually a greater difference between the two ratios.

3.8 PERFORMANCE CHARACTERISTICS

As mentioned earlier, the major use of the equivalent circuit of a transformer is in determining its characteristics. The characteristics of most interest to power engineers are voltage regulation and efficiency. *Voltage regulation* is a measure of the change in the terminal voltage of the transformer with load. From Fig. 3.12 it is clear that the terminal voltage V_1 is load-dependent. Specifically, we define voltage regulation as

$$\text{percent regulation} = \frac{V_{\text{no-load}} - V_{\text{load}}}{V_{\text{load}}} \times 100 \tag{3.23}$$

With reference to Fig. 3.12, we may rewrite (3.23) as

$$\text{percent regulation} = \frac{V_1 - aV_2}{aV_2} \times 100$$

There are two kinds of efficiencies of transformers of interest to us, known as *power efficiency* and *energy efficiency*. These are defined as follows:

$$\text{power efficiency} = \frac{\text{output power}}{\text{input power}} \tag{3.24}$$

$$\text{energy efficiency} = \frac{\text{output energy for a given period}}{\text{input energy for the same period}} \tag{3.25}$$

Generally, energy efficiency is taken over a 24-hour (h) period and is called *all-day efficiency*. In such a case (3.25) becomes

$$\text{all-day efficiency} = \frac{\text{output for 24 h}}{\text{input for 24 h}} \tag{3.26}$$

For our purposes, we will use the term *efficiency* to mean power efficiency from now on. It is clear from our discussion of nonideal transformers that the output power is less than the input power because of losses, the losses being I^2R losses in the windings, and hysteresis and eddy-current losses in the core. Thus, in terms of these losses, (3.24) may be more meaningfully expressed as

$$\begin{aligned}\text{efficiency} &= \frac{\text{input power} - \text{losses}}{\text{input power}} \\ &= \frac{\text{output power}}{\text{output power} + \text{losses}} \\ &= \frac{\text{output power}}{\text{output power} + I^2R\text{ loss} + \text{core loss}}\end{aligned} \tag{3.27}$$

Obviously, I^2R loss is load-dependent, whereas the core loss is constant and independent of the load on the transformer. The next examples show voltage regulation and efficiency calculations of a transformer.

Example 3.9 Determine the efficiency and voltage regulation of the transformer of Example 3.6 operating on rated load and a 0.8 lagging power factor.

Solution

$$\text{output} = 150 \times 0.8 = 120 \text{ kW}$$

$$\begin{aligned}\text{losses} &= I_1^2 R_1 + I_c^2 R_c + I_2^2 R_2 \\ &= (63.65)^2 \times 0.2 + (0.2427)^2 \times 10{,}000 + (625)^2 \times 0.002 \\ &= 2.18 \text{ kW}\end{aligned}$$

where I_1, I_2, and I_c are determined in Example 3.6.

$$\text{input} = 120 + 2.18 = 122.18 \text{ kW}$$

$$\text{efficiency} = \frac{120}{122.18} = 98.2\%$$

From Example 3.6, we have $V_{\text{no-load}} = 2455$ V. Thus, (3.23) yields

$$\text{percent regulation} = \frac{V_1 - aV_2}{aV_2} \times 100$$

$$= \frac{2455 - 2400}{2400} \times 100 = 2.3\%$$

Example 3.10 The transformer of Example 3.9 operates on full-load 0.8 lagging power factor for 12 h, on no-load for 4 h, and on half-full load unity power factor for 8 h. Calculate the all-day efficiency.

Solution

$$\text{output for 24 h} = (150 \times 0.8 \times 12) + (0 \times 4) + (150 \times ½ \times 8)$$

$$= 2040 \text{ kWh}$$

The losses for 24 h are

$$\text{core loss} = (0.2427)^2 \times 10{,}000 \times 24 = 14.14 \text{ kWh}$$

I^2R loss on full load for 12 h

$$= 12[(63.65)^2 \times 0.2 + (625)^2 \times 0.002] = 19.1 \text{ kWh}$$

I^2R loss on ½-load for 8 h

$$= 8\left[\left(\frac{63.65}{2}\right)^2 \times 0.2 + \left(\frac{625}{2}\right)^2 \times 0.002\right] = 3.18 \text{ kWh}$$

$$\text{total losses for 24 h} = 14.14 + 19.1 + 3.18 = 36.42 \text{ kWh}$$

$$\text{input for 24 h} = 2040 + 36.42 = 2076.42 \text{ kWh}$$

$$\text{all-day efficiency} = \frac{2040}{2076.42} = 98.2\%$$

3.9 APPROXIMATE EQUIVALENT CIRCUITS

Let us review the results of Example 3.6. Notice that the no-load current $I_0 = 1.6$ A, whereas the load current $I_1 = 63.6$ A. Also, $E_1 = 2427$ V and $V_1 = 2455$ V. In a well-designed commercial transformer, $I_0 << I_1$ and $E_1 \approx V_1$. As a result, we can move the shunt (no-load) circuit to the primary terminals (extreme left, as in Fig. 3.14). The circuit shown in Fig. 3.14 is known as the *approximate equivalent circuit* of the transformer. The principal advantage of using the approximate equivalent circuit is the reduction in the amount of complex arithmetic computations. In order to compare

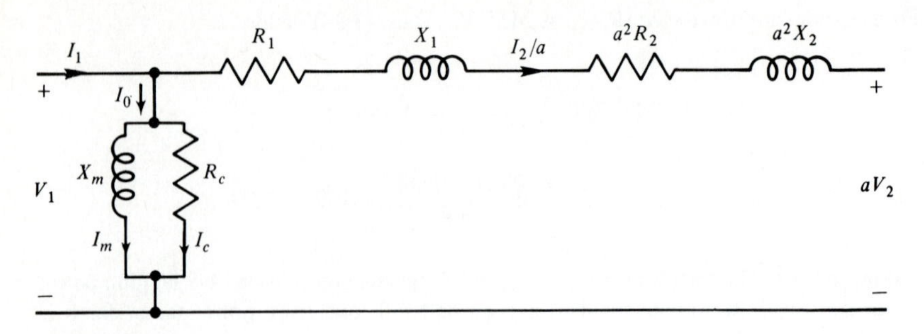

FIGURE 3-14
An approximate equivalent circuit referred to the primary.

the calculations involved in the approximate equivalent circuit, let us solve the problem of Example 3.6 again as follows.

Example 3.11 An approximate equivalent circuit of a transformer is shown in Fig. 3.14. Using this circuit, repeat the calculations of Example 3.6 and compare the results. Draw a phasor diagram showing all the voltages and currents of the circuit of Fig. 3.14.

Solution

Using Example 3.6, we have

$$aV_2 = 2400 \angle 0^0 \text{ V}$$

$$\frac{I_2}{a} = 50 - j37.5 \text{ A}$$

$$R_1 + a^2R_2 = 0.4 \ \Omega$$

$$X_1 + a^2X_2 = 0.9 \ \Omega$$

Hence

$$V_1 = (2400 + j0) + (50 - j37.5)(0.4 + j0.9)$$

$$= 2453 + j30 = 2453 \angle 0.7° \text{ A}$$

$$I_c = \frac{2453 \angle 0.7°}{10 \times 10^3} = 0.2453 \angle 0.7° \text{ A}$$

$$I_m = \frac{2453 \angle 0.7°}{1550 \angle 90°} = 1.58 \angle -89.3° \text{ A}$$

$$I_0 = 0.2453 - j1.58 \text{ A}$$

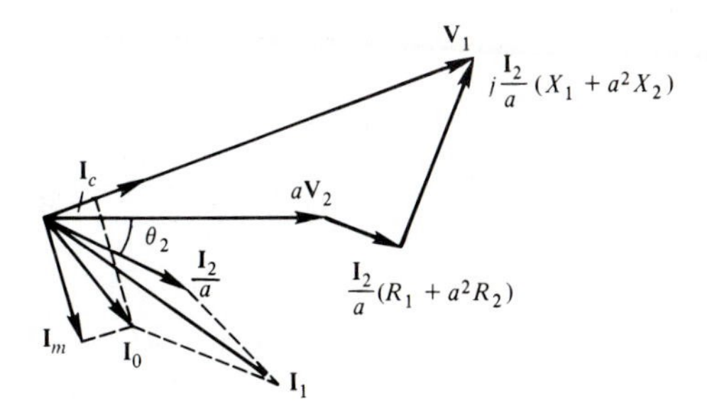

FIGURE 3-15
Phasor diagram corresponding to the equivalent circuit of Fig. 3.14.

$$I_1 = 50.25 - j39.08 = 63.66 \angle -37.9° \text{ A}$$

The phasor diagram is shown in Fig. 3.15.

$$\text{percent regulation} = \frac{2453 - 2400}{2400} \times 100 = 2.2\%$$

$$\text{efficiency} = \frac{120 \times 10^3}{120 \times 10^3 + (63.66)^2(0.4) + (0.2453)^2(10 \times 10^3)}$$

$$= 0.982 = 98.2\%$$

Notice that the approximate circuit yields results that are sufficiently accurate.

The circuit shown in Fig. 3.14 is the approximate equivalent circuit of the transformer referred to the primary. As with the exact equivalent circuit, we may obtain the approximate equivalent circuit referred to either of the two windings. The next example illustrates the procedure.

Example 3.12 The ohmic values of the circuit parameters of a transformer, having a turns ratio of 5, are $R_1 = 0.5\ \Omega$; $R_2 = 0.021\ \Omega$; $X_2 = 0.12\ \Omega$; $R_c = 350\ \Omega$, referred to the primary; and $X_m = 98\ \Omega$, referred to the primary. Draw the approximate equivalent circuits of the transformer, referred to *(a)* the primary and *(b)* the secondary. Show the numerical values of the circuit parameters.

Solution

The circuits are shown in Fig. 3.16 *(a)* and *(b)*, respectively. The calculations are as follows:

(a)

$$R' \equiv R_1 + a^2R_2 = 0.5 + (5)^2(0.021) = 1.025\ \Omega$$

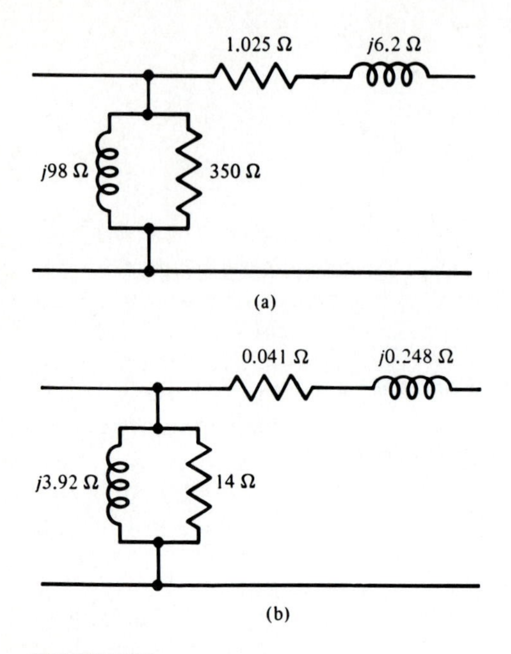

FIGURE 3-16
Example 3.12.

$$X' \equiv X_1 + a^2 X_2 = 3.2 + (5)^2(0.12) = 6.2\ \Omega$$

$$R_c' = 350\ \Omega$$

$$X_m' = 98\ \Omega$$

(b)

$$R'' \equiv \frac{R_1}{a^2} + R_2 = \frac{0.5}{25} + 0.021 = 0.041\ \Omega$$

$$X'' \equiv \frac{X_1}{a^2} + X_2 = \frac{3.2}{25} + 0.12 = 0.248\ \Omega$$

$$R_c'' = \frac{350}{25} = 14\ \Omega$$

$$X_m'' = \frac{98}{25} = 3.92\ \Omega$$

3.10 EQUIVALENT CIRCUITS FROM TEST DATA

The preceding two sections have clearly indicated the utility of transformer equivalent circuits. The parameters—the resistances and the reactances—of the equivalent circuits may be obtained from the following tests.

Open-Circuit (or No-Load) Test. In this test, one winding is open-circuited, and a voltage—usually, rated voltage at rated frequency—is applied to the other winding. The voltage, current, and power at the terminals of this winding are measured. The open-circuit voltage of the second winding is also measured, and from this measurement a check on the turns ratio can be obtained. It is usually convenient to apply the test voltage to the winding that has a voltage rating equal to that of the available power source. This means that in step-up voltage transformers, the open-circuit voltage of the second winding will be higher than the applied voltage, sometimes much higher. Care must be exercised in guarding the terminals of this winding to ensure safety for test personnel and to prevent these terminals from getting close to other electric circuits, instrumentation, grounds, and so forth.

In presenting the no-load parameters obtainable from test data, it is assumed that voltage is applied to the primary and that the secondary is open-circuited. The no-load power loss is equal to the wattmeter reading in this test; core loss is found by subtracting the ohmic loss in the primary, which is usually small and may be neglected in some cases. Thus, if P_0, I_0, and V_0 are the input power, current, and voltage, respectively, the core loss P_c is given by

$$P_c = P_0 - I_0^2 R_1 \tag{3.28}$$

The primary induced voltage is expressed in phasor form by

$$E_1 = V_0 \angle 0^\circ - (I_0 \angle \theta_0)(R_1 + jX_1) \tag{3.29}$$

where $\theta_0 \equiv$ no-load power-factor angle = $\cos^{-1}(P_0/V_0I_0)$. Other circuit quantities are found from

$$R_c = \frac{E_1^2}{P_c} \tag{3.30}$$

$$I_c = \frac{P_c}{E_1} \tag{3.31}$$

$$I_m = \sqrt{I_0^2 - I_c^2} \tag{3.32}$$

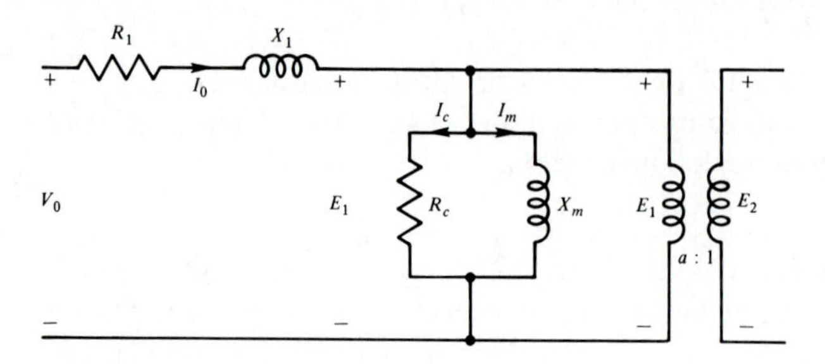

FIGURE 3-17
Equivalent circuit for open-circuit test.

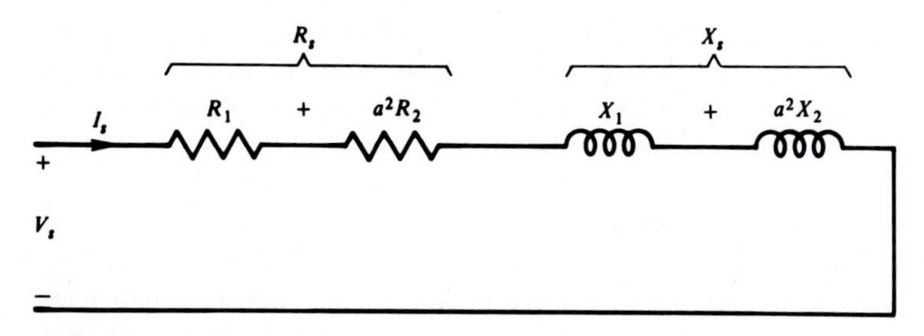

FIGURE 3-18
Equivalent circuit for short-circuit test.

$$X_m = \frac{E_1}{I_m} \tag{3.33}$$

$$a \approx \frac{V_0}{E_2} \tag{3.34}$$

The equivalent circuit for the open-circuit test is shown in Fig. 3.17.

Short-Circuit Test. In this test, one winding is short-circuited across its terminals, and reduced voltage is applied to the other winding. This reduced voltage is of such a magnitude as to cause a specific value of current—usually, rated current—to flow in the short-circuited winding. Again, the choice of the winding to be short-circuited is usually determined by the measuring equipment available for us in the test. However, care must be taken to note which winding is short-circuited, for this determines the reference winding for expressing the impedance components obtained by this test. Let the secondary be short-circuited and the reduced voltage be applied to the primary.

With a very low voltage applied to the primary winding, the core-loss current and magnetizing current become very small, and the equivalent circuit reduces to that of Fig. 3.18. Thus if P_s, I_s, and V_s are, respectively, the input power, current, and voltage under short circuit, then, referred to the primary,

$$Z_s = \frac{V_s}{I_s} \tag{3.35}$$

$$R_1 + a^2 R_2 \equiv R_s = \frac{P_s}{I_s^2} \tag{3.36}$$

$$X_1 + a^2 X_2 \equiv X_s = \sqrt{Z_s^2 - R_s^2} \tag{3.37}$$

Given R_1 and a, R_2 can be found from (3.36). In (3.37) it is usually assumed that the leakage reactance is divided equally between the primary and the secondary; that is,

$$X_1 = s^2 X_2 = \tfrac{1}{2} X_s \tag{3.38}$$

Notice that the open-circuit test is performed with the instrumentation on the low-voltage side, whereas the instrumentation is on the high-voltage side for the short-circuit test. Thus we have R_c and X_m referred to the secondary (low-voltage side) and R_1, R_2, X_1, and X_2 referred to the primary (high-voltage side). The equivalent circuit, however, must be referred to one side only. The next two examples show how we interpret the data obtained from the two tests.

Example 3.13 A certain transformer, with its secondary open, takes 80 W of power at 120 V and 1.4 A. The primary winding resistance is 0.25 Ω and the leakage reactance is 1.2 Ω. Evaluate the magnetizing reactance, X_m, and the core-loss equivalent resistance, R_c.

Solution

The no-load power factor angle is

$$\theta_0 = \cos^{-1} \frac{80}{1.4 \times 120} = -61.6°$$

The primary induced voltage is, from (3.29),

$$E_1 = 120 \angle 0° - 1.4 \angle -61.6° \, (0.25 + j1.25)$$

$$\approx 118.29 \text{ V}$$

Therefore, from (3.28) and (3.30) through (3.33),

$$R_c = \frac{(118.29)^2}{80 - (1.4)^2(0.25)} = 176\ \Omega$$

$$I_c = \frac{118.29}{176} = 0.672\ \text{A}$$

$$I_m = \sqrt{(1.4)^2 - (0.672)^2} = 1.228\ \text{A}$$

$$X_m = \frac{118.29}{1.228} = 96.3\ \Omega$$

Example 3.14 The results of open-circuit and short-circuit tests on a 25-kVA 440/220-V 60-Hz transformer are as follows.

Open-circuit test: Primary open-circuited, with instrumentation on the low-voltage side. Input voltage, 220 V; input current, 9.6 A; input power, 710 W.

Short-circuit test: Secondary short-circuited, with instrumentation on the high-voltage side. Input voltage, 42 V; input current, 57 A; input power, 1030 W.

Obtain the parameters of the exact equivalent circuit (Fig. 3.10), referred to the high-voltage side. Assume that $R_1 = a^2R_2$ and $X_1 = a^2X_2$.

Solution

From the short-circuit test:

$$Z_{s1} = \frac{42}{57} = 0.737\ \Omega$$

$$R_{s1} = \frac{1030}{(57)^2} = 0.317\ \Omega$$

$$X_{s1} = \sqrt{(0.737)^2 - (0.317)^2} = 0.665\ \Omega$$

Consequently,

$$R_1 = a^2R_2 = \frac{1}{2}R_{s1} = 0.158\ \Omega, \quad R_2 = 0.0395\ \Omega$$

$$X_1 = a^2X_2 = \frac{1}{2}X_{s1} = 0.333\ \Omega, \quad X_2 = 0.0832\ \Omega$$

From the open-circuit test:

$$\theta_0 = \cos^{-1}\frac{710}{(9.6)(220)} = \cos^{-1} 0.336 = -70°$$

$$E_2 = 220\ \angle 0° - (9.6\ \angle{-70°})\ (0.0395 + j0.0832) \approx 219\ \angle 0°\ \text{V}$$

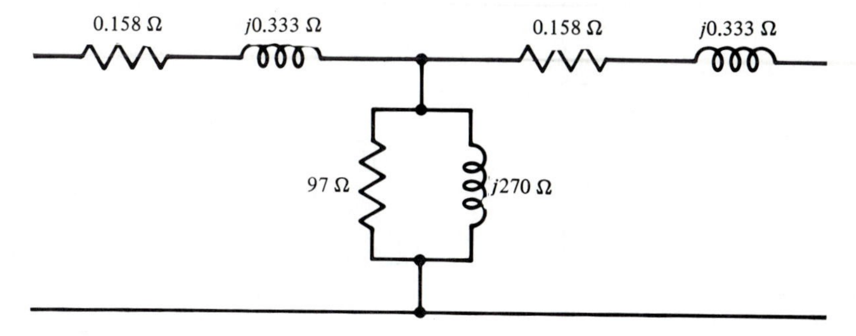

FIGURE 3-19
Example 3.14.

$$P_{c2} = 710 - (9.6)^2(0.0395) \approx 710 \text{ W}$$

$$R_{c2} = \frac{(219)^2}{710} = 67.5 \ \Omega$$

$$I_{c2} = \frac{219}{67.5} = 3.24 \text{ A}$$

$$I_{m2} = \sqrt{(9.6)^2 - (3.24)^2} = 9.03 \text{ A}$$

$$X_{m2} = \frac{219}{9.03} = 24.24 \ \Omega$$

$$X_{m1} = a^2 X_{m2} = 97 \ \Omega$$

$$R_{c1} = a^2 R_{c2} = 270 \ \Omega$$

The equivalent thus obtained is shown in Fig. 3.19.

3.11 TRANSFORMER POLARITY

Polarities of a transformer identify the relative directions of induced voltages in the two windings. The polarities result from the relative directions in which the two windings are wound on the core. For operating transformers in parallel, and for various transformer connections, it is necessary that we know the relative polarities. Polarities can be checked by a simple test, requiring only voltage measurements with the transformer on no-load. In this test, rated voltage is applied to one winding, and an electrical connection is made between one terminal from one winding and one from the other, as shown in Fig. 3.20. The voltage across the two remaining terminals (one

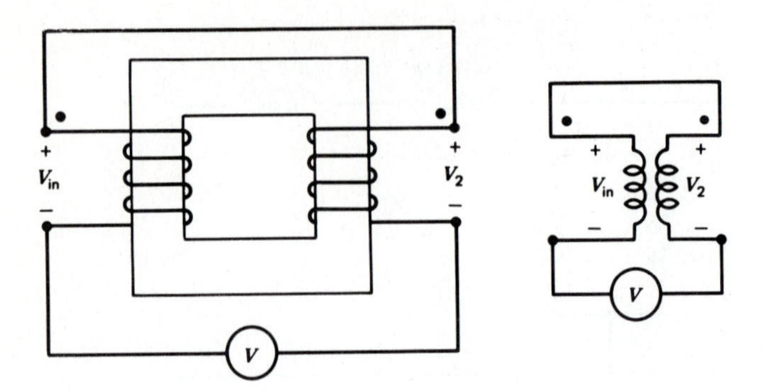

FIGURE 3-20
Polarity test on a transformer.

from each winding) is then measured. If this measured voltage is *larger* than the input test voltage, the polarity is *additive*; if smaller, the polarity is *subtractive*.

A standard method of marking transformer terminals is as follows. The high-voltage terminals are marked H1, H2, H3, . . ., with H1 on the right-hand side of the case when facing the high-voltage side. The low-voltage terminals are designated X1, X2, X3, . . ., and X1 may be on either side, adjacent to H1 or diagonally opposite. The two possible locations of X1 with respect to H1 for additive and subtractive polarities are shown in Fig. 3.21. The numbers must be so arranged that the voltage difference between any two leads of the same set, taken in order from smaller to larger numbers, must be of the same sign as that between any other pair of the set taken in the same order. Furthermore, when the voltage is directed from H1 to H2, it must simultaneously be directed from X1 to X2. Additive polarities are required by the American National Standards Institute (ANSI) in large (>200 kVA) high-voltage (>8660 V) power transformers. Small transformers have subtractive polarities (which reduce voltage stress between adjacent leads).

3.12 AUTOTRANSFORMERS

In contrast to the two-winding transformers considered so far, the autotransformer is a single-winding transformer having a tap brought out at an intermediate point. Thus, as shown in Fig. 3.22, *ac* is the single winding (wound on a laminated core) and *b* is the intermediate point where the tap is brought out. The autotransformer may be used as either a step-up or a step-down operation, like a two-winding transformer. Considering a step-down arrangement, let the primary applied (terminal) voltage be V_1, resulting in a magnetizing current and a core flux, ϕ_m. Let the secondary be open-circuited. Then the primary and secondary voltages obey the same rules as in a two-winding transformer, and we have

$$\frac{V_1}{V_2} = \frac{E_1}{E_2} = \frac{N_1}{N_2} = a \tag{3.39}$$

with $a > 1$ for step-down.

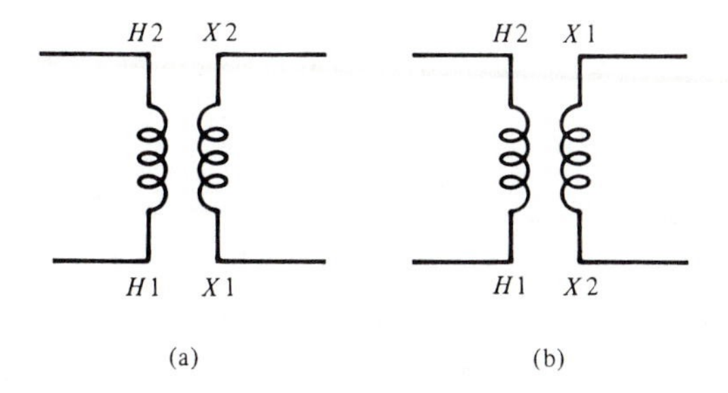

FIGURE 3-21
(a) Subtractive and *(b)* additive polarities of a transformer.

Furthermore, ideally,

$$V_1 I_1 = V_2 I_2 \tag{3.40}$$

and

$$\frac{V_1}{V_2} = \frac{I_2}{I_1} = a \tag{3.41}$$

Neglecting the magnetizing current, we must have the mmf balance equation as

$$N_2 I_3 = (N_1 - N_2) I_1$$

or

$$I_3 = \frac{N_1 - N_2}{N_2} I_1 = (a - 1) I_1 = I_2 - I_1 \tag{3.42}$$

which agrees with the current-flow directions shown in Fig. 3.22.

The apparent power delivered to the load may be written as

$$P = V_2 I_2 = V_2 I_1 + V_2 (I_2 - I_1) \tag{3.43}$$

In (3.43) the power is considered to consist of two parts:

$$V_2 I_1 \equiv P_c \equiv \text{conductively transferred power through } bc \tag{3.44}$$

$$V_2 (I_2 - I_1) \equiv P_i \equiv \text{inductively transferred power through } ab \tag{3.45}$$

These powers are related to the total power by

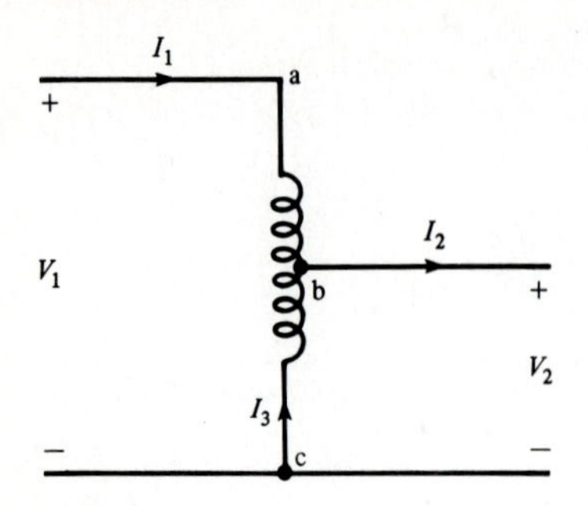

FIGURE 3-22
A step-down autotransformer.

$$\frac{P_i}{P} = \frac{I_2 - I_1}{I_2} = \frac{a - 1}{a} \tag{3.46}$$

and

$$\frac{P_c}{P} = \frac{I_1}{I_2} = \frac{1}{a} \tag{3.47}$$

where $a < 1$.

For a step-up transformer the power ratios are obtained as follows:

$$P = V_1 I_1 = V_1 I_2 + V_1 (I_1 - I_2)$$

implying that the total apparent power consists of two parts:

$P_c = V_1 I_2 =$ conductively transferred power

$P_i = V_1(I_1 - I_2) =$ inductively transferred power.

Hence

$$\frac{P_i}{P} = \frac{I_1 - I_2}{I_1} = 1 - a \tag{3.48}$$

and

$$\frac{P_c}{P} = \frac{I_2}{I_1} = a \tag{3.49}$$

where $a < 1$.

Example 3.15 A 10-kVA 440/110-V two-winding transformer is reconnected as a step-down 550/440-V autotransformer. Compare the volt-ampere rating of the autotransformer with that of the original two-winding transformer, and calculate P_i and P_c.

Solution

Refer to Fig. 3.22. The rated current in the 110-V winding (or in *ab*) is

$$I_1 = \frac{10,000}{110} = 90.91 \text{ A}$$

The current in the 440-V winding (or in *bc*) is

$$I_3 = I_2 - I_1 = \frac{10,000}{440} = 22.73 \text{ A}$$

which is the rated current of the winding *bc*. Thus the load current is

$$I_2 = I_1 + I_3 = 90.91 + 22.73 = 113.64 \text{ A}$$

Check: For the autotransformer

$$a = \frac{550}{440} = 1.25$$

and

$$I_2 = aI_1 = 1.25 \times \frac{10,000}{110} = 113.64 \text{ A}$$

which agrees with I_2 calculated above. Hence the rating of the autotransformer is

$$P_{\text{auto}} = V_2 I_2 = V_2 a I_1 = 440 \times 113.64 = 50 \text{ kVA}$$

Thus the inductively supplied apparent power is

$$P_i = V_2(I_2 - I_1) = \frac{a - 1}{a} P = \frac{1.25 - 1}{1.25} \times 50 = 10 \text{ kVA}$$

which is the volt-ampere rating of the two-winding transformer.

The conductively supplied power is

$$P_c = \frac{P}{a} = \frac{50}{1.25} = 40 \text{ kVA}$$

Example 3.16 Repeat the problem of Example 3.15 for a 440/550-V step-up connection.

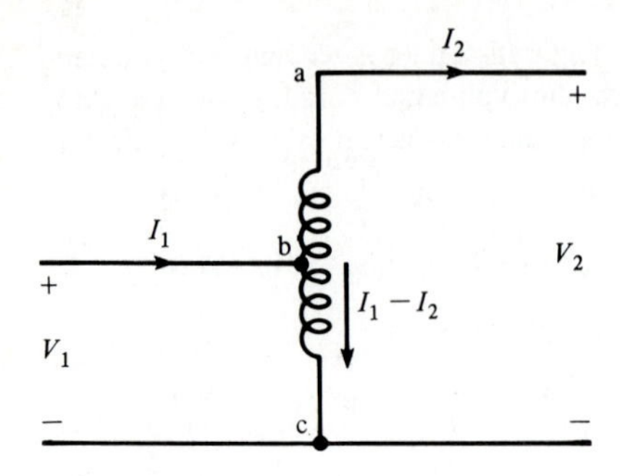

FIGURE 3-23
A step-up autotransformer.

Solution

The step-up connection is shown in Fig. 3.23. The rating of the winding *ab* is 110 V and the load current I_2 flows through *ab*. Hence

$$I_2 = \frac{10{,}000}{110} = 90.91 \text{ A}$$

The output voltage is

$$V_2 = 550 \text{ V}$$

Thus the volt-ampere rating of the autotransformer is

$$V_2 I_2 = 550 \times \frac{10{,}000}{110} = 50 \text{ kVA}$$

which is the same as in Example 3.16.

Thus the power transferred conductively is

$$V_1 I_2 = 440 \times 90.91 = 40 \text{ kVA}$$

and the power transferred inductively is

$$50 - 40 = 10 \text{ kVA}$$

Consequently, a two-winding transformer connected as an autotransformer will have a volt-ampere rating $a/(a - 1)$ times its rating as a two-winding transformer.

3.13 TRANSFORMER CAPACITANCE

Capacitance exists among many physical parts of all transformers. The principal elements of capacitance are between turns of the windings, between layers of the windings, between the core and the turns, between turns and housing (or case, if metallic), and between the terminals and external leads. These capacitive elements generally are distributed throughout the transformer volume, but the combined effect of the distributed capacitance can often be treated in terms of lumped capacitances and can be combined with the equivalent circuit previously discussed. The effects of capacitance are significant only in higher frequency and pulse applications, although the voltage gradients between turns attributable to interturn capacitance must be considered in designing the insulation system for high-voltage transformers at power frequencies.

A lumped circuit containing three capacitances is commonly used to describe the distributed capacitance of a transformer . This circuit is shown in Fig. 3.24*(a)*. The ideal transformer can be eliminated from the equivalent circuit, except for the mutual capacitance term C_{12}. This variation is shown in Fig. 3.24*(b)*.

These equivalent circuits containing capacitances are suitable for modeling transformers at frequencies where the distributed capacitance is significant. Methods for measuring the lumped equivalent capacitances C_1, C_2, and C_{12} are given in IEEE Standard No. 390-1975. Analysis of the equivalent circuits containing capacitance using "hand-calculation" techniques is extremely tedious, and computer circuit analysis programs are almost a necessity.

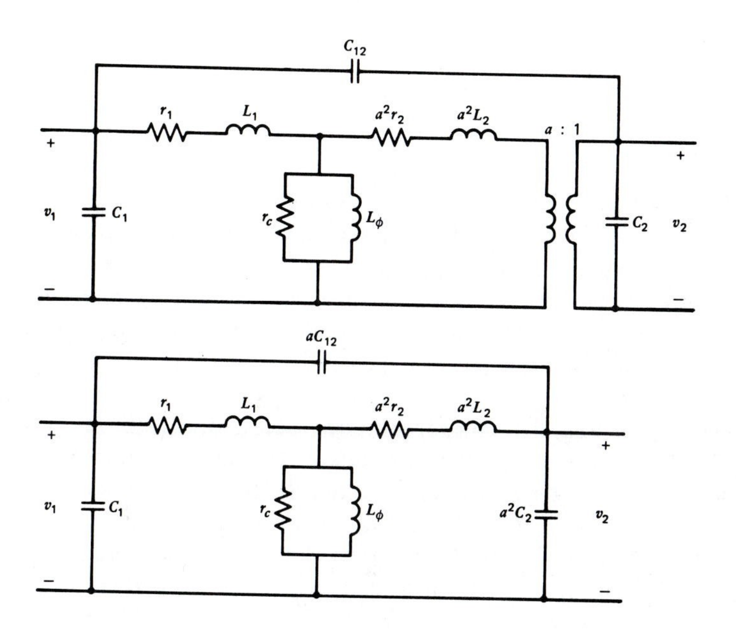

FIGURE 3-24

(a) Equivalent circuit including lumped capacitance to represent distributed capacitances. C_1 = winding 1 (primary) distributed capacitance; C_2 = winding 2 (secondary) distributed capacitance; C_{12} = bridging and direct leakage capacitance between input and output. *(b)* Modification of the circuit of Fig. 3.24*(a)* to eliminate the ideal transformer.

3.14 INRUSH CURRENT

Much of this chapter has been devoted to the steady-state theory and analysis of transformers. However, there is one aspect of transformer transient characteristics that deserves special mention: the potentially high current that may appear briefly upon initial energizing of a transformer from a sinusoidal voltage source. This initial current peak, termed *inrush* current, is a function of the instant at which the transformer is connected to the voltage source which, for practical purposes, is uncontrollable. It may result in a current peak many times the rated transformer current or it may not be observable. It is the bane of the machinery laboratory student (and instructor) and is a very significant concern during the energizing of large power transformers. It can be simply explained as follows.

Inrush current is modified somewhat by the load on the transformer. It is most observable at no-load, and this condition will be used to illustrate this phenomenon. Neglecting core loss and primary resistance, the relationship between voltage and flux at no-load in a transformer is:

$$V_m = N\frac{d\phi}{dt} = V_m \sin(\omega t + \alpha) \tag{3.50}$$

where α = the argument of the sine term at $t = 0$. The solution to this linear, first-order differential equation can be found to be

$$\phi = \Phi_m[\cos\alpha - \cos(\omega t + \alpha)] \tag{3.51}$$

where $\Phi_m = V_m/\omega N$.

The flux Φ consists of two components, the transient and steady state. Since we have neglected core loss and primary resistance, the transient term, $\Phi_m \cos\alpha$, contains no decrement factor. In any practical transformer, of course, there is such a term and the transient component decays exponentially with a time constant that is a function of the transformer resistance and core loss. In small transformers this time constant is usually of the order of a few milliseconds; in large power transformers, it may be 1 sec or more.

Remembering that the transient component does decay, we look at the implications of (3.47). First, it is seen that the magnitude of the transient component is a function of the instant at which the transformer is connected to the voltage source, which is given by the angle, α, in (3.47). If α is $\pi/2$ (which means that the source voltage at $t = 0$ is at its positive maximum), the transient term is zero and there is no inrush current. If α is zero, the transient term is maximum and a severe current peak can occur during the first few cycles after the transformer is energized. This is the worst case and is illustrated in Fig. 3.25. It is seen that the two flux components are additive after one-fourth of a period and, at $\omega t = \pi$, the flux in the transformer core would be $2\Phi_m$, neglecting decay in the transient term. Actually, this first peak of flux may be more than twice the normal flux maximum, since there is often residual magnetism in the core when initially energized. If the residual flux is in the proper direction, it will add to the flux calculated from (3.47).

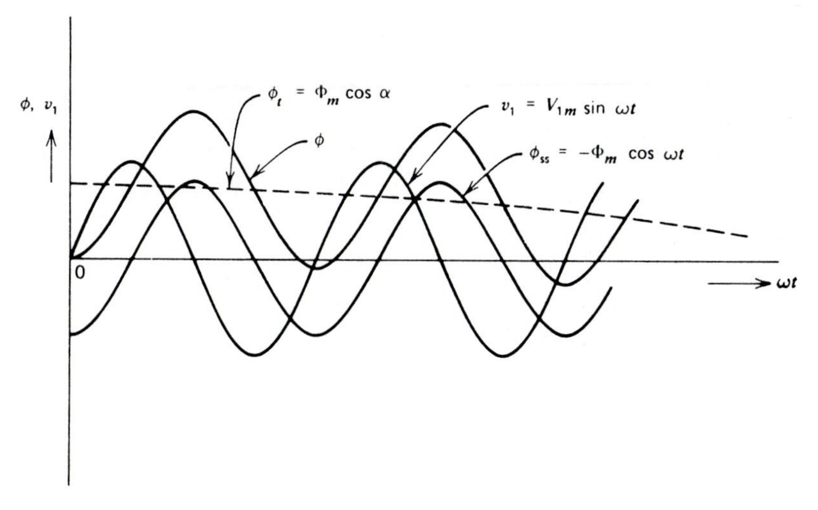

FIGURE 3-25
Flux relationships following initial energizing of a transformer switched on at $\alpha = 0$, Equation (3.51.)

The current peak associated with this flux peak is proportionately much larger due to the effect of core saturation. In most transformers the core and windings are designed so that, under rated voltage, the maximum steady-state flux is around the knee of the saturation curve. Therefore, the preceding condition described during the initial energizing of a transformer, which requires a flux of twice or more times the maximum steady-state flux, may demand an exciting current many hundreds of times the normal excitation current because of core saturation. This possibility must be considered in choosing the source for the circuit and for protection in any circuit involving a transformer. IEEE Standard No. 389-1978 gives a convenient approximation for relating the first inrush current peak (in amperes) to transformer design parameters:

$$I_{pK} = \frac{b(NA_c)(2B_m + B_r - B_s)}{\mu_0 N^2 A_s} \tag{3.52}$$

where

b = coil length along coil axis, m

A_s = total area of the space enclosed by the mean turn of the excited coil, m^2

A_c = net cross-sectional area of the magnetic core, m^2

N = number of turns of the excited coil winding

B_r = residual flux density at $t = 0$, T

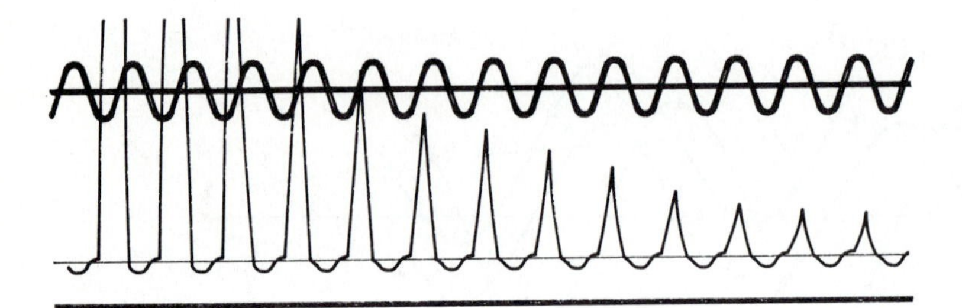

FIGURE 3-26
Starting-transient current wave for a transformer.

B_m = design maximum, steady-state flux density, T

B_s = saturation flux density, T

An oscillogram of the waveform of the inrush current for a transformer is shown in Fig. 3.26.

3.15 PULSE TRANSFORMERS

Transformers that are normally operated from a discontinuous source of excitation are known as pulse transformers. This mode of excitation results in some differences in both the design and analysis of pulse transformers compared to transformers designed for operation from a continuous excitation source, such as a sinusoidal voltage. A few analysis techniques will be summarized in this section, and several problems at the end of this chapter will permit the reader to explore this subject further.

The average power of a pulse transformer is generally very low, but the peak power may be very large. The earliest transformers specially designed for pulse excitation were developed for radar applications in World War II, and these were of very high peak-power capability. Similar pulse transformers are used in linear accelerators and similar equipment. Most recently, medium- and low-peak power pulse transformers have been developed for many electronic and control applications, such as in the gate-firing circuits of power thyristors. Pulse excitation usually consists of a rectangular voltage pulse or a short burst of high-frequency sine pulses, but many other pulse shapes are commonly used. Both the magnitude and the repetition rate of the pulse determine the average transformer power. Pulse transformer parameters are defined in terms of a standard pulse waveform, shown in Fig. 3.27. This waveform can represent either a voltage or current waveform resulting from a rectangular pulse excitation and, therefore, is described by a generalized symbol, A. The principal pulse parameters related to this wave are also shown in Fig. 3.27. It is seen that these parameter definitions follow those used in circuit analysis and control theory. The actual shape and magnitude of the output voltage or current pulses resulting from excitation from a rectangular pulse in a specific transformer can be found by analysis of the complete equivalent circuit of Fig. 3.24.

As has been noted, such analysis requires computer simulation techniques and is tedious and time consuming. It has been observed, however, that considerable simplification of the complete equivalent circuit is possible during various portions of the pulse. These simplifications are shown in Fig. 3.28, with all parameters assumed to be expressed on secondary (output winding) terms. All of these parameters have been defined previously except for C_D, which is the distributed capacitance of the secondary winding, and C_L, which is the load and terminating cable capacitance.

Both the average and peak volt-ampere and power ratings must be considered in the design and use of pulse transformers, and these may be several orders of magnitude apart. Another rating of significance in pulse transformers is the *voltage-time product* rating. This rating is defined as the maximum voltage-time integral of a rectangular voltage pulse that can be applied to a winding before core saturation effects cause the resultant exciting current pulse waveform to deviate from a linear ramp by a given percentage. The definition of the permeability of the magnetic core is also modified for pulse application to be "the value of amplitude permeability when the rate of change flux density is held substantially constant over a period of time during each cycle"; or

$$\mu_p = \frac{1}{\mu}\frac{\Delta B}{\Delta H} \tag{3.53}$$

where

ΔB = change in flux density during the stated time interval
ΔH = associated change in magnetic field strength

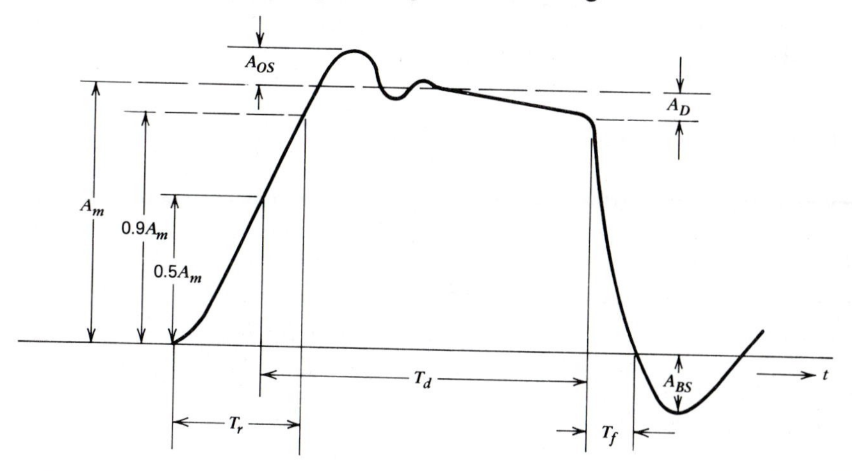

FIGURE 3-27
Standard voltage or current pulse shape used in the analysis and specification of pulse transformers. A_m = pulse amplitude, A_{OS} = overshoot, A_D = droop or tilt, A_{BS} = backswing, T_d = pulse duration, T_r = pulse rise time, and T_f = pulse fall time.

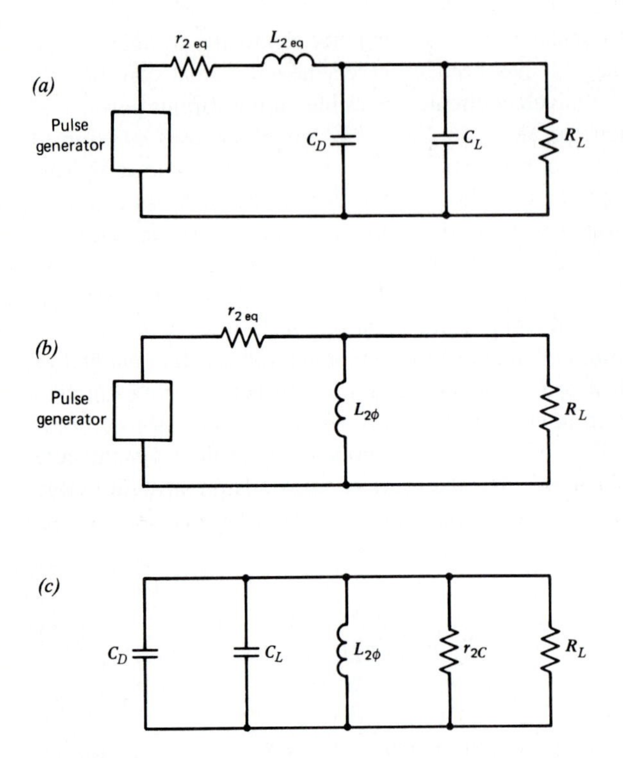

FIGURE 3-28
(a) Simplified equivalent circuit for evaluating the lead edge of a transformer pulse (rise time) with a resistive load. *(b)* Simplified equivalent circuit for evaluating the top of a transformer pulse. *(c)* Simplified equivalent circuit for evaluating the trailing edge (fall time and backswing) of a transformer pulse.

3.16 THREE-PHASE TRANSFORMER CONNECTIONS

In many power, rectifier, and motor control applications, polyphase circuits are used to reduce the losses in conductors and machines. The most common polyphase system is the three-phase system, but two-phase, six-phase, and even twelve-phase systems are used in various applications. Most power generation, transmission, and distribution systems, and many ac motor control systems are three-phase systems. The power is generated by three-phase generators and is transmitted by three-phase transmission lines. Obviously, these generators and transmission lines must be linked by three-phase transformers. The primary and secondary windings of single-phase transformers may be interconnected to obtain three-phase transformer banks.

Some of the factors governing the choice of connections are as follows:

1. Availability of a neutral connection for grounding, protection, or load connections.

2. Insulation to ground and voltage stresses.

3. Availability of a path for the flow of third-harmonic (exciting) currents and zero-sequence (fault) currents.

4. Need for partial capacity with one unit out of service.

5. Parallel operation with other transformers.

6. Operation under fault conditions.

7. Economic considerations.

Keeping these factors in mind, we now consider some of the three-phase transformer connections.

Wye/Wye Connection

The wye/wye connection is shown in Fig. 3.29, where the terminal markings show subtractive polarities. Primary terminals are designated by *ABC*, whereas *abc* is used to indicate the secondary terminals. The phase relationships between the various voltages are shown in Fig. 3.29*(b)*.

The equilateral triangles superimposed on the phasor diagrams aid in the construction of the phasor diagrams. For instance, *OA*, *OB*, and *OC* denote the phase voltages of the primary and *O'a*, *O'b*, and *O'c* correspond to the phase voltages of the secondary. To determine the time-phase relationships of line voltages, we observe that the primary phase sequence is *ABC*. The phase of the voltage between *A* and *B* is found by tracing the circuit *AOB*, where *AO* is traversed in the positive direction and *OB* in the negative direction. The phasor relationship between the voltages around the circuit *AOB* is

$$V_{BA} + V_{OB} - V_{OA} = 0$$

or

$$V_{BA} = V_{OA} - V_{OB} = V_A$$

Thus we reverse *OB* and geometrically add it to *OA* to obtain *BA*, and hence V_{BA}. Other voltages are determined in a similar fashion.

From Fig. 3.29*(b)* it is clear that there is no phase shift between the primary and secondary voltages. Furthermore, if V is the voltage between lines, then $V/\sqrt{3}$ is the voltage across the phase, on the terminals of a wye-connected transformer. Thus, compared to a delta connection (discussed later), the wye-connected transformer will have fewer turns, will require windings of larger cross section, and will have less dielectric stress on the insulation. On the other hand, the main disadvantage of wye/wye-connected transformers is that such transformers have *roving neutrals* when supplying unbalanced loads. By "roving neutral" we mean that the potential of point

O in Fig. 3.29*(b)* is not fixed with respect to the lines, and may take any position within the triangle if the transformer supplies an unbalanced load. Fig. 3.30 shows two positions of O under two unbalanced loading conditions. One way to prevent the shifting of the neutral is to connect the primary neutral of the transformer to the neutral of the generator. In such a case, however, if there is a third harmonic component in the generator voltage waveform, there will be a third harmonic in the secondary voltage, and there will be corresponding triple frequency currents in the secondary circuits.

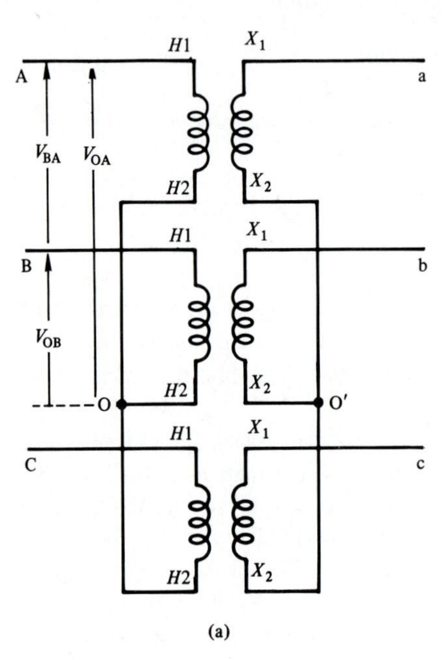

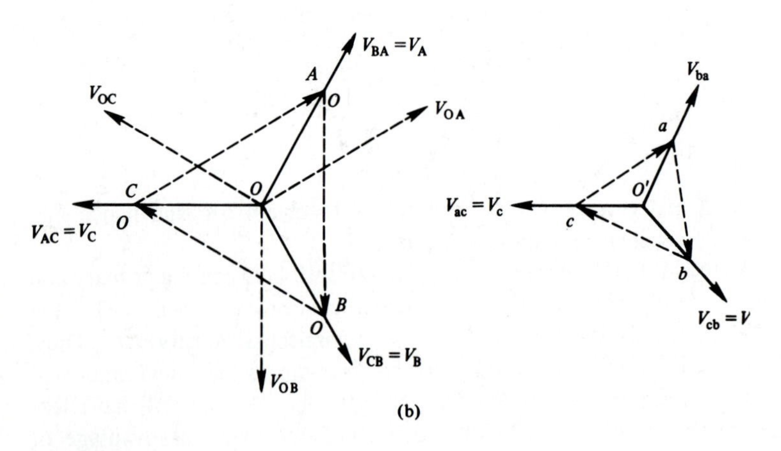

FIGURE 3-29
(a) Three-phase wye/wye connection; *(b)* phasor diagrams.

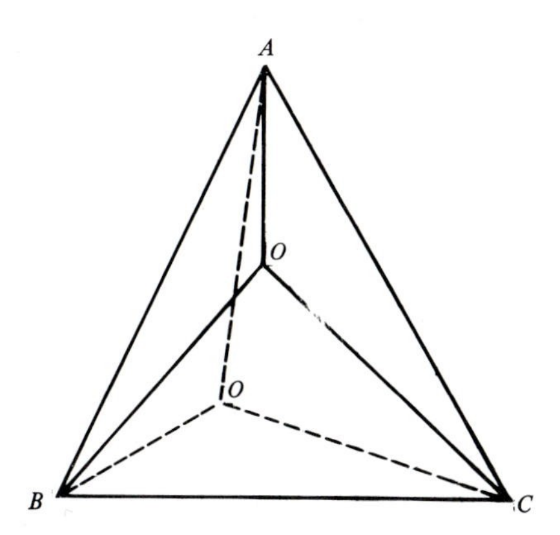

FIGURE 3-30
Roving neutral under unbalanced loads.

Delta/Delta Connection

Fig. 3.31 shows three transformers connected to make a three-phase delta/delta system. The phasor diagram showing the phase relationships of various voltages is also included in the figure. The transformers have subtractive polarities. Clearly, in such a connection, the individual windings of the transformers must be designed for full line voltages. This arrangement, however, has the advantage that one transformer can be out of service without an interruption in system operation except that the capacity with two transformers is proportionately reduced. The secondary voltages also tend to be slightly unbalanced with the loss of one transformer. The delta/delta connection provides a path for the flow of the third-harmonic magnetizing current, the current in each coil being in phase with that of the other. Hence third-harmonic currents and voltages do not appear on the line.

Delta/Wye and Wye/Delta Connections

Delta/wye and wye/delta connections of three-phase transformers are shown in Figs. 3.32 and 3.33, respectively. The polarities in both cases are subtractive. Notice the 30° phase shift between the line and phase voltages in the two cases.

This shift in the delta/wye connection is opposite to that in the wye/delta connection. These connections are particularly suited for high-voltage systems. The delta/wye connection is used for stepping up and the wye/delta for stepping down the voltage. The wye connection on the high-voltage side permits the grounding of the neutral. The delta connection offers a path for the flow of the third-harmonic currents.

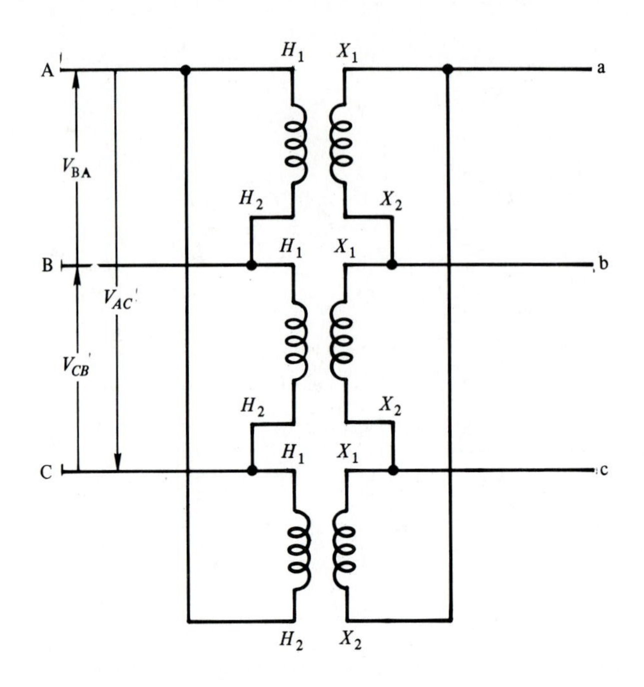

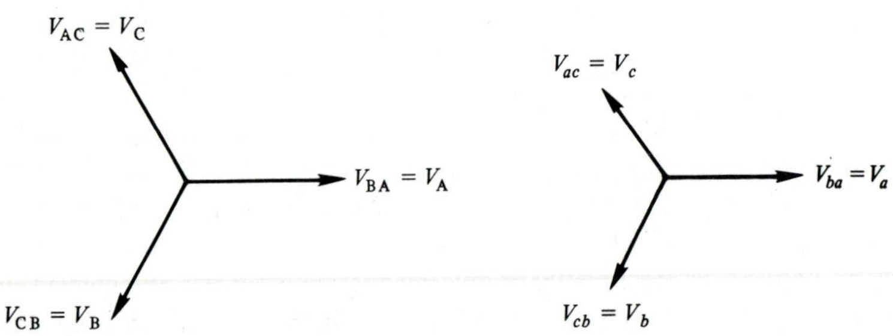

FIGURE 3-31
Delta/delta connection with subtractive polarities.

3.17 SPECIAL TRANSFORMER CONNECTIONS

In Section 3.16 we have studied three-phase transformer connections, where the primaries and/or the secondaries are connected in wye and/or delta. Such connections involved three transformers. There are certain three-phase transformer connections that require only two transformers. We will discuss three such connections here.

Open-Delta Connection

Recall from Section 3.16 that in the delta/delta connection, if one of the transformers is removed, the remaining two transformers can provide a three-phase system. In

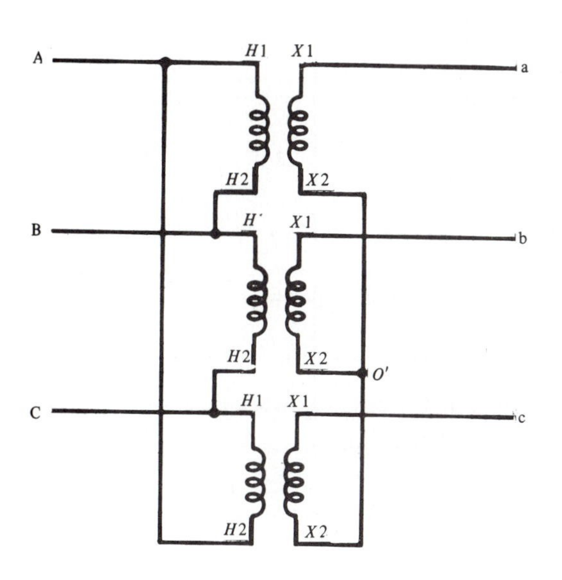

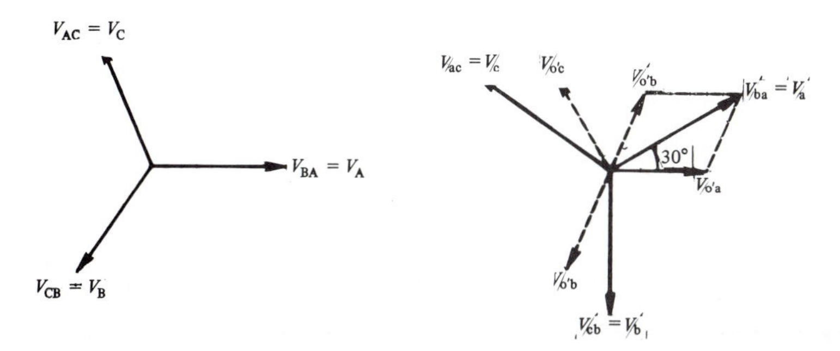

FIGURE 3-32
Delta/wye connection.

practice, such a connection is known as the *open-delta* or *V connection*, and is shown in Fig. 3.34. Obviously, the rating of an open-delta transformer bank is less than that of a delta/delta bank (assuming that each transformer of the two banks has the same rating). If each transformer of a delta/delta system is rated at V volts and I amperes, the line current for the system will be $\sqrt{3}\, I$. But for an open-delta system the line current cannot exceed I. Consequently, the combined rating of two transformers connected in open-delta form amounts to $(I/\sqrt{3}\, I)$ or 58 percent of the rating of the original three transformers.

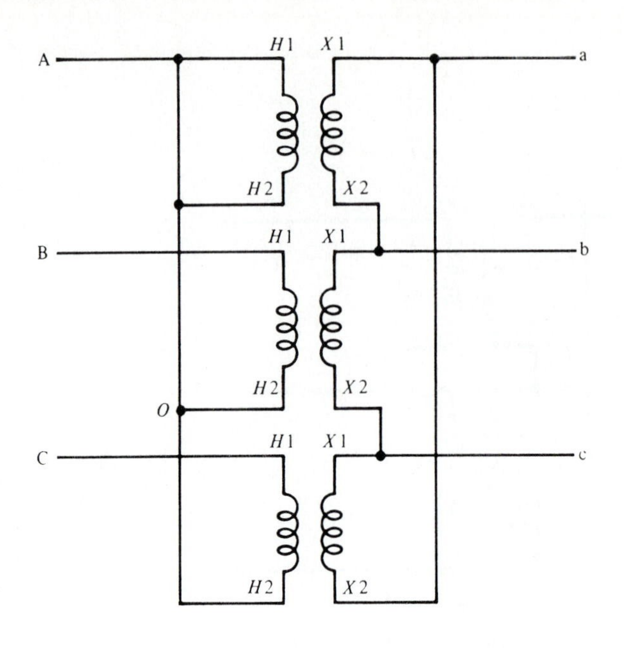

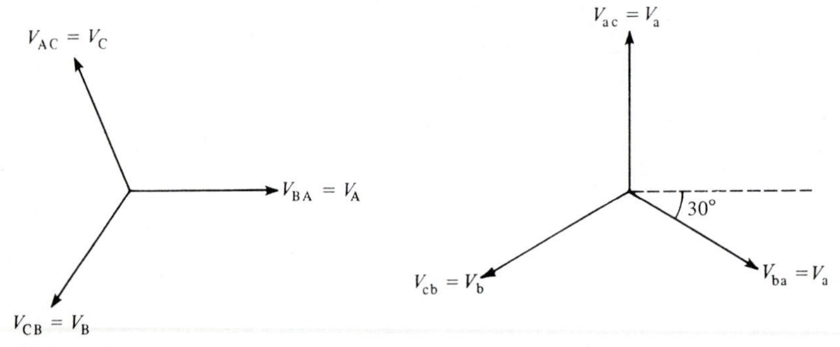

FIGURE 3-33
Wye/delta connection.

T Connection

The T connection also offers a method of using only two transformers in a three-phase system. This connection is shown in Fig. 3.35*(a)*. One transformer, having a center tap at *O*, is connected to terminals *A* and *B* of the three-phase system. This transformer is called the *main transformer*. The second transformer, called the *teaser transformer*, has one terminal connected to point *O* of the main transformer and the other to *C* of the three-phase source, as indicated in Fig. 3.35*(a)*. From the phasor diagram in Fig. 3.35*(b)*, it may be seen that the voltage V_{OC} across the teaser transformer = $(\sqrt{3}/2)V_{AB}$ = 0.866 V_{AB}. The ratio of the transformer of the two transformers being the same, we obtain a three-phase balanced system at the secondary. It is important that relative polarities be properly maintained, as given in

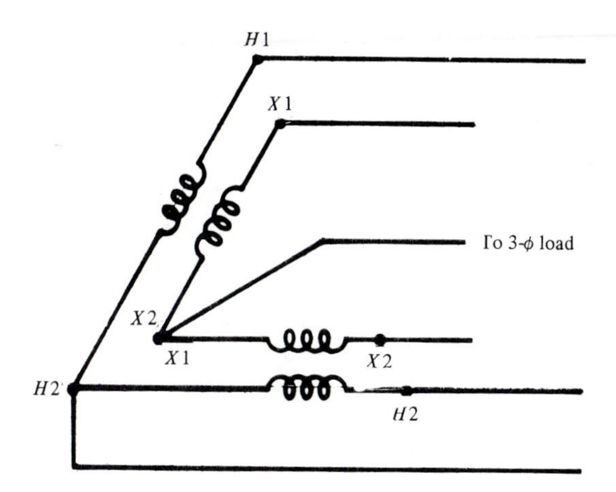

FIGURE 3-34
Open-delta connection.

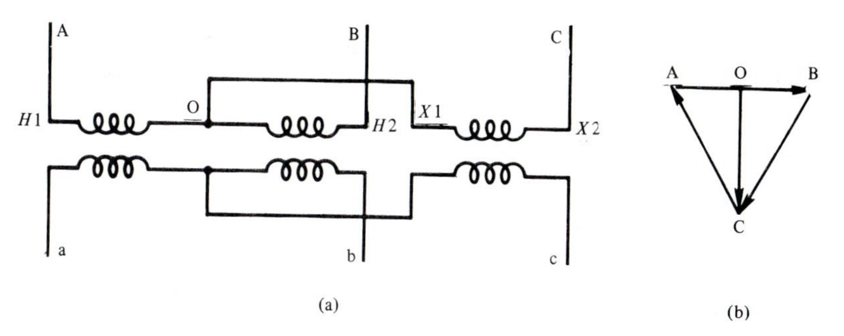

FIGURE 3-35
(a) T connection; *(b)* phasor diagram.

Fig. 3.35*(a)*. With the correct polarities, the currents in each of the halves of the main transformer flow in opposite directions.

Scott Connection

The Scott connection is a special connection used to obtain a two-phase system from a three-phase system, and vice versa. The connection is shown in Fig. 3.36. Transformer I has a midpoint tap and has a voltage rating V_2. Transformer II has a tap at a point corresponding to 86.6 percent of V_2.

There are numerous other transformer connections for various applications, but these are not considered here.

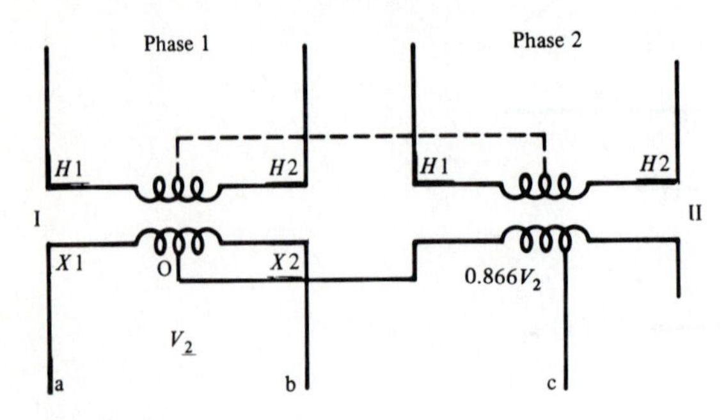

FIGURE 3-36
Scott connection.

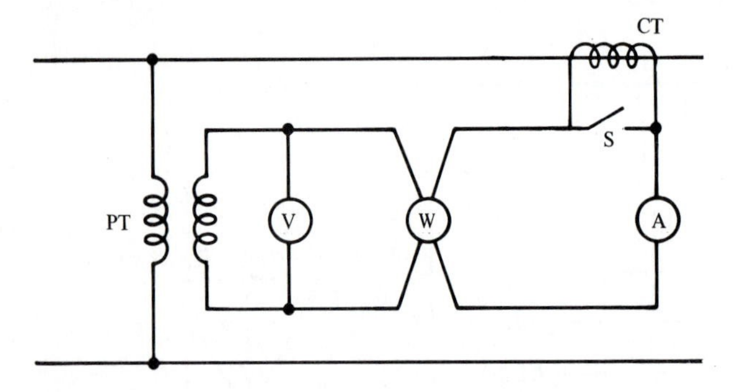

FIGURE 3-37
Instrument-transformer connections.

3.18 INSTRUMENT TRANSFORMERS

Instrument transformers are of two kinds: current transformers (CTs) and potential transformers (PTs). These are used to supply power to ammeters, voltmeters, wattmeters, relays, and so on. Instrument transformers are used for (1) reducing the measured quantity to a low value which can be indicated by standard instruments (a standard voltmeter may be rated at 120 V and an ammeter at 5 A); and (2) isolating the instruments from high-voltage sources for safety. A connection diagram of a CT and a PT with an ammeter, a voltmeter, and a wattmeter is shown in Fig. 3.37. The load on the instrument transformer is called the *burden*. Depending on the burden, instrument transformers are rated from 25 to 500 VA. However, a PT or a CT is much (two to six times) bigger than a power transformer of the same rating.

An ideal instrument transformer has no phase difference between the primary and secondary voltages (or currents), which are independent of the burden. Like the ideal power transformer discussed earlier, the voltage ratio of an ideal PT is exactly equal to its turns ratio. The current ratio of an ideal CT is exactly equal to the inverse of the turns ratio. In practice, however, load-dependent ratio and phase-angle errors are present in instrument transformers.

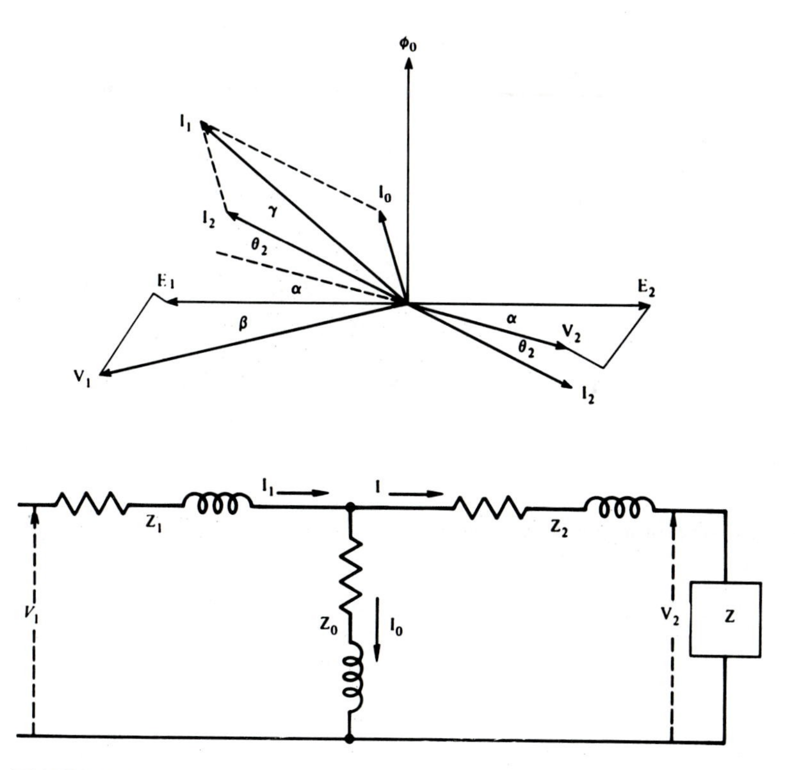

FIGURE 3-38
Phasor diagram and equivalent circuit of an instrument transformer.

The principle of operation of an instrument transformer is no different from that of an ordinary power transformer. Thus they have similar equivalent circuits and phasor diagrams, as shown in Fig. 3.38. It is clear from this diagram that the secondary impedance drop causes a phase displacement α, and the primary impedance drop a phase displacement β; the exciting current I_0 causes a further phase displacement γ, so that the angle between the primary voltage and current is $(\theta_2 + \alpha + \beta + \gamma)$, compared with an angle θ_2 between the secondary voltage and current. Thus the transformer introduces a phase-angle error $(\alpha + \beta + \gamma)$. Moreover, V_1 and V_2 will be only approximately in the ratio of the number of turns. The significance of these errors is indicated in Fig. 3.39. Calibration curves of this type are furnished by the manufacturer. In order to nullify or reduce the errors, instrument transformers are designed with (1) small leakage reactances and low resistances which reduce angles α and β; (2) low flux densities and good transformer iron, which reduces the exciting current I_0 and therefore angle γ; and (3) less than a nominal turns ratio, which compensates for the ratio error. Compensating impedances may also be provided, so that the burden can be kept constant as instruments are put in or out of the circuit. For a constant burden, the instruments may be calibrated, or corrected, against the load.

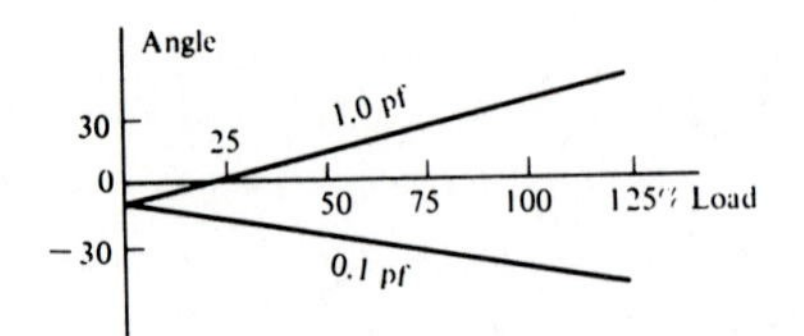

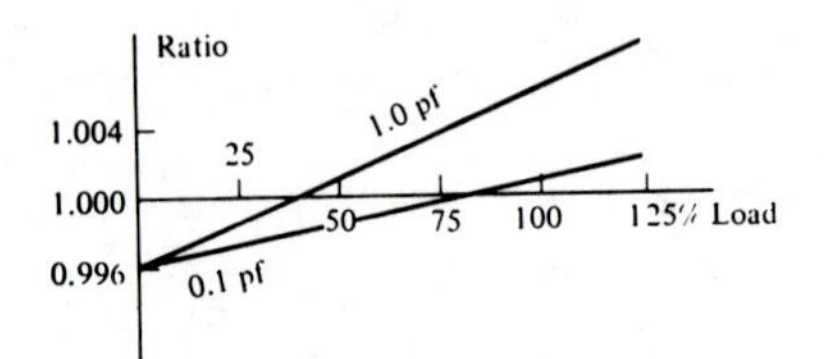

FIGURE 3-39
Phase-angle and ratio errors of an instrument transformer.

Provision must be made to short-circuit the secondary of a current transformer before removing any instruments, for otherwise, dangerously high voltages may occur. It is clear from the equivalent circuit of Fig. 3.38 that if the secondary is open-circuited, the primary current, of fixed magnitude as determined by the load, must flow through the exciting impedance and act entirely as a magnetizing current to the transformer. This results in high flux density and correspondingly high voltage. Oscillograms of such a voltage show it to be peaked by a dominating third harmonic, as would be expected.

PROBLEMS

3.1. A symmetrical square wave of voltage is applied to the primary winding of a transformer with the secondary winding open (no-load). Determine and sketch the waveform of the magnetic flux in the core. Neglect primary resistance and core losses.

3.2. Find the relationship among the root-mean-square voltage applied to the primary, the maximum flux in the core, and the square-wave frequency for the transformer of Problem 3.1. This equation will be analogous to the relationship for sinusoidal excitation, Equation (3.7).

3.3. Repeat Problems 3.1 and 3.2 for an applied voltage whose waveform is a symmetrical triangular wave.

3.4. A transformer is to be assembled using a magnetic core constructed of M-19 transformer steel laminations (see Fig. 2.5). The *net* cross-sectional area of the core is 10 cm^2. The transformer will be excited from a sinusoidal voltage source of 100 V (rms) at 400 Hz. Determine the number of primary winding turns required so that the maximum core flux density at rated applied voltage is at the knee of the saturation curve for M-19 steel (1.2 T). Neglect winding resistance and core loss.

3.5. The sinusoidal flux in the core of a transformer is given by the expression $\phi_1 = 0.0012 \sin 77t$ Wb; this flux links a primary winding of 150 turns. Determine the root-mean-square induced voltage in the primary winding.

3.6. The flux in the core of a transformer is expressed as $\phi = 0.002 \sin 377t + 0.00067 \sin 1131t$. Determine the induced voltage in an 80-turn coil linking this flux.

3.7. The voltage, $v = 100 \sin 377t - 20 \sin 1885t$, is applied to a 200-turn transformer winding. Derive the equation for the flux in the core, neglecting leakage flux and winding resistance. Determine the root-mean-square values of the voltage and the flux.

3.8. In a certain transformer tested with open secondary (no-load), a core loss of 150 W is measured when the no-load current is 2.5 A and the induced voltage is 400 V (voltage and current are root-mean-square values). Neglect the winding resistance and leakage flux. Determine *(a)* the no-load power factor, *(b)* the root-mean-square magnetizing current, and *(c)* the root-mean-square core-loss component of current.

3.9. A transformer is rated 10 kVA (kilovolt-amperes), 400-V primary and 200-V secondary voltages. Neglecting the magnetizing and core-loss volt-amperes, determine the turns ratio and the winding current ratings.

3.10. The transformer of Problem 3.9 has 1000 turns in its primary winding; the conductor used in the primary winding is copper AWG No. 16. If the secondary winding is designed to have exactly the same "copper loss" (ohmic resistance loss), determine the size of the wire to be used in the secondary winding. (See Appendix II for wire table.)

3.11. The mean magnetic length of the core of Problem 3.4 is 40 cm. Determine the root-mean-square magnetizing current with rated voltage applied using the turns calculated in Problem 3.4.

3.12. The alternating flux in the core of a transformer increases as the square of time from zero at the beginning of the flux period to 2.4×10^{-3} Wb at $t = 10$ ms. Then it decreases to zero at $t = 20$ ms along a curve that is symmetrical to the rising flux curve about the one-fourth period time. The negative half-period is symmetrical to the positive half-period about the time axis.
(a) Sketch the curve showing the voltage applied to a 500-turn winding that will result in this flux characteristic.
(b) Determine the maximum root-mean-square, half-period average and form factor of this applied voltage wave.

3.13. The transformer of Example 3.7 is excited from a current source (sine-current excitation). The root-mean-square value of the exciting current is 2.0 A. Using a graphical procedure similar to that used in Example 3.7, determine the waveform over a half-period of the flux in the core. Sketch the approximate half-period waveform of the primary induced voltage.

3.14. For the transformer of Problem 3.4, estimate the peak inrush current from (3.52). Use the turns calculated in Problem 3.4 and assume a coil length of 10 cm, a mean turn area (A_t) of 12 cm^2, and a residual density (B_r) of 0.3 T.

3.15. The nonsinusoidal voltage, $v = 120 \sin 377t - 60 \sin 1885t$, is applied to the 200-turn winding of a transformer. What is the equation of the time variation of the flux in the core? Determine the maximum value of the flux. Sketch the voltage and flux waveforms.

3.16. From Fig. 3.7 it is clear that the primary and the secondary of the transformer are linked or coupled with each other by the core or mutual flux ϕ_c. If we denote by ϕ_p and ϕ_s the total fluxes due to the primary and the secondary mmf's, respectively, then we may define a *coefficient of coupling*, k, by

$$k = \frac{\phi_{12}}{\phi_p} = \frac{\phi_{21}}{\phi_s}$$

where ϕ_{12} and ϕ_{21} are the core fluxes produced by the primary and the secondary windings, respectively. Using coupled-circuit approach, and defining inductance = flux linkage/ampere, show that

$$k = L_m / \sqrt{L_p L_s}$$

where L_m, L_p, and L_s are, respectively, the mutual, the primary, and the secondary inductances.

3.17. Using the theory developed in Problem 3.16, show that the steady-state secondary induced voltage is given by

$$E_2 = jX_m I_m$$

where X_m is the magnetizing reactance and I_m is the magnetizing current.

3.18. In an ideal transformer, with $\mu_{core} \simeq \infty$, show that the power transfer from the primary (source) to the secondary (load) occurs through the window (air space) and *not* through the iron of the core. (*Hint:* Use Poynting vector formulation.)

3.19. A transformer is rated 100 kVA, 11,000/2200 V, and 60 Hz. The no-load test on the low-voltage winding gives 2200 V, 2A, 100 W, and 60 Hz. If this test were performed on the high-voltage winding at rated voltage, determine the current and power that would be measured.

3.20. The voltage $v = 100 \sin 377t$, is applied to a transformer winding in a no-load test. The resulting current is found to be $i = 5 \sin(377t - 60°) + 2 \sin(1131t - 120°)$ A. Determine the core loss and the root-mean-square of the exciting current.

3.21. A no-load test on a certain transformer gives the following data: 120 V, 2.3 A, 75 W, and 60 Hz. Neglecting the winding resistance and leakage reactance, determine the value of the magnetizing reactance, X_m, the core-loss equivalent resistance, R_c, and the no-load power factor.

3.22. If the winding resistance and reactance of the transformer in Problem 3.21 are 0.4 Ω and 1.5 Ω, respectively, recalculate X_m and R_c including the effects of these winding impedances.

3.23. A transformer rated 220/440 V, 25 kVA, and 60 Hz is tested as follows. A no-load test, performed on the 220-V winding, gives 220 V, 10 A, 700 W, and 60 Hz; the short-circuit test, performed on the 440-V winding, gives 37 V, rated current,1000 W, and 60 Hz. Determine the impedances of the complete equivalent circuit (Fig. 3.10) in terms of the 440-V winding. State what assumptions were made in obtaining these impedances.

3.24. Express the complete equivalent circuit of Problem 3.23 in terms of the 220-V winding.

3.25. Voltage regulation of a transformer is defined as the secondary voltage at no-load, minus the secondary voltage at a given load condition, divided by the load voltage. In percentage this becomes

$$\text{voltage regulation} = 100\,\frac{V_2(\text{no-load}) - V_2(\text{load})}{V_2(\text{load})}$$

This calculation is usually made neglecting the exciting components, that is, using the approximate equivalent circuit of Fig. 3.14. For the transformer of Problem 3.23, determine the voltage regulation for the following.
(a) Rated load on the 440-V winding at a power factor of 0.85 lagging.
(b) Rated load on the 440-V winding at a power factor of 0.85 leading.
(c) One-half rated load on the 440-V winding at a power factor of 0.85 lagging.

3.26. A certain transformer is rated 1000 kVA, 11,000/2200 V, and 60 Hz. The short-circuit test on the 11,000-V winding gives 1000 V, rated current, and 9 kW. Determine the equivalent series resistance, reactance, and impedance in terms of both windings.

3.27. A transformer rated 500 kVA, 2400/120 V, and 60 Hz has a no-load loss (at rated voltage) of 1600 W and a short-circuit loss (at rated current) of 7500 W. Determine the efficiency of this transformer under the following conditions of load.
(a) Rated current at 0.8 power factor lagging.
(b) Three hundred kilowatts at 0.8 power factor leading.
(c) One hundred kilowatts at 0.8 power factor lagging.

3.28. Three identical transformers are each rated 200 kVA, 13,200/2300 V, and 60 Hz. The high-voltage windings are connected in delta and the low-voltage windings in wye. Determine the rated voltages and currents of the lines and phase windings on both sides of this polyphase connection.

3.29. A balanced, three-phase load of 300 kVA at 460 V is to be supplied from a 2300-V, three-phase system by a delta bank of transformers. Specify the current, voltage, and kilovolt-ampere rating of the windings of each transformer.

3.30. Two transformers, of the type described in Problem 3.28, are connected in open delta on both primary and secondary.
(a) Determine the load kilovolt-amperes that can be supplied from this transformer connection.
(b) A delta-connected, three-phase load of 300 kVA, 0.866 lagging power factor, 2300 V is connected to low-voltage terminals of this open-delta transformer. Determine the transformer currents on the 13,200-V side of this connection.

3.31. The transformer of Problem 3.23 is connected as an autotransformer to transform 600 V to 220 V.
(a) Determine the autotransformer ratio a.
(b) Determine the volt-ampere rating of the autotransformer.
(c) With a load of 25 kVA, 0.866 lagging power factor connected to the 220-V terminals, determine the currents in the load and the two transformer windings.

3.32. A load of 12 kVA at 0.7 lagging power factor is to be supplied at 110 V from a 120-V supply by an autotransformer. Specify the voltage and current ratings of each section of the autotransformer.

3.33. A transformer with additive polarity is rated 15 kVA, 2300/115 V, 60 Hz. Under rated conditions, the transformer has an excitation loss of 75 W and a short-circuit loss of 250 W. The transformer is to be connected as an autotransformer to transform 2300 V to 2415 V. With a load of 0.8 lagging power factor, what volt-ampere load can be supplied without exceeding the current rating of any winding? Determine the efficiency at this load.

3.34. A transformer is rated 10 kVA, 7200/120 V, 60 Hz. The following test data were measured on this transformer: short circuit (high-voltage winding): 220 V, 1.39 A, 200 W; open circuit (low-voltage winding): 120 V, 2.5 A, 76 W. Determine the constants of the equivalent circuit in high-voltage terms.

3.35. Determine the voltage transformation ratio, a', for the transformer of Problem 3.23.

3.36. An ideal transformer is rated 2400/240 V. A certain load of 50 A, unity power factor is to be connected to the low-voltage winding and must have exactly 200 V across it. With 2400 V applied to the high-voltage winding, what resistance must be added in series with the transformer if it is *(a)* located in the low-voltage winding, *(b)* located in the high-voltage winding?

3.37. A 60-Hz transformer is rated 10 kVA, 440/110 V. This transformer is to be operated from a 50-Hz supply with the high-voltage winding used as the primary. If the maximum value of flux in the transformer core is to be the same as when operated at 60 Hz:
(a) What root-mean-square voltage should be applied to the primary winding?
(b) What will be the secondary open-circuit root-mean-square voltage with the voltage calculated in part *(a)* applied to the primary?
(c) What will be the new kilovolt-ampere rating at 50 Hz if the flux and temperature rise are to be the same as at 60 Hz? (Assume the core loss change is negligible between 60 and 50 Hz.)

3.38. The transformer core of Problem 3.4 has a mean magnetic length of 0.25 m. For the conditions of Problem 3.4 and assuming that the losses of M-19 are those given in Fig. 2.7 (true for 0.019 laminations), determine the core loss in the core. M-19 has a density of 0.286 lb/in^3.

3.39. The core described in Problems 3.4 and 3.38 is to be used in a 400-Hz, 1000-VA, 100/24-V transformer. Design primary and secondary windings using current densities of 200 cmil (circular mils) per ampere.

3.40. Assume that the core cross section of the core used in Problem 3.39 is square with a *gross* area of 11 cm^2 . Determine the stacking factor. Estimate the total length of the two windings designed in Problem 3.39.

(a) Calculate the ohmic power loss in the two windings.

(b) Using the core loss calculated in Problem 3.38, calculate the transformer loss at rated volt-amperes (which should be 1000 at the output). Use the equivalent circuit of Fig. 3.14 for calculating ohmic losses.

CHAPTER

4

INDUCTION MACHINES

The induction motor is the most common of all motors. It has been called the "workhorse" of industry. An induction machine consists of a stator and a rotor mounted on bearings and separated from the stator by an airgap. Electromagnetically, the stator consists of a core made up of punchings (or laminations) carrying slot-embedded conductors. These conductors are interconnected in a predetermined fashion and constitute the armature windings.

Alternating current is supplied to the stator windings, and the currents in the rotor windings are induced by the stator currents; that is, in general, the rotor of an induction machine is not supplied by an external source of power. The rotor of the induction machine is cylindrical and carries either (1) conducting bars short-circuited at both ends, as in a *cage-type* machine (Fig. 4.1), or (2) a polyphase winding with terminals brought out to slip rings for external connections, as in a *wound-rotor* machine (Fig. 4.2). A wound-rotor winding is similar to that of the stator. Sometimes the cage-type machine is also called a *brushless* machine and the wound-rotor machine a *slip-ring* machine. The stator and the rotor, in its three different stages of production, are shown in Fig. 4.3. The motor is rated at 2500 kW, 3 kV, 575 A, two-pole, and 400 Hz. A finished cage-type rotor of a 3400-kW, 6-kV motor is shown in Fig. 4.4, and Fig. 4.5 shows the wound rotor of a three-phase slip-ring 15,200-kW, four-pole induction motor. A cutaway view of a completely assembled motor with a cage-type rotor is shown in Fig. 4.6.

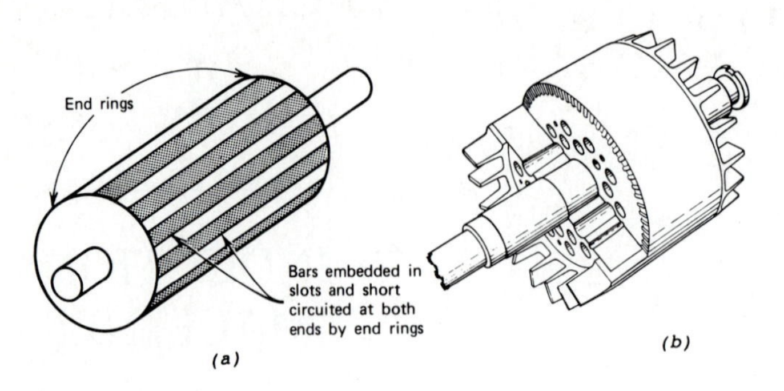

FIGURE 4-1
A cage-type rotor. *(a)* Schematic; *(b)* part section.

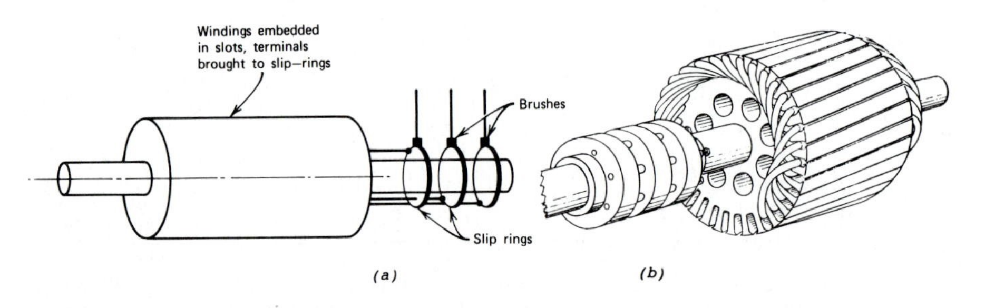

FIGURE 4-2
A wound rotor. *(a)* Schematic; *(b)* isometric view.

An induction machine operates on the basis of interaction of induced rotor currents and the airgap fields. If the rotor is allowed to run under the torque developed by this interaction, the machine will operate as a motor. On the other hand, the rotor may be driven by an external source beyond a speed such that the machine begins to deliver electrical power and operates as an induction generator (instead of as an induction motor, which absorbs electrical power). Thus we see that the induction machine is capable of functioning as a motor as well as a generator. However, almost invariably it is used as a motor and not too often as a generator.

Before we consider the induction motor in detail, it is worthwhile to study its stator construction and the magnetic field produced by the stator (or armature) windings.

4.1 MMFs OF ARMATURE WINDINGS

We have stated earlier that the stator of an induction machine has a winding that is distributed around the periphery of the stator. Thus slot-embedded conductors, covering the entire surface of the stator and interconnected in a predetermined manner, constitute the armature winding of the induction machine.

FIGURE 4-3
Rotor for a 2500-kW, 3-kV, two-pole, 400-Hz motor in different stages of production.

Often, more than one independent winding is on the stator. An arrangement of a three-phase stator winding is shown in Fig. 4.7. Notice that the stator windings are distributed in the slots over the entire periphery of the stator. Each slot contains two coil-sides. For instance, slot 1 has coil-sides of phases A and B, whereas slot 2 contains two layers (or two coil-sides) of phase A only. Such a winding is known as a double-layer winding. Furthermore, it is a four-pole winding laid in 36 slots, and we thus have three slots per pole per phase.

In order to produce the four-pole flux, each coil should have a span (or *pitch*) of one-quarter of the periphery. In practice, the pitch is made a little less and, as shown in Fig. 4.7, each coil embraces eight teeth. The coil pitch is about 89% of the pole pitch, and the winding is, therefore, a *fractional-pitch* (or *chorded*) winding.

For the present we will consider the mmf's produced by the armature windings. First, assume a single, full-pitch coil having N turns, as shown in Fig. 4.8, where the slot opening is negligible. Clearly, the machine has two poles (Fig. 4.8*(a)*). From Ampere's law, we have $\oint H \bullet dl = Ni$, which is the same for all lines of force. In other words, the mmf has a constant value of Ni between the coil sides, as shown in Fig. 4.8*(b)*. Traditionally, the magnetic effects of a winding in an electric machine are considered on a per-pole basis. Thus, if i is the current in the coil, the mmf per pole is $Ni/2$, which is plotted in Fig. 4.8*(c)*. The reason for such a representation is that Fig. 4.8*(c)* also represents a flux-density distribution, but to a different scale. Obviously, the flux-density over one pole (say the north pole) must be opposite to that over the other (south) pole, thus keeping the flux entering the rotor equal to that leaving the rotor surface. Comparing Figs. 4.8*(b)* and 4.8*(c)*, we notice that the representation of the mmf curve with positive and negative areas (Fig. 4.8*(c)*) has the advantage that it gives the flux-density distribution, which must contain positive and negative areas. The mmf distribution shown in Fig. 4.8*(c)* may be resolved into its harmonic components by Fourier analysis. The period of the fundamental component is the same (2τ) as that of the rectangular mmf wave. The amplitude of the fundamental wave is $4/\pi$ times the amplitude of the rectangular wave. Therefore the fundamental component of the mmf distribution is given by

$$F(x,\, t) = \frac{4}{\pi}\left(\frac{Ni}{2}\right)\cos\frac{\pi}{\tau}x \tag{4.1}$$

If i is sinusoidal alternating current, so that $i = I_m \sin \omega t$, (4.1) becomes

$$F(x,\, t) = \frac{\sqrt{2}}{2}\left(\frac{4}{\pi}\right) NI \cos\frac{\pi x}{\tau} \sin\omega t \tag{4.2}$$

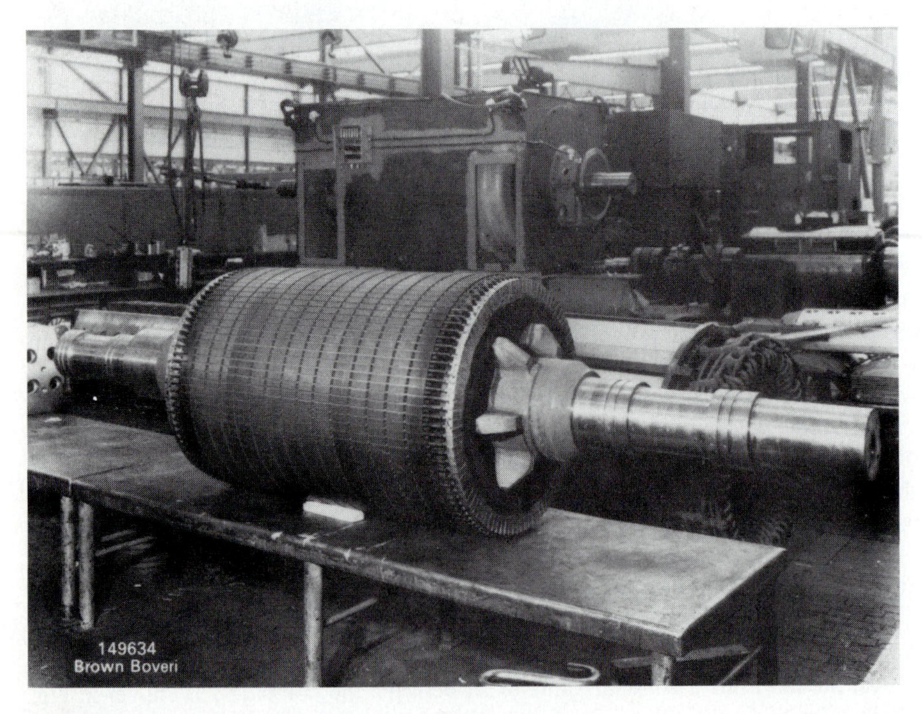

FIGURE 4-4
Complete rotor of a 3400-kW, 6-kV, 990-rpm motor.

FIGURE 4-5
Rotor of a three-phase, 15,200-kW, 2.4-kV slip-ring induction motor.

which simplifies to

$$F(x, t) = 0.9\ NI \cos\frac{\pi x}{\tau}\ \sin\omega t \tag{4.3}$$

where I is the root-mean-square value of the current. Notice that the time variation of the flux resulting from the mmf is alternating and the flux is stationary in space.

In an electric machine we seldom have a single N-turn coil (Fig. 4.8*(a)*) as a source of the armature mmf. Instead, we have windings that are distributed over the entire periphery of the machine, such as the one shown in Fig. 4.7. Besides utilizing all the space available, by distributing the winding we reduce the harmonic content in the mmf distribution, although the magnitude of the fundamental component will be less than that for a concentrated winding. Ideally, we attempt to distribute the winding so that the resulting mmf distribution is purely sinusoidal.

In practice, as a first approximation we assume a sinusoidal mmf distribution. For the study of the induction motor, we will assume such a distribution. Thus we let the mmf (or flux density) space distribution produced by three identical coils that are displaced from each other by 120° (in time and space) be given by

FIGURE 4-6
Cutaway of an induction motor.

$$F_a = F_m \sin\omega t \cos\frac{\pi x}{\tau}$$

$$F_b = F_m \sin(\omega t - 120°) \cos\left(\frac{\pi x}{\tau} - 120°\right) \tag{4.4}$$

$$F_c = F_m \sin(\omega t + 120°) \cos\left(\frac{\pi x}{\tau} + 120°\right)$$

Notice that the three coils are excited by a three-phase source. Because we have shown the three-phase windings to consist of three (independent) N-turn coils, $F_m = 0.9NI$ for each coil, from (4.3) and (4.4). The space and time variations of the resultant mmf is then the sum of the three mmf's of (4.4). Observing that $\sin A \cos B = \frac{1}{2}\sin(A - B) + \frac{1}{2}\sin(A + B)$ and adding F_a, F_b, and F_c we obtain the resultant mmf as

$$F(x, t) = 1.5 F_m \sin\left(\omega t - \frac{\pi x}{\tau}\right) \tag{4.5}$$

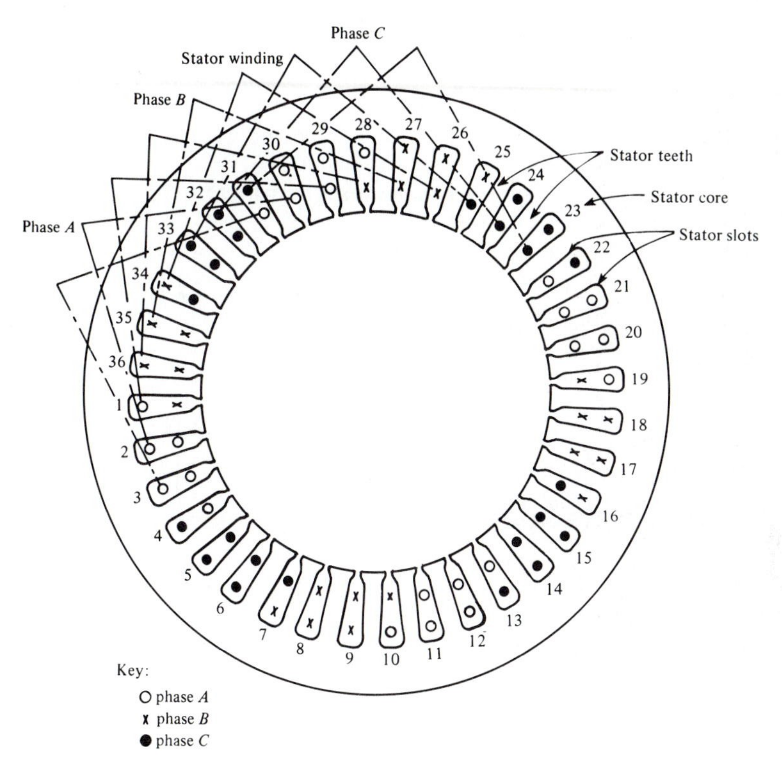

FIGURE 4-7
Stator windings.

Fig. 4.9 shows the position of the resultant mmf at three different instants $t_1 < t_2 < t_3$. Notice that as time elapses, a fixed point P moves to the right, implying that the resultant mmf is a traveling wave of mmf having a constant amplitude. The magnetic field produced by this mmf in an electric machine is then known as a *rotating magnetic field*. We may arrive at the same conclusion by considering the resultant mmf at various instants, as shown in Fig. 4.10. From these diagrams it is clear that as we progress in time from t_1 to t_3, the resultant mmf rotates in space from θ_1 to θ_3. The existence of the rotating magnetic field is essential to the operation of an induction motor.

To determine the velocity of the traveling field given by (4.5), imagine an observer traveling with the mmf wave from a point P. To this observer, the magnitude of the mmf wave will remain constant (independent of time), implying that the right side of (4.5) would appear constant. Expressed mathematically, this would be

$$\sin\left(\omega t - \frac{\pi x}{\tau}\right) = \text{constant}$$

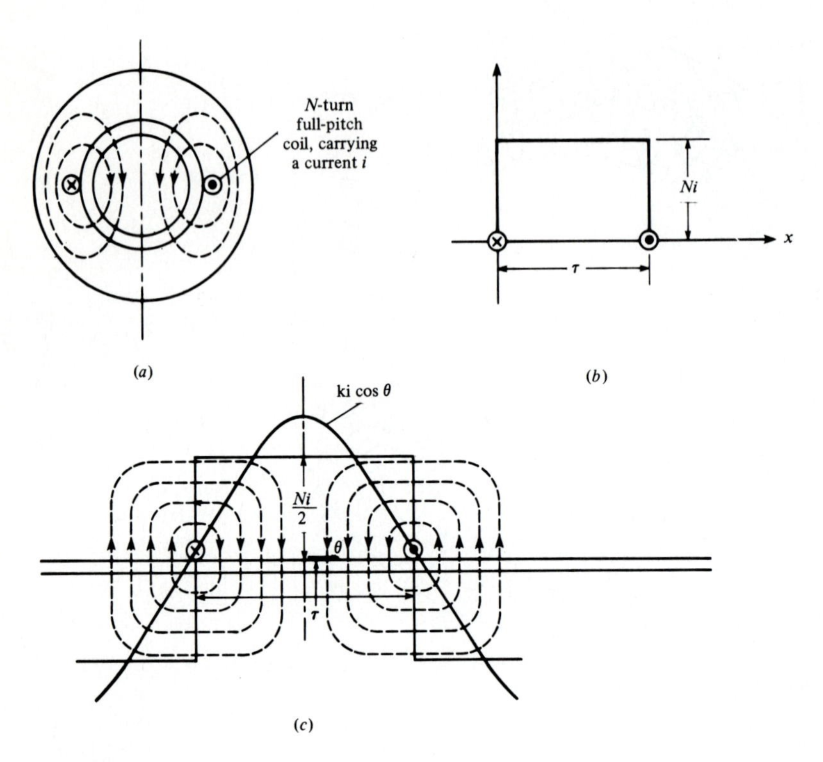

FIGURE 4-8
Flux and mmf produced by a concentrated winding. (*a*) Flux lines produced by an *N*-turn coil; (*b*) Mmf produced by the *N*-turn coil; (*c*) mmf per pole.

or

$$\omega t - \frac{\pi x}{\tau} = \text{constant} \tag{4.6}$$

Differentiating both sides of (4.6) with respect to t, we obtain

$$\omega - \frac{\pi}{\tau}\dot{x} = 0$$

or

$$\dot{x} = \frac{\omega\tau}{\pi} = 2f\tau = \frac{2\tau}{T} = \frac{\lambda}{T} \text{ m/s} \tag{4.7}$$

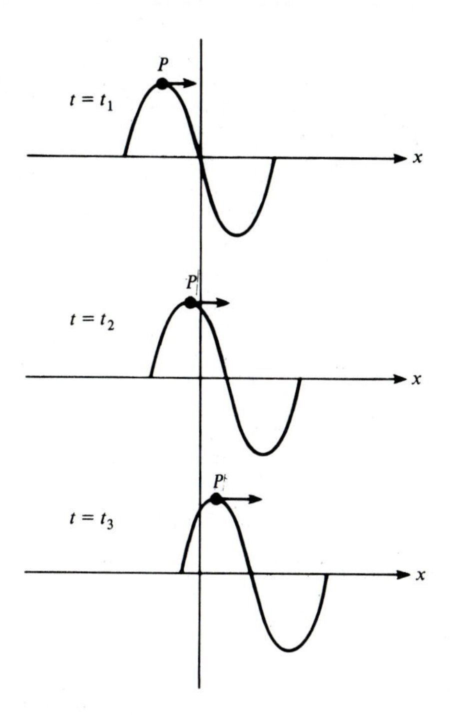

FIGURE 4-9
The function $\sin[\omega t - (\pi x/\tau)]$ at different time intervals $t_1 < t_2 < t_3$.

where τ = pole pitch, $\omega = 2\pi f$, f is the frequency of input currents, and T is the corresponding period (i.e., $f = 1/T$). From (4.7) we conclude that the mmf (or flux) wave travels, during one cycle of the current, a distance twice the pole pitch (or wavelength, λ). Therefore, for a given pole pitch and frequency, the velocity of the traveling field is constant and is known as the *synchronous velocity*.

The production of a rotating magnetic field is graphically depicted in Fig. 4.10. Fig. 4.10*(a)* shows a system of three-phase currents (or mmf's). The corresponding fluxes and the resultant flux are shown in Fig. 4.10*(b)*.

Notice from (4.1) that in one complete rotation in the airgap, corresponding to 360°, we obtain one complete cycle of the mmf. One cycle of mmf is said to correspond to 360 electrical degrees (in contrast to the 360 mechanical degrees in one rotation). For instance, in Fig. 4.11(*a*), the winding around the stator periphery is so arranged that we obtain four cycles of mmf. In other words, we obtain 1440 electrical degrees by going through 360 mechanical degrees. The "cycles of mmf's" are precisely designated by the number of poles, *P*. *One cycle of mmf corresponds to two poles and 360 electrical degrees*, and one complete rotation around the machine periphery corresponds to 360 mechanical degrees. Hence for a *P*-pole machine, the general relationship between the electrical degree θ and the mechanical degree θ_m is

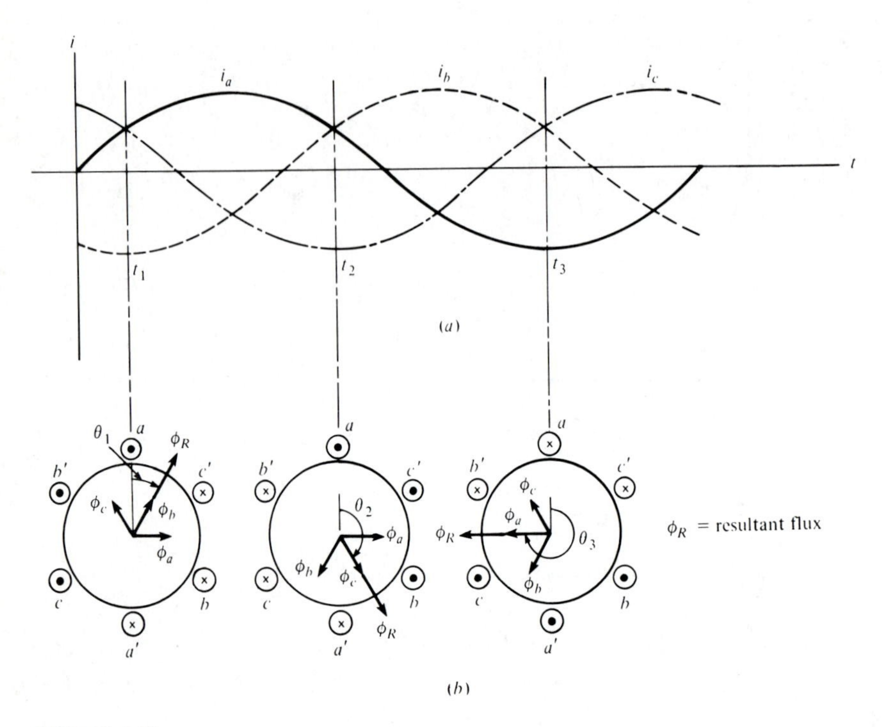

FIGURE 4-10
Production of a rotating magnetic field by a three-phase excitation. (*a*) Time diagram. (*b*) Space diagram.

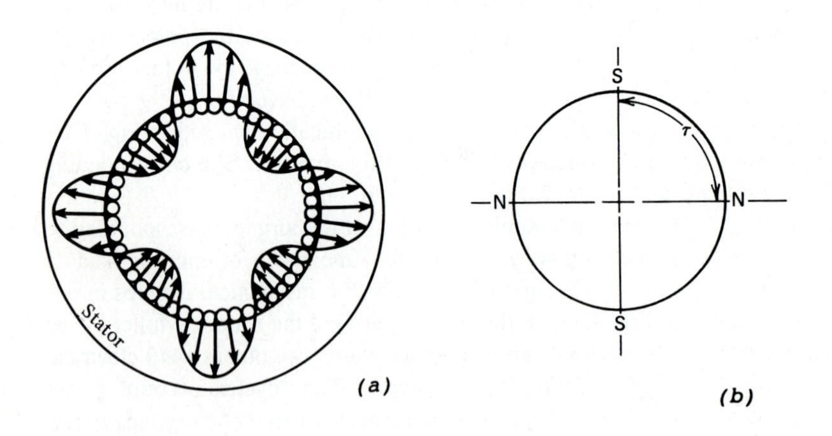

FIGURE 4-11
Definition of (*a*) number of poles P and (*b*) pole pitch τ.

$$\theta = \frac{P}{2}\,\theta_m \tag{4.8}$$

We can now relate the synchronous velocity (meters per second) to speed in revolutions per minute (rpm) by observing that one revolution around the airgap of the machine corresponds to a linear distance $P\tau$, where P is the number of poles and τ = pole pitch (Fig. 4.11*(b)*). The distance traveled by the wave in 1 min is $60(2f\tau)$. Thus the speed of the traveling field, which now corresponds to a rotating field in revolutions per minute, n, is given by

$$n = \frac{60(2f\tau)}{P\tau} = \frac{120f}{P}\ \text{rpm} \tag{4.9}$$

Again, this is the fixed speed, called synchronous speed, and is usually designated by n_s. So we rewrite (4.9) as

$$n_s = \frac{120f}{P}\ \text{rpm} \tag{4.10}$$

4.2 ACTION OF A POLYPHASE INDUCTION MOTOR

We recall from Section 4.1 that a three-phase stator excitation produces a rotating magnetic field in the air-gap of an induction motor, and the field rotates at a synchronous speed given by (4.10). As the magnetic field rotates, it "cuts" the rotor conductors. By this process, voltages are induced in the conductors. The induced voltages give rise to rotor currents. The pattern of rotor currents in a cage rotor is shown in Fig. 4.12(*a*), and its fundamental component is given in Fig. 4.12(*b*). These

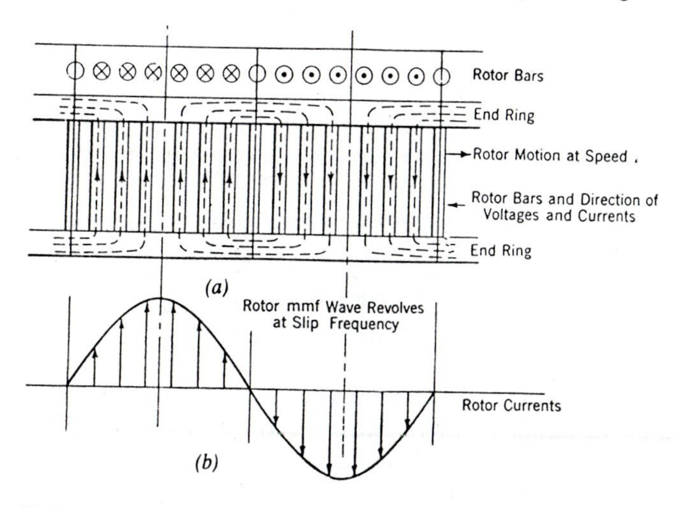

FIGURE 4-12
(*a*) Rotor currents in a cage rotor; (*b*) fundamental component of rotor mmf.

currents interact with the airgap field to produce a torque. The torque is maintained as long as the rotating magnetic field and the induced rotor currents exist. Consequently, the rotor starts rotating in the direction of the rotating field. (*Note*: It may be readily verified, by applying the principle of conservation of energy, that the rotor will not run on its own torque in a direction opposite to that of the rotating field.) The rotor will achieve a steady-state speed, n, such that $n < n_s$. Clearly, when $n = n_s$, there will be no induced currents and hence no torque. The condition $n > n_s$, corresponds to the generator mode.

An alternate approach to explaining the operation of the polyphase induction motor is by considering the interaction of the (excited) stator magnetic field with the (induced) rotor magnetic field. The stator excitation produces a rotating magnetic field, which rotates in the airgap at a synchronous speed. The field induces polyphase currents in the rotor, thereby giving rise to another rotating magnetic field, which also rotates with respect to the stator at the same synchronous speed as that of the stator. Thus we have two rotating magnetic fields, rotating at a synchronous speed with respect to the stator but stationary with respect to each other. Consequently, according to the principle of alignment of magnetic fields, the rotor experiences a torque. (It might be said to be dragged along by the stator magnetic field.) The rotor rotates in the direction of the rotating field of the stator.

The torque production in an induction motor may also be explained by a direct application of the Lorentz force equation (see Chapter 2).

4.3 SLIP AND FREQUENCY OF ROTOR CURRENTS

The actual mechanical speed, n, of the rotor is often expressed as a fraction of the synchronous speed, n_s, as related by *slip*, s, defined as

$$s = \frac{n_s - n}{n_s} \tag{4.11}$$

The slip may also be expressed as percent slip as follows:

$$\text{percent slip} = \frac{n_s - n}{n_s} \times 100$$

At standstill, the rotating magnetic field produced by the stator has the same relative speed with respect to the rotor windings as with respect to the stator windings. Thus the frequency of the rotor currents, f_2, is the same as the frequency of stator currents, f_1. At synchronous speed, there is no relative motion between the rotating field and the rotor, and the frequency of rotor current is zero. At other speeds, the rotor frequency is proportional to the slip, s. This may be demonstrated as follows.

First, from (4.10) we know that the frequency, f_r, of the current induced in a conductor rotating at n rpm in a field having P poles is given by

$$f_r = \frac{Pn_r}{120} \tag{4.12}$$

In an induction motor the relative speed, n_r, between the rotor conductors and the rotating field produced by the stator is

$$n_r = n_s - n \tag{4.13}$$

Thus, from (4.12) and (4.13), the frequency of rotor current is given by

$$f_r = \frac{P}{120}(n_s - n) \tag{4.14}$$

From (4.10) and (4.11) we have

$$(n_s - n) = sn_s = s\left(\frac{120f}{P}\right) \tag{4.15}$$

Substituting (4.13) and (4.15) in (4.12) yields

$$f_r = sf \tag{4.16}$$

which is known as slip frequency. In (4.16), f_r = frequency of the rotor currents and f = frequency of the stator (input) currents (or voltages).

We now summarize the preceding discussions.

1. The stator rotating magnetic field rotates at the synchronous speed ω_s (with respect to a stationary observer).
2. The rotor mmf produces a rotating magnetic field which also rotates at the synchronous speed and in the same direction as the field produced by the stator mmf. Thus the rotating fields produced by the stator and rotor are stationary with respect to each other.
3. The rotating field produced by the rotor rotates at a speed $(\omega_s - \omega_m)$ with respect to the rotor, where ω_m is the actual mechanical speed of the rotor.
4. Currents (and voltages) induced in the rotor are of slip frequency.

4.4 THE ROTOR EQUIVALENT CIRCUIT

Recognizing that the frequency of rotor currents is the slip frequency, we may express the per-phase rotor leakage reactance x_2, at a slip s, in terms of the standstill per-phase reactance X_2:

$$x_2 = sX_2 \tag{4.17}$$

Next, we observe that the magnitude of the voltage induced in the rotor circuit is also proportional to the slip.

A justification of this statement follows from transformer theory (Chapter 3) because we may view the induction motor at standstill as a transformer with an air-gap. For the transformer we know that the induced voltage, say E_2, is given by

$$E_2 = 4.44 f N \phi_m \tag{4.18}$$

But, at a slip s, the frequency becomes sf, according to (4.16); substituting this value of frequency into (4.18) yields the voltage e_2 at a slip s as

$$e_2 = 4.44 s f N \phi_m = s E_2 \tag{4.19}$$

We conclude, therefore, that if E_2 is the per-phase voltage induced in the rotor at standstill, the voltage e_2 at a slip s is given by

$$e_2 = s E_2 \tag{4.20}$$

Using (4.17) and (4.20), we obtain the rotor equivalent circuit shown in Fig. 4.13*(a)*. The rotor current I_2 is given by

$$I_2 = \frac{s E_2}{\sqrt{R_2^2 + (sX_2)^2}} \tag{4.21}$$

which may be rewritten as

$$I_2 = \frac{E_2}{\sqrt{(R_2/s)^2 + X_2^2}} \tag{4.22}$$

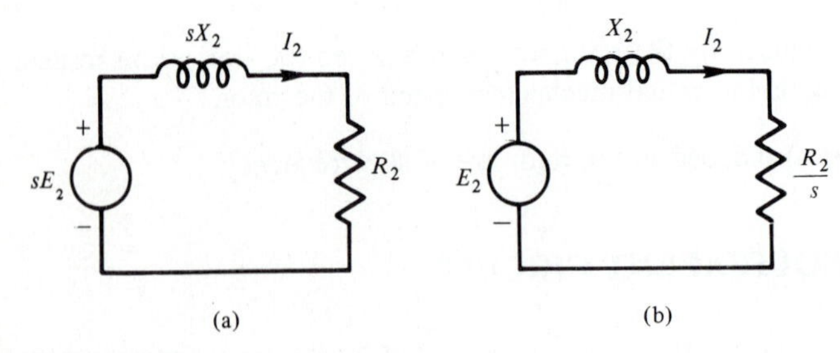

FIGURE 4-13
Two forms of rotor equivalent circuit.

resulting in the alternate form of the equivalent circuit shown in Fig. 4.13(*b*). Notice that the circuits shown in Fig. 4.13 are drawn on a per-phase basis. To this circuit we may now add the per-phase stator equivalent circuit to obtain the complete equivalent circuit of the induction motor, which will be discussed in Section 4.5.

4.5 DEVELOPMENT OF THE COMPLETE EQUIVALENT CIRCUIT

We recall that in an induction motor, only the stator is connected to the ac source. The rotor is not generally connected to an external source, and rotor voltage and current are produced by induction. In this regard, the induction motor may be viewed as a transformer with an airgap, having a variable resistance in the secondary. Thus we may consider that the primary of the transformer corresponds to the stator of the induction motor, whereas the secondary corresponds to the rotor on a per-phase basis. Because of the airgap, however, the value of the magnetizing reactance, X_m, tends to be relatively low compared with that of a transformer. As in a transformer (discussed in Chapter 3), we have a mutual flux linking the stator as well as rotor, represented by the magnetizing reactance and various leakage fluxes. For instance, the total rotor leakage flux is denoted by X_2 in Fig. 4.13. Although the leakage fluxes are subdivided into various components, such as end-connection leakage flux, slot leakage flux, tooth-top flux, and so forth, they will not be considered here. We note, however, that an appropriate leakage reactance component is assigned to each leakage flux component, and such components do not exist in a transformer.

Returning to the analogy of a transformer and considering that the rotor is coupled to the stator as the secondary of a transformer is coupled to its primary, we may draw the circuit shown in Fig. 4.14. To develop this circuit further, we need to express the rotor quantities as referred to the stator. For this purpose we must know the transformation ratio, as in a transformer (see Chapter 3).

Care must be exercised in defining the transformation ratio. The voltage transformation ratio in the induction motor must include the effect of the stator and rotor winding distributions. Therefore the ratio of the rotor to stator voltage becomes

$$\frac{E_2}{E_1} = \frac{k_{w2}N_2}{k_{w1}N_1} \tag{4.23}$$

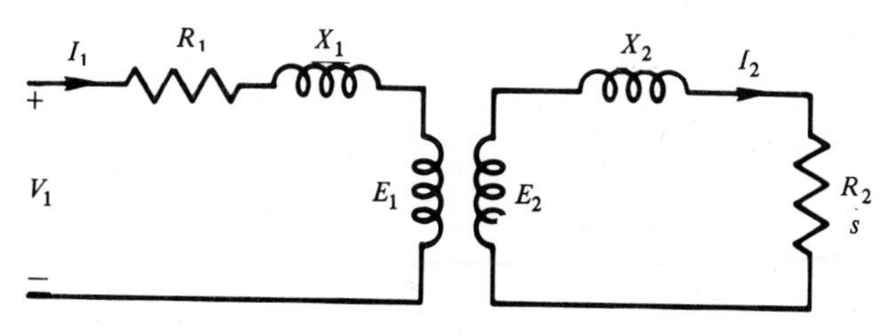

FIGURE 4-14
Stator and rotor as coupled circuits.

where k_{w1} is the winding factor of the stator having N_1 series-connected turns per phase and k_{w2} is the rotor winding factor having N_2 series-connected turns per phase. The winding factor, k_w, of a distributed winding is defined as a product of the distribution factor, k_d, and the pitch factor, k_p. Before proceeding any further with the concept of the transformation ratio of an induction motor, we define the distribution and pitch factors by the following two examples.

Example 4.1 Derive a general expression for the distribution factor for an ac armature winding. Indicate how the voltage equation is modified by the distribution factor.

Solution

We recall from Chapter 3 that the voltage, E, induced in an N-turn coil (all turns) linking a flux ϕ, alternating at a frequency f, is given by

$$E = 4.44 f \phi N \tag{4.24}$$

If these N turns are distributed in a number of slots, such as those shown in Fig. 4.15, the voltages induced in the coils will be displaced from each other in phase by the slot angle α, defined by

$$\alpha = \frac{180P}{Q} = \frac{180}{mq} \tag{4.25}$$

where

m = number of phases
q = number of slots per pole per phase
P = number of poles
Q = total number of slots

The net voltage available at the terminals of the N turns would then be the phasor sum of the voltages induced in each coil. Fig. 4.15 shows such a phasor addition, from which the ratio

$$k_d = \frac{\text{resultant voltage}}{\text{sum of individual coil voltage}} = \frac{E_r}{qE_e} \tag{4.26}$$

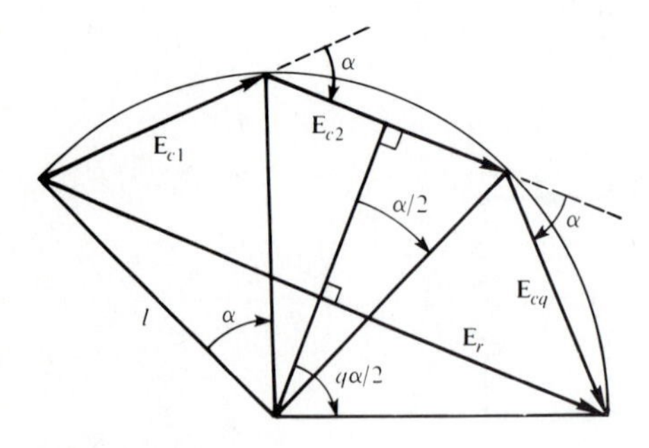

FIGURE 4-15
Determination of k_d.

and k_d is known as the *distribution factor*. From Fig. 4.15 we obtain

$$k_d = \frac{E_r}{qE_e} = \frac{2\alpha \sin q(\alpha/2)}{2\alpha q \sin(\alpha/2)} = \frac{\sin q\alpha/2}{q \sin\alpha/2} \tag{4.27}$$

The voltage equation (4.24) is modified by (4.27) as follows:

$$E = 4.44 k_d f \phi N \tag{4.28}$$

The distribution factors for some three-phase windings are given in Table 4.1.

Example 4.2 Recall from Fig. 4.7 that the coil pitch is not equal to the pole-pitch. Such a winding was termed as a fractional-pitch winding. The voltage induced in a fractional-pitch coil is reduced by a factor known as *pitch factor*, compared to the voltage induced in a full-pitch coil. Derive an expression for the pitch factor.

Solution

In a sinusoidally distributed flux density we show a full-pitch and a fractional-pitch coil in Fig. 4.16. The coil span of the full-pitch coil = pole-pitch = τ. Let the coil span of the fractional-pitch coil be β, as shown. The flux linking the fractional-pitch coil will be proportional to the shaded area (Fig. 4.16), as compared to the flux linking the full-pitch coil (i.e., proportional to the entire area under the curve). The ratio of the shaded area to the total area is, therefore, the *pitch factor, k_p*. Thus

$$k_p = \int_{(\tau - \beta)/2}^{(\tau + \beta)/2} \sin\frac{\pi x}{\tau} dx \bigg/ \int_{x=0}^{\tau} \sin\frac{\pi x}{\tau} dx = \sin\frac{\pi\beta}{2\tau} \tag{4.29}$$

Accounting for the pitch factor, the emf equation (4.28) further modifies to

$$E = 4.44 k_d k_p f \phi N = 4.44 k_w f \phi N \tag{4.30}$$

where $k_w = k_d k_p$ and was termed the winding factor in (4.23).

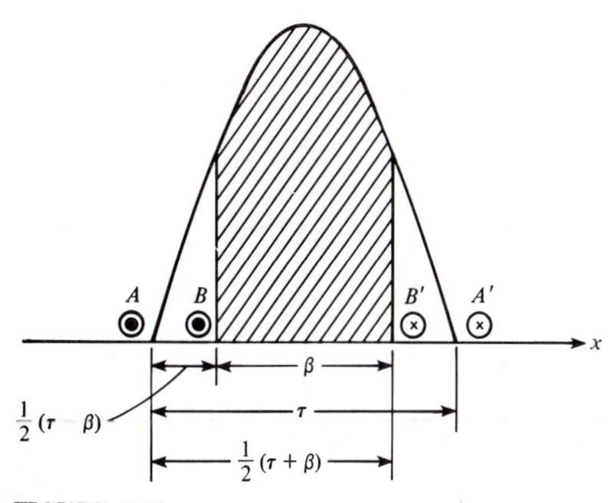

FIGURE 4-16
Determination of k_p.

TABLE 4.1
Distribution factors for three-phase windings

Slots per pole per phase	2	3	4	5	6	8	∞
k_d	0.966	0.960	0.958	0.957	0.957	0.956	0.955

Example 4.3 A cage-type rotor consists of Q bars and has P poles. Determine the distribution, pitch, and winding factors. Also, determine the number of phases and the number of turns per phase per pole.

Solution

number of phases on the rotor, m_2

$$= \frac{\text{number of pairs of bars}}{\text{number of pairs of poles}} = \frac{Q/2}{P/2} = \frac{Q}{P}$$

$$\text{number of turns per phase per pole, } N_{2/P} = \frac{\text{pairs of bars}}{\text{phase per pole}}$$

$$= \frac{Q}{2m_2P} = \frac{Q}{2\frac{Q}{P}P} = \frac{1}{2}$$

$$\text{pitch factor, } k_p = 1$$

(Assume an integral number of bars per pole. Even if not true, in a practical machine, a bar is positioned near one pole pitch, making the value of $k_p \sim 1$.)
Now,

$$q = \text{slots per pole per phase} = \frac{Q}{Pm_2} = \frac{Q}{PQ/q} = 1$$

Thus, from (4.27),

$$k_d = \frac{\sin(\alpha/2)}{\sin(\alpha/2)} = 1$$

and

$$k_w = k_p k_d = 1$$

To summarize, we have defined the distribution and pitch factors, k_d and k_p, respectively, in Examples 4.1 and 4.2. We have shown, in Example 4.3, that for a cage-type rotor, $k_d = k_p = 1$ and the number of series turns per phase per pole $N_2 = ½$.

For a wound rotor these quantities are determined in a manner similar to that for the stator.

Returning now to our discussion of the transformation ratio, we write for the amplitudes of stator and rotor mmf's F_1 and F_2 from (4.3)

$$F_1 = 0.9 m_1 k_{w1} \frac{N_1 I_1}{P} \tag{4.31}$$

and

$$F_2 = 0.9 m_2 k_{w2} \frac{N_2 I_2}{P} \tag{4.32}$$

where m_1 and m_2 are the number of phases on the stator and rotor, respectively, $m_2 = m_1$ in a wound rotor, and other symbols are as defined earlier. For a cage-type rotor, the number of phases m_2 = number of rotor bars per pole-pair = $Q_2/(P/2)$, Q_2 being the number of rotor bars. Referring the rotor quantities to the stator implies that the rotor has, in effect, the same mmf as the stator. That is, a rotor current I_2', referred to the stator and flowing in N_1 turns having m_1 phases, produces the same mmf as did the original F_2. From (4.32), therefore, we have

$$0.9 m_2 k_{w2} \frac{N_2 I_2}{P} = 0.9 m_1 k_{w1} \frac{N_1 I_2'}{P}$$

or

$$I_2' = \frac{m_2 k_{w2} N_2}{m_1 k_{w1} N_1} I_2 \tag{4.33}$$

Furthermore, rotor volt-amperes per phase referred to the stator must be the same as the original rotor volt-amperes. Thus

$$m_1 E_2' I_2' = m_2 E_2 I_2 \tag{4.34}$$

We substitute (4.23) and (4.33) into (4.34) to obtain the rotor voltage referred to the stator, E_2', as

$$E_2' = \frac{k_{w1} N_1}{k_{w2} N_2} E_2 = E_1 \tag{4.35}$$

The next equation that must be fulfilled is that the rotor I^2R losses be invariant. Expressed mathematically, this means that

$$m_1(I_2')^2R_2' = m_2I_2^2R_2 \tag{4.36}$$

where R_2' is the rotor resistance per phase referred to the stator and R_2 is the actual rotor resistance per phase. Equations (4.33) and (4.36) yield

$$R_2' = \frac{m_1}{m_2}\left(\frac{k_{w1}N_1}{k_{w2}N_2}\right)^2 R_2 \tag{4.37}$$

Finally, we require that the magnetic energy stored in the standstill rotor leakage reactance also remain unchanged; that is,

$$\frac{1}{2}m_1L_2'(I_2')^2 = \frac{1}{2}m_2L_2I_2^2 \tag{4.38}$$

where L_2 = rotor inductance per phase, referred to the rotor. Multiplying both sides of (4.38) by the stator angular frequency ω and substituting (4.33), we get

$$X_2' = \frac{m_1}{m_2}\left(\frac{k_{w1}N_1}{k_{w2}N_2}\right)^2 X_2 \tag{4.39}$$

which is the rotor leakage reactance referred to the stator.

To summarize, the rotor current, voltage, resistance, and reactance, when referred to the stator, must be multiplied by the factors contained in (4.33), (4.35), (4.37), and (4.39), respectively.

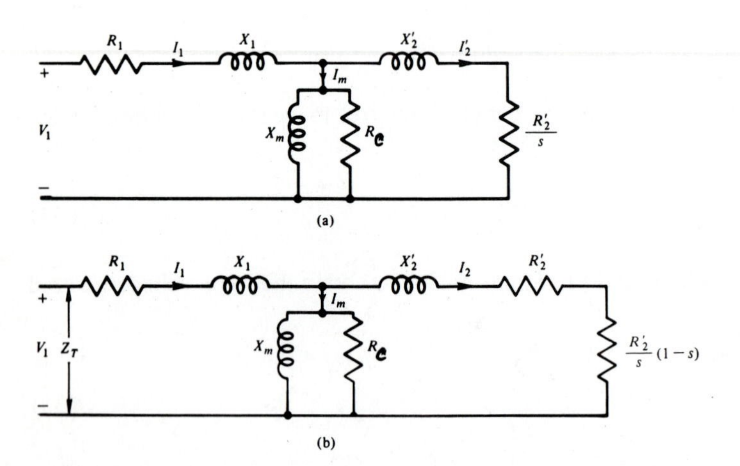

FIGURE 4-17
Two forms of equivalent circuits of an induction motor.

Having demonstrated the similarity between an induction motor and a transformer, and recognizing the essential differences, we can now refer the rotor quantities to the stator. Thus we obtain the exact equivalent circuit (per-phase) shown in Fig. 4.17*(a)* from the circuit given in Fig. 4.13. For reasons that will become immediately clear, we split R_2'/s as

$$\frac{R_2'}{s} = R_2' + \frac{R_2'}{s}(1 - s)$$

to obtain the circuit shown in Fig. 4.17. Here, R_2' is simply the per-phase standstill rotor resistance referred to the stator and $R_2'(1 - s)/s$ is a dynamic resistance that depends on the rotor speed and corresponds to the load on the motor. Notice that all the parameters shown in Figs. 4.17*(a)* and 4.17*(b)* are standstill values and the circuit is the per-phase exact equivalent circuit referred to the stator. We will show the usefulness of the circuit later, but first we consider an example to show the calculations of the factors used in referring rotor quantities to the stator.

Example 4.4 The stator winding of a cage-type induction motor shown in Fig. 4.7 has 24 turns per phase. Calculate the factor by which the rotor standstill resistance must be multiplied to refer it to the stator.

Solution

We notice from Fig. 4.7 that the stator has three phases, or $m_1 = 3$. Also, it has four poles, or $P = 4$; the number of slots per pole per phase, $q = 36/4 \times 3 = 3$, and the slot angle $\alpha = 180/3 \times 3 = 20$, from (4.34). Therefore, from (4.36) or Table 4.1,

$$k_{d1} = \frac{\sin(3 \times 20/2)}{3\ \sin(20/2)} = 0.96$$

Again, from Fig. 4.7, $\tau = 9$ slots and $\beta = 8$ slots. Thus, from (4.29) we get

$$k_{p1} = \sin\frac{8\pi}{18}$$

$$= \sin 80° = 0.985$$

The winding factor for the stator is

$$k_{w1} = k_{d1} k_{p1} = 0.945$$

For the rotor we have

$$k_{w2} = 1, \quad N_2 = \frac{P}{2} = 2, \quad \text{and} \quad m_2 = \frac{Q_2}{P} = \frac{28}{4} = 7$$

Substituting these values into (4.37), we obtain

$$R_2' = \frac{3}{7}\left(\frac{0.945 \times 24}{1 \times 2}\right)^2 R_2 = 55R_2$$

or the required factor = 55.

4.6 PERFORMANCE CALCULATIONS FROM EQUIVALENT CIRCUITS

The major usefulness of the equivalent circuit of an induction motor is in its performance calculations. Of course, here we assume that all the circuit parameters are known and either the input or output conditions or a combination of the two are specified. For instance, the input voltage and operating slip may be given, and we may be required to determine input current, power factor, efficiency, and so forth. We emphasize that all calculations are made on a per-phase basis, assuming a balanced operation of the machine. The total quantities are obtained by using appropriate multiplying factors, as shown in Example 4.4.

To illustrate we refer to Fig. 4.17. We redraw this circuit in Fig. 4.18, where we also show approximately the power flow and various power losses in one phase of the machine. Notice that we have neglected the core losses, most of which are in the stator. We will include core losses only in efficiency calculations. Therefore the power crossing the airgap, P_g, is the difference between the input power, P_i, and the stator $I_1^2R_1$ loss; that is,

$$P_g = P_i - I_1^2R_1 \quad \text{W/phase} \tag{4.40}$$

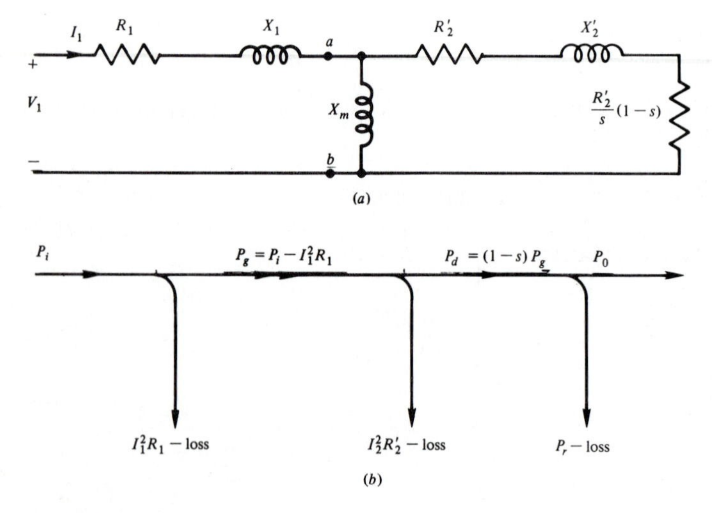

FIGURE 4-18
Power flow in an induction motor.

Clearly, this power is dissipated in the resistor R_2'/s (Fig. 4.17). Therefore

$$P_g = I_2'^2 \frac{R_2'}{s} \tag{4.41}$$

If we subtract the rotor $I_2'^2 R_2^2$ loss from P_g, we obtain the developed electromagnetic power, P_d, so that

$$P_d = P_g - I_2'^2 R_2' = (1 - s) P_g \tag{4.42}$$

This is the power that appears across a resistor having an ohmic value $R_2'[(1 - s)/s]$, which corresponds to the load. The rotational power loss, P_r, may be subtracted from P_d to obtain the shaft output power, P_o. Thus

$$P_o = P_d - P_r \tag{4.43}$$

Also

$$P_i = V_1 I_1 \cos\phi_1 \tag{4.44}$$

and the efficiency, η, is the ratio P_o/P_i.

We now illustrate this procedure by Example 4.5.

Example 4.5 The parameters of the equivalent circuit, Fig. 4.17*(a)*, for a 220-V, three-phase, 4-pole, *Y*-connected, 60-Hz induction motor are

$$R_1 = 0.2\ \Omega \qquad R_2' = 0.1\ \Omega$$

$$X_1 = 0.5\ \Omega \qquad X_2' = 0.2\ \Omega$$

$$X_m = 20.0\ \Omega$$

The total iron and mechanical losses are 350 W. For a slip of 2.5%, calculate input current, output torque, and efficiency.

Solution

Because the iron losses are known (350 W), we make an approximation by neglecting the resistance R_c. Thus, from Fig. 4.18*(a)*, the total impedance is

$$Z_t = R_1 + jX_1 + \frac{jX_m\left(\dfrac{R_2'}{2} + jX_2'\right)}{\dfrac{R_2'}{2} + j(X_m + X_2')}$$

$$= 0.2 + j0.5 + \frac{j20(4 + j0.2)}{4 + j(20 + 0.2)}$$

$$= (0.2 + j0.5) + (3.77 + j0.95) = 4.23\ \angle 20°\Omega$$

$$\text{phase voltage} = 220/\sqrt{3} = 127\ \text{V}$$

$$\text{input current} = 127/4.23 = 30\ \text{A}$$

$$\text{power factor} = \cos 20° = 0.94$$

$$\text{total input power} = \sqrt{3} \times 220 \times 30 \times 0.94$$

$$= 10.75\ \text{kW}$$

$$\text{total power across the airgap} = 3 \times 30^2 \times 3.77$$

$$= 10.18\ \text{kW}$$

$$\text{total power developed} = 0.975 \times 10.18 = 9.93\ \text{kW}$$

$$\text{total output power} = 9.93 - 0.35 = 9.58\ \text{kW}$$

$$\text{total output torque} = \text{output power}/\omega_m$$

$$= (9.58/184) \times 1000 = 52\ \text{N-m}$$

where

$$\omega_m = 0.975 \times 60 \times \pi = 184\ \text{rad/s}$$

$$\text{efficiency} = \frac{9.58}{10.75} = 89.1\%$$

Using a procedure similar to that given in Example 4.5, we can calculate the performance of the motor at other values of the slip, ranging from 0 to 1. The characteristics thus calculated are shown in Fig. 4.19.

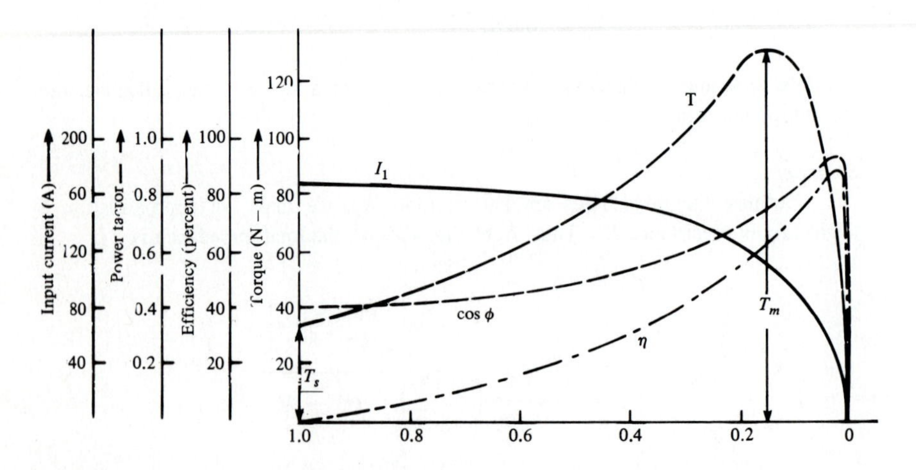

FIGURE 4-19
Characteristics of an induction motor. T_m = maximum torque; T_s = starting torque, Key: —, input current, A; — — —, power factor; — · — · —, efficiency, percent; and - - -, torque, newton-meters.

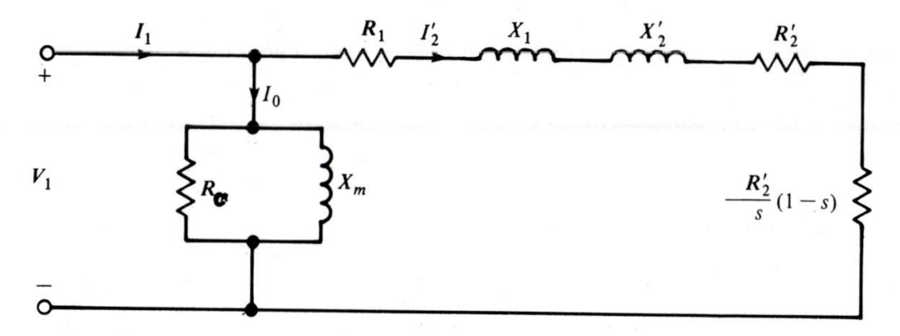

FIGURE 4-20
An approximate equivalent circuit of an induction motor.

4.7 THE EQUIVALENT CIRCUIT FROM TEST DATA

Example 4.5 illustrates the usefulness of an equivalent circuit. However, we did not actually use the exact circuit shown in Fig. 4.17. Instead, to simplify the calculations, we neglected the shunt-branch resistance R_c. In many calculations, for practical purposes the induction machine is represented by the approximate equivalent circuit shown in Fig. 4.20. In order to calculate the performance of the machine, its parameters must be known. The parameters of the circuit shown in Fig. 4.20 can be obtained from the following two tests.

1. *No-Load Test.* In this test rated voltage is applied to the machine and it is allowed to run on no-load. Input power, voltage, and current are measured. These are reduced to per-phase values and denoted by W_0, V_0, and I_0, respectively. Notice that W_0 = no-load per-phase power after the per-phase rotational loss on no-load has been subtracted from the measured no-load power. (See Example 4.6 for explanation.) When the machine runs on no-load, the slip is close to zero and the circuit to the right of the shunt branch is taken to be an open circuit. Thus the parameters R_c and X_m are found from the following equations:

$$R_c = \frac{V_0^2}{W_0} \tag{4.45}$$

$$X_m = \frac{V_0}{I_0 \sin\phi_0} \tag{4.46}$$

where

$$\phi_0 = \cos^{-1}\frac{W_0}{V_0 I_0} \tag{4.47}$$

2. *Blocked-Rotor Test.* In this test the rotor of the machine is blocked ($s = 1$) and a reduced voltage is applied to the machine so that the rated current flows through the stator windings. The input power, voltage, and current are recorded and

reduced to per-phase values. These are denoted, respectively, by W_s, V_s, and I_s. The iron losses are assumed to be negligible and the shunt branch of the circuit shown in Fig. 4.20 is considered to be absent. The parameters are thus found from

$$R_e = R_1 + a^2 R_2 = \frac{W_s}{I_s^2} \tag{4.48}$$

$$X_e = X_1 + a^2 X_2 = \frac{V_s \sin\phi_s}{I_2} \tag{4.49}$$

where

$$\phi_s = \cos^{-1}\frac{W_s}{V_s I_s} \tag{4.50}$$

In (4.48) and (4.49) a is a constant and is analogous to the transformation ratio of a transformer. It takes into account the effect of rotor resistance and reactance as referred to the stator, discussed in Section 4.5. The tests described here are approximate. The stator ac resistance per phase, R_1, can be directly measured and, knowing R_e from (4.48), we can determine $R_2' = a^2 R_2$, the rotor resistance referred to the stator. There is no simple method of determining the leakage reactances X_1 and X_2 separately. The total value of the leakage reactance is given by (4.49) and, approximately, we may assume $X_1 = X_2$. Consider now an example to illustrate the calculations involved in the determination of the machine constants from test data.

Example 4.6 The results of the no-load and blocked-rotor tests on a three-phase, Y-connected induction motor are as follows.

No-load test:	line-to-line voltage	=	220 V
	total input power	=	1000 W
	line current	=	20 A
	friction and windage loss	=	400 W
Blocked-rotor test:	line-to-line voltage	=	30 V
	total input power	=	1500 W
	line current	=	50 A

The stator resistance between two terminals = 0.16 Ω. Calculate the parameters of the approximate equivalent circuit shown in Fig. 4.20.

Solution

$$V_0 = \frac{220}{\sqrt{3}} = 127 \text{ V}$$

$$I_0 = 20 \text{ A}$$

$$W_0 = \frac{1}{3}(1000 - 400) = 200 \text{ W}$$

Thus, from (4.45) to (4.47),

$$R_c = \frac{127^2}{200} = 80.5 \, \Omega$$

$$\phi_0 = \cos^{-1}\frac{200}{20 \times 127} = 86°$$

$$X_m = \frac{127}{20 \times 0.99} = 6.4 \, \Omega$$

Now

$$V_s = \frac{30}{\sqrt{3}} = 17.32 \text{ V}$$

$$I_s = 50 \text{ A}$$

$$W_s = \frac{1500}{3} = 500 \text{ W}$$

Thus, from (4.48) to (4.50),

$$R_e = \frac{500}{50^2} = 0.2 \, \Omega$$

$$\phi_s = \cos^{-1}\frac{500}{17.32 \times 50} = 54°$$

$$X_e = 17.32 \times \frac{0.8}{50} = 0.277 \, \Omega$$

Knowing the circuit constants, we can calculate the machine performance, as in Example 4.5.

4.8 PERFORMANCE CRITERIA OF INDUCTION MOTORS

Examples 4.5 and 4.6 show the usefulness of the equivalent circuit and the method of determining its parameters from test data in order to calculate the performance of the motor. The performance of an induction motor may be characterized by the following factors.

1. Efficiency.

2. Power factor.

3. Starting torque.

4. Starting current.

5. Pull-out (or maximum) torque.

Notice that these characteristics are shown in Fig. 4.19. In design considerations, heating because of I^2R losses and core losses and means of heat dissipation must be included. It is not within the scope of this book to present a detailed discussion of the effects of design changes and, consequently, parameter variations, on each performance characteristic. Here we summarize the results as trends. For example, the efficiency is approximately proportional to $(1 - s)$. Thus the motor would be most compatible with a load running at the highest possible speed. Because the efficiency is clearly dependent on I^2R losses, R_2' and R_1 must be small for a given load. To reduce core losses, the working flux density (B) must be small. But this imposes a conflicting requirement on the load current (I_2') because the required torque, which is determined by the load, is dependent on the product of B and I_2'. In other words, an attempt to decrease the core losses beyond a limit would result in an increase in the I^2R losses for a given load.

It may be seen from the equivalent circuits (developed in Section 4.5) that the power factor can be improved by decreasing the leakage reactances and increasing the magnetizing reactance. However, it is not wise to reduce the leakage reactances to a minimum, since the starting current of the motor is essentially limited by these reactances. Again, we notice the conflicting conditions for a high power factor and a low starting current. Also, the pull-out torque would be higher for lower leakage reactances.

A high starting torque is produced by a high R_2'; that is, the higher the rotor resistance, the higher would be the starting torque. A high R_2' is in conflict with a high efficiency requirement.

We may arrive at some of these conclusions by considering the rotor circuit only as shown by Example 4.7.

Example 4.7 From the rotor equivalent circuit shown in Fig. 4.13, (1) find R_2 for which the developed torque would be a maximum. (2) What is the slip at this maximum torque? (3) Determine R_2 for a maximum starting torque. (4) What is the effect of X_2 on the torque?

Solution

From Fig. 4.13, the developed power P_d per phase is given by

$$P_d = I_2^2 \frac{R_2}{s}(1 - s) = T_e \omega_m$$

But the mechanical speed ω_m is related to the synchronous speed by

$$\omega_m = (1 - s)\omega_s$$

These two equations yield the expression for the electromagnetic torque, T_e, as

$$T_e = \frac{I_2^2 R_2}{s\omega_s} \tag{4.51}$$

But the rotor current I_2 is given by

$$I_2 = \frac{sE_2}{\sqrt{R_2^2 + (sX_2)^2}} \tag{4.52}$$

From (4.51) and (4.52) we have

$$T_e = \frac{E_2^2}{\omega_s}\left(\frac{sR_2}{R_2^2 + s^2X_2^2}\right) \tag{4.53}$$

For a maximum T_e we must have $\partial T_e/\partial R_2 = 0$, which together with (4.53) gives

$$\frac{\partial T_e}{\partial R_2} = \frac{sE_2^2}{\omega_s}\left[\frac{(R_2^2 + s^2X_2^2) - R_2(2R_2)}{(R_2^2 + s^2X_2^2)^2}\right] = 0$$

or

$$R_2^2 + s^2X_2^2 - 2R_2^2 = 0$$

(1)

$$R_2 = sX_2$$

and

(2)

$$s = \frac{R_2}{X_2}$$

At starting, $s = 1$.

(3)

$$R_2 = X_2$$

(4) For a given rotor resistance, the starting torque would be maximum of $X_2 = 0$, from (4.53).

Alternatively, we would arrive at the same conclusions as in parts (1), (2), and (3) by setting $\partial T_e/\partial s = 0$. This yields

$$R_2(R_2^2 + s^2X_2^2) - sR_2(2sX_2^2) = 0$$

from which $R_2 = sX_2$, $s = R_2/X_2$, and $R_2 = X_2$ at $s = 1$.

Clearly, this analysis is only approximate. However, we can arrive at similar conclusions by using the exact equivalent circuit as shown in Fig. 4.17.

4.9 SPEED CONTROL OF INDUCTION MOTORS

Because of its simplicity and ruggedness, the induction motor finds numerous applications. However, it suffers from the drawback that, in contrast to dc motors, its speed cannot be easily and efficiently varied continuously over a wide range of operating conditions. We will briefly review the various possible methods by which the speed of the induction motor can be varied either continuously or in discrete steps. However, we will not consider all these methods in detail here.

The speed of the induction motor can be varied by (1) varying the synchronous speed of the traveling field, or (2) varying the slip. Because the efficiency of the induction motor is approximately proportional to $(1 - s)$, any method of speed control that depends on the variation of slip is inherently inefficient. On the other hand, if the supply frequency is constant, varying the speed by changing the synchronous speed results only in discrete changes in the speed of the motor. We will now consider these methods of speed control in some detail.

Recall from (4.10) that the synchronous speed, n_s, of the rotating field in an induction machine is given by

$$n_s = \frac{120f}{P}$$

where P is the number of poles and f is the supply frequency, which indicates that n_s can be varied by (1) changing the number of poles P, or (2) changing the frequency f. Both methods have found applications, and we consider here the pertinent qualitative details.

POLE-CHANGING METHOD. In this method the stator winding of the motor is so designed that by changing the connections of the various coils (the terminals of which are brought out), the number of poles of the winding can be changed in the ratio of 2:1. Accordingly, two synchronous speeds result. We observe that only two speeds of operation are possible. Fig. 4.21 shows one phase interconnection for a 2:1 pole ratio. If more independent windings (e.g., two) are provided—each arranged for pole changing—more synchronous speeds (e.g., four) can be obtained. However, the fact remains that only discrete changes in the speed of the motor can be obtained by this technique. The method has the advantage of being efficient and reliable because the motor has a cage-type rotor and no brushes.

VARIABLE-FREQUENCY METHOD. We know that the synchronous speed is directly proportional to the frequency. If it is practicable to vary the supply frequency, the synchronous speed of the motor can also be varied. The variation in speed is continuous or discrete according to continuous or discrete variation of the supply frequency. However, the maximum torque developed by the motor is inversely proportional to the synchronous speed. If we desire a constant maximum torque, both supply voltage and supply frequency should be increased if we wish to increase the synchronous speed of the motor. The inherent difficulty in the application of this method is that the supply frequency, which is commonly available, is fixed. Thus the method is applicable only if a variable-frequency supply is available. Various schemes have been proposed to obtain a variable-frequency supply. With the advent of solid-state devices with comparatively large power ratings, it is now possible to use static inverters to drive the induction motor. (See Chapter 8.)

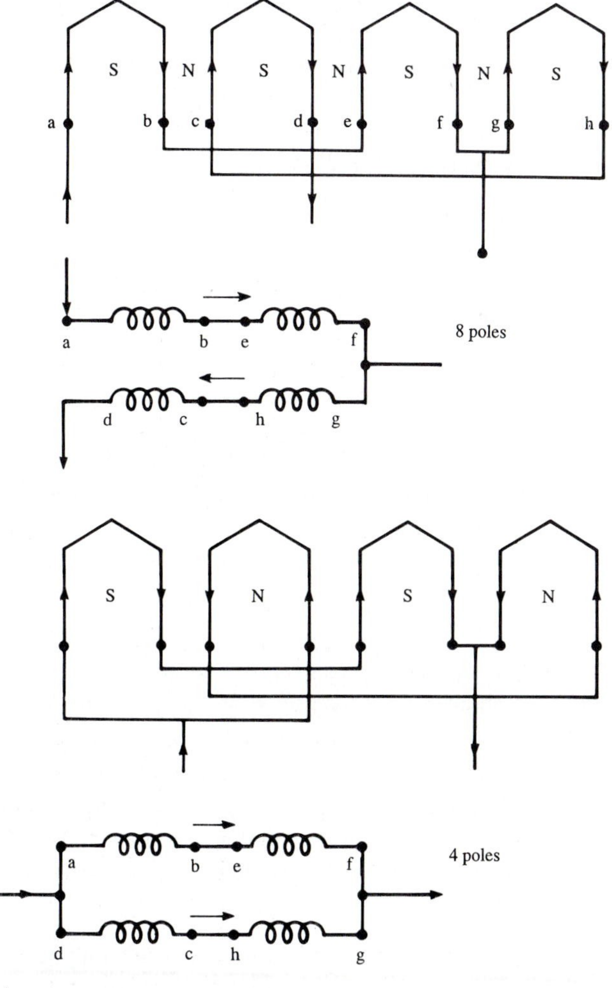

FIGURE 4-21
Pole-changing (2:1).

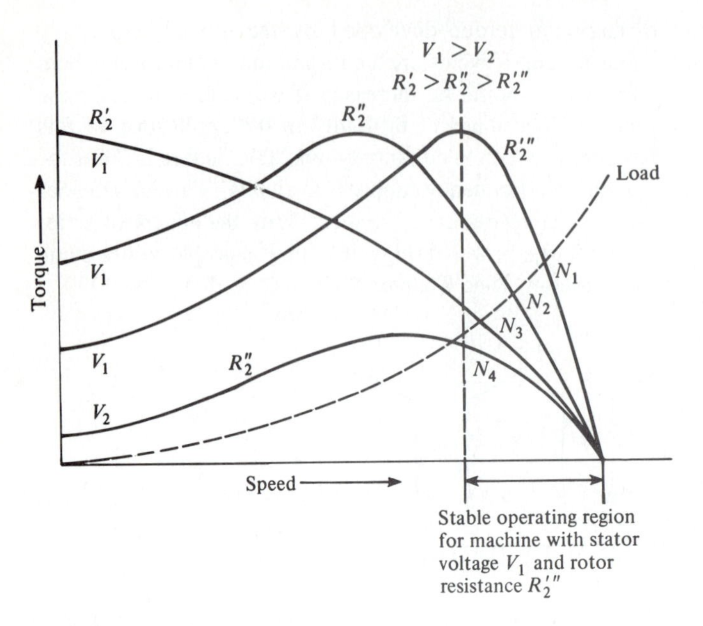

FIGURE 4-22
Speed control by changing the slip.

VARIABLE-SLIP METHOD. Controlling the speed of an induction motor by changing its slip may be understood by reference to Fig. 4.22. The dashed curve shows the speed-torque characteristic of the load. The curves with solid lines are the speed-torque characteristics of the induction motor under various conditions (such as different rotor resistances—R_2', R_2'', and R_2'''—or different stator voltages—V_1, V_2). We have four different torque-speed curves and, therefore, the motor can run at any one of four speeds—N_1, N_2, N_3, and N_4—for the given load. Note that the stable operating region of the motor is to the right of the peak torque. In practice, the slip of the motor can be changed by one of the following methods.

VARIABLE-STATOR-VOLTAGE METHODS. Since the electromagnetic torque developed by the machine is proportional to the square of the applied voltage, we obtain different torque-speed curves for different voltages applied to the motor. For a given rotor resistance, R_2, two such curves are shown in Fig. 4.22 for two applied voltages V_1 and V_2. Thus the motor can run at speeds N_2 or N_4. If the voltage can be varied continuously from V_1 to V_2, the speed of the motor can also be varied continuously between N_2 and N_4 for the given load. This method is applicable to cage-type as well as wound-rotor-type induction motors.

VARIABLE-ROTOR-RESISTANCE METHOD. This method is applicable only to the wound-rotor motor. The effect on the speed-torque curves of inserting external resistances in the rotor circuit is shown in Fig. 4.22 for three different rotor resistances R_2', R_2'', and R_2'''. For the given load, three speeds of operation are possible. Of course, by continuous variation of the rotor resistance, continuous variation of the speed is possible.

CONTROL BY POWER ELECTRONICS. Other than the inverter-driven motor, the speed of the wound-rotor motor can be controlled by inserting the inverter in the rotor circuit or by controlling the stator voltage by means of solid-state switching devices such as silicon-controlled rectifiers (SCRs or thyristors). The output from the SCR feeding the motor is controlled by adjusting its firing angle. The method of doing this is similar to the variable-voltage method outlined earlier. However, it has been found that control by an SCR gives a wider range of operation and is more efficient than other slip-control methods. For details, see Chapter 8.

4.10 STARTING OF INDUCTION MOTORS

Most induction motors—large and small—are rugged enough that they could be started across the line without incurring any damage to the motor windings, although about five to seven times the rated current flows through the stator at rated voltage at standstill. However, in large induction motors, large starting currents are objectionable in two respects. First, the mains supplying the induction motor may not be of a sufficiently large capacity. Second, because of a large starting current, the voltage drops in the lines may be excessive, resulting in reduced voltage across the motor. Because the torque varies approximately as the square of the voltage, the starting torque may become so small at the reduced line voltage that the motor might not even start on load. Thus we formulate the basic requirement for starting: the line current should be limited by the capacity of the mains, but only to the extent that the motor can develop sufficient torque to start (on load, if necessary).

Example 4.8 An induction motor is designed to run at 5 percent slip on full load. If the motor draws six times the full-load current at starting at the rated voltage, estimate the ratio of starting torque to the full-load torque.

Solution

The torque at a slip s is given by (4.51), which is repeated below:

$$T_e = \frac{I_2^2 R_2}{s\omega_s}$$

At full load, with $I_2 = I_{2f}$, the torque is

$$T_{ef} = \frac{I_{2f}^2 R_2}{0.05\,\omega_s}$$

At starting, $I_{2s} = 6I_{2f}$ and $s = 1$, so that

$$T_{es} = \frac{(6I_{2f})^2 R_2}{\omega_s}$$

Hence

$$\frac{T_{es}}{T_{ef}} = \frac{(6I_{2f})^2 R_2}{\omega_s} \left(\frac{0.05\,\omega_s}{I_{2f}^2 R_2} \right) = 1.8$$

Example 4.9 If the motor of the Example 4.8 is started at a reduced voltage to limit the line current to three times the full-load current, what is the ratio of the starting torque to the full-load torque?

Solution

In this case we have

$$\frac{T_{es}}{T_{ef}} = 3^2 \times 0.05 = 0.45$$

Notice that the starting torque has reduced by a factor of 4, compared to the case of full-voltage starting. In many practical cases, the line current is limited to six times the full-load current and the starting torque is desired to be about 1.5 times the full-load torque.

PUSHBUTTON STARTERS. There are numerous types of pushbutton starters for induction motors now commercially available. In the following, however, we will consider briefly only the principles of the two commonly used methods. We consider the current-limiting types first. Some of the most common methods of limiting the stator current while starting are:

1. *Reduced-Voltage Starting.* At the time of starting, a reduced voltage is applied to the stator and the voltage is increased to the rated value when the motor is within 25 percent of its final speed. This method has the obvious limitation that a variable-voltage source is needed and the starting torque drops substantially. The wye/delta method of starting is a reduced-voltage starting method. If the stator is normally connected in delta, reconnection to wye reduces the phase voltage, resulting in less current at starting. For example, at starting, if the line current is about five times the full-load current in a delta-connected stator, the current in the wye connection will be less than twice the full-load value. But at the same time, the starting torque for a wye connection would be about one-third its value for a delta connection. The advantage of wye/delta starting is that it is inexpensive and requires only a three-pole (or three single-pole) double-throw switch (or switches), as shown in Fig. 4.23.

2. *Current Limiting by Series Resistance.* Series resistances inserted in the three lines are sometimes used to limit the starting current. These resistances are shorted out when the motor has gained speed. This method has the obvious disadvantage of being inefficient because of the extra losses in the external resistances during the starting period.

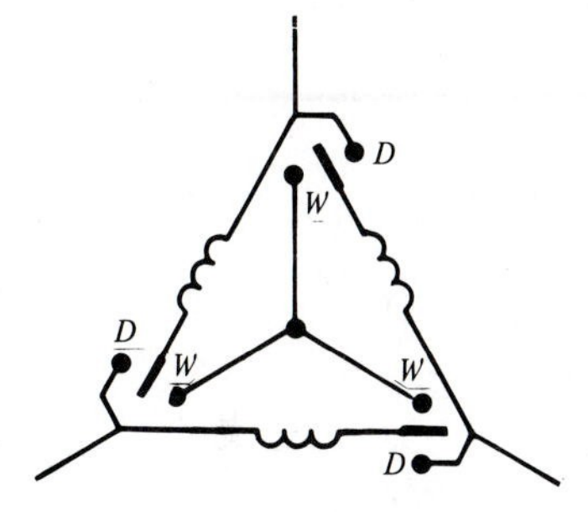

FIGURE 4-23
Wye/delta starting. Switches on W correspond to wye and switches on D correspond to the delta connection.

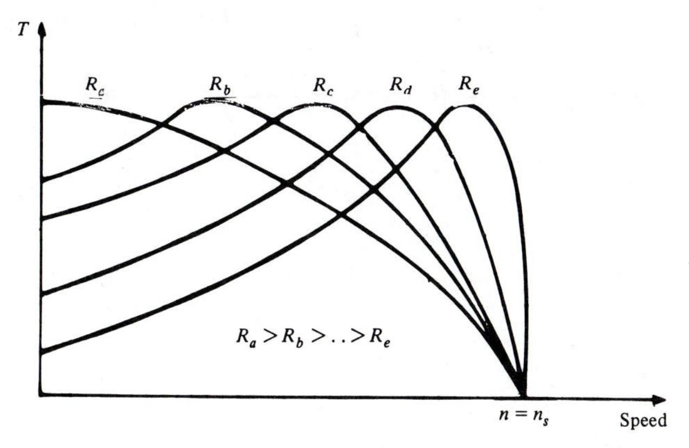

FIGURE 4-24
Effect of changing rotor resistance on the starting of a wound-rotor motor.

Turning now to the starting torque, we recall from Section 4.9 that the starting torque is dependent on the rotor resistance. Thus a high rotor resistance results in a high starting torque. Therefore, in a wound-rotor machine (see Fig. 4.24), external resistance in the rotor circuit may be conveniently used. In a cage rotor, deep slots are used, where the slot depth is two or three times greater than the slot width (see Fig. 4.25). Rotor bars embedded in deep slots provide a high effective resistance and a large torque at starting. Under normal running conditions with low slips, however, the rotor resistance becomes lower and efficiency high. This characteristic of rotor bar resistance is a consequence of *skin effect.* Because of the skin effect, the current will have a tendency to concentrate at the top of the bars at starting, when the frequency of rotor currents is high. At this point, the frequency of rotor currents will be the same as the stator input frequency (e.g., 60 Hz). While running, the frequency of rotor currents (= slip frequency = 3 Hz at 5 percent slip and 60 Hz) is much lower. At this level of operation, skin effect is negligible and the current almost uniformly distributes throughout the entire bar cross section.

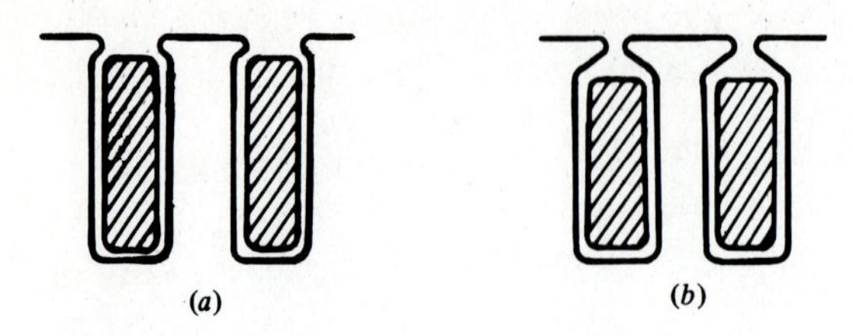

FIGURE 4-25
Deep-bar rotor slots: (*a*) open; (*b*) partially closed.

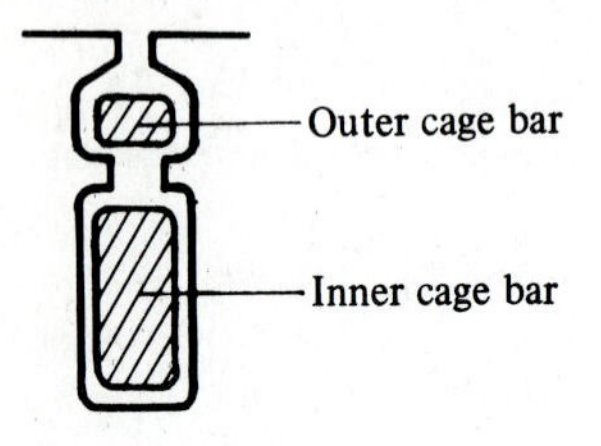

FIGURE 4-26
Form of a slot for a double-cage rotor.

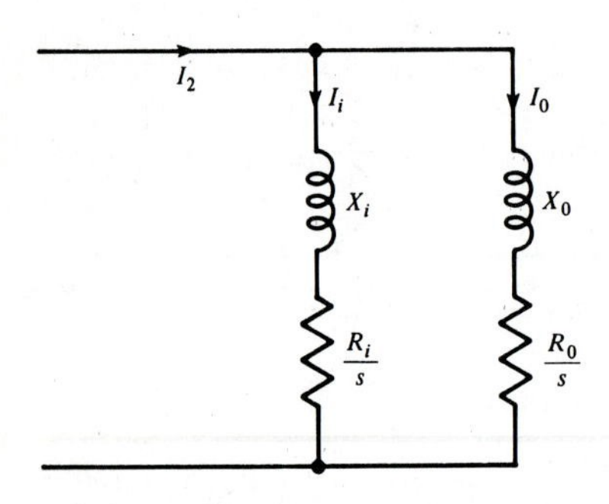

FIGURE 4-27
Equivalent circuit of a double-cage rotor.

The skin effect is used in an alternative form in a *double*-cage rotor (Fig. 4.26), where the inner cage is deeply embedded in iron and has low-resistance bars and has a high reactance. The outer cage has relatively high resistance bars close to the stator, and has the low reactance of a normal single-cage rotor. At starting, because of the skin effect, the influence of the outer cage dominates, thus producing a high starting torque. While running, the current penetrates to full depth into the lower cage—because of insignificant skin effect and lower reactance—which results in an efficient steady-state operation. Notice that under normal running conditions both cages carry current, thus somewhat increasing the power rating of the motor. The rotor equivalent circuit of a double-cage rotor then becomes as shown in Fig. 4.27. Approximately, the cages may be considered to develop separate torques, and the sum of these torques is the total torque. By appropriate design, the inner- and outer-cage resistances and leakage reactances may be modified to obtain a wide range of performance characteristics. Compared to a normal single-cage motor, the inner-cage leakage reactance in a double-cage motor lowers its power factor at full load. Furthermore, the high resistance of the outer cage increases the full-load I^2R loss and decreases the motor efficiency. But for a duty cycle of frequent starts and stops, a double-cage motor is likely to have a higher energy efficiency compared to a single-cage motor.

Example 4.10 A motor employs a wye/delta starter which connects the motor phases in wye at the time of starting and in delta when the motor is running. The full-load slip is 4 percent and the motor draws nine times the full-load current if started directly from the mains. Determine the ratio of starting torque to full-load torque.

Solution

When the phases are switched to delta, the phase voltage, and hence the full-load current, is increased by a factor of √3 over the value it would have had in a wye connection. Then it follows from the last equation in Example 4.9 that

$$\frac{T_s}{T_{\mathrm{FL}}} = \left(\frac{9}{\sqrt{3}}\right)^2 (0.04) = 1.08$$

Example 4.11 To obtain a high starting torque in a cage-type motor, a double-cage rotor is used. The forms of a slot and of the bars of the two cages are shown in Fig. 4.26. The outer cage has a higher resistance than the inner cage. At starting, because of the skin effect, the influence of the outer cage dominates, thus producing a high starting torque. An approximate equivalent circuit for such a rotor is given in Fig. 4.27. Suppose that, for a certain motor, we have the per-phase values, at standstill:

$$R_i = 0.1\,\Omega \qquad R_o = 1.2\,\Omega \qquad X_i = 2\,\Omega \qquad X_o = 1\,\Omega$$

Determine the ratio of the torques provided by the two cages at *(a)* starting and *(b)* 2 percent slip.

Solution

(a) From Fig. 4.27, at $s = 1$:

$$Z_i^2 = (0.1)^2 + (2)^2 = 4.01\,\Omega^2$$

$$Z_o^2 = (1.2)^2 + (1)^2 = 2.44\,\Omega^2$$

$$\text{power input to the inner cage} \equiv P_{ii} = I_i^2 R_i = 0.1 I_i^2$$

$$\text{power input to the outer cage} \equiv P_{io} = I_o^2 R_o = 1.2 I_o^2$$

$$\frac{\text{torque due to the inner cage}}{\text{torque due to the outer cage}} \equiv \frac{T_i}{T_o} = \frac{P_{ii}}{P_{io}} = \frac{0.1}{1.2}\left(\frac{I_i}{I_o}\right)^2 = \frac{0.1}{1.2}\left(\frac{Z_o}{Z_i}\right)^2$$

$$= \frac{0.1}{1.2}\left(\frac{2.44}{4.01}\right) = 0.05$$

(b) Similarly, at $s = 0.02$:

$$Z_i^2 = \left(\frac{0.1}{0.02}\right)^2 + (2)^2 = 29\ \Omega^2$$

$$Z_o^2 = \left(\frac{1.2}{0.02}\right)^2 + (1)^2 = 3601\ \Omega^2$$

$$\frac{T_i}{T_o} = \frac{0.1}{1.2}\left(\frac{3601}{29}\right) = 10.34$$

4.11 ENERGY-EFFICIENT INDUCTION MOTORS

Over the last 10 years the cost of electrical energy has more than doubled. For instance, it has been reported that the annual energy cost to operate a 10-hp induction motor 4000 h per year increased from $850 in 1972 to $1950 in 1980. The escalation of oil prices in the mid-1970s led the manufacturers of electric motors to seek methods to improve motor efficiencies. In order to improve motor efficiency, a motor's loss distribution must be studied. For a typical standard three-phase 50-hp motor, the loss distribution at full-load is given in Table 4.2. In this table we also show the average loss distribution in percent of total losses for standard induction motors. The per unit loss in Table 4.2 is defined as loss/(hp x 746).

In improving the efficiency of the motor, we must design to achieve a balance among the various losses and at the same time meet other specifications, such as breakdown torque, locked-rotor current and torque, and power factor. For the motor designer, a clear understanding of the loss distribution is very important. Loss reductions can be made by increasing the amount of the material in the motor. Without making other major design changes, a loss reduction of about 10 percent at full load can be achieved. Improving the magnetic circuit design using lower loss electrical grade laminations can result in a further reduction of losses by about 10 percent. The cost of improving the motor efficiency increases with output rating (hp) of the motor. Based on the improvements just mentioned to increase the motor efficiency, Fig. 4.28 shows a comparison of the efficiencies of energy-efficient motors with those of standard motors.

Several of the major manufacturers of induction motors have developed product lines of energy-efficient motors. These motors are identified by their trade names, which include

E-Plus (Gould Inc.).
Energy Saver (General Electric).
XE-Energy Efficient (Reliance Electric).
Mac II High Efficiency (Westinghouse).

TABLE 4.2
Loss distribution in standard induction motors

	50-hp Motor			
Loss Distribution	**Watts**	**Percent Loss**	**Per-Unit Loss**	**Average Percent Loss for Standard Motors**
Stator I^2R loss	1,540	38	0.04	37
Rotor I^2R loss	860	22	0.02	18
Magnetic core loss	765	20	0.02	20
Friction and windings loss	300	8	0.01	9
Stray load loss	452	12	0.01	16
Total losses	3,917	100	0.10	
Output (W)	37,300			
Input (W)	41,217			
Efficiency (%)	90.5			

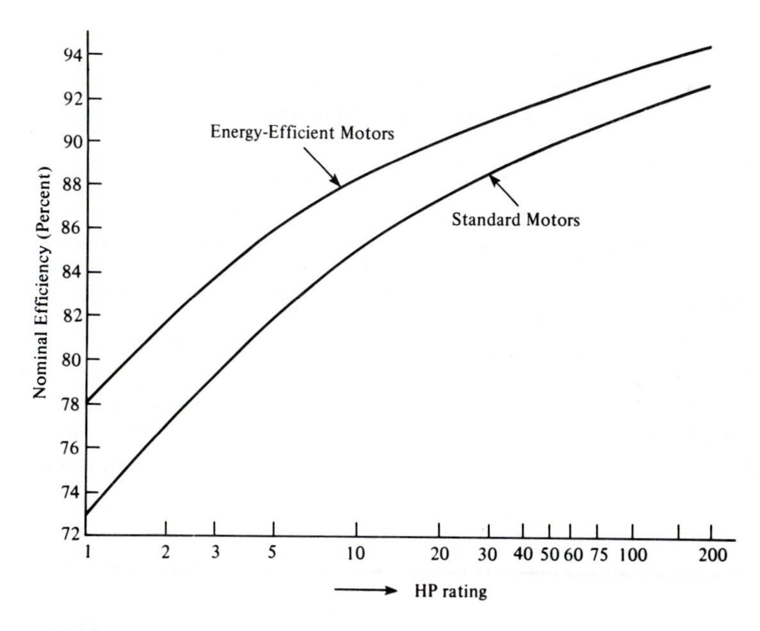

FIGURE 4-28
Comparison of nominal efficiencies of energy-efficient and standard motors.

Because energy-efficient motors use more material, they are bigger than standard motors.

4.12 INDUCTION GENERATORS

Up to this point, we have studied the behavior of an induction machine operating as a motor. We recall from the preceding discussions that for motor operation the slip lies between zero and unity, and for this case we have a conversion of electrical power into mechanical power. If the rotor of an induction machine is driven by an auxiliary means such that the rotor speed, n, becomes greater than the synchronous speed, n_s, we have, from (4.11), a negative slip. A negative slip implies that the induction machine is now operating as an induction generator. Alternatively, we may refer to the rotor portion of the equivalent circuit, such as that of Fig. 4.18*(a)*. If the slip is negative, the resistor representing the load becomes $R_s[1 - (-s)]/(-s)$, which results in a negative value of the resistance. Because a positive resistance absorbs electrical power, a negative resistance may be considered as a source of power. Hence a negative slip corresponds to a generator operation.

To understand the generator operation, we consider a three-phase induction machine to which a prime mover is coupled mechanically. When the stator is excited, a synchronously rotating magnetic field is produced and the rotor begins to run, as in an induction motor, while drawing electrical power from the supply. The prime mover is then turned on (to rotate the rotor in the direction of the rotating field). When the rotor speed exceeds synchronous speed, the direction of electrical power reverses. The power begins to flow into the supply as the machine begins to operate as a generator. The rotating magnetic field is produced by the magnetizing current supplied to the stator winding from the three-phase source. This supply of the magnetizing current must be available as the machine operates as an induction generator. For induction generators operating in parallel with a three-phase source capable of supplying the necessary exciting current, the voltage and the frequency are fixed by the operating voltage and frequency of the source supplying the exciting current.

An induction generator may be self-excited by providing the magnetizing reactive power by a capacitor bank. In such a case an external ac source is not needed. The generator operating frequency and voltage are determined by the speed of the generator, its load, and the capacitor rating. As for the dc shunt generator, for the induction generator to self-excite, its rotor must have sufficient remnant flux. The operation of a self-excited induction generator may be understood by referring to Fig. 4.29*(a)*. On no-load, the charging current of the capacitor, $I_c = V_1/X_c$, must be equal to the magnetizing current, $I_m = V_1/X_m$. Because V_1 is a function of I_m, for a stable operation the line $I_cX_c = I_mX_c$ must intersect the magnetization curve, which is a plot of V_1 versus I_m, as shown in Fig. 4.29*(b)*. The operating point P is thus determined, and we have

$$V_1 = I_m X_c \tag{4.54}$$

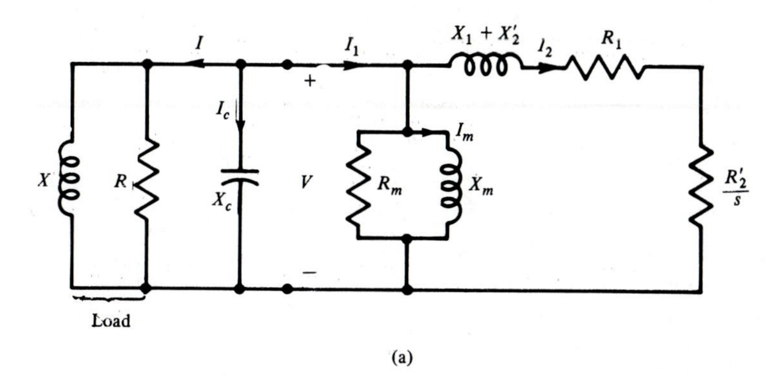

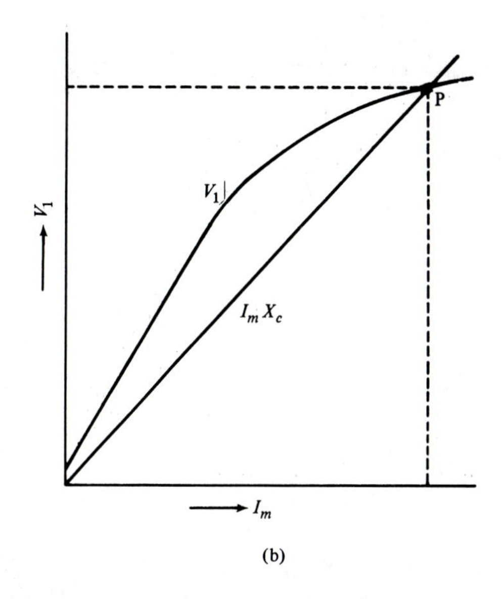

FIGURE 4-29
(a) Equivalent circuit of a self-excited induction generator; *(b)* determination of stable operating point.

Since $X_c = 1/\omega C = 1/2\pi fC$, we rewrite (4.54) as

$$I_m = 2\pi f C V_1 \tag{4.55}$$

From (4.55) the operating frequency is given by

$$f = \frac{I_m}{2\pi C V_1} \tag{4.56}$$

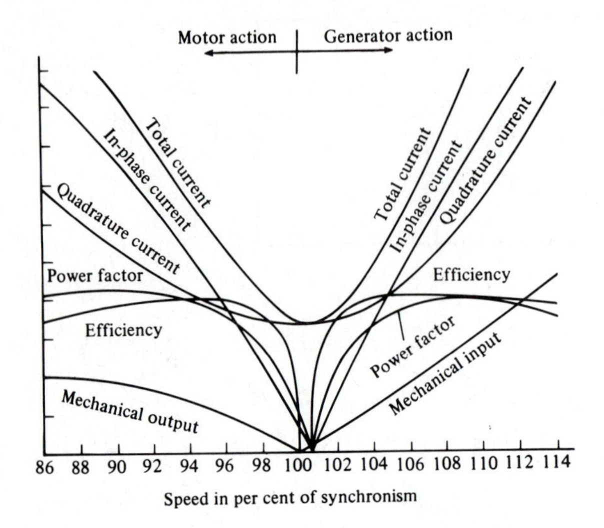

FIGURE 4.30
Motor and generator characteristics of an induction machine.

On load, the generated power $V_1 I_2' \cos \varphi_2'$ provides for the power loss in R_m and the power used by the load R. The reactive currents are related to each other by

$$\frac{V_1}{X_c} = \frac{V_1}{X} + \frac{V_1}{X_m} + I_2' \sin \varphi_2' \qquad (4.57)$$

which determines the capacitance for a given load.

Operating characteristics of an induction machine, for generator and motor modes, are shown in Fig. 4.30. Unlike in a synchronous generator, for a given load, the output current and the power factor are determined by the generator parameters. Therefore, when an induction generator delivers a certain power, it also supplies a certain in-phase current and a certain quadrature current. However, the quadrature component of the generator current generally does not have a definite relationship to the quadrature component of the load current. The quadrature current must be supplied by the synchronous generators operating in parallel with the induction generator.

Induction generators are not suitable for supplying loads having low lagging power factors. In the past, induction generators have been used in variable-speed constant-frequency generating systems. Large induction generators have found application in hydroelectric power stations. Induction generators are promising for windmill applications and are rapidly finding application in this area.

4.13 SINGLE-PHASE INDUCTION MOTORS

In the preceding sections we considered the polyphase—three-phase—induction machine operating under balanced conditions. Now let us consider a three-phase induction motor running at light load. If one of the supply lines is disconnected, the motor will continue to run, although at a different speed. Such an operation of a three-phase induction motor may be considered as the operation of a single-phase motor.

Let us now consider the three-phase motor at rest and fed by a single-phase source. Obviously, the motor will not start because we have a pulsating magnetic field in the airgap instead of a rotating magnetic field, which is required for torque production, as discussed earlier. Thus we conclude that a single-phase induction motor is not self-starting but will continue to run if started by some means. This implies that to make it self-starting, the motor must be provided with an auxiliary means of starting. In a later section we will examine the various means of starting the single-phase induction motor.

Not considering the starting mechanism, the essential difference between the three-phase and single-phase induction motors is that the single-phase induction motor has a single stator winding that produces an airgap field that is stationary in space but alternating in time. The three-phase induction motor has a three-phase winding that produces a time-invariant rotating magnetic field in the airgap. The rotor of the single-phase induction motor is almost always a cage-type rotor and is similar to that of a polyphase induction motor. The rating of a single-phase motor of the same size as a three-phase motor would be smaller, as expected, and single-phase induction motors are rated most often as fractional horsepower motors. These are the most widely used motors in household appliances, fans, and so forth.

PERFORMANCE ANALYSIS. The operating performance of the single-phase induction motor is studied on the basis of the following theories: (1) cross-field theory, and (2) double-revolving-field theory. We will use the revolving-field theory in the following in analyzing the single-phase induction motor, although both theories yield identical results. It is interesting, however, to study qualitatively the mechanism of torque production in a single-phase induction motor, as viewed by cross-field theory. Fig. 4.31*(a)* shows a single-phase induction motor, the stator winding of which carries a single-phase excitation and the rotor is at a standstill. The stator mmf is shown as F_1. Since the rotor is stationary, it acts like a short-circuited transformer. As a consequence, the mmf due to the rotor currents simply opposes the stator mmf. The resulting field will be stationary in space but will pulsate in magnitude. Next, let the motor be started (by some means) and let it run at some speed, as shown in Fig. 4.31(*b*). Rotor conductors are now rotating in the magnetic field produced by the stator. Consequently, rotational voltages will be induced in the rotor conductors. Applying the right-hand or $\bar{J} \times \bar{B}$ rule yields the directions of the voltages induced, and hence the current flow, in the rotor conductors, as given in Fig. 4.31(*b*). We now have two mmf's: F_1 due to the stator and F_2 due to the rotation of the rotor. These mmf's produce their respective airgap fields. These mmf's are displaced from each other in space, as shown in Fig. 4.31*(b)*. To determine the time displacement between the two mmf's, notice that the rotor-induced voltage (because of rotation) is in time phase with the stator mmf. However, the rotor circuit being highly inductive, the rotor current lags the rotor voltage by almost 90°. Hence there is a time displacement (of about 90°) between the stator and rotor mmf's. Thus we fulfill the condition for the production of a rotating magnetic field, and the rotor continues to develop a torque as long as the rotor is running in the rotating magnetic field.

We see from the above that the magnetic field produced by the stator of a single-phase motor alternates through time. The field induces a current—and consequently an mmf—in the rotor circuit, and the resultant field rotates with the rotor. A single-phase induction motor may be analyzed by considering the mmf's, fluxes, induced voltages (both rotational and transformer), and currents that are separately produced by the stator and by the rotor. Such an approach leads to the cross-field theory. However, we can also analyze the single-phase motor in a manner similar to that for the polyphase induction motor. We recall from Section 4.2 that the polyphase induction motor operates on the basis of the existence of a rotating magnetic field. This approach is based on the concept that an alternating magnetic field is equivalent to two rotating magnetic fields rotating in opposite directions. When this concept is expressed mathematically, the alternating field is of the form

$$B(\theta, t) = B_m \cos\theta \sin\omega t \tag{4.58}$$

Then (4.58) may be rewritten as

$$B_m \cos\theta \sin\omega t = \frac{1}{2} B_m \sin(\omega t - \theta) + \frac{1}{2} B_m \sin(\omega t + \theta) \tag{4.59}$$

In (4.59), the first term on the right-hand side denotes a forward rotating field, whereas the second term corresponds to a backward rotating field. The theory based on such a resolution of an alternating field into two counterrotating fields is known as the *double-revolving-field theory*. The direction of rotation of the forward rotating field is assumed to be the same as the direction of the rotation of the rotor. Thus if the rotor runs at n rpm and n_s is the synchronous speed in rpm, the slip, s_f, of the rotor with respect to the forward rotating field is the same as s, defined by

$$s_f = s = \frac{n_s - n}{n_s} = 1 - \frac{n}{n_s} \tag{4.60}$$

But the slip, s_b, of the rotor with respect to the backward rotating flux is given by

$$s_b = \frac{n_s - (-n)}{n_s} = 1 + \frac{n}{n_s} = 2 - s \tag{4.61}$$

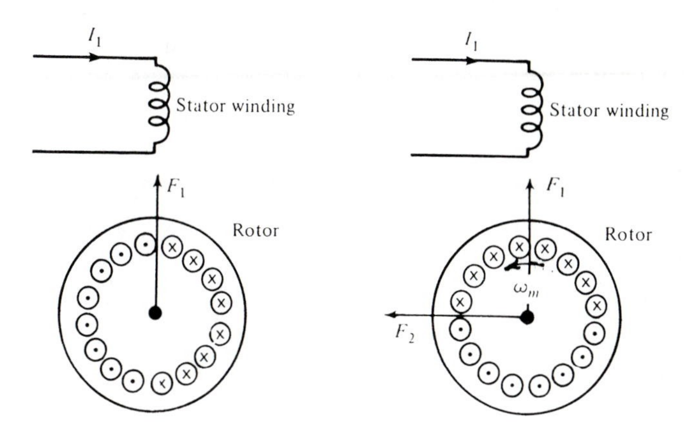

FIGURE 4-31
Representation of a single-phase induction motor: *(a)* standstill; *(b)* running.

We know from the operation of polyphase motors that, for $n < n_s$, (4.60) correspondsto a motor operation and denotes the braking region. Thus the two resulting torques have an opposite influence on the rotor.

The torque relationship for the polyphase induction motor is applicable to each of the two rotating fields of the single-phase motor. We notice from (4.59) that the amplitude of the rotating fields is one-half of the alternating flux. Thus the total magnetizing and leakage reactances of the motor can be divided equally so as to correspond to the forward and backward rotating fields. The approximate equivalent circuit of a single-phase induction motor, based on the double-revolving-field theory, becomes as shown in Fig. 4.32*(a)*. The torque-speed characteristics are qualitatively shown in Fig. 4.32*(b)*. The following example illustrates the usefulness of the circuit.

Example 4.12 For a 230-V single-phase induction motor, the parameters of the equivalent circuit, Fig. 4.32*(a)*, are: $R_1 = R_2' = 8\ \Omega$, $X_1 = X_2' = 12\ \Omega$, and $X_m = 200\ \Omega$. At a slip of 4 percent, calculate *(a)* the input current, *(b)* the input power, *(c)* the developed power, and *(d)* the developed torque (at rated voltage). The motor speed is 1728 rpm.

Solution

From Fig. 4.32*(a)*:

$$Z_f = \frac{(j100)\left(\dfrac{4}{0.04} + j6\right)}{j100 + \dfrac{4}{0.04} + j6} = 47 + j50\,\Omega$$

$$Z_b = \frac{(j100)\left(\dfrac{4}{1.96} + j6\right)}{j100 + \dfrac{4}{1.96} + j6} = 1.8 + j5.7\,\Omega$$

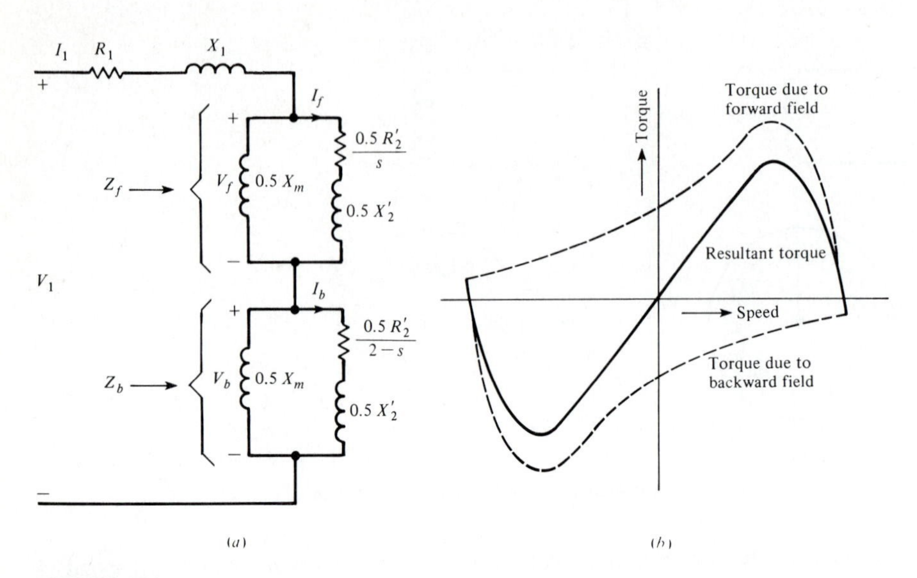

FIGURE 4-32
(a) Equivalent circuit, and *(b)* torque-speed characteristics, based on double-revolving-field theory of the single-phase induction motor.

$$Z_1 = R_1 + jX_1 = 8 + j12\,\Omega$$

$$Z_{\text{total}} = 56.8 + j67.7 = 88.4\angle 50^\circ\,\Omega$$

(a)

$$\text{input current} = I_1 = \frac{230}{88.4} = 2.6\text{ A}$$

$$\text{power factor} = \cos 50^\circ = 0.64 \text{ lagging}$$

(b) input power = (230)(2.6)(0.64) = 382.7 W

(c) Proceeding as in Example 4.5 we have

$$P_d = [I_1^2 \operatorname{Re}(Z_f)](1 - s) + [I_1^2 \operatorname{Re}(Z_b)]\,[1 - (2 - s)]$$

$$= I_1^2[\operatorname{Re}(Z_f) - \operatorname{Re}(Z_b)]\,(1 - s) = (2.6)^2(47 - 1.8)(1 - 0.04)$$

$$= 293.3\text{ W}$$

(d)

$$\text{torque} = \frac{P_d}{\omega_m} = \frac{293.3}{2\pi(1728)/60} = 1.62\text{ N-m}$$

Example 4.13 To reduce the numerical computation, Fig. 4.32*(a)* is modified by neglecting $0.5X_m$ in Z_b and taking the backward-circuit rotor resistance at low slips as $0.25R_2'$. With these approximations, repeat the calculations of Example 4.12 and compare the results.

Solution

$$Z_f = 47 + j50\ \Omega$$

$$Z_b = 2 + j6\ \Omega$$

$$Z_1 = 8 + j12\ \Omega$$

$$Z_{\text{total}} = 57 + j68 = 88.7\angle 50°\ \Omega$$

(a)

$$I_1 = \frac{230}{88.7} = 2.6\text{ A}$$

(b)

$$\cos\varphi = 0.64 \text{ lagging}$$

$$\text{input power} = (230)(2.6)(0.64) = 382.7\text{ W}$$

(c)

$$P_d = (2.6)^2(47 - 2)(1 - 0.04) = 292.0\text{ W}$$

(d)

$$\text{torque} = \frac{292.0}{2\pi(1728)/60} = 1.61\text{ N-m}$$

Following is a comparison of the results of Examples 4.12 and 4.13.

Example Number	Input Current (A)	Input Power (W)	Power Factor	Developed Torque (N-m)
4.12	2.6	382.7	0.64	1.62
4.13	2.6	382.7	0.64	1.61

This comparison indicates that the approximation suggested in Example 4.13 is adequate for most cases.

In the next example we show the procedure for efficiency calculations for a single-phase induction motor.

Example 4.14 A single-phase 110-V 60-Hz four-pole induction motor has the following constants in the equivalent circuit, Fig. 4.32*(a)*: $R_1 = R_2' = 2\ \Omega$, $X_1 = X_2' = 2\ \Omega$, and $X_m = 50\ \Omega$. There is a core loss of 25 W and a friction and windage loss of 10 W. For a 10 percent slip, calculate *(a)* the motor input current and *(b)* the efficiency.

Solution

$$Z_f = \frac{(j25)\left(\frac{1}{0.1} + j1\right)}{j25 + \frac{1}{0.1} + j1} = 8 + j4\,\Omega$$

$$Z_b = \frac{(j25)\left(\frac{1}{1.9} + j1\right)}{j25 + \frac{1}{1.9} + j1} = 0.48 + j0.96\,\Omega$$

$$Z_1 = 2 + j2\,\Omega$$

$$Z_{\text{total}} = 10.48 + j6.96 = 12.6\angle 33.6^\circ\,\Omega$$

(a)

$$I_1 = \frac{110}{12.6} = 8.73\text{ A}$$

(b) developed power = $(8.73)^2(8 - 0.48)(1 - 0.10) = 516$ W

output power = $516 - 25 - 10 = 481$ W

input power = $(110)(8.73)(\cos 33.6^\circ) = 800$ W

$$\text{efficiency} = \frac{481}{800} = 60\%$$

STARTING OF SINGLE-PHASE INDUCTION MOTORS. We already know that because of the absence of a rotating magnetic field, when the rotor of a single-phase induction motor is at standstill, it is not self-starting. The two methods of starting a single-phase motor are either to introduce commutator and brushes, such as in a repulsion motor, or to produce a rotating field by means of an auxiliary winding, such as by split phasing. We consider the latter method next.

From the theory of the polyphase induction motor, we know that in order to have a rotating magnetic field, we must have at least two mmf's which are displaced from each other in space and carry currents having different time phases. Thus, in a single-phase motor, a starting winding on the stator is provided as a source of the second mmf. The first mmf arises from the main stator winding. The various methods to achieve the time and space phase shifts between the main winding and starting winding mmf's are summarized below.

SPLIT-PHASE MOTORS. This type of motor is represented schematically in Fig. 4.33*(a)*, where the main winding has a relatively low resistance and a high reactance. The starting winding, however, has a high resistance and a low reactance and has a centrifugal switch as shown. The phase angle α between the two currents I_m and I_s is about 30 to 45°, and the starting torque T_s is given by

$$T_s = KI_mI_s \sin \alpha \tag{4.62}$$

where K is a constant. When the rotor reaches a certain speed (about 75 percent of its final speed), the centrifugal switch comes into action and disconnects the starting winding from the circuit. The torque-speed characteristic of the split-phase motor is of the form shown in Fig. 4.33*(b)*. Such motors find applications in fans, blowers, and so forth, and are rated up to ½ hp.

A higher starting torque can be developed by a split-phase motor by inserting a series resistor in the starting winding. A somewhat similar effect may be obtained by inserting a series inductive reactance in the main winding. This reactance is short-circuited when the motor builds up speed.

CAPACITOR-START MOTORS. By connecting a capacitor in series with the starting winding, as shown in Fig. 4.34, the angle α in (4.62) can be increased. The motor will develop a higher starting torque by doing this. Such motors are not restricted merely to fractional-horsepower ratings, and may be rated up to 10 hp. At 110 V, a 1-hp motor requires a capacitance of about 400 μF, whereas 70 μF is sufficient for a 1/8-hp motor. The capacitors generally used are inexpensive electrolytic types and can provide a starting torque that is almost four times that of the rated torque.

As shown in Fig. 4.34, the capacitor is merely an aid to starting and is disconnected by the centrifugal switch when the motor reaches a predetermined speed. However, some motors do not have the centrifugal switch. In such a motor, the starting winding and the capacitor are meant for permanent operation and the capacitors are much smaller. For example, a 110-V ½-hp motor requires a 15-μF capacitance.

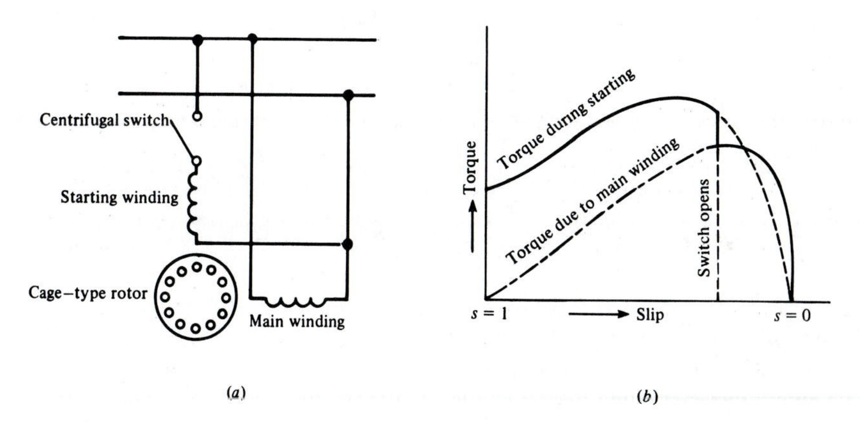

FIGURE 4-33
(a) Connections for a split-phase motor. *(b)* A torque-speed characteristic.

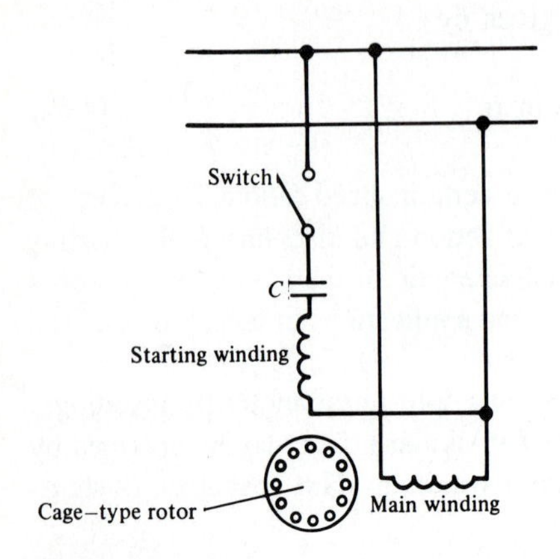

FIGURE 4-34
A capacitor-start motor.

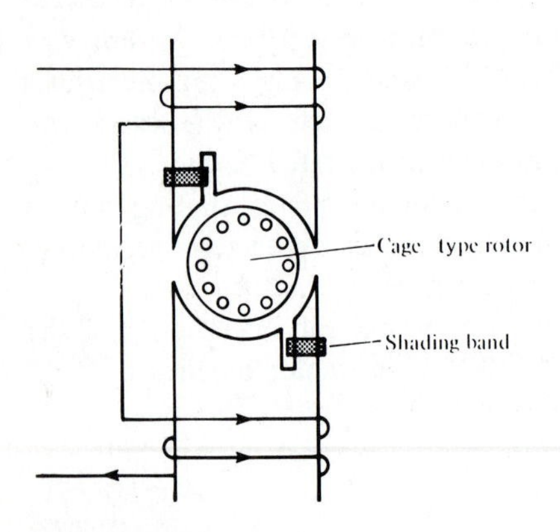

FIGURE 4-35
A shaded-pole motor.

A third kind of capacitor motor uses two capacitors: one that is left permanently in the circuit together with the starting winding, and one that gets disconnected by a centrifugal switch. Such motors are, in effect, unbalanced two-phase induction motors.

SHADED-POLE MOTORS. Another method of starting very small single-phase induction motors is to use a shading band on the poles, as shown in Fig. 4.35, where the main single-phase winding is also wound on the salient poles. The shading band is simply a short-circuited copper strap wound on a portion of the pole. Such a motor is known as a shaded-pole motor. The purpose of the shading band is to retard (in time) the portion of flux passing through it in relation to the flux coming out of the rest of the pole face. Thus the flux in the unshaded portion reaches its maximum before that located in the shaded portion. And we have a progressive shift of flux from the direction of the unshaded portion to the shaded portion of the pole, as shown in Fig. 4.35. The effect of the progressive shift of flux is similar to that of a rotating flux, and because of it, the shading band provides a starting torque. Shaded-pole motors are the least expensive of the fractional-horsepower motors and are generally rated up to 1/20 hp.

4.14 TRANSIENTS IN INDUCTION MACHINES

Transient conditions in induction machines occur under three circumstances: (1) at standstill, e.g., at the time of starting the machine; (2) at constant speed; and (3) at variable speed. These transients take place because of switching, plugging, overspeeding, reversing, and sudden applications of load on the machine. Induction machines supplied with a variable-frequency source also undergo transients. In addition to these causes, of course, there are other possible reasons for transient conditions. All in all, an understanding of the transients is quite important, and we shall therefore develop some quantitative analysis to obtain the transient performance of the machine. Unfortunately, no general method is available which can conveniently be used for all transient cases. In the following we shall consider certain special cases.

Speed Build-Up and Starting

The speed-build-up characteristics of an induction motor can be very simply obtained graphically if we assume that the torque developed by the motor is given by a steady-state torque-speed curve such as shown in Fig. 4.36*(a)*. The load torque is also shown in Fig. 4.36*(a)*. Note that ΔT is the accelerating torque, so that the speed differential equation is

$$\Delta T = J\dot{\omega}_m \tag{4.63}$$

and the speed-build-up time is given by

$$t = \int_0^{\omega_m} \frac{J}{\Delta T}\, d\omega_m \tag{4.64}$$

Knowing J, we plot $J/\Delta T$ in Fig. 4.36*(b)* where ΔT is obtained from Fig. 4.36*(a)*. Performing the graphical integration of the $(J/\Delta T)$-ω_m curve of Fig. 4.36*(b)* we obtain the curve shown in Fig. 4.36*(c)*, which shows the speed-build-up characteristic of the motor. It should be pointed out that the above analysis is quite approximate, and often the departure from the actual speed-torque characteristic is quite substantial, as discussed below. We term this approach a *time-domain formulation*, since all computation is done in real time.

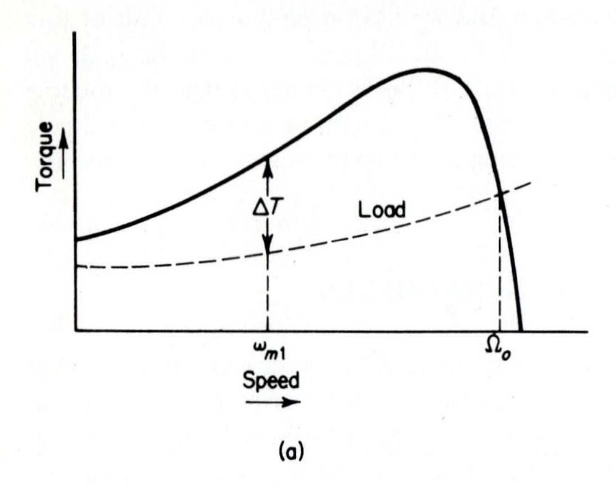

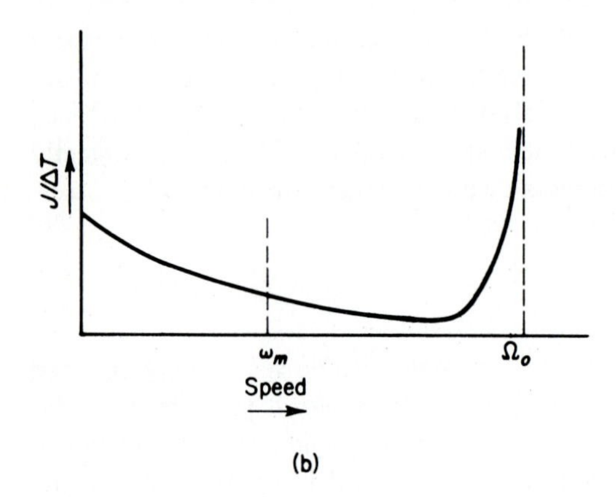

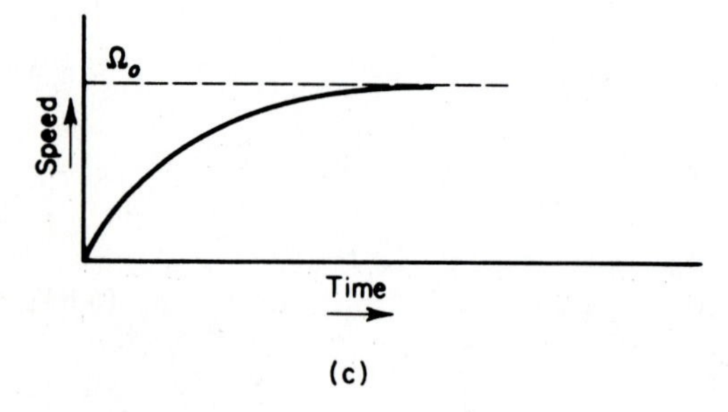

FIGURE 4-36
Speed build-up in an induction motor.

Consider a two-pole induction motor having a load of the form

$$J\dot{\omega}_m + b\omega_m + c$$

where J is the moment of inertia, b is the friction coefficient, c is a constant, and ω_m is the rotational speed. For this motor we let $[L]$ and $[R]$ represent the motor inductances and resistances in matrix form. Then if $[v]$ and $[i]$ are, respectively, the various phase voltages and currents represented as matrices, we may write the voltage and torque equations as

$$[v] = [R][i] + \frac{d}{dt}\{[L][i]\} \tag{4.65}$$

and

$$J\dot{\omega}_m + b\omega_m + c = \frac{1}{2}[\tilde{i}]\frac{\partial}{\partial\theta}[L][i] \tag{4.66}$$

These may be alternatively written as

$$\frac{d}{dt}[i] = -[L]^{-1}\left\{[R] + \frac{d}{dt}[L]\right\}[i] + [L]^{-1}[v] \tag{4.67}$$

$$\frac{d}{dt}\omega_m = \frac{1}{2J}\left\{[\tilde{i}]\frac{\partial}{\partial\theta}[L]\right\}[i] - \frac{b}{J}\omega_m - \frac{c}{J} \tag{4.68}$$

which may be combined to yield

$$[\dot{x}] = [A][x] + [B][v'] \tag{4.69}$$

which is the state equation. The various matrices in (4.69) are

$$[x] = \begin{bmatrix} [i] \\ \omega_m \\ \theta \end{bmatrix} \tag{4.70}$$

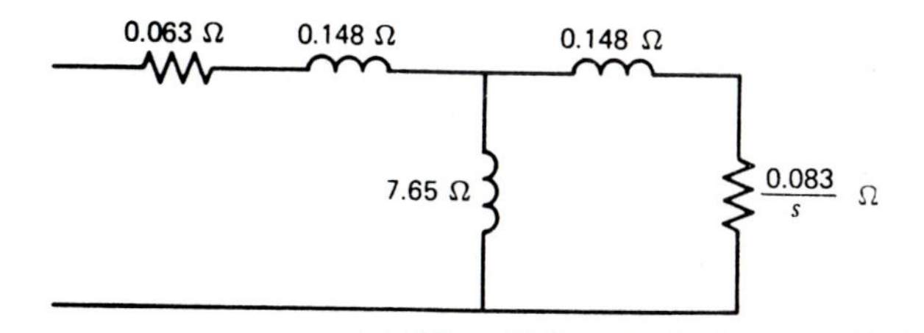

FIGURE 4-37
An approximate equivalent circuit of an induction motor.

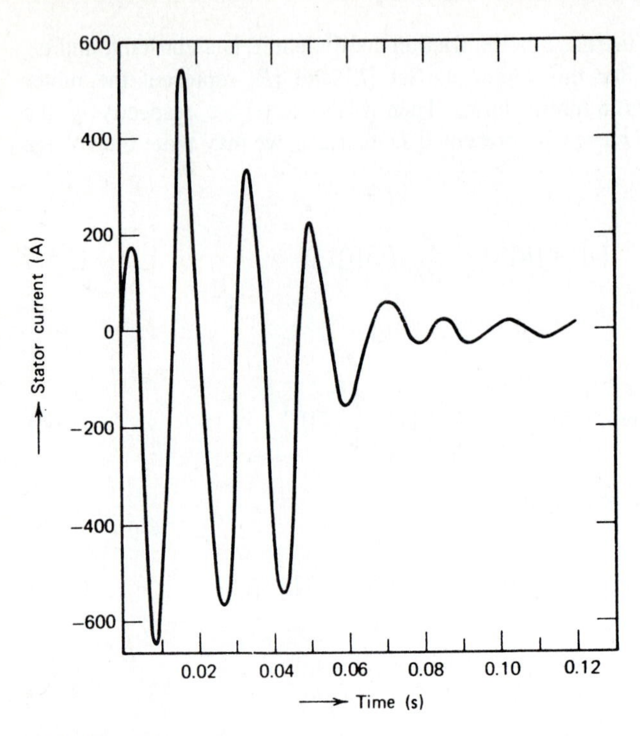

FIGURE 4-38
$i(t)$ during speed build-up.

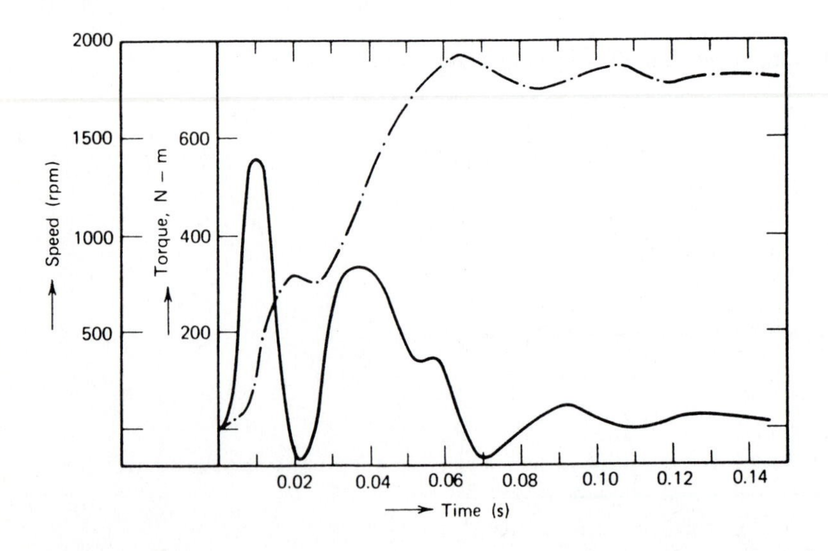

FIGURE 4-39
Speed and torque build-up. ____, torque: ___ . ___ . ___, speed.

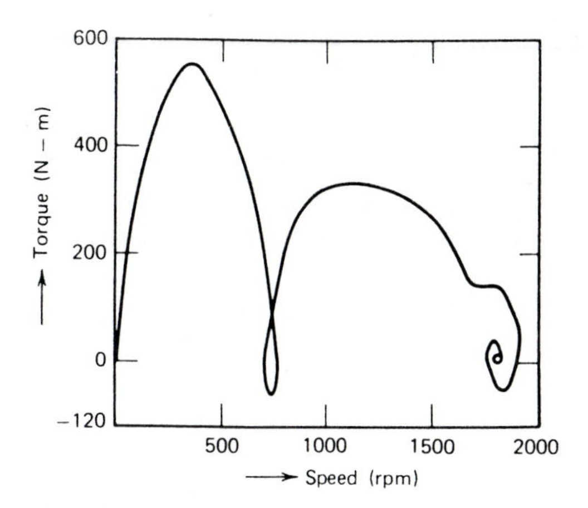

FIGURE 4-40
Transient torque-speed characteristics.

$$[v'] = \begin{bmatrix} [v] \\ 1 \\ 0 \end{bmatrix} \tag{4.71}$$

$$[B] = \begin{bmatrix} [L]^{-1} & 0 & 0 \\ 0 & -\dfrac{c}{J} & 0 \\ 0 & 0 & 0 \end{bmatrix} \tag{4.72}$$

and

$$[A] = \begin{bmatrix} -[L]^{-1}\left\{[R] + \dfrac{d}{dt}[L]\right\} & 0 & 0 \\ \dfrac{1}{2J}\{F[i]\} & -\dfrac{b}{J} & 0 \\ 0 & 0 & 0 \end{bmatrix} \tag{4.73}$$

where

$$\dot{\theta} = \omega_m \quad \text{and} \quad F[i] = \dot{\theta}\,(\partial/\partial\theta)\,[L][i]$$

Thus, the problem of transients in the machine is reduced to solving the state equation with time-varying coefficients. This equation can be solved numerically using a standard subroutine such as Runge-Kutta-Gauss-Seidel (RKGS) subroutine.

The procedure is extensively available in the literature.* Using this approach, for an induction motor whose equivalent circuit is shown in Fig. 4.37 and which has a load torque

$$T_L = (0.06\,\dot{\omega}_m + 0.03\,\omega_m + 6) \qquad \text{N-m}$$

the starting transients are shown in Fig. 4.38 through 4.40. The motor, assumed to be wye-connected, is initially at rest and is suddenly connected across a 200-V three-phase balanced source.

It is to be noted that Fig. 4.40 has been obtained by combining Figs. 4.38 and 4.39.

PROBLEMS

4.1. A six-pole 60-Hz induction motor runs at 1152 rpm. Determine the synchronous speed and the percent slip.

4.2. A six-pole, induction motor is supplied by a 50-Hz synchronous generator. If the speed of the induction motor is 750 rpm, what is the frequency of rotor current?

4.3. A four-pole, 60-Hz induction motor runs at 1710 rpm. Calculate *(a)* the slip in percent, *(b)* the frequency of rotor currents, and *(c)* the speed of the rotating magnetic field produced by *(i)* the stator, and *(ii)* the rotor, with respect to the stator, in revolutions per minute and in radians per second.

4.4. A two-pole, 60-Hz wound rotor induction motor has 127 V/phase across its stator. The voltage induced in the rotor is 3.81 V/phase. Assuming that the stator and the rotor have equal effective numbers of turns per phase, calculate *(a)* the motor speed, and *(b)* the slip.

4.5. A cage-type induction motor consists of 42 bars, each having a resistance of 4.12×10^{-5} Ω (including the resistances of the two end rings). The stator winding has the following data: six poles; three-phase; 36 slots; 144 turns/phase; 0.836 winding factor. Calculate the equivalent rotor resistance per phase that is referred to the stator.

4.6. The power crossing the airgap of an induction motor is 24.3 kW. If the developed electromagnetic power is 21.9 kW, what is the slip? The rotational loss at this slip is 350 W. Calculate the output torque, if the synchronous speed is 3600 rpm.

4.7. Using the circuit shown in Fig. 4.18*(a)*, obtain an expression of the slip at which the motor develops the maximum torque. Derive an expression for the maximum torque.

4.8. A three-phase, six-pole, 60-Hz induction motor develops a maximum torque of 180 N-m at 800 rpm. If the rotor resistance is 0.2 Ω per phase, determine the developed torque at 1000 rpm.

---

* See, for instance, I. Boldea and S. A. Nasar, *Electric Machine Dynamics*, Macmillan, 1986.

4.9. A 400-V four-pole, three-phase, 60-Hz induction motor has a wye-connected rotor having an impedance of $(0.1 + j0.5)\ \Omega$ per phase. How much additional resistance must be inserted in the rotor circuit for the motor to develop the maximum starting torque? The effective stator-to-rotor turns ratio is 1.

4.10. For the motor of Problem 4.9, what is the motor speed corresponding to the maximum developed torque without any external resistance in the rotor circuit?

4.11. A two-pole, 60-Hz induction motor develops a maximum torque of twice the full-load torque. The starting torque is equal to the full-load torque. Determine the full-load speed.

4.12. The input to the rotor circuit of a four-pole, 60-Hz induction motor, running at 1000 rpm, is 3 kW. What is the rotor copper loss?

4.13. The stator current of a 400-V, three-phase, wye-connected, four-pole, 60-Hz induction motor running at a 6 percent slip is 60 A at 0.866 power factor. The stator copper loss is 2700 W and the total iron and rotational losses are 3600 W. Calculate the motor efficiency.

4.14. An induction motor has an output of 30 kW at 86 percent efficiency. For this operating condition, stator copper loss = rotor copper loss = core losses = mechanical rotational losses. Determine the slip.

4.15. A four-pole, 60-Hz, three-phase, wye-connected induction motor has a mechanical rotational loss of 50 W. At 5 percent slip, the motor delivers 30 hp at the shaft. Calculate *(a)* the rotor input, *(b)* the output torque, and *(c)* the developed torque.

4.16. A three-phase, 230-V, 60-Hz, Y-connected, two-pole induction motor operates at 3% slip while taking a line current of 22 A. The stator resistance and leakage reactance per phase are $0.1\ \Omega$ and $0.2\ \Omega$, respectively. The rotor leakage reactance is $0.15\ \Omega$/phase. Calculate *(a)* the rotor resistance, *(b)* power crossing the airgap, and *(c)* developed power. Neglect X_m and R_c .

4.17. The per-phase constants of a three-phase, 600-V, 60-Hz, four-pole, Y-connected wound rotor induction motor are:

$$R_1 = 0.75\ \Omega \qquad X_1 = X_2' = 2.0\ \Omega$$

$$R_2' = 0.80\ \Omega \qquad X_m = 50.0\ \Omega$$

Neglect the core losses. *(a)* Calculate the slip at which the maximum developed torque occurs; *(b)* find the value of the maximum torque; *(c)* specify the range of speed for a stable operation of the motor, and *(d)* compute the starting torque and compare it with the maximum torque.

4.18. Repeat Problem 4.17, parts *(a)* and *(b)*, by considering the rotor circuit only. Assume that the entire 600 V may be taken as the applied line voltage to the rotor. Let turns ratio = 1.

4.19. A four-pole, 60-Hz, three-phase induction motor has a rotor resistance of $1.0\ \Omega$ and a leakage reactance of $2.5\ \Omega$/phase. Using the rotor circuit only, find the slip at which the motor develops a maximum torque. Also, determine the ratio of the maximum developed torque to the torque developed at 5% slip.

4.20. A three-phase, 60-Hz, four-pole induction motor runs at 1710 rpm. If the power crossing the airgap is 120 kW, what is the rotor copper loss? The motor has core and mechanical losses (at 1710 rpm) of 1.7 kW and 2.7 kW, respectively, and the stator copper loss is 3 kW. Calculate the motor efficiency and output torque.

4.21. A wound-rotor, six-pole, 60-Hz induction motor has a rotor resistance of 0.8 Ω and runs at 1150 rpm at a given load. The load on the motor is such that the torque remains constant at all speeds. How much resistance must be inserted in the rotor circuit to bring the motor speed down to 950 rpm? Neglect rotor leakage reactance.

4.22. A 400-V, three-phase, wye-connected induction motor has a stator impedance of $(0.6 + j1.2)$ Ω per phase. The rotor impedance referred to the stator is $(0.5 + j1.3)$ Ω per phase. Using the approximate equivalent circuit, determine the maximum electromagnetic power developed by the motor.

4.23. The motor of Problem 4.22 has a magnetizing reactance of 35 Ω. Neglecting the iron losses, at 3 percent slip calculate *(a)* the input current, and *(b)* the power factor of the motor. Use the approximate equivalent circuit.

4.24. On no-load a three-phase, delta-connected induction motor takes 6.8 A and 390 W at 220 V. The stator resistance is 0.1 Ω per phase. The friction and windage loss is 120 W. Determine the values of the parameters X_m and R_c of the equivalent circuit of the motor.

4.25. On blocked-rotor, the motor of Problem 4.24 takes 30 A and 480 W at 36 V. Using the data of Problem 4.24, determine the complete exact equivalent circuit of the motor. Assume that the per-phase stator and rotor leakage reactances are equal.

4.26. A 440-V, 25-Hz, two-pole, Y-connected motor has a magnetizing reactance of 10.5 Ω/phase and rotor leakage reactance of 0.12 Ω/phase. Using the rotor circuit only, determine the slip for maximum electromagnetic torque at a per-phase rotor resistance of *(a)* 0.03 Ω, *(b)* 0.06 Ω, and *(c)* 0.1 Ω. Show the effect of rotor resistance on the torque-speed characteristics of the motor. Neglect R_1 and X_1.

4.27. For the motor of Problem 4.17, determine the per-phase value of the resistance that must be inserted in the rotor circuit to obtain the maximum torque for the motor at starting.

4.28. A four-pole, 60-Hz, three-phase wound rotor induction motor has a rotor resistance of 0.2 Ω and a rotor leakage reactance of 2.0 Ω. Using the rotor circuit only, determine the value of the external resistance that must be inserted in the rotor circuit such that the motor develops the same torque at a 20% slip as at a 5% slip.

4.29. A 400-V, 60-Hz, three-phase, Y-connected, four-pole induction motor has the following per-phase equivalent circuit parameters [see Fig. 4.18*(a)*].

$$R_1 = 2R_2 = 0.2\ \Omega; \quad X_1 = 2.5X_2' = 0.5\ \Omega; \quad X_m = 20.0\ \Omega$$

The total mechanical and core losses at 1755 rpm are 800 W. At this speed, calculate *(a)* the output torque, and *(b)* the motor efficiency.

4.30. Develop a Thevenin equivalent circuit for the motor of Problem 4.29. What are the values of the Thevenin voltage and impedance? Using these values, compute the output torque and compare with the result of Problem 4.29.

4.31. Using the Thevenin circuit or otherwise, calculate the slip at which the motor of Problem 4.29 develops a maximum torque.

4.32. Sketch qualitatively the torque-speed characteristics of an induction motor, comparing it with normal characteristics and showing the effects of the following: *(a)* varying the frequency while keeping the applied voltage constant; *(b)* varying the applied voltage while keeping the frequency constant.

4.33. A 220-V, three-phase, 60-Hz, four-pole, Y-connected induction motor has a per-phase stator resistance of 0.25 Ω. The no-load and blocked-rotor test data on this motor are

no-load test:	stator voltage	= 220 V
	input current	= 3.0 A
	input power	= 600 W
	friction and windage loss	= 300 W
blocked-rotor test:	stator voltage	= 34.6 V
	input current	= 15.0 A
	input power	= 720 W

(a) Obtain the approximate equivalent circuit for the machine.
(b) If the machine runs as a motor with 5% slip, calculate the developed power, developed torque, and efficiency.
(c) Determine the slip at which maximum torque occurs, and calculate the maximum torque.

4.34. A double-cage rotor was mentioned in the text (see Fig. 4.27). Such a rotor may be represented by an approximate equivalent circuit having an impedance Z_o in parallel with an impedance Z_i, where Z_o and Z_i are the impedances of outer and inner cages, respectively. For a given motor $Z_o = 1 + j1$ and $Z_i = 0.1 + j2\ \Omega$. Using the rotor circuit only, determine approximately the ratio of the torques provided by the two cages at *(a)* starting, and *(b)* 2% slip.

4.35. For the motor of Problem 4.34, calculate the slip at which the torques contributed by the two cages become equal.

4.36. A single-phase, 110-V, 60-Hz, four-pole induction motor has the following equivalent circuit constants (see Fig. 4.32*(a)*).

$$R_1 = 2R_2 = 0.2\ \Omega; \quad X_1 = X_2 = 2.4\ \Omega; \quad X_m = 48\ \Omega$$

The motor has a core loss of 15 W and the friction and windage loss at 7% slip is 7 W. Calculate the output power and motor efficiency.

4.37. In the circuit of Fig. 4.32*(a)*, neglect X_m from Z_b and take $R_2 = \frac{1}{4}R_2$ in the backward field circuit. With these approximations, repeat the calculations of Problem 4.36 and compare the results.

CHAPTER 5

SYNCHRONOUS MACHINES

Synchronous machines are among the three most common types of electric machines, the other two being dc commutator machines and polyphase induction machines. The bulk of electric power for everyday use is produced by polyphase synchronous generators, which are the largest single-unit electric machines in production. For instance, synchronous generators with power ratings of several hundred megavolt-amperes (MVA) are fairly common. These are called synchronous machines because they operate at constant speeds and constant frequencies under steady-state conditions. Like most rotating machines, synchronous machines are capable of operating both as a motor and as a generator. They are used as motors in constant-speed drives and, where a variable speed drive is required, a synchronous motor is used with an appropriate frequency changer such as an inverter or cycloconverter (see Chapter 8). As generators, several synchronous machines often operate in parallel, as in a power station. While operating in parallel, the generators share the load with each other; at a given time, one of the generators may be allowed to "float" on the line as a synchronous motor on no-load. We will discuss such no-load operation of a synchronous motor later.

The operation of a synchronous generator is based on Faraday's law of electromagnetic induction, and in an ac synchronous generator the generation of emf's is by the relative motion of conductors and magnetic flux. The two basic parts of a synchronous machine are the magnetic field structure, carrying a dc-excited winding, and the armature. The armature often has a three-phase winding in which the ac emf is generated. Almost all modern synchronous machines have stationary armatures and rotating field structures. The dc winding on the rotating field structure is connected to an external source through slip rings and brushes. (Recall the construction of the slip-ring-type induction motor in Chapter 4.) Some field structures do not have brushes but, instead, have brushless excitation by rotating diodes. In some respects the stator carrying the armature windings is similar to the stator of a polyphase induction motor, which was studied in Chapter 4. In Section 5.2 we discuss some of the constructional features of synchronous machines.

5.1 OPERATION OF A SYNCHRONOUS GENERATOR

As mentioned earlier, the operation of the synchronous generator is based on Faraday's law of electromagnetic induction. The law states that an emf is induced in a circuit placed in a magnetic field if either (1) the magnetic flux linking the circuit is time varying, or (2) there is a relative motion between the circuit and the magnetic field such that the conductors comprising the circuit cut across the magnetic flux lines. The first form of the law, stated as (1), is the basis of operation of transformers. The second form, stated as (2), is the basic principle of operation of electric generators.

Consider a conductor of length l located at right angles to a uniform magnetic field of flux density B as shown in Fig. 5.1. Let the conductor be connected with fixed external connections to form a closed circuit. These external connections are shown in the form of conducting rails in Fig. 5.1. The conductor can slide on the rails, and thereby the flux linking the circuit changes. Hence, according to Faraday's law, an emf will be induced in the circuit, and a voltage, v, will be measured by the voltmeter. If the conductor moves with a velocity u (m/s) in a direction at right angles to B and l both, the area swept by the conductor in 1 second is lu. The flux in this area is Blu, which is also the flux linkage (since in effect, we have a single-turn coil formed by the conductor, the rails, and the voltmeter). In other words, flux linkage per unit time is Blu, which is, then, the induced emf, e. We write this in the equation form as

$$e = Blu \tag{5.1}$$

This form of Faraday's law is also known as the *flux-cutting rule.* Stated in words, an emf, e as given by (5.1), is induced in a conductor of length l if it "cuts" magnetic flux lines of density B by moving at right angles to B at a velocity u (at right angles to l). The mutual relationships among e, B, l, and u are given by the right-hand rule, as shown in Fig. 5.1. Recall from Chapter 2 that (5.1) is a specific form of the more general equation

$$v \equiv e = \oint (\bar{u} \times \bar{B}) \cdot d\bar{l} \tag{5.2a}$$

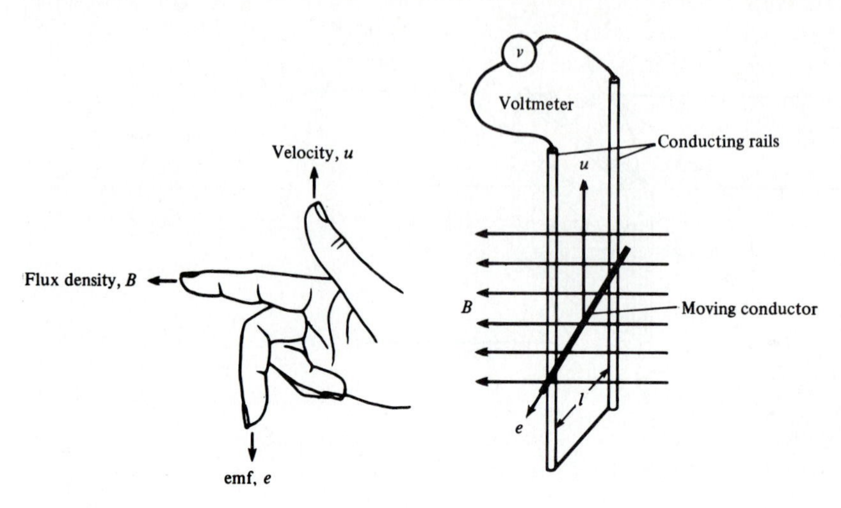

FIGURE 5-1
The right-hand rule.

Now, consider an N-turn coil, rotating at a constant angular velocity ω in a uniform magnetic field of flux density B, as shown in Fig. 5.2*(a)*. Let l be the axial length of the coil and r be its radius. The emf induced in the coil can be found by an application of (5.1). However, we must be careful in determining u in (5.2). Recall from the preceding section that u is the velocity at right angles to B. In the system under consideration, we have a rotating coil. Thus u in (5.1) corresponds to that component of velocity which is at right angles to B. We illustrate the components of velocities in Fig. 5.2*(b)*, where the tangential velocity $u_t = r\omega$ has been resolved into a component u_2, across (or perpendicular to) the flux. The latter component "cuts" the magnetic flux and is the component responsible for the emf induced in the coil. The u in (3.2) should then be replaced by u_2. The next term in (5.1) that needs careful consideration is l, the effective length of the conductor. For the N-turn coil of Fig. 5.2*(a)*, we effectively have $2N$ conductors in series (since each coil side has N conductors and there are two coil sides). If l_1 is the length of each conductor, then the total effective length of the N-turn coil is $2Nl_1$, which should be substituted for l in (5.1). The form of (5.1) for the N-turn coil then becomes

$$e = B(2Nl_1)u_2 \tag{5.2b}$$

From Fig. 5.2*(b)*, we have

$$u_2 = u_t \sin\theta = r\omega \sin\theta \tag{5.3}$$

If the coil rotates at a constant angular velocity ω, then

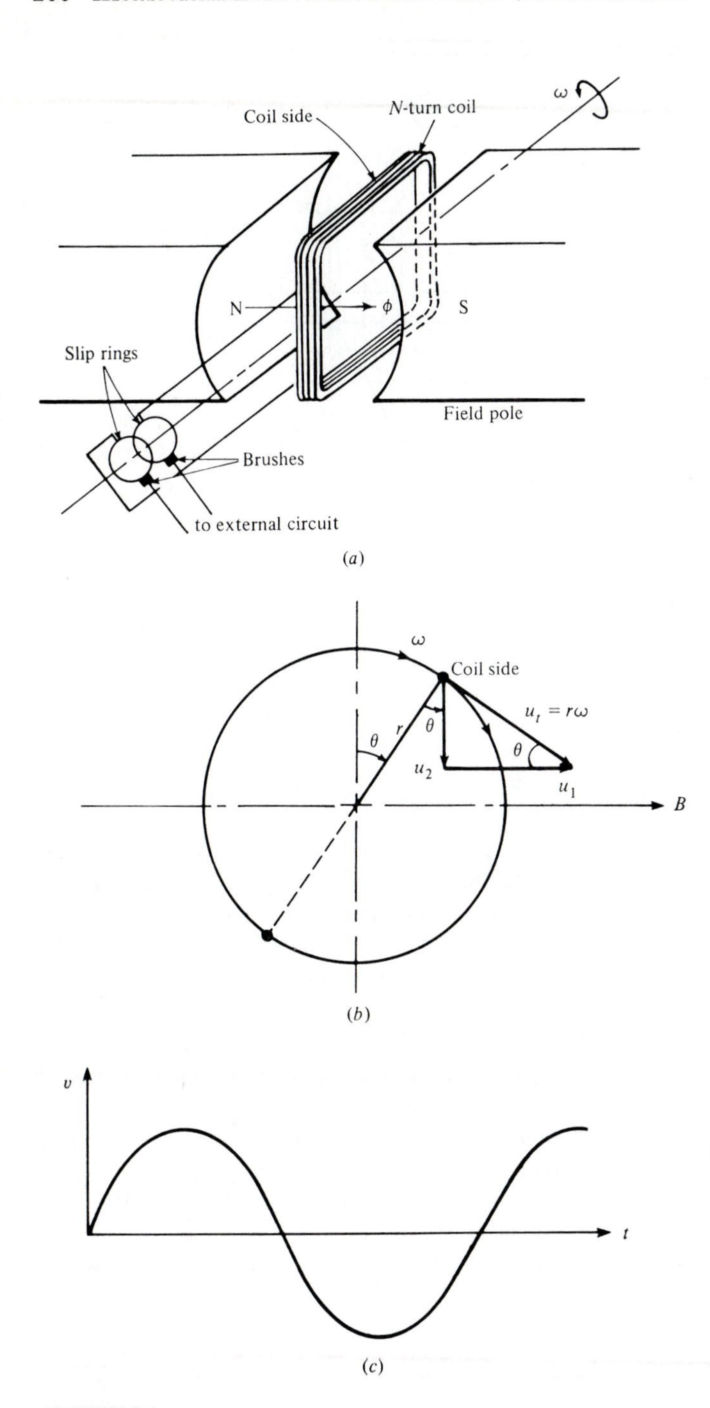

FIGURE 5-2
(a) An elementary generator; *(b)* resolution of velocities into "parallel" and "perpendicular" components; *(c)* voltage waveform at the brushes.

$$\theta = \omega t \tag{5.4}$$

Consequently, (5.2*b*) to (5.4) yield

$$e = 2BNl_1 r\omega \sin\omega t = E_m \sin\omega t \tag{5.5}$$

where $E_m = 2BNl_1 r\omega$. A plot of (5.5) is shown in Fig. 5.2*(c)*. The conclusion is that a sinusoidally varying voltage will be available at the slip rings, or brushes, of the rotating coil shown in Fig. 5.2*(a)*. The brushes reverse polarities periodically. Consequently, we obtain a sinusoidally varying ac voltage.

Alternatively, we may apply Faraday's law in the form of voltage-induced = time rate of change of flux linkage. Consider a three-phase round-rotor machine shown in Fig. 5.3*(a)*. We assume that the flux-density distribution due to the field excitation is uniform in the airgap. This implies that the flux density across the plane *aa'* [Fig. 5.3*(a)*] is given by

$$B_m = \frac{\phi}{Dl} \tag{5.6}$$

where

ϕ = total flux (per pole) produced by the field winding
D = diameter of the coil *aa'*
l = axial length of *aa'*

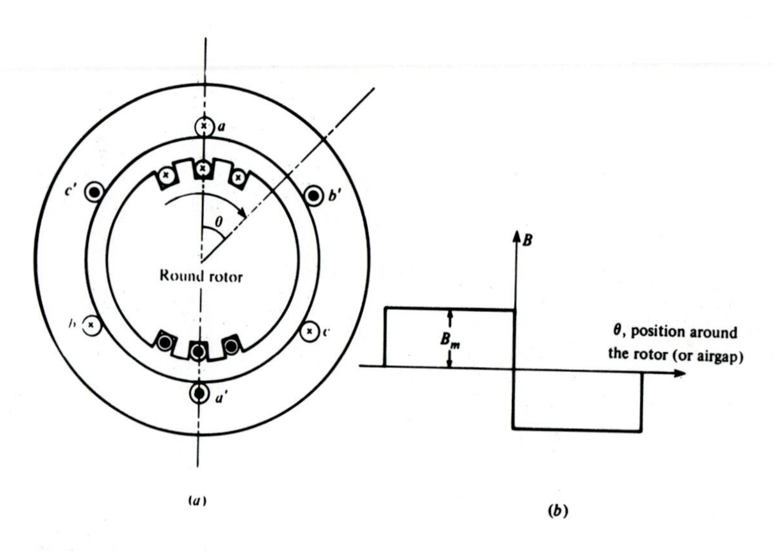

FIGURE 5-3
(a) A round or cylindrical-rotor synchronous machine. *(b)* Flux-density distribution due to the field excitation.

As the rotor is moved by an angle θ (Fig. 5.3), the flux linking the coil is $\lambda = N\phi \cos\theta$. If the rotor rotates at a constant velocity $\dot{\theta} = \omega$, the emf induced in aa' is

$$e_a = -\frac{d\lambda}{dt} = -\frac{d\lambda}{d\theta}\cdot\frac{d\theta}{dt} = N\phi \sin\theta\frac{d\theta}{dt} \tag{5.7}$$

Substituting $d\theta/dt = \omega = 2\pi f$ and $\theta = \omega t$ into (5.7) yields

$$\begin{aligned} e_a &= \omega N\phi \sin\omega t = 2\pi f N\phi \sin\omega t \\ &= E_m \sin\omega t \end{aligned} \tag{5.8}$$

where $E_m = 2\pi f N\phi$ and $f = \omega/2\pi$ is the frequency of the induced voltage. Because phases b and c are displaced from phase a by ±120° [Fig. 5.3*(a)*], the corresponding voltages may be written as

$$e_b = E_m \sin(\omega t - 120°) \tag{5.9a}$$

$$e_c = E_m \sin(\omega t + 120°) \tag{5.9b}$$

These voltages are sketched in Fig. 5.4.

Next we consider the salient rotor machine of Fig. 5.5*(a)*. We assume that the flux-density distribution, produced by the field winding, at the stator surface is sinusoidal, as shown. In the element dx the flux $d\phi = B(\theta)(D/2)d\theta l$, where $B(\theta)$ is the flux-density distribution, D is the inner diameter of the stator, and l is its axial length. Hence the total flux per pole ϕ is given by

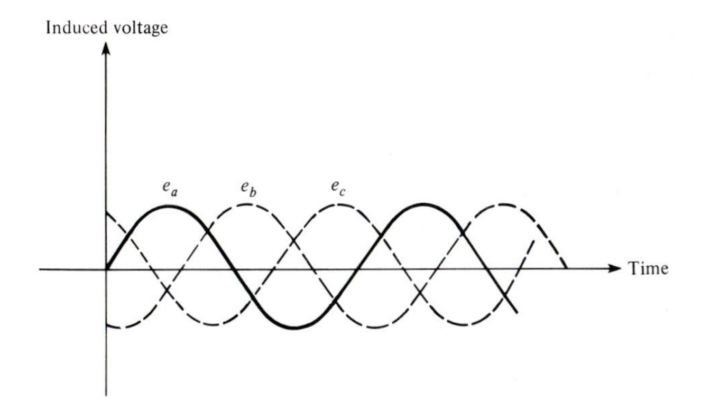

FIGURE 5-4
A three-phase voltage produced by a three-phase synchronous generator.

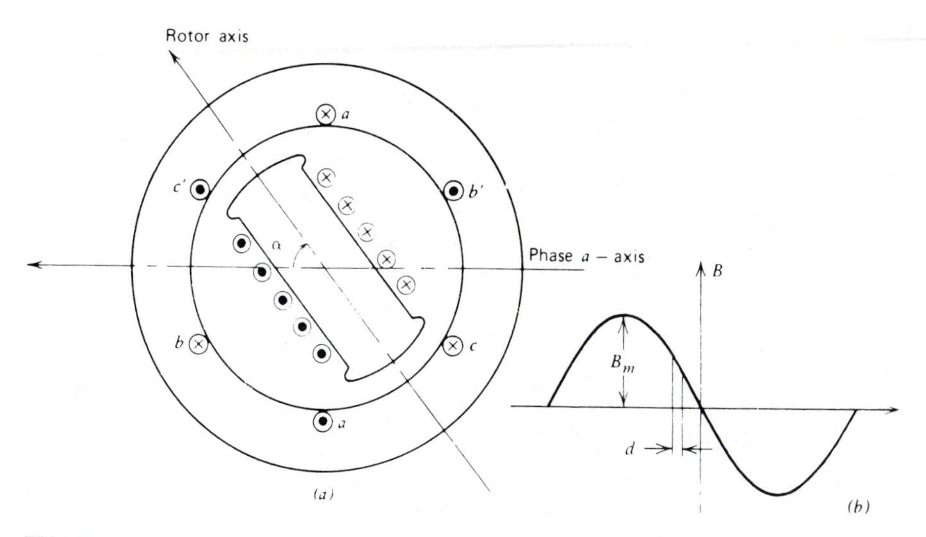

FIGURE 5-5
(a) A salient-rotor synchronous machine. *(b)* (Ideal) sinusoidal flux density distribution in the airgap.

$$\phi = \int_0^{\pi} \frac{D}{2} l B_m \sin\theta \, d\theta = B_m D l \tag{5.10}$$

From (5.10),

$$B_m = \frac{\phi}{Dl}$$

which is identical to (5.6). Hence the emf induced in aa' will be the same as in (5.7) and (5.8).

Notice that (5.10) implies that the flux per pole produced by the field excitation is the same as given by (5.6), and we may say that this flux is oriented along the rotor axis. Thus the flux linkage with phase a may be written as $\lambda = N\phi \cos\theta$. Now, proceeding as in (5.7) to (5.9*b*), we obtain identical results for the salient-rotor machine.

5.2 CONSTRUCTIONAL FEATURES

From the preceding discussion we conclude that a synchronous machine must have (in principle) a load (or armature) winding and a source for the magnetic flux. This source of flux may be either a permanent magnet or an electromagnet (providing for the field excitation). Some of the factors that dictate the form of construction of a synchronous machine follow.

FIGURE 5-6
Cutaway of a salient-pole synchronous machine. (Courtesy General Electric Company.)

1. *Form of Excitation.* We recall from the preceding remarks that the field structure is usually the rotating member of a synchronous machine and is supplied with a dc-excited winding to produce the magnetic flux. This dc excitation may be provided by a self-excited dc generator mounted on the same shaft as the rotor of synchronous machine. Such a generator is known as the *exciter*. The direct current thus generated is fed to the synchronous machine field winding, shown in Fig. 5.6. In slow-speed machines with large ratings, such as hydroelectric generators, the exciter may not be self-excited. Instead, a pilot exciter, which may be self-excited or may have a permanent magnet, activates the exciter. The maintenance problems of direct-coupled dc generators impose a limit on this form of excitation at about a 100-MW rating.

   An alternative form of excitation is provided by silicon diodes and thyristors, which do not present excitation problems for large synchronous machines. The two types of solid-state excitation systems are:

   (a) Static systems that have stationary diodes or thyristors, in which the current is fed to the rotor through slip rings.

   (b) Brushless systems that have shaft-mounted rectifiers that rotate with the rotor, thus avoiding the need for brushes and slip rings. Fig. 5.7 shows a brushless excitation system.

FIGURE 5-7
Rotor of a 3360-kVA, 6-kV brushless synchronous generator, with rotating diodes. (Courtesy Brown Boveri Company.)

2. *Field Structure and Speed of Machine.* We have already mentioned that the synchronous machine is a constant-speed machine. This speed, known as synchronous speed, n_s, is given by (4.10). The equation is repeated here for convenience:

$$n_s = \frac{120f}{P} \tag{5.11}$$

Thus a 60-Hz, two-pole synchronous machine must run at 3600 rpm, whereas the synchronous speed of a 12-pole, 60-Hz machine is only 600 rpm. The rotor field structure consequently depends on the speed rating of the machine. Therefore turbogenerators, which are high-speed machines, have *round* or *cylindrical rotors* (see Figs. 5.8 and 5.9). Hydroelectric and diesel-electric generators are low-speed machines and have *salient pole rotors*, as depicted in Figs. 5.6, 5.10, and 5.11. Such rotors are less expensive to fabricate than round rotors. They are not suitable for large, high-speed machines, however, because of the excessive centrifugal forces and mechanical stresses that develop at speeds around 3600 rpm.

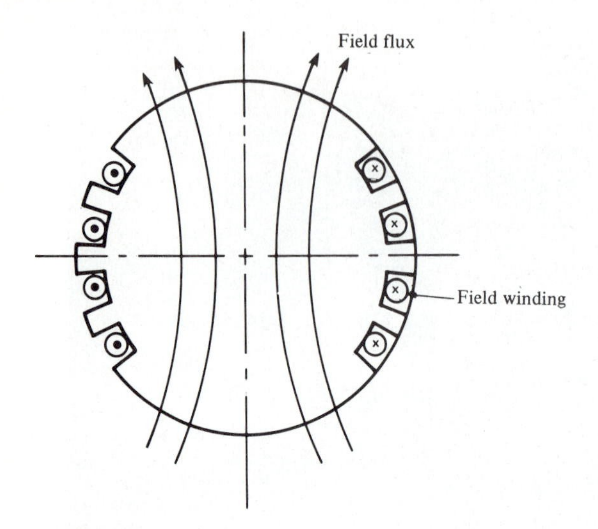

FIGURE 5-8
Field winding on a round rotor.

FIGURE 5-9
Turbine rotor with direct water cooling during the mounting of damper hollow conductors. (Courtesy Brown Boveri Company.)

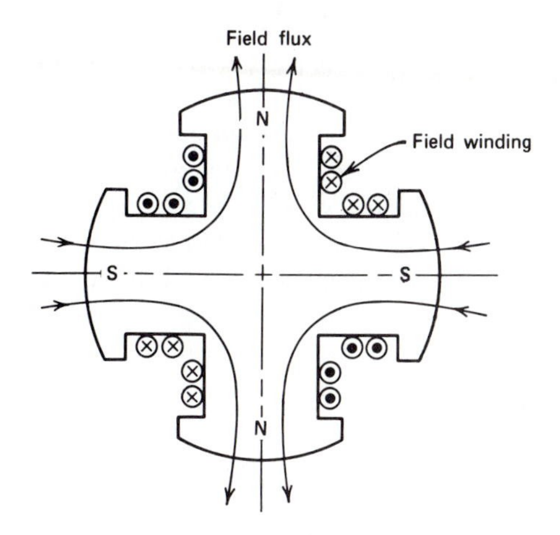

FIGURE 5-10
Field winding on a four-pole salient rotor.

Another feature in the construction of a synchronous machine stems from the mounting of the rotor. For example, a round-rotor, turbine-driven machine (Fig. 5.9) or a salient-rotor, diesel, engine-driven machine (Fig. 5.6) has a horizontally mounted rotor. A waterwheel-driven machine (Fig. 5.12) invariably has a vertically mounted, salient pole rotor.

3. *Stator.* The stator of a synchronous machine is similar to that of a polyphase induction motor (see Figs. 4.7 and 5.13). There is essentially no difference between the stator of a round-rotor machine and that of a salient-rotor machine. The stators of waterwheel generators, however, usually have a large-diameter armature compared to other types of generators (Fig. 5.14). The stator core consists of punchings of high-quality laminations having slot-embedded lap windings.

4. *Cooling.* Because synchronous machines are often built in extremely large sizes, they are designed to carry very large currents. A typical armature current density may be of the order of 10 A/mm^2 in a well-designed machine. Also, the magnetic loading of the core is such that it reaches saturation in many regions. The severe electric and magnetic loadings in a synchronous machine produce heat that must be appropriately dissipated. Thus the manner in which the active parts of a machine are cooled determines its overall physical structures. In addition to air, some of the coolants used in synchronous machines include water, hydrogen, and helium. Fig. 5.9 shows a turbine rotor with direct water cooling during the mounting of damper hollow conductors. The cooling arrangement of the stator of a turbine generator is shown in Fig. 5.13.

5. *Damper Bars.* So far we have mentioned only two electrical windings of a synchronous machine: the three-phase armature winding and the field winding. We also pointed out that, under steady-state, the machine runs at a constant speed, that

is, at the synchronous speed. However, like other electric machines, a synchronous machine undergoes transients during starting and abnormal conditions. During transients, the rotor may undergo mechanical oscillations and its speed deviates from the synchronous speed, which is an undesirable phenomenon. To overcome this, an additional set of windings, resembling the cage of an induction motor, is mounted on the rotor. This winding is called damper winding and is shown in Figs. 5.6, 5.9, 5.11, and 5.15. When the rotor speed is different from the synchronous speed, currents are induced in the damper windings. The damper winding acts like the cage rotor of an induction motor, producing a torque to restore the synchronous speed. Also, the damper bars provide a means of starting the machine, which is otherwise not self-starting.

In summary, Figs. 5.6 to 5.15 show the various structural features of different types of synchronous machines. In principle, the machine has three electrical windings—the armature, the field, and the damper—located on the stator and the rotor, as shown in Fig. 5.6.

5.3 SYNCHRONOUS MOTOR OPERATION

We have mentioned earlier that a synchronous machine is capable of operating either as a motor or as a generator. In Section 5.1 we have discussed the operation of a synchronous generator. In this section we will consider the qualitative aspects of the action of a synchronous machine as a motor.

From the discussions of Section 5.2 we notice a resemblance between the salient pole rotor of a synchronous machine and the rotor of a reluctance motor, discussed in Section 5.9. From the energy-storage and energy-conversion principles developed in Chapter 2, we can show how a polyphase synchronous machine operates as a motor. We know that the stator of a three-phase synchronous machine is similar to that of a three-phase induction motor. We have demonstrated in Chapter 4 that a three-phase excitation, such as that found in the stator of an induction motor, produces a rotating magnetic field in the airgap of the machine. Referring to Fig. 5.5*(a),* we will have a rotating magnetic field in the airgap of the salient pole machine when its stator (or armature) windings are fed from a three-phase source. The rotor will then have a tendency to align with the field at all times in order to present the path of least reluctance. Thus, if the field is rotating, the rotor will tend to rotate with the field. From Fig. 5.3*(a)*, we see that a round rotor will not tend to follow the rotating magnetic field because the uniform airgap presents the same reluctance all around the airgap and the rotor does not have any preferred direction of alignment with the magnetic field. This torque, which we have in the machine shown in Fig. 5.5*(a)* but not in that depicted in Fig. 5.3*(a)*, is called the *reluctance torque*. It is present by virtue of the variation of the reluctance around the periphery of the machine.

Next, let the field winding of the machine be fed by a dc source that produces the rotor magnetic field of definite polarities. By the principle of alignment of fields (Chapter 2), we conclude that when the rotor is excited, it will tend to align with the stator field and will tend to rotate with the rotating magnetic field. We observe that for an excited rotor, a round rotor, and a salient rotor, both will tend to rotate with the rotating magnetic field, although the salient rotor will have an additional reluctance torque because of the saliency. In a later section we will derive expressions for the electromagnetic torque in a synchronous machine attributable to field excitation and to saliency.

FIGURE 5-11
Salient rotor of a 152.5-MVA, 13.8-kV synchronous machine. (Courtesy Brown Boveri Company.)

FIGURE 5-12
Mounting the rotor of a hydroelectric generator. (Courtesy Brown Boveri Company.)

FIGURE 5-13
End-winding region of a 722-MVA, 22-kV turbine generator. (Courtesy Brown Boveri Company.)

FIGURE 5-14
Mounting stator conductors in slots of one stator half of a synchronous machine. (Courtesy Brown Boveri Company.)

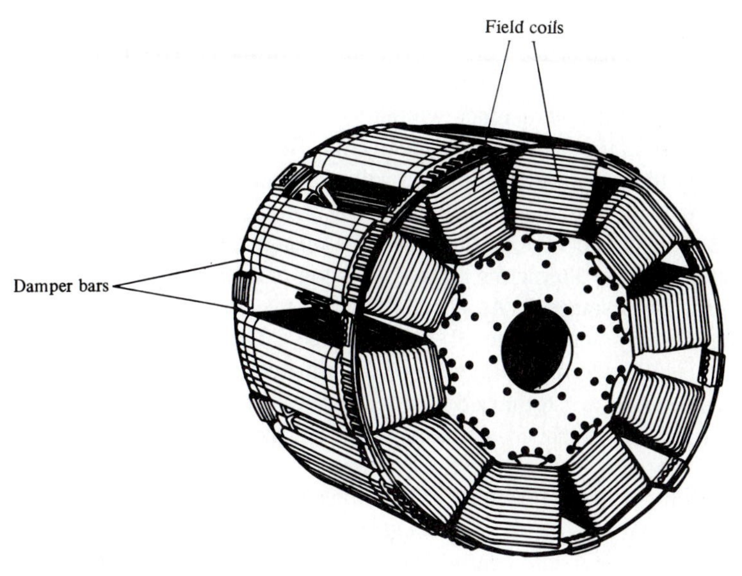

FIGURE 5-15
A salient rotor showing the field windings and damper bars (shaft not shown).

So far we have indicated the mechanism of torque production in a round-rotor and in a salient-rotor machine. To recapitulate, we might say that the stator magnetic field has a tendency to "drag" the rotor along, as if a north pole on the stator "locks in" with a south pole of the rotor. However, if the rotor is at a standstill, the stator poles will tend to make the rotor rotate in one direction and then in the other as they rotate and sweep across the rotor poles. Therefore, a synchronous motor is not self-starting. In practice, as mentioned earlier, the rotor carries damper bars that act like the cage of an induction motor and thereby provide a starting torque. Once the rotor starts running and almost reaches the synchronous speed, it locks into position with the stator poles. The rotor pulls into step with the rotating magnetic field and runs at the synchronous speed; the damper bars go out of action. Any departure from the synchronous speed results in induced currents in the damper bars, which tend to restore the synchronous speed. Machines without damper bars, or very large machines with damper bars, may be started by an auxiliary motor.

5.4 ANALYSIS OF SYNCHRONOUS MACHINES

We will discuss the operating characteristics of synchronous motors later, but first we consider some of the steady-state operating characteristics of synchronous machines from a quantitative viewpoint. (Recall that in Sections 5.1 and 5.3, respectively, we have presented briefly a qualitative discussion of generator and motor operation.) For convenience, we will consider generator operation separately from motor operation.

The analysis of a round-rotor machine is somewhat different from the procedure for a salient pole machine. In all cases, the analysis makes use of the machine parameters, which must be identified before proceeding with the analytical details.

Since, for now, we are considering only the steady-state behavior of the machine, circuit constants of the field and damper windings need not be considered. The presence of the field winding will be denoted by the flux produced by the field excitation. Turning to the armature winding, we will represent it on a per-phase basis (as we did with the induction motor in Chapter 4). Obviously, the armature winding has a resistance. But the ohmic value of this resistance must include the effects of the operating temperature and the alternating currents flowing in the armature conductors (causing skin effect, for instance). As a consequence, the value of the armature resistance becomes larger compared to its dc resistance. The larger value of the resistance is known as the *effective resistance* of the armature and is denoted by R_a. An approximate value of R_a is 1.6 times the dc resistance.

Next, we consider the reactances pertaining to the armature winding. First, the leakage reactance is caused by the leakage fluxes linking the armature conductors only because of the currents in the conductors. These fluxes do not link with the field winding. As in an induction motor, for convenience in calculation, the leakage reactance is divided into (1) end-connection leakage reactance, (2) slot-leakage reactance, (3) tooth-top and zigzag leakage reactance, and (4) belt-leakage reactance. All of these components are not significant in every synchronous machine. In most large machines the last two reactances are a small portion of the total leakage reactance.

Flux paths contributing to the end-connection leakage and slot-leakage reactances are shown in Fig. 5.16*(a)* and *(b)*, respectively. We denote the total leakage reactance of the armature winding per phase by X_a. Now, to proceed with the analysis, let us first consider a round-rotor synchronous generator. We will also introduce the concept of *synchronous reactance*, the most important parameter in determining the steady-state characteristics of a synchronous machine.

PERFORMANCE OF A ROUND-ROTOR SYNCHRONOUS GENERATOR. At the outset we wish to point out that we will study the machine on a per-phase basis, implying a balanced operation. Thus let us consider a round-rotor machine operating as a generator on no-load. Let the open-circuit phase voltage be V_0 for a certain field current I_f. Here, V_0 is the internal voltage of the generator. We assume that I_f is such that the machine is operating under unsaturated condition. Next, we short-circuit the armature at the terminals, keeping the field current unchanged (at I_f), and measure the armature phase current I_a. In this case, the entire internal voltage V_0 is dropped across the internal impedance of the machine. In mathematical terms,

$$V_0 = I_a Z_s \tag{5.12}$$

and Z_s is known as the *synchronous impedance*. One portion of Z_s is R_a and the other a reactance, X_s, known as *synchronous reactance*; that is,

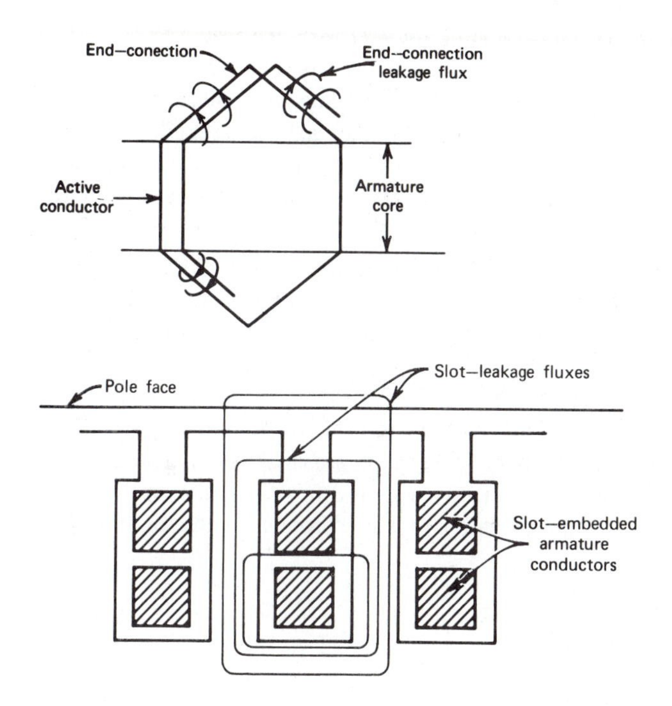

FIGURE 5-16
(a) End-connection leakage flux path. *(b)* Slot-leakage flux paths.

$$Z_s = R_a + jX_s \tag{5.13}$$

In (5.13) X_s is greater than the armature leakage reactance X_a, mentioned earlier. Where does the additional reactance come from? We will answer this question in the following discussion.

Let the generator supply a phase current I_a to a load at unity power factor and at a terminal voltage V_t V/phase. This is shown in the phasor diagram (Fig. 5.17), where V_a is the phasor sum of V_t and the drop due to the armature resistance and armature leakage reactance. Notice that we now have two mmf's—F_a attributable to the armature current and F_f attributable to the field current present in the machine. To find the mmf F_r that produces the voltage V_a, refer to the open-circuit characteristic of the generator, in Fig. 5.18, which shows F_r corresponding to V_a. The flux produced by an mmf is in phase with the mmf. As dictated by $e = Nd\phi/dt$, however, the voltage induced by a certain flux is behind the mmf by 90°. Therefore, we lay F_r ahead of V_a by 90° and F_a in phase with I_a, as depicted in Fig. 5.17. The mmf F_a is known as the *armature reaction mmf.* Sufficient mmf must be supplied by the field to overcome F_a such that we have a net F_r, to produce V_a. Because of F_a, an equal and opposite component, F_a, of the field mmf is laid off from F_r as shown in Fig. 5.17. The phasor

F_f is then the total field mmf in the machine. Corresponding to this mmf, the open-circuit voltage of the generator is V_0, as found from Fig. 5.18. This open-circuit voltage is known as the nominal induced emf and is also shown in Fig. 5.17. From the geometry of the phasor diagram, it is clear that the triangles *OST* and *OQR* are similar. We also see that *QR* is perpendicular to *OP* and must pass through *P* because $OP = V_t + I_a R_a$. Thus *QRP* is a continuous straight line, and we have

$$V_0 = V_t + I_a(R_a + jX_s) \tag{5.14}$$

where X_s, the synchronous reactance, agrees with the definitions of (5.12) and (5.13). The "extra" reactance, in addition to X_a (Fig. 5.17), is introduced by the armature reaction. Therefore the synchronous reactance is the sum of the armature leakage reactance and armature reaction reactance.

In an actual synchronous machine, except in very small ones, we almost always have $X_s >> R_a$, in which case $Z_s \cong jX_s$. We will use this restriction in most of the analysis. Among the steady-state characteristics of a synchronous generator, its voltage regulation and power-angle characteristics are the most important ones. As for a transformer, we define the voltage regulation of a synchronous generator at a given load as

$$\text{percent voltage regulation} = \frac{V_0 - V_t}{V_t} \times 100 \tag{5.15}$$

where V_t is the terminal voltage on load and V_0 is the no-load terminal voltage. Clearly, for a given V_t, we can find V_0 from (5.14) and, hence, the voltage regulation as illustrated by Examples 5.1 and 5.2.

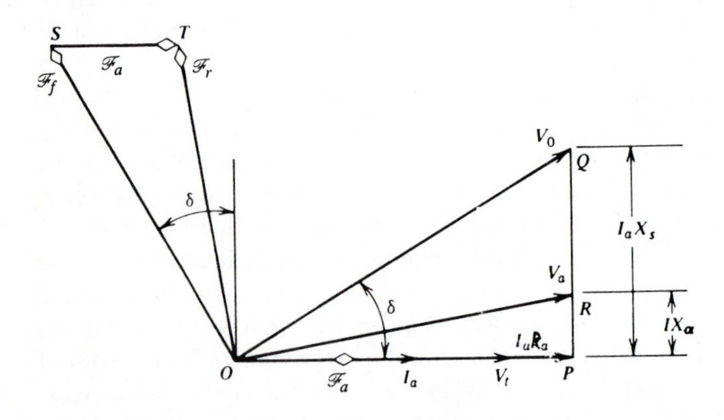

FIGURE 5-17
Phasor diagram for a round-rotor generator at unity power factor (to define X_s).

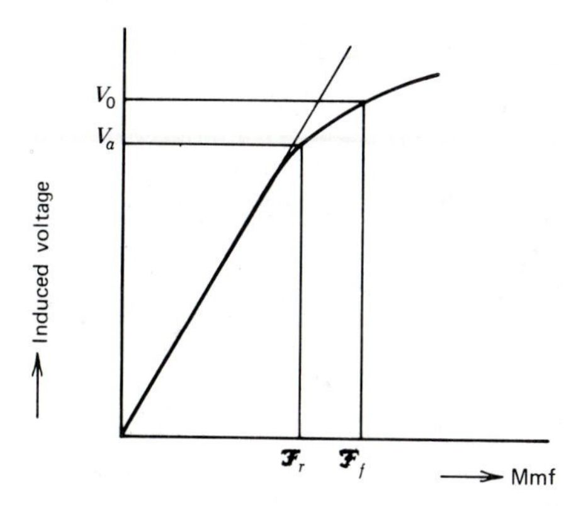

FIGURE 5-18
Open-circuit characteristics of a synchronous generator.

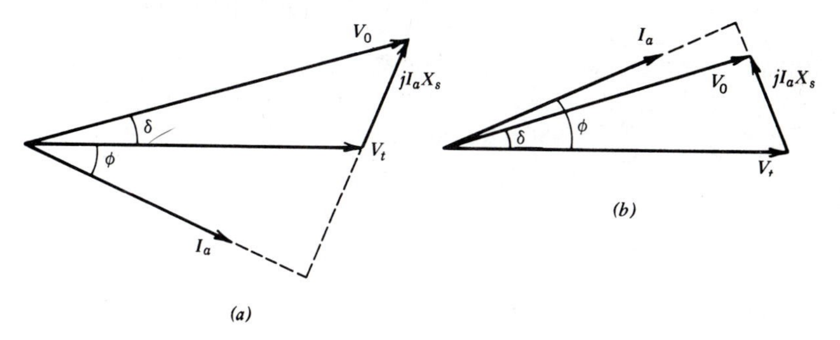

FIGURE 5-19
Phasor diagrams. *(a)* Lagging power factor. *(b)* Leading power factor.

Example 5.1 Calculate the percent voltage regulation for a three-phase, wye-connected, 2500-kVA, 6600-V turboalternator operating at full load and 0.8 power factor lagging. The per-phase synchronous reactance and the armature resistance are 10.4 and 0.071 Ω, respectively.

Solution

Clearly, we have $X_s >> R_a$. The phasor diagram for lagging power factor, neglecting the effect of R_a, is shown in Fig. 5.19*(a)*. The numerical values are as follows:

$$V_t = \frac{6600}{\sqrt{3}} = 3810 \text{ V}$$

$$I_a = \frac{2500 \times 1000}{\sqrt{3} \times 6600} = 218.7 \text{ A}$$

From (5.19) we have

$$V_0 = 3810 + 218.7(0.8 - j0.6)j10.4 = 5485 \angle 19.3° \text{ V}$$

and

$$\text{percent regulation} = \frac{5485 - 3810}{3810} \times 100 = 44\%$$

Example 5.2 Repeat the preceding calculations with 0.8 power factor leading.

Solution

In this case we have the phasor diagram shown in Fig. 5.19*(b)*, from which we get

$$V_0 = 3810 + 218.7(0.8 + j0.6)j10.4 = 3048 \angle 36.6° \text{ V}$$

and

$$\text{percent voltage regulation} = \frac{3048 - 3810}{3810} \times 100 = -20\%$$

We observe from Examples 5.1 and 5.2 that the voltage regulation is dependent on the power factor of the load. Unlike what happens in a dc generator (Chapter 6) the voltage regulation for a synchronous generator may even become negative. The angle between V_0 and V_t is defined as the *power angle*, δ. To justify this definition, we reconsider Fig. 5.19*(a)*, from which we obtain

$$I_a X_s \cos\phi = V_0 \sin\delta \tag{5.16}$$

However, $P_d = V_t I_a \cos\phi$. Hence, in conjunction with (5.16), we get

$$P_d = \frac{V_0 V_t}{X_s} \sin\delta \tag{5.17}$$

which shows that the internal power of the machine is proportional to sin δ. Equation (5.17) is often said to represent the power-angle characteristic of a synchronous machine.

PERFORMANCE OF A ROUND-ROTOR SYNCHRONOUS MOTOR. Except for some precise calculations, we may neglect the armature resistance as compared to the synchronous reactance. Therefore the steady-state per-phase equivalent circuit of a synchronous machine simplifies to the one shown in Fig. 5.20*(a)*. Here, we have shown the terminal voltage V_t, the internal excitation voltage V_0, and the armature current I_a going "into" the machine or "out of" it, depending on the mode of operation—"into" for motor and "out of" for generator. With the help of this circuit and (5.17) we will study some of the steady-state operating characteristics of a synchronous motor. In Fig. 5.20*(b)* we show the power-angle characteristics as given by (5.17). Here positive power and positive δ imply the generator operation, while a negative δ corresponds to a motor operation. Because δ is the angle between V_0 and V_t, V_0 is ahead of V_t in a generator, whereas in a motor, V_t is ahead of V_0. The voltage-balance equation for a motor is, from Fig. 5.20*(a)*,

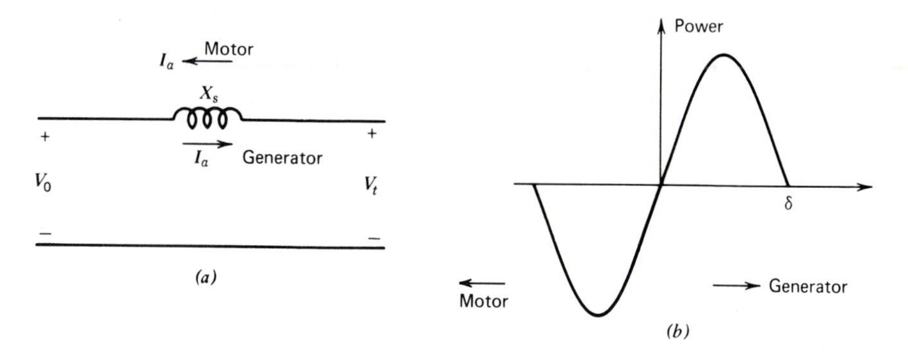

FIGURE 5-20
(a) An approximate equivalent circuit. *(b)* Power-angle characteristics of a synchronous machine.

$$V_t = V_0 + jI_aX_s \tag{5.18}$$

If the motor operates at a constant power, then (5.16) and (5.17) require that

$$V_0 \sin\delta = I_aX_s \cos\phi = \text{constant} \tag{5.19}$$

We recall that V_0 depends on the field current, I_f. Consider two cases: (1) when I_f is adjusted so that $V_0 < V_t$ and the machine is *underexcited*; and (2) when I_f is increased to a point that $V_0 > V_t$ and the machine becomes *overexcited*. The voltage-current relationships for the two cases are shown in Fig. 5.21*(a)*. For $V_0 > V_t$ at constant power, δ is greater than the δ for $V_0 < V_t$, as governed by (5.19). Notice that an underexcited motor operates at a lagging power factor (I_a lagging V_t), whereas an overexcited motor operates at a leading power factor. In both cases the terminal voltage and the load on the motor are the same. Thus we observe that the operating power factor of the motor is controlled by varying the field excitation, hence altering V_0. This is a very important property of synchronous motors. The armature current at a constant load, as given by (5.19), for varying field current is also shown in Fig. 5.21*(a)*. From this we can obtain the variations of the armature current I_a with the field current, I_f (corresponding to V_0), and this can be done for different loads, as shown in Fig. 5.21*(b)*. These curves are known as the V curves of the synchronous motor. One of the applications of a synchronous motor is in power factor correction, as demonstrated by Examples 5.3 and 5.4.

Example 5.3 A three-phase, wye-connected load takes 50-A current at 0.707 lagging power factor at 220 V between the lines. A three-phase, wye-connected, round-rotor synchronous motor, having a synchronous reactance of 1.27 Ω/phase, is connected in parallel with the load. The power developed by the motor is 33 kW at a power angle of 30°. Neglecting the armature resistance, calculate (1) the reactive kilovolt-amperes of the motor, and (2) the overall power factor of the motor and the load.

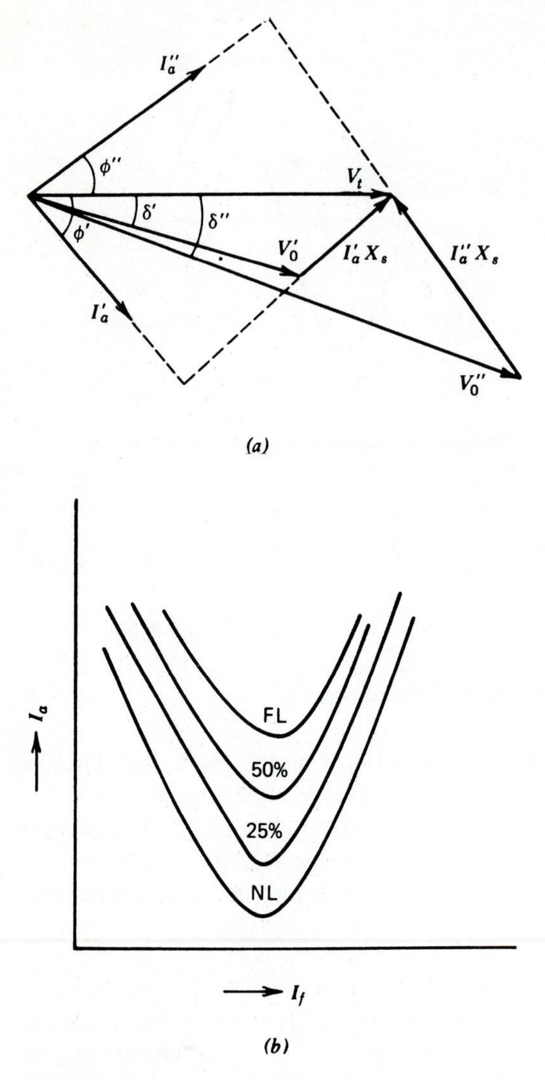

FIGURE 5-21
(a) Phasor diagram for motor operation. (V_0', I_a', ϕ', and δ') correspond to underexcited operation. (V_0'', I_a'', ϕ'', and δ'') correspond to overexcited operation. *(b)* *V* curves of a synchronous motor.

Solution

The circuit and the phasor diagram on a per-phase basis are shown in Fig. 5.22. From (5.17) we have

$$P_d = \frac{1}{3} \times 33{,}000 = \frac{220}{\sqrt{3}} \frac{V_0}{1.27} \sin 30°$$

which yields $V_0 = 220$ V. From the phasor diagram, $I_a X_s = 127$ or $I_a = 127/1.27 =$

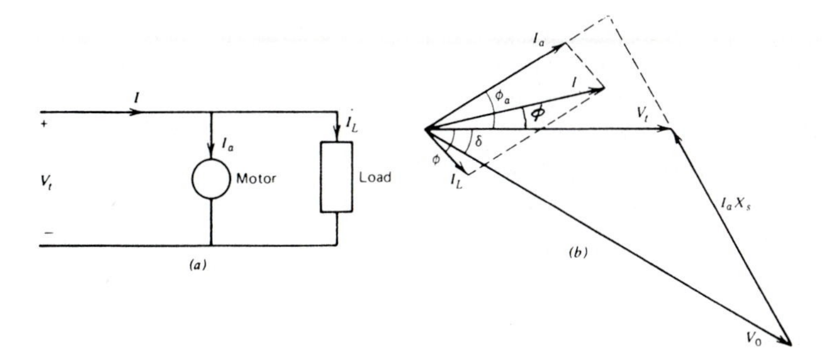

FIGURE 5-22
(a) Circuit diagram. *(b)* Phasor diagram.

100 A and $\phi_a = 30°$. The reactive kilovolt-amperes of the motor = $\sqrt{3}V_tI_a \sin \phi_a = \sqrt{3} \times (220/1000) \times 100 \times \sin 30° = 19$ kVAr.

The overall power factor angle is given by

$$\tan\phi = \frac{I_a \sin\phi_a - I_L \sin\phi_L}{I_a \cos\phi_a + I_L \cos\phi_L} = 0.122$$

or $\phi = 7°$ and $\cos \phi = 0.992$ leading.

Example 5.4 For the generator of Example 5.1, calculate the power factor for zero voltage regulation on full load.

Solution

Let ϕ be the power factor angle. Then

$$I_a Z_s = 218.7 \times 10.4 \angle\phi + 89.6 = 2274.48 \angle\phi + 89.6 \text{ V}$$

For voltage regulation to be zero $|V_0| = |V_t|$. Hence $|3810| = |3810 + 2274.48[\cos(\phi + 89.6) + j \sin(\phi + 89.6)]|$.

$$3810^2 = [3810 + 2274.48\cos(\phi + 89.6)]^2 + [2274.48 \sin(\phi + 89.6)]^2$$

from which

$$\phi = 17.76° \qquad \text{and} \qquad \cos\phi = 0.95 \text{ leading}$$

SALIENT POLE SYNCHRONOUS MACHINES. In the preceding discussion we have analyzed the round-rotor machine and made extensive use of the machine parameter, which we defined as synchronous reactance. Because of saliency, the reactance measured at the terminals of a salient-rotor machine will vary as a function of the rotor position. This is not so in a round-rotor machine.

To overcome this difficulty, we use the *two-reaction theory* proposed by André Blondel. The theory proposes to resolve the given armature mmf's into two mutually perpendicular components, with one located along the axis of the rotor salient pole, known as the *direct* (or *d*) axis and with the other in quadrature and known as the *quadrature* (or *q*) axis. The *d*-axis component of the mmf, F_d, is either magnetizing or demagnetizing; the *q*-axis component, F_q, results in a cross-magnetizing effect. Thus, if the amplitude of the armature mmf is F_a, then

$$F_d = F_a \sin\psi \tag{5.20}$$

and

$$F_q = F_a \cos\psi \tag{5.21}$$

where ψ is the phase angle between the armature current I_a and the internal (or excitation) voltage V_0. In terms of space distribution, the mmf's and ψ are shown in Fig. 5.23. The effects of F_d and F_q are that they give rise to voltages. To illustrate, we consider a salient pole generator having a terminal voltage V_t, supplying a load of lagging power factor (cos ϕ), and drawing a phase current I_a. For the operating condition (given field current), we also know the no-load voltage, V_0, from the no-load characteristics. These characteristics (V_t, V_0, I_a, and ϕ) are shown in Fig. 5.24. To construct this diagram, we choose V_0 as the reference phasor and neglect the armature resistance. Now, on load because of armature reaction alone, V_0 will be reduced to V_a as determined by V_d and V_q, which are caused by F_d and F_q, respectively. From V_a we obtain V_t by subtracting the armature leakage reactance drop I_aX_a. Ignoring V_a, we might say that the difference between V_0 and V_t is attributable to the armature reaction reactance and the armature leakage reactance, that is, the synchronous reactance. We can resolve I_a into its *d*- and *q*-axis components, I_d and I_q, respectively, so also the I_aX_s drop shown as I_dX_d and I_qX_q in Fig. 5.24. Thus the phasor diagram is complete. (See Example 5.5.) We can associate physical meanings to the reactances X_d and X_q. These are, respectively, the *direct-axis* and *quadrature-axis* reactance of a salient pole machine. These reactances can be measured experimentally, as we will discuss in a later section. The preceding discussions are valid on a per-phase basis for a balanced machine. We now show the details of some of the calculations by Example 5.5.

Example 5.5 A 20-kVA, 220-V, 60-Hz, wye-connected, three-phase salient pole synchronous generator supplies rated load at 0.707 lagging power factor. The phase constants of the machine are $R_a = 0.5\ \Omega$ and $X_d = 2X_q = 4.0\ \Omega$. Calculate the voltage regulation at the specified load.

$$V_t = \frac{220}{\sqrt{3}} = 127\ \text{V}$$

$$I_a = \frac{20{,}000}{\sqrt{3} \times 220} = 52.5\ \text{A}$$

$$\phi = \cos^{-1}0.0707 = 45^\circ$$

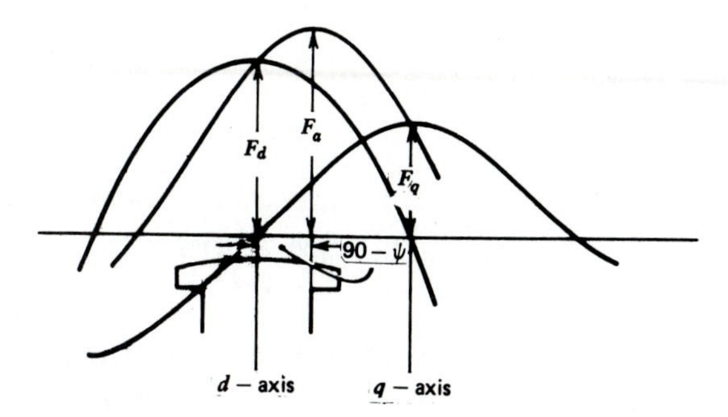

FIGURE 5-23
Armature mmf and its d and q components.

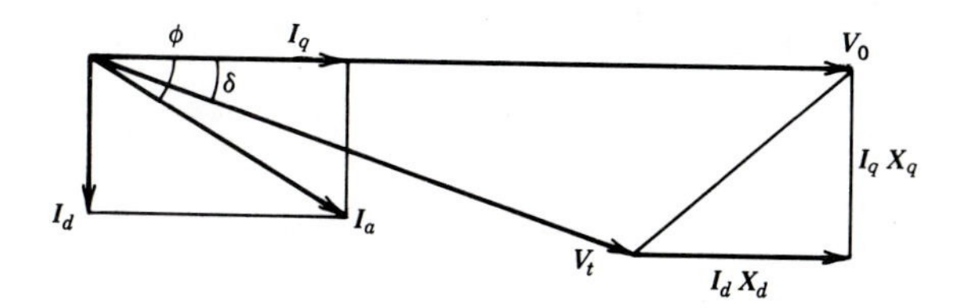

FIGURE 5-24
Phasor diagram of a salient pole machine.

From Fig. 5.24 we have (neglecting R_a):

$$I_d = I_a \sin(\delta + \phi)$$

$$I_q = I_a \cos(\delta + \phi)$$

$$V_t \sin\delta = I_q X_q = I_a X_q \cos(\delta + \phi)$$

or

$$\tan\delta = \frac{I_a X_q \cos\phi}{V_t + I_a X_q \sin\phi}$$

$$= \frac{52.5 \times 2 \times 0.707}{127 + 52.5 \times 2 \times 0.707} = 0.37$$

or

$$\delta = 20.6°$$

$$I_d = 52.5 \sin(20.6 + 45) = 47.5 \text{ A}$$

$$I_d X_d = 47.5 \times 4 = 190.0 \text{ V}$$

$$V_0 = V_t \cos\delta + I_d X_d$$

$$= 127 \cos 20.6 + 190 = 308 \text{ V}$$

and

$$\text{percent regulation} = \frac{V_0 - V_t}{V_t} \times 100\% = \frac{308 - 127}{127} \times 100 = 142\%$$

Example 5.5 shows how the phasor diagram (Fig. 5.24) can be used to determine the voltage regulation of a salient pole synchronous generator. In fact, the phasor diagram depicts the complete steady-state performance characteristics of the machine. For example, to obtain the power-angle characteristics of a salient pole machine, operating either as a generator or as a motor, we refer to Fig. 5.24. From this figure we have, neglecting R_a and the internal losses,

$$\text{power output} = V_t I_a \cos\phi = \text{developed power} = P_d \tag{5.22}$$

and

$$\begin{aligned} I_q X_q &= V_t \sin\delta \\ I_d X_d &= V_0 - V_t \cos\delta \end{aligned} \tag{5.23}$$

Also,

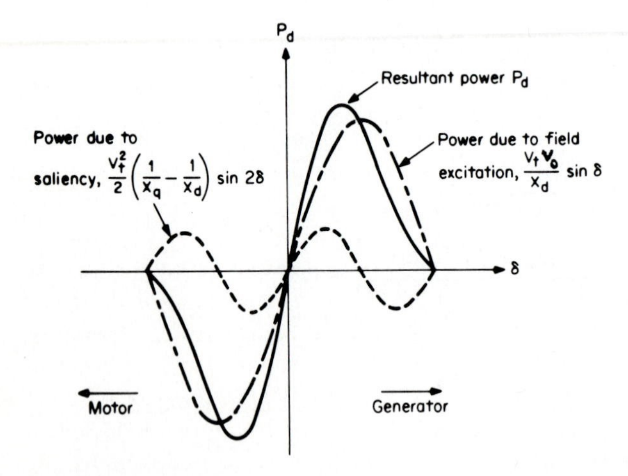

FIGURE 5-25
Power-angle characteristics of salient pole machine.

$$I_d = I_a \sin(\delta + \phi)$$
$$I_q = I_a \cos(\delta + \phi) \tag{5.24}$$

Substituting (5.24) into (5.23) and solving for $I_a \cos\phi$ yields

$$I_a \cos\phi = \frac{V_0}{X_d}\sin\delta + \frac{V_t}{2X_q}\sin 2\delta - \frac{V_t}{2X_d}\sin 2\delta \tag{5.25}$$

Finally, combining (5.25) and (5.22) gives

$$P_d = \frac{V_t V_0}{X_d}\sin\delta + \frac{1}{2}V_t^2\left(\frac{1}{X_q} - \frac{1}{X_d}\right)\sin 2\delta \tag{5.26}$$

This variation of the developed power, P_d, as a function of the power angle δ is shown in Fig. 5.25. Notice that the resulting power is composed of power due to saliency—the second term in (5.26)—and of power due to field excitation—the first term in (5.26). Clearly, when $X_d = X_q$, the machine has no saliency and only the first term in (5.26) is nonzero, which represents the power-angle characteristics of a round-rotor machine. On the other hand, if there is no field excitation, implying $V_0 = 0$, the first term in (5.26) reduces to zero. We then have the power-angle characteristics of a reluctance machine as given by the second term. As in a round-rotor machine (discussed earlier), the power-angle characteristics given by (5.26) reflect the generator as well as motor operation. The term δ is positive for the former and negative for the latter.

5.5 MACHINE LOSSES

Synchronous machine losses are similar to losses in other types of rotating machines. Some losses associated with large synchronous machines include the following.

Windage and Friction Loss

Windage loss is caused by air friction in the regions surrounding the rotor, primarily in the airgap. In large central station alternators, alternator efficiency is a significant factor in determining the cost of energy production of the central station. Therefore, many alternator design choices are made on the basis of operating efficiency rather than of initial cost. Most alternators rated 15,000 kW or larger used by utilities are housed in a sealed housing and the free spaces are filled with hydrogen rather than air in order to reduce windage losses. Hydrogen also serves as an improved cooling medium over air. Closed hydrogen systems are preferred, with the gas recirculated to the windings and air spaces after cooling in surface coolers. System pressures are in the neighborhood of 45 lb/in^2. The system is purged with CO_2 to avoid explosion hazards. Care must be exercised in the handling and storage of hydrogen, but modern practice has resulted in extremely high reliability of the hydrogen systems.

Friction losses are due to alternator bearing losses and slip-ring losses. An approximate combined friction and windage loss for air-cooled alternators is given by

$$P_{FW} = K\left(\frac{U}{10{,}000}\right)^{2.5} D\sqrt{L} \ \text{kW} \tag{5.27}$$

where

U = rotor peripheral speed, ft/min
D = rotor diameter, in
L = rotor magnetic length, in
K = 0.08 to 0.11 for slow-speed salient pole machines
= 0.06 to 0.08 for higher-speed salient pole machines
= 0.06 to 0.07 for cylindrical-rotor machines

Core Loss

In synchronous machines, core loss is a very complex function of (1) the flux densities in the various portions of the magnetic circuit, (2) the fundamental frequency, (3) the magnitude and frequency of the various harmonics of airgap flux density, (4) the lamination thickness, (5) the temperature, and (6) the type of magnetic material used in the magnetic circuit. Core loss is generally ascribed to the magnetic loss at the no-load condition of operation. An increase in core loss (and in certain other losses) as a function of load current is defined as a *stray-load loss*. Core loss is made up of two major components: (1) armature core loss due to time-varying magnetic flux in the armature and (2) pole-face loss on the surface of both the field structure and armature. Additional (usually minor) losses also exist in the shaft, housing, and other structural members owing to induced eddy currents.

Core loss can be reduced by constructing magnetic members of thinner laminations and by using low-loss types of magnetic materials, such as iron-nickel alloys, oriented silicon steel, or amorphous magnetic materials, although there is generally a cost penalty in all of these methods. Pole-face losses can be reduced by minimizing the slot openings, but this generally increases a winding's assembly cost.

Stray-Load Loss

This loss has been defined just above as the increase in core loss as a function of load current. It results from losses induced by armature-leakage fluxes and other variations in airgap flux distribution. In very large machines, eddy-current losses induced in armature conductors of large cross section are also included in the stray-load loss. To minimize this component of stray-load loss, armature conductors are frequently constructed of laminated members,such as bundled conductors, strip conductors, square conductors, or Litz wire. Stray-load loss is a difficult loss component to measure accurately, and a value of 1 percent of the power output is often assumed as a typical value for stray-load loss.

Armature Conductor Loss

This loss is made up of the ohmic (or dc) loss and the "effective" (or ac) loss in the armature conductors. Effective loss is due to the nonuniform distribution of flux linkages over the conductor's cross section, often called *skin effect*, and is a function of the conductor's cross-sectional area and the frequency of the armature current. In large conductors, skin effect can result in a significant increase in armature copper loss even at 60 Hz. For this reason, armature conductors are frequently laminated or segmented.

The armature conductor loss varies as a function of conductor temperature in the typical manner. Conductor loss must be associated with a specific temperature when rating or specifying a machine. Typical temperatures used in synchronous machine ratings are temperature rises above ambient of 80, 105, and 130°C.

Excitation Loss

This loss should generally include the entire loss of the excitation system, including the voltage regulator and the synchronous machine's field conductor loss. As a minimum, the field conductor loss must be included when specifying machine efficiency. The practice in this regard varies among manufacturers and users of synchronous machines. The field conductor loss is an ohmic loss and can be handled as the ohmic portion of the armature conductor loss.

5.6 TRANSIENTS IN SYNCHRONOUS MACHINES

In the preceding sections we focused our attention on the steady-state behavior of synchronous machines. In this section we will briefly review some cases involving transients in synchronous machines. Of particular interest are (1) the sudden short circuit at the armature terminals of a synchronous generator, and (2) mechanical transients caused by a sudden load change on the machine. There are numerous other cases involving transients in synchronous machines, but these will not be considered at this point.

We know from earlier considerations that the performance of a machine, for a given operating condition, can be determined if the machine parameters are known. For instance, we have already expressed the steady-state power-angle characteristics of a salient pole synchronous machine in terms of the d-axis and q-axis reactances. Similarly, the constants by which transient behavior of a synchronous machine is known are the transient and subtransient reactances and pertinent time constants. We now define these quantities while relating them to the study of an armature short circuit.

Sudden Short Circuit at the Armature Terminals

At the outset, we assume no saturation and neglect the resistances of all the windings—the armature, field, and damper windings. Thus only the inductances remain, implying that the flux linking a closed circuit (or winding) cannot change instantaneously, as dictated by the constant flux linkage theorem. Stated differently, the sum of the flux linkages is constant for each winding. With these assumptions in mind, we consider a round-rotor machine (Fig. 5.26) and focus our attention on phase a and the field winding. Let the field current be I_f at $t = 0$. Prior to $t = 0$, we assume the armature to be open. At $t = 0$, the armature winding is suddenly short-circuited, at which instant the mmf axis of phase a is at right angles to the mmf axis of the field winding. This occurs so that there is no mutual coupling between the two windings. Clearly, at $t = 0$, the flux that links the armature is $\lambda_a = 0$. However, the flux that links the field windings is $\lambda_f = L_f I_f$, where L_f = field-winding inductance. We may divide L_f such that

$$L_f = L_l + L_{ad} \tag{5.28}$$

where L_l = field-leakage inductance and L_{ad} = mutual inductance between the field and armature windings. Thus L_{ad} also corresponds to the armature reaction reactance. We may now rewrite λ_f, using (5.28), as

$$\lambda_f = (L_l + L_{ad})I_f = L_{ad}(1 + \tau_f)I_f$$

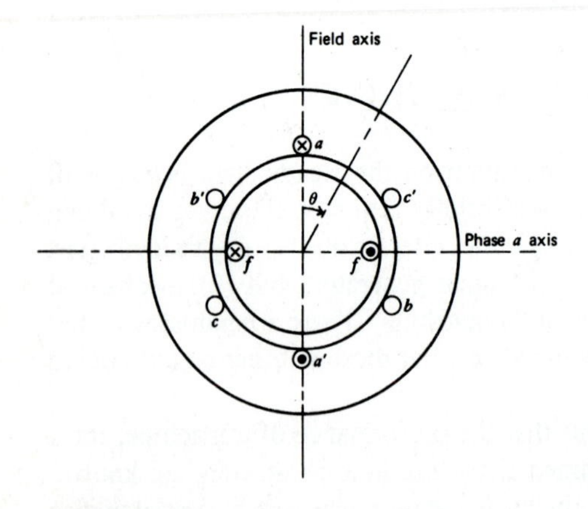

FIGURE 5-26
Three-phase round-rotor machine (only phase a and field winding carry currents).

where $\tau_f = L_f/L_{ad}$ = field-leakage coefficient.

After a time t, let the rotor rotate through angle θ (see Fig. 5.26), in which case i_a and $(i_f + I_f)$ will flow through the armature and field windings to maintain the flux linkages with these windings. Therefore, for the armature winding, we have

$$\lambda_a = i_a L_{ad}(1 + \tau_a) + (i_f + I_f) L_{ad} \sin\theta = 0 \tag{5.29}$$

where $\tau_a = X_1/\omega L_{ad}$ = armature leakage coefficient. Likewise, for the field winding, we have

$$\begin{aligned} \lambda_f &= (i_f + I_f) L_{ad}(1 + \tau_f) + i_a L_{ad} \sin\theta \\ &= I_f L_{ad}(1 + \tau_f) \end{aligned} \tag{5.30}$$

Solving (5.29) and (5.30) for i_a and i_f yields

$$i_a = \frac{[(1 + \tau_f)\sin\theta] I_f}{\sin^2\theta - (1 + \tau_a)(1 + \tau_f)}$$

$$i_f = -\frac{(\sin^2\theta) I_f}{\sin^2\theta - (1 + \tau_a)(1 + \tau_f)}$$

The maximum values of these currents occur at $\theta = \pi/2$. In this case we have

$$\begin{aligned} (i_a)_{\max} &= -\frac{(1 + \tau_f) I_f}{\tau_a + (1 + \tau_a)\tau_f} \\ (i_f)_{\max} &= \frac{I_f}{\tau_a + (1 - \tau_a)\tau_f} \end{aligned} \tag{5.31}$$

Multiplying the numerator and denominator of (5.31) by ωL_{ad} yields

$$(i_a)_{\max} = -\frac{(1 + \tau_f) V_0}{X_1 + (1 + \tau_a) X_f} \tag{5.32}$$

where $V_0 = L_{ad} I_f$ = (internal) induced voltage, $X_f = \omega L_{ad} \tau_f$ = field leakage reactance, and $X_1 = \omega L_{ad} \tau_a$ = armature leakage reactance. The circuit corresponding to (5.32) is shown in Fig. 5.27, and the input reactance of this circuit is the direct-axis transient reactance, x_d'.

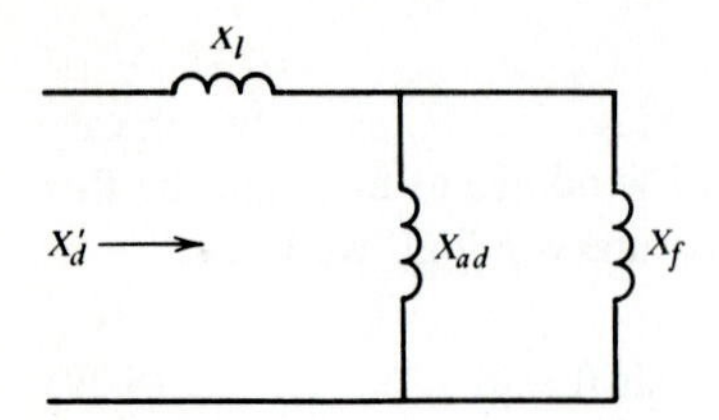

FIGURE 5-27
Equivalent circuit for the transient reactance.

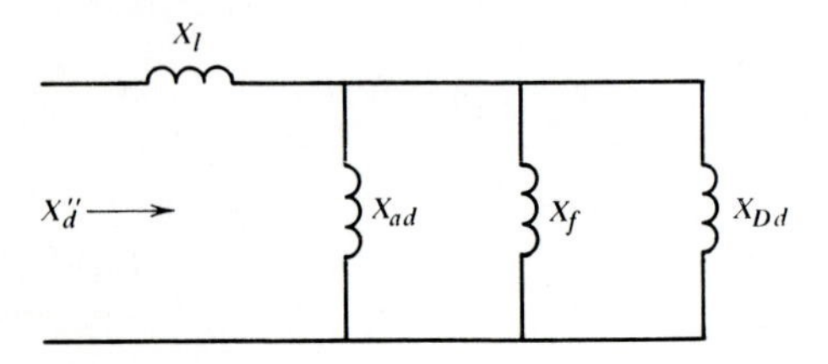

FIGURE 5-28
Equivalent circuit for the subtransient reactance.

Next, to include the effect of damper windings in the direct axis, we observe that the effect of the winding is indistinguishable in impact from that of the field winding, except for the current carried. Thus, in the d-axis, we now have the armature field and damper windings all in parallel, as shown in Fig. 5.28, where x_{Dd} = damper winding leakage reactance. From Fig. 5.28 we have the subtransient reactance, x_d'', given by

$$x_d'' = x_1 + x_{Dd} \frac{\tau_f}{\tau_{Dd} + \tau_f(1 + \tau_{Dd})} \tag{5.33}$$

where

$$\tau_{Dd} = \frac{x_{Dd}}{x_{ad}}$$

Having defined x_d' to account for the presence of the field winding and x_d'' to denote the effect of the damper winding, we can now consider the presence of the various resistances in determining the short circuit current in the armature. To obtain an explicit solution for the short circuit current, including the effects of x_d' and x_d'', is a cumbersome exercise and beyond our present scope. However, we can arrive at some useful and important conclusions about the transient-current waveform by the following reasoning. We assume that a sudden short circuit at the armature terminals occurs when the steady-state current passes through zero. In such a case the current waveform would be like that shown in Fig. 5.29. Here, the rate of decrease of the consecutive peaks is determined by the time constant of the windings. The current decays over time because of the presence of resistances. First, the damper winding

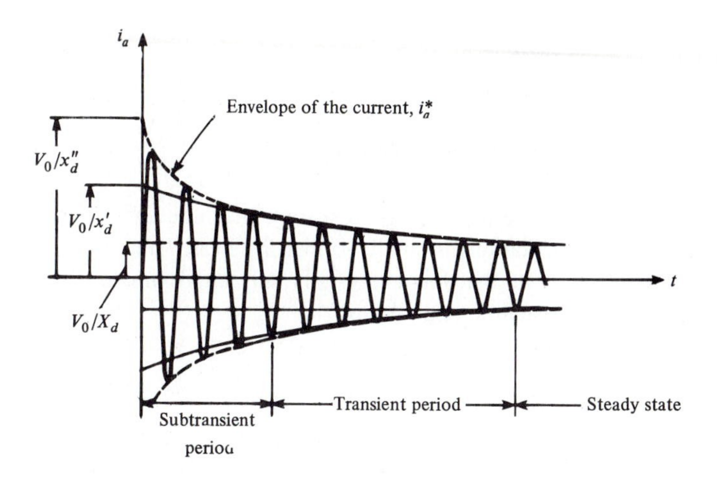

FIGURE 5-29
Armature current versus time, in a short-circuited generator.

TABLE 5.1
Per-unit synchronous machine reactances

Reactance	Salient Pole Machine	Round-Rotor Machine
X_d	1.0 to 1.25	1.0 to 1.2
X_q	0.65 to 0.80	
x_d'	0.35 to 0.40	0.15 to 0.25
x_d''	0.20 to 0.30	0.10 to 0.15
x_q''	0.20 to 0.30	0.10 to 0.15

has the smallest time constant. Thus the maximum current is V_0/x_d''. The effect of the damper lasts for only the first few cycles. Next, the maximum current is determined by x_d' and is V_0/x_d'. Finally, the steady-state current is limited by the synchronous reactance X_d.

Typical per-unit values of various reactances of synchronous machines are given in Table 5.1. In Table 5.1 the per-unit values are based on the machine rating.

Mechanical Transients

The mechanical equation of motion of a synchronous generator is

$$J\ddot{\Theta}_m + b\dot{\Theta}_m + T_e = T_m \tag{5.34}$$

where

T_e = torque developed by the machine

T_m = externally applied torque

J = moment of inertia of the rotating system (including the load or prime mover)

b = friction coefficient, including electrical damping.

To illustrate, let us consider a two-pole, cylindrical-rotor machine and assume that the frequency of mechanical oscillations is small, so that the steady-state power-angle characteristics can be used. Note that this analysis is only approximate. The per-phase power developed by the machine is given by

$$P_d = \frac{V_0 V_t}{X_s} \sin\delta = T_e \omega_m \tag{5.35}$$

which also represents the electrical torque on a different scale. Let the changes in Θ_m, T_e, and T_m caused by a sudden load change be represented by $\Delta\Theta_m$, ΔT_e, and ΔT_m respectively, so that (5.34) modifies to the following:

$$(Jp^2 + bp)\Delta\Theta_m + \Delta T_e = \Delta T_m \tag{5.36}$$

The change in the electrical torque is, from (5.35),

$$\Delta T_e = \frac{V_0 V_t \sin(\Delta\delta)}{\omega_m X_s} \tag{5.37}$$

where ω_m = mechanical velocity of the rotor and is the same as synchronous speed under steady-state conditions. In (5.37), for constant voltages, only δ changes for load changes. For small variations $\sin(\Delta\delta) \cong \Delta\delta$. Also $\Delta\Theta_m = \Delta\delta$; the number of poles is two. Therefore (5.37) becomes

$$\Delta T_e = K_e \Delta\delta \tag{5.38}$$

where $K_e = (V_0 V_t / \omega_m X_s)$. From (5.36) and (5.38) we have

$$(Jp^2 + bp + K_e)\Delta\delta = \Delta T_m \tag{5.39}$$

which is a linear, second-order differential equation in terms of the power angle δ. If we compare this with the second-order differential equation of a mechanical system, the natural frequency of oscillation and the damping ratio are, respectively,

$$\omega_n = \sqrt{\frac{K_e}{J}} \tag{5.40}$$

$$\zeta = \frac{b}{2\sqrt{K_e J}} \tag{5.41}$$

Example 5.6 A 30-hp, 220-V, three-phase, wye-connected, 60-Hz, 3600-rpm, cylindrical-rotor machine on no-load is brought up to the rated speed by an auxiliary motor and is then suddenly connected to a 220-V, three-phase source with the proper phase sequence. Study the mechanical transient from the following data:

synchronous reactance per phase = 2.0 Ω

excitation voltage V_0 of (5.35) = 150 V/phase

moment of inertia of rotating parts = 1.5 kg-m^2

damping torque b of (5.36) = 5 N-m/rad/sec

Solution

Denoting $\Delta\delta$ by δ', we find that the equation of motion is, from (5.39),

$$J\frac{d^2\delta'}{dt^2} + b\frac{d\delta'}{dt} + K_e\delta' = 0 \tag{5.42}$$

In (5.42) K_e is known as the *synchronizing torque*. For motor operation, K_e (for the three-phase machine) is given by

$$K_e = \frac{V_0 V_t}{\omega_m X_s} \times 3 \tag{5.43}$$

For the given machine, $V_0 = 150$ V, $V_t = 220/\sqrt{3} = 127$ V, $X_s = 2.0$ Ω, and $\omega_m = 120\pi$ rad/sec. Substituting these into (5.43), we obtain

$$K_e = \frac{150 \times 127 \times 3}{120\pi \times 2} = 75.8 \text{ N-m/rad}$$

Equation (5.42) therefore becomes

$$(1.5p^2 + 5p + 75.8)\delta' = 0 \tag{5.44}$$

From (5.40), (5.43), and (5.44) we have

$$\omega_n = \sqrt{\frac{75.8}{1.5}} = 7.1 \text{ rad/sec}$$

$$\zeta = \frac{5}{2\sqrt{75.8 \times 1.5}} = 0.23$$

In terms of cycles per second, the natural frequency of oscillation is obtained from

$$2\pi f_n = \omega_n$$

or

$$f_n = \frac{\omega_n}{2\pi} = \frac{7.1}{2\pi} = 1.1 \text{ cps}$$

In most machines,

$$0.2 < f_n < 2$$

$$\zeta \cong 0.2$$

Knowing ζ and ω_n, we can obtain the mechanical behavior from the equation

$$\frac{\delta'}{\delta'_{ss}} = 1 - \frac{1}{\sqrt{1 - \zeta^2}} e^{-\zeta\omega_n t} \sin\left(\sqrt{1 - \zeta^2}\, \omega_n t + \phi\right) \tag{5.45}$$

where

$$\phi = \tan^{-1} \frac{\sqrt{1 - \zeta^2}}{\zeta}$$

and δ_{ss}' = steady-state power angle. Note that (5.45) is the solution to (5.42) for $\zeta < 1$.

5.7 DETERMINATION OF MACHINE REACTANCES

The synchronous reactance X_s of a cylindrical-rotor machine can be obtained from the open- and short-circuit tests on the machine. The open-circuit saturation curve and the steady-state armature current are shown in Fig. 5.30 on a per-phase basis. For a 2-A field current, the short-circuit current is 25 A, whereas the open-circuit voltage is 57 V. Consequently, the synchronous impedance is 57/25 = 2.48 Ω. Neglecting armature resistance, $Z_s \cong X_s = AC/BC = 2.48\ \Omega$. As shown in Fig. 5.30, X_s varies with saturation.

For the salient pole machine, it is necessary that we know the values of both X_d and X_q. The physical significance of these reactances was discussed earlier, and these are the maximum and minimum values of the armature reactance, respectively, for different rotor positions. These reactances are determined by the slip test. In this test the machine is excited by a three-phase source (for a three-phase machine) and driven mechanically at a speed slightly different from the synchronous speed. The field winding is unexcited and open-circuited. Oscillograms are taken of the armature current, armature voltage, and the induced field voltage. These are shown in Fig. 5.31. The ratio of maximum to minimum armature current yields the ratio X_d/X_q. For instance, from the diagram we find that $X_d/X_q = 1.6$. Knowing X_d from the open- and short-circuit tests described for the cylindrical-rotor machine, we can calculate X_q. There are other methods available for the determination of these reactances.

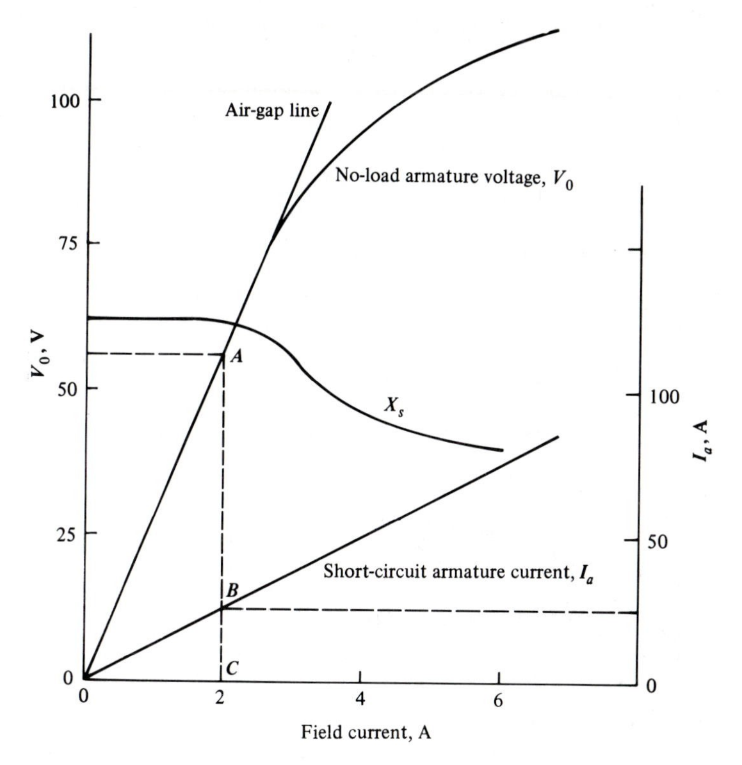

FIGURE 5-30
Test data for determining X_s.

The transient and subtransient reactances, x_d' and x_d'', are determined by recording the three-phase currents when a sudden short circuit is applied to the machine running on no-load and at rated speed. Generally, the currents of the various phases are not symmetrical about the time axis. However, x_d' and x_d'' can be determined either (1) by eliminating the dc component, or (2) from an oscillogram such as the one shown in Fig. 5.29.

5.8 TORQUE TESTS

A number of torque parameters are of interest in the design and application of synchronous machines, including the following.

1. *Locked-Rotor Torque.* Torque developed at zero speed with the rotor locked and prevented from turning, with rated voltage and frequency applied.
2. *Pull-Out Torque.* The maximum possible sustained torque developed at synchronous speed for 1 min with rated voltage and frequency.

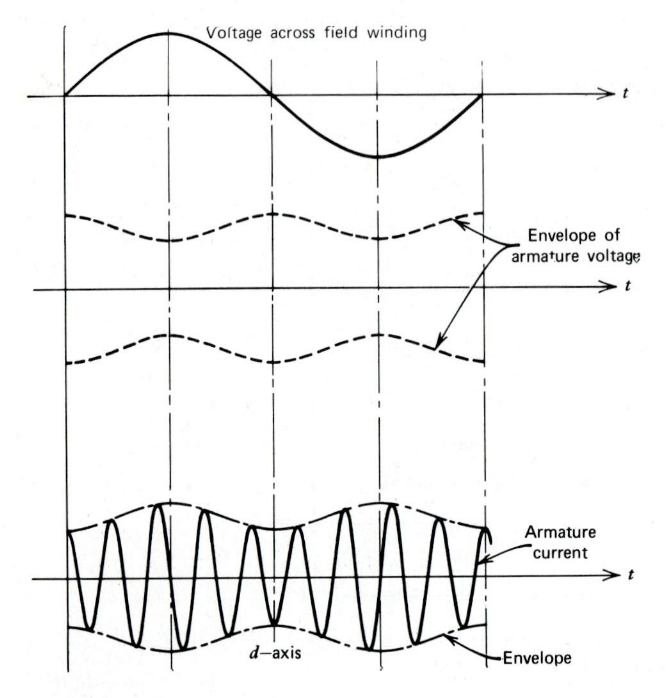

FIGURE 5-31
Oscillograms from slip test.

3. *Pull-In Torque.* In a synchronous motor, the maximum constant torque under which the motor will pull into synchronism at rated voltage and frequency when its excitation is applied.

4. *Pull-Up Torque.* The minimum torque developed at rated voltage and frequency during start-up at subsynchronous speeds.

5. *Speed-Torque Characteristic.* The variation of developed torque with rated voltage and frequency applied to a synchronous motor as a function of speed, with zero excitation applied.

Speed-Torque Characteristic. This characteristic is used to describe the operation of a synchronous motor when it is operating as an induction motor with no field excitation. In general, tests to obtain this characteristic in a synchronous machine are identical to those for induction motor speed-torque characteristics. Several of the torque parameters are readily obtained from the speed-torque characteristic of a synchronous machine. These are shown in the typical characteristic illustrated in Fig. 5.32. It should be noted, however, that there are some differences in winding impedances and in winding connections in a synchronous motor, as compared with an

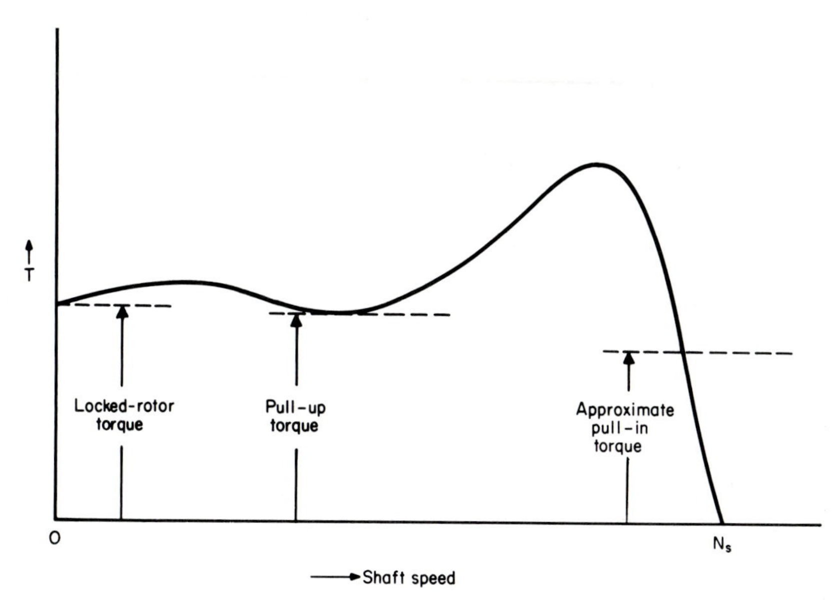

FIGURE 5-32
Speed-torque characteristics of a synchronous machine.

induction motor of similar horsepower: (1) The damper winding generally has a much higher resistance than a typical squirrel-cage winding of an induction motor of similar horsepower and frequency rating; thus, on the basis of the damper winding alone, the speed-torque characteristics of a typical synchronous motor appear more as those of a class C or D induction motor. (2) In addition to the damper winding, the speed-torque characteristic of a synchronous motor is significantly influenced by the field winding; most synchronous machines are designed for "closed-field" starting, that is, with the field winding short-circuited during start-up; the field winding, as "seen" from the armature winding, appears as a low-resistance winding and therefore—in itself—gives characteristics more like the class A or B induction motors. Thus, with the field short-circuited during start-up, the speed-torque characteristic of a synchronous machine is a composite of the characteristics resulting from the damper and field windings. Whether the field is shorted or open during starting is determined by the particular application of the synchronous machine. Fig. 5.33 shows the differences in the speed-torque characteristic of a relatively small synchronous machine between the conditions of the field open and the field shorted.

Pull-In Torque. This is a characteristic unique to synchronous machines. In general, it is a difficult parameter to measure or calculate. It is of most interest in the application of machines that are started and brought up to speed by means of the torque of the damper and field windings. Pull-in torque is the torque available to pull the rotor into synchronism with the armature rotating magnetic field.

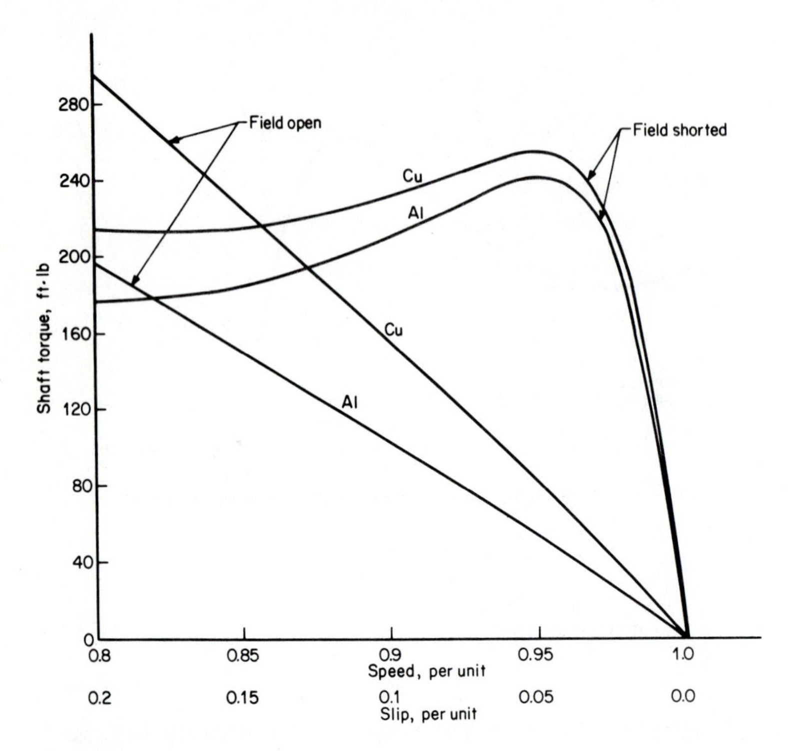

FIGURE 5-33
Synchronous motor speed-torque curves illustrating effect of field winding on motor torque. Effect of rotor-bar material also shown.

There is no recognized method for determining pull-in torque. Rather, the nominal pull-in torque is used to express this parameter. The nominal pull-in torque is arbitrarily defined as the torque developed by an induction motor at 95 percent of synchronous speed, as illustrated in Fig. 5.32.

Pull-Out Torque. This is another characteristic unique to synchronous machines; it describes the maximum torque capabilities of a synchronous machine. When a synchronous machine is subjected to a torque greater than pull-out, whether as a motor or a generator, the machine pulls out of synchronous speed, and a variety of conditions may result.

Pull-out torque may be obtained experimentally. This is not feasible for machines of power ratings beyond the capacity of the power supplies generally available in the laboratory. An approximate value of pull-out torque from machine constants is

$$T_{po} = \frac{KI_{f1}V}{I_{fsi}\eta\cos\theta} \tag{5.46}$$

where

T_{po} = pull-out torque, per unit

V = specified terminal voltage, per unit

I_{fsi} = specified per-unit field current

I_{fsi} = field current, per unit, corresponding to base armature current on the short-circuit saturation curve

$\cos\theta$ = rated power factor

η = efficiency at machine rating, per unit.

The factor K in (5.46) is to allow for *reluctance torque* and for positive sequence I^2R losses. This factor may be obtained from the machine manufacturer. It may also be estimated from the loading condition of the machine as follows:

$$K = \sin\delta + I_{fsi} V \frac{x_d - x_q}{2 I_{f1} x_d x_q} \tag{5.47}$$

where

x_d = direct-axis synchronous reactance, per unit

x_q = quadrature-axis synchronous reactance, per unit

δ = power angle, rad.

5.9 SMALL SYNCHRONOUS MOTORS

The three-phase synchronous machines discussed so far are assumed to be large (of the order of several hundred kilowatts, or even larger) because they have been considered in terms of possible applications in electric power systems. However, there are numerous applications that require synchronous motors of small (i.e., fractional horsepower) ratings. Most often such motors are designed to operate on a single-phase supply and do not require a dc excitation or the use of a permanent magnet. In this regard the fractional-horsepower synchronous motor is considerably different from its three-phase counterpart, which has a relatively large rating. The two types of small synchronous motors are the reluctance motor and the hysteresis motor. These motors are used in clocks, timers, turntables, and so forth.

The Reluctance Motor

First, we consider an elementary reluctance machine as shown in Fig. 5.34. The machine is singly excited; that is, it carries only one winding on the stator. The exciting winding is wound on the stator and the rotor is free to rotate. The rotor and stator are shaped so that the variation of the inductance of the windings is sinusoidal with respect to the rotor position. The space variation of the inductance is of double frequency; that is,

$$L(\theta) = L'' + L' \cos 2\theta$$

where the symbols are defined as in Fig. 5.34. For an excitation

$$i = I_m \sin \omega t$$

we now determine the instantaneous and average torques.

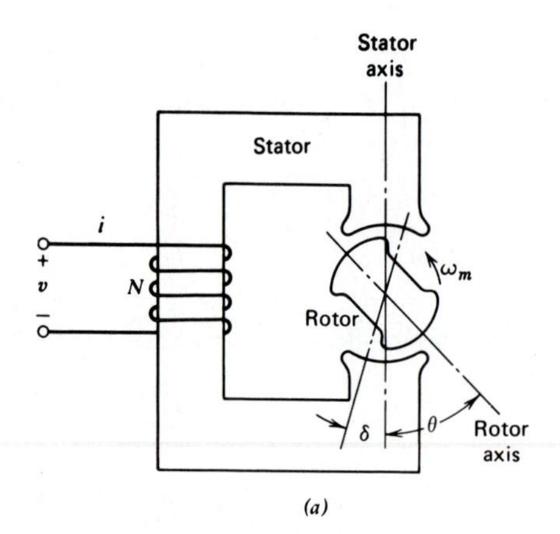

(a)

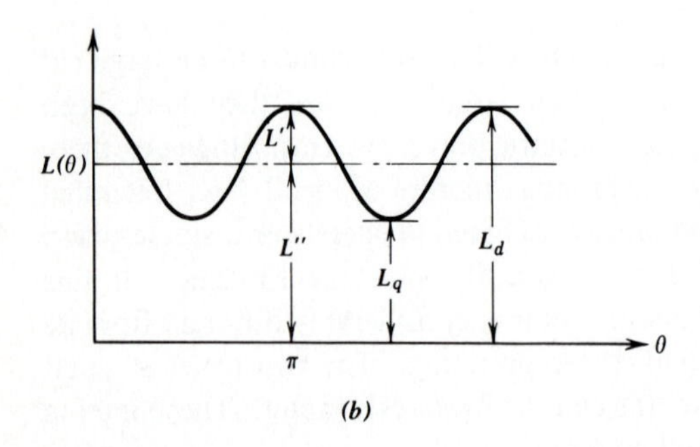

(b)

FIGURE 5-34
(a) A reluctance machine. *(b)* Inductance variation.

The magnetic energy stored is

$$W_m = \frac{1}{2}L(\theta)i^2$$

and the flux linkage is

$$\lambda(\theta) = L(\theta)i$$

where i is the independent variable.

Now (from Chapter 2) the torque is given by

$$T_e = -\frac{\partial W_m}{\partial \theta} + i\frac{\partial \lambda}{\partial \theta}$$

which, expressed in terms of inductance and current, becomes

$$T_e = -\frac{1}{2}i^2\frac{\partial L}{\partial \theta} + i^2\frac{\partial L}{\partial \theta} = \frac{1}{2}i^2\frac{\partial L}{\partial \theta} \tag{5.48}$$

For given current and inductance variations,

$$T_e = -I_m^2 L' \sin 2\theta \sin^2 \omega t$$

If the rotor is now allowed to rotate at an angular velocity ω_m, so that at any instant

$$\theta = \omega_m t - \delta$$

(where δ is the rotor position at $t = 0$, when the current i is also zero), then, in terms of ω and ω_m, the expression for instantaneous torque becomes

$$T_e = -\frac{1}{2}I_m^2 L'\{\sin 2(\omega_m t - \delta) - \frac{1}{2}[\sin 2(\omega_m t + \omega t - \delta) + \sin 2(\omega_m t - \omega t - \delta)]\}$$

To obtain this final form, we have used the following trigonometric identities:

$$\sin^2 A = \frac{1}{2}(1 - \cos 2A)$$

and

$$\sin C \cos D = \frac{1}{2}\sin(C + D) + \frac{1}{2}\sin(C - D)$$

From this expression for the torque, it can be concluded that the time-average torque is zero, since the value of each term integrated over a period is zero. The only case for which the average torque is nonzero is when $\omega = \omega_m$. At this particular frequency, the magnitude of the average torque becomes

$$T_{av} = \frac{1}{4} I_m^2 L' \sin 2\delta$$

Or, from Fig. 5.34*(b)*, since $L' = \frac{1}{2}(L_d - L_q)$,

$$T_{av} = \frac{1}{8} I_m^2 (L_d - L_q) \sin 2\delta \tag{5.49}$$

Thus, for example, at $I_m = 4$ A, $L_d = 0.2$ H, and $L_q = 0.1$ H, the maximum average torque is 0.2 N-m.

A number of conclusions can be drawn from the preceding analysis. The machine develops an average torque only at one particular speed corresponding to the frequency $\omega = \omega_m$, which is known as the synchronous speed. The reluctance machine is therefore a synchronous machine. The torque developed by the machine is called the reluctance torque, which will be zero if $L_d = L_q$. The torque varies sinusoidally with the angle δ, called the torque angle. Angle δ is a measure of torque. The inductances L_d and L_q are the maximum and minimum values of inductance and are called the direct-axis inductance and quadrature-axis inductance, respectively. The maximum torque occurs at the point at which $\delta = 45°$, which is called the pull-out torque. Any load requiring a torque greater than the maximum torque results in unstable operation of the machine.

A reluctance motor starts as an induction motor but normally operates as a synchronous motor. The stator of a reluctance motor is similar to that of an induction motor (single-phase or polyphase). Thus, to start a single-phase motor, almost any of the methods discussed in Chapter 4 may be used. A three-phase reluctance motor is self-starting when started as an induction motor. After starting, to pull it into step and then to run it as a synchronous motor, a three-phase motor should have low rotor resistance. In addition, the combined inertia of the rotor and the load should be small. A typical construction of a four-pole rotor is shown in Fig. 5.35. Here, the aluminum in the slots and in spaces where teeth have been removed serves as the rotor of an induction motor for starting.

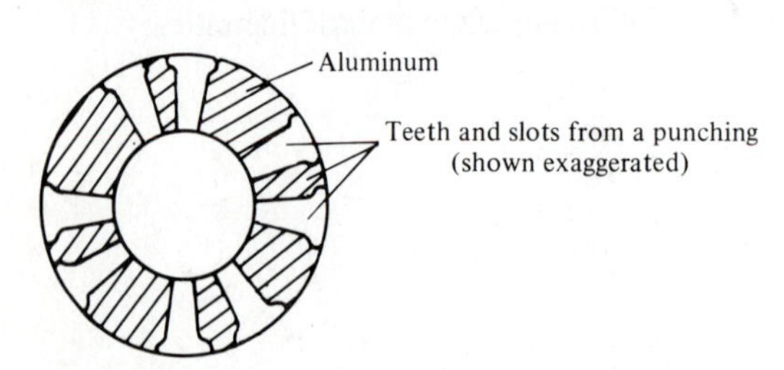

FIGURE 5-35
Rotor of a reluctance motor.

The Hysteresis Motor

Like the reluctance motor, a hysteresis motor does not have a dc excitation. Unlike the reluctance motor, however, the hysteresis motor does not have a salient rotor. Instead, the rotor of a hysteresis motor has a ring of special magnetic material, such as chrome, steel, or cobalt, mounted on a cylinder of aluminum or some other nonmagnetic material, as shown in Fig. 5.36. The stator of the motor is similar to that of an induction motor, and the hysteresis motor is started as an induction motor.

To understand the operation of the hysteresis motor, we may consider the hysteresis and eddy-current losses in the rotor. We observe that, as in an induction motor, the rotor has a certain equivalent resistance. The power dissipated in this resistance determines the electromagnetic torque developed by the motor, as discussed in Chapter 4. We may conclude that the electromagnetic torque developed by a hysteresis motor has two components—one by virtue of the eddy-current loss and the other because of the hysteresis loss. We know that the eddy-current loss can be expressed as

$$p_e = K_e f_2^2 B^2 \tag{5.50}$$

where

K_e = a constant

f_2 = frequency of the eddy currents in the rotor

B = flux density

In terms of the slip s, the rotor frequency f_2 is related to the stator frequency f_1 by

$$f_2 = sf_1 \tag{5.51}$$

Thus (5.50) and (5.51) yield

$$p_e = K_e s^2 f_1^2 B^2 \tag{5.52}$$

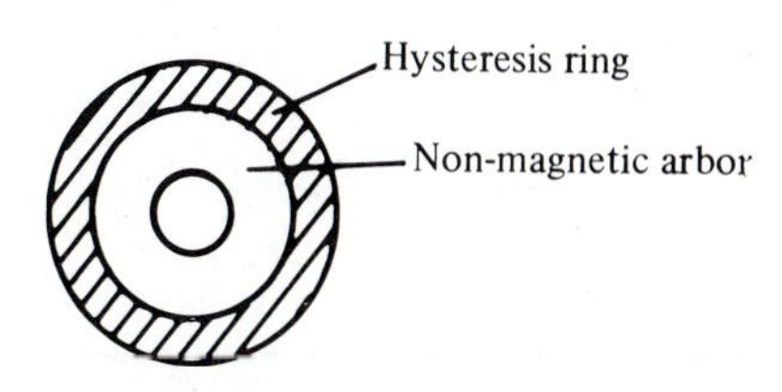

FIGURE 5-36
Rotor of a hysteresis motor.

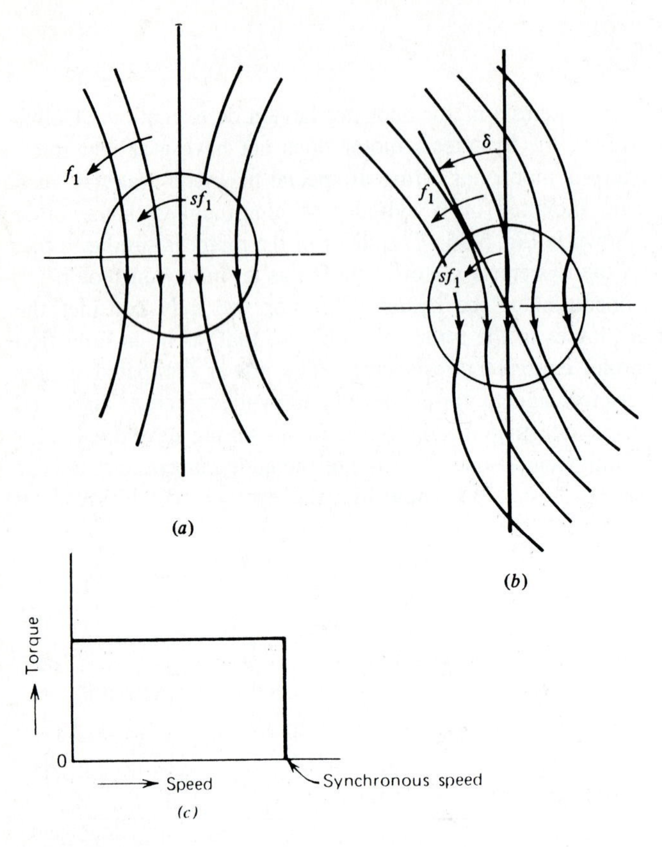

FIGURE 5-37
(a) Iron rotor, with no hysteresis in a magnetic field. *(b)* A rotor with hysteresis in a magnetic field. *(c)* Torque characteristics of a hysteresis motor.

And the torque T_e is related to p_e by (see Chapter 4)

$$T_e = \frac{p_e}{s\omega_s} \tag{5.53}$$

so that (5.52) and (5.53) give

$$T_e = K's \tag{5.54}$$

where $K' = K_e f_1^2 B^2/\omega_s$ = a constant.

Next, for the hysteresis loss, p_h, we have

$$p_h = K_h f_2 B^{1.6} = K_h s f_1 B^{1.6} \tag{5.55}$$

and for the corresponding torque, T_h, we obtain

$$T_h = K'' \tag{5.56}$$

where $K'' = K_h f_1 B^{1.6}/\omega_s$ = a constant.

Notice that the component T_e, as given by (5.54), is proportional to the slip and decreases as the rotor picks up speed. It is eventually zero at synchronous speed. This component of the torque aids in the starting of the motor. The second component, T_h, as given by (5.56), remains constant at all rotor speeds and is the only torque when the rotor achieves the synchronous speed. The physical basis of this torque is the hysteresis phenomenon, which causes a lag of the magnetic axis of the rotor behind that of the stator. In Figs. 5.37*(a)* and *(b)*, respectively, the absence and the presence of hysteresis are shown measured by the shift of the rotor magnetic axis. The angle of lag δ, shown in Fig. 5.37*(b)*, causes the torque arising from hysteresis. As mentioned, this torque is independent of the rotor speed (shown in Fig. 5.37*(c)*) until the breakdown torque.

PROBLEMS

5.1. A 60-Hz synchronous generator feeds an eight-pole induction motor running with a slip of 2%. What is the speed of the motor? At what speed must the generator run if it has *(a)* two poles, and *(b)* six poles?

5.2. The flux-density distribution produced by the field mmf of a two-pole salient rotor synchronous machine is sinusoidal, having an amplitude of 0.75 T. If the rotor runs at 3600 rpm, calculate the frequency and amplitude of the voltage induced in a 250-turn coil on the armature. The axial length of the armature is 12 cm and its inner diameter is 10 cm.

5.3. Let the field winding of a two-pole synchronous machine be excited by an ac source such that the airgap flux-density distribution is $B(\theta, t) = B_m \cos \omega_1 t \cos \theta$. The armature has a bore $2r$ and length l. Obtain an expression for the voltage induced in an N-turn coil on the armature if the rotor (or field) rotates at ω_2 rad/s. Study the special case when $\omega_1 = \omega_2 = \omega$.

5.4. A 30-kVA, 230-V, wye-connected, round-rotor synchronous motor operates at full-load at a leading power factor of 0.707. A three-phase, wye-connected inductive load having an impedance of $(4 + j3)$ Ω/phase is connected in parallel with the motor. Calculate *(a)* the overall power factor of the motor and the inductive load, *(b)* active and reactive power for the motor and for the load, and *(c)* line current for the motor and inductive load combination.

5.5. In the text we have obtained an expression for the power-angle characteristics of a round-rotor synchronous machine, neglecting the armature resistance. If this resistance has a value R_a, derive the modified expression for the power-angle characteristics of the machine.

5.6. A 5-kVA, 230-V, three-phase, wye-connected, round-rotor synchronous generator has R_a = 0.6 Ω and X_s = 1.2 Ω/phase. Calculate *(a)* the voltage regulation, and *(b)* the developed power at full-load and 0.8 lagging power factor. Notice that R_a is not negligible in this case.

5.7. For the generator of Problem 5.6, determine the power factor such that the full-load terminal voltage is the same as the no-load terminal voltage.

5.8. A synchronous generator produces 50 A of short-circuit armature current per phase at a field current of 2.3 A. At this field current the open-circuit voltage is 250 kV per phase. If the armature resistance is 0.9 Ω per phase, and the generator supplies a purely resistive load of 3 Ω per phase at a 130-V phase voltage and 50 A of armature current, determine the voltage regulation.

5.9. A 1000-kVA, 11-kV, three-phase wye-connected synchronous generator has an armature resistance of 0.5 Ω and a synchronous reactance of 5 Ω. At a certain field current the generator delivers rated load at 0.9 lagging power factor at 11 kV. For the same excitation, what is the terminal voltage at 0.9 leading power factor full-load?

5.10. An 11-kV, three-phase, wye-connected generator has a synchronous impedance of 6 Ω per phase and negligible armature resistance. For a given field current, the open-circuit voltage is 12 kV. Calculate the maximum power developed by the generator. Determine the armature current and power factor for the maximum power condition.

5.11. A 400-V, three-phase, wye-connected synchronous motor delivers 12 hp at the shaft and operates at 0.866 lagging power factor. The total iron, friction, and field copper losses are 1200 W. If the armature resistance is 0.75 Ω per phase, determine the efficiency of the motor.

5.12. The motor of Problem 5.11 has a synchronous reactance of 6 Ω per phase and operates at 0.9 leading power factor while taking an armature current of 20 A. Calculate the induced voltage.

5.13. A 1000-kVA, 11-kV, three-phase, wye-connected synchronous motor has a 10 Ω synchronous reactance and a negligible armature resistance. Calculate the induced voltage for *(a)* 0.8 lagging power factor, *(b)* unity power factor, and *(c)* 0.8 leading power factor, when the motor takes 1000 kVA (in each case).

5.14. The per-phase induced voltage of a synchronous motor is 2500 V. It lags behind the terminal voltage by 30°. If the terminal voltage is 2200 V per phase, determine the operating power factor. The per-phase armature reactance is 6 Ω. Neglect armature resistance.

5.15. A 150-MVA, 12.6-kV, three-phase, wye-connected, round-rotor synchronous generator has negligible armature resistance. The no-load magnetization curve is obtained from the following data:

Field current, A	100	200	300	400	500	600	700
Armature line-to-line voltage, kV	1.9	3.8	5.8	7.8	9.8	11.3	12.6

A 350-A field current results in an armature current of 4000 A under short-circuit.
(a) What is the unsaturated synchronous reactance?
(b) Determine the (saturated) synchronous reactance at 700-A field current.
(c) Calculate the voltage regulation at full-load 0.8 lagging power factor, using the value of X_s obtained in part *(b)*.

5.16. The per-phase power developed by a round-rotor synchronous machine having a synchronous impedance Z_s and an armature resistance R_a is given by

$$P_d = \frac{V_0 V_t}{Z_s}\cos(\delta - \phi) - V_0^2 \frac{R_a}{Z_s^2}$$

where V_0 = induced voltage, V_t = terminal voltage, δ = power angle, and ϕ = power factor angle. Determine the power angle for maximum developed power. If P_m is the maximum developed power, what is the corresponding value of V_0 (in terms of P_m, V_t, R_a, and Z_s)?

5.17. A 12.6-kV, three-phase synchronous motor, having a synchronous reactance of 0.9 Ω/phase and negligible armature resistance, operates at 0.866 leading power factor while taking a line current of 1575 A. Calculate the excitation (or induced) voltage and the power angle.

5.18. The load on the motor of Problem 5.17 is such that the power angle is 30°. Determine the armature current and the motor power factor.

5.19. A 400-V, three-phase, wye-connected synchronous motor has a synchronous impedance $Z_s = (0.15 + j3)$ Ω/phase. At a certain load and excitation, the motor takes 22.4-A armature current and operates at 0.8 leading power factor. Calculate the excitation voltage and the power angle.

5.20. The field excitation of the motor of Problem 5.19 is so adjusted that the motor takes a minimum armature current of 17.94 A. Determine the excitation voltage. Neglect the effect of armature resistance.

5.21. The excitation voltage of the motor of Problem 5.19 is 280 V/phase, while the armature current is 25 A. Neglecting the effect of armature resistance, obtain the operating power factor.

5.22. If the field current of the motor of Problem 5.19 is adjusted such that $V_0 = V_t$ and the armature current for this condition is 20 A, determine the power developed by the motor. Neglect R_a.

5.23. Draw the phasor diagram of a salient pole synchronous generator supplying a load having a leading power factor. From this diagram obtain an expression for the power angle δ in terms of the machine constants X_d and X_q, the armature current I_a, and the load power factor angle ϕ. Neglect armature resistance.

5.24. A salient pole synchronous generator is wye-connected and operates at 220 V (line-to-line) at a power angle of 30°. The machine constants per phase are $X_d = 5\ \Omega$, $X_q = 3\ \Omega$, and R_a is negligible. If the generator develops a total of 16-kW power, calculate the voltage regulation for the given operating conditions.

5.25. A 20-kVA, 220-V, 60-Hz, wye-connected, three-phase salient pole synchronous generator supplies rated load at 0.707 leading power factor. The per-phase constants of the machine are armature resistance $R_a = 0.05\ \Omega$; direct-axis reactance $X_d = 4.0\ \Omega$; and quadrature-axis reactance $X_q = 2.0\ \Omega$. Calculate the developed power and percent voltage regulation at the specified load. Notice that R_a is very small.

5.26. A synchronous motor is delivering 0.5 N-m torque at 1800 rpm. The load torque is suddenly reduced to zero, and the torque angle is observed to oscillate initially over a 12° range with a period of 0.3 s. After 7.5 s, the oscillations have decreased to 4.42°. Calculate the synchronizing torque, J, and b for the zero-load condition.

CHAPTER 6

DC COMMUTATOR MACHINES

In this chapter a family of rotating electromagnetic devices, dc commutator machines, is introduced. While analyzing dc commutator machines we will make use of many of the principles developed in earlier chapters. The material in this chapter would usually be entitled only "dc machines." We have inserted the word commutator for a number of reasons. First, the commutator is the distinguishing characteristic of the devices discussed in this chapter. Without the commutator, these machines would be indistinguishable from many other types of machines. The commutator is a mechanical rectifier/inverter that permits connection to a dc source and, in the process, provides this machine configuration with some useful characteristics as a motor, a generator, and a control device.

Second, the commutator type is not the *only* type of dc machine. In fact, it is truly not a dc machine at all—if by "dc" we mean a device whose currents and voltages are unidirectional for a given condition of speed and torque. Electronics-oriented readers would probably object to the intrusion of a "mechanical" rectifier/inverter as part of a device used so frequently in electronics and control systems and ask, "Why not supply the rectifier/inverter function with solid-state devices?" Such devices are fairly common today and are known as brushless dc machines. If one is still wondering what is a truly dc machine, it is a device known as the *homopolar machine*, which is a grandchild of the Faraday disc generator, developed by Michael Faraday in the 1830s. Therefore, we use "dc commutator machines" to typify a specific configuration of dc machines.

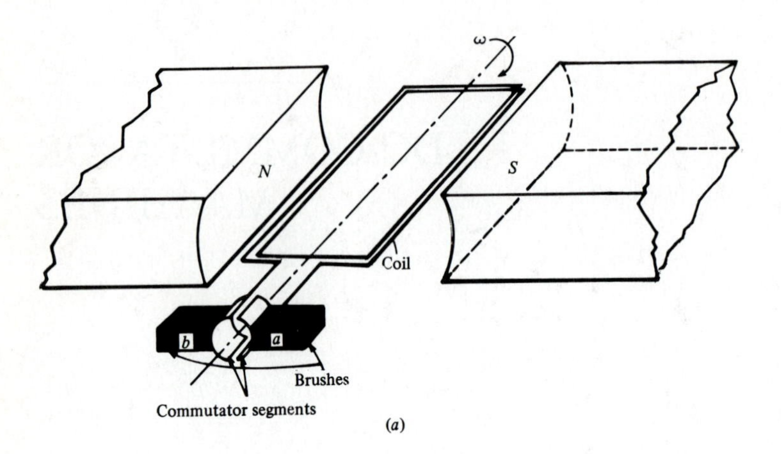

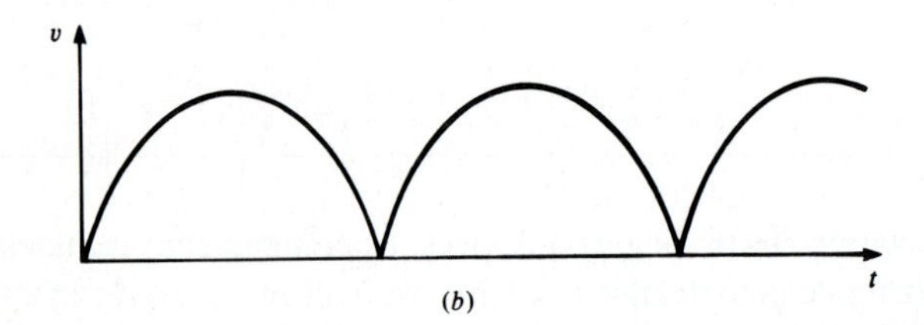

FIGURE 6-1
(a) An elementary dc generator; *(b)* output voltage at the brushes.

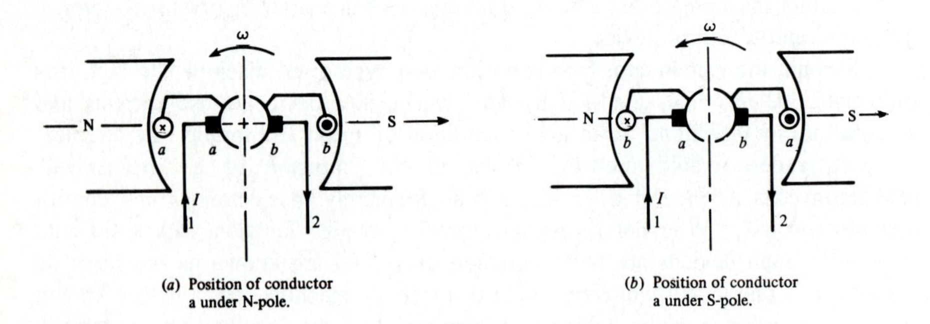

FIGURE 6-2
Production of a unidirectional torque and operation of an elementary dc motor.
(*a*) Position of conductor *a* under N pole. (*b*) Position of conductor *a* under S pole.

6.1 OPERATION OF DC GENERATOR

Just like the synchronous generator, the operation of a dc generator is based on Faraday's law of electromagnetic induction. See Section 5.1 and recall from (5.5) and Fig. 5.2 that the voltage available at the sliprings of a synchronous generator is alternating and, ideally, sinusoidal. In order to obtain a unidirectional voltage at the coil terminals, we replace the slip rings of Fig. 5.2*(a)* by the *commutator segments* shown in Fig. 6.1*(a)*. It can be readily verified, by applying the right-hand rule, that in the arrangement shown in Fig. 6.1*(a)* the brushes will maintain their polarities regardless of the position of the coil. In other words, brush *a* will always be positive, and brush *b* will always be negative for the given relative polarities of the flux and the direction of rotation. Thus the arrangement shown in Fig. 6.1 forms an elementary *heteropolar* dc machine. The main characteristic of a heteropolar dc machine is that the emf induced in a conductor, or a current flowing through it, has its direction reversed as it passes from a north-pole to a south-pole region. This reversal process is known as *commutation* and is accomplished by the commutator-brush mechanism, which also serves as a connection to the external circuit. The voltage available at the brushes will be of the form shown in Fig. 6.1*(b)*. In this respect, commutation is the process of rectification of the induced alternating emf in the coil into a dc voltage at the terminals.

FIGURE 6-3
A commercial large dc machine. (Courtesy of General Electric Company.)

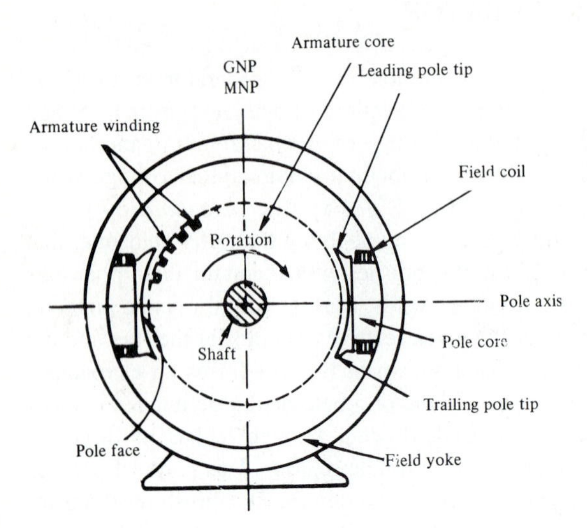

FIGURE 6-4
Parts of a dc machine.

The production of a unidirectional torque, and hence operation as a dc motor, can be verified by referring to Fig. 6.2. Conductors *a* and *b* are the two sides of a coil. Thus *a* and *b* are connected in series. Notice from Fig. 6.2*(a)* that the current enters into conductor *a* through brush *a* and leaves conductor *b* through brush *b*.

Applying the left-hand rule, the conductors will experience a force to produce a counterclockwise rotation. After half a rotation, the conductors interchange their respective positions, as shown in Fig. 6.2*(b)*. Now the current in conductor *b* is again in such a direction that the force on the coil will tend to rotate it in a counterclockwise direction. In other words, the torque is unidirectional and is independent of conductor position.

6.2 CONSTRUCTIONAL DETAILS

Before we consider some conventional dc machines, it is best that we study briefly the constructional features of their various parts and discuss the usefulness of these parts. We observe from the preceding section that the basic elements of a dc machine are the rotating coil, a means for the production of flux, and the commutator-brush arrangement. In a practical dc machine the coil is replaced by *armature winding* mounted on a cylindrical magnetic structure. The flux is provided by the *field winding* wound on field poles. Thus, the function of the field system is to supply energy to establish a magnetic field in the magnetic circuit. The use of an electrical field winding gives the great diversity and variety of performance characteristics that typify dc commutator machines. Permanent magnet excitation of the field system is often less costly and occupies less space than electric excitation and eliminates the need for a separate electrical source of energy.

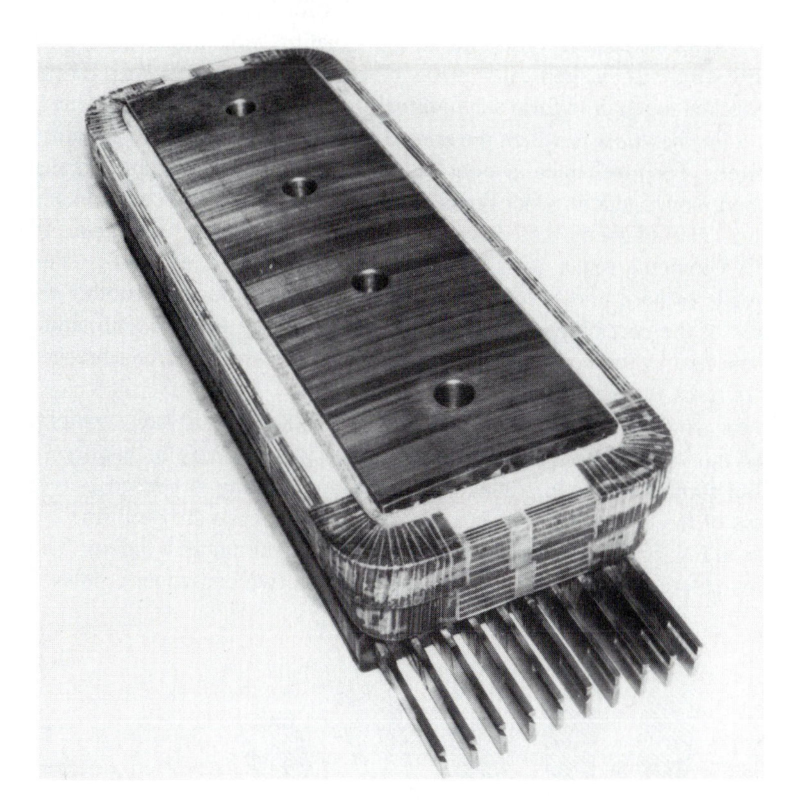

FIGURE 6-5
Field pole of a 2550-kW dc motor. (Courtesy of Brown Boveri Company.)

The armature winding is often referred to as the power winding of a dc commutator machine, since the machine's electromagnetic torque is a function of the armature winding current and, as mentioned earlier, the armature winding terminals are connected to the external power source through the *commutator/brush system.* The system acts as a mechanical switching device between the external armature circuit and the armature windings within the machine, in which currents and induced voltages are time-varying and reversing in polarity. As a bilateral device, the commutator/brush system is somewhat analogous to the action of an antiparallel pair of rectifiers.

Generally, the armature winding is placed on the rotating member—the *rotor*—and the field winding is on the stationary member—the *stator* of the dc machine. A common large dc machine is shown in Fig. 6.3, which, together with the schematic of Fig. 6.4, shows most of the important parts of a dc machine. The *field poles*, mounted on the stator, carry the field windings. Some machines carry more than one independent field winding on the same core. The cores of the poles are built of sheet-steel laminations. An assembled field pole is shown in Fig. 6.5. Because the field windings carry direct current, it is not necessary to have the pole cores laminated.

It is, however, necessary for the pole "faces" to be laminated because of their proximity to the armature windings. (Use of laminations for the cores as well as for the pole faces facilitates assembly.) The rotor or the armature core, which carries the rotor or armature windings, is generally made of sheet-steel laminations. These laminations are stacked together to form a cylindrical structure. On its outer periphery, the armature (or rotor) has *slots* in which the armature coils that make up the armature winding are located. For mechanical support, protection from abrasion, and greater electrical insulation, nonconducting slot liners are often wedged in between the coils and the slot walls. The magnetic material between the slots comprises the *teeth*. A typical slot/tooth geometry for a large dc machine is shown in Fig. 6.6. The commutator is made of hard-drawn copper segments insulated from one another by mica. The details of the commutator assembly are given in Fig. 6.7. The armature windings are connected to the commutator segments over which the carbon brushes slide and serve as leads for electrical connection.

The armature winding may be a *lap winding* [Fig. 6.8*(a)*] or a wave winding [Fig. 6.8*(b)*], and the various coils forming the armature winding may be connected in a series-parallel combination. In practice, the armature winding is housed as two layers in the slots of the armature core. In large machines the coils are preformed in the shapes shown in Fig. 6.9 and are interconnected to form an armature winding. The coils span approximately a pole pitch, the distance between two consecutive poles.

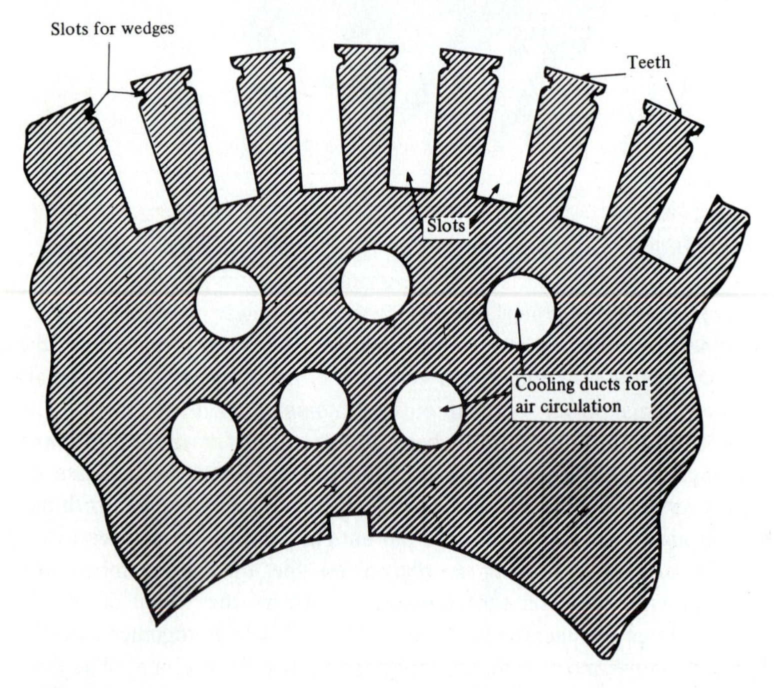

FIGURE 6-6
Portion of an armature lamination of a dc machine showing slots and teeth.

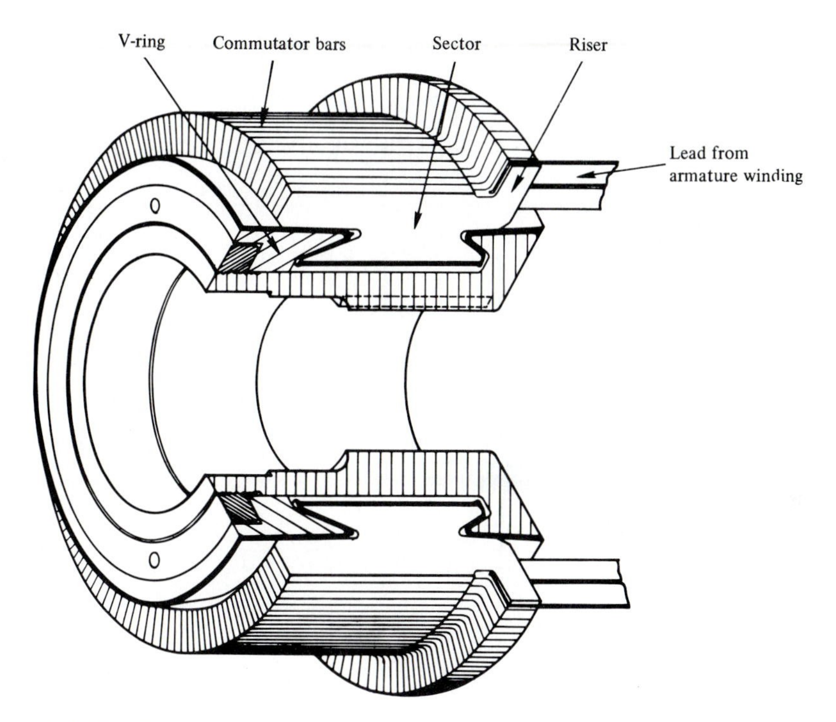

FIGURE 6-7
Details of commutator assembly.

We have just mentioned that the various coils forming an armature winding are interconnected in a series-parallel combination. Such an interconnection may be simulated by considering that the armature winding is analogous to a "battery" consisting of a number of inter-connected "cells," where the cells correspond to individual coils thus making the winding. Thus, Fig. 6.10 shows two possible cell arrangements to make a battery. Notice that in Fig. 6.10*(a)* we have four paths in parallel whereas Fig. 6.10*(b)* has two parallel. Correspondingly, in a lap winding the number of paths in parallel, a, is equal to the number of poles, P, whereas in a wave winding the number of parallel paths is always two.

Complete layouts of double layer lap and wave windings are given in Appendix IV. From the diagrams given there, the parallel paths (for currents) may be traced and the statement that $a = P$ in a lap winding and $a = 2$ in a wave winding may be verified.

An assembled armature of a large dc machine is shown in Fig. 6.11.

In addition to the armature and field windings, commutating poles and compensating windings are also found on large dc machines. These are shown in Fig. 6.12 and are used essentially to improve the performance of the machine, as we shall see later.

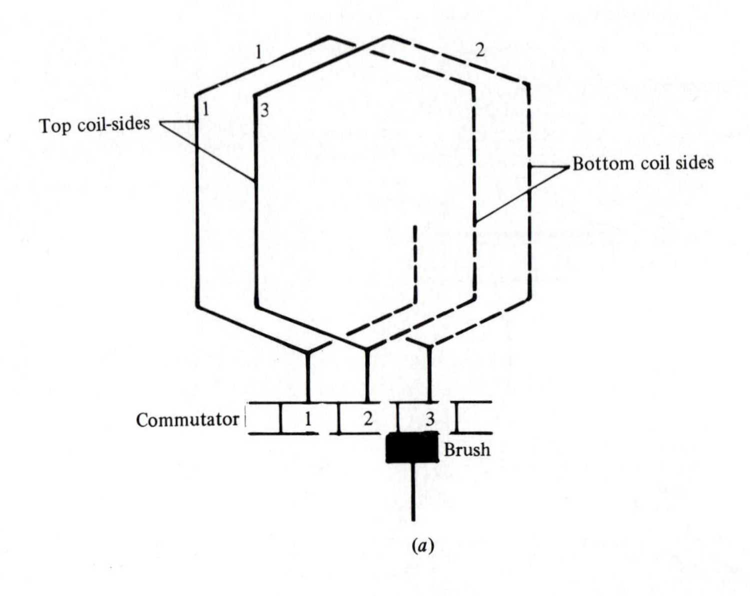

(*a*)

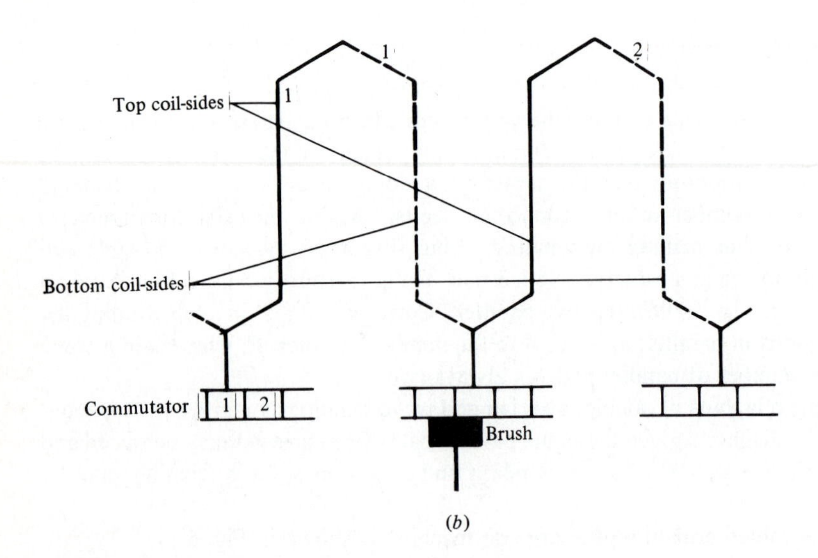

(*b*)

FIGURE 6-8
Elements of *(a)* lap winding, *(b)* wave winding. Odd-numbered conductors are at the top, and even-numbered conductors are at the bottom of the slots.

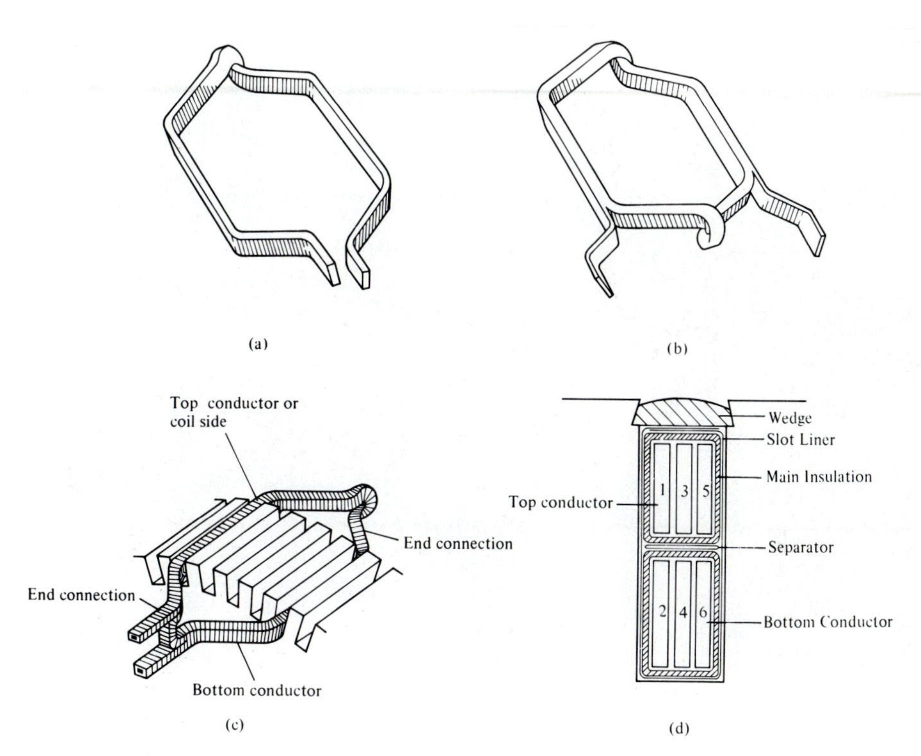

FIGURE 6-9
(a) Coil for a lap winding, made of a single bar; *(b)* multiturn coil for wave winding; *(c)* a coil in a slot; *(d)* slot details, showing several coils arranged in two layers.

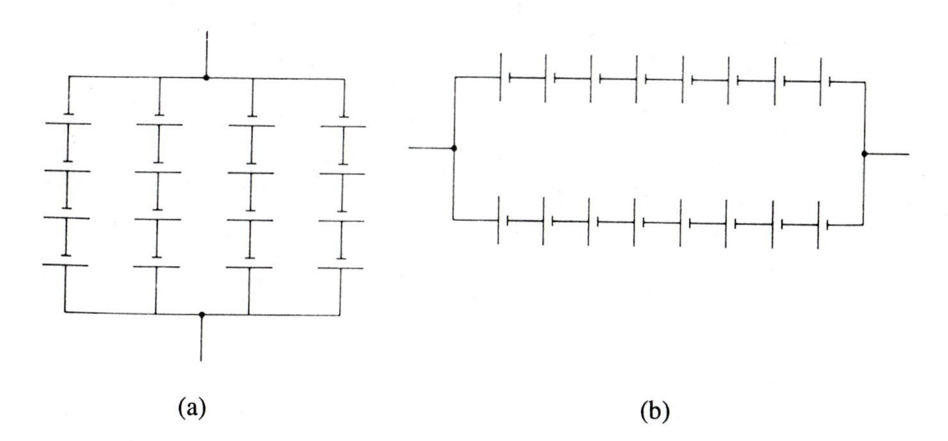

FIGURE 6-10
A battery having *(a)* four parallel paths and *(b)* two parallel paths. Notice that in each case we have a total of 16 cells.

FIGURE 6-11
Armature of a 2000-kW, 450-rpm dc generator. (Courtesy of Brown Boveri Company.)

The mechanical and structural design of rotating machines is a very challenging subject in itself and an area that is continuously changing with the advent of improved magnetic, electrical, and insulating materials, the use of improved heat transfer techniques, and the development of new manufacturing processes. For instance, the use of windings that eliminate the commutator is becoming more common in small dc machines. In these machines the brushes or equivalent current collectors make contact with the winding itself. The most common form of such machines is known as a printed-circuit machine, named from the early manufacturing technique that used electrochemical etching to form the armature windings. Similar armatures are now made by means of stamping techniques. An example of such an armature is shown in Fig. 6.13. Axial airgap configurations are the most common for printed or stamped armatures, but radial gap designs are also possible.

6.3 CLASSIFICATION ACCORDING TO FORMS OF EXCITATION

Conventional dc machines that have a set of field windings and armature windings can be classified, on the basis of mutual electrical connections between the field and armature windings, as follows:

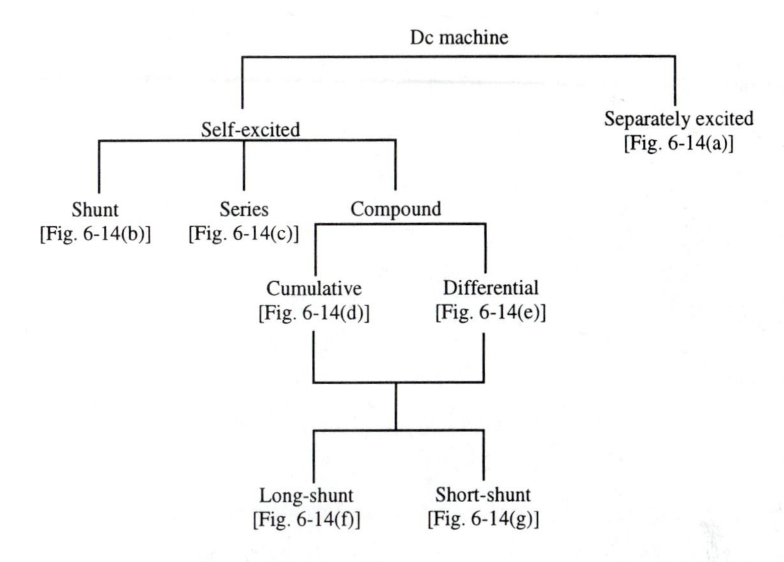

Fig. 6.14 illustrates schematically these different methods of connection of the field and armature circuits in the dc commutator machines. The circular symbol represents only the *active* portion of the armature circuit, that is, the generated voltage. There are also passive circuit elements in the armature circuit representing the resistance and inductance of the armature winding and of other windings connected in series with the armature for the purpose of improving commutation. The squares at the sides of the circle in the armature symbol indicate that the connection to the armature is through the commutator/brush system.

These interconnections of field and armature windings essentially determine the machine operating characteristics. In a separately excited machine, shown in Fig. 6.14*(a)*, there is no electrical interconnection between the field and the armature windings. On the other hand, the field winding is connected to the armature winding in a self-excited machine. A parallel connection between the field and the armature windings results in the shunt machine of Fig. 6.14*(b)*. The series machine has the field and the armature windings connected in series as in Fig. 6.14*(c)*. A compound machine has both shunt and series field windings in addition to the armature winding. If the relative polarities of the shunt and series field windings are additive, as illustrated in Fig. 6.14*(d)*, we obtain a cumulative compound machine. Notice from Fig. 6.14*(d)* that the two fields are shown to produce magnetic fluxes in the same direction. In a differential compound machine, the series field is in opposition to the shunt field, implying that the respective resulting fluxes are in opposition, as shown in Fig. 6.14*(e)*. A differential or cumulative compound machine may have a long-shunt connection, in which case the shunt field is across the armature-series field combination, as given in Fig. 6.14*(f)*. Fig. 6.14*(g)* shows a short-shunt connection, where the shunt field is directly across the armature. Finally, in a permanent magnet machine, we do not have the field winding, and the necessary magnetic flux is provided by the permanent magnet. Such dc machines are generally of fractional-horsepower ratings.

FIGURE 6-12
Stator of a 1030-kW dc motor showing interpoles and compensating bars. (Courtesy of Brown Boveri Company.)

6.4 PERFORMANCE EQUATIONS

The three quantities of greatest interest in evaluating the performance of a dc machine are (1) the induced electromotive force (emf), (2) the electromagnetic torque developed by the machine, and (3) the speed corresponding to (1) and/or (2). We will now derive the equations that enable us to determine these quantities.

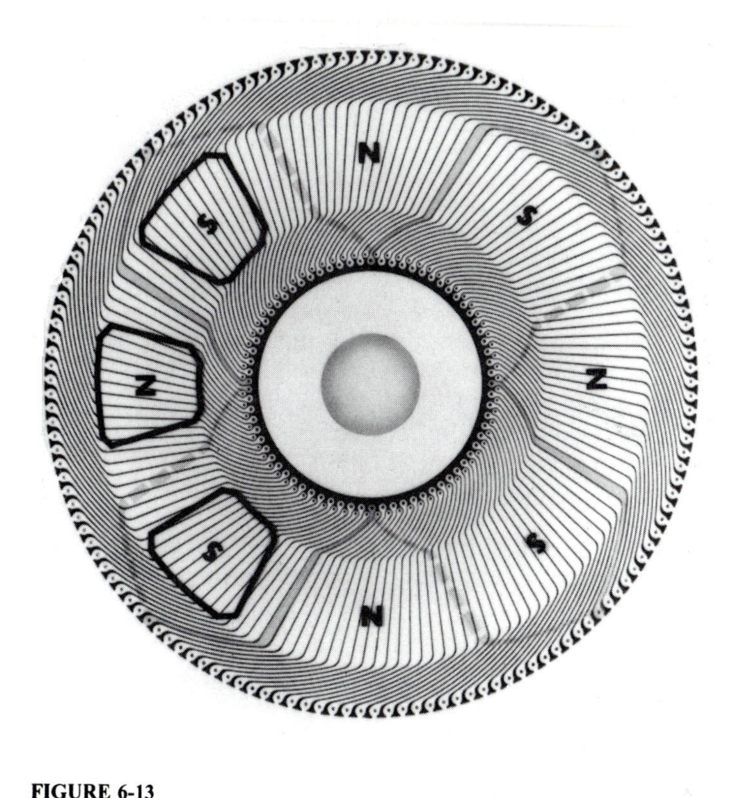

FIGURE 6-13
Printed-circuit winding layout for a disc-type motor. (Courtesy PMI Motors, Division of Kolmorgen Corporation.)

EMF Equation

The emf equation yields the emf induced in the armature of a dc machine. The derivation follows directly from Faraday's law of electromagnetic induction, according to which the emf induced in a moving conductor is the flux cut by the conductor per unit time.

Let us define the following symbols:

Z = number of active conductors on the armature
a = number of parallel paths in the armature winding
P = number of field poles
ϕ = flux per pole
n = speed of rotation of the armature, rpm

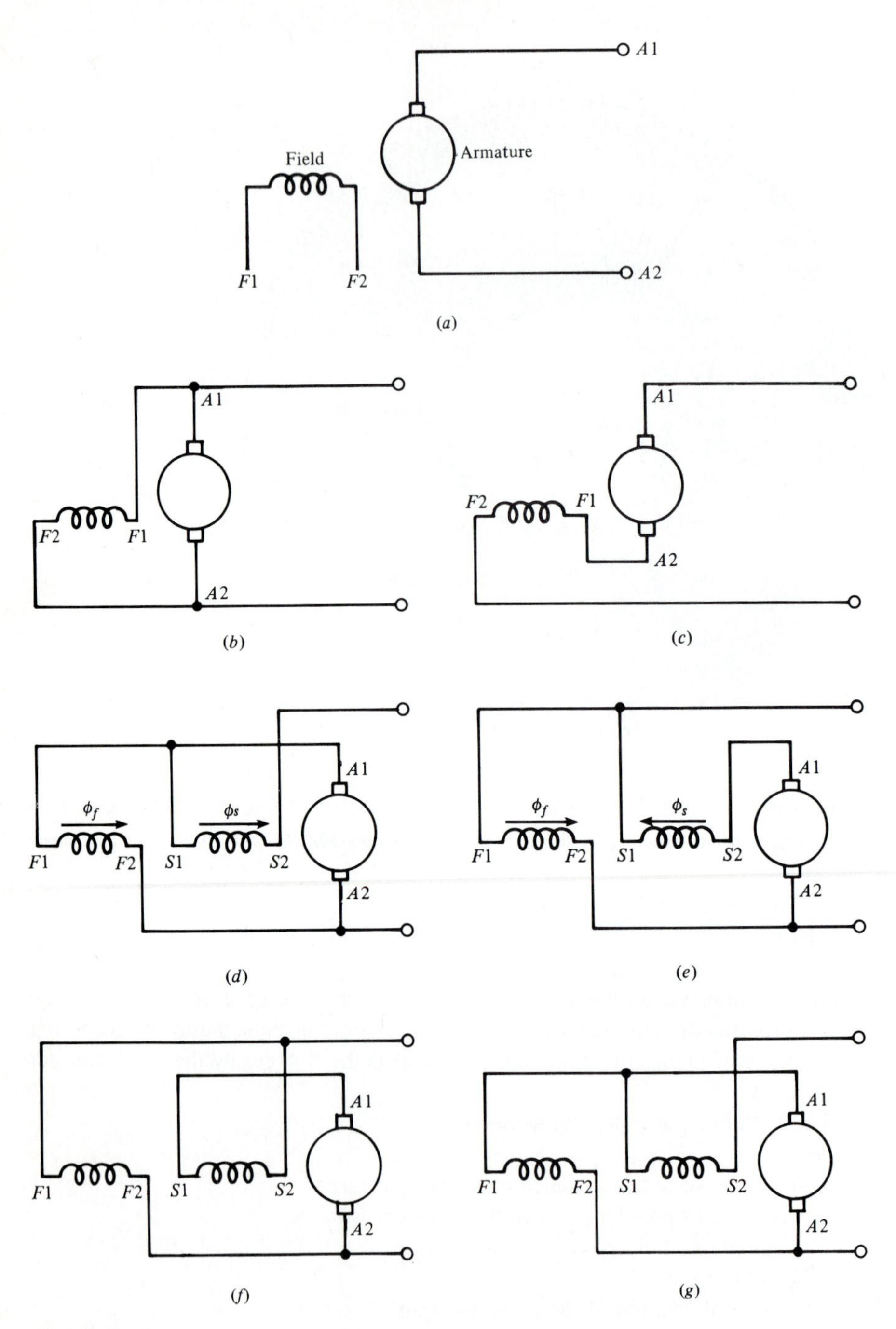

FIGURE 6-14
Classification of dc machines. *(a)* Separately excited; *(b)* shunt; *(c)* series; *(d)* cumulative compound; *(e)* differential compound; *(f)* long-shunt; *(g)* short-shunt.

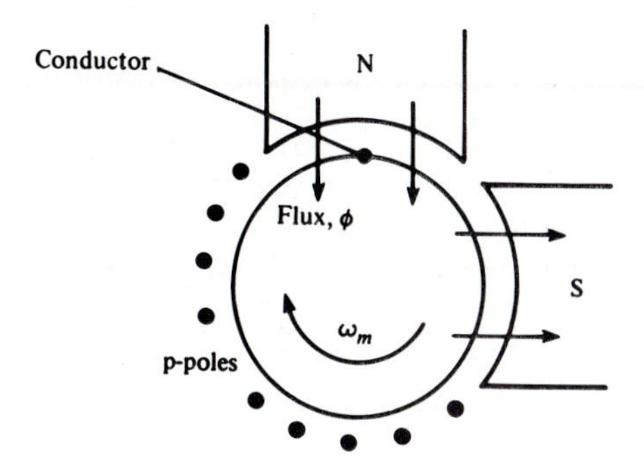

FIGURE 6-15
A conductor rotating at a speed ω_m in the field of P poles.

Then, with reference to Fig. 6.15,

$$\text{flux cut by one conductor in one rotation} = \phi P$$

$$\text{flux cut by one conductor in } n \text{ rotations} = \phi n P$$

$$\text{flux cut per second by one conductor} = \frac{\phi n P}{60}$$

$$\text{number of conductors in series} = \frac{Z}{a}$$

$$\text{flux cut per second by } Z/a \text{ conductors} = \frac{\phi n P}{60}\left(\frac{Z}{a}\right)$$

Hence

emf induced in the armature winding

$$= E = \frac{\phi n Z}{60} \times \frac{P}{a} \quad \text{V}$$
$$= \frac{ZP}{2\pi a}(\phi \omega_m) = k_a \phi \omega_m \quad \text{V} \tag{6.1}$$

where $k_a = ZP/2\pi a$ is known as the armature constant. Equation (6.1) is known as the *emf equation* of a dc machine. In (6.1) we have separated P/a from the other terms because P and a are related to each other for the two types of windings—lap and wave.

An Alternate Derivation

Recall that in Fig. 5.2 *one cycle of emf corresponds to two poles and 360 electrical degrees*, and one complete rotation around the machine periphery corresponds to 360 mechanical degrees. Hence for a *P-pole* machine, the general relationship between the electrical degree θ and the mechanical degree θ_m is

$$\theta = \frac{P}{2}\theta_m \tag{6.2}$$

Obviously, for the two-pole machine ($P = 2$) shown in Fig. 5.2 or 6.1, $\omega = \omega_m$. But, for a *P-pole* machine,

$$\omega = \frac{P}{2}\omega_m \tag{6.3}$$

Considering the output voltage at the brushes, shown in Fig. 6.1*(b)*, the average dc voltage becomes

$$E = \frac{1}{\pi}\int_0^{\pi} \omega N\phi \sin\omega t \; d(\omega t) = \frac{2}{\pi}\omega N\phi \tag{6.4}$$

since $2l_1 rB = \phi$ in (5.5). In terms of ω_m, from (6.3) and (6.4) we have

$$E = \frac{PN}{\pi}\phi\omega_m = 2PN\phi\frac{n}{60} \tag{6.5}$$

Finally, since $N = Z/2a$, (6.5) becomes

$$E = \phi\frac{nZ}{60}\left[\frac{P}{a}\right] = k_a\phi\omega_m \tag{6.6}$$

which is the same as (6.1), where $k_a = ZP/2\pi a$.

Example 6.1 Determine the voltage induced in the armature of a dc machine running at 1750 rpm and having four poles. The flux per pole is 25 mWb, and the armature is lap-wound with 728 conductors.

Solution

Since the armature is lap-wound, $P = a$, and (6.1) becomes

$$E = \frac{\phi n Z}{60} = \frac{25 \times 10^{-3} \times 1750 \times 728}{60} = 530.8 \text{ V}$$

Torque Equation

The mechanism of torque production in a dc machine has been considered earlier (see Fig. 6.2), and the electromagnetic torque developed by the armature can be evaluated by referring to Fig. 6.16, which represents a generic ideal electric machine. By virtue of its energy conversion property we have

$$vi = T_e \omega_m \tag{6.7}$$

where v and i are, respectively, the voltage and the current at the electrical port, and T_e and ω_m are, respectively, the torque (in newton-meters, N-m) and angular rotational velocity (in radians per second, rad/s) at the mechanical port. We wish to reiterate that (6.7) is valid for an ideal machine in that the machine is lossless.

Now, for a dc machine, for torque production we must have a current through the armature, as this current interacts with the flux produced by the field winding. Let I_a be the armature current and E the voltage induced in the armature. Thus the power at the armature electrical port (see Fig. 6.16) is EI_a. Assuming that this entire electrical power is transformed into mechanical form, we rewrite (6.7) as

$$EI_a = T_e \omega_m \tag{6.8}$$

where T_e is the electromagnetic torque developed by the armature and ω_m is its angular velocity in rad/s. The speed n (in rpm) and ω_m (in rad/s) are related by

$$\omega_m = \frac{2\pi n}{60} \tag{6.9}$$

Hence from (6.6) to (6.9) we obtain

$$\frac{\omega_m}{2\pi} \phi Z \frac{P}{a} I_a = T_e \omega_m$$

which yields

$$T_e = \frac{ZP}{2\pi a} \phi I_a = k_a \phi I_a \tag{6.10}$$

which is known as the *torque equation*. An application of (6.10) is illustrated by the next example.

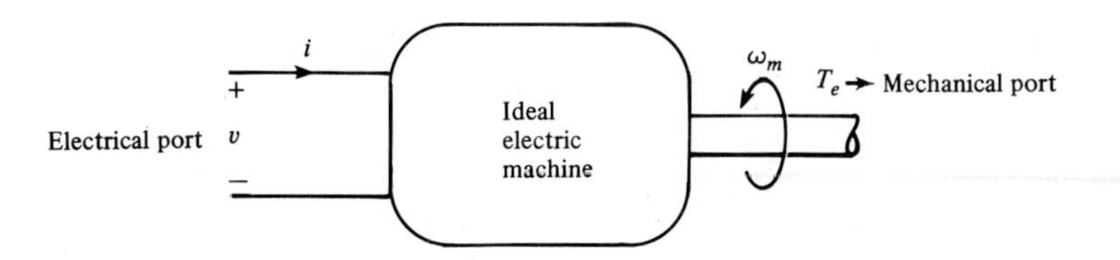

FIGURE 6-16
A general representation of an electric machine.

Example 6.2 A lap-wound armature has 576 conductors and carries an armature current of 123.5 A. If the flux per pole is 20 mWb, calculate the electromagnetic torque developed by the armature.

Solution

For lap winding, we have $P = a$. Substituting this and other given numerical values in (6.10) yields

$$T_e = \frac{576}{2\pi} \times 0.02 \times 123.5 = 226.4 \text{ N-m}$$

Example 6.3 If the armature of Example 6.2 rotates at an angular velocity of 150 rad/s, what is the induced emf in the armature?

Solution

To solve this problem we use (6.8) rather than (6.1). From Example 6.2, we have $I_a = 123.5$ A and $T_e = 226.4$ N-m. Hence (6.8) gives

$$E \times 123.5 = 226.4 \times 150$$

or

$$E = 275 \text{ V}$$

Speed Equation and Back EMF

The emf and torque equations discussed above indicate that the armature of a dc machine, whether operating as a generator or as a motor, will have an emf induced while rotating in a magnetic field, and will develop a torque if the armature carries a current. In a generator, the induced emf is the internal voltage available from the generator. When the generator supplies a load, the armature carries a current and develops a torque. This torque opposes the prime mover (such as a diesel engine) torque.

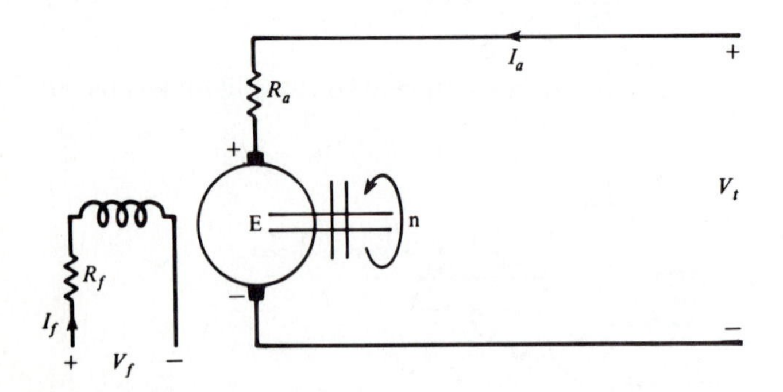

FIGURE 6-17
Equivalent circuit of a separately excited motor.

In motor operation, the developed torque of the armature supplies the load connected to the shaft of the motor, and the emf induced in the armature is termed the *back emf.* This emf opposes the terminal voltage of the motor.

Fig. 6.17 shows the equivalent circuit of a separately excited dc motor running at speed n while taking an armature current I_a at voltage V_t. From this circuit we have

$$V_t = E + I_a R_a \tag{6.11}$$

Substituting (6.1) in (6.11), putting $k_1 = ZP/60a$, and solving for n, we get

$$n = \frac{V_t - I_a R_a}{k_1 \phi} \tag{6.12}$$

This equation is known as the *speed equation,* as it contains all the factors that affect the speed of a motor. In a later section we shall consider the influence of these factors on the speed of dc motors. For the present, let us focus our attention on k_1 and ϕ. The term k_1 replacing $ZP/60a$ is a design constant in the sense that once the machine has been built, Z, P, and a cannot be altered. The magnetic flux ϕ is controlled primarily by the field current I_f (Fig. 6.17). If the magnetic circuit is unsaturated, ϕ is directly proportional to I_f. Thus we may write

$$\phi = k_f I_f \tag{6.13}$$

where k_f is a constant. We may combine (6.12) and (6.13) to obtain

$$n = \frac{V_t - I_a R_a}{k I_f} \tag{6.14}$$

where $k = k_1 k_f$ = a constant. This form of the speed equation is more meaningful because all the quantities in (6.14) can be conveniently measured. [In contrast, in (6.12), it is very difficult to measure ϕ.]

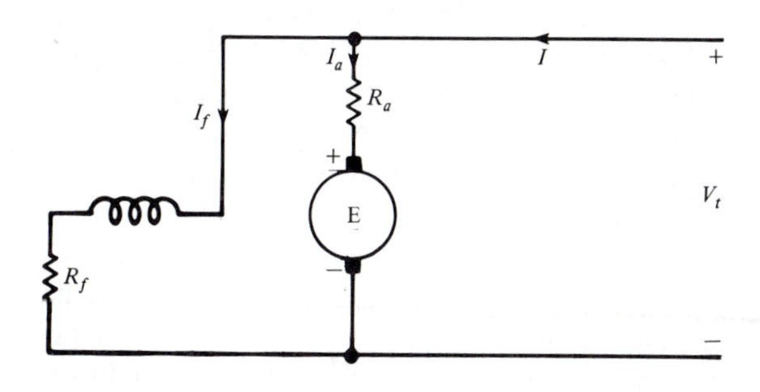

FIGURE 6-18
Equivalent circuit of a shunt motor.

Example 6.4 A 250-V shunt motor has an armature resistance of 0.25 Ω and a field resistance of 125 Ω. At no-load the motor takes a line current of 5.0 A while running at 1200 rpm. If the line current at full-load is 52.0 A, what is the full-load speed?

Solution

The motor equivalent circuit is shown in Fig. 6.18. The field current, $I_f = 250/125 = 2.0$A.

At no load: Armature current, $I_a = 5.0 - 2.0 = 3.0$ A
Back emf, $E_1 = V_t - I_a R_a = 250 - 3 \times 0.25 = 249.25$ V
Speed, $n_1 = 1200$ rpm (given)

At full-load: Armature current $= 52.0 - 2.0 = 50.0$ A
Back emf, $E_2 = 250 - 50 \times 0.25 = 237.5$ V
Speed, n_2 = unknown.

Now,

$$\frac{n_2}{n_1} = \frac{E_2}{E_1} = \frac{237.5}{249.25}$$

Hence

$$n_2 = \frac{237.5}{249.25} \times 1200 = 1143 \text{ rpm}$$

6.5 THE SATURATION CURVE

The emf equation (6.6) gives the armature-induced voltage as a function of motor speed, field pole magnetic flux, and winding configuration. This voltage is entirely independent of whether the machine is being operated as a motor or as a generator; that is, it is not dependent on the direction of current flow in the armature. In motor operation this voltage is often termed "back emf."

The graphical representation of (6.6) is one of the chief tools in the analysis of electrically excited dc commutator machines and is known as the *magnetization* or *excitation curve*. The curve is obtained by driving the machine at a constant mechanical speed, ω_m, and varying the field current (which varies ϕ) while measuring the induced voltage, E, across the open-circuited armature. Typical magnetization curves at two different mechanical speeds are shown in Fig. 6.19. This experimental procedure is termed the no-load test and is described later.

Note that the units used in the curves of Fig. 6.19 relate to the B-H curves described in Chapter 2. Although the units of the latter are usually associated with the characteristics of magnetic materials, they are equally applicable to magnetic circuits with airgaps, such as the commutator motor. The units of Fig. 6.19 are related to B and H as follows:

$$E = k_a \omega_m (A_p B_p) \tag{6.15}$$

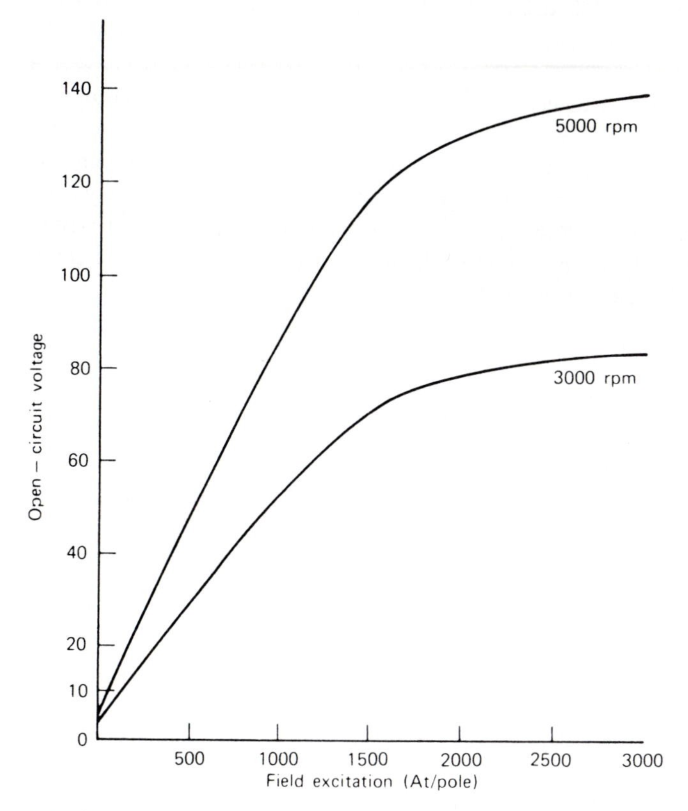

FIGURE 6-19
Typical magnetization curve of dc commutator machine.

$$\text{At/pole} = (NI)_p = H_g l_g + H_p l_p + \frac{H_{ay} l_{ay} + H_{sy} l_{sy}}{2} \tag{6.16}$$

where

$(NI)_p$ = equivalent ampere turns per pole used in the test
A_p = pole area facing airgap in square meters
B_p = average flux density on A_p in teslas
H = magnetic field intensity in amperes per meter
l = length of magnetic member in meters

subscripts

g, p, ay, sy = gap, pole, armature yoke, stator yoke, respectively.

The form of (6.6) in terms of flux is not always convenient. For this reason, (6.6) is often expressed in terms of field current rather than in terms of field flux as

$$E = k_t I_f \omega_m \tag{6.17}$$

The relationship between k_t of this equation and the armature constant, k_a, of (6.6) is the relationship between the magnetic flux in the pole and the exciting current or field current.

Equation (6.6) is a linear equation, but (6.17) is not, because of the nonlinear relationship between ϕ and I_f. The quantity k_t is nonlinear and varies with the slope of the magnetization curve of Fig. 6.17. Since this characteristic can be experimentally obtained for a dc commutator machine, the constant k_t is relatively easy to obtain as compared to the k_a in (6.6). Also, many machines, especially control-type machines, are operated in the linear region of the magnetization curve; in such cases the nonlinearities of (6.17) can be ignored.

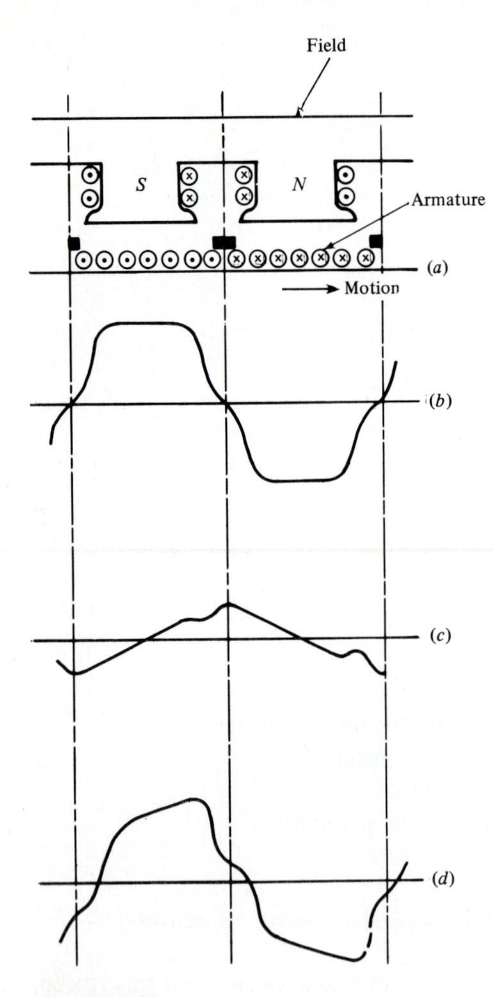

FIGURE 6-20
Airgap fields in a dc machine. *(a)* A two-pole machine, showing armature and field mmf's; *(b)* flux-density distribution due to field mmf; *(c)* flux-density distribution due to armature mmf; *(d)* resultant flux-density distribution [curve *(b)* + curve *(c)*].

6.6 ARMATURE REACTION

In the discussions so far we have assumed no interaction between the fluxes produced by the field windings and by the current-carrying armature windings. In reality, however, the situation is quite different. Consider the two-pole machine shown in Fig. 6.20*(a)*. If the armature does not carry any current (that is, when the machine is on no-load), the airgap field takes the form shown in Fig. 6.20*(b)*. The geometric neutral plane and magnetic neutral plane (GNP and MNP, respectively) are coincident. (*Note*: Magnetic flux lines enter the MNP at right angles.) Noting the polarities of the induced voltages in the conductors, we see that the brushes are located at the MNP for maximum voltage at the brushes. Notice from Fig. 6.10 that the tappings must be located at aa' in order to get the maximum voltage from the battery. We now assume that the machine is on "load" and that the armature carries current. The direction of flow of current in the armature conductors depends on the location of the brushes. For the situation shown in Fig. 6.20*(b)*, the direction of the current flow is the same as the direction of the induced voltages. In any event, the current-carrying armature conductors produce their own magnetic fields, as shown in Fig. 6.20*(c)*, and the airgap field is now the resultant of the fields due to the field and armature windings. The resultant airgap field is thus distorted and takes the form shown in Fig. 6.20*(d)*. The interaction of the fields due to the armature and field windings is known as *armature reaction*. As a consequence of armature reaction, the airgap field is distorted and the MNP is no longer coincident with the GNP. For maximum voltage at the terminals, the brushes have to be located at the MNP. Thus one undesirable effect of armature reaction is that the brushes must be shifted constantly, since the shift of the MNP from the GNP depends on the load (which is presumably always changing). The effect of armature reaction can be analyzed in terms of cross-magnetization and demagnetization, as shown in Fig. 6.21*(a)*. We just mentioned the effect of cross-magnetization resulting in the distortion of the airgap field and requiring the shifting of brushes according to the load on the machine. The effect of demagnetization is to weaken the airgap field. All in all, therefore, armature reaction is not a desirable phenomenon in a machine.

The effect of cross-magnetization can be neutralized by means of compensating windings, as shown in Fig. 6.21. These are conductors embedded in pole faces, connected in series with the armature windings and carrying currents in an opposite direction to that flowing in the armature conductors under the pole face (Fig. 6.21). Once cross-magnetization has been neutralized, the MNP does not shift with the load and remains coincident with the GNP at all loads. The effect of demagnetization can be compensated for by increasing the mmf on the main field poles. By neutralizing the net effect of armature reaction, we imply that there is no "coupling" between the armature and field windings.

6.7 REACTANCE VOLTAGE AND COMMUTATION

In discussing the action of the commutator earlier (Fig. 6.1, for instance) we indicated that the direction of flow of current in a coil undergoing commutation reverses by the

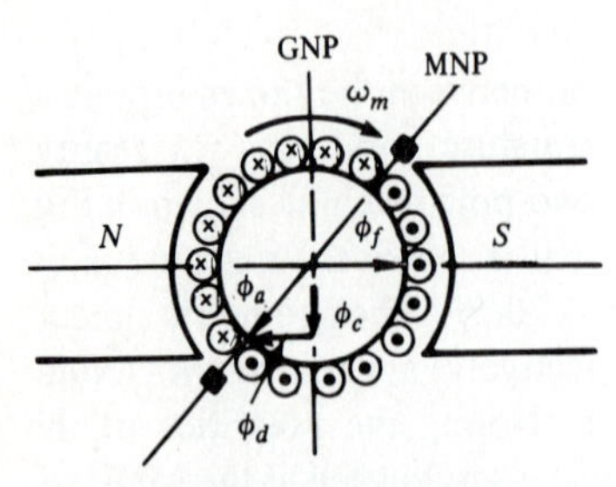

ϕ_a = Flux due to armature MMF
ϕ_c = Flux due to cross-magnetization
ϕ_d = Flux due to demagnetization
ϕ_f = Flux due to field MMF

(a)

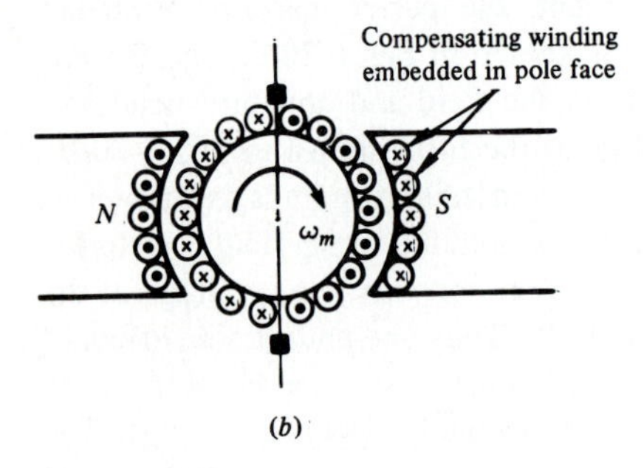

(b)

FIGURE 6-21
(a) Armature reaction resolved into cross and demagnetizing components. *(b)* Neutralization of cross-magnetizing component by compensating winding.

time the brushes move from one commutator segment to the other. This is represented schematically in Fig. 6.22. The flow of current in coil *a* for three different instants is shown. We have assumed that the current fed by a segment is proportional to the area of contact between the brush and the commutator segment. Thus, for satisfactory commutation, the direction of flow of current in coil *a* must completely reverse [Fig. 6.22*(a)* and *(c)*] by the time the brush moves from segment 2 to segment 3. The ideal situation is represented by the straight line in Fig. 6.23 and may be termed straight-line commutation. Because coil *a* has some inductance L, the change of current ΔI in a time Δt induces a voltage $L\ (\Delta I/\Delta t)$ in the coil. According to Lenz's law, the direction of this voltage, called *reactance voltage*, is opposite to the charge (ΔI) that is causing it. As a result, the current in the coil does not completely reverse by the time the brush moves from one segment to the other. The balance of the "unreversed" current jumps over as a spark from the commutator to the brush, and thereby the commutator wears out because of pitting. This departure from ideal commutation is also shown in Fig. 6.23.

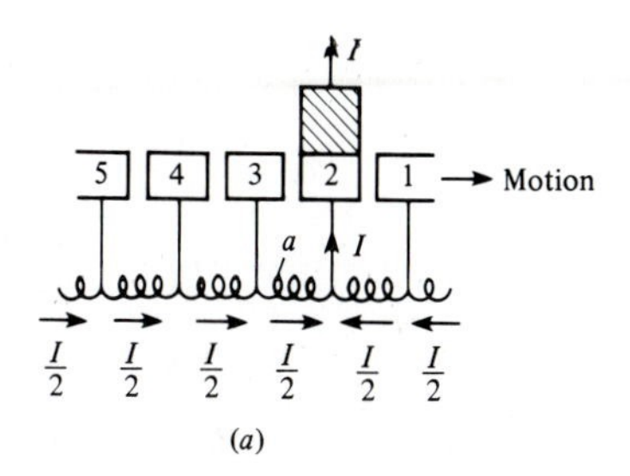

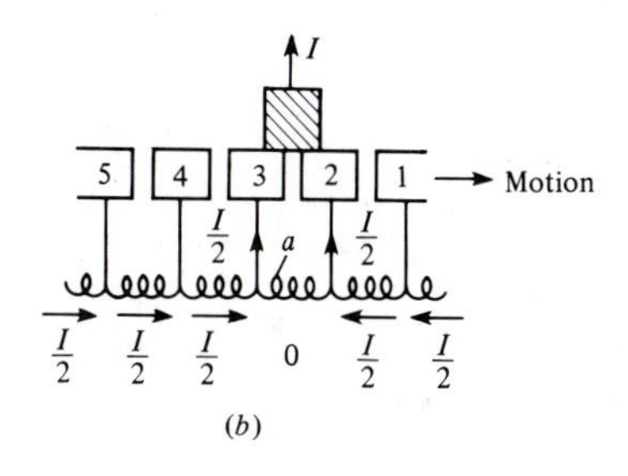

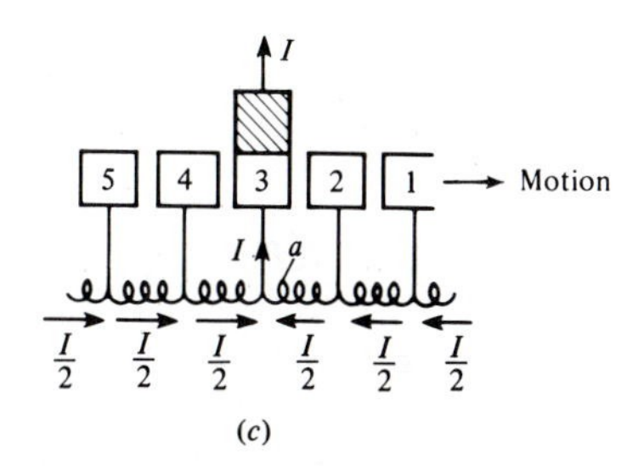

FIGURE 6-22
Coil *a* undergoing commutation.

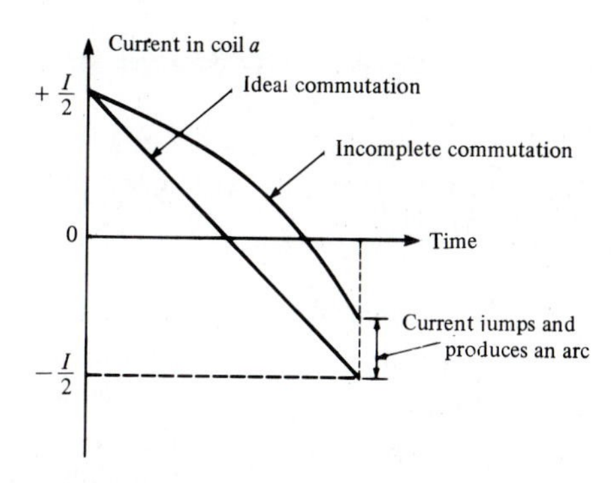

FIGURE 6-23
Commutation in coil *a*.

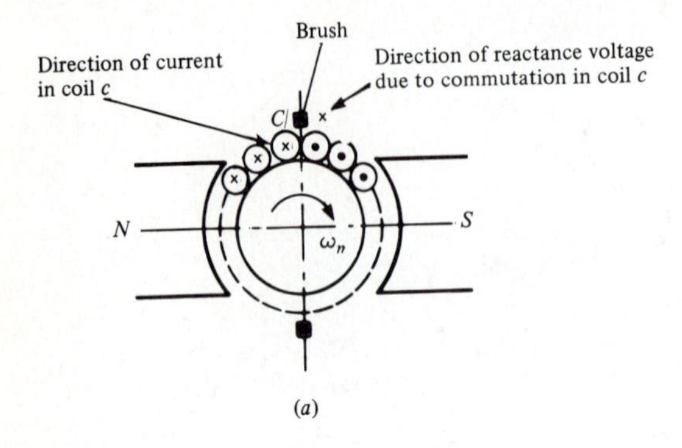

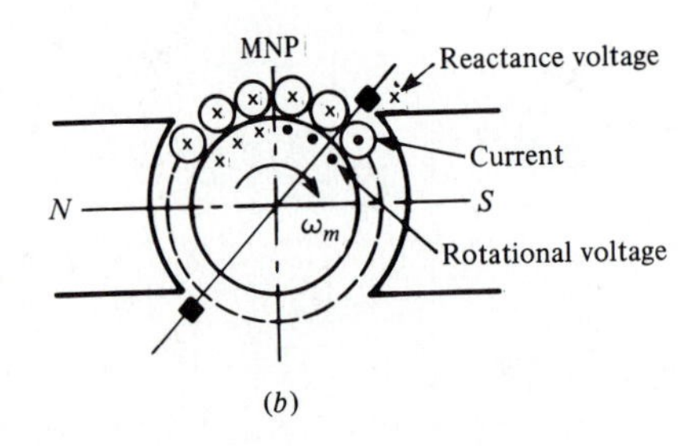

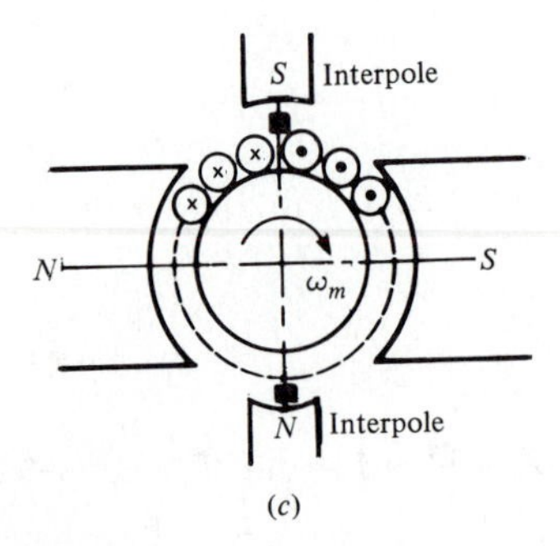

FIGURE 6-24
Reactance voltage and its neutralization. *(a)* Reactance voltage and current in coil *c*, rotational voltage = 0; *(b)* reactance voltage, rotational voltage, and current in coil *c; (c)* interpoles.

The directions of the (speed-) induced voltage, current flow, and reactance voltage are shown in Fig. 6.24*(a)*. Note that the direction of the induced voltage depends on the direction of rotation of the armature conductors and on the direction of the airgap flux. It is determined from the right-hand rule. Next, the direction of the current flow depends on the location of the brushes (or tapping points). Finally, the direction of the reactance voltage depends on the change in the direction of the current flow and is determined from Lenz's law. For the brush position shown in Fig. 6.24*(a)*, observe that the reactance voltage retards the current reversal. If the brushes are advanced in the direction of rotation (for generator operation), we may notice, from Fig. 6.24*(b)*, that the reactance voltage is in the same direction as the (speed-) induced voltage, and therefore the current reversal is not opposed. We may further observe that the coil undergoing commutation, being near the tip of the south pole, is under the influence of the field of a weak south pole. From this argument, we may conclude that commutation improves if we advance the brushes. But this is not a practical solution. The same—perhaps better—results can be achieved if we keep the brushes at the GNP, or MNP, as in Fig. 6.24*(a)*, but produce the "field of a weak south pole" by appropriately winding and connecting an auxiliary field winding, as shown in Fig. 6.24*(c)*. The poles producing the desired field for better commutation are known as *commutating poles* or *interpoles*.

6.8 LOSSES AND EFFICIENCY

The mechanical power at the shaft of a rotating electromagnetic machine is

$$P_m = T_s \omega_m \quad \text{W} \tag{6.18}$$

This power, in a motor, is the output power available, that is, the power that can be supplied to an external mechanical load, such as a pump, generator, or overhead crane. In a generator this is the required input power. The electrical power at the terminals of a dc commutator machine is

$$P_e = VI + P_f \tag{6.19}$$

where V and I are the voltage and current, respectively, of the external electrical system connected to the machine, and P_f is the power supplied to a separately excited field winding, if the machine is operated with such a field. In a motor P_e is the *input power*; in a generator the VI portion of P_e is the *output power*. The ratio of P_m and P_e is the machine power efficiency.

The difference between the input and output consists of various components of the internal machine losses, described with the aid of the energy-flow diagram in Fig. 6.25. In this diagram, power (or energy) flow can be in either direction, depending on whether the machine is being operated as a motor or as a generator. Many of the loss components are typical of any electromagnetic rotating machine, but the descriptions given next are in terms of the dc commutator machine.

1. *Mechanical Losses.* There are three components of the mechanical loss in a dc commutator machine.
   (a) **Bearing Friction.** This component varies with the type and quality of bearing used in the machine, with the torque loading (axial and radial) on

the bearing, and with the speed of rotation. An approximate relationship of the loss as a function of machine rotor speed, however, is given by

$$P_b = K_b \times 10^{-6}(\text{rpm})^{5/3} \quad \text{W} \tag{6.20}$$

where K_b is a constant, depending on bearing size, type, and method of lubrication. For two similar bearings per machine and oil mist lubrication, K_b is between 1.1 and 1.7.

(*b*) **Windage.** This component is due to the air motion in the airgap and interpolar regions caused by the rotation of the armature. A useful formula for estimating the windage loss of a radial airgap machine, assuming turbulent flow, is

$$P_w = \pi C_d \rho R^4 \omega_m^3 l \quad \text{W} \tag{6.21}$$

where R and l are the radius and length, respectively, of the cylindrical rotor, ρ is the air density, and C_d is a skin-friction coefficient evaluated from

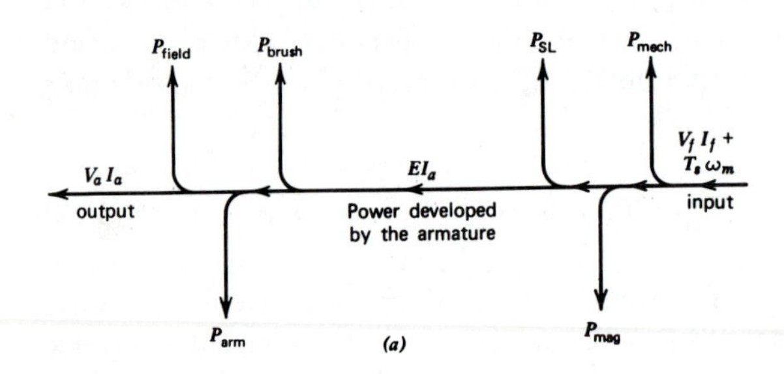

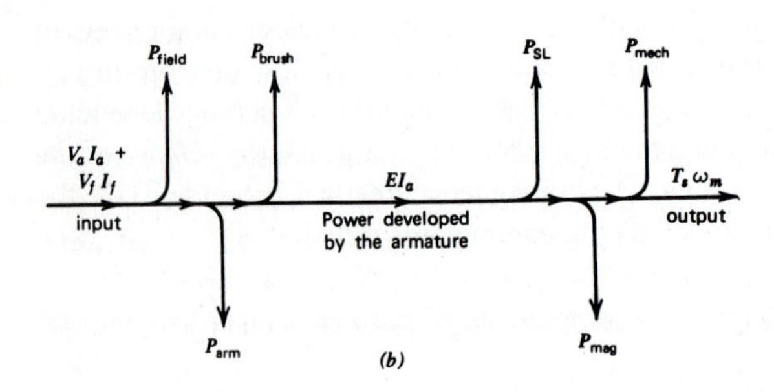

FIGURE 6-25
Power flow in a dc commutator machine. (*a*) Generator. (*b*) Motor.

$$\frac{1}{\sqrt{C_d}} = 2.04 + 1.768 \ln\left(\mathrm{Re}\sqrt{C_d}\right) \tag{6.22}$$

where Re is the Reynolds number for the radial airgap.

(c) **Brush Friction.** This component is unique to rotating machines which require a commutator/brush system. It is often the largest component of mechanical loss and is one of the undesirable features of this class of machines.

2. *Magnetic Losses or Core Losses.* These losses occur in magnetic sections of dc commutator machines in which there is a time-varying magnetic field. The nature of these losses has been described in Chapter 2. Magnetic losses occur primarily in the armature of dc commutator machines, since the magnetic field in the pole and yoke sections is relatively constant. The armature magnetic material is continuously rotating through the main field flux and armature field flux and therefore is usually constructed of laminations to minimize eddy currents. Pole and yoke sections are often constructed of solid magnetic material. However, there is a source of eddy currents and associated loss located in the regions of the poles adjacent to the airgap. This area is called the "pole face."

3. *Winding Losses.* These losses are often called copper losses. They are ohmic losses in the armature, field, interpole, and compensating windings. These losses are readily calculated if the winding resistances are known and can be measured by the test methods discussed in Section 6.9. The armature, interpole, and compensating pole resistances are usually categorized together as the total armature circuit resistance.

4. *Brush Electrical Loss.* The carbon graphite material used as the principal material in brushes has a different resistance characteristic than that of metallic conductors, a negative-resistance temperature characteristic similar to that of a semiconductor. Also, the brush loss results in part from the contact resistance between brush and commutator, which also cannot be described as a conventional ohmic resistance. For these reasons the brush voltage drop and power loss are often treated separately from the armature circuit voltage and loss. The brush loss depends on many factors, including the composition of the brushes, contact pressure, temperature of the brushes, and condition of the commutator surface. For pure graphite brushes, it is often assumed that the voltage across the brush contact resistance is constant as a function of armature current with a value of 2.0 V. The brush loss is therefore $2I_a$ W. However, in practice, the voltage-drop value varies considerably with the composition of the brushes and must be determined for each machine. In small machines the brush drop and loss are usually included with the total armature circuit parameters.

5. *Stray-Load Loss.* This is a somewhat anomalous loss used to account for loss increase caused by loading the machine. It is attributable mainly to an increase in magnetic losses caused by the changes in flux distribution due to the armature

reaction field. This loss is more pronounced in large machines. The stray-load loss is difficult to measure accurately; a value of 1% of the machine output in large machines is often assigned to account for this loss.

6.9 EXPERIMENTAL DETERMINATION OF MACHINE PARAMETERS

To predict the performance of a specific machine or to match the capabilities of a machine to a certain application, it is necessary to know the values of some of the machine's internal parameters. Of particular importance, especially in control applications, are the inductance and resistance of the armature and of field circuits. These can generally be obtained by conventional bridge methods or voltage-current measurements as used for any other type of circuit. It is often desirable to measure resistance and inductance of the armature circuit exclusive of the brushes. This can be done by locating the instrument probes underneath the brushes on the commutator circuit.

The armature circuit resistance is also frequently obtained from a standard machine test known as the *blocked rotor test.* In this test the machine's rotor is mechanically blocked and prevented from rotating by fastening a brake arm to the rotor shaft. A scale can then be used to measure the torque developed in the machine. In a shunt or separately excited machine, the field should be energized from a source of variable voltage in order to vary the field current. The armature circuit is supplied from a source of relatively low voltage, since the only impedance in the armature circuit at standstill is the armature circuit resistance.

Another useful machine test is the *no-load test.* The machine to be evaluated is driven as a generator from an external source at a constant speed and at zero armature current (armature circuit open-circuited). The measurement of armature voltage as a function of field current produces the magnetization curve data, as shown in Fig. 6.19. The torque constant, k_t, can also be obtained from this data, as seen from (6.17). The product of the input torque (in newton-meters) and the shaft speed (in radians per second) is the input power supplied to the test machine. The product is a measure of the no-load loss of the machine at the particular values of speed and field current. The no-load loss is the sum of mechanical losses and magnetic losses. The no-load test is repeated for several values of speed that encompass the operating speed range of the test machine. The resulting family of curves describes the variation of mechanical and magnetic losses over the range of speeds and excitations (field currents) used in the test.

In some series machines and many low-power machines, the field winding terminals are not accessible, and the test, as just described, cannot be performed. Also, many laboratories and test facilities are not equipped with an appropriate drive machine to operate the test machine as a generator. In situations where the no-load test cannot be performed, some of the same information can be obtained by operating the test machine as an unloaded motor. Magnetization curves can be obtained by varying the armature voltage as field current is varied in order to maintain constant speed and measuring armature current and voltage, field current, and speed.

The range of field current values that can be obtained at a constant speed in this manner is smaller than for the no-load generator method. Since the motor is unloaded, the armature current will be small and the effects of armature reaction are negligible. The induced voltage, E, is found by subtracting the small armature resistance voltage drop from the measured armature terminal voltage. The no-load loss is found by subtracting the armature I^2R loss from the armature power input, which is equal to the product of armature current and armature terminal voltage. Care must be exercised, when operating a motor with no external load, that the maximum safe operating speed of the motor is not exceeded during testing at low values of field current.

The loss components described in Section 6.8 can be determined by variations of the no-load and other tests.

1. *Mechanical loss* of a test machine can be determined by driving the test machine unloaded and unexcited from an external machine whose output is known or calculable. Mechanical loss is usually determined over the range of speeds for which the test machine is to be operated.

2. *Magnetic losses* are determined by repeating the test of step 1 with the test machine unloaded but excited over the range of excitation of interest. Magnetic loss is the difference between the test machine input between this test and the previous test of step 1.

3. *Ohmic losses* are determined from resistance measurements or from the blocked rotor test. Brush losses are often separated from armature winding loss by calculation or measurement of the brush and contact resistance.

4. *Stray-load loss* is determined by driving the test machine over a range of speeds of interest with the armature shorted and with excitation at a low value—large enough to cause rated armature current (or other specified values) to flow. Stray-load loss is the difference between test machine input in this test and the sum of ohmic and mechanical losses at appropriate speed and current levels measured in previous steps.

6.10 PREDICTION OF PERFORMANCE CHARACTERISTICS

The data obtained from the preceding tests can be used to predict the performance of the machine for either motor or generator operation under almost any operating condition. This process is made simpler if the results are presented graphically.

a. Mechanical losses versus speed.

b. Magnetic loss and induced armature voltage versus speed for the chosen values of field current (a family of curves).

c. Stray-load loss (if significant) versus armature current for the chosen values of speed (a family of curves).

In using these data it will often be necessary to interpolate among the curves for magnetic loss, induced voltage, and stray-load loss.

We will illustrate the method for a separately excited motor. The power-loss diagram of Fig. 6.25 will help to keep track of the loss components during the power summations used in the method. To start the performance prediction, it is necessary to assume either the power input or output. This choice will generally be made on the basis of how the machine is used and controlled. For a motor with both field and armature control, it is common practice to assume the output parameters.

1. Assume output torque T_s and shaft speed ω_m; the product of T_s (in newton-meters) and ω_m (in radians per second) gives the mechanical output power.

2. For the particular speed assumed, ω_m, look up the mechanical loss, P_{mech}, from the curve of item a.

3. Assume a value of field current I_f. Since the method is going to be repeated for a number of different values of field current, it is not too critical which value is assumed. However, the field current should be within the range of field currents used in the test data and one that is expected to be used during the actual operation of the machine. For the assumed values of field current and speed, look up the induced voltage, E, and the magnetic loss, P_{mag}, from the curves of item b.

4. Estimate or guess a value of stray-load loss, P_{SL}, from the curves of item c.

5. The electromagnetic or developed power of the machine is

$$P_e = T_s \omega_m + P_{\text{mech}} + P_{\text{mag}} + P_{\text{SL}} \tag{6.23}$$

6. The armature current for the assumed output and field current is

$$I_a = \frac{P_e}{E} \tag{6.24}$$

7. At this point, an iterative step can be inserted for accuracy. Now that the armature current is known, the stray-load loss can be obtained from the curve of item c. Look up this value and compare it with the estimate made in step 4. If the value is considerably different, insert the new value into (6.23) and determine a new value of armature current in (6.24). This process can be repeated several times until the values of P_{SL} and I_a are consistent with the measured data in item c.

8. Calculate the armature circuit loss, P_{Cu}, and the brush loss, P_b (if separately treated).

9. Calculate the field winding loss, P_f.

10. The motor efficiency is

$$\eta = \frac{T_s \omega_m}{T_s \omega_m + P_{\text{mech}} + P_{\text{mag}} + P_{\text{SL}} + P_{Cu} + P_b + P_f} \tag{6.25}$$

11. The required power input for the assumed output conditions is, of course, the denominator of (6.25).

12. The required armature terminal voltage, neglecting armature reaction, is

$$V = E + I_a R_a \tag{6.26}$$

where R_a has been corrected for an assumed temperature rise and includes brush drop. Brush drop may be handled independently, as discussed in Section 6.8.

13. It is possible to include the demagnetizing effects of armature reaction instead of using (6.26) if the demagnetizing armature ampere-turns are known. This is a difficult parameter to define numerically, since it varies with saturation in the machine's magnetic circuit. Because demagnetizing effects of armature reaction can be obtained only through extensive load testing, this parameter tends to defeat the intent of the technique being described, which is aimed at predicting performance with the use of only no-load and short-circuit test data. Therefore, if armature reaction demagnetizing effects are to be illustrated along with the method under discussion, estimated values of this parameter should be used. Otherwise, complete load testing can be used.

To introduce the effects of armature reaction into this method, the abscissa of the curves of item b must be expressed in ampere-turns. Then in step 3, the assumed field current can be expressed as an assumed value of ampere-turns. Add to this the assumed value of armature reaction demagnetizing ampere-turns, *AR*. Look up a new value of E from the curves of item b, which we will call E'. Use E' in (6.26) to determine V. Note that E is still used to determine magnetic losses in step 3 and armature current in step 6, since the magnetic loss and induced voltage are a function of flux (not mmf). However, the true field current required now must be

$$I_f' = \frac{N_f I_f + AR}{N_f} \tag{6.27}$$

This new value of field current, I_f', must be used in calculating the field winding loss in step 9.

14. Repeat steps 3 to 13 for other values of field current (or field ampere-turns, if armature reaction is to be included).

15. Repeat steps 1 to 13 for other values of output torque and speed.

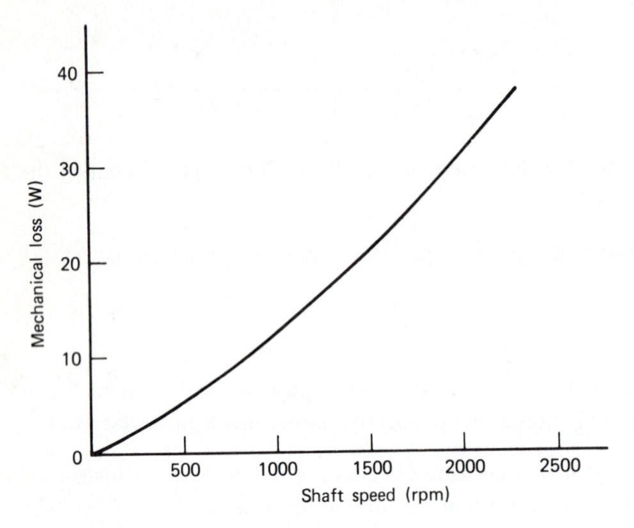

FIGURE 6-26
Mechanical loss for a 1-hp machine.

This simple performance prediction technique has taken many words to describe but actually requires a relatively short time to calculate, even by hand if the curves of items a to c are readily available. However, it is still desirable to program the preceding process on a computer whenever feasible, since it is readily amenable to computerized calculation.

Example 6.5 A separately excited motor is rated 1 hp, 220 V, 4.6 A, 1750 rpm, and 350 mA field current; the armature resistance is 4.8 Ω, including temperature correction; and the field resistance is 630 Ω. Laboratory tests have resulted in the loss information and magnetization curve shown in Figs. 6.26 to 6.28. Determine the required input armature voltage, input power, and efficiency when the motor is operated to supply rated torque at rated speed.

Solution

The rated torque, although not required for the solutions to this problem, can be found as 3.0 lb-ft or 4.06 N-m. The power output in watts is 746. Follow the procedure outlined in steps 1 to 15. From Fig. 6.26 (item a), the mechanical loss at rated speed is 26 W. From Fig. 6.27 (item b), the magnetic loss at rated field current of 350 mA is 22 W. From Fig. 6.28 (item c), assume that the armature current will be a rated value, or 4.6 A, giving a stray-load loss at a rated speed of 17 W. The electromagnetic power is therefore 746 + 26 + 22 + 17 = 811 W. From Fig. 6.27, the induced voltage at 350 mA is 192 V, giving an armature current of 811/192 = 4.25 A. This is close enough to the assumed value used in determining the stray-load loss so that an iteration is not required. The armature voltage drop is 4.25 x 4.8 = 20.4 V; the armature copper loss is 4.25^2 x 4.8 = 87 W; and the field copper loss is 0.35^2 x 630 = 77 W. Power input is 811 + 87 + 77 = 975 W. Efficiency is 746/975 = 0.77 or 77%.

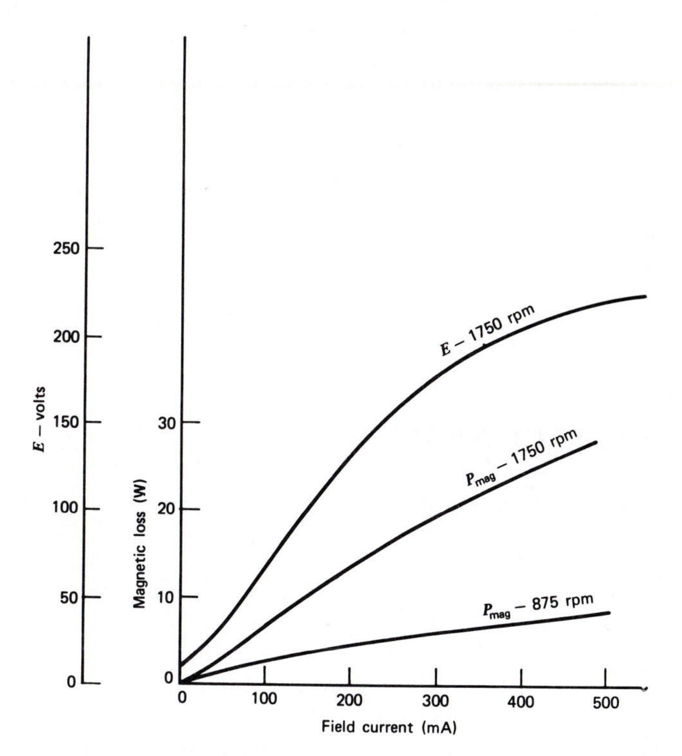

FIGURE 6-27
Magnetic loss and induced voltage for a 1-hp machine.

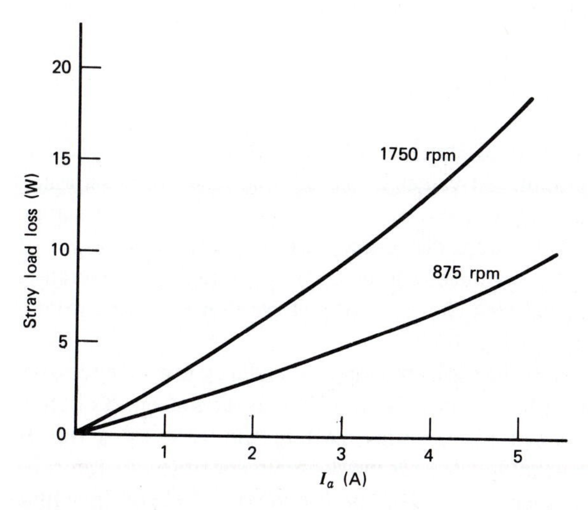

FIGURE 6-28
Stray-load loss for a 1-hp machine.

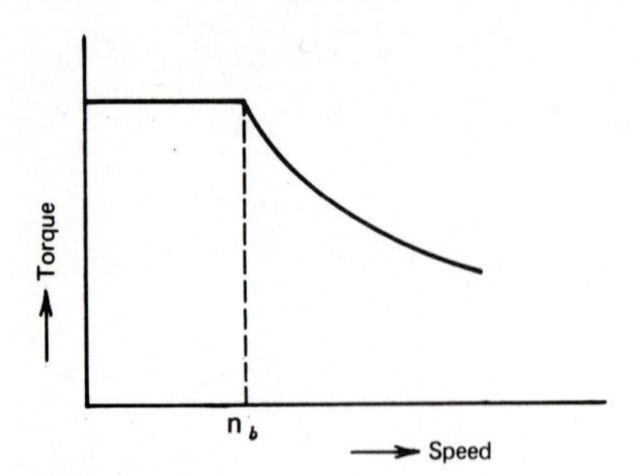

FIGURE 6-29
Speed-torque characteristic of a series motor with constant applied voltage.

6.11 DC MACHINE CHARACTERISTICS

In the preceding section we have discussed the method of predicting the machine characteristics. We now consider the characteristics of commonly used dc machines.

Series Motors

The series motor has been the principal electrical traction device for many years and is still widely used in all types of electric vehicles, electric trains, streetcars, industrial overhead cranes, and automotive starter motors. The highest torque per unit of input current can be achieved by the series motor. This characteristic is evidenced from an observation of the torque equation (6.10).

If the field winding is connected in series with the armature winding, the field and armature currents become identical, and the torque equation can be expressed as

$$T = K_m I_a^2 \tag{6.28}$$

where K_m is a new constant related to the armature constant of (6.10) and the characteristics of the magnetic circuit. At values of current that result in unsaturated conditions of the magnetic circuit, the torque of a series motor varies with the square of the armature current. Above saturation, the torque varies with the first power of armature current. Equation (6.28) is applicable in the unsaturated region of operation of the magnetic circuit. The general characteristics of a series motor excited from a constant voltage source are shown in Fig. 6.29.

An undesirable characteristic of the series motor is its tendency toward excessive speeds at light loads. This can best be observed from (6.14). If the machine is lightly loaded, the armature current is small. In a series machine the series field flux is therefore small, which results in a small value for the denominator of (6.14) and calls for a large value of mechanical speed, ω_m. This is a common cause of "machine runaways" in the typical college laboratory.

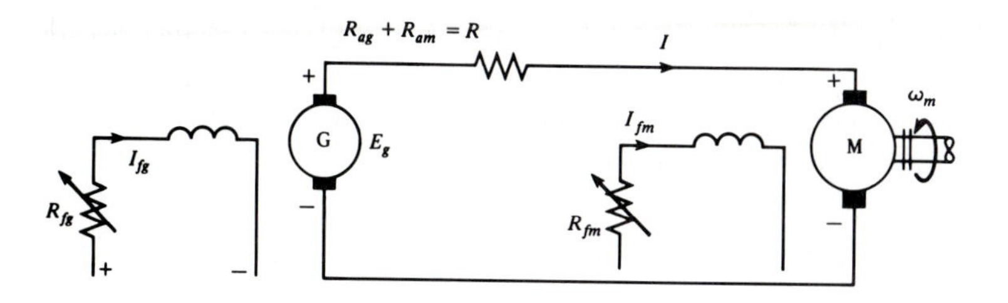

FIGURE 6-30
Schematic diagram of a Ward-Leonard system.

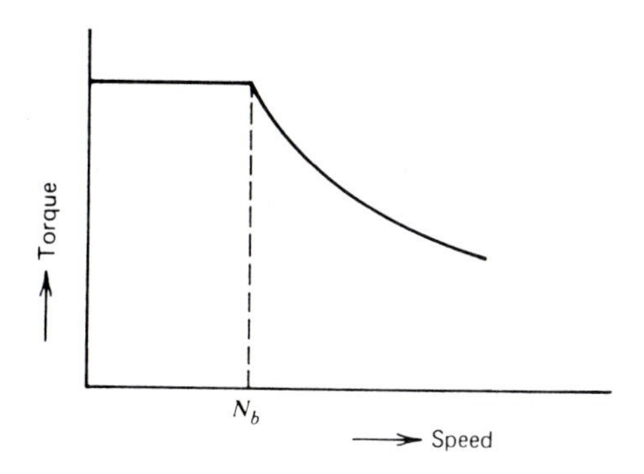

FIGURE 6-31
Torque-speed characteristics of a traction motor.

Separately Excited Motors

The separately excited machine has the capability of both independent armature control and field control, which—in terms of the operating parameters of a machine—imply independent speed and torque control. An early approach to exploiting this characteristic was the Ward-Leonard system, which is still widely used in high-power applications requiring variable speed and variable torque control, such as in steel and aluminum rolling mills. A schematic representation of the Ward-Leonard system is shown in Fig. 6.30. The separately excited configuration is especially adaptable to feedback control. For example, the effects of demagnetizing armature reaction and armature and brush resistance drop can be compensated for by increased field excitation.

A common speed-torque characteristic used in many variable-speed applications is shown in Fig. 6.31. This characteristic consists of two sections: a constant torque section at low speeds up to the "base speed," n_b, and a constant power section at speeds above the base speed. The separately excited machine is frequently operated in this mode using Ward-Leonard or various types of electronic control systems. In the constant-torque region, operation is usually at a constant value of field excitation (the maximum excitation in this region determines the maximum torque capability). Control of torque is achieved by controlling the armature voltage (or current).

In the constant-power regions at higher speeds, armature current is held constant and torque and speed are varied by varying the field current, or by "field weakening." These two modes of operation are understood by reference to (6.8). For example, constant electromagnetic power implies

$$P = EI_a = K_t I_f \omega_m I_a = \text{constant} \tag{6.29}$$

If the armature current is held constant in (6.29), the product, $I_f\omega_m$, must also be maintained constant, indicating an inverse relationship between speed and field current. This is the actual situation in the constant-power region of Fig. 6.31 and results in the parabolic characteristic above the base speed, since torque decreases with I_f (or inversely with ω_m). The speed, torque, and required excitation at the base speed determine the size of a motor designed to meet this type of characteristic. Traction motors are usually rated in terms of conditions at base speed. Since a motor rating is based on *continuous* operation, other ratings are of interest in traction motor applications due to the widely variable power requirements of traction applications. Typical traction motors have intermittent or "1-minute" torque and current ratings that are from three to five times the continuous rating.

Separately Excited Generators

The principal applications of separately excited generators today are as excitation sources for large synchronous alternators in power-generating stations, as the control generator in Ward-Leonard systems, and as auxiliary and emergency power supplies. The separately excited generator has all of the control flexibility of the separately excited motor. One important characteristic used in specifying a generator and its excitation control is *voltage regulation.* Voltage regulation is a measure of the variation in output voltage of a generator with load and is defined as

$$\text{percent regulation} = \frac{V(\text{no-load}) - V(\text{load})}{V(\text{load})} \times 100 \tag{6.30}$$

In a separately excited generator, load is synonymous with armature current.

Voltage regulation is usually specified in terms of the "full-load" or rated-load condition, although (6.30) is applicable to any other load. The variation of voltage with load in a separately excited generator arises from two causes: armature resistance voltage drop and the demagnetizing effects of armature reaction. With a fixed value of field excitation, a typical voltage characteristic might resemble that shown in Fig. 6.32. The drop in voltage with increasing load can be eliminated by appropriately increasing the field excitation. This is usually done by means of a feedback control system using armature voltage as the feedback signal.

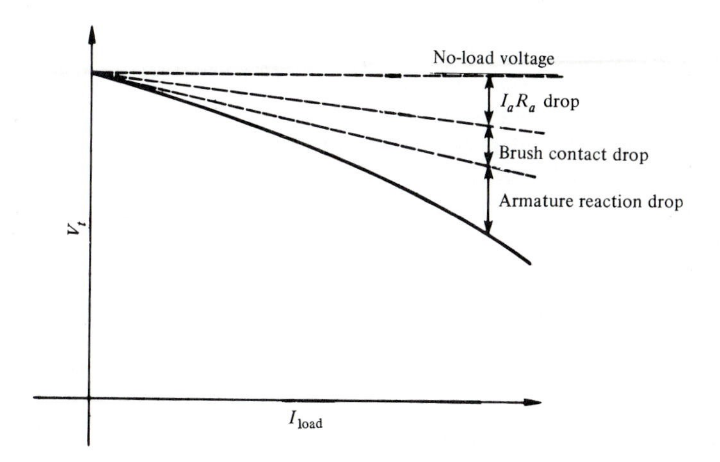

FIGURE 6-32
Load characteristic of a separately excited generator.

The main causes of the voltage drop in generators are

1. *Armature Resistance Drop.* This is an I_aR_a drop due to the resistance of the armature.

2. *Brush Contact Drop.* The mechanical contact between the brushes and the commutator offers an electrical resistance. Consequently, when a current flows through the brush, a voltage drop occurs. Usually, this voltage drop is taken as a constant (of 2 V).

3. *Armature Reaction Voltage Drop.* From Section 6.6 we recall that armature reaction has a demagnetizing component, which opposes the main field mmf, resulting in reduction of flux. The reduced flux will, in turn, reduce the armature induced emf and hence the terminal voltage.

4. Cumulative effects of 1 to 3 in self-excited (shunt, series, and compound) generators further lower the terminal voltage.

In view of the above, let us refer to Fig. 6.32, which shows the load characteristic of a separately excited dc generator. Notice that the terminal voltage on load differs from the no-load voltage by the three voltage drops mentioned in 1 to 3.

Self-Excited Shunt Machines

The shunt machine is shown in Fig. 6.14*(b)*. This configuration is used primarily in low-power, fixed-speed applications, where additional circuitry is inserted into the field circuit. This configuration becomes a variation of the separately excited motor configuration [Fig. 6.14*(a)*]. As a generator, the shunt configuration is known as a self-excited generator and has an interesting characteristic known as voltage build-up. This phenomenon is the process of the armature supplying its own field excitation and is useful in generator applications where no external source of excitation is available, such as on isolated farms or in camping areas.

Voltage build-up requires that some residual magnetism exist in the magnetic circuit of the generator to get the process started. Also, the field circuit must be connected to the armature circuit in such a manner that, for the given direction of armature rotation, the armature voltage causes a current to flow in the field circuit in a particular direction. The direction of current flow is such that the magnetic field resulting from this current *aids* the residual magnetism. The process may be observed with the aid of Fig. 6.33.

The magnetization curve of a typical machine is shown along with a "field resistance line," which is merely the plot of voltage versus current for the resistance of the field circuit of the self-excited shunt generator. Assuming that the resistance of the shunt field is linear, this plot is a straight line. The build-up process is as follows. Assume that the field circuit is initially disconnected from the armature circuit. The armature rotates at a given speed, which results in the residual induced voltage E_r. When the shunt field circuit is connected to the armature circuit, there is initially zero current in the field circuit as a result of the effect of the field circuit inductance.

The voltage, E_r, appearing across the armature circuit eventually causes the current I_1 to flow in the field circuit. The rate of rise of I_1 depends on the time constant of the field circuit. However, with I_1 exciting the field, the armature voltage—according to the magnetization curve—builds up to E_1. The increased armature voltage eventually increases the field current to I_2, which, in turn, builds up the armature voltage to E_2.

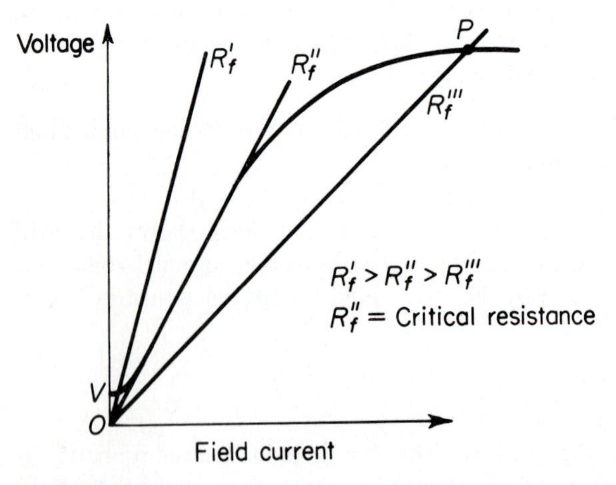

FIGURE 6-33
Voltage build-up in a self-excited generator.

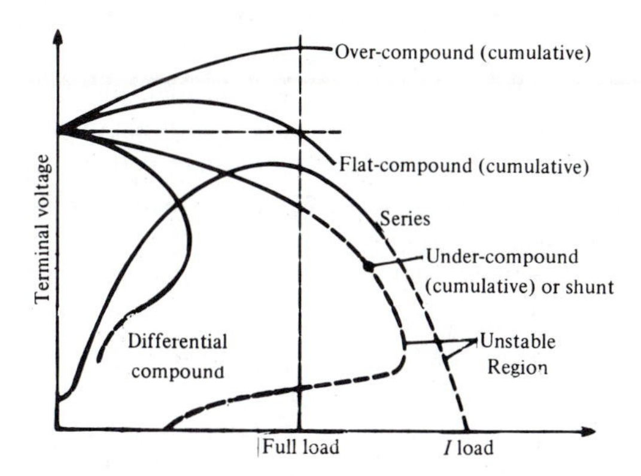

FIGURE 6-34
Load characteristics of dc generators.

This process is continued until the magnetization curve and resistance line cross, at which point voltage build-up ceases. Now stable operation as a shunt generator supplying load current can continue. Note that as the resistance of the field circuit is increased, the field resistance line in Fig. 6.33 moves counterclockwise. At some value of resistance, the resistance line is approximately coincident with the linear portion of the magnetization curve. This results in an unstable voltage condition as far as supplying power is concerned, and the corresponding value of the field circuit resistance is known as the *critical field* resistance. To summarize, the conditions for voltage build-up in a shunt generator are the presence of residual flux, field-circuit resistance less than the circuit resistance, and appropriate relative polarity of the field winding, which produces a flux that aids the residual flux.

No-load and load characteristics of dc generators are usually of interest in determining their potential applications. Of the two, load characteristics are of greater importance. As the names imply, no-load and load characteristics correspond, respectively, to the behavior of the machine when it is supplying no power (open-circuited, in the case of a generator) and when it is supplying power to an external circuit.

The only no-load (or open-circuit) characteristics that are meaningful are those of the shunt and separately excited generators. We have discussed above the no-load characteristic of a shunt generator as a voltage build-up process. The load characteristics of self-excited generators are shown in Fig. 6.34. The shunt generator has a characteristic similar to that of a separately excited generator, except for the cumulative effect mentioned earlier. If the shunt generator is loaded beyond a certain point, it breaks down in that the terminal voltage collapses. In a series generator, the load current flows through the field winding. This implies that the field flux, and hence the induced emf, increases with the load until the core begins to saturate magnetically. A load beyond a certain point would also result in a collapse of the terminal voltage of the series generator. Compound generators have combined characteristics of shunt and series generators. In a differential compound generator, shunt and series fields are in opposition. Hence the terminal voltage drops very rapidly with the load. On the other hand, cumulative compound generators have shunt and series fields aiding each other. The two field mmf's could be adjusted such that the terminal voltage on full load is less than the no-load voltage, as in an under-compound generator, or the full load voltage may be equal to the no-load voltage, as in a flat-compound generator. Finally, the terminal voltage on full load may be greater than the no-load voltage, as in an over-compound generator.

As motors, shunt machines are used in industrial and automotive applications where precise control of speed and torque are not required.

The speed torque relationship for a typical shunt motor is shown in Fig. 6.35. The declining speed versus torque characteristic is caused by the armature resistance voltage drop and armature reaction. At some value of torque, usually a value about 2.5 times the rated torque of the motor, armature reaction becomes excessive, causing a rapid decrease in field flux and a rapid decline in developed torque down to a stall condition. This abrupt drop in developed torque, which often leads to unstable operation at high overloads, can be smoothed out considerably by adding a series field winding connected in series with the armature circuit. Such a configuration is called a stabilizing winding although, contrary to its name, it results in a greater variation in speed in the normal operating torque range. This characteristic is also shown in Fig. 6.35. Such a configuration is termed a *compound motor*. Shunt and compound motors are used where a reasonably constant speed versus torque characteristic is required and the cost of field control is not justified.

Example 6.6 A 50-kW, 250-V, short-shunt compound generator has the following data: $R_a = 0.06\ \Omega$ and $R_f = 125\ \Omega$. Calculate the induced armature emf at rated load and terminal voltage. Take 2 V as the total brush-contact drop.

Solution

The equivalent circuit of the generator is shown in Fig. 6.36, from which

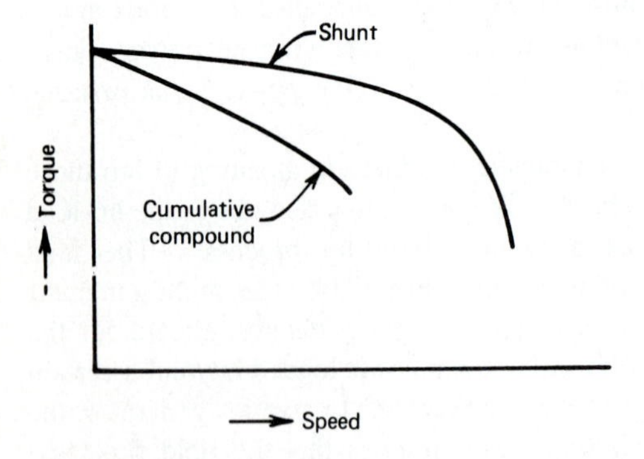

FIGURE 6-35
Torque-speed characteristics of shunt and compound motors.

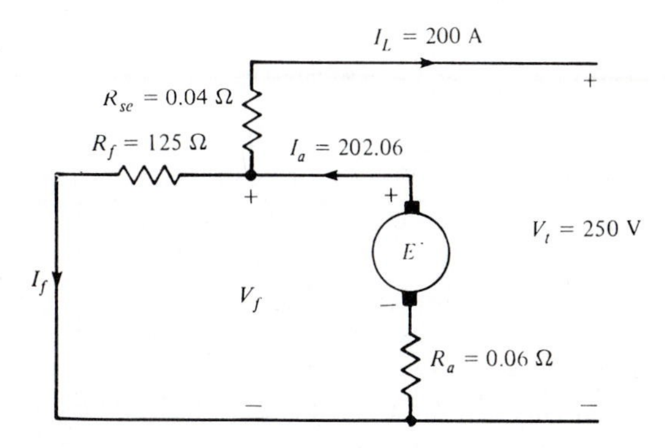

FIGURE 6-36
Example 6.6.

$$I_l = \frac{50 \times 10^3}{250} = 200 \text{ A}$$

$$I_l R_{se} = (200)(0.04) = 8 \text{ V}$$

$$V_f = 250 + 8 = 258 \text{ V}$$

$$I_f = \frac{258}{125} = 2.06 \text{ A}$$

$$I_a = 200 + 2.06 = 202.06 \text{ A}$$

$$I_a R_a = (202.06)(0.06) = 12.12 \text{ V}$$

$$E = 250 + 12.12 + 8 + 2 = 272.12 \text{ V}$$

6.12 STARTING AND CONTROL OF DC MOTORS

In addition to certain operational conveniences, the basic requirements for satisfactory starting of a dc motor are (1) sufficient starting torque, and (2) armature current, within safe limits, for successful commutation and for preventing the armature from overheating. The second requirement is obvious from the speed equation, according to which the armature current, I_a, is given by $I_a = V/R_a$ when the motor is at rest (n or $\omega_m = 0$). A typical 50-hp, 230-V motor having an armature resistance of 0.05 Ω, if connected across 230 V, draws 4600 A of current. This current is evidently too large for the motor, which might be rated to take 180 A on full load. Commonly, double the full-load current is allowed to flow through the armature at the time of starting. For the motor under consideration, therefore, an external resistance

$$R_{ext} = \left(\frac{230}{2 \times 180}\right) - 0.05 = 0.59\ \Omega$$

must be inserted in series with the armature to limit I_a within double the rated value.

From the speed equation of a dc motor, it follows that the speed of the motor can be varied by varying (1) the field-circuit resistance to control I_f and hence the field flux, (2) the armature circuit resistance, and (3) the terminal voltage. Let us now consider the scope of each of these three methods. In method 1, an external variable resistance is connected in the field circuit. When this resistance is set at zero, the field current is limited only by the field resistance. Corresponding to this field current, we obtain a motor speed n_1. Now, as the externally inserted resistance increases, the field current decreases and the motor speed increases (in accordance with the speed equation) to a corresponding speed n_2, where $n_2 > n_1$. In other words, by using an external resistance in series with the motor field, we can only increase the motor speed (from a minimum speed n_1). By inserting a resistance in series with the armature, as in method 2, we can reduce the voltage across the armature and hence decrease the motor speed. Because the armature current is of a relatively large magnitude, compared to the field current, method 2 is a wasteful method (as shown by Example 6.8). Method 3 can be efficiently used either to increase or decrease the speed of the motor, but is feasible only if a variable voltage source is available. Method 3 in conjunction with method 1 constitutes the *Ward-Leonard* system, which is capable of providing a wide variation of speed in both forward and reverse directions. This method is presented in a simplified form in Example 6.9.

Example 6.7 A 230-V shunt motor has an armature resistance of 0.05 Ω and a field resistance of 75 Ω. The motor draws 7 A of line current while running light at 1120 rpm. The line current at a certain load is 46 A. *(a)* What is the motor speed at this load? *(b)* At this load, if the field-circuit resistance is increased to 100 Ω, what is the new speed of the motor? Assume the line current to remain unchanged.

Solution

(a) On no-load,

$$n_0 = 1120 \text{ rpm (given)}$$

$$I_f = \frac{230}{75} = 3.07 \text{ A}$$

$$I_a = 7 - 3.07 = 3.93 \text{ A}$$

The speed equation gives

$$1120 = \frac{230 - 3.93 \times 0.05}{3.07k}$$

or $k = 0.0668$.

On load (with $R_f = 75\ \Omega$):

$I_f = 3.07$ A

$I_a = 46 - 3.07 = 42.93$ A

$$n = \frac{230 - 42.93 \times 0.05}{3.07 \times 0.0668} = 1111 \text{ rpm}$$

(b) On load (with $R_f = 100\ \Omega$):

$$I_f = \frac{230}{100} = 2.3 \text{ A}$$

$I_a = 46 - 2.3 = 43.7$ A

$$n = \frac{230 - 43.7 \times 0.05}{2.3 \times 0.0668} = 1483 \text{ rpm}$$

Example 6.8 Refer to part *(a)* of Example 6.7. The no-load conditions remain unchanged. On load, the line current remains at 46 A, but a 0.1 Ω resistance is inserted in the armature. Determine the speed of the motor and the power dissipated in the 0.1 Ω resistance.

Solution

In this case we have (from Example 6.7)

$I_f = 3.07$ A

$I_a = 42.93$ A

$k = 0.0668$

Thus, with $R_a = 0.05 + 0.1 = 0.15\ \Omega$,

$$I_a^2(0.1) = 42.93^2 \times 0.1 = 184.3 \text{ W}$$

$$n = \frac{230 - 42.93 \times 0.15}{3.07 \times 0.0668} = 1090 \text{ rpm}$$

Example 6.9 The system shown in Fig. 6.30 is called the *Ward-Leonard* system for controlling the speed of a dc motor. Discuss the effects of varying R_{fg} and R_{fm} on the motor speed.

Solution

Increasing R_{fg} decreases I_{fg} and hence E_g. Thus the motor speed will decrease. The opposite will be true if R_{fg} is decreased.

Increasing R_{fm} will increase the speed of the motor, as shown in Example 6.7. Decreasing R_{fm} will result in a decrease of the speed.

6.13 CONTROL MOTORS

Control motors are motors of relatively low power rating (usually less than a few hundred watts) with fixed-field excitation. Fixed-field excitation adds another variation to the basic torque equation (6.10):

$$T = K_c I_a \tag{6.31}$$

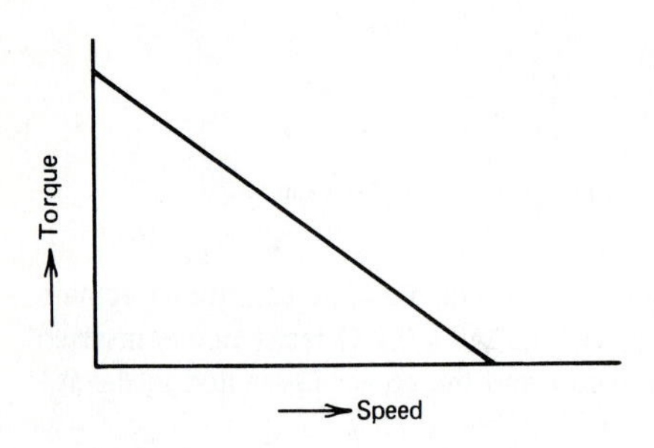

FIGURE 6-37
Torque-speed characteristic of a control motor.

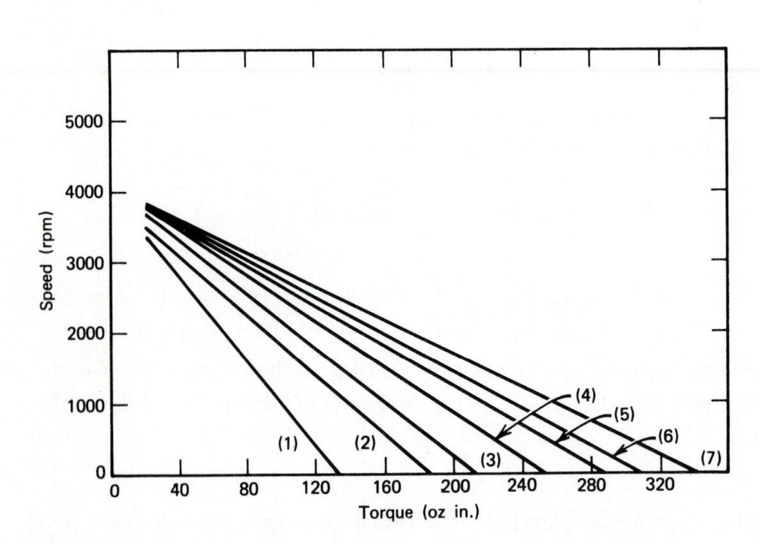

FIGURE 6-38
Speed-torque characteristics of a series of 12-V permanent magnet motors.

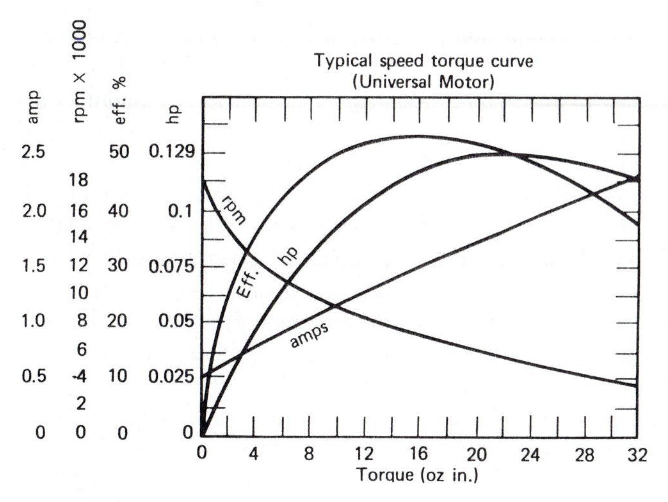

FIGURE 6-39
Characteristics of a universal motor, rated 1/8 hp and 115 V at 60 Hz. (Courtesy The Dunmore Co., Racine, Wisc.)

where K_c is a new constant in newton-meters per ampere. Control motors have relatively high armature current to provide a linear relationship between voltage and torque, which is the basis for applications as a control device. The desired speed-torque relationship of a control motor can be shown to be a degeneration of the shunt motor torque characteristic illustrated in Fig. 6.35. The shunt motor torque characteristic results from a large increase in armature resistance, as shown in Fig. 6.37. Figs. 6.37 and 6.38 illustrate the physical characteristics of several types of control motors. The fixed excitation is obtained by means of either permanent magnets or wound field.

An important parameter in control motors is the inertia of the rotating element, which generally should be minimized to permit fast response to step changes in the input control function. A useful parameter in evaluating control motors is the *torque/inertia* ratio, which is usually calculated using the motor stall torque (the torque at which the speed-torque characteristic meets the ordinate in Fig. 6.37).

6.14 UNIVERSAL MOTORS

A series motor is able to operate on alternating as well as direct current. Since armature and field windings are in series, a reversal of current results in a reversal of flux and, therefore, does not cause a reversal of electromagnetic force or torque, as can be observed from (6.10). Hence there is no change in direction of rotation of the armature. Motors that are designed to operate on either alternating or direct current are known as *universal motors.* Universal motors are widely used in household appliances, drills and other electric tools, electric lawn mowers, and similar applications. Low-frequency (16⅔, 25, or 30 Hz) motors of this type are also used in electric railway applications.

A series motor of a given frame size will generally have a lower power rating on alternating than on direct current. This is due to several causes: higher ohmic losses in the windings because of higher root-mean-square current, higher commutation losses because of the need for higher-resistance brushes to achieve adequate commutation, and the need for a slightly different winding design. A series motor designed for ac excitation is designed with fewer series field turns than a motor of equivalent size designed for dc excitation. Also, because of more severe commutation problems, an ac series motor usually requires a compensating winding (see Section 6.7). The reason for more severe commutation problems on alternating than on direct current is that there is a component of voltage in the coil undergoing commutation in addition to those listed in Section 6.7. This is a voltage due to a "transformer action" voltage induced in the coil caused by the time-changing flux linking the coil. The design of the compensation means to reduce sparking and commutator wear discussed in Section 6.7 is, therefore, more necessary in ac or universal series motors.

Typical characteristics of a 1/8-hp universal motor that can be operated from dc to 60-Hz sources is shown in Fig. 6.39. An interesting characteristic of ac series motors is that the input power factor decreases with increasing load (power output), which is the opposite to the condition for the other common ac motors (induction motors). The verification of this characteristic is left as a problem at the end of this chapter. Also, universal motors can be braked dynamically in the same manner as dc series motors, although additional precautions must be made to ensure that the field flux is not going through a zero crossover at the instant braking is initiated.

Example 6.10 A dc series motor operates at 750 rpm with a line current of 80 A from a 230-V source. The series field is wound for 15 turns/pole and has a total resistance of 0.11 Ω. The *total* armature-circuit resistance is 0.14 Ω. The magnetic circuit characteristic is given by the curve in Appendix III. Neglect armature reaction and determine the motor speed when the line current is 20 A.

Solution

From (6.12) note that motor speed is proportional to the induced voltage, E, and inversely proportional to flux per pole. At 80 A, E_{80} = 230 - 80(0.14 + 0.11) = 210 V (see Fig. 6.40); at 20 A, E_{20} = 230 - 20(0.14 + 0.11) = 225 V. Field excitation is at 80 A, NI_{80} = 15 × 80 = 1200 ampere-turns; at 20 A, NI_{20} = 15 × 20 = 300 ampere-turns; from Appendix III, ϕ_{80} = 4.3 mWb, ϕ_{20} = 1.4 mWb. Therefore the speed at 20 A = 750 (225/210)(4.3/1.4) = 2468 rpm.

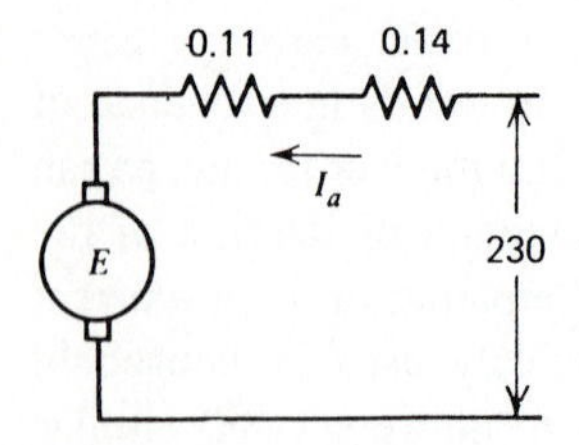

FIGURE 6-40
Equivalent circuit for motor of Example 6.10.

Example 6.11 A separately excited dc machine is rated as a generator at 25 kW, 230 V, 1200 rpm. The *total* armature circuit resistance is 0.12 Ω; the no-load magnetization curve at 1200 rpm is shown in Fig. 6.41. The total resistance of the field circuit is 100 Ω, and there are 1000 turns/pole in the field winding. At 1200 rpm, the mechanical loss is 300 W and the magnetic loss is 200 W at a field excitation of 2500 ampere-turns/pole. Neglect armature reaction and stray-load loss. Determine armature terminal voltage, efficiency, input torque, and voltage regulation when the machine is operated as a generator at rated load and 1200 rpm with an excitation of 1500 ampere-turns/pole.

Solution

The rated armature current is 25,000/200 = 125 A; the electromagnetic voltage is, from Fig. 6.41, $E = 230$ V; the terminal voltage, $V = 230 - 125 \times 0.12 = 215$ V. Power output = 215 × 125 = 26,875 W. The armature ohmic loss = $125^2 \times 0.12 = 1875$ W; field loss = $2.5^2 \times 100 = 625$ W. Power input = 26,875 + 1875 + 625 + 300 + 200 = 29,875 W. Efficiency = 26,875/29,875 = 0.90. Speed is $(1200/60)(2\pi) = 125.66$ rad/s; input torque = (29,875 - 625)/125.66 = 232.77 N-m. If the load were removed, the no-load voltage would be 230 V; voltage regulation is, therefore, from (6.30), (230 - 215) × 100/215 = 7%.

Example 6.12 The machine of Example 6.11 is operated as a separately excited motor with a field excitation of 2000 ampere-turns/pole on the field. When 250 V is applied to the armature, the armature current is 100 A. Determine the motor speed and electromagnetic torque. Neglect armature reaction, magnetic loss, and stray-load loss.

Solution

The induced voltage is $E = 250 - 100 \times 0.12 = 238$ V. However, if the speed were 1200 rpm, Fig. 6.41 shows that E would be 212 V at 2000 ampere-turns. Therefore the motor speed must increase proportionally to match the applied value of E, which is 238 V. This gives the speed as 1200(238/212) = 1347 rpm = 141 rad/s. The electromagnetic torque is (238 × 100)/141 = 168.8 N-m.

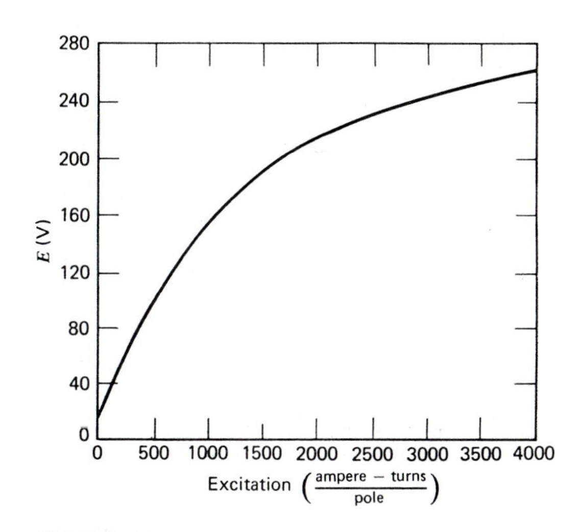

FIGURE 6-41
No-load voltage at 1200 rpm.

Example 6.13 Repeat Example 6.12 if armature reaction results in a demagnetizing effect of 250 ampere-turns/pole.

Solution

From Fig. 6.41, with excitation at 2000 - 250 = 1750 ampere-turns/pole, E = 202 V. The speed therefore increases to 1200 (238/202) = 1414 rpm = 148.1 rad/s. Electromagnetic torque is (238 x 100)/148.1 = 160.7 N-m.

6.15 DC MACHINE DYNAMICS

As in the case of developing steady-state relationships, the dynamic equations will vary somewhat depending on the configuration of the specific machine (series or shunt), the type of voltage sources and loads, and so forth. Most of the following discussion is based on the separately excited configuration illustrated schematically in Fig. 6.42. In terms of setting up the dynamic equations, this configuration is the most general and can be modified for use in other configurations. Lowercase symbols are used to represent instantaneous quantities. Thus, from Fig. 6.42, by summing voltages in the armature circuit, we obtain

$$v_a = e + i_a R_a + L_a \frac{di_a}{dt} \tag{6.32}$$

$$e = K_t i_f \omega_m \tag{6.33}$$

where L_a is armature inductance in henries. Equation (6.32) has been written for motor operation. To describe generator operation, the signs of i_a and di_a/dt are changed. By summing voltages in the field circuit, we obtain

$$v_f = i_f R_f + L_f(i_f) \frac{di_f}{dt} \tag{6.34}$$

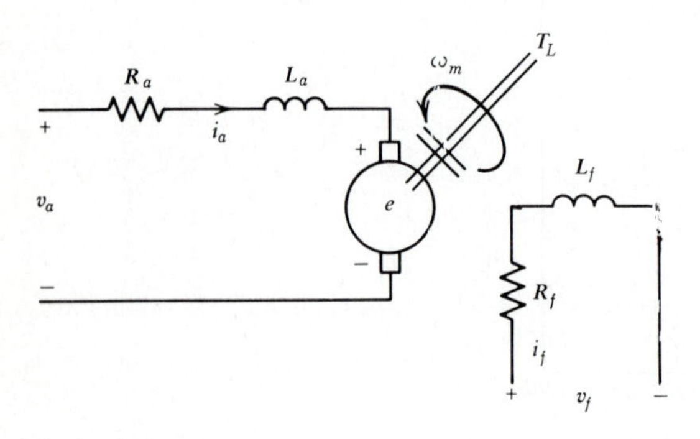

FIGURE 6-42
Schematic diagram for dynamics analysis.

The field circuit inductance, L_f (i_f), is shown as a nonlinear function of i_f to give generality to the set of equations. This nonlinear function is related to the magnetization curve of the machine or the flux versus ampere-turn characteristic of the magnetic circuit of the machine. Summation of torques acting on the motor shaft gives

$$T_d = T_L + D\omega_m + J\frac{d\omega_m}{dt} \tag{6.35}$$

$$T_d = K_t i_f i_a \tag{6.36}$$

where D is a viscous damping coefficient representing rotational loss torque in newton-meter-seconds; J is the moment of inertia of the total rotating system, including machine rotor, load, couplings, and shaft, in kilograms per square meter or N-m-s^2; and T_L is the load torque, in newton-meters.

Equation (6.34) is helpful in understanding the mechanism by which a dc commutator machine is loaded. For this purpose, it is convenient to rearrange (6.35) as follows:

$$T_d - T_L = D\omega_m + J\frac{d\omega_m}{dt} \tag{6.37}$$

Let us consider motor operation. Motor start-up is possible only if the electromagnetic torque, T_d, is larger than the load torque, T_L. At start-up, the speed is zero and the difference between these two torques determines the initial acceleration of the machine. Direct current commutator motors are capable of developing very high starting torques (as the torque at zero speed is called) and are used frequently because of this capability. As long as the left side of (6.37) is larger than the right side, the motor will continue to accelerate.

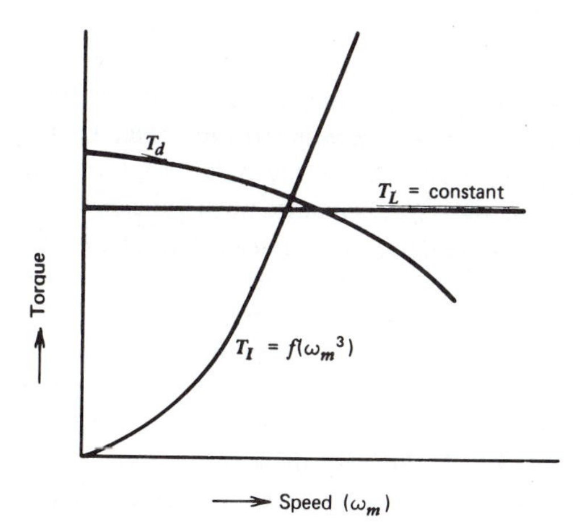

FIGURE 6-43
Motor and load torque characteristics.

An equilibrium condition is reached when T_d - T_L = $D\omega_m$, at which time the acceleration is zero and the motor operates at a constant speed. This situation is shown in Fig. 6.43 for two different types of load torques: constant torque, which might represent a cutting tool or lathe, and a torque varying approximately with cubic power of speed, which might represent a fan load. If the left side of (6.37) becomes negative, which can occur if the load torque is increased or the electromagnetic torque is decreased from their values at the equilibrium speed, the right side must also go negative. This situation is accomplished by means of a negative acceleration, that is, deceleration, causing the speed to decrease to a new equilibrium condition. The process of loading and speed variation is generally a stable process in dc commutator machines, except during conditions of very low field excitation or of excessive armature reaction with low value of load torque, which may result in excessive speed.

At start-up or stall, the back emf of the motor is zero and the armature current is equal to the applied voltage divided by the armature resistance. This condition results in an armature current far in excess of the rated current for the commutator/brush assembly and the armature winding. Such a situation can be tolerated only for a matter of seconds in most machines before permanent damage is done to some armature components. In many machines using armature voltage control, this situation is avoided by means of control circuitry, which includes armature current feedback signals. In machines with simpler control schemes, an armature resistance is usually added at start-up to limit armature current to safe values. Direct current commutator motor starters consist of discrete steps of resistance and have facilities for switching in various values of resistance to control starting current and torque.

Dynamic and Regenerative Braking

A most useful characteristic of dc commutator machines is the capability of a smooth transition from motor to generator operation, and vice versa. In most traction applications, much energy is wasted in the process of braking or decelerating the vehicle. This energy ends up as heat in the brake shoes or brake bands. There are three schemes of eliminating the wear of braking components in traction applications where dc commutator machines are used. One is *dynamic braking*. The kinetic energy of a moving vehicle is dumped into a resistance by means of the generator action of a dc commutator machine In *regenerative braking*, also most valuable as an energy-saving scheme, the kinetic energy of the vehicle is stored in a battery, flywheel, or other type of energy storage system by means of the generator action of the machine. A third type of electrical braking, sometimes employed in industrial dc commutator motors such as overhead crane or light truck motors, is known as *plugging*. This involves the sudden reversal of the connections of either the field or armature winding during motor operation, which causes a torque of the opposite direction to be developed at a large magnitude. In this case the kinetic energy of the moving system is dissipated in the armature resistance. This braking scheme requires an overdesigned commutator/brush assembly and heavy armature windings. Electrical braking of any type becomes less effective as speed decreases because the torque decreases as the speed approaches zero. Therefore backup mechanical or hydraulic brakes are usually required with electrical braking.

In regenerative braking only a portion of the stored kinetic energy is available for charging a battery or flywheel, since a portion of the kinetic energy is dissipated in the windage and friction losses of the motor and vehicle and in the electrical losses of the motor during deceleration. Therefore the effectiveness of regenerative braking as a means of conserving energy in traction applications is governed by the energy efficiency of the electrical components in the drive train. It has been shown that about 35% of the energy put into an automobile vehicle during typical urban driving is theoretically recoverable by means of regenerative braking. When this energy is reduced by the amount necessary to overcome the electrical system losses and the mechanical losses of the vehicle, about 10 to 15% of the energy input to the vehicle can be recovered and stored in batteries, flywheels, or other devices. The exact value of the recoverable energy depends on the type of driving, the terrain, the efficiency of the drive train, gear ratios in the drive train, and so forth.

Dynamic braking, that is, the dissipation of kinetic energy in a resistance external to a traction motor, is readily described analytically. For this case, (6.32) becomes

$$e = i(R_a + R_b) + L_a \frac{di_a}{dt} \tag{6.38}$$

where R_b is the resistance added to the armature circuit to dissipate the kinetic energy of braking. Equation (6.37) remains unchanged, with the understanding that its left side is negative. Equations (6.33) and (6.34) are unchanged by motor or generator operation. The solution of these equations representing dynamic braking is given as a problem at the end of this chapter.

Important parameters in the analysis of a machine dynamics are the inertia term, J, and the viscous damping term, D. A simple experimental method for obtaining these parameters is known as the *retardation* test, which consists of accurately measuring rotor speed as a function of time while the rotor is allowed to coast to a standstill with zero armature of field excitation. Speed is best measured by obtaining a voltage proportional to speed from a small tachometer and recording this voltage on a strip recorder.

A typical retardation speed characteristic is shown in Fig. 6.44. The parameters can be obtained from this test by means of (6.37). During a coastdown with no armature or field excitation and no load torque, (6.37) becomes

$$-D\omega_m = J\frac{d\omega_m}{dt} \tag{6.39}$$

Multiplying both sides of (6.39) by the mechanical speed ω_m will result in two power terms:

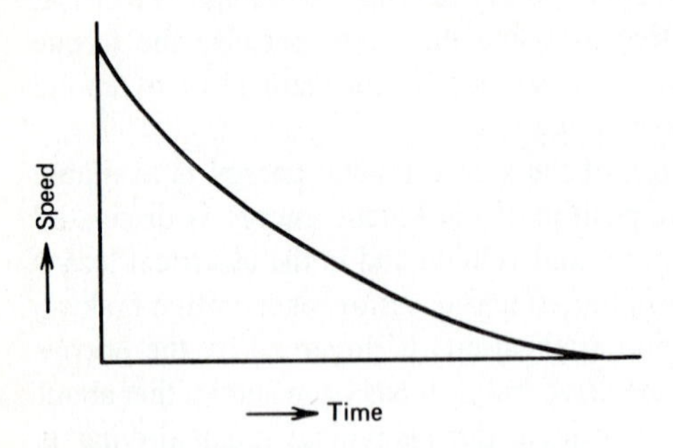

FIGURE 6-44
Speed versus time in a retardation test.

$$-D\omega_m^2 = J\omega_m \frac{d\omega_m}{dt} \tag{6.40}$$

The left side of (6.40) is the mechanical loss of the machine and the right side is the rate at which kinetic energy is supplied to the rotating system.

The inertia is found by picking a certain value of speed, determining the mechanical losses at this speed by the method outlined in Section 6.9, determining the speed-time slope at this value of speed from a retardation test, and solving for J from (6.40). Note that the retardation curve around the chosen speed only is required. A value for the linear mechanical damping coefficient can be obtained for the chosen speed by dividing the mechanical losses by the square of mechanical speed. The value of the damping coefficient will change considerably over the speed range of a machine.

PROBLEMS

6.1. A motor is operated with an airgap flux of 0.012 Wb/pole and an armature current of 100 A. Determine the electromagnetic developed torque, if the torque constant k_a of (6.6) is 45.8.

6.2. A separately excited motor has a nameplate rating of 50 hp, 440 V, and 3000 rpm.
(a) Determine the rated torque in pound-feet and in newton-meters.
(b) Assuming that the machine is 85% efficient at its rated output, what is the armature current at rated output with rated voltage applied?

6.3. Show that the constants, k_a, in (6.6) and (6.10) are dimensionally equivalent.

6.4. The motor characterized by curve (4) of Fig. 6.38 has a stall (0 rpm) current of 42 A. Determine the speed constant, K_s. What is the no-load (zero current or zero torque) induced voltage of this motor?

6.5. The motor of Problem 6.4 has an armature resistance of 0.4 Ω and is operated from a 12-V source. When the motor draws 10 A, at what value of speed and torque is the motor operating?

6.6. Test data taken on a shunt machine at 2400 rpm give the following points for the magnetization curve:

Field current (A)	0	1.0	2.0	3.0	4.0	5.0
Armature voltage (V)	4	75	133	160	172	182

(a) Express this graphically.
(b) If the machine is to be operated as a self-excited shunt generator, determine the shunt field resistance necessary to obtain a no-load voltage of 150 V at 2400 rpm.
(c) Determine the "critical" shunt field resistance for self-excited voltage build-up.

6.7. A separately excited machine has the magnetization characteristic as shown in Fig. 6.19. The machine has an armature resistance of 0.1 Ω and a field circuit resistance of 40 Ω; there are 500 turns/pole in the field winding. With 120 V applied to the field circuit:
(a) What is the armature no-load or open-circuit voltage at 5000 rpm?
(b) If the machine is operated as a generator with an armature current of 100 A, what is the armature terminal voltage? (Neglect the armature reaction.)
(c) If the armature reaction at 100 A results in a demagnetizing mmf of 220 ampere-turns/pole, determine the armature terminal voltage at 100 A.

6.8. A separately excited 120-V motor develops a torque of 20 lb-ft when the armature current is 8 A.
(a) Determine the armature current when the developed torque is 60 lb-ft, assuming no change in field current and neglecting armature reaction.
(b) If the speed is 6% higher with the 20 lb-ft torque than with the 60 lb-ft torque, determine the armature resistance.

6.9. A long-shunt compound generator has the magnetization curve given Fig. 6.19. The shunt field has 700 turns/pole, the series field 5 turns/pole. Armature circuit resistance (including brushes) is 0.08 Ω, and series field resistance is 0.005 Ω. At 5000 rpm, determine the terminal voltage when the shunt field current is 2.0 A and the armature current is 150 A. Neglect the armature reaction.

6.10. The permanent magnet motor whose speed-torque characteristic is given here as curve (7) of Fig. 6.38 is frequently used to drive air blowers for air-conditioning purposes. Assume a blower speed-torque characteristic described by the equation $T = 0.15 \times 10^{-6}\omega_m^3$, where T is in newton-meters and ω_m is in radians per second. Determine the steady-state speed at which this motor will drive the blower. (*Note.* This problem can be solved by either graphical or analytical methods. If graphical techniques are used, it will be more convenient to replot curve (7) of Fig. 6.38 with speed as the abscissa. Also keep in mind the difference in units between the data of curve (7) and that of the preceding equation.)

6.11. A series motor has the magnetic characteristic shown in Fig. A3.1, Appendix III. The series field consists of 6 turns/pole and has a resistance of 0.008 Ω. The armature circuit (including brushes) has a resistance of 0.017 Ω. The armature constant k_a is 32. At a certain load, the current is 400 A with 24 V applied.

(a) Determine the speed and torque at this value of armature current.
(b) Repeat part *(a)* for currents of 200, 100, and 50 A at 24 V.
(c) Plot the resulting speed-torque characteristic.

6.12. The mechanical and magnetic losses for the motor described in Problem 6.11 have been determined by test and are shown in the curves of Fig. 6.45. Determine the efficiency of this series motor at the four values of current used in Problem 6.11. Neglect stray-load losses.

6.13. Consider the machine described in Problem 6.7.
(a) Determine the initial (t = 0) armature current if the armature is directly connected across a 120-V supply.

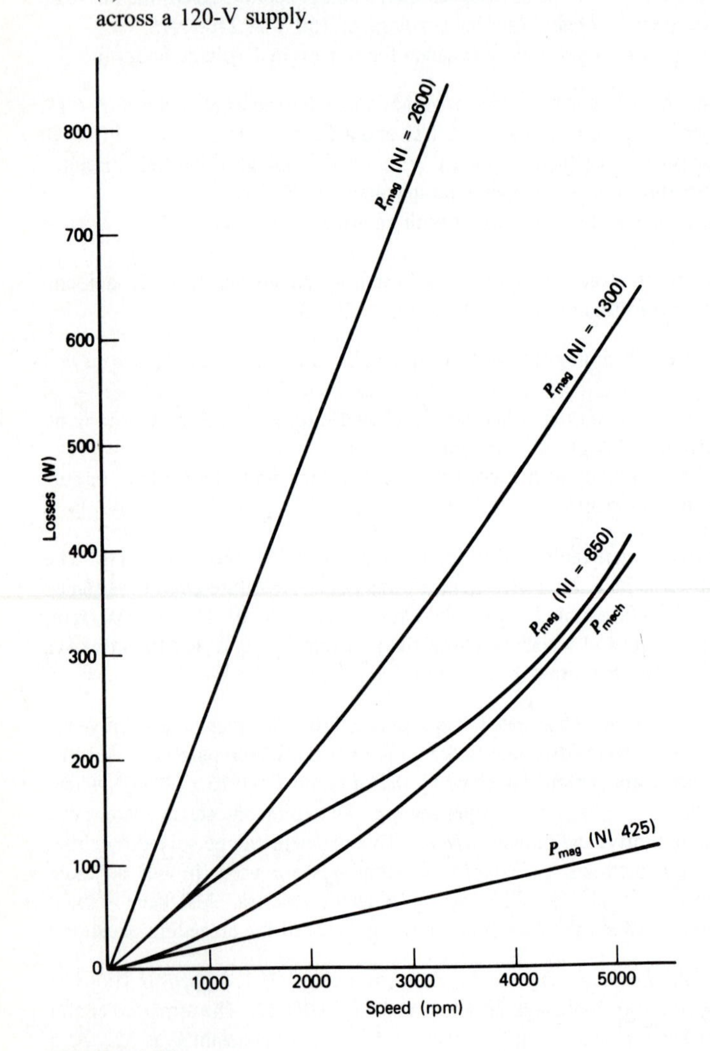

FIGURE 6-45
Problem 6.12. Series motor losses.

(b) Determine the resistancc to be added in series with the armature to limit the armature current to 200 A.

6.14. For the motor of Example 6.5, calculate the required armature terminal voltage, input power, and efficiency when:

(a) Operated as a motor to supply one-half rated torque at one-half rated speed.

(b) Operated as a generator to supply rated terminal voltage at rated armature current and rated speed.

6.15. A 400-V shunt motor delivers 15 kW of power at the shaft at 1200 rpm while drawing a line current of 62 A. The field and armature resistances are 200 Ω and 0.05 Ω, respectively. Assuming 1 V of contact drop per brush, calculate *(a)* the torque developed by the motor and *(b)* the motor efficiency.

6.16. A 400-V series motor, having an armature circuit resistance of 0.5 Ω, takes 44 A of current while running at 650 rpm. What is the motor speed for a line current of 36 A?

6.17. A 220-V shunt motor having an armature resistance of 0.2 Ω and a field resistance of 110 Ω takes 4 A of line current while running on no-load. When loaded, the motor runs at 1000 rpm while taking 42 A of current. Calculate the no-load speed.

6.18. The machine of Problem 6.17 is driven as a shunt generator to deliver a 44-kW load at 220 V. If the machine takes 44 kW while running as a motor, what is its speed?

6.19. A 220-V shunt motor having an armature resistance of 0.2 Ω and a field resistance of 110 Ω takes 4 A of line current while running at 1200 rpm on no-load. On load, the input to the motor is 15 kW. Calculate *(a)* the speed, *(b)* the developed torque, and *(c)* the efficiency at this load.

6.20. A 400-V series motor has a field resistance of 0.2 Ω and an armature resistance of 0.1 Ω. The motor takes 30 A of current at 1000 rpm while developing a torque T. Determine the motor speed if the developed torque is 0.6 T.

6.21. A shunt machine, while running as a generator, has an induced voltage of 260 V at 1200 rpm. Its armature and field resistances are 0.2 Ω and 110 Ω, respectively. If the machine is run as a shunt motor, it takes 4 A at 220 V. At a certain load the motor takes 30 A at 220 V. However, on load, armature reaction weakens the fields by 3 %. Calculate the motor speed and efficiency at the specific load.

6.22. The machine of Problem 6.21 is run as a motor. It takes 25 A of current at 800 rpm. What resistance must be inserted in the field circuit to increase the motor speed to 1000 rpm? The torque on the motor for the two speeds remains unchanged.

6.23. The motor of Problem 6.21 runs at 600 rpm while taking 40 A at a certain load. If a 0.8 Ω resistance is inserted in the armature circuit, determine the motor speed provided that the torque on the motor remains constant.

6.24. A 220-V shunt motor delivers 40 hp on full-load at 950 rpm, and has an efficiency of 88 percent. The armature and field resistances are 0.2 Ω and 110 Ω, respectively. Determine *(a)* the starting resistance such that the starting line current does not exceed 1.6 times the full-load current, and *(b)* the starting torque.

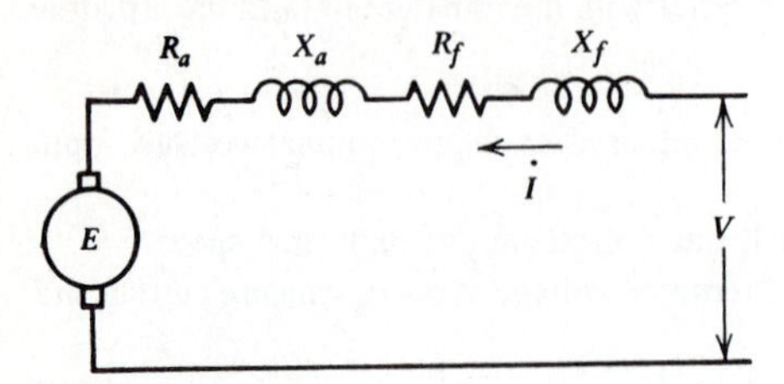

FIGURE 6-46
Problem 6.26.

6.25. A 220-V series motor runs at 750 rpm while taking 15 A of current. What is the motor speed if it takes 10 A of current? The torque on the motor is such that it increases as the square of the speed.

6.26. A universal motor is rated 1/8 hp, 5000 rpm, 1.8 A, 115 V ac or 60 Hz. Compute the rated torque of the motor, neglecting windage and friction. When the motor is supplying rated power output at rated speed, the input is measured to be 115 V, 60 Hz, 1.84 A, 182 W. Calculate input power factor and the efficiency. Draw a qualitative phasor diagram for this machine at the given load condition; the equivalent circuit shown in Fig. 6.46 will aid in drawing this diagram. It can be assumed that the electromagnetic voltage E times the motor current equals the electromagnetic power, which, assuming no mechanical losses, equals the output power.

Another test was made on this motor, giving the following measurements: 115 V, 60 Hz, 2.5 A, 214 W input; 35 oz-in. output torque at 3400 rpm. Calculate input power factor and efficiency. Draw the phasor diagram for this condition of load, and compare with that of the previous load.

6.27. Assuming linear damping coefficient, D, as in (6.37), and letting $T_L = 0$, solve for motor speed as a function of time during a dynamic braking operation. See (6.32) to (6.36). Assume that a braking operation is initiated at $t = 0$ when $i_a = I_0$ and $\omega_m = \Omega_0$. Also assume that V_f and i_f are maintained constant during the brake operation.

6.28. A separately excited dc motor is initially at rest, while the field winding carries a steady current I_f. A voltage V_t is connected across the armature at $t = 0$. The armature resistance is R_a and the total load on the motor is pure inertia J. Determine the total heat dissipated in the armature resistance over the period the motor reaches its steady-state speed. Compare the result with the kinetic energy stored in the inertia J.

CHAPTER 7

PERMANENT MAGNET MACHINES

Permanent magnet (PM) machines compose a well-known class of rotating and linear electric machines used in both the motoring and generating modes. PM machines have been used for many years in applications where simplicity of structure and a low initial cost were of primary importance. More recently, PM machines have been applied to more demanding applications, primarily as the result of the availability of low-cost power electronic control devices and the improvement of permanent magnet characteristics. In general, modern PM machines are competitive both in performance and cost with many types of field-wound dc machines and single-phase synchronous machines; the combinations of high-energy permanent magnets and solid-state power semiconductors are the principal ingredients of a relatively new class of machines commonly known as "brushless dc machines" and more appropriately termed "self-synchronous machines." The purpose of this discussion is to describe the theory and practice of electric machines of many types which use permanent magnets. In this connection it is a good idea to review Section 2.10, which deals with magnetic circuits containing PMs.

Energy converters using permanent magnets come in a variety of configurations and are described by such terms as motor, generator, alternator, stepper motor, linear motor, actuator, transducer, control motor, tachometer, brushless dc motor, and many others. In the following discussion, most of the presentations will refer to

1. Conventional (commutator) dc motor/generator
2. Synchronous alternator
3. Brushless dc motor
4. Digital machines

It should be noted that the latter two configurations are synchronous machines adapted for use with batteries and other dc energy sources by means of electronic switching. There are many varying physical configurations for PM machines including axial and radial airgap configurations. In general, the use of PM excitation provides the machine designer with a greater degree of freedom in physical configuration than most other classes of rotating electric machines, with the exception of the reluctance class of machines.

7.1 UNIQUE FEATURES OF PM MACHINES

How do PM machines relate to other classes of electric motors and generators? PM machines fall into a generalized classification known as "doubly excited" machines, which have two sources of excitation—usually known as the armature and the field (or excitation). In conventional synchronous and dc commutator machines, both of these excitation sources are electrical windings connected to an external source of electrical energy. In PM machines, the excitation or field winding is replaced by a permanent magnet and, of course, no external source of electrical energy is required. In other respects, a PM machine may be directly comparable to conventional synchronous or dc commutator machines, and armature windings and magnetic circuit may be identical in PM machines to those in conventional machines. There is no comparison or analogy between PM machines and singly excited machines such as the induction motor or hysteresis machines. However, PM machines generally have the structural simplicity of singly excited machines and are, therefore, often compared with singly excited machines in terms of cost, ease of assembly, size, and volume.

There are several major areas of difference between PM machines and other doubly excited machines which will be discussed in more detail subsequently.

1. *Control.* A major characteristic of conventional synchronous and dc commutator machines is the ability to control various external machine characteristics, such as terminal voltage and/or power factor. This feature is absent in a PM machine, although there have been certain attempts to try to exercise somewhat the same type of control by varying the machine airgap (in axial airgap machines) or by magnetically shunting the field magnets. Because of this limitation, PM machines have not been considered for central station alternator applications. However, for relatively small power applications, PM alternators can be effectively paralleled with conventional power systems by means of properly designed semiconductor switching devices. Such applications include wind power and power generated from waste heat.

2. *Cost.* Considering the simplicity of the permanent magnet versus a field winding excitation, it is often assumed that PM machines are inherently less expensive than wound-field machines. This is not necessarily true, and each case must be evaluated on its own merits and specifications. In general, where ceramic magnets can be used and where relatively little voltage control is required, a PM machine will be less costly than its wound-field counterpart.

3. *Volume and Weight.* A PM machine will generally offer a volume and weight savings compared to its equivalent wound-field configuration.

4. *Flexibility of Size and Shape.* One of the chief merits of PM machines as compared to conventional machines is that the PM machine can be constructed in many nonstandard sizes and shapes, which often compensates for a cost penalty, especially in automotive and aerospace applications. Machines using PMs with very high residual flux density can be constructed in what is called the "ironless stator" configuration; that is the magnetic material in the armature can be eliminated, resulting in a large weight savings. The same type of PMs also permit a machine to have a larger airgap. Both of these potential design features can have beneficial effects on various machine performance characteristics, such as cogging and ripple in the airgap flux density, as well as simplifying machine assembly and lowering initial cost.

5. *Demagnetization.* One limitation on a PM machine that has no counterpart in conventional machines is that the PM may be demagnetized by excessive armature reaction, by excessive temperatures, or, in some cases, by excessive mechanical shock. This limitation must be considered in all PM machine design.

7.2 PERMANENT MAGNET MATERIALS

In Sections 2.2 and 2.10 (Chapter 2) we have alluded to permanent magnets and some of their characteristics. We now consider these topics in more detail.

Permanent magnet materials are metallic alloys and metallic oxide materials which can exhibit a high condition of domain alignment in the absence of an external magnetic field. Permanent magnets used in rotating electric machines are of two general classes: ferromagnetic materials and ferrimagnetic materials. Ferrimagnetic permanent magnets, often called hard ferromagnetic materials, are crystalline structures formed from metallic alloys, usually containing one of the three natural magnetic metals, iron, nickel, or cobalt. Ferromagnetic materials, often called hard ferrites, are oxides of iron and one other metal, usually barium or strontium. Various thin- and thick-film forms of these oxides are also used in magnetic tapes, discs, and other recording media. Iron oxides (Fe_3O_4) occur naturally on the earth. An ancient stone of this material magnetized by the earth's magnetic field gave rise to the name magnetism.

In general, all magnetic materials exhibit varying degrees of permanent magnetism, often called remnance. To place a magnetic material in a state of zero magnetism requires a process known as "demagnetization," by means of an externally applied magnetic field. However, hard magnetic materials or permanent magnets are those magnetic materials which retain a much higher level of magnetism than "soft" magnetic materials in the absence of an external magnetic field. This property results from the internal microscopic structures of permanent magnet (PM) materials, which also cause differences in most other physical properties of hard and soft magnetic materials. PM materials generally are hard and somewhat brittle and have lower tensile and bending strength than soft magnetic materials. Their Curie temperatures and other thermal characteristics also differ from those of soft magnetic materials.

A great many types of PM materials have been developed in the 20th century, particularly in the latter half of this century, and there are many trends and indications that new PM materials will continue to be developed in the years ahead. As a result, the PM machine designer has a large choice of PM materials for consideration in most designs and often the ability to optimize a design for minimum size, weight, cost, or other design specification by means of PM material selection. Most commercially used magnets fall within six general categories, which helps to reduce the characteristics of which one should have at least a slight cognizance. Within each of these categories, there is still a myriad of variations in many parameters, but these are primarily due to the differences among materials manufacturers and to manufacturing techniques. Therefore, when precise magnetic parameters are required in a specific application, precise specifications should be made by the designer and agreed upon by the manufacturer. The characteristics presented in this text might be considered "generic" or typical characteristics of the stated material, and not the exact characteristics of a material supplied by a certain manufacturer.

Alnico Magnets

Alnico PM materials are metallic alloys of aluminum, nickel, cobalt, and iron, and were among the first high-energy PMs to be developed. Alnico magnets are generally characterized by relatively high residual flux density (B_r) and relatively low coercive force (H_c). The latter characteristic is undesirable from the electric machine standpoint. Certain grades of Alnico, such as grade 8HC, have been developed to remedy this weakness, but at the expense of lowered residual density. The characteristics of several popular grades of Alnico are summarized in Fig. 2.18 (Chapter 2). Alnico magnets are manufactured in "generic grades" from 1 through 9, with many variations within each grade. The grades generally represent the chronological order in which the properties designated by these grades were developed commercially.

Ceramic Magnets

Ceramic magnets are similar to other types of materials we commonly refer to as ceramics in physical properties, hence the popular name. However, ceramic PMs are properly defined as ferrite oxides of barium or strontium and exhibit the property known as ferrimagnetism. Due to the types of materials and manufacturing processes used, ceramic magnets are generally the lowest cost magnets available in terms of cost per unit of energy product. Ceramic magnets are by far the most widely used of any PM types in almost all applications, including rotating machines.

Ceramic magnets are characterized by relatively low residual flux densities (B_r) and relatively high coercive forces (H_c). Because of the latter characteristic, ceramic magnets are able to withstand armature reaction fields without demagnetization and are well suited for electric machine applications. Although ceramic magnets have generally poor mechanical and structural characteristics, they are the lightest in density of the common magnet types. This is often a distinct advantage in machine applications and tends to compensate for the increased pole-face area required due to the low residual flux density. Also, ceramic magnets have the lowest recoil permeability of common magnets, which is a stabilizing factor in machine application.

Samarium-Cobalt Magnets

Samarium-cobalt magnets gave an "order-of-magnitude leap" in energy product over ceramic magnets and most other types of magnets. Samarium-cobalt magnets have residual flux densities comparable with alnicos and coercive forces three to five times those of ceramic magnets. These magnets also have generally improved physical characteristics as compared to both alnicos and ceramics. From a technical standpoint, they are ideal for rotating electric machine applications.

Demagnetization characteristics of certain samarium-cobalt magnets are shown in Fig. 7.1.

Neodymium-Iron-Boron (NdFeB) Magnets

Neodymium-iron-boron (NdFeB) PM materials appear to offer the greatest promise for a PM material with greatly improved characteristics over those of ceramic magnets. This material has been shown in the laboratory to have the highest energy product of any PM material, and commercial versions of these laboratory samples are available with energy products above those of samarium-cobalt. Perhaps more important, NdFeB magnets hold the promise of relatively low cost in production quantities.

NdFeB has the highest coercive force available in commercial magnets and therefore is ideally suited for machine applications. Also, its residual flux density is relatively high, comparable to the best of the alnicos. As stated above, its energy product is the highest available today. Limitations of this material include very poor temperature characteristics. The low operating temperature requires the use of a larger size for an application required to operate at elevated temperatures, and therefore many of the reduced size and weight advantages of NdFeB are lost.

A trade name of NdFeB magnets is Magnaquench. Their characteristics are shown in Fig. 7.2.

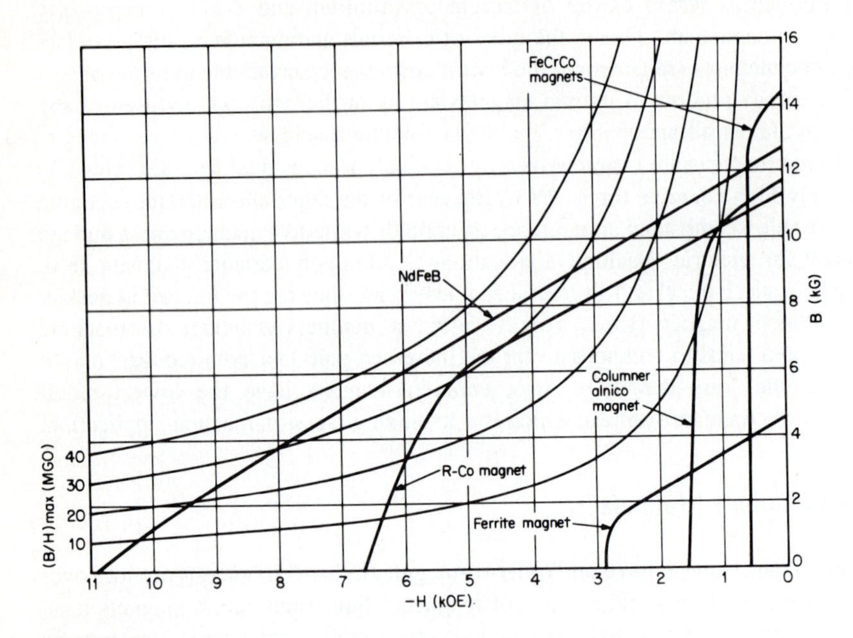

FIGURE 7-1
Magnetic characteristics of certain samarium-cobalt-rare earth magnets.

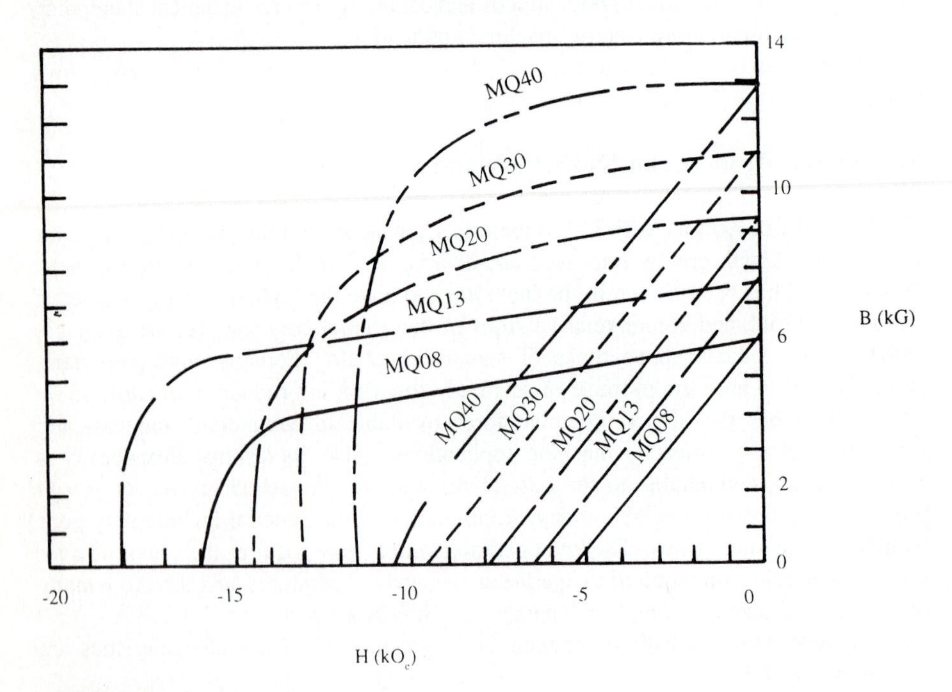

FIGURE 7-2
Magnetic characteristics of Magnaquench NdFeB magnets.

7.3 POWER LIMITATIONS OF PM MACHINES

The principal difference between a PM machine of any type and other types of rotating electric machines is the means by which machine excitation is developed. It is well known that the magnetic "output" parameters of PMs—flux density and field intensity—are less than those obtainable in conventional electrical excitation systems using soft iron and electrical conductors. Also, there is a definite limit in field intensity beyond which the magnet becomes demagnetized and the machine is essentially inoperable until the magnets are remagnetized. There is no similar limitation in the electrical excitation system. Most permanent magnet materials—especially those that have been used in rotating machine applications—also have much different physical characteristics than do the materials used in electrical excitation systems, which may impose mechanical and structural limitations on PM machines. Therefore, there are limitations on the power capabilities of PM machines due to both the above-mentioned magnetic and physical limitations of PM materials. It has generally been assumed—and with some justification—that PM machines are limited to relatively low power levels. The output rating of 10 hp has often been cited, although this limit has often been based more upon economic considerations than upon any inherent power limitation. Also, the introduction of economically feasible rare-earth magnets has greatly increased the power capability of PM machines. A number of PM machines with ratings around 100 hp have been built in recent years, and the use of PMs in machines rated well above 10 hp will become more common in future years. In any case, however, the power limit due to the characteristics of a permanent magnet is not rigid and is susceptible to many design features and innovations.

The mechanical characteristics of most PMs are considerably inferior to those of conventional soft iron and conductor materials used in electrically excited machines. These characteristics must be considered in both the operation and the handling or shipping of PM machines. In some cases, the mechanical characteristics also impose a power density limitation in a machine design by requiring additional magnet size or additional structural components to provide sufficient mechanical strength for a given application. Ceramic magnets are hard and brittle with very low tensile strength and are generally not suitable for use on a rotor. Cast alnico has low tensile strengths in the range of 7700 psi to 22,000 psi (lbs/in^2); rare earth—Co is approximately 18,000 psi; and NdFeB is approximately 35,000 psi. Ceramic and cast alnico are brittle and easily chipped during handling and mounting. The other materials are more ductile. It has been noted in the previous section that power output of a PM machine can be apparently increased by merely increasing the PM volume. However, it should be noted that the relatively poor mechanical characteristics of PM materials may impose limitations upon the power capability of a PM machine as the size of the PM is increased. Additional structural weight is generally required to properly support the PM materials as the size or power output of a PM machine is increased above 10 hp or so.

Other mechanical properties besides tensile strength must also be considered. The brittleness of ceramic magnets and cast alnico has already been noted. Corrosion problems with NdFeB magnets have been reported and must be considered in designs using these magnets. Most magnets are subject to possible demagnetization due to severe mechanical shocks, and this must be considered in the shipping, handling, and applications of PMs.

Finally, the characteristics of all permanent magnets are generally degraded by increasing temperature. The mechanisms by which the properties, such as residual flux density and coercive force, are diminished as a function of temperature at which the PM is operating are beyond the scope of this text. PM degradation as a function of temperature is described by means of temperature coefficients of flux density and field intensity and also by the material Curie temperature at which magnetism ceases.

The temperature characteristics of PMs must be considered in the design and applications of all PM machines. One can conclude from consideration of PM temperature characteristics that if a machine is to be operated at higher temperature levels, the size of the PM to give a power output equivalent to that at room temperature must be increased. Thermal shock of a machine's PM can easily result in demagnetization of the PM, and this must be considered in machine design. Unfortunately, the PM that shows the most promise for commercial application, NdFeB, has relatively poor temperature characteristics.

7.4 PERMANENT MAGNET DC MACHINES

DC commutator machines have long used PM excitation, and these machines are among the most common of machine types. In the early years of such PM applications, the flexibility of dc PM motors was limited by the lack of field control. However, field control has become of less significance with the advances in power semiconductors available for armature control and also by the use of microprocessors and other computer control devices in the supervision of motor performance under a variety of conditions. As has been noted earlier, PM dc commutator motors are generally less expensive, smaller in size, have reduced assembly costs in large quantities, and have higher efficiencies than their counterpart dc wound-field motors. Fig. 7.3 is a cutaway view of a commercial dc PM commutator motor.

PM dc commutator machines are generally designed and analyzed in a manner similar to that for any other dc commutator machine. The armatures of PM-excited machines are similar to those of wound-field machines and are designed by the same rules of winding layout as are dc machines in general. The choices of such parameters as number of slots and poles, slot geometry, commutator/brush layout, parallel paths, rotor yoke design, and winding layout (lap, wave, concentric, number of winding layers, etc.) are, in general, identical for both types of machines. The determination of the airgap flux density and, of course, the mechanical design of the field structure and stator yoke, are different for PM and wound-field machines, the former tending to favor a larger number of poles. The analysis of PM machines is similar to that of shunt machines with constant field excitation. A summary of the basic equations used in dc commutator machine design and analysis is given below.

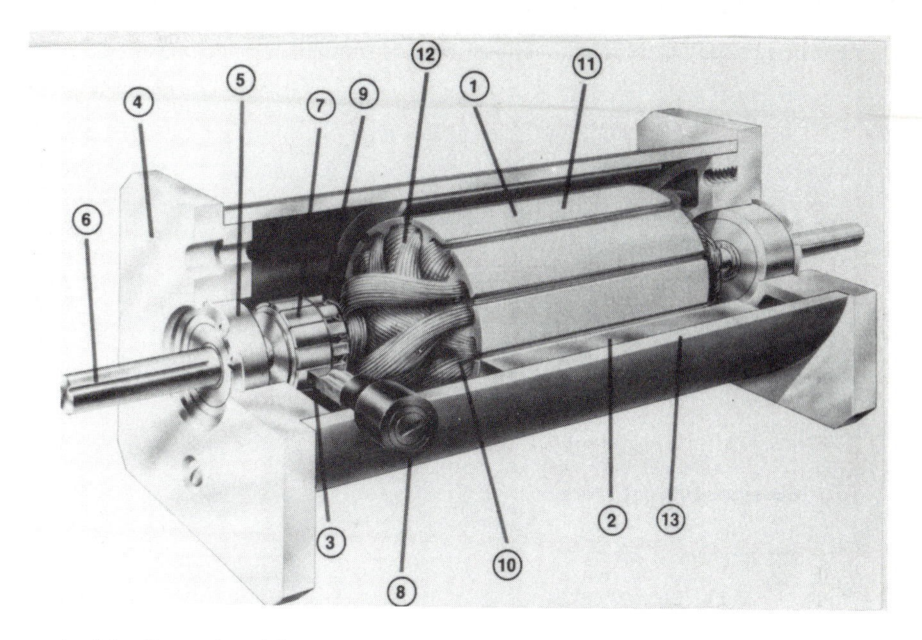

①	Insulation: Epoxy resin coated armature lamination structure. Underwriters' Laboratories, Inc. recognized insulation system available upon request.
②	Magnets: Ceramic, alnico or rare earth selection of permanent magnets.
③	Brushes: Replaceable, long-life brushes. Brush material composition selected on basis of application.
④	End Caps: Aluminum for efficient heat transfer and precision tolerances.
⑤	Bearings: Double shielded ball bearings lubricated for life, ABEC 1 standard. Other bearings optional.
⑥	Shaft: Stainless steel—wide selection of standard diameters and extensions.
⑦	Commutator: Diamond turned copper for smooth finish.
⑧	Brushholders: Molded thermoplastic.
⑨	Welds: Welded armature connections.
⑩	Magnet Wire: Class 155°C magnet wire.
⑪	Laminations: High permeability armature laminations.
⑫	Impregnant: Epoxy resin impregnated for added rigidity and insulation.
⑬	Housing: Totally enclosed carbon steel housing.

FIGURE 7-3
Construction features of a PM motor.

The basic steady-state equations describing dc commutator machine operation are

$$T_d = \frac{ZP}{2\pi a}(\phi I_a) = K\phi I_a \tag{7.1}$$

$$E = K\phi\omega_m \tag{7.2}$$

$$V = E \pm I_a R_a \tag{7.3}$$

$$\phi = \pi D l B_g / P \tag{7.4}$$

where

T_d = developed torque, N-m

E = developed emf, V

I_a = armature current, A

ϕ = flux per pole, Wb

V = terminal voltage, V

R_a = armature (including brush) resistance, Ω

ω_m = armature rotational speed, rad/s

D = stator bore, m

l = stator stack length, m

K = winding factor = $ZP/(2\pi a)$

Z = total armature conductors

P = number of poles

a = number of armature parallel paths.

Combining (7.1) and (7.4) gives

$$T_d = Z(D l B_g I_a)/(2a) \tag{7.5}$$

Letting $ZI_a/(2a) = X\pi D$ gives

$$T_d = \pi D^2 l B_g X = 4\,(\text{vol})\, B_g X \tag{7.6}$$

where X = electrical loading in armature, A/m; and vol = magnetic volume = $\pi D^2 l/4$.

For a specified developed torque and electrical loading, (7.6) can be expressed as

$$(\text{vol}) B_g = T_d / X = \text{constant} \tag{7.7}$$

Equation (7.7) is valid for any type of dc commutator machine for which the preceding equations are valid, but it is particularly useful in the evaluation and design of PM machines in which B_g is a function of the type of PM used for excitation. Note that (7.7) ignores thermal limitations of a machine, which often may be the primary limitation on reducing the volume of a machine.

The determination of the airgap density, B_g, is a complex task in most machines and requires introducing many dimensional parameters. It is further complicated in PM machines by the nature of PM variations caused by armature reaction, airgap length, leakage factors, etc. The analysis of PM characteristics in a motor environment is primarily a graphical procedure. However, these graphical procedures can usually be modeled for computer simulation and used in relatively conventional machine design and analysis computer programs. Rare-earth and ferrite magnetic characteristics are relatively simple to model, since these characteristics can be approximated by straight lines with relatively good accuracy. Alnico magnet characteristics are highly nonlinear, as is seen from the curves shown in Fig. 2.18 (Chapter 2).

Characteristics

In the absence of variable field excitation, the torque-speed characteristic of a PM dc motor tends to be linear, as shown in Fig. 7.4 (which is the same as Fig. 6.38). A phenomenon not usually encountered in conventional field-excited dc machines is cogging. Cogging is a phenomenon associated with PM machines and refers to the reluctance torque developed between the PM and the soft-iron surfaces facing the PM. It is more of a problem in brushless dc and ac machines due to the fact that these classes of machines operate at higher speeds and often are mounted on magnetic, or hydrodynamic, air bearings, which are quite susceptible to the torque induced by cogging. In certain classes of machines, known as "hybrid PM" machines, this reluctance torque is designed to become a useful component of the machine's output torque. Since cogging is due to the PM, it exists under any condition of operation of the machine, including that when no electrical excitation is supplied to the machine. It can usually be felt when turning the rotor of the machine by hand in the electrically unexcited condition.

In a PM commutator machine, cogging usually is a factor in the design of the machines. Cogging in this class of machines is due to the reluctance torque between the PM and certain groupings of teeth on the rotor or armature. Obviously, cogging is reduced by increasing the number of teeth on the rotor so that there is relatively

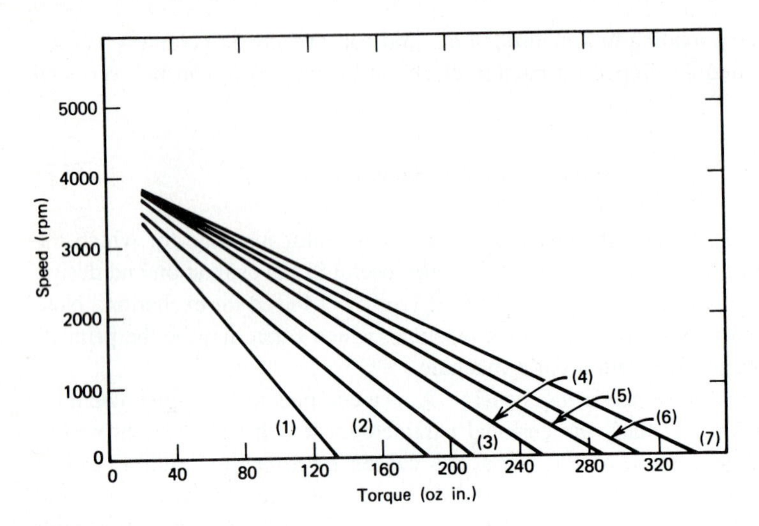

FIGURE 7-4
Torque-speed characteristic of a PM dc motor.

little difference in reluctance between an "aligned" and unaligned position. Reducing the slot-opening width reduces the airgap reluctance variation and, hence, coggingtorques. Cogging is also reduced through the use of an odd number of slots rather than an even number, for the same reason. However, note that the use of an odd number of slots requires a fractional-slot winding, which may involve increased design time. On the other hand, fractional-slot windings may be perfectly acceptable to the designer as a means of accommodating stack laminations. A second method for reducing cogging torque is that used frequently in induction machines to reduce torque pulsations, known as "skewing." A skewed armature winding skews the armature conductors away from parallelism with the shaft, as shown in Fig. 7.5. Skewing is a common practice in the construction of the rotors of many types of machines.

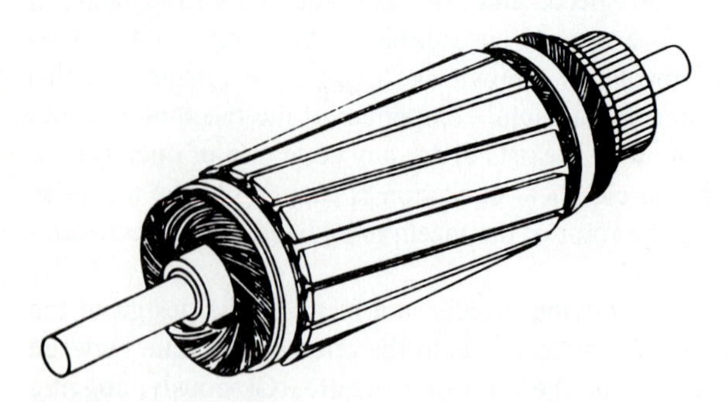

FIGURE 7-5
Skewed rotor, with windings placed on the rotor surface.

Cogging is virtually eliminated in several PM machine configurations in which reluctance torque is eliminated. The most obvious of these is the "ironless armature" configuration, in which the soft iron portions of the armature are replaced by air or other nonmagnetic materials. This technique is used frequently in aerospace machines using platinum-cobalt and rare-earth PMs. This construction is possible only with the use of PMs with very high coercive force, and is generally suitable only for brushless dc and ac PM machines. Other geometries eliminating or greatly reducing cogging are the rotating coil and printed circuit configurations.

7.5 PERMANENT MAGNET SYNCHRONOUS MACHINES

The vast array of synchronous machine configurations in the medium and low power ranges can generally be classified into two groups: *conventional* and *brushless*. PM machines fall into the latter group. PM synchronous machines generally have the same operating and performance characteristics as synchronous machines in general: operation at synchronous speed, a single- or polyphase source of alternating current supplying the armature windings, a power limit above which operation at synchronous speed is unstable, reversible power flow, damper (cage) windings for starting and stability purposes, a torque angle between armature and field phasors, etc. A PM machine can have a configuration almost identical to that of the conventional synchronous machine with the absence of slip rings and a field winding. This absence, of course, is responsible for the one major difference between a PM machine and a conventional synchronous machine: lack of power factor or reactive power control and its association with terminal voltage regulation. This characteristic will be discussed in detail later in this chapter.

The absence of brushes and slip rings is also a principal merit of the PM synchronous machine. However, in this regard, the PM synchronous machine is in "competition" with a number of other brushless configurations, such as the inductor, Lundell, reluctance, and rotating rectifier configurations, all of which are highly developed and in wide use today. Therefore, being brushless is not necessarily a major advantage of a PM machine over other types of synchronous machines. It is rather the low cost, simplicity of rotor construction, and reduced manufacturing assembly procedures and costs of a PM machine that make it very competitive in the marketplace. Only reluctance machines are simpler in construction and in assembly procedures than PM machines, but reluctance machines generally develop less torque per unit of current and per unit of weight. Therefore, on a basis of power output per unit weight (and, in general, per unit volume), the PM synchronous machine is superior to all other brushless synchronous machines—especially with the commercial feasibility of rare-earth magnets. In terms of cost per unit of power output, it is lower than all but possibly certain reluctance machine configurations.

Machine Components and Definitions

Fig. 7.6 illustrates a circumferential cross-section of a very simple PM synchronous machine. For a PM machine to be brushless, the magnets must be on the rotor, and this construction—although not essential for the operation of the machine—will be assumed throughout our discussions of PM synchronous machines. As illustrated in Fig. 7.6, the principal components of PM synchronous machines are described in the following.

1. *Stator*. This is the stationary member of the machine and includes a number of elements:
   (*a*) **Stator laminations**. Shown in Fig. 7.6 for a cylindrical, radial airgap machine; this type of lamination, in general, is similar to those of other types of synchronous and induction machines and is formed from punched soft magnetic steel stampings. Stator laminations for pancake or axial airgap machines are often formed by winding continuous strips of soft steel. Various parts of the laminations are the teeth, slots which contain the armature windings, and the stator yoke (the outer portion of the lamination shown in Fig. 7.6) which completes the magnetic path. Lamination thickness depends upon the frequency of the armature source voltage and cost and core-loss considerations. Laminations are bonded together in various manners to form the "stator stack." The stator stack length, l, is a principal parameter in determining the electromagnetic size of a cylindrical, radial airgap synchronous machine. In pancake and axial airgap machines, the equivalent size parameter is the stack diameter, D_{st}.
   (*b*) **Armature winding**. In the most common types of PM machines, especially those used in power applications, armature windings are conventionally single- or polyphase windings similar to those of other ac machines. Windings are generally double-layer (two coil sides per slot) and lap-wound;

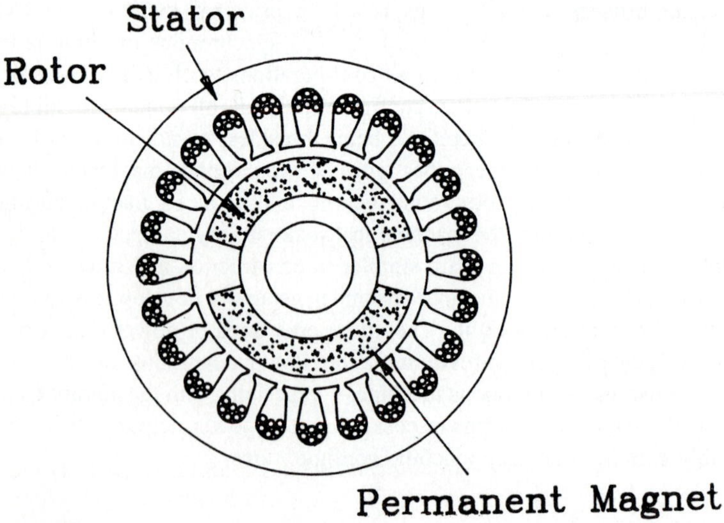

Stator
Rotor
Permanent Magnet

FIGURE 7-6
Circumferential cross section of a two-pole PM synchronous machine.

single-layer, concentric windings are also common. In contrast to dc armature windings, ac windings are "open ended." Individual coils are connected together to form *phase groups*, the basic element of an ac winding, and designated as the number of slots per pole per phase. Phase groups are connected together in series/parallel combinations to form wye, delta, zigzag, two-phase, or simple single-phase windings. AC windings are generally short-pitched to reduce harmonic voltage generated in the windings. Windings are usually formed from standard magnet wire of the type used in most machines and transformers. Coils, phase groups, and phases must be insulated from each other in the end-turn regions (the coil extensions beyond the stator stack), and the required dielectric strength of the insulation will depend upon the voltage rating of the machine. Conductors in the slots are often potted, both to increase insulation strength and for structural support. In smaller PM synchronous machines, the conventional winding methods are often not used, and solenoidal or *random* windings are used to reduce manufacturing costs. In many axial airgap configurations, solenoidal windings are also used.

2. *Airgap*. The airgap, the region between the stator and the rotor, is an essential parameter of any rotating machine. In a PM machine, the airgap serves an additional role in that its length largely determines the operating point of the PM in the no-load operating condition of the machine. The influence of the airgap upon machine performance is a combination of the effects of the actual physical airgap and the stator slot openings. This is treated by a technique using *Carter* coefficients as in the case of dc PM machines. The airgap length also influences the mechanical and structural design of a machine. In general, as the airgap length is increased, the machine tolerances on both rotor and stator surfaces are relaxed, which may often lower manufacturing costs. Also, longer airgaps reduce machine windage losses. However, longer airgaps require PMs with higher coercive force and also reduce the operating flux density in the airgap for any magnet.

3. *Rotor*. The principal elements of the rotor of a PM synchronous machine are:
   (a) **Pole structure**. The PMs form the poles equivalent to the wound-field poles of conventional synchronous machines. PM poles are inherently "salient," of course, and there is no equivalent to the cylindrical rotor pole configurations used in many conventional synchronous machines. Many PM synchronous machines may be cylindrical or "smooth rotor" physically, but electrically the PM is still equivalent to a salient pole structure. There are many ways of arranging the PMs physically, some of which will be further discussed later. Relatively few synchronous PM machine rotors have the PMs directly facing the airgap, as shown in Fig. 7.6. Rather, the PMs are part of a pole structure which may include a number of other structural and magnetic components. In the simplest configurations, these additional pole pieces may be only "pole shoes" or "flux concentrators," that is, magnetic

circuits for increasing the flux density between the magnet surface and the airgap. In more complex designs, the total pole structure may include complex magnetic circuits for changing the flux direction or blocking flux leakage. These concepts are further discussed later.

(b) **Damper winding**. This is the typical cage arrangement of conducting bars, similar to induction motor rotor bars and to damper bars used on many other types of synchronous machines. It is not essential for all PM synchronous machine applications but is found in most machines used in power applications. The principal purpose is to dampen oscillations about synchronous speed, but the bars are also used to start synchronous motors in many applications. The design and assembly of damper bars in PM machines are similar to those in other types of synchronous machines.

(c) **Rotor yoke**. This is the magnetic portion of the rotor to provide a return path for the PMs and also to provide structural support in some designs. The yoke is often a part of the pole structure discussed above and difficult to distinguish as a separate entity in many cases.

(d) **Shaft and bearing system**.

4. *Housing*. The stator is supported by the housing, which is a nonmagnetic structural member that generally contains the entire machine. There are several means of "housing" a machine, from relatively open housings to hermetically sealed housings. The housing usually serves several other functions, such as a thermal conductor for conducting heat generated within the machine to external thermal sinks; as a means of supporting the bearing system; and as a means of supporting the mechanical interfaces with other machines or load devices and/or to a fixed mounting surface.

Rotor Configurations

Synchronous machines are classified according to their rotor configuration. The distinguishing feature of PM synchronous machines is, of course, the presence of a set of PMs in the rotor. As has been noted earlier, there is a great variety of PM synchronous machine configurations, and much of this variety is due to the means of locating the PMs within the rotor structure. Most of the research and advance development in PM synchronous machines today is involved with searching for the optimum rotor structure. Rotor optimization may be based upon weight minimization or cost minimization or sometimes both. But in terms of machine parameters, an optimum rotor should maximize the airgap flux density and minimize the leakage flux between magnets that does not contribute to the energy conversion process. Structural integrity and ease of assembly are also concerns in the optimization process. Another consideration affecting rotor design is the number and location of the damper or squirrel-cage bars. And, finally, the optimum rotor design is a function of the type of PM being used.

As a starting point in designing a rotor for a PM synchronous machine, we can start with the conventional salient pole synchronous machine rotor and merely replace

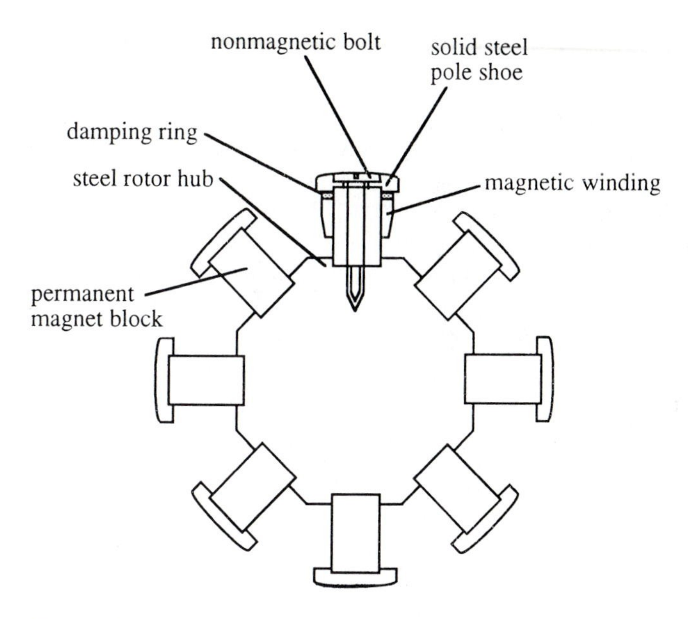

FIGURE 7-7
Salient pole PM generator rotor.

the soft-iron field poles—or a radial section of them—with PMs. Fig. 7.7 is an example of such a design which is fairly typical of rotor designs using alnico magnets. From this starting point, almost every conceivable location and orientation of PM is theoretically possible, and a great many have been evaluated. There are four general types of rotors found in PM synchronous machines, although some configurations may be hard to categorize into a single type. The category names are based upon the relative location of the PMs in the rotor, but they also differentiate the direction of the magnetic flux emanating from the PMs. The four general rotor types are:

1. *Peripheral.* The PMs are located on the rotor periphery, and PM flux is *radial.* Fig. 7.6 is definitely a peripheral type.

2. *Interior.* The PMs are located in the interior of the rotor, and flux is generally *radial.*

3. *Claw-Pole or Lundell.* The PMs are generally disc-shaped and magnetized *axially.* Long, soft-iron extensions emanate axially from the periphery of the discs like "claws" or Lundell poles. There is a set of equally spaced claws on each disc which alternate with each other, forming alternate north and south poles.

4. *Transverse.* In this configuration, the PMs are generally between soft-iron poles and the PM flux is *circumferential.* Fig. 7.8 illustrates a simple transverse

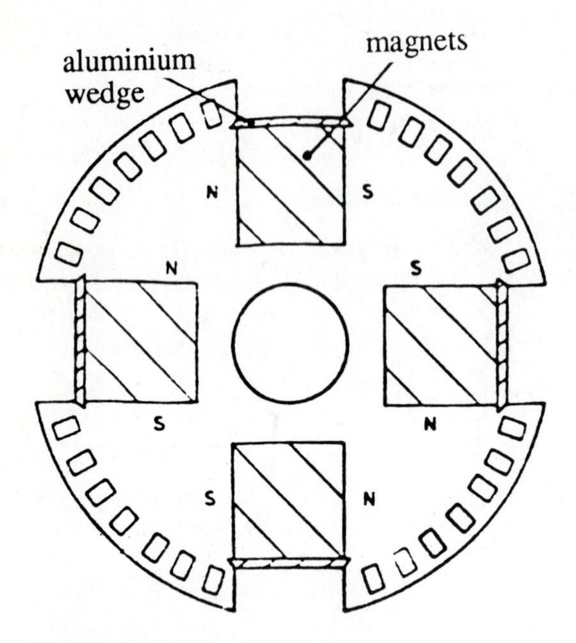

FIGURE 7-8
Transverse type rotor with damper bars.

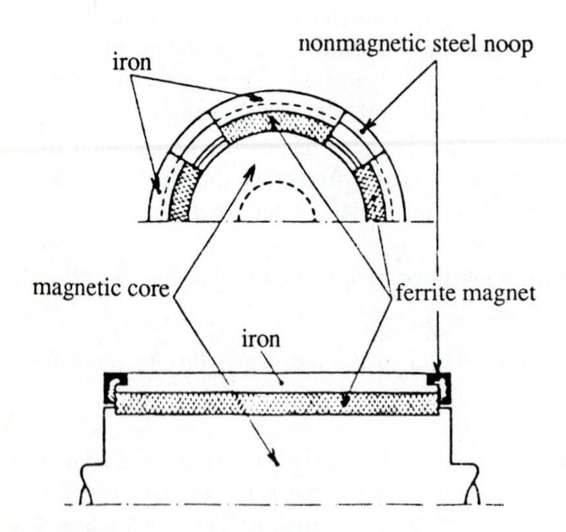

FIGURE 7-9
An interior-magnet rotor.

configuration; the rectangles in the soft-iron poles denote damper bars. Magnetically, this configuration is similar to a reluctance machine rotor, since the permeability of the PMs is very low, almost the same as that of a nonmagnetic material. Therefore, reluctance torque—as well as torque resulting from the PM flux—is developed. The transverse configuration is, therefore, also termed a *hybrid* configuration since the torque developed by the machines is a combination of two electromagnetic processes.

Fig. 7.9 illustrates the simplest form of interior type rotor, where the magnet has simply been moved to the interior of the rotor. To reduce leakage between the two PMs, a nonmagnetic slit is introduced into the lamination to provide a "flux barrier" with a different purpose thereby reducing leakage between the two magnets.

Theory and Steady-State Analysis

PM synchronous machines are closely related to conventional synchronous machines in performance and configuration, and the stators of most synchronous machines are identical. Therefore the theoretical analysis of PM synchronous machines as viewed from the stator terminals follows closely conventional synchronous machine theory. The following development follows conventional synchronous machine theory. The following equations and analysis are on a per-phase basis, as is customary in machine analysis. Subscripts have the following meaning:

d = direct axis component

q = quadrature axis component

m = a mutual reactance component

r = rotor component

ℓ = leakage reactance or resistance component

i = internal component (behind leakage impedance)

o = excitation voltage

c = core loss component

Uppercase symbols are used to represent rms values. Standard symbols are used for impedances, voltages, currents, and power. In addition, θ = power factor angle; δ = power (or torque) angle; a = number of parallel paths in armature; m = number of phases; Z = total armature conductors; V = terminal voltage; I = terminal current/phase; and K_w = winding factor, which is the product of distribution, pitch, and skew factors.

The following analysis proceeds in terms of components along the direct axis (pole axis) and the quadrature axis (interpole axis). The analysis is based upon the interior configuration but is also applicable to the peripheral and claw-pole configurations and, with some modification, to the transverse configuration. Note that in the transverse type, the magnetic circuit is similar to that of a reluctance machine and reluctance torque may be a major component of the total electromagnetic torque. In transverse types, the PMs are usually located in the interpole region, that is, on the quadrature axis.

We first consider a synchronous motor neglecting core losses and operating under steady state. Trigonometric considerations of the phasor diagram for an underexcited synchronous motor (Fig. 7.10) result in the following:

$$\begin{aligned} V \sin \delta &= I_q X_q - I_d R_\ell \\ V \cos \delta &= E_o + I_d X_d + I_q R_\ell \end{aligned} \tag{7.8}$$

Similar phasor diagrams may be drawn for synchronous alternators by reversing the sign of the reactance and resistance drops of Fig. 7.10; also, of course, the overexcited case can be considered by assuming (for a motor) a leading power factor. Solving for the component currents gives

$$I_d = \frac{V(X_q \cos\delta - R_\ell \sin\delta) - E_o X_q}{X_d X_q + R_\ell^2} \tag{7.9a}$$

$$I_q = \frac{V(R_\ell \cos\delta + X_d \sin\delta) - E_o R_\ell}{X_d X_q + R_\ell^2} \tag{7.9b}$$

The total input power to m phases is due to the current components in phase with the terminal voltage, V,

$$P_{\text{in}} = mV(I_q \cos\delta - I_d \sin\delta) \tag{7.10}$$

which can be modified to eliminate the functions of the torque angle, δ, as

$$P_{\text{in}} = m\left[I_q E_o + I_d I_q (X_d - X_q) + I^2 R_\ell\right] \tag{7.11}$$

The electromagnetic power developed at the airgap can be expressed as

$$P_e = m\left[I_q E_o + I_d I_q (X_{md} - X_{mq})\right] \tag{7.12}$$

which may also be expressed in terms of conventional theory as

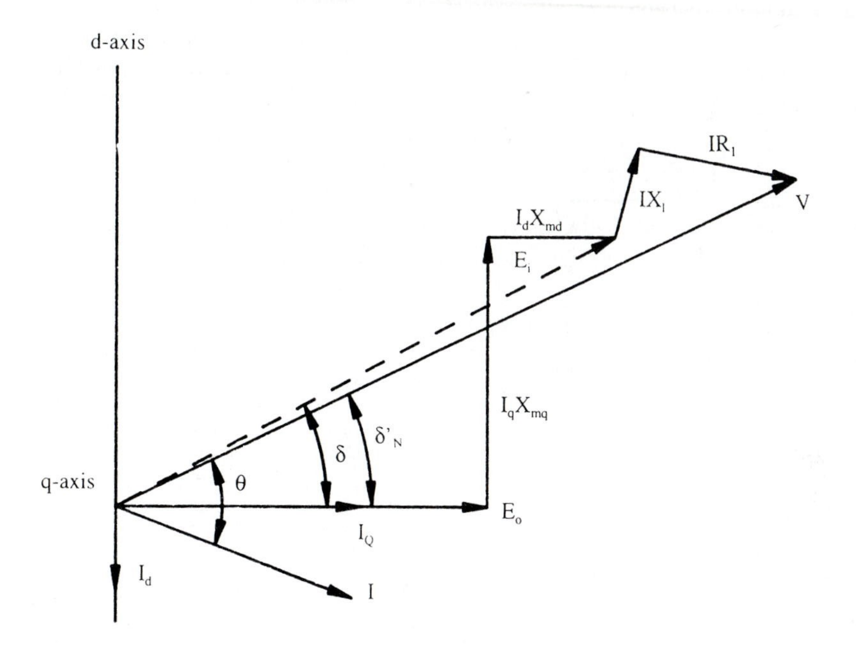

FIGURE 7-10
Phasor diagram for PM synchronous motor.

$$P_e = m\left[I_q E_{iq} - I_d E_{id}\right] \tag{7.13}$$

Another expression (in terms of the "interior torque angle," δ_i) is given as

$$P_e = m\left[\frac{E_o E_i}{X_{md}} \sin\delta_i + \frac{E_i^2 (X_{md} - X_{mq})}{2X_{md}X_{mq}} \sin 2\delta_i\right] \tag{7.14}$$

Note that

$$\begin{aligned} X_d &= X_{md} + X_\ell \\ X_q &= X_{mq} + X_\ell \end{aligned} \tag{7.15}$$

As a good approximation to (7.14), the terminal—rather than internal—values may be used, which means substituting V for E_i, X_d for X_{md}, δ for δ_i, etc.

When core losses are included, the above equations are modified, and these modifications may have considerable effect upon machine performance. Core loss is included by adding a resistor in parallel with the internal or mutual impedances of the machine. For a motor, the equivalent core-loss resister, R_c, would appear as shown in Fig. 7.11. It may be shown that the inclusion of R_c modifies the equivalent machine reactances as follows:

$$\varepsilon = \frac{\alpha_d \alpha_q}{1 + \alpha_d \alpha_q}; \quad \alpha_d = \frac{X_{md}}{R_c}; \quad \alpha_q = \frac{X_{mq}}{R_c} \tag{7.16}$$

$$\begin{aligned} \bar{X}_d &= X_d - \varepsilon X_{md} \\ \bar{X}_q &= X_q - \varepsilon X_{mq} \end{aligned} \tag{7.17}$$

$$\bar{R}_\ell = R_\ell - \varepsilon R_c \tag{7.18}$$

The phasor diagram for a motor in which core losses are included is shown in Fig. 7.12. The principal difference between this diagram and that of Fig. 7.10 is that the new current, $\bar{I}$, flows through the leakage impedances.

Power expressions for the case when core losses are included are similar to those given above, with the new current, $\bar{I}$, and its components substituted for I, I_d, and I_q. For example, (7.10) can be used directly with the substitute currents. Thus,

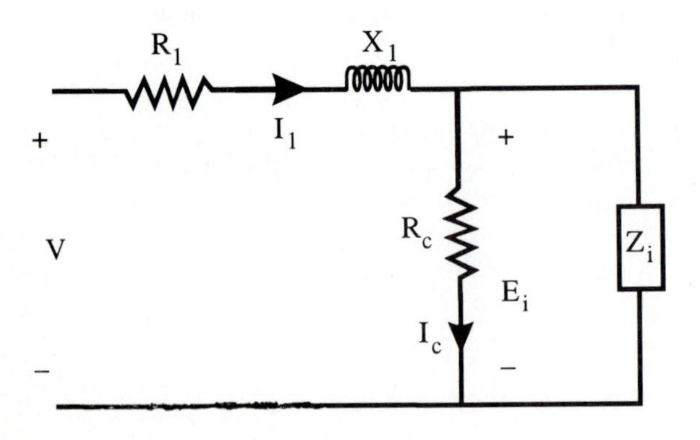

FIGURE 7-11
Equivalent circuit of a PM synchronous motor.

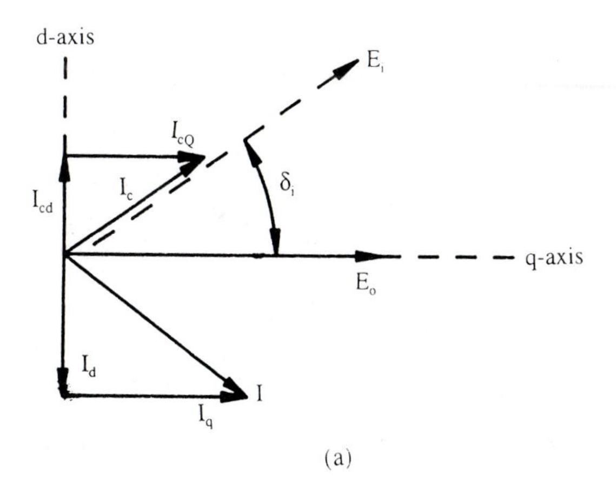

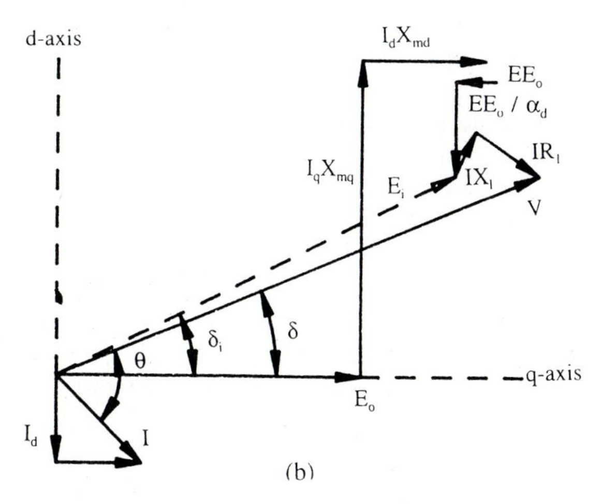

FIGURE 7-12
(a) Phasor diagram showing current, I_c, associated with core losses. *(b)* Complete phasor diagram including the effects of core losses.

$$P_{in} = m\left[\bar{I}_o E_q + \bar{I}_d \bar{I}_q (\bar{X}_d - \bar{X}_q) + \bar{I}^2 R_\ell + \varepsilon E_o (\bar{I}_d / \alpha_d - \bar{I}_q)\right] \tag{7.19}$$

The equations for internal electromagnetic power, P_e, are unchanged by the inclusion of core losses. In terms of power components, a simpler expression for P_e is

$$P_e = P_{in} - P_c - mI_l^2 R_\ell \tag{7.20}$$

where P_c is the assumed core loss (the loss in R_c in Fig. 7.11). In a motor, the output shaft power is

$$P_s = P_e - P_{wf} \tag{7.21}$$

where P_{wf} is the motor windage and friction loss. And, finally, the output shaft torque, in N-m, is

$$T_s = \frac{P_s}{4\pi f / P} \tag{7.22}$$

There are several interesting points to note in the above analysis. Although this analysis follows closely that of conventional synchronous machines, there are several important differences. In many types of PM machines, including the common interior types, $X_d < X_q$, and in most other types the two reactances are fairly close to each other in magnitude. Both of these conditions are in contrast to the conventional salient pole synchronous machine. The principal cause of these reactances' relative values is the very low permeability of the PM itself, which is located in the direct axis in interior types and many other configurations. The implications of these reactance values should be noted with reference to (7.14). The second term of this equation is normally known as "reluctance torque," and, in many types of PM machines, this term may be negative. It is only in the transverse types, which are basically reluctance machines, that this term contributes appreciable positive torque. In most other types of PM machines, the first term of (7.14) is the major contributor to torque development. As a corollary to this reactance relationship, the maximum power developed by an interior type—and some other types—of PM machines occurs at a torque angle, δ, greater than 90°, as would be expected, whereas maximum power occurs at torque angles less than 90° in conventional salient pole machines.

PM Synchronous Motor Starting and Torque Considerations

The synchronous machine itself, of course, develops no starting torque and has little capability for bringing itself up to synchronous speed. A phenomenon known as "hunting"—a condition of mechanical oscillations about synchronous speed—occurs when a synchronous machine is slightly out of synchronous speed as a result of a variety of causes, usually related to transient surges of load power or input line voltage or synchronization problems in paralleled machine systems. The standard remedy for the inherently unstable synchronous machine is to add an induction machine winding to the rotor, often called a damper or Amortisseur winding, which tends to damp the above-mentioned oscillations and enhances the capability of the machine to pull back into the synchronous speed condition, from which it can again operate as a truly synchronous machine. A by-product of the use of a damper winding is that a synchronous machine can start from zero speed as an induction motor. Therefore, most synchronous machines of power rating above a few hundred watts—including PM machines—have an electrical winding on the rotor similar to the cage winding of most induction machines. Very small synchronous motors use other methods of starting, such as the shaded-pole or split-phase techniques.

There are several ramifications of the use of cage windings on PM synchronous machines. Some of these are:

1. The PM flux interacts with the stator currents during any asynchronous operation of the rotor above zero speed and produces a drag torque which *decreases* the induction motor torque available for starting and accelerating the motor.

2. Besides supplying starting and stabilizing torques, the cage winding acts as a means of shielding the PMs from the high demagnetizing field caused by the armature currents during starting and other asynchronous operations. In rotors without cage windings, aluminum shield or copper plating is sometimes used for a similar purpose. However, there is a pulsating component of torque, associated with which are pulsating components of current that may cause demagnetization of portions of the PMs not protected by the cage bars.

3. The mechanical layout and design of the rotor is, of course, altered when cage bars are used.

7.6 APPLICATIONS OF PERMANENT MAGNET MACHINES

Permanent magnet (PM) machines are rapidly finding numerous applications. Here we outline some of the potential and current applications of PM machines.

Power Alternators

Permanent magnets have supplied excitation in alternator applications for many years. Essentially, a PM alternator is similar in configuration to a conventional synchronous alternator with the electrical excitation system replaced by permanent magnets. There are several benefits of this arrangement, including the elimination of the brush/slip ring system—particularly advantageous for aerospace applications—and greater design flexibility. The principal disadvantage is the loss of field control and, hence, reactive power control. Therefore, PM alternators are generally not suitable for parallel operation except in cases of fixed load power and power factor. Also, when alnico and other early magnet materials were used, magnet demagnetization due to load dump or other armature current surges was likely to occur, and provisions for frequent magnetization of the magnets were required. With ceramic and rare-earth magnets, demagnetization is less of a problem but still a possibility which must be considered. This is probably the principal reason that PM alternators have not been widely adopted in automotive applications. Recently, many innovative configurations have been developed which divert the demagnetizing armature reaction from the path of the PMs during current surges. Demagnetization due to thermal and mechanical shocks must also be considered in the application of PM alternators.

FIGURE 7-13
PM brushless axial-airgap alternator with strip-wound coil.

In aerospace applications, the PM alternator and polyphase motor are in competition with several other brushless synchronous machines, namely the inductor, Lundell, rotating rectifier, and various reluctance configurations. In applications where cost, high efficiency, and minimal volume are less significant than reliability, machines with electrical excitation systems or using line excitation are more frequently used. PM machines have therefore found niches in specialized configurations where unusual size or shape or high power in a small volume is necessary. Examples of such designs include tiny (7 W) turboalternators used as parasitic loads on high-speed turbines (up to 320,000 rpm) in cryogenic environments, the rotor consisting of a solid cylinder of chrome steel magnetized for two poles. Another unusual design used in an electric torpedo application is shown in Fig. 7.13; this is a pancake, "ironless stator" design with a short lifetime design but achieving efficiencies well over 90% and a power density of 1.2 kW/lb.

A more conventional application of PM alternators is as exciters or pilot exciters for conventional synchronous machines. This environment is good for PM application and involves few of the electrical, thermal, or mechanical shocks common to aerospace applications. In general, PM exciters offer good cost, size, and maintenance characteristics. For a 70-kVA PM pilot exciter, a comparison between an alnico field and a rare-earth (cobalt) field on the exciter with the same stator design in each case shows that with rare-earth magnets, output rating increased by 100%; voltage regulation decreases from 17% to 10.6%; and the magnet mass density (kg/kVA) decreases by almost 70%. Of further interest is the fact that the cobalt content of the required rare-earth magnets was only 60% that of the required Alnico magnets.

Automotive Applications

By far the largest user of PM machines today is the automotive industry. General Motors Corp. is the largest consumer of permanent magnets in the world, and the other large worldwide automotive companies follow closely behind. Most of the many motors used on modern automobiles and trucks today are PM-excited. Since these motors are produced in about as high quantities as any other motor, it follows that the economics of PM field excitation favor the PM. The one rotating machine in passenger cars that is not PM-excited is the alternator, due mainly to transient demagnetization considerations. However, PM alternators are used in other automotive applications such as auxiliary power supplies in trucks and off-road vehicles. Possible future applications of PM machines in vehicles include electric power steering, electric brakes, and certain types of valve controls. Most present-day vehicular motors use ceramic magnets. The gradual introduction of rare-earth magnets may increase the use of PM machines in automotive applications.

The starter motor is a somewhat unique rotating machine. It is perhaps the most widely produced rotating machine in the world and has been cost-reduced down to the last cent. During the electric vehicle developments of the 1970s, it was noted that the passenger car starter motor has the lowest cost density (cost/unit weight) of any known motor. The starter motor was the principal motivation for the development of arc-shaped ceramic magnets, which are still used on most starter motors. One of the most exciting recent developments is the design and construction of a starter motor using NdFeB magnets, which are flat rather than arc-shaped. This motor reduced starter motor weight by about 50% over the previously used ceramic magnet starters in the same vehicle. Although the starter motor is not used for a continuously operated application, it does exist in a hostile environment which includes a temperature range of -40°F to 125°F, much contamination and corrosion, and the ability to crank the engine to which it is mated until the battery energy is unable to crank the engine. It is a reliable commutator motor and even uses brush-shifting as a means of reducing the deleterious effects of armature reaction.

Vehicular Electric Drive Motors

A relatively recent development in applying rare-earth magnets to advanced rotating machines relates to a 40-hp, brushless motor using rare-earth-cobalt PMs. This motor has been extensively tested and shows much promise for electric drive applications, including nonautomotive. Full electronic control systems using power semiconductors have been developed and tested for motor control over a wide speed and torque range. Synchronous PM motors with electronic control, including feedback, generally require no cage winding on the rotor for starting or stabilizing, as discussed in Section 7.5.

Applications in Textile and Glass Industries

A growing application of variable speed drives similar to the automotive application noted above is in the textile and glass industries. In these industries, early drives were generally of the dc commutator type with various types of control. Later, induction motors were used in some applications. More recently, both reluctance and PM motors have been used with power semiconductor controllers. In the synthetic fiber and glass industries, it is often important that the speed of different machines be identical and exactly related to the supply frequency. This capability of synchronous machines gives them an advantage in such applications, and the low cost of reluctance and PM synchronous machines and the fact that they are brushless have resulted in increasing application in synchronized, variable-speed drives. A further advantage of the PM drive is its high efficiency and high power factor, especially at partial loading, compared to both reluctance and induction drives.

With the widespread acceptance and decreasing costs of modern power semiconductor controllers, PM machines will offer the advantages of minimum motor weight, high efficiency, greatly reduced rotor thermal loading, and high rotor reliability in almost all variable-speed drive applications.

Small Appliances

Small appliance motors have traditionally been of the ac series or "universal" configuration for appliances powered from the house 120-V system. In many types of appliances, such as mixers and hand-held drills, the rapidly declining speed as a function of load torque, which is characteristic of series motors, is a desirable feature. The relatively large friction in small appliance systems prevents excessive speed at light loading. Multiple speed ranges are achieved by series resistors, or, in larger motors, by winding reconnection, or in some cases by semiconductor phase control. The disadvantages of an ac series motor are its brush/commutator system and the resultant electrical and acoustical noise. Commutation problems are generally more severe in ac series machines than in dc machines of similar size, and more compensating circuitry is required. Also, brushes are usually the chief cause of failure or excessive maintenance on such appliances.

There are few, if any, reasons for the use of conventional dc PM motors in ac-powered appliances such as mixers, vacuum cleaners, and drills. More control circuitry will be required, stability at overloads may be a problem, and the brush/commutator is still required with added electromagnetic interference (emi) due to the required rectification. However, in certain applications where a more constant speed with load is desirable, the conventional PM motor may offer a reduced cost and reduced maintenance due to improved commutation. If the rectification is provided by thyristors or transistors, a controlled variable-speed characteristic can be simply obtained.

A rapidly growing market for PM motors in small appliances is the battery-powered, portable appliance application. Portable tools have become popular in many activities, and portable kitchen appliances are also being introduced. Whereas conventional dc-type PM motors can be used for battery-powered applications with very simple control schemes, these systems must be designed to prevent demagnetization of the PMs and unstable speed conditions at heavy loads. Brushless dc PM motors appear to offer the best solution for battery-powered applications and, since considerable electronic circuitry is required for operation of the brushless motor itself, very good speed/torque control can also be achieved with essentially the same power circuitry at little additional cost. Motor maintenance and downtime are reduced to a minimum with brushless dc motors.

Control Motors

PM machines are used in many types of dc control and instrumentation applications, including control motors, instrument drives, servomotors, resolvers, torque motors, and tachometers. Both commutator and brushless PM machines are extensively applied in control systems, and their use will continue to expand. Commutator motors generally have a cost advantage over wound-field motors in the sizes used in control applications and can easily be designed to give various control characteristics. Brushless motors eliminate maintenance problems and can give more flexible control characteristics. Both classes of machines have great flexibility in size, shape, and geometry. Rare-earth magnets are used in many control motors where high torque per unit volume is important.

PMs are also used in many sensing applications, especially the sensing of speed and position. Speed sensors called "reluctance sensors" are essentially reluctance generators with PM excitation. Wiegand speed sensors also require PM excitation. Magnetic encoders are PM-excited. PMs are also used in many damping functions in control and instrumentation applications, one of the oldest being the damping of the watthour-meter disk.

PM control motors and related electromagnetic devices already play a major role in control applications, and as the cost structure and technical characteristics of PMs improve, this role will certainly expand.

Computer and Robotics Applications

One of the largest applications of PM motors, primarily brushless motors, is in the computer industry—for obvious reasons: reduced noise levels, quiet operation, the ability of precise speed and torque control, and flexibility of shape and geometry. Most disk drives use PM brushless motors, called spindle motors. This has become one of the highest volume applications for spindle motors, which are generally of the "outer-rotating" type, i.e., with the PM rotor outside of the stationary (stator) winding system. The principal requirements for spindle motors are high starting torque, very precise speed regulation, and near-zero drift in characteristics with time and temperature. PM motors are also used in most other computer applications where controlled motion is required, such as printers and tape drives. Many PM motors used in motion control are digital or stepper motors of PM or hybrid configuration.

Another rapidly growing application of PM motors is in the areas of robotics and other motion control functions in automated manufacturing. The merits of PM motors discussed above in connection with computer applications apply equally to robotics applications, and most existing applications use this class of machines, usually in brushless configuration. DC brushless PM motors integrate well into the high degree of feedback required for precise motion control. Feedback control is generally digital, and this is also compatible with the controls required for dc brushless motor operation. There will surely be much future expansion in this application of PM machines.

Printed-Circuit Motors

The term, "printed-circuit motor" is both a trade name of a manufacturer and a generic name given to a certain type of PM motor. A printed-circuit armature is shown in Fig. 7.14 (which is the same as Fig. 6.13). The name was originally adopted to describe the process by which armature conductors were formed, that of photochemical printing similar to that used in some types of PC boards. More recently, most printed-circuit motor armature windings are formed by stamping processes. The other unique feature of printed-circuit motors is the elimination of the conventional brush/commutator system. Connection to the armature conductors is made directly by a conductor which is part of the external leads of the motor, and this contact also supplies the current commutating function. This design feature saves space in the overall motor package and is amenable to axial airgap geometry, which is the most common construction of printed-circuit motors. PM excitation also conserves space in the overall package.

PM printed-circuit motors are used in many control applications and some power applications, the latter in power ratings up to 5 hp. In control applications, printed-circuit motors have very high acceleration capabilities compared to cylindrical motors and can operate at higher speeds (up to 12,000 rpm). The disc shape offers advantages in many applications, and printed-circuit motors can often give an overall

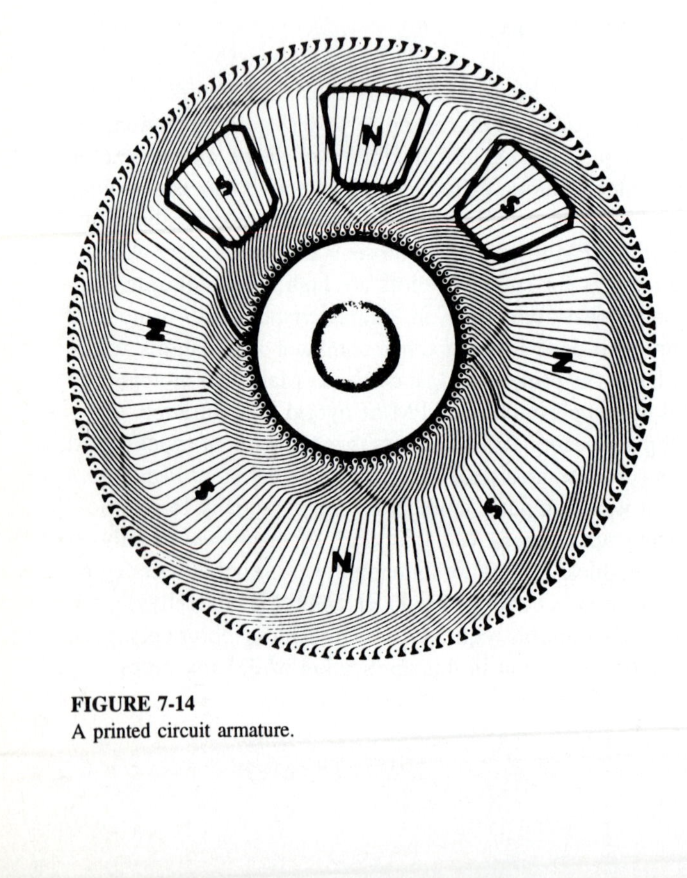

FIGURE 7-14
A printed circuit armature.

reduction in volume compared to cylindrical control motors. Most printed-circuit motors are of the "ironless armature" construction discussed previously and are, therefore, frequently lighter in weight than cylindrical control motors of similar power rating.

Adjustable-Speed Drives

There has been greatly increased interest in adjustable-speed drives (ASDs) in recent years, not only for traction applications, but also as energy-conserving drives for centrifugal pumps, blowers, and compressors, and in many other applications. The commercial and military application of ASDs has become economically feasible, mainly due to the tremendous performance improvements and declining cost structure of power semiconductors. Many types of motors are suitable for ASDs and are being applied today. The principal merits of PM synchronous motors in ASDs are high efficiency and high-speed capabilities. Since drive train efficiency is very important in realizing yearly energy savings when converting from a fixed-speed to variable-speed drive for, say, a 1000-hp pump, motor efficiency is significant. An improvement of a few percentage points in motor efficiency by using a PM motor can result in annual energy savings worth hundreds of dollars.

Both surface and interior type PM motors are being studied in ASD applications. ASD electronic control generally consists of a dc link inverter with PWM or six-step control (see Chapter 8), similar to that used in the more common induction motor drives. However, instead of the complex slip control with current and slip frequency feedback, the PM drive requires only rotor position feedback (with current limit to protect the power semiconductors). Position feedback permits electronic synchronization of the inverter frequency and can achieve very stable and self-starting motor operation. Hence, the damper or squirrel-cage winding is not required on PM machines with electronic control. This results in a simpler and less costly rotor design for such motors.

PROBLEMS

7.1. For the magnetic circuit of Fig. 2.20 (Chapter 2) we have $l_g = 1$ mm, $l_m = H = 5$ cm, and core cross section = 9 cm^2 (for the entire circuit). The magnet is made of Alnico V (see Fig. 2.4). Calculate the airgap flux density. Assume that the iron portion of the circuit is infinitely permeable, and neglect leakage and fringing.

7.2. Refer to the magnetic circuit of Fig. 2.20, and the data given in Problem 7.1. Obtain a relationship between B_m, the magnet flux density, and H_m, the field intensity in the magnet. Hence define the load line, and graphically determine the airgap flux density.

7.3. In (2.35) (Chapter 2) we have derived an expression for the magnet volume. Now, the magnetic circuit of Fig. 2.20 is modified by shaping the "poles" and thus reducing the cross section of the airgap to 6 cm^2. Determine the minimum magnet volume to obtain a 0.8-T airgap flux density. Use the data given in Problem 7.1.

7.4. The permanent magnet material used in a motor has a profound effect on the characteristics of the motor. For example, a ceramic (ferrite) magnet results in a motor that operates at a relatively low airgap flux density but that can sustain relatively high levels of reaction without demagnetization. Compare the ceramic motor with one using Alnico VI (Fig. 2.18 and Table 2.4) in the following steps.

(a) Plot the intrinsic demagnetization curve for Alnico VI. What is H_{ci} for Alnico VI? The normal demagnetization curve will also be necessary.

(b) It is seen that Alnico VI must be operated at a much higher permeance ratio than the ceramic material. Since the permeance ratio is a function of the magnetic permeance of the external magnetic circuit, mainly the airgap length, what does this imply concerning the length of practical airgaps usable in an Alnico VI motor as compared to those in ceramic motors?

(c) Assume a permeance ratio (B_d/H_d) of 50 for Alnico VI. What permeance ratio is used in the case of a ceramic magnet?

(d) What armature reaction (in terms of a field intensity, H_a) can be tolerated in the Alnico VI motor before the flux density drops to 0.8 B_r?

7.5. A permanent magnet dc motor develops a maximum torque of 2 N-m at an airgap flux density of 0.6 T. The axial length of the motor is 6 cm and the bore is 10 cm. Calculate the required electrical loading of the armature.

7.6. If the motor of Problem 7.5 has 2 poles and 144 active conductors, calculate the armature current. Also, determine the speed in rpm at which the motor develops a maximum torque, if the armature induced voltage at maximum torque is 36 V.

7.7. In a three-phase, 230-V, wye-connected PM synchronous motor, the *d*-axis and *q*-axis reactances are equal, each having a per phase value of 1.1 Ω. The armature resistance and the core losses are negligible. Calculate the power developed by the motor if it operates at a 45° power angle, and internal induced voltage/phase is 127 V.

7.8. Repeat Problem 7.7 if the armature resistance is 0.2 Ω/phase and the internal induced voltage/phase is 100 V.

CHAPTER 8

POWER ELECTRONICS

In Chapters 4 through 7 we have discussed the three basic types of dc and ac machines. Most methods of motor control involve switching operations—switches may be required to be opened or closed to achieve the desired goal. Modulation of power by turning switches on or off can be accomplished by mechanical switches, such as contactors, or by solid-state electronic switches, such as transistors and thyristors. Because power levels in electric motors and power systems are high compared to those in conventional electronic circuits (such as amplifiers, oscillators, etc.), the study of electronic circuits pertinent to electric machines and power systems is known as power electronics. Thus the scope of power electronics includes the applications of solid-state switches* to the control and modulation of power in electric motors and electric power systems.

There is a great variety of solid-state components and systems used to control electric motors. In terms of analysis and applications, no other aspect of electric machines has undergone such dramatic changes in recent years or holds greater potential for improving machine characteristics in the future than does the solid-state control of electric machines. Similarly, in application to the protection of electric power systems against faults and in high-voltage dc transmission, solid-state switches hold great promise.

* We will restrict ourselves to solid-state devices, as gas-filled and vacuum tubes are becoming obsolete.

In this chapter we discuss the various solid-state devices used in power electronics. This discussion will be from a circuit viewpoint, and the physics of semiconductors will not be included. Next, we review waveform analysis because invariably the output waveforms from solid-state switching devices are nonsinusoidal. This is followed by several dc and ac motor control schemes, including a brief review of thyristor commutation (or turn-off) techniques.

8.1 POWER SOLID-STATE DEVICES

Many types of solid-state devices exist which are suitable for power electronics applications. For a specific application, the choice depends on the power, voltage, and current requirements; environmental considerations such as ambient temperature; circuit considerations; and overall system cost. Some of the devices commonly used in power electronics circuits are presented in the following. In most cases, the maximum ratings such as voltage, current, and time of response of the device are also given. However, these capabilities are seldom achievable simultaneously in a single device. In practice, a device is chosen primarily either for its voltage or current rating. The devices listed here are *pn*-junction devices, having two layers as in a silicon rectifier; three layers, as in a power transistor; or four layers, as in a thyristor. Let us now consider these devices in some detail. [*Note*: In the symbolic representation of the device, we have A = anode; G = gate; K = cathode; B = base; C = collector; E = emitter.]

Silicon Rectifier

Silicon rectifiers are high-power diodes capable of operating at high junction temperatures. The principal parameters of a silicon rectifier are the repetitive peak reverse voltage (PRV) or blocking voltage, average forward current, and maximum operating junction temperature (≃ 125°C). The terminal *i-v* characteristic of a typical

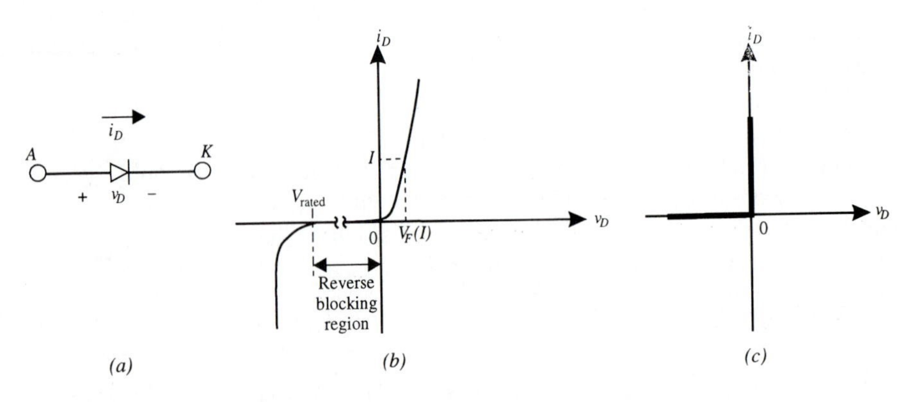

FIGURE 8-1
Silicon rectifier: *(a)* symbol, *(b)* *i-v* characteristics, *(c)* idealized characteristics.

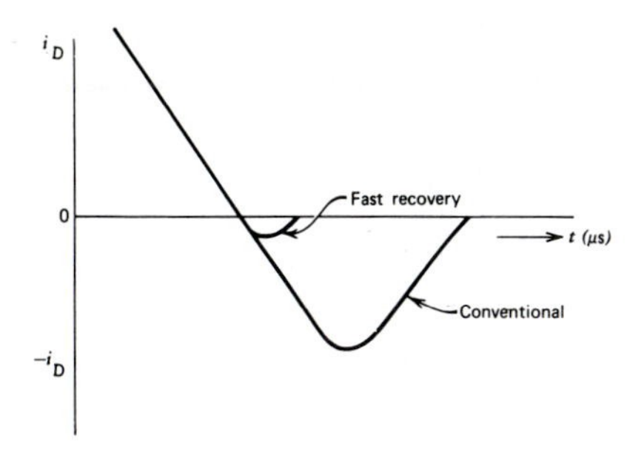

FIGURE 8-2
Comparison of conventional and "fast-recovery" silicon rectifiers in the reverse-recovery region.

silicon rectifier is shown in Fig. 8.1, which also shows the switching characteristic of an ideal diode. The silicon rectifier has a forward voltage drop of about 1 V at all current levels within its rating.

After the forward current in a silicon rectifier has ceased, a reverse current flows for a very short time. The rectifier assumes its full reverse blocking after the reverse current goes to zero. This characteristic of a diode is known as the reverse *recovery performance,* and the time interval during which the reverse current flows is known as the *recovery time*, which is of the order of a few microseconds. The recovery time determines the rate at which blocking voltage can be reapplied to the diode and thus governs its frequency of operation. For applications requiring very short recovery times (of the order of several hundred nanoseconds), fast-recovery devices have been developed. Fig. 8.2 compares the recovery characteristics of conventional and fast-recovery silicon rectifiers. Maximum voltage and current ratings of rectifier diodes are in the range of 5000 V and 7500 A.

Two of the major applications of silicon rectifiers in power electronics are as freewheeling diodes (providing a bypass for the flow of current) in motor controllers and, in general, as rectifiers.

Silicon-Controlled Rectifiers or Thyristors

A silicon-controlled rectifier (or *SCR*), also known as a *thyristor*, is a four-layer *pnpn* semiconductor switch. Unlike the diode, which has only two terminals—anode and cathode—the thyristor has three terminals—anode, cathode, and gate. The reverse characteristic of the thyristor is similar to that of silicon rectifier just discussed. However, the forward conduction of a thyristor can be controlled by utilizing the gate. Normally, a thyristor will not conduct in the forward direction unless it is "turned on" by applying a triggering signal to the gate. However, the full conduction in a thyristor is not instantaneous. We define the turn-on time, t_{on}, when the anode current reaches 90 percent of its final value. Once the thyristor starts to conduct, it continues to do so until turned off by external means. The turn-off of the thyristor is known as *commutation*. A thyristor is symbolically represented in Fig. 8.3*(a)* and has the *i-v*

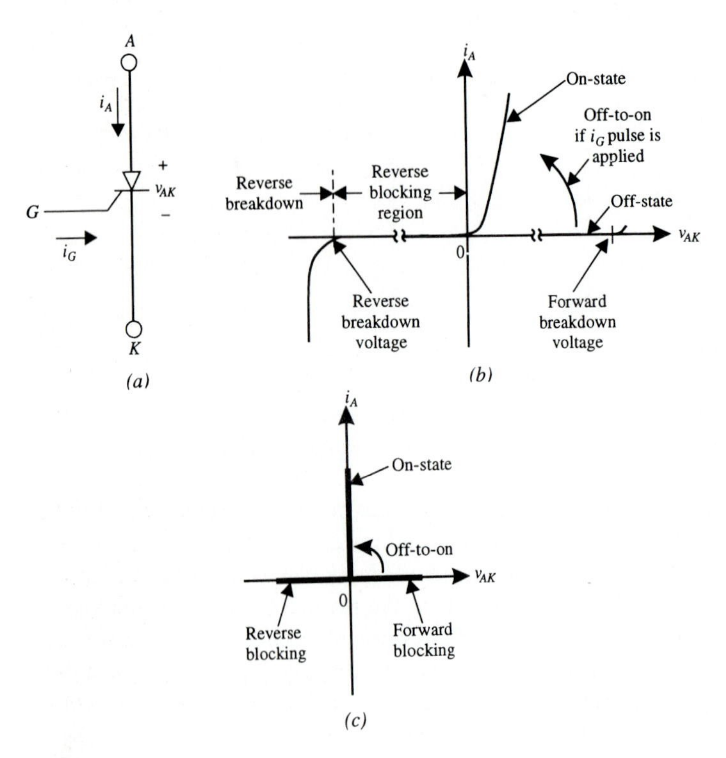

FIGURE 8-3
Thyristor: *(a)* symbol, *(b)* *i-v* characteristics, *(c)* idealized characteristics.

characteristics as shown in Fig. 8.3*(b)* and *(c)*. The maximum rating of the thyristor is in the range of 5000 V and 3000 A. The principal parameters characterizing the thyristor are as follows:

1. Repetitive peak reverse voltage (PRV).
2. Maximum value of average on-state current; this parameter is related to the heating within the semiconductor.
3. Maximum value of rms on-state current; this is the current rating of metal conductor portions of the device, such as the anode pigtail in stud devices.
4. Peak one-cycle on-state current; this is the surge current limit.
5. Critical rate of rise of forward blocking voltage. There are usually two ratings: initial (when the device is turned on) and reapplied (following commutation).
6. Turn-off time—the off-time required following commutation before forward voltage can be reapplied.

7. Maximum rate of rise of anode current during turn-on; too high *di/dt* may result in local hot-spot heating, a main cause of device failure.

8. Maximum operating junction temperature.

Thermal management of the thyristor is extremely critical in all applications. Most of the parameters just mentioned vary considerably as a function of device temperature. Therefore, much of the engineering required in the application of thyristors is in the design of its heat sink, mounting method, and auxiliary cooling (if required). The use of thyristors in series or parallel electrical connections generally assists in meeting thermal requirements.

Triacs

The triac, often called a bidirectional switch, is approximately equivalent to a pair of back-to-back or antiparallel thyristors fabricated on a single chip of semiconductor material. Triggered conduction may occur in both directions, that is, the triac is a quasi-bilateral device. Triac applications include light dimming, heater control, and ac motor speed control. The parameters listed earlier as important in the application of thyristors generally apply also to the triac. However, it should be noted that the triac is a three-terminal device with only one gate, which has an effect on its time response compared with that of two distinct thyristors connected in antiparallel position. The turn-off time of a triac is in the same order of magnitude as that of a thyristor. This implies that a time period approximately equal to the turn-off time must be observed before applying reverse voltage to a triac. In an antiparallel pair, however, reverse voltage can be immediately applied after cessation of forward current in one thyristor. Triacs are not available in as high voltage and current ratings as thyristors at the present time and therefore are used in control of motors of relatively low power ratings.

Symbolically, a triac is shown in Fig. 8.4. Triacs have a rating in the range of 1000 V and 2000 A, with a response time of 1 μs.

Diverse Thyristors

In addition to the triac, other forms of thyristors include the following:

1. *Gate turn-off thyristor* (GTO). This thyristor can be turned off at a high temperature, and a normal commutation circuit is therefore not required. This type

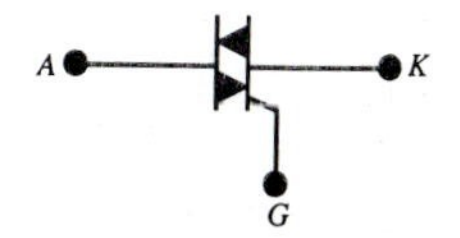

FIGURE 8-4
Representation of a triac.

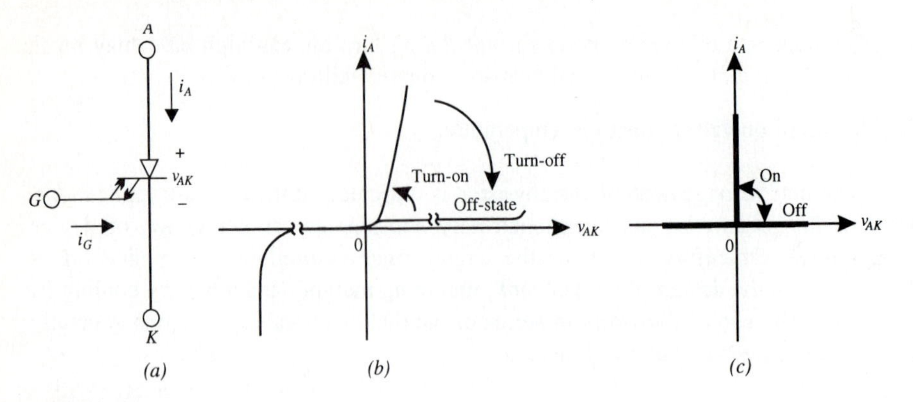

FIGURE 8-5
Gate turn-off thyristor (GTO): *(a)* symbol, *(b)* *i-v* characteristics, *(c)* idealized characteristics.

of thyristor has a high blocking voltage rating and is capable of handling large currents.

2. *Gate-assisted thyristor* (GAT). This thyristor requires large power for triggering. It has a small turn-off time and is specially suited for series-type inverters.

3. *Light-activated thyristor* (LAT or LASCR). This thyristor is turned on by photons of light. Such thyristors find application in high-voltage dc transmission.

Symbolic representation of the gate turn-off thyristor is shown in Fig. 8.5. These thyristors have voltage ratings in the range of 400 to 1000 V with a current rating of about 200 A and a response time of 0.2 to 2.0 μs.

Thyristors, when turned on, have *i-v* characteristics similar to that shown in Fig. 8.5*(b)* and *(c)*.

Power Transistors

When used in motor control circuits, power transistors are almost always operated in a switching mode. The transistor is driven into saturation, and the linear gain characteristics are not used. The common-emitter configuration is the most common because of the high power gain in this connection. The collector-emitter saturation voltage, $V_{CE(SAT)}$, for typical power transistors is from 0.2 to 0.8 V. This range is considerably lower than the on-state anode-to-cathode voltage drop of a thyristor. Therefore, the average power loss in a power transistor is lower than that in a thyristor of equivalent power rating. The switching times of power transistors are also generally faster than those of thyristors, and the problems associated with turning off or commutating a thyristor are almost nonexistent in transistors. However, a power transistor is more expensive than a thyristor of equivalent power capability. In addition, the voltage and current ratings of available power transistors are much lower than those of existing thyristors. It has already been stated that the maximum ratings of a power semiconductor are generally unobtainable concurrently in a single device. This is particularly true of power transistors. Devices with voltage ratings of 1000 V or above have limited current ratings of 10 A or less. Similarly, the devices with higher current ratings, 50 A and above, have voltage ratings of 200 V or less. For handling motor control requiring large current ratings at 200 V or below, it has been common to parallel transistors of lower current rating. This requires great care to ensure equal sharing of collector currents and proper synchronization of base currents among the paralleled devices.

Ratings of significance for motor control application include:

1. Breakdown voltage, specified by the symbols BV_{CEO}, collector-to-emitter breakdown voltage with base open, and BV_{CBO}, collector-to-base breakdown voltage with emitter open.

2. Collector saturation voltage, $V_{CE(SAT)}$.

3. Emitter-base voltage rating, V_{EBO}.

4. Maximum collector current, I_C, average and peak.

5. Forward current transfer ratio, H_{FE}, the ratio of collector to base current in the linear region.

6. Power dissipation.

7. Maximum junction temperature, typically 150 to 180°C.

8. Switching times; rise time, t_r; storage time, t_s; and fall time, t_f. Sometimes these switching times are related to a maximum frequency of switching.

The thermal impedances and temperature coefficients are also important parameters. In parallel power transistors, the variation of device characteristics with temperature becomes especially significant. The I_C-V_{BE} characteristic is extremely temperature-sensitive.

Some of the commonly used power transistors are as follows.

Bipolar Junction Transistor (BJT)

The symbol and the *i-v* characteristics of a bipolar junction transistor (BJT) are shown in Fig. 8.6. The BJT requires a sufficiently large base current such that

$$I > \frac{I_C}{h_{FE}} \tag{8.1}$$

where h_{FE} (≈5 to 10) is the dc gain of the BJT to turn it fully on. In this case the voltage V_{CE} is about 1 to 2 V. A BJT is a current-controlled device and requires a

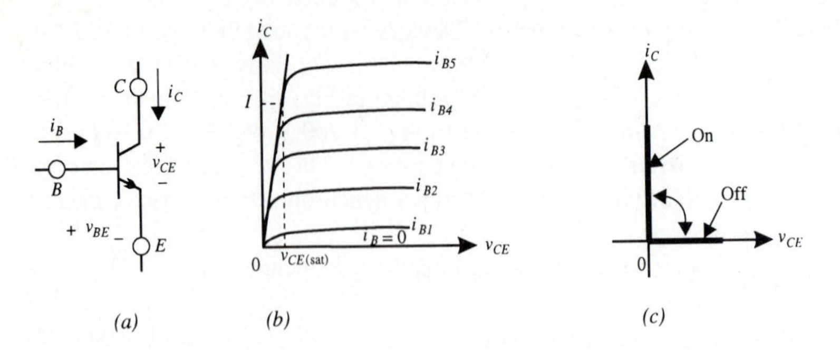

FIGURE 8-6
Bipolar junction transistor: *(a)* symbol; *(b)* *i-v* characteristics; *(c)* idealized characteristics.

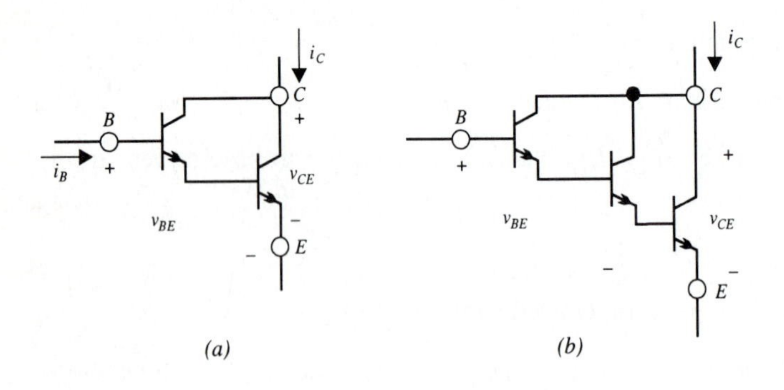

FIGURE 8-7
Darlington configurations: *(a)* Darlington; *(b)* triple Darlington.

continuous base current to operate in the on-state. A BJT may have a rating of about 1400 V and 200 A.

Power Darlington

This designation generally refers to the well-known Darlington-connected transistor pair fabricated on a single chip. The same characteristics are, of course, achievable through the use of two discrete transistors, albeit usually in a larger, more complex, and more costly package. The principal merit of the Darlington device is its high current gain. The operating parameters and failure modes discussed earlier for transistors are also applicable to the Darlington.

Darlington amplifiers are a recent entry into the area of motor controls but have met with considerable acceptance due to their potential for reducing the size, cost, and weight of motor controllers. These devices are used both in choppers for dc commutator motor control and in inverters for ac motor control, generally for lower-power applications. Recently, larger devices have been developed with ratings as high as 200 A and 100 V or 100 A and 450 V and have been applied to the control of traction motors used in lift trucks and industrial electric vehicles. Current gains as high as 1600 A have been achieved at these high current levels.

Two Darlington configurations are shown in Fig. 8.7.

Metal-Oxide-Semiconductor Field-Effect Transistor (MOSFET)

The symbol of a MOSFET and its characteristics are shown in Fig. 8.8. With a sufficiently large and continuous gate-source voltage, the MOSFET turns on. MOSFETs have ratings of over a 1000 V and 100 A. However, the large current rating comes at similar voltages.

Insulated Gate Bipolar Transistor (IGBT)

The symbol for an IGBT and its *i-v* characteristics are shown in Fig. 8.9. IGBTs have some of the advantages of the MOSFET, the BJT, and the GTO combined. However, MOSFETs have higher switching speeds. The ratings of IGBTs are up to 1200 V and 100 A.

Fig. 8.10 shows a family of power semiconductors: rectifiers, SCRs, and transistors. Disc type (or hockey puck) devices—one mounted and with heat sink—are shown in Fig. 8.11. Heat sinks and various components used in power electronics are shown in Fig. 8.12.

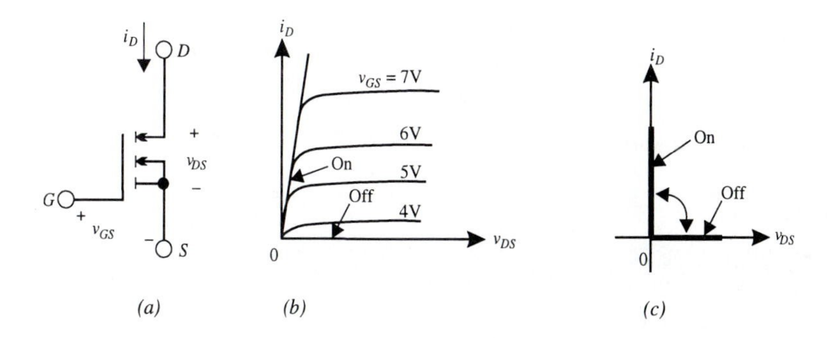

FIGURE 8-8
N-channel MOSFET: *(a)* symbol; *(b)* *i-v* characteristics; *(c)* idealized characteristics.

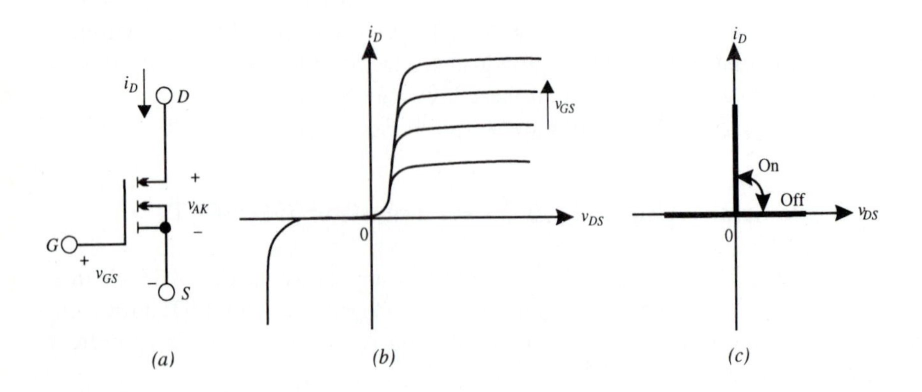

FIGURE 8-9
IGBT: *(a)* symbol; *(b)* *i-v* characteristics; *(c)* idealized characteristics.

FIGURE 8-10
A family of power semiconductors—general-purpose diodes. (Courtesy of Westinghouse Electric Corporation.)

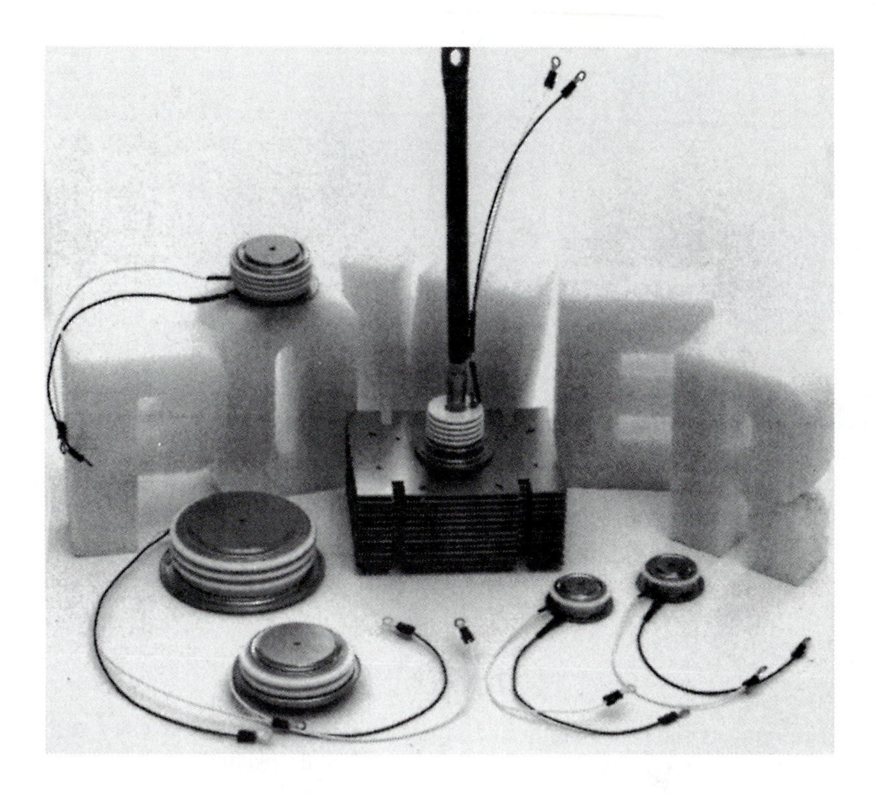

FIGURE 8-11
Phase-controlled thyristors and mounting with heat sink. (Courtesy of Westinghouse Electric Corporation.)

8.2 RMS AND AVERAGE VALUES OF WAVEFORMS

A characteristic of electronic control systems in motor controls, power system protection, and high-voltage dc transmission is that pertinent voltage and current waveforms are nonsinusoidal and often discontinuous. Furthermore, these waveforms change as a function of the level of operation. Some of the consequences of the above-mentioned nonlinearities in power electronic system are as follows.

1. The measurement of voltages and currents must be performed with instruments capable of accurately indicating the type of waveforms being measured. Thermocouple instruments are adequate for measuring power components in most electronic motor control systems. Oscilloscopes are usually essential for waveform analysis of both the power and control signal parameters.

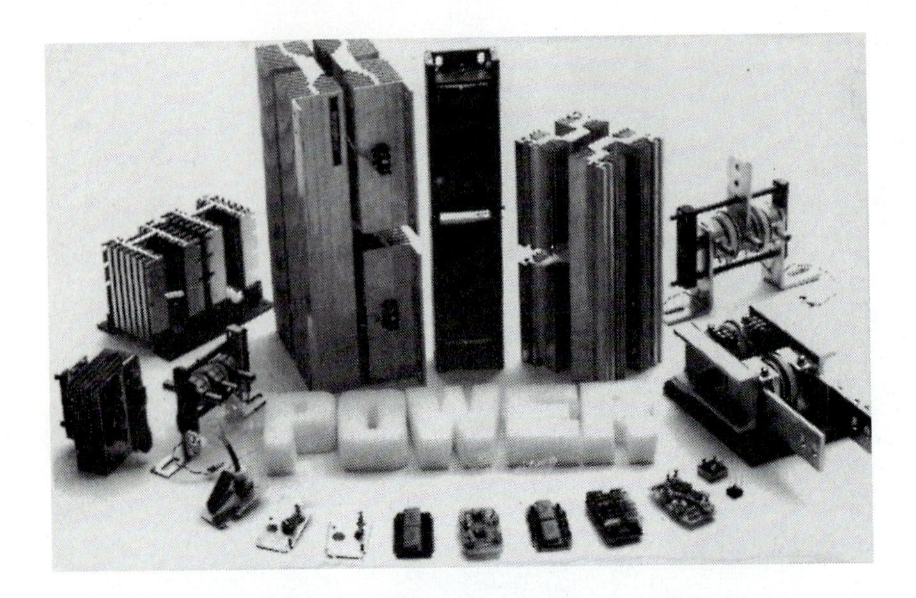

FIGURE 8-12
Heat sinks and other components for power semiconductor circuits. (Courtesy of Westinghouse Electric Corporation.)

2. Loss measurements should be performed, if possible, with the motor excited as it is to be used, rather than with standard sinusoidal or dc excitation. Magnetic material loss data are obtained with sinusoidal excitation and are often incorrect for other types of excitation. The measurement of core losses is difficult when the waveforms are like those described above. In that case, special wattmeters, such as electronic multipler, Hall-effect, or thermal-type instruments, should be used.

3. Standard circuit theory based on single-frequency sine-wave parameters is inadequate in the analysis of electronic motor control circuits. It is frequently necessary to evaluate the instantaneous time variation of both power and control signal currents and voltages. Fourier methods are also useful, as noted above.

4. The standard numerical values for the relationships between average, rms, and maximum values of current and voltages are seldom applicable.

5. The range of frequencies of the voltage and current components in an electronic motor control system is always much greater than the fundamental frequency applied to the motor. This is readily apparent if one considers the Fourier components of a nonsinusoidal periodic function. The fundamental frequency results from the switching action of power semiconductors in the control system and is usually in the power or low audio range of frequencies, seldom more than

3000 Hz. The range of frequencies in various currents and voltages may easily be 100,000 Hz or higher, however. This fact must be recognized when the choice of instrumentation used in the laboratory is made, when considering audible and electromagnetic noise interference that results from the control system, when designing filters, and when protecting the control logic circuitry used to switch the power devices.

The calculation of average and rms values of voltage and current is quite important in electronic control systems for calculating motor power and torque, for heating of wires and other components, and for sizing components and instrumentation. To make these calculations, it is often necessary to return to the definitions of average and rms values, which are respectively defined as

$$A_{\text{ave}} = \frac{1}{T_0}\int_0^{T_0} a\,dt \tag{8.2}$$

$$A_{\text{rms}} \equiv A = \left(\frac{1}{T_0}\int_0^{T_0} a^2\,dt\right)^{1/2} \tag{8.3}$$

where a represents instantaneous value of the parameter and T_0 is the period over which the average (or rms) value is evaluated. In motor control circuits involving power semiconductors, T_0 is usually the "on-time" duration. The fundamental frequency of the signal referred to above is defined by

$$f_p = \frac{1}{T_p} \tag{8.4}$$

where T_p is the length of a full period.

We illustrate the calculations of rms and average values of waveforms by the following examples.

Example 8.1 An electronic motor controller has a chopped half-wave rectified sinusoidal output voltage waveform, as shown in Fig. 8.13. Determine the average and rms values of the output voltage.

Solution

The output voltage, from Fig. 8.13, is given by

$$\begin{aligned} v(t) &= V_m \sin\frac{\pi t}{T_0}, & 0 < t < T_0 \\ &= 0, & T_0 < t < T_p \end{aligned}$$

Hence from (8.1) and (8.2) we obtain

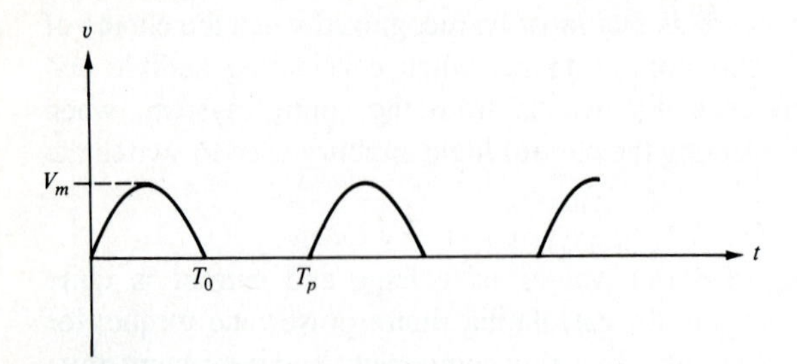

FIGURE 8-13
Chopped sine wave.

$$V_{\text{ave}} = \frac{1}{T_p}\left(\int_0^{T_0} V_m \sin\frac{\pi t}{T_0}dt + 0\right) = \frac{2V_m}{\pi}\left(\frac{T_0}{T_p}\right)$$

and

$$V_{\text{rms}} = \left[\frac{1}{T_p}\left(\int_0^{T_0} V_m^2 \sin^2\frac{\pi t}{T_0}dt + 0\right)\right]^{1/2} = V_m\sqrt{\frac{T_0}{2T_p}}$$

Example 8.2 A silicon rectifier is connected to an inductive load shown in Fig. 8.14*(a)*. With the parameters shown, determine the average value of the load current over a period $T = 2\pi/\omega$.

Solution

The instantaneous load current is given by

$$L\frac{di}{dt} + Ri = v_0 \tag{8.5}$$

which has a solution of the form

$$i = \begin{cases} \frac{V_m}{Z}[\sin(\omega t - \varphi) + e^{-(R/L)t}\sin\varphi] & 0 < \omega t < \beta \\ 0 & \beta < \omega t < 2\pi \end{cases} \tag{8.6}$$

where $Z = \sqrt{R^2 + (\omega L)^2}$ and $\tan\varphi = \omega L/R$. Notice that, from Fig. 8.14*(b)*, β/ω is the time when the diode stopped conducting. To determine the average current we may use (8.6) directly or (8.5) indirectly. Choosing the latter, we get

$$\frac{L}{T}\int_0^T \frac{di}{dt}dt + R\left(\frac{1}{T}\int_0^T i\,dt\right) = \frac{1}{T}\int_0^T v_0\,dt \tag{8.7}$$

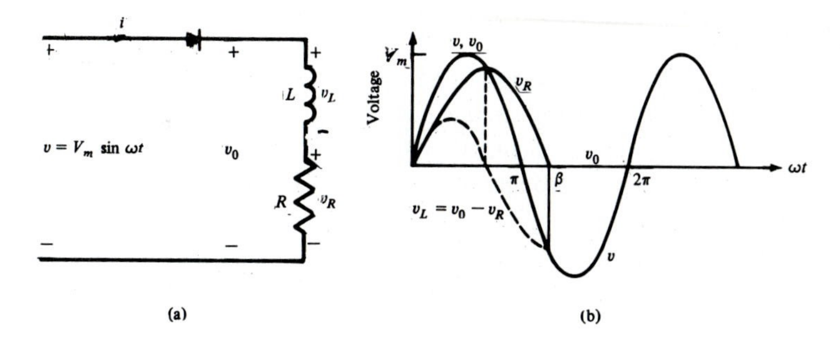

FIGURE 8-14
(a) An *R-L* circuit fed from a half-wave rectifier; *(b)* voltage waveforms.

The first integral in (8.7) equals $[i(t) - i(0)](L/T) = 0$ since it is periodic, of period T. The second integral is simply RI_{ave} and (8.5) becomes

$$RI_{\text{ave}} = \frac{1}{T}\int_0^{\beta/\omega} V_m \sin\omega t\, dt = \frac{\omega}{2\pi}\frac{V_m}{\omega}(1 - \cos\beta)$$

Hence

$$I_{\text{ave}} = \frac{V_m}{2\pi R}(1 - \cos\beta) \tag{8.8}$$

8.3 THYRISTOR COMMUTATION TECHNIQUES

We mentioned in Section 8.1 that a thyristor can be turned on by injection of energy into it through the gate connection. Once the anode current exceeds a certain minimum value, the thyristor latches on, and the gate loses control over the anode current. Subsequent anode current is determined by the external circuit between the anode and cathode, until the thyristor is brought into the blocking state or turned off. Commutation of a thyristor refers to the process of turning it off. The thyristor can be turned off when the forward anode current is reduced to zero and held at zero for a period at least equal to the turn-off time. The three basic methods of commutation are as follows.

1. *Line Commutation.* In this case, the source is ac and in series with the thyristor. The anode current goes through zero in a cycle. If the current remains zero for a period greater than the turn-off time, the thyristor will be turned off, until turned on again by some external means.

2. *Load Commutation.* Owing to the nature of the load, the anode current may go to zero and thereby turn off the thyristor. This type of commutation is useful mainly in dc circuits.

3. *Forced Commutation.* Forced commutation is achieved in systems energized from dc sources by an arrangement of energy storage elements (capacitors and inductors) and by additional switching devices (usually thyristors). In systems energized from ac sources, forced commutation is brought about by means of the cyclic potential reversal of the power source.

Line and load commutation, items 1 and 2, are sometimes grouped as one, known as natural or starvation commutation. We now derive the conditions for commutation by the following examples.

Example 8.3 An *R-L* circuit is fed from an ac source in series with a thyristor, as shown in Fig. 8.15*(a)*. The thyristor is fired at an angle α. Derive the condition for line commutation, and determine the thyristor conduction period.

Solution

The voltage equation is

$$L\frac{di}{dt} + Ri = V_m \sin\omega t \tag{8.9}$$

where the symbols are defined in Fig. 8.15*(a)*. The solution to (8.9) is of the form (see also Example 8.2)

$$i = \frac{V_m}{Z}\sin(\omega t - \varphi) + ke^{-(R/L)t} \tag{8.10}$$

where $Z = \sqrt{R^2 + (\omega L)^2}$ and $\tan\varphi = \omega L/R$. To evaluate k, we use the condition $i = 0$ at $\omega t = \alpha$ in (8.13) to obtain

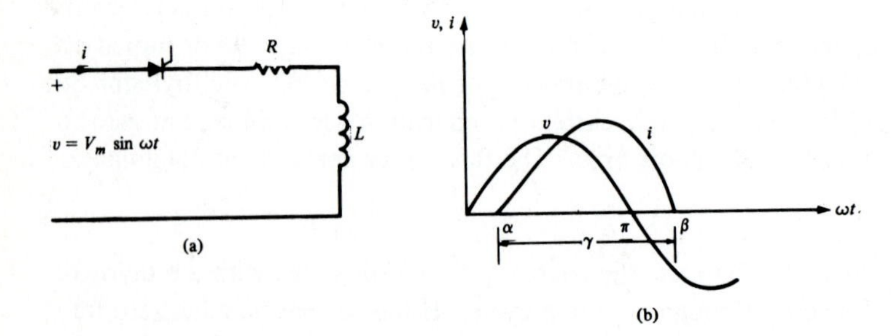

FIGURE 8-15
(a) An *R-L* thyristor series circuit; *(b)* voltage and current waveforms.

$$i = \frac{V_m}{Z}[\sin(\omega t - \varphi) - \sin(\alpha - \varphi)e^{R(\alpha-\beta)/\omega L}] \tag{8.11}$$

To obtain the condition for line commutation we notice that $i = 0$ at $\omega t = \beta$, as shown in Fig. 8.15*(b)*. Thus the required condition for commutation is given by the transcendental equation

$$\sin(\beta - \varphi) = \sin(\alpha - \varphi)e^{R(\alpha-\beta)/\omega L} \tag{8.12}$$

The conduction angle γ is then given by

$$\gamma = \beta - \alpha \tag{8.13}$$

Example 8.4 From Example 8.3 it is clear that line commutation is possible only in ac systems. Similarly, in a dc system, load commutation is not possible for an *R-L* circuit. However, load commutation in an *R-L-C* circuit fed from a dc source is possible if the circuit current could be made oscillatory. For the circuit shown in Fig. 8.16, derive the equation governing the time of commutation.

Solution

The voltage equation for the given circuit is

$$L\frac{di}{dt} + Ri + \frac{1}{C}\int_0^t i\,dt + v_c(0) = V \tag{8.14}$$

where $v_c(0)$ is the charge on the capacitor at $t = 0$. Defining

$$\zeta = \frac{R}{2L} = \text{damping ratio} \tag{8.15}$$

and

$$\omega_0 = \frac{1}{\sqrt{LC}} = \text{resonant frequency} \tag{8.16}$$

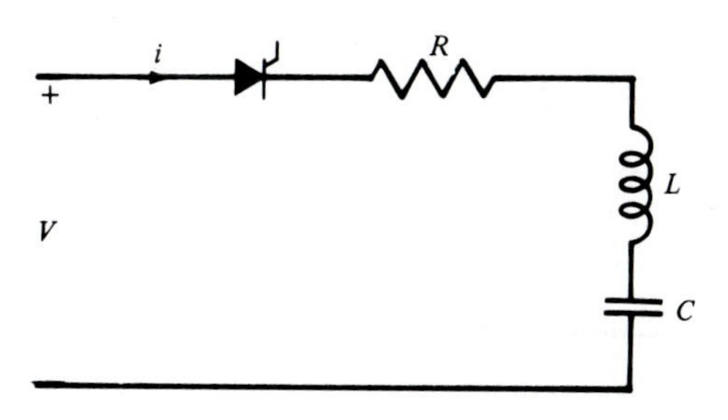

FIGURE 8-16
R-L-C circuit driven by a thyristor.

for $\zeta < \omega_0$, the solution to (8.14) becomes

$$i(t) = e^{-\zeta t}(A \cos\omega_r t + B \sin\omega_r t) \tag{8.17}$$

where A and B are arbitrary constants and

$$\omega_r = \sqrt{\omega_0^2 - \zeta^2} = \text{ringing frequency} \tag{8.18}$$

Because of the inductance, the current in the circuit cannot change instantaneously. Hence $i(0) = 0$, which when substituted in (8.17) yields $A = 0$, and (8.17) then becomes

$$i(t) = Be^{-\zeta t} \sin\omega_r t \tag{8.19}$$

For commutation, this current must go zero, requiring that

$$\omega_r t = \pi \tag{8.20}$$

Example 8.5 For the circuit shown in Fig. 8.16, $V = 96$ V, $L = 50$ mH, $C = 80$ μF, and $R = 40\ \Omega$. The initial charge on the capacitor is zero. At what time will the thyristor turn off?

Solution

From (8.15) and (8.16) we have

$$\zeta = \frac{40}{2 \times 50 \times 10^{-3}} = 400$$

and

$$\omega_0 = \frac{1}{\sqrt{50 \times 10^{-3} \times 80 \times 10^{-6}}} = 500 \text{ rad/s}$$

From (8.18)

$$\omega_r = \sqrt{500^2 - 400^2} = 300 \text{ rad/s}$$

Finally, (8.20) yields

$$300t = \pi$$

or

$$t = \frac{\pi}{300} = 10.47 \text{ ms}$$

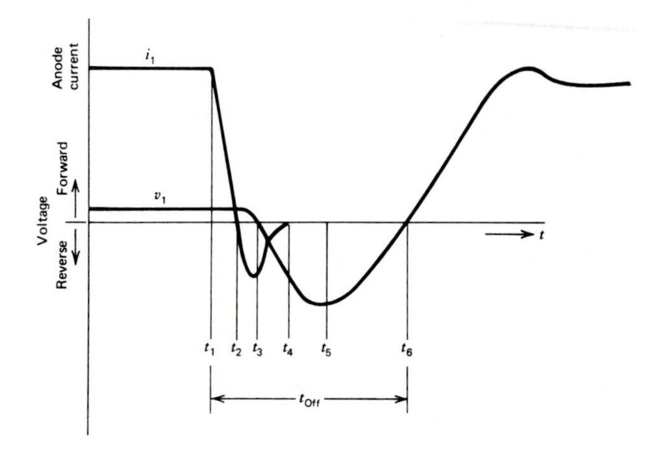

FIGURE 8-17
Thyristor voltage and current during commutation.

Forced Commutation

Example 8.5 shows that thyristor commutation in an *R-L* circuit fed from a dc source can be accomplished by including a series capacitor. The capacitor must be rated to carry full-load current. Such a scheme may not be economical in all instances, and as an alternative, forced commutation is used. Various forms of forced-commutation techniques include

1. Series-capacitor commutation.
2. Parallel-capacitor commutation.
3. Parallel capacitor-inductor commutation.
4. External pulse commutation.

The requirements that must be fulfilled by the commutation circuit are that the thyristor current be reduced to zero, the reverse-bias voltage be applied to the thyristor for an interval greater than the turn-off time, and any stored energy be properly discharged. Numerous circuits fulfilling these requirements have been developed. In the following we illustrate the principles rather than discussing the details of various commutation circuits.

For the sake of illustration we consider Figs. 8.17 and 8.18. In Fig. 8.17 the thyristor is conducting initially. Commutation is initiated at time t_1 by introducing a negative voltage into the external anode-cathode circuit. The anode-cathode voltage drop (v_1) remains at the low on-state (1.0 to 2.0 V) until the anode current (i_1) goes to zero at time t_2. At t_2, v_1 begins to decrease; v_1 goes negative at t_3 and i_1 goes to a negative maximum at t_5 and back to zero at t_6. Until time t_6, the anode must be maintained at a negative (reverse-biased) potential. Now, referring to Fig. 8.18, when T_1 is turned on, it carries only the charging current, which decays to less than the holding current when C is charged to the value V. At a later time (> T_{off}, the turn-off time), T_2 is turned on and C is discharged through L_c and T_2. The series circuit (RLC) is underdamped such that the voltage across C is greater than V. This reverse bias assists in turning off T_1. When T_1 is turned on, the current in the RLC series circuit is governed by

$$L\frac{di}{dt} + Ri + \frac{1}{C}\int i\,dt = V \tag{8.21}$$

In most cases of practical interest, $i = 0$ at $t = 0$. Subject to this condition, and for the underdamped case ($R^2 < 4L/C$), the solution to (8.21) becomes

$$i = \frac{V_0}{Z_0} e^{-\zeta t} \sin \omega_r t \tag{8.22}$$

where $V_0 = V - V_{c0}$

V_{c0} = voltage across C at $t = 0$ (an arbitrary constant)

Z_0 = characteristic impedance = $\sqrt{(L/C) - (R^2/4)}$

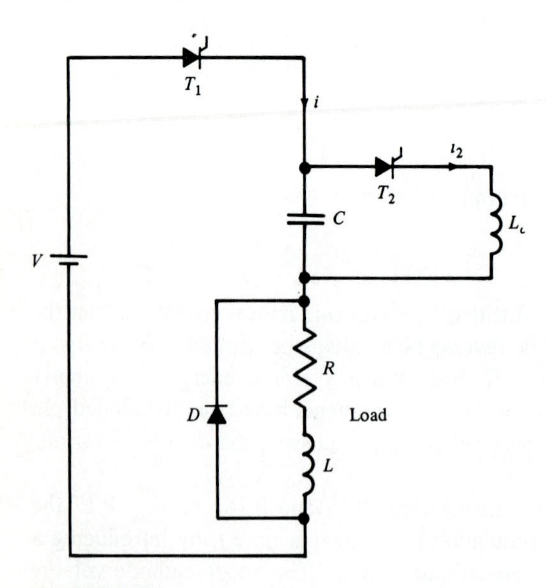

FIGURE 8-18
Series commutation circuit.

ζ = attenuation constant = $R/2L$

ω_r = ringing frequency = $Z_0 L$

and other symbols are as defined in Fig. 8.18. Defining $\tan \varepsilon = \omega_r/\zeta$, the voltage across the capacitor, v_c, is given by

$$v_c = V - \frac{V - V_{c0}}{\sin \varepsilon} e^{-\zeta t} \sin(\omega_r t + \varepsilon) \tag{8.23}$$

The pulse width, T_0, of the current given by (8.22) is

$$T_0 = \frac{\pi}{\omega_r} \tag{8.24}$$

The time t_m after initial turn-on of T_1, when the current pulse reaches its maximum, is

$$t_m = \frac{\varepsilon}{\omega_r} \tag{8.25}$$

From (8.22) and (8.25), the corresponding maximum current becomes

$$I_m = \frac{V_0}{Z_0} e^{-\zeta \varepsilon/\omega_r} \sin \varepsilon \tag{8.26}$$

When the current is building up, L starts storing energy, which reaches its maximum value at $t \simeq t_m$. This energy is subsequently returned to C. At the end of the sine pulse ($t = T_0$), $i = 0$, L is fully discharged, and the voltage across C is given by (with $V_{c0} = 0$)

$$v_c(T_0) = V(1 + e^{-\zeta T_0}) \tag{8.27}$$

Because $\zeta < 1$ and $T_0 < 1$, $\zeta T_0 << 1$ for the underdamped case, (8.27) implies that at the end of the pulse the capacitance voltage becomes almost twice the source voltage. Hence the net voltage appearing between the anode-cathode terminals of T_1 is the difference between V and $v_c(T_0)$, and thereby T_1 is reverse biased and is turned off, provided the reverse bias is maintained for a period greater than the thyristor turn-off time.

The capacitance voltage polarity is next reversed by turning on T_2. As a result, a second sine pulse of current flows through the CT_2L_c circuit. At the end of this second pulse the capacitance voltage is reversed in polarity, with the upper plate of C

second pulse the capacitance voltage is reversed in polarity, with the upper plate of C (Fig. 8.18) now negative with respect to the lower plate. The second pulse needed to reverse capacitance voltage can be initiated if the second pulse period results in a reverse-biased condition that is maintained across T_1 for a time interval, T_Q, slightly greater than the turn-off time of T_1. This time interval (T_Q) is obtained from (for the circuit of Fig. 8.18*(e)*

$$\sin(\psi - \omega_r T_Q) = \frac{V}{V_0} \sin\psi \tag{8.28}$$

where

$$\tan\psi = \frac{\sin\varepsilon}{\sqrt{L/C} - \text{load impedance}}$$

The time T_Q is the circuit commutation time. Notice that T_2 is also turned off by a series commutation process. Finally, the steady-state voltage across C at the instant T_1 is turned on is

$$V_{c0} = V \tanh \zeta T_0 \tag{8.29}$$

As mentioned earlier, a large variety of thyristor commutation circuits are used in practice. It is beyond our scope to discuss these circuits here. The preceding analysis has been presented to illustrate the principles of operation of the series-commutation circuit. We now turn to the applications of solid-state switching devices to the control of dc and ac motors.

8.4 GENERAL CONSIDERATIONS OF MOTOR CONTROL

The purpose of a motor control system is to control one or more of the motor output parameters, that is, shaft speed, angular position, acceleration, shaft torque, and mechanical output power. The control of temperature at various points in the motor is also a frequent objective of motor control systems. Since it is the mechanical parameters of the motor that are being controlled by the input electrical parameters, the peculiar characteristic of the individual machine—that is, the characteristics that relate input electrical quantities to output mechanical quantities—are vitally important in the design and analysis of electronic control. It is therefore customary to treat the control of dc commutator motors separately from that of synchronous and induction motors. The nature of the load and power supply will also influence the nature of the control. Fig. 8.19 illustrates the basic arrangement of electronic motor control. This figure illustrates a total motor system including load and power source. The feedback loops are shown by dotted lines, since many motor control schemes are "open-loop" schemes. The load box represents a very general concept of load and may be a pure inertia load. The power source box is also generalized and is meant to cover *all* power or excitation sources required by the motor, such as the field and armature power supplies in a dc commutator motor. Fig. 8.19 has no analytical value and should not be confused with the block diagram describing control signal flow, such as those in control theory. Here we simply illustrate the general topological layout of the principal elements of a general motor control system. The primary concern of this chapter will be with the box labeled "controller."

The materials and structural features of power semiconductors used in motor control are much different from those of electromechanical devices, which, as would be expected, result in differing operating and environmental characteristics. It is essential to recognize these differences in the design and use of electronic motor control in which the two different types of components are used in a common environment and are subject to the same voltage and current values. The principal differences may be summarized as follows.

1. Motors have large thermal capacities and can sustain thermal overloads for time periods measured in minutes. Semiconductors have very short thermal time constants, lasting often less than 1 s. They also have poor natural thermal conduction paths and require heat sinks in most applications.

2. The current overload characteristics of the two types of devices are quite different, partly as a result of the thermal differences just noted. Semiconductor devices have relatively little overcurrent capability.

3. Many semiconductors are limited by the rate of change of current, as will be explained for thyristors. There is no equivalent limitation in a motor.

4. Thyristors are also limited by a rate of change of voltage characteristic, which is not a factor in motor operation.

5. Semiconductor circuits are very susceptible to electromagnetic interference, both conductive and inductive, whereas motors are in no way similarly affected.

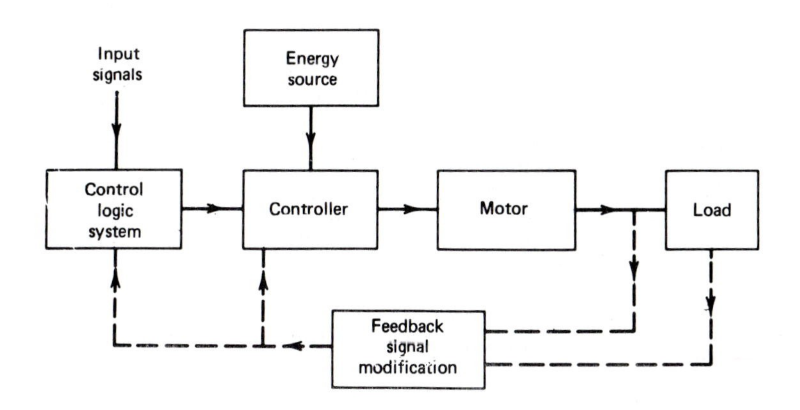

FIGURE 8-19
General signal flow in electronic motor control.

6. Semiconductors are much more sensitive to shock and vibration than motors. However, with proper mounting and packaging, semiconductors can be applied in most levels of vibration.

7. Motors have traditionally been operated in ways that result in continuous current and voltage waveforms, such as steady direct current or sinusoidal alternating current. When controlled by semiconductor systems, waveforms are usually less regular and often discontinuous, with steeply rising wave fronts. Such waveforms have various implications concerning motor losses and excitation characteristics and in some cases suggest modification of the motor design.

8. Electromagnetic motors are inherently inductive devices and large magnitudes of inductive energy storage are associated with their normal operation. The management of inductive energy during the rapid switching action of semiconductor devices, to prevent voltage spikes and excessive *dv/dt* that may damage the devices, is a principal design problem in electronic motor control.

8.5 CONTROL OF DC MOTORS

A motor controller should be designed as an integrated part of the drive system, which includes the motor, the controller, and the source. Clearly, the objective of the design is to achieve maximum efficiency at minimum cost and weight of the overall system. The controller serves as a link between the power source and the motor. Hence the controller determines the efficiency of energy transfer from the source to the motor. Consequently, the efficiency of the controller is an important factor in governing the performance of the drive system. In addition to high efficiency, some of the other desirable features of a motor controller are

1. Quick and smooth response to the operator's signals.
2. Operation to result in minimum internal losses of the source and the motor.
3. Protection against overloads, to protect itself as well as the motor.

The torque, speed, and regeneration characteristics (under steady-state conditions) of dc motors are governed essentially by the equations derived in Chapter 6. For convenience, we now repeat these pertinent equations:

$$E = \frac{\varphi n Z}{60}\frac{p}{a} = k_a \varphi \Omega_m \tag{8.30}$$

$$T_e = k_a \varphi I_a \tag{8.31}$$

$$\Omega_m = \frac{V_t - I_a R_a}{k_a \varphi} \tag{8.32}$$

where

E = voltage induced in the armature, V
I_a = armature current, A
R_a = armature resistance, Ω
φ = flux per pole, Wb
Z = number of armature conductors
a = number of parallel paths
p = number of poles
Ω_m = armature speed, rad/s
n = armature speed, rpm
V_t = armature terminal voltage, V
$k_a = Zp/2\pi a$, a constant

These equations indicate the great flexibility of controlling the dc motor. For instance, the speed of the motor may be varied by varying V_t, R_a, or φ (that is, the field current). Control of dc motors is governed by (8.30) to (8.32), and the various practical schemes are manifestations of these equations in one form or another.

From the governing equations (8.31) and (8.32) it is clear that the motor torque and speed can be controlled by controlling φ (that is, the field current), V_t, and R_a, and changes in these quantities can be accomplished as follows. In essence, the method of control involves field control, armature control, or a combination of the two. Electronically, field and/or armature control in a dc motor is achieved by modulating the voltage across the field and/or armature of the motor by means of thyristor circuits. The choice of the voltage control method depends on the nature of the available supply—whether ac or dc. For an ac source, phase-controlled rectifiers are employed; for a dc source, choppers are used. But before we discuss these, it is worthwhile to consider the analysis of a dc motor supplied by a half-wave rectifier.

Half-Wave Rectifier with DC Motor Load

The circuit is shown in Fig. 8.20*(a)*, where R and L are, respectively, the armature-circuit resistance and inductance, and e' is the motor back emf, assumed constant. The circuit analysis leads to the following expression for the current:

$$i = \begin{cases} 0 & 0 < \omega t < \alpha \\ \dfrac{V_m}{Z}[\sin(\omega t - \theta) + Be^{(R/L)t}] - \dfrac{e'}{R} & \alpha < \omega t < \beta \\ 0 & \beta < \omega t < 2\pi \end{cases} \tag{8.33}$$

where

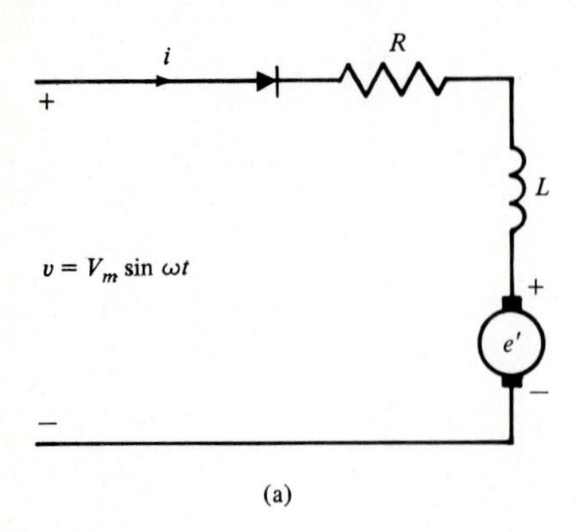

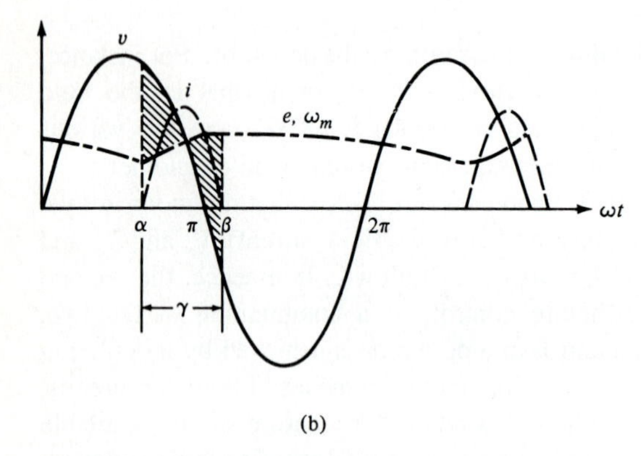

FIGURE 8-20
(a) A dc motor supplied by an ac source through a diode; *(b)* voltage and current waveforms.

$$Z = \sqrt{R^2 + (\omega L)^2}, \quad \tan\theta = \frac{\omega L}{R}$$

and where

$$B \equiv \left[\frac{e'}{V_m \cos\varphi} - \sin(\alpha - \theta)\right] e^{\alpha R/\omega L} \tag{8.34}$$

is such as to make i continuous at $\omega t = \alpha$. It is seen from (8.33) that the diode starts conducting at $\omega t = \alpha$; the firing angle, α, is determined by the condition

$$\sin\alpha = \frac{e'}{V_m} \tag{8.35}$$

As shown in Fig. 8.20*(b)*, conduction does not necessarily stop when v becomes less than e'; rather, it ends at $\omega t = \beta$, when the energy stored in the inductor during the current build-up has been completely recovered. The extinction angle, β, may be determined from the continuity of (8.33) at $\omega t = \beta$; we find that

$$\sin(\beta - \theta) + Be^{-\beta\cot\varphi} = \frac{\sin\alpha}{\cos\theta} \tag{8.36}$$

as the transcendental equation for β, in which B is known from (8.34). The average value of the current over one period of the applied voltage is found to be

$$I_{\text{ave}} = \frac{1}{R}V_{R\text{ave}} = \frac{V_m}{2\pi R}(\cos\alpha - \cos\beta - \gamma\sin\alpha) \tag{8.37}$$

where $\gamma \equiv \beta - \alpha$ is the conduction angle. Fig. 8.20*(b)* shows the waveforms.

Thyristor-Controlled DC Motor

In the example above, the dc motor load was not controlled by the half-wave rectifier; the back emf remained constant, implying that the motor speed was unaffected by the cyclic firing and extinction of the diode. To achieve control, we use a thyristor instead of the diode, as shown in Fig. 8.21*(a)*. The corresponding waveforms are illustrated in Fig. 8.21*(b)*. The motor torque (or speed) may be varied by varying α. Explicitly, for the armature we integrate

$$v_m = Ri + L\frac{di}{dt} + e \tag{8.38}$$

over the conduction period $\alpha/\omega < t < \beta/\omega$, during which v_m coincides with the line voltage v. The result is

$$V_m' = \frac{V_m(\cos\alpha - \cos\beta)}{\gamma} = RI' + E' \tag{8.39}$$

where a prime indicates an average over the conduction period. Over a full period of line voltage, the average armature current is given by

$$I_{\text{ave}} = \frac{\gamma}{2\pi}I' \tag{8.40}$$

and the average torque is given by

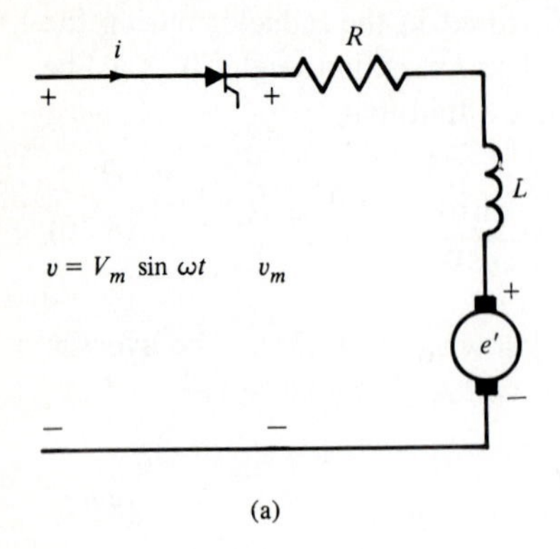

(a)

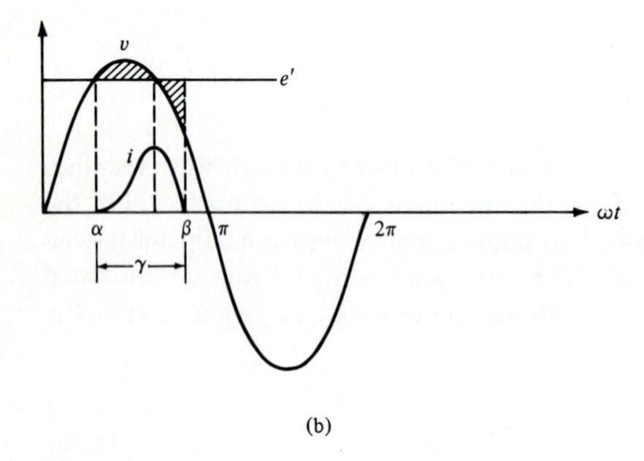

(b)

FIGURE 8-21
(a) A dc motor supplied by an ac source through a thyristor; *(b)* voltage, current, back emf, and speed waveforms.

$$T_{\text{ave}} = kI_{\text{ave}} = \frac{k\gamma}{2\pi} I' \tag{8.41}$$

Equations (8.39) to (8.41) govern the steady-state behavior of the thyristor-controlled dc motor.

Chopper Control

Using power semiconductors, the dc chopper is the most common electronic controller for the dc motor. In principle, a chopper is an on/off switch connecting the load to and disconnecting it from the dc source, thus producing a chopped voltage across the load. Symbolically, a chopper as a switch is represented in Fig. 8.22*(a)*, and a basic

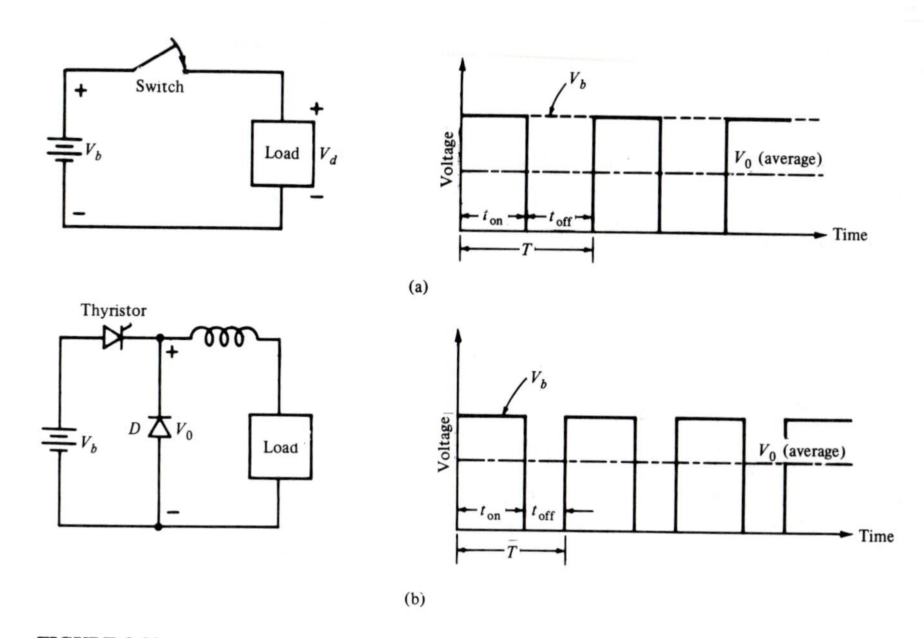

FIGURE 8-22
(a) Symbolic representation and *(b)* basic circuit of a chopper and output waveforms.

chopper circuit is shown in Fig. 8.22*(b)*. In the circuit shown in Fig. 8.22*(b)*, when the thyristor does not conduct, the load current flows through the freewheeling diode *D*. From Fig. 8.22, it is clear that the average voltage across the load, V_0, is given by

$$V_0 = \frac{t_{on}}{t_{on} + t_{off}} V_b = \frac{t_{on}}{T} V_b = \alpha V_b \tag{8.42}$$

where the various times are shown in the figure, T is known as the chopping period, and $\alpha = t_{on}/T$ is called the duty cycle. Thus the voltage across the load varies with the duty cycle.

There are three ways in which the chopper output voltage can be varied, and these are illustrated in Fig. 8.23. In the first method, the chopping frequency is kept constant and the pulse width (or on-time, t_{on}) is varied, and the method is known as pulse-width modulation. The second method, called frequency modulation, has either t_{on} or t_{off} fixed, and a variable chopping period, as indicated in Fig. 8.23*(b)*. The preceding two methods can be combined to obtain pulse-width and frequency modulation shown in Fig. 8.23*(c)*, which is used in current limit control. In a method involving frequency modulation, the frequency must not be decreased to a value that may cause a pulsating effect or a discontinuous armature current, and the frequency should not be increased to such a high value as to result in excessive switching losses. The switching frequency of most chopper applications range from 50 to 500 pulses per

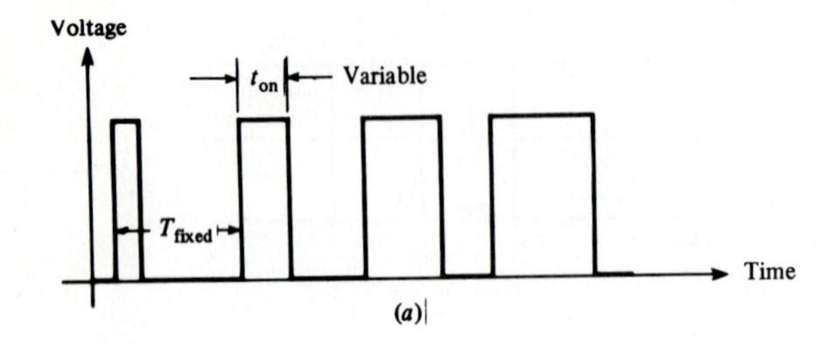

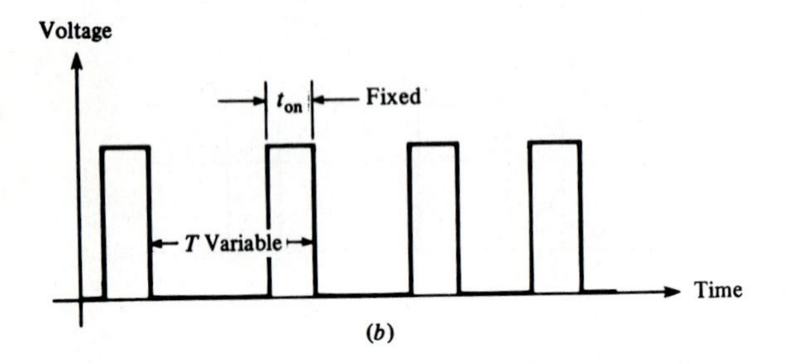

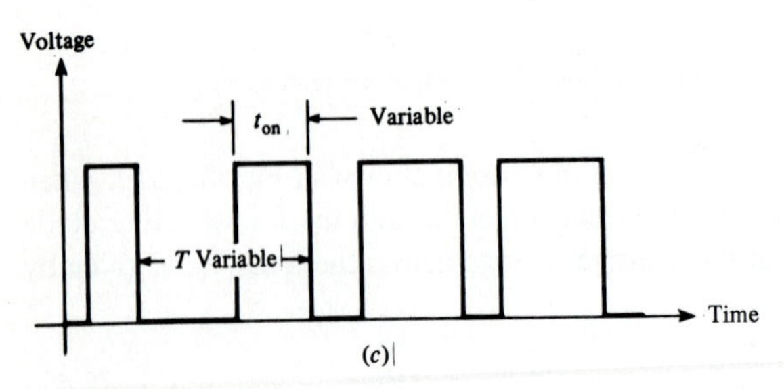

FIGURE 8-23
(a) Constant frequency, variable pulse width; *(b)* variable frequency, constant pulse width; *(c)* variable frequency, variable pulse width.

second. The drawback of high-frequency choppers is that the current interruption generates a high-frequency noise.

We mentioned earlier that when current begins to flow through a thyristor it remains in the conductive state until turned off by external means. In a chopper we may have load commutation, in which case the load current flowing through the thyristor goes to zero and thereby turns off the thyristor. The other method of commutation is forced commutation, whereby the thyristor is turned off by forcing the current to zero. This is achieved in a dc system by an arrangement of energy storage elements (capacitors and inductors) and by additional switching devices, usually thyristors. In systems energized from ac sources, forced commutation is brought about by means of the cyclic potential reversal of the power source. Several commutation circuits for a dc motor are shown in Fig. 8.24. In these circuits commutation is forced by reversing the voltage across the thyristor T_1 and diverting the load current. This is achieved by storing energy in the capacitor C during the "on" period and discharging C across the thyristor in the reverse polarity to turn it off.

In order to model a chopper-controlled separately excited motor, we consider a simplified circuit, and the corresponding voltage and current waveforms, as given in Fig. 8.25. Observe that when the thyristor turns off the applied voltage, v_m drops from V_1 to zero. However, armature current continues to flow through the path completed by the freewheeling diode until all the energy stored in L has been dissipated in R. Then v_m becomes equal to the motor back emf and stays at that value until the thyristor is turned off, whereupon it regains the value V_1.

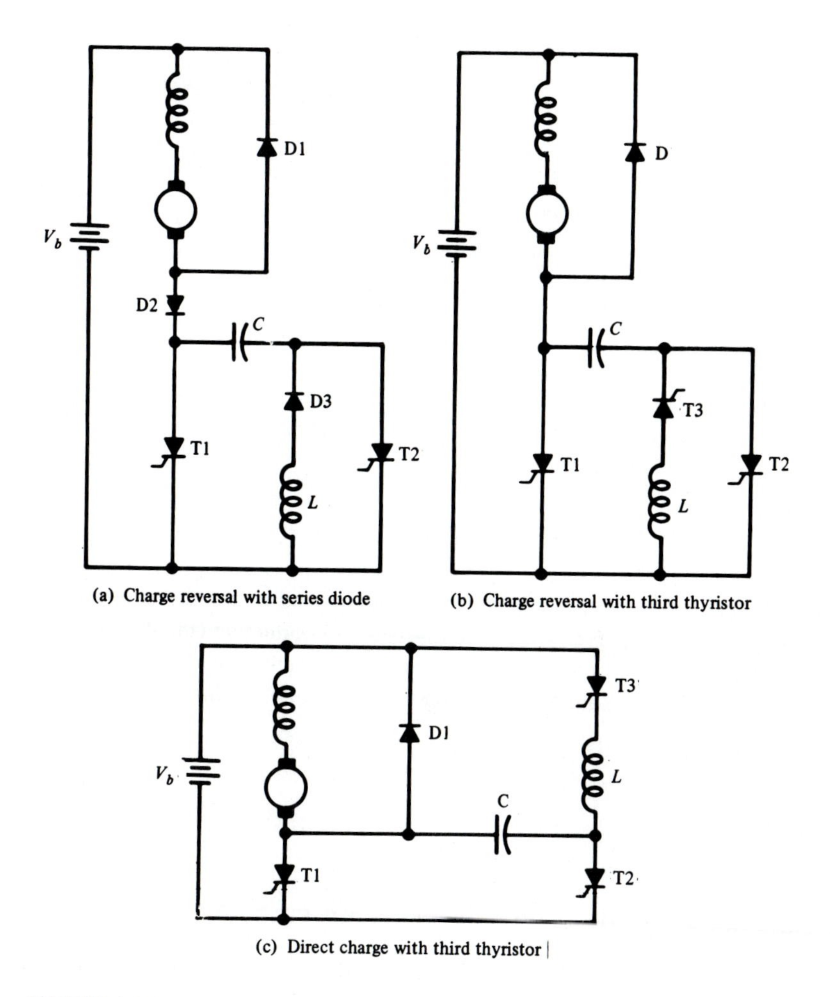

FIGURE 8-24
Commutation circuits for dc motors.

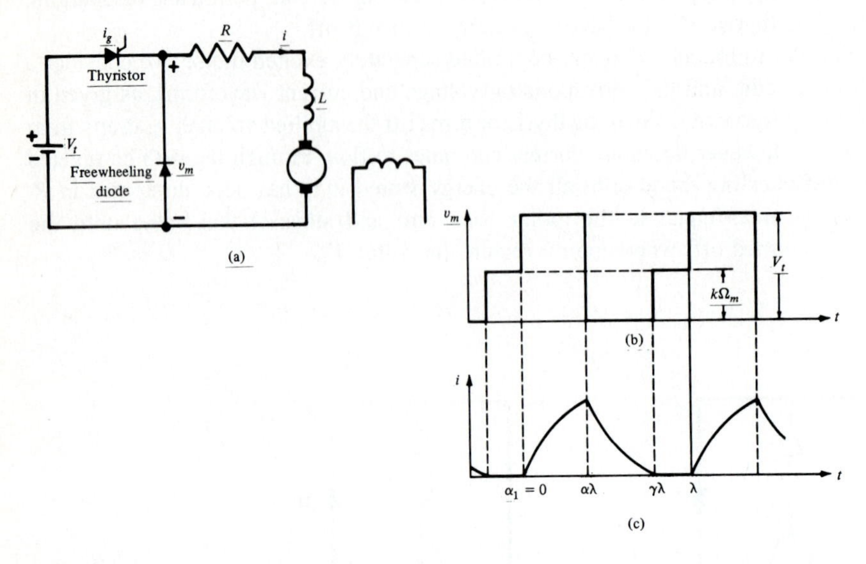

FIGURE 8-25
(a) A simplified circuit of a chopper-driven motor; *(b)* voltage waveform; *(c)* current waveform.

If the speed pulsations are small, the motor back emf may be approximated by its average value, $k\Omega_m$, yielding

$$L\frac{di}{dt} + Ri + k\Omega_m + v_m = \begin{cases} V_1 & 0 < t < \alpha\lambda \\ 0 & \alpha\lambda < t < \gamma\lambda \\ k\Omega_m & \gamma\lambda < t < \lambda \end{cases} \tag{8.43}$$

as the electrical equation of the system. Here λ is the period of the thyristor signal, α is the fraction of the period over which the thyristor is conductive (the duty cycle), and γ is the fraction of the period over which armature current flows.

Equation (8.43) has the solution, subject to the initial condition $i(0) = 0$,

$$i = \begin{cases} \dfrac{V_1 - k\Omega_m}{R}(1 - e^{-t/\tau}) & 0 < t < \alpha\lambda \\ \dfrac{k\Omega_m}{R}[e^{(\gamma\lambda - t)/\tau} - 1] & \alpha\lambda < t < \gamma\lambda \\ 0 & \gamma\lambda < t < \lambda \end{cases} \tag{8.44}$$

where $\tau = L/R$ is the armature time constant. Since $i = 0$ at $t = \alpha\lambda$, the following equation for γ is obtained:

$$\gamma = \frac{\tau}{\lambda} \ln\left(1 + \frac{e^{\alpha\lambda/\tau} - 1}{\Omega^*}\right) \tag{8.45}$$

in which $\Omega^* = k\Omega_m/V_t$ is the normalized (dimensionless) average rotational speed of the motor. Now, it is apparent that when α is sufficiently large and Ω^* sufficiently small, (8.45) gives $\gamma > 1$, which is impossible. Thus we must distinguish between two modes of operation of the machine.

Mode I is defined by all (α, Ω^*) combinations satisfying

$$1 > \frac{\tau}{\lambda} \ln\left(1 + \frac{e^{\alpha\lambda/\tau} - 1}{\Omega^*}\right)$$

In this mode, γ is given by (8.45) and the armature current, (8.44), vanishes over a fraction $(1 - \gamma)$ of the basic cycle.

Mode II is defined by all (α, Ω^*) combinations satisfying

$$1 \leq \ln\left(1 + \frac{e^{\alpha\lambda/\tau} - 1}{\Omega^*}\right)$$

If the equality holds, (8.44) is valid with $\gamma = 1$; that is, the armature current becomes zero only at the period points. If the inequality holds, (8.44) is no longer valid. The governing differential equation, (8.43), and the boundary conditions must be changed to admit a strictly positive solution, for which again $\gamma = 1$.

The average torque—average speed characteristic of the motor can now be derived. Integration of (8.43) over one period of the thyristor signal gives

$$RI_{\text{ave}} + k\Omega_m = \alpha V_1 + k\Omega_m(1 - \gamma) \tag{8.46}$$

On the other hand, the torque equation of the motor,

$$J\dot{\omega}_m + b\omega_m + t_0 = ki$$

where t_0 is the load torque, b is the rotational friction coefficient, and J is the moment of inertia, integrates to give

$$b\Omega_m + t_{0\text{ave}} = kI_{\text{ave}} \tag{8.47}$$

Eliminating I_{ave} between (8.46) and (8.47), we obtain the desired relation between $T^* \equiv T_{0\text{ave}}/(kV_t/R)$, the normalized (dimensionless) average torque, and $\Omega^* \equiv k\Omega_m/V_1$:

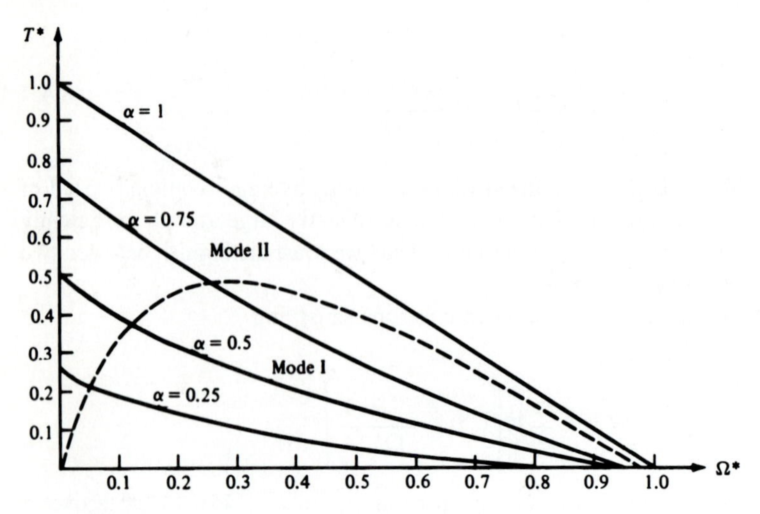

FIGURE 8-26
Normalized torque-speed characteristics.

$$T^* = \alpha - \left(\frac{bR}{k^2} + \gamma\right)\Omega^*$$

that is,

$$T^* = \alpha - \left[\frac{bR}{k^2} + \frac{\tau}{\lambda}\ln\left(1 + \frac{e^{\alpha\lambda/\tau} - 1}{\Omega^*}\right)\right] \qquad \text{in mode I}$$

$$= \alpha - \left(\frac{bR}{k^2} + 1\right)\Omega^* \qquad \text{in mode II} \qquad (8.48)$$

Fig. 8.26 shows the torque-speed curves for several values of α. Observe the linearity of the curves in the region corresponding to mode II, which is separated from the mode I region by the dashed curve.

The analysis of a chopper-driven series motor (Fig. 8.27) is slightly more involved than the analysis of the shunt motor presented earlier, because the torque of a series motor varies nonlinearly with the armature current, and magnetic saturation must be taken into account. In such a case the computations are too tedious unless a digital computer is used. The procedure is as follows.

1. Store the saturation curve φ versus I_{ave} in the computer. It suffices to use the three-segment approximation shown in Fig. 8.28.

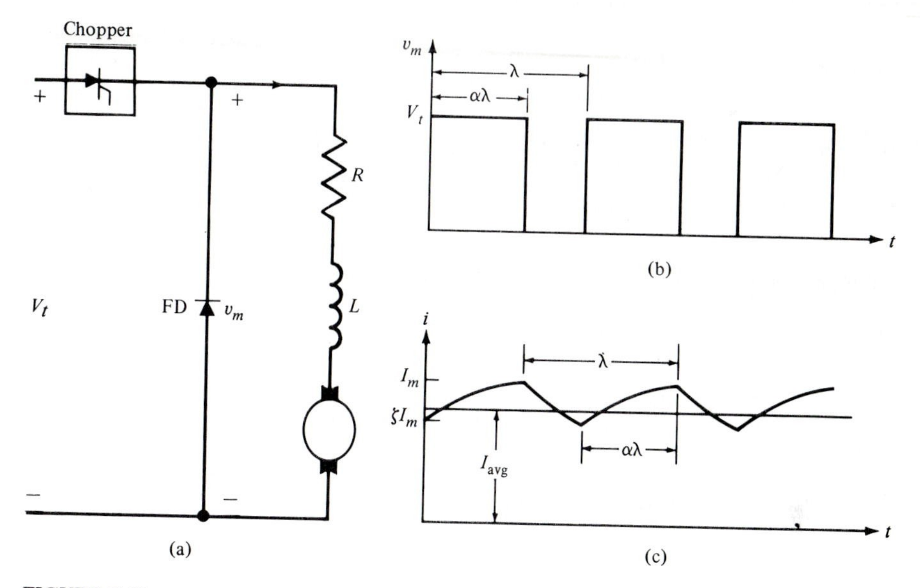

FIGURE 8-27
(a) A chopper-driven series motor; *(b)* motor voltage and duty cycle; *(c)* motor current.

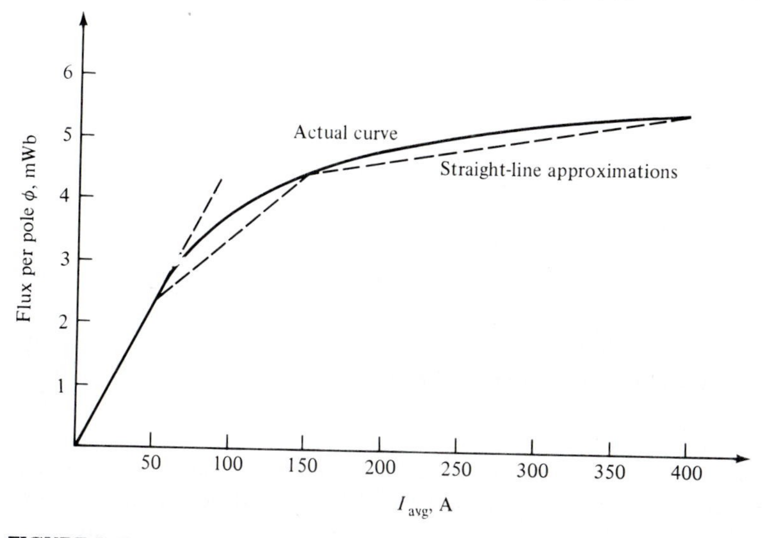

FIGURE 8-28
Straight-line approximation of the saturation curve of a series motor.

2. Integration of the circuit equation over a period λ gives the average motor speed as

$$\Omega_m = \frac{\alpha V_t - I_{ave} R}{k_a} \tag{8.49}$$

Setting $\Omega_m = 10$ rad/s, iteratively choose $[I_{ave}, \varphi(I_{ave})]$ pairs from step 1 until (8.49) is satisfied; the final I_{ave} is the average armature current for the speed 10 rad/s.

3. Repeat step 2 for a set of evenly spaced values of Ω_m and thus obtain $I_{ave}(\Omega_m)$.

4. For each Ω_m, compute the corresponding developed power from

$$P_d = \alpha V_t I_{ave} \Omega_m - R[I_{ave} \Omega_m]^2 \tag{8.50}$$

thus obtaining the power-speed characteristic.

5. For each Ω_m, compute, from step 4,

$$T_{ave} = \frac{P_d}{\Omega_m} \tag{8.51}$$

generating the torque-speed characteristic.

8.6 CONTROL OF AC MOTORS

We recall from Chapters 4 and 5 that the two major categories of ac motors are induction motors and synchronous motors. The synchronous speed, n_s, of an ac motor is given by

$$n_s = \frac{120f}{P} \tag{8.52}$$

where f is the supply frequency (Hz) and P is the number of poles. The most efficient method of varying the speed of an ac motor is by varying the supply frequency. If the available source of power is dc, the variable frequency is obtained by means of an inverter. On the other hand, if the source is ac, it may be converted into dc and then inverted, or the variable frequency may be obtained using a cycloconverter. These methods of speed control of ac motors are illustrated in terms of functional block diagrams in Fig. 8.29*(a)* and *(b)*.

We will discuss the components of these block diagrams later in this chapter. With a variable-frequency controller, the torque-speed characteristics of an induction motor for several frequencies are shown in Fig. 8.30*(a)*. The dashed-line envelope defines two distinct operating regions of constant maximum torque and constant output power. In the constant maximum torque region, the ratio of applied motor voltage to supply frequency (V/Hz) is held constant by increasing motor voltage directly with frequency, as shown in Fig. 8.30*(b)*. The volts/hertz ratio defines the airgap magnetic flux in the motor and can be held constant for constant torque within the excitation frequency range and corresponding motor speeds extending to frequency f_0. Beyond frequency f_0, the motor voltage cannot be increased to maintain a constant volts/hertz

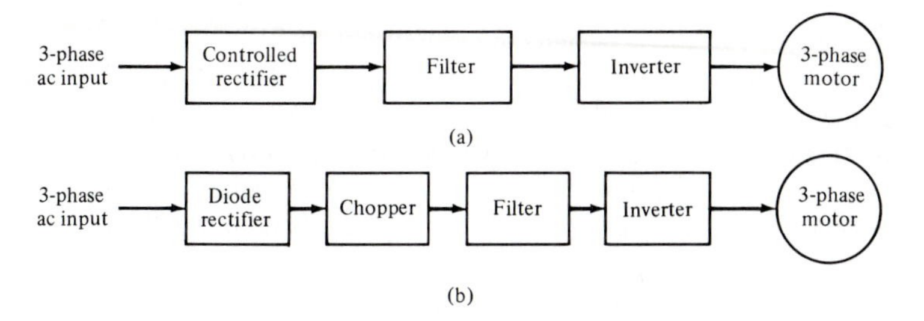

FIGURE 8-29
Two schematics of ac motor speed control by frequency variation.

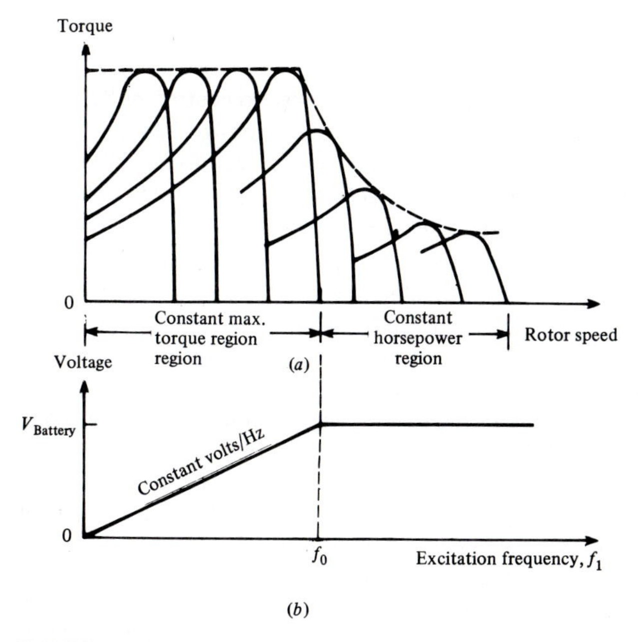

FIGURE 8-30
(a) Induction motor speed-torque characteristics; *(b)* voltage characteristics for variable-frequency operation.

ratio due to the limitations of a finite source voltage. For motor speeds in the range where f_r is greater than f_0, the supply voltage is held constant at its maximum level, and the supply frequency is increased to provide the motor speed demanded. This is the constant maximum horsepower region of operation, since the maximum developed torque decreases nonlinearly with speed.

The envclope illustrated in Fig. 8.30*(a)* characterizes induction motor operation in a typical electric traction system. Any speed-torque combination within the envelope can be provided by appropriate voltage and frequency control. For heavy loads, such as vehicle acceleration, the motor is operated in the constant-torque region to provide the torque demanded. For high-speed operation at vehicle cruising speeds, the motor operates in the constant-horsepower region at a frequency to satisfy the demanded speed. Voltage control is usually accomplished in the constant-torque region by pulse-width (duty-cycle) modulation. The motor operates with a fixed-voltage, variable-frequency square wave in the constant-horsepower region. Voltage and frequency control are used in both driving and braking operating modes of the vehicle. Regenerative braking is accomplished by reducing the supply frequency below the rotor frequency, that is, f_0 less than f_r. Under these conditions, the motor acts as a generator, developing a negative braking torque. Both the degree of braking and the level of power returned to the battery are controlled by voltage and frequency control.

We now turn our attention to some of the solid-state controllers for ac motors.

Inverters

There is a wide variety of inverter circuits which may be used to control the speed of an ac motor. For the present, however, we consider the full-bridge inverter circuit shown in Fig. 8.31, which also shows the voltage and current waveforms. When T_1 and T_3 are conducting, the battery voltage appears across the load with the polarities shown in Fig. 8.31*(b)*. But when T_2 and T_4 are conducting, the polarities across the load are reversed. Thus we get a square-wave voltage across the load, and the frequency of this wave can be varied by varying the frequency of gating signals. If the load is not purely resistive, the load current will not reverse instantaneously with the voltage. The antiparallel connected diodes shown in Fig. 8.31*(a)* allow the load current to flow after voltage reversal.

The principle of the bridge inverter mentioned above can be extended to form a three-phase bridge inverter of Fig. 8.32*(a)*. The gating signals and the output voltages are shown in Fig. 8.32*(b)*. The fundamental components of the line-to-line voltages will form a balanced three-phase system. The antiparallel connected diodes are used to allow flow of currents out of phase with the voltage.

Some commonly used inverters are discussed below.

Adjustable Voltage Inverter

In an adjustable voltage inverter (AVI), the output voltage and frequency both can be varied. The voltage is controlled by including a chopper between the battery and the inverter, whereas the frequency is varied by the frequency of operation of the gating signals. In an AVI, the amplitude of the output decreases with the output frequency and the ratio V/f remains essentially constant over the entire operating range. The AVI output waveform does not contain 2nd, 3rd, 4th, 6th, 8th, and 10th harmonics. Other harmonic contents as a fraction of the total rms output voltage are

fundamental = 0.965
5th harmonic = 0.1944

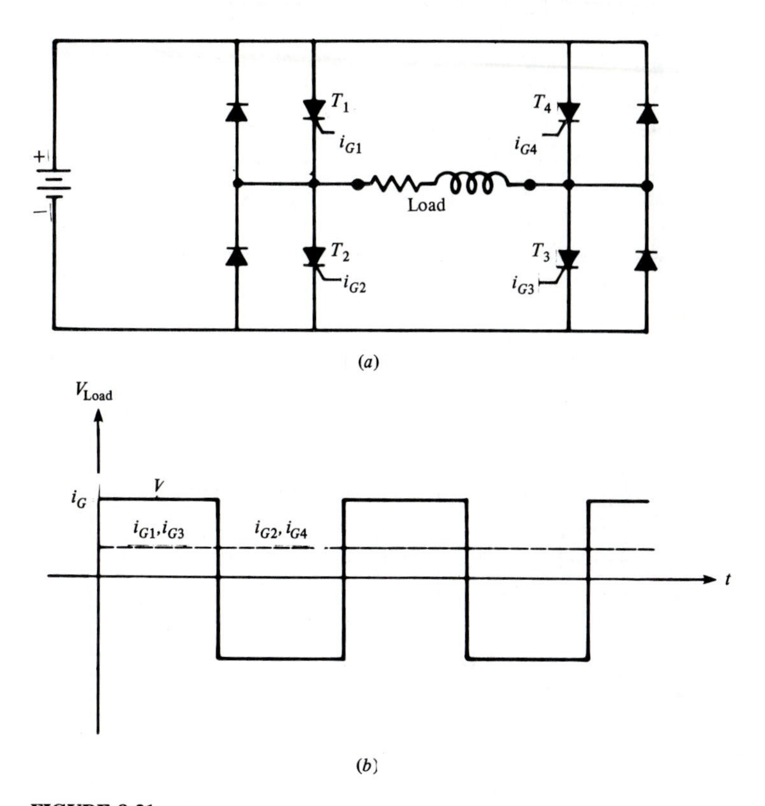

FIGURE 8-31
(a) Single-phase full bridge inverter; *(b)* load voltage waveform.

7th harmonic = 0.138
11th harmonic = 0.087

The losses produced by these harmonics tend to heat the motor, and its efficiency decreases by about 5 percent compared to a motor driven by a purely sinusoidal voltage.

Pulse-Width-Modulated and Pulse-Frequency-Modulated Inverters

The voltage control in pulse-width-modulated (PWM) and pulse-frequency-modulated (PFM) inverters is obtained in a manner similar to that for a chopper shown earlier in Fig. 8.23. In a PWM inverter, the output voltage amplitude is fixed and equal to the battery voltage. The voltage is varied by varying the width of the pulse on-time relative to the fundamental half-cycle period, as illustrated in Fig. 8.33*(a)* and *(b)*. Notice that the full-voltage output wave is similar to that of the AVI output shown in

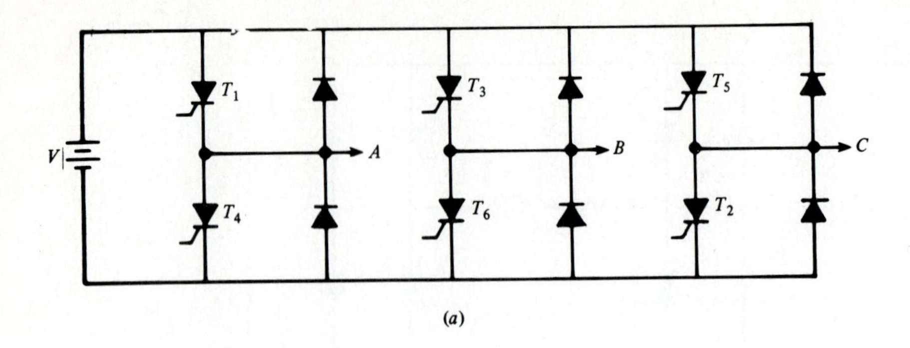

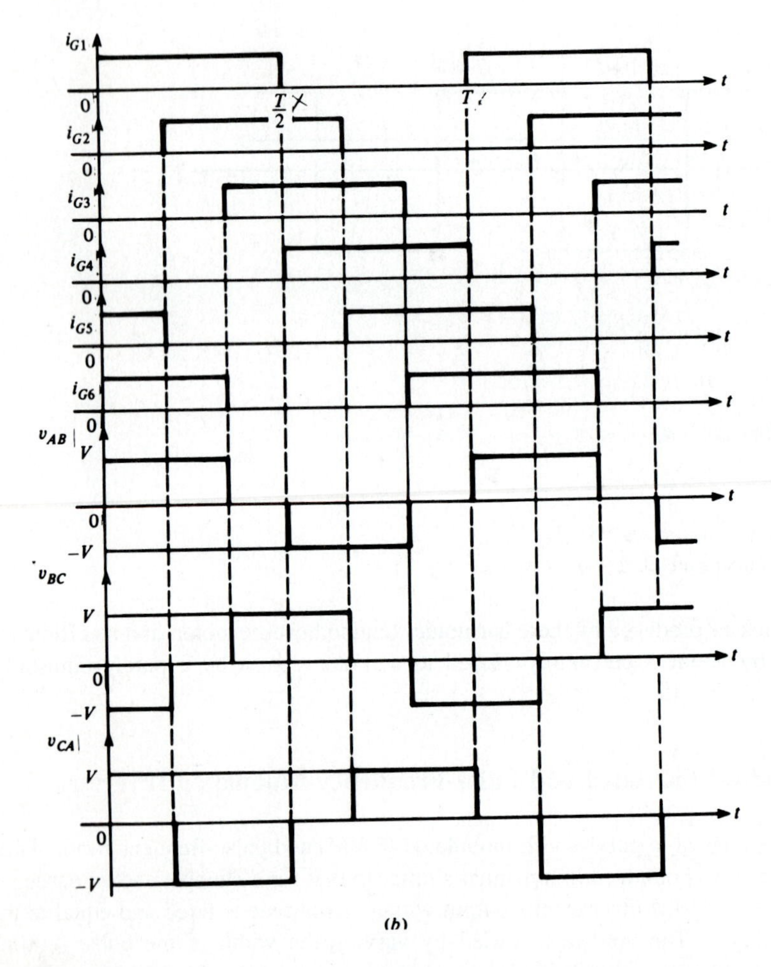

(b)

FIGURE 8-32
(a) Three-phase bridge inverter; *(b)* gating current and output voltage waveforms.Fig. 8.32*(b)*.

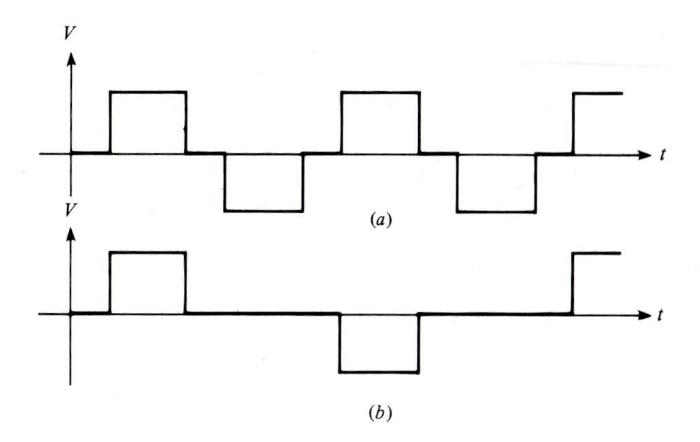

FIGURE 8-33
(a) Full voltage output and *(b)* half-full voltage output of a single-pulse PWM inverter.

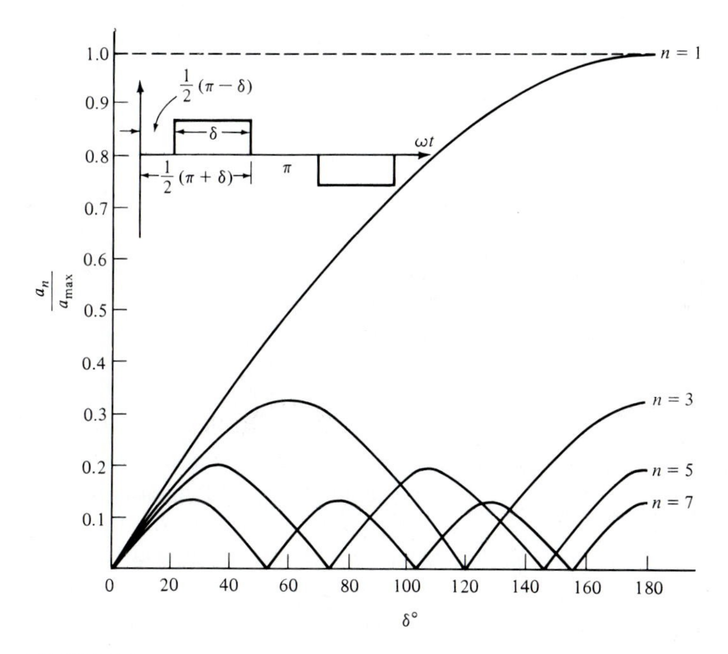

FIGURE 8-34
Harmonic content of a single pulse.

However, as the voltage is reduced (Fig. 8.33*(b)*) the harmonics vary rapidly in magnitude. Fig. 8.34 shows the harmonic content of the output voltage with single-pulse modulation.

Low-frequency harmonics can be reduced by pulse frequency modulation (PFM), in which the number and width of pulses within the half-cycle period is varied. PFM

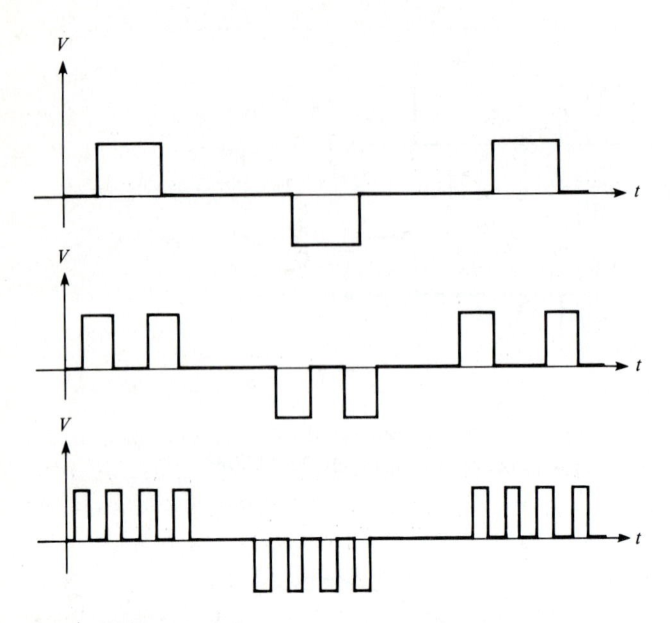

FIGURE 8-35
Output from a PFM inverter.

waveforms for three different cases are shown in Fig. 8.35. Harmonics from the output of a PFM inverter can be reduced by increasing the number of pulses per half-cycle. But this requires a reduction in the pulse width, which is limited by the thyristor turn-off time and the switching losses of the thyristor. Furthermore, an increase in the number of pulses increases the complexity of the logic system and thereby increases the overall cost of the PFM inverter system.

In comparing an AVI system with a PWM (or PFM) inverter system, we observe that whereas the AVI system is efficient but expensive, the PWM inverter is relatively inexpensive but inefficient. Both types of inverter system are suitable for induction motors. However, for the control of a synchronous motor, a thyristor inverter requires a motor voltage greater than the dc link voltage in order to turn off the thyristors. Also, no diodes are required in the ac portion of the circuit, thus avoiding uncontrolled currents. If a synchronous motor is controlled by a transistor inverter, the motor internal voltage must be less than the battery voltage; otherwise, the diodes in the inverter circuit will conduct, resulting in uncontrollable currents and high losses.

Note on Motor Performance (Fed from Inverters)

The inverter output voltage and current waveforms are rich in harmonics. These harmonics have detrimental effects on motor performance. Among the most important effects are the production of additional losses and harmonic torques. The additional losses that may occur in a cage induction motor owing to the harmonics in the input current are summarized below.

1. *Primary I^2R Losses.* The harmonic currents contribute to the total rms input current. Skin effect in the primary conductors may be neglected in small wire-wound machines, but it should be taken into account in motor analysis when the primary-conductor depth is appreciable.

2. *Secondary I^2R Losses.* When calculating the additional secondary I^2R losses, the skin effect must be taken into account for motors of all sizes.

3. *Core Losses Due to Harmonic Main Fluxes.* These core losses occur at high frequencies, but the fluxes are highly damped by induced secondary currents.

4. *Losses Due to Skew-Leakage Fluxes.* These losses occur if there is relative skew between the primary and secondary conductors. At 60 Hz the loss is usually small, but it may be appreciable at harmonic frequencies. Since the time-harmonic mmf's rotate relative to both primary and secondary, skew-leakage losses are produced in both members.

5. *Losses Due to End-Leakage Fluxes.* As in the case of skew-leakage losses, these losses occur in the end regions of both the primary and secondary and are a function of harmonic frequency.

6. *Space-Harmonic MMF Losses Excited by Time-Harmonic Currents.* These correspond to those losses which, in the case of the fundamental current component, are termed high-frequency stray-load losses.

In addition to these losses and harmonic torques, the harmonics act as sources of magnetic noise in the motor.

Example 8.6 A single-phase half-bridge inverter is shown in Fig. 8.36*(a)*. The thyristor triggering currents are shown in Fig. 8.36*(b)*. Sketch the load voltage and current waveforms. Solve for the steady-state load current and determine the maximum value of the current.

Solution

The circuit equation is

$$L\frac{di}{dt} + Ri = v = \begin{cases} \dfrac{V}{2} & 0 < t < T/2 \\ -\dfrac{V}{2} & T/2 < t < T \end{cases} \tag{8.53}$$

Expressing v in a Fourier series, we have

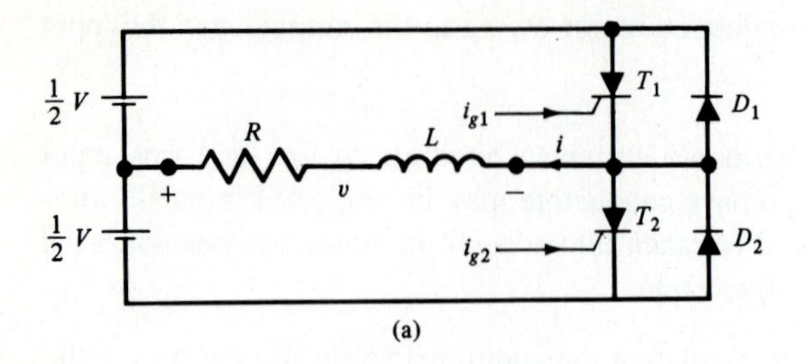

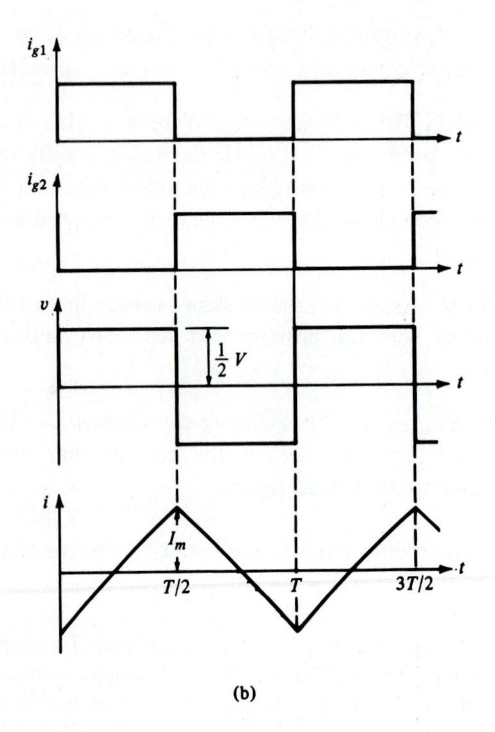

FIGURE 8-36
(a) Single-phase, half-bridge inverter. *(b)* Voltage and current waveforms.

$$v = \frac{2V}{\pi} \sum_{n \text{odd}} \frac{1}{n} \sin n\omega t \tag{8.54}$$

We solve (8.53) by superposition, to obtain the steady-state current:

$$i = \frac{2V}{\pi} \sum_{n \text{odd}} \frac{1}{nZ_n} \sin(n\omega t - \theta_n) \tag{8.55}$$

where

$$Z_n = \sqrt{R^2 + (n\omega L)^2}, \quad \tan\theta_n = \frac{n\omega L}{R}$$

At the instants of commutation ($t = 0, T/2, T, 3T/2, 2T, \ldots$),

$$\begin{aligned} i = \pm I_m &\equiv \pm \frac{2V}{\pi\omega L} \sum_{n\,\text{odd}} \frac{1}{n^2 + (R/\omega L)^2} \\ &= \pm \frac{V}{2R} \tanh \frac{R\pi}{2\omega L} \end{aligned} \tag{8.56}$$

When the plus sign holds in (8.56), forced commutation is required; it may not be necessary when the minus sign holds.

Voltage and current waveforms are shown in Fig. 8.36*(b)*.

Cycloconverters

The cycloconverter is a control device used on variable-speed motors supplied by an ac power source. It is a means of converting a source at fixed (peak) voltage and fixed frequency to an output with variable voltage and variable frequency. The source frequency must be at least three to four times the maximum frequency of the output. A single-phase bridge cycloconverter is shown in Fig. 8.37*(a)*, and the various waveforms are shown in Fig. 8.37*(b)*, which also indicates the firing sequence of the thyristors. In Fig. 8.37*(b)*, α_p denotes the minimum delay time for the positive group of converters, and α_n denotes the maximum delay time for the negative group. The variation of the delay controls the output voltage, in that it determines how many half-cycles of line voltage go to make up one half-cycle of the load voltage fundamental.

The cycloconverter has the advantage that it does not require a dc link, and consequently has high efficiency. Further advantages of a cycloconverter are the possibility of voltage control within the converter, and the fact that it has line commutation. Major applications of cycloconverters are for low-speed motors rated in the range 300 to 20,000 hp (or 15,000 kW). Such a motor-drive system is capable of providing low-voltage starting, reversing, and regenerative braking.

PROBLEMS

8.1. Find the average rms values of the waveforms shown in Fig. 8.38.

8.2. Show that the solution given by (8.6) in Example 8.2 is correct.

8.3. Solve (8.14) for the three cases $\zeta > \omega_0$, $\zeta = \omega_0$, and $\zeta < \omega_0$.

8.4. Verify that (8.22) is a valid solution to (8.21) for the conditions stated in the text.

8.5. The half-wave rectifier shown in Fig. 8.39 may be used as a filtered power supply. The input is 220 V at 60 Hz.

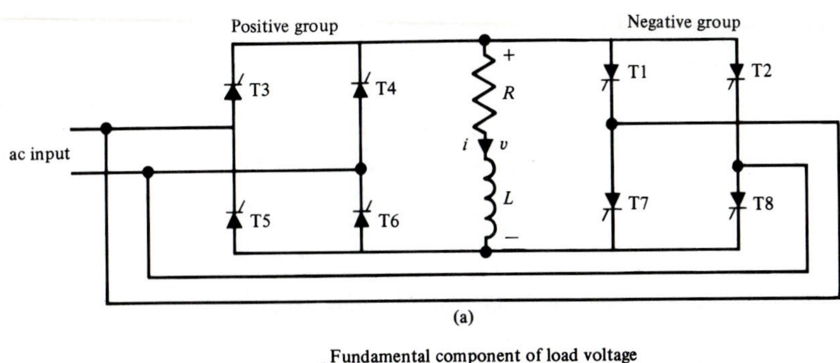

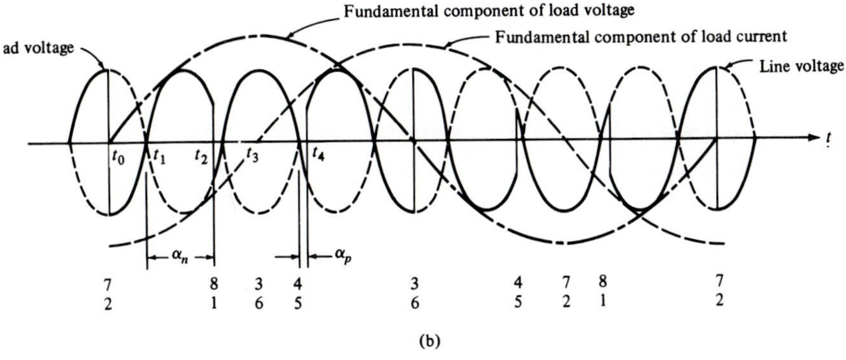

FIGURE 8-37
(a) A single-phase cycloconverter; *(b)* voltage and current waveforms.

(a) Determine the value of C to limit the ripple in v_0 (peak to peak) to less than 5%.
(b) For this value of C, calculate the average value of the output voltage.

8.6. A half-wave rectifier for battery charging may be modeled by the circuit shown in Fig. 8.40. Given values are $v_s = 14 \sin 100t$, $V_0 = 10$ V, and $R = 0.2\ \Omega$.
(a) Sketch v_0.
(b) Find an expression for $i_o(t)$ for $0 < \omega t \leq T$.
(c) Calculate the average and rms values of i_o.
(d) Determine the power delivered to the battery.

8.7. For the circuit shown in Fig. 8.41, the rms value of the input voltage is 10 V. *(a)* What is the minimum value of the firing angle to turn on the thyristor? *(b)* For a firing angle of 75°, determine the extinction angle.

8.8. Consider the circuit of Fig. 8.7, where $V = 48$ V, $L = 20$ mH, $C = 60\ \mu$F, and $R = 50\ \Omega$. If the initial charge on the capacitor is zero, at what time will the thyristor turn off?

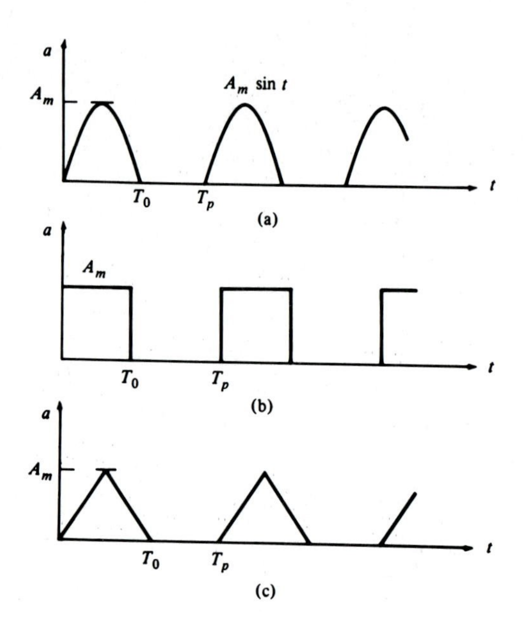

FIGURE 8-38
Problem 8.1

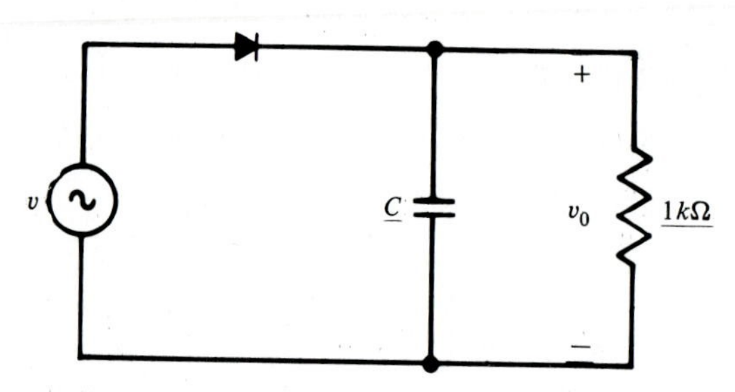

FIGURE 8-39
Problem 8.5.

8.9. The input voltage to a chopper, supplying a 2-Ω resistive load, is 50 V. For a duty cycle of 75%, determine the average output voltage and current.

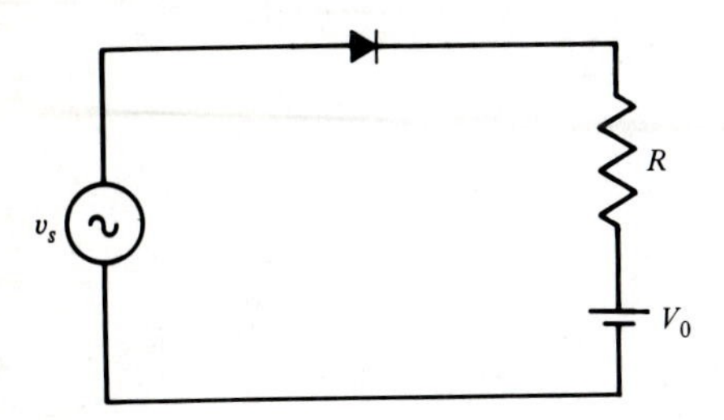

FIGURE 8-40
Problem 8.6.

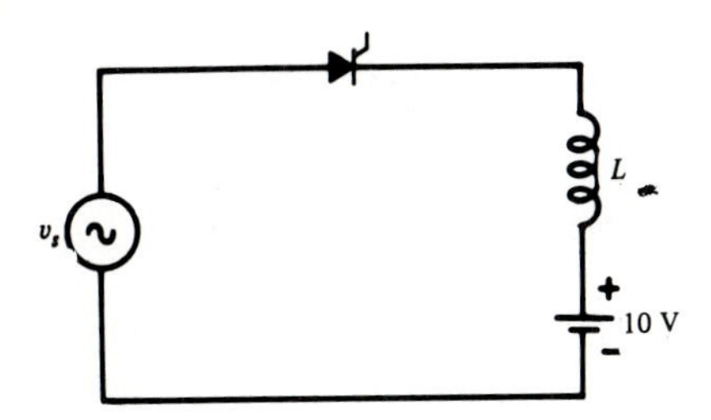

FIGURE 8-41
Problem 8.7.

8.10. A chopper drives a separately excited dc motor having the following data:

armature resistance, $R = 0.2\ \Omega$

armature inductance, $L = 2$ mH

energy conversion constant, $k = 1.27$ V-s/rad

input voltage = 300 V

chopping period, $T = 0.05$ s

duty cycle, $\alpha = 0.6$

load/motor friction coefficient, $b = 0.11$ N-m-s/rad

load torque, $T_L = 893$ N-m

Determine *(a)* the speed at which conduction becomes marginally continuous and *(b)* the average motor current at the point of discontinuous conduction.

8.11. The following data pertain to a chopper-driven dc series motor: average input current to the motor = 160 A; armature-circuit resistance = 0.08 Ω; armature-circuit inductance = 15 mH; input voltage to the chopper = 300 V; duty cycle = 0.8; chopping period = 0.005 s; motor energy conversion constant = 0.017 V-s/A-rad; motor-load friction coefficient = 0.35 N-m-s/rad. Assuming continuous conduction, find *(a)* the average speed, *(b)* the rms value of the armature current, and *(c)* the mechanical output power.

8.12. For a dc series motor driven by a chopper, the following data are given: supply voltage = 440 V; duty cycle = 30%; armature circuit inductance = 0.04 H. Determine the chopper frequency if the maximum allowable change in the armature current is 8 A.

APPENDIX I

UNIT CONVERSION

Symbol	Description	One: (SI Unit)	Is Equal to: (English Unit)	(CGS Unit)
B	Magnetic flux density	tesla (T) (= 1 Wb/m^2)	6.452×10^4 $lines/in^2$	10^4 G
H	Magnetic field intensity	ampere per meter (A/m)	0.0254 A/in	0.004π Oe
ϕ	Magnetic flux	weber (Wb)	10^8 lines	10^8 Mx
D	Viscous damping coefficient	newton-meter-second (N-m-s)	0.73756 lb-ft-sec	10^7 dyn-cm-sec
F	Force	newton (N)	0.2248 lb	10^5 dyn
J	Inertia	kilogram-square meter ($kg\text{-}m^2$)	23.73 $lb\text{-}ft^2$	10^7 $g\text{-}cm^3$
T	Torque	newton-meter (N-m)	0.73756 ft-lb	10^7 dyn-cm
W	Energy	joule (J)	1 W-sec	10^7 ergs

AWG Number	Diameter (in.)	Area, d^2 (circular mils; 1 mil = 0.001 in.)	Pounds per 1000 ft Bare Wire	Length (ft/lb)	Resistance at 77°F (Ω/1000 ft)
6	0.1620	26250.	79.46	12.59	0.3951
7	0.1443	20820.	63.02	15.87	0.4982
8	0.1285	16510.	49.98	20.01	0.6282
9	0.1144	13090.	39.63	25.23	0.7921
10	0.1019	10380.	31.43	31.82	0.9989
11	0.09074	8234.	24.92	40.12	1.260
12	0.08081	6530.	19.77	50.59	1.588
13	0.07196	5178.	15.68	63.80	2.003
14	0.06408	4107.	12.43	80.44	2.525
15	0.05707	3257.	9.86	101.4	3.184
16	0.05082	2583.	7.82	127.9	4.016
17	0.04526	2048.	6.20	161.3	5.064
18	0.04030	1624.	4.92	203.4	6.385
19	0.03589	1288.	3.90	256.5	8.051
20	0.03196	1022.	3.09	323.4	10.15
21	0.02846	810.1	2.45	407.8	12.80
22	0.02535	642.4	1.95	514.2	16.14
23	0.02257	509.5	1.54	648.4	20.36
24	0.02010	404.0	1.22	817.7	25.67
25	0.01790	320.4	0.970	1031.0	32.37
26	0.01594	254.1	0.769	1300.0	40.81
27	0.01420	201.5	0.610	1639.0	51.47
28	0.01264	159.8	0.484	2067.0	64.90
29	0.01126	126.7	0.384	2607.0	81.83
30	0.01003	100.5	0.304	3287.0	103.2
31	0.00893	79.70	0.241	4145.0	130.1
32	0.00795	63.21	0.191	5227.0	164.1
33	0.00708	50.13	0.152	6591.0	206.9
34	0.00631	39.75	0.120	8310.0	260.9
35	0.00562	31.52	0.095	10480.0	329.0
36	0.00500	25.00	0.076	13210.0	414.8

APPENDIX II

DATA FOR PURE COPPER WIRE

AWG Number	Diameter (in)	Area, d^2 (circular mils; 1 mil = 0.001 in)	Pounds per 1000 ft Bare Wire	Length (ft/lb)	Resistance at 77°F (Ω/1000 ft)
	1.151	1000000.	3090.	0.3235	0.0108
	1.029	800000.	2470.	0.4024	0.0135
	0.963	700000.	2160.	0.4628	0.0154
	0.891	600000.	1850.	0.5400	0.0180
	0.814	500000.	1540.	0.6488	0.0216
	0.726	400000.	1240.	0.8060	0.0270
Stranded	0.574	250000.	772.	1.30	0.0341
0000	0.4600	211600.	640.5	1.56	0.490
000	0.4096	167800.	507.9	1.97	0.0618
00	0.3648	133100.	402.8	2.48	0.0871
0	0.3249	105500.	319.5	3.13	0.0983
1	0.2893	83690.	253.3	3.95	0.1239
2	0.2576	66370.	200.9	4.96	0.1563
3	0.2294	52630.	159.3	6.28	0.1970
4	0.2043	41740.	126.4	7.91	0.2485
5	0.1819	31100.	100.2	9.98	0.3133

AWG Number	Diameter (in)	Area, d^2 (circular mils; 1 mil = 0.001 in)	Pounds per 1000 ft Bare Wire	Length (ft/lb)	Resistance at 77°F (Ω/1000 ft)
6	0.1620	26250.	79.46	12.59	0.3951
7	0.1443	20820.	63.02	15.87	0.4982
8	0.1285	16510.	49.98	20.01	0.6282
9	0.1144	13090.	39.63	25.23	0.7921
10	0.1019	10380.	31.43	31.82	0.9989
11	0.09074	8234.	24.92	40.12	1.260
12	0.08081	6530.	19.77	50.59	1.588
13	0.07196	5178.	15.68	63.80	2.003
14	0.06408	4107.	12.43	80.44	2.525
15	0.05707	3257.	9.86	101.4	3.184
16	0.05082	2583.	7.82	127.9	4.016
17	0.04526	2048.	6.20	161.3	5.064
18	0.04030	1624.	4.92	203.4	6.385
19	0.03589	1288.	3.90	256.5	8.051
20	0.03196	1022.	3.09	323.4	10.15
21	0.02846	810.1	2.45	407.8	12.80
22	0.02535	642.4	1.95	514.2	16.14
23	0.02257	509.5	1.54	648.4	20.36
24	0.02010	404.0	1.22	817.7	25.67
25	0.01790	320.4	0.970	1031.0	32.37
26	0.01594	254.1	0.769	1300.0	40.81
27	0.01420	201.5	0.610	1639.0	51.47
28	0.01264	159.8	0.484	2067.0	64.90
29	0.01126	126.7	0.384	2607.0	81.83
30	0.01003	100.5	0.304	3287.0	103.2
31	0.00893	79.70	0.241	4145.0	130.1
32	0.00795	63.21	0.191	5227.0	164.1
33	0.00708	50.13	0.152	6591.0	206.9
34	0.00631	39.75	0.120	8310.0	260.9
35	0.00562	31.52	0.095	10480.0	329.0
36	0.00500	25.00	0.076	13210.0	414.8

APPENDIX III

COMPUTER TECHNIQUES IN THE ANALYSIS OF ELECTRIC MACHINES

Analog and digital computer techniques have been applied to the analysis and design of transformers, rotating electric machines, and control systems from the earliest days of the "computer era." Computer techniques are used today in every aspect of the study of machines, transformers, and other electromechanical systems. This appendix is included to acquaint the reader with a few of the many computer techniques and specific programs used for this purpose.

A.1 MAGNETIC CHARACTERISTICS

A principal impediment to accurate systems is the nonlinear relationship of the magnetic circuit. Computer methods offer an invaluable alternative to the graphical, analog, and manual techniques used in early machine analysis. Some approaches for handling the magnetic characteristics include the following.

1. An actual B versus H (or ϕ versus NI or E versus NI) characteristic can be stored in a data file.

2. The curve can be represented by two or more straight-line segments. Fig. A3.1 is a typical flux versus mmf characteristic for an automobile starter motor. Two straight-line segments are shown for the major slopes of the curve. The equations for these segments are

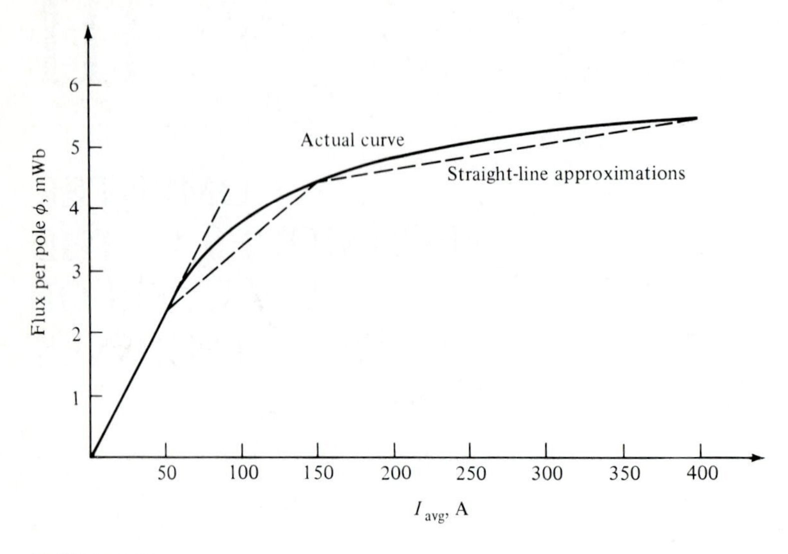

FIGURE A3.1
Flux versus ampere-turns curve and straight-line approximation.

$$\phi = 6.786 \times 10^{-6} (NI)_f W, \qquad (NI)_f < 700$$

$$\phi = 4.2 \times 10^{-3} + 4.1 \times 10^{-7} (NI)_f W, \qquad (NI)_f > 700$$

A third segment could be introduced to depict more accurately the curve in the region of the knee of the ϕ versus NI characteristic.

3. There are various methods for representing a curve, such as Fig. A3.1, by means of a power series in NI (or H). In general, a minimum of three terms is required for such a representation. In many computer systems that may be available to the reader, software programs for regression analysis can give the power series expressions for specific B versus H characteristics.

4. Many analytical expressions for the B versus H characteristics have been proposed. One very useful analytical expression that has been used in finite element methods of magnetic circuit analysis is

$$H = \left(k_1 e^{k_2 B^2} + k_3\right) B$$

where k_1, $k_2 m$, and k_3 are determined by causing the above equation to pass through three experimental points on a measured B versus H characteristic.

5. Finite elements. There are many computer programs for the general analysis of magnetic circuits used in electromechanical devices. Principal use of the finite-element technique is the derivation of equipotential lines and flux plots, calculation of energy, inductance, and force.

APPENDIX IV

DC MACHINE ARMATURE WINDINGS

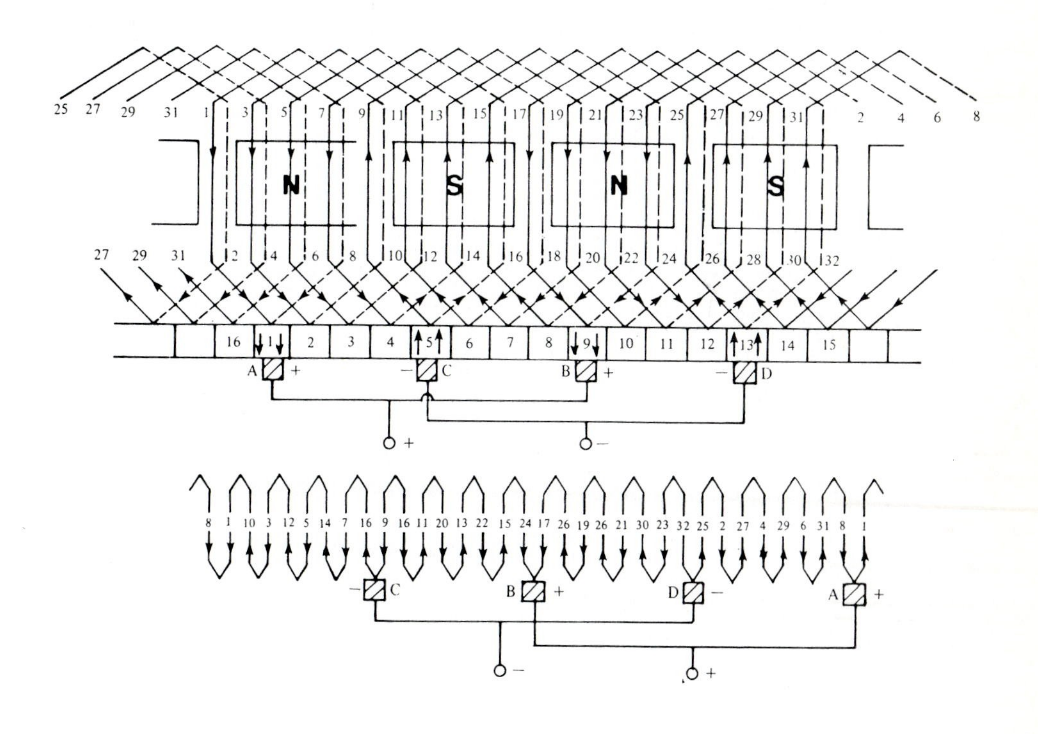

FIGURE A4-1
A four-pole double-layer lap winding.

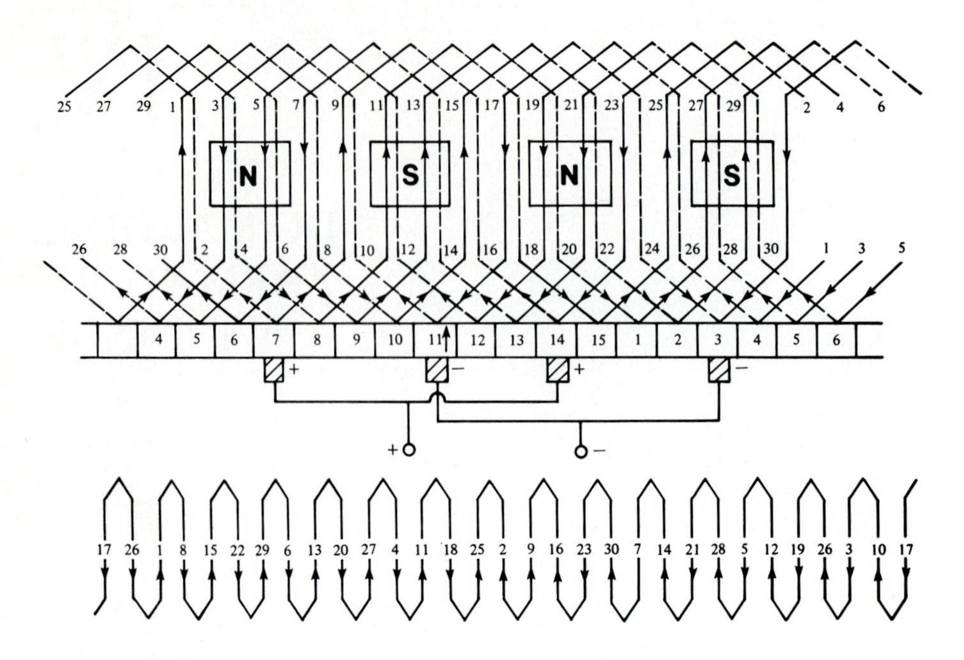

FIGURE A4-2
A four-pole double-layer wave winding.

INDEX

WIDENER UNIVERSITY
WOLFGRAM
LIBRARY
CHESTER, PA

THE NEW PRINCETON HANDBOOK OF POETIC TERMS

T. V. F. BROGAN
EDITOR

PRINCETON, NEW JERSEY
PRINCETON UNIVERSITY PRESS
1994

DEER PARK PUBLIC LIBRARY
44 LAKE AVENUE
DEER PARK, NY 11729

Copyright © 1994 by Princeton University Press

Published by Princeton University Press
41 William Street, Princeton, New Jersey 08540
In the United Kingdom:
Princeton University Press, Chichester, West Sussex
All Rights Reserved

THIS BOOK IS DEDICATED TO THE
UNIVERSITY OF NOTRE DAME

Power corrupts, and absolute power corrupts absolutely
—First Baron Lord Acton

Preparation of this volume was made possible in part by generous grants from the National Endowment for the Humanities and by other major foundations and private donors who value the acquisition, preservation, and dissemination of knowledge as well as the furtherance of all discourse which is reasoned, pluralist, and humane, toward the improvement of our common life.

Library of Congress Cataloguing-in-Publication Data
The New Princeton Handbook of Poetic Terms
T. V. F. Brogan, editor
p. cm.
Based on the New Princeton Encyclopedia of Poetry and Poetics
Includes bibliographical references
ISBN: 0-691-03671-3 (hardback edition)
ISBN: 0-691-03672-1 (paperback edition)
1. Poetry—Dictionaries. 2. Poetics—Dictionaries.
3. Poetry—History and Criticism.
I. Brogan, T. V. F. (Terry V. F.)
II. New Princeton Encyclopedia of Poetry and Poetics.
PN1021.N39 1994 808.1'03—dc20 93-43944

This book is based on the
New Princeton Encyclopedia of Poetry and Poetics
Alex Preminger and T. V. F. Brogan, editors,
Frank J. Warnke,† O. B. Hardison, Jr.,† and Earl Miner, associate editors

Princeton University Press books are printed on acid-free paper and meet the guidelines for permanence and durability of the Committee on Production Guidelines for Book Longevity of the Council on Library Resources.

Composed in ITC New Baskerville and custom fonts
Designed and produced by Leximetrics, Inc., South Bend, Indiana
Printed in the United States of America

18.95 BB ✓

3 5 7 9 10 8 6 4

PREFACE

This book is an offspring of the recently published *New Princeton Encyclopedia of Poetry and Poetics*, refashioned in smaller compass and intended for a different audience. Whereas the full *Encyclopedia* sought to provide readers with detailed, accurate, and timely information on every significant subject concerning poetry—not only the practice of poetry in every national literature of the world with any discernible poetic tradition but also the theory of poetry in all its manifestations both Western and Eastern from ancient times up to the last decade of the twentieth century—the present volume is meant to be, rather, a more convenient compilation that is easier to read, more affordable in price, and—not least—easier to hold and carry. Readers who want full and scholarly surveys of topics both major and minor will naturally turn to the *Encyclopedia*, but many other readers will equally naturally want more concise accounts of major terms, the kinds of accounts, that is, which will help them in their study of literature without becoming ends in themselves. This book is for them. Its purpose, in short, is to provide ready accounts of poetic and (some) prosodic terms likely to be encountered in modern literary study. No scholarly expertise is assumed; all entries are written for the general reader. We want particularly to make the volume accessible to students and general readers, and we aim to deliver more substantive accounts than are usually available in most literary dictionaries and other handbooks.

Moreover, we focus particulary on the most important terms for discussing poetry itself; we do not present detailed surveys of national poetries or poetic movements or the manifold schools of literary criticism: specialized volumes on these subjects we intend to publish separately in the near future. Many entries, of course, apply far more widely than to poetry—"Imagination," for example, is a subject fundamental to the cognition of literature itself, and an example too which will serve for many herein. Nevertheless, one central thread of Western poetics since preclassical times has been the privileging of poetry, or verse, over drama and, latterly, prose as the literary medium par excellence. Hence the focus of the book on poetry. Finally, most entries in this book—in all, 182 entries by 130 scholars—are completely comparative in scope; in a number of entries, such as "Allegory" and "Love Poetry," for example, we have provided a cross-cultural

perspective on subjects which some readers might tend to conceive in Western terms. but which look quite different from the perspectives of half a dozen other cultures. Updated, substantial bibliographies for further reading have been provided throughout.

Abbreviations used in the *New Princeton Encyclopedia* have been retained. As in several continental reference works, the word or words in the title of an entry are abbreviated to the first letter or letters when they occur in the text of the entry. Thus "versification" is abbreviated "v.," "metaphor" is abbreviated "m.," and so forth. Common abbreviations ("Gr." for "Greek," "c." for "century," etc.) are used freely. Finally, authors and works referred to frequently are abbreviated both in text and bibliographies simply by last name; the references will be found in the List of Bibliographical Abbreviations beginning on p. ix. Author initials at the ends of entries are expanded in the List of Contributors beginning on p. xiii.

The pursuit of knowledge must always seem esoteric in a dumbed-down world: the forces of violence and vulgarity presently so strong in America, together with changes in gender roles and cultural assimilation which have been as sweeping as they have been swift, have brought us into a temporary state of cultural vertigo. In this din, voices which value the humanizing effects of art and the necessity of toleration in discourse, voices which call again for all things which enrich the life of the spirit, and voices which would not throw away the long-cherished artifacts of where we have come from simply because how we got here now seems less than ideal, must seem, in a supercharged climate of politicization, voices from the past. Possibly. To some others, however, they may seem the only means by which we may *learn to change* and thus avoid repeating the ideological intolerances of the past under a new name for some momentarily shining idea. In a time when many ordinary people are homeless, intolerance in the academy whether from the right or left seems an especially brutal luxury. We must learn from the mistakes of the past and learn, now, how to go forward together, respecting our mutual differences while yet affirming the value of the common weal.

In the course of the work on which the present volume was based, one academic, too enchanted perhaps by the obscurantism of the would-be philosopher of National Socialism, pronounced the patient accumulation of scholarship "mere clerical work." This, we may feel, is evidence of no progress. Yet if, against such fascisms of theory and practice, this little book should persuade even a few readers that poetry matters, that there is much still to be discovered, and that there are important questions long ago asked but still awaiting answer, then the work will have been worth all that blear-eye'd wisdom born of midnight oil.

PREFACE

Professors Edward R. Weismiller and Fabian Gudas, senior scholars who both continue to work under physical limitations of increasing severity, offered the fruits of their remarkable erudition so as to make this book, like its parent volume, as good as human hands and minds could make it. The late O. B. Hardison, Jr., too, reminded me of what matters, and how deeply. I have not known how to thank any of them, beyond what poor words can convey and what a pair of eyes can acknowledge. What thanking is there? I do not know that thanking is of much use, for I learned from these men that what one gives is not what is convenient nor even what one wants but only what is necessary. In some future time and on some other plane of existence, we may hope, we may all speak together, I with them as with you, of the mysteries of these things, and of this place we call life.

My editor at Princeton University Press, Robert Brown, has watched over my work with patience beyond generosity; in some darkened, blue- haze jazz club, one night, I will make verbal what these lines cannot fully say. Janet Stern too watched over my pages expertly.

The following authors, publishers, and agents granted us permission to use brief extracts from the publications listed below. We have taken pains to trace all the owners of copyrighted material used in this book; any inadvertent omissions pointed out to us will be gladly corrected in future editions:

Charles Bernstein for five lines of "Sentences My Father Used" from *Controlling Interests*, reprinted by permission of Charles Bernstein and ROOF Books.

Robert Bly for four lines of "Six Winter Privacy Poems" from *Sleepers Joining Hands*, reprinted by permission of Robert Bly.

Carcanet Press Ltd. for five lines from "Portrait of a Lady" from *Collected Poems of William Carlos Williams, 1909–1939*, vol. 1, copyright 1938.

Faber and Faber, Ltd., for six lines of Canto VII from *The Cantos of Ezra Pound*, copyright 1934 by Ezra Pound, reprinted by permission of the publishers.

Farrar, Straus & Giroux for an excerpt from "In the Waiting Room" from *The Complete Poems, 1927–1979* by Elizabeth Bishop, copyright 1940, 1971, renewal copyright 1968 by Elizabeth Bishop, copyright 1979, 1983 by Alice Helen Methfessel.

Granada and HarperCollins Publishers for eight lines of "Buffalo Bill 's" from *Tulips and Chimneys* by e e cummings, edited by George James Firmage, copyright 1923, 1925, renewal copyright 1951, 1953 by e e cummings, copyright 1973, 1976 by the Trustees for the E. E. Cummings Trust, copyright 1973, 1976 by George James Firmage.

Henry Holt and Company for eight lines from "Come In" from *The Poems of*

Robert Frost, ed. Edward Connery Lathem, copyright 1923, 1947, 1969 by Holt, Rinehart and Winston; copyright 1942, 1951 by Robert Frost; copyright 1970, 1975 by Lesley Frost Ballantine.

Indiana University Press for two lines from *Martial: Selected Epigrams*, translated by Rolfe Humphries.

Jonathan Cape, Ltd., for eight lines from "Come In" from *The Poems of Robert Frost*, ed. Edward Connery Lathem, copyright 1923, 1947, 1969 by Holt, Rinehart and Winston; copyright 1942, 1951 by Robert Frost; copyright 1970, 1975 by Lesley Frost Ballantine.

Liverwright Publishing Corporation for eight lines of "Buffalo Bill 's" from *Tulips and Chimneys* by e e cummings, edited by George James Firmage, copyright 1923, 1925, and renewed 1951, 1953 by e e cummings, copyright 1973, 1976 by the Trustees for the E. E. Cummings Trust, copyright 1973, 1976 by George James Firmage.

New Directions Publishing Corporation for six lines of Canto VII from *The Cantos of Ezra Pound*, copyright 1934 by Ezra Pound; and for five lines of "Portrait of a Lady" from *Collected Poems of William Carlos Williams, 1909–1939*, vol. 1, copyright 1938 by New Directions Publishing Corporation.

Ohio University Press for Epigram no. 68 from *Collected Poems and Epigrams of J. V. Cunningham*, copyright 1971.

Taylor & Francis, Ltd., for six lines of "Walker Skating" by Brian Morris from *Word & Image*, vol. 2, copyright 1986.

BIBLIOGRAPHICAL ABBREVIATIONS

Abrams	M. H. Abrams, *The Mirror and the Lamp: Romantic Theory and the Critical Tradition*, 1953.
Analecta hymnica	*Analecta hymnica medii aevi*, ed. G. M. Dreves, C. Blume, and H. M. Bannister, 55 v., 1886–1922.
Auerbach	E. Auerbach, *Mimesis: The Representation of Reality in Western Literature*, tr. W. R. Trask, 1953.
Beare	W. Beare, *Latin Verse and European Song*, 1957.
Bec	P. Bec, *La Lyrique française au moyen âge (XIIe–XIIIe siècles): Contribution à une typologie des genres poétiques médiévaux*, 2 v., 1977–78.
Bowra	C. M. Bowra, *Greek Lyric Poetry from Alcman to Simonides*, 2d ed., 1961.
Brogan	T. V. F. Brogan, *English Versification, 1570–1980: A Reference Guide with a Global Appendix*, 1981.
Brooks	C. Brooks, *The Well Wrought Urn*, 1947.
CBEL	*Cambridge Bibliography of English Literature*, ed. F. W. Bateson, 4 v., 1940; v. 5, *Supplement*, ed. G. Watson, 1957.
CBFL	*A Critical Bibliography of French Literature*, gen. ed. D. C. Cabeen, 1–; 1947–; revisions and supplements, gen. ed. R. A. Brooks, 7 v., 1968–.
Chambers	F. M. Chambers, *An Introduction to Old Provençal Versification*, 1985.
CHCL	*Cambridge History of Classical Literature*, v. 1, *Greek Literature*, ed. P. E. Easterling and B. M. W. Knox, 1985; v. 2, *Latin Literature*, ed. E. J. Kenney, 1982.
CHEL	*Cambridge History of English Literature*, ed. A. W. Ward and A. R. Waller, 14 v., 1907–1916.
CHLC	*Cambridge History of Literary Criticism*, v. 1, *Classical Criticism*, ed. G. A. Kennedy, 1989.
Corbett	E. P. J. Corbett, *Classical Rhetoric for the Modern Student*, 3d ed., 1990.
Crane	*Critics and Criticism, Ancient and Modern*, ed. R. S. Crane, 1952.
Crusius	F. Crusius, *Römische Metrik: Ein Einfürung*, 8th ed., rev. H. Rubenbauer, 1967.
Culler	J. Culler, *Structuralist Poetics: Structuralism, Linguistics, and the Study of Literature*, 1975.
Curtius	E. Curtius, *European Literature and the Latin Middle Ages*, tr. W. R. Trask, 1953.
DAI	*Dissertation Abstracts International.*
Dale	A. M. Dale, *The Lyric Meters of Greek Drama*, 2d ed., 1968.
de Man	P. de Man, *Blindness and Insight: Essays in the Rhetoric of Contemporary Criticism*, 2d ed., 1983.
Derrida	J. Derrida, *Of Grammatology*, tr. 1976.
DHI	*Dictionary of the History of Ideas*, ed. P. P. Wiener, 6 v., 1968–74.
Dronke	P. Dronke, *Medieval Latin and the Rise of European Love Lyric*, 2d ed., 2 v., 1968.
Eliot, *Essays*	T. S. Eliot, *Selected Essays*, rev ed., 1950.
Elwert	W. T. Elwert, *Französische Metrik*, 4th ed., 1978.
Elwert, *Italienische*	W. T. Elwert, *Italienische Metrik*, 2d ed., 1984.
Empson	W. Empson, *Seven Types of Ambiguity*, 3d ed., 1953.
Faral	E. Faral, *Les Arts poétiques du XIIe et du XIIIe siècles*, 1924.
Fisher	*The Medieval Literature of Western Europe: A Review of Research, Mainly 1930–1960*, ed. John H. Fisher, 1965.
Fowler	A. Fowler, *Kinds of Literature: An Introduction to the Theory of Genres and Modes*, 1982.
Frye	N. Frye, *Anatomy of Criticism*, 1957.
Gasparov	M. L. Gasparov, *Sovremennyj russkij stix: Metrika i ritmika*, 1974.
GRLMA	*Grundriss der romanischen Literaturen des Mittelalters*, ed. H. R. Jauss and E. Köhler, 9 v., 1968–.
Group Mu	Group Mu (J. Dubois, F. Edeline, J.-M. Klinkenberg, P. Minguet, F. Pire, H. Trinon), *A General Rhetoric*, tr. P. B. Burrell and E. M. Slotkin, 1981.
Halporn et al.	J. W. Halporn, M. Ostwald, and T. G. Rosenmeyer, *The Meters of Greek and Latin Poetry*, 2d ed., 1980.
Hardie	W. R. Hardie, *Res metrica*, 1920.
Hollander	J. Hollander, *Vision and Resonance: Two Senses of Poetic Form*, 2d ed., 1985.
Hollier	*A New History of French Literature*, ed. D. Hollier, 1989.
Jakobson	R. Jakobson, *Selected Writings*, 8 v., 1962–88.

ABBREVIATIONS

Jarman and Hughes — *A Guide to Welsh Literature*, ed. A. O. H. Jarman and G. R. Hughes, 2 v., 1976–79.

Jeanroy — A. Jeanroy, *La Poésie lyrique des troubadours*, 2 v., 1934.

Jeanroy, *Origines* — A. Jeanroy, *Les Origines de la poésie lyrique en France au moyen âge*, 4th ed., 1965.

Kastner — L. E. Kastner, *A History of French Versification*, 1903.

Keil — *Grammatici latini*, ed. H. Keil, 7 v., 1855–80; v. 8, *Anecdota helvetica: Supplementum*, ed. H. Hagen, 1870.

Koster — W. J. W. Koster, *Traité de métrique grecque suivi d'un précis de métrique latine*, 4th ed., 1966.

Lausberg — H. Lausberg, *Handbuch der literarischen Rhetorik*, 2d ed., 2 v., 1973.

Le Gentil — P. Le Gentil, *La Poésie lyrique espagnole et portugaise à la fin du moyen âge*, 2 v., 1949–53.

Lewis — C. S. Lewis, *The Allegory of Love*, 1936.

Lord — A. B. Lord, *The Singer of Tales*, 1960.

Lote — G. Lote, *Histoire du vers français*, 3 v., 1949–56.

Maas — P. Maas, *Greek Metre*, tr. H. Lloyd-Jones, 3d ed., 1962.

Manitius — M. Manitius, *Geschichte der lateinischen Literatur des Mittelalters*, 3 v., 1911–31.

Mazaleyrat — J. Mazaleyrat, *Éléments de métrique française*, 3d ed., 1981.

Meyer — W. Meyer, *Gessamelte Abhandlungen zur mittellateinischen Rhythmik*, 3 v., 1905–36.

MGG — *Die Musik in Geschichte und Gegenwart: Allgemeine Enzyklopädie der Musik*, ed. F. Blume, 16 v., 1949–79.

MGH — Monumenta germaniae historica.

Michaelides — S. Michaelides, *The Music of Ancient Greece: An Encyclopaedia*, 1978.

Migne, *PG* — *Patrilogiae cursus completus, series graeca*, ed. J. P. Migne, 161 v., 1857–66.

Migne, *PL* — *Patrilogiae cursus completus, series latina*, ed. J. P. Migne, 221 v., 1844–64.

Miner et al. — E. Miner, H. Odagiri, and R. E. Morrell, *The Princeton Companion to Classical Japanese Literature*, 1986.

Morier — H. Morier, *Dictionnaire de poétique et de rhétorique*, 3d ed., 1981.

Morris-Jones — J. Morris-Jones, *Cerdd Dafod*, 1925, rpt. with Index, 1980.

Murphy — J. J. Murphy, *Rhetoric in the Middle Ages: A History of Rhetorical Theory from St. Augustine to the Renaissance*, 1974.

Navarro — T. Navarro, *Métrica española: Reseña histórica y descriptiva*, 6th ed., 1983.

New CBEL — *New Cambridge Bibliography of English Literature*, ed. G. Watson and I. R. Willison, 5 v., 1969–77.

New Grove — *New Grove Dictionary of Music and Musicians*, ed. S. Sadie, 20 v., 1980.

Nienhauser et al. — W. H. Nienhauser, Jr., C. Hartman, Y. W. Ma, and S. H. West, *The Indiana Companion to Traditional Chinese Literature*, 1986.

Norberg — D. Norberg, *Introduction a l'étude de la versification latine médiévale*, 1958.

Norden — E. Norden, *Die antike Kunstprosa*, 9th ed., 2 v., 1983.

OED — *Oxford English Dictionary*, 1st ed.

Omond — T. S. Omond, *English Metrists*, 1921.

Parry — M. Parry, *The Making of Homeric Verse*, ed. A. Parry, 1971.

Parry, *History* — T. Parry, *A History of Welsh Literature*, tr. H. I. Bell, 1955.

Patterson — W. F. Patterson, *Three Centuries of French Poetic Theory: A Critical History of The Chief Arts of Poetry in France (1328–1630)*, 2 v., 1935.

Pauly-Wissowa — *Paulys Realencyclopädie der classischen Altertumswissenschaft*, ed. A. Pauly, G. Wissowa, W. Kroll, and K. Mittelhaus, 24 v. (A-Q), 10 v. (R-Z, Series 2), and 15 v. (Supplements), 1894–1978.

Pearsall — D. Pearsall, *Old English and Middle English Poetry*, 1977.

PLAC — *Poetae latini aevi carolini*, ed. E. Dümmler (v. 1–2), L. Traube, P. von Winterfeld, and K. Strecker, 5 v., 1881–1937.

Raby, *Christian* — F. J. E. Raby, *A History of Christian-Latin Poetry From the Beginnings to the Close of the Middle Ages*, 2d ed., 1953.

Raby, *Secular* — F. J. E. Raby, *A History of Secular Latin Poetry in the Middle Ages*, 2d ed., 2 v., 1957.

Ransom — *Selected Essays of John Crowe Ransom*, ed. T. D. Young and J. Hindle, 1984.

Reallexikon — *Reallexikon der deutschen Literaturgeschichte*, 2d ed., ed. W. Kohlschmidt and W. Mohr (v. 1–3), K. Kanzog and A. Masser (v. 4), 4 v., 1958–84.

Reallexikon I — *Reallexikon der deutschen Literaturgeschichte*, ed. P. Merker and W. Stammler, 1st ed., 4 v., 1925–31.

Richards — I. A. Richards, *Principles of Literary Criticism*, 1925.

Saisselin — R. G. Saisselin, *The Rule of Reason and the Ruses of the Heart: A Philosophical Dictionary of Classical French Criticism, Critics, and Aesthetic Issues*, 1970.

Sayce — O. Sayce, *The Medieval German Lyric, 1150–1300: The Development of Its Themes and Forms in Their European Context*, 1982.

Scherr — B. P. Scherr, *Russian Poetry: Meter, Rhythm, and Rhyme*, 1986.

Schipper — J. M. Schipper, *Englische Metrik*, 3 v., 1881–1888.

Schipper, *History* — J. M. Schipper, *A History of English Versification*, 1910.

Schmid and Stählin — W. Schmid and O. Stählin, *Geschichte der griechischen Literatur*, 7 v., 1920–48.

Scott — C. Scott, *French verse-art: A study*, 1980.

Sebeok — *Style in Language*, ed. T. Sebeok, 1960.

Sievers — E. Sievers, *Altgermanische Metrik*, 1893.

Smith — *Elizabethan Critical Essays*, ed. G. G. Smith, 2 v., 1904.

Snell — B. Snell, *Griechische Metrik*, 4th ed., 1982.

Spongano — R. Spongano, *Nozioni ed esempi di metrica italiana*, 2d. ed., 1974.

Stephens — *The Oxford Companion to the Literature of Wales*, ed. M. Stephens, 1986.

Terras — *Handbook of Russian Literature*, ed. V. Terras, 1985.

Thieme — H. P. Thieme, *Essai sur l'histoire du vers français*, 1916.

Trypanis — C. A. Trypanis, *Greek Poetry From Homer to Seferis*, 1981.

Weinberg — B. Weinberg, *A History of Literary Criticism in the Italian Renaissance*, 2 v., 1961.

Wellek — R. Wellek, *A History of Modern Criticism, 1750–1950*, 8 v., 1955–92.

Wellek and Warren — R. Wellek and A. Warren, *Theory of Literature*, 3d ed., 1956.

West — M. L. West, *Greek Metre*, 1982.

Wilamowitz — U. von Wilamowitz-Moellendorf, *Griechische Verskunst*, 1921.

Wilkins — E. H. Wilkins, *A History of Italian Literature*, rev. T. G. Bergin, 1974.

Wimsatt — *Versification: Major Language Types*, ed. W. K. Wimsatt, Jr., 1972.

Wimsatt and Brooks — W. K. Wimsatt, Jr., and C. Brooks, *Literary Criticism: A Short History*, 1957.

GENERAL ABBREVIATIONS

Two sets of abbreviations are used systematically throughout this volume in order to conserve space. First, general terms (below) are abbreviated both in text and bibliographies. Second, within each entry, the headword or -words of the title of the entry are abbreviated by the first letter of each word, unless the word is a general abbreviation, which takes precedence. Thus in the entry, "Egyptian Poetry," that phrase is regularly abbreviated E. p., while in "English Poetry," the head phrase is abbreviated Eng. p. Both general and headword abbreviations may also show plural forms, e.g. ms. for "metaphors" in the entry "Metaphor" (and "Meter," etc.) and lits. for "literatures" in any entry.

Af.	African
Am.	American
anthol.	anthology
Ar.	Arabic
Assoc.	Association
b.	born
bibl.	bibliography
c.	century
ca.	*circa*, about
cf.	*confer*, compare
ch.	chapter
Cl.	Classical
comp.	comparative
contemp.	contemporary
crit.	criticism
d.	died
devel.	development
dict.	dictionary
diss.	dissertation
ed.	edition, editor, edited by
e.g.	*exempli gratia*, for example
Eng.	English
enl.	enlarged
esp.	especially
et al.	*et alii*, and others
ff.	following
fl.	*floruit*, flourished
Fr.	French
Ger.	German
Gesch.	Geschichte
Gr.	Greek
Heb.	Hebrew
hist.	history, histoire
IE	Indo-European
i.e.	*id est*, that is
incl.	includes, including
interp.	interpretation
Intro.	Introduction
Ir.	Irish
It.	Italian
Jh.	Jahrhundert
jour.	journal
lang.	language
Lat.	Latin
ling.	linguistics
lit.	literature
lit. crit.	literary criticism
lit. hist.	literary history
ME	Middle English
med.	medieval
MHG	Middle High German
mod.	modern
ms.	manuscript
NT	New Testament
OE	Old English
OF	Old French
OHG	Old High German
ON	Old Norse
OT	Old Testament
p., pp.	page, pages
Port.	Portuguese
Proc.	Proceedings
pros.	prosody
Prov.	Provençal, i.e. Occitan
pub.	published
q.v.	*quod vide*, which see
qq.v.	*quae vide*, both (all of) which see
Ren.	Renaissance
rev.	revised
Rev.	Review
rhet.	rhetoric
rpt.	reprinted
Rus.	Russian
Sp.	Spanish
supp.	supplement(ed)
tr.	translation, translated
trad.	tradition
v.	volume(s)

THE CONTRIBUTORS

A.B.F. Albert B. Friedman, Rosecrans Professor of English, Claremont Graduate School
A.B.L. Albert Bates Lord, late Arthur Kingsley Porter Professor Emeritus of Slavic and Comparative Literature, Harvard University
A.G.E. Alfred Garvin Engstrom, Alumni Distinguished Professor Emeritus of French, University of North Carolina
A.L.S. A. Lytton Sells, late Professor of French and Italian, Indiana University
A.M.D. Andrew M. Devine, Professor of Classics, Stanford University
A.P. Anne Paolucci, President, Council on National Literatures
A.PR. Alex Preminger, scholar and editor
A.T.C. A. Thomas Cole, Professor of Greek and Latin, Yale University
A.W. Andrew Welsh, Associate Professor of English, Rutgers University
A.W.H. Albert W. Halsall, Professor of French, Carleton University
B.A.N. B. Ashton Nichols, Assistant Professor of English, Dickinson College
B.B. Beth Bjorklund, Associate Professor of German, University of Virginia
B.BO. Betsy Bowden, Associate Professor of English, Rutgers University
B.G. Bernard Groom, late Professor of English, McMaster University
B.H.S. Barbara Herrnstein Smith, Braxton Craven Professor of Comparative Literature and English, Duke University
B.S.M. Barbara Stoler Miller, Samuel R. Milbank Professor of Oriental Studies, Columbia University
C.A.K. Charles A. Knudson, late Professor of English, University of Illinois
C.B.L. Claudia Brodsky Lacour, Associate Professor of Comparative Literature, Princeton University
C.K. Christopher Kleinhenz, Professor and Chairman of the Department of Italian, University of Wisconsin
C.P. Camille Paglia, Associate Professor of Humanities, Philadelphia College of Performing Arts, University of the Arts
C.S. Clive Scott, Lecturer in French, University of East Anglia
D.H. Dick Higgins, Research Associate, State University of New York at Purchase
D.HO. Daniel Hoffman, Poet in Residence and Felix E. Schelling Professor of English, University of Pennsylvania
D.L.R. David Lee Rubin, Professor of French, University of Virginia
D.W. Donald Wesling, Professor of English Literature, University of California at San DIego
E.A.B. Edward A. Bloom, Professor of English, Brown University
E.A.P. Erskine A. Peters, Professor of English, University of Notre Dame
E.B. Eleanor Berry, author
E.B.J. Elise Bickford Jorgens, Professor of English, Western Michigan University
E.BO. Eniko Bollobaś, scholar and diplomat, Hungary
E.C. Eleanor Cook, Professor of English, Victoria College, University of Toronto
E.D. Edward Doughtie, Professor of English, Rice University
E.H.B. Ernst H. Behler, Professor of German and Chairman, University of Washington
E.M. Earl Miner, Professor of English and Comparative Literature, Princeton University
E.R.W. Edward R. Weismiller, Professor Emeritus of English, George Washington University
F.F. Frances Ferguson, Professor of English and the Humanities, Johns Hopkins University
F.G. Frederick Garber, Professor of Comparative Literature, State University of New York
F.GU. Fabian Gudas, Professor Emeritus of English, Louisiana State University
F.J.W. Frank J. Warnke, late Professor of Comparative Literature, University of Georgia
F.T. Frances Teague, Associate Professor of English, University of Georgia
F.W. Frederic Will, author
G.F.E. Gerald F. Else, late Professor of Classical Studies, University of Michigan
G.T.W. George T. Wright, Professor Emeritus of English, University of Minnesota
H.B. Henryk Baran, Associate Professor of Russian, State University of New York at Albany
H.P. Henry Paolucci, Professor of Government and Politics, St. John's University
H.R.E. Helen Regueiro Elam, Associate Professor of English, State University of New York at Albany
H.S. Hiroaki Sato, author and translator
J.A. John Arthos, Distinquished University

Professor Emeritus of English, University of Michigan

J.A.W. James A. Winn, Professor of English, University of Michigan

J.B.V. John B. Vickery, Professor of English and Associate Executive Vice-Chancellor, University of California at Riverside

J.C.S. Joseph C. Sitterson, Jr., Associate Professor of English, Georgetown University

J.E. James Engell, Professor of English and Comparative Literature, Harvard University

J.E.C. J. E. Congleton, Dean Emeritus, School of Humanities and Sciences, University of Findlay

J.G.F. Joseph G. Fucilla, late Professor of Romance Languages, Northwestern University

J.K.N. J. K. Newman, Professor of the Classics, University of Illinois at Urbana-Champaign

J.L. John Lindow, Professor of Scandinavian, University of California

J.L.F. John L. Foster, Professor of English, Roosevelt University

J.M.F. John Miles Foley, Professor of English and William H. Byler Distinguished Chair in Humanities, University of Missouri

J.R.B. James Richard Bennett, Professor of English, University of Arkansas

J.S.M. Julie S. Meisami, Lecturer in Persian, University of Oxford

J.V.B. Jaqueline Vaught Brogan, Professor of English, University of Notre Dame

J.W.H. James W. Halporn, Professor of Classical Studies and Comparative Literature, Indiana University

J.WH. Jon Whitman, Senior Lecturer, Hebrew University of Jerusalem

J.W.J. James William Johnson, Professor of English, University of Rochester

K.S.C. Kang-i Sun Chang, Professor of Chinese Literature, Yale University

L.D.S. Laurence D. Stephens, Professor of Classics, University of North Carolina

L.J.Z. Lawrence J. Zillman, Professor Emeritus of English, University of Washington

L.M. Lawrence Manley, Associate Professor of English, Yale University

L.N. Leonard Nathan, Professor of Rhetoric, University of California at Berkeley

L.NE. Lowry Nelson, Jr., Professor of Comparative Literature, Yale University

L.P. Laurence Perrine, Frensley Professor Emeritus of English, Southern Methodist University

M.A.C. Mary Ann Caws, Distinguished Professor of English, French, and Comparative Literature, City University of New York

M.B. Michael Beard, Professor of English, University of North Dakota

M.D. Michael Davidson, Associate Professor of Literature, University of California at San Diego

M.E.B. Merle E. Brown, late Professor of English, University of Iowa

M.L.R. M. L. Rosenthal, Professor Emeritus of English, New York University

M.O'C. Michael Patrick O'Connor, author

M.S. Marianne Shapiro, Visiting Professor of Comparative Literature, Brown University

M.ST. Martin Steinmann, Jr., Professor Emeritus of English, University of Illinois at Chicago

M.STE. Martin Stevens, City University of New York Distinguished Professor of English, City University of New York

M.T.H. Marvin T. Herrick, late Professor of English, University of Illionis

M.V.F. Michael V. Fox, Professor of Hebrew, University of Wisconsin

N.F. Norman Friedman, Professor of English, Queens College, City University of New York

O.B.H. O. B. Hardison, Jr., late University Professor of English, Georgetown University

P.G.A. Percy G. Adams, Lindsay Young Professor Emeritus of English and Comparative Literature, University of Tennessee

P.H.F. Paul H. Fry, Professor of English, Yale University

P.M. Pamela McCallum, Associate Professor of English, University of Calgary

P.M.D. Peter M. Daly, Professor of German, McGill University

P.RO. Philip Rollinson, Associate Professor of English, University of South Carolina

P.S. Peter Sacks, Associate Professor of English, John Hopkins University

P.S.C. Procope S. Costas, late Professor of Classics, Brooklyn College, City University of New York

P.W. Philip Weller, scholar

R.A. Robert Alter, Professor of Hebrew and Comparative Literature, University of California at Berkeley

R.A.H. Roger A. Hornsby, Professor Emeritus of Classics, University of Iowa

R.A.S. Roy Arthur Swanson, late Professor of Classics, University of Wisconsin at Milwaukee

R.C. Ruby Cohn, Professor of Comparative Drama, University of California at Davis

R.F. Roger Fowler, Professor of English and Linguistics, University of East-Anglia

R.GR. Roland Greene, Professor of Comparative Literature, University of Oregon

R.H.W. Ruth Helen Webb, Frances A. Yates Research Fellow, Warburg & Courtauld Institute

R.L.H. Robert L. Harrison, author

R.O.E. Robert O. Evans, Professor Emeritus of

CONTRIBUTORS

English and Comparative Literatue, University of Kentucky

R.P.F. Robert P. Falk, Professor Emeritus of English Literature, University of California at Los Angeles

R.P.S. Raymond P. Scheindlin, Provost and Professor of Medieval Hebrew Literature, The Jewish Theological Seminary of America

R.V.H. Robert von Hallberg, Professor of English, University of Chicago

R.W. Richard Wendorf, Director, Houghton Library, Harvard University

S.A.L. Stephen A. Larrabee, late author

S.E.H. Stephen E. Henderson, Professor of Afro-American Studies, Howard University

S.F. Solomon Fishman, late Professor of English, University of California at Davis

S.F.F. Stephen F. Fogle, late Professor of English, Adelphi Suffolk College

S.J.K. Sholom J. Kahn, Professor Emeritus of English, Hebrew University

S.LY. Sverre Lyngstad, Distinguished Professor of English, New Jersey Institute of Technology

S.M.G. Sally M. Gall, author

S.P.R. Stella P. Revard, Professor of English, Southern Illinois University at Edwardsville

T.B. Timothy Bahti, Associate Professor of Comparative Literature, University of Michigan

T.G. Thomas Gardner, Associate Professor of English, Virginia Polytechnic Institute and State University

T.J.R. Timothy J. Reiss, Professor and Chairman of the Department of Comparative Literature, New York University

T.O.S. Thomas O. Sloane, Professor of Rhetoric, University of California

T.R.A. Timothy R. Austin, Associate Professor of English, Loyola University of Chicago

T.V.F.B. T. V. F. Brogan, scholar and editor

U.T.H. Urban T. Holmes, Jr., late William R. Kenan Professor of Romance Languages, University of North Carolina

W.B. Willard Bohn, Professor of French, Illinois State University

W.B.F. Wolfgang Bernhard Fleischmann, late Professor of Comparative Literature, Montclair State College

W.B.P. William Bowman Piper, Professor of English, Rice University

W.D.P. William D. Paden, Professor of French, Northwestern University

W.H.R. William H. Race, Associate Professor of Classical Studies, Vanderbilt University

W.L.H. William L. Hanaway, Professor of Persian, University of Pennsylvania

W.M. Wallace Martin, Professor Emeritus of English, University of Toledo

W.T. Wesley Trimpi, Professor of English, Stanford University

W.V.O'C. William Van O'Connor, late Professor of English, University of California at Davis

W.W.P. Walter Ward Parks, Associate Professor of English, Louisiana State University

THE NEW PRINCETON HANDBOOK OF POETIC TERMS

A

ACCENT (Gr. *prosôdia*, Lat. *accentus*, "song added to speech," i.e. the melody of lang.).

I. TERMINOLOGY
II. DEGREES AND TYPES
III. HISTORY

In the general sense, the emphasis or prominence that some syllables in speech bear over others, regardless of how achieved (the intonational means are via pitch, stress, or length); in the specific sense, a synonym for stress, a dimension of speech not reflected in most Western systems of orthography. Technically, "stress" in linguistics denotes intensity of articulatory force, resulting from greater musculatory exertion in forming a sound.

I. TERMINOLOGY. Prominence of a syllable may be produced by stress, but it may also be produced by obtruded (raised or lowered) pitch (fundamental frequency) or increased length (duration) or (most often) by a combination of these factors, which have been found to correlate highly with one another (stressed syllables are often higher in pitch and increased in length as well); together the three phenomena constitute an "intonational contour." Up until the 1950s, it was common to speak of a "pitch a." in tonic (pitch-based) langs. such as Ch., Japanese, and ancient Gr., as opposed to the "stress a." of the mod. Germanic langs., incl. Eng. But subsequent work (e.g. Bolinger), which was in fact in accord with a long line of theorizing by prosodists (e.g. Bright), suggested that pitch was after all the crucial element of "a." even in stress langs. At the same time, Allen has suggested that, contrary to centuries of trad., stress not pitch was the basis of a. in ancient Gr. At present one may say only that the phonetic phenomena concerning a. are still not fully understood. Since stress is not readily isolated from other factors, it is still not clear whether stress is after all a real entity in itself or simply the resultant of perceived prominence caused mainly by pitch and length, but the latter view looks more likely.

Usage of the terms "stress" and "a." consequently shows a welter of variation; Crystal reserves "stress" for word-stress and "a." for prominence at the phrase and sentence levels ("stress belongs to the lexicon, and a. to the utterance"), but many others speak of "sentence stress." Bolinger, emphasizing pitch, distinguishes (word) *stress*, (pitch) *a.* (on stressed syllables), and *intonation* (pitch changes at the sentence level). Liberman and Prince, by contrast, posit stress as a relation between syllables in a hierarchy above the lexical level, rather than assigning it as a feature of the segment (so Jespersen 1901). But these are technical problems in linguistics. In nontechnical usage, the simplest thing to say is that "a." in the sense of emphasis is the more general term, "stress" the more precise, and that "stress" can denote intensity as opposed to pitch or length. Most of the conceptual disputes in metrics over the past two millennia resulted from confusions about the nature of "quantity" vs. "a.," deeply interrelated acoustic phenomena whose subtlety and complexity were not much penetrated until the advent of modern linguistics.

II. DEGREES AND TYPES. Monosyllables and short polysyllables, particularly substantives such as nouns, normally bear a stress; this is "word-stress." In longer polysyllables (which form by compounding shorter words, prefixes, and suffixes), one stress dominates all the others, reducing them to unstressed syllables or to a secondary degree of stress. How many degrees of stress are thus created, and perceived, is a disputed question. The Ren. grammarians, following those in Med. Lat., who in turn followed the ancient Greeks, held that there were three, which they called "acute," "circumflex," and "grave." The structural linguists of the mid 20th c. (i.e. Trager-Smith) claimed there were four (primary, secondary, tertiary, weak, denoted 1-2-3-4, respectively, and based on loudness), as in the noun phrase *elevator operator*, stressed 1-4-3-4 2-4-3-4, though there is some evidence that in ordinary speech auditors do not hear that many degrees. The extra degree added by Trager-Smith is for sentence stress and corresponds to Chomsky and Halle's nucleus; inside a given word there are still only three degrees. As in the formation of polysyllabic words, so in the formation of phrases one stress comes to dominate all the others within its domain. So too at the level of the sentence: this process of dominance, of forming stress hierarchies, is general. All other things being equal, the Nuclear Stress Rule dictates that within any major syntactic constituent, the rightmost major stress will be primary, weakening all others to the left. At the sentence level, its effect is to make the last major element the strongest.

Word-stress is relatively fixed and is coded into the lexicon (it can be found in the dictionary), being controlled by the phonological and morphological rules of the lang. But phrase- and sentence-stress is more variable, depending in part on syntactic rules but also on context and on the emphases that the speaker wishes to make in that particular context. Word-stress thus controls

whether a syllable *can* be stressed, but semantic intent and other factors control whether it *will.* Stress at the phrasal or sentence level was traditionally called "rhetorical a."—so Alexander Gil (1619): "accentus est duplex Grammaticus aut Rhetoricus" (ed. Danielsson and Gabrielson [1972], 124). One result of it is that a speaker can say the same word or sentence several different ways by moving the main a. and mean quite different things. "Próject" is a noun, but "projéct" is a verb. One can say "I never said I loved you," for example, so as to mean (stressing "loved") "I may have said I *liked* you but I never said I *loved* you" or (stressing the second "I") "I said *Jack* loved you" or (stressing "you") "I said I loved *Ann*, not you, Diane."

Such cases as this, where mere shift of stress deeply alters meaning, are instances of the phenomenon known as "contrastive stress," which Shakespeare exploits to effect in Sonnet 130 and which Donne employs frequently (Melton's "arsis-thesis variation"). This is an important device in poetry. If, to take another example, we alter Ben Jonson's line and say, "love me only with thine eyes," which words shall we stress—*love* me with thine eyes (don't give me hateful looks)? love *me* (only me, not anyone else)? only with thine *eyes* (not with any other part of you)? Semantic clues from the wider context of the poem may help us decide. But a more important determinant in metrical poetry is *meter*, which is not determined by but must align with and so helps us determine the intended or most probable stressings (meanings) of the line. Meter regulates the placement of stresses in sentences according to a fixed scheme via a process called *modulation*, in which stress values can be demoted or promoted. Knowing the meter will therefore help one to elucidate sense; conversely, grasping the sense of lines in an unfamiliar poem can help one correctly identify its meter. It is one of the great limitations of free verse (q.v.) that it abandons both these hermeneutic instruments.

III. HISTORY. To give a history of a. in the poetries of the West is really to give a history of our progressive understanding, or clarification, of a. in the poetries in the West. In Gr., a. was, according to both the ancient authorities and most modern scholars (though see Allen), determined by pitch. The quantitative meters of Gr. poetry, however, were based on syllabic length. When Ennius appropriated Gr. meters for Lat. poetry, there may have been an indigenous native trad. of stress-based meters in Italic—the Saturnian—which was thereafter suppressed from Lat. artverse, surviving only in popular verseforms such as soldiers' songs; or it may have served a more important role we do not now well understand (see Beare). In any event, sometime about the 3d–4th cs. A.D., in some massive linguistic transformation still not understood, a. based on pitch was replaced by a. based on intensity or stress in Lat., with the result that a. replaced syllabic length as the basis of verse composition. Subsequent Old Germanic versification is heavily accentual.

With the decline of Rome, understanding of the (artificial) rules of quantity were progressively lost, and Med. Lat. "metrical verse" (based on quantity) begins to be replaced by "rhythmical verse" (based on a.)—first and most importantly in a popular form of poetry, the hymn (q.v.), expressly developed by Augustine, Ambrose, and Hilary for easy singing and memorization by the illiterate masses. Quantitative verse in Lat. continued to be written in quantity at least up to the 12th c., but only as a scholarly exercise.

With the emergence of the European vernaculars, each Romance poetry had to invent a prosody for itself. The general view at present is that the metrical and stanza forms were taken over from Med. Lat., while the rules for accentuation and syllabification varied somewhat in each lang., their prosodies reflecting this fact also (Norberg, Pulgram). Accents do not have the force in Fr. that they do in the Germanic langs. such as Eng. In Eng., word-stress greatly outweighs word-boundary: the stressed syllable is lengthened. But in Fr. the last syllable in the word is characteristically lengthened, while word stress is reduced in favor of phrasal stress. These phrasal stresses at line-end and before the caesura, called *accents toniques,* are bolstered on occasion by *accents d'appui* ("prop" as., made by promotion of weakly stressed syllables) and *accents contre-toniques* (as. on the antepenult of polysyllabic words without mute -e). The Fr. *accent oratoire* corresponds to Eng. "rhetorical a." In sum, Fr. prosody (meter) is phrasal in nature, based on rhythmic groupings in verse identical to those in prose. Eng. prosody, by contrast, developed smaller, nonphrasal groups—feet—each based on one stress. Romance words which came into Eng. (esp. in ME) underwent "recession of a.," by which Eng. speakers shifted left to the beginning of the word the rightward word-stress of the Fr.

In the Ren. reaction against medievalism, the Humanists looked back to Classicism, and esp. the prestigious trad. of Cl. prosody: consequently, quantitative meters were successively attempted in the 16th c. in every vernacular, first It., then Fr., Eng., and Ger. But length is simply not phonemic in the mod. langs., so any metric based on it is doomed to be artificial. Stress, by contrast, is phonemic; and when independent thinking about poetics also emerged in the Ren., many critics and prosodists recognized this fact—though only dimly, and with no accurate terminology with which to describe what they heard and felt when they read poetry. Most of the Eng. Ren. prosodists (collected in Smith) grasped, intuitively, the existence of stress, but they were unable clearly to distinguish stress from pitch (as indeed many still do not), and the terms which they had available—those descended from antiquity—were of little help. The Ren. grammarians often speak of a

raising and lowering of the voice; Lily treats a. s.v. "Tonus," and identifies three varieties: grave, circumflex, and acute. An accurate terminology only begins to appear in the 18th c., with its renewed interest in the origin of lang. It is abetted materially by the rise of Philology in the 19th c., esp. in Germany, though here too most of the terminology continues to be explicitly Classical. In Eng. one finds fresh air in the discussions in the London Philological Society in the latter 19th c. (Ellis, Sweet). But it is chiefly in the 20th c. that linguistics, now fully emerged as an autonomous discipline, begins to develop instruments and techniques for analyzing precisely the phenomena of acoustic phonetics. As for integrating stress within a unified theory, structural linguistics described stress patterns by assigning levels or degrees of stress (see above); generative phonology, following upon (and imitating) the transformational grammar of Chomsky, attempted to find rules which would generate correct stressings but exclude all others and which would mesh with the rules of syntax (see Liberman and Prince, Selkirk). "Generative metrics" sought to apply such rules to scansion (see Beaver) but failed to grasp the nature of convention (esp. metrical) in poetry. Modern linguistics has swept away much of the detritus of two millennia of confusion, error, and received opinion. But if concepts such as pitch and stress have become clearer, they also have become far more complex. The state of linguistic theory at present may well suggest that our understanding of these intricate phenomena is still far from complete.

J. Hart, *The Opening of the Unreasonable Writing of Our InglishToung* (1551); J. Steele, *Prosodia rationalis*, 2d ed. (1779); A. J. Ellis, *On Early Eng. Pronunciation*, 2 v. (1868–69), "On the Physical Constituents of A. and Emphasis," *TPS* (1873–74); Smith; Schipper, *History*, ch. 8; Thieme, 390—lists Fr. works to 1912; P. Habermann, "Akzent," "Takt," *Reallexikon I*; B. Danielsson, *Studies in the Accentuation of Polysyllabic Loan-Words in Eng.* (1948); G. L. Trager and H. L. Smith, Jr., *An Outline of Eng. Structure* (1951); Beare—still essential; Norberg; P. Habermann and W. Mohr, "Hebung und Senkung," *Reallexikon* 1.623–29; D. L. Bolinger, *Forms of Eng.* (1965), pt. 1, *Intonation and Its Parts* (1986), ch. 2; S. Chatman, *A Theory of Meter* (1965), chs. 3, 4, App.; J. McAuley, *Versification* (1966); E. J. Dobson, *Eng. Pronunciation 1500–1700*, 2d ed., 2 v. (1968)—historical phonology; P. Garde, *L'A.* (1968); N. Chomsky and M. Halle, *The Sound Pattern of Eng.* (1968)—generative phonology; J. C. Beaver, "Contrastive Stress and Metered Verse," *Lang&S* 2 (1969), "The Rules of Stress in Eng. Verse," *Lang.* 47 (1971); D. Crystal, *Prosodic Systems and Intonation in Eng.* (1969), *Dict. of Linguistics and Phonetics*, 2d ed. (1985); I. Lehiste, *Suprasegmentals* (1970); M. Halle and S. J. Keyser, *Eng. Stress* (1971)—better avoided; W. S. Allen, *A. and Rhythm* (1973), esp. chs. 5, 7, 12, 15, 16—stress in Gr.; E. Pulgram, *Latin-Romance Phonology: Prosodics and Metrics* (1975); S. F. Schmerling, *Aspects of Eng. Sentence-Stress* (1976); M. Liberman and A. S. Prince, "On Stress and Ling. Rhythm," *LingI* 8 (1977)—metrical trees; *Studies in Stress and A.*, ed. L. Hyman (1977); I. Fónagy and P. Léon, *L'A. français contemporain* (1979); D. R. Ladd, *The Structure of Intonational Meaning* (1980); Scott, 24–28, 39–55; Brogan, 54–57, 110–19; Mazaleyrat, ch. 4; Morier, s.v. "A." and "Contre-a.," and also "Consonne," "Voyelle," and "Syllabe," some of the latter very long articles; E. Fudge, *Eng. Word-Stress* (1984); E. O. Selkirk, *Phonology and Syntax* (1984)—grid theory; M. Halle and J.-R. Vergnaud, *An Essay on Stress* (1987); A. C. Gimson, *Intro. to the Pronunciation of Eng.*, 4th ed. (1989); D. Oliver, *Poetry and Narrative in Performance* (1989).

T.V.F.B.

ACROSTIC (Gr. "at the tip of the verse"). In an a. the first letter of each line or stanza spells out either the alphabet (an *abecedarius*) or a name—usually of the author or the addressee (a patron, the beloved, a saint)—or the title of the work (e.g. Plautus; Ben Jonson's *The Alchemist*). More rarely, the initials spell out a whole sentence—the oldest extant examples, seven Babylonian texts dating from ca. 1000 B.C., are of this sort (they use the first syllable of each ideogram). By far the most common form is that which reveals, while it purports to conceal, the name of the author, as in the a. Cicero says Ennius wrote, or Villon's a. to his mother which spells "Villone." The spelling is usually straightforward but may be in anagram for the sake of concealment. If the medial letter of each line spells out the name, the poem is a *mesostich*, if the final letter, a *telestich*; if both initials and finals are used, the poem is a double a. (two of the Babylonian as. are such), if all three, a triple a. Finals may also read from the bottom up. It is significant that recognition of an a. depends on perceiving not the aural rhythms or the sense of the text but its visual shape.

From the forms just enumerated it is but one step to *carmina quadrata* and the fantastic word-square *intexti* of Hrabanus Maurus—with a modern analogue in Edward Taylor's poem to Elizabeth Fitch—and from there but one step more to *carmina figurata* or true pattern poetry (q.v.), with which as. are commonly found in ancient texts (e.g. the Gr. Anthol.). In the East, as. are found in both Chinese (ring-poems, wherein one can begin reading at any character) and Japanese poetry (*kakushidai* and *mono no na*).

As. are the kind of mannered artifice that will be popular in any Silver Age poetry; they flourished in Alexandria and in the Middle Ages, being written by Boniface, Bede, Fortunatus, Boccaccio, Deschamps, and Marot among many. Commodian has a book of 80 as., the *Instructiones*; Aldhelm's *De laudibus virginitatis* is not only a double a. made out of its first line, but the last line is the first line

read backwards, making a box. The longest a. in the world is apparently Boccaccio's *Amorosa visione*, which spells out three entire sonnets. In the Ren., Du Bellay excepted the ingenious a. and anagram from his sweeping dismissal of medieval Fr. verseforms. Sir John Davies wrote a posy of 26 as. to Queen Elizabeth (*Hymnes Of Astraea*, 1599). The modern disparagement begins as early as Addison (*Spectator*, 60), but as. were very popular among the Victorians. In Poe's valentine poem to Frances Sargent Osgood, her name is spelled by the first letter of the first line, the second letter of the second line, and so on. Outside verse, this process of elevating initials into a "higher" script produces *acronyms* (snafu, gulag), now very common. In the early Christian church the symbol of the fish is such an acronym: the initials of the five Gr. words in the phrase Jesus-Christ-God's-Son-Saviour spell out the Gr. word for fish, *ichthys*. See also ANAGRAM.

K. Krumbacher, "Die Acrostichis in der griechischen Kirchenpoesie," *Sitzungsberichte der königlich-bayerische Akad. der Wiss., philos.-philol.-hist. Klasse* (1904), 551–691; A. Kopp, "Das Akrostichon als kritische Hilfsmittel," *ZDP* 32 (1900); H. Leclercq, "Acrostiche," *Dictionnaire d'archéologie chrétienne*, ed. F. Cabrol (1907); E. Graf, "Akrostichis," Pauly-Wissowa; Kastner; R. A. Knox, *Book of As.* (1924); F. Dornseiff, *Das Alphabet in Mystik und Magie*, 2d ed. (1925); Lote, 2.305; *Reallexikon*; W. G. Lambert, *Babylonian Wisdom Lit.* (1960), ch. 3; R. F. G. Sweet, "A Pair of Double As. in Akkadian," *Orientalia* 38 (1969); T. Augarde, *Oxford Guide to Word Games* (1984); Miner et al., pt. 4.

T.V.F.B.

ALEXANDRINE. In Fr. prosody, a line of 12 syllables. From the 16th c. up to *vers libre* (q.v.), the a. has been the standard meter of Fr. poetry, in which it has had an importance comparable to that of the (quantitative) hexameter in Gr. and Lat. or the iambic pentameter in Eng.; it has been used esp. in narrative and dramatic forms.

The earliest Fr. as. appear in *Le Pèlerinage de Charlemagne*, a *chanson de geste* (q.v.) of the early 12th c. which abandons the traditional decasyllable of the med. Fr. epics for the longer line. But the a. probably takes its name from a slightly later poem, the *Roman d'Alexandre* (ca. 1170; in couplets) of Lambert le Tort, one of many romances of the late 12th to late 14th c. based on the legendary exploits of Alexander the Great. The term "a." itself first appears in the anonymous *Les Regles de la seconde rhetorique* (ca. 1411–32): the name of the hero of the cycles is transferred metonymically to their form (cf. "Sapphic"; "Hudibrastic"). Having fallen into disuse in the later Middle Ages when replaced by prose, the meter was revived in the 16th c. by J. A. de Baïf and was widely used by Ronsard and other members of the *Pléiade*. After being perfected by the great dramatists of the 17th c., esp. Corneille and Racine, it became the principal meter for all serious Fr. poetry. A certain regularity, characteristic of even the earlier a. verse, was intensified and standardized by the theory and practice of the 17th-c. poets, and the cl. form of the a. known as the *tétramètre* emerged. Most significant for the line's rhythmic structure was the fixed medial caesura, with its accompanying accent on the sixth syllable, which, dividing the line into two hemistichs, made it an apt vehicle for dramatic polarization, paradox, parallelism, and complementarity:

> Lui céder, c'est ta gloire, et la vaincre, ta honte
> (To yield to it is your glory, and to suppress it your shame)
> (Corneille, *Cinna*)

> J'embrasse mon rival, mais c'est pour l'étouffer
> (I will embrace my rival, but only to suffocate him)
> (Racine, *Britannicus*)

This medial articulation, its common tetrametric shape, and its syntactic integrity suited the a. for periodic and oratorical utterance in the grand manner, making it ideal for tragedy. But Molière was equally able to exploit the a. for comedy, not only for mock-heroic purposes, but also because it was sufficiently flexible and leisurely to handle lang. at much lower temperatures (the *a. familier*). After its 17th-c. heyday, the a. became increasingly subject to conventionalized diction and mechanical rhymes and syntax until the advent of the Fr. romantics, who rejuvenated it by generating a new diction and by increasing the incidence both of enjambment (q.v.) and of the effaced caesura and the tripartite form of the a. known as *trimètre*; thus Hugo:

> J'ai disloqué / ce grand niais / d'alexandrin
> (I have dislocated this great simpleton, the alexandrine)

The evolution toward a line which is more a freely structured sequence of 12 syllables than a rule-governed system of rhythmic segmentation was carried to its conclusion in the *vers libéré* of Verlaine, Rimbaud, and Mallarmé; the *alexandrin libéré* with its frequent obliteration of the caesura, deemphasized rhymes, more extreme kinds of enjambment, and masking of rhythmic contours by intensified acoustic activity, brought regular Fr. verse to the brink of *vers libre* (q.v.). Since symbolism, the a. has continued to explore the whole gamut between Malherbian rigidity and liberated evanescence.

The a. has had a great importance in the poetry of several other langs., particularly Dutch, in which it was the most widely used meter from the early 17th c. until around 1880. It is a common meter in 17th-c. Ger. poetry—widely used by the

school of Opitz because of the sanction lent it by *Pléiade* practice, and imaginatively exploited by Andreas Gryphius because of its formal appropriateness to his highly antithetic style. From the a. also developed the *cuaderna via,* the important 14-syllable Sp. meter, as well as the It. meter analogous to it. The Eng. a. is composed of iambic feet with six stresses rather than the fluid four (occasionally three or two) of the Fr. a. Spenser adopted the a. as the ninth line of the Spenserian stanza (q.v.): the slight increase in length contrasts with the eight preceding pentameter lines, giving emphasis and closure to the stanza. Several longer Eng. poems—Drayton's *Polyolbion,* Browning's *Fifine at the Fair*—are written entirely in as., but in general the Eng. a. never took hold for continuous use; virtually all as. after 1600 are allusively Spenserian. Strictly speaking, the term "a." is appropriate to Fr. syllabic meters, and it may be applied to other metrical systems only where they too espouse syllabism as their principle, introduce phrasal accentuation, or rigorously observe the medial caesura, as in Fr. (e.g. Bridges' *Testament of Beauty,* though Bridges' a. has no fixed caesura).

E. Träger, *Der Französische Alexandriner bis Ronsard* (1889); H. Thieme, *The Technique of the Fr. A.* (1898); T. Rudmose-Brown, *Étude comparée de la versification française et de la versification anglaise* (1905); V. Horak, *Le Vers a. en français* (1911); G. Lote, *L'A. d'après la phonétique expérimentale,* 3 v. (1911–12); A. Rochette, *L'A. chez Victor Hugo* (1911); Thieme, 360 ff., 374 ff.; P. Verrier, *Le Vers français,* 3 v. (1931–32), v.2–3; J. B. Ratermanis, "L'Inversion et la structure de l'a.," *FS* 6 (1952); M. Burger, *Recherches sur la structure et l'origine des vers romans* (1957); Elwert; Mazaleyrat; P. Lusson and J. Roubaud, "Mètre et rythme de l'a. ordinaire," *LF* 23 (1974); J. Roubaud, *La Vieillesse d'Alexandre* (1978); Morier; Scott; B. de Cornulier, *Théorie du vers* (1982); C. Scott, *A Question of Syllables* (1986).

A.PR.; C.S.; T.V.F.B.

ALLEGORY.

I. WESTERN
- A. *Text and Concept*
- B. *Text and History*
- C. *Text and Structure*
- D. *Text and Consciousness*

II. NONWESTERN

I. WESTERN. A. (Gr. *allos,* "other," and *agoreuein,* "to speak") is a term that denotes two complementary procedures: a way of composing lit. and a way of interpreting it. To compose allegorically is to construct a work so that its apparent sense refers to an "other" sense. To interpret allegorically ("allegoresis") is to explain a work as if there were an "other" sense to which it referred. Not only do both of these procedures imply each other; the two forms of a. increasingly stimulate one another as they develop into full-scale literary and interpretive movements in their own right.

Because a. is so various in its operations, turning from one sense to another in widely divergent texts and times, it resists any attempt at strict and comprehensive definition. It can be argued, for example, that any composition may have some "other" sense. The question of which kinds of composition are specifically "allegorical" has received diverse responses since ancient definitions of the term in Gr. and Roman rhetorical treatises. By the late 1st c. A.D., the term is applied by Quintilian (*Institutio oratoria* 8.6.44–59) to a brief trope (Virgil's shepherd "Menalcas" signifying Virgil himself), a sustained metaphor (Horace's "ship of state" ode), and an ironic form of discourse, in which a speaker says something other than what he intends (Cicero's mocking praise of a man's "integrity" in order to blame him). At times the term has been associated with even more general tensions attributed to literary lang.—from the aspiration of works to express the "inexpressible" (F. Schlegel, *Gespräch über die Poesie,* 1800) to the liability of narratives to betray conflicts in their own value systems (de Man 1979). Perhaps the dominant attitude in current classifications is that there are degrees of allegorical composition, depending on the extent to which a text displays two divided tendencies. One tendency is for the elements of the text to exhibit a certain fictional autonomy. The other tendency is for these elements to imply another set of actions, circumstances, or principles, whether found in another text or perceived at large. The story of a journey, for example, may develop into an a. of the Exodus by its evocation of a wilderness and a promised land. The extent of such an a. diminishes, however, insofar as the "journey" is either reduced to a short-lived figure of speech or elaborated into an explicit paraphrase of the biblical account itself. Insofar as a composition is allegorical, it tends to signal the ambivalence or allusiveness of its lang. and to prescribe the directions in which a reader should interpret it.

While interpretive a. is implicit in the design of works composed allegorically, it is not limited to works intended to be allegorized. Allegoresis has a momentum of its own. By the time of the Jewish biblical exegete Philo (1st c. B.C.–1st c. A.D.) and the near-contemporary Gr. Homeric commentator Heraclitus the Allegorizer, the term itself appears with reference to both brief passages and sustained narratives in the service of explanations ranging from the spiritual (the planting of a garden in Eden as the planting of virtues in the soul [Philo, *De plantatione*]) to the physical (the angry Apollo sending a plague as the fiery sun provoking disease [Heraclitus, *Homerika problēmata*]). Recent crit. sometimes applies the term to all interp., viewed as the reformulation of a text in other terminology. The normal application of the word, however, does not include many interpretive procedures: philological notes on an ancient text, scholastic "questions" about a biblical expression,

historical background for a modern novel. It is arguable whether even pointing the moral of a tale is necessarily to "allegorize" it. To say, for example, that the labors of Hercules display the triumph of fortitude over hardship seems different in degree, if not in kind, from saying that they represent the victory of reason over passion. The latter interp. entails the kind of shift in categories of understanding that is now normally associated with "allegorical" interp. Because such categories themselves change over time, the degree to which an interp. is considered allegorical varies with the degree of difference perceived between the apparent sense of a text and the other-sense assigned to it. Given these changes in perspective, interp. that is considered allegorical does not necessarily offer a less functional reading than do other interpretive attempts to reconstitute a work written in a different place or time. Like them, it needs to be tested against those elements of the work which it excludes. In the end, allegorical interp. may help to preserve a text by applying it to contemporary circumstances or even by creating the terms under which understanding itself can take place.

As allegorical interp. and composition turn into broad movements, their approaches to the relation between one sense and another change in emphasis. Although these approaches overlap in character and time, a number of major relationships may be selected for special consideration.

A. *Text and Concept.* In its earliest systematic forms, interpretive a. tends to treat the other-sense as a conceptual framework that provides the only "true" significance of a text. Developing in the 6th and 5th cs. B.C. with Gr. efforts to interpret Homer and Hesiod according to certain "scientific" or moral categories and to defend the poets against charges of immorality, such strategies intensify in Stoic exegesis of the last centuries B.C. and 1st c. A.D. A story like Homer's battle of the gods (*Iliad*, Bk. 20), for example, turns into the physical interaction of the elements, with Hera as air and Poseidon as water. Interp. of this kind, initially called not a. but *hyponoia*, the "under-sense," coincides with broader movements in antiquity from mythological to "logical" thought. It tends to divide the text, along with the gods, into an apparent form and an underlying significance. By late antiquity, Neoplatonic interpreters are correlating different figures of a narrative with different levels of an orderly universe; to Plotinus in the 3d c. A.D., the Hesiodic succession from Uranus to Cronos to Zeus signifies the cosmic emanation from "One" to "Mind" to "Soul." Such strategies, which continue to be adapted long after antiquity, help to turn the personalities and features of mythological stories into conceptual principles which can be deployed in a philosophical or ethical argument.

Related tendencies develop early in the compositional trad. In ancient Gr. drama, figures like Aeschylus' Themis ("Justice") and Euripides' Lyssa ("Madness") are in one sense daemonic personalities in their own right, in another sense dramatic projections of human concerns. Eventually, writers of epic lit. itself give their divine figures multiple dimensions. Thus, Virgil associates Juno so closely with the physical element of air and the emotional principle of passion that she scarcely appears in the *Aeneid* without stirring up a storm. Already by the era of the ancient Gr. stage, however, a complementary procedure is developing alongside this deployment of dramatic characters as abstract principles: the deployment of abstract principles as dramatic characters. "Personification" (Gr. *prosōpopoiia*), which long after this period comes to be considered central to Western a., in fact refers originally to the dramatic practice of "staging" personalities, figures whose masks (Lat. "personae," Gr. *prosōpa*) reveal inner meanings. The practice itself develops well outside drama, from Prodicus' 5th-c. B.C. "Choice of Heracles," in which Virtue and Vice confront a man at a crossroads, to Martianus Capella's 5th-c. A.D. *Marriage of Mercury and Philology*, in which abstract figures interact with the gods in a tale of human ascent through different levels of the cosmos. Allegorical writing, however, operates not just by the compositional process of turning conceptual systems into narrative ones, or by its antique interpretive counterpart, the process of turning narrative systems into conceptual ones. By late antiquity, a. is developing other ways of relating its divided reference points to each other.

B. *Text and History.* One of these ways is to associate both the provisional sense and the projected sense of a text with the progression of actual events over time. In the interpretive trad., this strategy begins to develop systematically with the exegesis of a text treated not as a mythological tale but as a sacred history: the story of Scripture. Tendencies to allegorize aspects of Scripture appear early in Jewish interp., from Rabbinical treatments of certain biblical passages as predictions of later events to Philo's elaborate allegorization of the text according to ethical and philosophical designs. When in the 1st c. A.D., however, the Christian Paul calls the biblical account of two wives of Abraham and their sons an a. (*Gal.* 4:24), he gives this divided relationship a different kind of historical orientation, interpreting it as the relationship between an Old Covenant and a New one. Early Christian exegetes increasingly explain the narrative transition from Old to New Testament as an historical transition from the "types" or "figures" of an early order to their eventual fulfillment in a later one. In Augustine's influential phrase, the a. of Scripture takes place "in facto," in historical events. In late antiquity and the Middle Ages, the exegesis of changing events sometimes turns into a more static form of "typology": the study of recurrent patterns. Such tendencies appear not only in the classification of natural phenomena according to spiritual categories (the lion, for example, as Christ or the Devil),

but in the interp. of selected Scriptural passages according to "four senses": (1) "literal" or "historical" facts recorded in the text; (2) their "allegorical" fulfillment in Christ and/or the Church (a specific use of "allegorical"); (3) their moral or "tropological" application to the individual; and (4) their "anagogic" ("leading up") significance in the other world. A widespread example of the method occurs in a critical piece once attributed to Dante, the letter to Can Grande, which aligns a Scriptural reference to the Exodus with (1) the departure of the Israelites from Egypt; (2) the redemption of Christ; (3) the conversion of the soul from sin to grace; and (4) the departure of the soul to eternal glory. Yet formal strategies of this kind have counter-implications. By the late Middle Ages and the Ren., exegetes are increasingly adapting mythological stories to the multifaceted order of the Christian cosmos. Thus, to the author of the 14th-c. *Ovide moralisé*, Jupiter's metamorphosis into creatural form suggests the Incarnation of the Christian God; to Boccaccio, poetic "fable" and Scriptural "figure" so broadly overlap that ancient mythology at large becomes an early form of theology, gradually revealed over time. In this kind of continuum, although mythological tales themselves remain at some remove from hist., they are increasingly correlated with the world "in facto."

Nothing more strikingly displays the influence of the element of time upon the compositional trad. than the first full-scale Western personification a., Prudentius' early Christian *Psychomachia* (5th c. A.D.). The poem has two parts; the author uses the brief preface to recall what he terms the "figura" of Old Dispensation events; he uses the main plot to play out this "figure" in a New Dispensation battle between personified Virtues and Vices. Here the technique of personification has more than a conceptual function, as in the "Choice of Heracles"; now it has a chronological force, turning the a. toward eschatology. Yet attempts to dramatize the movement from figure to fulfillment *after* the time of Scripture have an ambiguous relation to their Scriptural source. As later figurations of the divine story in their own right, these compositions look back to Scripture as a prior hist. that authorizes their own efforts, while they simultaneously look forward to a future hist. that they themselves cannot fully compose. The effort to engage and transfigure this state of suspension, similar in some respects to the suspended typology of "patterns" in certain uses of fourfold exegesis, helps to organize Dante's late medieval *Divine Comedy*. Here even the conceptual dimensions of the other world turn into individual souls forever playing out their particular histories, while the wayfarer himself converts to his divine author by historically recreating the Christian model of descent and ascent. This kind of pressure to pass from the sense of a story to the other-sense of hist. takes more apocalyptic forms near the end of the Middle Ages in Langland's *Piers Plowman*, where Scriptural quotations keep disrupting and reorienting the narrative, as if God himself needed to compose the tale and fulfill it "in facto."

C. *Text and Structure*. The internal strains produced by such shifts between fiction and fact increasingly direct allegorists toward the problems of their own formal designs. Particularly in Ren. and neoclassical approaches to a., the question of how one sense of a text relates to another becomes a formal issue of how the imaginative lang. of a work relates to its controlling structure. Insofar as writers in this period try to establish strict rules of structural decorum, they tend to restrict a., with its divided reference points, to special cases. Thus, 16th-c. It. critics influenced by Aristotle's *Poetics* stress that a story must be "unified" and "credible" to be true; those who reserve a place for a. treat it primarily as a specialized device for giving marvelous tales this kind of plausibility. In the exegesis of sacred lit., Protestants largely reject the "four senses" of Scripture on behalf of its "plain" sense, spiritually illuminated; at the same time, by emphasizing the need to display that illumination in the inner world of the soul, such Reformers bring their own typology of the spirit into biblical interp. In the critique of ancient mythology at large, 17th- and 18th-c. critics such as Vico tend to "rationalize" myth as an early historical phenomenon yet revalue it as a "picturesque" testimony to the primal, imaginative impulses in human development. In such interpretive movements, a. no sooner disappears than it begins to reappear in changed forms.

In the Western compositional trad., allegorists long before the It. Ren. make the formal control of diverse figures a thematic concern. The 12th-c. *Cosmographia* of Bernard Silvestris, for example, elaborately organizes its cosmological *dramatis personae* around their divine source, while the 13th-c. *Romance of the Rose* carefully orchestrates its psychological figures around the human personality. With the expansive works of Ariosto and Tasso, however, the problem of structural coherence takes the form of a general tension between the diversity of marvels and the unity of truth; Tasso himself, assisted by the *Poetics*, self-consciously analyzes the problem in a critique of his own *Jerusalem Delivered*. The part/whole relation often takes less fluid but visually arresting forms in the influential emblem books and emblematic art works of the 16th c., with their allusive characters and objects conspicuously posed as distinct facets of a philosophic design. In lit. of this period a particularly intricate treatment of such tensions is Spenser's *Faerie Queene*, where the narrative sequence and the moral progress both develop as characters interact with diverse allegorical aspects of themselves. Thus, a Spenserian knight scarcely abandons the consolidating figure of Una before he befriends the divisive figure of Duessa, while Una herself is deceived by the arch-image-maker Archimago, disguised as the knight she has

lost. With Bunyan's 17th-c. *Pilgrim's Progress*, allegorical romance of this kind develops into more programmatic fiction; diversity of incident is subordinated to the inner reflections and Scriptural prooftexts of the Puritan pilgrim. By the 18th c., the question of reconciling imaginative expression with structural control takes forms ranging from Swift's *Tale of a Tub*, which elaborately parodies certain strategies of Christian exegesis while repeatedly digressing from its own story, to the more stylized lyric, which often reduces allegorical abstractions to picturesque, though animated, expressions of a rational design.

D. *Text and Consciousness*. The effort to give imaginative expression a logic of its own provokes a critique of the very basis of allegorical writing: its differentiation between the apparent sense of a text and some other sense. The critique develops esp. in theorists of the romantic period, who argue that in a., an image can be separated from the abstract concept that it signifies. By contrast, they maintain, in the preferred forms of "symbol" and "myth" (qq.v.), an image inherently participates in the whole that it reveals. The specific terms and emphases of such theorists vary. For Goethe, "symbolic" objects of experience include a "certain totality"; for Coleridge, a growing plant is a "symbol" of the organic world to which both the plant and the perceiving mind belong; for Schelling, a mythological god is inseparable from the general principle it embodies. Such challenges to a. are gradually qualified by both practical and theoretical concerns. In practice, the decision about which images are separable from the thoughts and emotions they evoke often proves difficult even for romantic critics to make, esp. in works like the *Divine Comedy* and the *Faerie Queene*. In theory, romantic assumptions about "symbolic" harmonies of mind, nature, and lang. increasingly yield to attitudes more reminiscent of a. Thus, in early 20th-c. psychological and anthropological theories stimulated by Freud, Jung, and Frazer, myths and symbols themselves turn into hidden, explanatory structures underlying disparate strains in human activity at large. In aesthetic theory of this period, Walter Benjamin charges that the romantic notion of the "symbol" tends to obscure the disruptive effect of time on any effort to unify an ephemeral object with an eternal idea. By contrast, he argues, a. allows nature to be seen not "in bud and bloom" but in its inevitable "decay," leaving allegorical fragments of a desired whole. In the later 20th-c. poststructuralist crit. of Paul de Man and others, this disparity in hist. is largely treated as a disparity in lang., in which a sign evokes an origin with which it can never coincide. In such crit., the very effort to construct a whole like the "self" or the "world" depends upon, and is undermined by, the differentiation of senses associated with allegorical writing.

The critical theory of a distinction between "symbol" or "myth" and a. has its counterpart in the writing of romantic poetry itself. Here the distinction often takes the form of a tension between the poet's imaginative vision and the analytic framework it implies. In dramatic myths like Blake's *Jerusalem* and Shelley's *Prometheus Unbound*, the plot is partly an a. of the very psychological and social conditions which the poet seeks to transcend with his prophetic vision. In the more reflective lyrics of Wordsworth, Hölderlin, and, later, Baudelaire and other Fr. poets associated with the turn toward symbolist poetry, the visionary is a character who at once invites and resists the possibility that the imaginative landscape which frames him is an a. of his own consciousness. When more than one consciousness is placed in a "symbolic" scene, as in Am. romantic narratives like Hawthorne's *Scarlet Letter* or Melville's *Moby Dick*, enigmatic objects like the letter "A" and the white whale nearly allegorize the conflicting mentalities of the very characters (and readers) who seek to interpret their "significance." In such writing, standards of significance tend to shift from a realm of authoritative, public norms to a realm of limited, private perceptions. The dilemma of orientation implied by such a shift develops in several 20th-c. works where a. is both suggested and deflected, as in Joyce's *Ulysses*, where a day of urban wandering distantly recalls the primal myth of the *Odyssey*, or in Kafka's *The Castle*, where a human personality struggles inconclusively to establish the other-sense of the institutions which impinge upon it.

By definition, allegorical writing is a particularly elusive procedure. No account of selected changes in its operation, including the one presented here, can fully show the interplay of its compositional and interpretive forms. The very elusiveness of a., however, generates a certain equilibrium of its own. For if a. is always criticizing the systems on which it appears to depend, it also keeps creating new systems in turn. In the process, a. not only complicates the strategies of composition and interp., it also helps to coordinate the sense of the one with the sense of the other.

Lewis; Auerbach; Curtius; F. Buffière, *Les Mythes d'Homère et la pensée grecque* (1956); E. D. Leyburn, *Satiric A.: Mirror of Man* (1956); Frye; E. Auerbach, "Figura," *Scenes from the Drama of European Lit.* (1959); E. Honig, *Dark Conceit: The Making of A.* (1959); H. de Lubac, *Exégèse médiévale: les quatre sens de l'écriture*, 2 pts. in 4 v. (1959–64); J. Daniélou, *From Shadows to Reality: Studies in the Biblical Typology of the Fathers* (1960); A. C. Hamilton, *The Structure of A. in the Faerie Queene* (1961); D. W. Robertson, Jr., *A Preface to Chaucer* (1962); A. Fletcher, *A.: The Theory of a Symbolic Mode* (1964); O. Seel, "Antike und Frühchristliche Allegorik," *Festschrift für Peter Metz*, ed. U. Schlegel and C. Z. von Manteuffel (1965); R. Tuve, *Allegorical Imagery* (1966); A. C. Charity, *Events and Their Afterlife: The Dialectics of Christian Typology in the Bible and Dante* (1966); R. Hahn, *Die Allegorie in der antiken Rhetorik* (1967); H. R. Jauss and U. Ebel, "Entstehung und Strukturwandel der allegoris-

chen Dichtung," *GRLMA*, v. 6.1 (1968); R. Hollander, *A. in Dante's "Commedia"* (1969); M. Murrin, *The Veil of A.* (1969), *The Allegorical Epic* (1980); D. C. Allen, *Mysteriously Meant* (1970); J. MacQueen, *A.* (1970); M.-R. Jung, *Études sur le poème allégorique en France au moyen âge* (1971); U. Krewitt, *Metapher und tropische Rede in der Auffassung des Mittelalters* (1971); P. Piehler, *The Visionary Landscape* (1971); G. Clifford, *The Transformations of A.* (1974); P. Dronke, *Fabula: Explorations into the Uses of Myth in Med. Platonism* (1974); J. Pépin, *Mythe et allégorie: les origines grecques et les contestations judéo-chrétiennes*, enl. ed. (1976), *La Tradition de l'allégorie de Philon d'Alexandrie à Dante* (1987); J. A. Coulter, *The Literary Microcosm: Theories of Interp. of the Later Neoplatonists* (1976); W. Benjamin, *The Origin of Ger. Tragic Drama*, tr. J. Osborne (1977); H.-J. Klauck, *Allegorie und Allegorese in synoptischen Gleichnistexten* (1978); S. A. Barney, *As. of Hist., As. of Love* (1979); B. K. Lewalski, *Protestant Poetics and the 17th-C. Religious Lyric* (1979); M. Quilligan, *The Lang. of A.: Defining the Genre* (1979); P. de Man, *As. of Reading* (1979), "The Rhet. of Temporality" in de Man; H. Brinkmann, *Mittelalterliche Hermeneutik* (1980); P. Rollinson, *Cl. Theories of A. and Christian Culture* (1981); *A., Myth, and Symbol*, ed. M. W. Bloomfield (1981); H. Adams, *Philosophy of the Literary Symbolic* (1983); L. Barkan, *The Gods Made Flesh: Metamorphosis and the Pursuit of Paganism* (1986); R. Lamberton, *Homer The Theologian: Neoplatonist Allegorical Reading and the Growth of the Epic Trad.* (1986); *Midrash and Lit.*, ed. G. H. Hartman and S. Budick (1986); J. Whitman, *A.: The Dynamics of an Ancient and Med. Technique* (1987); S. M. Wailes, *Medieval As. of Jesus' Parables* (1987); *Allegoresis: The Craft of A. in Med. Lit.*, ed. J. S. Russell (1988); *Med. Literary Theory and Crit. ca. 1100–ca. 1375: The Commentary Trad.*, ed. A. J. Minnis and A. B. Scott (1988); G. Teskey, "A.," *The Spenser Encyc.*, ed. A. C. Hamilton et al. (1990); D. Stern, *Parables in Midrash* (1991). J.WH.

II. NONWESTERN. Most of the distinctions that can be drawn about Western a. can be found elsewhere—e.g. that between a. as a compositional procedure and allegoresis as an interpretive procedure, as also the problem of distinguishing between the two. Practice outside the West is not uniform, however. For example, sustained poetic narrative a. is typical almost solely of Sanskrit and other later Indic lits. There, the sacred and profane may be not so much allegorically related as aspects of each other. The boundaries are clearer in the teaching of Buddha, as in the *Lotus Sūtra*, in which there are seven major parables, and as in the concepts of expedients (*upāya*), adaptive means of teaching that gave credit to fictionality in Asia. And these religious, prose sources have their counterparts in mythical and other religious lyricism from the Middle East to East Asia.

In Chinese poetry, a. is often said to make historical and political references ("contextualization") in contrast to medieval Western as., where referents are usually philosophical or religious rather than historical facts. Of course, Daoist and Buddhist poets did write religious as., but they did not determine the main currents of Chinese poetics. As a mode of hermeneutics, allegorical interp. is as old as the Chinese literary trad. itself. The Confucian classics, for example, depict diplomats using tags from poetry in lieu of direct statement. Commentators since the 2d c. B.C. have developed a trad. of reading the *Classic of Poetry* (China's first anthology of poems), and subsequently, poetry in general, as referring obliquely to political situations. This ancient habit has provided the foundation for Chinese allegorical (*jituo*) writing. At the center of Chinese a. is a poetics of implicit imagery or symbolism, the goal of which is to create an impression of distancing while yet guiding readers to the intended meaning. Such imagistic a. flourished esp. during periods of dynastic transition—i.e. in periods of cultural crisis—because it lent itself so readily to expressing, indirectly but effectively, a poet's secret loyalty to the fallen dynasty.

Chinese practice had effect also in Korea and Japan, though chiefly on poetry written in Chinese. In Japan, a. is confined almost solely to lyric, although secular narrative echoes of Buddhist scriptures and allegoresis may be found. A. in lyrics is of two kinds: the incomplete or "open," in which nonallegorical features are also present, and the complete or "closed," in which nothing in the poem itself signals a.; external evidence is required. The incomplete kind is found in poems devoted to love (a very large category) and to congratulations or complaints (as for failure to gain promotion). The complete kind marks many poems on Buddhist topics: commonly the a. would be missed entirely without a headnote to direct interp. In both kinds natural imagery is normally the signifier of other meaning. There is also frequent use of figural imagery, e.g. the moon as a figure of enlightenment or the sky as a figure for the doctrine of Emptiness. In highly allusive poetic passages, e.g. in nō, and in allegoresis, the line between the allegorical and the nonallegorical may be difficult to draw, as in other lits.

In Ar. and Persian lit., a. is both a structural principle and a hermeneutic method. Ar. literary a. developed out of allegorical exegesis of the Koran, which distinguished between the literal (*ẓāhir*) and inner (*bāṭin*) meaning of Scripture, and was later extended to encompass other texts, until ultimately all creation was viewed as a system of signs pointing to concealed meanings. But allegorical exegesis was ultimately rejected by the orthodox; in consequence, both it and much literary a. are heterodox and esoteric, and incorporate many Iranian and Hellenizing elements. Mystical a. informs both prose treatises and lyric poems such as those of Ibn al-Fāriḍ (d. 1195?) and Ibn al-ʿArabī (d. 1240), e.g. *Tarjumān al-ashwāq* (The

Interpreter of Desires), to which the poet later added an allegorical commentary. Moral a., which developed independent of exegesis and was influenced by trs. from Middle Persian, is found primarily in animal fables such as the *Kalila and Dimna* and the "trial of man by the animals" included in the *Rasāʾil* (Epistles) of the Ikhwān al-Ṣafāʾ.

In Persian lit., a. pervades all literary types, prose and verse, narrative and lyric, and serves a wide range of purposes. The lyric and *maṣnavī* works of Farīd al Dīn ʿAṭṭār (d. 1220), the ghazals (q.v.) and *Maṣnavī maʿnavī* (Spiritual Masnavi) of Jalāl al-Dīn Rūmī (d. 1273), the ghazals of ʿIrāqī (d. 1289) and the romances of Jāmī (d. 1492) provide examples of mystical a. Moral a. informs Sanāʾī's (d. 1135) *Sayr al-ʿibād ilā al-maʿād* (The Believers' Journey Toward the Eternal Return). Panegyric *qaṣīdas* often employ topical a.; for example, Manūchihrī (d. 1040–41) uses the extended metaphor of Spring's battle against Winter to allude to contemp. political events. Ḥāfiẓ (d. 1389) combines topical, moral, and philosophical a. in many of his ghazals. Mystical allegoresis was widely employed as a means of legitimizing works whose content was regarded as being of dubious orthodoxy; this is particularly true in the case of Ḥāfiẓ, whose ghazals are interpreted according to a lexical code equating poetic images with mystical concepts. Such reading must be regarded as highly suspect.

R. H. Brower and E. Miner, *Japanese Court Poetry* (1961); P. Nwyia, *Exégèse coranique et langage mystique* (1970); J. W. Clinton, *The Divan of Manūchivī Dāmaghānī* (1972); A. Hamori, *On the Art of Med. Ar. Poetry* (1974); L. Hurvitz, *Scripture of the Lotus Blossom of the Fine Dharma* (1976); A. H. Plaks, *Archetype and A. in the Dream of the Red Chamber* (1976); E. Gerow, *Indian Poetics* (1977); M. Kiyota, *Shingon Buddhism: Theory and Practice* (1978); A. Schimmel, *The Triumphal Sun: A Study of the Works of Jalāl al-Dīn Rūmī* (1978); J. S. Meisami, "Allegorical Techniques in the *Ghazals* of Hāfez," *Edebiyat* 4 (1979), *Med. Persian Court Poetry* (1987); Fowler; K. S. Chang, "Symbolic and Allegorical Meaning in the *Yüeh-fu pu-t'i* Poem Series," *HJAS* 46 (1986); P. Yu, *The Reading of Imagery in the Chinese Poetic Trad.* (1987); L. Zhang, "The Letter or the Spirit: *The Song of Songs*, Allegoresis, and the *Book of Poetry*," *CL* 39 (1987).

J.WH.; E.M.; K.S.C.; J.S.M.

ALLITERATION. The repetition of the sound of an initial consonant or consonant cluster in stressed syllables close enough to each other for the ear to be affected. The term is sometimes also used for the repetition of an initial consonant in unstressed syllables, as in Poe's "lost Lenore," where the weak second *l* affects the ear less than the long *o* followed by *r*, but this less direct patterning is arguably not of the same class as stress-enhanced a. From Cl. times up to the 20th c., a. as a figure in rhet. meant the repetition of the initial *letter* of words; and in the absence of the relatively recent term *assonance* (q.v.), a. formerly included the echo of initial vowels, even of initial letters in weak syllables—a fact that explains Churchill's 18th-c. attack on those who unsuccessfully "pray'd / For apt alliteration's Artful aid." But it must constantly be borne in mind that since a. is a device of the *sound* stratum of poetry, the vagaries of spelling systems must be discounted: it is sounds not letters that count, as in Dickinson's "The cricket sang / And set the sun," or Dryden's "Thy force, infus'd, the fainting Tyrians prop'd; / And haughty Pharoah found his Fortune stop'd" (*Ab. & Ac.*, 842–43), where the [f] phone begins six stressed syllables, one in a medial syllable, one spelled *Ph.*

In the alliterative meter of OHG and OE poetry, esp. *Beowulf*, a. usually binds three (at least two) of the four stressed syllables in the line structurally, i.e. by fixed metrical rule, as in "Oft Scyld Scefing sceathena threatum," but the same vowel can also begin the stressed syllables (today called assonance) and any vowel can "alliterate" with any other. In poetry after OE, by contrast, a. is neither linear nor structural; in fact it is often carried through several lines. Today, then, a. is one of the three most significant devices of phonic echo in poetry.

One must remember that unintentional a. will occur less often than unintentional assonance, since in most IE langs. consonants outnumber vowels. Corollary facts are (1) that the semi-vowels *h*, *y*, and *w* in Eng. are universally said to alliterate rather than to assonate and (2) that each of certain consonant clusters (e.g. *st*, *sp*, and *sc* [sk]) has traditionally been said to alliterate only with itself—this practice derives from OE. Housman, aware of this trad., lets his poet in "The Grecian Galley" alliterate seven *st* clusters in three lines.

Almost every major poetry in the world except Israeli, Persian, and Arabic seems to have made considerable use of a., which has been more popular and persistent than rhyme. Although there is disagreement about the nature or importance of stress in Gr. and Lat. poetry, it is certain that Cl. poets understood the uses of a. Aeschylus echoed initials to emphasize key words and phrases, to point up a pun, to aid onomatopoeia, or simply to please the ear (Stanford, *Aeschylus*). Lucretius revelled in phonic echoes, incl. a., while Virgil, a quieter poet, wrote many lines like these two, heavy with assonance, in which eight important syllables also bear initial [k] or [l]: "cuncta mihi Alpheum linquens locusque Molorchi / cursibus et crudo decernet Graecia caestu."

More than most others, the Romance langs. have neglected a. in favor of assonance, the reason perhaps being that It., Sp., and Port. poetry have traditionally preferred short lines with vowel (often followed by consonant) echoes in the final two syllables. Even though it is said that Fr. poetry

by the 17th c. considered a. "mauvaise" (Thieme), the Fr. cl. alexandrine (q.v.) did make some use of it, as in Boileau's "Des traits d'esprit semés de temps en temps pétillent." But most Fr. poets from Hugo to St.-John Perse have found other devices more appealing to the ear.

Even the Oriental tone langs. have been as fond of a. as of other phonic echoes. Chinese, a monosyllabic lang. with short-line poems based on parallel structure and patterning of rising vs. falling tones, insists on rhyme and plays much with repeated vowels and consonants, as with initial *ch* in the opening lines of "Fu on Ascending a Tower" by Wang Ts'an (ca. A.D. 200 [Frankel]). Also tonal, but polysyllabic, Japanese poetry has from ancient times employed much a. (Brower).

All of the Celtic langs., but esp. Ir. and Welsh, have been renowned for their elaborate schemes of phonic ornamentation. Early Welsh and Ir. "cadenced" verse was seldom rhymed but used much a. for linking words and lines. By the 6th c. this verse gave way to "rhyming" verse in which each line has stressed word pairs that alliterate or assonate, all to go with the "generic rhyming" that links lines of a stanza. By the 14th c., Welsh bards had evolved the elaborately decorated *cynghanedd* based on exact rhymes in couplets and complex echoes of vowels and consonants, incl. a. (Dunn).

Of the poetry in Germanic langs., modern Ger. has employed a. consistently while Eng. has liked it perhaps even more. Goethe used a. much as Shakespeare did, in conjunction with other phonic devices and normally with restraint. He might, however, open a lyric (e.g. "Hochzeitlied") with heavy a. or help elevate Faust's most important speech to Mephistopheles with "Dann will ich gern zu Grunde gehn!" Heine preferred rhymed short-line verse and needed a. less, but he often tended to heavy ornament (e.g. "Ein Wintermarchen"). Gottfried Benn and Rilke were perhaps more sound-conscious than most other 20th-c. Ger. poets, and while they too preferred assonance, Benn could write dozens of lines like this one in "Untergrundbahn": "Druch all den Frühling kommt die fremde Frau," a line no more ornate than some by Rilke.

Eng. poets continued to use a. widely, if not in such a mannered way, after the 14th-c. revival of alliterative verse in England and Scotland. Spenser makes use of it in the *Shepheardes Calender*, and in *The Faerie Queene* the first 36 lines have 7 trisyllabic alliterations—by some standards very heavy. In *MND* (5.1.147–48) Shakespeare makes fun of a., but he employs it heavily in the early long poems (e.g. *Lucrece*)—less so in the lyrics, sonnets, and plays—though a. is often conspicuous in the plays. Milton, who loved the sound of words even before his blindness, preferred assonance, but he also liked a., e.g. in "L'Allegro" ("And to the stack, or the barn door, / Stoutly struts his Dames before") and in *P.L.* ("Moping melancholy, / And moonstruck madness" [2.424–25]). Eng. poets from 1660 to 1780 worked hard with phonic echoes in order to vary not only their couplets but their blank verse and octosyllables. Dryden, Pope, Gay, and Thomson used a. best to tie adjectives to nouns, to balance nouns or verbs or adjectives, to stress the caesura and end rhyme, to join sound to sense, and to decorate their lines. After 1780, poets used a. less; still, almost every poet from Blake to Hopkins to Housman used it well, sometimes excessively, as with this typical burst in Browning's "The Bishop Orders His Tomb": "And stretch my feet forth straight as stone can point." Tennyson is reported to have said, "When I spout my lines first, they come out so alliteratively that I have sometimes no end of trouble to get rid of the a." (H. Tennyson, *Memoir* [1897]).

In the 20th c., Frost, Pound, and Williams are relatively unornamental, though Frost could sometimes alliterate heavily—"Nature's first green is gold, / Her hardest hue to hold." Others—Stevens, Eliot, Yeats—are less sparing with a., as Eliot is in the Sweeney poems and "Prufrock," though sometimes they indulge sudden excesses, as with Stevens's "Winding across Wide Waters." But there is a larger group of recent poets who, learning from Hopkins and Hardy, have depended very much on phonic echoes, esp. in unrhymed verse. Among these are Owen and Thomas in Britain and Jeffers, Moore, Crane, Roethke, Lowell, and Wilbur in the U.S. Jeffers, typical of this group, echoes not only initial consonants but also important vowels in almost every line—"Sun-silt rippled them / Asunder," "Rail-squatters ranged in nomad raillery." Although Thomas, with his charged lang. and syntax, has much a., it is no heavier than his other phonic echoes and seldom includes more than two syllables. See now SOUND.

R. E. Deutsch, *The Patterns of Sound in Lucretius* (1939); W. B. Stanford, *Aeschylus in His Style* (1942), *The Sound of Gr.* (1967); N. I. Hérescu, *La Poésie latine* (1960); L. P. Wilkinson, *Golden Lat. Artistry* (1963); J. D. Allen, *Quantitative Studies in Prosody* (1968); Wimsatt, esp. the essays by Frankel, Brower, Lehmann, and Dunn; J. A. Leavitt, "On the Measurement of A. in Poetry," *CHum* 10 (1976); P. G. Adams, *Graces of Harmony* (1977); J. T. S. Wheelock, "Alliterative Functions in the *Divina Comedia*, *LeS* 13 (1978); Brogan, esp. items C111–C170, L65, L406 (Jakobson), L150, L474, L564, L686, and pp. 479–585. P.G.A.

ALLUSION (Ger. *Anspielung, Zitat*). A poet's deliberate incorporation of identifiable elements from other sources, preceding or contemporaneous, textual or extratextual. A. may be used merely to display knowledge, as in some Alexandrian poems; to appeal to those sharing experience or knowledge with the poet; or to enrich a poem by incorporating further meaning. A. differs from repetition (q.v.) by recalling portions only of the original; from parody and imitation (qq.v.) in not being necessarily systematic (al-

though Rochester's "A. to Horace" is an imitation); from source borrowing because it requires readers' knowledge of the original borrowed from; and from use of topoi, commonplaces, and proverbs in having, in each instance, a single identifiable source.

A. is also distinguishable from intertextuality (q.v.), from precedented lang., and from plagiarism. It was said in praise of Tu Fu's late poems that every word and phrase was precedented. In Japanese poetry-matches and in poetry written on other formal occasions, some judges held that the lang. should be restricted to that used in the first few royal collections. Intertextuality is involuntary: in some sense, by using any given real lang., one draws on the intertexts from which one has learned the words, and neither the poet nor the reader is aware of the connections. Precedented poetic lang. is somewhat similar, except that the words are in principle traceable to identifiable sources. But without the special meaningfulness of connection required for a., the awareness is solely of a poetic lang. traceable to numerous possible sources. The test for a. is that it is a phenomenon that some reader or readers may fail to observe. Plagiarism is deliberate and text- or words-specific, but the plagiarizer presents what is stolen as personally original, and there is none of the interaction required by a.

Although poetic a. is necessarily manifested in words, what it draws on in another work need not be verbal. The words of the alluding passage may establish a conceptual rather than a verbal connection with the passage or work alluded to. Nonverbal a. was recognized by medieval Japanese critics and poets as belonging to conception rather than to words and verbal phrasing. This was of course thought particularly true of a poem alluding to a nonpoetic source. If a. to another poem was conceived of as working a foundation poem (*honkadori*), a. to a prose predecessor was taken as working a foundation story (*honzetsu*). There is obviously no possibility of verbal connection in a poet's a. to a historical situation or to a work in another art (as Auden's to Breughel's "Icarus" in "Musée des beaux arts"). A. presumes, then, either intentionality or whatever other name a voluntary, deliberate effort goes under.

A. assumes: (1) prior achievements or events as sources of value; (2) readers sharing knowledge with the poet; (3) incorporation of sufficiently familiar yet distinctive elements; and (4) fusion of the incorporated and incorporating elements. A. is not restricted to poetry, and has analogues in other arts, religious writings, and other possible uses of echo. It usually presumes a close relation between poet and audience, a social emphasis, a community of knowledge, and a prizing of tradition, even fears of the loss of valued trad. Sometimes it offers coded messages. A. is also a matter of interpretation (q.v.), since readers may disagree as to the deliberateness or intentionality of a poet's incorporation of elements. Debates turn on whether Jonson's *Cataline* alludes to the Gunpowder Plot; whether Milton alludes to Interregnum or Restoration events in specific passages of *Paradise Lost* and *Paradise Regained*; and whether a Ch. poem on a woman abandoned (e.g. Tu Fu, "Jiaren") alludes to the poet's disfavor under a new regime. Examples of a. in *Paradise Lost* show a variety of effects and issues. The claim to write "Things unattempted yet in Prose or Rhime" (1.16) puts Ariosto, *Orlando Furioso* 1.2, into Eng., but the result is not clear. Milton's "Thick as Autumnal Leaves" simile (1.303–4) is not an a. but a topos, as in Isaiah 34:4, Homer, *Iliad* 6.146, Virgil, *Aeneid* 5.309–10, and Dante, *Inferno* 3.112–15. There is no question but that Milton recalls a famous passage, *Aeneid* 6.126–31 (the descent to Avernus is easy, reascent barred to all but those favored by Jupiter and of bright virtue), for allusive correction of the fallen angels, as in 2.81, but less certainly in 1.633 and 2.432–33.

Kinds: (1) *Topical* a. normally refers to recent events and need not be textual. (2) *Personal* a. involves facts drawn from the poet's life, necessarily apparent or to some degree familiar, as in much of Horace, in the Petrarchan lyric sequence (q.v.), and in Shakespeare and Donne punning on their own names; it should be distinguished from East Asian assumptions of literary nonfictionality and from romantic direct use of personal experience. (3) *Formal* a. is exemplified in Ch. poetry by the use of another poet's rhymes (the use of a common form such as the sestina or sonnet is insufficiently specific without additional evidence, e.g. an acrostic). (4) *Metaphorical* a., found chiefly in periods valuing trad. (e.g. Rome, pre-modern China, Korea, Japan, and 16th- to 20th-c. England), uses the incorporated elements as signifiers to generate signification in the new context (for example, Dryden's a. to *Aeneid* 5–6 in his poem on Oldham expresses, through metaphor, the relation both between himself and Oldham and between Roman and Eng. cultural values). (5) *Imitative* a. varies, whether specific (e.g. Johnson in *London* to Juvenal, *Satire* 3, so sustained as to constitute imitation; this form, the picking up of a specific verbal string, typically a phrase, from an antecedent text of some canonical authority, is perhaps the most common variety), generic (Dryden to epic in *Absalom and Achitophel*), parodic (Philips to Milton in *The Splendid Shilling*), or synthetic (Pope's *Rape of the Lock*, specific to Milton and others, generic to epic, and parodic). (6) *Structural* a. gives form to a new work by recalling the organization of an older—in this it resembles other kinds, but clearly differs when the a. is to works in other genres or arts (e.g. the *Odyssey* used by Joyce in *Ulysses* and music by Eliot in *Four Quartets*). In spite of Lessing's strictures in *Laokoön*, many poets have also alluded, sometimes in descriptions, sometimes less visually, to other arts than verbal ones (e.g. Auden in "Musée des Beaux Arts" to

Breughel's *Icarus*; see VISUAL ARTS AND POETRY).

Rules governing a. are rare, but in some Japanese crit., verbal a. is thought to require roughly a certain proportion of echoed words, whereas particularly famous poems (e.g. Narihira, "Tsuki ya aranu," *Kokinshū* 15.747) are to be used for conceptual rather than verbal a. Often a. resembles allegory, figuration, imitation, intertextuality, or parody (qq.v.). Another distinction lies in subject matter: thus Ch. a. is unusually given to a. that is historical-political. Religious a. is the prevailing variety in many cultures and in certain periods; Islamic lit., for example, is dominated as no other by its holy book.

Although there are numerous studies of a. in individual authors, no comprehensive study exists. A fully comparative study would be very valuable.

R. D. Havens, *The Influence of Milton on Eng. Poetry* (1922); G. Smith, *T. S. Eliot's Poems and Plays: Sources and Meanings* (1956); S. P. Bovie, "Cl. As.," *CW* 52 (1958); R. Brower, *Alexander Pope: The Poetry of A.* (1959); R. H. Brower and E. Miner, *Japanese Court Poetry* (1961); D. P. Harding, *The Club of Hercules* (1962); A. Hamori, *On the Art of Med. Ar. Poetry* (1974); Z. Ben-Porat, "The Poetics of Literary A.," and A. L. Johnson, "A. in Poetry," both in *PTL* 1 (1976); C. Perri et al., "A. Studies: An Internat. Annot. Bibl., 1921–1977," *Style* 13 (1979); J. Hollander, *The Figure of Echo* (1981); Morier; G. B. Conte, *The Rhet. of Imitation*, ed. and tr. C. Segal (1986); P. Yu, *The Reading of Imagery in the Ch. Poetic Trad.* (1987); E. Stein, *Wordsworth's Art of A.* (1988); E. Cook, *Poetry, Word-play, and Word-war in Wallace Stevens* (1988); *Intertextuality, A., and Quotation: An Internat. Bibl. of Crit. Studies*, ed. U. J. Hebel (1989); R. Garner, *From Homer to Tragedy: The Art of A. in Gr. Poetry* (1989). E.M.

AMBIGUITY. An utterance is ambiguous if it is open to more than one interp. in the context (linguistic and pragmatic) in which it occurs. Such linguistic a. is obvious even in the ordinary use of a lang., and it has been the subject of sophisticated treatment by logicians, linguists, and rhetoricians since Cl. times. In 20th-c. lit. crit., a. has become a key concept particularly for critics interested in the complex problem of distinguishing the lang. of poetry from the precise, literal, univocal lang. required by scientific and other expository prose. The popularity of a. began in 1930 with the publication of William Empson's *Seven Types of A.* His detailed explications of the as. he found in poems became a model for the methods of "close reading" advocated esp. by the New Criticism.

Empson says that in ordinary usage, a. is associated with equivocal, doubtful, or confused meaning; with carelessness, evasiveness, or deceit. But the linguistic mechanisms producing a. can also generate desirable results only some of which have been fully appreciated (e.g. by the punster or the allegorist). Empson proposed to use "a." for "any verbal nuance, however slight, which gives room for alternative reactions to the same piece of lang." For him, "a." becomes an umbrella term for the multiple readings of a passage made possible by the use of polysemous or homonymous words; by double or deviant syntax; by unconventional or omitted punctuation; by sound structure; by incomplete or misleading contexts in which words or sentences appear; by descriptive or affective connotation; by the use of vague or very general words; by trope, allusion, paradox, or symbol; by tautology, contradiction, or irrelevance; or by irony, puns, or portmanteau words. The multiple readings may have a variety of relations with each other: they may be of equal relevance and importance, or one may be a mere overtone or innuendo; they may supplement each other, may be fused into a larger unity of meaning, or may be mutually exclusive; they may produce polyphony or dissonance. The subtlety, density, and depth of meaning made possible by the use of these devices demonstrate that "the machinations of a. are among the very roots of poetry."

Empson says that he has arranged his seven types as "stages of advancing logical disorder" generated by a poet's use of one or more of the above mechanisms. He sees these degrees as objective linguistic facts that appear in a poem. But his classification also depends in part on psychological considerations: (1) the degree of perplexity caused in the reader or the degree of effort required to "work out" the latent meanings and their relations; and (2) the degree of tension or conflict in the mind of the author that is reflected by the presence of a. His first few types show how the multiple meanings of an expression tend to support and enrich each other or resolve themselves into a single meaning, while in his later types the multiple meanings diverge, until in his seventh type the variety of meanings reflect contradictions and unresolvable alternatives (best illustrated in his extended analysis of Hopkins' "The Windhover").

Not all critics have been happy with the extension of ordinary usage that Empson had proposed for "a." or with the clarity, logic, or usefulness of his classification of as. Beardsley, Kooij, and Rimmon make a case for restricting the term "a." to utterances whose multiple interpretations are mutually exclusive and thus lead to indecisiveness of meaning. Other terms have been suggested for lang. conveying conjoined multiple meanings: "resourcefulness" (I. A. Richards); "plurisignation" (Wheelwright); "extralocution" (Nowottny). Kaplan and Kris, wishing to retain Empson's insights but to make his theoretical scheme more precise, have suggested "disjunctive a." for as. in which the separate meanings are interpreted as mutually exclusive; "additive a.," "conjunctive a.," and "integrative a." for those kinds in which the separate meanings are conjoined in one way or another; and "projective a." for cases in which the "responses vary altogether with the interpreter"—that is,

cases in which different interps. of a text are due not to properties of the text but to the varying subjectivities of reader response. Other critics have debated whether "a." should not be extended even further to cover the linguistic phenomena (anticipated in Empson's seventh type) that deconstructionist critics call "undecidability," "unreadability," "dissemination," or "misreading." In a century of discord and alienation, human uncertainty and ambivalence, some readers have developed a taste for the opaque and inscrutable, the labyrinth and the abyss. They delight in opportunities for "creative" reading; they study the interplay, or "free play," of possible meanings previously unsuspected, or at least unvalued, in the masterpieces of the past. And some poets, particularly since Rimbaud, are composing "enigma" texts, poems deliberately written to frustrate the expectations of traditional readers for clarity and coherence. Because of the nature of lang. and the complexities and unpredictability of human experience, clarity and coherence, it is said, are illusions. Lit. can provide no answers; at best, it can only raise questions. The relations between a "poetics of a." and a "poetics of indeterminacy" remain to be worked out (see esp. Perloff).

One extension that does seem advisable is to regard a. as a potential problem (or resource) not just of lang. but of the use of a sign in any semiotic context. Signs, linguistic and nonlinguistic, are involved in the interp. of most of the strata of a literary work: the thoughts, attitudes, motivations of the characters represented; the nature of their actions, verbal and nonverbal; the degree of reliability of the narrator; the values and attitudes of the implied or real author toward the characters and their actions; the theme (q.v.) or thesis of the work; and the nature of the organizing principles of the work. If there are gaps or contradictions in the system of signs which the reader must use as clues to arrive at an interp. on any of these levels, the reader may be left with a number of plausible interps. or a sense of bewilderment.

Linguistic a., at least the kind that is caused by polysemy, has also become an important concern in metacriticism. In his earliest writings I. A. Richards pointed out the various senses in which certain terms (e.g. "beauty," "meaning," "poetry," "sincerity," "fiction") were used in aesthetics and lit. crit. In his writings of the 1930s and 1940s he recommended the linguistic discipline of "multiple definition" to help readers get over the "habit of behaving as though if a passage means one thing it cannot at the same time mean another and incompatible thing."

The documentation of the polysemy of many other terms used in poetics has interested other scholars; an excellent example of such a study is Richard McKeon's "Imitation and Poetry" (*Thought, Action, and Passion* [1954]). In this essay McKeon lists a veritable multitude of senses in which "imitation" (q.v.) has been used in poetics. Besides such exercises in historical semantics, efforts have been made to account for this polysemy and explore its consequences. One position is that the polysemy of the key terms of poetics, like that of most words used in humanistic studies, is inevitable. The objects of humanistic concern are extraordinarily complex and, more important, change as the result of new philosophical or political orientations. A theorist can handle only a limited "aspect" of a subject matter, and the terms that she inherits from the past undergo shifts of reference which can lead to confusion and misunderstanding. What can save the theorist is an abandonment of dogmatism and a commitment to pluralism. The concepts of poetics should be considered "open" or "essentially contested" concepts. The various advantages of this position have been pointed out by a number of critics, incl. Richards, Abrams, and Booth.

I. A. Richards and C. K. Ogden, *The Meaning of Meaning* (1923); I. A. Richards, *The Philosophy of Rhet.* (1936), *How to Read a Book* (1942); W. B. Stanford, *A. in Gr. Lit.* (1939); A. Kaplan and E. Kris, "Aesthetic A.," *PPR* 8 (1948); E. Olson, "William Empson," *MP* 47 (1950); W. Empson, *The Structure of Complex Words* (1951), *Seven Types of A.*, 3d ed. (1953); J. D. Hubert, *L'Esthétique des "Fleurs du Mal"* (1953); M. C. Beardsley, *Aesthetics* (1958); W. Nowottny, *The Lang. Poets Use* (1962); T. Tashiro, "A. as Aesthetic Principle," *DHI*; P. Wheelwright, *The Burning Fountain*, 2d ed. (1968); J. G. Kooij, *A. in Natural Lang.* (1971); M. H. Abrams, "What's the Use of Theorizing about the Arts?" *In Search of Literary Theory*, ed. M. W. Bloomfield (1972); J. R. Kincaid, "Coherent Readers, Incoherent Texts," *CritI* 3 (1977); S. Rimmon, *The Concept of A.* (1977); M. Weitz, *The Opening Mind* (1977); W. C. Booth, *Critical Understanding* (1979); M. Perloff, *The Poetics of Indeterminancy* (1981); D. N. Levine, *The Flight from A.* (1985).

F.GU.

ANAGRAM. In its original, narrower sense, an a., a rearrangement of the letters (sounds) of a word or phrase to produce another word or phrase, is an explicit, consciously constructed device. It is related to the palindrome, in which a textual sequence reads the same way backwards as forwards, and to paronomasia (see PUN). Examples of the a. are found occasionally in ancient writers and quite frequently in the 17th c., primarily in occasional verse: thus each text in Mary Fage's *Fame's Roll* [of British royalty and nobility] (1637) is prefaced by the name of its subject and by an a. of it (THOMAS HOWARD // OH, DRAW MOST, HA!). The a. has also been widely used by authors to create pseudonyms (e.g. Rabelais, John Bunyan, Lewis Carroll) and occasionally used in acrostics (q.v.).

More recently, the a. has come to be conceived by many critics as a substantive component of poetic discourse occurring in a number of poetic trads. This view, which subsumes conscious in-

stances of the a., originates with the partial publication in the 1960s of Ferdinand de Saussure's notebooks (for 1906–9) containing his researches on as. Saussure studied the role of several types of sound structure in texts of various provenience, incl. IE poetry, Lat. poetry (Saturnian verse), such authors as Virgil and Lucretius, and Vedic hymns. These structures range from paired repetitions of sounds, either individually or in groups, to the looser reproduction in the phonic material of the work of elements of a *key-word* (*mot-thème*, later *hypogram*) central to the plot of the text. Frequently this word is the name of the hero or deity to whom the text (such as a hymn) is dedicated.

In the Saussurean model, the functioning of the a. presupposes both a poet capable of sophisticated operations on verbal material (the segmentation of a key-word into elements and their dissemination in the text) and a reader able to recognize the presence of an a. and to reconstitute the hidden whole. To the extent that this conception is valid, it has profound implications for lit. hist. and theory. For a time, Saussure regarded the a. as a fundamental principle of IE poetry; subsequently, however, he came to question both his methodology and conclusions, in part because there is no explicit testimony by ancient authors about such use of the a. But Saussure's ideas have been further developed by both linguists and literary theorists. From linguistics has come added evidence concerning the principle of word-analysis in the IE poetic trad. (e.g., in archaic Ir. verse).

In literary theory, focus on the a. has stimulated a deeper look at the nature of the sound-meaning nexus in poetry; this in turn has led some researchers to emphasize the role of the a. as a semantic category which, under certain conditions, generates added meaning in the poem, semanticizing elements of the lowest levels of the text. Semiotic theory, which has explored the role of trad. as a factor in the autonomous development of sign systems, has provided a framework within which the issue of authorial intention may be removed as a point of contention. New evidence concerning the presence of as. in the work of modern poets has been produced by members of the Moscow-Tartu school: Toporov and Ivanov among others have demonstrated numerous as. in Rus. poetry of the early 20th c., where certain poets and schools attached a special ontological significance to the role of names and developed systems of composition resulting in complex, esoteric, and often encrypted messages.

Future research will need to address the problem of the objective verification of as. and also the question of whether the a. operates on a subliminal as opposed to a denotative or connotative level. Resolving these problems will involve finding answers to such questions as: (1) the conditions under which semantic information "overflows" onto the lower levels of the text, increasing the interconnections between formally discrete elements (recent studies have suggested that as. are most likely to appear in the work of poets who use paronymy extensively); and (2) the genres in which as. are most likely to arise.

H. B. Wheatley, *Of As.* (1862); F. de Saussure, *Cours de linguistique generale* (1916); P. Habermann, "A.," *Reallexikon*; A. Liede, *Dichtung als Spiel*, 2 v. (1963); J. Starobinski, *Les Mots sous les mots* (1971, tr. 1979); P. Wunderli, *F. de Saussure und die Anagramme* (1972); M. Meylakh, "À propos des anagrammes," *L'Homme* 16,4 (1976); V. V. Ivanov, "Ob anagrammax F. de Sossjura," in F. de Saussure, *Trudy po jazykoznaniju* (1977); A. L. Johnson, "Anagrammatism in Poetry," *PTL* 2 (1977); S. Baevskij and A. D. Koševev, "Poetika Bloka: anagrammy," *Blokovskij sbornik 3* (1979); D. Shepheard, "Saussure's Vedic As.," *MLR* 77 (1982); S. C. Hunter, *Dict. of As.* (1982); T. Augarde, *Oxford Guide to Word Games* (1984); V. N. Toporov, "K issledovaniju anagrammatičeskix struktur (analizy)," *Issledovanija po strukture teksta*, ed. T. V. Civ'jan (1987); *Očerki istorii jazyka russkoj poezii XX veka. Poetičeskij jazyk i idiostil'. Obščie voprosy. Zvukovaja organizacija teksta*, ed. V. P. Grigor'ev (1990). H.B.

ANAPHORA (Gr. "a carrying up or back"); also *epanaphora*. The repetition of the same word or words at the beginning of successive phrases, clauses, sentences, or lines. Conversely, *epistrophe* (also called *epiphora*, e.g. Shakespeare, *Merchant of Venice* 3.3.4) repeats words at the ends of clauses, lines, or stanzas; so Tennyson repeats "the days that are no more" at the end of each stanza of "Tears, Idle Tears." Synonyms for *epistrophe* are *epiphora* and *antistrophe* (in the rhetorical sense). Combining both figures is *symploce* (*3 Henry 6* 2.5.103–8). Sonnet 62 of Bartholomew Griffin's *Fidessa* is a Ren. tour de force of symploce, each of the 14 lines beginning "Most true that" and ending "love." Discussions of symploce date from Alexander, the Gr. rhetorician of the 2d c. A.D. (*Peri schematon* 2; rpt. in C. Walz, *Rhetores graeci*, 9 v. [1832–36] 8.464–65) and the *Rhetorica ad Herennium* (4.13). Repetition of the *same* word at the beginning and end of each phrase, clause, or line is *epanalepsis* (*King John* 2.1.329–30). Linking of the end of one member to the beginning of the next is *anadiplosis*, a form of concatenation. Demetrius (1st c. A.D.; *On Style* 268) and Longinus (*On the Sublime* 20.1–2) use the term *epanaphora*, and virtually all postclassical authorities treat a. as its exact synonym.

A. has been a favored device in the poetry of many cultures, particularly in the form where the repeated words or phrases begin lines: this structure enhances the sense of the line even as it foregrounds the larger enumerative sequence. A. may thus be seen as one form of parallelism (q.v.) which uses the repetitions to bring the metrical and syntactic frames into alignment. It has been used particularly in religious and devotional poetry—where in part, it may be thought, the device

echoes the iterations of ecstatic experience—and is common in the Middle Ages; Martianus Capella (5th c.) remarks on the 21 repetitions of *sol* in a Lat. poem. In the Bible many examples foreground "The Lord" or "He," e.g. Psalms 23 and 29. It is conspicuous too in the lyric, esp. the sonnet (q.v.). Daniel (*Delia*) uses as. regularly, preferring triple repetitions, Shakespeare more sparingly—only 16 times in the 154 Sonnets, with 6 of these triple—though the 10-line a. of "And" in sonnet 66 is striking; this sort of extended, plurilinear a. is instanced in Petrarch, no. 145; in Ronsard, *Amours* 1.54; in Spenser *Amoretti* 9 and 15; and in *Hekatompathia* 75 and 77. A. is also common in Shakespeare's plays, particularly the early plays, e.g. *Richard 2* 2.1.40–42 and 4.1.207 ff.; *Richard 3* 4.4.92–104; and *Othello* 3.3.345–47. Sidney is enchanted with it—witness more than 40 examples in *Astrophil and Stella.*

In verse where the bonds of strict meter have loosened or partly disappeared, a. is still very much of value. It is Whitman's most frequent device, serving to give structure to his extended catalogues and syntactically involuted constructions; no poet has exploited the device more than Whitman. In T. S. Eliot most instances are no longer than in most of the Ren. sonneteers, but it dominates section V of *The Waste Land.* In recent years anaphoric structures have been the subject of considerable research in linguistics. But the definitive historical and theoretical study of a. in poetry still remains to be written.

Sr. M. Joseph, *Shakespeare's Use of the Arts of Lang.* (1947); C. Schaar, *An Elizabethan Sonnet Problem* (1960); A. Quinn, *Figures of Speech* (1982), 83–87; Norden; J. Aoun, *A Grammar of A.* (1985); B. Vickers, *In Defence of Rhet.* (1988), *Cl. Rhet. in Eng. Poetry*, 2d ed. (1989); Corbett. T.V.F.B.

ANTHOLOGY (*Florilegium, Chrestomathie*). Etymologically a "bouquet," from Gr. *anthos* (flower) and *legein* (to gather, pick up). In the hist. of Western lit. the most influential a. is unquestionably the Bible, a heterogeneous collection of Jewish (Heb.) and Christian (Gr.) texts written over more than a millennium. The OT includes historical, genealogical, legal, proverbial, wisdom, and prophetic texts codified in its present form by rabbinical council in 90 A.D. The NT includes epistolary and apocalyptic texts and a hybrid genre related to but not biography (the Gospels) codified about the 4th c. The term "Apocrypha" is used sometimes to refer to the 7 texts accepted by Catholics but rejected by Protestants, but more often for all texts whose canonical status is disputed.

In Gr., the a. was originally a collection of epigrams (q.v.), generally composed in elegiac distichs (q.v.) and treating specific subjects or poets. Compilations were made at least as early as the 4th c. B.C. About 90 B.C., Meleager of Gadara collected a *Garland* of short epigrams in various meters, but chiefly elegiac, and on various subjects; some fifty poets from Archilochus (7th c. B.C.) to his own era (incl. himself) were represented. About A.D. 40, Philippus of Thessalonica collected a *Garland* of exclusively elegiac epigrams by poets since Meleager. Approximately a century later Straton of Sardis put together some hundred epigrams on a single subject (homosexual love). About A.D. 570, the Byzantine anthologist Agathias collected a *Circle* of epigrams in various meters; he included selections, arranged by subject, from both *Garlands* as well as a large selection of contemporary epigrams.

Constantinus Cephalas, a Byzantine Greek of the 10th c., compiled an a. combining and rearranging the collections of Meleager, Philippus, Straton and Agathias. In all, it included 15 divisions: Christian epigrams, descriptions of statuary, temple inscriptions, prefaces (by Meleager, Philippus, and Agathias), erotic poems, dedicatory poems, epitaphs (including Simonides' famous lines on the Spartan dead at Thermopylae), epigrams by St. Gregory of Nazianzus, epideictic epigrams, moral epigrams, social and satirical epigrams, Straton's collection, epigrams in special meters, riddles, and miscellaneous epigrams. This a., or the *Gr. A.*, as it is now called, was edited, revised, and expurgated in 1301 by the monk Maximus Planudes. His was the only available text until 1606, when the great Fr. scholar Claude Saumaise (Salmasius) discovered a single manuscript of Cephalas in the Elector Palatine's library at Heidelberg. Thereafter this latter text came to be known as the *Palatine A.*; it supplanted the *Planudean A.* but retained a Planudean appendix as a 16th division. The first ed. of the *Palatine A.* was published by Friedrich Jacobs in 13 vols. from 1794 through 1814.

The influence of the *Gr. A.* in the modern world dates from Janus Lascaris' Florentine ed. of Planudes in 1494. Trs. of the epigrams into Lat. and later into the vernacular langs. multiplied consistently until about 1800, when enthusiasm for the unpointed Gr. epigram was supplanted by one for the pointed epigram as perfected by Martial (1st c. A.D.). Interest in the *Gr. A.* in the 20th c. is evidenced by a number of eds., selections, and trs., e.g. that by Kenneth Rexroth.

Other important Cl. as. include the 5th-c. Stobaeus' Gr. *Eclogae* (Selections) and *Anthologion* and the *Anthologia latina* (ed. Riese, Bücheler, and Lommatzsch, 1894–1926, and incl. otherwise uncollected Lat. verse and a 6th-c. compilation which contained the *Pervigilium Veneris*). Medieval as. were created and preserved mainly by the clerical orders, and survive in influential ms. collections such as the OE *Proverbs of Alfred* and the Eng. lyric collection called the Harley Ms. (British Museum Ms. Harley 2253; ca. 1330); among other medieval *florilegia*, esp. notable are the *Carmina Cantabrigiensia* ("Cambridge Songs"; 11th c.) and the *Carmina Burana* (collected at the Ger. monastery at Benediktbeuren in Bavaria in the 13th c.). Ren.

collections of proverbs drew inspiration from Erasmus' *Adagia* (early 16th c.).

As. took on new importance in the Ren. with a vogue inaugurated in England by the collection assembled by Richard Tottel and now called Tottel's Miscellany (originally *Songes and Sonettes, written by the ryght honorable Lorde Henry Haward late Earle of Surrey, and others*, 1557; ed. Rollins, rev. ed., 2 v., 1965). After Tottel the vogue for the "miscellanies," as they were called (accent on the second syllable), grew to a flood in the last quarter of the century, incl. Clement Robinson's *Very Pleasaunt Sonettes and Storyes in Myter* (1566; surviving only as *A Handefull of Pleasant Delites*, 1584; ed. Hyder E. Rollins, 1924); Richard Edwards' *The Paradyse of Daynty Devises* (1576; ed. Rollins, 1927); Thomas Proctor's *A Gorgious Gallery of Gallant Inventions* (1578; ed. Rollins, 1926); *The Phoenix Nest* (1593; ed. Rollins, 1931); *Englands Helicon* (1600, 1614); Francis Davison's *A Poetical Rapsody* (1602; ed. Rollins, 2 v., 1931); and Nicholas Breton's *Brittons Bowre of Delights* (1591; ed. Rollins, 1933) and *The Arbor of Amorous Devices* (1597; ed. Rollins, 1936).

Other significant European as. are the massive *Flores poetarum* compiled early in the 16th c. by Octavianus Mirandula and used throughout Europe until the 18th c.; Jan Gruter's *Delitiae* (It., Fr., Belgian, and Ger. poems in Lat.; 1608–14); J. W. Zincgref's *Anhang unterschiedlicher aussgesuchter Gedichten* (1624); Thomas Percy's *Reliques of Ancient Eng. Poetry* (1765), an a. of the popular ballads which proved very influential in the 18th-c. revival of antiquarian interest in primitive poetry; Oliver Goldsmith's *The Beauties of Eng. Poetry* (1767); Thomas Campbell's *Specimens of the British Poets* (1891); and Francis Palgrave's *Golden Treasury of the Best Songs and Lyrical Poems in the Eng. Lang.* (1861–; 5th ed. supp. J. Press, 1987), the most important Victorian a. of lyric poetry.

The popularity of as. in the 20th c. has if anything increased, with the expansion of the institutions of higher education, esp. in America. Important as. include the *Oxford Book of Eng. Verse*, successively edited by Sir Arthur Quiller-Couch (1900, 1939) and Helen Gardner (1972); *The New Poetry* (1917) by Harriet Monroe and Alice C. Henderson, which influenced the high modernists; Sir Herbert Grierson's *Metaphysical Lyrics and Poems* (1921), which inaugurated the vogue for metaphysical poetry; W. B. Yeats's *Oxford Book of Modern Verse* (1936); Brooks and Warren's *Understanding Poetry* (1938; 4th ed. rev. extensively, 1976), which codified for pedagogy and practical crit. the principles of the New Criticism; Donald Hall's *The New Am. Poetry* (1960), which opened up the formalist canon; and the seemingly ubiquitous Norton As. of Lit. in their several manifestations (World, Eng., Am.) and As. of Poetry and of Lit. by Women—which, for all practical purposes, in each successive revision, effectually constitute the canon (q.v.). In the later 1980s these began to be seriously challenged by competitors offering much more radically revisionary as. See now EPIGRAM; LYRIC SEQUENCE.

J. O. Halliwell-Phillipps, *Early Eng. Miscellanies* (1855); *An OE Miscellany*, ed. R. Morris (1872); F. Lachère, *Bibliographie des receuils collectifs de poésies publiés de 1597 à 1700* (1901); A. Wifstrand, *Studien zur griechischen Anthologie* (1926); A. E. Case, *A Bibl. of Eng. Poetical Miscellanies, 1521–1750* (1935); J. Hutton, *The Gr. A. in Italy to the Year 1800* (1935), *The Gr. A. in France and in the Writers of the Netherlands to the Year 1800* (1946); *The Harley Lyrics*, ed. G. L. Brook (1948); A. S. F. Gow, *The Gr. A.: Sources and Ascriptions* (1958); *Anthologia graeca*, ed. H. Beckby, 4 v. (1965); R. F. Arnold, *Allgemeine Bücherkunde*, 4th ed. (1966); *Die deutschsprachige Anthologie*, ed. J. Bark and D. Pforte, 2 v. (1969–70); *New CBEL* 2.327 ff.; E. W. Pomeroy, *The Elizabethan Miscellanies, Their Devel. and Conventions* (1973); *The Gr. A.*, sel. and tr. P. Jay (1973); Pearsall 94 ff.; R. McDowell, "The Poetry A.," *HudR* 42 (1990). T.V.F.B.; R.A.S.

APOSTROPHE (Gr. "to turn away"). A figure of speech which consists of addressing an absent or dead person, a thing, or an abstract idea as if it were alive or present. Examples include Virgil's "Quid non mortalia pectora cogis, / auri sacra fames!" (*Aeneid* 3.56); Dante's "Ahi, serva Italia, di dolore ostello" (*Purgatorio* 6.76); Shakespeare's "O judgment! thou art fled to brutish beasts" (*Julius Caesar* 3.2.106); Wordsworth's "Milton! thou should'st be living at this hour" ("London, 1802"); and Tennyson's "Ring out, wild bells" (*In Memoriam* 106). The traditional epic invocation of the Muse is an a. A. is particularly useful in sonnet sequences (q.v.) for addresses to the beloved, to other parties, and to abstractions. Fully 134 of Shakespeare's 154 sonnets contain an a., 100 of which are direct addresses (to the lady, the friend); these are almost always to a specific person, and only very rarely to a personified abstraction, e.g. Time (only twice), Love (thrice), or the Muse (twice). The term originally referred to any abrupt "turning away" from the normal audience to address a different or more specific audience, whether present or absent (Quintilian, *Institutio oratoria* 4.1.63–70, 9.3.24). Corbett reminds us that a. is a figure of *pathos*, i.e. "calculated to work directly on the emotions" (460), but other critics have seen it as a metapoetic device by means of which a speaker "addresses" his own utterance. Shelley's "Ode to a Skylark" begins with the conventional kind of a.: "Hail to thee, blithe spirit"; Robert Burns' "Address to the Unco Guid," on the other hand, sarcastically apostrophizes the speaker's moral opponents: "O ye wha are sae guid yoursel, / Sae pious and sae holy" (1–2). The a. in Charles Calverley's "Ballad" parodies the device's artificiality: "O Beer. O Hodgson, Guinness, Allsop, Bass. / Names that should be on every infant's tongue." Ezra Pound uses a. metapoetically in "Coda": "O my songs, / Why do you

look so eagerly and so curiously into people's faces, / Will you find your lost dead among them?" Culler attempts to identify a. "with lyric itself" in order to deconstruct the notion of lyric timelessness, seeing the lyric as *écriture*, though this approach ignores the explicit emphasis on orality in Quintilian and others.—Sr. M. Joseph, *Shakespeare's Use of the Arts of Lang.* (1947), 246–47; C. Perelman and L. Olbrechts-Tyteca, *The New Rhet.* (tr. 1969); M. R. McKay, "Shakespeare's Use of the A.," *DAI* 30 (1970): 4459A; J. T. Braun, *The Apostrophic Gesture* (1971); Lausberg; J. Culler, "A.," *The Pursuit of Signs* (1981), ch. 7; L. M. Findlay, "Culler and Byron on A. and Lyric Time," *SiR* 24 (1985); P. Lang, *The Apostrophic Moment in 19th- and 20th-C. Ger. Lyric Poetry* (1988); Corbett.

L.P.; A.W.H.; T.V.F.B.

ARCHAISM. Deliberate use of old, old-fashioned, or obsolete words, esp. in poetry, in order to evoke the mood of an earlier time—i.e. to appropriate that ethos for a text—or else, alternatively, to recapture a connotation or denotation not borne by any modern word. As. appear in both Virgil and Lucretius; in Fr. poetry of the Ren., as. were promoted in the doctrines of the *Pléiade.* In Eng. poetry, so trad. held, a. found its deepest source in Spenser's *Faerie Queene* (1589–96), in which "strange inkhorn terms" seem to be used for the sake of their association with the chivalry and romances of the past. But modern scholarship contradicts the received view: Strang for example says that Spenser's as. "are superficial and limited; the essential character of his poetic lang. is its modernity." As. are less common in Spenser than was often thought: in fact there are only about a hundred archaic words in all of the *Faerie Queene,* and half of these are used but once. But Spenser salts his diction regularly: two-thirds of the 55 stanzas of Book I, Canto 1, contain at least one archaic locution. Further, the *effect* of a. may be achieved by a number of lexical strategies besides resuscitating words found in an older author, as Spenser did with Chaucer: these include the addition of archaic prefixes or suffixes to ordinary contemporaraneous words, capture of words still in use but rapidly passing out of fashion, use of rare and unusual (but not obsolete) words, use of dialectal words, and the coining of words that look or sound archaic but in fact never existed. Syntactic strategies such as hyperbaton (q.v.)—common in Spenser—also give the effect of a.

All this suggests that one should proceed with caution in appraising a., for its strategies are various and not easily disentangled from innovation. Too, many words that today may seem archaic might not in fact have been so, or seemed only slightly so, to Spenser's own contemporaries: it requires great erudition to assess with any accuracy exactly how much—and how—a text from the past evoked, from its audience, its own past. What is essential for the critic is the recognition that the desired effect on the audience is that the archaic words seem "strange but not obscure" (McElderry).

Sometimes a. also serves metrical ends when the older form of a word, having a different number of syllables, more readily fits into the meter than its modern equivalent or descendant. Readers of Eng. poetry are familiar with such words as *loved, wished,* etc., used in poetry as disyllables, namely *lovéd, wishéd.* Yet this should not be thought mere exigency. With older texts, one must exercise some caution in weighing the metrical value of as., since it is often difficult to know precisely when a term went out of contemporaneous usage and became an a.: many words, we know, survived for a considerable time in alternative, syllabically variant forms (now called *doublets*) derived from pronunciation (esp. dialect), e.g. "heaven" as a monosyllable or disyllable.

When archaic words are preserved in verse but not in prose or common speech, and are adopted by poets beyond the usage of some one poet in whose work as. are very prominent, and in that of his followers, such words come to form a select body of "poetic diction" built up by and as trad., and the diction of poetry then comes to be commonly conceived *as* a. This is true for the "School of Spenser" in the 17th c. and the Spenserian imitators of the 18th, and of Pre-Raphaelites such as William Morris in the 19th. (Conversely, some words, such as *morn* for "morning" and *eve* for "evening," become so common that, in time, they lose much of their sense of a. and are felt to be simply "poetic.") Inexorably such movements provoke reaction, e.g. romanticism to the rigidities of the diction of 18th-c. poetry, and modernism to those of Victorian. However, these reactions are far from simple, as witnessed by the as. common in Coleridge. Nor have archaic words ceased to be used in the 20th c., despite the seeming contravention of Pound's dictum to "Make It New": Pound himself drew heavily upon the diction of the past for his own poetry as well as his translations and imitations, and Eliot in the first movement of *East Coker* deploys as. extensively so as to evoke the "country mirth" of "those long since under earth," making "time past" "time 2present." Elsewhere Eliot says that "last year's words belong to last year's lang."; and if the poet must find new words for new experience, so too must he or she suscitate old words for a sense of the past. See now EPITHET; LEXIS.

G. Wagner, *On Spenser's Use of As.* (1879); R. S. Crane, "Imitation of Spenser and Milton in the Early 18th C.," *SP* 15 (1918); T. Quayle, *Poetic Diction* (1924); E. F. Pope, "Ren. Crit. and the Diction of the *Faerie Queene,*" *PMLA* 41 (1926); B. R. McElderry, Jr., "A. and Innovation in Spenser's Poetic Diction," *PMLA* 47 (1932); W. L. Renwick, "The Critical Origins of Spenser's Diction," *MLR* 17 (1922); O. Barfield, *Poetic Diction,* 2d ed. (1952); N. Osselton, "A.," and B. M. H. Strang,

"Lang., Gen.," *The Spenser Encyc.*, ed. A. C. Hamilton et al. (1990). T.V.F.B.

ASSONANCE (Lat. *assonare*, "to answer with the same sound"; Ir. *ammus*). In a trad. stemming from OF and OSp., and until the 20th c. with its more varied versification, a. served the same function as rhyme, i.e. formalized closures to lines in the poems of most Romance langs. The OF *Chanson de Roland* (early 12th c.) attempts in each *laisse* to have not only the same vowel in the final stressed syllable of each line but also the same vowel in the following weak syllable (if there is one). A similar system prevailed in Sp. and Port., as can be seen not only in the great epic *Os Lusíadas* by Camões (16th c.) but also in his lyrics, sonnets, and pastorals. In It., the final words in four typical consecutive lines of Dante's *Divina commedia* end *viaggio-vide-selvaggio-gride* (91–94). In all of these, it is the vowel that cannot be parted with, the result being, by increasing constraint, a., rhyme, or identity.

Along with these complex end echoes, however, Romance poets have employed much internal a., partly perhaps because their langs., like Lat., are so rich in vocables. Almost any Lat. or Gr. poet can be shown to have loved vowel play (Wilkinson; Stanford). Dryden quotes this line from the *Aeneid* to demonstrate Virgil's "musical ear"—"Quo fata trahunt retrahuntque, sequamur"—which stresses *a* four times against a background of three unstressed *u* sounds. In modern times, Fr. has far more end rhymes than other Romance langs., but because so many Fr. final consonants are not pronounced, the rhyme is sometimes really a. The cl. alexandrine (q.v.) ends in a vowel (i.e. a.) perhaps 20 percent of the time. Furthermore, internal a. is as marked an acoustic device of the alexandrine as of later Fr. poetry, as in *Phèdre*, which has a number of lines with a. of three syllables, this one (1.3.9) with four—"Tout m'afflige et me nuit et conspire a me nuire." Verlaine with his belief in "musique avant toute chose" is often credited with inspiring poets of his day to employ sonorous sounds ("rime or assonate; otherwise, no verse" [1888]); a. was most important in his own versification.

Far less a subject of traditional interest than alliteration (q.v.) but more noticed than consonance, a. has often been thought a mere substitute for rhyme. It is true that the Germanic langs. had an early history of alliterative poetries—e.g. *Beowulf*—and continue to employ alliteration, often heavily. True, too, the Celtic langs., notably Welsh, developed schemes of sound correspondence, esp. *cynghanedd*, that have complex alliterative patterns. But it is equally true that in nearly all langs., incl. Ir. and Welsh, poets have continually employed a. internally if not as a substitute for end rhyme. The term "a." was borrowed by the Eng. from Sp. and Fr. to describe the various vocalic substitutes for rhyme in the Romance poetries. But end a. is only a small part of the total phenomenon called "a." Consequently, a. is best defined as the repetition of the sound of a vowel or diphthong in nonrhyming stressed syllables near enough to each other for the echo to be discernible. Note the following facts about this definition:

First and most important, a. is an *aural* device. Because alliteration is normally initial in words, it usually receives aid from the reader's eye. But a., since it is normally medial, can expect little such help. Only the ear can respond to Shakespeare's "Time's scythe," Eliot's "recovers / My guts," or MacLeish's "cough / In waltz-time."

Second, because a. is sonant, and since vowels and diphthongs change their pronunciation, sometimes radically, over time and from region to region, while consonants change relatively little, a reader needs to know how the vowels were pronounced by the poet being read. In Eng., because the Great Vowel Shift was largely completed by Shakespeare's time, the verse of Chaucer and other ME poets sounds different from that of later times. Furthermore, a number of vowels continued to change pronunciation after Shakespeare's day; by the 18th c., Eng. poets were still naturally rhyming *Devil-civil, stem-stream, pull-dull-fool, feast-rest, tea-obey*, and *join-fine.*

Third, because there are more consonants than vowels in the IE langs., those langs. will have more accidental a. than accidental alliteration. The consonants in Eng. number about 24, but Eng. has at least 18 different vocalic sounds that poets can employ for rhyme or a. In addition to simple assonances such as *back-cast, rose-float*, and *feat-seek*, or the echoes of diphthongs such as *fine-bride* and *proud-cowl*, poets assonate *er, ir, ur*, and sometimes *ear* in words like *serve, furred, earn, dirge*, or *beard* with *miracle* and *cheer*. Given so many choices, if one can assume that Pope intended the trisyllabic a. in "silver-quivering rills," that Wilbur intended it in "ladders and hats hurl past," and that Lowell intended it in "mast-lashed master," one is perhaps willing to accept as intended the disyllabic a. in Pope's "a lazy state," in Wilbur's "hollow knocks" and "optative bop," or in Lowell's "gagged Italians" ("Falling Asleep over the *Aeneid*").

Ger. poets seem to have found a. even more appealing than alliteration, from Brentano, who has whole poems with a polysyllabic a. in nearly every line, to Goethe, Heine, and modern poets. Goethe in his lyrics and plays has more a. than other phonic echoes, as in Faust's contract with Mephistopheles, which in its last three lines has a six-syllable a. of [aɪ] as well as other echoes: "Dann bist du d*ei*nes Dienstes fr*ei*, / Die Ühr mag stehn, der Z*ei*ger fallen, / Es s*ei* die Z*ei*t fur mich vorb*ei*!" And Rilke, one of the most sonorous of poets, opens "Wendung" with "L*a*nge err*a*ng ers im *A*nschauen."

Rus. poets too have employed a. consistently, esp. Pasternak, Blok, and Bryusov, the latter two frequently substituting it for end rhyme (Plank; Donchin).

Finally, the notion that a. is no longer used for

end rhyme must be dispelled. Blok replaced rhyme with a. in whole series of poems, as did Rilke; Verlaine used a. instead of rhyme perhaps half the time, and Eng. poets such as Thomas in Britain and Jarrell and Crane in the U.S. have often made such a substitution. It should also be pointed out that poets other than Sp. and It. have dabbled with polysyllabic a., the echoes coming only in the weak final syllables. But in Eng. such echoes are rare, difficult, and relatively ineffective. In fact, just as computers would be of little help in finding a., any reader should hesitate before including obviously weak syllables in a study of vowel repetitions.

R. E. Deutsch, *The Patterns of Sound in Lucretius* (1939); W. B. Stanford, *Aeschylus in His Style* (1942), *The Sound of Gr.* (1967); B. Donchin, *The Influence of Fr. Symbolism on Rus. Poetry* (1958); N. I. Hérescu, *La Poésie latine* (1960); W. Kayser, *Gesch. des deutschen Verses* (1960); P. Delbouille, *Poésie et sonorités,* 2 v. (1961, 1984); D. L. Plank, *Pasternak's Lyrics* (1968); L. P. Wilkinson, *Golden Lat. Artistry* (1970); J. Gluck, "A. in Ancient Heb. Poetry," *De Fructu oris sui: Essays in Honor of Adrianus von Selma,* ed. I. H. Eybers et al. (1971); Wimsatt; P. G. Adams, *Graces of Harmony* (1977); R. P Newton, *Vowel Undersong* (1981); Brogan, esp. items C129-30, C150, C152–53, K2, K116, L233, L322, L406, L518.

P.G.A.

B

BALLAD. The "folk," "popular," or "traditional" b. is a short narrative song preserved and transmitted orally among illiterate or semiliterate people. As one of the if not the most important forms of folk poetry, bs. have been composed, sung, and remembered in every age. But wider use of the term "b." recommends that we draw distinctions between the following three types of bs.: (1) the b. as a species of oral poetry (q.v.), which is orally composed and transmitted, popular in register, and mainly rural in provenience: most of the famous examples are late medieval, but survived as folkverse via oral transmission down to the 19th c., when they were discovered and collected; (2) 16th-c. assimilations and adaptations of the oral b., and particularly its meter, to urban contexts and religious uses: these are the broadside or street b. and the hymns, as in the metrical psalters, esp. Sternhold and Hopkins; and (3) literary imitations of the popular b., beginning in the Ren. but flourishing particularly in the later 18th-c. revival of antiquarianism and folk poetry which culminated in romanticism, esp. in Goethe and Wordsworth.

Short narrative songs have been collected in all European countries, and though each national balladry has its distinctive characteristics, three constants seem to hold for all genuine specimens. (1) Bs. focus on a single crucial episode. The b. begins usually at a point where the action is decisively directed toward its catastrophe. Events leading up to this conclusive episode are told in a hurried, summary fashion. Little attention is given to describing settings; indeed, circumstantial detail of every sort is conspicuously absent. (2) Bs. are dramatic. We are not told about things happening; we are shown them happening. Every artistic resource is pointed toward giving an intensity and immediacy to the action and toward heightening the emotional impact of the climax. Protagonists are allowed to speak for themselves, which means that dialogue (q.v.) bulks large. At strategic moments, dialogue erupts into the narrative. Such speeches are sparingly tagged; frequently we must deduce the speaker from what is being said. (3) Bs. are impersonal. The narrator seldom allows his or her own attitude toward the events to intrude. Comments on motives are broad, general, detached. There may be an "I" in a ballad, but the singer does not forget his or her position as the representative of the public voice. Bias there is in bs., of course, but it is the bias of a party, community, or nation, not an individual's subjective point of view.

Stylistically, the b. is a distinctive genre. Like folksong, but unlike all the literary genres, the b. is an oral phenomenon and, as a consequence, preserves traces of the archaic modes of preliterature. Plot is the central element: all other artistic possibilities are subordinated to it. The lang. is plain and often formulaic, a small stock of epithets and adjectives serving as formulas (q.v.) for extensive repetition. There are few arresting figures of speech and no self-conscious straining after novel turns of phrase. And because the emphasis is on a single line of action precipitately developed, there is no time for careful delineation of character or exploration of psychological motivation. The heavy amount of repetition and parallelism (qq.v.) characteristic of the bs. only *seems* to be mere ornamental rhet. Repetition in heightened passages is, as Coleridge explained brilliantly, the singers' effort to discharge emotion that could not be exhausted in one saying. Much repetition is mnemonic: in a story being recited or sung, crucial facts must be firmly planted in the memory since the hearer cannot turn back a page to recover a fact that slipped by in a moment of inattention.

Much repetition, too, is cumulative in effect: this device is known as "incremental repetition."

Between the balladries of Western and Eastern Europe there are marked differences. Except for certain Romanian bs., rhyme and assonance are unusual in the Slavic territories and in the Balkans, incl. Greece. The Ukrainian *dumy*, which assonate, are another exception. Finnish bs. employ alliteration as a binding principle. In Western countries, rhyme or assonance is general. All bs. are essentially short narrative poems with a greater or lesser infusion of lyrical elements, and the strength of the lyrical quality is another discriminant in the classification of balladries. As a general rule, strophic (stanzaic) bs. tend to be more lyrical than nonstrophic. Rhyme, assonance, refrains (q.v.), and short-lined meters further suggest lyricism, as do singability and dance animation. Even the *viser* of Denmark, lengthy heroic bs., are kept lyrically buoyant by rhyme, assonance, stanza breaks, and by a refrain technique which shows these bs. were meant to be performed in ring or chain dances. A leader sang the narrative stanza as the ring moved; the dancers paused to sing the refrain, an exultation irrelevant to the progress of the story. Least lyrical of European bs. are the Serbian men's songs on historical or martial topics, written in a heavy pentameter line, and the Sp. *romances*, which seem, like the *viser*, to have been cut down from epic lays, and which, though held together by assonance, are not strophic. Lithuania, Poland, France, Italy, and Scotland possess the most lyrical story-songs. As to metrical schemes, British and Scandinavian bs. use the 4–3–4–3 stress verse of the Common Meter quatrain; an octosyllabic line is the staple of Sp., Bulgarian, Rumanian, and much Ger. b. poetry; France and allied b. territory (Catalonia, Northern Italy, Portugal) take as standard a verse of 12 or 16 syllables broken into hemistichs which rhyme or assonate with one another; the scansion of the Rus. *bylina* is free and highly irregular, the musical phrasing governing the organization of the verse.

The charms of British balladry were first brought to the attention of the lettered and learned world in the 18th c. During the period 1790–1830 many important recordings from oral trad. were made. F. J. Child, a professor at Harvard, made the definitive thesaurus of British popular bs. (1882–98), printing 305 bs., some in as many as 25 versions. In the 20th c., about 125 of Child's bs. have been recovered in the United States and Canada from the descendants of 18th-c. immigrants from the British Isles. *Judas*, the oldest b. in Child's collection, is recorded in a late 13th-c. manuscript; it deals with the betrayal of Christ. There are a few other religious bs., mostly concerned with the miracles of the Virgin, but these are far outnumbered by the pieces dealing with the pagan supernatural, like *Tam Lin*, *The Wife of Usher's Well*, and *Lady Isabel and the Elf-Knight*. In America the supernatural elements have been deleted or rationalized, as they have also been in recent British trad. The commonest b. theme is tragic love of a sensational and violent turn (*Earl Brand*, *Barbara Allen*, *Childe Maurice*). Incest and other domestic crimes are surprisingly common. The troubles on the English-Scottish Border in the 15th and 16th cs. inspired a precious body of sanguine heroic bs. (*Johnny Armstrong*, *Hobie Noble*, *Edom O'Gordon*), many of which are partly historical. Most other historical bs. are the work of minstrels and transparently urge the causes of their noble patrons. Propaganda seems to inform the Robin Hood bs., which exalt the virtues of the yeoman class.

The origin of the British popular b. was once hotly argued among b. scholars. A school known as "individualists" (John Meier, Louise Pound) asserted that all bs. are the work of individual poets and are "popular" merely in having been taken up by the folk. "Communalists" (F. B. Gummere, W. M. Hart, G. L. Kittredge) insisted that the prototypical b. was concocted in assemblies of the folk in the exultations of choral dance. Current opinion concedes that the traits of "balladness" may be explained by the communal theory, but holds that all extant bs. are originally the work of individuals. As the individualists failed to understand, however, the work of an individual poet does not become a b. until it is accepted by the folk and remodeled by b. conventions in the course of its tour in oral trad.

Perhaps the liveliest recent issue in b. theory is the question of whether the Eng. and Scottish bs. were composed extemporaneously in the same oral-formulaic way that Milman Parry and Albert B. Lord argue for the oral-formulaic composition of epics (see Jones and Friedman 1961). David Buchan based his structural analysis of b. narration on a wholehearted acceptance of the bs. as oral-formulaic compositions. His arguments were attacked by Friedman (in Porter 1983), but Buchan's analyses have been found valuable even by scholars who reject the theory they subtend (see Andersen et al. 1982).

Native Am. bs., influenced partly by the British traditional bs. but mainly by the broadsides, exist alongside the imported bs. The oldest perhaps is *Springfield Mountain*, the story of a Yankee farmboy fatally bitten by a snake, which may be pre-Revolutionary. Better known are outlaw songs like *Jesse James* and *Sam Bass*, the lumberjack classic *The Jam on Gerry's Rock*, the cowboys' *The Streets of Laredo*, and the popular favorites *Casey Jones*, *John Henry*, *Frankie and Albert* (*Frankie and Johnny*), *Tom Dooley*, and *Big John*.

Bs. have had an enormous influence on Eng. and Ger. literary poetry, though Entwistle exaggerates when he claims that "the debt of [romantic] lit. to the b. has been comparable to that of the Ren. to the Gr. and Lat. classics." The early 18th-c. imitations, Thomas Tickell's *Lucy and*

Colin, for example, patterned themselves on broadsides and dealt mainly with village tragedies. Wordsworth and Hardy pursued a similar vein in much of their b.-colored poetry. Bishop Percy, the editor of the famous *Reliques* (1765), sentimentalized and prettified the b. style in *The Hermit of Warkworth*, but some of his completions of genuine fragments were reasonably faithful. His reconstruction of *Sir Cauline* lent many touches to Coleridge's *Christabel*. Scott's poetic career began with a translation of Bürger's *Lenore*, a Ger. literary b., and he concocted several counterfeit bs. in his early years, but he eventually came to feel that b. impersonality and the stylization of b. lang. were contrary to the aesthetic of composed poetry. In the b.-like poems of Coleridge, Keats, Rossetti, Meredith, and Swinburne, we see the popular b. style being crossed with borrowings from medieval romances and from minstrel poetry to yield poems of richer texture, more circumstantial and more contrivedly dramatic than are the bs. *La Belle Dame Sans Merci* and *The Ancient Mariner*, perhaps the finest of the literary bs. or b. imitations, could never pass for the genuine article but are greater poems than any b. Yeats's bs. (*Moll Magee, Father Gilligan* etc.) are bs. only in stanza structure and simplicity; they do not employ stock lang. or the rhet. peculiar to traditional song.

Herder's encouragement of folksong collecting resulted in Arnim and Brentano's *Des Knaben Wunderhorn* (1805–8), a collection which exerted an almost tyrannical influence over the Ger. lyric throughout the 19th c. Many of the best shorter poems of Heine, Mörike, Chamisso, Eichendorff, Uhland, and Liliencron purposely resemble *Volkslieder*. The narrative element in these poems is usually so completely overwhelmed by the lyrical that they would hardly be considered bs. by Eng. standards. A.B.F.

T. Percy, *Reliques of Ancient Eng. Poetry*, 3 v. (1765); W. Wordsworth, Preface to *Lyrical Bs.*, 3d ed. (1802); W. Scott, *Minstrelsy of the Scottish Border* (1803); F. J. Child, *The Eng. and Scottish Popular Bs*, 10 v. (1882–98); F. B. Gummere, *The Popular B.* (1907)—use with caution; R. Menéndez Pidal, *Poesía popular y poesía tradicional* (1922); G. Grieg, *Last Leaves of Traditional Bs.* (1925)—post-Child Scottish coll.; W. P. Ker, *Form and Style in Poetry* (1928); P. Barry et al., *British Bs. from Maine* (1929); G. H. Gerould, *The B. of Trad.* (1932); J. and A. Lomax, *Am. Bs. and Folk Songs* (1934); L. K. Goetz, *Volkslied und Volksleben der Kroaten und Serben* (1936–37); S. B. Hustvedt, *A Melodic Index of Child's B. Tunes* (1936); W. Kayser, *Gesch. der deutschen Ballade* (1936); W. J. Entwistle, *European Balladry* (1939); M. J. C. Hodgart, *The Bs.* (1950); M. Dean-Smith, *A Guide to Eng. Folk Song Collections* (1954); *The B. Book*, ed. M. Leech (1955); *Viking Book of Folk-Bs.*, ed. A. B. Friedman (1956); A. B. Friedman, "The Late Med. Ballade and the Origin of Broadside Balladry," *MÆ* 27 (1958), *The B. Revival* (1961); *Traditional Tunes of the Child Bs.*, 4 v. (1959–72), *The Singing Trad. of Child's Popular Bs.* (1976), both ed. B. H. Bronson; D. K. Wilgus, *Anglo-Am. Folksong Scholarship since 1898* (1959); *The Critics and the B.*, ed. M. Leach and T. P. Coffin (1960); J. H. Jones, "Commonplace and Memorization in the Oral Trad. of the Eng. and Scottish Popular Bs.," *JAF* 74 (1961), with foll. reply by Friedman; A. K. Moore, "The Literary Status of the Eng. Popular B.," *ME Survey*, ed. E. Vasta (1965); *Deutsche Balladen*, ed. H. Fromm, 4th ed. (1965); C. M. Simpson, *The British Broadside B. and its Music* (1966); J. B. Toelken, "An Oral Canon for the Child Bs.," *JFI* 4 (1967); L. Vargyas, *Researches into the Med. Hist. of Folk B.* (tr. 1967), *Hungarian Bs. and the European B. Trad.*, 2 v. (tr. 1983); D. C. Fowler, *A Lit. Hist. of the Popular B.* (1968); B. H. Bronson, *The B. as Song* (1969); *Die deutsche Ballade*, 3d ed. (1970), *Moderne deutschen Balladen* (1970), both ed. K. Bräutigam; D. W. Foster, *The Early Sp. B.* (1971); D. Buchan, *The B. and the Folk* (1972); G. M. Laws, Jr., *The British Literary B.* (1972); S. M. Bryant, *The Sp. B. in Eng.* (1973); B. Nettl, *Folk and Traditional Music of the Western Continents* (1973); J. S. Bratton, *The Victorian Popular B.* (1975); *Penguin Book of Bs.*, ed. G. Grigson (1975); J. Reed, *The Border B.* (1975); M. R. Katz, *The Literary B. in Early 19th-C. Rus. Lit.* (1976); *B. Studies*, ed. E. B. Lyle (1976); T. P. Coffin, *The British Traditional B. in America*, 2d ed. (1977); *The European Med. B.*, ed. O. Holzapfel (1978); *Bs. and B. Research*, ed. P. Conroy (1978); B. R. Jonsson et al., *The Types of the Scandinavian Med. B.* (1978); W. Hinck, *Die deutsche Ballade vom Bürger bis Brecht*, 3d ed. (1978); W. Freund, *Die deutsche Ballade* (1978); A. Bold, *The B.* (1979); *Balladenforschung*, ed. W. Müller-Seidel (1980); R. de V. Renwick, *Eng. Folk Poetry: Structure and Meaning* (1980); *The B. as Narrative*, ed. F. Andersen et al. (1982); *The B. Image*, ed. J. Porter (1983); R. Wildbolz, "Kunstballade," and R. W. Brednich, "Volksballade," *Reallexikon* 1.902–9, 4.723–34; C. Freytag, *Ballade* (1986); *Sp. Bs.*, ed. and tr. R. Wright (1987)—anthol.; *The Pepys Bs.*, ed. W. G. Day, 5 v. (1987); P. Cachia, *Popular Narrative Bs. of Mod. Egypt* (1988)—Arabic; *Hispanic Balladry Today*, ed. R. H. Webber (1989); W. E. Richmond, *B. Scholarship: An Annot. Bibl.* (1989); C. Preston, *A Concordance to the Child Bs.* (1990); W. B. McCarthy, *The B. Matrix* (1990); *The B. and Oral Lit.*, ed. J. Harris (1991). T.V.F.B.

BALLADE. The most important of the OF fixed forms and the dominant verseform of OF poetry in the 14th and 15th cs. The most common type comprises 28 lines of octosyllables, i.e. three 8-line stanzas rhyming *ababbcbC* and a 4-line *envoi* rhyming *bcbC*. As the capital letter indicates, the last line of the first stanza serves as the refrain, being repeated as the last line of each stanza and the *envoi*. In the complexity of its rhyme scheme, restriction of its rhyme-sounds, and use of the refrain, the b. is one of the most exacting of the

fixed forms. Some variants of the standard b. employ 10-or (less often) 12-line stanzas with, respectively, 5- and 6-line *envois.* The *envoi*, which frequently begins with the address "Prince," derived from the medieval literary competition (*puy*) at which the presiding judge was so addressed, forms the climatic summation of the poem.

Although the b. may have developed from an Occitan form, it was standardized in northern Fr. poetry in the 14th c. by Guillaume de Machaut, Eustache Deschamps, and Jean Froissart. It was carried to perfection in the 15th c. by Christine de Pisan, Charles d'Orléans, and, most of all, François Villon, who made the b. the vehicle for the greatest of early Fr. poetry. Such works as his "B. des pendus" and his "B. des dames du temps jadis" achieved an unequaled intensity in their use of refrain and *envoi.* The b. continued in favor up to the time of Marot (early 16th c.), but the poets of the *Pléiade*, followed by their neoclassical successors in the 17th c.—with the exception of La Fontaine—had little use for the form and regarded it as barbaric. Both Molière and Boileau made contemptuous allusions to the b.

The b. of the vintage Fr. period was imitated in England by Chaucer and Gower, though now in decasyllables. Chaucer uses it for several of his early complaints and takes the single octave (q.v.) from it for the *Monk's Tale* stanza. Beyond their practice, it never established itself firmly. In the later 19th c., the so-called Eng. Parnassians (Edmund Gosse, Austin Dobson, Andrew Lang) and poets of the Nineties (W. E. Henley, Richard Le Gallienne, Arthur Symons) revived the form with enthusiasm, inspired by the example of Banville (*Trente-six Bs. joyeuses à la manière de Villon*, 1873), who gave a new impetus to the b. equally among fellow Parnassian and decadent poets in France (François Coppée, Verlaine, Jean Richepin, Maurice Rollinat). But the modern b., with the possible exception of a few pieces by Swinburne and Pound's Villonesque adaptations ("Villonaud for this Yule" and the freely constructed "A Villonaud: Ballad of the Gibbet"), has not aimed at the grandeur and scope of Villon: it has been essentially a vehicle for light verse (q.v.), e.g. by Chesterton, Belloc, and, more recently, Wendy Cope.

The double b. is composed of six 8- or 10-line stanzas; the refrain is maintained, but the *envoi* is optional (e.g. Villon's "Pour ce, amez tant que vouldrez"; Banville's "Pour les bonnes gens" and "Des sottises de Paris"; Swinburne's "A Double Ballad of August" and "A Double Ballad of Good Counsel"). The "b. à double refrain," which has Marot's "Frère Lubin" as its model, introduces a second refrain at the fourth line of each stanza and again at the second line of the *envoi*, producing, most characteristically, a rhyme scheme of *abaBbcbC* for the stanzas and *bBcC* for the *envoi*; Dobson, Lang, and Henley number among the later 19th-c. practitioners of this variant.

J. Gleeson White, *Bs. and Rondeaus* (1887); G. M. Hecq, *La B. et ses derivées* (1891); Kastner; H. L. Cohen, *The B.* (1915), *Lyric Forms from France* (1922); P. Champion, *Histoire poétique du XVe siècle*, 2 v. (1923); G. Reaney, "Concerning the Origins of the Rondeau, Virelai, and B.," *Musica Disciplina* 6 (1952), "The Devel. of the Rondeau, Virelai, and B.," *Festschrift Karl Fellerer* (1962); A. B. Friedman, "The Late Medieval B. and the Origin of Broadside Balladry," *MÆ* 27 (1958); J. Fox, *The Poetry of Villon* (1962); Morier. A.PR.; T.V.F.B.; C.S.

BLANK VERSE.

I. IN ITALIAN AND ENGLISH
II. IN OTHER LANGUAGES

I. IN ITALIAN AND ENGLISH. B. v. first appeared in It. poetry of the Ren. as an unrhymed variant of the *endecasillabo*, then was transplanted to England as the unrhymed decasyllable or iambic pentameter. Though these lines are thought to have derived metrically from the Cl. iambic trimeter, they were designed to produce, in the vernaculars, equivalents in tone and weight of the Cl. "heroic" line, the line of epic, the hexameter (q.v.). The unrhymed *endecasillabo*, while popular and important, never became a major It. meter; in England, however, b. v. became, under the influence of Shakespeare and Milton, the staple meter of Eng. dramatic verse and a major meter of nondramatic verse as well. Fr. poet-critics noted the work of the It. poets and made experiments of their own, but these never took hold in a lang. where word-accent was weak; hence Fr. poetry never developed a significant b.-v. trad. The Ger., Scandinavian, and Slavic trads. are discussed in section II below.

In *England*, b. v. was invented by Henry Howard, Earl of Surrey (1517–47), who sometime between 1539 and 1546 translated two books of the *Aeneid* (2 and 4) into this "straunge meter." Surrey knew Gavin Douglas's Scottish tr. (written, in rhymed couplets, ca. 1513; pub. 1553); as much as 40% of Surrey's diction is taken directly from Douglas. But the lexical borrowing is localized; in general, Douglas is prolix, Surrey laconic—even more so than Virgil. More significantly, Surrey develops—in part, perhaps, to offset absent rhyme—an extensive network of sound patterning.

Surrey also surely knew Luigi Alamanni's *Rime toscane* (pub. 1532) and other famous It. works in *versi sciolti* (i.e. *versi sciolti da rima* "verse freed from rhyme") such as Trissino's tragedy *Sophonisba* (1515) and epic *Italia liberata dai Goti* (pub. 1547), or Liburnio's 1534 tr. of Virgil or the 1539 tr. by the de'Medici circle. But the It. *versi* "freed" from rhyme are hendecasyllables, technically of 11 syllables but permissibly also of 9 or 10, and with only one required stress in each hemistich, whereas Surrey clearly intended to produce, in Eng., an alternating (iambic) rhythm in a strict 10-syllable line. Behind Surrey of course stand Chaucer and Wyatt: Chaucer had mastered the Eng. decasyllable over a century earlier, but the

rhythm of his verse was lost to Eng. ears in the 15th c. by virtue of phonological changes. From Wyatt's sonnets and sonnet trs. Surrey could have learned a great deal, but there is much irregularity in Wyatt, and all of his verse is rhymed. The precise nature of Surrey's metrical accomplishment is still not entirely understood, and the two features of the b.-v. line—the metrical structure and the forgoing of rhyme—are quite different issues. If Surrey sought to appropriate features of It. verseform to Eng., they would of course be naturalized in Eng. rhythms. It remains to be shown precisely how the rhythms in Surrey's lines derived from ones native to the It. *endecasillabo*.

As b. v. developed in Eng., generic considerations also became important. Clearly verse without rhyme is esp. suited to long works, permitting an idea to be expressed at whatever length is appropriate, not imposing on the lang. a repeated structure of couplet or stanza which would tend to produce conformity in syntactic structures as well. The omission of rhyme promoted continuity, sustained articulation, enjambment, and relatively natural word order. It permitted, on the other hand, the deliberate use of syntactic inversion, an effect comparable to the hyperbaton (q.v.) of Cl. verse. Relatively natural word order made b. v. a fitting vehicle for drama; inversion, suspension, and related stylistic devices suited it to epic.

"Blank" as used of verse (the earliest *OED* citation is by Nashe in 1589) suggests a mere absence (of rhyme), not that liberation from a restrictive requirement implied in the It. term. Eng. defenders of b. v. repeatedly asserted that rhyme acts on poets as a "constraint to express many things otherwise, and for the most part worse than else they would have exprest them" (Milton). That rhyme has its virtues and beauties needs no argument here. That it blocks off more avenues of thought than it opens, for the skilled poet, is doubtful. Rhyme does, however, tend to delimit—even to define—metrical structures; it has its clearest effect when the syllables it connects occur at the ends also of syntactic structures, so that meter and syntax reinforce one another. One associates rhyme with symmetries and closures. Omission of rhyme, by contrast, encourages the use of syntactic structures greater and more various than could be contained strictly within the line, and so makes possible an amplitude of discourse, a natural-seeming multiformity, not easily available to rhymed verse.

Though b. v. appeared in Eng. first in (translated) epic, attempts at Eng. heroic verse after Surrey use, as Hardison remarks, "almost every form *but* b. v." The form achieved its first great flowering in drama. After Surrey the dramatic and nondramatic varieties have significantly different histories, suggesting that they differ in nature more than critics once thought (see Wright; Hardison).

In nondramatic verse, the influence of Petrarchism, manifested in the vogue for sonneteering (see SONNET SEQUENCE), ensured that rhyme held sway in Eng. nondramatic verse up to Milton. The heavy editorial regularization by the editor of Tottel's Miscellany (1557) set the trend up to Sidney, who showed that metrical correctness and natural expressiveness were not mutually exclusive (see Thompson), a demonstration extended even further by Donne—but again, in rhyme.

The first Eng. dramatic b. v., Thomas Norton's in the first three acts of *Gorboduc* (1561), is smooth but heavily end-stopped, giving the impression of contrivance, of a diction shaped—and often padded out—to fit the meter. Sackville's verse in the last two acts of the play is more alive, more like the verse of Surrey's *Aeneid*. But the artificial regularity of Norton's verse came to characterize Eng. dramatic poetry until Marlowe came fully into his powers. Marlowe showed what rhetorical and tonal effects b. v. was capable of; his early play *Dido* echoes lines from Surrey's *Aeneid*, and Shakespeare's early works show what he learned from Marlowe. T.V.F.B.; E.R.W.

Shakespeare. Shakespeare's b. v., the major verseform of his plays throughout his career, is marked by several features, some of them shared with or derived from earlier Eng. poets (e.g. Lydgate, Sidney, and Marlowe) but developed with unprecedented coherence. (1) B. v. is always mixed with other metrical modes (e.g. rhymed verse, songs) and (except for *Richard II* and *King John*) with prose; two plays offer more rhymed verse than b. (*Love's Labour's Lost* and *A Midsummer Night's Dream*), and seven plays (two histories and five middle-period comedies) are largely or predominantly written in prose. Shifts within a scene from one mode to another are often subtle, gradual, and hard to hear; reserving different metrical registers for different social classes, however, can be heard, and helps to identify sets of characters—as in *MND*. (2) Resourceful use of common Elizabethan conventions of metrical patterning and esp. of metrical variation gives many individual lines great flexibility, variety, melody, and speechlike force. (3) Frequent use of lines deviant in length or pattern (short, long, headless, broken-backed, and epic-caesural) extends the potentialities of expressive variation beyond what was commonly available to Ren. writers of stanzaic verse. (4) Shrewdly deployed syllabic ambiguity, esp. by devices of compression, makes many lines seem packed. (5) Lines become increasingly enjambed: sentences run from midline to midline, and even a speech or a scene may end in midline. Conversely, metrically regular lines may comprise several short phrases or sentences and may be shared by characters ("split lines," q.v.). In the theater, consistently enjambed b. v., unlike Marlowe's endstopped "mighty" line, sounds more like speech but also tests the audience's awareness of the meter. Esp. in the later, more troubled plays, the audience, like the characters it is scrutinizing, follows an uncertain path between comprehension and bafflement.

Deploying all these means in speeches, scenes, and plays that differently approach, complicate, and resolve a varied array of characters and issues, the b. v. Shakespeare uses to negotiate the sometimes tempestuous marriage of phrase and line is fully equal in suppleness and energy to the plays' expressive syntax, imagery, and wordplay. Besides carrying the characters' emotional utterances and conveying (with appropriate intensity) their complex states of mind, this b. v. may figure, through its rich dialectic of pattern and departure-from-pattern, a continuing tension between authority and event, model and story, the measured structures of cosmic order and the wayward motions of erratic individual characters. G.T.W.

After Shakespeare, and esp. after Donne's (rhymed) *Satyres*, the dramatic b. v. line grew looser in form. Feminine endings, infrequent in all early b. v., became common; in Fletcher they often carry verbal stress. Later, true feminine endings become common even in nondramatic rhymed verse. Milton uses feminine endings in *Paradise Lost* and *Paradise Regained* only with great restraint. They occur rather seldom in the early books of *PL*, but much more frequently after Eve's and Adam's fall, and are thereby appropriated to the speech of fallen mortals.

In dramatic b. v., extrametrical syllables begin to appear within the line—first, nearly always following a strong stress and at the end of a phrase or clause at the caesura; later, elsewhere within the line. In some late Jacobean and Caroline drama, the line has become so flexible that at times the five points of stress seem to become phrase centers, each capable of carrying with it unstressed syllables required by the sense rather than by the metrical count of syllables. But such lines were to call down the wrath of later critics; and the closing of the theaters in 1642 saw also the end of a brilliant—if almost too daring—period of experimentation with the structure of the dramatic b. v. line. When verse drama was written again, after the Restoration, and under the impetus of the newly popular Fr. model, the line was once again a strict pentameter, but now, and for the next century, rhymed. In any event, after the closing of the theaters, b. v. was never to be of major importance in the drama again. The attempt to renew verse drama (incl. b.-v.) in the 20th c.—as in Eliot and Yeats—never won either popular or critical acclaim.

It was *Milton* who returned b. v. to its earliest use as a vehicle of epic, and to strict, though complex, metrical order; his influence was so powerful that the form bore his impress up to the 20th c. He did, of course, write b.-v. drama—*A Mask* (*Comus*) and *Samson Agonistes*, the first much influenced by Shakespeare. In both, b. v., though it is unquestionably the central form, is intermingled with other forms—lyric in *A Mask*, choric in *SA*—to such effect that we think of both works as being in mixed meters. In *SA* Milton is trying to produce the effect of Gr. tragedy (though with a biblical subject); but the b. v. is similar enough to Milton's nondramatic b. v. that discussion of the two may be merged.

We know that Milton was profoundly familiar with the *Aeneid* in Lat., perhaps in It. tr., and in some Eng. versions; he was equally familiar with It. epic and romance. As a theorist of form he wanted to make of b. v. in Eng. the instrument which the Humanist poets had been attempting to forge since Trissino. For his subject in *PL* he needed a dense, packed line, as various as possible in movement within the limits set by broadly understood but absolute metricality; at the same time he needed a syntax complex and elaborate enough to overflow line form, to subordinate it to larger forms of thought, appropriately varied in the scope of their articulation. He needed "the sense variously drawn out from one line into another." The idea was not new, yet no one had managed enjambment in Eng. as skillfully as Milton learned to manage it. Before him—except in Shakespeare's later plays—the congruence of syntax with line form had been too nearly predictable.

Even more than Shakespeare's, Milton's b. v. line differs from his rhymed pentameter—not in basic meter, but in management of rhythm, and not only through enjambment, but also through constant variation of the placement of pause within the line. In the course of enjambing lines, Milton writes long periodic sentences, making liberal use of inversion, parenthesis, and other delaying and complicating devices. Whether or not Milton knew Surrey's *Aeneid*, there are real similarities between the two men's work. Like Surrey, Milton at times uses Italianate stress sequences that disturb the duple rhythm and in some instances all but break the meter. Also important, given his refusal of rhyme, is a more extensive (if irregular) deployment of the varied resources of sound patterning. Numerous forms of internal rhyme occur, as well as final assonance and half-rhyme; and whole passages woven together by patterns of alliteration, assonance, and half-rhyme.

Milton's influence on subsequent nondramatic b. v. in Eng. was, as Havens showed, enormous. Yet it is a fact that all b. v. after Milton became essentially a romantic form—no longer epic, and no longer dramatic, but the vehicle of rumination and recollection. The line of descent leads through Wordsworth (*Michael*; *The Prelude*; *The Excursion*) to Tennyson (*Ulysses*; *Tithonus*; *Idylls of the King*) and Browning (*The Ring and the Book*). By the mid 18th c., the forces of metrical regularity had begun to weaken, and for the first time genuine extra syllables begin to appear in the nondramatic line which cannot be removed by elision, producing triple rhythm. The Eng. pentameter both dramatic and nondramatic between 1540 and 1780 all but disallowed triple rhythms; strict count of syllables was deemed central to line structure. Real, irreducible triple rhythms prior to 1780 were associated with music; they occur fairly com-

monly in song lyrics and ballads, but not in accentual-syllabic meter. After 1780, however, triple rhythms gradually invade poems in duple rhythm. In part this is because the romantics were devoted to the work of their 16th- and 17th-c. predecessors, and often used the diction of Spenser, Shakespeare, and Milton, though without entirely understanding earlier metrical conventions. There was also a revival of interest in the ballads, which varied the basic duple rhythms with irreducible triple rhythms, a practice soon evident in Blake's *Songs* and even occasionally in Wordsworth.

The rhythmic and syntactic flexibility of the line once called "heroic" had come undeniably closer to that of speech. Conservative trends would still appear: Tennyson uses many fewer triple rhythms than Browning, and most of those he does use seem relaxations of traditional metrical compressions; his choices of rhythm and diction together give an impression of formality, of dignity. Browning, on the other hand, in his dramatic monologues (see MONOLOGUE) achieves an astonishing range of effects, each suited to the persona of the character he is presenting. In the 20th c., comparable shadings of metrical effect can be found in the b. v. of Robinson, Frost, Eliot, Stevens, Jeffers, and Lowell. Lines of b. v. may still be used to resolve deliberately inchoate materials into form; and they may still arise naturally and as though involuntarily among lines of differing rhythm. But the advent of free verse sounded the death-knell of this meter which was once and for long a powerful, flexible, and subtle form, the most prestigious and successful modern rival to the greatest meter of antiquity. E.R.W.; T.V.F.B.

S. Johnson, *Rambler*, nos. 86–96 (1751); Schipper; J. B. Mayor, *Chapters on Eng. Metre*, 2d ed. (1901); R. Bridges, *Milton's Pros.* (1921); R. D. Havens, *The Influence of Milton on Eng. Poetry* (1922); A. Oras, *B. V. and Chronology in Milton* (1966); R. Beum, "So Much Gravity and Ease," *Lang. and Style in Milton*, ed. R. D. Emma and J. T. Shawcross (1967); R. Fowler, "Three B. V. Textures," *The Langs. of Lit.* (1971); E. R. Weismiller, "B. V.," "Versif.," and J. T. Shawcross, "Controversy over B. V.," *A Milton Encyc.*, ed. W. B. Hunter, Jr., 8 v. (1978–80); Brogan, 356 ff.; H. Suhamy, *Le Vers de Shakespeare* (1984); M. Tarlinskaja, *Shakespeare's Verse* (1987); G. T. Wright, *Shakespeare's Metrical Art* (1988); J. Thompson, *The Founding of Eng. Metre*, 2d ed. (1989); O. B. Hardison, Jr., *Pros. and Purpose in the Eng. Ren.* (1989).

E.R.W.; T.V.F.B.; G.T.W.

II. IN OTHER LANGUAGES. A. *German.* Outside of England, whence it derived, b. v. celebrated its greatest triumph in Germany. Although 17th-c. Ger. poets knew Shakespeare and Milton, their early attempts at b. v. were inconsequential. After the dominance of the alexandrine (q.v.) and its rhyme had passed, however, b. v. was reprised in the second half of the 18th c. with much greater success. Wieland's tragedy *Lady Johanna Gray* (1758) is the first instance, but Lessing's *Nathan der Weise* (1779) established the most influential precedent. Goethe followed suit by rewriting his earlier prose version of *Iphigenie auf Tauris* (1787) and *Torquato Tasso* (1790) in b. v. Similarly, Schiller rewrote *Don Carlos* (1787), and Kleist used b. v. exclusively for his dramas. A. W. Schlegel's tr. of Shakespeare (1797–1810), continued by Tieck, did much to solidify the convention, which had gained prestige by its association with Weimar Classicism. Throughout the 19th c., b. v. remained the standard meter for Ger. verse drama, as exemplified by the works of Grillparzer, Grabbe, Hebbel, and even Hauptmann (in contrast to his naturalist prose dramas); and b. v. appeared with a final flourish in the lyric dramas of Hofmannsthal.

This broad spectrum points up the main advantage of b. v., its adaptability to diverse styles. At one pole stands Goethe, who labored to perfect a lyrical form of b. v. with end-stopped lines, balanced caesuras, and smooth rhythmic phrasing; at the opposite pole stands Lessing, whose heavily enjambed lines are often divided into short phrases and shared by different speakers, making the verse speechlike. Kleist too overrides the line end; his strong contrasts and rapid shifts impel the verse with a driving thrust. Schiller at first assumes an intermediate position, later developing in the direction of Goethe. But the great flexibility of b. v. has also been seen as a disadvantage, for the form itself lacks a distinctive profile and has been thought monotonous at times. The elder Goethe maintained that b. v. "reduced poetry to prose," which recalls Dr. Johnson on Milton, a sentiment echoed after Goethe by other poets and critics such as Heusler. But b. v. can accommodate many expressive possibilities, much depending on precisely how the metrical pattern is realized.

The relative lack of contour to b. v. in Ger. may explain why it was not common in traditional Ger. lyric (though Schiller's *Das verschleierte Bild zu Sais* is a notable exception). Still, b. v. was picked up by the early modern poets, such as Meyer, George, Hofmannsthal, and Rilke, who gave it a distinctly formal tone, thus distancing themselves from the rhymed folk poetry of the 19th c. Two of the 10 elegies in Rilke's *Duineser Elegien* are in b. v. In contrast to the liberties allowed in the dramatic form, b. v. in the lyric is very strict, similar to its counterpart, rhymed iambic pentameter, which like the sonnet form derived from the It. *endecasillabo,* was more common than b. v. throughout the lyric trad. In the 20th c., Ger. lyric b. v. was largely supplanted by *freie Rhythmen* and free verse, as dramatic b. v. was by prose, though a ghostly form of traditional meter often remains in the background—just behind the arras, as T. S. Eliot said.

B.B.

B. *Scandinavian.* The Scandinavian countries received b. v. from Italy, England, and Germany, the chief models being the dramatic verse of Goethe and Schiller. With *Balders Død* (1773),

Johannes Ewald introduced hendecasyllabic b. v. in Denmark, evidently of It. extraction. Preferring the decasyllable with masculine endings, Oehlenschläger created a medium capable of a wider range of dramatic effects; his treatment of b. v. in his numerous historical tragedies is much like that of the later Schiller. Oehlenschläger's followers in the drama, Ingemann and Hauch, naturally adhered to the verseform of their master; more individual is the b. v. used later by Paludan-Müller (in his dramatic poems on mythological subjects) and by Rørdam. In Swedish poetry, b. v. was introduced with Kellgren's narrative fragment *Sigvarth och Hilma* (1788) in the It. form acquired from Ewald. Excellent Eng. b. v. appeared as early as 1796 in a few scenes of a projected historical play by the Finn Franzén; but not until 1862 did Wecksell, another Finn, produce the only significant b.-v. drama in Swedish, *Daniel Hjort.* Unlike the situation in Denmark and Norway, where b. v. appears infrequently in narrative, didactic, satirical, and reflective poetry, in Sweden b. v. has a long and still living trad. in these genres, with contributions by such poets as Tegnér, Stagnelius, Sjöberg, Malmberg, Edfelt, and Ekelöf. The form of Swedish b. v. is generally conservative, not unlike that of Goethe. Norwegian poets started using b. v. about the same time, but the first significant works are several farces and dramas beginning in 1827 by Wergeland, whose metrical usage, modeled on Shakespeare, is very free. Neither Andreas Munch's imitations of Oehlenschläger (beginning in 1837) nor Ibsen's *Catilina* (1850) displays a distinctive form of b. v. The rhymed verse intermittently used in *Catilina* is far superior to the b. v., which may explain Ibsen's subsequent preference for rhyme. Bjørnson, on the other hand, composed excellent b. v. in his saga dramas in a free form which points to Shakespeare and Schiller as the chief models.

C. *Slavic.* In Russia, b. v. appeared with Žukovskij's tr. of Schiller's *Die Jungfrau von Orleans* (1817–21). It was subsequently used by Puškin in *Boris Godunov* (1825; pub. 1831) and his "Little Tragedies," and by Mey, Ostrovsky, and A. K. Tolstoy in their historical dramas. Tolstoy's *Czar Fyodor Ioannovich* (1868) has been a popular success up to recent years. B. v. also appears in narrative and reflective poetry. In Russia, as elsewhere, b. v. varies within a certain range, from the conservative line-structured form with constant caesura of *Boris Godunov* to the more loosely articulated verse of "little tragedies" like *Mozart and Salieri* and *The Covetous Knight.* In Poland, the hendecasyllabic b. v. of It. origin used by Kochanowski in his tragedy *Odprawa posłów greckich* (The Dismissal of the Gr. Envoys, 1578) failed to inaugurate a trad. Rhymed verse became standard for Polish drama, and neither J. Korzeniowski's many b.-v. plays (beginning in 1820) nor his persistent theoretical advocacy of the medium greatly modified the situation. Among individual works in b. v. may be mentioned *Lilla Weneda* (1840), one of Słowacki's best tragedies, Norwid's comedy *Miłośc* (Chaste Love at the Bathing Beach), and J. Kraszewski's epic trilogy *Anafielas* (1840). S.LY.

F. Zarncke, *Über den fünffüssigen Jambus . . . durch Lessing, Schiller, und Goethe* (1865); A. Sauer, *Über den fünffüssigen Jambus vor Lessings* Nathan (1878); L. Hettich, *Der fünffüssigen Jambus in den Dramen Goethes* (1913); O. Sylwan, *Den svenska versen från 1600-talets början*, 3 v. (1925–34), *Svensk verskonst från Wivallius till Karlfeldt* (1934); A. Heusler, *Deutsche Versgesch.* (1925–29), v. 3; K. Taranovski, *Ruski dvodelni ritmoci*, 2 v. (1953); B. Unbegaun, *Rus. Versif.* (1956); R. Haller, "Studie über den deutschen Blankvers," *DVLG* 31 (1957); P. Habermann, "Blankvers," *Reallexikon*; R. Bräuer, *Tonbewegung und Erscheinungsformen des sprachlichen Rhythmus; Profile des deutschen Blankverses* (1964); L. Schädle, *Der frühe deutsche Blankvers* (1972); B. Bjorklund, *A Study in Comp. Pros.* (1978); Brogan; Scherr. B.B.; S.LY.

BLUES are more than sad songs about lost love and loneliness. They exist as instrumental and vocal music, as psychological state, as lifestyle, and as philosophical stance. As folk or popular poetry, b. lyrics are distinguished by graphic imagery and themes drawn from a wide range of group and personal experiences. Thematically, they address suffering, struggle, and sexuality. They confront life head-on, with an indomitable will expressed in sarcasm, wit, and "signifying" (q.v.) humor, and they move on an expressive spectrum ranging from the grim and the bawdy to the tragic. These features and concerns are rooted in the Black Am. oral trad.—in secular rhymes and "toasts," in ritual verbal combat (the "dozens," q.v.) and ironic teasing in tales and jokes. Their dynamic pattern of "confrontation, improvisation, affirmation, and celebration" (Murray) evokes the history and texture of the Black Am. presence.

B. must be seen then in their dynamic contextuality with Black oral trad. and Am. popular culture, which they adopt and transform through critique and improvisation. Thus there are b. versions of ballads like "John Henry," as well as allusions to standard hymns, spirituals, popular songs, and topical events. In effect, around the turn of the 20th c. the b. emerged out of the work songs and field hollers as a durable, elastic artform that could efficiently digest and render the contradictory elements in Am. life. In the classic b. form, the first line makes a statement which is repeated with a variation in the second. The third line provides an ironic contrast or extension. Thus all kinds of combinations are possible, incl. call and response patterns between the voice and the accompanying instrument (guitar, harmonica) or band.

Early b. may be classified as city or country, with the former being recorded first. They may also be called classic or rural. Of the first great classic singers the best known were Gertrude "Ma" Rainey and Bessie Smith. Among the rural singers

were Charley Patton, Ed "Son" House, and the legendary Robert Johnson.

Formally, there are many different kinds of b., but they may be usefully discussed within the country/city framework. Both forms share features like the three-line song lyric and the so-called blue notes, in which the third and seventh notes of the scale are flattened. These intervals give rise to haunting melodies and harmonies and add a distinctive coloration to any song regardless of its verbal content. Thematically, the classic b. were written for entertainment but retain strong connections with earlier rural styles. The personal lyrical voice from which the form arose remained a feature of the classic style. In fact, the two styles developed simultaneously during the first two decades of this century, nourishing and strengthening one another. In contrast to the classic style with its emphasis on conflict and sexuality, the country b. responded to everything, and often retains elements of other songforms like the ballad and popular song.

Although, taken as a whole, the b. are tragicomic in both a thematic and structural sense, they often sustain a tragic tone, as in Robert Johnson's "Hellhound on My Trail," with its image of a man driven by fate stumbling through a b. storm; in this depiction of Native Son as Orestes, Johnson reveals himself as one of the most poetic of b. singers.

The b. entered the literary canon as "folk poetry" in anthologies such as *The Negro Caravan*, and as "literary poetry" or poems in the folk manner in collections such as Langston Hughes' *The Weary Blues*, where the title poem is not, strictly speaking, b. at all but a lyric framework with narrative elements. The poem dramatizes the b. experience as performance, as social ritual, and as existential experience. Explicit in the poem is the relationship of performer to audience, to his art, and to his experience of the b. Questions of this sort were raised fairly early by b. commentators such as Hughes, Sterling A. Brown, and others. Recent scholarship demonstrates the usefulness of the b. in understanding the unique qualities of Am., esp. Black Am., life and art. Thus the b. are relevant to presentations and analyses of dance, fiction, cinema, dress, and other modes—from Ellison's *Invisible Man* to the paintings of Romare Bearden and Jacob Lawrence to the novels of Gayle Jones and Alice Walker. Indeed, Houston A. Baker has projected a challenging theory of Black lit. based on his understanding of the b., while elsewhere discussions of the Black Aesthetic are being subsumed in the B. Aesthetic—P. Oliver, *The Meaning of the B.* (1963); A. Murray, *Stomping the B.* (1976); J. T. Titon, *Early Downhome B.* (1977); M. Taft, *B. Lyric Poetry*, 2 v. (1983, 1988)—anthol. and concordance; H. A. Baker, Jr., *B., Ideology and Afro-Am. Lit.* (1984); M. L. Hart et al., *The B.: A Bibl. Guide* (1989). S.E.H.

C

CALLIGRAMME. The c. derives its name from Guillaume Apollinaire's *Calligrammes* (1918) and represents one type of visual poetry (q.v.). Its antecedents incl. the ancient forms of pattern poetry (q.v.), Gr. *technopaigneia* and Lat. *carmina figurata*, as well as medieval religious verse, Ren. emblems (q.v.), and, latterly, Mallarmé's *Un Coup de dés*. It thus occupies an intermediary position between the older forms of pattern poetry and the more recent concrete poetry (q.v.). In the context of the early 20th century, the c. reflects the influence of cubist painting, It. futurism, and the ideogram (q.v.; the first examples were called *idéogrammes lyriques*) and thus inaugurates a period of radical experimentation by fusing art and poetry (see VISUAL ARTS AND POETRY) to form a complex interdisciplinary genre or intermedium.

Like other forms which dissolve the traditional barriers between visual and verbal, the c. mediates between the two fundamental modes of human perception, sight and sound. Whereas the futurists preferred abstract visual poetry, Apollinaire's compositions are largely figurative, portraying objects such as hearts, fountains, mandolins, and the Eiffel Tower, as well as animals, plants, and people. Still, a significant number are abstract, making generalization difficult, though on the whole the abstract cs. are later works. Of the 133 works arranged in realistic shapes, the visual image duplicates a verbal image in about half the cases. Some of the poems are autonomous, but most are grouped together to form still-lifes, landscapes, and portraits, an important innovation in the visual poetry trad. which complicates the play of visual and verbal signs. As Apollinaire himself remarks, "the relations between the juxtaposed figures in one of my poems are just as expressive as the words that compose it." Although some of the cs. are solid compositions, most feature an outlined form that emphasizes the genre's connection with drawing. Unlike most of his predecessors, however, Apollinaire weds visual poetry to lyric theme. Some of the cs. are in verse, but many others are in prose. One group, composed in 1917, continues Apollinaire's earlier experiments with *poésie critique* and functions as art crit. Although its

critical reception has been mixed, the c. was widely imitated in France, Italy, Spain (esp. Catalonia), Mexico, the United States, and elsewhere.

A.-M. Bassy, "Les Schématogrammes d'Apollinaire," *Scolies* 3–4 (1974–75); G. Apollinaire, *Cs.*, ed. and tr. A. H. Greet and S. I. Lockerbie (1980); C. Debon, *Guillaume Apollinaire après Alcools, I* (1981); *Que Vlo-Ve?*, 1st ser., 29–30 (1981)—an important colloquium; D. Seaman, *Concrete Poetry in France* (1981); J. G. Lapacherie, "Écriture et lecture du c.," *Poétique* 50 (1982); G. H. F. Longrée, *L'Expérience idéo-calligrammatique d'Apollinaire* (1984); W. Bohn, *The Aesthetics of Visual Poetry, 1914–1928* (1986); Hollier, 842 ff. W.B.

CANON (Ger. *kanon*, "rule" or "standard"; also "list" or "catalog" [see Childs 24–25]). The two etymological senses of the term c. are conflated in its most common use: an authoritative list of books forming the Judeo-Christian Bible—authoritative here meaning inspired by God or legitimized by the believing community. In secular lit. crit., c. has three senses: (a) rules of crit.; (b) a list of works by a single author; and (c) a list of texts believed to be culturally central or "classic." Whatever the practical difficulties in determining specific rules or authorship, usage in the first two senses is clear, e.g. "the cs. of taste" or "the c. of Shakespeare's works." The third sense—the subject of the remainder of this entry—is more problematic, since it supposes that secular texts can be authoritative in a manner analogous to sacred texts. Authority here is located not in a single text or author but in a particular group of texts whose formulators see that group as esp. valuable to its culture. Although the word c. is not used in the Bible itself, the formation and continuing existence of the various biblical cs. are instructive for studying secular cs. since there the issue of authority is necessarily foregrounded.

The word c. was first applied to the Bible in the 4th c. A.D. The first early medieval non-biblical c. was patristic, followed by the medieval *auctores*, lists for school study of authors both Christian and pagan (Hardison). Except for the inevitable Christian-pagan tension—often alleviated by allegorical accommodation of pagan to Christian—this Lat. c. remained relatively uncontroversial until the rise of the vernacular langs. in the Ren., beginning in Italy, motivated c. change (Curtius, Auerbach, Weinberg). Subsequent 17th- and 18th-century c. formulations from national and comparative perspectives alike usually linked cs. with cultural values explicitly (Dryden, S. Johnson), as did Matthew Arnold and T. S. Eliot in late 19th- and early 20th-c. England (see also Sainte-Beuve).

If we exclude accident (e.g. the fact that the destruction of the Alexandrian Library in effect helped canonize works in the surviving mss.), the central factors in c. formation are ideological, religious, and aesthetic values in particular and complex combinations (Kermode 1985, Mitchell, van Hallberg). These factors may be intertwined but are not always consciously or explicitly acknowledged to be so. As a modern example, much New Critical emphasis on the "tradition" seeks to ground it in exclusively aesthetic value. Brooks is "concerned with the poems as poems"; poems "in the main stream of the trad.," and aesthetically most valuable, "giv[e] us an insight which preserves the unity of experience and which . . . triumphs over the apparently contradictory and conflicting elements of experience by unifying them into a new pattern" (Brooks 215, 214, 192). But "unity of experience" is more than an aesthetic value.

Biblical and secular canonical crit. alike study not only c. formation but the subsequent life of a c. already formed. Secular crit. that takes for granted a c. practices, whether self-consciously or not, a hermeneutics of accommodation, mediating between the classic text and the modern reader (Kermode 1975). And late 20th-c. crit., despite its theoretical self-consciousness, often takes cs. for granted. Marxist crit. mostly reads canonical texts, as does deconstruction, although the latter draws from more diverse psychoanalytic, philosophical, and critical cs. (e.g. Freud, Nietzsche, Derrida, de Man) as often as literary ones—thus opening the c. horizontally, to include nonliterary texts, but not vertically. Reader-response crit. is most often interested in educated readers responding to canonical texts (cf. Sammons). Implicit challenges to Western literary cs. have come from structuralism and semiotics, disciplines which consider any literary text—if not any cultural product at all—appropriate for study. But the most explicit challenges to Western literary cs. have come from Fr. and Am. feminist crit., curricular crit. (Graff), working-class writing in England, and third-world writing, which often view existing cs. as ideologically oppressive in effect if not always in motive (Eagleton).

There have probably always been cs. (cf. Barr), formalized or not, and they have always had—or been capable of being given, which is not the same—ideological relevance. Such relevance however is not easily predictable. How and in what contexts a c. is read, taught, or learned may or may not reinforce a particular ideological rationale for constituting that c. For example, the belief that classics ultimately express harmony rather than discord, and thus serve or at least complement empire (e.g. Roman, British), does not explain why other texts expressing harmony are not canonized, or why such classics cannot also be read as undermining "unity of experience" and expressing discord rather than harmony (W. Johnson). Likewise, the relation of ideology to life cannot be taken for granted (Abercrombie). No generalization about these relationships has been adequate to cover all instances of c. formation and effect.

J. Dryden, Preface to *Fables Ancient and Modern* (1700); S. Johnson, *Lives of the Poets* (1781); C.

Sainte-Beuve, "What is a Classic?" (1850), "On the Literary Trad." (1858); M. Arnold, *Essays in Crit.* (1865, 1888); C. Brooks, *Mod. Poetry and the Trad.* (1939); Brooks; Eliot, *Essays*, and *On Poetry and Poets* (1956); Curtius, 247–72; E. Auerbach, *Literary Lang. and Its Public in Late Lat. Antiquity and in the Middle Ages* (1958; tr. 1965); Weinberg; W. Johnson, "The Problem of the Counter-classical Sensibility and Its Critics," *Calif. Studies in Cl. Antiquity* 3 (1970); *Med. Lit. Crit.*, ed. O. B. Hardison, Jr., et al. (1974); F. Kermode, *The Classic* (1975), *Forms of Attention* (1985); J. Sammons, *Literary Sociology and Practical Crit.* (1977); N. Abercrombie et al., *The Dominant Ideology Thesis* (1980); Fowler; J. Barr, *Holy Scripture* (1983); T. Eagleton, *Literary Theory* (1983); *The Politics of Interp.*, ed. W. J. T. Mitchell (1983); B. Childs, *The New Testament as C.* (1984); *Cs.*, ed. R. von Hallberg (1984); J. Weber, *Kanon und Methode* (1986); G. Graff, *Professing Lit.* (1987); J. Guillory, "C.," *Critical Terms for Literary Study*, ed. F. Lentricchia and T. McLaughlin (1990); W. V. Harris, "Canonicity," *PMLA* 106 (1991); C. Kaplan and E. C. Rose, *The C. and the Common Reader* (1991); C. Altieri, *Cs. and Consequences* (1991); P. Lauter, *Cs. and Context* (1991).

J.C.S.

CANZONE. In *De vulgari eloquentia* Dante defines the *c.* as the most excellent It. verseform, the one which is the worthy vehicle for those "tragic" compositions which treat the three noblest subjects: martial valor, love, and moral virtue. Noting the intimate link between poetry and music (the term *canzone* comes from *cantio,* a song), Dante remarks that "although all that we put in verse is a 'song' (*cantio*), only *canzoni* have been given this name" (2.3.4). As a result, the term *c.* has come to be applied to quite a number of verseforms with differing metrical patterns. Among the better known types is the *c. epico-lirica,* whose center of diffusion was originally the Gallo-It. dialect area: it belongs to the Celtic substratum and is akin to compositions of the same genre in France and Catalonia. More indigenous to the It. soil is the *c. a ballo* or *ballata* and other popular compositions such as the *frottola* and *barzelletta,* the *canto carnascialesco,* and the *lauda sacra.* At various times these types have been used by the *poeti d'arte,* but the type exclusively employed for refined artistic expression is the so-called *c. petrarchesca,* which bears strong traces of Occitan influence. The It. *strambotto* and *ballata* and Ger. Minnesang are also said to have conditioned its architectonic structure.

In Italy the first forms of the *c.* are found among the poets of the Sicilian School. It is employed extensively by Guittone d'Arezzo and his followers and by the poets of the *dolce stil nuovo,* but acquires fixed patterns and perfection in Petrarch's *Canzoniere,* hence the qualifying adjective *petrarchesca.* Its greatest vogue in Italy occurred during the age of Petrarchism, which lasted until the death of Torquato Tasso (1595). While the Petrarchan *c.* was employed in England by William Drummond of Hawthornden and in Germany by A. W. von Schlegel and other Ger. romantic poets, Spain and Portugal were really the only countries outside Italy where it was used to a considerable extent.

One structural feature of the *c.* proved to be of lasting importance for the subsequent devel. of a number of lyric genres in the European vernaculars, esp. Occitan, OF, and MHG. This is the organization of the poem into a structure that is simultaneously tripartite and bipartite. In the terminology Dante gives in *De vulgari eloquentia* (2.10), the poem is divided into a *fronte* (Lat. *frons*) and a *sirma* (Lat. *cauda*)—head and tail—but the *frons* itself is further subdivided into two *piedi* (*pedes,* feet) which are metrically and musically identical, making the essential structure AA/B. Subsequent Ger. terms for these partitions are the *Aufgesang* divided into *Stollen* and the *Abgesang.* This structure appears in analogues as late Chaucerian *rhyme royal* (q.v.), the Venus and Adonis stanza, and the Shakespearean sonnet, where the octave, unlike the sestet, neatly divides into quatrains. (In some cases the *sirma* may be divided into two identical parts called *volte* [Lat. *versus*].) In the It. c. there is, further, usually a single *commiato* at the close of the poem serving as a valediction. Stanzaic length is indeterminate, varying from a minimum of 7 to a maximum of 20 verses. The lines are normally hendecasyllables with some admixture of heptasyllables, and this mix of 11s and 7s became distinctive of the form. After Tasso, under the strong influence of the Fr. *Pléiade,* this type was supplanted by new forms labeled *canzoni*—the Pindaric and Anacreontic odes, whence the inexact use of *ode* (q.v.) as a translation for *c.* Chiabrera played a leading role in their diffusion. He also revived the *canzonetta* originally employed by the poets of the Sicilian School. This became the favorite type used by Metastasio and the Arcadian school. Toward the close of the 17th c., Alessandro Guidi acclimated the *c. libera,* which reached its highest devel. at the hands of Leopardi.

P. E. Guarnerio, *Manuale di versificazione italiana* (1893); O. Floeck, *Die Kanzone in der deutschen Dichtung* (1910); F. Flamini, *Notizia storica dei versi e metri italiani* (1919); R. Murari, *Ritmica e metrica razionale ital.* (1927); V. Pernicone, "Storia e svolgimento della metrica," *Problemi ed orientamenti critici di lingua e di letteratura italiana,* ed. A. Momigliano, v. 2 (1948); E. Segura Covarsí, *La canción petrarquista en la lírica española del Siglo de Oro* (1949); E. H. Wilkins, "The C. and the Minnesong," *The Invention of the Sonnet* (1959); I. L. Mumford, "The *C.* in 16th-C. Eng. Verse," *EM* 11 (1960); L. Galdi, "Les origines provencales de la métrique des canzoni de Petrarque," *Actes romanes* 74 (1966); M. Pazzaglia, *Il verso e l'arte della C. nel "De vulgari eloquentia"* (1967), *Teoria e analisi metrica* (1974); E. Köhler, "'Vers' und Kanzone," *GRLMA,* v. 2.1.3; Spongano; Wilkins; M. Fubini,

Metrica e poesia, 3d ed. (1975); D. Rieger, *Gattungen und Gattungsbezeichungen der Trobadorlyrik* (1976); J. A. Barber, "Rhyme Scheme Patterns in Petrarch's *Canzoniere*," *MLN* 92 (1977); C. Hunt, *"Lycidas" and the It. Critics* (1979); G. Gonfroy, "Le reflet de la canso dans *De vulgari eloquentia* et dans les *Leys d'amors*," *CCM* 25 (1982); F. P. Memmo, *Dizionario de metrica italiana* (1983), s.v. "C." et seq.; Elwert, *Italienische*, sect. 79. J.G.F.; C.K.

CAROL. The medieval carol was a lyric of distinct verseform which derived from and for long preserved a close connection with the dance. The Fr. form, the *carole*, was a dance-song similar in structure and movement to the early Eng. carol and probably its ancestor. But by 1550, dancing at religious festivals had fallen into desuetude, leaving the element of song. Hence, since the 16th c., the word has come to mean any festive religious song, whatever its metrical or stanzaic form, sung to a tune which in pace and melody follows secular musical traditions rather than those of hymnody. In America the carol is now almost invariably associated with Christmas; this is less true in England, where Easter carols are also widely sung. The Fr. *noël* (from Lat. *natalis*), a joyous song of the Nativity, is the counterpart of the Christmas carol; it has been an established song type since the 15th c.

Medieval carols are composed of uniform stanzas which are preceded by an initial refrain, the *burden*, usually a rhymed couplet, which is then repeated after every stanza. Often the stanza is a quatrain of three 4-stress lines in monorhyme followed by a shorter tag line rhyming with the burden. The following slightly modernized example is from a 15th.-c. carol of moral advice:

> Man, beware, beware, beware,
> And keep thee that thou have no care.
>
> Thy tongue is made of flesh and blood;
> Evil to speak it is not good;
> By Christ, that died upon the rood,
> So give us grace our tongues to spare.

More commonly the stanza rhymes *abab*; it may also be extended to 5, 6, or 8 lines and bound together by a variety of rhyme schemes. The burdens, too, are sometimes extended by one or two lines. Perhaps the single most notable variation from the norm is having the tag line identical with a refrain line, an integration that is common when a dance-song ceases to be danced.

In the round dances at which carols were originally performed, the stanza was probably sung by the leader of the dance and the burden by the chorus as they executed an accompanying dance figure. Modern children's games like "Now We Go 'Round the Mulberry Bush" and "A Tisket, A Tasket" represent corrupt descendants of the medieval round dances—then an adult pastime. From the violent denunciations of caroling that fulminated from medieval clerics, it is clear that caroling was regarded as a wicked pagan survival, though perhaps it was more than the songs that the clerics were concerned about: many of the older specimens are highly erotic and some very explicit. Doubtless the reason why caroling waxed strongest at Christmas and Easter was that these Christian festivals supplanted earlier winter and spring fertility revels. Of some 500 medieval carols extant in ms., about 200 deal directly or indirectly with Advent and the Nativity. But political, moral, and satirical carols are also abundant.

Carols were popular not only among the folk but in courtly circles as well; most of those extant show learned influences such as Lat. tags. Differing ms. versions of the same carol do not exhibit the kinds of variation that one would expect if carols had been orally transmitted in the manner of folksong.

With the Reformation, the medieval carol began to die out, mainly due to the more sober fashion of celebrating Christmas and other religious holidays. The formal carol was thus gradually replaced by festive songs learned from broadsides, chapbooks, and devotional songbooks. Some carols of this new kind, like *The Seven Joys of Mary, I Saw Three Ships, God Rest You Merry, Gentlemen*, and *The Virgin Unspotted*, are regularly described as "traditional," a term which means only that such pieces were long popular and are anonymous, not necessarily that they are folksongs or oral-formulaic.

The carols which supplanted the medieval carols were themselves beginning to wither in popularity when musical antiquaries like Gilbert Davies (*Some Ancient Christmas Carols*, 1822) and William Sandys (*Christmas Carols, Ancient and Mod.*, 1833) collected and revived them. J. M. Neale and Thomas Helmore in 1852 introduced the practice, since followed in most British and Am. carol books, of plundering Fr., Basque, Dutch, Sp., It., Ger., and Scandinavian collections for tunes to which Eng. words could be adapted. After 1870 the rural counties of England were scoured for folk carols, and the collectors' discoveries have been impressive. By being made available in excellent arrangements by Cecil J. Sharp, Vaughan Williams, and other folk-music experts, the folk carols have been artificially revitalized among educated people. In America, folk carols are comparatively rare; the only ones widely reported in this century, *The Seven Joys of Mary, Jesus Born in Bethlehem*, and *The Twelve Days of Christmas*, are all British imports.

A. E. Gillington, *Old Christmas Carols of the Southern Counties* (1910); C. J. Sharp, *Eng. Folk Carols* (1911); R. V. Williams, *Eight Traditional Carols* (1919); *Oxford Book of Carols*, ed. P. Dearmer (1928); P. Verrier, *Le Vers français*, v. 1 (1931), "La Plus Vieille Citation de carole," *Romania* 58 (1932); Sir R. R. Terry, *Two Hundred Folk Carols* (1933); *F. C. Brown Coll. of North Carolina Folklore*, v. 2 (1952), 199–212—abundant references to other sources on Am. folk carols; R. H. Robbins, "ME Carols as Processional Hymns," *SP* 56 (1959), "Greene's Revised Carols," *Review* 1 (1979); E. Routley, *The Eng. Carol*

(1959); R. L. Greene, "Carols in Tudor Drama," *Chaucer and ME Studies*, ed. B. Rowland (1971), *The Early Eng. Carols*, 2d ed. (1977)—the definitive collection of medieval carols, with superb and lengthy critical intro.; F. McKay, "The Survival of the Carol in the 17th C.," *Anglia* 100 (1982).

A.B.F.; T.V.F.B.

CATACHRESIS (Gr. "misuse"). The misapplication of a word, esp. to produce a strained or mixed metaphor. Most often it is not a ridiculous misapplication, as in doggerel verse, but rather a deliberate wresting of a term from its proper signification for effect. Sometimes it is deliberately humorous. Quintilian calls it a necessary misuse (*abusio*) of words and cites *Aeneid* 2.15–16: "equum divina Palladis arte / ædificant" (They build a horse by Pallas' divine art). Since *aedificant* literally means "they build a house," it is a c. when applied to a horse. A disapproving view of c. might call it a mixed metaphor, which is a fault, but many examples, e.g. "the sword shall devour," make sense and are striking. Two celebrated examples are found in Shakespeare—"To take arms against a sea of troubles" (*Hamlet* 3.1.60)—and in Milton—"Blind mouths! that scarce themselves know how to hold a sheep-hook" (*Lycidas* 119–20). A very effective c. is Shakespeare's "'Tis deepest winter in Lord Timon's purse" (*Timon of Athens* 3.4.15), which suggests comparison with some of the strained metaphors and conceits (qq.v.) of metaphysical poetry and modern poetry, e.g. "The sun roars at the prayer's end" (Dylan Thomas, *Visions and Prayer*). In any event, the semantic transference produced by c. generates new meaning insofar as it is accepted as a successful effect; Joseph (146) says Shakespeare uses it to achieve "sudden concentrations of meaning" together with "compression, energy, and intensity."—Sr. M. Joseph, *Shakespeare's Use of the Arts of Lang.* (1947); Lausberg; Morier; A. Quinn, *Figures of Speech* (1982), 94–96; S. R. Levin, "C.: Vico and Joyce," *P&R* 20 (1987).

M.T.H.; T.V.F.B.

CATALOG, catalog verse. A list of persons, places, things, or ideas in poetry which have some common denominator such as heroism, beauty, or death; it is also extended as the structure and meaning of entire poems, called c. v. The c. device is of ancient origin and found in almost all lits. of the world. In antiquity the c. often had a didactic or mnemonic function, as may be seen in the genealogical lists commonly found in ancient lits., e.g. in the Bible (Genesis 10) and in Old Germanic verse. In some Polynesian and Abyssinian verses, lists of places seem intended to furnish geographical information. But frequently c. v. has a more clearly aesthetic function, such as indicating the vastness of a war or battle or the power of a prince or king. This is the primary purpose of the c. of heroes in epic lit., for example the heroes of the Trojan war found in *Iliad* 2, the Argonauts in Apollonius Rhodius' *Argonautica* 1, the heroes in *Aeneid* 7, and the list of fallen angels in Milton's *Paradise Lost*, Book 2. Closely allied is the treatment of God's power in the "Benedicite omnia opera domini" in the *Book of Common Prayer*.

In medieval and Ren. poetry, the c. device is often used for itemizing topics such as the beauty of women. Such usage seems to follow from ancient example, e.g. Ovid's c. of trees in *Metamorphoses*, and can even be found in modern song writing, e.g. Cole Porter's "They couldn't compare with you." At other times catalogs seem used for the sake of play, or because the poet enjoyed the sound of particular words, e.g. the list of jewels in Wolfram von Eschenbach's *Parzifal*. In 19th- and 20th-c. European and Am. poetry, other functions of the c. have emerged. Whitman, for example, employs long lists to demonstrate the essential unity of the universe amid its seemingly endless multeity. Modern poets such as Rilke and George as well as Auden and Werfel have in varying degrees followed Whitman in this use.

A particularly important type is the c. of women, which in the West derives from Hesiod and Homer, and is practiced thereafter by Virgil, Ovid, Juvenal, Plutarch, Boccaccio, and Chaucer. McLeod argues that the c. of the "good woman" type (e.g. Boccaccio, Chaucer) helped preserve gender stereotypes.

T. W. Allen, "The Homeric C.," *JoHS* 30 (1910); D. W. Schumann, "Enumerative Style and its Significance in Whitman, Rilke, Werfel," *MLQ* 3 (1942), "Observations on Enumerative Style in Mod. Ger. Poetry," *PMLA* 59 (1944); S. K. Coffman, "'Crossing Brooklyn Ferry': A Note on the C. Technique of Whitman's Poetry," *MP* 51 (1954); L. Spitzer, "La enumeración caótica en la poesía moderna," *Linguística e historia literaria* (1955); H. E. Wedeck, "The C. in Lat. and Med. Lat. Poetry," *M&H* 13 (1960); S. A. Barney, "Chaucer's Lists," *The Wisdom of Poetry*, ed. L. D. Benson and S. Wenzel (1982); M. L. West, *The Hesiodic C. of Women* (1985); N. Howe, *The OE C. Poems* (1985); C. Goodblatt, "Whitman's Cs. as Literary Gestalt," *Style* 24 (1990); G. McLeod, *Virtue and Venom: Cs. of Women from Antiquity to the Ren.* (1991).

R.A.H.; T.V.F.B.

CHANSONS DE GESTE is the term by which the OF epic poems relating the deeds of Charlemagne and his barons, or other feudal lords of the Carolingian era, were known in their own time, principally the 12th and 13th cs. "Geste" has, aside from the original sense of "deeds," additional senses of "history" and "historical document," and by further extension it comes to mean "family, lineage." About 120 poems survive, in whole or in part, in existing mss., some of them in several redactions. They celebrate heroic actions, historical or pseudo-historical, and the chivalric ideals of a Christian, monarchical, and feudal France. There is no critical consensus as to whether the ideologi-

cal preoccupations are primarily those of Carolingian times or of the period of the Crusades and 12th-c. France. The history is at best considerably overlaid with legend, and many of the epics are largely or wholly fictitious, reworking themes made popular by earlier works from oral tradition. This is particularly true where the taste for the romantic and the fantastic nurtured by the medieval romance and by folklore was carried over into the invention of plots for the epics: this hybrid type is best illustrated by poems like *Huon de Bordeaux, Renaud de Montauban*, and *Le Chevalier au Cygne*.

I. CYCLES. Several more or less well-defined groups of poems may be distinguished. The most cohesive of these (24 poems) is the cycle of Guillaume d'Orange, to be identified with the historical Count Guillaume de Toulouse, contemporary of Charlemagne, and in which are recounted his deeds and those of his six brothers, his nephews, particularly Vivien and Bertrand, his father Aymeri de Narbonne, and others of his line. The principal poems of this cycle are *Le Couronnement de Louis, Le Charroi de Nîmes, La Prise d'Orange, Le Couvenant Vivien, La Chanson de Guillaume, Aliscans, Le Moniage Guillaume*, and *Aymeri de Narbonne*.

The so-called cycle of Charlemagne is less extensive and less unified. To it are assigned the poems treating of Charlemagne's wars (*La Chanson de Roland, Aspremont, Les Saxons*) or youth (*Mainet*), or earlier royal heroes such as the son of Clovis the Merovingian (*Floövant*). One of these is the partly comic *Pélerinage de Charlemagne*, which includes the description of a highly fanciful visit to the court of the Emperor of Constantinople.

The third main group has as its common element the theme of a feudal lord's revolt, usually provoked by an act of injustice, against his *seigneur*, who is in several cases Charlemagne, as in *Girart de Roussillon* (the only *c. de g.* surviving in a dialect of the *langue d'oc*), *La Chevalerie Ogier de Danemarche, Renaud de Montauban*, known also as *Les Quatre fils Aymon*, and *Huon de Bordeaux*. In the oldest poem of this group, the 11th-c. *Gormont et Isembart*, surviving only in a fragment, the renegade Isembart fights against his lord King Louis III. *Raoul de Cambrai* relates the sombre violence of feudal warfare following a forcible dispossession. The unforgiving bitterness of struggle between two great families is the subject of a minor cycle, *Les Lorrains*, of which the principal poems are *Garin le Lorrain* and *Hervis de Metz*.

II. LA CHANSON DE ROLAND is the masterpiece of the genre and the earliest surviving example, composed around 1100, transcribed probably around 1150 in continental France, and preserved in an Anglo-Norman manuscript of the mid-12th c. (Oxford version). Later versions lengthen the poem and insert additional episodes. The historical event on which the poem is based is the annihilation of Charlemagne's rear guard under Count Roland while recrossing the Pyrenees after an expedition against Saragossa in 778. In the poem the attackers are referred to as Saracens, whereas they were in all probability Basques, and a succession of councils is related, leading to the decision to leave Spain. A traitor Ganelon is introduced, as Roland's stepfather, who urges the Saracens to attack the rear guard in revenge for Roland's having designated him for the perilous embassy to the enemy camp. The disaster is assured when Roland overconfidently refuses to call back Charlemagne by sounding his horn when attacked, in spite of the urgings of his companion Oliver. After the defeat at Roncevaux, the poem ends with another battle in which Charlemagne is victorious, and with the trial and execution of Ganelon. The *Chanson de Roland* is as remarkable for the ideals it exalts—unstinting devotion to God and to feudal lord, and to the fatherland, "douce France"—as for its vigorous and incisive portrayals of great characters, closely knit structure, elevation of tone, and firm, concise lang.

III. FORM, VERSIFICATION, STYLE. The usual line is a decasyllable with caesura after the fourth syllable; unstressed syllables (mute -e) after the accented fourth or tenth syllables, when they appear, are not counted. In the earlier epics, octosyllables or alexandrines are sometimes used (*Gormont et Isembart, Pélerinage de Charlemagne*). The strophic form is that of the *laisse*, a variable number of lines bound together by assonance in the earlier poems, by rhyme in the later ones. In the *Chanson de Roland*, the *laisses* average 14 lines in length; in later poems they tend to be much longer. In length, the *c. de g.* range from about a thousand lines (*Pélerinage de Charlemagne*) to over 35,000; the *Chanson de Roland* has four thousand in the Oxford version. As their name would indicate, the *c. de g.* were sung, and musical notation has been preserved for some.

In coherence of composition, the epics vary greatly, from well-knit poems like the *Chanson de Roland* or the *Pélerinage de Charlemagne* to rambling and even self-contradictory successions of episodes. The style is vigorous in the best poems but diffuse and filled with clichés and formulaic diction in the poorer ones. The formulaic diction provides some evidence for the view that the epics were improvised orally by *jongleurs*, itinerant minstrel-performers of the later Middle Ages. It is certain that the epics contain genuine elements of oral poetry as defined in the terms of the oral-formulaic theory of Parry and Lord, but the issues are complicated by the presence of later, literary elements in the written texts.

IV. ORIGINS. The debate over the origins and the prehistory of the *c. de g.* was for nearly a century the outstanding controversy in Fr. medieval lit. hist. In the 19th c., it was customary to consider the surviving *c. de g.* as deriving ultimately from poems inspired by contemporary historical events, composed anonymously, and altered and expanded constantly in the course of oral transmission, over two, three, or even four centuries. At the

beginning of the 20th c., however, Bédier denied the continuity-of-transmission theory, arguing instead that the historical content of the poems amounted only to such information as could have been discovered by jongleur-poets in sanctuaries along the pilgrimage routes in the 11th and 12th cs. (incl. learned and written sources), that the poems were composed relatively late, and that the composers were specific individuals little interested in "tradition."

But Bédier's "individualism," as it has been called, has increasingly been attacked by scholars who have revived the older view and buttressed it with new arguments. It is principally around the *Chanson de Roland* that controversy wages, according to the degree of originality the critic is willing to ascribe to the author of the Oxford version. Traditionalists see him as a mere arranger of a poem with a long prehistory of collective elaboration; an intermediate group finds his sources in Med. Lat. hagiography or epic; and "individualists" minimize his debt to hypothetical predecessors and credit him with the largest possible measure of creativeness.

V. DIFFUSION. The *c. de g.* quickly became popular outside northern France and Norman England, being translated notably into MHG and Old Norse and Icelandic. In Italy they were made accessible in an Italianized Fr. before serving as the models for wholly It. poems. They were known in Spain, where, however, national heroes were preferred for epic (the *Cid*). In France, the 15th and 16th cs. knew the epic legends through prose adaptations before these in turn were forgotten through two and a half centuries of Classicism.

L. Gautier, *Épopées françaises*, 2d ed., 4 v. (1878–92), supp. by his *Bibliographie des c. de g.* (1897)—still valuable despite the obsolescence of many of its views; J. Bédier, *Légendes épiques*, 4 v. (1908–13); *CBFL*, v. 1; Lote; I. Siciliano, *Les Origines des c. de g.*, 2d. rev. ed. (1951), *Les C. de g. et l'épopée* (1968); J. Horrent, *La Chanson de Roland dans les litts. française et espagnole au moyen âge* (1951); R. Bossuat, *Manuel bibliographique de la litt. française du moyen âge* (1951, supp. 1955, 1961); M. Delbouille, *Sur la genèse de la* Chanson de Roland (1954); P. Le Gentil, *Chanson de Roland* (1955); J. Rychner, *La C. de g.: Essai sur l'art épique des jongleurs* (1955)—oral transmission with improvisation; A. Junker, "Stand der Forschung zum Rolandslied," *GRM* (1956); M. de Riquer, *Les C. de g. françaises*, 2d ed. (tr. 1957); *La Technique littéraire des c. de g.* (1959); R. Menéndez Pidal, *La Chanson de Roland et la trad. épique des Francs*, 2d. rev. ed. (1960)—the neotraditionalist view; Lord; W. B. Calin, *The Epic Quest: Studies in Four OF C. de G.* (1966), *A Muse for Heroes: Nine Cs. of the Epic in France* (1983); Parry; J. Duggan, *The Song of Roland: Formulaic Style and Poetic Craft* (1973); G. J. Brault, "The Fr. C. de G.," *Heroic Epic and Saga*, ed. F. J. Oinas (1978), ed., *The Song of Roland* (1979); N. Daniel, *Heroes and Saracens* (1984); *Au carrefour des routes d'Europe: La C. de g.: Xe Congres international de la Société Rencesvals pour l'étude des épopées romanes* (1985); B. Guidot, *Recherches sur la c. de g. au XIIIe siècle* (1986); R. F. Cook, *The Sense of the* Song of Roland (1987); H.-E. Keller, *Autour de Roland* (1989); Hollier.

C.A.K.; T.V.F.B.

CHIASMUS (Gr. "a placing crosswise," from the name of the Gr. letter X, "chi"; but the term is first attested only in 1871, in Eng., not appearing as a Lat. term until 1903). Any structure in which elements are repeated in reverse, so giving the pattern *ABBA*. Usually the repeated elements are specific words, and the syntactic frames holding them (phrases, clauses) are parallel in construction, but may not necessarily be so. The c. may be manifested on any level of the text or (often) on multiple levels at once: phonological (sound-patterning), lexical or morphological (word repetition), syntactic (phrase- or clause-construction), or semantic/thematic. The fourth of these requires one of the preceding three, and the second usually entails the third. C. of sound will be found in the first sentence of Coleridge's "Kubla Khan"; of word repetition (the commonest form) in Shakespeare's *Julius Caesar*: "Remember March, the ides of March remember" (4.3.18); of syntax, Milton's "That all this good of evil shall produce, / And evil turn to good" (*PL* 12.470–71); of sense or theme, Coleridge's "Frost at Midnight," which sets up the thematic sequence frost–secret ministry–lull–birdsong–birdsong–lull–secret ministry–frost (Nänny). Other poems for which chiastic structures have been claimed are Chaucer's *Parlement of Foules* and Tennyson's *In Memoriam*. C. is thus but another term for an envelope pattern, and will be seen too as one form of inversion within repetition. In cl. rhet., c. is the pattern of a sentence consisting of two main clauses, each modified by a subordinate clause, in which sentence each of the subordinate clauses could apply to each of the main clauses, so that the order of these four members could be altered in several ways without change in the meaning of the whole (Hermogenes [2d c. A.D.], *Peri heureseōn* 4.3, in Walz 3.157).

Lausberg defines c., usefully, as the criss-cross placing of sentence members that correspond in either syntax or meaning, with or without exact verbal repetition. Group Mu classifies c. among "metataxes," i.e. figures which "focus attention on syntax," and describes it as a syntactic device which effects a "complete" suppression and addition of semes and taxemes. They add that such crossed symmetry "emphasizes both meaning and grammar." Mahood shows how a simple inverted parallelism in c. can achieve economically a relatively complex effect. In *Richard II*, Bolingbroke receives the following answer when he asks the King if he is content to abdicate: "Ay, no; no, Ay: for I must nothing bee: / Therefore no, no, for I resign to thee" (4.1.202–3). Mahood comments

that Richard's reply, "besides suggesting in one meaning (Ay, no; no, Ay) his tormenting indecision, and in another (Ay—no; no I) the overwrought mind that finds an outlet in punning, also represents in the meaning 'I know no I' Richard's pathetic play-acting, his attempt to conjure with a magic he no longer believes. Can he exist if he no longer bears his right name of King?" (87). A less complex modern example which nonetheless shows c. affecting both semantics and syntax is Hopkins' "This seeing the sick endears them to us, us too it endears" ("Felix Randal," 9–10), which concludes with an ellipsis of the grammatical object. And Adrienne Rich uses the fully parallel, repetitive potentiality of c. in "Planetarium": "A woman in the shape of a monster / a monster in the shape of a woman" (1–2). Quinn defines c. by comparing it to other repetitive figures like *epanodos* (amplification or expansion of a series), *antimetabole,* and *palindrome,* a sentence which spells the same words forwards or backwards. These figures all manifest rhetoric's multi-levelled and synchronous functions in the text.—C. Walz, *Rhetores graeci,* 9 v. (1832–36); Sr. M. Joseph, *Shakespeare's Use of the Arts of Lang.* (1947); M. M. Mahood, *Shakespeare's Wordplay* (1957); Lausberg; A. Di Marco, "Der C. in der Bibel," *Linguistica Biblica* 37–39 (1976); *C. in Antiquity: Structures, Analyses, Exegesis,* ed. W. Welch (1981); Group Mu, 45, 80, 124–25; A. Quinn, *Figures of Speech* (1982), 93–95; H. Horvei, *The Chev'ril Glove* (1984); R. Norrman, *Samuel Butler and the Meaning of C.* (1986); M. Nänny, "C. in Lit.: Ornament or Function?" *W&I* 4 (1988); Corbett. T.V.F.B.; A.W.H.

CLOSURE refers most broadly to the manner in which texts end or the qualities that characterize their conclusions. More specifically, the term "poetic c." is used to refer to the achievement of an effect of finality, resolution, and stability at the end of a poem. In the latter sense, c. appears to be a generally valued quality, the achievement of which is not confined to the poetry of any particular period or nation. Its modes and the techniques by which it is secured do, however, vary in accord with stylistic, particularly structural, variables.

Closural effects are primarily a function of the reader's perception of a poem's total structure; i.e. they are a matter of his/her experience of the relation of the concluding portion of a poem to the entire composition. The generating principles that constitute a poem's formal and thematic structure characteristically arouse continuously changing sets of expectations, which elicit various "hypotheses" from a reader concerning the poem's immediate direction and ultimate design. Successful c. occurs when, at the end of a poem, the reader is left without residual expectations: his/her developing hypotheses have been confirmed and validated (or, in the case of "surprise" endings, the unexpected turn has been accommodated and justified retrospectively), and s/he is left with a sense of the poem's completeness, which is to say of the integrity of his/her own experience of it and the appropriateness of its cessation at that point.

C. may be strengthened by certain specifically *terminal* features in a poem, i.e. things that happen at the end of it. These include the repetition and balance of formal elements (as in alliteration and parallelism [qq.v.]), explicit allusions to finality and repose, and the terminal return, after a deviation, to a previously established structural "norm" (e.g. a metrical norm). Closural failures (e.g. anticlimax) usually involve factors that, for one reason or another, leave the reader with residual expectations. They may also arise from weak or incompatible structural principles or from a stylistic discrepancy between the structure of the poem and its mode of c. Weak c. may, however, be deliberately cultivated: much modernist poetry shares with modernist works in other genres and artforms a tendency toward apparent "anticlosure," i.e. the rejection of strong closural effects in favor of irresolution, incompleteness and, more generally, a quality of "openness."—B. H. Smith, *Poetic C.: A Study of How Poems End* (1968); P. Hamon, "Clausules," *Poétique* 24 (1975); *Concepts of C.,* spec. iss. of *YFS* 67 (1984). B.H.S.

COMEDY.

I. DEFINITION
II. ANCIENT
III. RENAISSANCE AND MODERN EUROPEAN
IV. RELATION TO TRAGEDY

I. DEFINITION. Like tragedy (q.v.), the Western trad. of comic theater is considered to have begun with the ancient Greeks. Such a claim is however less clear than that made on behalf of tragedy, if only because no known culture appears to lack some form of comic performance. This fact has inspired various speculative theses concerning laughter—like reason and speech—as one of the defining characteristics of humanity. As in the case of tragedy, therefore, we need to distinguish with some care between speculative generalizations about the "comic spirit," and that more precise historical description needed to analyze the function of c. in society. We must also discriminate between such description and attempts to analyze the "psychology" of laughter, because the event of c. and the eruption of mirth are by no means the same. (I should add that although the term "c." has been applied to any literary genre that is humorous, joyful, or expresses good fortune, what follows will concern above all the theater, even if some observations have a broader application.)

The name "c." comes from Comus, a Gr. fertility god. In ancient Greece "c." also named a ritual springtime procession presumed to celebrate cyclical rebirth, resurrection, and perpetual rejuvenation. Modern scholars and critics have thus taken c. to be a universal celebration of life, a joyous outburst of laughter in the face of either an

incomprehensible world or a repressive socio-political order. Carnival, festival, folly, and a general freedom of action then indicate either an indifference to and acceptance of the first, or a resistance to the second. But scholars have taken such notions yet further: if tragedy represents the fall from some kind of "sacred irrationality," c. on the contrary becomes the triumphant affirmation of that riotous unreason, marking some ready acceptance of human participation in the chaotic forces needed to produce Life. The comic protagonist's defeat is then the counterpart to the tragic protagonist's failure, both versions of some ritual cleansing by means of a scapegoat—in this case one representing life-threatening forces. Such speculations have been advanced in one form or another by classicists (F. M. Cornford, Jane Harrison, Gilbert Murray), philosophers (Mikhail Bakhtin, Susanne Langer), and literary critics (C. L. Barber, Northrop Frye), not to mention anthropologists and even sociologists.

How much these theories help us understand what *cs.* are is another, and perhaps a different, question. For in the last resort such arguments depend on the assumption that beneath all and any particular c. is some kind of profound universal "carnival," a common denominator of the human in all times and places. Recalling Nietzsche's *Gay Science*, Jean Duvignaud has thus spoken of "laughter that for a fleeting moment pitches humans before an infinite freedom, eluding constraints and rules, drawing them away from the irremediable nature of their condition to discover unforeseeable connections, and suggesting a common existence where the imaginary and real life will be reconciled" (229–30). But theories of this sort depend upon the idea that one can obtain the deepest comprehension of cs. by removing them from their distinct historical moment and social environment. They forget that such carnival and such laughter are themselves the creations of a particular rationality, just as Dante's *Divine Comedy* universalized a particular theology. Even so seemingly fantastic a theater as that of Aristophanes (ca. 445–385 B.C.) is misconstrued by a theory that neglects c.'s essential embedding in the social and political intricacies of its age and place (Athens during the Peloponnesian Wars).

Setting aside these broad metaphysical speculations, then, we must look at accounts of laughter as a human reaction to certain kinds of events. By and large, these may be divided into two theories. The one asserts that laughter is provoked by a sense of superiority (Hobbes' "sudden glory"), the other that it is produced by a sudden sense of the ludicrous, the incongruous, some abrupt dissociation of event and expectation. The theory of superiority is the more modern one, developed mainly by Hobbes, Bergson, and Meredith. It presumes our joy in seeing ourselves more fortunate than others, or in some way more free. Bergson's notion that one of the causes of laughter is an abrupt perception of someone as a kind of automaton or puppet, as though some freedom of action had been lost, is one version of this.

The theory of c. as the ludicrous or as the dissociation of expectation and event has a longer pedigree. It begins with Aristotle and has come down to us via Kant, Schopenhauer, and Freud. In the *Poetics*, Aristotle mentions another work on c., now lost; what remains are a few comments. In *Poetics* 5 Aristotle remarks that c. imitates people "worse than average; worse, however, not as regards any and every sort of fault, but only as regards one particular kind, the Ridiculous, which is a species of the Ugly. The Ridiculous may be defined as a mistake or deformity not productive of pain or harm to others" (tr. Bywater). Similar remarks exist in his *Rhetoric* and in a medieval Gr. ms. known as the *Tractatus Coislinianus*, Aristotelian in argument and possibly even an actual epitome of Aristotle's lost writing on c. (ed. Janko, 1984). Save for suggesting some detail of dissociative word and action, this text adds little to what may be gleaned from extant texts of Aristotle. It does make a parallel between c. and tragedy, however, by saying that catharsis also occurs in c., "through pleasure and laughter achieving the purgation of like emotions." The meaning of such a phrase is not at all clear, although it suggests c. as an almost Stoic device to clean away extremes of hedonism and to root out any carnivalesque temptations.

Although both theories involve the psychology of laughter, the superiority theory seems less particularly applicable to c. than that of incongruity, for the latter seeks both to indicate *devices* specifically provocative of laughter and to explain their *effect* on a spectator. The "Aristotelian" analyses suggest several matters. First, their kind of laughter requires oddness, distortion, folly, or some such "version of the ugly," but without pain. Such laughter thus depends on sympathy. Second, although this theory is kinder than that of superiority, it too has its part of cruelty, just because of the touch of ugliness. Third, theories of superiority and of incongruity both take laughter as *means*, as commentary upon or correction of what we may call the real or even "local" world—unlike metaphysical theories, which make mirth an end in itself and an escape into some "universality." Fourth, both these theories (which supplement rather than oppose one another) require the laugher to be aware of some disfiguring of an accepted norm. C. and laughter imply a habit of normality, a familiarity of custom, from which the comic is a deviation. It may indeed be the case that c., like tragedy, shows the construction of such order, but above all it demonstrates why such order must be conserved.

II. ANCIENT. The fourth theory would at least partly explain why comic competition was instituted at the Athenian Dionysia some 50 years *after* that for tragedy (in 486 B.C.). Aristotle has told us

the first competition was won by Chionides, who with Magnes represented the first generation of writers of c. Around 455 a comic victory was won by Cratinus, who with Crates formed a second generation. Many titles have survived and some fragments, but these constitute nearly the sum total of extant facts about Athenian c. until Aristophanes' victory with *Acharnians* in 425. We know that in this competition Cratinus was second with *Kheimazomenoi*, and Eupolis third with *Noumeniai*. These names tell us little, but we may perhaps assume that Aristophanic c. was fairly typical of this so-called Gr. "Old C.": a mixture of dance, poetry, song, and drama, combining fantastic plots with mockery and sharp satire of contemp. people, events, and customs. Most of his plays are only partly comprehensible if we know nothing of current social, political, and literary conditions.

Aristophanes did not hesitate to attack education, the law, tragedians, the situation of women (though it is clearly an error to take him for a "feminist" of any kind), and the very nature of Athenian "democracy." Above all he attacked the demagogue Cleon, the war party he led, and the war itself. This says much about the nature of Athenian freedoms, for Aristophanes wrote during the struggle with Sparta, when no one doubted at all that the very future of Athens was in question. Aristophanes' last surviving play (of 11, 44 being attributed to him) is *Plutus* (388), a play criticizing myth, but whose actual themes are avarice and ambition. Quite different in tone and intent from the preceding openly political plays, *Plutus* is considered the earliest (and only extant) example of Gr. "Middle C."

The situation of c. was, however, quite different from that of tragedy, for another powerful trad. existed. This was centered in Sicilian Syracuse, a Corinthian colony, and claimed the earliest comic writer, Epicharmus, one of the authors at the court of Hieron I in the 470s. We know the titles of some 40 of his plays. Other comic poets writing in this Doric trad. were Phormis and the slightly younger Deinolochus, but the Dorians were supplanted by the Attic writers in the 5th c. and survive only in fragments. The best known composer of literary versions of the otherwise "para-" or "sub-" literary genre of comic mime was another Sicilian, Sophron, who lived during the late 5th c. From the 4th c. we have a series of vase paintings from Sicily and Southern Italy which suggest that c. still throve there. The initiative had largely passed, however, to the Gr. mainland. *Plutus* is an example of that Middle C. whose volume we know to have been huge. Plautus' Lat. *Amphitruo* (ca. 230 B.C.) seems to be a version of another one, and, if so, one characteristic was the attack on myth. (Aristophanes' earlier *Frogs* [405], attacking Euripides and Aeschylus, tried in the underworld, may well be thought a forerunner.)

By the mid 4th c., so-called "New C." held the stage. Among its poets the most celebrated and influential was unquestionably Menander (ca. 342–290 B.C.). His "teacher" was a certain Alexis of Thurii in southern Italy, so we can readily see how the "colonial" influence continued, even though Alexis was based in Athens. He is supposed to have written 245 plays and to have outlived his pupil. We know of Philemon from either Cilicia or Syracuse, of Diphilos from Sinope on the Black Sea, and of Apollodorus from Carystos in Euboea—worth mentioning as illustrating the great spread of c. Until the 1930s, however, only fragments seemed to have survived. Then what can only be considered one of the great literary discoveries turned up a papyrus containing a number of Menander's cs., complete or almost so. These plays deal not with political matters or crit. of myth, but with broadly social matters (sometimes using mythical themes). The situations are domestic, the c. is of manners, the characters are stock.

The widespread familiarity of comic forms helps explain why c. was soon diffused once again over the Gr. and Roman world. By the mid 3rd c., not only had itinerant troupes spread from Greece throughout the Hellenistic world, but already by 240, Livius Andronicus, from Tarentum in southern Italy, had adapted Gr. plays into Lat. for public performance. Like Gnaeus Naevius and later Quintus Ennius, this poet composed both tragedy and c. From the 3rd c. as well dates Atellan farce (named from Atella in Campania), using stock characters and a small number of set scenes, and featuring clowns (called Bucco or Maccus), foolish old men, and greedy buffoons. These farces were partly improvised, on the basis of skeletal scripts, much like the *commedia dell'arte* of almost two millennia later. The influence of Etrurian musical performance, southern It. drama, Gr. mime, New C., and Atellan farce came together in the cs. of Titus Maccius Plautus, who wrote in the late 3rd c. (he is said by Cicero to have died in 184). By him 21 complete or almost complete cs. have survived. A little later Rome was entertained by the much more highbrow Publius Terentius Afer (Terence), by whom six plays remain extant. These two authors provided themes, characters, and style for c. as it was to develop in Europe after the Ren. (though farces, *sotties*, and comic interludes were widely performed in the Middle Ages).

III. RENAISSANCE AND MODERN EUROPEAN. As in the case of tragedy, c. was rediscovered first in Italy. While humanist scholars published and then imitated both Plautus and Terence, vernacular art developed alongside such efforts. The early 16th c. saw the publication of much school drama in both Lat. and It., while just a little later there developed the *commedia dell'arte*, whose influence was to be enormous. This was a c. of improvisation, using sketchy scripts and a small number of stock characters—Harlequin, Columbine, Pantaloon, the Doctor and others—placing these last in various situations. These plots were as frequently derived from antiquity as they were from folk art.

Later on, these two forms of c. tended to feed one another; the popular *Comédie italienne* of late 17th-c. France was one outcome. The *Commedia*'s influence was equally visible in Marivaux (1688–1763) and Goldoni (1707–93), though in the case of the first, the *Italienne* was just as important. The *Commedia* survives vividly in our own time in the theater of the San Francisco Mime Troupe, which has put the old characters to work in the service of powerful political satire.

Spain vied with Italy in its devel. of c., starting with the late 15th-c. *Celestina* of Fernando de Rojas, written in Acts and in dialogue but never really intended for performance. By the late 16th c., Spain's theater was second to none in Europe. Lope de Vega (1562–1635), Calderón (1600–81), and a host of others produced a multitude of romantic and realistic c. dealing mainly with love and honor. They provided innumerable plots, themes, and characters for comic writers of France and England. These two countries started rather later than the South, but, like them, benefited from both an indigenous folk trad. and the publication of Lat. c. The influence of It. humanist c. was significant in both nations during the 16th c., and that of Spain particularly in France in the early 17th c.

In France, humanist c. gave way in the late 1620s to a romantic form of c. whose threefold source was the prose romance and novella of Spain, Italy, and France, Sp. c. (esp. that of Lope and Cervantes), and It. dramatic pastoral. The first influential authors in this style were Pierre Corneille (1606–84) and Jean de Rotrou (1609–50). They were followed by many, incl. Cyrano de Bergerac, Thomas Corneille, and the poet whom many consider the greatest writer of c. of all times, Jean-Baptiste Poquelin Molière (1622–73). He wrote an enormous variety, in verse and prose, ranging from slapstick farce to something approaching bourgeois tragic drama. *Comédie ballet*, c. of situation, of manners, of intrigue, and of character all flowed from his pen. He did not hesitate to write on matters that provoked the ire of religious *dévots* or of professional bigots, nor did he shirk the criticism of patriarchy, and many of his plays have political overtones. Having begun his theatrical career as leader of a traveling troupe, Molière made full use of folk trad., of provincial dialect, of *Commedia* and of farce, as well as of Cl. example. Many of his characters have become familiar types in Fr. trad. (e.g. the "misanthrope," "tartuffe," "don juan"); many of his lines have become proverbial. While his plays do contain the now familiar young lovers, old men both helpful and obstructive, wily servants both female and male, sensible wives and mothers (whereas husbands and fathers are almost always foolish, headstrong, cuckolded, or downright obstructive), they bear chiefly upon such matters as avarice, ambition, pride, hypocrisy, misanthropy, and other such extreme traits. What interests Molière is how such excess conflicts directly with the well-regulated and customary process of ordered society.

Having followed a similar trajectory to that of its southern neighbors in the first half of the 16th c., England created a comic trad. unique in variety and longevity. The extraordinarily diverse c. of Shakespeare (1564–1616) and the so-called c. of humors favored by Jonson (1573–1637) seemed about to create two distinct comic trads. Shakespeare wrote in almost every mode imaginable: aristocratic romance, bitter and problematic farce, c. of character, slapstick farce, and the almost tragic *Troilus and Cressida.* If any c. may be analyzed with some "metaphysical" theory it is no doubt Shakespeare's, with its concern for madness and wisdom, birth and death, the seasons' cycles, love and animosity. Yet Shakespeare's c. remained unique, and he had no successor in this style. Jonson's more urbane c. of types and of character, satirizing manners and morals, social humbug and excess of all kinds, and falling more clearly into the forms already seen, was soon followed by the quite remarkable flowering of Restoration c., with a crowd of authors, incl. Dryden, Wycherley, Congreve, Behn, and Centlivre, among many others. They produced a brittle c. of manners and cynical wit whose major impression is one of decay and an almost unbalanced self-interest. They were in turn succeeded by a widely varied 18th-c. c. from the staunch complacency of Steele through the political satire of Gay to the joyous and mocking cynicism of Goldsmith, Inchbald, and Sheridan. This trad. was pursued through the late 19th and early 20th cs. by a series of great Irish dramatists: Shaw, most notably, then Wilde, Yeats, Synge, and O'Casey.

During this period France was equally productive, but with few exceptions failed to attain the quality represented by the names just mentioned. At the turn of the 17th c., Regnard produced serious and significant social satire, as did Lesage (esp. in *Turcaret*, 1706). Marivaux dominated the first half of the 18th c., as Voltaire did the middle and Beaumarchais the end. If any new form appeared it was doubtless the *comédie larmoyante*, a sentimental drama whose main (and stated) purpose was to draw the heartstrings; in a way, it did for c. what the later melodrama did for tragedy. In the 19th c. Musset produced his delicate c. of manners, while Dumas *fils* and others strove to produce a c. dealing with society's ills. This culminated on the one hand in Scribe's "well-made play," on the other in the "realist" drama of Zola and Antoine at the end of the century.

In other European lands, authors tended to be isolated: in late 19th-c. Norway, Ibsen; in early 20th-c. Russia, Chekhov; slightly later in Italy, Pirandello. To mention them so briefly is to be unjust, for they were all major creative figures. In many ways they foreshadowed that breakdown of traditional c. that marks the mid to late 20th c. Laughter tends to become mingled problematically with that sense of discomfort in the world and

unease in the self which is perhaps a principal sign of our age. Among representative authors one might mention such as Witkiewicz, Mrozek, and Gombrowicz from Poland; Brecht, Dürrenmatt, and Handke from Germany, Switzerland, and Austria; Adamov, Ionesco, Arrabal, and Beckett in France; Čapek, Fischerova, Havel, and Kohut in Czechoslovakia; Pinter, Arden, Bond, Stoppard, Benton, Hare, and Churchill in England; and Hellman, Albee, Baraka, and Simon in America. All have been writing plays that sport ironically with the political, social, and metaphysical dimensions of the human condition. Usually such issues are no longer held separate, and all are fair game for an ambiguous, perplexed, and uncertain derision. Such theater is now widely distributed, as strong in Lat. America as in Czechoslovakia, in Italy or Spain as in Nigeria. It is almost as though c. had lost a sense of that social norm to which we referred at the outset, as if it were increasingly imbued with an inescapable sense of the tragic.

IV. RELATION TO TRAGEDY. C. had from the start a rather ambiguous relation to tragedy, and it was never difficult to see in Aristophanes' *Thesmophoriazusai* an inversion of Euripides' *Bacchae*, for example. A celebrated passage at the end of Plato's *Symposium* has Socrates obliging Agathon and Aristophanes to agree that c. and tragedy have the same source. Chekhov's *The Cherry Orchard* has been played as both c. and tragedy; so has *The Merchant of Venice*. Even the elements compounding the confrontation may be identical, as in *Macbeth*, Jarry's *Ubu Roi*, or Ionesco's *Macbett*. When the comic protagonist acquires attributes of typicality or of some absolute, then c. may take on overtones of tragedy. A critic of Molière's *Tartuffe* (1667) remarked that whatever "lacks extremely in reason" is ridiculous: anything contrary to a predictable reaction or an expected and habitual situation is absurd. This is of course straight from the Aristotelian trad., but the emphasis on excess is significant. It shows just how close c. always was to tragedy, explaining such cs. as *Dom Juan* or *Le Misanthrope*. Both focus on an idealism either misplaced or preposterous. Dom Juan's ideal self is misplaced because it serves a violent and injurious sexuality; Alceste's self-righteous scorn becomes comic when he refuses even the most innocent concession, and his responses become inappropriate to his urbane surroundings. Yet if he lowered his tone to suit his milieu he would fall short of his ideal: the dilemma is that of dissonance between the ideal and the situation where it is expressed—incongruity again. The excessive ideal in this case contradicts society's needs and fails its norm.

Tragedy appears to require a world view such that a recognized human quantity may be pitted against a known but inhuman one (variously called Fate, the gods, the idea of some Absolute, etc.), permitting the "limits" of human action and knowledge to be defined. C. seems rather to oppose humans to one another, within essentially *social* boundaries. And if, as both the superiority and the incongruity theories hold, c. is essentially a social phenomenon, then wherever humans are will be somehow conducive to it; whereas tragedy seems to signify a moment of passage from one sociocultural environment to another. That *social* nature of c. may be why its characters seem to us so down-to-earth, pragmatic, and familiar. Even where a theater's real (external) social context is very different we can still recognize creatures of a *social* order. That is also why cs. are in league with their audience, obtaining their spectators' sympathy for what are given as the dominant social interests. Volpone menaces that order, as do Shylock, Tartuffe, and Philokleon (Aristophanes, *Wasps*). Volpone and Shylock are defeated in the name of the Venetian Republic, as is Tartuffe in that of the King, and Philokleon in that of a city longing for peace. In Plautus' *Epidicus*, the eponymous slave—archetypal outsider for 3rd-c. Rome—is absorbed into and becomes a part of the social system. In Pirandello's *Six Characters in Search of an Author*, the actors remain at loose ends because they are unable to situate either themselves *or* a social order. Similarly, Beckett's two tramps remain despairingly expectant at the end of *Waiting for Godot*. C. has always emphasized the conservation of an order it may well have helped construct. When we can no more grasp or even envisage that order, then derisive irony may make us laugh, but it also leaves us painfully disturbed.

G. Meredith, *An Essay on C.* (1877; ed. W. Sypher, 1980); F. Nietzsche, *The Gay Science* (1882); H. L. Bergson, *Laughter* (1912; ed. W. Sypher, 1980); F. M. Cornford, *The Origin of Attic C.* (1914); S. Freud, *Wit and Its Relation to the Unconscious* (1916); L. Cooper, *An Aristotelian Theory of C.* (1922); M. A. Grant, *The Ancient Rhetorical Theories of the Laughable* (1924); J. Harrison, *Themis* (1927); K. M. Lea, *It. Popular C.*, 2 v. (1934); J. Feibleman, *In Praise of C.* (1939); M. T. Herrick, *Comic Theory in the 16th C.* (1950), *It. C. in the Ren.* (1960); G. E. Duckworth, *The Nature of Roman C.* (1952); W. Sypher, *C.* (1956); Frye; S. Langer, *Philosophy in a New Key*, 3d ed. (1957); A. Artaud, *The Theater and Its Double*, tr. M. C. Richards (1958); C. L. Barber, *Shakespeare's Festive C.* (1959); E. Welsford, *The Fool* (1961); J. L. Styan, *The Dark C.* (1962); A. Nicoll, *A Hist. of Eng. Drama, 1660–1900*, 6 v. (1952–59), *The World of Harlequin* (1963); *Theories of C.*, ed. P. Lauter (1964); N. Frye, *A Natural Perspective* (1965); H. B. Charlton, *Shakespearean C.* (1966); W. Kerr, *Tragedy and C.* (1967); M. M. Bakhtin, *Rabelais and His World*, tr. H. Iswolsky (1968); E. Olson, *The Theory of C.* (1968); E. Segal, *Roman Laughter: The C. of Plautus* (1968); L. S. Champion, *The Evolution of Shakespeare's C.* (1970); G. M. Sifakis, *Parabasis and Animal Choruses: A Contribution to the Hist. of Attic C.* (1971); W. M. Merchant, *C.* (1972); K. J. Dover, *Aristophanic C.* (1972); M. C. Bradbrook, *The Growth and Structure of Elizabethan C.*, 2d ed.

(1973); R. B. Martin, *The Triumph of Wit: A Study of Victorian Comic Theory* (1974); A. Rodway, *Eng. C.: Its Role and Nature from Chaucer to the Present Day* (1975); M. Gurewitch, *C.: The Irrational Vision* (1975); F. H. Sandbach, *The Comic Theatre of Greece and Rome* (1977); A. Caputi, *Buffo: The Genius of Vulgar C.* (1978); E. Kern, *The Absolute Comic* (1980); R. Nevo, *Comic Transformations in Shakespeare* (1980); R. W. Corrigan, *C.: Meaning and Form*, 2d ed. (1981); Trypanis; Fowler; K. H. Bareis, *Comoedia* (1982); D. Konstan, *Roman C.* (1983); E. L. Galligan, *The Comic Vision in Lit.* (1984); R. Janko, *Aristotle on C.* (1984); J. Duvignaud, *Le Propre de l'homme: histoire du comique et de la dérision* (1985); K. Neuman, *Shakespeare's Rhetoric of Comic Character* (1985); E. W. Handley, "C.," *CHCL*, v. 1; R. L. Hunter, *The New C. of Greece and Rome* (1985); W. E. Gruber, *Comic Theaters* (1986); T. Lang, *Barbarians in Gr. C.* (1986); T. B. Leinward, *The City Staged: Jacobean C., 1603–1613* (1986); E. Burns, *Restoration C.: Crises of Desire and Identity* (1987); H. Levin, *Playboys and Killjoys* (1987); L. Siegel, *Laughing Matters: Comic Trads. in India* (1987). T.J.R.

COMPLAINT (Gr. *schetliasmos*) is an established rhetorical term by the time of Aristotle's *Rhetoric* (1395a9); in later rhetoricians it designates that portion of a lament (q.v.) or farewell speech in which the speaker rails against cruel fate (Menander Rhetor; Cairns). Catullus and the Roman elegists often employ a c. (Lat. *querela*) in their love poetry, directing it against a reluctant mistress, sometimes in the form of a *paraclausithyron*, a lament by the beloved's door (Copley). In the Middle Ages and Ren., three sometimes overlapping strains of c. are evident: (1) satiric poems that expose the evil ways of the world (*contemptus mundi*, e.g. Alain de Lille, *De planctu naturae*, and Spenser's collection, *Cs., Containing sundrie small Poems of the Worlds Vanitie*); (2) didactic, that relate the fall of great persons (e.g. Boccaccio, *De casibus virorum illustrium*; Chaucer, *The Monk's Tale*; and the Ren. collection *The Mirror for Magistrates*); and (3) amatory, incl. both short poems written in the plaintive Petrarchan mode (e.g. by Wyatt and Surrey in Tottel's Miscellany [1557]) and more ambitious monologues (e.g. Daniel, *The C. of Rosamond*, and Shakespeare, *A Lover's C.*). In addition, elegies like Milton's "Lycidas" retain the traditional function of the c. in laments. Other prominent examples of c. incl. Chaucer, "A C. unto Pity" (Chaucer has many early, experimental cs., on the model of Fr. and It., as *ballades* or in *ottava rima*); Cowley, "The C."; and Young's long discursive poem *C., or Night Thoughts*. In Occitan poetry the c. takes the form of the *enueg*, whence Ezra Pound found the model for the 36th of his *Cantos*.

H. Stein, *Studies in Spenser's Cs.* (1934); *The Mirror for Magistrates*, ed. L. B. Campbell (1938); F. O. Copley, *Exclusus amator* (1956); J. Peter, *C. and Satire in Early Eng. Lit.* (1956); H. Smith, *Elizabethan Poetry* (1964); F. Cairns, *Generic Composition in Gr. and Roman Poetry* (1972); R. Primeau, "Daniel and the *Mirror* Trad.," *SEL* 15 (1975); *Menander Rhetor*, ed. D. A. Russell and N. Wilson (1981); W. A. Davenport, *Chaucer: C. and Narrative* (1988). W.H.R.

CONCEIT. A complex and arresting metaphor, in context usually part of a larger pattern of imagery, which stimulates understanding by combining objects and concepts in unconventional ways; in earlier usage, the imagination or fancy in general. Derived from the It. *concetto* (concept), the term denotes a rhetorical operation which is specifically intellectual rather than sensuous in origin. Its marked artificiality appeals to the power of reason to perceive likenesses in naturally dissimilar and unrelated phenomena by abstracting from them the qualities (or logical "places") they share.

We may distinguish two types, the Petrarchan and the metaphysical. In the Petrarchan c., physical qualities or experiences are metaphorically described in terms of incommensurate physical objects, e.g. "When I turn to snow before your burning rays . . . (Petrarch, *Canzone* 8). The Petrarchan c. greatly influenced poetics in Italy, France, and England—it was widely used by the Elizabethan sonneteers and by Tasso—before giving way to the more codified styles of Petrarchism and, later, Marinism, whose catalogues of incongruous images function less like forms of c. than like mnemonic epithets.

By contrast, distinctly conceptual vehicles of comparison characterize the metaphysical c. (the kind of c. usually intended in critical discussions of the term), for here internal qualities are conveyed by vehicles with which they share no physical features. The "metaphysical poets" were so dubbed by Dr. Johnson because, in his view, their poems "imitated" nothing, "neither . . . nature nor life" (*Life of Cowley*). Ever since Donne, the first and foremost of the metaphysicals, took erotic and, later, religious love for his themes, the ingeniousness of the metaphysical c. has been associated with esp. intense sensual and spiritual experience. Aiming at describing the incomparable nature of that experience by way of figural comparisons, Donne's imagery veered from the concrete toward objects of essentially abstract content (the book, the mind, the tear, the picture) and of cosmological proportions (the sun, the heavens, the sphere), and such rhetorical figures as hyperbole, catachresis, and paradox (qq.v.). The antimimetic logic of the metaphysical c. allowed for extended metaphor sometimes bordering on allegory (q.v.), but usually returning to restate the opening terms of its analogy, as in Donne's famous figure, in *A Valediction: Forbidding Mourning*, of lovers' souls as compass legs:

> If they be two, they are two so
> As stiff twin compasses are two,

Thy soul the fixed foot, makes no show
To move, but doth, if th'other do.
. . . .
Thy firmness makes my circle just,
And makes me end, where I begun.

In modern poetry and poetics the c. has been associated chiefly with T. S. Eliot, who championed his own interp. of the polyvalent "sensibility" of the metaphysicals. But its chief resurgence after the metaphysicals occurred in Fr. Romantic and Symbolist poetry, as in the combination of metaphor and personification in Baudelaire's "Spleen": "I am a graveyard abhorred by the moon, / Where long worms drag themselves on like remorse / Relentlessly devouring my most cherished dead." The uncommon, often unbecoming, objectifications of Baudelaire's cs. both disturb and please by describing artificially what the mind can imagine and recognize but the imitation of nature cannot express, significant relations found in the least expected of places.

This capacity for forging improbable relations—once called *wit* (q.v.)—is central to the c., for the sense evoked by a c. is not simply surprise or, in Dr. Johnson's terms, wonder at the perversity which created it. Succinct in its immediate effect, while allowing for ever greater amplification and development, the c. imitates the fiction-making capability of metaphor itself.

R. M. Alden, "The Lyrical C. of the Elizabethans," *SP* 14 (1917), "The Lyrical Cs. of the Metaphysical Poets," *SP* 17 (1920); K. M. Lea, "Cs.," *MLR* 20 (1925); G. Williamson, *The Donne Trad.* (1930); C. Brooks, "A Note on Symbol and C.," *Am. Rev.* 3 (1934); M. Praz, *Studies in 17th-C. Imagery* (1939); G. E. Potter, "Protest Against the Term C.," *PQ* 20 (1941); Tuve; T. E. May, "Gracián's Idea of the 'Concepto,'" *HR* 18 (1950); J. A. Mazzeo, "A Critique of Some Mod. Theories of Metaphysical Poetry," *MP* 50 (1952); D. L. Guss, *John Donne, Petrarchist* (1966); K. K. Ruthven, *The C.* (1969); E. Miner, *The Metaphysical Mode from Donne to Cowley* (1970); F. Warnke, *Versions of Baroque* (1972); A. Fowler, *Conceitful Thought* (1975); C. Brodsky, "Donne: The Imaging of the Logical C.," *ELH* 49 (1982); A. A. Parker, "'Concept' and 'C.': An Aspect of Comp. Lit. Hist.," *MLR* 77 (1982).

F.J.W.; C.B.L.

CONCRETE POETRY is visual poetry, esp. of the 1950s and 1960s, in which (1) each work defines its own form and is visually and, if possible, structurally original or even unique; (2) the piece is without any major allusion to any previously existing poem; and (3) the visual shape is wherever possible abstract, the words or letters within it behaving as ideograms. This dissociates the prototypical c. poem from the freer-form visual poems of futurism and dada as well as from the less consciously programmatic visual poetry of the 1920s through the 1940s, epitomized in e e cummings' "l(a." The term "c. p." was adopted in 1955 when the Swiss poet Eugen Gomringer (b. 1924) met Decio Pignatari (b. 1927), who had founded, with his fellow Brazilians Augusto and Haroldo de Campos (b. 1931 and 1929), the Noigandres Group. The term was adapted from the Swiss sculptor Max Bill (b. 1908), who called some of his works *konkretionen* and whose secretary Gomringer was at that time; however, it had already been used independently in 1953 by the Swedish poet and artist Öyvind Fahlström to describe his own visual poems, which are unlike those of the later group. The main manifesto of the Noigandres group, "Pilot Plan for C. P." (1958), is reprinted in Solt (1968) and may be taken as describing the early concrete program.

Over the next four years the group expanded to include such Germans as Claus Bremer, Dieter Roth, and Franz Mon, the Swiss Daniel Spoerri, the Austrians Gerhard Rühm and Ernst Jandl, the Scot Ian Hamilton Finlay, and the expatriate Am. Emmett Williams. By the middle 1960s, when the largest-scale anthols. of c. p. appeared, those edited by Williams (1967) and by Mary-Ellen Solt (1969), the movement included several dozen more poets in countries as various as Spain, Italy, Argentina, and Japan. By that point the rigid original definition had been stretched and some degree of textual allusion and visual mimesis was allowed, though with the growing fashion for c. p., not everyone who called himself a "c. poet" was accepted by everyone else. But the influence of the anthols. and the dissemination to a larger audience encouraged many other poets who were by no means avant-garde to experiment with visual poetry (still known as "c." in most cases). This was the case with such Am. poets as May Swenson, M. L. Rosenthal, and John Hollander. Thus c. p. became not the style of a group but a possible form for the mainstream. The most conspicuous result of this popularization was a widespread confusion about the term "c. p." which has persisted to the present day.

From the late 1960s onward, some younger visual poets formed new groups, calling their works "post-c." (Cavan McCarthy), "visuelle Gedichte" (K. P. Dencker) or, ultimately, *poesia visiva* (sometimes attributed to Luciano Ori but actually first used by Eugenio Miccini as early as 1962); because "visual poetry" (q.v.) describes the general field of works which fuse visual and literary art, *poesia visiva* is used untranslated, and has become the commonest term in the Eng.-speaking world to describe the visual poetic forms that have developed since c. p. *Poesia visiva* typically combines photography and graphic techniques with letters, and these do not always make sense semantically, so that with *poesia visiva* one truly enters the world of visual art more than lit. As for the original c. poets, many of them turned to sound poetry (q.v.) after the 1960s and today combine the media of c. and sound poetry into activities for radio, television, and cinema.

R. H. Fogle, *The Imagery of Keats and Shelley* (1949), ch. 5; R. L. Beloof, "E. E. Cummings: The Prosodic Shape of His Poems," *DAI* 14 (1954); *An Anthol. of C. P.*, ed. E. Williams (1967); *C. P.: A World View*, ed. M.-E. Solt and W. Barnstone (1969); R. P. Draper, "C. P.," *NLH* 2 (1971); "Konkrete Poesie I, II," *Text & Kritik* 25, 30 (1974–75); *Theoretische Positionen zur konkreten Poesie*, ed. T. Kopfermann (1974); A. Marcus, "Intro. to the Visual Syntax of C. P.," *VLang* 8 (1974); H. Hartung, *Experimentelle Literatur und konkrete Poesie* (1975); L. Gumpel, *"C." P. from East and West Germany* (1976)—conflates terms; A. and H. de Campos and D. Pignatari, *Teoria da poesia concreta* (1975); J. L. McHughes, "The Poesis of Space," *QJS* 63 (1977); *Visual Lit. Crit.* (1979), *Am. Writing Today* (1982), both ed. R. Kostelanetz; D. W. Seaman, *C. P. in France* (1981); W. Steiner, "Res poetica," *The Colors of Rhet.* (1982); G. Janecek, *The Look of Rus. Lit.* (1984); *Poetics of the Avant-Garde / C. P.*, Spec. Iss. of *PoT* 3, 3 (1982); *Verbe et image: Poésie visuelle*, ed. D. Higgins and K. Kempton, Spec. Iss. of *Art contemporain* 5 (1983); E. Sacerio-Gari, "El despertar de la forma en la poesía concreta," *RI* 50 (1984); *The Ruth and Marvin Sackner Archive of C. and Visual P. 1984* (1986)—catalog and bibl.; K. McCullough, *C. P.: An Annot. Internat. Bibl.* (1989); C. Bayard, *The New Poetics in Canada and Quebec* (1989). D.H.

CONVENTION. By "c." is meant any rule that by implicit agreement between a writer and some of his or her readers (or of his audience) allows him certain freedoms in, and imposes certain restrictions upon, his or her treatment of style, structure, genre, and theme and enables these readers to interpret his work correctly. Combining social and objective functions, literary cs. are intersubjective; they hold (like linguistic cs.) the normative force which underlies the possibility of communication at all.

Unlike users of linguistic cs., readers who are party to literary cs. may be very few indeed, else a writer could never create a new c. (e.g. free verse or sprung rhythm), revive an old c. (such as alliterative meters in modern languages), or abandon an old c. (the pastoral elegy). Readers who are ignorant of—or at least out of sympathy with—the c. must to some extent misinterpret a work that exemplifies it, and when the number of such readers becomes large, writers may abandon the c.—though of course works that exemplify it remain to be interpreted. Dr. Johnson in his judgment of *Lycidas* is an instance of a reader who misinterprets a work because he is ignorant of or out of sympathy with its cs. (*Life of Milton*).

Cs. govern the relations of matter to form, means to ends, and parts to wholes. Some examples of cs. of style are the rhyme scheme of the sonnet and the diction of the ballad; of structure, beginning an epic *in medias res* and the strophic structure of the ode; of genre, representing the subject of a pastoral elegy as a shepherd; of theme, attitudes toward love in the Cavalier lyric and toward death in the Elizabethan lyric. The function of any particular c. is determined by its relationship to the other cs. which together form the literary system. At any point in the hist. of its transmission, this system provides for the finite articulation of infinite literary possibilities.

Cs. both liberate and restrict the writer. Because cs. usually form sets and are motivated by traditional acceptance, a writer's decision to use a certain c. obliges him or her to use certain others or risk misleading the reader. The cs. of the epic, for example, allow a writer to achieve effects of scale but compel her or him to forgo the conversational idiom of the metaphysical lyric.

To break with cs. (or "rules") is sometimes thought a merit, sometimes a defect; but such a break is never abandonment of all cs.—merely replacement of an old set with a new. Wordsworth condemns 18th-c. poetry for using poetic diction (Preface to the 1800 *Lyrical Ballads*); F. R. Leavis condemns Georgian poetry for adhering to "19th-c. cs. of 'the poetical.'" The institutional character of c. supports theories of literary change based on the dialectic of trad. and innovation. In the New Criticism, "conventional 'materials'" are "rendered dramatic and moving" by the individual work (Brooks 98), while in Rus. Formalism, dominant cs. change in function as they are displaced by the nascent ones they contain.

The social nature of cs. also distinguishes them from universals. R. S. Crane proposes that the sense of "c." be restricted to denote "any characteristic of the matter or technique of a poem the reason for the presence of which cannot be inferred from the necessities of the form envisaged but must be sought in the historical circumstances of its composition" (198). In other words, those features that all works in a certain genre must by definition share are not (in Crane's sense) cs. He would not count an unhappy ending as a c. of tragedy but does count the chorus of Gr. tragedy as such. Structuralism, by contrast, concedes the conventionality of all norms but seeks to identify the universal relational laws by which they are organized. The pervasiveness of c. accounts in poststructuralism for the absorption of subjectivity and authorial intention into *écriture*, the autonomous productivity of writing as an institution—(Barthes, 1953, 17–29) and for the concept of a plural, "writable" text whose meaning is unintelligible in traditional, "readerly" terms and must be written or invented by its readers (1970, 11–12). In the New Historicism, the social basis of literary cs. allows for their resituation in relation to the cs. of nonliterary discourse and of nondiscursive practices and institutions. Because cs. mediate between nature and its representation, as well as between authors and readers, different theories of c. (and different historical periods) will lay stress on their relation

to universals or to the poetic system, to the individual or to society.

J. L. Lowes, *C. and Revolt in Poetry* (1922); F. R. Leavis, *New Bearings in Eng. Poetry* (1932); Brooks; R. S. Crane, *The Langs. of Crit. and the Structure of Poetry* (1953); R. Barthes, *Le Degre zero de l'écriture* (1953), *S/Z* (1970); R. M. Browne, *Theories of C. in Contemp. Am. Crit.* (1956); N. Frye, "Nature and Homer," *Fables of Identity* (1963); S. R. Levin, "The Cs. of Poetry," *Literary Style: A Symposium*, ed. S. Chatman (1971); V. Forrest-Thomson, "Levels in Poetic C.," *JES* 2 (1972); J. Culler, *Structuralist Poetics* (1976), "C. and Meaning: Derrida and Austin," *NLH* 13 (1981); L. Manley, *C., 1500–1750* (1980), "Concepts of C. and Models of Critical Discourse," *NLH* 13, 1 (1981)—spec. iss. on c.; M. Steinmann, "Superordinate Genre Cs.," *Poetics* 10 (1981); Fowler; *NLH* 14, 2 (1983)—spec. iss. on c.; C. E. Reeves, "The Langs. of C.," *PoT* 7 (1986), "'Conveniency to Nature': Literary Art and Arbitrariness," *PMLA* 101 (1986). M.ST.; L.M.

COUPLET (Lat. *distich*, Fr. *rime plate* and [obs.] *vers commun*, Ger. *Reimpaar*; from Lat. *copula; Welsh cywydd*). Two contiguous lines of verse which function as a metrical unit and are so marked (usually) either by rhyme or syntax or both. Since the advent of rhymed verse in the European vernaculars in the 12th century, the c. has counted as one of the principal units of versification in Western poetry, whether as an independent poem of a gnomic or epigrammatic nature, as a subordinate element in other stanzaic forms—two of the principal stanzaic forms of the later Middle Ages and the Ren., *ottava rima* and rhyme royal [qq.v.], both conclude with a c., as does the Shakespearean sonnet—or as a stanzaic form for extended verse composition, narrative or philosophical. In each of these modes the tightness of the c. and closeness of its rhyme make it esp. suited for purposes of formal conclusion, summation, or epigrammatic comment. In dramatic verse, the c. occurs in the cl. Fr. drama, the older Ger. and Dutch drama, and the "heroic plays" of Restoration England. It also fills an important function in Elizabethan and Jacobean drama as a variation from the standard blank verse (q.v.), its principal use being to mark for the audience, aurally, the conclusion of a scene or a climax in dramatic action. The c. occupies a unique and interesting position in the typology of verseforms. Standing midway between stichic verse and strophic, it permits the fluidity of the former while also taking advantage of some of the effects of the latter.

In the medieval Fr. epic, the older assonanted *laisse* gives way to rhyme in cs. around the 12th century. The earliest examples, such as the *Cantilene de Sainte Eulalie* and the *Vie de Saint Léger* are in closed couplets (Meyer 7). It is Chrétien de Troyes (fl. after 1150) who seems to have introduced *enjambement*, which by the 13th century is common (e.g. Raoul de Houdenc). After the *Chanson de Roland* (ca. 1100), first the decasyllabic then the alexandrine c. is the dominant form of OF narrative and dramatic poetry. In the hands of the masters of Fr. Classicism—Corneille, Moliére, Racine, La Fontaine—this is end-stopped and relatively self-contained, but a freer use of enjambment is found among the romantics. Under Fr. influence, the alexandrine c. became the dominant metrical form of Ger. and Dutch narrative and dramatic verse of the 17th and 18th centuries. Subsequently, a more indigenous Ger. c., the tetrameter c. called *Knittelvers*, was revived by Goethe and Schiller. The term "c." is sometimes used in Fr. prosody with the meaning of stanza, as in the *c. carré* (square c.), an octave composed of octosyllables.

In Eng. poetry, though the octosyllabic or iambic tetrameter c. has been used well, by far the most important c. form has been isometrical and composed of two lines of iambic pentameter. As perfected by Dryden and Pope, the so-called "heroic c." (q.v.) is "closed"—syntax and thought fit perfectly into the envelope of rhyme and meter sealed at the end of the c.—and in this form dominates the poetry of the neoclassical period: "Know then thyself, presume not God to scan," says Pope in the *Essay on Man*, "The proper study of Mankind is Man." When meter and syntax thus conclude together, the c. is said to be "end-stopped."

The c. is "open" when enjambed, i.e. when the syntactic and metrical frames do not close together at the end of the c., the sentence being carried forward into subsequent cs. to any length desired and ending at any point in the line. This form of the c. is historically older and not much less common than the closed form in Eng. poetry as a whole. It was introduced by Chaucer and continued to be produced (e.g. Spenser, *Mother Hubberd's Tale* [1591]; Nicholas Breton, *Machivils Instructions* [1613]) well into the 17th c. during the very time when the closed or "heroical" c. was being established. Puttenham and other Ren. critics labeled this older form "riding rhyme" (origin of this term unknown). It was further explored in the 19th century (e.g. Browning), proving then as ever esp. suited for continuous narrative and didactic verse. Long sentences in enjambed cs. constitute the rhymed equivalent of the "verse paragraph" (q.v.) in blank verse. In a medial form between closed and open, each two-line sentence ends at the end of the *first* line of the c., making for systematic counterpointing of syntax against meter.

Not all Eng. cs. are isometric, however. Poets as diverse as George Herbert and Robert Browning have developed c. forms which rhyme lines of unequal length: " . . . With their triumphs and their glories and the rest. / Love is best!" (Browning, "Love Among the Ruins"). A related heterometric form of c. which developed in Gr. and Lat., a hexameter followed by a pentameter, came to be known for its generic usage as the *elegiac distich*. Passed down through the Middle Ages as a semipopular form (e.g. the *Distichs of Cato*), it

survived into modern times.

Schipper; P. Meyer, "Le C. de deux vers," *Romania* 23 (1894); C. H. G. Helm, *Zur Rhythmik der kurzen Reimpaare des XVI Jahrhunderten* (1895); C. C. Spiker, "The Ten-Syllable Rhyming C.," *West Va. Univ. Philol. Papers* 1 (1929); P. Verrier, *Le Vers français*, 3 v. (1931–32), v. 2; G. Wehowsky, *Schmuckformen und Formbruch in der deutschen Reimpaardichtung des Mittelalters* (1936); E. N. S. Thompson, "The Octosyllabic C.," *PQ* 18 (1939): F. W. Ness, *The Use of Rhyme in Shakespeare's Plays* (1941), esp. ch. 5, App. C; B. H. Smith, *Poetic Closure* (1968), 70 ff.; J. A. Jones, *Pope's C. Art* (1969); W. B. Piper, *The Heroic C.* (1969); Brogan, 389 ff.

T.V.F.B.; W.B.P.

D

DECORUM. The Cl. term *d.* (Gr. *to prepon*) refers to one of the criteria to be observed in judging those things in our experience whose excellence lends itself more appropriately to qualitative than to quantitative measurement. This definition depends upon the ancient distinction between two types of measure stated by Plato in *The Statesman* (283d–85a) and adopted by Aristotle in his ethical, political, rhetorical, and literary treatises. Excess and deficiency, Plato says, are measurable not only in terms of the quantitative largeness or smallness of objects in relation to each other but also in terms of each object's approximation to a norm of "due measure." All the arts owe the effectiveness and beauty of their products to this norm, *to metron*, whose criteria of what is commensurate (*to metrion*), decorous (*to prepon*), timely (*to kairon*), and needful (*to deon*) all address themselves to the "mean" (*to meson*) rather than to the fixed, arithmetically defined, extremes. This distinction is central to literary d. because lit. shares its subject matter and certain principles of stylistic representation with the disciplines of law, ethics, and rhet., which also are concerned with the qualitative analysis and judgment of human experience.

With respect to subject matter, Aristotle says that poetry treats human actions (*Poetics* 2, 4, 6–9) by revealing their universal significance in terms of the "kinds of thing a certain kind of person will say or do in accordance with probability or necessity" (Else, 9.4). The kinds of thing said or done will indicate certain moral qualities (*poious*) of character which, in turn, confer certain qualities (*poias*) upon the actions themselves (6.5–6). Extending the terms of Plato's *Statesman* (283d–85b, 294a–97e) to ethics, law, and rhet., Aristotle defines the concept of virtue necessary to evaluate such qualities of character as a balanced "disposition" (*hexis*) of the emotions achieved by the observation of the "mean" (*mesotes*) *relative to each situation* (*EN* 2.6). Likewise in law, equity is a "disposition" achieved through the individual application of a "quantitatively" invariable code of statutes for (amounts of) rewards and penalties to "qualitatively" variable and unpredictable human actions. As the builder of Lesbos bent his leaden ruler to a particular stone, so the judge, in drawing upon the universal law of nature (*to katholou*), might rectify the particular application of civil law by making a special ordinance to fit the individual case (*EN* 5.7, 10). Equity brings these universal considerations to bear by looking to the qualitative questions of the legislator's intentions and of what kind (*poios*) of man the accused has generally or always been in order to mitigate his present act (*Rhet.* 1.13.13–19).

With respect to stylistic representation, all verbal (as opposed to arithmetical) expression is qualitative, and rests content if it can establish a likely similarity (rather than an exact equivalence) between things (Plato, *Crat.* 432ab, *Tim.* 29cd). Diction achieves d. by a proper balance between "distinctive" words, which contribute dignity while avoiding obtrusiveness, and "familiar" words, which, while avoiding meanness, contribute clarity (Aristotle, *Rhet.* 3.2–3, *Poetics* 21–22). Lang. as a whole achieves d. if the style expresses the degree of emotion proper to the importance of the subject and makes the speaker appear to have the kind of character proper to the occasion and audience. The greater the subject, the more powerful the emotions the speaker or writer may decorously solicit. The more shrewdly he estimates the "disposition" which informs the character of his listeners, the more he can achieve the "timely" (*eukairos*) degree of sophistication appropriate to the occasion (*Rhet.* 3.7; cf. Plato, *Phaedrus* 277bc).

Influenced by the Middle Stoicism of Panaetius and Posidonius, Cicero did most to define and transmit the relation of literary d. to ethics and law, suggested by such words as *decor* and *decet*, in his philosophical and rhetorical treatises (*De Off.* 1.14, 93–161; *Orat.* 69–74, 123–5). From his reformulation of Platonic, Aristotelian, and Isocratean attitudes, there flowed an immense variety of literary and artistic applications of the concept by grammarians, poets, rhetoricians, historians, philosophers, theologians, and encyclopedic writers on specialized disciplines. Most exemplary, perhaps, of the vitality and versatility of this trad. is St. Augustine's adaptation of it to Christian oratory and exegesis in *De doctrina cristiana*, where, while virtually rejecting the Cl. *doctrine* of "levels

of style"—to which d. was often reduced—he gave the Cl. *principle* of d. new life and applications (see Auerbach).

Like the "mean" and "disposition" of ethics and equity, d. is an activity rather than a set of specific characteristics of style or content to be discovered, preserved, and reproduced. It is a fluid corrective process which must achieve and maintain, instant by instant, the delicate balance between the formal, cognitive, and judicative intentions of literary discourse. Its constant resistance to imbalance, taking different forms in different periods, never offers a "solution" to be found or expressed by literary "rules" with which later neoclassical theorists often try to identify it. As soon as they reduce it to a doctrine concerning either a given kind of subject or a given kind of style or a fixed relation of the one to the other, d. ceases to exist, because its activity is a continuous negotiation between the two. For the forms that particular reductions or recoveries have taken, the reader must look to those literary controversies endemic to the specific period, genre, or issue of interest.

E. M. Cope and J. E. Sandys, *The Rhet. of Aristotle*, 3 v. (1877); G. L. Hendrickson, "The Peripatetic Mean of Style and the Three Stylistic Characters," *AJP* 25 (1904), "The Origin and Meaning of the Ancient Characters of Style," *AJP* 26 (1905); J. W. H. Atkins, *Lit. Crit. in Antiquity*, 2 v. (1934); W. Jaeger, *Paideia*, 3 v. (1943)—cultural background; E. De Bruyne, *Études d'esthétique médiévale*, 3 v. (1946)—med. background; M. T. Herrick, *The Fusion of Horatian and Aristotelian Lit. Crit. 1531–1555* (1946); Curtius, ch. 10; Norden; F. Quadlbauer, "Die Genera Dicendi bis Plinius d.J.," *WS* 71 (1958), "Die antike Theorie der Genera Dicendi im lateinischen Mittelalter," *Akad. D. Wissens. zu Wien, Sitzungsberichte* 241.2 (1962); H. I. Marrou, *Histoire de l'éducation dans l'antiquité*, 5th ed. (1960)—cultural background; Weinberg; C. O. Brink, *Horace on Poetry*, 3 v. (1963–82); G. F. Else, *Aristotle's* Poetics: *The Argument* (1963); E. Auerbach, "*Sermo Humilis*," *Literary Lang. and its Public in Late Lat. Antiquity and in the Middle Ages* (tr. 1965), ch. 1; M. Pohlenz, "*To prepon*," *Kleine Schriften*, ed. H. Dorrie (1965); A. Patterson, *Hermogenes and the Ren.* (1970); T. McAlindon, *Shakespeare and D.* (1973); Murphy; W. Edinger, *Samuel Johnson and Poetic Style* (1977)—18th-c. background; S. Shankman, *Pope's* Iliad (1983), ch. 3; W. Trimpi, *Muses of One Mind* (1983)—see index; T. A. Kranidas, *The Fierce Equation* (1965); K. Eden, *Poetic and Legal Fiction in the Aristotelian Trad.* (1986). W.T.

DESCRIPTIVE POETRY.

I. CLASSICAL AND LATE ANTIQUITY
II. THE MODERN PERIOD

I. CLASSICAL AND LATE ANTIQUITY. D. p. in the West concerns the theory and practice of description in versified texts. D. p. is not a genre in itself, but d. portrayal has always played an important role in verse art, in a continuous trad. reaching back to the oldest surviving poetic texts. Its tenacity is epitomized in Horace's well-known phrase *ut pictura poesis* (*Ars poetica* 361; q.v.), which, divorced from its context, became a Ren. dictum encapsulating the aspiration of poetry to portray its subject matter quasi-visually. Ultimately, description should be considered as a strategy and an instancing of the doctrine of imitation (q.v.; see also REPRESENTATION AND MIMESIS). Plato's condemnation (*Republic* 10) of the powerful mimetic arts of epic and tragic poetry as dangerously misleading "copies of copies" was historically superseded by Aristotle's defense of imitation as satisfying a natural instinct for knowledge and instruction, hence an effective means of learning. It is impossible in this space to give an exhaustive account of the uses of description in poetry throughout the ages; rather, this article aims simply to point out some of the diverse functions of description in Western poetry, recognizing that its presence in Eastern lits. is equally widespread and profound.

In practice, d. technique has been applied mainly: (1) to the female body, as in the biblical Song of Solomon (1.5,10,15; 4.1–7, 10–15; 6.4–7,10; 7.1–9; but at 1.13–14 and 5.10–16 the man's body is eulogized) and the Fr. *blason*; (2) to man's surroundings and habitat, whether natural or cultivated landscape or crafted architecture; and (3) to isolated objects, whether utensils or representational works of art.

In antiquity, a theory of description (*ekphrasis* [q.v.], *descriptio*) was elaborated for practical use by Cl. rhetoricians, but it was equally applicable, with certain modifications, to poetry. Gr. and Lat. rhetorical handbooks frequently cited examples from poetry, esp. epic, to illustrate their points. *Ekphrasis*, one of the elementary rhetorical exercises (*progymnasmata*), was intended to train pupils in the art of "bringing a subject before the eyes [of the audience]." Events and actions unfolding in time were as appropriate to *ekphrasis* as static entities such as cities, landscapes, and buildings. The modern distinction between the narrative thread and interpolated d. 'digressions' does not apply in antiquity. Quintilian (8.3.68–70, quoted by Erasmus, *De copia* 2.5) shows how the summary phrase "a sacked city" can be expanded to show in action the unfolding process of its devastation in pictorial detail. This example would apply not only to prose oratory but also to versified poetry, as can be seen from the fact that the quotations in 6.2.32–33 are all from the *Aeneid*.

Vividness or *enargeia* is the effective quality deriving from poetic description. Its use was didactically codified in rhetorical rather than poetic treatises, yet both arts engage the subjective faculties of the performer (poet, actor, or orator) who delivers the text with those of the audience. The stimulation of inward vision in the imagination (q.v.) and the arousal of concomitant feelings are closely linked. The practical aim is to evoke vividly

the scene of an action or the situation of an object. Pseudo-Longinus praises passages in tragedy where the poet seems to have been actually present at the scene (*De sublimitate* 15). The immediacy of the effect of subjective representation is more important than the strict truth of its contents. Veracity—as opposed to verisimilitude—must therefore be guaranteed by other means: professional historiographers and forensic lawyer-orators were therefore advised to scrutinize their material with caution; nevertheless, they too were to appeal to the visual imagination and to the emotions of the audiences. In poetry the intended effect was the qualitatively more striking one of *ekplexis*—astonishment, amazement—which permitted, even encouraged, the depiction of fantastic, supernatural, and mythological creatures such as the Furies, allegorical personifications (e.g. the figure of Discord in *Iliad* 4.442, which is praised at the expense of the description of Strife in the pseudo-Hesiodic *Shield of Herakles* in *De sublimitate* 9), and powerful events such as the battle of the Titans in the *Gigantomachia*.

Aristotle (*Rhetoric* 1411b–1412a) always uses the adverbial phrase "before the eyes" (*pro ommathon*) as a concrete synonym for the abstract notion of "vivid(ness)." And he goes further in asserting that this quality produces an effect of "actuality": the Gr. term is *energeia*; and lexical confusion with *enargeia* may stem from this source. The aim of such representation remains that of vitality. And Aristotle recommends the use not only of adjectives of visual or tactile qualities but even more of verbs of motion (esp. present participles) and adverbial phrases denoting—representing—active movement. He draws his examples from Euripides and Homer rather than from prose oratory.

Such prescriptive epitomes for powerful description were a didactic codification of practice within a living trad. The Homeric poems were composed long before any extant rhetorical handbook; yet they are full of vividly described events (battles, Odysseus' adventures) and places (the palace and gardens of Alcinous and the cave of the nymphs in *Odyssey* 7, 13). The most famous and most discussed d. passage in Homer is indubitably the ekphrasis of the Shield of Achilles in *Iliad* 18. The account of the object is presented in a narrative form: Homer depicts Hephaestus actually making the object and describes the scenes on the shield as if they were unfolding in time. Poetry thus invests static spatial objects with vitality by transfusing into them its own rhythmic, temporal succession.

Such use of description for vividly heightened narrative and for pictorial set pieces is to be found throughout the history of epic (q.v.) in authors such as Apollonios Rhodios, Nonnos, and Virgil. Virgil adopts Homer's device of the shield (*Aeneid* 8) but takes the finished object as the point of departure for his description. He imbues it with significance by taking each scene portrayed as the occasion for a prophetic excursus on the future glory of Rome. Different kinds of allegorical exegesis were applied to Homer's descriptions by Late Antique interpreters (e.g. Porphyry, *On the Cave of the Nymphs*). Allegorical significance was detected in Virgil's (d.) pastoral (q.v.) poetry by Servius and other commentators.

The highly polished, smaller-scale genres such as the epigram, pastoral, and epyllion) developed in the Hellenistic age also made use of concentrated description, usually of places and scenes, as in Theocritus' *Idylls*. The physical characteristics and attributes of the pastoral context are explicitly described either by the author or in the speech of the characters. Here, descriptions constitute a nostalgic evocation of a state of rustic innocence remote from urban life, mores, and lit. But the bucolic *locus amoenus* might equally stand for quiet leisure, necessary for sophisticated poetic learning, a simple, agreeable metaphor for *otium*. In Roman poetry, this theme is developed in accounts of the Golden Age such as in Virgil's fourth *Eclogue*, Horace's sixteenth *Epode*, and, with irony, in Catullus 64.

Isolated objects such as a rustic cup (Theocritus, *Idylls* 1.27–56), a ship ("Phaselus ille," Catullus 4; Horace, *Odes* 1.3) or a water droplet caught in a crystal (Claudian, *Epigrams* 4.2.12, a set of nine different descriptions of the same thing) can be found alongside poems addressed to (and evocative of) real places ("O funde noster," Catullus 44; "O Colonia," Catullus 17) or imagined locations, like the Palace of the Sun and the dwellings of Sleep and Fame in Ovid's *Metamorphoses*.

In descriptions of weather, straight meteorological eulogy (e.g. of Spring, Horace, *Odes* 1.4) may be developed as political allegory (e.g. storms as a metaphor for civic turbulence: Horace, *Odes* 1.2), as observations on the social economics of agriculture (*Odes* 2.15), or as a parable of Fortune and her consolations (*Odes* 1.9); this practice is still evident in the 19th c.

The use of description in dramatic action finds its natural function in evoking absent scenes (remote in time and space), and esp. in messenger speeches, which describe a (usually violent) action which has taken place offstage. Where this device includes reported speech, it is related to the *progymnasma* called *prosopopoiia* or *ethopoiia*, which consists of a recreation of the words of an individual in a certain situation, an exercise which would have trained an author in the composition of pithy, highly characteristic, direct speech. Pseudo-Longinus cites passages from tragedy to illustrate the dramatic effect of description and to underline the need for the poet to use the visual imagination in the composition process (*De sublimitate* 15). In Euripides' *Hecuba* (444–83), the chorus of captive Trojan women contrast the beautiful images of their memories of the sea, the sun, and the winds with the misery of their present plight and surroundings. Talthybius then enters to

relate (518–82) the sacrifice—or murder—of Polyxena by Neoptolemos, son of Achilles. His speech is full of vivid d. detail; and his representation of Polyxena's physical beauty, her decorous *pudeur,* and radiant nobility (557–70) evokes pathos by the intensity of its contrast with the barbarity of the act.

The importance of description in intensifying the rhetorical effect and dramatic impact of messenger speeches is attested by Erasmus (*De copia* 2.5; rev. 1532 ed.) as part of his wide-ranging analysis of description as a powerful device of oratory in all manner of prose and verse genres. Erasmus' treatment includes four divisions: *descriptio rei, personae, loci, temporis.* He cites examples from lyric, dramatic and epic poetry, as well as from prose speeches, historiography, and "natural science." It remains the most comprehensive classicizing Ren. survey. Homer is again proposed as the supremely vivid poet.

Didactic poetry (q.v.) such as Hesiod's *Works and Days,* Lucretius' *De rerum natura,* and Virgil's *Georgics* uses d. passages and personification to invest the subject matter with immediacy and vitality. Such appeals to the visual imagination may also have reinforced the poets' didactic intentions, since ancient memory techniques relied on visualization to store information in the form of images.

In the early centuries A.D., parallel developments can be seen in the use of description in poetry and rhet. Under the Roman Empire, rhet. changed from being a practical instrument used in courtrooms—and thus directly concerned with real human actions—to a highly sophisticated art form, a showcase for the orator's skill and ingenuity. Judicial rhet. is replaced by epideictic. School exercises such as *ekphrasis* became an end in themselves rather than a means. Speeches were often composed for particular occasions to praise objects such as cities, temples, or civic landmarks. Statius' *Silvae*—a collection of poems describing villas, statues, society weddings—illustrates the use of description in poetry for similar ends.

In Byzantium, which continued the trad. of ancient epideictic, versified poetry often replaced prose rhet.: long *ekphraseis* were composed in poetic meter using a literary, quasi-Homeric lang. Paul the Silentiary's verse *ekphrasis* of Hagia Sophia was publicly recited, like an epideictic speech, at the second consecration of the Church in 563. Like prose *ekphraseis,* these metrical works reproduce a viewer's sense impressions of the building (glittering gold, variegated marble) and expound its spiritual significance, and thus translate the inner response to the subject matter or its figural decoration of biblical scenes. These inward reactions are described in a radically subjective manner: they do not normally provide accurate information as to the exact appearance of the object. Nor are they exact in the archeological sense, permitting the detailed and precise reconstruction of an original. This is a legacy of the emotive function of description in rhet., where *vraisemblance* is more important than truth to nature.

It is the meshing of images (*phantasiai*) with emotions (*pathe*) which underlies the affective power of description. Ultimately, these literary and psycho-physiological effects all aspire to the mimetic condition, which revivifies static linguistic constructions so that they reproduce subjectively, via artistic means, the vitality and physical immediacy of natural experience.

II. THE MODERN PERIOD. D. technique is crucial to the effect of Ren. epic. In Tasso's *Gerusalemme liberata,* the interpenetration of vivid imagery, pictorial intensity, and psychological discernment powerfully emphasizes the seriousness of Rinaldo's desertion and reconversion. The fantastic, mythic-encyclopedic vision of the "Mediterranean" (Canto 15), the enchanted palace and gardens of Armida (15, 16), and the picturesque review of the Egyptian armies (17) are all resolved into the transfiguration of Rinaldo (Canto 18).

Ancient pastoral was revived in Ren. Neo-Lat. eclogues (Petrarch, Boccaccio) and in long poems such as Sannazzaro's *Arcadia.* In England, Sidney also wrote an *Arcadia,* and Spenser used d. technique in evoking the surroundings of the "Bower of Bliss" episode in *The Faerie Queene.* Milton composed pastoral poems in both Lat. (*Damon*) and Eng. (*Lycidas*) and used the pastoral setting for his depiction of Eden in *Paradise Lost* (4.689 ff.). But the whole epic is permeated by startling d. and visual style: Satan's call-to-arms ("Pandaemonium," 1.283 ff.) of his legions and their terrestrial advance (4.205 ff.), the portrayal of Raphael (5.246 ff.), and Michael's representation to Adam of a vision of the future course of hist. down to the Flood (Bks. 11, 12) are all splendid examples, as is Satan's temptation of Christ in *Paradise Regained* (4.25 ff.).

In the 17th c., Boileau (*Art poétique* 3.258 ff.) recommended a rich and sensuous elegance in description so as to avoid tedium, inertia, and cold melancholy and to relieve the austerity of the severe high style in epic and romance. Here too, description should enliven narrative. Boileau's acknowledged master is Homer, whose poems are "animated by a felicitous warmth of style" derived from his rich stock (3.297) of graceful and pleasing pictures (3.288) of stimulating beauty. The vibrancy of the images must be matched nuance for nuance by the figures of style: in this way the poet can, like Homer, alchemically transmute his imagined visions into poetic gold ("Everything [Homer] touched was turned to gold; / Everything receives in his hands a new grace" [3.298–99]). In the mind, vividness produces animation, which in turn generates vital warmth.

Racine's messenger speeches (*récits*) too are animated through description (Phèdre 5.6.1498–1592; *Andromaque* 5.3.1496–1524; *Iphigénie* 5.6. 1734–90). The premonitory dream of Athalia

(*Athalie* 2.5.485–540) colors a supernatural vision with enough appropriate, carefully selected physical detail to make the representation lifelike.

Perhaps the finest example, in this period, of description functioning as laudatory rhet. is La Fontaine's long (though incomplete) *Le Songe de Vaux* (1659–61), which describes the architecture and gardens of Fouquet's Château of Vaux-le-Vicomte and the spectacular celebrations prepared there for Louis XIV in 1661. The poem was intended as a monument in verse to Fouquet's worldly glory and munificence. The close relation in antiquity between *ekphrasis/descriptio* and *encomion/laus* is preserved in 17th-c. Fr. Classicism. La Fontaine's work is a combination of commissioned panegyric and theatrical, topographical fantasy. Of equal interest is the cycle of odes by Racine, *Le Paysage*, which describes a journey on foot to a Jansenist convent. It is a eulogy of the retired, contemplative religious life, couched in the pastoral terms of the idyll.

In Ger., the most significant text on poetic description to appear is Lessing's *Laokön* (1766, 1788), which, although it had immediate polemical intent in a dispute with Winckelmann, presents a pivotal analysis of the effects of poetry in comparison with painting and sculpture. Lessing quotes the epigrammatic statement attributed by Plutarch to Simonides, "painting is dumb poetry and poetry speaking painting," to show that, although there may be similarity or even identity of mental and spiritual effect ("vollkommene Ähnlichkeit der Wirkung") between two different arts, their technical systems of construction and expression remain irreducibly different. Lessing exemplifies his theoretical exposition by citing examples from Virgil and Homer, but he draws on a formidable range of Cl. sources, as well as citing Ariosto, Milton, and Pope. The *Laokön* is arguably the fundamental Western text for any appraisal of poetic description. Lessing's two main topics concern the necessity of matching moral qualities with physical, and of linking, through art, the different motions of time, events, the eye, the mind, and the feelings.

Goethe rarely presents such full description as can be found in *Ilmenau* (which like *Zueignung* continues a moral allegory) or *Ein zärtlich jugendlicher Kunimer*, but *Harzreise im Winter* shows more typically how vivid sense perception, subjective emotion, and moral reflection are everywhere interwoven in the fabric of Goethe's poetry. A couplet from the seventh *Roman Elegy* ("Now the sheen of the brighter ether illumines the stars; Phoebus the god calls forth forms and colours") shows how closely his poetic eye informs the "Theory of Colours" (*Farbenlehre*). And the light-hearted parable *Amor als Landschaftsmaler* puts in a humorous light Goethe's intuitive sense that Eros is the motivating force in Art as well as in Nature.

In the Eng. 18th c., Thomson's *The Seasons* (1727–30) initiated a literary fashion for poems describing the working conditions and settings of agricultural life, the pictorial elements of landscape, anecdotal details of rustic existence, and man's relation to nature. But the specific genre of topographical poetry was not recognized until 1779, when Samuel Johnson characterized it as "local poetry, of which the fundamental subject is some particular landscape . . . with the addition of such embellishments as may be supplied by historical retrospection, or incidental meditation" (*Life of Denham*); this prescription of landscape as subject matter restricts the wide historical application of the inherent power of vivid description of all kinds of objects. Hence, the genre needed to be sustained by strength of style. Dr. Johnson's observations were made in regard to Sir John Denham's *Cooper's Hill* (1642); and indeed, he found Denham's style to possess compact strength and concision. Denham had made paraphrastic translations of Virgil in 1636, and some bucolic fragments of these found their way into *Cooper's Hill* (Bush 165–69).

Gottfried van Swieten modeled his libretto for Haydn's *Die Jahreszeiten* on Thomson's poem. The poetry and music of this oratorio embody the loftier aspirations of the topographical genre toward an integrated, quasi-anthropological study of rural life in its confrontation with nature, but one which did not deny the sophisticated emotional response expected in pastoral. It blends the Enlightenment strain of sentimental, moralizing narrative with a noble feeling for the emotional and ethical grandeur of nature.

In the hands of Wordsworth, the technical limits and psychological distinctions of sense perception respected by earlier writers are dissolved. The early poems *An Evening Walk* and *Descriptive Sketches* (1793) show an attention to natural detail close to 18th-c. practice. But the tendency in the Tintern Abbey poem, the Intimations Ode, and *The Prelude* (e.g. the opening of Books 6 and 12) is to dissolve the separate boundaries of sense and cognition so as to achieve an utterly subjective union with a undifferentiated pantheistic Nature. This aspiration was made explicit by Wordsworth himself when he claimed to have "completely extricated the notions of time and space" and to have made an attempt "to evolve all the five senses; that is, to deduce them from one sense, and to state their growth and the causes of their difference, and in this evolution to solve the process of life and consciousness" (Renwick 173). This entailed the breakdown of the differentiated perceptual system within which carefully selected details provided the concrete material of vivid description. Instead, the emotions, which would otherwise be aroused by the mental apprehension of the d. details, are directly expressed in wholly subjective terms. Only passing references are made to the material, constituent elements of the natural world as normally perceived. This utterly inward response to nature took to an extreme Johnson's

recommendation to include in d. p. the fruits of "incidental meditation"; and in so doing it spelled the end of the topographical poem as such.

The directness of Wordsworth's encounter with nature is obvious: so many of his poems, from the *Descriptive Sketches* onwards, were composed in ecstatic response to a physical confrontation with landscape. His response is rapturous and utterly involving, but he rarely provides the common natural details of description for the reader to absorb and process mentally, and so to translate into congruent images.

In Germany, Heine's *Die Harzreise* (1824) and *Die Nordsee* (1825–26) show that the confrontation of a traveler on foot with the natural landscape in the hills or on the coast could still form the basis for an extended cyclical poem. But the concrete representation of landscape tends to fragment into a series of vignettes which cohere only by virtue of the underlying current of the poet's subjective response. The telling d. detail used by Hölderlin and Mörike always serves the ends of emotion or of metaphysical speculation. Mörike's fantasy island landscape of Orphid and Hölderlin's baffling *Hälfte des Lebens* are two instances of d. evocation utterly dissimilar in approach.

Wordsworth's revitalization of poetic diction (see LEXIS) and his practical reform of the aims, techniques, and limits of poetic description are linked. Similar parallels can be drawn between the proliferation of 19th- and 20th-c. poetic movements (e.g. symbolism, imagism, surrealism) intent on restating the qualitative virtues of poetic lang. in resistance to the pressures of quantitative science and expository prose, and the practice of individual poets who are also critics (Baudelaire, Mallarmé, Eliot, Seferis, Auden, Rilke, Bonnefoy). For all of these varied poets, the d. values of image and metaphor can be seen as an essential quality in poetic perception. Trace elements of sensation, drawn from experience and recombined to form images, are presented so as to stimulate vivid awareness of the external world or to express particular, sometimes extreme inner states. The result is a kind of "psychological description" (as detected by De Robertis in Leopardi's *Canti*) which portrays an "interior landscape," the *Weltinnenraum* of Rilke or the spiritual pilgrimage towards Bonnefoy's *arrière-pays*. Even the ekphrastic poems-about-paintings of Bruno Tolentino ("About the Hunt," "Those Strange Hunters") contain more of the reflective "embellishments . . . of incidental meditation," i.e. of imaginative appraisal of the subject matter, than an accurate verbal reconstitution of their subjects' physical construction. For Cavafy too, the act of radically subjective appropriation is the kernel of vivid poetic narrative: "You won't encounter [Odysseus' adventures] unless you carry them within your soul, unless your soul sets them up in front of you" (*Ithaka*). And Ungaretti's "Choruses Descriptive of Dido's States of Mind" (*La Terra promessa*) attempt a direct physiognomy of the soul rather than of the external symptoms of emotion.

The modern rural descriptions and nature poems of R. S. Thomas, Seamus Heaney, Ted Hughes, and Charles Tomlinson show neither the sophisticated pastoral ease of the urbanite at leisure nor the pantheist fervor of an idealizing imagination such as Wordsworth's, but rather a grim confrontation with the pitiless harshness, even brutality, of the natural world they observe. They describe unsentimentally the struggle for survival within the animal kingdom and the dogged, heroic resistance of real farmers to nature's destructive power. Again, the moral question of the fearless recognition of the true existential plight of man in his habitat is more important in influencing the description of his natural surroundings then the creation of aesthetic pleasure.

Altogether, description on a broad canvas, elaborate, explicit, and comprehensive in scope, has probably been less common in 19th- and 20th-c. poetry than concentration on intensity of d. detail in the imagery of lyric. And while the persistence of certain conventional themes, which create the fictional assumption of a travel journal or pilgrimage, or of a poetic monument to a piece of architecture or sculpture, or of a versified record of a geographical or topographical survey, may give the initial impression that there is a specifically d. genre, there is in fact always another governing intention (moralizing, didactic, persuasive, emotive) which is served, rather than conditioned, by the technique of description, which is rather to be considered as a rhetorical-poetic strategy to be applied in genres established on other grounds, as occasion demands.

E. W. Manwaring, *It. Landscape in 18th-C. England* (1925); M. Cameron, *L'Influence des 'Saisons' de Thomson sur la poésie d. en France (1759–1810)* (1927); R. A. Aubin, *Topographical Poetry in 18th-C. England* (1936); D. Bush, *Eng. Lit. in the Earlier 17th C.* (1945); G. De Robertis, *Saggio sul Leopardi* (1960); R. L. Renwick, *Eng. Lit. 1789–1815* (1963); J. Arthos, *The Lang. of Description in 18th-C. Poetry* (1966); A. Roper, *Arnold's Poetic Landscapes* (1969); J. W. Foster, "A Redefinition of Topographical Poetry," *JEGP* 69 (1970); P. Hamon, "Qu'est-ce qu'une description?" *Poétique* 2 (1972); G. Genette, *Figures III* (1972); J. B. Spencer, *Heroic Nature: Ideal Landscape in Eng. Poetry from Marvell to Thomson* (1973); H. H. Richmond, *Ren. Landscapes: Eng. Lyrics in a European Trad.* (1973); R. Williams, *The Country and the City* (1973); E. Guitton, *Jacques Delille (1738–1813) et le poème de la nature en France de 1750 à 1820* (1974); J. Gitzon, "British Nature Poetry Now," *MidwestQ* 15 (1974); D. Lodge, "Types of Description," *The Modes of Mod. Writing* (1977); J. G. Turner, *The Politics of Landscape: Rural Scenery and Society in Eng. Poetry, 1630–1660* (1979); "Towards a Theory of Description," ed. J. Kittay, spec. iss. of *YFS* 62 (1981); B. Peucker, "The Poem as Place," *PMLA* 96 (1981)—the lyric;

R. Barthes et al., *Littérature et réalité* (1982); A. Reed, *Romantic Weather* (1983); C.-G. Dubois, "Itinéraires et impasses dans la vive représentation au XVIe siècle," *Litt. de la ren.* (1984); D. E. Wellerby, *Lessing's* Laocöon: *Semiotics and Aesthetics in the Age of Reason* (1984); J. A. W. Heffernan, *The Re-Creation of Landscape* (1984); J. Holden, "Landscape Poems," *DenverQ* 20–21 (1986); J. Applewhite, *Seas and Inland Journeys: Landscape and Consciousness from Wordsworth to Roethke* (1986); E. W. Leach, *The Rhet. of Space: Literary and Artistic Representations of Landscape in Republican and Augustan Rome* (1988). R.H.W.; P.W.

DIALOGUE is a term used primarily in two ways, to refer to an exchange of words between speakers in lit., and to refer to a literary usage that presents the speech of more than one character without specific theatrical intentions. In the first sense, d. certainly lies at the center of all writing for the stage, although there are also important uses of monologue (q.v.) in dramatic writing. Like dramatic d., poetic d. consists of conversation in which characters and readers achieve varying levels of understanding by listening to a verbal exchange between speakers. The mode allows an author to reveal a wider range of ideas, emotions, and perspectives than would be possible with a single voice. Verbal exchanges between characters, whether in drama, fiction, or poetry, produce heightened dramatic tension and constitute d., while collections of monologues, such as Browning's *The Ring and the Book*, do not. Literary d. demands the interaction occasioned by immediate response, though the exchange need not be limited to two speakers. Monologues broken by interruptions and poems in which one speaker introduces another main voice are not legitimate examples of d. The dialogic technique dramatizes all forms in which it appears and captures the rhythms and nuances of spoken lang.

The best known Cl. examples are the Socratic ds. of Plato. Though written in prose, they were apparently based on 5th-c. B.C. dramatic mimes by Sophron and Epicharmus. Cl. authors included verse d. in pastoral, satiric, and philosophical poems. Lucian modeled his prose *Ds. of the Dead* on Plato's works but used the form primarily for satiric and comic purposes. The satires of Horace became models for poetic exchanges between speakers. Virgil's eclogues present short pastoral d. in verse. D. in poetry often remained purely literary, although it also encouraged recitation, as in philosophical debates and musical forms such as the duet and oratorio. Lyric ds. intended to be sung were used widely in antiphonal music, particularly hymns and litanies. The form lends itself to powerful emotional or philosophical exchanges, as in love lyrics and intellectual debates.

Verse d. flourished in the Middle Ages in "debates" and "flyting." The dualistic philosophical and theological temper of the age fostered d., as in *The Owl and the Nightingale.* A typical subject was the debate between the Soul and the Body. Med. romances and allegories used d. in ways that produce confusion as to whether these poems were meant soley for reading or were also meant to be staged. Many Oriental poems fit this description as well. The 13th-c. *Roman de la rose* employed multiple speakers for satiric and dramatic purposes, while the Fr. *débat* was a form of poetic contest that influenced the devel. of the drama. Villon's d. between heart and body is a notable example. Popular ballads often included more than one speaker in order to heighten dramatic tension or suspense, as in "Lord Randall" and "Edward."

The Ren. saw the rise of specifically philosophical d. in verse, related to the devel. of prose ds. by Elyot, Thomas More, and Ascham. Tasso called the d. *imitazione di ragionamento* (imitation of reasoning), claiming that it reconciled drama and dialectic. In the 16th c., John Heywood's *D. of Proverbs* and Margaret Cavendish's d. poems were effective examples of the form. Songs by Shakespeare, Sidney, and Herrick included questions and answers, emphasizing the close connection between d. and the interrogative mode. Spenser made effective use of multiple voices in *The Shepheards Calender,* developing the d. trad. of Virgil's eclogues. Notable Ren. examples of d. poems incl. Samuel Daniel's *Ulysses and the Siren* and Andrew Marvell's *A D. Between the Soul and the Body.* Courtly masques, like those by Jonson and Milton, combined d. with music and lyrics. Allegorical poems like Dryden's *The Hind and the Panther* provide an occasion for d., as do 18th-c. direct-speech poems like Pope's *Satires.* Many dialect poems adopt multiple speakers, as in the Scottish ballads of Robert Burns and the Dorset eclogues of William Barnes.

Although romantic emphasis on subjectivity increased the monologic aspect of lyric poetry during the 19th c., verse d. appears in poems and dramatic experiments by Byron, Shelley, and Keats. Ger. and Scandinavian writers of songs and romantic ballads made widespread use of the form. The Victorians sought to objectify romantic poetry by dramatizing the single-voiced lyric, but Arnold and later Hardy used d. effectively in their poems; Browning and Tennyson both wrote dramas in verse. Revived interest in prose d. appears in Wilde's *The Critic as Artist* and later, in France, in the d. experiments of Valéry.

The role of d. in 20th-c. poetry is often connected with the blurring of generic distinctions. While Yeats's *D. of Self and Soul* is a clear example of a traditional verse d., other works by Yeats, Beckett, Dylan Thomas, and Robert Frost suggest the role of poetic d. in a wider variety of dramatic and nondramatic forms. Implied d. in poems by Geoffrey Hill and Ted Hughes, for example, objectifies the subject and produces heightened dramatic tension in the lyric. Like the Victorian dramatization of the monologue, d. poems de-em-

phasize the determinate force of lang., stressing the conditional aspects of all utterances. The term "double lyric" has been used to refer to lyrics that contain a suppressed but significant dialogic element. D. thus remains a significant concern in poetry as traditional distinctions between poetry, prose, and drama are questioned and explored.

In recent years, emphasis on the "dialogic" aspect of all lang. has suggested that even the most seemingly monologic utterances are dependent on the interplay of meanings found only in d. This view has arisen largely as a result of increased interest in the work of the Rus. critic M. M. Bakhtin. Although Bakhtin's own writings emphasize the novel, his analysis of the role of the "dialogic" has implications for the understanding of all literary meaning. Bakhtin argues that every utterance is subject to reinterpretation from the perspective of d. On these terms, even apparent monologues always have a conditional dialogic element. Such a view has important consequences for the interp. of lyric and narrative poetry, where the first-person speaker or narrator often tries to achieve authority for a particular description of emotions or version of events. For Bakhtin, the unsaid, the partially said, and the equivocally said are as potentially meaningful as the clearly said. Likewise, all lang. uses are a product of conflicting social forces. On these terms, every verbal utterance contributes dialogically to the shaping of the world, and the interactions between such utterances produce an endless interplay of voices in d.

R. Hirzel, *Der Dialog: ein literarischer-historischer Versuch*, 2 v. (1895); E. Merrill, *The D. in Eng. Lit.* (1911); E. R. Purpus, "The D. in Eng. Lit. 1660–1725," *ELH* 17 (1950); R. Wildbolz, *Der philosophisches D. als literarisches Kunstwerk* (1952); J. Andrieu, *Le D. antique: Structure et presentation* (1954); C. Diesch, "Gespräch," and H. Schelle, "Totengespräch," *Reallexikon*; W. J. Ong, *Ramus, Method, and the Decay of D.* (1958); A. Langen, *Dialogisches Spiel* (1966); G. Bauer, *Zur Poetik des Dialogs* (1969); F. M. Keener, *Eng. Ds. of the Dead* (1973); J. Mukařovskij, *The Word and Verbal Art* (tr. 1977), ch. 2; M. E. Brown, *Double Lyric* (1980); D. Marsh, *The Quattrocento D.* (1980); M. M. Bakhtin, *The Dialogic Imagination* (tr. 1981); T. Todorov, *Mikhail Bakhtin: The Dialogical Principle* (1981, tr. 1984); W. G. Müller, "Das Ich im Dialog mit sich selbst," *DVLG* 56 (1982); A. K. Kennedy, *Dramatic D.: The Duologue of Personal Encounter* (1983); *Tasso's Ds.* [incl. his *Discourse on the Art of the D.*], tr. C. Lord and D. A. Trafton (1983); *Le D. au temps de la ren.*, ed. M. T. Jones-Davies (1984); *Bakhtin: Essays and Ds. on His Work*, ed. G. S. Morson (1986); J. R. Snyder, *Writing the Scene of Speaking: Theories of D. in the Late It. Ren.* (1989); *The Interp. of D.*, ed. T. Maranhão (1989); G. S. Morson and C. Emerson, *Mikhail Bakhtin: Creation of a Prosaics* (1990); M. S. Macovski, *D. and Lit.* (1992); D. H. Bialostosky, *Wordsworth, Dialogics, and the Practice of Crit.* (1992).

B.A.N.

DIDACTIC POETRY.

I. CONCEPT AND HISTORY
II. ANTIQUITY
III. MEDIEVAL, RENAISSANCE, AUGUSTAN
IV. 19TH AND 20TH CENTURIES

I. CONCEPT AND HISTORY. *Didaktikos* in Gr. relates to teaching, and implies its counterpart: learning. "All men by nature desire knowledge" and all experience (embodied in lang., says Croce); hence all lit. (in the broadest sense) can be seen as "instructive."

Given such interacting complexities, our problem becomes one of "historical semantics" (Spitzer): of modulations and transformations (Fowler) of the d. concept, and of attempts to distinguish it from near neighbors such as allegory, archetype, myth, symbol, fable, and satire. The category of unconscious or unintended teaching could lead us astray into infinite mists and abysses. (What do we learn from "Jabberwocky?" That "nonsense" can make an attractive poem.) Aware of such proliferating contexts, we try to focus here chiefly (but not exclusively) on poetry which clearly *intends* "useful teaching," embodying Horace's "instruction *and* delight" in the genre of "d. poems."

Basic categories relate to the contents ("themes" [q.v.]) of poems, inseparable from their forms. In *De vulgari eloquentia*, Dante specified the worthiest objects as Safety, Love, and Virtue—his corresponding poetic themes being War, Love, and Salvation (or Morality: "direction of the will"). To these we add knowledge (Science and Philosophy), Beauty (Aesthetics), Efficiency (e.g. "How to" run a farm or write a poem), and Information ("Thirty days hath September"). Modern discussions have tended to emphasize the problematics of Knowledge and Morality. Beardsley treated "the D. Theory of lit." as seeing "a close connection between truth and value" (426–32). He mentions *De rerum natura*, but no one reads Lucretius' poem today for information on materialism and atomic theory; Beardsley wants rather to clarify the relations among "predications," cognition, and how readers submit poems to experiential "testing"—with reference to "philosophical, economic, social, or religious" doctrines. Arnold Isenberg confronted the "strong case" of "the d. poem or essay" in an intricate analysis (265–81).

Another necessary preliminary is Fowler's concept of "mode," illuminating the continuing life of genres. Fowler writes: "modal terms never imply a complete external form" and "tend to be adjectival"—as in "d. essay" or "d. lyric." He finds it remarkable that "several important literary kinds, notably georgic, essay, and novel, are not supposed to have corresponding modes. Can it be that these modal options have never been taken up? By no means" (108). Once we accept modal extensions of didacticism, its presence and power throughout the hist. of poetry become obvious.

II. ANTIQUITY. A product of lang. and hist., poetry also transcends these via archetypes and translations; and d. modes probably preceded the invention of alphabets and writing. Indeed, for oral trads. (religious and secular) rhythm and metaphor have probably always been used to aid memory and enliven ritual; such speculations about prehistoric poetry seem to be confirmed by surviving fragments and modern anthropology. Religious scriptures in all langs. tend to be "poetic," mingling freely epic hist. (narrative), hymn and psalm, and prophetic vision and preaching. In the Judeo-Christian Bible, for example, there are elements of poetic drama, philosophy, practical wisdom (proverbs, q.v.), and parable—in d. modes.

It was the Greeks who first, in Europe, distinguished clearly between poetry of imitation (Homer) and versified science (e.g. the lost poem by Empedocles); within mimetic uses of melodious lang., between "manners" of narration and drama; and within the latter, analyzed the tragic catharsis (Aristotle)—with comedy, epic, and other "kinds" in the background. Plato notoriously banished the poets from his ideal Republic because they taught lies about the gods, and aroused and confused men's passions. Two main tendencies of the d. in verse were created by Hesiod (8th c. B.C.): in *Theogony*, knowledge about the gods, their origins and stories, problems of culture—moving towards philosophic abstractions; and in *Works and Days*, "how to" farm and the like—towards practical and specific information. Each kind (details are fragmentary) leaned towards a different style and meter (e.g. Aratus' "Phainomena" used hexameters); and in later Alexandria, the latter emphasis (erudition, technical information) became popular (e.g. Nicander of Colophon).

The Romans derived from the Greeks not only gods and ideas but most of their poetic genres (translating and adapting Empedocles and Aratus, for example). Four major Lat. poets wrote masterpieces which transformed d. poetry creatively: (a) Lucretius' *De rerum natura* became the prototypical "philosophic" poem (Santayana) in the Empedocles and *Theogony* line, invoking the values of sense-experience and materialistic metaphysics. (b) Virgil's *Georgics*, derived from Hesiod, became the popular "how to" poem: running a farm, living with the seasons, keeping bees, and so forth. (c) Horace, on his Sabine farm, not only wrote memorable odes, satires, and epistles, but also wrote letters of practical advice to poets (*Epistulae ad Pisones*, i.e. the *Ars poetica*), some of his ideas and phrases becoming proverbial: e.g. "the labor of the file," "in the midst of things," "from the egg," and "make Greece your model." Finally, (d) Ovid's versified advice related chiefly to "sex and society" (*Ars amatoria*), a Lat. primer in matters of love. Manilius wrote a five-book poem on astrology (ed. A. E. Housman); and there were others.

In sum, Lat. poets gave priority to instructing citizens and artists in a variety of subjects. Satire developed beyond light-hearted Horace to bitter Juvenal ("The Vanity of Human Wishes"). And in Asia, esp., poetry was central to Confucian teaching. Hindu philosophy is embodied in Vedic hymns (with the *Upaniṣads*); Buddhism, in epic poems (Mahābhārata and Rāmāyaṇa); and Persian "teaching" is embodied for Westerners in Omar Khayyam's *Rubáiyát* (quatrains) and the *Avesta* of Zoroaster. The bible of Islam, the *Qur'ān*, is quintessential Ar. poetry.

III. MEDIEVAL, RENAISSANCE, AUGUSTAN. Christian lit. in Europe (indeed, all religious poetry) was almost entirely d. Even in mimetic (narrative and dramatic) modes, its central purpose was to impart religious doctrines and values (see Curtius, ch. 3 and passim). For example, Martianus Capella, *On the Marriage of Mercury and Philology* (Menippean satire), *Le Roman de la rose*, and Spenser's *Faerie Queene* (allegory), Dante, Chaucer—d. elements are everywhere. Vagabond scholars mixed orthodox doctrine and symbols with irreverent satire and "pagan" feeling. Medieval lit. is dense with rhymed chronicles, encyclopedias, devotional manuals and saints' lives, popularized excerpts from church doctrine, and collections of aphorisms. Early modes of theater in western Europe—mysteries and moralities, Passion plays—aimed to combine indoctrination with celebration and entertainment. This is also true of the late masque, where the allegorical teaching is increasingly subordinated to spectacle and dance. Milton's *Comus* (1634), for example, ends with a moral: "Love Virtue."

The emerging Ren. and Reformation saw Lat. poetry enriched by the vernacular langs. In Ger., for example, d. works incl. Luther's Bible, *Bescheidenheit* by "Freidank" (1215–16), *Der Renner* by Hugo von Trimberg (ca. 1300), *Narrenschiff* by Sebastian Brant (1494), *Narrensbeschwörung* and *Gauchmatt* by Thomas Murner (1512–14). In Sp., such authors as Francisco Pacheco, Lopé de Vega, and Cervantes in *Don Quixote* and elsewhere wrote didactically at times. In It., literary theory flourished, reaffirming Cl. ideals and finding Virgil a "better teacher" than Cato (but Fracastoro wrote that "Teaching is *in a measure* the concern of the poet, but *not* in his peculiar capacity" [1555; emphases added]). Such satires as those of Parini and Ariosto mingled d. elements with satire and autobiography; and the *Georgics* were imitated. In France, Rabelais and Montaigne virtually created a lit. which excelled in *raison* as well as *esprit*; masterpieces by Racine and Molière shaped the nation's emerging trad. and education; Diderot preferred the *Georgics* to Virgil's other poetry; A. Chenier's enthusiasm for "modern" science parallels that of Goethe and the Eng. romantics; and Jacques de Lille translated Virgil and wrote *Les Jardins*.

But the d. line (and problem) was most clearly developed, perhaps, in England, as in Sir Philip Sidney's *Defense of Poesie* (1583, passim—but, despite Plato, the poet is for Sidney not a liar: "he nothing affirms and therefore never lieth"). Mil-

ton's great epic attempt was "to justify the ways of God to man"; and only in England did the Puritans close the theaters and behead a king in order to establish a Godly commonwealth. Though "engaged" tendencies in verse developed esp. in England and France, they were evident throughout much of Europe. Notoriously, the Restoration in England (one of whose classics is Dryden's *Virgil*) and the Neoclassic century which followed was a long period of transitions leading to romanticism. Dryden wrote political, historical, satiric, religious, and theological poems, and translated all four of the great Lat. d. poets; Pope wrote *An Essay on Crit.* (1711) and other imitations of Horace, philosophic and "Moral" essays in verse, and *The Dunciad* (1742–43)—a d. and mock-epic-satiric masterpiece; and it is hardly necessary to insist on the didacticism of Swift and Dr. Johnson. As Chalker puts it, important poems in the georgic trad. "were often remote from any practical purpose, although others were d. *in intention*" (emphasis added).

Not only the Lat. education of gentlemen who became poets, but the emergence of Newtonian science (see Nicolson) and The Royal Society (Cowley's poems) bore poetic fruit. New concepts of nature and light (optics) mingled with the descriptive *paysage moralisé* in complex ways, transforming georgic trad. most effectively in James Thomson's *The Seasons* (1730; see Cohen). We recall Dr. Johnson's distinction: the poet "does not number the streaks of the tulip." Grierson and Smith's history (1944) shows the dominance of d. and satiric modes, as well as the representative "timid revolt" of Gray in his Odes—yet few poems are more blatantly moralistic than Gray's "Elegy." The scientific (informative and theoretic) tendencies bore strange and influential fruit in Erasmus Darwin's poems (e.g. *The Botanic Garden*, 1791). And we recall the variety of d. elements in Crabbe, Goldsmith, and Burns: the Am. and Fr. Revolutions did make a difference.

IV. 19TH AND 20TH CENTURIES. After 1800, d. aims and methods underwent radical transformations. The main shift was one from an Augustan use of Cl. and Ren. models to a growing variety of philosophies and ideologies, followed by later compromises which made "Victorian" almost synonymous with moralizing in prose and verse—as in Carlyle, Emerson, Tennyson, Ruskin, some of the pre-Raphaelites, and Arnold. Few poets are more obviously d. than Blake, for example—but his mode became one of vision and prophecy, of what we now see as "myth-making." Wordsworth and Coleridge wrestled with didacticism, esp. in odes and poems of meditation ("The Growth of a Poet's Mind"); and though Coleridge lived to regret it mildly, he did conclude *The Rime of the Ancient Mariner* with a moral.

In Shelley's *Defense*, Milton's "bold neglect of a *direct* moral purpose" (emphasis added) is seen as proof of that poet's genius; but neither Blake nor Shelley was unaware that to seek "to justify the ways of God to man" was to be essentially d. True, Shelley wrote in the Preface to *Prometheus Unbound* that "d. poetry is my abhorrence"—meaning poetry whose teaching "can be equally well expressed in prose" (I. A. Richards' "separable content"—implying superficial moralizing); but surely in *Prometheus* and elsewhere Shelley was, and understood himself to be, a prophet-teacher, one of the "unacknowledged legislators" of the world. Thus Wimsatt and Brooks characterize his critical position correctly as a "rhapsodic didacticism" and "a didacticism of revolution." And it was Keats who saw Shakespeare as having lived "a life of allegory," Keats who wrote that "Beauty is truth, truth beauty."

One may generalize that the opposites of the Horatian pleasure-use polarity tend to meet; and that a strong anti-didacticism usually emerges in opposition to a boringly conservative culture (cf. Blackmur's "intolerable dogma"). Thus, when Poe attacked "the heresy of the D." (a position later adopted by the Fr. symbolists and others), it was because, like Shelley, he was surrounded by inept poetasters spouting clichés; and Whitman (whose didacticism has been compared to that of Blake and Wordsworth) picked up Poe's (and Emerson's) idea of working "indirectly" and symbolically. Even the fin-de-siècle proponents of art for art's sake were themselves moralists in rebellion against Arnold's "Barbarians" and "Philistines." But the older, moralizing trad. of Bryant, Longfellow, Whittier, and others in America and England had a strength of its own. In the 20th c., the best poets (e.g. Yeats, Auden) have been ardently "engaged" in various social, political, philosophical, and religious controversies.

In sum, modern didacticism assumed protean forms that could no longer be forced into the genre classifications of the Augustans. We can follow this process clearly through Wellek's *Hist. of Modern Crit.*: thus when A. W. Schlegel discusses "the d. philosophical poem," he falls into the trap (for Wellek) of defining all poetry as "esoteric philosophy"—excusable, however, when one means, as the romantics did, "a poetic philosophy, a thinking in symbols as it was practiced by Schelling or Jakob Böhme." In this, Wellek is disparaging inferior poems "held together *merely* by logic." Similarly, Wellek remarks that "with the years Wordsworth's point of view became . . . more and more *simply* d. and instructive" (emphases added). Still, Wellek quotes Wordsworth (1808): "every great poet is a teacher: I wish either to be understood as a teacher, or as nothing." This debate about the changing nature of didacticism took a variety of shapes in England, America, and Europe. Yet what long poem in Eng. had a more clearly d. intent (i.e. interp. of hist., fate) than *The Dynasts*—Hardy's *War and Peace*? Since the substance of such teachings has become increasingly complex, fragmented, and problematic, we find the emphasis now falling on conflict, dialectics, quest, doubt, psychol-

ogy, existential immediacy, and pluralism rather than on any fixed doctrine.

For instance, in a poem such as Karl Shapiro's *Essay on Rime*, and in much of T. S. Eliot, in the Pound of *The Cantos*, and in Stevens' "meditations," we witness modal transformations of the d. Robert Frost's inveterate didacticism finds expression by transformations of Cl. eclogue and pastoral (e.g. "Build Soil," "The Lesson for Today"; see Empson). Modern degradations of didacticism, of course, occur in the use of poetry for propaganda or even advertising. Anglo-Am. poetry is esp. rich in works with historical, regional, geographic, philosophic, and political substance. The "how to" motif is still evident when Pound writes an *ABC of Reading* or John Hollander a versified textbook on prosody. One thinks of such old-new genres as utopian (and dystopian) narrative and science fiction; and there is a strong d. element in modern satire (R. Campbell, A. M. Klein). A recent study by H. J. Blackham devotes a chapter to "Modern Instances" and concludes with "The Message" (cf. Scholes).

One concludes that the d. *mode* is very much alive in modern lit. (possibly more so in prose than in verse); and that many of the traditional d. genres have undergone complex transformations and modulations. We find not only "modern georgics" (Frost, MacNeice), but epigrams (Ogden Nash), parodies without number, parables (Kafka), and other genres aiming at some sort of didacticism. Critical theory, however, has tended either to skirt the issues or to convert the d. mode into related categories. S.J.K.

R. Eckart, *Die Lehrdichtung: ihr Wesen und ihre Vertreter*, 2d ed. (1909); G. Santayana, *Three Philosophical Poets* (1910); W. Kroll, "Lehrgedicht," Pauly-Wissowa; I. A. Richards, "Doctrine in Poetry," *Practical Crit.* (1929); H. M. and N. K. Chadwick, *The Growth of Lit.*, 3 v. (1932–40); T. S. Eliot, *The Uses of Poetry and the Uses of Crit.* (1933), *On Poetry and Poets* (1957), esp. "The Social Function of Poetry"; W. Empson, *Some Versions of Pastoral* (1935); G. Solbach, *Beitrag zur Beziehung zwischen deutschen und italienischen Lehrdichtung im Mittelalter* (1937); H. J. C. Grierson and J. C. Smith, *A Critical Hist. of Eng. Poetry* (1944); M. H. Nicolson, *Newton Demands the Muse* (1946); R. A. B. Mynors, "D. P.," *Oxford Cl. Dict.* (1949); Curtius; Wellek; D. Daiches, *Critical Approaches to Lit.* (1956), chs. 3–4; Wimsatt and Brooks; M. C. Beardsley, *Aesthetics* (1958); G. Pellegrini, *La Poesia didascalia inglese nel settecento italiano* (1958); E. Neumann, "Bispêl," W. Richter, "Lehrhafte Dichtung," and H. Zeltner, "Philosophie und Dichtung," *Reallexikon*; B. F. Huppé, *Doctrine and Poetry* (1959); W. G. Lambert, *Babylonian Wisdom Lit.* (1960); R. Cohen, *The Art of Discrimination: Thomson's* The Seasons *and the Lang. of Crit.* (1964); H. de Boor, *Fabel und Bispêl* (1966); L. L. Albertsen, *Das Lehrgedicht* (1967); L. Lerner, *The Truthtellers* (1967); R. Scholes, *The Fabulators* (1967); J. Chalker, *The Eng. Georgic* (1969)—good bibl.; C. S. Lewis, *Sel. Literary Essays* (1969); B. Sowinski, *Lehrhafte Dichtung im Mittelalter* (1971); A. Isenberg, "Ethical and Aesthetic Crit.," *Aesthetics and the Theory of Crit.* (1973); C. Siegrist, *Das Lehrgedicht der Aufklarung* (1974); B. Boesch, *Lehrhafte Literatur* (1977); B. Effe, *Dichtung und Lehre* (1977)—typology of Cl. d. genres; Hesiod, *Works and Days*, ed. M. L. West (1978), "Prolegomena"—excellent comparative survey; R. L. Montgomery, *The Reader's Eye* (1979); *GRLMA*, v. 6, *La Littérature didactique, allegorique et satirique*; *Europäische Lehrdichtung*, ed. H. G. Rötzer and H. Walz (1981); R. Wellek, "Crit. as Evaluation," *The Attack on Lit. and Other Essays* (1982); M. B. Ross, *The Poetics of Moral Transformation: Forms of Didacticism in the Poetry of Shelley* (1983); "Lit. and/as Moral Philosophy," spec. iss. of *NLH* 15 (1983); K. Muir, *Shakespeare's D. Art* (1984); L. Sternbach, *Subhasida: Gnomic and D. Lit.* (1984)—Indian; H. J. Blackham, *The Fable as Lit.* (1985); W. Spiegelman, *The D. Muse* (1990).

T.V.F.B.; S.J.K.

DOZENS is a game of exchanging, in contest form, ritualized verbal insults, which are usually in rhymed couplets and often profane. The term is probably a literate corruption of the vernacular *doesn'ts*, relating to forbidden lang. activity. The game is practiced now mostly among adolescent Afro-Am. males, though its origin is thought to lie in the verbal insult contests of West Africa. The d. is a subcategory of *signifying* (q.v.). Sometimes referred to as woofing, sounding, cutting, capping, or chopping, the ritual is most often called "playing the d." The subjects of the insults are frequently the relatives of the verbal opponent, esp. his mother; the insults are frequently sexual. A mildly phrased example of the d. technique and style would be: "I don't play the d., the d. ain't my game, / But the way I loved your mama is a crying shame."

The players of the d. must display great skill with rhyme, analogy, wit, humor, and rhythm to maintain the momentum and vivacity necessary to win the accolades of their audience. The d. is partly an initiation ritual that teaches a player how to hold his equilibrium by learning to master the power of words and humor. The exchange takes place as a verbal duel in which words and humor are chosen to sting, so that the opponent will be goaded to either greater lexical creativity or defeat. But another purpose is to make the opponent aware of his or her intellectual rank within the group. The notion of the d. as only a game is a rule generally adhered to, in that the players are expected not to base their insults upon their opponent's real life, but if this rule is broken, what might have been a game can turn into an outright battle. Langston Hughes draws upon the trad. of the d., particularly in his long poetic work *Ask Your Mama* (1961), as do Richard Wright and Ralph Ellison.—R. D. Abrahams, *Deep Down in the Jungle* (1970); *Rappin' and Stylin' Out*, ed. T. Kochman (1972); L. W. Levine, *Black Culture and Black Con-*

sciousness (1977); G. Smitherman, *Talkin and Testifyin* (1977); H. L. Gates, Jr., *The Signifying Monkey* (1988). E.A.P.

DRAMATIC POETRY. In our time of explosion or implosion of literary genres, we should recall that some two millennia of Western poetics repose on a tacit separation of drama from lyric and narrative. Traditionally, the lyric expressed personal emotion; the narrative propelled characters through a plot; the dramatic presented an enactment. All three genres have been hospitable to, may have originated in, verse, which is the simplest synonym for poetry. The contemporary verse dramatist Ntozake Shange has proclaimed: "In the beginning was the beat," but the beat is sounded differently in different langs. Moreover, the beat marks the rhythmic foot, but in poetry the verse line is in tension with the grammatical sentence, and in d. p. that tension may be high.

Western critics have interpreted the phrase d. p. in three main ways: (1) lyrics or short poems that imply a scene; (2) plays that are valorized with the adjective "poetic"; and (3) dramas whose dialogue is calculatingly rhythmed—in rhythms that are often regularized into meters and that are usually presented as discreet lines on the page. Avoiding the impressionism of the first two categories, the following overview focuses on d. p. as verse drama, which embraces such icons of our culture as *Oedipus*, *Faust*, and *Hamlet*. And although the survey is limited to published texts, we must also recall that virtually all d. p. was conceived for performance. In Martin Esslin's telling image, "a dramatic text is a blueprint for mimetic action."

Milestone documents of Eng. lit. hist. vary in their terminology for d. p.: although Sidney's *Apologie for Poetrie* (1581) cites only dramas written in verse, he refers to them as comedy, tragedy, or "play matter"; Dryden's very title subsumes his concern—*Essay of Dramatic Poesy* (1668); Shelley's *Defense of Poetry* (1820) also defends poetic drama; in a number of essays at the turn of the 20th c. W. B. Yeats champions "verse-drama"; and T. S. Eliot, probably the major modern champion of d. p., writes interchangeably of d. p., poetic drama, and verse drama. His spokesman A in his "Dialogue on D. P." (1928) makes the astonishing claim that "the craving for poetic drama is permanent in human nature."

The lineage of d. p. in the West extends back to cl. Greece, where d. p. was a chief form of celebration of the god Dionysus in religious and civic festivals. Although Gr. names are still retained for Eng. verse feet—iamb, trochee, dactyl, anapest, spondee—the accentual nature of the Eng. lang. betrays its Gr. source, which counted syllable quantity rather than stress. The duration and not the accent of the syllable determines Gr. metrics, which lays down intricate laws of scansion. In both tragedy and comedy, spoken dialogue alternates with choral song; together, they compose the oldest form of d. p. in the Western world. Typically, a Chorus of commoners interacts with and comments upon the heroic characters of Gr. tragedy. The cl. Gr. Chorus is rich in metrical variety, which was in some way expressive of the music, now lost, but the dialogue of the drama is largely confined to the iambic trimeter, which modern scholars, following Aristotle, believe to resemble the rhythm of everyday Gr. speech. The comic trimeter is more amenable than the tragic to metrical substitution, but the speech of both genres occasionally runs to trochaic or anapestic tetrameter. It is, however, the cl. Gr. Chorus whose metrical diversity defeats the translator. From Aeschylus to Euripides in tragedy, as from Aristophanes to Menander in comedy, the choral role shrinks, and dialogue usurps most of the drama. At its epitome, however, the Chorus can provide a cosmic context, a reflective interlude, or an emotional association for the episodes involving the actors. The range of resonance is reflected in the wide range of choral meters in any single drama.

For all the embarrassments in modern staging of Choruses, they lie at the heart of the deepest Gr. tragedies, and they are important enough to figure in the titles of Aristophanes' Old Comedy (e.g. *The Clouds*, *The Wasps*, *The Birds*). A half-century separates Aeschylus' *Oresteia* from Euripides' *Bacchae*, but both tragedies achieve dramatic intensity through their respective Choruses of women (played, like all Gr. roles, by men). The fearsome Erinyes of the final play of Aeschylus' trilogy eschew metrical diversity for iambic and trochaic rhythms of great complexity, and Euripides' nonGreek Bacchantes display a panoply of meters in their Dionysian frenzies.

By Roman times the Chorus virtually disappears from comedy, but not from tragedy. Livius Andronicus in the 3rd c. B.C. is credited with establishing the basic meter for both tragedy and comedy—a free iambic trimeter which modern scholars call the iambic senarius. In tragedy Seneca is strict yet versatile in his adherence to the trimeter, sometimes distorting syntax to attain regularity, and he builds to dramatic climaxes with line-for-line repartee, called stichomythia. Punctuating the declamations of his characters with four reflective choral passages, Seneca also influenced the structural form of European Ren. drama, in its proclivity for five-act structure.

In Roman comedy the iambic senarius is the meter of choice for Terence, but the favorite of free-ringing Plautus is the trochaic septenarius, which he sprinkles with metrically inventive songs, so that scholars compare his farces to modern musical comedy. Medieval mss. of Plautus and Terence mark scenes as DV (*diverbium*) for spoken verse and C (*canticum*) for all other meters, but scholars disagree on the gradations between recitative and song. Indisputable is Terence's predilection for quiet control, whereas Plautus is more flamboyant. In Ben Jonson's words: "neat Terence, witty Plautus."

DRAMATIC POETRY

Although an occasional dramatic sport appears in prose, verse is basic to two generative influences on Ren. drama—the Roman legacy and Med. Lat. liturgical plays. Lat. is the lang. and Lat. meters the verseform of both these strands, but it is very different Lat. Terence served as the basic Cl. Lat. reading text in Ren. schools, whereas Med. Lat. was increasingly undermined by the various vernaculars. Verse nevertheless remained the norm for medieval European drama, both in Lat. and in the new native meters, such as the nine-line stanzas of the Wakefield Master (Stevens) or the octosyllabic couplets of anonymous Fr. farce. The Englishman animated his *Second Shepherd's Play* with strong alliterative stress as well as internal rhymes in short-lined stanzas. The Wakefield Master could distribute his nine lines among several characters, for both the colloquial conversation of the shepherds and the majestic pronouncements of the angel and of Mary, while across the Eng. Channel his near (15th-c.) contemporary, the anonymous author of *Master Pathelin,* one of the most scintillating comedies of the Fr. lang., could etch characters in colloquial Fr. or bend his line into the several dialects with which the astute Master Pathelin evaded his creditors and delighted his audiences—all in impeccable octosyllables.

The Ren. witnessed not the rebirth but the birth of sustained prose drama (e.g. Machiavelli's *Mandragola*), and yet verse or d. p. continued to be the dramatic norm. After some fluctuation in the meters of d. p. in the several langs. of Western Europe, each country settled into its own preferred rhythm. In Italy, opera emerged, based on the assumption that the dialogue of Gr. tragedy was sung rather than spoken; but when music dictates the meter of dramatic dialogue, relegating lang. to a secondary function, the work does not fall into the province of d. p. and hence will not be considered here. Nor will such later forms as operetta and musical comedy.

On the Elizabethan stage Marlowe's mighty line towered above its competitors. Bold and reverberant in the mouths of his over-reachers, the pentameter was also whipped by Marlowe into stichomythic exchanges between his Faustus and Mephistopheles. Whether in tirade or duet, Marlowe's blank verse (q.v.) is regular, resonant, and end-stopped. In moments of dramatic stress, however, Marlowe sometimes breaks the basic meter, as in Faustus' anguished question to himself: "And canst thou not be saved?" Although Marlowe follows what was becoming custom in assigning prose to lower-class characters, he has no scruples against mixing prose and verse for the aristocratic onstage audiences of Faustus' paltry stunts. Unlike Shakespeare, who will shift his rhythm to announce creatures of another world, Marlowe limits both Good and Bad Angels to blank verse—in a virtual extension of the hero's inner conflict, to which Shakespeare's soliloquies would rise matchless.

Except for the prose *Merry Wives of Windsor,* all of Shakespeare's plays are grounded in blank verse, in which he achieved greater suppleness than any other dramatic poet, while also showing mastery of rhyme and prose. As G. R. Hibbard wrote: "Perhaps the major artistic problem that confronted Shakespeare . . . was that of bringing his exuberant delight in words and figures of speech into harmony with the dependence of drama on action and character." It is an understatement to say that Shakespeare solved this problem masterfully. He could infuse the pentameter with the questions of Hamlet, the exclamations of Lear, the imprecations of Timon of Athens. He could break the line in play or in passion (Wright); he could run the pentameters into soliloquies of labyrinthine syntax. Although Dorothy Sipe has shown that the late plays also follow the pentameter norm, the mature Shakespeare nevertheless departed from ten-syllable strictness at moments of high tension. The most famous pentameter in the Eng. lang. is a hendecasyllable: "To be, or not to be; that is the question"; Lear rages on the heath: "Then kill, kill, kill, kill, kill, kill!"; Macbeth begins and ends his soliloquy of grief with short lines: "She should have died hereafter," "Signifying nothing." The model of later writers of blank verse, Shakespeare is also deft with rhyme (Ness), inventing stanzas for many of the songs in his d. p. Couplets served him traditionally to cue an entrance, close a scene, or announce a change of mood; his rhymes are heard in prologues, epilogues, choruses, parodies, the sonnets of *Love's Labor's Lost,* the octosyllables of *Measure for Measure,* the dirges of *Cymbeline* and *The Tempest.* In Shakespeare's mature plays his rhythmic virtuosity always functions dramatically (Wright).

Despite T. S. Eliot's distaste for *Hamlet,* its rhythms shape the tragedy. In the very first scene the pentameters are distributed among the several soldiers to convey their disquiet (see SPLIT LINES). In the second scene an entirely different rhythm permeates Claudius' lines in the same meter—glib and regular. Hamlet's first extended blank-verse speech (of grief) rises to the emphasis of a concluding couplet. His first soliloquy reveals his torment by exclamations and self-interruptions; the breaks in iambic fluidity predict: "But break, my heart. . . ." Then the mercurial prince shifts to easy familiarity for his greeting to Horatio—still in blank verse. The same meter questions the Ghost but snaps abruptly in the throes of the oath of secrecy. No other Shakespearean hero speaks so often in prose, a signal for the many levels of Hamlet's "antic disposition." Once the king's conscience is caught, Hamlet bursts exultantly into a ballad-rhyme, but he cannot manage the rhyme for a second ballad, as Horatio points out. Nevertheless, Hamlet utters seven blank-verse soliloquies at key dramatic points of the play's first four acts, and in them as nowhere else in d. p., the mind is caught live in "blooded thought." Ophelia in her madness zigzags from

prose to bawdy song to elegy, and the clown sings as he digs her grave, offending Hamlet's sense of decorum. Other characters are sharply differentiated through their blank verse—sanctimonious Polonius, troubled Gertrude, duplicitous Claudius. Hamlet's final words—"the rest is silence"—lay to rest not only the prince but the most dramatically varied rhythms (largely in a single meter) in the Eng. lang.

Shakespeare's virtuosity was unique on the Elizabethan stage, but the age produced unparalleled skill in d. p., whether the panoramic comedies of Ben Jonson, the dark tragedies of Webster and Tourneur, the incisive satire of Marston. Yet by the time of the closing of the theaters in 1642, the energy of Eng. blank-verse drama had ebbed; and with the Restoration came prose comedy and heroic tragedy, the latter written in rhyming pentameter couplets. It is puzzling that the same audiences apparently responded to these polar forms; perhaps they were attracted to the frankness of each artifice.

Certainly the artifice of the verse was part of the appeal of d. p. in rhyme-rich Sp. drama of the Golden Age (Morley). In these plays, ruled by an ineluctable poetic justice, theatrical variety was confined to incident and rhythm. Lope de Vega was both prime theoretician and practitioner. In his *Arte nuevo de hacer comedias* (1609), he specifies the appropriate verse form for the specific dramatic situation: romance for narration, redondilla for dialogue, sonnet for monologue; he himself was reputed to pour forth every conceivable rhyme and meter in his 400-odd plays.

Calderón's noteworthy contribution to the Sleeper Awakened topos, *Life Is a Dream*, is a typical Golden Age drama in its metrical diversity, where no character "owns" any particular meter. Calderón's scenes of ceremony are couched traditionally in long, irregularly rhyming lines (hendecasyllables), but far more frequent are shorter lines (octosyllables) linked by assonance, which serve both monologue and dialogue, both action and reflection. In quatrains and quintillas, these short lines heighten scenes of pathos, and in ten-line stanzas, or decimas, they structure the two soliloquies of the protagonist Segismundo, who learns through suffering to accept the dreamlike quality of life. During the course of the tragicomedy, shifts of meter and placement of rhyme reflect the evanescence of events: thus the play opens on Rosaura's distraught *silva*, but soon Segismundo yearns for liberty in a *decima*; the passionate reactions of this protagonist brim with questions, exclamations, and verbless phrases—in shifting meters. Several scenes bristle with *redondillas* rising at climactic moments to the *octava* (eight octosyllabic lines rhyming abababcc). The *romance* accomplishes narrative summary in the play's few quiet moments. Although this bald listing of forms may suggest a ragged effect, the pattern of the whole moves inexorably toward an almost abstract acceptance of fate, wherein the many rhythms and wayward feelings are finally subdued to a destined order.

In Fr. d. p. of the 17th c., conventions impose order. Physical action is banished from the stage, and the often passionate dialogue is largely confined to alexandrine couplets with fixed rules for pauses. The strict five-act form is built with tense and static scenes, which in turn are built with balanced couplets, but the balance varies with the dramatic moment. In what Jacques Scherer has called "the tyranny of the tirade," monologues and messengers' speeches can run on for a hundred lines, with liberal enjambment: the best-known example is Théramène's account of the death of Hippolytus in Racine's *Phèdre*. In comedy, on the other hand, piquancy is gained by dividing the rhymed lines between two characters, and the alexandrine itself can be split between characters, as in the opening scene of Molière's *Misanthrope*. Thus distributed or, more usually, flowing unimpeded, alexandrine couplets ruled the cl. Fr. stage. Corneille's heroes sometimes sound as though they are impatient with these constraints, but Racine's tragedies gain in power from the very rigor of expression of his erotically disposed heroines. Phèdre's complaint: "Tout m'afflige et me nuit, et conspire á me nuire" becomes a quiet scream through the balance of the stressed /i/ sounds. Or in a celebrated alexandrine: "Venus toute entière à sa proie attachée" conveys Phèdre's helplessness within the bonds of her passion, and yet she does not even mention the subjective first person; the variety of vowels before the caesura dissolves abruptly into five merciless /a/ sounds. The contrast, tension, and symmetry of the whole play is reflected in such an alexandrine.

In Molière's character comedies the alexandrine functions to comic point—the catechism of *The School of Wives*, the lubricious hypocrisies of *Tartuffe*, the minatory manners of *The Misanthrope*. In spite of Molière's wit, however, the alexandrine is primarily associated with tragedy, and although it has not translated well into any other lang., it was imitated in several countries of 18th-c. Europe. Geriatric alexandrines continued to dominate Fr. d. p. until the 20th c., but their stranglehold upon the d. p. of other langs. was broken by the advent of bardolatry as a tenet of romanticism.

In Germany, where romanticism ruled deepest and longest, translations of Shakespeare established blank verse for d. p., with Lessing's *Nathan der Weise* (1779) serving as a model, and with Goethe committing his cl. *Iphigenie auf Tauris* to that meter. Perhaps the single most ambitious work of European d. p. is Goethe's *Faust*, whose composition extended over half a century; whose scenes take place in Heaven, Hell, and many domains of the earth, incl. the theater; whose characters are drawn from commoners, aristocrats, mythological figures, and other-worldly spirits; and whose verse performs acrobatic feats of rhyme.

Scholars differ as to whether it is a single unified work at all, since various parts have been played both separately and together, requiring actors well attuned to d. p. In the apparently disjunctive scenes of Part I (but not in the cl. five-act structure of Part II) Goethe relies mainly on two free but traditional Ger. verse forms, the iambic madrigal with its free-ranging rhymes, and Knittelvers, or four-beat rhyming couplets with wide latitude for unaccented syllables. Deliberately unShakespearean in its virtual rejection of blank verse, the vast panorama of *Faust* called forth the poet in Goethe, subsuming the Cl. heritage (in the Helen scenes), the Med. (the famous *Walpurgisnacht*), the Ren. (Faustus' insatiable desire for experience), and the romantic (the Gretchen story). The verse-forms reflect that range: in Part II, for example, Faustus addresses Helen in a redondilla, shifting then to blank verse enhanced by anaphora, word play, and parallelism. Helen herself first speaks a Ger. approximation of quantitative verse but soon moves forward to rhyme in a scene which, while lacking the immediacy of Faustus' first view of Gretchen, nevertheless contains its own drama.

Goethe read Marlowe's *Faustus* (in tr.) only after he had published Part I and was well into Part II. Although both poets follow the 16th-c. *Faustbuch* in their dramatization of a sage who acquires power by selling his soul to the devil, Marlowe damns him, whereas Goethe saves him. Moreover, the Marlovian devil's "Why, this is hell, nor am I out of it," is far more orthodox than Goethe's nihilistic spirit ("Ich bin der Geist, der stets verneint"), who is unable to erode Faust's exhilaration in pure experience. Both dramas have thrived in the 20th-c. theater, which is not intimidated by the Herculean roles, the disjunctive story-lines, and—in Goethe's case—the rhythm shifts and myriad rhymes.

Before the posthumous publication of Goethe's *Faust* Part II in 1832, many of the romantic poets of Europe tried to write d. p.—Tieck and Grabbe in Germany; Musset, Hugo, and Dumas in France; Wordsworth, Coleridge, Keats, Byron, Shelley in England, soon to be followed by Tennyson, Browning, Arnold, and Swinburne. The Eng. were particularly vulnerable to Shakepeare's misleadership, and many 19th-c. hacks loaded frail pentameters with unspeakable images. Unfortunately, competent poets produced dull verbiage, then as now largely relegated to the closet, although Byron, Browning, and Tennyson did show a certain flair for d. p. and were sporadically staged. With the dominance of prose on the 19th-c. stage, the very sound of verse triggered the expectation of a lofty subject and, less automatically, a tragic ending. This association between d. p. and tragedy carries over to the present.

At the turn of the 20th c., two popular versifiers, one on each side of the Eng. Channel, attracted enthusiastic audiences to their respective theaters—Edmond Rostand in France and Stephen Phillips in England. In spite of the rapid decline of Phillips' popularity, the Am. Maxwell Anderson later lingered on the same blank-verse path, and after World War II Christopher Fry in England sowed it with a richer lexicon. Even such skillful Jacobean pastiche as Djuna Barnes' *Antiphon* or Lawrence Durrell's *Irish Faustus* is anachronistic in the 20th c. Analogous to these speleologists of blank verse are Fr. archaeologists of the alexandrine, notably Cocteau, despite his celebrated distinction between poetry in the theater and poetry of the theater (the latter in prose).

Prose gradually displaced d. p. on the 18th-c. stage, but dramatic prose is beyond the scope of this survey, which is limited to the d. p. of Western Europe. Passing mention must nevertheless be made of the importance of d. p. in the very different lits. of Russia and Poland. Poland in particular, the Nowhere of Jarry's *Ubu roi*, declared adherence to the nationalist aspirations of Mickiewicz's *Forefather's Eve* (1832) and Slovacki's *Kordian* (1834). The new century, still sustaining hope for a Polish nation, was ushered in with Wyspianski's *The Wedding* (1901).

Although Rus. drama begins with prose comedy in the 17th c., plays in syllabic verse soon followed. Theater histories celebrate Sumarokov as the first major Rus. playwright, influenced alike by Voltaire and Shakespeare. In the 18th c., Catherine the Great tried her hand at various genres of d. p. in imitation of the Fr., and the neoclassical influence lingered into the 19th c., particularly in comedy, of which the outstanding example is *Woe from Wit* by Griboedov (1824). A year later Russia's greatest romantic poet, Aleksandr Pushkin, completed his neo-Shakespearean history play *Boris Godunov*, followed by the so-called "little tragedies." Verse satire and Shakespearean tragedy continued into the 20th c., as well as short symbolist d. p. by Tsvetaeva and Gumilyov. Although the adjective "poetic" often precedes the name of Chekhov, his plays are confined to prose, as are those of the antirealists Evreinov and Majakovsij, and even the great poet Pasternak.

The Norwegian Ibsen may serve as exemplar of the late 19th-c. shift from historical d. p. to modern realism. After writing his Faustian *Brand* (1866) and *Peer Gynt* (1867) in several forms of rhyming verse (and, like *Goethe's Faust*, these plays were not conceived for the theater), Ibsen shifted to prose for his comparably cosmic *Emperor and the Galilean* (1873), which was intended for the theater. Thereafter Ibsen never looked back to d. p. By the turn of the 20th c., realistic prose dominated the stage throughout Europe and the Americas, and yet reactions against realism followed hard upon its inauguration—mainly in the prose of such dramatists as Villiers de l'isle Adam and Maeterlinck, but also in verse plays. In Austria the young Hugo von Hofmannsthal leaned lightly on Shakespeare as he dabbled in magic, fantasy, and lyrical longing; later his firmer verse would mold libretti for the music of Richard Strauss. Not dissimi-

lar to the earlier work of Hofmannsthal were the early plays of W. B. Yeats, hovering in Irish mist and myth; later his taut lines in freer verse rhythms, sometimes interspersed with prose, would mold *Four Plays for Dancers* on the model of the Japanese Ghost Nō.

At the end of the 19th c., Fr. Paul Claudel began a half-century of dramas that flouted Fr. traditional d. p.; his plots sprawled, his characters spanned countries and centuries, and his long verse line heeded neither tonic accent nor syllable count but took its inspiration from the Bible. Called *verset* rather than *vers* in Fr., this flowing line is sometimes anglicized as "versicle"—particularly by critics who deny that verse can be free. As monolithic in their faith as Sp. Ren. dramas, Claudel's plays only slowly reached the stage, but they pointed the way for a few Fr. poets who later ventured into theater—Henri Pichette, Jules Supervielle, and, most recently, Novarina. The Francophone poet of Martinique, Aimé Césaire, mixed prose and verse in his *Tragedy of King Christophe* as in his adaptation of Shakespeare's *Tempest*; like Claudel, Césaire writes long lines that are rhythmed by rhetorical parallelisms. Although Jacques Vauthier resembles Césaire in his attraction to Shakespearean and Jacobean imagery, the Frenchman writes mainly in prose, but his aptly titled *Sang* swells in Claudelian versicles for a tragedy derived from the Jacobean *Revengers Tragedy*.

Ger. writers are as aware as Eng. of Shakespeare's shadow over d. p. Gerhart Hauptmann versifies several plays based on Gr. tragedy or Shakespearean drama. In sharp contrast, blank-verse passages serve Bertolt Brecht mainly as parody in *St. Joan of the Stockyards* and *The Resistible Rise of Arturo Ui*. Brecht wrote only one play entirely in verse—the opera libretto for *Mahagonny*. Into most of his plays, however, Brecht injects lyrics (called "songs" in Ger.) to teach delightfully. His characters often achieve a degree of self-awareness in their songs, from which they are screened in their prose dialogue, and Brecht means his audience to share that awareness. Brecht's "cool" drama was in part a reaction against the emotions of expressionism, and indeed the expressionist Walter Hasenclever often inserts blank-verse scenes of suffering into his basic prose plays. In postwar East Germany several playwrights reverted to blank verse to disengage themselves from Socialist Realism—Volker Braun, Peter Hacks, Harmut Lange, Stefan Schütz.

More consistent and original is the German Heiner Müller, most of whose plays mix prose and verse: "It is really a problem of the impact of history—even of the tempo of historical devel. When there is stagnation, you have prose. When the process has a more violent rhythm, you have verse." Also in Germany, Rolf Hochhuth seeks emotional distance in his documentary dramas by breaking the dialogue into irregular verse lines, with emphasis falling at the end of the line. Peter Weiss introduced Knittelvers for parodic effect, notably in his *Marat/Sade*, but in other plays (*The Investigation, Hölderlin*) Weiss breaks speeches into irregular verse lines determined by syntax, possibly to achieve a Brechtian estrangement.

The most consistent contemporary writer of Ger. d. p. is the Austrian Thomas Bernhard. His short unpunctuated lines clarify his syntax, but he spurns such sonic enhancements as rhyme, assonance, alliteration. In both the dialogue and long monologues of Bernhard's dramas, merciless repetition drenches the stage in a fetid atmosphere that warns the audience away from realism. In spite of their nominally contemporary settings, Bernhard's d. p. probes into the unlocalized cruelty of human relationships. Simple in plot but complex in character, these plays thrive on incisive stage images that are etched the more sharply for the mordant verse lines.

It is, however, the Eng. lang., forever haunted by Shakespeare's ghost, that has hosted most 20th-c. efforts at d. p. Each generation seems to sound its own clarion call for a revival of poetic drama: at the turn of the c., Phillips and Yeats; after World War I, James Elroy Flecker, Gordon Bottomley, Lascelles Abercrombie, Laurence Binyon, John Drinkwater, John Masefield, John Middleton Murry, William Archer, T. Sturge Moore, and dozens of authors who managed a single verse play—almost inevitably in blank verse, with subjects wrenched from myth, history, or what passed for peasant life. Masefield conducted a personal crusade for verse drama, not only hazarding other meters than blank verse, but also building a theater for d. p. at his home in Boars Head, Oxford. At that time too appeared the first of T. S. Eliot's essays on d. p.—"Rhet. and Poetic Drama" (1919), "Euripides and Professor Murray" (1920), and "A Dialogue on D. P." (1928). Also in 1928 began the parade of poetic pilgrims to Canterbury—notably Eliot with his *Murder in the Cathedral* (1935). Eliot's verse plays with contemporary settings did not, however, appear till a generation later, and by that time Auden and Isherwood had turned verse to parody in *The Dog Beneath the Skin* (1936); they enfolded passages of various meters into the basic prose of *The Ascent of F6* (1936) and *On the Frontier* (1938). Stephen Spender resurrected blank verse for *The Trial of a Judge* (1938), and Louis MacNeice hazarded different meters for his radio plays.

After World War II came another Eng. movement toward d. p.—emanating from the commitment of Eliot. In contrast to Yeats, who desired a small, elite audience, Eliot accommodated himself to the commercial theater of his time, and he invented a verse line commensurate with what he took to be the speech of his time. Earlier (1926) he had verbalized jazz rhythms in the Sweeney fragments he never wielded into a whole, but for *Family Reunion* (1939) he conceived a "rhythm close to contemporary speech . . . a line of varying

length and varying number of syllables, with a caesura and three stresses." Although Eliot did not know it at the time, *Family Reunion* was to mold the pattern for his subsequent three contemporary verse plays: 3-stress line, underground Gr. plot, foreground drawing-room drama, and character's discovery of religious vocation. It is in the latter two aspects that Eliot influenced Christopher Fry, Anne Ridler, Norman Nicholson, and Ronald Duncan, who were welcomed either at the Canterbury Festival or the Mercury Theatre of London. Free of these pretensions is *Under Milk Wood*, a prose-verse blend by Dylan Thomas, not quite completed at the time of his death in 1953; commissioned for radio, it has played joyously on stage and screen. By the date of Eliot's last, least lyrical play, *The Elder Statesman* (1958), angry prose had invaded the London stage, spurring the critic Kenneth Tynan to write: "Messrs. Eliot and Fry suggest nothing so much as a pair of energetic swimming instructors giving lessons in an empty pool."

Even on the island of London theater, the pool was not quite so dry as Tynan claimed. Not only did Eliot and Fry attract West End audiences to their d. p. (an attraction that has afterwards flared up sporadically), but Anne Ridler embarked on a new career as librettist: "Libretto-writing, as W. H. Auden said, gives the poet his one chance nowadays of using the high style." Ronald Duncan maintained such a "high style" in his several verse modes—satiric couplets, free rhymes, intricate Italianate stanzas. His *Don Juan* and *Death of Satan* played at the Royal Court Theatre in the same year as *Look Back in Anger*—1956—and that was his misfortune; Duncan's modernized myths paled against the heat of Osborn's vitriolic prose. In Duncan's own bitter opinion: "the convention of Shaftesbury Avenue duchesses fiddling with flower vases was replaced by Jimmy Porters picking their noses in public." Less graphically, Christopher Fry asked: "Do you think that when speech in the theatre gets closer to speech in the street we necessarily get closer to the nature of man?"

Of those playwrights who implied an affirmative answer, several nevertheless revived verse for satiric thrusts. Adrian Mitchell came to theater through his witty trs. of Peter Weiss' Knittelvers in Peter Brook's production of *Marat/Sade*. This launched Mitchell into the deft songs by which he structures—if that is the word—his own wayward plays, such as *Tyger* and *Mind Your Head*. Festive in nature, Michell's d. p. loses ebullience on the printed page, and more recently he has adapted himself to television, where associational fantasy dilutes his earlier satiric bite. More faithful to the theater is David Edgar, who deromanticizes *Romeo and Juliet* into prose, but who twirls couplets in his Watergate parody, *Dick Deterr'd* (1974), which shadows Shakespeare's *Richard III*. Edgar experiments seriously with d. p. in his ambitious *O Fair Jerusalem* (1975), a play within a play. The characters of the frame play, set in 1948, speak in undistinguished modern prose, but those (the same actors) in the inner drama, set in 1348, speak mainly in verse of varying meters. Caryl Churchill, another playwright who is, like Edgar, mainly a social realist, has also turned to verse for her less-than-serious *Serious Money* (1987), a parody of the London Stock Exchange. Where Edgar's Dick is a comic-book villain, however, Churchill's (pointedly named) Scilla wavers uncertainly between caricature and character. A similar uncertainty of tone afflicts Steven Berkoff's several verse plays, set in modern London, in which obscenities and misquotations sprinkle flaccid pentameters. More purposeful is the trenchant, sometimes savage antiwar satire in Charles Wood's *H* and *Veterans*. With Pirandellian and filmic dissolves of character into role, Wood parodies the mindless posturing, outworn rhet., and simple greed that propel armies to destroy and be destroyed. His verse lines are short and unmetrical; yet cliché, repetition, wrenched syntax, and subtle rhyme endow the dialogue with a military precision which paradoxically undermines war. On the contemp. London stage the most popular mode of d. p. is parody, but the serious and honorable exception is John Arden, who blends from ten to fifty percent verse into his mainly prose plays. Moreover, he deploys verse like a composer of opera for moments of high dramatic tension. Partial to the ballad, Arden also writes long stretches of dialogue in a kind of sprung rhythm, rhyming freely.

In the wide reaches of the British Commonwealth, dramatic poets have plumbed serious d. p. The heroic verse of the Australian poet Douglas Stewart contrasts with the understated colloquial rambles of his countryman Ray Mathew. During the 1940s, Stewart centered several dramas on the failed yet noble dreams of his protagonists, and a decade later Mathew quietly celebrated the humilities of provincial Australian life. In rhythm-rich Africa several playwrights have mined d. p. in the Eng. lang. Recently awarded a Nobel Prize, Wole Soyinka couches most of his dramas in densely imaged prose, but *A Dance of the Forests* (1960) shifts to verse for tense scenes between the gods. Verse is also the medium for Soyinka's adaptation of *The Bacchae* (1973), sounding out in a multi-rhythmed verbal percussion that does honor to Euripides. Soyinka's fellow Nigerian John Pepper Clark sets his (shorter) plays in the village life of his country, rendering native dialogue in short, colloquial, repetitive verse lines. Outside Africa but linked to its culture is the versatile d. p. of the Trinidadian Derek Walcott. From *The Sea at Dauphin*, a Synge-like tragedy of men who live by the sea, to *Pantomime*, a drama that plays pointedly with race relations, and encompassing the metaphysical *Dream on Monkey Mountain*, the adapted *Joker of Seville*, and the satiric *Remembrance*, Walcott tunes his loose and supple verse line to the particular dramatic situation. Walcott's *Joker of Seville* is only one of many accomplished verse trs. of the

classics by which today's Anglophone theater has been enriched—Tony Harrison's *Oresteia* and *Mysteries*, Ted Hughes' *Seneca-Oedipus*, Robert David MacDonald's Molière, Adrian Mitchell's Calderón, and a lonely Am. poet—Richard Wilbur's Molière and Racine.

Several Am. poets dabbled in a verse play or two—e e cummings, Richard Eberhart, Archibald MacLeish, Wallace Stevens, William Carlos Williams. Alone among the poets MacLeish contrived a Broadway success in Bible-based *JB*. In view of the subsequent counter-cultural impact of the Becks' Living Theater, it is rarely recalled that they started humbly as a Poets' Theater, producing several plays of William Carlos Williams. The Living also produced the aleatory plays of Jackson MacLow, where compulsive repetition structures near-abstractions. It may be that this divorce from lexical language, which gives extraordinary freedom to the director, was one of the causes of a revival of interest in Gertrude Stein's undramatic short dramas dating from the 1920s and 1930s. In contrast, the decidedly down-to-earth Beat Poets of the 1950s, who read their poems aloud to jazz, were impervious to the interactive dialogue of drama; the single exception is Michael McClure, whose *Beard* hammers at free-verse patterns. More durable are Robert Lowell's verse dramatizations of stories by Hawthorne and Melville into a trilogy that criticizes the United States as *Old Glory*. Kenneth Koch's short-lined short plays are more playful demystifications of Am. history.

A very few contemporary Am. dramatists have dipped into d. p. Arthur Miller wrote his first version of *The Crucible* in verse, and Tennessee Williams, who published a volume of poetry, resorted to free-verse lines in his Lorca-like *Purification*, although Lorca himself adhered to prose after his first play. Edward Albee's *Counting the Ways* is a slight verse piece that departs from a sonnet by Christina Rossetti, but David Mamet's *Reunion* gains poignancy through his short verse lines—a rhythm he also introduces into parts of his *Water Engine*. James Merrill has also written d. p.

By the late 20th c., generic divisions have been undermined, and two major poet-playwrights have been instrumental in such subversion—Bertolt Brecht and Samuel Beckett. Although they have published highly dramatic lyrics, their plays are largely prose. Nevertheless, Brecht's songs heighten memorable moments of his drama; passages of Beckett's plays sound like verse, the leaves stanzas of *Godot*, for example; two Beckett plays even break into lines of formal verse—*Come and Go* and *Rockaby*. The rhythms of *Play* are scored as for music, and the three voice-strands of *That Time* are segmented into free-verse stanzas. The adjective "poetic" is affixed more often to Beckett than to any other playwright of the second half-century, but Brecht and Beckett both were aware of the extra-verbal means of achieving highly dramatic moments, and both verbal masters therefore summoned music and lighting to enhance their polyvalent words. Born into a technological era, later playwrights nevertheless have learned from the lyrical intimacies of Beckett as from the choral dialectics of Brecht to enrich their own prose plays—prose by the arbitrary criteria I imposed at the outset of this survey.

I conclude on a dramatic poet who is poised between accomplishment and promise—the black Am. woman Ntozake Shange. Her *for colored girls who have considered suicide / when the rainbow is enuf* began as individual lyrics, but in performance, she says, "I came to understand these twenty-odd poems as a single statement, a choreopoem" which was nurtured by the tragedies of black women. On stage, a slight pause punctuates Shange's syntactic verse-units; on the page, a slash separates the phrases. Popular both on stage and page, all Shange's work has forged a language and a rhythm of her own in this recent, resonant outpost of d. p.

L. Abercrombie, "The Function of Poetry in the Drama," *Poetry Rev.* 1 (1912); S. G. Morley, *Studies in Sp. Dramatic Versification of the Siglo de Oro* (1918); T. S. Eliot, *The Sacred Wood* (1920), *Selected Essays* (1950), *Poetry and Drama* (1951); N. Díaz de Escovar and F. de P. Lasso de la Vega, *Historia del teatro español*, 2 v. (1924); W. B. Yeats, *Essays* (1924); H. C. Lancaster, *Hist. of Fr. Dramatic Lit. in the 17th C.*, 9 v. (1929–42); C. V. Deane, *Dramatic Theory and the Rhymed Heroic Play* (1931); J. S. Kennard, *It. Theatre*, 2 v. (1932); H. Granville-Barker, *On Poetry in Drama* (1937); F. W. Ness, *The Use of Rhyme in Shakespeare's Plays* (1941); M. E. Prior, *The Lang. of Tragedy* (1947); B. Brecht, *Kleines Organon für das Theater* (1948); J. L. Barrault, *Reflections on the Theatre* (1951); F. Fergusson, *Idea of a Theater* (1953); G. F. Else, *Aristotle's* Poetics*: The Argument* (1957); S. Atkins, *Goethe's* Faust (1958); R. Lattimore, *Poetry of Gr. Tragedy* (1958); D. Donoghue, *The Third Voice: Mod. British and Am. Verse Drama* (1959); R. Peacock, *The Poet in the Theatre*, rev. ed. (1960); M. Bieber, *Hist. of the Gr. and Roman Theater*, 2d ed. (1961); C. Fry, "Talking of Henry," *20th C.* 169 (1961); W. H. Auden, "Notes on Music and Opera," *The Dyer's Hand* (1962); C. H. Smith, *T. S. Eliot's Dramatic Theory and Practice* (1963); M. C. Bradbrook, *Eng. Dramatic Form: A Hist. of its Devel.* (1965); Dale; J. Scherer, *La Dramaturgie classique en France* (1968); D. Sipe, *Shakespeare's Metrics* (1968); J. Cocteau, "Préface de 1922," *Les Mariés de la Tour Eiffel* (1969); R. Williams, *Drama from Ibsen to Genet* (1971); R. Duncan, Preface to *Collected Plays* (1971); K. J. Worth, *Revolutions in Mod. Eng. Drama* (1972); H. H. A. Gowda, *D. P. from Med. to Mod. Times* (1972); A. Nicoll, *A Hist. of Eng. Drama, 1660–1900*, 6 v. (1952–59), *Eng. Drama 1900–1930* (1973); K. Tynan, "Prose and the Playwright," *A View of the Eng. Stage* (1975); A. P. Hinchliffe, *Mod. Verse Drama* (1977); F. H. Sandbach, *The Comic Theatre of Greece and Rome*

(1977); H. Müller, "Heinar Müller on Verse," *New Ger. Critique Supplement* (1979); J. Baxter, *Shakespeare's Poetic Styles* (1980); D. Breuer, *Deutsche Metrik und Versgeschichte* (1981); C. Freer, *The Poetics of Jacobean Drama* (1981); Trypanis, esp. chs. 3,4,6,9—Greek; G. R. Hibbard, *The Making of Shakespeare's D. P.* (1981); O. Mandel, "Poetry and Excessive Poetry in the Theatre," *CentR* 26 (1982); T. Rosenmeyer, *The Art of Aeschylus* (1982); A. Brown, *A New Companion to Gr. Tragedy* (1983); G. Hoffmann, "Das moderne amerikanische Versdrama," *Das amerikanische Drama*, ed. G. Hoffmann (1984); M. Carlson, *Theories of the Theatre* (1984); *CHCL*, v. 1, chs. 10–12; C. W. Meister, *Dramatic Crit.: A Hist.* (1985); C. J. Herington, *Poetry into Drama* (1985); K. Pickering, *Drama in the Cathedral* (1985); Terras; R. M. Harriott, *Aristophanes: Poet and Dramatist* (1986); M. Esslin, *The Field of Drama* (1987); M. Stevens, *Four ME Mystery Cycles* (1987), ch. 2—on the Wakefield Master; H. Blau, *The Eye of Prey* (1987); G. T. Wright, *Shakespeare's Metrical Art* (1988); A. W. Pickard-Cambridge, *The Dramatic Festivals of Athens*, 2d. ed. rev. and corr. (1989).

R.C.

E

ECLOGUE. A short pastoral (q.v.) poem, usually quasi-dramatic, i.e. a dialogue (q.v.) or soliloquy. The setting, the *locus amoenus* ("pleasant place"), is Arcadian (later Edenic)—rural, idyllic, serene—and the highly finished verse smooth and melodious. But its ends are far from primitive or serene: a poet writes an e., says Puttenham, "not of purpose to counterfeit or represent the rusticall manner of loves and communication, but under the vaile of homely persons and in rude speeches to insinuate and glaunce at greater matters." Such matters might include praise of a person, beloved, worthy, or dead, or disquisition on the nature of poetry or the state of contemp. poetry, or criticism of political or religious corruption. Elevation of rustic life implied denigration of urban.

Derived from Gr. *eklegein* ("to choose"), "e." originally denoted a selection, i.e. a notable passage from a work or a choice poem, as in Statius (*Silvae*, Bks 3–4, Pref.). The term *ecloga* was applied by the grammarians to Horace's Epistles 2.1 and *Epodes*, but most importantly to Virgil's bucolic poems, this application possibly by Virgil himself. From this association "e." became a common designation in Carolingian and Ren. usage for a pastoral poem following the traditional technique derived from the idylls of Theocritus. Though there are precedents in both Cl. and Ren. lit. of city and piscatory es., most es. are pastorals. The term, however, signifies nothing more than form. The spelling *aeglogue* (or *eglog*), popularized by Dante, was based on a false etymology which derived the word from *aix* (goat) and *logos* (speech) and was construed to signify, as Spenser's "E. K." argued, "Goteheards tales."

E. was not a major genre in the Middle Ages, the most famous Christian version being the *Ecloga Theoduli* (9th c.?). It was revived perhaps by Dante (*E.* 1 [1319]), then by Petrarch and Boccaccio (both in Lat.), coming into full flower under the culture of the Ren. Humanists of the 15th and 16th cs. Widely studied and of dominating influence were the es. of Baptista Mantuanus Spagnuoli (mentioned with affection by Shakespeare). Barnabe Googe was the first writer to attempt the e. in Eng. (*Es., Epitaphs, and Sonnets* [1563], ed. J. M. Kennedy [1989]), but the finest flower is certainly Spenser's *Shepheardes Calender* (1579). Pope's es., though called *Pastorals*, epitomize the Neoclassical e. and rococo art. Later in the 18th c., new matter was poured into the mold, yielding a variety of es.—urban, exotic, political, war, school, culinary, Quaker—the most celebrated of which is Swift's *A Town Eclogue. 1710. Scene, The Royal Exchange.*

M. H. Shackford, "A Definition of the Pastoral Idyll," *PMLA* 19 (1904); W. W. Greg, *Pastoral Poetry and Pastoral Drama* (1906); W. P. Mustard, Intro. to *The Piscatory Es. of Jacopo Sannazaro* (1914), and "Notes on *The Shepheardes Calender*," *MLN* 35 (1920); R. F. Jones, "E. Types in Eng. Poetry of the 18th C.," *JEGP* 34 (1925); M. K. Bragg, *The Formal E. in 18th-C. England* (1926); A. Hulubei, *L'Eglogue en France au XVIe siècle* (1938); D. Lessig, *Ursprung und Entwicklung der spanischen Ekloge* (1962); P. Cullen, "Imitation and Metamorphosis," *PMLA* 84 (1969); G. Otto, *Ode, Eklogue und Elegie im 18. Jh.* (1973); R. Borgmeier, *The Dying Shepherd* (1976)—the 18th-c. Eng. trad.; P. Alpers, *The Singer of the* Es. (1979); Fowler; R. E. Stillman, *Sidney's Poetic Justice* (1986); *Es. and Georgics of Virgil*, tr. D. R. Slavitt (1990).

J.E.C.; T.V.F.B.

EKPHRASIS, ecphrasis (Gr. "description," pl. *ekphraseis*). E. first appears in rhet. writings attributed to Dionysius of Halicarnassus (*Rhetoric* 10.17). Later it becomes a school exercise, where it is defined as "an expository speech which vividly (*enargōs*) brings the subject before our eyes" (Theon, 2d c. A.D.—Spengel). Among the topics listed by Theon are descriptions of people, actions, places, seasons, and festivals. Priscian's Lat. tr. of the *Progymnasmata* ("School Exercises") attributed to Hermogenes (2d c. A.D.) made instruc-

tions for writing ekphraseis (*descriptiones*) available to the West. The medieval poetic *Artes* of Matthew of Vendôme and Geoffrey of Vinsauf devote considerable attention to the composing of descriptions (Faral, Arbusow). A tendency to limit e. to descriptions of works of art is discernable in the later *progymnasmata* of Nicolaus Sophistes (5th c. A.D.) and in the prose works of Libanius, the Philostrati, and Callistratus, but only in modern times does it bear that exclusive meaning.

Still, in this limited sense, as a description of a work of art, e. has had a long and complex history. The *locus classicus* in epic is Homer's description of Achilles' shield (*Iliad* 18.483-608), imitated by the more mannered depiction of Heracles' shield in the *Scutum* (139-320) attributed to Hesiod, and rivaled by Virgil's description of Aeneas' shield (*Aeneid* 8.626-731) (Kurman). A late epic example is the e. of Dionysus' shield in Nonnus' *Dionysiaca* (25.384-567).

During the Hellenistic period, notable ekphraseis occur in epic (Jason's cloak in Apollonius Rhodius, *Argonautica* 1.730-67) and in pastoral (the rustic cup in Theocritus, *Idyll* 1.27-56, and Europa's basket in Moschus' *Europa* 43-62). Roman examples indebted to Hellenistic ekphraseis include the description of the bed cover in Catullus 64 and the portrayal of Mars and Venus in Lucretius, *De rerum natura* (1.33-40), which appears to be a description of a Hellenistic painting. The ekphraseis of the tapestries of Minerva and Arachne in Ovid's *Metamorphoses* (6.70-128) provide an instructive contrast between Cl. and Hellenistic artistic styles (Race). Spenser provides an extended imitation of Arachne's tapestry in *The Faerie Queene* 3.11.29-46 (DuBois), and Shakespeare includes a lengthy e. on the siege of Troy in *The Rape of Lucrece* (1366-1561; Hulse).

Another strain of e. consists of epigrammatic descriptions and interpretations of paintings and statues. Although first composed in the Cl. period as inscriptions for actual statues, they eventually became a minor genre with many representatives in the *Gr. Anthol.* (Friedländer). Ren. examples include Marino's *La Galeria*, Marvell's "Gallery" (Hagstrum), and emblem poetry. An offshoot is the Ren. *blason*, a description of the beloved (Wilson) which often follows the prescription of the *progymnasmata* for describing statues by proceeding from the head to the foot, a practice also recommended by Geoffrey of Vinsauf (*Poetria nova* 562-99). In addition, the popular *Tabula of Cebes* encouraged the combination of description with moral allegory, as in Le Moyne's *Peintures morales* (Hagstrum).

Buildings and their surroundings have been popular subjects for ekphraseis, beginning with the description of Alcinous' palace and garden at *Odyssey* 7.84-132. Aeneas describes the murals on the temple of Juno in Carthage and the doors to the temple of Apollo (*Aeneid* 1.446-93, 6.20-30); Ovid describes the palace of the Sun at *Metamorphoses* 2.1-18. In the Byzantine period churches became a popular subject for description, most notably Paul the Silentiary's description of Santa Sophia. A Ren. example is Milton's description of Pandaemonium in Book I of *Paradise Lost*. The Flavian poet Statius wrote descriptions of villas (*Silvae* 1.3 and 2.2) that influenced Ben Jonson's "To Penshurst," the first in a long line of poems in the 17th and 18th cs. (e.g. by Carew, Herrick, Marvell, and Pope) describing estates and country houses (Hibbard).

A related theme is the depiction of a lovely spot ("pleasance"), traceable to Homer's description of the wild area around Calypso's cave at *Odyssey* 5.63-75. It is also found in Gr. lyric (Sappho, fr. 2) and drama (Sophocles, *Oedipus at Colonus* 668-93), and becomes so frequent in Lat. poetry that it has acquired a topical name, *locus amoenus* (Curtius, Arbusow, Schönbeck), with examples in Ren. epic (Spenser's Bower of Bliss and Milton's Eden) and lyric (Marvell's "The Garden").

The most famous romantic e. is Keats' "Ode on a Grecian Urn." In the 20th c., poets have produced many ekphraseis of paintings, particularly of Brueghel's works. W. H. Auden set the fashion with "Musée des Beaux Arts" and was followed by many others, including William Carlos Williams in his sequence, *Pictures from Brueghel* (Fowler). Auden's "Shield of Achilles" is an ironic adaptation of the Homeric e. See now DESCRIPTIVE POETRY; VISUAL ARTS AND POETRY.

L. Spengel, *Rhetores graeci* 3 v. (1853-56); P. Friedländer, *Johannes von Gaza und Paulus Silentiarius* (1912); Faral; C. S. Baldwin, *Med. Rhet. and Poetic* (1928); H. Rosenfeld, *Das deutsche Bildgedicht* (1935); Curtius; G. R. Hibbard, "The Country House Poem of the 17th C.," *JWCI* 19 (1956); J. H. Hagstrum, *The Sister Arts* (1958); G. Downey, "E.," *Reallexikon für Antike und Christentum* (1959); G. Schönbeck, *Der Locus Amoenus von Homer bis Horaz* (1962); L. Arbusow, *Colores rhetorici* (1963); D. B. Wilson, *Descriptive Poetry in France from Blason to Baroque* (1967); *Three Med. Rhet. Arts*, ed. J. J. Murphy (1971); G. Kranz, *Die Bildgedicht in Europa* (1973), ed., *Gedichte auf Bilder* (1975)—anthol.; G. Kurman, "E. in Epic Poetry," *CL* 26 (1974); M. Albrecht-Bott, *Die bildende Kunst in der italienischen Lyrik der Ren. und des Barkok* (1976); S. C. Hulse, "'A Piece of Skilful Painting' in Shakespeare's *Lucrece*," *ShS* 31 (1978); E. L. Bergmann, *Art Inscribed* (1979)—e. in Sp. Golden Age poetry; P. DuBois, *Hist., Rhet., Description and the Epic* (1982); Fowler; W. H. Race, *Cl. Genres and Eng. Poetry* (1988); M. Krieger, *E.* (1991). W.H.R.

ELEGIAC DISTICH, elegiac couplet (Gr. *elegeion*). In Gr. poetry, a distinctive meter, consisting of a hexameter followed by a pentameter, which developed in the archaic period for a variety of topics but which came to be associated thereafter with only one, i.e. loss or mourning, hence elegy in the modern sense. It first appears in the 7th–6th cs.

B.C. in the work of the Gr. "elegiac poets" (meaning only that they wrote in this meter; the name was not applied until the 4th c.)—i.e. Archilochus, Callinus, Tyrtaeus, Theognis, and Mimnermus—who used it for flute songs at banquets and competitions, war songs, dedications, epitaphs and inscriptions, and laments on love or death. It seems to embody reflection, advice, and exhortation—essentially "sharing one's thoughts." Elegiac meter is a species of the epode (sense 2), and perhaps was originally used in melic poetry, but in any event the shorter second line gives the distinctive and satisfying effect of end-shortening or catalexis. Threnodies, ritual laments or cries uttered by professional poets at funerals, may also have been related.

Outside of the "elegiac" context, whether on love or death, the e. d. was specifically the meter of epigrams, esp. after the 4th c., when literary imitations of verse inscriptions were cultivated by the Alexandrian poets. It is this fixation of meter to genre which lasted the longest. Ennius introduced it into Lat., and later, the skill of Martial ensured its passage into the Middle Ages. In Lat. the love-elegy emerges as a major genre. The Augustan elegiac poets (Tibullus, Propertius, Ovid) tended to make the sense end with the couplet, whereas in Gr. poetry and Catullus it was often continuous. Among the refinements which in time became regular in the Roman elegists, esp. Ovid, was the restriction of the final word in the pentameter to a disyllable. In the opening d. of the *Amores*, Ovid jokes that though he intended to write of epic things (hence in hexameters), Cupid stole a foot from his second line, making it hypometric, whence the form.

In the Middle Ages the e. d. was associated with leonine verse, where it acquired rhyme. In the Ren., poets attempted imitations of Gr. quantitative meters, incl. the e. d., such efforts being revived again in the 18th and 19th cs. Examples are found in Eng. by Spenser, Sidney, Coleridge, Clough, Kingsley, and Swinburne; in Ger. by Klopstock, Schiller, Goethe, and Hölderlin; and in It. by D'Annunzio. Coleridge's tr. of Schiller's e. d. is well known: "In the hexameter rises the fountain's silvery column, / In the pentameter aye falling in melody back." But when the e. d. was naturalized into the accentually based prosodies of the vernacular meters, it was imitated as isometric couplets, as, in Eng., in the work of Grimald and in Marlowe's Ovid, whence it exerted influence on the devel. of the heroic couplet. Heterometric (iambic) imitations incl. Swinburne's "Hesperia."

R. Reitzenstein, "Epigramme," Pauly-Wissowa; K. Strecker, "Leoninische Hexameter und Pentameter im 9. Jahrhundert," *Neues Archiv für ältere deutsche Geschichtskunde* 44 (1922); C. M. Bowra, *Early Gr. Elegists*, 2d ed. (1938); P. Friedländer, *Epigrammata: Gr. Inscriptions in Verse from the Beginnings to the Persian Wars* (1948); Hardie, ch. 3; M. Platnauer, *Lat. E. Verse* (1951); L. P. Wilkinson, *Golden Lat. Artistry* (1963); T. G. Rosenmeyer, "Elegiac and Elegos," *Calif. Studies in Cl. Antiquity* 1 (1968); D. O. Ross, *Style and Trad. in Catullus* (1969); M. L. West, *Studies in Gr. Elegy and Iambus* (1974); West; A. W. H. Atkins, *Poetic Craft in the Early Gr. Elegists* (1984). T.V.F.B.; A.T.C.

ELEGY (Gr. *elegeia*, "lament").

I. PASTORAL ELEGY
II. HISTORY
III. FUNCTIONS

In the modern sense of the term, the e. is a short poem, usually formal or ceremonious in tone and diction, occasioned by the death of a person. Unlike the dirge, threnody, obsequy, and other forms of pure lament or memorial, however, and more expansive than the epitaph, the e. frequently includes a movement from expressed sorrow toward consolation. In the larger historical perspective, however, it has been most often a poem of meditation, usually on love or death. Discussion of the origin and devel. of the e. in Western poetry is complicated by major shifts in definition and has been limited in the past by a failure to distinguish between the elegiac as a mode, or motive, and the several species of e. which fall wholly or partly within that mode (Bloomfield).

I. PASTORAL ELEGY. One of the oldest and most influential species of the genre, the pastoral e. has sometimes been thought a subdivision of the pastoral. Since its inception in Cl. lit., it has directed itself toward ceremonial mourning for an exemplary figure, originally associated with forces of fertility or (poetic) creativity. From Theocritus' First Idyll to Yeats's "Shepherd and Goatherd," examples of the pastoral e. will be found among the various elegies listed below. Its practitioners incl. Virgil, Petrarch, Sannazaro, Marot, Spenser, Milton, Goethe, Shelley, and Arnold, among others.

While the pastoral e.'s period of postclassical dominance extended from the Ren. through the 17th c., several of its conventions have informed nonpastoral elegies throughout the history of the genre. Such conventions incl. a procession of mourners, extended use of repetition and refrain, antiphony or competition between voices, appeals and questionings of deities and witnesses, outbreaks of anger or criticism, offerings of tribute and rewards, and the use of imagery such as water, vegetation, sources of light, and emblems of sexual power drawn from a natural world depicted as either injured victim or site of renewal.

The pastoral e. is most notably illustrated in Eng. by Spenser's "November" in the *Shephearedes Calender*, Milton's *Lycidas*, a monody on the death of Edward King, Shelley's *Adonais* (1821) on the death of Keats, and Arnold's *Thyrsis* (1867) on the death of Clough. The gap between *Lycidas* and *Adonais* is due to the fact that, while the e. was continued in the 18th c., the pastoral e. was often derided: examples between Milton and Shelley—

Oldham's on Rochester, Dryden's "Amyntas," Pope's "Winter," Thomson's "Damon"—are relatively weak.

II. HISTORY. In Gr. poetry, the term referred not to a genre or subject matter but to a specific verseform, the elegiac distich. In Cl. Gr., the terms *elegion* and (pl.) *elegia* simply denote verse in the elegiac meter regardless of content (so Aristotle). However, *elegos*, pl. *elegoi* (Lat. *elegi*), referred to a sad song (such as those of the 6th-c. Peloponessian poet Echembrotos) sung to the accompaniment of a flute, i.e. a sung lament about death (so Euripides and Apollonius Rhodius 2.782) regardless of meter. As West says, "in archaic Greece it was the occasion, not the meter, that conferred the name" on a poetic class—e.g. paean, dithyramb—but by the 5th c., classifying poems by meter was more acceptable (cf. the *Poetics*), and since elegiac poems had no one function or occasion, the name for the meter came to represent the kind of songs usually written in that meter. The pastoral laments, such as those of Theocritus for Daphnis, of Bion for Adonis, or (reputedly) of Moschus for Bion—laments which seem in subject matter to be prototypes of the modern funeral e.—were classed by the Greeks as "bucolic poetry," later as "idylls," and later still, "pastoral"; by the Ren., "pastoral e." (see PASTORAL), while mainly applied to works of lament, could also simply denote a pastoral written in verse. Cl. Gr. e. was virtually extinct by the first quarter of the 4th c. B.C., but was revived again at the beginning of the Alexandrian age, ca. 300 B.C.

In Lat., the erotic or love e., in the hands of Gallus, Catullus, Propertius, Tibullus, Ovid, and their successors, becomes in effect a new genre, although derived from Gr. materials. It is distinguished from other genres by meter (still the distich), theme, and tone (often complaint). In Ovid the meter and the tone are extended across several subgenres, from the *Heroides*, *Ars amatoria*, and *Remedia amoris* to the *Tristia* and *Epistulae ex ponto*. Virgil's Fifth and Tenth Eclogues, which like Theocritus' First Idyll combine the themes of love and death, also served as models, extending the trad. of pastoral lament. Nemesian's First Eclogue follows Virgil, as do the eclogues of Calpurnius; Radbertus' *Egloga* (9th c.) consolidates the use of pastoral e. for Christian and allegorical purposes, reinforcing thereby the Virgilian influence on the later, Neo-Lat. elegies of Petrarch's Second and Eleventh Eclogues (mid 14th c.), Boccaccio's Fourteenth Eclogue, Poliziano's elegy on Albeira di Tomasso, Sannazaro's piscatory eclogues, and Castiglione's "Alcon" (1505). These poems were crucial to the shaping of the vernacular e. in It., Fr., and Eng.

In Italy, Sannazaro's *Arcadia* (1502, 1504) includes vernacular It. elegies in the direct trad. of Cl. Gr. and Lat. models. Alamanni uses a different Cl. model for each of the four elegies on the death of Cosimo Rucellai (ca. 1518), and Trissino's *Dafne* (1527) continues this vein. By 1588, Tasso's "Il Rogo Amoroso" uses the pastoral conventions, but with a gloomier sense of historical and biographical pressures. Prosodically, one of the forms by which the Cl. quantitative meters were adapted to the vernacular was *terza rima* (q.v.), a popular vehicle for the subject matter of the e. in the hands of such writers as Tasso and Ariosto, the latter of whose *Rime* and *Satire* received the label of elegies only after his death. The elegiac strain of extended lyrics expressing melancholy and tender sentiments was represented in the baroque period by Filicaia and in the 19th c. by Leopardi and Carducci, whose memorial poem for Shelley is written in elegiac distichs. The elegiac meter has been subsequently used by D'Annunzio, one of whose collections is *Elegie romane* (1892). In the 20th c., impressive elegies have been written by Montale and Pasolini.

In Spain, the e. began as an imitation of It. models, as in Garcilaso de la Vega's *First Eclogue* (ca. 1535) on the death of his lady, and in some of his other poems modeled upon the work of Tasso. Lope de Vega (1562–1635) used octaves and other stanzas in imitation of Tasso for his elegiac verse. In the 20th c., the prevailing tone of Juan Ramón Jiménez in his *Arias tristes* and *Elegías* is melancholy and elegiac. The work of Federico García Lorca, in *Elegía a Doña Juana La Loca* and *Llanto por Ignacio Sánchez Mejías* shows a more direct obsession with the presentiments of death.

In France, the first attempt at copying the e. form from the ancients was by imitating the elegiac distich in alexandrine (q.v.) couplets, first by alternating masculine and feminine rhymes, later by alternating decasyllabic with octosyllabic lines, an experiment of Jean Doublet in his *Élégies* (1559). Ronsard in his *Élégies, Mascarades, et Bergeries* (1565) abandoned the attempt to reproduce the Cl. meter and returned to the subject matter of the Cl. elegists, a treatment also adopted by Louise Labé and Malherbe. In so doing, Ronsard followed the initiative of Marot, whose celebrated lament for Louise of Savoy (1531) faithfully employed the conventions of the pastoral e., thereby setting a precedent not only for Fr. elegists but also for Spenser's introduction of the pastoral e. into England. Other 16th-c. Fr. elegies incl. Baïf's Second Eclogue, Scève's "Arion," and Belleau's "Chant pastoral" on the death of Du Bellay.

With the decline of convincing pastoral, the e. broadens its content: Boileau's *Art poétique* insists on themes either of love or death; and in the 18th c. the genre comes to deal with the tender and the melancholy rather than with grief or mourning. The climax of this tendency is reached with André Chénier at the end of the century. After Lamartine's *Méditations* (1820), the elegiac mode in Fr. poetry becomes confused with others, notably *tombeaux* and poems of homage such as those by Baudelaire, Verlaine, and Valéry.

In Germany, the subject matter of the e. has been so little restricted that Sir Edmund Gosse could say that the e. as a poem of lamentation does

not exist in the Ger. lang. There had been a number of attempts during the Ren. to write elegies in Lat., but it remained for Opitz, in the early 17th c., to write elegies in the vernacular. His equivalent for the Cl. meter was alexandrine couplets with alternating masculine and feminine rhymes. In the 18th c., Klopstock, with greater metrical freedom, turned to sentimental subjects, general sadness, and the troubles of love, as well as the memorializing of actual death. The influence of Eng. "graveyard poetry," as well as of Young, Goldsmith, Gray, Ossian, and others, was also felt heavily in Germany. The *Römische Elegien* of Goethe, although imitative of the Lat. elegiac distich, should probably be classed as idylls, his chief e. being *Metamorphosen der Pflanzen.* Schiller in his essay, "On Naive and Sentimental Poetry," distinguishes the elegiac from the satiric and idyllic by saying that the elegiac longs for the ideal while the satiric rails against the present situation and the idyllic represents the ideal as actually existent. His notion of the elegiac is illustrated in his own works *Die Götter Griechenlands, Die Sehnsucht,* and *Der Pilgrim.* Hölderlin's elegies also deal with the impossibility of attaining an ideal and the longing for the golden days of Hellenism and youth. Mörike and Geibel produced the only Ger. elegies of note in the remainder of the 19th c. In the 20th c., Rainer Maria Rilke's "Requiem" (1909) and ten *Duino Elegien* (1912–1922) constitute an important extension of the genre in its more symbolic and speculative modes and have been widely influential both inside and outside Germany. More recent work by Trakl and Celan bring the genre to new expressiveness; Celan's "Todesfuge" is one of the most powerful elegies ever written for the victims of the Holocaust.

In England, the OE poems *The Wanderer* and *The Seafarer* are elegiac, while *Beowulf* contains a scene of professional mourning for the hero. In ME, *Pearl* and Chaucer's *Book of the Duchess* are closer to elegies proper. In the Ren., except for Surrey's fine memorials to Wyatt and Clere, and apart from some work by the Tudor poets, the Eng. e. remains unimaginative until the pastoral revivals by Spenser and Sidney. Looking back via Marot to Virgil and Theocritus, Spenser's "November" in the *Shepheardes Calender* (1579), together with Sidney's elegies in the *Old Arcadia,* inaugurate the pastoral e. in England and influence an array of subsequent poets. Spenser's "Daphnaida" (1590) looks back to Chaucer, while "Astrophel," written on the death of Sidney, is the keystone in the first anthology of Eng. elegies, prompting numerous others. In the 17th c., crucial contributions are made by Jonson and Donne. Donne brings a newly direct, argumentative manner to his *Anniversaries* for Elizabeth Drury, while the more restrained, neoclassical style of Jonson's stoicism forges a link between the Lat. epitaph and the 18th c.

As for form, after the few Ren. attempts to imitate the Cl. elegiac distich in quantitative verse, the e. was written in iambic couplets, but the e. never developed a fixed form in Eng.: Donne's elegies are in pentameter couplets; Milton's *Lycidas* is in an intricate *canzone*-like schema of irregularly deployed stanzas, rhymes, and short lines (which may echo the choruses of Tasso and Guarini or may simply be the apt vehicle for grief); thereafter, the "elegiac stanza" was used with some frequency up to romanticism, alongside anapestic tetrameters and couplets.

The term *elegie* was used in the 16th and early 17th cs. for poems with a variety of content, including Petrarchan love poetry as well as laments (q.v.), and carries simply the sense "meditation." The Cl. term thus takes over the native trad. of the complaint (q.v.). The connection between death and e. may have been strengthened by Donne, who chose "A Funeral E." to title the last section of his *First Anniversary* (1611), though most of his twenty-odd elegies (in pentameter couplets) are about love. But it remained for Milton in his pastoral e. *Lycidas* (1637) to establish definitively the e. as a separate genre in Eng., the concerns of which are lament for the dead and the search for consolation in the contemplation of some permanent principle.

In the 17th c., within the larger corpus of Eng. meditative and reflective verse, the distinction between "elegiac" verse and the e. proper was sharpened, although the boundary is by no means sharp—many odes, for example, are elegiac in effect. Notable Eng. elegies or poems in the elegiac mode—excepting here the Eng. pastoral elegies mentioned above—incl. Jonson's Cary-Morison ode, Henry King's "An Exequy," Cowley's e. for William Hervey, Denham's e. on Cowley, Marvell's Horatian ode, Dryden's elegies for Oldham and Mrs. Anne Killigrew, works by the Americans Anne Bradstreet and Edward Taylor, Pope's "E. on the Death of an Unfortunate Lady" and pastoral e. "Winter," Gray's *E. Written in a Country Churchyard* (1751), Young's more diffuse *Night Thoughts,* Johnson's "Vanity of Human Wishes" and e. for Dr. Levet, Swift's parodic "Verses on the Death of Dr. Swift," Wordsworth's Lucy poems and elegiac "Immortality Ode," Tennyson's *In Memoriam* (1850); Whitman's "When Lilacs Last in the Dooryard Bloom'd" (for Lincoln), Hopkins' "Wreck of the Deutschland," Swinburne's "Ave atque vale" (for Baudelaire), E. B. Browning's "Bereavement," Hardy's "A Singer Asleep" (1910, for Swinburne) and elegies of 1912–13 for his first wife, Owen's elegies from the First World War, Yeats's "In Memory of Major Robert Gregory," W. H. Auden's "In Memory of W. B. Yeats" (employing three meters) and "In Memory of Sigmund Freud" (both 1939) as well as others for Toller and Henry James, Dylan Thomas's anti-elegiac e., "A Refusal to Mourn the Death, by Fire, of a Child in London," Geoffrey Hill's book-length elegiac homage to Charles Péguy, and Seamus Heaney's elegies for the victims of violence in Northern

Ireland. Among many 20th-c. Am. elegies of note are Tate's "Ode to the Confederate Dead," Stevens' "The Owl in the Sarcophagus," Lowell's "The Quaker Graveyard in Nantucket" and "For the Union Dead," Bishop's e. for Lowell, Sexton's for Plath, and Berryman's for Delmore Schwartz, as well as innovative works such as Ginsberg's *Kaddish,* Bidart's "E." and "Confessional," Clampitt's "Procession at Candlemas," Olds's poems for her father, Grossman's elegies in *Of the Great House,* and Merrill's epic *The Changing Light at Sandover,* a work whose many elegiac moments renovate the most ancient motives and conventions of the genre.

III. FUNCTIONS. Traditionally the functions of the e. were three, to lament, praise, and console. All are responses to the experience of loss: lament, by expressing grief and deprivation; praise, by idealizing the deceased and preserving her or his memory among the living; and consolation, by finding solace in meditation on natural continuances or on moral, metaphysical, and religious values. Often involving questions of initiation and continuity, inheritance and vocation, the e. has been a favored form not only for mourning deceased poets but also for formulating ambitions and shaping poetic genealogies. As such it is a genre deeply implicated in the making of literary history. Closely related are the epitaph, ode, and satire, eulogy and epideictic poetry, and meditative poetry, often serving many of the same functions on related occasions. Milton takes the occasion for virulent denunciation of ecclesiastical or political corruption. Sacks emphasizes how "repetition itself and the submission to codes are crucial elements in the work of mourning" (326). Many elegies of the past century, however, have reflected the modern skepticism about idealization and consolation. Auden adds criticism to his praise of the dead: Yeats was "silly"; Freud "wasn't clever at all" and was "wrong" and sometimes "absurd." Still, if 20th-c. poets have undercut the traditional conventions of e., they have used the "residue of generic strategies" to do so—as well as to achieve what comfort they can—and the number of works in the elegiac mode makes it clear that in poetry the 20th c. has been a "distinctly elegiac age" (Sacks).

H. Potez, *L'Élégie en France avant le Romantisme* (1897); M. Lloyd, *Elegies, Ancient and Modern* (1903); J. H. Hanford, "The Pastoral E. and Milton's *Lycidas*," *PMLA* 25 (1910); G. Norlin, "The Conventions of the Pastoral E.," *AJP* 32 (1911); J. W. Draper, *The Funeral E. and the Rise of Romanticism* (1929); P. Aiken, *The Influence of the Lat. Elegists on Eng. Lyric Poetry 1600–1650* (1932); C. M. Bowra, *Early Gr. Elegists* (1938); R. Wallerstein, *Studies in 17th-C. Poetic* (1950); M. Platnauer, *Lat. Elegiac Verse* (1951); J. Wiegand, "Elegie," *Reallexikon*; G. Luck, *The Lat. Love E.* (1960); O. B. Hardison, Jr., *The Enduring Monument* (1962); F. Beissner, *Gesch. der deutschen Elegie,* 3d ed. (1965); V. B. Richmond, *Laments for the Dead in Medieval Narrative* (1966); *Milton's* Lycidas, ed. S. Elledge (1966); A. F. Potts, *The Elegiac Mode* (1967); C. M. Scollen, *The Birth of the Elegie in France 1500-1550* (1967); G. Otto, *Ode, Ekloge und Elegie im 18. Jh.* (1973); M. L. West, *Studies in Gr. E. and Iambus,* 2 v. (1974); D. C. Mell, Jr., *A Poetics of Augustan E.* (1974); P. Ariès, *Western Attitudes to Death* (1tr. 1974); M. Alexiou, *The Ritual Lament in Gr. Trad.* (1974); J. E. Clark, *Élégie: The Fortunes of a Cl. Genre in 16th-C. France* (1975); K.-W. Kirchmeir, *Romantische Lyrik und neoklassizistische Elegie* (1976); E. Z. Lambert, *Placing Sorrow* (1976)—pastoral e.; R. Borgmeier, *"The Dying Shepherd"* (1976)—the 18th-c. Eng. eclogue trad.; C. Hunt, Lycidas *and the It. Critics* (1979); T. Ziolkowski, *The Cl. Ger. E. 1795–1950* (1980); Trypanis, ch. 7; Fowler; R. J. Ball, *Tibullus the Elegist* (1983); *The OE Elegies: New Essays in Crit.,* ed. M. Green (1983); A. W. H. Adkins, *Poetic Craft in the Early Gr. Elegists* (1985); *Three Elegies of Ch'u,* tr. G. R. Waters (1985); G. W. Pigman, III, *Grief and Eng. Ren. E.* (1985); *CHCL,* v. 1, ch. 5; P. M. Sacks, *The Eng. E.* (1985); M. W. Bloomfield, "The E. and the Elegiac Mode," *Ren. Genres,* ed. B. K. Lewalski (1986); C. M. Schenk, "Feminism and Deconstruction: Re-Constructing the Elegy," *TSWL* 5 (1986), *Mourning and Panegyric* (1988); R. L. Fowler, *The Nature of Early Gr. Lyric* (1987), ch. 3; P. Veyne, *Roman Erotic E.* (tr. 1988); D. Kay, *Melodious Tears* (1990)—Eng. funeral e.; A. L. Klinck, *The OE Es.: A Critical Ed. and Genre Study* (1992).

T.V.F.B.; P.S.; S.F.F.

EMBLEM. A memorable combination of texts and images into a composite picture, the characteristic e. has three parts (see Schöne): a short motto (lemma, *inscriptio*) introduces the theme or subject, which is symbolically bodied forth in the picture itself (icon, *pictura*); the picture is then described and elucidated by an epigram (*subscriptio*) or short prose text. The three parts each contribute in different ways to the dual functions of representation and interp. that particularly characterize the e.

As a genre (see Daly), the e. is related to various exegetical, literary, and iconographic trads. upon which e. writers drew to differing degrees. Illustrated forerunners include broad sheets, dance-of-death sequences, *biblia pauperum,* and illustrated fables. Even more important are the traditions of the *impresa,* heraldry, Ren. hieroglyphics, the *Greek Anthology,* and medieval Bible allegory and symbolism (see Praz; Schöne; Daly).

When in 1531 the Venetian Andrea Alciato (Alciati, Alciatus) published the first collection of es., *Emblematum liber,* he could not have known that his unlikely little book would launch a new genre that would become immensely popular. Alciato became known as "the father and prince of es." The *Emblematum liber* itself has gone through over 170 eds.; Francis Quarles's *Emblemes* (1635) have appeared at least 50 times; Herman Hugo's *Pia*

desideria (1624) enjoyed over 42 Lat. eds., to say nothing of the translations; and Otto van Veen's *Moralia horatiana* (1607) runs to at least 34 eds. It was estimated that some 600 authors produced over 1000 books that appeared in more than 2000 eds. However, since we now know that the Jesuits alone account for over 600 titles (in over 1700 eds. and trs.), the corpus must be much larger.

The function of the e. is didactic in the broadest sense: it is intended to convey knowledge or truth in a brief yet compelling form that would persuade the reader and impress itself upon memory.

In origin humanistic and erudite, the form in the vernacular served a variety of audiences with a panoply of interests ranging from love to war, from social comment to alchemistic mysteries, from encyclopedic knowledge to pure entertainment. Es. were widely employed for the presentation of spiritual and moral instruction and political and religious propaganda.

The first e. book published with Eng. texts is a tr.: Jan van der Noot's *A Theater for worldlings* (1569); the first Englishman's publication of a coll. of trs., incl. some original es., is Geffrey Whitney's *A Choice of Emblemes* (1586). The most successful Eng. emblem book is Francis Quarles's collection of spiritual *Emblemes* (1635), modeled on Jesuit sources. Later, the Victorian period witnessed a revival of interest in the form (see Green and Höltgen). Rosemary Freeman's study, though outdated in conception, remains the only monograph on the Eng. e. Daniel Russell's study of Fr. es. and devices is methodologically sound and will be recognized as a standard work on the subject.

As an art form and a mode of allegorical thinking (see Jöns), the e. was enormously influential during the 16th and 17th cs. It shaped verbal images in poetry and lit., such as Donne's lovers as "stiff twin compasses," Crashaw's image of Christ's bleeding wounds as "thornes . . . proving rose," the Christian vision of salvation in the extended image of metal worked in a furnace in Southwell's "The Burning Babe," Spenser's "golden bridle" (an e. of temperance), Ben Jonson's allegorical masque figures (e.g. Reason), Chapman's taper (*Bussy D'Ambois*), Shakespeare's red and white roses (*I Henry IV*), Falstaff's cushion used as a crown, and Shylock's whetted knife.

The book is only one medium through which the emblematic combination of text and picture was disseminated. Emblematic designs are found in painting and portraiture, interior decoration, carving, jewelry, tapestry, and embroidery. Es. were also used as theatrical properties in dramas, pageants, and street processions (see Schöne; Daly). Poets, preachers, writers, and dramatists frequently employed es. and e.-like structures in the spoken as well as the written word (see Green; Praz; Schöne; Jöns; Daly). The e. thus helped shape virtually every form of verbal and visual communication of the 16th and 17th cs.

H. Green, *Andrea Alciati and His Books of Es.* (1872); R. Freeman, *Eng. E. Books* (1948); W. Heckscher and K. Wirth, "E., Emblembuch," *Reallexikon zur deutschen Kunstgesch.*, v. 5 (1959); R. Clements, *Picta poesis: Literary and Humanistic Theory in Ren. E. Books* (1960); M. Praz, *Studies in 17th-C. Imagery*, 2d ed. (1964); D. Jöns, *Das "Sinnen-Bild": Studien zur allegorischen Bildlichkeit bei Andreas Gryphius* (1966); A. Schöne, *Emblematik und Drama im Zeitalter des Barock*, 2d ed. (1968); H. Homann, *Studien zur Emblematik des 16. Jahrhunderts* (1971); *Ausserliterarische Wirkungen barocker Emblembücher*, ed. W. Harms and H. Freytag (1975); *Emblemata: Handbuch zur Sinnbildkunst des XVI. und XVII. Jahrhunderts*, ed. A. Henkel and A. Schöne, 2d ed. (1976); K. Höltgen, *Francis Quarles (1592–1644)* (1977), *Aspects of the E.* (1986); P. Daly, *Lit. in the Light of the E.* (1979), *E. Theory* (1979), ed., *Andrea Alciato and the E. Trad.* (1989), et. al., ed., *The Eng. E. Trad.*, v. 1 (1988), *The Eng. E. and the Continental Trad.*, (1988); A. Young, *Henry Peacham* (1979), *The Eng. Tournament Imprese* (1988); D. S. Russell, *The E. and Device in France* (1985); *Andreas Alciatus: The Lat. Es.: Es. in Tr.*, ed. P. Daly et al., 2 v. (1985); J. S. Dees, "Recent Studies in the Eng. E.," *ELR* 16 (1986); H. Diehl, *An Index of Icons in Eng. E. Books* (1986). P.M.D.

ENJAMBMENT (Fr. *enjambement*; Ger. *Versüberschreitung* or *Brechung*; Sp. *encabalgamiento*). Nonalignment of (end of) metrical frame and syntactic period at line-end: the overflow into the following poetic line of a syntactic phrase (with its intonational contour) begun in the preceding line without a major juncture or pause. The opposite of end-stopped.

In reading, the noncoincidence of the frames of syntax and meter in e. has the effect of giving the reader "mixed messages": the closure of the metrical pattern at line-end implies a stop (pause), no matter how infinitesimal, while the obvious incompletion of the syntactic period says, *go on*. The one scissors the other. These conflicting signals, in heightening readerly tension, also thereby heighten awareness, so that in fact one is made more aware of the word at line-end than its predecessors: rhyme itself, by enhancing closure, will diminish this effect, while absence of rhyme, in blank and free verse, increases it—toward the same end.

In lang., the hierarchy of domains in the syntagm is: sound–syllable–word–phrase–clause (sentence). The joints between elements both within each domain and between domains are bridged by a bonding force; as one moves up the chain, the bonding force between domains weakens and cuts are easier to make. Within domains, however, much greater variability obtains, particularly inside phrases: the bond between determiners and following nouns is stronger than between adjectives and nouns, so that enjambing the former is a little more severe than the latter. In *Troilus and Criseyde* Chaucer has " . . . help me from / the deth" (1.535–36) and, later, "have" and "ben

youres al" separated by an entire line (3.100–2). Complex verb phrases offer even more prospects. E. below the phrasal level is radical, with submorphemic examples being rare, but famous—as in Hopkins' intrepid e. ending the first line of "The Windhover": "I caught this morning morning's minion, king- / dom of daylight's dauphin." But splitting a word across line-end can also be used to comic effect, as in Byron. No complete taxonomy of effects has yet been given.

It is often assumed that e. is a phenomenon of rhymeless stichic verse and most suitable for dramatic verse or for extended narrative, such as blank verse (q.v.). However, e. and rhymed verse are not at all incompatible. Rhymed stanzaic verse normally shows an overwhelming coincidence of major syntactic boundaries with borders of (some) metrical frames, esp. stanzaic borders, but in many stanza forms a sentence often occupies most or all of a stanza, and the types of syntactic cuts made at line-end vary widely. Even in couplets, where rhyme is tight, the most constrained form, the Eng. closed or "heroic" couplet (q.v.), was a late development: older and more pervasive are the enjambed or "open" couplets, where rhyme-pairing is preserved while syntactic spans often run to length—as in the opening sentence of the *Canterbury Tales*, which is 18 lines long. In rhymeless verse—particularly Milton—the long syntactic period is known as the "verse paragraph" (q.v.). Free verse (q.v.), abandoning rhyme and meter both, finds e. almost compulsory.

Though the Cl. hexameter (q.v.) is mainly stopped (as is Sanskrit verse), Homer uses e. (Kirk). It is the norm in Old Germanic alliterative verse, incl. OE—e.g. *Heliand* and *Beowulf*—where rhyme is unknown and lines are bound together by alliteration used to mark the meter. Prior to the 12th c., e. is rare in OF poetry: it is the *trouvère* Chrétien de Troyes (fl. after 1150) who seems to have been the first poet in the European vernaculars to break his verses systematically. By the 15th c. it is widely used, and even more so in the 16th, esp. by Ronsard (who coined the term) and the *Pléiade*. In the 17th c. it was impugned by Malherbe and later Boileau. These neoclassical authorities, however, allowed its use in certain circumstances—in decasyllabic poetry and in the less "noble" genres such as comedy and fable. Occasionally e. occurs even in tragedy. Since André Chénier (1762–94), it has been accepted in all genres. The device was exploited to the full by Hugo, whose famous e. at the beginning of *Hernani*—"Serait-ce déjà lui? C'est bien à l'escalier / Dérobé" (Is he already here? It must be by the secret staircase)—had all the force of a manifesto. E. was a fundamental characteristic of the *vers libéré* of the later 19th c. and the *vers libre* (q.v.) which emerged from it.

In Fr. prosody e. has been a subject of controversy: since Fr. rhythms are in essence phrasal, line-terminal accents naturally tend to coincide with significant syntactic junctures. Consequently the terminology for analyzing types of constructions in e. is more developed in Fr. prosody than in Eng. Modern Ger. poets who have valued e. incl. Lessing (*Nathan*), Goethe, and Rilke.

In Eng., e was used widely by the Elizabethans for dramatic and narrative verse. The closed couplet drove e. from the scene in the 18th c., but the example of Milton to Wordsworth revived it for the romantics, who saw it as the metrical emblem for liberation from neoclassical rules. Keats's *Endymion* supplies some strong examples of e., as does Byron's narrative verse. Hopkins' "sprung rhythm" introduces "roving over," that is, metrical as well as syntactic e., so that "the scanning runs on without a break from the beginning . . . of a stanza to the end." But it is Milton who is the great master of e. effects in Eng. poetry: indeed, in justifying his choice of blank verse for *Paradise Lost* in its preface ("The Verse"), Milton points to e., i.e. "sense variously drawn out from one verse into another," as one of the three central features of his verse. Here *variously* is a crucial term for describing the manifold lengths and positionings of the Miltonic sentence for manifold effects, many of them at line-end. Perhaps chief among these are those instances in which the reader arrives at line-end and makes a prediction about how the next line will complete the phrase—only to have that expectation thwarted. The momentary shock of our *error*, we may believe, is the Miltonic exhibition of our postlapsarian nature:

> . . . Now conscience wakes despair
> That slumber'd, wakes the bitter memory
> Of what he was, what is, and what must be
> Worse; of worse deeds worse sufferings must
> ensue. (*PL* 4.23–26)

One had thought the series of parallel phrases complete at the end of line 25. One was wrong.

Most generally, e. is simply the most conspicuous example of that dialectics of closure and flow which is the whole substance of formal imposition in poetry, and its study one of the central avenues by which the devices of prosody lead directly to the phenomenology of reading. See now LINE.

K. Borinski, "Die Überführung des Sinnes über den Versschluss und ihr Verbot in der neueren Zeit," *Studien zur Literaturgeschichte* (1893); Kastner, ch. 5; F. Klee, *Das E. bei Chaucer* (1913); F. Wahnschaffe, *Die syntaktische Bedeutung der mittelhochdeutschen Enjambements* (1919); P. Habermann, "Brechung," "Enjambement," "Hakenvers," *Reallexikon I*; J. Levý, "Rhythmical Ambivalence in the Poetry of T. S. Eliot," *Anglia* 77 (1959); K. Taranovski, "Some Problems of E. in Slavic and West European Verse," *IJSLP* 7 (1963); G. S. Kirk, "Studies in Some Technical Aspects of Homer's Style," *YClS* 20 (1966), 105 ff.; R. Cremante, "Nota sull'e.,'" *LeS* 2 (1967); M. Parry, "The Distinctive Character of E. in Homeric Verse," in Parry; J. Gruber, *Laura und das Trobar car* (1976)—e. in

Occitan and It. lyric; Elwert; H. Golomb, *E. in Poetry* (1979); J. G. Lawler, *Celestial Pantomime* (1979); Mazaleyrat; Scott; Morier, s.v. "Enjambement," "Contre-E.," "Contre-rejet"; J. Hollander, "'Sense Variously Drawn Out,'" in Hollander; S. Cushman, *Wm. Carlos Williams and the Meanings of Measure* (1985), ch. 1; D. Godet, "À propos de quelque es. anarchiques," *Versants* 9 (1986); C. Higbie, *Measure and Music: E. and Sentence Structure in the* Iliad (1991). T.V.F.B.; C.S.

ENVOI, envoy (Occitan *tornada*; Ger. *Geleit*). A short concluding stanza found in certain Fr. poetic forms such as the *ballade* and the *chant royal*. In the *ballade* the e. normally consists of 4 lines, in the *chant royal* of either 5 or 7, thus repeating the metrical pattern as well as the rhyme scheme of the half-stanza which precedes it. The e. also repeats the refrain which runs through the poem (e.g. Villon's "Mais ou sont les neiges d'antan?"). In its typical use of some form of address, such as "Prince," the e. shows a trace of its original function, which was to serve as a kind of postscript dedicating the poem to a patron or other important person. However, its true function during the great period of the OF fixed forms was to serve as a concise summation of the poem. For this reason the Occitan troubadours called their envoys *tornadas* (returns). Among the Eng. poets, Scott, Southey, and Swinburne employed envoys. Chaucer wrote a number of *ballades* in which he departs from the customary form by closing with an e. which is equal in length to a regular stanza of the poem, usually his favorite rhyme royal (q.v.).—R. Dragonetti, *La Technique poétique des trouvères dans la chanson courtoise* (1960), pt. 1, ch. 4. A.PR.

EPIC.

I. HISTORY
- A. *Ancient*
- B. *Roman*
- C. *Middle Ages*
- D. *Renaissance*
- E. *Eighteenth Century*
- F. *Modern*

II. THEORY
- A. *Classical and Alexandrian Greek*
- B. *Classical Latin and Medieval*
- C. *Renaissance to Modern*

I. HISTORY. An e. is a long narrative poem that treats a single heroic figure or a group of such figures and concerns an historical event, such as a war or conquest, or an heroic quest or some other significant mythic or legendary achievement that is central to the traditions and belief of its culture.

E. usually develops in the oral culture of a society at a period when the nation is taking stock of its historical, cultural, and religious heritage (see ORAL POETRY). E. often focuses on a hero, sometimes semi-divine, who performs difficult and virtuous deeds; it frequently involves the interaction between human beings and gods. The events of the poem, however, affect the lives of ordinary human beings and often change the course of the nation. Typically long and elaborate in its narrative design, episodic in sequence, and elevated in lang., the e. usually begins "in the midst of things" (*in medias res*) and employs a range of poetic techniques, often opening with a formal invocation to a muse or some other divine figure, and frequently employing elaborate formulaic figures (see FORMULA), extended similes (usually termed epic or Homeric similes; see SIMILE), and other stylized descriptive devices such as catalogues of warriors, detailed descriptions of arms and armor (see EKPHRASIS), and descriptions of sacrifices and other rituals. Recurrent narrative features include formal combat between warriors, prefaced by an exchange of boasts; accounts of epic games or tournaments; and fabulous adventures, sometimes with supernatural overtones and often involving display of superior strength or cunning. E. incorporates within it not only the methods of narrative poetry (q.v.), but also of lyric and dramatic poetry (qq.v.). It includes and expands upon panegyric and lament. With its extended speeches and its well-crafted scenic structure, it is often dramatic and is perhaps with the choral ode the true ancestor of ancient drama.

A. *Ancient*. The earliest surviving es. are the Babylonian *Gilgamesh* saga, the Sanskrit *Mahābhārata* and *Rāmāyana*, and the Cl. Gr. *Iliad, Odyssey*, and fragments from the E. Cycle. The Gilgamesh e. (ca. 2000 B.C.) is contained on 12 tablets and runs to approximately 3000 lines in its surviving form. Gilgamesh, a hero of almost superhuman strength, like the Gr. Hercules or the Hebraic Samson, rescues the city of Uruk from misfortune and completes other deeds, among them the destruction of a bull sent by the goddess Ishtar. The e. also includes digressive material on the Creation, the Flood, and the story of Paradise similar to that in Babylonian and Hebraic scripture. Gilgamesh is tempted by demons (incl. one disguised as a serpent) and after the death of Eabani enters the underworld to seek (like the Gr. heroes Hercules, Theseus, and Odysseus) knowledge of the fortunes of the dead. His relationship to Eabani is one of the earliest of e. warrior-friendships, and his grief at his friend's death has its counterpart in Achilles' mourning for Patroclus or Roland's for Oliver.

The Sanskrit es., the *Mahābhārata* and the *Rāmāyana*, while dating in written form from between 400 B.C. and 400 A.D., were probably composed orally and revised repeatedly by different poets over a period of centuries, beginning perhaps as early as 2000 B.C. The *Mahābhārata*, divided into from 18 to 24 books and composed of over 110,000 couplets or *slokas* of 16 syllables each, is the longest e. in existence. The *Rāmāyana* is over 24,000 couplets long. Like the *Iliad*, the *Mahābhārata* tells of a dynastic war between rival princes; like the *Odyssey*, the *Rāmāyana* concerns

the wanderings of a hero who must rescue his wife and regain his kingdom. The central plot of both es. is embellished by a vast amount of digressive material that includes didactic and doctrinal lore on religion, morals, law, and philosophy, as well as narrative tales, adventures, anecdotes, and fables. Like the Homeric poems among the Hellenic and the Bible among the Hebraic peoples, the Sanskrit es. acquired the status of sacred texts for the Hindi people. The struggle between the Bharatas and the Pandavas in the *Mahābhārata* may be construed allegorically as the war between Hindi gods and demons. Rama, the hero of the *Rāmāyana*, who on one level may be taken as the incarnation of the god Vishnu, achieves an allegorical victory over Ravana that may be similarly taken as India's victory over the drought-demon. Both es., however, include recognizable e. types: the king, the tragically doomed warring princes, the almost superhuman strongman. The immense Bhima of the *Mahābhārata* and Hanumat, Rama's ally in the *Rāmāyana*, fight like Hercules with a club and are, like Ajax and Achilles, violent and impetuous. Although performing deeds on a more massive scale than those of the Homeric heroes, Bhima subscribes to an heroic code in which the making and keeping of vows and the defense of honor are supreme virtues.

Both the *Iliad* and the *Odyssey*, and in the Middle Ages the battle es. of the Germanic, Anglo-Saxon, and Romance cycles, are intimately connected with the heroic codes of their societies. All agree that a violation of honor must be requited and revenge for a slain friend, dishonored spouse, or wronged family member undertaken. Achilles' terrible anger comes upon him in the *Iliad* because Agamemnon has violated his honor or *timé* by wrongfully seizing the girl Briseis; a comparable anger moves him later when he seeks to kill Hector, to avenge the death of his friend Patroclus. Homer tells us, moreover, that the Trojan War came about not merely because Paris had stolen Menelaus' wife, Helen, but because in so doing he had violated the sanctions of guest-friendship. Similarly, in the *Odyssey* Penelope's suitors have violated the laws of hospitality and so Odysseus is fully justified in dispensing justice and killing them all. The Bharatas in the Sanskrit e. pursue the war against their cousins both because they have cheated them in a game and because they have refused to honor their promise and relinquish the kingdom they wrongfully hold.

Both the Sanskrit and the Homeric es. deal with material that is in part based on historical events. Homer, an Asiatic Greek (fl. ca. 750 B.C.), recounts events that involved the great Mycenean society of the Gr. mainland (ca. 1200 B.C.). The excavations of Heinrich Schliemann from 1870 to 1890 confirmed the existence of Troy, actually 9 Troys, one of them (Troy 6) probably that treated by Homer. Composed from material almost certainly handed down orally by rhapsodes, the poems were not written down until the 6th c. B.C. nor organized into the 24 books each now contains until the Alexandrian period. They are both composed in dactylic hexameter (q.v.), the meter that became from Homer's example the standard for all Gr. and Lat. e., and both make extensive use of the metrical and verbal formulas that are distinctive of oral poetry (q.v.). The formulas include epithets (such as "swift-footed" for Achilles) and set phrases arranged in half-lines, whole lines, or sets of lines that can be repeated by the poet to fill the metrical shapes required for his material. Formulaic density varies from scene to scene, tending to predominate in battle scenes or scenes of ceremonial action such as the offering of sacrifices or the launching of ships. Homeric verse is also characterized by the use of simile (q.v.), sometimes brief, sometimes extended, a device much imitated by later Gr., Roman, and Continental and Eng. Ren. poets.

Both the *Iliad* and the *Odyssey* open *in medias res* ("in the midst of things") and pursue to conclusion the action set in motion at the beginning: in the *Iliad*, the resolution of Achilles' anger, and in the *Odyssey*, Odysseus' resumption of kingship in Ithaca. Much episodic material is included in each. Achilles and Odysseus are fully mortal and fallible even though each is attended by a protecting goddess. Lesser characters in each e. serve as foils to the heroes; the Trojan hero Hector is Achilles' inevitable opponent, and Telemachus, Odysseus' son, matures to become his father's assistant against the suitors. While Achilles and Odysseus are endowed with almost superhuman strength at the height of the action, the former filling the river Xanthus with dead bodies, the latter slaughtering over 100 suitors with his mighty bow, Homeric heroes are not supreme strongmen, like Hercules or Gilgamesh, nor incarnations of gods like the Hindi heroes. They do, however, act out a predetermined fate. As Rama in the *Rāmāyana* is ordained to be victorious in his fight with Ravana, so Achilles in the *Iliad* anticipates early death and Odysseus in the *Odyssey* perseveres, knowing he is fated to return safely.

Contemp. or later es. in Gr. survive only in fragments. The *Cypria*, the *Aethiopis*, the *Sack of Ilium*, the *Little Iliad*, the *Nostoi*, and the *Telegony* tell other parts of the Troy cycle; the *Thebais*, the *Titanomachy*, the *Oedipoedia*, and the *Epigoni* recount stories from other cycles. These es. provided later Gr. and Roman poets with material and alternative models. Hesiod's *Theogony* and *Works and Days* (ca. 700 B.C.), though not strictly es., are composed in hexameters and are in Homeric dialect; moreover both contain material that attaches itself to the e. trad. The *Theogony* not only tells of the generation of the gods, but also of Zeus's battle with the Titans and with Typhoeus; the *Works and Days*, which recounts the generation and ages of man, is the first extant didactic poem and influenced later Roman semi-es.: the *Georgics*, *De*

rerum natura, and the *Metamorphoses*.

Little remains of the Alexandrian e. of the 4th and 3rd cs. B.C. Even the *Hecale* of Callimachus (b. ca. 330 B.C.), Apollonius of Rhodes' teacher, exists only as a set of fragments that tells how Theseus, on his way to fight the bull of Marathon, converses with the old woman Hecale on the eve of battle. The *Argonautica* of Apollonius of Rhodes, written in hexameters in four books comprising 5834 lines, is the only complete e. of the period and the earliest example of written or purely literary e. Although deliberately imitating Homeric lang. and devices, the *Argonautica* marks a decided departure from the ethos of Homeric e. Its hero, Jason, is neither a super-warrior like Achilles nor a man of endurance like Odysseus; he performs his most impressive acts through magic and presides over a crew of heroes whose deeds frequently surpass his own. Apollonius heightens the marvelous that had been a feature of the apologue of the *Odyssey* and creates with Medea, the virgin-witch, a psychological study of a woman in love. He hands down to the Romans an e. that differs markedly from the oral e. of Homer.

B. *Roman*. The historical e. in Rome dates from Naevius' *Bellum punicum*, now lost, and Ennius' *Annales*, a work in 18 books of which only fragments survive. Ennius (239–169 B.C.), reputed father of Roman e., replaced the older Saturnian meter, used by Livius Andronicus in his tr. of the *Odyssey* (3d c. B.C.), with Lat. hexameters on Gr. metrical principles and employed the conventions of Homeric e. for a semilegendary history of Rome from Aeneas' landing in Italy to his own time. He bequeathed to Virgil (70–19 B.C.) and to later Roman poets the notion that hexameter was the only proper verseform and national destiny the only possible theme for the Roman e. Virgil himself limited his account of Roman history to key prophetic sequences in Books 1 and 6 of the *Aeneid* and to various apostrophes and allusions to future Rome throughout the e. Imitating the *Odyssey*, he recounts in the first six books of the *Aeneid* the story of Aeneas' successful voyage to his new homeland in Italy and in the final six books, imitating the *Iliad*, of Aeneas' ascendance over Turnus, an Achilles-type opponent, in the It. wars. In Virgil's hands Aeneas becomes an exemplar of loyalty to family and comrades and of Roman *pietas*, a stoic leader whose endurance conquers adversity and whose determination brings peace to warfaring Italy, just as Augustus, to whom the e. was dedicated, had brought an end to civil war. Providential destiny replaces the Homeric *moira* or fate, a force to which not only Aeneas but also Dido, Turnus, and the doomed young warriors of the It. wars must bow. Within a tightly controlled frame, episodic material plays a proscribed role. Hence the story of Troy's downfall in Book 2 is contrasted to Rome's eventual rise. The account in Book 4 of Dido, a character who recalls the Homeric Calypso, Circe, and Nausicaa as well as the abandoned Hypsypyle and love-sick Medea of Apollonius' *Argonautica*, must be more than a digressive episode of ill-fated love; it must foreshadow the future enmity of Carthage, the nation Rome had to destroy to secure its position in the ancient world. Portent, oracle, and sign, used by Apollonius as devices to move the Argo on its forward and return journeys, become in Virgilian e. assurances of Rome's future success. Similarly, the archetypal voyage to the underworld, so prominent in ancient oral e., becomes for Virgil the means for Aeneas to look ahead to the future leaders of Rome. Further, while Virgil may imitate, in Venus' fetching of arms for Aeneas, Thetis' rearming of Achilles in the *Iliad*, he places upon Aeneas' shield scenes that lead inexorably to Augustus' triumph at Actium.

Virgil's most accomplished contemporaries—the older Lucretius (ca. 99–ca. 55 B.C.) and the younger Ovid (43 B.C.–17 A.D.)—both of whom create ambitious long poems in hexameters, follow different e. models from Virgil. Like Hesiod, Lucretius in *De rerum natura* (Of the Nature of Things) writes a philosophical and didactic poem dealing with the creation and the composition of the universe and problems of human existence from the vantage of Epicurean philosophy. Ovid, taking his inspiration from Callimachus' *Aetia* and the Theocritean *epyllion*, creates in the *Metamorphoses* a series of tales, some drawn from the ancient e. trad., some from Alexandrian romance, that involve transformations and are told briefly with interlocking narrative material. With particular interest in the erotic and in the marvelous, which had been prominent in Alexandrian e. and *epyllia*, Ovid introduces into the e. trad. material that has enormous impact not only on the later writers of mythological e.—Valerius Flaccus in his *Argonautica* (1st c. A.D.) and Claudian in *De raptu proserpina* (late 4th c. A.D.)—but also on Ren. writers from Dante through Milton. With Lucretius, who makes cosmic and philosophical material the province of e., and with Virgil, who reshapes Homeric e., Ovid is the most important influence on e. writers of the Ren.

Virgil's influence on writers in the e. trad. is both immediate and long-lasting. Even for poets such as Lucan (39–65 A.D.), who abandons the mythological machinery of the *Aeneid* and accepts the boundaries of recent history for his 10-book unfinished e., the *Pharsalia* (or Civil War), Virgil is the inevitable model. Echoes of the dynastic struggle between Aeneas and Turnus resonate in the e. duel between a relentless Caesar and a tragic Pompey. To a greater extent, Silius Italicus (26–101 A.D.), whose 17-book *Punica* on the Second Punic war is the longest Roman e., follows Virgil, imitating his mythological framework and battle scenes, projecting an underworld vision for Scipio modeled on Aeneas', and even constructing a shield for Hannibal embossed, like Aeneas', with scenes from hist. Although he chooses a subject

from Gr. rather than Roman hist., Statius (45–96 A.D.) in the *Thebais* depends heavily on the machinery of Virgilian e. Employing supernatural messengers, portents, and dreams, he increases the episodic content of e. and the number of set pieces, describing battle scenes, chariot races, and a journey to the Hall of Sleep, embellished with allegorical imagery. The aim of the later Roman e. poet was to sustain a highly crafted lang. of imitation, based on standard features in Virgil. The high reputation as e. poets that Statius and Lucan enjoyed through the Ren. indicates their success.

C. *Middle Ages.* In the Christian era, e. undergoes a dramatic change. The mythological subject still survives in Claudian and in Nonnus' *Dionysiaca* (5th c.), a compendium in hexameters of stories about Dionysus and the other gods that extends to 48 books. While the first Christian es., such as Juvencus' gospel paraphrases, are little more than versified scripture, Marius Victor's *Alethia* and Avitus's *Poemata* adapt to biblical narratives from Genesis and Exodus the Roman epic's methods of treating historical material. Substituting Christian providence for Roman destiny, they tell how God's people attain the goal of salvation rather than the triumph of empire. Es. such as Prudentius' *Psychomachia* (4th c.) and Dracontius' *Carmen de deo* (5th c.) substitute the battle of the soul against evil for traditional e. combat. Saints begin to assume the place of the e. hero, with the hero of faith superseding the heroic warrior. Not only in Lat. but also in the considerable body of vernacular lit., the religious e. begins to dominate. The Anglo-Saxon *Genesis* A and B, following Lat. e., give heroic proportion to stories from Genesis, Satan as an e. hero making his first appearance in the latter. The OE *Christ*, attributed to Cynewulf, is one of the first poems to give heroic treatment to Christ's birth, death and resurrection, and return at Judgment Day, while *Andreas* and *Elene* tell of the saint's devotion to Christ and search for the true cross. The Christian e., begun in the Middle Ages, revives in the Ren. with renewed interest in hexaemera and es. on biblical subjects.

During the Middle Ages, pre-Christian material from the ancient oral and heroic culture also began to find written form, and in the process pagan and Christian often were synthesized. Like the Gr., the Romance and Germanic e. has its genesis in the heroic *lai* and its roots in a feudal society headed by a king who was surrounded by his vassals or *comitatus* bound by oath to support him in battle in return for protection and gifts. Anglo-Saxon e. pictures the center of the society as the hall, where king and warriors meet and where the bard sings of the heroic past. Many e. events are based upon documented historical events, such as the raid of Hygelac in *Beowulf* or the slaughter of the Burgundians at the court of Attila, narrated in the *Nibelungenlied* and in other Germanic es. The OF *Chanson de Roland* and the Sp. *Poema de mio Cid* recount versions of specific episodes in Charlemagne's struggle against the Saracens in the 8th c. and Castile's repulse of the Moors in the 10th c.

Most Germanic poetry, though emerging from a Christian society, is heavily overlaid with pre-Christian elements. A sense of fatality stronger than Christian providence is conspicuous. Although *wyrd* or fate may relent and save an undoomed man if, as the *Beowulf* poet says, his courage is good, the hero usually fights a losing battle, knowing the odds are against him. He is controlled, moreover, by a strict code of honor that requires at all costs loyalty to lord and friends. The earl Byrthnoth in the Anglo-Saxon *Battle of Maldon*, Roland in the *Chanson de Roland*, and Hagen in the *Nibelungenlied*, having made tragic errors in judgment, all prefer honor to life.

Older stories of pre-Christian gods and heroes who fight monsters with superhuman strength survive side-by-side with stories of dynastic strife. Sigurd, the hero of the Icelandic Elder Edda (ca. 900–1300), is the son of a god; he fights a dragon, wins treasure, and rides through a wall of fire to conquer an Amazonian goddess, deeds attributed in the *Nibelungenlied* (13th c.) to Siegfried, a prince of the Netherlands, who behaves more in the chivalric than the supernaturally heroic style. The Irish hero Cuchulain, also a god's son, can singlehandedly withstand an army. Beowulf, the hero of the 8th-c. OE poem, combines supernatural strength with feudal loyalty and Christian faith. Against a background of dynastic intrigue that dooms the court-society in Denmark and Sweden, Beowulf defeats the monster Grendel in his youth and a marauding dragon in old age.

Germanic e. deals mainly with the tragedy that results from conflicting loyalties to family, spouse, feudal lord, and comrades-in-arms. In the Hildebrand saga (ca. 800) a father, come as champion of an enemy army, is challenged by a son ignorant of his identity and is forced by the code of honor to answer the challenge. Walther, hero of the OE fragment *Waldere*, the fragmentary Ger. e. *Walther*, as well as the Lat. *Waltherius*, having been kept hostage with his friend Hagen in the court of Attila, is challenged by Hagen on his return home. In the various versions of the Nibelungen saga, Hagen murders Siegfried to protect the honor of his liege-lord Gunther, then falls victim to the long-plotted revenge of Siegfried's wife and Gunther's sister Kriemhild, who, like Hagen, must exact an honorable retribution. Assuming the role of the avenging male, unlike most women in Germanic poetry, Kriemhild in the *Nibelungenlied*, like Achilles and Aeneas in Cl. e., denies mercy to Hagen when she sees him wearing Siegried's sword, the fatal token of his victory over Siegfried.

Early Romance e., like Germanic e., was composed orally for a royal society by poets called minstrels or jongleurs, and is formulaic. Anglo-Saxon and Germanic e. employs a heavily alliterative line, often with 4 stresses and marked with a prominent medial caesura. The most common

meter for Romance epic, the one used by the anonymous poet of the *Chanson de Roland* is decasyllabic with a regular caesura after the fourth syllable. It is organized into rhyming units called laisses, which average about 14 lines but can vary from 5 to 100.

Romance e., like its counterpart the medieval romance, is episodic in structure, arranging adventures in panoramic sequence, often juxtaposing farcical and heroic action, and frequently incl. fantastic and supernatural elements. Although it sometimes involves a love affair, Romance e. focuses on the hero's achievements as a warrior rather than as a lover. With its roots in the *chansons de geste* (q.v.), Romance e. was divided into three main cycles: the matter of France, which told of the deeds of Charlemagne and Roland, and includes the most famous Fr. e., the *Chanson de Roland* (ca. 1100); the matter of Britain, which dealt with King Arthur and his knights of the Round Table; and the matter of Rome, which encompassed material from antiquity, esp. stories of the siege of Troy, and includes the *Roman d'Alexandre* (ca. 1160), which used and gave the verse line, the alexandrine (q.v.), its name. Apart from the *Chanson de Roland*, the most important chansons are the *Chanson de Guillaume*, celebrating the exploits of the historical Duke of Orange, and the *Voyage de Charlemagne*.

The Fr. poet of the *Chanson de Roland* reshapes the historical events of Roland's defeat at Roncevaux (ca. 778) into an e. of betrayal and courage. Divided into three parts, the second of which concludes with the death of Roland, the *Chanson* narrates the events that lead to the betrayal of Roland, describing how the warrior-hero and his peers fight off the Saracens. As in the *Iliad*, descriptions of the aristeias of each warrior follow in rapid succession. Roland's loyalty to his liege-lord is only surpassed by that to his friends, whose deaths he inadvertently causes by his reluctance to summon aid. Nevertheless, Oliver, like Patroclus, dies blessing his friend, having sacrificed his life to defend Roland's honor. As a battle e. the *Chanson* has much in common both with ancient and with med. Germanic e. The warriors of the Fr. *geste* follow the traditional pattern of exchanging vows and boasts with their opponents before combat. Special weapons are prized that are closely identified with the hero and which he alone can wield, such as the huge spear that Achilles possesses or the magnificent sword granted Beowulf to kill Grendel's mother. Like Sigurd or Siegfried before him, and the Cid after him, Roland has a named sword, sent from heaven and made of holy relics, that he would rather destroy than hand over to his enemies. Like Achilles and Sigurd, he has a horse who grieves for his death. The trad. of named swords and loyal horses carries on into Ren. chivalric es., where they often assume allegorical significance.

The great Sp. e., the anonymous *Poema de mio Cid* (ca. 1140), tells the adventures of a hero and his comitatus. As the oldest surviving text of the medieval Castilian e. trad., it is remarkable in that it attains literary form less than 100 years after the events its celebrates: the exploits of Don Roderigo of Castile, dubbed by the Moors he conquers El Cid. Written in *mester de juglaria* (irregular heptasyllabic hemistichs) and divided into three sections or cantars, the e. celebrates a vassal exemplary in his loyalty to his king and to his warring allies but, unlike many of the tragic heroes of Germanic e. and the Fr. Roland, both valiant and successful, defeating his enemies and defending insults to his honor, until finally becoming reconciled to his king and, through his daughters' marriages to princes, a future patriarch himself. Though much shorter than the *Aeneid* or the Sanskrit es., like them it is a national e. that concludes with a prince's successful founding of a royal dynasty.

D. *Renaissance.* Ren. e. begins in Italy with an attempt to reconcile Romance e. to the Christian and Cl. trads. From Dante (1265–1321) in the late Middle Ages on, European e. poetry, espousing the literary values of Virgil and his Cl. predecessors, tries to make e. poetry meaningful to the Christian present and to the national interests of the different Eng. and Continental cultures. For Dante this means uniting, in the *Divine Comedy* (written in the vernacular), the Cl. model of the e. journey quest with the Christian model of the allegorical dream vision. Making himself an e. pilgrim and journeying soul, he travels down to survey Hell, then up the pseudogeographical mount of Purgatory to the Earthly Paradise, and finally through the spheres of a Ptolemaic universe to Heaven, where he sees the smile of a triune deity. His journey, unlike those of Odysseus or Aeneas, seeks neither homeland nor political ideal but the soul's place with God and a justification of his native Florence's nonpolitical, nonsectarian destiny. The three realms that Dante surveys are inhabited by historical and mythical characters, as well as contemp. It. sinners and saints. Encountering each successive individual or group as he moves through the circles of Hell, Purgatory, and Paradise (to which he devotes 34, 33, and 33 cantos respectively in his 100-canto poem), Dante approaches closer to numerological perfection and to defining what it would mean to dwell in a society of just and sane people. Not unlike the Cl. Lucretius and the medieval Boethius (whose philosophical poem *The Consolation of Philosophy* [6th c.] had impact on the writers of intellectual e.), Dante investigates the material and intellectual composition of the universe in order to understand its history and man's destiny.

Dante's immediate successors in Italy—Petrarch (1304–74) and Boccaccio (1313–75)—and in England—Chaucer (ca. 1343–1400) and Langland (ca. 1332–1400)—take a different course with e. material. Petrarch, writing in Lat.,

attempts to revive in *Africa* the Roman historical e. Boccaccio and Chaucer in the *Teseida* and *The Knight's Tale* (from the *Canterbury Tales*), respectively, and also in *Filostrato* and *Troilus and Criseyde*, attempt to unite e. and medieval romance. They narrate stories that involve the traditional heroes of the Theban and Trojan War cycles and that climax in traditional events such as e. tournaments and heroic battle. The narrative thrust of these chivalric es., however, involves the winning of a lady's affection, which will remain an important ingredient in the romantic es. of the next two centuries. Boccaccio and Chaucer eschew the terza rima (q.v.) of Dante's *Commedia* and choose longer stanzaic patterns. Boccaccio develops ottava rima (q.v.), which becomes the meter of choice for writers of romance e.—Pulci, Boiardo, Ariosto, Tasso, and Camoëns. From Boccaccio Chaucer develops rhyme royal (q.v.), which in turn becomes the choice of the Scottish Chaucerians in the 15th c. and leads to Spenser's devel. of the Spenserian stanza (q.v.) for *The Faerie Queene*. Except for Ren. e. in Lat., which uses Virgilian hexameters, e. after Boccaccio typically employs one stanzaic pattern or another, until the mid 17th c., when Milton chooses unrhymed iambic pentameter as the meter for *Paradise Lost*. Langland's *Piers Plowman* (late 14th c.), like the *Divine Comedy*, takes the allegorist's way with a dream vision that surveys mankind as a fair field of folk, considers the alternate temptations of Lady Meed and the Seven Deadly Sins, and unravels the instructions of Holy Church. But despite its pronounced influence on future Eng. allegorists such as Spenser, *Piers Plowman* belongs more to the alliterative trad. that finds its final flowering in the 14th c. than to the Ren.

It. poets of the 15th and 16th cs. look back to the *chansons de geste* and the matter of France for a new chivalric e. in which love and marvelous adventures became as important as heroic battle. Taking seriously the trad. that Homer was the author of the so-called comic e., *Margites*, Luigi Pulci (1432–84) determines in *Morgante Maggiore* (1483) to transform the adventures of Orlando into humor by creating as Orlando's associates two comic giants—Morgante and his semi-giant companion to Margutte. Without being able to change the tragic end to the story—Orlando's betrayal by Ganelon and his death at Roncevaux—Pulci lightens the effect by having Rinaldo, accompanied by the comic devil Astarotte, attempt to save Orlando while burlesque devils and angels snatch at the souls of dying Saracens and Christians. *Orlando Innamorato*, the incomplete e. pub. in two installments in 1486 and 1494 by Matteo Boiardo (1441–94), follows soon after. Influenced by the treatment of love in Arthurian chivalric romances, Boiardo reshapes the character and story of Roland, making his love for the princess Angelica of Cathay the principal theme. Pitting Christian against Saracen knights in battle, in tournaments, and in love, Boiardo makes the golden Angelica the object heroes vie for as once they sought glory or the Golden Fleece in the traditional es. of war and quest. Boiardo heightens the element of the marvelous, moreover, with magic lances and rings, enchanted castles and fountains, for Angelica inherits the role of sorceress Medea once had. War is not entirely banished, however, for the pagan enemies of Charlemagne's peers—the Tartar Agricane and the Moor Gradasso—are fictional counterparts of the real threat that the Turk continued to pose for 15th-and 16th-c. Europe. With his sequel to Boiardo's e., *Orlando Furioso*, pub. first in 1516 and in a definitive edition of 46 cantos in 1532, Ludovico Ariosto (1474–1533) resumed the story of Orlando's pursuit of Angelica. Upon her marriage to a humble-born warrior, Angelica departs, mid-way through the e., leaving Orlando, like the maddened Ajax, to wreak havoc until, restored to his senses, he resumes his proper role as Charlemagne's leading warrior. Ariosto takes an ironic view of the world of men and women, undercutting the very heroism his e. purports to praise and cynically demonstrating the folly of the love that in chivalric romances moves men to brave deeds and women to virtue. Not even Ruggiero and Bradamante, characters inherited from Boiardo and the alleged founders of the dynasty of the d'Estes (for whom both Boiardo and Ariosto wrote), escape Ariosto's satire, for Ruggiero must be rescued from the sorceress Alcina by his warrior-fiancée.

After Ariosto, e. assumes a more serious purpose, with the national and Christian e. coming to the fore in Camões, Ronsard, Tasso, and Spenser. In *Os Lusiadas* Luis de Camões (1524–80) attempted to do for Portugal what Virgil in the *Aeneid* had done for Rome—to write an e. celebrating the national character that both told the hist. of the nation and adhered to the design of traditional Cl. e. Like the contemp. Sp. poet Alonso de Ercilla y Zuñiga (1533–94), who told in the *Araucana* of Spain's conquest of the South Am. Indians, the Araucanians, Camões chose for his e. a national adventure from recent history. Recounting in the *Lusiads* the voyage of Vasco da Gama to India to find a new trade route and establish an empire to rival Spain's, Camões created a contemp. quest e. He adopts the mythological machinery of the *Aeneid*: Vasco's voyage to the East proceeds under the protective eye of a semi-Christian Venus, Aeneas' patroness, and overcomes obstacles posed by a malevolent Bacchus, representing the disorderly pagan anarchy of the East. Vasco's voyage is, as Camões demonstrates through Virgilian digressions and prophecies, the outcome of Portugal's own destiny, in which, having repulsed the Moors and defended its freedom against Spain, Portugal can assume its proper role as the leading defender of Christianity. Formally, the *Lusiads* is in 10 cantos of ottava rima, following It. e. rather than Cl.; the ninth canto, the episode

of the Island of Loves, indicates in another way the influence of Ariosto and Boiardo. Pierre de Ronsard (1524–85), although he only completed four books of the *Franciade* (pub. 1572), had an aim similar to Camões—to glorify the future hist. of his nation by telling of its founding by the eponymous Trojan ancestor, Francus, and so to produce a latter-day *Aeneid* for the kings of France. The Croatian poets Brno Krnarutić (1520–72) in *The Capture of the City of Sziget*, Juraj Baraković (1548–1628) in *Vita Slovinska*, and Ivan Gundulić (1588–1638) in *Osman* also produced es. inspired by similar national interests.

Torquato Tasso (1544–95), combining Virgilian and chivalric e., produced in *Gerusalemma liberata* (1582) a Christian e. conscious of its Counter-Reformation mission to urge the liberation of Jerusalem from the Turks, as once the heroes of the poem had liberated the city at the end of the First Crusade. Substituting the pious Godfrey of Bologne for the chivalric Charlemagne of older e. and the impetuous Rinaldo for Orlando as the leading warrior, Tasso considers, as Ariosto and Boiardo had before him, the alternate demands of war and love. Rinaldo, a composite of the Cl. and chivalric hero, is, like Achilles, headstrong and quarrelsome and, like Orlando, led by his love for a woman to neglect his duty. Other characters are divided by their love for pagans and their loyalty to Christianity. Angelic messengers take the parts that pagan deities played in the *Aeneid*. Although written, like the es. of Ariosto and Boiardo, for the d'Estes, and like them in ottava rima (and 20 cantos in length), *Gerusalemma liberata* is more nationalistic and Christian in design. As such, it more nearly resembles the *Faerie Queene* (1590, 1596), the allegorical work that Edmund Spenser (ca. 1552–99) designed as the national e. for England to celebrate the virtues of the perfect knight Arthur and to glorify Elizabeth, the greatest princess of Christendom. Although only six Books of the original 24 projected were completed, each of the six celebrates a different virtue. The *Faerie Queene* adapts material from e., romance, and allegorical trad., not only using Virgil, Tasso, Ariosto, Dante, Chaucer, and Langland as direct models, but also paying attention to what his contemporaries Samuel Daniel and Edward Hall were attempting in producing chronicles and poetic histories for England. Hence in Book I, in recounting the adventures of the Red Crosse Knight, Spenser is also recounting allegorically both how Protestantism triumphed in England after Henry VIII and how the individual Christian pilgrim, like the typical allegorical hero, aspires to overcome sin in the battle each Christian must wage. Spenser's female knight Britomart, whose adventures occupy Books 3–5, is both one of the proposed ancestresses of Elizabeth Tudor (as Ariosto's Bradamante was of the d'Estes) and the personification of Elizabeth's virtue, Chastity.

Unlike Tasso and Spenser, many Christian writers of the 16th and 17th cs. turned directly to the Bible for their plots and characters, adapting Old Testament narrative, as medieval writers had done, taking Christ as the example of the ideal hero, or elaborating extra-biblical material on Satan and his angels. The Croation writer Marko Marulić (1450–1524) wrote a *Judita* (1521–23) in his native Croatian and a Lat. e. on David, *Davidijada*. Later in the 16th c., Gabriello Chiabrera (1552–1638) produced several short biblical es. on David, Judith, the Deluge, and other subjects, using *versi sciolti*, perhaps the Continental precursor to Milton's blank verse. The most famous e. of the century, however, was the *Christiad* (1535) by Girolamo Vida (ca. 1490–1566), composed in six Books in Lat. hexameters, closely modeled on Virgil's, and an important influence on Continental and Eng. writers through Milton. Vida in making Christ the hero of his poem looks forward to the Christian es. of the 17th c. The Fr. Huguenot Guillaume Du Bartas, reviving the hexaemeral trad. in his enormously successful Creation e., *La Semaine* (1578), sparked other Creation es. throughout Europe well into the next century, among them Tasso's *Le Sette Giornate del Mondo Creato* and es. by the Ger. Mollerus (1596), the It. Murtola (1608), and the Sp. Acevedo (1615). A different type of religious e. recounted Satan's rebellion and fall, the most notable being Erasmo de Valvasone's *L'Angeleida* (1590) and Odorico Valmarana's *Daemonomachiae* (1623). The Eng. Spenserians, Giles and Phineas Fletcher, produced es. as diverse as Giles' celebration of *Christ's Victory and Triumph* (1610) and Phineas' dual-lang. Lat. and Eng. *The Locusts* (1627), on the Gunpowder Plot, a subject that also moved the young Milton's first attempt at e. Probably the most promising of the Eng. biblical es. before Milton's, however, was the *Davideia* (1656), Abraham Cowley's ambitious but unfinished work in pentameter couplets that tried to give Cl. scope to the history of David.

John Milton (1608–74), opting for a topic popular in Ren. biblical drama over the Arthurian material he originally considered, composed in *Paradise Lost* (1667, 1674) the first Cl.-style e. on Man's fall to include not only the story of the temptation of Adam and Eve but also accounts of Creation, the battle of the angels, and an extended prophetic sequence on biblical hist. Like Dante, Milton turned away from Cl. notions of the heroic, eschewing as supreme virtue both battle courage and desire for empire (qualities he attributes to the anti-hero Satan), and instead lauding the patience and heroic martyrdom that the Son of God willingly espouses as true merit. He makes Adam and Eve his central characters and the central event their tragic yielding to temptation. Looking back on the didactic semi-es. of Hesiod, Lucretius, and the Virgilian *Georgics*, Milton makes instruction a large component of the e. He modifies the basic structure of the Cl. e., retaining, however, its

12 books, beginning *in medias res,* and including narrative digressions; the invocation to the Muse, the extended simile, and mythic allusion he adapts to the purposes of a Christian e. The sequel, *Paradise Regained,* in four books only and often termed a "brief e.," is plainer in style; Milton chooses, like Vida, Jesus as his hero, and using the techniques of Platonic dialogue and Jacobean debate tells the story of Satan's temptation of Jesus in the wilderness.

E. *Eighteenth Century.* Interest in Cl. e. continues strong in the latter 17th c. and into the 18th. Treatises on e., such as R. Le Bossu, *Traité du poème epique* (1675) and *Treatise of the Epick Poem,* tr. "W. J." (1695), appear; major poets, particularly in England, translate Homer and Virgil; and poets continue to imitate Cl. and Romance e. Voltaire produced a *Henriade* (1723) on the Fr. religious wars, and in Russia M. Lomonosov and M. Kheraskov also wrote nationalistic epics, the former on Peter the Great, and the latter on Ivan the Terrible's capture of Kazan, the *Rossiada* (1779). In England, though minor writers such as Richard Blackmore compose e. after e., no writer of talent after Milton wrote a successful Cl. e. Many successful trs. of Homer and Virgil appear, however, the first having been that of Homer by George Chapman (1559–1634). John Dryden (1631–1700) produced a neoclassical tr. of the *Aeneid,* as did Alexander Pope (1688–1744) and William Cowper (1731–1800) of the Homeric poems. Mock e. (q.v.), however, abounded, from Nicholas Boileau's *Le Lutrin* (1673–83) to Dryden's *Absalom and Achitophel* (1681) to Pope's *Rape of the Lock* (1714), a form employed by Neo-Lat. poets in the previous century and given Cl. approval by the existence of the type in such poems as the 6th-c. B.C. *Batrachomyomachia* (Battle of the Frogs and Mice), attributed in the Ren. to Homer. Pope's most ambitious mock-e. poem, *The Dunciad* (1738, 1742), consigned untalented dabblers in heroic verse to an underworld ruled by the Goddess Dulness. Lyric poets use some techniques of e. in long descriptive poems, such as, in England, James Thomson in *The Seasons* (1726–28); on the Continent, Albrecht von Haller in *Die Alpen* (1732), a preromantic panegyric in 490 hexameters on the beauty of mountains; and in Russia, S. Bobrov in *Tavrida* (1798) or *Khersonida* (1804).

The rediscovery and revival of interest in oral e. (both genuine and spurious) also dates from this period, however, as does serious evaluation of the conditions of composition that produced the Homeric poems (see F. A. Wolf, *Prologomena ad Homerum* [1795]). Even though the most famous of these so-called recovered ancient es.—*Fingal, an Ancient E. Poem in Six Books*—was a hoax, attributed to the Celtic bard Ossian but actually written by James Macpherson (1736–96), the interest generated led to the study of ballads and folksongs. Celtic es. and medieval ballad material began to be gathered in Scotland, Wales, and Ireland. The Irish hero Cuchulain emerges from the Celtic e., the *Taín,* which in its reconstructed version is prosimetric. In Finland the poet-scholar Elias Lönnrot (1802–84) recovered for his nation the e. known as the *Kalevala* (1835, 1840), reconstructed from orally transmitted material.

F. *Modern.* After 1800 few poets attempt to revive Cl. e., though the immense cultural prestige of e., lingering, attracted many of them to displaced or related types of long verse compositions. John Keats (1795–1821) leaves his Miltonically-inspired *Hyperion* unfinished. William Wordsworth (1770–1850) in his long seminarrative poem in blank verse, *The Prelude* (1805, 1850), creates an autobiographical account of his devel. as a poet that qualifies in intellectual scope as a philosophical e. Similarly, Elizabeth Barrett Browning (1806–61) in *Aurora Leigh* (1857) could lay claim to have written a novel in verse or a feminist e. George Gordon Lord Byron (1788–1824), having attempted different types of narrative verse in his Oriental tales and in *Childe Harold's Pilgrimage* (1812, 1816–18), turned to comic e. in ottava rima in the style of Pulci and Ariosto in *Don Juan* (1822), which even in its unfinished state still reaches 16 cantos. In Russia, Aleksandr Puškin (1799–1837) followed Byron's lead, particularly in his exotic oriental tale *The Gypsies* (1824). His romantic e. *Ruslan and Lyudmila* (1820) is in the style of Ariosto but incl. folklore elements; the historical es.—*Poltava* (1829) and *The Bronze Horseman* (1833)—follow earlier Rus. models, and finally the new novelistic e., *Evgeny Onegin* (1833), his most accomplished poem. In England both Alfred Lord Tennyson (1809–92) and William Morris (1834–96) turn back to the Romance and Germanic e. of the Middle Ages. Tennyson attempted in *Idylls of the King* (1888) in a series of 12 linked Books to study the rise, decline, and fall of an heroic society, allegorically representing Camelot as his own England in the waning 19th c. Morris devotes himself to retelling Cl. legend in *The Life and Death of Jason* (1867) and Icelandic saga in *Sigurd and Volsung* (1876), using a rhymed couplet rather than the blank verse Tennyson had employed. In America Henry Wadsworth Longfellow turns to the legends of the Am. Indians for *Hiawatha* (1855), which he writes in the meter of the Finnish *Kalevala.* Walt Whitman (1819–92), publishing the first ed. of *Leaves of Grass* in the same year (1855), heralds the new e. Adding to and reorganizing in successive editions the lyric, semi-narrative poems that make up the sections of *Leaves of Grass,* Whitman tells the story of America by representing himself as the collective ego of its people. He moves from birth through life to death, touching on the triple realms of physical, historical, and spiritual reality, making—as Dante had—his own personality the center of his poem and making the key figures of Am. hist.—Columbus, Washington, and above all Lincoln—heroic foils to his own persona.

On the Continent, those 19th-c. works most e. in character and scope are either dramatic or novelistic in form. Goethe (1749–1832) collaborated with Friedrich Schiller on an essay "Über epische und dramatische Dichtung" (1797), which attempts to differentiate between e. and dramatic action. Although Goethe generally opposes the confusion of these genres, he created in *Faust*, which in form is dramatic, a poem that takes the e. quest as its theme. Reaching back to the medieval tale of the man who sold his soul to the devil, Goethe enacts on both the human and cosmic level the struggle of man for salvation, a theme that Christian e., and the *Divine Comedy* in particular, has made preeminant in the Western world. In his cycle of music dramas, *The Ring of the Nibelung*, Richard Wagner (1813–83), drawing his material, like William Morris, from the Elder Edda, creates a musical e. in dramatic form. His cycle, like Milton's e., encompasses Creation and rebellion, the struggle of the human heroes Siegmund and Siegfried with their divine progenitor Wotan, the tragic death of Siegfried, and the noble sacrifice of Brunhilde to redeem the ring and make possible a new heaven and new earth. *Les Trojens* by Hector Berlioz (1803–1869), based on Virgil's *Aeneid*, could make a similar claim to musical e. Like the e.-drama or the music drama, the novel often possesses not only the narrative sweep of the e., but also its bardic consciousness. Efforts at e. scope in the novel are made by Manzoni in Italy (*I Promessi Sposi*, 1827), Balzac and Flaubert in France; Tolstoy in Russia; Scott, Dickens, Thackeray, and George Eliot in England; and Melville in America.

In the 20th-c., Homeric e. continues to inspire both writers of verse and prose. The modern version of the Odysseus story can be found both in the psychological novel *Ulysses* (1922), by the Irish writer James Joyce (1882–1941), which narrates one day in the life of a Dubliner, and in the modern narrative poem, *The Odyssey, a Modern Sequel* (1938) by the Gr. poet Nikos Kazantzakis (1885–1957), which recounts the further adventures of Odysseus. While historical novelists and directors of so-called "e." films carry on in one way the trad. of e. narration, 20th-c. poets seek for new underpinnings for the long poem, often using the techniques of lyric, prose fiction, or drama to give their poetry the authority and breadth of statement that e. seems to demand. In *Anabasis* (1924), a brief e. in 10 lyric interludes of varying length, St.-John Perse adopts the persona of a princely hero of the past to describe his founding of an ancient city and undertaking of an expedition, the poem celebrating the heroic spirit in man and his thirst for adventure and conquest. At the same time, Boris Pasternak (1890–1960) in *Lieutenant Schmidt* (1927) and *Spektorsky* (1931) revives the novelistic poem made popular by Puškin a century earlier. Both World Wars have had their effect on writers of long poems. H. D. (1886–1961) in *Trilogy* (1944–46), using the lyric voice to record her experiences of wartime London, tackles the universal problem of war. William Carlos Williams (1883–1963) in *Paterson* (1944–63) deals with the history and inner life of a city and its inhabitants. In these poems, as in Ezra Pound's *Cantos*, the personal voice of the poet is central and shapes the experiences narrated. The *Cantos*, over 100 in number and published over the course of 50 years, are this century's most serious attempt at verse e. Pound (1885–1972) defines e. as a poem containing history, in which one man speaks for a nation. Placing himself at the center of his poem, like Dante and Whitman, as an e. pilgrim or voyager, Pound encompasses time from the mythic past to the present and civilizations as disparate as Homeric Greece, Imperial China and Rome, Med. France, Ren. Italy, and Federalist America. His lang. can be vulgar or exalted; his vision moves from the common to the mystical and paradisical. Focusing often on the violent episodes in history, on the villains and heroes of the past, Pound attempts to draw heroic exemplars—from Odysseus to the 15th-c. It. Sigismundo Malatesta to Confucius to John Adams and Mussolini—to teach man how to live within the world he must inhabit.

As 20th-c. poets such as Pound extend the boundaries of the e. as genre (see MODERN LONG POEM), 20th-c. critics, from C. M. Bowra to E. M. W. Tillyard to C. S. Lewis, search for a classification scheme or typology, contrasting the primary or oral e. with the secondary or written literary product. Investigations of oral e. in the line of those begun in Yugoslavia by Milman Parry and continued by Albert B. Lord extend now to the study of orally transmitted material in Albania, Turkey, Russia, Africa, Polynesia, New Zealand, and the Americas. And the study of the roots of Chinese and Japanese e., which began in the West only in the present century, is flourishing most strongly at present, as is the study of the numerous, emergent, flourishing African e. trads.

E. W. Hopkins, *The Great E. of India* (1901); W. P. Ker, *E. and Romance*, 2d ed. (1908); W. M. Dixon, *Eng. E. and Heroic Poetry* (1912); R. Heinze, *Virgils epische Technik*, 3d ed. (1915); R. C. Williams, *The Theory of the Heroic E. in It. Crit. of the 16th C.* (1917); C. G. Osgood, *The Trad. of Virgil* (1930); C. M. Bowra, *Trad. and Design in the* Iliad (1930), *From Virgil to Milton* (1945), *Heroic Poetry* (1952); R. Menéndez Pidal, *Historia y epopeya* (1934), *La epopeya castellana a través de la literatura española* (1945); H. Massé, *Les Épopées persanes* (1935); C. S. Lewis, *A Preface to* Paradise Lost (1942); E. Mudrak, *Die nordische Heldensage* (1943); H. T. Swedenberg, *The Theory of the E. in England, 1650–1800* (1944); G. Highet, *The Cl. Trad.* (1949); D. Knight, *Pope and the Heroic Trad.* (1951); J. Crosland, *The OF E.* (1951); U. Leo, *Torquato Tasso* (1951); Auerbach; Curtius; G. R. Levy, *The Sword from the Rock* (1953); E. M. W. Tillyard, *The Eng. E. and its Background* (1954); K. H. Halbach, "Epik des Mittelalters," and H. Maiworm, "Epos der

Neuzeit," *Deutsche Philologie im Aufriss*, ed. W. Stammler, v. 2 (1954); A. Heusler, *Nibelungensage und Nibelungenlied*, 5th ed. (1955); R. A. Sayce, *The Fr. Biblical E.* (1955); Raby, *Secular*; C. Whitman, *Homer and the Homeric Trad.* (1958); Lord; Weinberg; G. S. Kirk, *The Songs of Homer* (1962), *Homer and the Oral Trad.* (1976); D. Bush, *Eng. Lit. in the Earlier 17th C.* 2d ed. (1962), chs. 11–12; D. M. Foerster, *The Fortunes of E. Poetry* (1962); T. M. Greene, *The Descent from Heaven: A Study in E. Continuity* (1963); J. Clark, *A Hist. of E.P.* (1964); P. Hägin, *The E. Hero and the Decline of Heroic Poetry* (1964); B. Wilkie, *Romantic Poets and E. Trad.* (1965); A. Cook, *The Classic Line* (1966); W. Calin, *The E. Quest* (1966), *A Muse for Heroes: Nine Cs. of the E. in France* (1983); G. Williams, *Trad. and Originality in Roman Poetry* (1968); N. K. Chadwick and V. Žirmunskij, *Oral Es. of Central Asia* (1969); R. Finnegan, *Oral Lit. in Africa* (1970); Parry; P. Merchant, *The E.* (1971); M. P. Hagiwara, *Fr. E. Poetry in the 16th C.* (1972); D. Maskell, *The Historical E. in France, 1500–1700* (1973); *Parnassus Revisited: Mod. Crit. Essays on the E. Trad.*, ed. A. C. Yu (1973); *E. and Romance Crit.*, ed. A. Coleman, 2 v. (1973); J. J. Duggan, The Song of Roland: *Formulaic Style and Poetic Craft* (1973); Wilkins; Pearsall; *Heroic E. and Saga*, ed. F. J. Oinas (1978)—surveys 15 cultures; *Europaïsche Heldendichtung*, ed. K. von See (1978)—34 essays on 9 national poetries; I. Okpeuho, *The E. in Africa* (1979); *Das römische Epos*, ed. E. Burck (1979); G. Nagy, *The Best of the Achaeans* (1979); K. Simonsuuri, *Homer's Original Genius* (1979)—18th-c. concepts; F. C. Blessington, Paradise Lost *and the Cl. E.* (1979); J. E. Miller, Jr., *The Am. Quest for a Supreme Fiction* (1979); *Trads. of Heroic and E.P.*, ed. A. T. Hatto (1980); M. A. Bernstein, *The Tale of the Tribe: Ezra Pound and the Mod. Am. Verse E.* (1980); Trypanis; W. T. H. Jackson, *The Hero and the King* (1982); *Les Épopées romanes, GRLMA* v. 3.1–2; Fowler; J. D. Niles, Beowulf: *The Poem and Its Trad.* (1983); *CHCL*, v. 1, chs. 2, 4; J. Opland, "World E.: On Heroic and E. Trad. in Oral and Written Lit.," *Comp Crit* 8 (1986); C. Martindale, *John Milton and the Transformation of Ancient E.* (1986); J. K. Newman, *The Cl. E. Trad.* (1986); *L'Épopée*, ed. J. Victorio (1988); J. Walker, *Bardic Ethos and the Am. E. Poem* (1989); J. P. McWilliams, Jr., *The Am. E.: Transforming a Genre, 1770–1860* (1989); J. M. Foley, *Traditional Oral E.* (1991); J. B. Hainsworth, *The Idea of E.* (1991). S.P.R.

II. THEORY. A. *Classical and Alexandrian Greek.* The rich e. trad. of the Greeks (Homer, Hesiod, Arctinus, Antimachus) was at first criticized more from an ethical than a literary standpoint. Xenophanes (fr. 11, Diels-Kranz), for example, objected to Homer and Hesiod's depiction of gods as thieves, adulterers, and deceivers. In the *Ion* Plato pokes fun at Homer's interpreters for not having a rational understanding of their topics. In the *Republic* he combines the ethical and literary when he attacks Homer for teaching the young morally pernicious ideas by the pernicious method of imitation (q.v.). Here for the first time the notion of art as imitation, basic to Gr. aesthetics and already familiar as a positive concept to Pindar (*Pythian* 12.21), is developed into an argument that the whole poetic enterprise is corrupt for providing not the truth but a copy of a copy of the truth. Since for Plato, Homer is also the founder of tragedy (q.v.), both genres stand condemned. Much later this position would be adopted by Tolstoy (*What is Art?* 1898), with which the Neoplatonic arguments of Iris Murdoch (*The Fire and the Sun*, 1977) may be compared.

In the *Poetics* Aristotle accepts the Platonic theory of imitation but draws a radically different conclusion. Not only is imitation natural to man, but as intensified by tragedy it can produce, by means of pity and fear, a *katharsis tōn pathēmatōn.* The lost second book of the *Poetics* may have claimed some parallel experience for comedy, according to the surviving work called the *Tractatus Coislinianus.* That for Aristotle this puzzling but evidently drastic effect is also valid for e. follows from his acceptance of another Platonic insight, that e. is inherently dramatic. Since, unlike Plato, Aristotle argued that drama is the superior genre, the best e. must be Homer's because it is already "dramatic" (an adjective which Aristotle may have coined). Since Aristotle believes that drama (Gr. "that which is done") is so important, it follows that the action has primacy. Not the soul of the hero but the mythic interaction of personae is the soul of the play. "Hero" is never used by Aristotle in its modern sense of "principal character," and indeed he criticizes es. which assert their unity merely by hanging a collection of disparate adventures onto a well-known name. Aristotle demands organic unity from the e., and also requires that it be *eusynopton* ("easily grasped in its totality") and that it not exceed the length of dramas shown at one sitting, a period of time usually calculated at six or seven thousand lines. Apollonius Rhodius's *Argonautica*, written in Alexandria (3d c. B.C.), exactly fills the bill.

Aristotle objected to the confusion of e. and historical narrative, perhaps because he had a strict notion of the historian's duty to do nothing but record. He ignores the fact that already Thucydides, for example, had used mythical models to interpret events. In disallowing the versification of Herodotus, however, Aristotle is obviously dismissing the (unmentioned) *Persica* of Choerilus of Samos (late 5th c.), and by implication the whole genre of historical, eulogistic e. dedicated to a patron to which Choerilus' work would give rise. By contrast, in allowing for "poetic" or imaginative prose, Aristotle opened the door to the modern e. novel. The definition of e. attributed to his pupil and successor as Head of the Lyceum in Athens, Theophrastus, as "that which embraces divine, heroic, and human affairs" also shifts the emphasis from form to content. Aristotle's modi-

fication in the *Rhetoric* of his views on vocabulary is related. The *Poetics* had laid down that e. vocabulary should be clear but elevated, marked by the use of poetic words or "glosses." The later *Rhetoric*, however, under the impulse of Euripides, permits the use of words from the lang. of everyday life. Aristotle saw in Homer more than the pre-tragic. He accepted as Homeric the comic *Margites* ("The Crazy Man": cf. *Orlando Furioso*), now lost, a silly account in mixed meters of a bumpkin who is not even sure how to proceed on his wedding night. Such openness completely contradicts Plato's severity and that of many later "Aristotelians."

The scholia or interpretive comments written in the margins of Homeric mss. prove that a practical crit. was worked out after Aristotle which preserved and extended his insights. Here the *Iliad* and *Odyssey* are regarded as tragedies, though the notion was not lost that Homer was also the founder of comedy. The contrast between e. and history is maintained. The analogy with painting is approved. "Fantasy" is praised (cf. Dante, *Paradiso* 33.142) both as pure imagination and as graphic visualization of detail. Nonlinear presentation of the story may be made and at a variety of linguistic levels.

Already Callimachus in Alexandria (3d c. B.C.) shared Aristotle's objections to the versification of Herodotus. In rejecting the eulogistic e., he worked out, in his own e. *Hecale*, a different kind of Homer-*imitatio* from the straight comparison of the modern champion to a Homeric counterpart of the type apparently sought by Alexander the Great from the poetaster Choerilus of Iasos. Too little of Callimachus' e. *Galateia* now survives to make judgment possible, but his *Deification of Arsinoe* in lyric meter set the precedent for the appeal by a junior deity to a senior for elucidation that would become a topos of the Lat. eulogy (e.g. in Claudian and Sidonius Apollinaris) and even be adapted by Dante in the first canzone of the *Vita nuova*. Since this topos had Homeric precedent (*Iliad* 1.493 ff.), Callimachus was indicating which parts of the Homeric legacy were imitable, and how. In his elegiac *Aetia*, Callimachus advanced Hesiod as the figurehead of his new approach to e., since Homer was already pre-empted by the theorists and poets of the rival school. He used Hesiod also to defend the e. of his contemporary, Aratus of Soli (epigram 27), assailed by critics because it was didactic. In later lit., Hesiodic or didactic e. will therefore be a clue to Callimachean allegiance.

The *Argonautica* of Callimachus's pupil Apollonius of Rhodes (3d c. B.C.) is a further demonstration of the Alexandrian theory of e. Out of their element in the heroic ambience, the characters collectively and Jason individually are often gripped by *amēkhaniē* ("helplessness"). Unified by verbal echoes of the red-gold icon of the Fleece, the poem underplays the conventionally heroic in showing both the futility of war and the degradation of the hero who, dependent on a witch's Promethean magic, eventually becomes the cowardly murderer of Apsyrtus. Homeric allusions point the lesson, and the reader knows from Euripides that eventually the marriage of Jason and Medea will end not in triumph but in disaster.

B. *Classical Latin and Medieval.* The Callimachean and Apollonian e., i.e. the kind written by scholar-poets, treated the lit. of the past by allusion and reminiscence in a polyphonic way. Catullus, the bitter foe of historical e., also uses this technique in his poem 64. Similarly, in the *Aeneid*, Virgil uses Homer as a sounding board rarely for simple harmony but to secure extra and discordant resonance for his modern symphony. Thus Dido is at once the *Odyssey*'s Calypso, Circe, Nausicaa, Arete, and the *Iliad*'s Helen and Andromache. Aeneas is Odysseus, Ajax, Paris, Agamemnon, Hector, Achilles. There is no end to the sliding identities and exchanges ("metamorphoses") of the characters. The poet who had quoted from Callimachus in *Eclogue* 6 in order to introduce an Ovidian poetic program, and had progressed to e. through the Hesiodic *Georgics*, may therefore be properly regarded as Callimachean. But he is also Aristotelian, both in the dramatic nature of his poem and in its tragic affinities. In Italy, Aeneas gropes toward victory over the bodies of friend and foe alike. Vengeful Dido by the technique of verbal reminiscence and recurrent imagery is never absent from the poem. Although the e. serves therefore in its way to exalt the origins of Rome, "shadows" (*umbrae*) is characteristically its last word.

The first Alexandrians were scholar-poets, and the third Head of the Library, Eratosthenes (276?–196? B.C.), still exemplified this ideal. But the early divergence of the two vocations led to a split between creative and critical sensibilities. Horace's *Ars poetica* (22? B.C.), written by a poet, is the last trustworthy theorizing about e. from antiquity. Described by an ancient commentator as a versification of the prose treatise of the Alexandrian scholar Neoptolemus of Parium, though clearly in its emphasis on the *vates* (391–407) more than this, it recommends by its form a musicality in which arrangement and correspondence, interlace and arabesque, will replace the pedestrian logic of prose. In content, it both allows for the closeness of tragic and comic lang. and enjoins the *callida iunctura* ("cunning juxtaposition") by which context will redeem commonplace words.

A Roman and an Augustan, Horace moves beyond Callimachus when he urges that the poet, without in any way betraying Gr. refinement, must also be a *vates*, engaged with his society and with the reform of public morals. This is an aspect of Cl. e. subsumed by Dante in his allusion to *Orazio satiro* (*Inferno* 4.89).

Finally, Horace, who spoke of Homer's "listener" (*auditor*), took for granted a feature of ancient e. theory now often overlooked. E. did not cease to be oral with Homer. The power of Virgil's

"acting ability" (*hypocrisis*), the "sweetness and marvelous harlotries" of his voice are attested in the *Life* written by Donatus (4th c. A.D.). This Aristotelian closeness to drama again implies polyphonic composition. There cannot be a single, univocal, "right" interp. of the action. Although in the 12th c. Geoffrey of Vinsauf in his theory of *pronuntiatio* seems to preserve some memory of this orality, it was competely ignored by those Ren. "Aristotelians" who, unlike their master but like Plato (who poked fun at dramatic and oral presentation of e., and of whom Callimachus said that he was incapable of judging poetry), set a statically conceived e. at the head of all the genres.

The crit. of the *Aeneid* for its *communia verba* by Agrippa, recorded by Donatus in his *Life* of Virgil (44), shows that there had developed even under Augustus the theory that e., as the sublimest of genres, demanded the sublimest of lang. Petronius' implied criticisms of Lucan (*Satyricon* 118 ff.) prove the persistence of this notion. At the end of Cl. antiquity, the same theory received a fatally deceptive application. *Rhetorica ad Herennium* (early 1st c. B.C.), influential in the Middle Ages, distinguished three styles, high, middle, and low, and conveniently Virgil had written three major poems: obviously the *Eclogues* must exemplify the low style, the *Georgics* the middle, and the *Aeneid* the high. This was the doctrine that eventually found its medieval canonization in the *Rota virgilii* ("Virgil's Wheel") devised by John of Garland. By this, names, weapons, even trees that could be mentioned in the different styles were carefully prescribed. One part of the deadly consequences of this doctrine was that the opening towards comedy in the e. (the *Margites*) was lost. Yet even the late antique commentator Servius had remarked of *Aeneid* 4: *paene comicus stilus est: nec mirum, ubi de amore tractatur* ("the style is almost comic, and no surprise, considering the theme is love").

On the Gr. side, pseudo-Longinus' *De sublimitate* (*On the Sublime*), variously dated between the 1st and 3d cs. A.D. or later, directly attacks Alexandrian theories of e., denying the Callimachean doctrine of purity of style and rejecting Apollonius in favor of Homer. The author supports his theory of grandeur with Homeric quotations which are in two cases conflated from different passages. Homer is indeed "grand," but in its inadequate appreciation of comedy and irony, and preference for magnificent lang. producing *ekplēxis* ("knockout"), the treatise represents a late and misleading simplification of Aristotle's subtle analysis.

Stung by Platonic crit. of Homer's "lying," Gr. Stoic philosophers in particular had developed a method of interpreting the Homeric narrative in symbolic terms intended to rescue its moral and theological credibility. In his efforts to reclaim all the genres for the new religion, the Christian Lat. poet Prudentius (348–post 405) wrote an e. *Psychomachia* in which the contending champions were no longer flesh-and-blood heroes but abstract qualities of the soul. It was only a short step from this to allegorizing Virgil's *Aeneid*, the most important example of this being Fulgentius (late 5th c.). This method seems in fact to rob the poem of its human immediacy, esp. as taken to extremes by Bernardus Silvestris in the 12th c. (*Commentum super sex libros eneidos virgilii*). The *Ovide moralisé* of Pierre Bersuire (d. 1362) is even less redeemable, since Ovid's interest in morality is slight at best. Highly praised allegorical es. have been written, however, of which the most important is probably Spenser's incomplete *Faerie Queene* (1590, 1596; see ALLEGORY).

The critical failure of later antiquity meant in effect that any e. theory that was to make sense had to be recoverable from the practice of major poets. This fact lent even greater significance to the already towering figure of Dante, since it was he who, as the author of an e. *Comedy* in the vernacular, broke decisively with both medieval prescription and practice, as seen at a relatively naive level in the anonymous *Waltharius* (930?) and with more sophistication in the *Alexandreis* of Walter of Châtillon (1180?). By anticipation Dante also rejects the classicizing poetics of the 16th c. His confrontation with academic orthodoxy in his exchange with Giovanni del Virgilio shows that, like Virgil, he used an eclogue to legitimize the claims of ordinary vocabulary and to reject the grand historical e. eulogy. These arguments are bolstered in Letter X to Can Grande, traditionally ascribed to Dante, with a quotation from Horace bordering on a line (97) which we are expressly told by the late commentator Porphyrio was borrowed from Callimachus.

C. *Renaissance to Modern.* The Ren. critics (Marco Girolamo Vida, *De arte poetica* [1527]; F. Robortelli, *In librum aristotelis de arte poetica explicationes* [1548]; Antonio Minturno, *De poeta* [1559], *De arte poetica* [1563]) were too often prescriptive rather than descriptive. Armed with the *Poetics* (tr. into Lat. by Giorgio Valla in 1498, 1536; Gr. text 1508; It. tr. 1549) and eventually with an amalgam of Aristotle and Horace, they advanced to war down the "unclassical" in e., but were largely the unconscious victims of old ideas. J. C. Scaliger, the most gifted scholar among them (*Poetices libri septem* [1561]), uses the evidence of lang. to decry Homer and exalt Virgil: the display by Homer of *humilitas, simplicitas, loquacitas,* and *ruditas* in his style must make him inferior to the Roman. If Virgil echoes Homer's description of Strife in his picture of Fama in Book 4 of the *Aeneid*, that is an excuse for loading Homer with abuse. The Alexandrian notion of the polyphonically echoed model has been replaced by that of annihilation, as it had been already in pseudo-Longinus.

Like their medieval forbears, therefore, these critics demanded sublimity in lang. and seriousness in theme, denoted by lofty character and moral uplift (Tommaso Campanella, *Poetica*

[1595?]); and just as ruthlessly as Agrippa with Virgil they set about Dante, Ariosto, and Tasso for their unclassical backsliding. There was much argument over "probability," which for Aristotle was identical with persuasiveness. Now it came to mean what the rationalist critic would regard as probable, and thus could be used to clip the wings of poetic imagination. In fact, the fidelity of It. e. to the Cl. heritage may be observed both positively, in the minuteness with which it echoes the Alexandrian code both in detail (e.g. the *recusatio* or "refusal" topos) and, more largely, in its assumption of the romantic or pastoral element found also in the es. of Apollonius Rhodius and Virgil; and negatively, in the failure of Petrarch's *Africa*, which had sought to resurrect the historical, eulogistic e. in the manner of Ennius. Similarly, Ronsard's *Hymnes* (1555–) in the style of Callimachus were successful, while his e. *Franciade* (1572), which had been preceded by his rather Horatian *Abrégé d'art poétique* (1565), remained unfinished. In the 17th c., Fr. critics adapted and propagated the It. recension of Aristotle's rules, emphasizing the unities, decorum, and verisimilitude, although the first tr. of the *Poetics* into Fr. did not appear until 1671. André Dacier's edition and commentary (*La Poétique d'Aristote contenant les règles les plus exactes pour juger du poème héroique et des pièces de théâtre, la tragédie et la comédie* [1692]) became standard. This and other critical works (René Rapin, *Réflexions sur la poétique d'Aristote* [1674]; René le Bossu, *Traité du poème épique* [1675]) were regarded as normative throughout Europe. The es. they have inspired have however been universally regarded as failures. In Eng., John Milton (*Paradise Lost*, 1667, *Paradise Regained*, 1671) was greatly influenced by both It. example and precept. He knew Mazzoni's *Della Difesa della Commedia di Dante* (1587) and the theoretical work of Tasso (see Weinberg), whose old patron, Count Manso, he had met during his travels in Italy (1637–39). He quotes from Ariosto's *Orlando Furioso* at the beginning of *Paradise Lost* (1.16=*O.F.* I.2.2). His *Epitaphium damonis* (1639) toys with the idea of a historical epic on Arthurian legends, and his notebooks still preserved in Cambridge show that he considered a dramatic treatment of the same topics and of the story of Adam. In returning to epic he fixed on this theological theme, to which the closest Cl. parallel would be Hesiod's *Theogony*. It enabled him to set out his profoundest beliefs in the origin of the moral order of the universe, the human condition, and the Christian promise of atonement. His assumption in the poem of a difficult lang., criticized by Dr. Johnson, is part of the struggle to convey truths larger than life. His *Paradise Regained* uses a simpler style to depict Christ's rebuttal of the temptations of Satan. This successful resistance sums up the Christian victory.

Some Ren. treatises on poetics had argued in favor of the apparently unclassical, "romance" e. (Tasso, *Discorsi dell'arte poetica* [1567–70]), while others (Sidney, *Defence of Poesie* [1595]) had allowed modern authors to become "classical" in time. A wider breach in the "rules" seemed to open when, thanks to the study of Shakespeare in England (e.g. by Dr. Johnson) and in Germany (e.g. by the Schlegels), the deficiencies of normative crit. were recognized and replaced, slowly, by the inductive study of how great poetry is related to its national roots. At first the movement affected drama more than e., but the notion of short popular "lays" later assembled by editorial labor into longer e. poems was advanced by F. A. Wolff in his *Prolegomena ad Homerum* (1795). This theory, however, had the unfortunate consequence of breaking up the transmitted text of Homer into many sometimes incongruous layers (Karl Lachmann, *Betrachtungen ueber Homers Ilias* [1837]). Though it received a fresh boost from the investigations of Milman Parry and Albert B. Lord into Serbo-Croatian oral poetry (see Parry; Lord), nowadays it is seen by some Classicists as reducing Homer's aristocratic masterpieces to an uncouth common denominator. The problem of understanding the Cl. e. trad. is to be solved not by denying its literariness but only by sustaining an awareness of how complex that trad. really was.

Since the Ren., potential writers of e., in seeking new ways to deploy the poetic imagination at its fullest, have been confronted by a bewildering variety of choices and advice. If they choose verse for their mode, they accept all the risks of a genre not now organically related to their culture. This dilemma was adumbrated by Schiller in his *Über Naive und sentimentalische Dichtung* (1795–96) and was subsequently developed by Marxist critics like György Lukacs. Such writers also run the danger of ignoring Aristotle's insight (*Poetics*, ch. 26) that drama is superior to e., a view which Berthold Brecht (by his theory of e. drama) still illustrates. More traditional poets either find themselves compelled to treat of lofty themes in a way that is bound to be lifeless and unconvincing, because it mistakes the polyphonic and potentially even comic nature of the genre, or to ignore the decorum of Cl. e. in an extravagant effort to impress. If they turn away from mythical loftiness (Goethe's *Hermann und Dorothea* [1797]; Wordsworth's *Prelude* [1805, 1850]), they run the opposite risk of seeming to have forgotten their commitment to heroic sublimity. An author like St.-John Perse (*Anabase* [1924]) uses Cl. rhet. to produce a musical evocation of history that appealed to T. S. Eliot. Its relation to the historical e. condemned by Aristotle and Callimachus has not yet been explored. N. Kazantzakis (*Odyssey* [1938], *The Last Temptation of Christ* [1960]) treats both Gr. and Christian myths with flamboyant extravagance, in which, however, the Cl. elements seem to have exploded into unrecognizability.

The fragmentation of the Cl. trad. in our time has sometimes therefore led to impatient rejection of its relevance. Each artist responds freshly to new

challenges. Even so, a unifying e. theory may seek to accommodate their individual achievements. Aristotle had allowed for imaginative writing in prose; and though Tolstoy was no friend of the Classics, even *War and Peace*, modestly described by him as a companion piece to the *Iliad*, and with its corrosive irony at the expense of the officially great, may be seen as a blow against the eulogistic e. favored by Callimachus's enemies and as a recovery of Homer's tragicomic picture of the human condition. In another vein, Tolstoy's *Anna Karenina*, with its "endless labyrinth of connections that is the essence of art," is a good introduction to the complexities of the narrative technique of the *Aeneid*. But the Cl. amalgam may be broken into its constituent elements (Pound's *Cantos*), given an Aristotelian push towards the drama (Eliot's *Murder in the Cathedral*), expanded to explore the human comedy and the illusions of eros (Proust), or deployed with all the tragic resonance of a Virgil (Thomas Mann's *Doktor Faustus*). Joyce's *Ulysses* and *Finnegans Wake*, with their allusive learning, verbal echoes, musicality, irony, and parody, are more Ovidian essays in the Alexandrian manner. The e. film may represent the most grandiose modern foray into the genre, recreating its own engagement with contemp. society, popular in appeal, yet, for at least one of its modern theorists, S. M. Eisenstein, profoundly traditional.

D. Comparetti, *Vergil in the Middle Ages*, tr. E. F. M. Benecke (1895); Faral; J. E. Spingarn, *Hist. of Lit. Crit. in the Ren.*, 2d ed. (1925); H. Strecker, "Theorie des Epos," *Reallexikon I* 4.28–38; E. Reitzenstein, "Zur Stiltheorie des Kallimachos," *Festschrift Richard Reitzenstein* (1931), 23–69, esp. 41 ff.; Lewis; M.-L. von Franz, *Die aesthetischen Anschauungen der Iliasscholien* (1943); C. M. Bowra, *From Vergil to Milton* (1945), *Heroic Poetry* (1952); M. T. Herrick, *The Fusion of Horatian and Aristotelian Lit. Crit., 1531–1555* (1946); Auerbach; Curtius; Frye, 318 ff.; Lord; Weinberg; D. M. Foerster, *The Fortunes of E. Poetry: A Study in Eng. and Am. Crit. 1750–1950* (1962); F.-J. Worstbrock, *Elemente einer Poetik der Aeneis* (1963); G. N. Knauer, *Die Aeneis und Homer* (1964); S. M. Eisenstein, *Nonindifferent Nature* [1964], tr. H. Marshall (1987); A. Lesky, *A Hist. of Gr. Lit.* (tr. 1966); K. Ziegler, *Das hellenistische Epos*, 2d ed. (1966); J. K. Newman, *Augustus and the New Poetry* (1967), *The Cl. E. Trad.* (1986); E. Fränkel, *Noten zu den Argonautika des Apollonios* (1968); S. Koster, *Antike Epostheorien* (1970); Parry; *Cl. and Med. Lit. Crit.*, ed. A. Preminger et al. (1974); *Homer to Brecht: The European E. and Dramatic Trad.*, ed. M. Seidel and E. Mendelson (1976); R. Häussler, *Das historische Epos der Griechen und Römer bis Vergil* (1976), *Das historische Epos von Lucan bis Silius und seine Theorie* (1978); G. S. Kirk, *Homer and the Oral Trad.* (1976); R. O. A. M. Lyne, *Further Voices in Vergil's* Aeneid (1987); D. Shive, *Naming Achilles* (1987). J.K.N.

EPIGRAM. A form of writing which makes a satiric or aphoristic observation with wit, extreme condensation, and, above all, brevity. As a poetic form, the e. generally takes the shape of a couplet or quatrain, but tone defines it better than verse-form. The etymology of the term—Gr. *epigramma*, "inscription"—suggests the brevity and pithiness of the form. Pithy poems in the *Gr. Anthol.*—first assembled in 60 B.C. (see ANTHOLOGY) and the major source of es. in the Western trad.—run a gamut of tones from biting to sharp to gentle, and thus introduce the wider sense of the term (a sense which still lets us distinguish the e. from the proverb [q.v.] and apothegm, which are impersonal and gnomic in tone).

Japanese, Chinese, and Af. trads. exhibit their own versions of e., though the term is hard to translate into any non-Western lang. The Japanese, like the Greco-Roman trad., breaks into two strands—*haikai* (see RENGA; HAIKU) and *senryū*—both of which display a loosely comic thrust. Bashō, the renowned 17th-c. *haikai* poet, typifies the skill: "The salted bream / Look cold, even to their gums, / On the fishmonger's shelf" (tr. E. Miner). The more satiric *senryū*, which came into being in the 18th c., runs to the following realism: "If you ask directions / From a man pounding rice / First he wipes his sweat" (tr. E. Miner).

An 18th-c. Chinese epigrammatist, Cheng Hsieh, cuts from the same block:

> With wild hair, you block out your characters.
> Deep in the mountains, you engrave your poems.
> Don't even mention this man's "bone and marrow"—
> Who could even imitate his "skin"?

From the vastness of Af. repartee and wit one might pluck—as a different species of e.—John Pepper Clark's "Ibadan" (1965):

> Ibadan,
> running splash of rust
> and gold—flung and scattered
> among seven hills like broken
> china in the sun.

In the *Greek Anthology*, the romantic or elegiac inscription poem is prominent, amidst other occasional and light verse, but in 1st-c. A.D. Rome it yielded place to the more acid *vers de société* of Martial and Catullus:

> Once there was perfume in this little jar.
> Paphylus sniffed; it turned to vinegar.
> (Martial 7.94, tr. R. Humphries)

This sharp-tongued trad. inspired the Ren. e. both in Lat. and in the vernacular, though by the 18th-c. the double mode of the trad. was apparent again. In the 17th c., the fine Ger. epigrammatist Friedrich von Logau wrote *Sinngedichte* (1654), which inspired Lessing. Among the Eng. the e. is practiced by Heywood, Davies, Harington, More,

and Jonson, who called his *Es.* (1616) "the ripest of my studies." The form was a favorite of Herrick's: nearly 200 of the 272 poems in his *Noble Numbers* are es., as "Upon a child that died":

> Here she lies a pretty bud,
> Lately made of flesh and blood:
> Who, as soone, fell fast asleep,
> As her little eyes did peep.
> Give her strewings; but not stire
> The earth, that lightly covers her.

While a harder hitting Donne, treating the same loss, wrote: "By childrens' births, and death, I am become / So dry, that I am now mine own sad tombe" ("Niobe").

Goethe's *Zahme Xenien* belong to both streams of the trad.; as do the critical studies of the e. and its history by Lessing and Herder. In France, where the vernacular e. was initiated by Marot and St. Gelais in the early 16th c., the satiric and personal e. reached perfection in Boileau, Voltaire, and Lebrun. By the 19th c., generic purities fade, yet there remain splendid examples of the major e. trads., as well as paths into new forms. Shelley can weep in the form: "Rome has fallen, ye see it lying / Heaped in undistinguished ruin: / Nature is alone undying" ("Fragment: Rome and Nature").

Max Beerbohm can manage the old rapier in his "Epitaph for G. B. Shaw":

> I strove with all, for all were worth my
> strife.
> Nature I loathed, and, next to Nature,
> Art.
> I chilled both feet on the thin ice of
> Life.
> It broke, and I emit one final fart.

Nietzsche's aphorisms suggest what possibilities lay inside the e. as it met a century concerned with both the natural voice and the ancient form, and eager to honor both freshly.

In the 20th c. the stinging e. survives in Yeats, Pound, Roethke, and J. V. Cunningham, and abroad in the work of such Continental poets as Erich Kästner, Christian Morgenstern, and René Char. At the same time, an almost new genus, the "lyrical e.," makes a fresh move back toward the softer grounds of the *Greek Anthology*, as for example in Robert Bly's

> When I woke, new snow had fallen.
> I am alone, yet
> someone else is with me,
> drinking coffee, looking out at the
> snow.
> (*Sleepers Joining Hands*)

or J. V. Cunningham's

> Arms and the man I sing, and sing for
> joy,
> Who was last year all elbows and a boy.
> (*The Exclusions of a Rhyme*)

There seems to be a balance of self-identity with innovation inside the e. trad., a compacting from which fresh energy is constantly extruded.

COLLECTIONS: *Epigrammata: Gr. Inscriptions in Verse*, ed. P. Friedländer (1948); *The Lat. Es. of Thomas More*, ed. L. Bradner and C. A. Lynch (1953); *Edo Satirical Verse Anthols.*, ed. R. H. Blyth (1961); *Selected Es.*, tr. R. Humphreys (1963)—Martial; *Deutsche Epigramme aus fünf Jahrhunderten*, ed. K. Altmann (1966); *The Gr. Anthol.*, ed. P. Jay (1973); *Russkaya epigramma vtoroi poloviny XVII-nachalo XXV.*, ed. V. E. Vasil'ev et al. (1975); *The Faber Book of Es. and Epitaphs*, ed. G. Grigson (1977); *Es. of Martial Englished by Divers Hands*, ed. J. P. Sullivan and P. Whigham (1987).

HISTORY AND CRITICISM: Thieme, 369—Fr.; R. Reitzenstein, "Epigramm," Pauly-Wissowa; T. K. Whipple, *Martial and the Eng. E.* (1925); P. Nixon, *Martial and the Mod. E.* (1927); H. H. Hudson, *The E. in the Eng. Ren.* (1947); G. R. Hamilton, *Eng. Verse E.* (1965); *L'Epigramme grecque*, ed. A. Raubitschek (1967); G. Bernt, *Das lateinische Epigramm im Übergang von der Spätantike zum frühen Mittelalter* (1968); *Das Epigramm*, ed. G. Pfohl (1969); G. Highet, "E.," *Oxford Cl. Dict.* (1970); D. H. Garrison, *Mild Frenzy* (1978); P. Erlebach, *Formgesch. des engl. Es.* (1979); *Das dt. Epigramm des 17. Jhs.*, ed. J. Weisz (1979); F. Will, "E. or Lapidary Engraving," *Antioch Rev.* 40 (1982); Fowler; *CHCL*, v. 1, chs. 18, 20; P. Laurens, *L'Abeille dans l'ambre* (1989).

F.W.

EPITAPH (Gr. "writing on a tomb"). A literary work suitable for placing on the grave of someone or something which indicates the salient facts about or characteristics of the deceased. A shortened form of the elegy (q.v.), the e. may vary in tone from panegyrical to ribald, frequently addressing its message to the passerby, compelling him to read and reflect on the life of the person commemorated and, by implication, on his own life. In the West, the earliest extant es. are Egyptian, written on sarcophagi and coffins. They generally include the name, the person's descent, his office, and a prayer to some deity, e.g. "Royal chosen offering to Anubis, Director of the Balance . . . that he may give a good wrought Coffin in the Consecrated Enclosure . . . for the votary Osiris . . . deceased." Like the Egyptian material, Japanese tomb inscriptions, medieval Eng. brasses, and tombstone carvings in backcountry Am. graveyards—to suggest the cultural breadth of the activity—abound in efforts to draw favorable attention to the deceased.

Gr. and Roman es. in particular distinguish themselves: they are often highly personal and epigrammatic—note the convergence of epigram with e. (see below)—and their literary level is unusually high. They may be written in verse (usually the elegiac distich) or in prose. Their details may include the name of the person, his family, certain facts of his life, a prayer to the underworld (esp. in Roman es.), and a warning or imprecation

against defilement, though rarely are all these found in a single e. In Cl. es. we also find conventional topoi and motifs such as Fortuna and Fate, the thread of life, the removal from light, the payment of a debt, and a variety of consolations and lamentations.

The major collection of Cl. es. is Book 7 of the *Greek Anthology* The es. in this collection are of high poetic quality and cover the whole range of the form from satiric and comic to intensely serious. (The following, much imitated and here put literally, gives an idea of the range of modes implicit in such poems: "To Hope and to thee, O Fortune, a long farewell! I have found the haven. You and I have no more to do with each other; make sport of those who come after me.") They have influenced subsequent writers of es. from Roman times (Propertius, Horace, Martial, Ausonius) through the Ren. (Pontanus, Erasmus, More, Jonson) into the present (Pound, Yeats [esp. the tercet which concludes "Under Ben Bulben"], Auden, Edgar Lee Masters). Perhaps the single most famous e. from the *Greek Anthology* is that attributed to Simonides about the Spartan dead at Thermopylae: "Go, tell the Lacedaimonians, passerby, / That here obedient to their words we lie."

The Middle Ages used the Lat. e. both in prose and verse, often leonine verse. Following the themes and practices laid down by the Greeks and particularly the Romans, the Eng. used the form, developing it to an exceptionally high art in the 15th and 16th cs., e.g. William Browne's on the Countess of Pembroke and Milton's on Shakespeare. Both Dr. Johnson and Wordsworth wrote essays on the e. as an artform.

The e. has not always been used to commemorate the dead, however. It has been put to satirical use against an enemy who is alive, and it has even been aimed at an institution, e.g. Piron's e. on his rejection by the Fr. Academy. Of the relationship of e. to epigram, Puttenham held that the e. was "but a kind of epigram only applied to the report of the dead person's state and degree, or of his other good or bad partes." The Gr. preposition *epi* (upon) shares out to both these genres the quality of being impressed upon something, and of the concision implicit in carving into a hard material. Ancient assumptions about the nature of the act of writing are evident here.

S. Tessington, *Es.* (1857); *Carmina latina epigraphica,* ed. F. Bücheler et al., 2 v. (1894–1926)—Lat. epigraphical verse; W. H. Beable, *Es.* (1925); H. W. Wells, *New Poets from Old* (1940); *Epigrammata,* ed. P. Friedlander (1948); Frye; R. W. Ketton-Cremer, "Lapidary Verse," *PBA* 45 (1959); *Griechische Grabgedichte,* ed. W. Peek, (1960)—Gr. funerary inscriptions; R. A. Lattimore, *Themes in Gr. and Lat. Es.* (1962); G. Pfohl, *Gr. Poems on Stones: Es. from the 7th to the 5th Cs. b.c.* (1967), ed., *Das Epigramm* (1969); R. Brown, *A Book of Es.* (1967); E. Bernhardt-Kabisch, "The E. and the Romantic Poets: A Survey," *HLQ* 30 (1967); J. Sparrow, *Visible Words* (1969); G. Grigson, *The Faber Book of Epigrams and Es.* (1977); J. Bakewell and J. Drummond, *A Fine and Private Place: A Coll. of Es. and Inscriptions* (1977); D. D. Devlin, *Wordsworth and the Poetry of Es.* (1980); F. Will, "The Epigram or Lapidary Engraving," *Antioch Rev.* 40 (1982); K. Mills-Courts, *Poetry as E.* (1990); J. K. Scodel, *The Eng. Poetic E.* (1991). R.A.H.; F.W.

EPITHET. An adjective or adjectival phrase, typically attached to a proper name, used for distinctive purposes in poetry, these purposes most often having to do with allusion, connotation, repetition, and meter. In Western poetry the most important es. are undoubtedly those in Homer. The conspicuousness of the Homeric e. was recognized for centuries before Milman Parry demonstrated, in the early 20th c., that its utility for Homer is far greater than merely for lexical embellishment. Parry showed conclusively that in fact Homer uses only a small set of es. for each name, and indeed only a single phrase, with a distinct metrical pattern, for each grammatical case—Parry calls it a "formula" (q.v.). Homer uses these prefabricated metrical building-blocks to facilitate rapid composition of long narrative poems in an oral setting (see ORAL POETRY).

Es. also form sensation-complexes or *compounds* which enable poets in their most energetic moments to find or create, in Coleridge's phrase, "*one word* to express *one act* of imagination." The Homeric phrase "Pēlion einosiphullon" (Pelion with quivering leafage) and Aeschylus' "anērithmon gelasma" ("multitudinous laughter," i.e. of the sea) are brilliant examples of this: each produces the effect described by Wordsworth in his poem on the daffodils: "Ten thousand saw I at a glance." This is the imaginative power which Coleridge called "esemplastic." The "embracing" or "extensive" e. occurs frequently in both Gr. and Eng. The compound es. in Gr. beginning with *eury-*, *tele-*, and *poly-* have their Eng. analogues beginning with *wide-*, *far-*, and *many-*. The *poly-* or "many" group is particularly large. By the side of Gr. es. having the sense of "rich in flowers," "with many furrows," "with many trees," "with many ridges," "poluphloisbos" ("loud-roaring," Homer's e. for the sea), and many more of like character may be placed Milton's "wide-watered" (shore), Keats's "far-foamed" (sands), and Tennyson's "many-fountain'd" (Ida). Such es. are imaginative in their reduction of multiplicity to unity.

OE poetry possessed an elaborate diction which included periphrases or kennings ([q.v.] e.g. *swanrād,* "swan's-road," for "sea") and compound es. (e.g. *fāmigheals,* "foamy-necked"). Chaucer also employed compound es. such as *golden-tressed* (*Troilus and Criseyde* 5.8) and *laurer-crouned* (*Anelida and Arcite* 43), as well as picturesque metaphorical verbs such as *unneste* ("fly out of the nest" [*Troilus and Criseyde* 4.305]). But the Eng. poet who most cultivated es. as an enrichment of style is Edmund

Spenser. Occasionally he fills a whole line with them, as in his description of Britain as a "saluage wildernesse, / Unpeopled, unmanurd, unprou'd, unpraysd" (*Faerie Queene* 2.10.5). His compound es. are more numerous and memorable than Chaucer's, and are of various kinds: the morally serious (e.g. *hart-murdring* love [*F.Q.* 2.5.16]), the picturesque (*firie-footed* teeme [1.12.2]), the classical (*rosy-fingred* Morning [1.2.7]). Spenser's es. were the model and inspiration of much brilliant work in Marlowe, Shakespeare, Chatterton, Keats, Tennyson, and others.—M. Parry, "The Traditional E. in Homer" (1928), rpt. in Parry; B. Groom, *The Formation and Use of Compound Es. in Eng. Poetry from 1579* (1937), *The Diction of Poetry from Spenser to Bridges* (1955); N. Peltola, *The Compound E. and Its Use in Am. Poetry from Bradstreet through Whitman* (1956); H. J. Rose, "Es., Divine," *Oxford Cl. Dict.*, 2d ed. (1970); N. Austin, *Archery at the Dark of the Moon* (1975), ch. 1; P. Vivante, *The Es. in Homer* (1982). B.G.; T.V.F.B.

F

FABLE. A brief verse or prose narrative or description, whose characters may be animals ("The Cicada and the Ant") or inanimate objects ("The Iron Pot and the Clay Pot") acting like humans; or, less frequently, personified abstractions ("Love and Madness") or human types, whether literal ("The Old Man and the Three Young Men") or metaphorical ("The Danube Peasant"). The narrative or description may be preceded, followed, or interrupted by a separate, relatively abstract statement of the f.'s theme or thesis.

I. HISTORY. Despite suggestions that the *Panchatantra* (transcribed ca. 3d c. A.D.) is the fountainhead of the European f., the genre probably arose spontaneously in Greece with Hesiod's poem of the hawk and the nightingale (8th c. B.C.), followed by Archilochus' fragments on the fox and the eagle (7th c. B.C.). The first collection of Gr. fs. is attributed to Æsop (6th c. B.C.) and is known to us through Maximus Planudes' 14th-c. ed. of a prose text transcribed by Demetrius of Phalerum (4th c. B.C.). Phædrus and Babrius were the first to cast the f. into verse; their works attained such popularity that fs. became part of the regular school exercises.

Phaedrus (1st c. A.D.), the first fabulist we may reckon a poet, imitated Æsop in Lat. iambic senarii, but also invented many new fs., recounted contemp. anecdotes, and introduced political allusions. Babrius, writing in Gr. (2d c. A.D.), went further by inventing racy epithets and picturesque expressions while enlarging the formula of the genre in the direction of satire and the bucolic; his *Muthiamboi aisopeioi*, originally in 10 books, is in choliambics, the meter of the lampoon. A famous collection by Nicostratus, also of the 2d c., is now lost. Avianus (4th c. A.D.) paraphrased and expanded Babrian models, which he enriched with Virgilian and Ovidian phraseology for mock-heroic effect. *Romulus*, a 10th-c. prose tr. of Phaedrus and Babrius, was later versified and enjoyed celebrity into the 17th c. But the best medieval fabulist was Marie de France, who composed 102 octosyllabic fs. (ca. 1200), combining Gr. and Lat. themes with insight into feudal society, fresh observation of man and nature, and Gallic irony. The *Ysopets* (13th and 14th c.) were Fr. verse trs. of older Lat. fs.

The European beast epic, in particular the *Roman de Renart*, owes much to many fs. as told in antiquity and in their Med. Lat. forms (notably Babrius and Avianus), and to Marie. It in turn influenced many fs., esp. those in the *Ysopets* and in Robert Henryson's late 15th-c. *Moral Fabillis of Esope*—an innovative version in lowland Scots English. Beast epic differs from f. not only in magnitude and in its exclusively animal cast of characters but also in its single, mock-heroic modality. Bestiary differs from f. in its emphasis on the symbolic and allegorical meanings of the features or traits attributed to its subjects, animals both legendary and real.

The *Fs. choisies et mises en vers* of Jean de La Fontaine (1621–95) are both summative and innovative. In the first two collections (1668), the Fr. poet adapted subjects and techniques from trs. of Phaedrus, Babrius, and Avianus, incl. Gilles Corrozet's *Fs. du très ancien esope phrygien* (1542), which anticipated La Fontaine's use, in the same poem, of *vers mêlés*. The second collection and subsequent additions (1678–79; 1693) contained materials from Indic sources, incl. *Le Livre des lumières*, a 1644 tr. of fs. based on an 8th-c. Ar. version of the *Panchatantra*. Two features distinguish La Fontaine's *Fs.* from their predecessors: first, they make thematically significant use of pastiche and parody across a wide spectrum of modes and genres; and second, they are systematically philosophical, setting forth, extending, and revising an epicureanism derived from Lucretius and Gassendi.

La Fontaine was widely imitated during the 17th and 18th cs.: in France, by Eustache le Noble (1643–1711) and J.-P.-C. de Florian (1754–94); in England, by John Gay (1685–1732); in Spain, by Tomás de Iriarte (1750–91); and in Germany, by C. F. Gellert (1715–69). G. E. Lessing (1729–81)

modeled his fs. on Æsop.

In the first two decades of the 19th c., the Rus. Ivan Adreyevich Krylov (1769–1844) won wide acclaim for his trs. of La Fontaine and his original fs., still read for their satire and realism of matter and lang. The verse f. trad. was carried on in America by Joel Chandler Harris, who drew on Afro-Am. trads. in his Uncle Remus collections (1881–1906).

II. TYPES. All fs. are didactic in purpose but may be subdivided by technique into three categories, the assertional, the dialectical, and the problematic. Assertional fs. plainly and directly expound simple ideas through a harmonious union of precept and example. The Æsopic trad.—ancient and modern—is in the main assertional. Certain versions of the *Panchatantra* and other Indian collections are dialectical. They present assertional fs. in a sequence where each is clarified, nuanced, or even corrected by those that follow. Problematic fs. feature moral dilemmas or enigmatic presentation. Among the devices used to "problematize" a f. are omission of the thesis statement, unreliable or playful narration, subtle allusion to other literary works, verbal ambiguity, abstruse metaphors, and symbolism. Many fs. by La Fontaine, Bierce, and Thurber are because of one or more of these devices problematic.

L. Hervieux, *Les Fabulistes latins depuis le siècle d'auguste jusqu'à la fin du moyen-âge*, 2d ed., 5 v. (1893–99); A. Hausrath, "Fabel," Pauly-Wissowa 6.1704–36; *Krylov's Fs.*, tr. B. Pares (1921); F. Edgerton, *The* Panchatantra *Reconstructed* (1924); M. Stege, *Die Gesch. der deutschen Fabeltheorie* (1929); "Avianus—*Fabulae*," *Minor Lat. Poets*, ed. and tr. J. W. and A. M. Duff (1934); Marie de France, *Fs.*, ed. A. Ewert and R. C. Johnston (1942); C. Filosa, *La favola e la letteratura esopiana in Italia. . . .* (1952); *Fs. of Æsop*, tr. S. A. Handford (1954); M. Guiton, *La Fontaine: Poet and Counter-Poet* (1961); M. Noejgaard, *La F. antique*, 2 v. (1964–67); *Babrius and Phaedrus*, ed. and tr. B. E. Perry (1965)—invaluable intro. and appendices; H. de Boor, *Fabel und Bîspel* (1966); T. Noel, *Theories of the F. in the 18th C.* (1975); *La Fontaine: Sel. Fs.*, tr. J. Michie (1979); H. J. Blackham, *The F. as Lit.* (1985); P. Carnes, *F. Scholarship: An Annot. Bibl.* (1985); R. Danner, *Patterns of Irony in La Fontaine's Fs.* (1985); *La Fontaine: Fs.*, ed. M. Fumaroli (1985)—brilliant intros., revolutionary bibl., penetrating commentary; G. Dicke and K. Grubmüller, *Die Fabeln des Mittelalters und der frühen Neuzeit, ein Katalog der deutschen Versionen und ihrer lateinischen Entsprechungen* (1987); M.-O. Sweetser, *La Fontaine* (1987)—up-to-date commentaries and bibl.; C. D. Reverand, *Dryden's Final Poetic Mode: The Fs.* (1988); D. L. Rubin, *A Pact with Silence* (1991); A. Patterson, *Fs. of Power* (1991). D.L.R.; A.L.S.

FABLIAU. A comic tale in verse which flourished in the 12th and 13th cs., principally in the north of France. Although the name, which means "little fable" in the Picard dialect, suggests a short story, a few fabliaux run over a thousand lines. Most, however, are short and share a common thread of unabashedly bawdy humor, the subjects being mainly sexual or excretory. A few names, such as Rutebeuf, Philippe de Beaumanoir, Jean Bodel, and Gautier Le Leu, have been associated with the f., but most examples remain anonymous. Critical opinion in the past has been divided as to whether the fabliaux were bourgeois or courtly in origin. Modern scholars favor the view that their authors as well as their audiences were not limited to any one social class. On the other hand, the predominance of courtly and lyrical sentiments make it all but anachronistic to speak of the f. as a bourgeois genre, as Bédier did in his classic study (1st ed. 1893).

The characteristic f. style is simple, unsophisticated, and practical—the materialistic view of everyday life. Though the fabliaux have been accused of antifeminist sentiment, it is interesting to note that, in most instances where such a sentiment is expressed, it is appended to a tale which illustrates female ingenuity and superiority, often over a foolish husband. If there is a prevailing theme in the plots of the fabliaux, it is that of "the trickster tricked." The verseform is commonly octosyllabic couplets.

When critics complain of the obscenity of the f.—and historically many have done so—they usually seem to be offended more by the vulgarity of the lang. than by the prurience of the stories. True, in sexual matters the lang. is routinely explicit, yet the social attitudes underlying it are conventional, even conservative, and may indicate a medieval sensibility toward such matters that underlies and antedates both courtly purity and Christian puritanism.

In the 14th c. the vogue of the f. spread to Italy and England, where Chaucer borrowed freely from the genre for eight or nine of his *Canterbury Tales*, esp. the *Miller's Tale* and *Reeve's Tale*. The f. may fairly be said to be the genre of greatest interest to Chaucer in his mature work. Fr. trad. continued in the prose *nouvelle*, but the influence of the older form may be seen centuries later in the poetry of La Fontaine in France, C. F. Gellert in Germany, and I. A. Krylov in Russia.

W. M. Hart, "The F. and Popular Lit.," *PMLA* 23 (1908), "The Narrative Art of the OF F.," *Kittredge Anniv. Papers* (1913); J. Bédier, *Les Fabliaux*, 5th ed. (1925); J. Rychner, *Contribution à l'étude des fabliaux*, 2 v. (1960); *The Literary Context of Chaucer's Fabliaux*, ed. L. D. Benson and T. M. Andersson (1971); P. Nykrog, *Les Fabliaux*, rpt. with a "Post-scriptum 1973" (1973); P. Dronke, "The Rise of the Med. F.," *RF* 85 (1973); *The Humor of the Fabliaux*, ed. T. D. Cooke and B. L. Honeycutt (1974); *Gallic Salt*, tr. R. L. Harrison (1974); T. Cooke, *The OF and Chaucerian Fabliaux* (1978); *Cuckolds, Clerics, and Countrymen*, tr. J. Duval (1982); R. E. Lewis, "The Eng. F. Trad. and Chaucer's *Miller's Tale*," *MP* 79 (1982); P. Ménard, *Les Fabliaux* (1983); *Nouveau recueil complet des*

fabliaux, ed. W. Noomen and N. van den Boogaard, 10 v. (1983–); R. H. Bloch, *The Scandal of the Fabliaux* (1986); C. Muscatine, *The OF Fabliaux* (1986). A.PR.; R.L.H.

FIGURE, SCHEME, TROPE. F., s., and t. are parts of what is collectively called rhetorical or "figurative" lang. But the difficulty of distinguishing and relating these terms, or of giving a principled definition of any one of them, is well known, and writers from Quintilian (1st c. A.D.) to Fontanier (19th c.) and Todorov (20th) have customarily begun by discussing the limitations of all previous positions. The present article considers first some of the attempts at defining t. in terms of f., and f. in terms of t., then returns briefly to a consideration of s.

Discussions of f. and t. have traditionally used the philosophic and linguistic (semiotic) distinctions between word, or sign, and meaning to define t. as a delimited form of the more expansively conceived f. T. is defined by Quintilian (*Institutio oratoria*, ed. H. E. Butler, v. 3 [1920]) as "the artificial alteration of a word or phrase from its proper meaning to another," and f. as "a change in meaning or lang. from the ordinary and simple form"; for Fontanier (*Les Fs. du discours*, 2 v. [1827–30]), ts. are "all the fs. of discourse which consist of the *divergent meaning* of words, i.e. of a meaning more or less removed and different from their proper and literal meaning," and fs. are "the characteristics, forms and turns . . . by which discourse, in the expression of ideas, thoughts and feelings, stands more or less apart from what would have been their simple and common expression." Such classic formulations presuppose both a norm of "proper" meanings and "ordinary" usage from which ts. and fs. can then diverge (to which Todorov [1972] objects that, by this definition, "ordinary" lang. only becomes retrospectively—and inaccurately—viewed as lang. as it ought to be, "normal"), and also a qualitative distinction between changes in the meanings of words (ts.), and changes in the words and meanings of larger units of discourse (fs.). It would appear that there are only changes of meaning in the uses of individual words in such ts. as "rose" for a quality of affection or beauty in "My love is a rose" (metaphor), or "the White House" for the U. S. government (metonymy), or "hands" for seamen in "All hands on deck!" (synecdoche). But the words "rose," "the White House," or "hands" could not have any altered meaning as individual words (ts.) without an altered or at least specified context of *words and meanings* (fs.), following Austin's argument that individual words alone can never have any meaning at all without context (1961). T. could thus appear to be a compressed instance of f., and f. to be an expanded t.

With regard to ts. and fs. as nonliteral, altered, or improper uses of lang.'s words and meanings, this is an aspect of how rhetorical lang. itself has always been viewed. The varying characterizations of this "rhetorical" difference in lang.'s employment bring with them different considerations of t. and f. Plato's denigration of rhet. as manipulative and untruthful when measured against his model lang. of philosophy meant that ts. and fs. would primarily be characterized by falsehood (*Phaedrus*). Aristotle maintains Plato's view of rhet. as manipulative and less-than-philosophic but is appreciative of the pragmatic functions it serves; furthermore, he does not distinguish rhet. as "unlike" logic without at the same time noting the respects in which they are comparable or "like" (*homos*); he also characterizes his syllogistic reasoning as one form or f. of speech among others (*Rhetoric*). Shortly after Aristotle, rhetorical lang. receives its standard division into *inventio* (subjects, arguments, commonplaces), *dispositio* (arrangement of larger units of discourse such as exhortation, narration, peroration), and *elocutio* (choice and arrangement of specific words and phrases); fs. and ts. become simply parts of *elocutio*—which, after Ramus, we call "style" (q.v.)—constituting some of the verbal material of rhet.'s efforts at persuasion, incl. those of literary expression. Longinus similarly sees fs. in poetry as means for effecting emotional "elevation" of the audience (*Peri hypsos*). In Cicero (*De orat.*) and the anonymous *Rhetorica ad Herennium*, fs. of speech (*figurae dicendi*) appear to mean "modes of eloquence" and to designate different levels of style (the "grand" or "full," the middle, and the "simple" or "plain"; see STYLE). Dante refers to certain forms of allegory (q.v.), long classified as a f., as "truth hidden beneath a beautiful lie" (*De vulgari eloquentia*), and Fontanier writes of "fs. of expression" (or words; see below)—which include personification, allegory, allusion, paradox, and irony—as offering "an apparent illusory meaning in order to allow you better to find or grasp the real and true meaning."

This brief survey of some notable conceptions of rhetorical lang. indicates that the alterations of "ordinary" usage in ts. and fs. may be variously analyzed as ones that change the true to the false, that shape or mold the unstylized into the highly embellished and specifically targeted, and ones that turn via the false or artificial to the true. The general insistence upon ts. and fs.'s divergence from a quasi-natural or basic norm is apparently preserved in the terms themselves, t. being from the Gr. *tropein*, "to turn," "to swerve," f. from the Lat. *figura*, "the made," "the shaped," "the formed." But this view reaches a *terminus ad quem* in accounting for one of the briefest and most common of ts., catachresis (q.v), or the forced (Gr. "abusive") attribution of a new meaning to a given word. Catachresis is not divergence from a literal meaning—in the sense of opposition to a word's proper meaning—but rather its and its word's *extension* to a place where there is no other sign (as in "*leg* of a chair" or "*wing* of a building"), which

extension then itself becomes a necessary norm. Catachresis can thus be viewed either as the extension of literal lang. toward t., or as the entry of a t. (close to but more "basic" than metaphor) into "literal" lang. (Derrida).

Fs. function not only as extensions of ts. into larger units of discourse, but also as intrusions of rhet. into thought. In Cl. rhet., the traditional distinction between fs. of words and of thought (*figurae verborum* and *sententiarum*) attempts to address this problem. Quintilian claims that fs. of words always involve a changed aspect of lang. if not of thought (repetition, antithesis, asyndeton), while fs. of thought always involve an artificial or contrived thought and may either leave lang. unchanged or be composed of several fs. of words. Fontanier similarly distinguishes between ts. (altered meanings of individual words), fs. of "expression" or words (in which the form, order, choice, or assortment of words is affected—e.g. allusion, litotes, inversion, apposition, ellipsis, apostrophe), and fs. of thought (among which he counts prosopopoeia, concession, and description). But there could not be a f. of words such as litotes without its intruding upon the *thought*—that, for example, "not unlike" is weaker than or at least different from "like"; and there could not be an apostrophe to something absent or ordinarily incapable of being an object of audible address without this affecting the thought of the expression. (This difficulty is evident in Quintilian categorizing apostrophe as a f. of thought, but Fontanier as a f. of "expression" or words.) Irony may not alter a single word from an ordinary expression in order to be manifest in a context—"I really like that" may ironically mean that one does not like it at all—and thus Quintilian understands it as a f. of thought; but for the very reason that such a f. involves divergence from—indeed, opposition to—a customary meaning of words, Fontanier considers irony to be a f. of "expression." Fs. of thought, then, may simply be extended instances of or views upon fs. of words: in each case, either words change, or meanings change, or both. Modern linguistics as well as ancient rhet. is hard put to find any instance of sheer fs. of words, in which only sounds or arrangements of words are involved (e.g. repetition, opposition, parallelism, alliteration, assonance) that does not bring with it at least a minimal f.—alteration or artificing—of thought as well.

Ts. and fs. (with the possible exception of catachresis) may be said always to present two senses under the guise of a single verbal formulation: either they say something unexpected by way of a striking expression precisely because the expression or novel sense rebounds against an expected sense or usage which still attaches to the words in question; or they do not "say" it so much as they let it, contextually, be understood beneath (or even in spite of) the actual expression (as in allegory and irony). This effect of *doubling* a single word or expression's appearance or meaning may be construed as alteration of a first meaning by the addition of a second (Genette summarizes Fontanier's t. as "the *change* of meaning of a word") or as "the *substitution* of one expression for another" (Genette on Fontanier's f.). It allows lang.—which is necessarily thought at any given moment to have more or less customary and stable meanings and uses attached to it—to say and mean *more* or *something other* (to allegorize, from Gr. *allos-agoreuein*, "to speak otherwise"). Minimally, this suggests that ts. and fs. always involve at least the *relating* of other words, meanings, and usages to the ones at hand, or the *comparing* of various meanings for words (ts.) or of one arrangement or usage of words for another possible one (f.). In the 20th c., when figurative lang. has most often been studied as a part of poetics or literary stylistics, this view has become part of a psychology of literary response, as in I. A. Richards' position that a metaphor allows two or more ideas of different things to be carried by a single word or expression (the "vehicle"), its meaning (the "tenor") resulting from their "interaction" (cf. Empson).

Opposed to these long-standing views of t. and f. as swerves from normal lang. usage, with their concomitant effects understood psychologically as an audience's reaction to such divergence, there have also been treatments of f. and t. as ineluctable discursive structures, fundamental to the organization of verbal experience and consciousness, and to the production of lang. and understanding themselves. Paul de Man considers the fs. of allegory and irony to be constitutive of interrelated treatments of temporality, allegory projecting a consciousness of meaning in relation to time along an axis of past and future—thus constituting narrative—irony doing so along an axis of a present moment divided between empirical and linguistic or "fictional" selves—thus constituting a divided self-consciousness ("The Rhet. of Temporality"). Roman Jakobson uses the structural linguistic (Saussurian) understanding that all lang. displays two axes of selection (similarity, the "paradigmatic" axis) and combination (contiguity, the "syntagmatic" axis) to argue that metaphor is essentially *selective*, metonymy essentially *combinatory*, and that all lang., incl. all lit., orients itself toward one or another of these principles of verbal organization (see METAPHOR; METONYMY)

With ts. (or fs.) such as allegory, irony, metaphor, and metonymy thus being held to generate literary texts and "original" verbal utterances, it is no longer a matter of t. being understood as a species of f., and f. as a special (divergent) kind of lang., but rather one of t. and f. being used to name and analyze lang. in its fundamental structures. A lineage for this view can be found in such diverse writers as Vico, Hamann ("Über den göttlichen und menschlichen Ursprung der Sprache" [On the Divine and Human Origin of Lang.], 1772), Rousseau ("Essai sur l'origine des langues" [Essay on the Origin of Langs.], 1817), and Nietzsche,

who variously see the "origin" of lang. in t. or f. such as catachresis or metaphor. When Nietzsche says that one can only speak figuratively about the relations between lang., fs., and their meaning ("Über Wahrheit und Lüge im aussermoralischen Sinn" [1873]), he recalls Vico's effort to analyze historical "stages" of lang.-use with the names and structures of ts. and fs. (*Scienza nuova* [1725]; see also N. Frye, *The Great Code* [1982]) and anticipates de Man's view that attempts at historical understandings of lang., its origins, and its meanings (Rousseau's and Nietzsche's among them) necessarily show themselves to be ts. and fs., e.g. allegory, metonymy, metalepsis (de Man 1979, 1983; see also H. Bloom, *A Map of Misreading* [1975], on lit. hist. as the tropological turns, displacements, or alterations of an earlier poet's lang. by later ones). Erich Auerbach shows that, for a millennium, the late Cl. and medieval Christian doctrine of figural exegesis extended *figura* from a domain of eloquence narrowly construed into the interrelations of figurative lang. with identifications of historical meaning.

Independent of these problems of a metalanguage of history being addressed to lang., nonhistorical analytic or descriptive lang. about lang. also appears to involve t. and f. Quintilian, defining t. generically as "transfer of expression," uses the technical term for metaphor (Gr. *metapherein*, "to carry over," "to transfer") to describe the set of which metaphor would be a member; and Fontanier concedes that to speak of "fs. of discourse" at all is itself *metaphoric*, using a metaphor (or perhaps a catachresis) from the traits, form, and contours of the body for those of lang. Derrida has analyzed this persistence of f. and t. in philosophy's discussion of the very terms. Jakobson, observing that "similarity in lang. connects the symbols of a metalanguage with the symbols of the lang. referred to" and that "similarity [the metaphoric pole] connects a metaphorical term with the term for which it is substituted," concludes that this dichotomy "appears to be of primal significance and consequence for all verbal behavior and for human behavior in general."

Whether or not all metalanguage is metaphorical—which would mean that all rhetorical (and grammatical and logical) treatises and also all linguistics are figurative—it is undeniable that all commentaries upon fs. and ts. are metalinguistic, i.e. substitutive, and thus, in the view running from Quintilian to Jakobson, tropaic. From the medieval *trivium* to Fontanier and much modern linguistics, understandings of figurative lang. have traditionally subordinated rhet. to grammar and grammar to logic, but this has not diminished the recurrent power of analyses of fs. and ts. to uncover a disruptive questioning of the ostensible "logic" of explanations of f. and t., and of the tropological forces at the basis of both lang. and metalanguage (de Man 1986). The argument that fs. and ts. are constitutive of and coextensive with lang. has as one of its consequences that lit. becomes not merely an occasion for "foregrounding" lang. or bringing it to consciousness—a view found from Hegel (*Vorlesungen über die Ästhetik*, v. 13–15 [1835–38]) to some Rus. Formalists and Czech and Fr. structuralists (e.g. Riffaterre 1971, 1978)—but also a privileged vehicle or avenue for access to the problems lang. poses for understanding, with de Man "calling 'literary,' in the full sense of the term, any text that implicitly or explicitly signifies its own rhetorical mode and prefigures its own misunderstanding as the correlative of its rhetorical nature; that is, of its 'rhetoricity'" (1986).

S., in its relation to f. and t., enters here as a summary instance of the breadth and complexity of figurative lang. and its understanding. S. in antiquity often means fs. of words and of thought (*schémata lexeós* and *dianoia*), and Cicero, the *Ad Herennium*, Quintilian, and later Ren. rhetoricians frequently appear to use the term for fs. of speech or levels of style. But in Cicero's Lat. trs. of Aristotelian and other Gr. uses of s. for outward appearances, semblances, or perceptible forms and shapes, it can also broadly represent thought and lang. in their relations to reality, its true perception and understanding. Thus Aristotle writes of his *schémata syllogismou*, and of the *schéma tés ideas* (*Metaphys.*), and the atomists Democritus and Lucretius consider ss. to emanate from bodies and to be like images (Auerbach). Kant uses s. to name an order of thought and lang. that cannot be known perceptually or empirically, but in relating s. to such fs. as hypotyposis and symbol, he reintroduces its imagistic and figurative ("shaped" or "formed") aspects (*Critique of Pure Reason, Critique of Judgment*). When one asks colloquially after "the s. of things," this trad. of forms and meanings is summarized as one invokes a nonobvious, nonvisual pattern or sense which attaches to things or events, but which nonetheless can be brought to lang. and thought in the form of a verbal image or narrative.

S. is scarcely used any longer to speak of figurative lang. in rhet. or lit. But it is schematization, in the wider sense of the tropaic rendering of appearances and meanings, that is recalled when de Man analyzes the t. of prosopopeia—from Gr. "to make or give a person or face," and once categorized as a f. of thought or a s. (see PUN)—as both materially a textual f. and, in its catachretic projections, a means by which "face" (Fr. *figure*) and voice are inscribed for lang., cognition, and consciousness ("Hypogram and Inscription," "Anthropomorphism and T. in the Lyric"). Prosopopeia and f., s., and t. are not unique or proper to certain forms of the lyric such as the elegy, the ode, or ekphrastic poetry, nor to lit. in general. Rather, poetry and poetics, to the extent that they can be said to be constructed exclusively of lang., are sites among others where f., s., and t. have been deposited as materials of lang. and modes of thought.

I. A. Richards, *The Philosophy of Rhet.* (1936); K. Burke, "Four Master Ts.," *A Grammar of Motives*

(1945); W. Empson, *The Structure of Complex Words* (1951), *Seven Types of Ambiguity*, 3d ed. (1953); E. Auerbach, "Figura," *Scenes from the Drama of European Lit.* (1959); J. L. Austin, "The Meaning of a Word," *Philosophical Papers* (1961); G. Genette, *Fs.*, 3 v. (1966–72), selections tr. as *Fs. of Literary Discourse* (1982); R. Jakobson, "Two Aspects of Lang. and Two Types of Aphasic Disturbances," in Jakobson, v. 2; R. Lanham, *Handlist of Rhetorical Terms* (1968), 101–3; M. Riffaterre, *Essais de stylistique structurale* (1971), *Semiotics of Poetry* (1978); J. Derrida, "La Mythologie blanche," *Marges de la philosophie* (1972); T. Todorov, "Rhétorique et stylistique" and "F.," *Dictionnaire encyclopédique des sciences du langage* (1972); Lausberg; D. Rice and P. Schofer, *Rhetorical Poetics* (1973); M. and M. Shapiro, *Hierarchy and the Structure of Ts.* (1976); Group Mu, 45—schematic taxonomy; *Rhet. Revalued*, ed. B. Vickers (1982); P. de Man, "The Rhet. of Temporality," and "The Rhet. of Blindness," *Blindness and Insight*, 2d ed. (1983), "Anthropomorphism and T. in the Lyric," *The Rhet. of Romanticism* (1984), "Hypogram and Inscription" and "The Resistance to Theory," *The Resistance to Theory* (1986); R. J. Fogelin, *Figuratively Speaking* (1988).
T.B.

FORM. Few terms in literary study are more widely used than f., but the term is so variable and inclusive that it ends by being one of the most ambiguous that we have. At one level it can refer to the minutiae of the text, at another to the shape of the text in itself, at a third to the characteristics a text may share with others, at still another level to a transcendental or Platonic model from which the text imperfectly derives; and it may have several of these meanings within the work of a single theorist. Such usage can take in a wide variety of conditions that have to do with f.—its nature, or its location, or its relationship to content and to the world content proclaims. Yet that territory is not only broad but the site of issues so fundamental that they have had to be faced from the beginning of literary study. Recent theorizing on the nature of the literary text gives far less attention to f. than to the related concept of genre, quite possibly because so much modern study of f. has been associated with varieties of formalism. Still, some of the most potent contemporary theories of lit. have considerable (if implicit) significance for the study of this ancient issue.

Whatever the specific import given the concept of f., whatever the period in which it becomes part of literary discussion, the question of representation (q.v.) or mimesis always comes into play, if only, as in parts of our century, to reject the desirability and even possibility of such a process. This inevitable relation of f. to representation explains why Plato and Aristotle have been permanent topics in the study of literary f., the recurring dialectic of their differing concepts of f. always a compelling force. The basic Platonic position appears in the discussion of painter, craftsman, bed, and model in Book Ten of *The Republic*. For Plato a f. is an instance of absolute purity which, because it is purity, has to be distanced from the place of emergence of its necessarily imperfect echoes. The Platonic conception of fs. (*eidei*, ideas) has, therefore, two principal components: first the model's unsullied perfection and second the fact that it is external to its material manifestations. That the f. has its origin outside the text or object seems to have as one of its implications the possibility that f. is imposed upon matter, and as another the corollary that f. and content have to be viewed as separate elements brought together in the manifestation.

It is along these lines in particular that the conception of f. in Aristotle's *Poetics* has to be seen as an answer to Plato, as history's major alternative. F. in Aristotle is not fixed but emergent and dynamic, not external or transcendental but intrinsic and immanent, coexistent with the matter in which it develops towards its fullest realization. F. so conceived cannot be identified with structure but is, instead, the informing principle that works on matter and causes the text or object to become all it is. In Aristotelian terms f. (the "formal cause") is one of the four causes, the others being the material cause (the matter), the efficient cause (the maker), and the final cause (the end in the sense of both conclusion and purpose). This last involves another crucial distinction between the Platonic and Aristotelian modes, for in Plato f. is identified with origin, a place of beginnings against which the result is to be tested, while Aristotelian f. is based on entelechy, an end which brings about the full and complete being of the object (*telos* has to do not only with ending but with a consummation which is completeness). The implications for literary theory are considerable. René Wellek and others have argued that the concept of organic f., which rejects any separation of f. and content, began with Aristotle but was ignored until the 18th c., a history which may have much to do with the Platonic understanding of literary f. and its attendant implication that f. is originary, separable, and distinct. Aristotle's own practice clearly rejects originary thinking: in the *Poetics*, *anagnorisis* (recognition) and *peripeteia* (reversal) are elements both in the f. of a tragedy and in its content as well, as is the insistence that all texts should have a beginning, middle, and end. F. and content so conceived are not only inseparable but indistinguishable, and in later critics may be considered as precisely the same aspect taken from different points of view.

The purity of the Aristotelian distinction between f. and what is often called structure has rarely been kept, the result being the plethora of meanings the term always has. At its least encompassing level, f. is used to refer to metrical patterns as well as lexical, syntactic, and linear arrangements: for example, the f. of the heroic couplet or

the Spenserian stanza. At this level the term has much to do with matters of technique and style (conceived as ways of performing, of putting into f.) as well as the various implications that emerge from those ways: for example, what Pope's heroic couplets show of his sense of the world, what Byron's use of the Spenserian stanza in *Childe Harold's Pilgrimage* states about lit. hist. (These examples show again how f. is always linked to questions of representation, however different the tangents.) A more encompassing view considers f. as structure, put most simply as the overall "mode of arrangement" (Thomas Munro) of the text; that is, the way textual materials are organized so as to create shape. In a classically rhetorical mode, f. in this sense refers to scheme or argument. In a classically New Critical mode, it refers to the ironic patterns of tension within a text, patterns that create the overall shape of the text's unparaphraseable meanings (Brooks). Put in terms of structure, f. is esp. prone to problems of the separation of f. and content, a point of particular sensitivity to all manner of modern formalists. It is equally prone to that problem when taken at the same level but in a different context, that is, when identified with genre or kind. Clearly dissatisfied with the blurring of concepts that actually touch only rarely, Claudio Guillén has argued that genre is "an invitation to f." Theodore Greene's more involved distinction between specific f. (unique and essentially incomparable) and generic f. (shared and contextual) seems designed not only to establish practical categories but to satisfy Croceans, formalists, and literary historians at once. At this point f. can become an element or principle of classification, a segment of another order, of a more encompassing f.

The need to satisfy such a plurality of views indicates the variety of positions taken on f. since the middle of the 18th c. The concept of "inner f." was developed by Goethe and others, its immediate origins in Shaftesbury and Winckelmann, its ultimate source Neo-Platonic. The principle concern of inner f. is with the internal totality of the work and the correspondence of that wholeness to the external world, what Shaftesbury speaks of as a microcosmic mirroring of experience (in *Characteristics* he refers to "inward f."). That emphasis on wholeness found its most productive routing in A. W. Schlegel's distinction between organic and mechanical f., surely the most influential reading of the term in modern times. Schlegel speaks in his *Dramatic Lectures* of a f. which is essentially additive, imposed from without and therefore quite possibly arbitrary. This is the mode of neoclassical drama. The opposite is organic f., generated from within, coexistent with the material, complete only when the object itself is complete, working its way just as natural fs. do. This is the Shakespearian mode. Schlegel's pairing is repeated in the first of Coleridge's *Lectures on Shakespeare*, an influence which (carrying along the attendant ideas of variety in unity and the reconciliation of opposites) led to ways of conceiving f. still at work in our own day. Versions of it appear in Croce and others who are not necessarily Crocean, where f. is thought of not only as implying a new unity of all given materials but as ultimately the only means whereby expression becomes possible (see Greene and La Drière). Critics like Wimsatt and Brooks have argued, however, that Croce is actually concerned with lyric moments rather than wholes, a reading which would place him squarely in line with Poe. Far less useful than the concept of f. as expression is the notion of "significant f.," fathered by A. C. Bradley and furthered by Clive Bell and Roger Fry, which gets more attention in histories than it did in thinking about lit. Significant f. has little or nothing to do with representations of "life" but is, instead, concerned only with color, line, space, and those other elements which arouse aesthetic emotion. That Bell and Fry have their sources in decadence and aestheticism clarifies their limitations. Potentially much more fruitful, though as yet hardly tapped in general speculations on f., are recent explorations of the question of indeterminacy that are likely to become a major legacy of end-of-the-century theorizing. Fostered by a suspicion of total systems and the fs. that mimic them—suspicions that extend, in some cases, to any sort of absolute closure—such speculations are surely the most potent threat yet to two centuries of Schlegelian thinking about the shape of literary f. Whether one argues that this, too, involves questions of representation, or that we can never really know fixities such as closure, these speculations are likely to compel radical readjustments of a basic literary concept that has too long been taken with complacency.

For discussion of the theoretical bases of poetic form in poetics, see VERSE AND PROSE.

C. Bell, *Art* (1914); H. Wölfflin, *Kunstgeschichtliche Grundbegriffe* (1915), 7th Ger. ed. tr. M. D. Hottinger, *Principles of Art Hist.* (1976), ch. 3—"open" vs. "closed" f.; W. P. Ker, *F. and Style in Poetry* (1928); R. Schwinger, *Innere F.* (1935); B. Fehr, "The Antagonism of Fs. in the 18th C.," *ES* 18–19 (1936–37); T. H. Greene, *The Arts and the Art of Crit.* (1940); K. Burke, "Container and Thing Contained," *SR* 8 (1945), "Psychology and F." and "Lexicon Rhetoricae," *Counter-Statement*, 2d ed. (1953), *Philosophy of Literary F.* (1957); Brooks; I. A. Richards, "Poetic F.," *Practical Crit.*, 2d ed. (1948); C. La Drière, "F.," *Dict of World Lit.*, ed. J. T. Shipley (1953); S. Langer, *Feeling and F.* (1953); W. S. Johnson, "Some Functions of Poetic F.," *JAAC* 13 (1955); Wellek; Wimsatt and Brooks; W. A. Davis, "Theories of F. in Mod. Crit.," Diss., Univ. of Chicago (1960), *The Act of Interp.* (1978), ch. 1; R. Wellek, "Concepts of F. and Structure in 20th-C. Crit.," *Concepts of Crit.* (1963); J. Levý, "The Meanings of F. and the Fs. of Meaning," *Poetics, Poetyka, Poetika*, ed. R. Jakobson et al. (1966); W. Tatarkiewicz, "F. in the Hist. of Aesthet-

ics," *DHI*; T. Munro, *F. and Style in the Arts* (1970); C. Guillén, *Lit. as System* (1971); A. Fowler, "The Life and Death of Literary Fs.," and S. Chatman, "On Defining F.," *NLH* 2 (1971); T. E. Uehling, Jr., *The Notion of F. in Kant's Critique of Aesthetic Judgment* (1971); G. S. Brown, *Laws of F.* (1972); *Organic F.: The Life of an Idea*, ed. G. S. Rousseau (1972); K. K. Ruthven, *Critical Assumptions* (1979), ch. 2; J. D. Boyd, *The Function of Mimesis and Its Decline*, 2d ed. (1980); M. Perloff, *The Poetics of Indeterminacy* (1981); R. Duncan, "Ideas of the Meaning of F.," *Claims for Poetry*, ed. D. Hall (1982)—and others therein; Fowler; N. E. Emerton, *The Scientific Reinterpretation of F.* (1984); D. Wesling, *The New Poetries* (1985), ch. 2; Hollander. F.G.

FORMULA, together with the "theme" and "story-pattern," are the istructural units of oral composition delineated by the oral-formulaic theory. Originally defined by Milman Parry in 1928 as "a group of words which is regularly employed under the same metrical conditions to express a given essential idea," the f. has since been identified as the "atom" of oral-traditional phraseology in dozens of trads. The f. is to be distinguished from simple repetition on the basis of its consummate usefulness to the composing oral poet, who employs it not for rhetorical effect but because it is part of the traditional poetic idiom bequeathed to him by generations of bards, an idiom that combines the practical and immediate value of ready-made phraseology with the aesthetic advantage of enormous connotative force.

Although earlier scholars, chiefly the Ger. philologists of the 19th c., had referred in general terms to formulaic diction (esp. Sievers' *Formelverzeichnis* [1878] and the linguistic analyses of Düntzer), Parry was the first to describe the existence and morphology of this phraseological element with precision, to demonstrate its tectonic function, and to link the traditional phrase with composition in oral performance. His studies of Homeric epic (1928–32) showed how traditional composition was made possible by the bard's ability to draw on an inherited repertoire of poetic diction consisting of fs. and *formulaic systems*, each system comprising a set of fs. that share a common pattern of phraseology. Thus he was able to illustrate, for example, how prosêuda ("addressed") could combine with 11 different nominal subjects to constitute a metrically correct measure in the hexameter line (selections below):

prosêuda +
— ∪ ∪ — —
dia qeawn (divine of goddesses)
— ∪ ∪ — —
mantis amumwn (blameless seer)
— ∪ ∪ — —
dios ᵓAcilleus (divine Achilles)
— ∪ ∪ — —
dios ᵓOdusseus (divine Odysseus).

Parry understood the poet's task not as a search for original diction but as a fluency in this traditional idiom—that is, as the talent and learned ability to weave together the ready-made diction and, occasionally, new phraseology invented by analogy to existing fs. and systems to produce metrically correct hexameters. One further useful aspect of this special epic lang. he found to be its *thrift*, whereby more than one f. of the same metrical definition for the same character was but seldom allowed. Having concentrated in his 1928 theses principally on the noun-epithet fs. so typical of Homeric style, Parry then broadened his examination in 1930 and 1932 to consider all of Homer's diction. His analysis of the first 25 lines of the *Iliad* and *Odyssey*, which show graphically how Homer depended on fs. and systems to make his poems, remains a *locus classicus* for the interp. of ancient Gr. epic lang.

These deductions were augmented as a result of fieldwork in Yugoslavia carried out by Parry and Albert Lord. Lord demonstrated that the Yugoslav *guslar* employed a similar formulaic idiom in the making of South Slavic epic, using stock phrases like lički Mustajbeže ("Mustajbeg of the Lika" [a place-name]), as well as more flexible formulaic systems, to compose his poem in oral performance. With the benefit of a large corpus of material, including many songs from the same *guslar*, Lord was able to show that virtually every line of South Slavic oral epic could be understood as a f. or member of a formulaic system.

On this foundation other scholars have erected a series of studies of the f., some on ancient, medieval, and other ms. trads., others on recoverable modern trads.—e.g. Webber (1951) on formulaic frequency in the Sp. ballad; Magoun (1953) on the first 25 lines of *Beowulf*; Duggan (1973) on the OF *chansons de geste*, postulating a threshold of 20% straight repetition for probable orality; Lord (1986) on comparative studies; Culley (1986) on the Bible; Edwards (1986, 1987) on ancient Gr.; Olsen (1986, 1987) on OE; Parks (1986) on ME; and Zwettler (1978) on Arabic. Foley provides bibl. (1985) and detailed history (1988).

In recent years scholars have focused on the problem of interpreting a formulaic text: whether the idiom is so predetermined and mechanical as to preclude verbal art as we know it, whether the oral poet can select esp. meaningful fs. appropriate to certain narrative situations and thus place his individual stamp on the composition, or whether some other aesthetic must be applicable. Most have emphasized either the traditional and inherited or the spontaneous and individual. More account needs to be taken of the differences as well as similarities among fs., with attention to different trads., genres, and kinds of texts (e.g. mss. and acoustic recordings). Similarly, the referential fields of meaning commanded by fs. need more investigation, since fs. and other structural units convey meaning not by abstract denotation but by metonymic reference to

the trad. as a whole.

H. Düntzer, *Homerische Abhandlungen* (1872); E. Sievers, "Formelverzeichnis," in his ed. of *Heliand* (1878); R. Webber, *Formulistic Diction in the Sp. Ballad* (1951); F. P. Magoun, "The Oral-Formulaic Character of Anglo-Saxon Narrative Poetry," *Speculum* 28 (1953); Lord; Parry; J. Duggan, The Song of Roland: *Formulaic Style and Poetic Craft* (1973); M. Nagler, *Spontaneity and Trad.* (1974); M. Zwettler, *The Oral Trad. of Cl. Arabic Poetry* (1978); J. Foley, *Oral-Formulaic Theory and Research* (1985), *The Theory of Oral Composition* (1988), *Comparative Studies in Traditional Oral Epic* (1989); R. Culley, "Oral Trad. and Biblical Studies," *OT* 1 (1986); A. B. Lord, "Perspectives on Recent Work on the Oral Traditional F.," *OT* 1 (1986); W. Parks, "The Oral-Formulaic Theory in ME Studies," *OT* 1 (1986); M. Edwards, "Homer and Oral Trad.: The F.," *OT* 1–2 (1986–87); A. Olsen, "Oral-Formulaic Research in OE Studies," *OT* 1–2 (1986–87).

J.M.F.

FOURTEENER. A metrical line of 14 syllables which since the Middle Ages has taken two distinct forms used for a variety of kinds of poetry and thus exhibiting widely differing characteristics. When two fs. are broken by hemistichs to form a quatrain of lines stressed 4–3–4–3 and rhyming *abab* they become the familiar "eight-and-six" form of ballad meter called Common Meter or Common Measure; Coleridge imitated the looser form of this meter in "Kubla Khan," as did many other 19th-c. poets attempting literary imitations of "folk poetry." The eight-and-six pattern was used widely in the 16th c. for hymns; indeed, Sternhold and Hopkins adapted it from contemporary song specifically to take advantage of its popularity and energy for their metrical psalter (2d ed. 1549). It was also used for polished lyrics such as Herrick's "Gather ye rosebuds while ye may." Shakespeare uses it for comic effect in the "Pyramus and Thisbe" play performed by the "rude mechanicals" in *A Midsummer Night's Dream.* Closely related is Poulter's Measure, a line of 12 syllables followed by a f., which when broken into a quatrain gives the 3–3–4–3 pattern of Short Meter.

But the f. was also used extensively in the 16th c. for the elevated genres, in particular for tr. Lat. tragedy and epic. It is used for trs. of Seneca's tragedies by Jasper Heywood and others pub. by Thomas Newton in the *Tenne Tragedies* (1581) and by Arthur Golding for his tr. of Ovid's *Metamorphoses.* The most famous and impressive use of fs. in Eng. is doubtless George Chapman's tr. of Homer's *Iliad* (1616). The contrast between the effect of fs. in popular and learned genres is striking:

> The king sits in Dumferline town
> Drinking the blood-red wine;
> "O whare will I get a skeely skipper
> To sail this new ship of mine?"
> (*Sir Patrick Spens*)

> Achilles' banefull wrath resound, O
> Goddesse, that impos'd
> Infinite sorrowes on the Greekes, and
> many brave soules los'd
> From breasts Heroique—sent them
> farre, to that invisible cave
> That no light comforts; and their lims
> to dogs and vultures gave.
> (Chapman's Homer)

Opinions differ about the origins of the f. Schipper calls it a *septenarius*—more precisely, the trochaic tetrameter brachycatalectic—implying that it was derived from imitation of Med. Lat. poems in that meter. The native influence of ballad meter is also obvious. But another explanation is equally likely: the great Cl. epic meter, the dactylic hexameter, contains on average about 16 "times," and it may be that early Eng. translators thought the f. was the closest vernacular equivalent.—Schipper; P. Verrier, *Le Vers français,* 3 v. (1931–32), v. 2–3; J. Thompson, *The Founding of Eng. Metre* (1961); O. B. Hardison, Jr., *Prosody and Purpose in the Eng. Ren.* (1988).

O.B.H.; T.V.F.B.

FREE VERSE.

I. TYPES
II. ORDERING PRINCIPLES

F. v. is distinguished from meter by the lack of a structuring grid based on counting of linguistic units and/or position of linguistic features. Some of f. v.'s primary features (nonmetrical structuring, heavy reliance on grammatical breaks, absence of regular endrhyme) may be traced back into the roots of lyric, and f. v. is today a frequently used (if extremely various) set of prosodies for contemporary poetry around the world.

I. TYPES. Many of the same analytical terms and methods employed to describe metrical verse can be applied to f. v. Nonetheless, f. v. brings with it different relations of regularity to irregularity and of the auditory to the visual; and different kinds of repetition, equivalence, and voice.

The term, though not always the formal impulse and practice, is a derivative of metrical verse. It is a prosody of avoidance of meter, and at the same time it fuses other prosodies. The term itself is a manifesto and a contradiction-in-terms first used by Fr. poet Gustave Kahn in the late 1880s at the advent of European modernism. The Fr. term *vers libre* (q.v.) has its counterparts in the Eng. f. v., Ger. *freie Rhythmen,* and Rus. *svobodnyj stix, volnyj stix.* In Eng., f. v. was a term of derogation before it became a battle cry or, as today, a more neutral descriptor.

F. v. must be described by integrating historical and structural approaches. The types of f. v. are products of different literary-historical conditions. We could argue that f. v. reaches back to the oral roots of poetry, to the period preceding the devel.

of regular metricity. Sumerian, Akkadian, Egyptian, Sanskrit, and Heb. poetries all share one characteristic: in their texts, repetition and parallelism (qq.v.) create prosodic regularity, while meter in any of its types does not yet regulate the verse.

The first type of f. v., the form partly freed from the constraints of traditional meter, has its origins in Fr. *vers libéré* of the 17th c., when La Fontaine and his followers began to loosen the strictly syllabic lines of Fr. poetry by inventing "liberated" metrical forms. Such was the precursor of those prosodic strainings against meter which in the 19th c. became a major trend. Zhukovskij, Sumarokov, Pushkin in Russia; Klopstock, Goethe, Hölderlin in Germany; Batsányi, Kazinczy, Petőfi in Hungary; Rimbaud, Verhaeren, Apollinaire in France—all wrote with radical deviation from the meters common in their respective national poetries. In Eng., this partial liberation of and from meter existed here and there before Whitman (e.g. Smart's catalogue poems, Scott's "Verses in the Style of the Druids," Arnold's "Strayed Reveler"). This type and, if included, the premetrical form described briefly above, appear as f. v. only in retrospect.

The second type, avant-garde f. v., has no direct roots in the metrical trad., at least in the way the verse-line is built up. Meter is what this self-conscious, self-proclaimed f. v. is free of. This type was born when Walt Whitman's proudly nonmetrical *Leaves of Grass* (1855) flouted accent-and-syllable-counting regularity. Whitman brought back into poetry strong stress at unpredictable places, grammatical emphasis and parallelism, anaphora, and long lines. His oral-derived form is expansive, asymmetrical, mixing dialects and modes, and above all, personal. A related subtype, based in Whitman and possibly also in the Fr. verset or verse-paragraph (q.v.), is the long measure of Allen Ginsberg ("Howl") or Robert Bly ("The Teeth-Mother Naked at Last"), where the line is claimed to derive from the rhythms of breathing, and where it is often used in a list-structure that condemns or celebrates.

Writers of another subtype, notably William Carlos Williams and those in his trad. like Cid Corman, George Oppen, and Robert Creeley, strive for more delicate dynamics, wrenching tighter Whitman's long and sometimes prolix lines. These writers often use lineation out of phase with grammatical and/or visual units. Still others like the early 20th-c. imagist poets, Osip Mandelstam in his one poem (1923) in f. v., the early Wallace Stevens, Carl Sandburg, and D. H. Lawrence habitually use lines coincident with grammatical units.

Williams' theory of the variable foot and Charles Olson's theory of projective verse provide critical justifications for avant-garde f.-v. practice—both more polemical than definitive, both failing to account adequately for even these two poets' own practices.

Many poets now write entirely in f. v. measures, or they move easily from meter through liberated forms to f. v., depending on the needs of the given poem. An international list of avant-garde practitioners would include, among many others, Arthur Rimbaud, André Breton, Paul Éluard, Michel Déguy, Aimé Césaire, Anne-Marie Albiach (Fr.); Kuo Mo Jo, Yip Wai-lim (Chinese); Aleksandr Blok, Velemir Xlebnikov, Nikolai Zabolotskij, Olga Berggolts, Evgeny Evtušenko (Rus.); Pablo Neruda (Sp.—from Chile); Lajos Kassák, Milán Füst (Hungarian); and Gael Turnbull and Charles Tomlinson (Eng.).

II. SOME ORDERING PRINCIPLES OF AVANT-GARDE FREE VERSE. None of the principles exemplified briefly below is particular to f. v., but each has a specific usage and weight within the f. v. poem. The relations of regularity to irregularity may vary widely over the whole poem, for example. Much f. v. is, however, as regular as much metrical verse; the difference is that in f. v., regularity is based on linguistic or textual features not much noticed by most scholars of versification (e.g. intonation or length in typewriter ems).

Generalizations may not apply to all subtypes of 20th-c. practice. Broadly speaking, in f. v. one-time-only events tend to be as important as repetitive events in the poem's structure. A pleasure for the reader comes from unpredictability, when lang. is thickened by sound and visual effects that will or will not be echoed later on. F. v. enriches texture by grammatical and other means, by equivalence, symmetry, and repetition, though not to the same extent as in syllable-counting verse. The effect of breaking-off is typical of some subtypes (Williams), while the effect of deliberative or headlong continuity is typical of others (Lawrence, Ginsberg). In f.-v. prosodies as in all other kinds, technologies of typesetting and historically new visual media co-exist with, even modify and extend, earlier practices of reading.

A. *Equivalence subsets of line/rhyme/stanza.* Long lines, after Whitman, have in general been avoided by the Anglo-Am. writers of f. v., though notable exceptions would be David Jones (*In Parenthesis*), Robinson Jeffers, and A. R. Ammons. The short line is preferred for speed, or for slowness gained by isolating clots of phrase, as in Creeley, Denise Levertov, and Elizabeth Bishop. Lines, whether long or short, are equivalent by virtue of being lines and may also be phonological or syntactic units of the same type or equal typographically. Lines of short or medium length have been chosen by writers of extended poems in f. v. such as Pound (*Cantos*), Williams (*Paterson*), and Edward Dorn (*Slinger*). A carrying measure for a long poem after Whitman will usually not be oratorical, but will make the line turn more tightly. 20th-c. long poems often employ a variety of kinds of f. v., variously tighter and looser, as well as prose.

Alternation of long and short lines gives a conversational cast either to extended narratives like Charles Reznikoff's *Witness*, or to brief lyrics like

e e cummings':

> Buffalo Bill's
> defunct
> who used to
> ride a watersmooth-silver
> stallion
> and break onetwothreefourfive
> pigeonsjustlikethat
> Jesus
>
> he was a handsome man. . . .

No line is the same as any other; f. v. does that habitually, whereas meter does it rarely. Rhyme is missing in this passage, but rhyme is one resource of f. v.—used in unpredictable places (e.g. middles of words and lines) and is used as instance rather than as part of overall design:

> O Alessandro, chief and thrice warned, watcher,
> Eternal watcher of things,
> Of things, of men, of passions.
> Eyes floating in dry, dark air,
> E biondo, with glass-grey iris, with an even side-fall of hair
> The stiff, still features.
> [Pound, *Canto* VII].

Here line-end rhyme occurs, but only once, while rhyme-like repetition makes for a profusion of like sounds which contribute to the noble tone of praise. The accentual cadence, which gives the unique Poundian movement, seems regular but is not; with such variety of stress from line to line, it is pointless to speak of spondee or trochee.

In designs larger than the rhyme-mated pair, f. v. uses stanzas in innovative ways, notably in what has been called the "sight-stanza" (Berry) in Williams and others after him (Creeley, Paul Blackburn). Syntax is driven past the end of a tight, boxlike stanza. In f. v., as usually also in metrical verse, there is often higher-level phonological equivalence between lines and narrative equivalence between stanzas and other kinds of line-groups.

B. *Visual devices* (see VISUAL POETRY). Some f. v. uses visual form, such as heavy capitalization in Aleksej Kručonich (Rus.) and in Kenneth Patchen (Am.), as an overlay rhythm to gain articulation and emphasis. The disposition of the poem across the page in representational designs is another variant, as in Mallarmé's page-fold-out in the shape of a bird's feather in *Un Coup de dés* and in Apollinaire's *Calligrammes*. In these patterns, as in the less dramatic symmetries and fragmentings of more ordinary f. v., the whitespace on the page may be as important as the black print—as an image of pause, disjunction, the silence that surrounds and spaces the text. But it is well to remember that arrangement of the page may function expressively and rhetorically without its being in the least representational.

C. *The Verse period* (grammar in the poem; see SYNTAX). Like visual prosody, the mutual scissoring of sentence and line is an explicit feature of the prosody of f. v. that completely ignores the norm of meter. F. v. depends on a tension between grammar and line-length, stanza length. The cognitive act of sentencing is constantly counterpointing the physical-acoustic and visual features of the poem. The counterpointing patterns of meter, verbal rhythm, lineation, and rhyme exist in metrical verse too, of course, but there they are less determining. In avant-garde f. v., counterpoint becomes the primary principle because this type, esp., relies on the centrality of the line as the meeting ground of grammatical and prosodic processes.

When the line-breaks occur within phrases, cutting off function-words from lexical words, poetic speech may seem discontinuous. While it remains the choice of the performer whether the notated break is read aloud as such, intraphrasal line-breaking, in the most fragmented kind of f. v., de-automatizes our reading—words and groups of words, even lowly prepositions, tend to stand out more as material entities. However, in long-line verse, low-level breaks in the syntax tend to be subsumed in the process.

All poetry restructures direct experience by means of devices of equivalence; all poetry has attributes of a naturalizing and an artificializing rhetoric. However, more explicitly than the metrical poetry of the period from Chaucer to Tennyson, from Pushkin to Tsvetaeva, f. v. claims and thematizes a proximity to lived experience. It does this by trying to replicate, project, or represent perceptual, cognitive, emotional, and imaginative processes. Lived experience and replicated process are unreachable goals, but nevertheless this ethos is what continues to draw writers and readers to f. v. For discussion of other modern poetic forms see LYRIC SEQUENCE; MODERN LONG POEM; PROSE POEM; VERS LIBRE.

T. S. Eliot, "Reflections on *Vers Libre*," *New Statesman* 8 (1917); E. Pound, "A Retrospect," *Pavannes and Divisions* (1918); J. L. Lowes, *Convention and Revolt in Poetry* (1919); H. Monroe, "The F. V. Movement in America," *Eng. Jour.* 3 (1924); B. Hrushovsky, "On Free Rhythms in Mod. Poetry," in Sebeok; R. Duncan, "Ideas of the Meaning of Form," *Kulchur* 4 (1961); J. McNaughton, "Ezra Pound's Metres and Rhythms," *PMLA* 78 (1963); R. Kell, "Note on Versification," *BJA* 3 (1963); O. Brik, "Ritm i sintaksis," tr. in *Two Essays on Poetic Lang.* (1964); H. Gross, *Sound and Form in Mod. Poetry* (1964); D. Levertov, "Notes on Organic Form," *Poetry* 106 (1965); C. Olson, "Projective Verse," *Sel. Writings* (1966); P. Ramsey, "F. V.: Some Steps Toward Definition," *SP* 65 (1968); *Naked Poetry: Recent Am. Poetry in Open Forms*, ed. S. Berg and R. Mezey (1969)—with authors' statements; L. Ern, *Freivers und Metrik* (1970); R. Mitchell, "Toward a System of Grammatical Scansion," *Lang&S* 3 (1970); E. Fussell, *Lucifer in*

Harness (1973); J. Kwan-Terry, "The Prosodic Theories of Ezra Pound," *PLL* 9 (1973); J. Roubaud, *La Vieillesse d'Alexandre* (1978); H.-J. Frey, "Verszerfall," *Kritik des freien Verses*, ed. H.-J. Frey and C. Lorenz (1980); C. O. Hartman, *F. V.* (1980)—unreliable; Scott, ch. 7; M. Perloff, *The Poetics of Indeterminacy* (1981), *The Dance of the Intellect* (1985); Brogan; E. Berry, "Williams' Devel. of a New Prosodic Form," *WCWR* 7 (1981), "Syntactical and Metrical Structures in the Poetry of William Carlos Williams" *DAI* 42 (1982): 4449A; H. Meschonnic, *Critique du rhythme* (1982); C. Miller, "The Iambic Pentameter Norm of Whitman's F. V.," *Lang&S* 15 (1982); "F. V.," Spec. Iss. of *OhR* 28 (1982); "Symposium on Postmodern Form," *NER/BLQ* 6 (1983); R. Hass, *20th-C. Pleasures* (1984); D. Justice, "The F. V. Line in Stevens," *Antaeus* 53 (1984); O. Ovčarenko, *Ruskij Svobodny Stix* (1984); R. Cureton, "Rhythm: A Multi-level Analysis," *Style* 17 (1985); J. Hollander, "Observations on the Experimental," in Hollander; D. Wesling, *The New Poetries* (1985), ch. 5; S. Cushman, *W. C. Williams and the Meanings of Measure* (1985); E. Bollobás, *Trad. and Innovation in Am. F. V., Whitman to Duncan* (1986); C. Scott, *A Question of Syllables* (1986), ch. 6; T. Steele, *Missing Measures* (1990).
D.W.; E.BO.

G

GENRE. The term "g." is often used interchangeably with "type," "kind," and "form." Western theory on the subject of whether works of lit. can be classified into distinct kinds appears at the beginning of literary study and has sustained active controversies in every stage of lit. hist. Alternately extolled and condemned, praised for its potential for order and ignored as, finally, irrelevant, the concept takes its tone, in every age, from the particular theory that surrounds it. Theorists approached it prescriptively until about the end of the 18th c., descriptively thereafter; and it retains its viability (if not always its honor) through the plethora of modern comments about the nature of possibilities of lit. But built into its ways of working are difficulties that have ultimately to do with a version of the hermeneutic circle: how can we choose specific works from which to draw a definition of, say, epic unless we already know what an epic is? Though answered in various ways, the question continues to insinuate itself.

Classical poetics had no systematic theory about the concept of g. What thinking there is at the beginning of Western poetics originates from a distinction made by Plato between two possible modes of reproducing an object or person: (1) by description (i.e. by portraying it by means of words) or (2) by mimicry or impersonation (i.e. by imitating it). Since poetry according to the mimetic theory (see IMITATION; REPRESENTATION AND MIMESIS) was conceived as such a reproduction of external objects, these two modes became the main divisions of poetry: dramatic poetry or the theatre was direct imitation or miming of persons, and narrative poetry or the epic was the portrayal or description of human actions.

But since this simple division obviously left out too much, a third division was inserted between the two others (*Republic* 3.392–94): the so-called mixed mode, in which narrative alternates with dialogue, as is usually the case in epic poetry (which is rarely pure narrative). But no new principle of classification was thereby introduced, so no room was left for the genre of self-expression or the lyric, in which the poet expresses directly her or his own thoughts and feelings. The extensive use of Homer as a model gave clear if implicit preference to the epic, a point echoed in *Laws*, which comments effectually mark the beginning of the hierarchy of gs. The classification is as much moral as it is literary; Plato says subsequently that the guardians should imitate only the most suitable characters (395) but that there are impersonators who will imitate anything (397).

Prior to Plato, during the Attic age, we find a wide variety of terms for specific gs.: the epic or recited poetry; the drama or acted poetry, subdivided into tragedy and comedy; then iambic or satirical poetry, so called because written in iambic meter; and elegiac poetry, also written in a distinctive meter, the elegiac couplet, with its offshoots the epitaph and the epigram (qq.v.), all classed together because composed in the same meter. Then there was choral or melic poetry, as it was later called, poetry sung by a chorus to the accompaniment of a flute or stringed instrument. Melic poetry comes closest to our concept of the lyric, but it is not divorced from music and it excludes what we consider the essentially lyric gs. of the elegy and epigram. In addition, there was the hymn (q.v.), the dirge or *threnos*, and the dithyramb, a composition in honor of Dionysus which could be anything from a hymn to a miniature play. Songs of triumph or of celebration included the paean, the encomium, the epinikion, and the epithalamium. There was certainly plenty of material in Gr. poetry to make up a concept of lyric poetry, but the early Greeks apparently contented themselves with classifying by such criteria as metrical form.

The purely extrinsic scheme used for the nonce by Plato is taken up by Aristotle in *Poetics* ch. 3, where it becomes the foundation of his main classification of poetic gs. Aristotle gives no express recognition of the lyric there, much less in his statement that in the second of these gs. the poet "speaks in his own person": that is merely Aristotle's way of saying that the narrative is the poet's own discourse and not a speech by a fictitious character of drama. So the traditional tripartite division of poetic gs. or kinds into epic, dramatic, and lyric, far from being a "natural" division first discovered by the Gr. genius, is not to be found in either Plato or Aristotle. It was, rather, the result of a long and tedious process of compilation and adjustment, through the repetition with slight variations of certain traditional lists of poetic gs., which did not reach the modern formula of the three divisions until the 16th c.

Nevertheless, Aristotle's classifications of kind in the *Poetics* make him the source and arbiter of g. study (though often at only second or even third hand, and frequently warped) for nearly two millennia. Like Plato, Aristotle argues that poetry is a species of imitation. The medium of imitation concerns the instrumentality through which the various kinds are presented. The object of imitation, men in action, has both contentual and moral aspects, tragedy and comedy dealing with men as better or worse than they are; but the package is not nearly so neat because Cleophon, though a tragic poet, represents a middle way, men as they are—a significant point which shows how Aristotle's examples can complicate the issue appropriately. On the manner of imitation, Aristotle continues the general Platonic divisions according to the status of the speaker.

All of this supports the view that Aristotle is arguably the first formalist, the first exponent of organic unity; for him, mode, object, and manner, working together, not only make for the "character" of the kind but affect (and effect) all that each aspect does and is. Yet he is formalist and a good deal more, for his connection of g. with tone and moral stance led not only to later quarrels about decorum (q.v.) and mixed modes but also, and more profitably, about the ways in which texts seek to conceive and appropriate the world—that is, the difficult business of representation, incl. his implicit debate with Plato over its possibilities and value.

After Aristotle, it was Alexandrian scholarship that undertook the first comprehensive stock-taking of Gr. poetry and began the process of grouping, grading, and classifying gs. Lists or "canons" of the best writers in each kind were made, which led to a sharper awareness of g. The first extant grammarian to mention the lyric as a g. was Dionysius Thrax (2d c. B.C.), in a list which comprises, in all, the following: "Tragedy, Comedy, Elegy, Epos, Lyric, and Threnos," lyric meaning for him, still, poetry sung to the lyre. In Alexandrian lit., other gs. were added to the list, esp. the idyll and pastoral (qq.v.).

The Gr. conceptions of g. were themselves radically generic in the sense of putting the issues in their elemental forms. What followed—adulation, elaboration, correction, rejection—built on those ways of working. Yet it was clear to later Classicism that these treatises needed supplementary detail, their nearly exclusive emphases on epic and tragedy being insufficient to cover the complex topography of g. Further, with the model of the Greeks so potent that there was no thought of undoing their principles, it seemed best not only to elaborate but to clarify and purify, to establish principles of tact which were not only matters of taste (q.v.) but, ultimately, of the appropriate. The Middle Ages, and later, the 17th c., was a time for codification, which could slip easily into rules. Quintilian's *Institutio oratoria* argued for such practices, but most important was Horace's *Ars poetica* (a name given by Quintilian to the *Epistula ad Pisones*), a text of extraordinary influence because so many later students read the Greeks through Horace's letter. The attitudes of Horace were often taken as the classical ways.

Part of the irony is that his letter is not particularly original: its outstanding contribution is the principle of decorum Aristotle had referred to the interrelation of style with theme, but in Horace this combined with the demands of urbanity and propriety to become the principal emphasis. Tragedy does not babble light verses. Plays ought to be in five acts, no more and no less, with all bloodiness offstage. Plunge into epics *in medias res*, but echo the categories of the strong predecessors either by telling those events or having them acted out. At this distance, Horace comes through mainly as the exponent of a set of mind, one who surely had much to do with later equations of social and literary decorum. Given his emphasis on "the labor of the file," he is probably best seen as the ultimate craftsman, completer of the Cl. triumvirate on which g. study built for most of the rest of Western lit. hist.

Scholars are generally agreed that the Middle Ages offered little if any commentary of permanent value on the theory of g., and they usually cite Dante's remarks in *De vulgari eloquentia* (ca. 1305) as the major points of interest. In fact though, Dante's account shows a curious transformation of trad., esp. in his insistence that his poem is a comedy because it has a happy ending and is written in a middle style; this sense of "comedy" Dante found in Donatus, *De comoedia*, and Euanthius, *De fabula*. Dante argued for a quasi-Horatian decorum of g. and style. In a sense the *Commedia* culminates medieval mixtures of the grotesque and the sublime (q.v.), as in the mystery plays, but it also suggests, if unwittingly, an undoing of generic norms that was to cause much bitter controversy in subsequent approaches.

As though to counter such implicit subversion, the theorists of the It. Ren. focused intensely on

g.; the rediscovery of the *Poetics* around 1500 became an impetus to codifications such as had never been conceived even in the most rigorous late Cl. formulations. Part of the intensity came from the wide variety of gs. and mixed modes such as the *prosimetrum* practiced in the Middle Ages, leading to the blend of medieval romance and epic in figures such as Ariosto and Tasso. If there were 16th-c. defenders of these "mixed" works—among which tragicomedy was surely the most notorious (Guarini argued in his own defense, but the greatest of the kind were written in England)—there were codifiers such as Scaliger and Castelvetro who had considerable influence well into the 18th c.

Out of these theorists came that ultimate codification, the "unities" of time, place, and action, which was finally put to rest only in the 18th c. by critics such as Samuel Johnson. Though they were claimed to have their sources in Aristotle's categories, in fact these arguments distorted Aristotle and carried Horatian conservatism to reactionary lengths. Fr. Neoclassicism continued the codification, quite brilliantly in Boileau's *Art poétique*, more ambivalently in Corneille's *Discours*, the latter an apologia for his dramatic practice which is, at the same time, an act of support for the unities. Suggestions that Neoclassical generic hierarchies and standards of decorum have sociopolitical and philosophical implications are, for the most part, convincing: the potential analogies among these favorite subjects ensured their mutual support and offer still another instance of the relations of lit. and power. Yet as both lit. and society worked their way into romanticism, most of those hierarchies shifted: in lit. the lyric ascended to the top of the hierarchy, signaling the confirmation of the triad of lyric, epic (i.e. narrative), and drama which is set forth by Hegel and still dominates g. theory. Friedrich Schlegel (in his *Dialogue on Poetry* [1800] and essay on Goethe [1828]) argued for the abolition of generic classification, which would in effect eliminate g. Schlegel and others had in mind the example of Cervantes, expanding the concept of the novel to speak of it as a package that could carry all other genres within itself, e.g. ballads and romances within tales, as in Matthew Gregory Lewis's *The Nonk* and, memorably, Poe's *Fall of the House of Usher*. International romanticism explored such issues routinely. But when 19th-c. Darwinian biology found application to lit., it produced a rigidly evolutionary theory of g. in Brunetière and others, a dead-end whose main value was that it annoyed theorists like Croce, who considered gs. mere abstractions, useful in the construction of classifications for practical convenience, but of no value as aesthetic categories. Thereby it stimulated interest in g. theory in the 20th c., one of the great ages of speculation on the subject.

Croce became the case against which theorists tested themselves for much of the early 20th c. If g. classifications have a certain convenience, they nevertheless conflict with Croce's conception of the individual work of art as the product of a unique intuition. G., in this view, has a merely nominalist existence, a position echoed in varying ways by later theorists as significant and different as Jameson and, in one of his moods, the unclassifiable Frye—though the latter set up an elaborate system of classification which all commentators have taken as another way of talking about g. Todorov's structuralist attack on Frye resulted in a controversial proposal concerning historical and theoretical gs., but Frye remains the most important theorist of the subject since Croce. Scholars like Fowler have argued eloquently for looser, more historically based readings, recognizing the fact of change and the necessity for flexibility, while concepts like intertextuality (q.v.) obviously have much of importance to say about the workings of g. Formalists of various persuasions have worried about g. in terms of form-content relations. Drawing on the work of Karl Viëtor, Claudio Guillén distinguishes persuasively between universal modes of experience (lyric, epic, drama) and genres proper (tragedy or the sonnet). Other recent theorists argue for the institutional nature of g., for its function as a series of codes, and (less convincingly) as an element in a *lanque-parole* relationship, while Fowler and others, working out of Frye, stress the significance of the concept of "mode." Still, Jameson's argument that g. theory has been discredited by modern thinking about lit. seems now largely convincing. Recognition of the embodiment of lit. in the necessarily shifting conditions of culture has led a number of theorists to argue that a g. is whatever a particular text or time claims it to be. Skepticism about universals has clearly taken its toll, as have, in other ways, the arguments of Croce. Such skepticism has appeared among contemporary artists as well, e.g. the performance artist Laurie Anderson and the composer-writer-performer John Cage, who pull down all walls of distinction among gs. and media as well as what has been called "high art" and "low art." (Here as elsewhere sociocultural elements cannot be separated from other facets of the work.) Terms like "multimedia" and "intermedia" can be complemented by others such as "intergeneric," such practices denying, in varying degree, the validity of absolute distinctions, categories, and hierarchies. Theorizing about g. has not been so vigorous since the 16th c. The suggestiveness of the 20th c.'s quite variegated work makes it a period of extraordinary achievement in the history of this stubborn, dubious, always controversial concept. See now VERSE AND PROSE.

JOURNAL: *Genre* 1– (1968–).

STUDIES: F. Brunetière, *L'Evolution des gs.*, 7th ed. (1922); B. Croce, *Aesthetic*, tr. D. Ainslie, 2d ed. (1922); J. Petersen, "Zur Lehre von den Dichtungsgattungen," *Festschrift Aug. Sauer* (1925); K. Viëtor, "Probleme der literarische Gattungsgesch.," *DVLG* 9 (1931), "Die Gesch. der literarischen Gattungen," *Geist und Form* (1952); R. Bray,

Des Gs. littéraires (1937); K. Burke, "Poetic Categories," *Attitudes Toward History* (1937); I. Behrens, *Die Lehre von der Einteilung der Dichtkunst* (1940)—best account of devel. of g. classif. in Western lit.; J. J. Donohue, *The Theory of Literary Kinds*, 2 v. (1943–49)—ancient Gr. g. classif.; I. Ehrenpreis, *The "Types Approach" to Lit.* (1945); C. Vincent, *Théorie des gs. littéraires*, 21st ed. (1951); Abrams, chs. 1, 4, 6; E. Olson, "An Outline of Poetic Theory," in Crane; A. E. Harvey, "The Classif. of Gr. Lyric Poetry," *ClassQ* n.s. 5 (1955); Wellek and Warren, ch. 17; Frye; Wimsatt and Brooks; Weinberg, ch. 13; C. F. P. Stutterheim, "Prolegomena to a Theory of Literary Gs.," *ZRL* 6 (1964); B. K. Lewalski, *Milton's Brief Epic* (1966), Paradise Lost *and the Rhet. of Literary Forms* (1985), ed., *Ren. Gs.* (1986); F. Séngle, *Die literarische Formenlehre* (1967); W. V. Ruttkowski, *Die literarischen Gattungen* (1968)—bibl. with trilingual indices, *Bibliographie der Gattungspoetik* (1973); E. Staiger, *Grundbegriffe der Poetik* (1968), tr. J. C. Hudson and L. T. Frank as *Basic Concepts of Poetics* (1991); K. R. Scherpe, *Gattungspoetik im 18 Jh.* (1968); E. Vivas, "Literary Classes: Some Problems," *Genre* 1 (1968); H.-R. Jauss, "Litt. médiévale et théorie des gs.," *Poétique* 1 (1970); T. Todorov, *Intro. à la lit. fantastique* (1970), *Gs. in Discourse* (tr. 1990); M. Fubini, *Entstehung und Gesch. der literarischen Gattungen* (1971); C. Guillén, *Lit. as System* (1971), chs. 4–5; P. Hernadi, *Beyond G.* (1972); F. Cairns, *Generic Composition in Gr. and Roman Poetry* (1972); R. L. Colie, *The Resources of Kind: G. Theory in the Ren.* (1973); K. W. Hempfer, *Gattungstheorie* (1973); R. Cohen, "On the Interrelations of 18th-C. Literary Forms," and R. W. Rader, "The Concept of G. and 18th-C. Studies," *New Approaches to 18th-C. Lit.*, ed. P. Harth (1974); A. Jolles, *Einfache Formen*, 5th ed. (1974); G. W. F. Hegel, *Aesthetics*, tr. T. M. Knox (1975); G. Genette, "Gs., 'types,' modes," *Poétique* 32 (1977); K. Müller-Dyes, *Literarische Gattungen* (1978); "Theories of Literary G.," ed. J. Strelka, spec. iss. of *YCC* 8 (1978); Spec. Iss. on G., *Glyph* 7 (1980); Spec. Iss. on G. Theory, *Poetics* 10, 2–3 (1981); F. Jameson, *The Political Unconscious* (1981); Fowler—the major mod. study; H. Dubrow, *G.* (1982); W. E. Rogers, *The Three Gs. and the Interp. of Lyric* (1983); B. J. Bond, *Literary Transvaluation from Vergilian Epic to Shakespearean Tragicomedy* (1984); *Canons*, ed. R. von Hallberg (1984); *Discourse and Lit.: New Approaches to the Analysis of Literary Gs.*, ed. T. A. Van Dijk (1984); T. G. Rosenmeyer, "Ancient Literary Gs.: A Mirage?" *YCGL* 34 (1985); *Postmodern Gs.*, ed. M. Perloff (1989).

F.G.; G.N.G.O.; T.V.F.B.

GEORGIC. A didactic poem primarily intended to give directions concerning some skill, art, or science. In his "Essay on the G." (1697), which is the most important modern discussion of the genre, Addison specifically distinguished this kind of poetry from the pastoral (q.v.) and crystallized the definition of the g. by pointing out that this "class of Poetry . . . consists in giving plain and direct instructions." The central theme of the g. is the glorification of labor and praise of simple country life. Though this didactic intention is primary, the g. is often filled with descriptions of nature and digressions on myths, lore, or philosophical reflections suggested by the subject matter.

The g. begins as early as Hesiod's *Works and Days* (ca. 750 B.C.) and was used by many of the great ancients—Lucretius, Ovid, Oppian, Nemesianus, Columella. Some of the better known poems in the trad. are Poliziano's *Rusticus* (1483), Vida's *De bombyce* (1527), Alamanni's *La coltivazione* (1546), Tusser's *Five Hundred Points of Good Husbandry* (1573), Rapin's *Horti* (1665), and Jammes' *Géorgiques chrétiennes* (1912). Nevertheless, the finest specimens of the type remain the *Gs.* of Virgil. Virgil's purpose is to celebrate the virtue and dignity of honest toil by the small landowner. The *Gs.* thus differ from pastoral in that work is emphasized instead of leisure, and from epic in planting over killing. It is these virtues—hard work, cultivation—that made the poems appeal to Augustus, newly consolidating the Roman state as empire.

The *Gs.* cast a long shadow over the poetry of the late 17th and 18th cs.: Dryden called the *Gs.* "the best poem of the best poet"; and James Thomson was called the "Eng. Virgil" on account of his hugely successful *The Seasons* (1726; Fr. tr. 1759). Thomson's far-reaching influence was felt even on the Continent. Even where the term "g." does not appear in their title, scores of 18th-c. poems were written which imitate Virgil's gs. in form and content—poems on the art of hunting, fishing, dancing, laughing, preserving health, raising hops, shearing sheep, etc. But the neoclassical g. manifested considerable changes from its Cl. models. It emphasized heterogeneous information rather than detailed instructions; it exalted landscape over work processes; and it invited the reader to feast on the native beauties of rural life, not to wring a living from the recalcitrant soil. And its attempts to elevate a lowly subject by exaggerated periphrases led many imitators to grotesqueries of style. Indeed, at times the serious imitation, as in William Cowper's *The Task*, can hardly be separated from the burlesque, as in Gay's *Trivia; or the Art of Walking the Streets of London* (1716).

M. L. Lilly, *The G.* (1919); D. L. Durling, *G. Trad. in Eng. Poetry* (1935); R. Cohen, *The Art of Discrimination* (1964), *The Unfolding of* The Seasons (1970); D. B. Wilson, *Descriptive Poetry in France from Blason to Baroque* (1967); J. Chalker, *The Eng. G.* (1969); L. P. Wilkinson, *The Gs. of Virgil: A Critical Survey* (1969); R. Feingold, *Nature and Society* (1978); J. G. Turner, *Politics of Landscape: Rural Scenery and Society in Eng. Verse 1630–1660* (1979); M. C. J. Putnam, *Virgil's Poem of the Earth* (1979); *Virgil's Ascraean Song*, ed. A. J. Doyle (1979); G. B. Miles, *Virgil's Gs.: A New Interp.* (1980); A. Low, *The G. Revolution* (1985); A. Fowler, "The Beginnings of Eng. G.," *Ren. Genres*, ed. B. K. Lewalski (1986); D. O. Ross, Jr., *Virgil's*

Elements (1987); R. A. B. Mynors, *Virgil's Gs.: A Commentary* (1989); C. Perkell, *The Poet's Truth* (1989); *Handspan of Red Earth*, ed. C. L. Marconi (1991)—anthol. of Am. farm poems.

J.E.C.; T.V.F.B.

GHAZAL. A monorhymed lyric poem, shorter than a *qaṣīda*, common to Arabic, Persian, Turkish, Uzbek, Pashto, and Urdu lit. The rhyme scheme is *aa ba ca*, etc., and, after the 12th c., the poet mentions his name toward the end of the poem. Cl. gs. share a limited range of images and a fairly restricted vocabulary. The principal subject of the g. is earthly or mystical love, and the mood is melancholy, expressing sadness over separation from the beloved. The g. took its canonical form in Persian in the 11th–12th cs. The most famous of all g. poets were the Persians Saʿdi (d. 1292), Ḥāfeẓ (d.1389–90), and Ṣāʾeb (d. 1677–78); and the genre developed in other lits. under strong Persian influence. Attempts to modernize the Urdu g. have met with some success and it continues to flourish in that lit. but has been mostly superseded by modern forms in the other lits. Tr. of Ḥāfeẓ inspired the Ger. romantic poets, esp. Goethe, whose *West-östlicher Divan* imitated Persian models. Platen wrote a large number of gs. Recently there has been something of a revival of interest in translating gs., with significant efforts being made by Am. poets such as Adrienne Rich and James Harrison.—E. J. W. Gibb, *Hist. of Ottoman Poetry*, v. 1 (1900); D. Balke, *West-östliche Gedichtformen: Sadschal-Theorie und Gesch. des deutsches Ghasels*, Diss., Bonn (1952); A. Bausani, *Storia delle letterature del Pakistan* (1958); A. Pagliaro and A. Bausani, *Storia della letteratura persiana* (1960); *Encyclopaedia of Islam*, 2d ed., s.v. "G."; *Gs. of Ghalib*, ed. A. Ahmad (1971); *An Anthol. of Cl. Urdu Love Lyrics*, ed. D. J. Matthews and C. Shackle (1972); A. Schimmel, "The Emergence of the Ger. G.," *Studies in the Urdu Gazal and Prose Fiction*, ed. M. Memon (1979); W. Andrews, *Poetry's Voice, Society's Song* (1985).

W.L.H.

H

HAIKU (also called *hokku* and *haikai*). This Japanese poetic form was originally the opening section (hence *hokku*, "opening part") of *renga* (q.v.), which took shape in the 13th and 14th cs. as a sequential verseform that alternates up to fifty times 5-7-5- and 7-7-syllable parts composed in turn by two or more poets. The name *haiku* derives from the variety of *renga* known as *haikai*, "humorous." *Haikai* poets, most notably Matsuo Bashō (1644–94), rejected the poetic diction and lyricism of court poetry that prevailed in orthodox *renga* and found "humor" in describing the mundane. But they retained some basic features of orthodox *renga*, among them the inclusion of a *kigo*, a word or phrase that specifies the season of the composition.

Even though the *hokku* in units of 5-7-5 syllables was being written more or less independently by the 16th c., it was not completely severed from *renga* until the end of the 19th c. As it became an independent form, largely at the instigation of the reformist Masaoka Shiki (1867–1902), the name *hokku* was replaced by *haiku*. Some poets, such as Ogiwara Seisensui (1884–1976) and Ozaki Hōsai (1885–1926), went further and rejected the two basic requirements of the *hokku*, the syllabic pattern of 5-7-5 and the inclusion of a *kigo*. Some also departed from the one-line printing format, which was adopted as standard in the 19th c.—a practice which instilled in *haiku* writers a strong sense of the h. as a one-line poem. Such departures from the norm have not converted the majority, however.

Around 1900, h. began to attract the attention of Western poets, the Fr. being among the first to try to take up the form, calling it *haikai*. In the following decades, mainly under the influence of Fr. poets, a number of poets in the United States, in Latin America, and in other regions tried the form, writing their pieces mainly in three lines of 5-7-5 syllables. But they did not go much beyond experimentation. Imagism, a literary movement linked with h., produced Ezra Pound's technique of "superposition" and otherwise had a considerable influence on the poetry of the time, and esp. on reducing discursiveness in Western poetry.

Outside Japan, interest in writing h. gained solid ground after World War II, esp. among Am. poets, as a result of a sudden increase in the number and quality of trs. of Japanese lit. and studies of Japanese culture. The works of three scholars have been seminal: Blyth, Yasuda, and Henderson. Of the three, Blyth's volumes displayed an intimate knowledge of the subject in relating h. to Zen Buddhism, thereby awakening interest in this form among poets, such as those of the Beat Generation, who were pursuing Eastern philosophies and religions during the '50s. Taking a contrasting secular approach, Henderson greatly helped to make the practical aspects of h.-writing understood during the '60s through his roles as adviser to the poets and as guiding hand in the formation of the H. Society of America in 1968. In 1963 *Am. H.*, the first magazine devoted to h. in Eng., began publication, followed by at

least ten h. magazines over the next two decades. In 1974 *The H. Anthol.*, edited by Cor van den Heuvel, was published; a collection of h. by contemp. Am. and Canadian poets, it was the first notable anthol. in a non-Japanese lang. and was greatly enlarged in its 2d ed. (1986).

H. is now written in a great many langs., but receptivity to new ideas and experimentation seems strongest in North America. One indication of this is the writing of one-line h. that began in the mid-'70s largely in recognition of the one-line format employed by most Japanese h. writers and as a result of trs. of such Japanese h. in one line. This and other devels. make a definition of h. impractical, although at least the three-line format, if not the 5-7-5 syllable pattern, seems still to be regarded as the norm in many countries outside Japan.

E. Miner, *The Japanese Trad. in British and Am. Lit.* (1958); R. H. Blyth, *H.*, 4 v. (1949–52), *A Hist. of H.*, 2 v. (1963–64); K. Yasuda, *The Japanese H.* (1957); *Haikai Dai-jiten* [The Great Dict. of Haikai], ed. Ijichi Tetsuo et al. (1957); H. G. Henderson, *An Intro. to H.* (1958), *H. in Eng.* (1967); G. L. Brower, *H. in Western Langs.: An Annot. Bibl.* (1972); M. Ueda, *Mod. Japanese H.* (Tokyo, 1976), *Basho and His Interpreters* (1991); R. Étiemble, "Sur une bibl. du h. dans les langues européennes," *CLS* 11 (1974); *Gendai H. Dai-jiten* [The Great Dict. of Mod. H.], ed. Azumi Atsushi et al. (1977); R. Figgins, "A Basic Bibl. of H. in Eng.," *Bull. of Bibl.* 36 (1979); *From the Country of Eight Islands*, ed. and tr. H. Sato and B. Watson (1981); H. Sato, *One Hundred Frogs* (1983), *Eigo H.* [H. in Eng.] (1987); *H. Rev. '84*, ed. R. and S. Brooks (1984); J. T. Rimer and R. E. Morrell, *Guide to Japanese Poetry*, 2d ed. (1984); S. Kodoma, *Am. Poetry and Japanese Culture* (1984); S. Sommerkamp, "Der Einfluss des H. auf Imagismus und Jüngere Moderne," Diss., Univ. of Hamburg (1984); *Anthologie Canadienne: H.*, ed. D. Howard and A. Duhaime (1985); W. J. Higginson, *The H. Handbook* (1985).

H.S.

HEROIC COUPLET. A rhyming pair of Eng. heroic (that is, iambic pentameter) lines, most often used for epigrams, verse essays, satires, and narrative verse, and the dominant form for Eng. poetry from ca. 1640 to ca. 1790. The English created the h. c. in the 16th c. by imposing a regular iambic stress pattern and a regular caesura (falling normally after the fourth, fifth, or sixth syllable) on the old Chaucerian decasyllabic line and by imposing on the Chaucerian couplet (called "riding rhyme" by Puttenham and Gascoigne) a regular hierarchy of pauses—respectively, caesural, first-line, and end-of-couplet—adapted from the Lat. elegiac distich. This form, in which the end of the couplet regularly coincides with the end of a sentence, would later be called "closed," as opposed to the "open" type, in which there is no such coincidence of line and syntax.

The transformation of the Chaucerian line into what Puttenham and Dryden called the "heroic" line was evolutionary, spanning the 16th c. from ca. 1520 to 1600 (Thompson). But the imposition of the hierarchy of pauses on the loose couplet inherited from Chaucer was more dramatic, taking place almost explosively between 1590 and 1600, when numbers of Eng. poets translated and adapted Lat. poems in elegiac distichs—chiefly the *Amores* and *Heroides* of Ovid and the *Epigrammaton* of Martial—achieving or approximating a correspondence of couplet to distich. Marlowe's tr. of the *Amores* accomplishes a virtually unexceptionable equivalence:

> sive es docta, places raras dotata per
> artes;
> sive rudis, placita es simplicitate tua.
> est, quae Callimachi prae nostris rustica dicat
> carmina—cui placeo, protinus ipsa
> placet.
> est etiam, quae me vatem et mea
> carmina culpet—
> culpantis cupiam sustinuisse femur.
> molliter incedit—motu capit; altera
> dur est—
> at poterit tacto mollior esse viro.

> If she be learn'd, then for her skill I
> crave her,
> If not, because shees simple I would
> have her.
> Before *Callimachus* one preferres me
> farre,
> Seeing she likes my bookes why should
> we jarre?
> An other railes at me and that I write
> Yet would I lie with her if that I might.
> Trips she, it likes me well, plods she,
> what then?
> Shee would be nimbler, lying with a
> man.

Other poets significant for the devel. of the form were Grimald (who preceded the wave), Harrington, Marston, Heywood, Hall, Drayton, Jonson, and Donne.

The immediate result of this structural imitation was the affixing onto Eng. couplets of the Lat. distich's hierarchy of pauses and of a correspondingly balanced rhet. Note the balances above between the sharply demarcated lines of single couplets—"If . . . / If not . . . "—and the sharply divided halves of single lines—"If . . . then"; "Trips . . . plods." Eng. rhyme, which had the same closural power as the recurrent half-line pattern in the pentameter of the Lat. distich, allowed the Eng. measure, once it outgrew its dependence on the Lat. model, to become much more flexible and thus absorb the modifications by which successive Eng. poets from Donne to Crabbe transformed it into a medium of great expressive power.

At the same time, the h. c. also endured a second major devel., beginning with Marlowe's *Hero and Leander* (1593) and extending until about 1660, in which the formal elements, the rhyme and all the pauses, were subsumed and all but concealed. This process, which Marlowe himself practiced cautiously, was carried further by Chapman in his tr. of the *Odyssey* (1614) and in Chamberlayne's long romance, *Pharonnida* (1659). This "romance" or "open" couplet pointedly rejected the emphasis and definition established by the closed couplet; here rhyme and syntax do not parallel and reinforce each other, and the formal elements work against sense, creating mystery and remoteness instead of clarity and precision. But the romance couplet vanished from Eng. poetry before the Restoration, to be revived—briefly, and appropriately—by such romantic poets as Keats (*Endymion*), Hunt (*The Story of Rimini*), and Moore (*Lalla Rookh*).

Meanwhile, the closed couplet was refined and extended in the 17th c. by such poets as Beaumont, Sandys, and Falkland, then Waller, Denham, and Cowley, and thus handed on to Dryden (practicing 1660–1700) and Pope (practicing ca. 1700–1745). The h. c. as all these poets understood it is both described and exemplified in the famous lines from *Cooper's Hill* (1642) that Denham addressed to the Thames:

> O could I flow like thee, and make thy
> stream
> My great example, as it is my theme!
> Though deep, yet clear, though gentle,
> yet not dull,
> Strong without rage, without ore-flow-
> ing full.

As this widely known and widely imitated little passage illustrates, the closed couplet provides continuous support in its half-line, line, and couplet spans for certain rhetorical devices and corresponding intellectual procedures, particularly *parallelism* (q.v.; "deep . . . gentle . . . / Strong . . . full"), with which poets enforced systems of induction, and *antithesis* ("gentle . . . not dull"), with which they determined comparisons and refined analyses. They often intensified such patterns with the elliptical figures zeugma and syllepsis ("O could I flow like thee, and [could I] make thy stream") and with inversion ("Strong without rage, without ore-flowing full"). The argumentative and persuasive implications of such practices were esp. pertinent to an age empiricist in intellectual tendency and social in cultural orientation.

Dryden chiefly developed the oratorical possibilities of the h. c., addressing large, potentially turbulent groups: in theatrical prologues, men and women, or fops and gentlemen; in satires, Whigs and Tories, or Puritans and Anglicans. He projected himself not as an individual but as a spokesman, a representative of one faction or, if possible (as in the opening of *Astraea Redux*), of sensible Englishmen in general. In *Religio laici* he concludes his exposition of how Catholics and Protestants each in their own way abuse the Bible:

> So all we make of Heavens discover'd
> Will
> Is not to have it, or to use it ill.
> The Danger's much the same; on sev-
> eral Shelves
> If *others* wreck *us*, or *we* wreck our *selves*.

Notice, first, Dryden's use of the formal aspects of the closed couplet, the definitive use of caesura in the second and fourth lines, and the emphasis of the rhymes; and, second, the way he unifies the two sides, partially by using plural pronouns, himself taking the position of spokesman for both.

Pope developed the h. c. into a primarily social—as opposed to Dryden's political—instrument. He presented himself as a sensible gentleman addressing either an intimate friend—as in the opening lines of the Essay on Man—or an attentive society—as in the *Essay on Crit.*—or, as in his mature essays and epistles, both at once. In this conversational exchange with his friend Dr. Arbuthnot, the sensible (if sensibly exasperated) gentleman asks who has been hurt by his social satire:

> Does not one Table *Bavius* still admit?
> Still to one Bishop *Philips* seems a Wit?
> Still *Sapho*—'Hold! for God-sake—
> you'll offend:
> No Names—be calm—learn Prudence
> of a Friend:
> I too could write, and I am twice as tall,
>
> But Foes like these!—One Flatt'rer's
> worse than all;
> Of all mad Creatures, if the Learn'd
> are right,
> It is the Slaver kills, and not the Bite.

Notice, first, the caesura-defined parallel in line 5, and the formally illuminated antithesis, dividing *both* "Foes . . . [and] Flatt'rer" *and* the two conversationalists, in line 6; and, second, Pope's and his friend's awareness, although they are in intimate talk, that the walls have ears. Dryden in his most characteristic achievements aimed at political harmony or what he called "common quiet." Pope aimed, rather, at generally shared understanding or what he called "common sense."

The h. c., modified in various ways, supported the generation of great poets who followed Pope, esp. Johnson, Churchill, Goldsmith, and Crabbe. And even when Eng. culture shifted its focus from general understanding to individual expression—and quite properly revived the open couplet—several poets, chief among them Byron, still found the closed couplet useful for public satire. It continues in occasional use to this day. Browning ("My Last Duchess") and Arnold used the h. c. in the Victorian period; Yvor Winters has produced several

beautiful h. c. poems in our time; and Eliot composed an h. c. section for *The Waste Land*, which he wisely excised when Pound made him see that he could not compete with Pope.

Schipper; F. E. Schelling, "Ben Jonson and the Cl. School," *PMLA* 13 (1898); G. P. Shannon, "Nicholas Grimald's H. C. and the Lat. Elegiac Distich," *PMLA* 45 (1930); R. C. Wallerstein, "The Devel. of the Rhet. and Metre of the H. C.," *PMLA* 50 (1935); G. Williamson, "The Rhet. Pattern of Neo-classical Wit," *MP* 33 (1935–36); E. R. Wasserman, "The Return of the Enjambed Couplet," *ELH* 7 (1940); Y. Winters, "The H. C. and its Recent Revivals," *In Defense of Reason* (1943); W. C. Brown, *The Triumph of Form* (1948); W. K. Wimsatt, Jr., "One Relation of Rhyme to Reason," *The Verbal Icon* (1954); J. Thompson, *The Founding of Eng. Metre* (1961); J. H. Adler, *The Reach of Art* (1964); J. A. Jones, *Pope's Couplet Art* (1969); W. B. Piper, *The H. C.* (1969), ed., *An H. C. Anthol.* (1977); G. T. Amis, "The Structure of the Augustan Couplet," *Genre* 9 (1976); H. Carruth, "Three Notes on the Versewriting of Alexander Pope," *MQR* 15 (1976); Brogan, 389 ff. W.B.P.

HEXAMETER (Gr. "of six measures").

I. CLASSICAL
II. POSTCLASSICAL

I. CLASSICAL. The dactylic h. is the oldest Gr. verseform, the meter of Homeric and later epic poetry, Hesiodic and later didactic poems, Hellenistic epyllia, bucolic poetry, hymns, oracles, riddles, and verse inscriptions. It also forms the first verse of the elegiac distich and of the two "Pythiambic" *Epodes* of Horace. It was written in quantitative meter in Lat. through the Middle Ages, thereafter in accentual imitations and adaptations of quantitative meter in the vernaculars (see section II below). Linguistic evidence from the Homeric epics indicates its great antiquity; for a review of present thinking about its origin and prehistoric devel., see West 1973, Tichy 1981, and West 1982.

The Cl. Gr. and Lat. h. is a catalectic line which consists of six feet, five of which may be dactyls (– ⏑ ⏑) or spondees (– –), the sixth of which is always disyllabic—usually a spondee, but may show *brevis in longo* (– x).

Caesura: The caesura in the Cl. h. occurs at one of three places: after 2 1/2 feet (i.e. after the third longum in the line), after 2 3/4 feet (after the first breve in the third foot), or after 3 1/2 feet (after the fourth longum in the line): the first of these is known as "penthemimeral" (Gr. [after] "five half-feet") and the last as "hephthemimeral" (Gr. "seven half-feet"); the second has no name. The first two types (i.e. in the third foot) greatly outnumber the third (of which only 14 instances occur in the *Iliad* and 9 in the *Odyssey*) and, between these two, the second (which might be said to have a "feminine" ending) slightly predominates over the first (with "masculine" ending) in the ratio of 4:3 (West). Conversely, hs. divided into equal three-foot halves are extremely rare, a fact which *may* lend support to the view sometimes advanced that originally the h. was not made up of six feet—as would appear from the traditional method for scansion of it—but rather of two *hemiepes*—i.e. the sequence – ⏑ ⏑ – ⏑ ⏑ – (which is the first 7 syllables, up to the penthemimeral caesura) joined together in the middle.

Contraction of *brevia* (– for ⏑ ⏑) is most common in the first two feet and decreases from the fourth to the third (because of the preference for the feminine caesura) to the fifth, where it is quite rare. The h. is subject to a complex of bridges regulating word boundary after trochaic segments and contraction; these bridges become more stringent in the Hellenistic period and later. Word boundary after a dactylic fourth foot is frequent and esp. favored in pastoral poetry, whence its traditional name, "bucolic diaeresis"; it too becomes more frequent over time.

The Gr. h. was adapted to Lat. by Ennius (239–169 B.C.) and employed, with the significant addition of satire, over the same range of genres as the Gr. The major differences from the Gr. h. are motivated by the Lat. stress accent. Disyllabic and trisyllabic words predominate at the end of the verse, ensuring, by the rules of Lat. stress, coincidence of linguistic stress and metrical prominence in the fifth and sixth feet. To prevent a monotonous coincidence throughout the line, however, the masculine caesura after the *longum* of the third foot is strongly preferred over the feminine, and, in those lines with a feminine caesura, word end after the second *longum* increases in frequency.

Variant Forms: (1) *Versus spondaicus*, "spondaic verse," refers to a h. whose fifth foot contains a spondee rather than a dactyl, thus causing the verse to end in at least two spondees, as in *Aeneid* 2.68. Spondaic verses appear only occasionally in Homer but are common in his Alexandrian imitators (esp. in lines which end with quadrisyllabic words: Catullus 64.78–80 has three in succession); they are however rarer in Lat. poets, and virtually disappear after Ovid. (2) *Versus Pythius* was another name sometimes used for the h., e.g. by Aphthonius (3d c. A.D.), because it was the meter used in the Pythian (i.e. Delphic) oracles. Combinations of hs. and iambic dimeters or trimeters were called *pythiambics*. (3) *Meiurus* or *myurus* (Gr. "tapering," "mouse-tailed"; also called *teliambos*) refers to hs. in which the first syllable of the last foot is short instead of long. The classical example is: "Troes d' errhigesan, hopos idon aiolon ophin" (*Iliad* 12.208), translated by Terentianus Maurus as "attoniti Troes viso serpente pavitant," where in both cases the penultimate syllable in the line is naturally short. Mousetails were recognized by the ancients as a special verse.

Wilamowitz, pt. 2, ch. 10; Hardie, ch. 1; T. F. Higham, *Gr. Poetry and Life* (1936); E. G. O'Neill,

Jr., "The Localization of Metrical Word-Types in the Gr. H.," *YClS* 8 (1942); H. N. Porter, "The Early Gr. H.," *YClS* 12 (1951); C. G. Cooper, *Intro. to the Lat. H.* (1952); H. Drexler, *Hexameterstudien* (1953), *Einführung in die romische Metrik* (1967); E. Norden, "Die malerischen Mittel des Vergilischen Hs.," *Aeneis Buch VI*, ed. E. Norden, 4th ed. (1957); Maas, sects. 82–100; L. P. Wilkinson, *Golden Lat. Artistry* (1963); Koster; Crusius; G. E. Duckworth, *Vergil and Cl. H. Poetry* (1969); W. S. Allen, *Accent and Rhythm* (1973); M. L. West, "Indo-European Metre," *Glotta* 51 (1973), "Greek Poetry 2000–700 B. C.," *CQ* 23 (1973); G. Nagy, *Comparative Studies in Gr. and Indic Meter* (1974); B. Gentili, "Preistoria e formazione dell'esametro," *QUCC* 17 (1977); E. Liénard, *Répertoires prosodiques et métriques* (1978); *Lateinisches H.-Lexikon*, ed. O. Schumann, 6 v. (1979–83); Halporn et al.; E. Tichy, "Homerische *androtēta* und die Vorgesch. des daktylischen Hs.," *Glotta* 59 (1981); *H. Studies*, ed. R. Grotjahn (1981)—with annot. bibl. of statistical studies; Snell; West; W. G. Thalmann, *Conventions of Form and Thought in Early Gr. Epic Poetry* (1984); A. M. Devine and L. D. Stephens, *Lang. and Metre* (1984); M. W. Edwards, "Homer and Oral Trad.," *Oral Trad.* 1–2 (1986)—bibl.

A.M.D.; L.D.S.; T.V.F.B.; P.S.C.

II. POSTCLASSICAL. The quantitative Lat. h. continued to be written in late antiquity and during the Middle Ages, first by Christian writers such as Juvencus, Sedulius, Arator, and Avitus for epic, and later by clerics and secular poets for a variety of other kinds of poems as well (e.g. John of Salisbury's *Entheticus*; see Raby). But the loss of quantity in the Lat. lang. sometime around the 4th c. in favor of stress accent led to the rise of accentual, rhymed, medieval "rhythmical verse"; and though strict quantities continued to be written by scholastics well past the 12th c., for all practical purposes, quantity no longer mattered to popular poetry after the emergence of the hymn—written in iambic dimeters but based on stress—by Augustine, Ambrose, and Hilary in the 3d–4th cs. A particularly popular form of the h., esp. after the 9th c., is *leonine verse*—accentual hs. with internal rhyme.

When the Romance vernaculars emerged from vulgar Lat. as autonomous langs., poets sought to fashion meters equivalent to the h., but equivalent only insofar as the (quite different) phonological structure of each lang. permitted: thus emerged the *endecasillabo* (11 syllables) in It. and *alexandrine* (12) in Fr., and eventually the iambic pentameter (10) in Eng.—radically different in structure from, but meant as equivalents of, the Cl. h. In the Cl. langs., the h. might range from a theoretical minimum of 12 syllables (if all six feet were spondees) to a maximum of 18, though in practice, the most common form, five dactyls plus a spondee, has 17. But if the quantities are treated as temporal units, these 17 syllables comprise 24 "times" (one long equaling two shorts). Vernacular equivalents of the Cl. h. attempted to approximate these numbers: in Eng., the first analogue of the h. was the fourteener (14 syllables, 21 times), followed by the iambic pentameter (10 syllables, 15 times). The first blank verse (q.v.) in Eng. was Surrey's tr. of two books of Virgil's *Aeneid.* Most of the history of Western verse can be derived from the history of efforts to reproduce or find metrical equivalents for the Cl. h., meters which would rival its prestige and power as the great and sufficient meter for the greatest Western poetic form, the epic.

Early in the Ren., attempts were also made to return to the Cl. h., first written in Lat. then in the vernaculars, but meant as accentual imitations of Cl. quantitative verse, both in patterning of dactyls and spondees and also in claims for long and short syllables. These attempts first appear in It., in the *certame coronario* of 1441, where L. B. Alberti and Leonardo Dati introduced experimental *esametri italiani*, followed in the next century by Claudio Tolomei and other poets. In Fr., such verse first appears in the 16th c. and is associated with the movement known as *vers mesurés à l'antique*; similar efforts appear in Eng. (e.g. Harvey, Sidney, Stanyhurst, Webbe) and in Rus. (Maksim the Greek, *Maksimovshaja prosodija*, based on an artificial classification of long and short vowels).

In the 18th c., the quantitative experiments of Ger. poets—e.g. Klopstock's *Der Messias* (1748–73; cf. his *Vom deutschen H.*, 1779)), Voss's *Homer* and *Luise*, and Goethe's *Reineke Fuchs* and *Hermann und Dorothea* (1798)—inspired a new vogue which was widely influential both on the Continent and abroad for over a century. In Russia N. Gnedič's h. tr. of the *Iliad* at the end of the 18th c. was admired by Pushkin, who himself wrote much h. verse (Burgi). In England, in the 19th c., poets from Southey, Coleridge, Kingsley (*Andromeda*), Clough, Tennyson, Longfellow, Swinburne, and Bridges (*Ibant obscuri* [1916]) to a host of minor poets translated Homer or wrote other poems in several varieties of h. imitations 2(Omond). Even the first Eng. tr. of *Gilgamesh* (1927) was in hs.

Meyer—for Med. Lat.; K. Wölk, *Gesch. und Kritik des englischen Hs.* (1909); Saintsbury 3.394 ff.; R. Bridges, *Milton's Prosody* (1921); Omond—still the fullest discussion for Eng.; H. G. Atkins, *Hist. of Ger. Versif.* (1923); J. S. Molina, *Los hexámetros castellanos* (1935); G. L. Hendrickson, "Elizabethan Quantitative Hs.," *PQ* 28 (1949); Raby, *Christian* and *Secular*; R. Burgi, *A Hist. of the Rus. H.* (1954); W. Bennett, *Ger. Verse in Cl. Metres* (1963); D. Attridge, *Well-Weighed Syllables* (1974); Brogan—fullest bibl., supp. in *Verseform* (1988); N. Wright, "The Anglo-Lat. H.," Diss., Cambridge Univ. (1981); Z. Kopczyńska and L. Pszczołowska, "Heksametr polski," *Pamietnik Literacki* (1983)—for Polish; Navarro—Sp.; Scherr—Rus. T.V.F.B.

HYMN, an ancient Gr. liturgical and literary genre which assimilates Near Eastern and specifically

Heb. trad. in the Septuagint tr. (LXX) of the Old Testament, is introduced into Christian trad. in the Gr. New Testament, the Lat. Vulgate Bible, and the Gr. and Lat. Fathers, and has continued to flourish in Western culture in both liturgical and literary trads. down to the 20th c. Other trads. represented in Oriental and in early Am. cultures, as well as in those of ancient Akkad, Sumeria, and Egypt which influenced Heb. trad., are undeniably important but lie beyond the scope of this entry.

The Gr. noun *hymnos* refers to a song, poem, or speech which praises gods and sometimes heroes and abstractions. Hs. are in lyric measures, hexameters, elegaic couplets, and even prose in late antiquity; the longest critical discussion of the h. by the rhetorician Menander, *Peri epideiktikōn* (3d c. A.D.), makes no distinction between prose and poetry. Most of the Gr. lyric hs. (mainly fragments by Alcaeus, Simonides, Bacchylides, Timotheus, and Pindar) were probably sung in religious worship. Similarly, the hexameter *Orphic Hs.* with their markedly theurgic emphasis may have been designed for religious purposes, although such abstractions as Justice, Equity, Law, and Fortune as well as the Stars, Death, and Sleep are subjects of praise. Thucydides refers to one of the hexameter *Homeric Hs.* as a prelude or preface; these presumably were literary introductions to epic recitations. The famous philosophic h. by the Stoic Cleanthes (4th–3d c. B.C.) is probably literary rather than liturgical in intention, as are the *Hs.* by the Neoplatonic philosopher Proclus (5th c. A.D.), although there is a significant theurgic element in the latter. The six extant *Hs.* of the Gr. poet Callimachus, in hexameters and elegaic couplets, are certainly literary emulations of the *Homeric Hs.*, and the long prose hs. of the apostate Emperor Julian (4th c. A.D.) could not have been used in worship.

When the LXX translators came to the Heb. Old Testament they used the nouns *hymnos, psalmos* (song or tune played to a stringed instrument), *ōdē* (song), and *ainesis* (praise), and the related verbs (*hymneō*, etc.) to translate a variety of Heb. words for praise, song, music, thanksgiving, speech, beauty, and joy. Several Psalms are identified in their titles as hs. (e.g. 53, 60), incl. sometimes the explicit use of both words, *psalmos* and *hymnos* (6, 75). The Psalms are also identified with odes/songs in several titles (e.g. 51), sometimes in the expression "a song of a psalm" (*ōdē psalmou*, 65), at others as "a psalm of a song" (*psalmos ōdēs*, 67 and 74). Psalm 136.3 connects *ōdē* and *hymnos*, and in two similar New Testament passages all three terms are mentioned together with the implication of rough equivalence (*Eph.* 5.19, *Col.* 3.16).

While literary hs. were not a popular or important genre in pagan Roman adaptations of Gr. literary forms, Lat. Christians of the 4th c., influenced by the h.-singing of the Eastern Church, revived and introduced both literary and liturgical hs. into the Lat. West. This revival is also reflected in Jerome's Vulgate tr. of the Bible, where the transliterations *hymnus* and *psalmus* are regularly employed (along with a more accurate and consistent translation of the Old Testament Heb.). In his *Confessions* (9.7), Augustine remarks on the introduction of the singing of hs. and psalms in Milan by Ambrose. Isidore's influential *Etymologiae* (7th c.) identifies the Psalms of David as hs. (1.39.17) and asserts that properly speaking they must praise God and be sung, although in another work he notes that Hilary of Poitiers (also 4th c.) first composed hs. in celebrations of saints and martyrs (*De ecclestiacis officiis* 1.7).

Ambrose's short hs. (8 quatrains of iambic dimeter) are explicitly Trinitarian and were intended to counteract Arian doctrine, which had itself apparently been fostered by hs. of an Arian bent. Considerable suspicion remained in the 6th and 7th cs. over the liturgical use of hs. composed by men as opposed to the divinely inspired Psalms of the Bible. But the unquestioned orthodoxy of Ambrose himself probably helped the h. become accepted as a regular part of the liturgy along with the Psalmody. As the Middle Ages progressed, many brilliant new hs. were added to the Lat. hymnody, from the great passion h. *Vexilla regis prodeunt* by Fortunatus (7th c.) and the *Veni, creator spiritus* (tr. by Dryden) to Aquinas' hs. for Corpus Christi day and the Franciscan *Stabat mater.* Closely related to the h. is the medieval "prose" or sequence. Noteworthy are the works of Notker Balbulus (9th–10th cs.) and of Adam of St. Victor (12th c.), and the magnificent Franciscan *Dies irae.*

The Reformation renewed the suspicion of hs. among many Protestants. Although Luther composed hs. from paraphrases of the Psalms and other portions of the Bible, he also translated hs. from the Roman Breviary and wrote original ones. Calvin, on the other hand, who opposed the use of man-made hs. in congregational worship, retained the poet Marot to create a metrical psalmody or psalter in Fr., and the Calvinistic position strongly influenced the Anglican and Presbyterian churches in Great Britain in the 16th and 17th cs. The renderings of the Anglican Psalmody by Sternhold and Hopkins (2d ed., 1549), generally infelicitous but set in ballad meter, may have prompted many Eng. poets of the Elizabethan and Jacobean periods to experiment with trs. of the Psalms (Milton translated 17 of them in a variety of verseforms), but in popular terms the Sternhold-Hopkins psalter was enormously successful.

The Roman Catholic Church experienced a different problem. The great influence of Ren. humanism led to a succession of more or less unfortunate revisions of the hymnody beginning with that by Leo X in 1525 and ending with Urban VIII's in the early 17th c. The motive was to turn the late-Cl. and Med. Lat. of the hs. and sequences into good Cl. Lat.

It remained for the Eng. Nonconformist minis-

ter Isaac Watts to revive the writing of original hs. in the 18th c. and turn the primary focus of hs. away from literary artistry back to the rugged emotional expression of strong religious beliefs (*Hymns and Spiritual Songs*, 1707–09). As the 18th c. progressed, there occurred an enormous outpouring of hs., along with new metrical versions of psalms, in the works of the Wesleys, Newton, Toplady, Perronet, the poet Cowper, and many others, some of which (e.g. those by Charles Wesley) have great lyric power.

This outpouring continued in the 19th c. The works of Fanny Crosby and Katherine Hankey and the musical arrangements of W. H. Doane are pre-eminent among those of literally hundreds of writers and composers. New hs. are still being composed to this day (e.g. by Gloria and William Gaither, Andraé Crouch, Stuart Hamblen, Ralph Carmichael, and Doris Akers), and new and revised hymnals are still being arranged.

At about the same time that Ambrose introduced congregational h.-singing to the Christian church in Milan, the Christian poet Prudentius (4th–5th c. A.D.) wrote Christian literary hs. in the manner of the pagan poets of antiquity. Although some of Prudentius' creations are brief enough to be sung congregationally, many of his hs. are much too long for worship, and all have obvious literary and poetic pretensions. The fine lyric measures in two of his collections (*Peristephanon* and *Cathemerinon*) celebrate mainly saints and martyrs.

The Cl. literary h. was revived during the Ren. in both its Christian/Prudentian and secular forms. Pope Paul II's 140-line Sapphic "Hymnus de passione" (15th c.) could have been written by Prudentius, but Pontano's (d. 1503) Sapphic "Hymnus in noctem," from his *Parthenopei* (*Carmina*, ed. J. Oeschger, 1948) imitates Catullus' lyric celebration of Diana. Pontano also wrote Christian literary hs. in Lat. elegaics. Another Neo-Lat. poet, Marco Antonio Flaminio (1498–1550), who wrote probably the best liturgical hs. of the Ren. (in Ambrosian dimeters), also composed a celebrated literary lyric, "Hymnus in auroram."

Michele Marullo's 15th c. *Hymni Naturales* (*Carmina*, ed. A. Perosa, 1951) were found by many in the Ren. to be too thoroughly pagan in subject and tone as well as form. There were other celebrated Christian literary hs. in the Ren., esp. those by Vida in the early 16th c. (*Hymni*, 1536). The later vernacular poets follow the lead of the Neo-Latinists. Ronsard, along with Marullo the most prolific writer of literary hs., composes both explicitly Christian and secular/philosophical hs. along with classical/mythological celebrations. The first two of Spenser's *Four Hs.* celebrate love and beauty in a Cl. philosophical (and Petrarchan) way, while the latter two celebrate heavenly (i.e. Christian) love and beauty.

These two trads. continue in Neo-Lat. and the vernacular lits., most notably in Chapman's two long hs. in *The Shadow of Night* (1594), in Raphael Thorius' celebrated *Hymnus tabaci* (1625), in several of Crashaw's long Christian literary hs., James Thomson's "A H. on the Seasons," Keats's "H. to Apollo," and Shelley's hs. "to Intellectual Beauty," "of Apollo," and "of Pan."

While liturgical hs. tend simply to start and stop, the literary h. frequently has an Aristotelian coherence—a beginning, middle, and end. Typically it begins with an invocation and apostrophe. The main body will narrate an important story or describe some moral, philosophic, or scientific attribute. A prayer and farewell provide the conclusion. The history of literary hs. is marked by great stylistic variation and rhetorical elaboration depending on the writer's object of praise and his conception of the relationship of style to content. The definitive survey of this trad. remains to be written.

TEXTS AND ANTHOLOGIES: *Analecta hymnica*; *Mystical Hymns of Orpheus*, tr. T. Taylor (1896); *Lyra graeca*, ed. and tr. J. Edmonds, 3 v. (1922); *Early Lat. Hs.*, ed. A. Walpole (1922); *Proclus's Biography, Hs., and Works*, tr. K. Guthrie (1925); *Poeti Latini del Quattrocento*, ed. F. Arnaldi et al. (1964); *Hymni latini antiquissimi XXV*, ed. W. Bulst (1975); *Hs. for the Family of God*, ed. F. Bock (1976); *New Oxford Book of Christian Verse*, ed. D. Davie (1981).

STUDIES: G. M. Dreves, *Aurelius Ambrosius* (1893); R. Wünsch, "Hymnos," in Pauly-Wissowa; Meyer, esp. 2.119, 3.145 ff.; Manitius; *Dict. of Hymnology*, ed. J. Julian, 2d ed., 2 v. (1907), s.v. "Psalters" *inter alia*; L. Benson, *Hymnody of the Christian Church* (1927), *The Eng. H.* (1962); P. Von Rohr-Sauer, *Eng. Metrical Psalms From 1600 to 1660* (1938); O. A. Beckerlegge, "An Attempt at a Classification of Charles Wesley's Metres," *London Quarterly Rev.* 169 (1944); H. Smith, "Eng. Metrical Psalms in the 16th C. and Their Literary Significance," *HLQ* 9 (1946); R. E. Messenger, *The Med. Lat. H.* (1953); Raby, *Christian*; *Reallexikon* 1.736–41; Bowra; J. Szövérffy, *Die Annalen der lateinischen Hymnendichtung*, 2 v. (1964–65), *Iberian Hymnody: Survey and Problems* (1971), *Guide to Byzantine Hymnography* (1978), *Religious Lyrics of the Middle Ages: Hymnological Studies and Coll. Essays* (1983), *Concise Hist. of Med. Lat. Hymnody* (1985), *Lat. Hs.* (1989); C. Maddison, *Marcantonio Flaminio* (1966); H. Gneuss, *Hymnar und Hymnen im englischen Mittelalter* (1968); P. Rollinson, "Ren. of the Literary H.," *RenP* (1968); C. Freer, *Music for a King* (1971); C. E. Calendar, "Metrical Trs. of the Psalms in France and England," *DAI* 33 (1973): 6863A; J. Ijsewijn, *Companion to Neo-Lat. Studies* (1977); *Menander Rhetor*, ed. and tr. D. Russell and N. Wilson (1981); P. S. Diehl, *The Med. European Religious Lyric* (1985). P.RO.

HYPERBATON (Gr. "a stepping over," Eng. "inversion"). In rhet., the genus for several figures of syntactic dislocation, i.e. alteration of the normal (that is, prose) word order. Puttenham calls it the term for "all the auricular figures of disorder." Such dislocation may take several forms, may be

ranked according to increasing severity, and may range from a single word moved from its accustomed place, to a pair of words simply reversed, to more extreme instances of disarray (Lat. *transgressio*), often used to depict extreme emotion. In Cl. rhet. h. would therefore subsume such figures as *anastrophe* (Lat. *inversio* or *reversio*), the reversal of a word-pair—this is one common meaning of the Eng. term "inversion"; *hypallage*; and *hysteron proteron*. Further, "normal (prose) order" is itself a problematic concept: simple cases are easy to analyze, but more complex structures are not. In inflectional langs., such as Gr. and Lat., where word-class is marked by ending, word order is, while very free, still not entirely free; and in positional langs., such as Eng., where inflections are almost entirely absent and word order determines case, still there may often be several acceptable ways of ordering many sentence elements. If, for example, one wants to insert the word "however" into the sentence "John disputed Smith's findings," there are five slots, and only one of them (between "Smith's" and "findings") is ungrammatical; but presumably this sort of optional placement is not what was traditionally meant by h. and its relatives. In Cl. rhet., dislocation was usually viewed not as displacement but either as the separation of two words normally belonging together or by simple reversal of the order of two adjacent words. Quintilian (*Institutio oratoria* 8.6.65) thus explains h. as the interposition of a foreign sentence element which has to be "stepped over," and restricts the term *anastrophe* (Gr. "a turning upside down") to the inversion of only two words, e.g. *mecum* for *cum me* or *quibus de rebus* for *de quibus rebus*. In any event, "normal" implies not a rule but a norm, i.e. statistical predominance; and the study of poetic syntax from a normative and statistical point of view is very much a modern phenomenon. All prose is stylized to some degree, so it is a problem to establish a base or zero degree against which to measure deviance, though extreme instances are obvious enough. Too, usage varies from age to age, allowing differing degrees of freedom in word order; this must be factored in. Nevertheless, it is undeniable that the several species of h. were recognized as rhetorical figures and exploited in poetry. Dionysius of Halicarnassus' *On Literary Composition* (ca. 10 B.C.) is actually on the placing of words; Sappho, Pindar, Sophocles, Euripides, and above all, Homer, are analyzed in detail. The example of Homer had the effect that h. increasingly came to be viewed as a specific stylistic feature of the high or grand style, suitable for heroic subjects—i.e. war and love—whence it passed from the epic trad. into the Ren. It. lyric and the sonnet: it is Tasso's favorite figure. Longinus (*On the Sublime*, ch. 22) applies the term h. to inversion of ideas rather than words, i.e. reveral of the expected or logical order of elements in an argument, pointing out what extra force a surprising order imparts to ideas that thus seem "wrung from" the speaker rather than "premeditated." Sister Joseph remarks that it "might almost be said to characterize the style" of Shakespeare's *Tempest* (54).

If we examine in particular the main line of syntactic sequence in Eng., i.e. subject–verb–object, some interesting facts emerge. In Eng. poetry the most common form of displacement is SOV, moving the verb to the end of the clause and line, often for the rhyme word; this is conspicuous in the heroic couplet (q.v.), but end-of-line is a point of maximum visibility in all verse. Milton uses normal SVO order 71% of the time, compared to 68% for Pope, 85% for Shelley, 86% for Shakespeare, and 88% for Tennyson; there are only 3 inversions in all of *The Waste Land* (Jespersen 7.60). Milton's chief inversions are OSV and SOV., e.g. "Him th'Almighty Power / Hurl'd headlong flaming . . . " (OSV; *PL* 1.44–45), and "They looking back, all th'Eastern side behold / Of Paradise . . . " (SOV; 12.641–42). But he relishes more complex h. as well: "ten paces huge / He back recoiled" (6.193–94); and esp. Miltonic is the placing of a word between two others which both modify it or both of which it modifies, such as a noun between two adjectives or other nouns, e.g. "temperate vapours bland," "heavenly forms angelic," or "Strange horror seize thee, and pangs unfelt before" (2.703). It was Miltonic inversions which Keats said were the reason for his abandoning *Hyperion*.

H. is thus exploited in poetry for a variety of useful ends, some rhetorical, some prosodic, but even in the latter, these extend far beyond the exigencies of meter or rhyme to include a number of strategies for emphasis, such as isolating, pointing, and conjoining. Thus Milton uses hyperbatic word-order more systematically so as to delay the narrative progression of the verse, thereby bringing the reader's attention to the phonic and rhythmic nuances of each line *qua* line.

O. Jespersen, *A Mod Eng. Grammar*, 7 v. (1909–49); M. Redin, *Word-Order in Eng. Verse from Pope to Sassoon* (1925); Sr. M. Joseph, *Shakespeare's Use of the Arts of Lang.* (1947); N. A. Greenberg, "Metathesis as an Instrument in the Crit. of Poetry," *TAPA* 89 (1958); S. R. Levin, "Deviation—Statistical and Determinate—in Poetic Lang.," *Lingua* 12 (1963); T. K. Bender, "Non-Logical Syntax: Lat. and Gr. H.," *Gerard Manley Hopkins* (1966); G. M. Landon, "The Grammatical Description of Poetic Word-Order in Eng. Verse," *Lang&S* 1 (1968); Lausberg; G. L. Dillon, "Inversions and Deletions in Eng. Poetry," *Lang&S* 8 (1975), "Literary Transformations and Poetic Word Order," *Poetics* 5 (1976); J. M. Lipski, "Poetic Deviance and Generative Grammar," *PTL* 2 (1977); F. Björling, "On the Question of Inversion in Rus. Poetry," *Slavica Lundensia Litteraria* 5 (1977), 7–84—on adj.–noun h. in Pushkin; W. P. Bivens, III, "Parameters of Poetic Inversion in Eng.," *Lang&S* 12 (1979); Morier, s.v. "Inversion"; Group Mu, 82–84; A. Quinn, *Figures of Speech*

(1982); T. Eekman, "Some Questions of Inversion in Rus. Poetry." *Rus. Poetics* (1982); A. Clement, "The Technique of Hebraic Inversion in Eng. Lit.," *The Literary Criterion* 22 (1987); J. P. Houston, *Shakespearean Sentences* (1988); M. J. Hausman, "Syntactic Disordering in Mod. Poetry," *DAI* 49, 8A (1989): 2210; Corbett. T.V.F.B.

HYPERBOLE. (Gr. "overshooting," "excess"). A figure or trope, common to all lits., consisting of bold exaggeration, apparently first discussed by Isocrates and Aristotle (*Rhetoric* 1413a28). Quintilian calls it "an elegant straining of the truth" which "may be employed indifferently for exaggeration or attenuation" (*Institutio oratoria* 8.6.67) and exemplifies with Virgil's "Twin rocks that threaten heaven" (*Aeneid* 1.162). Shakespeare has "Not all the water in the rude rough sea / Can wash the balm off from an anointed King" (*Richard 2* 3.2.54–5). H. is thus any extravagant statement used to express strong emotion, not intended to be understood literally ("I've told you a million times not to exaggerate"). According to Puttenham, who understood the figure's ideological possibilities, h. is used to "advance or . . . abase the reputation of any thing or person" (*The Arte of Eng. Poesie* [1589]). Use of h. is common in Elizabethan drama from Marlowe (*Tamburlaine*) to Jonson, particularly for comic irony.

H., litotes (Gr. "plainness") and meiosis (Gr. "lessening") belong in the "super-category" in rhet. which some rhetoricians have called "figures of thought." The latter organize other tropes or figures such as metaphor, metonymy, and synechdoche (qq.v.) to make ideological statements. Group Mu points out (132–38) that such super-figures refer to an "ostensive situation," i.e. to the pragmatic context of the utterance, and that it is only reference to this situation which makes possible communication of either exaggeration or understatement. The metaphoric assertion "this cat is a real tiger" can only be tested for truth-value or for its degree of "literalness" (as opposed to its "overstatement") by inspection of the individual member of the feline genus in question. Only such conventional h. as is found in adynaton based on obvious impossibility conditions evades the necessity of reference to the ostensive situation. But h. certainly has a number of other functions as well.—A. H. Sackton, *Rhet. as a Dramatic Lang. in Ben Jonson* (1948); W. S. Howells, *Logic and Rhet. in England, 1500–1700* (1956); G. Genette, "Hs.," *Figures I* (1966); C. Perelman and L. Olbrechts-Tyteca, *The New Rhet.* (tr. 1969); B. Vickers, "The *Songs and Sonets* and the Rhet. of H.," *John Donne: Essays in Celebration*, ed. A. J. Smith (1972); Lausberg; Group Mu; Corbett. R.O.E.; A.W.H.

I

IDEOGRAM. In lang., a pictogram, a character in an alphabet which functions as an image for that to which it refers. In poetry, a field or matrix of juxtaposed words and images. The i. is the crux of Ezra Pound's "ideogrammic method," a new form of poetic structure which he hoped would allow Modernist poets to escape the traps of both late-Victorian vagueness, metaphor, and sentimentality, on the one hand, and the loss of immediacy created by the syntax of discursive, representational speech, on the other.

Pound's search for new poetic techniques led him, successively, into imagism and vorticism, but it was his discovery of the work of the Am. Orientalist Ernest Fenollosa (d. 1908) which provided the catalyst. Pound read Fenollosa's lecture notes (written ca. 1904) on the nature of Chinese pictograms in 1914, published trs. of Chinese poems based on Fenollosa's work as *Cathay* in 1915, and edited the notes into an essay which he published in 1919. Fenollosa's essay, *The Chinese Written Character as a Medium for Poetry*, may be fairly called the first and perhaps the most important *Ars poetica* of the 20th c.

In Chinese, said Fenollosa, the characters of the alphabet are, in greater or lesser degree, images or visual representations of the things they signify. Thus the i. for "East" is a synthesis of the is. for "sun" and "tree"—the sun in the trees. Here abstract concepts arise not out of abstract symbols but out of concrete particulars and the immediacy of experience. In the Western langs., by contrast, graphemes denote sounds which refer to objects purely by arbitrary convention. Fenollosa only contrasted Chinese and Western modes of thought; it remained for Pound to see how the pictogram could be formulated as a principle and developed into a poetic method.

The essence of the ideogrammic method is juxtaposition of separate objects or events so as to evoke a new matrix or constellation of meaning. The syntax is paratactic or asyndetic. But juxtaposition does not imply fragmentation or the mere "heaping together" of unrelated items. Rather, the poet still exercises selection and combination; separate items are juxtaposed so as to be re-integrated by the reader exactly as objects and events are integrated in ordinary perception of the external world: the reader's mind will organize the elements into a coherent pattern. The ideogram-

mic method thereby imitates perception itself, and in this respect, it is a mimetic strategy (see REPRESENTATION AND MIMESIS); is. are motivated signs. Further, emphasis shifts away from static description: now symbols are used to represent not objects but relations, motion, energy, and action. Now lang. has an immediacy of presentational form lacking in the discursive-descriptive-rationalist nature of Western writing.

Pound in fact misunderstood part of Fenollosa, but it was a terrifically fertile misprision. And Fenollosa himself was in error about Chinese: only 10% of its characters are pictographic; the vast majority are phonetic. Nevertheless, reading Fenollosa gave Pound the idea for a wholly new poetics, arguably a method which became more influential, in its several forms, than any other in the 20th c. Géfin claims that the principle of juxtaposition underlies much subsequent Am. poetry in the 20th c., incl. the Williams trad., Objectivism, free verse (q.v.), "open form," Olson's "composition by field," and Duncan's "collages." Pound seems to have intended *The Cantos* as a modern epic, though one organized not via narrative but via the ideogrammic method.—D. Perkins, *Hist. of Mod. Poetry*, 2 v. (1976, 1987); L. Géfin, *I.: Hist. of a Poetic Method* (1982). T.V.F.B.

IDYLL, idyl (Gr. *eidyllion* [diminutive of *eidos*, "form"], "short separate poem"; in Gr. a pastoral poem or i. is *eidylllion Boukolikon*). One of the several synonyms used by the Gr. grammarians for the poems of Theocritus, Bion, and Moschus, which at other times are called *bucolic*, *eclogue* (q.v.), and pastoral (q.v.). Theocritus' ten pastoral poems, no doubt because of their superiority, became the prototype of the i. In the 16th and 17th cs., esp. in France, there was frequent insistence that pastorals in dialogue be called *eclogues*, those in narrative, *is.* In Ger., *Idylle* is the ordinary term for pastoral. Two biblical books—*Ruth* and *The Song of Songs*—are sometimes called is., which may be taken to illustrate the latitude of the term. Note that Tennyson's *Is. of the King* 1859) is hardly pastoral. Perhaps Tennyson thought the use of the term was appropriate: each i. contains an incident in the matter of Arthur and his Knights which is separate (or framed) but at the same time is connected with the central theme; the contents treat the Christian virtues in an ideal manner and in a remote setting. But there is very little in Browning's *Dramatic Is.* (1879–80), which mainly explore psychological crises, to place them in the pastoral trad. The adjectival form, *idyllic*, is more regularly and conventionally applied to works or scenes which present picturesque rural scenery and a life of innocence and tranquillity, but this has little specific relation to any poetic genre.

M. H. Shackford, "A Definition of the Pastoral I.," *PMLA* 19 (1904); P. van Tieghem, "Les Is. de Gessner et le rève pastoral," *Le Préromantisme*, v. 2 (1930); P. Merker, *Deutsche Idyllendichtung* (1934); E. Merker, "Idylle," *Reallexikon* 1.742–49; J. Tismar, *Gestörte Idyllen* (1973); T. Lange, *I. und exotische Sehnsucht* (1976); R. Böschenstein-Schäfer, *Idylle*, 2d ed. (1977); K. Bernhard, *Idylle* (1977); V. Nemoianu, *Micro Harmony: Growth and Uses of the Idyllic Model in Lit.* (1977). T.V.F.B.; J.E.C.

IMAGERY.

I. 16TH TO 19TH CENTURIES
II. 20TH CENTURY
  A. *Mental Imagery*
  B. *Figures of Speech*
  C. *Cluster Criticism*
  D. *Symbol and Myth*
III. RECENT DEVELOPMENTS
IV. APPLICATION

Both the root meanings and broad implications of this term are akin to the word "imitate," and hence refer to a likeness, reproduction, reflection, copy, resemblance, or similitude. Its cognate terms, "icon" and "idol," often refer to objects of devotion and worship, manifesting the profound historical and cultural issues involved in i. The history of the term dates from at least the Old Testament, wherein are found the statement that humankind was made in God's image and the injunction against making graven images. The Judaic trad. took its stand firmly against the Many in favor of the One, Appearance in favor of Reality. Material representations of the Divine were seen as interfering with rather than enhancing humanity's realization of God.

Such dualisms are difficult to sustain, however, and later the Catholic Church perforce sanctioned the general use of images as an aid to worship. But the pendulum subsequently swung back with the Islamic rejection of the Trinity in the 7th c. A.D., the iconoclasm of the Catholic Church itself in the 8th and 9th c. A.D., and the Protestant rejection of images in the 16th and 17th c. It is during the Ren. and the Reformation that the history of lit. crit. in Eng. may be said to begin, and—although cl. critics such as Horace, Longinus, and Quintilian spoke of i. and its uses—here too begins the special consideration of i. in literary theory and practice (Frazer, Furbank, Legouis [in Miner], and Mitchell [1986]).

I. 16TH TO 19TH CENTURIES. The term i. itself, however, was not in general use during the Ren., where the emphasis fell on "figures"—of thought, of lang., and of speech—concepts largely derived from the Cl. rhet. of Greece and Rome. The underlying philosophy of this rhet. was that meaning is primary and manner of expression secondary. Thus the use of figures—e.g. metaphor, simile, allegory (qq.v.)—was viewed as ornamental and decorative, enlivening and enhancing a pre-existing meaning, and a premium was placed upon artificiality (artfulness) and technical virtuosity.

A primary example is Sir Philip Sidney's "Defence of Poesie" (1595). In characterizing a poem as a "speaking picture," a concept derived from

Horace, and in claiming that the poet "nothing affirms, therefore never lieth," Sidney meant that the poet deals with what ought to be (the ideal) rather than with what is (the real)—a Platonized adaptation of Aristotle's distinction between poetry (the probable) and history (the actual) (*Poetics*, ch. 9). For Sidney the power of images lay in their ability to move us toward virtue and away from vice by means of their strong emotional appeal.

But with the growth of skepticism and empiricism in the late 17th and early 18th cs. there was a reaction against rhet. in favor of plainness and truth in lang. (Frazer). This left a vacuum which was filled by "image." The epistemology of Hobbes and the associationist psychology of Locke led to a new way of looking at poetry in which the image was seen as the connecting link between experience (object) and knowledge (subject). An image was defined as the reproduction in the mind of a sensation produced in perception. Thus, if a person's eye perceives a certain color, he or she will register an image of that color in the mind—"image," because the subjective sensation experienced will be ostensibly a copy or replica of the objective phenomenon of color.

But of course the mind may also produce images when not receiving direct perceptions, as in the attempt to remember something once perceived but no longer present, or in reflection upon remembered experience, or in the combinations wrought out of perception by the imagination (q.v.), or in the hallucinations of dreams and fever. More specifically, in literary usage i. refers to images produced in the mind by lang., whose words may refer either to experiences which could produce physical perceptions were the reader actually to have those experiences, or to the sense-impressions themselves.

Thus, according to Frazer, the descriptive poetry of the 18th c. was new in that it focused more on the landscape itself than on that which it was supposed to represent. Creative power was assumed to reside in the "imagination," a storehouse of images in the empty mind. There was a shift, then, from the Ren. conception of the image as exemplum to the 18th-c. conception of the image as a basis for associative ruminating.

By the time of the romantic poets, however, the concept of "image" shifted once again. Given romantic transcendentalism, the world appearing as the garment of God, the abstract and general were seen as residing in the concrete and particular, and thus Spirit was felt to be immanent in Matter. Attempting to resolve the old dualism, they regarded Nature as Divine and nature poetry as a way to body forth the sacred. I., therefore, was elevated to the level of symbol (q.v.), to become thereby one of the central issues of modern criticism.

Meanwhile, several scientific and intellectual developments were underway during the latter 19th c. Max Müller put forth a theory of the growth of lang. (1861–64) which seemed to account for metaphorical imagery as an organic part of lang. rather than as an ornament or decoration. According to this theory, humanity, as it develops its conceptions of immaterial things, must perforce express them in terms of material things or images because its lang. lags behind its needs. So lang. as it expands grows through metaphor (q.v.) from image to idea. The word "spirit," for example, has as its root meaning "breath"; thus as the need to express an immaterial conception of soul or deity emerged, a pre-existent concrete word had to be used to stand for the new abstract meaning. Although Müller's thesis has been questioned (Barfield in Knights and Cottle), and although there may well be differences between ordinary and poetic metaphor, the implication that metaphor is created as a strategy for expressing the inexpressible is still influential today.

Sir Francis Galton reported in 1880 on his experiments in the psychology of perception. He discovered that people differ in their image-making habits and capacities. While one person may reveal a predominating tendency to visualize his reading, memories, and ruminations, another may favor the mind's "ear," another the mind's "nose," and yet another may have no i. at all. This discovery has also had some importance in modern crit. and poetry.

II. 20TH CENTURY. In what follows we shall trace out the modern devel. of four issues: (a) i. as a description of how part of the mind works, (b) i. as figure, (c) cluster criticism, and (d) symbol and myth.

A. *Mental Imagery.* Psychologists have identified seven kinds of mental images: visual (sight, then brightness, clarity, color, and motion), auditory (hearing), olfactory (smell), gustatory (taste), tactile (touch, then temperature, texture), organic (awareness of heartbeat, pulse, breathing, digestion), and kinesthetic (awareness of muscle tension and movement). These categories are preliminary to the other approaches to i., for they define the very nature of the materials.

Several valuable results have emerged from the application of these categories to lit. (Downey). In the first place, the concept of mental i. has encouraged catholicity of taste, for once it is realized that not all poets have the same interests and capacities, it is easier to appreciate different kinds of poetry. Much of Browning's i., for example, is tactile, and those who habitually visualize are unjust in laying the charge of obscurity at his door (Bonnell). Or again, the frequently-voiced complaint that Shelley's poetry is less "concrete" than Keats's suffers from a basic misconception, for Shelley's poetry contains just as much i. as Keats's, although it is of a somewhat different kind (Fogle). Second, the concept of mental i. provides a valuable index to the type of imagination with which a given poet is gifted. Knowing that Keats's poetry is characterized by a predominance of tactile and organic i., for example, or Shelley's by the

i. of motion is crucial to an accurate appraisal of the achievement of each poet. Third, the concept of mental i. is pedagogically useful, for a teacher or a critic may encourage better reading habits by stressing this aspect of poetry.

But the disadvantages of the mental-i. approach almost outweigh its advantages. For one thing, there is an insoluble methodological problem, in that readers are just as different from one another in their i.-producing capacities as poets, and therefore the attempt to describe the imagination of a poet is inextricably bound up with the imagination of the critic who analyzes it. Second, this approach tends to overemphasize mental i. at the expense of meaning, feeling, and sentiment (Betts). And third, in focusing upon the sensory qualities of images themselves, it diverts attention from the *function* of these images in the poetic context.

B. *Figures of Speech.* It is evident that each of the key issues in the study of i. overlaps with the others and broadens out to encompass many areas of human knowledge and culture. For practical purposes, however, we may treat the second issue—the nature and function of i. in metaphor—as if it were a separable topic.

The figures of speech in common currency today have been reduced to seven: synecdoche, metonymy, simile, metaphor, personification, allegory, and—a different but related device—symbol. Each is a device of lang. by virtue of which one thing is said (analogue) while something else is meant (subject), and either the analogue, the subject, or both may involve i.—although most characteristically it is the analogue. Subject and analogue (or "tenor" and "vehicle" as I. A. Richards termed them) may be related with respect to physical resemblance—as when Homer compares the charge of a warrior in battle to that of a lion on a sheepfold, in which case the study of mental i. provides useful distinctions—here we may visualize the furious movement of the attack and perhaps imagine the roaring sound of rage. But many figures relate two different things in other ways: a lady's blush, in Burns's "my love is like a red, red rose," her delicate skin, or her fragrance may find analogues in the color, texture, or odor of a rose, but her freshness and beauty are qualities suggestively evoked by the rose rather than images displayed in it. Burns's speaker is saying that his lady is to him as June is to the world, in the sense of bringing rebirth and joy. The similarity, then, is that his lady makes him feel as spring makes him feel: the ground of comparison is an emotion, and the i. functions in relation to that purpose rather than to a physical object.

Furthermore, the two things related may each be images (as in the Burns example), or feelings or concepts; or the subject may be an image and the analogue a feeling or concept; or the subject may be a feeling or concept and the analogue an image. Some critics have asserted that the fundamental subject or tenor of a figure is in reality the relationship itself and that therefore the analogy or vehicle includes the two things related.

Thus, although the term i. is commonly used to refer to all figures of speech, as well as to literal images, further distinctions are obviously necessary. The kinds of things related and the nature and function of their relationship, as we have seen, provide grounds upon which these distinctions may be made. It was once common to claim that proper practice precluded mixing one's analogues in any one figure (Jennings), but more recent critics have argued that no such rule is universally valid, esp. in poetry (Brooks): it all depends on whether the mixture of analogues is consistent with the context and what their effect upon the reader is. Shakespeare is notoriously full of mixed metaphors, as in Hamlet's famous reflection on whether he should "take arms against a sea of troubles." Here we must either imagine some sort of demented warrior slashing at the sea with his sword, or we must imagine two separate metaphors—"feeling overwhelmed by troubles as by the inundation of the sea, and doing something about them as when a warrior arms himself and marches out to meet the enemy"—conflated. With such mixtures, when the figures flash by in rapid sequence, it is counterproductive to try to stop and picture each in our minds; the point is that they *seem* fitting and effective in context.

Or again, it was once considered good form to teach students to visualize figures, but not only are most metaphors constructed on bases other than on mental i. but also much mental i. is other than visual. In fact, persistent visualizing will break down the relationship between subject and analogue in many figures entirely—as in the Burns simile, for example: if we consider the rose further, we could imagine that there may be a worm buried within it, that it has thorns, and that its life is all too brief (qualities which have certainly been exploited in many other poems), we will have pretty much destroyed the image and the poem.

On the other hand, much attention was once focused on that kind of figure in which the difference between subject and analogue seems unusually great, the "metaphysical image" (Wells, Rugoff, Tuve, Miller)—a term and concept which go back to Dr. Johnson's essay on Donne, though Holmes has argued that such disparate figures were anticipated by the Elizabethan dramatists. Thus, while the reader must beware of going too far with an image lest it break down into irrelevance, poets are licensed to challenge the reader with an image which at first appears far-fetched but which on closer inspection reveals its profounder core of significance.

Fourth, much was once made of the function of the "central" or "unifying" image in a poem, according to which the poet develops a sustained analogy which then serves as the structural frame of the poem. Ericson, for example, distinguishes among different sorts of i. patterns in a poem; in Herbert he finds that images may be extended, end-on-end, mixed, and so on, concluding that the

extended image seems to be the most effective. But here again we can only conclude, as with mixed metaphors, visualized figures, and metaphysical images, that the central or unifying image is merely one possibility among many, and as such it has no special normative value but depends largely upon meaning and function in a particular context.

When such distinctions regarding the meaning and function of i. were adapted as the basis for theories about the nature of poetic lang., or the quality of the artistic imagination, i. became one of the key terms of the New Criticism, which developed what it viewed as a radically "functional" theory of i., based on the assumptions that figures are the differentiae of poetic lang. and that poetic lang. is the differentia of the poetic art. This assumption was opposed to what the New Critics took to be the "decorative fallacy" of traditional rhet. and the "heresy" of modern positivist semantics, which claimed that logical statement is the only indicator of truth and all other forms of statement are untrue.

C. *Cluster Criticism.* The study of figurative i., therefore, anticipates and overlaps the subsequently developed study of symbolic i. In the latter approach the essential question is how *patterns* of i.—whether literal, figurative, or both—in a work reveal facts about the author or the poem, on the basic assumption that repetitions and recurrences (usually of images but on occasion of word patterns in general) are in themselves significant. Hence the method involved the simple application (and sometimes distortion) of some elementary statistical principles. These patterns may appear either within the work itself or among literary works and myths in general or both.

Assuming that repetitions are indeed significant, we must examine the nature of this significance. What exactly will counting image-clusters tell the critic? There developed at least five distinguishable approaches, each a bit more complex than the preceding: (1) texts of doubtful authorship can be authenticated (Smith); (2) some aspects of the poet's own experiences, tastes, temperament, and vision of life can be inferred (Spurgeon, Banks); 3) the causes of tone, atmosphere, and mood in a work can be analyzed and defined (Spurgeon); (4) the ways in which the structure of conflict in a play is embodied can be examined (Burke 1941); and (5) symbols can be traced out, either in terms of how image patterns relate to the author or how they relate to archetypes or both (Frye, Knight, Heilman).

The first two approaches relate to problems extrinsic to the work itself, although they seek internal evidence. The procedure involved counting all the images in a work or all the works of a given poet (and here the problems of what an image is, what kind it is, and whether it is literal or figurative had to be resolved anew by each critic doing the counting) and then classifying them according to the areas of experience these images represent: Nature—Animate and Inanimate, Daily Life, Learning, Commerce, and so on. Since these categories and their proportions represent aspects of the poet's perceptions and imagination, two inferences can be made on the basis of the resultant charts and figures: first, that the patterns are caused by the poet's personal experiences and therefore give a clue to the poet's personality and background; and second, that since they are unique—like fingerprints, no two minds have exactly the same patterns—they offer a means of determining the authorship of doubtful works. Perhaps the second assumption is sounder than the first, although both are dubious, for frequently images appear in a work because of literary and artistic conventions rather than because of the poet's personality or experience (Hornstein, Hankins).

The third and fourth approaches relate to problems intrinsic to the artistic organization of the work itself. Thus Kenneth Burke remarked, "One cannot long discuss imagery without sliding into symbolism. The poet's images are organized with relation to one another by reason of their symbolic kinships. We shift from the image of an object to its symbolism as soon as we consider it, not in itself alone, but as a function in a texture of relationships" (1937). Certain plays of Shakespeare, as Spurgeon showed, are saturated with one or another kind of similar images or clusters (usually figurative)—the i. of light and dark in *Romeo and Juliet*, or of animals in *King Lear*, or of disease in *Hamlet*—and it was reasoned that these recurrences, although they might not be consciously noticed by the ordinary reader or playgoer, are continually at work conditioning our responses as we follow the action of the play. Thus Kolbe, a pioneer of cluster crit. along with his predecessors Whiter and Spaulding, claimed in 1930: "My thesis is that Shakespeare secures the unity of each of his greater plays, not only by the plot, by linkage of characters, by the sweep of Nemesis, by the use of irony, and by the appropriateness of style, but by deliberate repetition throughout the play of at least one set of words or ideas in harmony with the plot. It is like the effect of the dominant note in a melody." Later critics added that clusters may form dramatic discords as well as harmonies.

From this argument it was but a small step to classifying images according to their relationship to the dramatic conflicts in the work. There are basically two sorts of clusters: the recurrence of the same image at intervals throughout the work, or the recurrence of different images together at intervals. If the same image recurs in different contexts, then it should serve to link those contexts in significant ways, and if different images recur together several times, then the mention of any one of them will serve to call the others to mind. Such notions of going below the surface to find associative patterns owed something, of course, to Freudian psychoanalysis as well.

D. *Symbol and Myth.* The next and fourth step was to reason once again from inside to outside the work, but this time ostensibly for the sake of greater insight into the work. According to Burke, a poem is a dramatic revelation in disguised form of the poet's emotional tensions and conflicts, and if, therefore, some idea of these tensions and conflicts can be formed, the reader will then be alerted to their symbolic appearance in the work. Thus Burke could make "equations" among Coleridge's image-clusters by comparing the poet's letters with the *Ancient Mariner* and could conclude that the albatross symbolizes Coleridge's guilt over his addiction to opium. This, he reasoned, illuminates the "motivational structure" of that particular poem.

It is not difficult, on the other hand, to equate image-clusters in a particular work with larger patterns found in other works and in myths instead of merely in the poet's personal life (a dream is the "myth" of the individual, a myth is the "dream" of the race), as does Frye, for example, and even Burke himself, for the "action" of which a poem is "symbolic" frequently resembles larger ritualistic patterns such as purgation, scapegoating, killing the king, initiation, and the like, though expressed on a personal level and in personal terms. Robert Penn Warren sees in the *Ancient Mariner* a symbol of the artist-archetype, embodied in the figure of the Mariner, torn between the conflicting and ambiguous claims of reason (symbolized by the sun) and the imagination (symbolized by the moon); thus the crime is a crime against the imagination, and the imagination revenges itself but at the same time heals the Mariner; the wandering is also a blessing and a curse, for the Mariner is the *poète maudit* as well as the "prophet of universal charity" (see Olson's review of this interp. in Crane). One may therefore find implicit images of guilt, purification, descent, and ascent running throughout a poem whose literal action may be of quite a different nature. Image clusters may form "spatial pattern" or even a "subplot" calling for special attention (Cook).

There are certain difficulties with cluster crit., however, which revolve primarily around the fact that it tends to reduce complex and individual works of lit. to simpler and more general paradigms. In emphasizing primarily certain aspects of a work for inspection, in finding similarities among all literary works, and in assuming that all recurrences have a certain kind of significance, this approach has been called seriously into question (Adams in Miner).

III. RECENT DEVELOPMENTS. More recently the literary study of i. has become at once more advanced and more problematic. There are a plethora of studies in speculative and experimental psychology, involving phenomenology, epistemology, and cognitive psychology, looking very closely at the question of what exactly mental i. is.

Block, for example, analyzes the recent objection that, although it seems as if we can see internal pictures, this cannot in fact be the case, for it would mean that we have pictures in the brain and an inner eye to see them with, and no such mental operations can be found. Images are, it is claimed, verbal and conceptual. Block concludes that there are two kinds of i., one which represents perception in roughly the same way pictures do, the other which represents as lang. does, i.e. conceptually, that some combine pictorial and verbal elements, and that the real question is what types of representation are possible.

Furbank, on the other hand, after giving a very useful history of the term "image" up to the imagists, reveals the ambiguities and confusions involved in using the terms "image" and "concrete" to describe qualities of the verbal medium. We should not confuse "abstract" with "general" and "concrete" with "specific." "Concrete," for example, does not necessarily mean "sensuous," and that which is "specific" may very well also be "abstract." The only actual *physical* element of poetry is found in the subvocal or silent actions of tongue and larynx as a poem is recited or read. Furbank has suggested that we drop the term i. altogether.

More inclusive is the work of Mitchell, who places the entire issue within the broad historical context of knowledge as a cultural product. First, he has provided a chart of the "family of images" to indicate the different meanings of i.: in art history are found the graphic images (pictures, statues, designs), in physics the optical images (mirrors, projections), in philosophy and theology the perceptual images (sense data, "species," appearances), in psychology and epistemology the mental images (dreams, memories, ideas, fantasmata), and in lit. crit. the verbal images (metaphors, descriptions, writing).

Mitchell goes on to develop an argument concerning the nature and function of i. which is deliberately constructed to steer a middle course between the old-fashioned realism of empiricists such as John Locke and the recently fashionable nominalism of poststructuralists and deconstructionists such as Jacques Derrida. Both our signs and the world they signify are a product of human action and understanding, and although on the one hand our modes of knowledge and representation may be arbitrary and conventional, they are on the other hand the constituents of the forms of life—the practices and trads. within which we must make epistemological, ethical, and political choices. The question is, therefore, not simply "What is an image?" but more "How do we transfer images into powers worthy of trust and respect?" Discourse, Mitchell concludes, does project worlds and states of affairs that can be pictured concretely and tested against other representations.

Approaching the problem from a socio-political-historical perspective, Weimann would remind us that, while for modern critics the meaning of a poem is secondary to its figures, for Ren. and metaphysical poets the meaning was primary.

Here he agrees with Tuve that modern interps., therefore, fail to do justice to the intentions and structures of earlier works. Metaphor, he says, is neither autonomous nor decorative; rather, it relates man and universe, and the link or interaction between tenor and vehicle is the core. History is part of the meaning of a metaphor. Shakespeare's freedom of linguistic transference reflects the social mobility of his era—an age of transitions and contradictions. That neoclassical critics did not value Shakespeare's over-rich i. but rather preferred plot over diction reflects their own view of man in society. This valuation was reversed during the romantic age, but even the romantics did not place form over meaning. Modernist poets and critics, Weimann argues, emphasizing the autonomy of a literary work and its spatial patterns, have removed lit. both from history and its audience. In seeking liberation from time and space, and from history into myth, this trad. has rejected the mimetic function of lit. in Cl. crit., as well as the expressive principle of romanticism. Metaphor is now seen as an escape from reality and has become severed from its social meaning. In reply, Weimann calls for an integration of the study of i. within a more comprehensive vision of lit. hist. Other commentators more sympathetic to the achievements of modernism have held that, far from severing form and meaning, it sees meaning as a function of form; and far from seeking an escape from history, it seeks to redeem history. No more time-bound works can be imagined than those two monuments of modernism, Joyce's *Ulysses* and Eliot's *The Waste Land.*

Finally, Yu's analysis of Chinese i. suggests ways in which the study of poetic i. can be enriched further by means of comparative studies. She sees a fundamental difference between "attitudes toward poetic i. in classical China" and "those commonly taken for granted in the West." Western conceptions are based on the dualism of matter and spirit and upon the twin assumptions of mimesis and fictionality: poetry embodies concretely a transcendent reality, and the poet is a creator of hitherto unapprehended relations between these two disparate realms. The Chinese assumption, by contrast, is nondualistic: there are indeed different categories of existence—personal, familial, social, political, natural—but they all belong to the same earthly realm, and the poet represents reality both literally and by joining various images which have already been molded for him by his culture. Thus the Chinese poetic and critical trad. reveals the conventional correspondences used by the poets in their efforts to juxtapose images so as to suggest rather than explain meaning. One might point out, however, that these principles sound very much like those of the Western modernist poetic previously discussed, itself influenced by Oriental philosophy and poetry—namely, the objective correlative and the juxtaposition of images. Yeats, Pound, Eliot, and the imagists, not to mention Whitman before them, were all striving, each in his own way, to heal the split of Western dualism and achieve, in Yeats's phrase, Unity of Being. Whether they did so by means of a vision of transcendence or one of immanence, there are remarkable similarities in technique among them, as well as between them and the Chinese poets and critics so illuminatingly analyzed by Yu.

IV. APPLICATION. We may now ask what i. does in an actual poem. Although i. has come to be regarded as an essentially poetic device, many good poems contain little or no i. When i. *is* present, however, it is best viewed as part of the larger integral fabric of the poem's form and as having a variety of possible functions.

I. may be, in the first place, the speaker's subject, whether that thing is present in the situation or recollected afterwards, or simply the subject of a predication. Such subjects are, roughly speaking, people, places, objects, actions, and events. In Frost's "Come In," for example, the speaker, narrating in the past tense, represents himself as having come to a place of boundaries—a not uncommon situation in Frost—between the woods and field, night and day, earth and heaven. He heard a thrush's song echoing through the darkening trees, and he felt the pull of sorrow. He decided, however, to turn away from this invitation and to gaze, rather, at the stars.

The literal i. of subject here comprises the woods, the thrush music, the shades of darkness and light, and the stars. As for mental i., we may if we choose visualize the scene and "hear" the song of the thrush. Frost does no more than sketch it all in, however: he does not create a descriptive set piece; he merely names the objects and actions. Thus, although he presents a physical action and setting, he is much more concerned with their mood and atmosphere than their physical details, and so he presents just enough i. to create that effect. Nevertheless, all of the lang. of the poem, except perhaps for the last two lines, is involved in this physicality. In the concluding quatrain,

> But no, I was out for stars:
> I would not come in.
> I meant not even if asked,
> And I hadn't been.

The first two lines refer to action and setting, while the last two are ratiocinative and nonphysical. It does make sense, then, despite the problems of cognitive psychology, the strictures of Furbank, and the subtleties of Mitchell, to distinguish between imaginal and nonimaginal statements. Further, since economy is a fundamental artistic principle, we often expect to find that literal i. is converted into a second subject, becoming the symbol (q.v.) of something else as a result of the speaker's reflective and deliberative activity. Mere scenery is rarely enough to justify its presence in a poem. As we proceed through Frost's poem, these expectations begin materializing around a

certain sort of symbolic structure: the sun has all but set, the speaker is hesitating between open field and enclosed woods, and he is pausing between going inside or staying outside. The speaker is not simply out for a pleasant evening walk; some more crucial issue is at stake.

Here is the penultimate stanza:

> Far in the pillared dark
> Thrush music went—
> Almost like a call to come in
> To the dark and lament.

Now we know something about the symbolic *significance* of the speaker's representation of this landscape: he is resisting the temptation of despair. Thus does the literal i. become symbolic, clustering around the poles of woods-inside-dark-birdsong-sunset-lamentation *versus* field-outside-dusk-stars-determination. The entire scene, therefore, and the speaker's response to it, are now symbolic of his inner struggle, which is his second or "real" subject—scene and act are now "vehicle" and inner struggle is the "tenor" of the basic "metaphor" which is the poem. And though most of the lang. of the poem is literal, in the third stanza the light of the sun is personified as having "died in the west," though it still "lived for one song more / In a thrush's breast." This figure is enfolded within another and more important one: the dying light of the sun is identified with the thrush's song. The thrush is singing, in other words, of the death of the day. All of these associations are reinforced by the i. of extremity which permeates the poem: "edge of the woods," "Too dark," "The last of the light," "Still lived for one song more," "I meant not even if asked." Everything seems just a bit desperate. But in the final two lines the speaker admits he realized the whole thing was a projection of his own despair: it was just a bird, and so we are led right out of the symbolic structure he has encouraged us to build—the same one he himself built. He has exorcised it. Are the "stars," then, an anti-symbolic symbol? Tension, irony, and ambiguity enough to satisfy any clever critic.

We could go even further, such is the skill and richness of this little lyric. Is there not the archetype of death-and-rebirth here? of the descent into the underworld and consequent ascent? of a ritual initiation in which the speaker-scapegoat tests himself against his own attraction-repulsion in relation to death? We could conclude with Mitchell that the i. in this poem is a representation of a cultural and literary convention rather than simply an unmediated piece of reality, and with Weimann that the poem's use of literal and figurative images embodies the social and historical realities of its time—1942. The literalism of the poem's i. reflects the modern suspicion of decorative figures and preference for implicit and symbolic i. The speaker's presentation of the experience as a matter of boundaries-hesitation-commit ment may be thought to reflect the nature of his society—its initial hesitation after the Depression in entering Word War II and its subsequent dedication to that task. In this view, the moods of Frost's speaker are not simply conventional poetic moods; they are, rather, the effects of a particular culture and history.

Thus, while we are not to regard the poem as embodying naked actuality, we are also not to ignore the fact that it springs from and refers back to an actuality, however mediated, and that the problem for the poet and the critic is to adjudicate among various representations of reality. The modernists did not believe that the autotelic nature of poetry either divorced it from history or revealed an "objective" truth about it. Realists such as Kermode and Weimann have exaggerated the influence of Fr. symbolism on 20th-c. Eng. and Am. poetry and poetics, while nominalists such as Kuhn, White, Rorty, and the deconstructionists (see Culler), have exaggerated the difficulty of matching mind to world through the agency of the Word. As Furbank says, a work is a self-contained "world" aesthetically so that it can refer to the world really. The entire enterprise has always been an attempt, whether successful or not, to define the way or ways in which Mind can come to know World other than simply factually, to balance our projections against what's out there, in the knowledge that we can never really finally say what it is. We know, on the one hand, that Mind mediates between ourselves and the World, and we seek, on the other, a validation-process for choosing among Mind's interpretations. The answer, as Frost's poem demonstrates—Stevens is another example, although without that literality—is to explore the boundary between them.

I. in poetry, therefore, may be seen to have the following uses. It may, in the first place, serve as a device for externalizing and making vivid the speaker's thoughts and feelings. It would have been much less effective had Frost's speaker simply stated, flatly, that he was tempted by despair and was seeking a way to resist it. In the second place, and consequently, since this scene serves to call up to the speaker's consciousness—and thereby becomes, as we have seen, the vehicle of—a problem which has been troubling him, it stimulates and externalizes further his mental activity. Mind and World are not simply being thought about, they are indeed confronting and interacting with one another—that is what the poem is "about," a *process* of thinking and feeling in relation to the environment rather than a set of "ideas" *about* it. Third, the poet's handling of i. serves to dispose the reader either favorably or unfavorably toward the various elements in the poetic situation. That Frost devotes four out of five stanzas on the thrush, for example, serves to indicate the magnitude of the temptation and hence the difficulty of turning away from it. I. may serve, in the fourth place, as a way of arousing and

guiding the reader's expectations. Frost sets up the nature of the problem implicitly from the beginning via "edge," "outside," "inside," and so on. And finally, i. may serve to direct our attention, as we "slide," in Burke's term, from i. to symbolism, to the poem's inner meanings—realizing, in this case, that the speaker is testing himself against his despair and coming out the other side.

JOURNALS: *Journal of Mental Imagery; The Literary Image; Word & Image.*

HISTORICAL CONSIDERATIONS: R. Frazer, "The Origin of the Term 'Image'," *ELH* 27 (1960); P. N. Furbank, *Reflections on the Word "Image"* (1970); *17th-C. I.*, ed. E. Miner (1971); W. J. T. Mitchell, *Iconology* (1986).

MENTAL IMAGERY: F. Galton, "Statistics of Mental I.," *Mind* 5 (1880); G. H. Betts, *The Distributions and Functions of Mental I.* (1909); J. K. Bonnell, "Touch Images in the Poetry of Robert Browning," *PMLA* 37 (1922); E. Rickert, *New Methods for the Study of Lit.* (1927); J. E. Downey, *Creative Imagination* (1929); R. H. Fogle, *The I. of Keats and Shelley* (1949); Wellek and Warren, ch. 15; R. A. Brower, *The Fields of Light* (1951); J. Press, *The Fire and the Fountain* (1955); M. Tye, *The I. Debate* (1992).

FIGURES OF SPEECH: M. Müller, *Lectures on the Science of Lang.* 2d ser. (1894); F. I. Carpenter, Metaphor and Simile in Minor Elizabethan Drama (1895); G. Buck, *The Metaphor* (1899); J. G. Jennings, *An Essay on Metaphor in Poetry* (1915); H. W. Wells, *Poetic I.* (1924); O. Barfield, *Poetic Diction* (1928), "The Meaning of the Word 'Literal,'" *Metaphor and Symbol*, ed. L. C. Knights and B. Cottle (1960); E. Holmes, *Aspects of Elizabethan I.* (1929); I. A. Richards, *The Philosophy of Rhet.* (1936), *Interpretation in Teaching* (1938); M. A. Rugoff, *Donne's I.* (1939); C. Brooks, *Mod. Poetry and the Trad.* (1939); R. Tuve, *Elizabethan and Metaphysical I.* (1947); C. D. Lewis, *The Poetic Image* (1947); H. Coombs, *Lit. and Crit.* (1953); C. Brooke-Rose, *A Grammar of Metaphor* (1958); D. C. Allen, *Image and Meaning* (1960); M. Peckham, "Metaphor" [1962], *The Triumph of Romanticism* (1970); E. E. Ericson, "A Structural Approach to I.," *Style* 3 (1969); D. M. Miller, *The Net of Hephaestus* (1971); R. J. Fogelin, *Figuratively Speaking* (1988).

CLUSTER CRITICISM: W. Whiter, *A Specimen of a Commentary on Shakespeare* (1794); W. Spaulding, *A Letter on Shakespeare's Authorship of* The Two Noble Kinsmen (1876); S. J. Brown, *The World of I.* (1927); F. C. Kolbe, *Shakespeare's Way* (1930); C. F. E. Spurgeon, *Shakespeare's I. and What it Tells Us* (1935); U. Ellis-Fermor, *Some Recent Research in Shakespeare's I.* (1937); K. Burke, *Attitudes Toward Hist.* (1937), *The Philosophy of Lit. Form* (1941); G. W. Knight, *The Burning Oracle* (1939), *The Chariot of Wrath* (1942), *The Crown of Life* (1961); M. B. Smith, *Marlowe's I. and the Marlowe Canon* (1940); L. H. Hornstein, "Analysis of I.," *PMLA* 57 (1942); E. A. Armstrong, *Shakespeare's Imagination* (1946); Brooks; R. B. Heilman, *This Great Stage* (1948); D. A. Stauffer, *Shakespeare's World of Images* (1949); T. H. Banks, *Milton's I.* (1950); W. H. Clemen, *The Devel. of Shakespeare's I.* (1951); F. Marsh, *Wordsworth's I.* (1952); J. E. Hankins, *Shakespeare's Derived I.* (1953); Frye; J. W. Beach, *Obsessive Images* (1960); G. W. Williams, *Image and Symbol in the Sacred Poetry of Richard Crashaw* (1963); D. A. West, *The I. and Poetry of Lucretius* (1969); S. A. Barlow, *The I. of Euripides* (1971); R. S. Varma, *I. and Thought in the Metaphysical Poets* (1972); W. E. Rogers, *Image and Abstraction* (1972); J. Doebler, *Shakespeare's Speaking Pictures* (1974); J. H. Matthews, *The I. of Surrealism* (1977); V. N. Sinha, *The I. and Lang. of Keats's Odes* (1978); R. Berry, *Shakespearean Metaphor* (1978); V. S. Kolve, *Chaucer and the I. of Narrative* (1984); J. Steadman, *Milton's Biblical and Cl. I.* (1984); J. Dundas, *The Spider and the Bee* (1985).

SYMBOL AND MYTH: O. Rank, *Art and Artist* (1932); R. P. Warren, "A Poem of Pure Imagination," *The Rime of the Ancient Mariner by Samuel Taylor Coleridge* (1946); Crane; P. Wheelwright, *The Burning Fountain* (1954), *Metaphor and Reality* (1962); Frye; F. Kermode, *Romantic Image* (1957); H. Musurillo, *Symbol and Myth in Ancient Poetry* (1961); K. Burke, *Lang as Symbolic Action* (1966); *17th-C. I.*, ed. E. Miner (1971); A. Cook, *Figural Choice in Poetry and Art* (1985).

RECENT DEVELOPMENTS: T. S. Kuhn, *The Structure of Scientific Revolutions* (1962); R. Weimann, *Structure and Society in Lit. Hist.* (1976); H. White, *Tropics of Discourse* (1978); R. Rorty, *Philosophy and the Mirror of Nature* (1979); *The Lang. of Images*, ed. W. T. J. Mitchell (1980); *I.*, ed. N. Block (1981); J. D. Culler, *On Deconstruction* (1982); P. Yu, *The Reading of I. in the Chinese Poetic Trad.* (1987). N.F.

IMAGINATION.

I. DISTINCTIONS FROM FANCY; GENERAL SCOPE
II. CLASSICAL AND MEDIEVAL
III. RENAISSANCE AND 17TH CENTURY
IV. 18TH CENTURY AND ROMANTICISM
V. LATER 19TH AND 20TH CENTURIES

I. DISTINCTIONS FROM FANCY; GENERAL SCOPE. I. derives from Lat. *imaginatio*, itself a late substitute for Gr. *phantasia.* During the Ren. the term "fancy"—connoting free play, mental creativity, and license—often eclipsed i., considered more as reproducing sense impressions, primarily visual images. By ca. 1700, empirical philosophy cast suspicion on fancy; i. seemed preferably rooted in the concrete evidence of sense data. Hobbes nevertheless retains "fancy" and is perhaps the last Eng. writer to use it to signify the mind's greatest inventive range. Dryden describes i. as a capacious power encompassing traditional stages of composition: invention, fancy (distribution or design), and elocution. Leibniz contrasts *les idées réelles* with *les idées phantastiques ou chimériques* (*Nouveaux essais*). Many, incl. Addison, use fancy and i. synonymously, but Addison calls his important *Spectator*

series (nos. 409, 411–21) "The Pleasures of the I."

More susceptible to prosodic manipulation in actual lines of verse, the term "fancy" retains a higher place in poetic diction than in crit. (e.g. Collins' "Young Fancy thus, to me divinest name"). But in England and particularly in Germany, writers increasingly distinguish the terms long before Coleridge's definitions in *Biographia literaria* (1817). By 1780, Christian Wolff, J. G. Sulzer, J. N. Tetens, and Ernst Platner make explicit distinctions. Coleridge recognizes this by claiming himself the first "of my countrymen" to distinguish fancy from i. But several Eng. writers record distinctions between 1760 and 1800.

Since 1800, and to some degree before as well, poets and critics have considered i. the chief creative faculty, a "synthetic and magical power" responsible for invention and originality (Coleridge). Writers have associated or identified i. with genius, inspiration, taste, visionary power, and prophecy. During the past three centuries, no other idea has proved more fruitful for poetics and critical theory, or for their intersection with psychology and philosophy. Before the watershed in the history of the idea during the 18th–19th cs., commentaries and poetics generally accord i. an important but ambivalent role: judgment or understanding must trim its vagaries and correct its wayward force.

Ancient poets, philosophers, and psychologists consider i. a strong and diverse power, but unregulated it produces illusion, mental instability (often melancholy), bad art, or madness. Yet i. nevertheless becomes the chief criterion of European and Am. romanticism. Even if we speak of romanticisms in the plural, i. is important to each one. Not only philosophers and psychologists, but critics and poets elevate it as the prime subject of their vital work. Wordsworth's *Prelude* proclaims i. as its main theme; Keats' *Lamia* and great odes debate in symbolic terms the function and worth of imaginative art. To a surprising extent, the transcendental or "critical" philosophy in Germany and America explores and elaborates the idea, spawning theories that champion the process and function of art as the final act and highest symbolic expression of philosophizing.

II. CLASSICAL AND MEDIEVAL. Aristotle advances in *De anima* the Cl. definition of i.: mental reproduction of sensory experience. In this elegant simplicity, i. registers sensory impressions that are immediately present in the act of perception. With sense data absent, i. becomes a form of memory, mother of the Muses. Aristotle discusses the role of i. in what Locke will call "the association of ideas," central to empirical psychology. Hume, Priestley, the associationists in general, Hazlitt, and to some degree Wordsworth will later view i. in terms of heightened associations of ideas (either unconscious in origin or determined by conscious choice) coexisting with feelings and passions, thus giving all associative operations a subjective and affective element. I. thus forms the basis of taste, by definition grounded in the perceiving subject, an argument Burke broaches in his *Enquiry into the Sublime and Beautiful* (1757) and later at the core of Kant's *Kritik der Urteilskraft* (Critique of Judgment, 1790).

But in the *Poetics*, where Aristotle pursues a structural or generic approach based on dramatic texts, he skirts the function of i. The audience's imaginative response projecting sympathy and identification (pity and fear) may be inferred, but Aristotle does not analyze i. as crucial to either reception or production.

Plato distrusts the artist's i. and assigns it only an essentially reproductive duty (copying a copy). In psychological terms, the soul passively receives an image, which reflects an idea. However, the Platonic *nous* (reason) carries a force similar to later conceptions of creative i. Sidney and Schelling, among others, will later interpret Plato's thought and image-laden writing to counter Plato's own condemnation of poetry and poets. For Plotinus, the i. (phantasy) is a plastic, constructive faculty which can change and alter experience, permit or realize a form of intellectual intuition or insight. Phantasy comes in two forms. The lower—linked to sense and the soul's irrational power—may harmonize with the higher, which reflects ideas and the rational. Plotinus and Proclus are among several who provide antecedents for Coleridge's "primary" and "secondary" i. As Plotinus considers nature an emanation of soul, and soul an emanation of mind, his thought adumbrates a large system of nature and sense, idea and soul, cosmos and mind, creator and creativity, all interconnected through imaginative power. Elements of this thought recur in the Hermetic philosophers, Jakob Boehme, Blake, Schelling, and Coleridge. Plotinus also provides broad implications for mimetic theory, and for a poetics that spills over into theodicy, where the external world is "the book of nature" or, as Goethe and others later express it (before the Rosetta Stone is deciphered), a divine "hieroglyphic." Using this general idea, Emerson and Thoreau develop their own angles of vision, as do the *Naturphilosophen* in their search for connections between the forces and laws discovered by natural science and poetry conceived as an inventive power creating its own related—but original—nature.

From the beginning, i. thus possesses roots both in empiricism based on sense experience and in transcendentalism. Its function may be seen in terms of natural and psychological phenomena, perception, acquisition of knowledge, creative production of art, and even, as Spinoza says, the prophets who receive "revelations of God by the aid of i." (*Tractatus theologico-politicus*, 1670). These manifestations are not mutually exclusive but interact, as in the thought of Giordano Bruno, who conceives of sense, memory, emotion, cognition, and the divine mind all linked through i.

Many writers, in fact, divide the power into levels or degrees, with adjectives denoting particular functions (e.g. re/productive, *erste/zweite/dritte potenz*, primary or secondary, creative, sympathetic, perceptive). *The Prelude* traces a maturation of the power through different stages. The explosion of Ger. analytical terms for i. in the 18th c. is staggering (*Einbildungskraft, Phantasie, I., Fassungskraft, Perceptionsvermögen, Dichtungsvermögen* or *Dichtkraft*—with variants, and more). Since antiquity, then, images produced by this power have been variously considered as materially real, or simply as appearances and pure illusion, or even as idealized forms. Poetry, dreams, divine inspiration (madness and prophecy), delusion and emotional disturbances—all involve one or more activities of i. It therefore harbors the greatest potential not only for insight and intuition but for deception and illusion.

Horace, long influential in Western poetics, barely mentions i. in his *Ars poetica*. This absence, coupled with Platonic distrust of i. and later neoclassical emphasis on verisimilitude and decorum in imitation (what much 18th-c. and romantic theory scorns as mere "copy"), diminishes the place of i. in poetic theory for centuries after Horace. Although Philostratus declares imitation inferior to phantasy and Quintilian connects i. (*visiones*) with raising absent things to create emotion in the hearer, not until the rediscovery in the 16th c. of the *Peri hypsous* (*On the Sublime*) attributed to Longinus do European critics turn more directly to the power he describes: "moved by enthusiasm and passion you seem to see the things whereof you speak and place them before the eyes of your hearers." Boileau translates Longinus; in England Dennis and Pope spread his views. The useful, if rough, distinction between "Aristotelian" and "Longinian" crit. stems largely from divergent emphases on i. and its corollaries for mimetic theory.

III. RENAISSANCE AND 17TH CENTURY. Pico della Mirandola's *De imaginatione sive phantasia* (ed. and tr. H. Caplan, 1971) collects from the Cl. and Med. trads., but stresses vision as the primary or archetypal sense, circumscribing the possibility of a larger poetics founded on i. Puttenham summarizes Ren. notions of i. in the production of poetry and emphasizes the link between poet as creator and a divine creator. Sidney and Scaliger call the poet a "creator" (Johnson calls him a "maker"), implying that such creation imitates or follows nature rather than copies it.

Hence a rigid dichotomy should not be established between imitation and creative i.; it is possible to ally them, as Coleridge does in *On Poesy or Art*. As Coleridge frequently says, to distinguish is not necessarily to divide. For even with Sidney (and later Leibniz, Reynolds, Shelley, and Carlyle), imitation is not so much a duplication of nature as an echo of divine creative power. The 19th-c. Platonist Joubert will claim that the poet "purifies and empties the forms of matter and shows us the universe as it is in the mind of God. . . . His portrayal is not a copy of a copy, but an impression of the archetype." Macaulay will call this the "imperial power" of poetry to imitate the "whole external and the whole internal universe." For Shaftesbury the poet is a "just Prometheus under Jove," and the variation on the Ovidian phrase—a god or daemon in us—becomes common in discussions of i.

During the *Sturm und Drang* and romantic eras, the theme of Prometheus further modifies the connection between divine creative energy and human poetry. In Blake's Christian vision these powers merge as the "divine-human i." Schelling calls genius "a portion of the Absolute nature of God." Poetry for Shelley elicits "the divinity in man." Coleridge considers the secret of genius in the fine arts to make the internal external and the external internal; in religious terms, Jesus is the living, communicative intellect in God and man—suggestively phrased in Emerson's "living, leaping Logos."

Ren. psychologists, among them Melancthon, Amerbach, and esp. Vives, advance increasingly sophisticated views of the mind wherein i. and the association of ideas play important roles. Sidney and Spenser consider imaginative power central to both theory and practice. Shakespeare uses i. (or fancy as its equivalent) suggestively, as in the Chorus of *Henry V*. In Eng. critical discussion of i., Theseus' lines in *A Midsummer Night's Dream* (5.1) become the most-quoted verses (*Paradise Lost* also figures prominently). Although close reading reveals Theseus' distrust of "strong" i. and its "tricks," identifying i. with madmen, lovers, and poets—whose tales may never be believed—Hippolyta reinterprets all the players and the night's action as a totalized myth which "More witnesseth than fancy's images, / And grows to something of great constancy."

Bacon's view of rhet. and poetry must be spliced together from his many writings. Two trends emerge. He splits rational or scientific knowledge from poetic knowledge, which is "not tied to the laws of matter." But this bestows greater freedom on i. to join and divide the elements of nature, to appeal to psychological satisfaction rather than verisimilitude, and to permit the imitation of nature frankly to differ from nature itself not only in the medium of presentation but in its appeal to moral value—and to what would later be called aesthetics and the sublime (q.v.). Later these premises resurface often, incl. Addison's "Pleasures of the Imagination," Hume's "Of the Standard of Taste," and Reynold's *Discourses* (esp. 6, 7, and 13), where i. rather than the matter-of-fact becomes "the residence of truth." In *The Advancement of Learning* Bacon mentions an "imaginative or insinuative reason," a concept Addison repeats. But Bacon concludes, "I find not any science that doth properly or fitly pertain to the i."

Hobbes fails to see the capacious possibilities of Bacon's scheme for crit. or the arts, but his contri-

butions to empirical psychology open other avenues and begin to supply the science Bacon found wanting. For Hobbes the "compounded i." forms "trains of ideas" or "mental discourse" which evolves into judgment or *sagacitas.* In "compounding" images and directing these associations to larger designs, the "contexture" of i. offers a picture of reality. This suggests what thinkers in the next century commonly express: i. subsumes judgment. In 1774 Alexander Gerard's *Essay on Genius* explicitly declares this.

Locke coins the phrase "association of ideas" in the *Essay Concerning Human Understanding* (4th ed., 1700). Originally he means idiosyncratic links between mental representations peculiar to an individual, a process Sterne explores in *Tristram Shandy.* But the phrase, like William James' "streams of consciousness," changes signification and soon stands for a pervasive habit of mind, both conscious and unconscious, akin to Hobbes' mental discourse and trains of compounded ideas. Locke receives little credit for recognizing i. as a strong faculty because he criticizes poetry and shows scant sympathy for art, and he distrusts figuration in lang. But the "tabula rasa" tag oversimplifies his epistemology and psychology. He states that "the mind has a power," an innate power (distinguished from innate ideas, which it never possesses) "to consider several" simple ideas "united together as one idea; and that not only as they are united in external objects, but as itself has joined them." As in Bacon, the active formation of complex ideas in "almost infinite variety" need not correspond to nature.

Replying to Locke's difficult arguments about innate ideas, Leibniz in his *Nouveaux essais* quotes Aristotle's *De anima*: "There is nothing in the mind not previously in the senses—except the mind itself." Here, as Santayana later notes, Leibniz uncovers the germ of Kant's critical philosophy. In Leibniz—and in Hobbes' and Locke's ascription of an active power forming complex ideas or trains of them—i. acquires a central position in a new, flexible faculty-psychology with its accompanying facultative logic.

IV. 18TH CENTURY AND ROMANTICISM. (See also paragraphs II.1,3,4 and III.1,3 above.) It is Joseph Addison who brings these concepts within the ken of critical theory and poetics. "The Pleasures of the I." arise from comparing our imaginative perception of nature with our perception of art, where art is itself an imitation of nature achieved through the artist's imaginative production which "has something in it like creation; it bestows a kind of existence.... It makes additions to nature." The interest rapidly becomes psychological and involves a projective faculty, active and passive, productive and reproductive. Like Pico, Addison emphasizes vision, but remarks that the pleasures of i. "are not wholly confined to such particular authors as are conversant in material objects, but are often to be met with among the polite masters of morality, crit., and other speculations abstracted from matter, who, though they do not directly treat of the visible parts of nature, often draw from them their similitudes, metaphors, and allegories. . . . A truth in the understanding is . . . reflected by the i.; we are able to see something like colour and shape in a notion, and to discover a scheme of thoughts traced out upon matter. And here the mind . . . has two of its faculties gratified at the same time, while the fancy is busy in copying after the understanding, and transcribing ideas out of the intellectual world into the material." The material and the intellectual or passionately moral worlds thus fuse through i.; matter and spirit find a common faculty and may be represented by a single image.

In part derived from Addison's crit., Mark Akenside's popular poem *The Pleasures of the I.* combines empirical and Platonic elements. I. "blends" and "divides" images; its power can "mingle," "join," and "converge" them—phrases which anticipate Coleridge's "secondary" i. Collins and Joseph Warton reject satiric and didactic modes of verse; Warton proclaims that "invention and i." are "the chief faculties of a poet." Edward Young links originality and genius with i.; Gerard, whom Kant later praises as "the sharpest observer of the nature of genius," says, "it is i. that produces genius." Gerard, Priestley, and Duff extend associationist and empirical psychology to give a fluid model of the mind, not rigid or compartmentalized. Hazlitt concludes: "the i. is an *associating* principle" assuring "continuity and comprehension of mind," a definition allied with his remark that poetry portrays the flowing and not the fixed. Emerson eventually gives a twist to this organic process by hinting at a Neoplatonic foundation in "The Poet": "the endless passing of one element into new forms, the incessant metamorphosis, explains the rank which the i. holds in our catalogue of mental powers. The i. is the reader of these forms."

In Italy, L. A. Muratori advances a mutually beneficial combination of intellect and i. to produce "artificial" or "fantastic" images applied metaphorically and charged with emotion. Vico's *Scienza nuova* (3d ed., 1744) mentions a recollective *fantasia* but, more importantly, examines how poetic i. creates the basis for culture through the production of myth and universal patterns that shape understanding of both nature and human nature. Largely ignored during his lifetime, Vico produced ideas that have influenced historiography, anthropology, and imaginative writers since the early 19th c.

While Johnson and Hume generally distrust the i., they stress its pervasive ability to supplant reason. I. for Hume becomes the central faculty, and for Johnson, in his morality and psychology, the central concern. Both keenly appreciate the role of the passions in strengthening imaginative activity and the association of ideas. They yoke the Cl. theory of passions, further developed in the Ren.,

to empirical psychology. Hume sees i. as a completing power acting on suggestiveness; he claims that the elements a writer creates "must be connected together by some bond or tie: They must be related to each other in the i., and form a kind of *unity* which may bring them under one plan or view." This anticipates the romantic stress on organic unity. As Coleridge defines "the secondary" i., it "dissolves, diffuses, dissipates, in order to re-create; or where this process is rendered impossible, yet still at all events it struggles to idealize and to unify."

The associationists—among them Kames, Alison, H. Blair, Gerard, Hazlitt (to some extent), and others—stress imaginative association as pervasive; it determines taste. Images and sense data receive modifying "colors" of feeling and passion, a notion Wordsworth studies and uses widely. The emphasis on feeling fused with perception or cognition may be compared with I. A. Richards' later discussion of "emotive" versus "intellectual" belief (*Practical Crit.*), or T. S. Eliot's "dissociation of sensibility."

In Burke's *Enquiry* the imaginative arts become the "affecting arts." Because they trigger the completing i. of reader or audience, "suggestion" and "obscurity" attain positive value in poetry and visual art. This helps explain the growing interest in literary forms and genres not rigidly fixed but in metamorphosis, also the fascination in Blake, Novalis, and others for aphorism, and the importance of the literary fragment in the early 19th c. The i. of the reader becomes regarded as an important critical concept, too, ranging from Dryden's and Locke's "assent" to Coleridge's "willing suspension of disbelief." J. G. Sulzer, in his *Allgemeine Theorie der Schönen Künste*, notes, "the i. of those who hear or see the artist's work comes to his aid. If through any of the latent qualities in the work this i. takes on a vivid effect, it will thereupon complete what remains by itself."

Sensing rapid expansion of the idea, Burke claims in his *Enquiry*, (6th ed., 1770) that to i. "belongs whatever is called wit, fancy, invention, and the like." Hazlitt later remarks, "this power is indifferently called genius, i., feeling, taste; but the manner in which it acts upon the mind can neither be defined by abstract rules, as is the case in science, nor verified by continual unvarying experiments, as is the case in mechanical performances." Critics analyze passages of poetry as reflecting the pervasive operation of i., one aspect of what Coleridge first calls "practical crit.," a forerunner of the New Criticism in the 20th c.

Adam Smith bases his *Theory of Moral Sentiments* (1759) on the sympathetic identification that i. allows us to extend to others. As James Beattie says, "the philosophy of Sympathy ought always to form a part of the science of Crit." Hazlitt later states: "passion, in short, is the essence, the chief ingredient in moral truth; and the warmth of passion is sure to kindle the light of i. on the objects around it." The application to crit. is found in Wordsworth, Coleridge, and Keats, culminating rhetorically (along with much else) in Shelley's *Defence*: "the great secret of morals is love; or a going out of our own nature, and an identification with the beautiful . . . not our own. A man . . . must imagine intensely and comprehensively; he must put himself in the place of another and of many others. . . . The great instrument of moral good is the i.; and poetry administers to the effect by acting upon the cause. Poetry enlarges the circumference of the i." Applications of what Coleridge calls "sympathetic i." are crucial to changes in dramatic crit. Before 1800, William Richardson, Thomas Whatley, and Maurice Morgann approach Shakespeare in this fashion, and Coleridge becomes the most brilliant exponent of such psychologically based crit., this too a forerunner of the psychological criticism of the 20th c.

The idea of imaginative sympathy also affects theories of poetic lang.: it should be "natural" and "spontaneous." Metaphor, personification, and figurative writing in general are now viewed less as ornament and more as essential to impassioned "natural lang." Theories of the primitive origin of lang. (Rousseau, Herder, Duff, Monboddo, and many recapitulations, incl. Shelley's and Hegel's) strengthen connections between i., poetry, early devel. of societies and their langs., and figurative speech in general. As Hazlitt says, exemplifying the connection between i. and the fascination with figurative lang., a metaphor or figure of speech requires no proof: "it gives *carte blanche* to the i." In the best poetry, images are "the building, and not the scaffolding to thought." What is true not only may but must be expressed figuratively, and is a specially valid form of knowledge. Hazlitt, neither a religious enthusiast nor transcendentalist, even says i. holds communion with the soul of nature and allows poets "to foreknow and to record the feelings of all men at all times and places."

Blake's emphasis falls heavily on i. as a power ultimately communicating and sharing with the holy power of creation, a vision that culminates in an Edenic state. "All things Exist in the Human I." "Man is All I.," so also "God is Man & exists in us & we in him." "The Eternal Body of Man . . . The I., that is, God himself," is symbolically "the Divine Body of the Lord Jesus." We are "Creating Space, Creating Time according to the wonders Divine of Human I."

Wordsworth's ideas of i. evolve from the Preface to the 2d ed. of *Lyrical Ballads* (1800), through its 1802 version, to a new preface and "Essay Supplementary" in 1815. In *The Prelude*, he describes the infant:

> . . . his mind
> Even as an agent of the one great mind,
> Creates, creator and receiver both,
> Working but in alliance with the works
> Which it beholds.—Such, verily is the first
> Poetic spirit of our human life.

Cf. Coleridge's definition of "primary" i.: "the

living Power and prime Agent of all human Perception, and . . . a repetition in the finite mind of the eternal act of creation in the infinite I AM." Coleridge actually began the *Biographia* as a Preface to his own volume of verse, in part answering and modifying Wordsworth, but expanded it to a critical exposition and autobiography of his *literary* life that pivots on a reply to Wordsworth's published ideas concerning fancy and i. Coleridge considers that Wordsworth too closely links fancy and i. and explains i. too exclusively via associationism.

In the 1815 Preface, Wordsworth tries to explain how i. creates as well as associates. The power becomes for him so vast it challenges all conception, as with the apostrophe in *The Prelude* after crossing the Alps, or later: "I., which, in truth, / Is but another name for absolute power / And clearest insight, amplitude of mind, / And Reason in her most exalted mood." Wordsworth also claims: "I. having been our theme, / So also hath that intellectual Love, / For they are each in each, and cannot stand / Dividually." He uses his idea of i. to establish new grounding in poetic lang. and vindicates it in his practice (though not always in perfect conformity with his theoretical statements). He chooses not to exert a systematic philosophical concern for the idea as Coleridge does. He even considers that the word i. "has been overstrained . . . to meet the demands of the faculty which is perhaps the noblest of our nature."

Keats explores the idea of sympathetic i. and "negative capability," a search for truth or reality without any compulsive reaching after fact. Many of his letters and greater poems are undogmatic speculations on the value, "truth," and nature of the imaginative inner life. Shelley, regarding reason more in its deductive and experimental mode, contrasts it sharply with i., and on this distinction builds his *Defence of Poetry*. The essay recapitulates developments of the previous half century. He constructs his argument through powerful images and analogies rather than by Keats's "consequitive reasoning." Shelley's impassioned and figurative prose actualizes its own subject. He emphasizes the unconscious power of i. over the conscious will in poetic composition.

Coleridge, familiar with the philosophical developments in the idea since Plato and Aristotle, fuses and transforms them into the most suggestive and fruitful critical observations of Eng. romanticism. Though he never completes a systematic work that includes extended discussion of i., his theoretical pronouncements and practical insights provide rich ground for surmise. He combines British empirical psychology, Platonism and Neoplatonism, scholastic and Hermetic philosophies, and Ger. transcendentalism. The result is more than an admixture: he crystallizes and connects issues of perception and constitutive or regulative ideas, of associationism, of theories of lang. and poetic diction, and of the function of i. into a *mimesis* that "humanizes" nature not only by reproducing natural objects and forms (the "fixities and definites" manipulated by fancy) but also by imitating the living process through which they exist and through which we feel and realize them. Unlike many contemp. Ger. writers (e.g. Fichte or Schelling), he applies the theory of i. at the level of phrase and individual image in the crit. of poetry.

Symbols for Coleridge are "living educts of the i., of that reconciling and mediatory power which, incorporating the reason in images of the sense, and organizing (as it were) the flux of the senses by the permanence and self-encircling energies of the reason, gives birth to a system of symbols, harmonious in themselves and consubstantial with the truths of which they are the conductors" (*The Statesman's Manual*). This definition is virtually identical to Bishop Lowth's discussion of "mystical allegory" in *De sacra poesi hebraeorum* (1753). Thus, even in the symbol as defined by Coleridge, we see elements of prophecy and its earlier critical devel. in the 18th c. Various romantic definitions of symbol, allegory, myth (qq.v.), and schema, whether from Kant, Schelling, Solger, Hegel, or others, all rely on i. Coleridge enlarges, reformulates, and applies the idea of i. in ways seminal for poetics. He may be regarded as the most important progenitor of the New Critical valuation of organicism and of unity.

The British devel. of the idea thus draws from both Platonic and Neoplatonic strains as well as from the line of 18th-c. empiricism initiated by Hobbes and Locke. With Coleridge, and to some extent Shelley and Carlyle, Ger. transcendental philosophy further enriches the Anglophone trad., just as British writing on the subject had vitalized Ger. thought in the 18th c. The Ger. background draws heavily on Spinoza, Locke, Leibniz, Wolff, Addison, Shaftesbury, the "Swiss Critics" Bodmer and Breitinger, Baumgarten, Hume, and the associationists (histories of associationism are written in 1777 by Michael Hissmann and in 1792 by J. G. E. Maass). In the later 18th c., Platner, Herder, and Sulzer extend discussion into psychology, lit., culture, and myth. Sulzer declares, "mythological poems must be considered as a lang. of i. . . . they make a world for themselves." Tetens, most eminent psychologist of the era, breaks the power of i. into different levels responsible for perception, larger associations and cognition, and ultimate poetic power. He profoundly influenced Kant, and Coleridge read him carefully.

The issue in Kant is central and vexing; his presentations of i. are multiple. Taking a less than clear-cut but prominent place in the *Critique of Pure Reason*, i. is "an active faculty of synthesis" operating on sense experience, "a necessary ingredient of perception itself" that mediates between senses and understanding. But "a transcendental synthesis of i." also exists, so that again, i. is both an empirical and a transcendental faculty or power. In the *Critique of Judgment* i. is vital to the

analytic of both beauty and the sublime. It is not simply reproductive and operating under "the laws of association," but also "productive and exerting an activity of its own." Kant tries to reconcile empirical views of the "reproductive" i. with those derived from i. in its "pure" or "productive" mode; this leads to a "tertium medium" joining the two, an idea Coleridge also suggests in the *Biographia.* In our perceptions, Kant claims, we "introduce into appearances that order and regularity which we name nature." In aesthetic theory he stresses the "free play" of i. and how, combined with taste, it draws an analogy between beauty and morality through the medium of the symbol.

Fichte proclaims "all reality is presented through the i." On the faculty of "*the creative i.* . . . depends whether we philosophize with or without spirit . . . because the fundamental ideas . . . must be presented by the creative i. itself. . . . The whole operation of the human spirit proceeds from the i., but an i. that can be grasped no other way than through i." He elevates i. as the most important epistemological faculty and bases philosophy on it. In part to resolve potential contradictions or dualities in Kant, and to escape Fichte's more abstract epistemology, Schiller writes his *Ästhetische Erziehung* (1794–95), where free play of i. becomes *Spieltrieb* (the "play drive"). With antecedents of i. as "play" or "free play" in Bacon, Kant, Wieland, and Lessing, Schiller explores i. as an aesthetic state of being that oscillates (recalling Fichte's "schweben der Einbildungskraft") between "form" and "sense" drives. This aesthetic i. renovates the soul and permits it to be "fully human," opening a line of Utopian thought. An extensive transcendental poetics develops with Schiller, the Schlegels, and Novalis. They all hold i. to be of supreme importance in the psychological and epistemological grounding of art.

Schelling constructs his *System des tranzscendental Idealismus* (1800) on the idea of i., which in that and other works he carries further than perhaps any other thinker. It is central to his philosophy of nature and of mind as one larger system insured by the unifying revelation of the work of art. The *Kunstprodukt* combines the force of nature with that of mind into something that had not previously existed in either. I. creates the myths that secure all cultural and spiritual significance. Finally, the artist's i. generates the most comprehensive symbols; in them knowledge realizes its highest manifestation. Art becomes necessary to complete philosophy; philosophy's highest goal is the philosophy of art, ultimately based on i. in God, artist, philosopher, and audience.

Hegel's *Aesthetik* (1835) utilizes i. or "Geist" (spirit) as a key element for his historical and critical views but does not much enlarge the theory of i. Goethe emphasizes i., though in unsystematic fashion. The hermeneutic trad. from Schleiermacher through Dilthey and down to Gadamer relies on i. as an instrument of knowledge and interp., with Gadamer roughly retaining the older distinction between i. and discursive reason as that between *Wahrheit* ("truth") and *Methode* ("method").

Romantic writers provide the idea with its most extreme formulation and make the highest claims for i. It stands directly related to—and necessary for—perception, memory, images, ideas, knowledge, worldviews, poetry, prophecy, and religion. Santayana proclaims Emerson the first modern philosopher to base a system on i. rather than reason. Schelling or Coleridge may have a stronger claim, though Coleridge's later thought increasingly grasps a fuller reason, both discursive and intuitive, as subsuming imaginative power. Even a brief review of romantic manifestations of i. threatens to expand infinitely.

Perhaps the greatest "romantic" claim is that i. resolves all contradictions and unifies the soul and being of creator and receiver, writer and reader, subject and object, and human nature and *Naturgeist* alike. I., says Coleridge, "reveals itself in the balance or reconciliation of opposite or discordant qualities: of sameness, with difference; of general, with the concrete; the idea, with the image; the individual, with the representative. . . ." Coleridge, Schelling, Schiller, and Shelley all say that i. or poetry calls upon the "whole" soul or individual. To characterize this activity Coleridge coins two words: in an 1802 letter "co-adunating" (from the Lat. "to join" or "to shape into one") and in the *Biographia* the famous "esemplastic" (with its analogous Gr. basis). Schelling posits a fanciful etymology reminiscent of Herder: "In-Eins-Bildung" ("making into one") for "Einbildungskraft." The traditional way of expressing this unifying action, recognized in Neoplatonic and Hermetic circles, was to say, as Wordsworth does, that i. "modifies" or throws "one coloring" over all its productions. As the resolution of contradictions or antinomies, i. could be seen as a metaphysical, psychological, and artistic principle—as Kant said, a "blind power hidden in the depths of the soul"—that unifies noumenal and phenomenal, sensory and transcendental, mind and spirit, self and nature (e.g. Fichte's *Ich* and *Nicht-Ich*), even freedom and necessity. The power thus resolves Cartesian dualism and all subject/object, ego/world, Aristotelian/Platonic or Neoplatonic divisions. It drives all dialectical process and provides genuine knowledge.

V. LATER 19TH AND 20TH CENTURIES. Such full-dress syntheses left scant room for further analysis or higher claims. Arguably, i. had come to stand for many ideas, not one. The associated terms and adjectives grew confusing, and G. H. Lewes remarked: "there are few words more abused."

Ruskin posits three modes of i., and an elaborate, schematic division of fancy and i. He stresses the intuitive grasp of art rather than its reasoning or analysis. Pater echoes Coleridge to some degree, as Wallace Stevens will, but neither adds a new dimension to theory of i. In Germany Theo-

dore Lipps' studies of *Einfühlung* or empathy are related to an imaginative grasp of living truth; in this connection Hopkins' "inscape" also suggests an intuition of object and the feelings it arouses. Oscar Wilde's simultaneous attacks on realism and on unrealistic romances in "The Decay of Lying" are playful defenses of what earlier had been called "feigning," now phrased in more provocative and paradoxical lang. than Sidney's or Shelley's.

Irving Babbitt, while unsympathetic to romanticism, nevertheless develops a theory of ethical i. similar to 18th-c. discussions. Croce's aesthetics and poetics lean heavily on a productive i. (*fantasia*) that acts under the influence of intuition and feeling to produce a unifying image. Though without extensive elaboration on the theory of i. as such, anthropological and psychological studies by Frazer, Lévi-Strauss, Freud, Jung, Eliade, and others have kept alive and deepened concerns voiced by Vico and Blackwell. The interest here, as to some extent with Cassirer's theory of the symbol, is more with the formation and importance of the individual image or myth than with all the diverse powers once attributed to i. Richards, through his scholarship and crit. of late 18th-c. figures and esp. through *Coleridge on I.*, combines elements of a romantic aesthetic with modern psychology and helps create the New Criticism, indebted to Coleridge and the concept of organic unity.

R. G. Collingwood, positing i. as a mediating faculty between sense and intellect, refashions romantic theory and attempts to give it consistent shape. I. provides real knowledge guaranteed by the work of art, which raises perception, feelings, creation, and expression to consciousness through a concrete act. Collingwood jettisons higher claims and focuses on a comprehensive aesthetic, but the currency and effect of his views—fundamentally out of step with postmodernism—diminishes in the second half of the 20th c. More recently, neurophysiologists have attempted to analyze areas of the brain, while developmental psychologists have empirically studied artists' and children's creative activities to determine the functions played by brain hemispheres, biochemistry, the environment, habit, and association in the imaginative process, incl. poetry. Results have been mixed, with no single theory or explanation emerging.

But the notion of i. as creating unity is generally opposed to the spirit of literary modernism and postmodernism. Nor is it central to phenomenological approaches. More stress falls on the power of the individual image, or images juxtaposed, e.g. in imagism or the metaphysical revival. Claims for art to save and enlighten, or to provide special knowledge, are reduced. Detractors of the New Criticism simplify and attack the concept of organic unity. However, though less massively than in romantic poetics, the idea of i. continues to play a role in postmodernism. Structuralism offers affinities with romantic organicism and an imaginative, intellectual reconstruction of reality, the creation of this "simulacrum" itself an imitation and not a copy. Deconstruction shuns any unifying power to resolve contradictions and builds itself—or rather deploys its various moves and strategies—in part by exploiting those contradictions or divisions. But while deconstruction seems a polar opposite of imaginative unity and organic synthesis, the romantic theory of i. itself thrives on such contradictions and polarities. In one sense i. and deconstruction are allied: in crit. or philosophy they distrust any formal system or set of rules for either analysis or creativity; they rejoin poetry and philosophy through the medium of words and through a consideration of the nature and the history of writing in general, and figurative lang. in particular.

I. Babbitt, *Rousseau and Romanticism* (1919); H. C. Warren, *Hist. of Association Psychology* (1921); G. Santayana, *Character and Opinion* (1921); B. Croce, *Aesthetic*, tr. D. Ainslie (1922), *Essays on Lit. and Lit. Crit.*, ed. and tr. M. E. Moss (1990); L. P. Smith, "Four Romantic Words," *Words and Idioms* (1925); Richards, esp. ch. 32; M. W. Bundy, *The Theory of I. in Cl. and Mod. Thought* (1927), "'Invention' and 'I.' in the Ren.," *JEGP* 29 (1930), "Bacon's True Opinion of Poetry," *SP* 27 (1930); R. Wellek, *Kant in England* (1931); A. S. P. Woodhouse, "Collins and the Creative I.," *Studies in Eng.*, ed. M. Wallace (1931), "Romanticism and the Hist. of Ideas," *Eng. Studies Today*, ed. C. Wrenn and G. Bullough (1951); I. A. Richards, *Coleridge on I.* (1934); S. H. Monk, *The Sublime* (1935); D. Bond, "'Distrust of I.' in Eng. Neoclassicism," *PQ* 14 (1935), "Neoclassical Psychology of the I.," *ELH* 4 (1937); R. G. Collingwood, *Principles of Art* (1938); C. D. Thorpe, *The Aesthetic of Hobbes* (1940); K. R. Wallace, *Bacon on Communication and Rhet.* (1943); W. J. Bate, "Sympathetic I. in 18th-C. Eng. Crit.," *ELH* 12 (1945), *From Classic to Romantic* (1946); W. J. Bate and J. Bullitt, "Distinctions between Fancy and I.," *MLN* 60 (1945); W. Stevens, *The Necessary Angel* (1951); Crane; Abrams; E. L. Fackenheim, "Schelling's Philosophy of the Literary Arts," *PQ* 4 (1954); Wellek, v. 1; M. H. Nicolson, *Science and I.* (1956); R. Cohen, "Association of Ideas and Poetic Unity," *PQ* 36 (1957); Wimsatt and Brooks; E. Tuveson, *The I. as a Means of Grace* (1960); E. D. Hirsch, Jr., *Wordsworth and Schelling* (1960); W. P. Albrecht, *Hazlitt and the Creative I.* (1961), chs. 1, 3, 5; B. Hathaway, *The Age of Crit.* (1962); R. Barthes, "The Structuralist Activity," (1963); F. Yates, *The Art of Memory* (1966); T. McFarland, *Coleridge and Pantheist Trad.* (1969), *Originality and I.* (1985); H. Mörchen, *Die Einbildungskraft bei Kant* (1970); T. Todorov, *Theories of the Symbol* (1977, tr. 1982); W. Wetherbee, *Platonism and Poetry in the 12th C.* (1972)—for i. and *ingenium*; R. Scholes, *Structuralism in Lit.* (1974), ch. 6; M. Warnock, *I.* (1976); E. S. Casey, *Imagining: a Phenomenological Study* (1976); J. Derrida, *Writing and Difference*, tr. A.

Bass (1978); J. Engell, *The Creative I.* (1981), ch. 1, 13; D. P. Verene, *Vico's Science of I.* (1981); de Man; E. Dod, *Die Vernünftigkeit der I.* (1985)—on Schiller and Shelley; M. Kipperman, *Beyond Enchantment* (1986)—Eng. poetry and Ger. idealism; C. G. Ryn, *Will, I. and Reason* (1986)—on Babbitt; R. Kearney, *The Wake of I.* (1988); *Coleridge, Keats, and the I.*, ed. J. R. Barth and J. L. Mahoney (1989); *Coleridge's Theory of I. Today*, ed. C. Gallant (1989).
J.E.

IMITATION.

I. FROM THE HISTORICAL PERSPECTIVE
II. FROM THE MODERN PERSPECTIVE

I. FROM THE HISTORICAL PERSPECTIVE. Until the 20th c., when it was restored to the critical vocabulary by Auerbach and the members of the Chicago School, "i." had been out of favor as a literary term since the 18th c. Its eclipse began with the neoclassical critical stirrings that led the way to romanticism, when "i." (Gr. *mimesis*; Lat. *imitatio*) was increasingly felt to be out of keeping with the new spirit of originality, spontaneity, and self-expression. Despite the fact that romantic poetics was preserved essentially intact well into the 20th c., interest in i. as a critical concept has been revived in recent years in connection with efforts to question the concept of representation in lang. and with a new interest in rhetoricity (see section II below).

The original connotation of *mimesis* (see REPRESENTATION AND MIMESIS) seems to have been dramatic or quasidramatic. Whether any theory of poetry as i. was developed before Plato is uncertain. Gorgias's notion of tragedy as a "beneficent deception" (*apate*) perhaps anticipated it in part; and Democritus certainly held that the arts in general arose out of i. of nature: singing, for example, from i. of the birds. The first place where we can actually grasp *mimesis* as a critical term is Plato's *Republic*, Books 3 and 10, though in Book 3, the context is political and pedagogical rather than literary. Plato's concern there is with the education of his elite corps of Guards, and he judges poetry strictly by that criterion. "I." is identified almost exclusively with the dramatic mode, i.e. with the direct impersonation of literary characters. This involves an identification of oneself with others which is perilous for the young; it may lead them to indiscriminate i. of low and unworthy persons. Hence poetry, but esp. the drama, must be banished from the professional education of the ideal ruler. In Book 10, Plato renews his attack on a broader front. I. is now identified as the method of *all* poetry, and of the visual arts as well. The poet, like the painter, is incapable of doing more than counterfeiting the external appearance of things; Truth, the realm of Ideas, is inaccessible to him. In this second discussion (perhaps written later), Plato's attention has shifted from the method of i. to its object, and i.—i.e. art—is condemned not merely for its moral effects but because it cannot break through the surface of Appearance to the reality it ought to reproduce, Ideas.

This crushing verdict upon "i." does not result, as we might expect, in banishing the term from Plato's world of discourse; on the contrary, it permeates his thinking more and more in the later dialogues. In the *Sophist* (236) he hints at the possibility of a "true i." which would reproduce the real nature and proportions of its object. Indeed Plato came to think of the whole complex relation of Becoming to Being, Particular to Idea, as a kind of i. Thus the *Timaeus* (27 ff.) presents the universe itself as a work of art, an "image" of the world of Ideas made by a divine craftsman. From this it is only a step to conceiving visual art, and then poetry, as a sensuous embodiment of the ideal. The Neoplatonists (see Plotinus, *Enneads* 5.8; cf. Cicero, *Orator* 2.8–9) took this step, but Plato himself did not. The condemnation of poetry in the *Republic* was never explicitly revised or withdrawn (it is substantially repeated in the *Laws*, Books 2 and 7), and the developments just mentioned remain hints (highly fruitful ones for later thought) rather than a new and positive doctrine of poetic i.

Aristotle accepts i. (*Poetics* 4) as a fundamental human instinct—an *intellectual* instinct—of which poetry is one manifestation, along with music, painting, and sculpture. His real innovation, however, and the cornerstone of his new theory of poetry, is a redefinition of *mimesis* to mean not a counterfeiting of sensible reality but a presentation of "universals." By "universals" he means not metaphysical entities like the Platonic Ideas but simply the permanent, characteristic modes of human thought, feeling, and action (9). It goes without saying, or at least Aristotle does not bother to say, that knowledge of such universals is not restricted to the philosopher. The poet can represent them, and his readers can grasp them, without benefit of metaphysical training. Poetic i. is of action rather than simply of men (i.e. characters). Tragedy (and, with certain reservations, the epic) is an i. of a single, complete, and serious action involving the happiness of an important human being. More specifically, the i. is lodged in the *plot* (*mythos*) of the poem; and by "plot" Aristotle means not merely a sequence but a *structure* of events, so firmly welded together as to form an organic whole. It follows that the poet's most important duty is to shape his plot. He cannot find it already given; whether he starts from mythical trad., history, or his own invention, he is a poet only so far as he is a builder (*poietes*, "maker") of plots. Thus "i." comes very close to meaning "creation." But the poet's creation is not of some "second nature" existing only in his fancy; it is a valid representation of the actions of men according to the laws of probability or necessity.

Aristotle's concept of i. was subtle and complex. His chief successors in crit. were men of another

stamp, more literary than philosophical in their view of poetry. So far as i. remained a key term in the Hellenistic age (actually we do not hear a great deal about it), it seems to have been conceived as meaning the portrayal of standardized human types: the hot-headed man, the braggart soldier, the wild Thracian, etc. Aristotle's "probability" (*to eikos, verisimile*), and the even more characteristic concept of "appropriateness" (*to prepon, decorum*, q.v.), are now tailored to the measure of particular social standards and conventions more than to any permanent principles of human nature. At the same time Aristotle's insistence on action gives way to more relaxed and eclectic views: the object of i. may be character, thought, or even natural phenomena. Anything can be imitated, in accordance with the laws of the genre (q.v.) one has chosen, and the object, whether fable, fact, or fiction, is tacitly assumed to have more or less the same status as a natural object.

Alongside the Aristotelian concept of i., thus denatured, another of very different provenience took on increasing importance in the Hellenistic and Roman periods. This was the relatively simple idea of imitating the established "classics" (the word is Roman, the concept Gr.), the great models of achievement in each genre. Its origin was rhetorical but it ended by spreading impartially over prose and poetry. The treatise of Dionysius of Halicarnassus *On I.* is lost except for fragments, but we can get some idea of the theme from the second chapter of Book 10 of Quintilian's *Institutio oratoria.* From these two authors, and more particulariy from "Longinus," we can see that the doctrine had its higher side. I. of the great writers of the past need not and should not be merely a copying of devices of arrangement and style, but a passionate emulation of their spirit. Dryden (*Essay of Dramatic Poesy*) puts it very well: "Those great men whom we propose to ourselves as patterns of our i., serve us as a torch, which is lifted up before us, to illumine our passage and often elevate our thoughts as high as the conception we have of our author's genius." Here i. is united with its apparent opposite, inspiration. Nevertheless, both in antiquity and in the Ren., i. in the sense of emulation of models meant chiefly stylistic i., and thus helped to fortify the prevalent understanding of poetry as an art of words.

The Ren. inherited three major concepts of i. from antiquity: (1) the Platonic: a copying of sensuous reality; (2) the Aristotelian: a representation of the universal patterns of human behavior embodied in action; and (3) the Hellenistic and rhetorical: i. of canonized literary models. But each of these was further complicated by a deviation or variant interpretation: (1) the Platonic by the Neoplatonic suggestion that the artist can create according to a true Idea; (2) the Aristotelian by the vulgarization of Aristotle's "universals" into particular social types belonging to a particular place or time; and (3) the rhetorical by its rather adventitious association with "enthusiasm" and the *furor poeticus* (e.g. Vida's *Ars poetica* 2.422–44). That this mixed inheritance did not lead to complete critical chaos was due partly to the chronological accident that the *Poetics* did not become known in Italy until well after 1500, partly to the incorrigible syncretism of the humanists, and partly to the plain fact that the chief literary creed and inspiration of the It. Ren. was rhetorical. Humanism was an imitative movement in its very root and essence: the i. of Cl., and particularly Cl. Lat., lit. was its life-blood. Thus the burning question in the 15th c., and well into the 16th, was not "What is i.?" or "Should we imitate?" but "Whom [i.e. which Cl. authors] should we imitate?" The fiercest battle was waged over prose style, i.e. whether Cicero should be the sole and all-sufficient model.

A genuine theoretical interest in the concept of poetic i. as such could not arise, however, until Aristotle's *Poetics* had come to light again and begun to be studied (Gr. text 1508; Lat. trs. 1498, 1536; It. tr. 1549). Vida's *Art of Poetry* (1527) is still innocent of this new trend. It preaches the i. of "nature" (2.455) but for no other real purpose than to inculcate the i. of the ancient poets, and above all Virgil, who followed her to the best advantage. Daniello (*Poetica*, 1536) knows Aristotle's definition of tragedy as i. but hardly knows what to make of it since he draws only a faltering distinction between poetry and history. Robortelli, in his commentary on the *Poetics* (1548), allows the poet to invent things that transcend nature. Fracastoro (*Naugerius, sive de poeta dialogus*, 1555) pieces out Aristotle's concept of i. with the Platonic idea of beauty, identifying the latter with the universal. Scaliger (1561) recommends the i. of Virgil because Virgil had created a "second nature" more beautiful than the first; and Boileau gave the problem its definitive formulation for neoclassical theory: the surest way to imitate nature is to imitate the classics. But the real difficulty and challenge of Aristotle's idea of i. had not been grasped, much less solved. The later Ren. was as unable as the earlier to make an effective distinction between poetry and history on the one hand, and between poetry and rhet., on the other, because it could not seize and define any true "universal" as the object of poetic i., except in vague Platonic (Neoplatonic) terms, and so fell back into regarding poetry as essentially a special way of discoursing about "things."

Although i. was implicitly accepted down through the 18th c. as the goal and method of the fine arts in general, incl. poetry and painting, it began to slip into disrepute after 1770, being increasingly felt to imply a derogation of the artist's integrity. Edward Young sneered at "the meddling ape, I.," and Coleridge opined that "To admire on principle, is the only way to imitate without loss of originality." The revival of "i." in the 20th-c. has very little to do with either the Cl. or the neoclassical trad.; it goes straight to Aris-

totle, not through intermediaries, and views i. above all as a structural concept, the principle of organization of poetic wholes. G.F.E.

II. FROM THE MODERN PERSPECTIVE, i. refers to two different but related concepts: representation of external reality and copying or adaptation of artistic or literary models. Periods in which critical theory is privileged have tended to devalue the concept of i. in either sense and to stress, as its opposite, originality, influence, and intertextuality. Some recent theorists dismiss i. because it seems to rely on a rhet. of "presence" (insisting on "sameness" by suppressing the "difference" that haunts all representations); others point out that both in practice and in theory i. invokes that very "difference" which it is purported to suppress.

In Plato's articulation of two types of mimesis—as artistic reproduction of the physical object, and as inward representation of the idea—one activity represents the external world by means of lang. and the visual arts. The other is a conceptual activity, at once implicated in lang. yet presumably transcending its problematic elements, though this transcendence is put into question in the *Cratylus* and the *Theaetetus*, where lang. appears to be so intimately connected with thought itself that what is known and how it is known become inextricably entangled. Plato distinguishes poetry from philosophy in terms of their respective possibilities of representing the Idea, but if the act of representing can contaminate the Idea, then the claims of philosophy cannot readily be separated from the performance of poetry. Aristotle responds to Plato's debunking of mimesis by linking it to the concept of form (q.v.); in the *Poetics*, mimesis is governed by the rules of its form rather than by the accuracy with which it represents the object.

This shift in the concept of i. in antiquity in fact represents a broader divergence between affective and formalist theories of art that has endured to this day. Affective theories, such as Plato's, which recognize the power of the text to move the reader, tend to disregard i. (either of nature or of texts) as an adequate source of that power. Formalist theories, on the other hand, regard i. as a cognitive function made possible by the closure of form. A dramatic action with a beginning and an end, portrayed within the form of art, makes it possible for a text to yield to its reader knowledge which is both textual and human. In the formalist mimetic context, the order of lit. is also the order of nature. Once this representational link is established, mimesis slips into an i. of trad. Thus Horace (*Ars poetica*) asserts the propriety of i. because he does not see nature and art as disparate orders. Cl. accomplishments become rules for the production of lit., and the secret of good writing is learning (*sapere*). This connection between learning and originality remains important from the Ren. through the neoclassical period. When Dryden calls Jonson "a learned Plagiary" he means that Jonson has a profound knowledge of the classics: "he has done his Robberies so openly" and so authoritatively that the line between plagiarism and originality becomes blurred. Dante too echoes the importance of learning in i. ("the more closely we imitate those great poets, the more correctly we write poetry"), yet he engages at the same time in a theory of i. and a practice of invention by which the model (Virgil) is displaced in the *Commedia* and Paradise created. I. attempts to bridge the distance from its models, or more aggressively to suppress that distance, and in the process it slips into invention, creating a web of intertextual relations which make the link of "copy" to source problematic.

Ren. critical theory routinely grounded itself by referring to the Cl. trad., but the i. of those models only served to highlight the cultural and aesthetic distance between Ren. writers and their originals. In this respect i. forced upon Ren. critics a consciousness of historical change, and in turn this consciousness of their difference from the past compelled them to transform i. into an invention of beginnings (Greene). Thus i., which aims at the recuperation and presence of the originals, opens up the question of originality and poetic identity and drifts insensibly into the concerns of influence and intertextuality to which at first glance it appears to be opposed. The stability which accompanies the notion of i. is in this respect disrupted by the gap which i. as a practice opens, and this tension breaks into major debates in the 16th c. as to what i. means, who the models should be, and how the relation of nature to the classics should be defined.

At issue in this dispute was also the adequacy of the ancients' representation of nature—a question that remains in the foreground through the 18th c. The reliance on Cl. models produced a tension between i. of nature as external reality and i. of nature as represented in those models. In drama in particular this tension translated itself into a distinction between art as the i. of nature and art as an i. in which nature is elevated "to a higher pitch." Dryden (*An Essay of Dramatic Poesy*) suggests that the exact representation of nature must give way to an i. in which plot, characters, and description "are exalted above the level of common converse." This notion refers us back to Aristotle's statement in the *Physics* that art completes what nature leaves undone, or "imitates the missing parts" (2.8). The neoclassical period reaffirmed Aristotle's idea that art imitates not just actual nature but nature's potential form. When properly accomplished, i. is expected to reflect at once Cl. models and the nature which those models imitated: "Nature and Homer" become, in Pope's *Essay on Crit.*, "the same."

In this way i. brings together two antithetical concepts: representation of reality, which assumes the priority of the object and demands of the work of art verisimilitude or accuracy of depiction, and formalism, which insists on the work of art "completing" its subject. Both concepts come to rest on

a metaphysics of presence (though the notion of "completing" the model puts into question the fullness or presence of the original), and both idealize the nature of the artistic act. I. is made possible by assuming "sameness" to be the governing condition of lit. and life, and "difference" to be a mere departure from this central fact (Miller). The "sameness" of nature and model which Pope points to is affirmed in 18th-c. aesthetics through the painted image, which, as de Man suggests (1983), is supposed to restore the object to view and in that sense guarantee the continuity of its presence (see UT PICTURA POESIS). The possibility that the model might be absent is repressed.

Also repressed is the possibility that lang. might constitute its own fictions rather than represent, submissively, an external reality. The tension between nature as external reality and as potential totality conceals a deeper schism between theories of i. and theories of metaphor (q.v.), and consequently between formalist and affective theories of reading. Aristotle's view of metaphor as improper naming (*Poetics* 21) tends to put into question both the mimetic order that is assumed to exist between nature and art and the totality of form. When Longinus (*On the Sublime*) takes up the question of art's relation to nature, he develops a theory of the fragment which insures the priority of metaphor over the possibility of mimesis. Fragments rather than total forms strike fire from the reader's mind, and the power of the fragment resides not in i. of nature but in a figuration in which the argument is concealed (15). Affective theories of crit. in this way develop a line of argument antithetical to that of mimetic theories of crit., yet the two approaches are implicated in one another in the course of critical trads. Sidney (*An Apology for Poetry*) refers to Aristotle's mimesis at once as "representing" and "counterfeiting." Nature comes to be perceived as form in formalist readings, but power comes to be perceived as the breaking of that form. It may be that the felt need for "defenses" of poetry in the Ren. was a result of this tension between incompatible critical models.

Romantic writings engage in precisely this exploration of the limits of form and the inadequacies of i. For the romantics the text is merely a fragment of a larger vision, but this vision has less to do with nature as subject for i. than it does with the workings of consciousness itself. Both the Cl. and neoclassical periods privileged reason as the determining structure of consciousness, and this assumption made i. the proper method of interaction between text and world. But in romantic writings concealment plays such a large role (in its various guises as the unconscious, the "life of things") that nature becomes closed to the workings of reason, form, and i., and imagination (q.v.) comes to take their place; the visionary perspective now opposes itself to the mimetic one. At the same time, the neoclassical acceptance of i. as a practice of writing is displaced in romantic texts by a stress on originality, a refusal to adopt inherited forms. At times this stress on newness is couched in the rhet. of i.—e.g. Wordsworth's desire to "imitate the very lang. of common men"—but romantic practice stresses a different rhet., built on the inescapable figuration that attends all knowledge "of ourselves, and of the universe." When Wordsworth looks down from a slow-moving boat into a lake he cannot tell apart what is at the bottom, what is reflected from the shore, and his own image: he finds he "cannot part / the shadow from the substance" (*Prelude* Bk. 4). The romantic subject is implicated in what he sees, and this implication undoes the possibility of a true, holistic, or transcendent representation of nature.

The 20th c. has witnessed both periods of formalism and antiformalism. The New Criticism repeated the gesture of associating form with a representation of idealized nature. In a different vein, Auerbach traces in *Mimesis* the history of the idea of representation in textual practice by assuming, in synecdochic fashion, the coherence between literary text and the culture it represents. But after 1960, in part as a result of the influx of Fr. crit. into America, the possibility of i. as representation has been put into question by the notion of textuality: Derrida asserts the impossibility of representation that is not always already lang., and de Man terms representation "an ambivalent process that implies the absence of what is being made present again," an absence which "cannot be assumed to be merely contingent" (*Blindness and Insight* 123). For deconstruction, mimesis is groundless repetition haunted by difference. The authority of representation is undermined in such theories by the perceived power of rhet. to constitute its object, that is, by the strange priority of effect over cause, thus reversing the traditional cause-and-effect relationship and taking the act of representation itself to be constitutive of the "origin" or cause which it supposedly mirrors. For deconstructive crit., "there are no originals; there are only copies," unavoidably repeated in their "failure to originate."

This upheaval in the structure and authority of i. is pursued on another front by Harold Bloom, who reverses the possibilities of i. of trad. by mapping out trad. as a battleground in which writers continually attempt to shake off or repress those precursors who would take away their freedom; in the process, these writers rewrite their precursors and become their cause.

Whether the term i. invokes Cl. models or nature, it rests on a metaphysics within which fulfillment is possible. But if the "original object" of i. is itself contaminated by the act of representing—if rhet. constitutes the object it purports to imitate—then i. shifts from the "re-presenting" of an "original" to a replication or doubling in which the original is displaced, inaccessible—imagined. The rhet. of presence and fulfillment opposes itself to the rhet. of desire, and loss; and the shift

in the concept of i. from the one to the other aligns mimesis with violence (Girard). I. has become an important topic once again in critical theory precisely because it highlights the problematic relation to the "object" and the conception of history itself. It may well be, of course, that the very theoretical concerns which recognize the unstable play of i. and the problem of the original from Plato to Derrida are themselves instances of a mimetic replication by which theory constructs and realigns its models. H.R.E.

I. Scott, *Controversies Over I. of Cicero* (1910); S. H. Butcher, *Aristotle's Theory of Poetry and Fine Art*, 4th ed. (1911), ch. 2—suggestive, but overmodernizes; E. Nitchie, "Longinus and the Theory of Poetic I. in 17th- and 18th-C. England," *SP* 32 (1915); R. D. Havens, *The Influence of Milton on Eng. Poetry* (1922); U. Galli, "La mimèsi artistica secondo Aristotele," *Studi Ital. di Filol. Class.*, n.s. 4 (1926)—comprehensive, also covers Plato; J. W. Draper, "Aristotelian 'Mimesis' in 18th-C. England," *PMLA* 36 (1926); J. Tate, "I. in Plato's *Republic*," *ClassQ* 22 (1928), "Plato and I.," *ClassQ* 26 (1932); H. Gmelin, "Das Prinzip des *Imitatio* in den romanischen Literaturen der Ren.," *Romanischen Forschungen* 46 (1932); H. O. White, *Plagiarism and I. During the Ren.* (1935); R. McKeon, "Lit. Crit. and the Concept of I. in Antiquity," *MP* 34 (1936)—rpt. in Crane, and "I. and Poetry," *Thought, Action, and Passion* (1954)—discriminates types; M. T. Herrick, *The Fusion of Horatian and Aristotelian Lit. Crit, 1531–1555* (1946); H. F. Brooks, "The 'I.' in Eng. Poetry, Esp. in Formal Satire, before the Age of Pope," *RES* 25 (1949); W. J. Verdenius, *Mimesis* (1949); K. Burke, "A Dramatistic View of I.," *Accent* 12 (1952); Crane; Abrams, ch. 1–2; R. C. Lodge, *Plato's Theory of Art* (1953); R. R. Bolgar, *The Cl. Heritage and Its Beneficiaries* (1954); H. Koller, *Die Mimesis in der Antike* (1954)—ambitious but unreliable; Auerbach; Wimsatt and Brooks, ch. 10; G. F. Else, *Aristotle's* Poetics: *The Argument* (1957), "I. in the 5th C.," *CP* 53 (1958); V. F. Ulivi, *L'Imitazione nella poetica del Rinascimento* (1959); R. A. Brower, *Alexander Pope: The Poetry of Allusion* (1959); R. W. Dent, *John Webster's Borrowing* (1960); Weinberg; B. Hathaway, *The Age of Crit.* (1962); A. J. Smith, "Theory and Practice in Ren. Poetry: Two Kinds of I.," *BJRL* 47 (1964); G. Sorban, *Mimesis and Art: Studies in the Origin and Early Devel. of an Aesthetic Vocabulary* (1966); H. D. Weinbrot, "Translation and Parody: Towards the Genealogy of the Augustan I.," *ELH* 33 (1966); J. Derrida, *De la grammatologie* (1967), "La Mythologie blanche," *Marges de la philosophie* (1972), *La Dissémination* (1972); W. K. Wimsatt, Jr., "I. As Freedom—1717–1798," *NLH* 1 (1970); R. Girard, *La Violence et le sacré* (1972); J. Steadman, *The Lamb and the Elephant: Ideal I. and the Context of Ren. Allegory* (1974); *Mimesis*, ed. S. Agacinski et al. (1975)—esp. Derrida, "Economimesis"; G. Braden, *The Classics and Eng. Ren. Poetry* (1978); P. de Man, *Allegories of Reading* (1979); D. A. Russell, "De imitatione," *Creative I. and Lat. Lit.*, ed. D. West and A. Woodman (1979); G. W. Pigman, III, "Versions of I. in the Ren.," *RenQ* 33 (1980); L. Manley, *Convention 1500–1750* (1980); P. Ricoeur, "Mimesis and Representation," *Annals of Scholarship* 2 (1981); T. Greene, *The Light in Troy* (1982)—essential; J. H. Miller, *Fiction and Repetition* (1982); D. Quint, *Origin and Originality in Ren. Lit.* (1983); M. Ferguson, *Trials of Desire: Ren. Defenses of Poetry* (1983); De Man; B. J. Bono, *Literary Transvaluation From Vergilian Epic to Shakespearean Tragicomedy* (1984); J. Weinsheimer, *I.* (1984); N. Hertz, "A Reading of Longinus," *The End of the Line* (1985); J. L. Mahoney, *The Whole Internal Universe: I. and the New Defense in British Crit., 1660–1830* (1985); F. Stack, *Pope and Horace: Studies in I.* (1985); C. Prendergast, *The Order of Mimesis* (1986); G. B. Conte, *The Rhet. of I.*, ed. and tr. C. P. Segal (1986); S. K. Heninger, Jr., *Sidney and Spenser: The Poet as Maker* (1989); P. Lacoue-Labarthe, *Typography: Mimesis, Philosophy, Politics* (1989); R. W. Dasenbrock, *Imitating the Italians* (1991). G.F.E.; H.R.E.

INDETERMINACY. A term drawn from poststructuralist crit. which suggests the impossibility of stabilizing a text's (or word's) meaning. In traditional lit. hist., meaning is either inherent in the work or produced by context. But when context itself is drawn into the conditions of textuality, all connections one may make to "determine" meaning are open to this originary instability of lang.—M. Perloff, *The Poetics of I.* (1981); G. Graff, "Determinacy/I.," *Critical Terms for Literary Study*, ed. F. Lentricchia and T. McLaughlin (1990); B. J. Martine, *I. and Intelligibility* (1992). H.R.E.

INFLUENCE. Traditionally, i. has been associated with imitation (q.v.) and most often understood as the result of learning and technique. Horace counsels writers to follow trad.; "Longinus" considers imitation of earlier writers as a means to sublimity; Dante urges writers to first drink of Helicon before taking up their singing instruments. I. is a beneficent and necessary corollary of creative genius, routinely addressed and acknowledged up until the late 18th c. romantic poetry and poetics instead stress originality, taking the latter to be the direct opposite of the former. A trad. of "affective" crit. then arises in both the poetry and crit. of the 19th c. which proceeds to reread earlier critical texts, such as Longinus, in terms of the struggle between, on the one hand, the techniques steeped in learning and imitation and, on the other, the sublime and wholly original power of genius. Genius, however, cannot be fully separated from learning, and the notion of i. develops as a way of negotiating the distance between the two. In the 4th c., i. had referred to the stars, and later came to be associated with the exercise of power. It signals an influx or flowing in, and thereby relates the imitation of earlier writers to the power that

such writers exert on the trad.

This concept of "tradition," of an ideal order of lit., has had a considerable impact on the hist. of i. as traced by critics in the 20th c. Northrop Frye, for example, repeats Eliot's ideal order by reading the lines of filiation (of i.) between one writer and another in terms of myth and archetype, knowledge of which makes possible an understanding of the "total form" of lit. Each work is a synecdoche for the whole. The myth constitutes an origin (*archê*) outside of time and lang., and each creative act becomes a "fable of identity." Both Frye and Eliot perceive lit. as a totality, present in space and expressed through form.

This model of i., articulated within the framework of traditional lit. hist., has produced many other solid scholarly works. R. D. Havens' *The I. of Milton on Eng. Poetry* (1922), for example, traces the 18th and 19th centuries' indebtedness to Milton by a meticulous inventory of repetitions and echoes of Milton's style. Havens refers, for instance, to Wordsworth's poems about Milton as well as his conscious stylistic borrowings from Milton, incl. his lofty diction and "organ tone." Several assumptions underwrite such studies as Havens': (1) the earlier poet functions as an undisputed and stable "source," a foundation or origin itself not open to question; (2) i. is conscious and explicit, not hidden; (3) the scholarly study of i. observes the boundaries of the discipline, eschewing interactions with other disciplines such as philosophy and psychoanalysis; and (4) literary trad. works in cumulative fashion and holds out the promise of further riches, so that i. (such as Milton's) is both powerful and beneficent.

It is W. J. Bate (1970) who for the first time emphasizes the difficulty that a rich trad. produces for each poet who comes after. He sees the past as a "burden," though he suggests that this burden is a recent phenomenon ("the product of the last 50 years"). Bate is explicit that the strength of the past in no way indicates the weakness of the future, a condition which only a self-fulfilling prophecy could bring about. An idealizing, humanistic strain informs *The Burden of the Past*, with the assertion of human freedom and human choice its paramount concern. For Bate the rich heritage of the past points the way to our own "identities," the way "to be ourselves."

These conceptions of i. rest on the traditional view of lit. hist. as a stable context with determinate cause and effect relations, recognizable sources, and a reliably straightforward chronology. Literary relations, in this scheme of things, are equally straightforward and determinable. Their determinacy is built on the premises of traditional crit., but these premises were put into question by the advent of theoretical crit. originating in France in the 1950s and '60s. "Textuality" and "intertextuality" (qq.v.) disrupt the stability of the "work" and its meaning, deny its "origin" and "originality," and open up infinite and random connections between texts.

The concept of the origin is problematic long before the advent of "textuality," however. Poststructuralist critics note that the history of Western metaphysics is haunted by the impossibility of establishing a system of knowledge authorized by a cause external to the system, and after poststructuralism the study of causes and effects between texts often goes by the name of "genealogy." The name suggests a genesis which is already a metaphor within the system rather than the cause of it. Gregory Jay defines genealogy as "charting, between writers and texts, those often unauthorized relationships that nonetheless belong to the literary lineage of an essay or poem" (1983). Genealogy rereads the simultaneous order as a figure for textuality itself. Relations with the history of texts may shift, but genealogy does not produce an origin: it already knows that beginnings do not trace a history but constitute instead a metaphor for such a history (Riddel).

Poststructuralist critics suggest that the absent origin punctuates a history of texts stretching from Plato to the present. Three writers within this history place particular emphasis on the problematics of the origin. That is, in the terms outlined above, three writers are repeatedly drawn out of a history of texts to function as a genealogical locus and exemplar for the problem: Giambattista Vico (1668–1744), Friedrich Nietzsche (1844–1900), and Sigmund Freud (1856–1939).

Vico's major work, *Scienza Nuova* (*The New Science*, 1725, tr. 1744) asks how human beings first began to think humanly, a question prefaced by an account of the difficulty Vico encountered in reaching back to inaccessible origins which he could "imagine only with great effort." The text dealing with origins enacts the problematic for which it seeks to account. Vico's response to the question he poses is that thought begins in metaphor. The first human beings begin to think humanly when, in response to fear, they create a world out of themselves and assign to it the names of gods. This is the "imaginative metaphysics" born of ignorance and fear and needing no authority other than its own creation. We might say that it is born out of need, and functions as defense. Vico traces a later "rational metaphysics" which emphasizes learning and which he describes as a fall from the first. Imaginative metaphysics is a metaphysics of power: it imagines the origins it needs in order to survive. Rational metaphysics recognizes the tropological nature of those origins and the groundlessness of all understanding.

This double attitude is echoed by Nietzsche in "Über Wahrheit und Lüge im aussermoralischen Sinn" ("On Truth and Lying in their Ultramoral Sense" [1873]), which notes two significant moments in thought: the creation of "truth" through metaphor ("truth is a mobile army of metaphors") and the forgetting of the fact of that creation. The first points to an epistemological

impasse, since we cannot know anything in its essence, only in the relations posited by lang. The second points to a will-to-power by which an artist creates out of the "delicate material of ideas." Nietzsche denigrates the "impulse to truth" crucial to a metaphysics of presence, since it promotes a false sense of understanding and security. Understanding for Nietzsche is built on metaphor and cannot claim a privileged point of reference beyond it. There is no metalanguage or external reference point from which to view the whole. Identity is a fiction, cause and effect relations an illusion, and origin an imagined thing; this is the frightening world without rational order, without bottom, that art reveals to us when it tears for a moment "the woof of ideas." It is the power of art to witness the abyss that underlies human thinking, and it is also the power of art to create the illusion of origin and poetic identity.

Nietzsche's "forgetting" is Freud's "repression." Though Freud is perceived as a scientist, and his initial gestures are in that direction, his theories lead him further away from the world of real causes and events and into the area of "phantasy." The case history of the Wolf Man proposes "cause" as phantasy. At first Freud surmises that the Wolf Man's trauma is caused by a "primal scene" in which the child, at one and a half years of age, watches his parents copulating. Yet it is not until the child is four years old that another event triggers the memory of the primal scene and renders it traumatic. Freud terms this phenomenon "Nachträglichkeit" (retroactive meaningfulness). In this case cause and effect so contaminate one another that the rational chronology of events (the kind of chronology that would be employed in "normal" lit. hist.) does not apply. The later even could well be said to be the cause of the earlier one, or rather the cause of its "meaning." Freud further suggests that even the "primal scene" is a phantasy, constructed out of the child's experience on a farm and the stories he was read. In place of origin, then, we have imagined construct. In the beginning is another metaphor or text.

In the wake of Vico, Nietzsche, and Freud, deconstruction assumes the problematic nature of the origin, and in so doing it undermines the logocentric and rationalist Western metaphysics upon which presence, continuity, and the notion of fulfillment are based. Following Nietzsche, Paul de Man emphasizes "the tropological structure that underlies all cognitions, incl. knowledge of the self" ("Autobiography as De-facement"). Far from making any totalization possible, rhet. suggests that even the ultimate self-understanding (autobiography) comes to rest on a figure. Through his discussions of irony (q.v.) and prosopopoeia, de Man points to the "rigorous unreliability" of literary lang. and to the "systematic undoing of all understanding." Irony is the recognition that conflict cannot be reduced to logic, and prosopopoeia (the giving of a face) is the *giving* of a face to that which is otherwise faceless. Such a rhetorical (ironic) reading undoes the possibility of lit. "hist." or of a "genetic" study of i., since neither history nor lit. would be generative processes (de Man speaks of the literary text's repeated failure to originate). Deconstruction makes impossible the reading of lit. in terms of periods, dismantles the notion of "lit. hist.," and dismisses the vision of a trad. as an aggregation of great works. Lit. hist. conceived along the lines of rhetorical figures and aporias would be a halting phenomenon, accounting "at the same time for the truth and falsehood of the knowledge lit. conveys about itself" (1983).

While the controversy over traditional lit. hist. vs. deconstruction was raging, Bloom published in 1973 *The Anxiety of I.*, followed by three subsequent books (1975, 1975, 1976) setting forth a theory of i. which opposed both the idealizing views of traditional lit. hist. and the deconstructive readings of Derrida and de Man. Three more books followed (1976, 1982, 1982), refining and extending these theories. Together the seven studies have made Bloom the major modern theorist on the subject. Bloom rejects the view of trad. as a "handing over" from precursor to later poet and argues for a reading of trad. as a series of struggles in which the later poet attempts to turn his lateness into earliness. The informing rhetorical figure for Bloom is metalepsis—the "trope-reversing-trope" which reverses cause and effect. Bloom rejects the conventional view of trad. as an inclusive order in which space is always made for the newcomers. Trad. is the force that takes potential originality away from a poet, always rendering the space available more limited and more exclusive. Originality in this view is a moment of freedom which the poet must achieve by creating discontinuity and thus turning loss into the illusion of origins. Bloom's story of influence-relations is one of power, violence, and appropriation, since these will be the marks of a poet's strength. He has no compunctions about dividing poets into strong and weak: weak poets accept the "handing over" and are silenced by it; strong poets react against it and their reaction is mapped out by Bloom into specific "revisionary ratios." The later poet "swerves" from the precursor by finding where the earlier poet went wrong, establishes discontinuity with the precursor, and rewrites the parent poem in such a way that, far from appearing to be influenced by the earlier, the later poet now appears to have created the earlier poet's work.

Bloom's approach to texts is not entirely unlike intertextual conceptions. For him, texts have neither closure nor meaning: meaning wanders between texts, and the meaning of a poem lies in its reaction to another poem. This wandering of meaning is a key concept in Bloom's theory of i., and when it becomes appropriation it merits the term "misreading." Misreading is the necessary condition of all strong poetry and all strong crit.,

for it describes a swerving away from the precursor's poem which in turn makes possible the strength of the belated poet. Misreading points to a practice which Bloom terms "antithetical crit." and which refuses to distinguish between poetry and crit. In the predicate that there are no right or wrong readings, only strong or weak readings, misreading opens up the space of persuasion, "the domain of the lie." Bloom attacks the humanistic and New Critical procedures which would make crit. an ancillary activity, and which, like Auerbach's "figura," would make the later reading a fulfillment of the earlier one. Strong crit. and strong poetry constitute a lie against time, granting the poet for a moment the illusion that he is self-begotten. Though he draws from poststructuralist theories of intertextuality, however, Bloom insists finally on "figures of willing" against "figures of knowing." Epistemology tells us that all origins are metaphors; Bloom's readings of Freud and Vico remind us that origins can be imagined. It is the power of great poetry to achieve the latter, and Bloom's seven books on i. are an account of the strategies by which a poet turns his belated "authority" into imaginative "priority," or the illusion of originality. Deconstruction transforms i. into intertextuality because the poet is himself an effect of a system of lang. in which he is inscribed. Bloom denies on the one hand a self that is not the construct of a poem, but on the other hand insists on the figurative creation of a poetic self. He argues that repression and forgetting (embedded in his theory of i.) offer a much more realistic account of the relations between one text and another and of the ways in which a text finally comes to read itself than deconstruction's emphasis on epistemology. In this way he insists on the importance of psychic strategies of reaction and defense as opposed to the epistemological concerns of deconstruction. Within a single poem, the condition of belatedness expresses itself as a series of disjunctive moments, and the "crossings" between these moments are defined as a repression of i., a leaping over time, and an image of voice. Voice (q.v.) is a term drawn from traditional crit., but Bloom's use of it is directed against purely rhetorical readings that focus on an impasse rather than a metaleptic leaping over that leads to self-begetting.

Self-begetting can be achieved in other ways. Geoffrey Hartman links quotation with the creation of a poetic self. A poet has to "overhear" and work through the ghostly echoes or allusions which haunt him, and in poets like Wordsworth quotation turns into self-begetting. Intertextuality (and its implication that the poem cannot complete itself) becomes *intra*textuality, with the poet finally quoting not Milton or the classics, but himself.

The arguments over i. over the past quarter century continue unabated. De Man refers to Bloom's scheme of a father-son struggle as a genetic model and hence a "rhetorical mystification." But while the issues between the priority of epistemology in deconstruction and the priority of the psyche in Bloom's theories will continue to affect the way in which i. is understood, it is clear that these concerns have radically transformed the study of i., turning it away from any simplistic kind of "source-study" toward more elaborate patterns of allusion.

Far from attempting to define a text's origin or cause, i. has shifted to a concern with the problematic wanderings between texts and has taken on the shape of theory. It is no longer possible to distinguish, in an age of theoretical self-consciousness, the objective literary relations between writers from the critical articulations that shape those relations, so that the study of i. engages the critical history through which i. is articulated. In a 1987 collection suggestively entitled *Re-membering Milton*, Robin Jarvis begins with a meditation on Havens' study and takes us to an "i.-theory" that argues the self-divided, androgynous, effeminate nature of poetic fathers and their texts, "shored against an otherness which inscribes them in the irrecuperable ebb of meaning." In contemp. theory the paternity of the text is an illusion, and i. and intertextuality address the complex wandering and dismemberments and the critical "re-memberings" of the orphaned text. In the realm of theory, there are no last words, only further articulations. In this respect the theories of the past 20 years have, in transforming the understanding of i., opened up space for other work to follow.

T. S. Eliot, "Trad. and the Individual Talent" (1917), rpt. in *Essays*; S. Freud, "Aus der Gesch. einer infantilen Neurose" (1918), rpt. in *Standard Works*; R. D. Havens, *The I. of Milton on Eng. Poetry* (1922); N. Frye, "The Archetypes of Lit.," *KR* 13 (1951); J. Derrida, *L'Écriture et la différence* (1967); W. J. Bate, *The Burden of the Past and the Eng. Poet* (1970); C. Guillén, "The Aesthetics of Literary I.," *Lit. as System* (1971); H. Bloom, *The Anxiety of I.* (1973), *A Map of Misreading* (1975), *Kabbalah and Crit.* (1975), *Poetry and Repression* (1976), *Figures of Capable Imagination* (1976), *Wallace Stevens* (1976), *Agon* (1982), *The Breaking of the Vessels* (1982); E. Said, *Beginnings* (1975); *The New Nietzsche*, ed. D. Allison (1977); P. de Man, "Shelley Disfigured," in *Deconstruction and Crit.* (1979), *Blindness and Insight*, 2d ed. (1983), "Autobiography as De-facement," *The Rhet. of Romanticism* (1984); J. Riddel, "Decentering the Image," *Textual Strategies*, ed. J. Harari (1979); F. Lentricchia, *After the New Crit.* (1980)—critique of Bloom; S. M. Gilbert and S. Gubar, *The Madwoman in the Attic* (1980); G. H. Hartman, "Words, Wish, Worth: Wordsworth," in *Deconstruction and Crit.* (above); J. Hollander, *The Figure of Echo* (1981); J. Culler, *On Deconstruction* (1982); J. Guillory, *Poetic Authority* (1983); G. S. Jay, *T. S. Eliot and the Poetics of Lit. Hist.* (1983); *The Anxiety of Anticipation*, spec. iss. of *YFS* 66 (1984); C. Baker, *The Echoing Green* (1984); P. Brooks, "Fictions of the Wolf Man,"

Reading for the Plot (1984); D. Fite, *Harold Bloom* (1985); J. P. Mileur, *Literary Revisionism and the Burden of Modernity* (1985); N. Lukacher, *Primal Scenes* (1986); *Re-membering Milton*, ed. M. Nyquist and M. Ferguson (1987); H. R. Elam, "Harold Bloom," *Mod. Literary Critics 1955 to the Present*, ed. G. Jay (1988); *Intertextuality, Allusion, and Quotation: An Internat. Bibl. of Crit. Studies*, ed. U. J. Hebel (1989); L. A. Renza, "I.," *Critical Terms for Literary Study*, ed. F. Lentricchia and T. McLaughlin (1990); *I. and Intertextuality in Lit. Hist.*, ed. J. Clayton and E. Rothstein (1991). H.R.E.

INSPIRATION (Lat. "breathing in"). Every poet recognizes that, during poetic composition, material emerges—words, images, figures, rhythms—from sources which lie beyond the pale of consciousness. Conscious effort and craft may indeed produce such material, and certainly they will shape all material, but at least some material seems to come into the mind from that place which we know only as *other.* In most cultures, poets, esp. those of religious conviction, have believed this source, this other, to be divine, the "breath" of the god or God blown into the recipient poet, who becomes the vehicle of godhead. In the West, the doctrine of i. has deep roots in both Hellenic and Hebraic cultures, from which streams it fed into Christianity and so was transmitted to the modern world, where it still holds sway: the concept of production by other-than-the-conscious-self has survived all radical alterations in attribution of its source.

At least as early as Homer, i. holds a central place in Gr. poetics, both as invocation to the gods or, more often, the Muses for the gift of memorable speech, and also as claim that when the god does take possession, the poet enters a state of transcendent ecstasy or frenzy, a "poetic madness" or *furor poeticus.* Throughout most of archaic Gr. thought, the creation of art is associated with ritual, religion, and substance-induced *ecstasis.* In *Odyssey* 22.347–48, the bard Phemius acknowledges that the god has put songs into his heart. Hesiod and Pindar also invoke divine i., as does Theocritus, though with the latter perhaps the invocation has already become conventional. Plato refers to i. and to the doctrine of poetic madness often (*Laws* 719c), sometimes at length (*Symposium* 197a; *Phaedrus* 244–45). In the *Ion,* Plato suggests, borrowing from Democritus, that just as iron filings become magnetized by the magnet, so the poet is inspired through divine power, which power is conveyed by him in turn to the professional rhapsodes and their audiences. Aristotle repeats Plato's view in the *Poetics* (cf. *Rhetoric* 3.7) but seems uninterested in pursuing it; and the Peripatetic *Problemata* (Bk. 30; ca. 250 B.C.) offers an alternative, organic explanation, the four "humors."

Virgil's invocation to the Muse at *Aeneid* 1 is well known, and Ovid also refers to i. (*Ars amatoria* 3.549; *Fasti* 6.5). Longinus, one of the key texts of antiquity, treats i. from the reverse direction: the litmus test of great lit. is sublimity, and in all works which are sublime, the reader is transported, carried out of herself, expired as it were into divinity. Cicero discusses i. in his *On Divination* (1.18.37), *On the Nature of the Gods* (2.66: "No man was ever great without divine i."), *On the Orator* (2.46.194), and *The Tusculan Disputations* (1.26). Modern Eng. trs. of these passages use the word *i.*, but Cicero's Lat. terms are *afflatus, instinctus,* and *concitatio*: *inspiratio* does not appear until the late Lat. period. *Afflatus* in particular long survived as a synonym for i.

I. has, if anything, even more central a position in Hebraic poetics, as the poets and prophets of the Old Testament freely acknowledged (*Joel* 2:28–30, *Ezekiel* 2:1–10), though for them possession was by no means always willed, desired, or ecstatic (*Jeremiah*). When Christianity emerged triumphant over the other Roman mystery cults, it appropriated the notion of the Muses to the Christian God as the source of i., making not the breath of the god but divine grace the inseminating force. The Church Fathers—Jerome in particular—often referred to David as the perfect poet-prophet, inspired by God. St. Paul's claim that "all scripture is given by i. of God" (*2 Timothy* 3:16) was reasserted by literalists for millennia thereafter. I. became a central tenet of Neoplatonism, which transmitted it through the Middle Ages. The emergence of science in the 17th c. and of psychology in the 18th, however, transferred the locus of i. from outer to inner: now the source is the subconscious mind—creative and energetic, but also appetitive and unstable. In this form, i. becomes, along with spontaneity, one of the principal tenets of romanticism, and is given the fixative, for the 20th c., by Freud. But the two loci, outer and inner, should not be taken as mutually exclusive, for even in the religious view the mind itself is a divine creation: the issue is only whether i. is directly given or mediate and proximal. Nor, similarly, should the two dimensions of poetic production—i. and craft, or seeing and making—be taken as mutually exclusive: the only question is whether the materials which appear are already fully formed and finished, or require subsequent shaping.

In its most extreme form, the doctrine of i. makes of the poet a passive receptacle and mere mouthpiece. This would reverse Auden's remark, elegizing Yeats, that poetry is only a mouth and makes nothing happen, for even there the saying itself is attributed to the poem; literalist i., by contrast, strips even the poet of creative capacity. Composition becomes now merely automatic writing, as Yeats reported of his wife, or Blake reported of himself. It is Blake who says, "I have written this Poem from immediate Dictation, 12 or sometimes 20 or 30 lines at a time, without Premeditation & even against my Will," and "I dare not pretend to be any other than the secretary the Authors are in

Eternity." And we may recall that Socrates in the *Thaetetus* (149–50) claims that even the philosopher is merely the midwife of truth, possessing no wisdom in himself. Subsequent examples incl. Bede's account of Caedmon (*Ecclesiastical History* 4.24); Dante, *Purgatorio* 1.1–20; Boccaccio, *Genealogy of the Gods* 14, 15.39, 15.99, etc.; J. C. Scaliger, *Poetics libri septem* 1.2; Bacon, *Advancement of Learning*; and Ben Jonson, *Discoveries*. The most significant example in Eng. poetry is found in Milton's invocations of the Muse in *Paradise Lost* (e.g. 9.24). Milton's Muse is a source of enlightenment comparable to the Protestant "inner light" and equated with the spirit from whom Moses received the Ten Commandments.

This sort of theory finds its strongest modern statement in romantic poetics: preromantic and romantic examples incl. Young's *Conjectures on Original Composition* ("genius" is "the god within"); Blake's letter to Thomas Butts of April 25, 1803; Wordsworth's conclusion to "The Recluse"; Coleridge's account of the origin of "Kubla Khan"; Poe's "Poetic Principle"; and Emerson's "The Poet." The doctrine will be found full blown [sic] in Shelley's *Defence of Poetry* (1821) and "Ode to the West Wind." But study of the notebooks and working habits of the romantic poets (incl. Shelley; cf. the preface to *Prometheus Unbound*) shows unequivocably that one should regard i. as more a part of the mythology of romanticism than of its praxis, for the romantics recognized the necessity of composition after—or without—i.: even Blake admitted that "without practice nothing can be accomplished." To say this is to acknowledge the importance of craft, of what Horace called "the labor of the file," i.e. the application of effort and skill in the acquisition of technique—practiced, learned, remembered, repeated. Thus the concepts of i. and making may come to be seen as antinomies, which mutually entail and illuminate each other; and in the heat of sustained and complex poetic composition, a good poet, asked to distinguish the two, might not be able to say. It is of interest to notice that, in the *Iliad* and *Odyssey*, on every occasion when Homer appeals to the Muses, the appeal is for matter of a factual nature rather than for form. That is, Homer always asks for i. concerning what to say, never how to say it.

Less extreme forms of i. would probably be acceptable to many poets, particularly if the locus is left unspecified or if the nature of the infusion is seen as less systematic, less direct (the Aeolian harp, the correspondent breeze), or simply occasional. And if the joint effect of romanticism and Freudian psychology was ultimately to shift the locus of i. from external divinity to internal psyche, the mystery of the process of production itself was not thereby lessened.

W. Dilthey, *Die Einbildungskraft des Dichters* (1887); Patterson, esp. 507–20; C. D. Baker, "Certain Religious Elements in the Eng. Doctrine of the Inspired Poet During the Ren.," *ELH* 6 (1939); E. A. Armstrong, *Shakespeare's Imagination: A Study of the Psychology of Association and I.* (1946); G. Grigson, *The Harp of Aeolus* (1947); J. C. Simopoulous, "The Study of I.," *JP* 45 (1948); R. E. M. Harding, *An Anatomy of I.*, 3d ed. (1948); G. Kleiner, *Die I. des Dichters* (1949); B. Ghiselin, *The Creative Process* (1952); J. Prévost, *Baudelaire* (1953); Abrams; C. M. Bowra, *I. and Poetry* (1955); M. C. Beardsley, "On the Creation of Art," *JAAC* 23 (1964–65); A. Anvi, "I. in Plato and the Heb. Prophets," *CL* 20 (1968); E. N. Tigerstedt, *Plato's Idea of Poetical I.* (1969); B. Vauter, *Biblical I.* (1972); P. Milward, *Landscape and Inscape* (1975); H. Mehnert, *Melancholia und I.* (1978); K. K. Ruthven, *Critical Assumptions* (1979), ch. 4; C. Fehrmann, *Poetic Creation* (tr. 1980); P. Murray, "Poetic I. in Early Greece," *JHI* 101 (1981); A. J. Harding, *Coleridge and the Inspired Word* (1985); *CHLC*.

T.V.F.B.

INTERPRETATION. This entry concerns the interp. of poetry written in modern langs., which did not become a central concern of crit. until the 20th c.

I. TERMINOLOGY: THEORY VS. PRACTICE. Interp. begins when we have difficulty understanding a poem. "Explanation"—achieved by obtaining extratextual information about an allusion, for example—may solve the problem. If the relevance of an allusion (q.v.) is not apparent or a metaphor (q.v.) is hard to fathom, "explication"—elucidation of the meaning through analysis—will be necessary. In Fr. crit., *explication de texte* includes explanation and explication. The Ger. *erklären*, to explain, is often opposed to *verstehen*, to understand, the former being characteristic of the natural sciences, the latter of the human sciences. *Explizieren* means to explicate, in the limited sense; *auslegen*, like *interpretieren*, means to explicate and to interpret. As in Ger., the Eng. term "interp." includes explanation and explication but goes beyond them: the purpose of analysis is to understand the meaning of the poem as a whole.

Some critics (e.g. Hirsch) hold that there is, in principle, a single correct interp. of a poem, notwithstanding the fact that it may have varied kinds of significance in connection with one or another situation. Critical pluralists argue that different theories of lit. can produce different interps. of a poem, all of which are legitimate (Pasternack 1975). "Syncretists" would draw together the useful parts of other theories to produce an encompassing interp. (Hermand and Beck). One school

of Ger. crit. has undertaken detailed studies of the evidence and logic that critics use in interp. The results are disappointing: critics' methods often prove to be haphazard and sometimes even contrary to their own theories. Attempts to create formal procedures for interp. based on principles taken from the philosophy of science, while revealing, have not proved influential (Göttner; Schmidt; Kindt and Schmidt; Pasternack 1979). Those who look on the multiplicity of interps. not as a defect but as a necessary result of literary study, one that cannot be explained or contained by pluralism or syncretism, are often called relativists or skeptics. But these terms are products of the very philosophical and critical trads. they would call into question.

Theories of interp. often seem remote from interpretive practice, which cannot be reduced mechanistically to a set of rules or a method. But arguments about interp., beginning with reasons, lead toward theoretical conclusions, and, as in philosophy, the quest for meaning becomes one for certainty, for knowledge grounded on indubitable premises. Some interpreters avoid this problem by rejecting theory (Knapp and Michaels), and some theorists seem determined to eschew praxis at all costs, concentrating instead on the study of literary conventions and textual structures. But this separation of theory and philosophy from interpretive praxis cannot be sustained, as de Man (1982) points out. In identifying a phrase as a metaphor, for example, we are unavoidably interpreting it, not merely describing it, and at the same time invoking poetics (the interp. depends on a theory of figures). This reciprocal relation between theory and praxis encompasses all aspects of interp. If we find a poem ambiguous, we can either appeal to a theory that provides methods of reducing ambiguity (q.v.), on the assumption that it is a fault in the poem or the interp.; or we can invoke a theory that explains why poems should be ambiguous, after which we might seek ambiguity where we had not noticed it before. The history of crit. shows that methods used in interpreting one type of poetry tend to be generalized into a theory that is then applied to other genres, in the same way that theories developed in other disciplines have been used to interpret poetry.

Two conceptual schemes discussed in other entries provide a useful means of classifying theories of interp. The first is Jakobson's communication model. On a horizontal axis, left to right, an "addresser" sends a "message" to an "addressee." The transmission depends on three factors named on a vertical axis over "message": the "context" (a situation and its realities), the "code" (in this case, lang.), and "contact," a means of communication (e.g., speech and hearing, or print and reading). This diagram is related to one in the entry for REPRESENTATION AND MIMESIS (see page 255): a poet produces a poem for an audience, and the poem refers, in a general sense, to reality.

Intrinsic theories of interp. derive the meaning of a poem (Jakobson's "message") from its lang., isolating it from the poet, variations of audience, and other factors. In 20th-c. poetics, such theories have been termed formalist or New Critical. Other theorists emphasize the connection between the poem and other terms in the communication model; while not necessarily rejecting poetics, their allegiances are with rhet., linguistics, semiotics, or the social sciences. *Intentional* theories concern the relation of the poem to the consciousness of the poet or "addresser." *Referential* theories, which emphasize the poem's connections with reality (the "context"), include traditional literary historians, biographers, and social and political critics, as well as psychoanalytic and myth critics (who hold that poems embody encoded references to reality). The codes and conventions important in interp. are emphasized in structuralist and semiotic theories; in research these are concerned with the history of ideas. *Reader-response* theories show how variations in the audience ("addressee") affect interp.

This classification is heuristic; for other purposes, it could be constellated differently. Some critics and philosophers see the communication model, based on the idea that we mentally encode and decode messages in specifiable contexts, as itself an erroneous interp. of how lang. works. The last section of this entry treats this view.

II. THEORETICAL ORIENTATIONS. In the 19th c., Arnold and Taine could refer to poetry as an interp. of life; in our century, the burden of interp. has shifted from the poets to the critics. For the emergence of significant theories of interp., three factors seem crucial. The 20th c. inherited a high esteem for the lyric from romanticism. The enigmatic texts of modernist poets, the progeny of symbolism, imagism, and surrealism, required new interpretive methods. A third factor contributing to the rise of intrinsic interp. was the growing importance of modern lits. in the universities. For an understanding of Gr., Lat., and Heb. texts, linguistic and historical knowledge had been paramount; modern poetry presented entirely different problems.

A. *Intrinsic Interp.* 19th-c. philosophy had provided a place for poetry in the scheme of things without indicating how it should be interpreted. Poetry, said Kant, unites "a concept with a wealth of thought to which no verbal expression is completely adequate"; it is peculiarly suited to express feeling and the inner life, according to Hegel, and can do so without reliance on representation of the outer world. As a mode of discourse, it is opposed to rhet.; it is not heard, but overheard (Mill); when produced by genius, it seems to flow from nature, not the poet's intentions (Kant). Similar ideas can be found in Coleridge. While not in agreement about what a poem is, these thinkers nevertheless all assume that it is *not* a "message" that the poet transmits to readers to inform or persuade them,

as in ordinary discourse. For modern Fr. and It. critics, Bergson and Croce provided philosophical grounds for a conception of poetry as a special use of lang.

Drawing on these philosophical resources, a broad range of 20th-c. critics developed methods of poetic analysis that are fundamentally intrinsic or formalist. The assumption that the meaning of poetry is entangled in its concrete and figural lang.—so ambiguous, paradoxical, ironic, and polysemous that it is not reducible to paraphrase—is most obviously applicable to modern poems. Because modern poets were influenced by a poetics that advocated impersonality and discouraged mimetic representation, early advocates of intrinsic interp. often found in practice what they held in principle: modern poems were self-sufficient entities, complex verbal structures creating meanings that could not be recovered from the poet's biography or historical circumstances. Mallarmé, Valéry, Rilke, the late Hölderlin, Eliot, and Wallace Stevens were poets ideally suited to intrinsic interp., as were some metaphysical poets, *Culteranismo*, Scève, and Dante's *canzoni*.

In order to apply intrinsic interp. to poetry written between the 17th and 20th cs., it was necessary, as Cleanth Brooks said, to read "the intervening poems . . . as one has learned to read Donne and the moderns." Confirmation of the intrinsic theory required the discovery of complexity in poems that appeared simple (Staiger). Originally a method for the interp. of obscure lyrics, the immanent approach now became a theory that determined how *all* poetry should be read: poems not displaying the features it esteemed were correspondingly devalued.

The faults of intrinsic crit., which have been more than adequately exhibited, should be balanced against its pedagogical achievements. As Richards showed, the wildly disparate interps. produced by students often resulted from an inability to recognize the basic conventions of poetry, quite apart from the ambiguity that could be imputed to it (Empson). Influential textbooks led to the improvement of exegetical skills (Brooks and Warren; Kayser), and cogent accounts of the logic underlying intrinsic interp. (Wimsatt; Burckhardt) provided later generations of critics with a theoretical awareness that helped them pinpoint the deficiencies of the theory.

B. *Intentional Interp.* Intrinsic interp. achieved such widespread acceptance that even politically conscious critics (Bakhtin; Sartre) did not challenge it; conceding the autonomy of poetry, they devoted their critical energies to socially relevant prose forms. Those who opposed intrinsic interp. in the 1950s, despite their differences, agreed on one thing: poems are not self-contained entities cut off from their authors, readers, and reality. Though distinctive in some structural or textual respects, in other respects they do not differ in kind from other forms of verbal communication.

To separate the poet from the poem, intrinsic interp. had argued that the speaker should always be construed as a dramatized persona, a poetic voice (q.v.), never to be identified with the flesh-and-blood author. Apart from the difficulties that this position presents with respect to poems that are obviously personal, it glosses over important theoretical issues. Unlike organisms, poems have meaning because someone intends them to mean something. To interpret them, it is helpful to consider the poet's intentions insofar as they are manifest or latent in his or her other writings and inferable from biographical study.

From this premise, *intentional* theorists diverge along several lines. Philosophers have distinguished nine meanings of "intention," three of which are important here. Some critics use the word in its everyday sense but add to it a concept from phenomenology: meaning *as such* is intentional (Hirsch; Juhl). A poem therefore means what the poet intended it to mean, in general and in every particular, though we are unlikely ever to know exactly what that meaning was. In a more fully phenomenological sense, intentionality is not just a matter of conceptual meaning; it is that property of consciousness whereby objects are constituted in all their experiential dimensions. Interp. from this point of view is a matter of assimilating all of a poet's writings in order to replicate his or her mode of experiencing the world—sensory and emotional as well as intellectual and volitional. In intrinsic crit., image and idea are the parts that constitute the unity of the poem; for "critics of consciousness," the poet's mind is the unity to be interpreted, and images taken from various poems become parts used to reconstitute that whole.

C. *Referential Interp.* When the relationship between the poet and the poem is the focus of interp., intention and meaning may become nearly synonymous. But the latter terms are often opposed to each other in ordinary usage. Like other mortals, poets may intend to conceal facts (representing as a general situation some incident that has a specific biographical basis) and may unintentionally convey meanings that they do not try to communicate. *Referential* or mimetic theories emphasize the importance of concrete circumstances in determining what a poem means. They recover from history and biography the facts or situations to which poems directly or indirectly refer. In a similar manner, we interpret what friends say in light of what we know of them; and to understand other cultures, we must know what they assume about the world as well as what they "intend" to say about it. To interpret a poem from the past, we must understand not just its words but its pastness; its distance from us becomes part of its meaning.

To these traditional methods of interp. the 20th c. has added others based on the assumption that references to reality in lit. can themselves be en-

coded. Todorov refers to these as symbolic or "finalist" theories. He includes Marxist interps. in this category, but their interpretive decoding usually connects poetry to the facts of history rather than to an individual or transhistorical psyche. Jameson bridges the gap between psyche and history by positing a "political unconsciousness" that contains an imaginary replication of one's socioeconomic circumstances.

The success of referential theories is inseparable from their weakness. The wholeness of meaning they reconstruct is that of a life, an epoch of history, or a racial unconscious; poems are merely parts useful in creating these wholes. That a poem is a product of its time is undeniably true; but it is easier to posit connections between the two than to confirm or refute them. Rather than providing a factual basis for interpretive inference, historiography produces different histories, which are themselves interps.; the same is true of biography and psychology. Some recent literary critics have tried to resolve this dilemma by treating all the events and writing of a period as a "social text." From this point of view, history is not the cause or source of lit.; acts and texts are both embodiments of cultural powers and practices (Greenblatt; Pechter). Alternatively, one may find conceptual configurations emerging in poetry before they are registered elsewhere in culture (Fineman).

D. *Structuralist and Semiotic Theories.* In the traditional opposition of intrinsic to extrinsic interp. and of language-bound meanings to facts independent of linguistic embodiment, what is lacking is attention to the mediating terms through which these antitheses interact: the conventions and codes that bind poet, poem, lang., and reality together in shared systems of signification. An emphasis on the distinctive *codes* of poetry is characteristic of Rus. Formalism and structuralism. Shklovskij conceived them as purely formal structures: "The purpose of the new form is not to express new content, but to change an old form which has lost its aesthetic quality." Jakobson isolated the "poetic function" in phonetic, syntactic, and grammatical features. Like the formalists, some structuralists and poststructuralists have argued that the purpose of literary study should not be to produce still more interps. but "to advance one's understanding of the conventions and operations of an institution, a mode of discourse" (Culler).

Semioticians argue that social and cultural codes are as important as linguistic and literary conventions in the interp. of poetry. Challenging traditional methods of extracting a theme from a poem, J. Mukařovský held that *all* of its features, including sound, are endowed with meaning, and that poetry refers not to reality as such, but to the "total context of so-called social phenomena, for example, philosophy, politics, religion, economics, and so on." The Prague School structuralists found in poems a dialectical tension between the poet and others of his era; between inherited and innovative forms; between lit. and society. Interp. thus becomes an inclusive project that moves back and forth between text and culture, correlating the semiotic configurations of the text with the cultural contexts in which they take on one or another significance. In its final phase, this school began to explore the ways in which interp. depends upon the assumptions, habits, and expectations of readers in different historical periods (see Galan).

E. *Reader-response Theories.* Phenomenological critics, structuralists, and semioticians all hold that a poem exists as an object of interp. only when it is read; in this sense they recognize the importance of the reader. But they are usually interested in finding what remains constant in varied instances of interp. rather than in determining why these vary. Riffaterre, for example, criticized Jakobson and Lévi-Strauss for disregarding the reader when identifying patterns in poetry, but he did not go on to argue that readers could legitimately find different meanings in poems. His semiotic approach assumes that there is only one valid interp. of a poem, that it involves not "meaning" (reference to reality) but "significance" (semantic coherence within the poem), and that all deviations from clear meaning are generated by a commonplace or cliché that serves as the poem's matrix. "All of the reader's possible reactions to the text" constitute for him the audience; this inclusiveness insulates his theory from individual and historical differences.

Reader-response critics explore the ways in which interp. depends upon the audience. Their answers to the following questions are crucial: (a) Who is the reader? For structuralists and semioticians, the reader is anyone who has acquired "competence" in understanding literary codes and conventions. From an empirical point of view, however, the reader is a particular person who may produce a unique interp., or simply disregard the exegetical labors undertaken by interpreters to enjoy the poem. Other critics posit an ideal reader or the kind of reader implied by the text as the appropriate interpreter of its content. (b) To what extent does the poem exercise control over the interp.? The phenomenologist's assertion that a text exists only as it is perceived and understood can correct a naive belief that the meaning lies "in" the words, independent of interp., but it can also lead to the conclusion that meaning is in the reader, independent of any corrigible experience of words. Interps. that cannot plausibly be related to the words of a poem may serve as evidence for this conclusion, but it has had little effect on the practices of interpretive communities, which produce tacit (and evolving) criteria governing how readings will be justified. (c) What is "interp."? As Ricoeur and Ray point out, critics at one extreme hold that it is an act, the process of experiencing a text, which may cancel as many meanings as it creates (Fish); at the other extreme is the traditional view that the purpose of interp. is to recon-

stitute a stable meaning after the experience of reading.

Jauss tries to accommodate these differences in his inclusive model of reader response. In his view response comprises three phases: understanding (the act of reading), interp. (reflective constitution of meaning), and application (relating the interp. to other knowledge and experience). He distinguishes the *Wirkung* or effect of a poem, which the structure of the text lays out for the reader, from its *Rezeption* or reception, the actualization of its meaning conditioned by the reader's social and personal existence. Despite their differences, readers from a particular period have much in common. The history of literary reception shows that poems survive because they remain open to varied interps. governed by needs and expectations that could not have been imagined when they were written. Other critics would add that interp. can thus be seen as a product of "interpretive communities," institutional practices, and social structures that will determine whether or not an innovative reading is acceptable (Fish; Kermode).

III. PHILOSOPHY, HERMENEUTICS, AND DECONSTRUCTION. The shift from intrinsic interp. to theories based on the communication model corresponded to a shift in interest from lyric poetry to narrative modes in literary study. Consideration of how literary codes, reality, and readers determine the meaning of texts seemed to provide a sound basis for interp. of lit. within a general theory of communication (e.g. Jakobson's); while disagreements about interp. persisted, it was at least possible to point out their sources within the framework of the model. Beginning in the 1970s, however, the model itself was called into question—either because it failed to account for specific features of literary lang. or because of its philosophical presuppositions.

If poets intend to convey clear messages, they are remarkably unsuccessful in doing so; the variety of interps. elicited by their works suggests that they have other aims. Bloom holds that their intention is to attain literary immortality and that their poems are misinterpretations of earlier poems, created to clear a space for their own meanings in the trad. In a similar manner, the "strong" reader or critic proposes a new interp. of a poem to assert priority over other readers and, indeed, over the poem itself.

The assumption that literary lang. is a deviation from ordinary lang., one in which conventions (the use of figures and sound effects) add ambiguity and pleasure to messages that could be transmitted without them, is likewise questionable. Intrinsic critics held that poetry unites meaning (codes) and being (reference); de Man argued that allegory reveals the gap between the two, and that rhet. cannot be contained within the tidy codes of linguistics and semiotics. Despite their differences, intrinsic and subsequent critics agree that poets explore the relationship between lang. and reference, rather than assuming it to be unproblematic.

The communication model presupposes that, on the basis of reliable knowledge about intentions, history, literary conventions, and lang., we can elucidate the less evident meanings contained in poems. But that purportedly reliable knowledge is *itself* an interp., based on philosophical and linguistic assumptions that are far from certain. Interp. is always caught in the hermeneutic circle: rather than using facts to determine meaning, we cannot help but let our anticipations of meaning determine what "facts" (themselves prior interps.) we will select to support it (see Fish 1980).

Linguists, philosophers, and critics who employ the communication model treat speech as the typical instance of lang. use. In a particular context, at a certain moment, someone says something for a specifiable purpose. When this model is applied to written poetry, interp. becomes a matter of recovering information we lack about the poet, the audience, and a unique moment of inscription. Writing is seen as an impoverished version of speech, since it has been torn from a presence that made it meaningful. It is possible, however, to construe writing as a distinct method of creating meaning, in which case interp. becomes a very different activity.

As Ricoeur says, writing severs lang. from the writer and "opens it to an indefinite range of potential readers in an indeterminate time." An interp. that would confine the meaning of a poem to the historical moment of its creation is at odds with the poet's intentions, and denial that poems written in earlier centuries have meaning in our world is at odds with our desires. If, in the case of writing, the "context" and "addressee" of the communication model continually change, interp. cannot be invariant. The different interps. of a poem we inherit from the past may themselves become part of its meaning for us (Jauss).

For deconstruction, the instability of meaning apparent in the case of writing results not just from the text's survival in changing circumstances but from the character of lang. itself. Philosophers have argued that problems of interp. result from literary uses of lang. (e.g. metaphor and polysemy) that could be eliminated in a systematic literalism. Deconstructive critics reply that philosophers also make use of these "supplementary" aspects of lang. in building systems that themselves contain paradoxes and contradictions. Literary lang. and the interpretive difficulties it poses can be seen as characteristic of lang. in general. As a self-reflexive unfolding of the figural and rhetorical constituents of meaning, poetry explores and exemplifies the impossibility of attaining a univocal interp. of words and the world. From this point of view, the critic's task is to discover the patterns underlying alternative readings of a poem and the points at which they intersect to

create an impasse or dilemma.

By discovering a diversity of interpretive possibilities in the lang. of poetry, deconstruction confers on the poetic text a richness denied it in intentionalist, referential, structuralist, and reader-response theories. Miller suggests that the impossibility of univocal interp. is implied in poems that he interprets—a paradoxical conclusion that can only be as correct as his interp. Many critics admit or assert that there is no single correct interp. of a poem; yet when themselves engaged in interp., they very often urge the superiority of theirs; and almost always cannot help but imply that the poem has one range of meanings rather than another, referring to its formal features to substantiate their arguments. Thus theory and praxis, while interdependent, remain dialectically at odds with each other in the critical acts of reading and writing.

I. A. Richards, *Practical Crit.* (1929); Brooks; Empson; W. K. Wimsatt, Jr., *The Verbal Icon* (1954); E. Staiger, *Die Kunst der Interp.* (1955); Wellek and Warren; M. C. Beardsley, *Aesthetics* (1958); A. Child, *Interp.: A General Theory* (1965); S. Sontag, *Against Interp.* (1966); *Immanente Ästhetik—Ästhetische Reflexion*, ed. W. Iser (1966); Jakobson; E. D. Hirsch, Jr., *Validity in Interp.* (1967), *The Aims of Interp.* (1976); symposium on Hirsch 1967 in *Genre* 1,3 (1968), with his reply in 2,1; S. Fish, *Surprised by Sin* (1972), *Is There a Text in This Class?* (1980); H. Bloom, *The Anxiety of Influence* (1973); H. Göttner, *Logik der Interp.* (1973); J. Hermand and E. T. Beck, *Interpretive Synthesis* (1975); G. Pasternack, *Theoriebildung in der Literaturwiss.* (1975), *Interp.* (1979); L. Pollmann, "Lyrikinterp. heute," *Sprachen der Lyrik*, ed. E. Köhler (1975); S. J. Schmidt, *Literaturwiss. als argumentierende Wiss.* (1975); W. Kindt and S. J. Schmidt, *Interpretationsanalysen* (1976); J. A. Coulter, *The Literary Microcosm* (1976)—Neoplatonic theories of interp.; J. Mukařovský, "Art as Semiotic Fact," *Semiotics of Art*, ed. I. R. Titunik et al. (1976); P. Ricoeur, *Interp. Theory* (1976); J. W. Meiland, "Interp. as a Cognitive Discipline," *P&L* 2 (1977); W. Kayser, *Das Sprachliche Kunstwerk*, 18th ed. (1978); M. Riffaterre, *Semiotics of Poetry* (1978); T. Todorov, *Symbolism and Interp.* (1978); F. Kermode, *The Genesis of Secrecy* (1979), *The Classic* (1983); H. Bloom et al., *Deconstruction and Crit.* (1979); P. de Man, *Allegories of Reading* (1979), Intro. to Jauss (below); S. Greenblatt, *Ren. Self-Fashioning* (1980); P. D. Juhl, *Interp.* (1980); *Text und Applikation*, ed. M. Fuhrmann et al. (1981); J. Culler, "Beyond Interp.," *The Pursuit of Signs* (1981); F. Jameson, *The Political Unconscious* (1981); Fowler; H. R. Jauss, *Toward an Aesthetic of Reception* (1982), *Aesthetic Experience and Literary Hermeneutics* (1982); S. Knapp and W. B. Michaels, "Against Theory," *CritI* 8 (1982); de Man; K. Mueller-Vollmer, "Zur Problematik des Interpretationsbegriffs," *Erkennen und Deuten*, ed. M. Woodmansee et al. (1983), and "Understanding and Interp.," *Lit. Crit. and Philosophy*, ed. J. P. Strelka (1983); W. E. Rogers, *The Three Genres and the Interp. of Lyric* (1983); S. Schmidt, "Selected Bibl. on Interp. (1970–1982)," *Poetics* 12 (1983); F. Galan, *Historic Structures* (1984); W. Ray, *Literary Meaning* (1984); J. H. Miller, *The Linguistic Moment* (1985); J. Fineman, *Shakespeare's Perjured Eye* (1986); E. Schauber and E. Spolsky, *The Bounds of Interp.: Linguistic Theory and Literary Text* (1986); E. Pechter, "The New Historicism and Its Discontents," *PMLA* 102 (1987); A. Barnes, *On Interp.: A Critical Analysis* (1988); J. M. Ellis, *Against Deconstruction* (1989); P. B. Armstrong, *Conflicting Readings: Variety and Validity in Interp.* (1990); S. Mailloux, "Interp.," *Critical Terms for Literary Study*, ed. F. Lentricchia and T. McLaughlin (1990); U. Eco, *The Limits of Interp.* (1990).

W.M.

INTERTEXTUALITY refers to those conditions of textuality (q.v.) which affect and describe the relations between texts, and in most respects is synonymous with textuality. It originates in the crisis of representation and the absent origin that would guarantee meaning, centrality, and reference. Without an ultimate referent that would make possible the self-presence and meaning of a text, texts are by definition fragments in open and endless relations with all other texts.

In traditional models of influence (q.v.), a text comes to rest on a prior text which functions as a stable source which is retrieved and made present by a study of allusion (q.v.), quotation, and reference. Relations between texts are thus straightforward and determinate. Their determinacy is the result of five premises of traditional crit.: (1) that lang. has the capacity to create stable meaning; (2) that such meaning exists within the confines of form; (3) that the artist is in control of meaning; (4) that a work has closure (q.v.), its tensions, ambiguities, and ironies coming to a point of resolution; and (5) that crit. is an ancillary activity, separate from lit. These premises tend toward totality (either in the mode of "the work itself" or Frye's mode of the "total form of lit."), and the concepts of stable meaning and the artist's control of it are basic to the humanistic trad. of learning, which affirms and emphasizes the notions of human self and human will.

In the late 1950s and '60s, however, as a result of the influx of Fr. crit. into America, the premises upon which the study of lit. had been carried out changed dramatically, and the conceptual approach to literary relations underwent an equally radical transformation. With structuralism and poststructuralism, the concept of "influence" was discarded in favor of the concept of "i." Seven major premises now come into play: (1) Lang. is not a transparent medium of thought or a tool in the service of communication; it is arbitrary and dense, and its very excessiveness leads to an infinite number of interps. (2) Texts are fragments, without closure or resolution. No text is self-sufficient; each text is fraught with explicit or invisible

quotation marks that dispel the illusion of its autonomy and refer it endlessly to other texts—like the entries in this *Encyclopedia*, referring parenthetically to other references, except that there is no way to contain all possible references in any encyclopedic "whole." (3) Given the above, no writer can ever be in control of the meaning of the text. I. does away with the concept of "author" in its conventional meaning (authority, property, intention), supplanting it with the concepts of "author-function" (Foucault) or "subject" (Lacan). (4) Meaning is supplanted by the notion of "signification" (a sign is composed of signifier and signified, but in poststructuralist thinking the signified is lost, leaving the signifier in search of a referent it can never find). Poststructuralism thus discards the humanistic version of human beings as creators of meaning, and proposes them instead as creatures (effects) of lang. (5) Crit. is no longer an ancillary activity, but is now considered part of the poem, creative of its meaning or signification. In formalism and humanism, the task of crit. is "explication," which distinguishes the reading subject from the literary object and defines lit. as a discipline and a mode of knowledge. I. stands in direct opposition to explication, with its explicit distinction between primary and secondary texts, and instead opens up literary, critical, and indeed many other texts to illimitable relations. (6) Disciplinary boundaries are erased: such fields as philosophy and psychoanalysis are all considered discursive practices and ultimately inseparable from lit. (7) Finally, poststructuralist crit. defies the rules of reason and identity and suggests instead the idea of contradiction. "Contradiction," says Adorno, "indicates the untruth of identity, the fact that the concept does not exhaust the thing conceived." Adorno's "contradiction" is very close to Derrida's "*différance*."

I. is marked by two key features: the absence of an origin and the function of randomness. Traditional ("logocentric") critics consider lit. as a privileged mode of communication, expressive of human nature in its highest form. For these critics a transcendent referent (the "transcendental signified") organizes all lang. and experience, making possible a "self" which expresses itself in terms of "voice" (q.v.). Poststructuralist crit. "steals" that ultimate referent. In the place of a privileged origin it finds a trace of something prior, which is lang. (writing) itself. Derrida speaks of the absence of a center or origin which would "arrest and ground the play of substitutions" because the sign which replaces the center "supplements" it and thus underscores its lack. *Différance* is the name for this unnamable absence of origin and for the "chain of differing and deferring substitutions" which it unleashes. In this respect textuality is precisely synonymous with i., in that it signals the impossibility of boundaries or borderlines that would adequately frame a "work" and its "meaning," and points instead to writing's "dissemination." Under these conditions, there could be no metalanguage, no privileged point that would make reference and knowledge possible, that would not "always already" be implicated in the tropological relations it would seek to describe. This submission to the legislature of lang. transforms the nature of intertextual relations and thereby the relations between crit. and lit. In the absence of an origin that would guarantee presence, meaning, and voice, there can be no originals—only copies. And without a univocal and transcendental referent, all texts refer to one another—translate one another—in infinite and utterly random ways.

Representation and reference, the mainstays of traditional humanism, underwrite a patriarchal "logocentric" order (Derrida terms it "phallogocentric") in which a work or a referent functions as a stabilizing ("seminal") source and provides the authority of meaning. Explicit in i. is the dismantling ("dissemination") of paternity. Barthes claims that "there is no father-author"; Derrida argues that "writing is an orphan." I. underwrites a critique of logocentrism and of patriarchy, substituting for patriarchal self-presence the feminizing "otherness" of intertextual "lapses."

These lapses which signal lost and irrecuperable meaning have altered the very shape of the "book." Notable as an effect of i. is the number of quotation marks deployed across texts, indicating "other" or alien contexts in which the words in question might be understood. The book, as a concept and as an entity, has undergone a similar transformation. While it still appears to us as words between two covers, the traditional rules by which prefaces functioned as openings, introductions introduced, and afterwords or epilogues closed are often missing. A preface may be termed a "pre-text" or "prefatory material"; a conclusion may well be a "postscript" or an "afterword" (or both, in sequence); "interchapters" or "interstices" point to the fragile relations between parts; and footnotes will sometimes constitute a parallel text, as in Derrida's "Living On . Borderlines," where the essay is entitled "Living On" and the long footnote threading its way through the essay is entitled "Borderlines." The latter does not function in submission to or as a subtitle to the former, so even the punctuation between them for routine reference becomes problematic. All of these strategies point to the impossibility of a text's wholeness or self-presence, and to the changed relations i. has wrought between reading and writing.

J. Derrida, *De la grammatologie* (1967, tr. 1976)—with essential preface by G. Spivak, "Signature, Event, Context," *Marges de la philosophie* (1972), *La Dissémination* (1972), *Glas* (1974, 1981, tr. 1986), "Living On . Borderlines," *Deconstruction and Crit.*, ed. H. Bloom (1979); J. Kristeva, *Séméiotikè* (1969), *La Révolution du langage poétique* (1974, tr. 1984); *Poétique* 27 (1976)—spec. iss. on i.; M. Riffaterre, *Semiotics of Poetry* (1978), *Text Production* (tr. 1983), *Fictional Truth* (1990); *Tex-*

tual Strategies, ed. J. Harari (1979)—excellent intro., bibl.; J. Culler, "Presupposition and I.," *The Pursuit of Signs* (1981), *On Deconstruction* (1982); *Untying the Text*, ed. R. Young (1981); *The Question of Textuality*, ed. W. Spanos and P. Bové (1982); *Displacement*, ed. M. Krupnick (1983); *The Anxiety of Anticipation*, spec. iss. of *YFS* 66 (1984); G. Atkins and M. Johnson, *Writing and Reading Differently* (1985); G. Jay and D. Miller, *After Strange Texts* (1985); *Intertextualität: Formen, Funktionen, anglistische Fallstudien*, ed. U. Broich et al. (1985); A. McHoul and D. Wills, "The Late(r) Barthes: Constituting Fragmenting Subjects," *Boundary 2* 14, 1–2 (1985–86); C. Chase, *Decomposing Figures* (1986); M. Nyquist, "Textual Overlapping and Dalilah's Harlot-Lap," *Literary Theory / Renaissance Texts*, ed. P. Parker and D. Quint (1986)—on textual "lapses"; S. Weber, *Demarcating the Disciplines*, ed., (1986), *Institution and Interp.* (1987); R. Dasenbrock, "Accounting for the Changing Certainties of Interpretive Communities," *MLN* 101 (1986); *Poems in Their Place: The I. and Order of Poetic Collections*, ed. N. Fraistat (1986); C. Norris, *Derrida* (1987); R. Goodkin, "Proust and Home(r): An Avuncular Intertext," *MLN* 104 (1989); *Autour de Racine: Studies in I.*, spec. iss. of *YFS* 76 (1989); *I. and Contemp. Am. Fiction*, ed. P. O'Donnell and R. Davis (1989)—esp T. Morgan, "The Space of I."; *I., Allusion, and Quotation: An Internat. Bibl. of Crit. Studies*, ed. U. J. Hebel (1989); G. Jay, "Paul de Man: Being in Question," *America the Scrivener* (1990); *I.: Theories and Practices*, ed. M. Worton and J. Still (1990); *Influence and I. in Lit. Hist.*, ed. J. Clayton and E. Rothstein (1991); *I.*, ed. H. F. Plett (1991). H.R.E.

INTUITION.

I. IN AESTHETICS AND POETICS
II. IN POETRY

I. IN AESTHETICS AND POETICS. The term "i." owes its importance in 20th-c. poetics to Benedetto Croce's use of it in his *Aesthetic as Science of Expression and General Linguistic* (1902, tr. D. Ainslie, 2d ed., 1922), where he identifies it with expression. Rejecting as naive the popular view of i. as a completely subjective phase of the cognitive process, Croce lays down the warning: "it is impossible to distinguish i. from expression in this cognitive process. The one appears with the other at the same instant, because they are not two, but one." Also: "to intuit is to express; and nothing else (nothing more but nothing less) than to express" (Ainslie 8–11). In 1915, Croce acknowledged that his own dissatisfaction with his 1902 account of i. had led to its "conversion" into the "further concept of pure or lyrical i." Under the influence of Giovanni Gentile (*Philosophy of Art*, tr. G. Gullace, 1972) as well as Vico and Hegel, Croce had delivered his Heidelberg lecture in 1908 on "Pure I. and the Lyrical Character of Art," where the term's meaning deepens to include the "successful union of a poetic image with an emotion."

By the time he is done, Croce has given an aesthetic theory of i. which is a theory of expression and of imagination (q.v.) as well. By identifying it with expression and imagination, Croce was able to give his use of i. a measure of variety and novelty it probably could not have sustained on its own. Croce nearly makes us forget that, before Kant took it up with fresh insight, the concept had already had a long history in Med. Lat. as *intuitus*, and an even longer history in its original Gr. form, *nous*.

It was C. S. Peirce who pointed out that the Lat. term *intuitus*—which Kant puts in parentheses after the Ger. equivalent, *Anschauung*—first occurs as a technical term in St. Anselm's *Monologium* (11th c.; tr. S. N. Deane 1903). Anselm, Peirce explains, had tried to draw a clear distinction between seeing things through a glass darkly, (*per speculum*) and knowing them "face to face," calling the former *speculation* and the latter *intuition*. In a famous passage of the *Monologium*, Anselm says: "to the supreme Spirit, expressing [*dicere*] and beholding through conception [*cogitando intueri*], as it were, are the same, just as the expression of our human mind is nothing but the i. of the thinker." Some students of Pound's *Cantos*, commenting on the so-called St. Anselm canto (105), have suggested that the entire *Monologium* may be read as an adumbration of what Eliot called the "objective correlative."

In the *Proslogium*, St. Anselm would later attempt to show that, in the concept of God intuited through Christian faith, enough is contained to "prove" discursively that anything so conceived must not only be thinkable but also exist. Pressing his argument (against the fool who has said in his heart "there is no God"), Anselm compares human i. of God with a painter's i. of a painting he has actually painted, as contrasted with his i. of the same before he has painted it. This so-called ontological "proof" of God's existence, later advanced by Descartes, Leibnitz, and Spinoza, prompted many later critics to attack Anselm. And foremost among them has been Immanuel Kant.

I. lies at the heart of Kant's entire intellectual paradigm. In his first critique, the *Critique of Pure Reason* (1781), Kant is at pains to distinguish two kinds of i.: first, a *receptive* kind, which enables the understanding (*Verstand*) to take in phenomenal sensations in the *a priori* forms of space and time, and as related to one another causally; and second, a *nonreceptive* kind, which Kant does not hesitate to characterize as a "productive imagination." It provides the understanding with a contents of supersensory things-in-themselves, not knowable through sensory experience. A proper name for it, says Kant, is "intellectual i." The chief supersensory ideas it provides are soul, world, and God. However grand their suggestiveness, those ideas have no empirical validity: they belong entirely to intellectual i.; the understanding cannot "prove" to itself whether they exist or not.

In Kant's second critique, the *Critique of Practical Reason* (1788), the entire context of his argument derives from one of the three supersensory ideas provided by intellectual i., namely the soul or human psyche. As the subjectivity of reason, the soul is made "aware" of itself not in the process of trying to know itself, but in that of being or willing itself. Kant's critique shows that, for the practical reason, the will can in no sense be a phenomenon related by laws of necessity to other phenomena; on the contrary, it is a noumenon or thing-in-itself, unaffected by anything external to it, and therefore completely free, answering only to itself for how it responds to the *a priori* "categorical imperative" by which its ends are determined intuitively.

But i. rises to a still higher level in Kant's third essay, the *Critique of Judgment* (1790; ed. K. Vorländer 1923, tr. J. C. Meredith, 1911, 1928). At the outset, Kant speaks of a "great gulf" yawning between the two previous critiques, "between the realm of the natural concept, as the sensible, and the realm of the concept of freedom, as the supersensible." What is needed, he says, is "a third mediating principle," an *a priori* intuitive synthesis of the opposed perspectives of nature (which mind comprehends only through its phenomenal impressions on the understanding) and freedom (where mind is at home with itself in its inner, noumenal reality). A table at the beginning of the third critique shows it is "art and beauty in general" that must "bridge the gap" between the two.

Only art or aesthetic experience, Kant explains, can provide the needed *a priori* intuitive synthesis, precisely because it is neither natural nor free in itself, and yet participates, at least apparently, in both. In nature, our mind seeks knowledge; in freedom, it wills or desires; in art, it neither seeks knowledge nor wills, but rather finds itself viewing things intuitively—Wordsworth in his *Excursion* (4.1295) will call it a "passionate i."—so as to link the realms of nature and freedom together in what amounts to a great metaphor or intuitive synthesis. Despite his Humean skepticism, Kant allows himself to speak of the "divine mind" as author of that "highest synthesis of the critical philosophy." Having identified it as "that reason which creates the content at the same time with its forms," Kant adds that it here appears at last as "intellectual perception or intuitive understanding."

It was Hegel who remarked that, in rising through his three critiques to the concept of intellectual i. or intuitive understanding as the governing principle of the experience of purposiveness and beauty in nature and art, Kant merits the kind of praise Aristotle accorded Anaxagoras in Gr. antiquity. When Anaxagoras first said that *nous* (understanding in general or reason) rules the world, "he appeared," says Aristotle, "like a sober man among drunkards." Awed, like Kant, by the spectacle of the night sky with its "undisturbed circling of countless worlds," Anaxagoras had concluded that it could only be the result of a mind or *nous*, a divine intuitive reason which sorts out the original constituents of things in order to join like to like in a well-ordered whole (*kosmos*). Still, as Aristotle notes, Anaxagoras wavers between assigning his *nous* completely to man's thinking soul (like Kant in his first critique) and proclaiming its total objectivity as the "soul-stuff" of the sensory world of nature.

In Plato and, even more, in Aristotle and the Neo-Platonists, the *nous* of Anaxagoras retains its original significance as the divine mind that gives order to all things. It comes finally to be conceived as a pure act of "thinking thinking about thinking" (*nóesis noéseos nóesis*). The human soul or psyche, like everything else that is, "imitates" or "participates" in that pure act in the measure of its natural potencies. The psyche receives intuitively—quite as Kant will later explain it—the principles of its theoretical, practical, and productive kinds of rational activity. Through rational making, men and women build the cities or states in which they are able, individually and collectively, to behave rationally. And it is only when their making and behaving have acquired habitual excellence that human beings can have the experience of sharing, however briefly, in the divine existence of thinking thinking about thinking, which is the height of *sophia* or wisdom. The good habit or virtue of making well is *techne* or art; of behaving well, it is *phronesis* or prudence; of explaining well, it is *episteme*, or excellence in discursive reasoning. *Nous* or intuitive reason is what ultimately activates all three; but when the three become one at the pole, like the meridians on the globe, it is *sophia*, the highest intuitive reason joined with the highest discursive reason, that absorbs the other two (the highest good and the highest beauty) in its highest truth.

Through St. Jerome, St. Augustine, and St. Thomas, the form of rational i. that activates practical reason will get a special name: *synderesis*. But it will be left to Dante to do the same for the productive reason, or fine art. Dante will dare to say that his whole *Commedia*, in which he sought to "write like God," was for him a metaphoric, intuitive vision, completed to perfection in his "creative imagination" (*alta fantasia*) before he applied himself to the task of writing it down. Love, he says, helped him to turn his vision into poetry by dictating words in his heart. Fortunately, he had native talent—*ingenium* or genius—and had acquired the traditional poetic skills (*ars*) through long practice (*usus*), to be able to do some justice to love's divine dictation; otherwise, as he says, his trying to write like God might have looked rather like the efforts of a goose to fly like an eagle.

For a rounded view of the importance of i. as an aesthetic concept that does full justice to its long history, one must turn to Hegel (*Aesthetics*, tr. T. M. Knox 1975). Hegel makes use of i. to distinguish the three great kinds of art that have characterized the civilizations of ancient Greece, the

Near and Far East, and European Christendom. The three are the Classical, with its ideal reciprocal adequacy of content and form; the Symbolic, which falls short of that ideal, presenting a reciprocal inadequacy of constituents, the results of which are sometimes ugly but sometimes tremulously sublime (q.v.) in Kant's sense of the term; and the Romantic, which transcends the classical reciprocal adequacy—as Shakespeare does in his characterizations of Falstaff and Hamlet, and Dante does at the close of his *Paradiso*, where, after telling us that his creative imagination finally experienced a vast intuitive power-shortage (*all'alta fantasia qui mancò possa*), he permits us to share with him, intuitively, the dizzying heights of art in the process of transcending itself as art.

The same insight permits Hegel to treat comprehensively the great epic, lyric, and dramatic genres or voices of poetry, and to predict that poets of the future in the Western world are apt to find the lyric voice more vital than the other two voices of poetry. As a living experience, epic poetry belongs to the beginnings of a people's national history, just as dramatic poetry does to the decades of its full maturity. Lyric poetry has no special time in a people's history, and therefore its voice is never silent. From that Hegelian vantage point, it is possible to explain why theorists of art since Kant who have counted on i. to tell them what is actually living for them in aesthetic experience have tended to emphasize the ideal of the lyric in poetry and of musical subjectivity among all the arts. That has been so from Schopenhauer and Nietzsche down through Henri Bergson (*The Creative Mind*, tr. 1946) and Jacques Maritain (*Creative I. in Art and Poetry*, 1953). René Wellek has seen in it a tendency to abolish the "whole concept of art as one of the distinct activities of man" by collapsing the rich legacy of traditional artistic distinctions into a crude "unity of experience" that makes all things subjectively and intuitively one. Still, Hegel's best insights permit us to say on Croce's behalf that, when he faithfully scrutinized his personal aesthetic experience, he found indeed that its depths resounded, for him at least, with a voice of singularly lyrical intuitive inspiration. A.P.; H.P.

II. IN POETRY. Theories of poetry as a form of i. are even today somewhat alien to our ways of thinking. G. N. G. Orsini observes that Am. literary critics are profoundly distrustful of their intuitive capacities and, one might add, of the intuitive power of poetry itself. Am. histories of aesthetics do contain clear descriptions of theories of poetry as i., but their clarity depends on their remoteness from that which they are describing. Even when most objective, they seem to suggest that such theories are simply untenable. Of course, the theories may in truth be untenable. But because they have emerged in cultures with much richer conditions of poetry and art than our own, we should reject them only with extreme caution.

In its most fully developed form, the theory of poetry as essentially intuitive is based on the belief that lines like the following are exemplary: "the dry sound of bees / Stretching across a lucid space" (Hart Crane, "Praise for an Urn") and "As a calm darkens among water lights" (Wallace Stevens, "Sunday Morning"). To be sure, any line of verse, according to this theory, is poetic only if it is intuitive. But the two lines just quoted seem to exhibit what is meant by poetic i. with unusual clarity and vividness. Even these lines, it is true, do not force the reader to respond to them intuitively. The lines may be taken to be imitative of certain natural events or as illustrations of visual illusions or even as instances of rhetorical catachresis (q.v.). But if the reader takes these lines in as they really are—or so the theory goes—then the reader will experience them as poetic intuitions, as immediate fusion of feeling and image. The reader is pulled into a new place and time and becomes one with its desperate beauty. If the reader mediates within this intuitive moment, it will enfold his whole world; Crane's world will become the reader's world, and he will see everything afresh, colored by a pain and lucidity as never before. In experiences such as this, life, lang., the world, and the word are irrefrangibly one. The experience one has of the poetry and one's knowledge of the experience are identical. The poetry creates the experience—as a fusion of world as experienced and of the person as experiencing—and gives knowledge of the experience as an identity of world and person in a single, seamless act of i.

This theory does not equate poetry with creative acts such as the primal creative act of God, for it views poetic i. not as a creation out of nothing but as the creation of a poetic maker and the world, of lang. and being, out of a material which is its prior condition. But this condition is utterly formless; it is only a hum and buzz, a pure flux of sensation. In effect, then, poetic i. discovers only what it creates: it knows that which it itself makes. Moreover, in this theory, poetry as we ordinarily think of it is only the highest form of that creative experience in which each of us becomes a human being, both sentient and verbal, living in a world.

Poetic i. differs from sensation because it is neither passive nor psychological; it is a oneness of person and world expressed in lang. In the other arts its lang. may be song or shape or gesture. It is knowledge, but of an immediate kind, and thus is prior to conceptual, judgmental, discursive knowledge. There is no claim in poetic i. that its world is either real or unreal or that that world and the experiencing person are distinct; because it is not a self-conscious experience, it does not even contain the claim that it is itself poetic i. Although it is possible to extract concepts and abstract ideas from a poem, in the poem experienced as a poem, these ideas are fused within the i.

Vico, who may be credited as the originator of this conception of poetry, argued that Homer con-

ceived of Achilles not as a courageous individual or as an example of courage or as courage itself, but as an utter fusion of all of these. In poetic i., in other words, individuality and universality are identical (as the "concrete universal"). Poetic i., moreover, is radically distinct from perception, which is the basis of empirical knowledge. If one perceives "the blue spot here and now," one observes it as part of a spatial and temporal and chromatic framework, a structure already composed by conceptual thought. It is, of course, possible to perceive rather than intuit poems, to consider their space and time as part of some large, conventional structure within which we live our days. But to do so is to miss the poems as poems. Space and time are abstractions by means of which we think and perceive the world. But poetic i. creates the world and with it our living sense of space and time. The crudeness or fineness of our very ideas of space and time is thus derivable from the quality of our poetic, intuitive experience. Finally, in its purest form, the concept of poetry as i. is at odds with the idea of poetry as self-expression. In a poetic i., self and world, subject and object, are immediately identical. This is the way the world begins. This is the way the self begins. On its basis alone we construct our distinctions, self and world, space and time, real and unreal, truth and error, even beauty and ugliness. M.E.B.

After Croce, theories of i. in the 20th c. have been pursued from directions he might never have anticipated, but with results which in some cases at least he would have found congenial. Expressionist theories of poetry, which are the last inheritors of romanticism and of which poetic i. is one, have had surprising strength through much of the 20th c. In Italy, the publication of Gentile's *Philosophy of Art* in 1932 effectively put an end to the currency of Crocean expressionism, but in Switzerland and France, the rise of the Geneva School of critics during the same decade produced a strong form of intuitivist crit. This, however, is based now on Bergson, existentialism, and (esp.) phenomenology. Here too the operating assumption is that a reader will, through inseeing or i., come into a rapport with the imaginative space of the text, and through it that of the poet, the authenticity of which is taken as guaranteed. On the basis of this guarantee one can then say that the ordinary category boundaries between subjective and objective are indeed dissolved, and along with them the usual concerns of critics with the gaps or spaces between the world and the world as embodied in words. In such i., one is literally seeing through words into the very life of things. Eng. translations of Geneva-School phenomenological crit. brought its methods into currency in America in the 1960s. The subsequent assaults on referentiality and determinate meaning that were associated with deconstruction, however, sought to dissolve all possibility of intuitive readerly rapport with the text, much less with the poet, as convenient but vain delusions built upon a fictive metaphysics. But in denying the possibility of stable meaning, deconstruction reduced its own position to merely one voice among all the rest, and not one on which any productive or even viable cultural practice could be built.

Much more productive were developments in reader-response criticism, which sought to reestablish links between individual readers and texts, at least, and perhaps even to restore links among readers themselves, and so guarantee intuitive authenticity, via the concept of readerly and social convention (q.v.). Nevertheless, intuitivist theories of poetry, like phenomenological philosophy, provides the most radical alternative to traditional Western dualist metaphysics. Most of the great Western poets have indicated implicitly or explicitly their belief in the power of i. to bypass the circuits of feeble human rationality and fickle human perception, going straight to the source. "If a sparrow come before my window," said Keats, "I take part in its existence and pick about the gravel." Whether such entering-in upon the conscious lives of other selves, other beings not human, and even events beyond all selfhood be dream or truth is a question that seems, finally, less important than the evident fact that it proceeds from a human capacity certain beyond cavil, and one which poetry above all arts, for some reason, makes central.—G. N. G. Orsini, *Benedetto Orsini* (1961); M. E. Brown, *Neo-Idealistic Aesthetics: Croce, Gentile, Collingwood* (1966); M. R. Westcott, *Psychology of I.* (1968). T.V.F.B.

INVECTIVE. A personal attack or satire, often scurrilous, a lampoon, formerly written mainly in verse. I. is to be differentiated from satire (q.v.) on the grounds that it is personal, motivated by malice, and unjust; thus John Dennis remarks that satire "can never exist where the censures are not just. In that case the Versifyer, instead of a Satirist, is a Lampooner, and infamous Libeller." I. is as old as poetry, and as widespread; in the West it appears (if not in the Homeric *Margites*) at least as early as Archilochus, who wrote an i. against Lycambes; other notable Gr. examples include those by Hipponax against Bupalus, by Anacreon against Artemon, and others by Xenophanes, Timon of Phlius, Sotades, Menippus, and (less virulently) Callimachus. Indeed, iambic meter itself (see IAMBIC) is in its earliest, Ionian form so called specifically because of its association with i., which has the specific characteristics both of a speaker giving vent to personal hatred and of common speech for its vehicle, to which iambic meter was thought by the ancients to conform. In Lat., i. is written, though in a wider variety of meters, chiefly by Catullus, Ovid (*Ibis*), Martial, and Varro. In the Middle Ages, Petrarch's i. against doctors, *I. contra medicum*, is notable; in the Ren., the scope of personal i. is expanded considerably by the

invention of printing, which provided broadsides, bills, and ballads particularly well suited for rapid and wide dispersal of political i. and satire. Eng. i. of this sort abounds particularly in the Restoration and 18th c.; indeed, the Eng. word "lampoon" (from the Fr. slang term *lamper*, "to guzzle, swill down") dates only from mid 17th c. Rochester's *History of Insipids, a Lampoon* (1680) is but one of many of his, and others'. Dryden, a master of i., nevertheless deplores it in his "Discourse concerning the Original and Progress of Satire" (1693) as both illegal and dangerous. After 1750, however, verse i., like other verse genres such as narrative poetry, rapidly gave ground to prose as the medium of choice, except in the (remarkably durable) trad. of the epigram, incl. scurrilous and vindictive epigrams, which have been produced in the 20th c. most notably by J. V. Cunningham.

J. Addison, *Spectator*, no. 23; *An Anthol. of I. and Abuse*, and *More I.*, both ed. H. Kingsmill (1929, 1930); C. Reichley, "Lampoon: Archilochus to Byron," *DAI* 32 (1971), 2703A; J. C. Manning, *Blue I.* (1973); *The Book of Insults, Ancient & Modern*, ed. N. McPhee (1978); G. Nagy, *The Best of the Achaeans* (1979), chs. 12–13; *Tygers of Wrath: Poems of Hate, Anger, and I.*, ed. X. J. Kennedy (1981); A. Richlin, *The Garden of Priapus* (1983); *The Devil's Book of Verse*, ed. R. Conniff (1983); *The Blasted Pine: An Anthol.*, ed. F. R. Scott and A. J. M. Smith (1965)—Canadian; R. M. Rosen, *Old Comedy and the Iambographic Trad.* (1988); H. Rawson, *Wicked Words* (1989); G. J. van Gelder, *The Bad and the Ugly: Attitudes towards I. Poetry* (Hija') *in Cl. Arabic Lit.* (1989); K. Swenson, *Performing Definitions* (1991); L. Watson, *Arae: Curse Poetry in Antiquity* (1991).

T.V.F.B.

IRONY (Gr. *eironeia*, Lat. *dissimulatio*, esp. through understatement).

A. *Classical I.* In Gr. comedy the *eiron* was the underdog, weak but clever, who regularly triumphed over the stupid and boastful *alazon.* In Theophrast's *Characters*, the ironist appears as a deceitful hypocrite pursuing his own advantage. The cl. image of i. as a lofty, urbane mode of dissimulation, practiced in conversation and public speech and without one's own advantage in mind, finds its origin in the Platonic Socrates (hence the term "Socratic i."). In front of his conversational partners who claim to know, Socrates professes not to know, but through insistent questioning proves them also not to know, thereby finding a common basis for their quest for knowledge. Hence Socrates dissimulates not for his own advantage but for the sake of truth. Aristotle (*Rhet.*, Bk. 3) presents i. as "a mockery of oneself": "the jests of the ironical man are at his own expense; the buffoon excites laughter at others" (1419b7). In the *Nicomachean Ethics* Aristotle conceives of i. as "the contrary to boastful exaggeration; it is a self-deprecating concealment of one's own powers and possessions; it shows better taste to deprecate than to exaggerate one's virtues" (1108.19–23). In the same work Aristotle discusses *eironeia* and *alazoneia*, understatement and boastfulness, as deviations from truth, holding that i. is a noble form because it deviates not for the sake of one's own advantage but from a dislike for bombast and a desire to spare others the feeling of inferiority. The prototype of this genuine i. is Socrates (1127b22–26).

Cl. rhetoricians distinguished several varieties of i. In i. proper, the speaker is conscious of double meaning and the victim unconscious; in *sarcasm* both parties understand the double meaning. Other forms incl. *meiosis* and *litotes* (understatement); *hyperbole* (overstatement); *antiphrasis* (contrast); *asteism* and *charientism* (forms of the joke); *chleuasm* (mockery); *mycterism* (the sneer); and *mimesis* (imitation [q.v.], esp. for the sake of ridicule). Cicero termed i. "that form of dissimulation which the Greeks named *eironeia*" (*Academica posteriora* 2.5.15) and also considered Socrates the prototype of this witty and refined art of conversation (*De officiis* 1.30.108). Quintilian assigned i. its position among the tropes and figures discussed in Books 8 and 9 of his *Institutio oratoria.* For Quintilian the common feature in all rhetorical forms of i. is that the intention of the speaker is different from what he says, that we understand the contrary of what he says (9.2.44).

In late antiquity, the Middle Ages, the Ren., and neoclassicism, the cl. delineation of i. was varied and enriched by including other rhetorical forms and elaborating a more complex system of figures, but the basic meaning remained the same. The Fr. *Encyclopédie* of 1765 summarizes the various nuances of i. found in numerous critical handbooks of the time by defining i. as "a figure of speech by which one indicates the opposite of what one says" (8.905). To this one should add that, according to cl. opinion, in order to distinguish i. from mere lying, the entire tenor of speaking, incl. intonation, emphasis, and gesture, is supposed to reveal the intended meaning. One should also recollect that authors (Boccaccio, Cervantes, Shakespeare) whom today we consider ironic in their literary creations were not viewed so in their time; this term remained confined to the field of rhet. until late in the 18th c.

B. *Romantic I.* The most significant change in meaning took place in 1797, when Schlegel observed in his *Fragments*: "there are ancient and modern poems which breathe throughout, in their entirety and in every detail, the divine breath of i." Schlegel's most constant description of i. in its literary and poetic forms is that of a consistent alternation of affirmation and negation, of exuberant emergence from oneself and self-critical retreat into oneself, of enthusiasm and skepticism. In Ger. romantic poetry (Tieck, Jean Paul, Hoffmann, Heine), i. became a conscious form of literary creation, although its prototype was seen, as is now fully recognized, in older European

authors such as Boccaccio, Cervantes, and Shakespeare. Particular points of ironic contrast, of creation and annihilation, were the relationships of illusion and reality, subjective and objective, self and world, the inauthenticity and authenticity of the self, the relative and the absolute. A new dimension was introduced when these relationships were viewed not only in terms of artistic playfulness (Schlegel) but also in terms of melancholy and sadness as the *mal du siècle* (Fr. romanticism), the transitoriness of life (Keats), or the perishing of the divine in this world (Solger).

C. *Tragic I.* was introduced by Connop Thirlwall in 1833, who based it on a distinction among three basic types of i.: verbal, dialectic, and practical. Verbal i. establishes, as in cl. rhet., a contrast between what is said and meant; dialectic i. relates to works of lit. and thought in which i. permeates the entire structure. Practical i., however, is the most comprehensive form, present throughout life in individuals as well as in the history of states and institutions, and constitutes the basis for tragic i. The contrast of the individual and his hopes, wishes, and actions, on the one hand, and the workings of the dark and unyielding power of fate, on the other, is the proper sphere of tragic i. The tragic poet is the creator of a small world in which he reigns with absolute power over the fate of those imaginary persons to whom he gives life and breath according to his own plan.

D. *Cosmic I.* I. took on a new and more comprehensive dimension with Hegel, who strongly opposed romantic i. because of its "annihilating" tendency, seeing in it nothing but poetic caprice. Yet in *The History of Philosophy*, Hegel sensed in the "crowding of world historical affairs," in the trampling down of the "happiness of peoples, wisdom of states, and virtue of individuals," in short, in his comprehensive view of history, an ironic contrast between the absolute and the relative, the general and the individual, which he expressed by the phrase, "general i. of the world." Later we find this phrase in Heine and Kierkegaard. In Kierkegaard, however, i. becomes an absolute and irreconcilable opposition between the subjective and objective. Heine uses terms such as "God's i." and "i. of the world" to express the disappearance of reasonable order in the world. The ultimate extension of Hegel's concept was made by Nietzsche, who asked what would happen if all our ultimate convictions would become "more and more incredible, if nothing should prove to be divine any more unless it were error, blindness, lie—if God himself should prove to be our most enduring lie."

E. In *Verbal I.* (see paragraph 2 above), one meaning is stated and a different, usually antithetical, meaning is intended. In understatement the expressed meaning is mild and the intended meaning often intense, e.g. Mercutio's comment on his death-wound, "No, 'tis not so deep as a well, nor so wide as a church door; but 'tis enough, 'twill serve." Overstatement, esp. common in Am. folk humor, effects the reverse. The i. of a statement often depends on context. If one looks out of his window at a rain storm and remarks to a friend, "Wonderful day, isn't it?" the contradiction between the facts and the implied description of them establishes the i. When Hamlet rejects the idea of suicide with the remark, "Thus conscience does make cowards of us all," his remark is ironic because *conscience* is a sacramental word associated with moral goodness, whereas *coward* is a pejorative. The same kind of i. is illustrated in Comus' seduction speech, where a true principle (natural fertility) is used to prove an untrue doctrine (libertinism). Thus, i. can arise from explicit or implicit contradiction, as when Marvell begins his proposition to his coy mistress with the remark that time is short, but ends with the observation that love can make time pass more quickly.

F. *Dramatic I.* is a plot device according to which (a) the spectators know more than the protagonist; (b) the character reacts in a way contrary to that which is appropriate or wise; (c) characters or situations are compared or contrasted for ironic effects, such as parody; or (d) there is a marked contrast between what the character understands about his acts and what the play demonstrates about them. Foreshadowing is often ironic: Hamlet's speech on the fall of the sparrow has one meaning in its immediate context and a somewhat different one in Hamlet's own "fall" at the end of the scene. Tragedy is esp. rich in all forms of dramatic i. The necessity for a sudden reversal or catastrophe in the fortunes of the hero (Aristotle's *peripeteia*) means that the fourth form of i. (form d) is almost inevitable. *Oedipus Rex* piles i. on i. Form (a) is present because of the fact that the audience becomes increasingly conscious as the play progresses that Oedipus is rushing blindly to his doom. Form (b) is present in Oedipus' insistence on pursuing his investigation to its bitter climax (the fact that his basic motivation is a desire for justice and public welfare is a further i.—his fall is in part caused by his nobility). Form (c) is illustrated in the parallel between blind Tiresias (who can "see" morally) and the figure of Oedipus when he, too, has gained "vision" after blinding himself. Form (d) is, of course, present in the contrast between what Oedipus hopes to accomplish and what he finally does accomplish.

G. *Poetic I.* An important new step in the crit. hist. of i. can be noticed with the literary theory elaborated by the New Criticism, esp. in the work of I. A. Richards, Cleanth Brooks, and Robert Penn Warren, equating i. with poetry as such. For Brooks in particular, i. is considered the "principle of structure" in literary works, a reconciling power fusing the ambiguity, paradox, multiplicity, and variety of meaning in a work into the unity, wholeness, and identity which constitutes its modes of being. On this basis, but by way of reversal, Paul de Man conceives of i. in terms of a

discrepancy between sign and meaning, an absence of coherence or gap among the parts of a work, and an inability to escape from a situation that has become unbearable. In this sense i. practically coincides with the notion of deconstruction. In all its forms of expression, cl. and modern, i. functions as an agent of qualification and refinement. But during the modern period esp., beginning with romanticism, i. has become inseparable from literary and poetic expression itself.

S. Kierkegaard, *The Concept of I.* (1841), tr. L. M. Capel (1965); J. A. K. Thomson, *I., an Historical Intro.* (1926); G. G. Sedgwick, *Of I., Esp. in the Drama* (1935); C. Brooks, *Mod. Poetry and the Trad.* (1939), *The Well Wrought Urn* (1947), "I. as a Principle of Structure" (1949), in *Literary Opinion in America*, ed. M. W. Zabel (1951); D. Worcester, *The Art of Satire* (1940); A. R. Thompson, *The Dry Mock: A Study of I. in Drama* (1948); R. P. Warren, "Pure and Impure Poetry," *Critiques and Essays in Crit.*, ed. R. W. Stallman (1949); R. B. Sharpe, *I. in the Drama* (1959); N. D. Knox, *The Word I. and its Context, 1500–1755* (1961); V. Jankelevitch, *L'Ironie* (1964); E. M. Good, *I. in the Old Testament* (1965); A. E. Dyson, *The Crazy Fabric: Essays in I.* (1965); N. D. Knox, "I.," *DHI*; B. Allemann, *Ironie und Dichtung*, 2d ed. (1969); B. O. States, *I. and Drama: A Poetics* (1971); E. Behler, *Klassische Ironie, Romantische Ironie, Tragische Ironie* (1972), *I. and the Discourse of Modernity* (1990); W. C. Booth, *A Rhet. of I.*, 2d ed. (1974); D. H. Green, *I. in the Med. Romance* (1979); D. Simpson, *I. and Authority in Romantic Poetry* (1979); A. K. Mellor, *Eng. Romantic I.* (1980); Morier; D. C. Muecke, *I. and the Ironic*, 2d ed. (1982); U. Japp, *Theorie der Ironie* (1983); L. R. Furst, *Fictions of Romantic I.* (1984); E. Birney, *Essays on Chaucerian I.*, ed. B. Rowland (1985); D. J. Enright, *The Alluring Problem: An Essay on I.* (1987); M. Yaari, *Ironie paradoxale et ironie poétique* (1988); C. D. Lang, *I./Humor: Crit. Paradigms* (1988); R. J. Fogelin, *Figuratively Speaking* (1988); F. Garber, *Self, Text, and Romantic I.* (1988), ed., *Romantic I.* (1989); S. Gaunt, *Troubadours and I.* (1989); L. Bishop, *Romantic I. in Fr. Lit.* (1990); D. Knox, *Ironia* (1990); J. A. Dane, *The Critical Mythology of I.* (1991). W.V.O'C.; E.H.B.

J

JONGLEUR (Fr. *jongleur*, Occitan and Catalan *joglar*, Sp. *juglar*, Port. *jogral*, It. *giullare*). A medieval minstrel in France, Spain, or Italy. Although the word dates only from the 8th c., js. seem to have existed in France from the 5th c. to the 15th. In the earlier period the name was applied indiscriminately to acrobats, actors, and entertainers in general, as well as to musicians and reciters of verse. From the 10th c, on, however, the term is confined to musicians and reciters of verse, evidencing the importance of the distinction between the composer and the performer of verse. Cf. the OE *scop*, the ON *skald*, and the Celtic *bard*.

The Occitan *joglar* performed lyrics composed by a troubadour (q.v.) whom he served as a singing messenger, perhaps accompanying himself with an instrument or alternating song with instrumental performance. The troubadour might name the j. in the *tornada*, with instructions to take the song to an addressee who could be expected to reward him for it. Thus a given song was usually entrusted to one j., but the troubadour might commission several js. with different songs during his career, occasionally revising a given song for a second performer. During the 12th and 13th c. the frequency of such mentions of js. declined steadily, suggesting that oral diffusion was being gradually replaced by written transmission.

The OF j. performed not only the lyrics of the *trouvères* (q.v.), but also *chansons de geste* (q.v.), medieval romances, saints' lives, and other narratives. The *chansons de geste* were originally performed to a simple melody, and according to the oral-formulaic theory were composed in performance by a technique similar to that of the 20th-c. illiterate Yugoslavian *guslar* (see Lord, Duggan); but other scholars maintain that the js. read the *chansons de geste* from a book even in the earliest times, as they certainly did later on, and as they must have normally read romances and hagiographic texts.

E. Faral, *Les Js. en France au moyen âge* (1910); Jeanroy; R. Morgan, "OF Jogleor and Kindred Terms," *RPh* 7 (1954); R. Menéndez Pidal, *Poesia juglaresca y origines de las literaturas románicas* (1957); Lord; J. Duggan, *The Song of Roland* (1973); W. A. Quinn and A. S. Hall, *J.: A Modified Theory of Oral Improvisation* (1982); W. D. Paden, "The Role of the Joglar," *Chrétien de Troyes and the Troubadours*, ed. P. S. Noble and L. M. Paterson (1984). W.D.P.

K

KENNING (pl. *kenningar*). A multi-noun substitution for a single noun, e.g. "din of spears" for battle. Although found in many poetries, the k. is best known from Old Germanic verse. Ks. are common in West Germanic poetry, and scholars have recognized a k. in the expression of "corpse-sea," i.e. "blood," on the Eggjum runic inscription from Western Norway, ca. 700 A.D. ON Eddic poetry also makes use of ks., but their greatest importance was in skaldic poetry.

In medieval Icelandic rhet., the verb *kenna* (*við*), "make known (by)," was used to explain these expressions: "din" (the "base word" in modern analysis) is "made known" as battle by "of spears" (the "determinant"). The determinant may be in the genitive case, as here, or may attach to the base word to form a compound ("spear-din"); the base word takes the morphological and syntactic form of the concept the k. replaces. In skaldic poetry the determinant could in turn be replaced by another determinant; if "flame of battle" means "sword," then "flame of the din of spears" makes an acceptable k. Snorri Sturluson, the 13th-c. poet and man of letters and the first to attempt a rhetoric of ks., called this example *tvíkennt* ("twice-determined") in the commentary to *Háttatal* in his *Edda*. If another determinant were added, to make a four-part k., he would call it *rekit*, "driven." Snorri cautioned against "driven" ks. with more than six parts.

The relationship between the base word and determinant(s) could be essentially metonymic, as in "Baldr's father" for Odin, or metaphoric, as in the examples above. The ks. of West Germanic poetry, frequently used in connection with variation, tend toward the first category, those of skaldic poetry toward the second. Many skaldic ks. rely on Norse mythology or heroic legend for the links between the parts; thus poetry is the "theft of Odin" because he stole it from the giants. In skaldic poetry the number of concepts for which ks. may substitute is limited to about one hundred, among which warrior, woman, weapons, and battle are well represented. Since the base words tend to be fairly stereotyped (ks. for "woman" often have the name of a goddess as base word, for example), the system is relatively closed, but ks. make up nearly all the nouns in skaldic poetry, and the verbs are not important. Having imposed on themselves this closed system, the skalds exploited it brilliantly. In skaldic poetry the sum of the ks. is, to be sure, greater than their parts, but the best skalds made every word count.

As the following ks. for "sea" from *Beowulf* show, ks. enabled poets to express various aspects of an underlying noun: *ganotes bæð* "gannet's bath": a shoreward salt-water area where the sea-fowl fishes, sports, and bathes; *floda begang* "expanse of the floods or currents": emphasizes the vast extent of the oceans, esp. between the lands of the Geats and the Danes; *lagustræt* "path of the sea": the sea on which men sail their ships from one port to another; *windgeard* "enclosure or home of the winds": the sea, its storms, a watery expanse marked off from and enclosed by land, to be compared with *middan-geard* "middle enclosure," the land surrounded by the sea. See also LEXIS.—R. Meissner, *Die Kenningar der Skalden* (1921); H. van der Merwe Scholtz, *The K. in Anglo-Saxon and ON Poetry* (1927); *Edda Snorra Sturlusonar*, ed. F. Jónsson (1931); C. Brady, "The Synonyms for 'Sea' in *Beowulf*," *Studies in Honor of A. M. Sturtevant* (1952); H. Lie, *"Natur" og "unatur" i skaldekunsten* (1957); R. Frank, *ON Court Poetry* (1978); E. Marold, *Kenningkunst* (1983); M. Clunies Ross, *Skáldskaparmál* (1986).

J.L.

L

LAMENT (Ger. *Klagelied, Totenklage, Trost*). In its largest sense, l. is an expression of grief at misfortune, esp. at the loss of someone or something, and occurs throughout poetry in most genres. There are ls. in Sumerian poetry (e.g. over the fall of Sumer and Ur) and in *Gilgamesh*, Homer (e.g. *Iliad* 22.477–514), and the Heb. Bible (e.g. Lam.; Isa. 47; Ps. 44; Job 3). Ls. are a common feature of Gr. tragedy in the form of a monody or *kommos* (Alexiou). In Gr. choral lyric poetry, Simonides developed the *threnos* or dirge, followed by Pindar, of whose book of dirges only fragments remain (Smyth). Theocritus introduced l. into pastoral (q.v.) poetry (*Idyll* 1), foll. by Bion ("L. for Adonis") and Virgil (*Eclogue* 5). Pseudo-Moschus' "L. for Bion" is the precursor of numerous ls. or

elegies for poets, incl. Virgil's l. for Gallus (*Eclogue* 10), Ovid's for Tibullus (*Amores* 3.9), Garnier's for Ronsard, Spenser's for Sidney, Carew's and Walton's for Donne, Milton's *Lycidas* for Edward King, Shelley's "Adonais" for Keats, Arnold's "Thyrsis" for Clough, and Auden's "In Memory of W. B. Yeats."

Standard topics in ls. incl. a summons to and list of mourners, praise of the deceased, depiction of the death, contrast between the past and the present, the *ubi sunt* topos, and complaints about the cruelty of fortune, the purposelessness of life ("What boots it?"), and the finality of death (Menander Rhetor; Hardison; Alexiou; Race). In pastoral ls., nature is often depicted as mourning.

Poetic ls. or elegies over deceased individuals, most often including various topics of consolation or "comforting" (Wilson), are legion in virtually all langs. A sampling includes: *Hebrew*: David's l. for Saul and Jonathan at 2 Sam. 1.19–27; *Greek*: *Iliad* 24.748–59, Hecabe's l. for Hector; *Latin*: Catullus 101 for his brother; Propertius 3.18 for Marcellus; and Geoffrey of Vinsauf, *Poetria nova* 368–430, for Richard the Lionhearted; *French*: Ronsard's *Epitaphes*; *English*: *Deor's L.* (OE); Jonson's "On My First Son"; Donne's *Epicedes and Obsequies*; Whitman's "When Lilacs Last in the Dooryard Bloom'd"; and Dylan Thomas's "Do Not Go Gentle Into That Good Night"; *Italian*: Foscolo's "In morte del Fratello Giovanni" and Carducci's "Piano Antico"; *German*: Schiller's "Nänie"; *Spanish*: Manrique's "Coplas por la muerte de su padre" and García Lorca's "Llanto por Ignacio Sánchez Mejías."

H. W. Smyth, *Gr. Melic Poets* (1900); *Wilson's Arte of Rhetorique*, ed. G. H. Mair (1909); R. Kassel, *Untersuchungen zur griechischen und römischen Konsolationslit.* (1958); R. Lattimore, *Themes in Gr. and Lat. Epitaphs* (1962); O. B. Hardison, *The Enduring Momument* (1962); N. K. Gottwald, *Studies in the Book of* Lamentations (1962); V. B. Richmond, *Ls. for the Dead in Medieval Narrative* (1966); G. Davis, "*Ad sidera notus*: Strategies of L. and Consolation in Fortunatus' *De gelesuintha*," *Agon* 1 (1967); H.-T. Johann, *Trauer und Trost* (1968); J. B. Pritchard, *Ancient Near Eastern Texts* (1969); P. von Moos, *Consolatio: Studien zur mittellateinischen Trostliteratur*, 2 v. (1971–72); M. H. Means, *The* Consolatio *Genre in ME Lit.* (1972); M. Alexiou, *The Ritual L. in Gr. Trad.* (1974); *Menander Rhetor*, ed. and tr. D. A. Russell and N. G. Wilson (1981); Terras; W. H. Race, *Cl. Genres and Eng. Poetry* (1988). W.H.R.

LANGUAGE POETRY emerged in the mid 1970s as both a reaction to and an outgrowth of the "New Am. Poetry" as embodied by Black Mountain, New York School, and Beat aesthetics. Within the pages of little magazines like *Tottel's, This, Hills*, and the Tuumba chapbook series, poets such as Ron Silliman, Barrett Watten, Charles Bernstein, Lyn Hejinian, Bruce Andrews, Bob Perelman, and Robert Grenier developed modes of writing that implicitly criticized the bardic, personalist impulses of the 1960s and explicitly focused attention on the material of lang. itself. This practice was supplemented by essays in poetics, pub. in journals like *L=A=N=G=U=A=G=E, Open Space, Paper Air*, and *Poetics Jour.*, or presented in "talk" series conducted at lofts and art spaces. The general thrust of this critical discourse has been to interrogate the expressive basis of much postwar Am. poetry, esp. the earlier generation's use of depth psychology, its interest in primitivism and mysticism, and its emphasis on the poetic line (q.v.) as a score for the voice. While l. p. has derived much from the process-oriented poetry of Charles Olson, Robert Creeley, and John Ashbery, it has been skeptical about the claims of presence and participation that underlie such practice.

The response of l. p. to expressivism has taken several forms, most notably a deliberate flattening of tonal register and extensive use of non-sequitur. Experimentation in new forms of prose, collaboration, proceduralism, and collage have diminished the role of the lyric subject in favor of a relatively neutral voice (or multiple voices). L. poets have endorsed Victor Shklovsky's notion of *ostranenie* or "making strange," by which the instrumental function of lang. is diminished and the objective character of words foregrounded. The poetry of Rus. Futurism and Am. objectivism has been influential. Far from representing a return to an impersonal formalism, l. p. regards its defamiliarizing strategies as a critique of the social basis of meaning, i.e. the degree to which signs are contextualized by use. In order to defamiliarize poetic lang., l. p. has had recourse to a variety of formal techniques, two in particular. The first involves the condensation and displacement of linguistic elements, whether at the submorphemic level (as in the poetry of David Melnick, P. Inman, or Steve McCaffery) or at the level of phrases and clauses (Charles Bernstein, Bruce Andrews, Barrett Watten). Bernstein, for example, condenses what appear to be larger syntactic units into brief, fragmentary phrases:

> Casts across otherwise unavailable
> fields.
> Makes plain. Ruffled. Is trying to
> alleviate his false: invalidate. Yet all is
> "to live out," by shut belief, the
> various, simply succeeds which.

Although the title of this poem, "Sentences My Father Used," implies some autobiographical content, there is little evidence of person. The use of sentence fragments, false apposition, and enjambment displaces any unified narrative, creating a constantly changing semantic environment.

A second prominent feature of l. p. has been extensive work in prose. In the most influential essay, "The New Sentence," Ron Silliman calls for the organization of texts on the level of sentences and paragraphs. The "new" sentence refers less to deformations of normal sentences as to alternate

ways of combining them within larger structures. Lyn Hejinian's *My Life*, for example, consists of 37 paragraphs of 37 sentences, each one of which leads to the next by the substitution or replacement of materials from the previous one or by mutiple forms of association.

The issues raised by such writing are not simply aesthetic but involve the social implications of literary reception. By blurring the boundaries between poetry and prose, everyday and literary lang., theory and practice, l. p. has attempted to establish a new relationship with the reader, one based less on the recuperation of a generically or stylistically encoded work and more on the reader's participation in a relatively open text. By thwarting traditional reading and interpretive habits, the poet encourages the reader to regard lang. not simply as a vehicle for preexistant meanings but as a system with its own rules and operations. However, since that system exists in service to ideological interests of the dominant culture, any deformation forces attention onto the material basis of meaning production within that culture. If such a goal seems utopian, it has a precedent in earlier avant-garde movements from symbolism to futurism and surrealism. Rather than seeking a lang. beyond rationality by purifying the words of the tribe or by discovering new langs. of irrationality, l. p. has made its horizon the material form rationality takes.

ANTHOLOGIES: *The L=A=N=G=U=A=G=E Book*, ed. B. Andrews and C. Bernstein (1984); *Writing / Talks*, ed. B. Perelman (1985); *In the Am. Tree*, ed. R. Silliman (1986); *"Lang." Poetries: An Anthol.*, ed. D. Messerli (1987).

CRITICISM: B. Watten, *Total Syntax* (1985); M. Perloff, *The Dance of the Intellect* (1985), ch. 10; S. McCaffery, *North of Intention: Crit. Writings 1973–1986* (1986); C. Bernstein, *Content's Dream: Essays 1975–1984* (1986); L. Bartlett, "What is 'L. P.'?" *CritI* 12 (1986); R. Silliman, *The New Sentence* (1987); J. J. McGann, "Contemp. Poetry, Alternate Routes," *CritI* 13 (1987); A. Ross, "The New Sentence and the Commodity Form: Recent Am. Writing," *Marxism and the Interp. of Culture*, ed. C. Nelson and L. Grossberg (1988); G. Hartley, *Textual Politics and the L. Poets* (1989). M.D.

LEXIS. The term "l." (Gr. "diction"), was accepted as Eng. usage by the *OED* in the 2d ed. (first citations 1950, 1957 [Frye]) and defined as "the diction or wording, in contrast to other elements, of a piece of writing." "L." is a more useful term than "diction" because more neutral. Even where "the diction of poetry" is distinguished from "poetic diction" (esp. in the 18th-c. sense), "diction" may elicit only the question of unusual lang. rather than all questions concerning the lang. of poetry. L. in poetics is further to be differentiated from l. in linguistics (*OED*, sense 1.2).

The primary rule for thinking about l. is that words in a poem always exist in relation, never in isolation: "there are no bad words or good words; there are only words in bad or good places" (Nowottny 32). Otherwise, classifying l. can be a barren exercise, just as concentrating on isolated words can be barren for a beginning poet. Consistency within the chosen area of l. is necessary for a well-made poem, and consistency is not necessarily easy to achieve (Johnston). Ascertaining the consistency of l. in a poem enables the reader to hear moves outside that range (e.g. Kenner on Wordsworthian l. in Yeats's "The Tower" [*Gnomon* (1958)]). Great skill in l. implies that a poet knows words as she or he knows people (Hollander 228), knows how "words have a stubborn life of their own" (Elton), and knows that words need to be "at home" with the "complete consort dancing together" (Eliot, *Little Gidding* 216–25, the best modern poetic description of l. "that is right").

Some useful categories for studying l. may be drawn from the *OED* (preface on "General Knowledge"), where vocabulary is classified as follows: (1) *Identification*, incl. usual spelling, pronunciation, grammatical part of speech, whether specialized, and status (e.g. rare, obsolete, archaic, colloquial, dialectal); (2) *Morphology*, including etymology, and subsequent word-formation, including cognates in other languages; (3) *Signification*, which builds on other dictionaries and on quotations; and (4) *Illustrative quotations*, which show forms and uses, particular senses, earliest use (or, for obsolete words, latest use), and connotations. Studies of l. might test these categories for any given poem. It should be remembered that in common usage "meaning" refers to definition under category (3), but "meaning" as defined by the *OED* includes all four categories. And "meaning" in poetry, fully defined, includes all functions of the word.

L. includes all parts of speech, not simply nouns, adjectives, and verbs. Emphasis on what is striking tends to isolate main parts of speech and imposes a dubious standard of vividness analogous to Arnold's critical method of reading "touchstones." Even articles matter (Browning sometimes drops them; cf. Whitman and Forster on passages to India). Verb forms matter (see James Merrill's *Recitative* [1986] 21 on the prevalence of first-person present active indicative). The metaphorical force of, for example, prepositions must be remembered as well as their double possibilities (e.g. "of," as in "the love of three oranges," a favorite device of Stevens). Different langs. offer different possibilities for plurisignation and ambiguity according to their grammatical structures.

Discussions of l. often tend more toward polemics than poetics. It may be impossible to separate the two, but the effort is essential (Nowottny is exemplary). Coleridge's dictum should be kept in mind: that every great and original author "has had the task of creating the taste by which he is to be enjoyed" (cited in Wordsworth 1815), a task that perforce includes polemics. Thus Eliot's attacks on the Keats-Tennyson line of l., esp. as

developed by Swinburne, are better read generically in terms of charm and riddle, as Frye does (1976; and see Welsh), rather than in Eliot's terms. Similarly, readers and critics must be vigilant so as not to read modern assumptions about l. back into older poetry. Thus, says Strang, "the reader of Spenser should approach the text as being in Spenser's lang., which is a very different matter from reading him as if he were writing modern Eng. with intermittent lapses into strange expressions which require glossing." Here critics must avail themselves of the results of historical scholarship on the contemporaneity or archaism (q.v.) of words—often a difficult assessment.

There are only a few general questions concerning l., and they have remained for centuries. The most fruitful may be the more particular ones. One longstanding general issue is whether a special l. for poetry exists, or should exist. This in turn depends on what poetry is thought to be, or what type of poetry is in question. Of discussions in antiquity, those by Aristotle, Dionysius of Halicarnassus, Horace, and Pseudo-Longinus are the most important. Aristotle's few remarks remain pertinent: poetic l. should be both clear and striking, "ordinary words" should be used for clarity, and "unfamiliar terms, i.e. strange words, metaphors" should be used to make l. shine and to avoid l. that is inappropriately "mean." There should be a mixture of ordinary and unfamiliar uses of lang. (*Poetics*, tr. Bywater [1909]). In the Middle Ages and early Ren., the issue of diction takes on particular importance as Med. Lat. gives way to the vernaculars. Dante's *De vulgari eloquentia* (Of Eloquence in the Vernacular, ca. 1303; tr. R. S. Haller [1973]) is the central text in the *questione della lingua*; Dante debates this same division at some length, setting l. in the context of the disputes on the suitability of the vernaculars for elevated expression, on the biblical origins of lang., and on prosody. He judges the best l. to be "illustrious," "cardinal," "courtly," and "curial" (i.e. well-balanced, as in a law-court). In *DVE* 2.7, Dante gives detailed criteria for those words that are suitable for "the highest style." Some are as specific as those constraining Valéry's well-known search for "a word that is feminine, disyllabic, includes P or F, ends in a mute syllable, and is a synonym for break or disintegration, and not learned, not rare. Six conditions—at least!" (Nowottny 2). Here Dante classifies words in the "noble vernacular" as "childish" (too elementary), "feminine" (too soft in sound, and unelevated), and "manly"; only the last type will serve for poetry. These last, in turn, are either "rustic" or "urban," and only the urban will serve for poetry. And of the urban words, some are "smooth-haired or even oily" while others are "shaggy or even bristly." Only words that are "smooth-haired" or "shaggy" are suitable for the high style in poetry, as, for example, in tragedy; words that are "oily" or "bristly" are "excessive in some direction." Dante sees that the main question, as so often, is appropriateness or decorum (q.v.), but not simply for a given type of writing. He also stresses appropriateness for the person using a given l., a criterion unfamiliar to many current writers. Only those with sufficient natural talent, art, and learning should attempt the most demanding l., says Dante, for l. has its own implicit demands. Even today it remains true that the use of l. reflects a poet's judgment, for good or ill.

The term "poetic diction" is strongly associated with 18th-c. poetry, largely because of Wordsworth's attacks on it in the "Preface" to *Lyrical Ballads* (3d ed., 1802; see also the "Appendix on . . . Poetic Diction" and "Essay, Supplementary to the Preface" [1815]). Wordsworth remarks that "there will be found in these volumes little of what is usually called poetic diction," by which he means the epithets, periphrases, personifications, archaisms, and other conventionalized phrases all too often employed unthinkingly in Augustan poetry. As against Thomas Gray, for example, who wrote that "the lang. of the age is never the lang. of poetry" (Letter to R. West, April 1742), Wordsworth advocated using the "real lang. of men," esp. those in humble circumstances and rustic life. But Wordsworth's many conditions governing such "real lang." in poetry must be kept clearly in mind (e.g. men "in a state of vivid sensation," the lang. adapted and purified, a selection only).

Coleridge, with his superior critical mind, saw that "the lang. of real life" was an "equivocal expression" applying only to some poetry, and there in ways never denied (*Biographia literaria*, chs. 14–22). He rejected the argument of rusticity, asserting that the lang. of Wordsworth's rustics derives from a strong grounding in the lang. of the Eng. Bible (the Authorized Version, 1611) and the liturgy or hymn-book. In any case, the best part of lang., says Coleridge, is not derived from objects but from "reflection on the acts of the mind itself." By "real," Wordsworth actually means "ordinary" lang., the "lingua communis" (cf. *OED*, Pref.), and even this needs cultivation to become truly "communis" (Coleridge cites Dante). Wordsworth's real object, Coleridge saw, was to attack assumptions about a supposedly necessary poetic diction. The debate is of great importance for l. It marks the shift from what Frye calls a high mimetic mode to a low mimetic one, a shift still governing the l. of poetry today. (In Fr. poetry, the shift comes a little later and is associated with Hugo [1827].)

Coleridge disagreed with Wordsworth's contention that "there neither is, nor can be any *essential* difference between the lang. of prose and metrical composition." Though there is a "neutral style" common to prose and poetry, Coleridge finds it "a singular and noticeable fact . . . that a theory which would establish the *lingua communis* not only as the best, but as the only commendable style, should have proceeded from a poet, whose diction, next to that of Shakespeare and Milton, appears to me

of all others the most *individualized* and characteristic." To be sure, some words in a poem may be in everyday use; but, he adds, "are those words *in those places* commonly employed in real life to express the same thought or outward thing? Are they the style used in the ordinary intercourse of spoken words? No! nor are the modes of connections; and still less the breaks and transitions" (*Biographica literaria*, ch. 20). In Coleridge's modification of Wordsworth's well-intentioned arguments, readers may still find essential principles applicable to questions of poetic l.

The 20th c. has, in one sense, taken up Wordsworth's words, steadily removing virtually every restriction on diction. The 20th c. now generally bars no word whatever from the l. of poetry, at least in the Germanic and Romance langs. Struggles over appropriate l. in the 19th c. included attacks on the romantics, Browning, and Whitman. Attempts by Bridges and others to domesticate Hopkins' extraordinary l. are well known. In the early 20th-c., Edwardian critics with genteel notions of poetry objected to Brooke writing about seasickness and Owen's disgust at the horrors of World War I (Stead). Wordsworth's "real lang. of men" was twisted by some into attacks on any unusual l. whatsoever—difficult, local, learned—a problem to this day, though now less from genteel notions than egalitarian ones inappropriately extended to the l. of all poetry. Yet the l. of poetry may still be associated with the lang. of a certain class—see Tony Harrison's poems playing Standard Eng. against working-class Eng. But if poetry now generally admits all types of l., it remains true that the l. of poetry—of the Bible, Shakespeare, and the ballads, for example—needs to be learned; otherwise most older poetry, as well as much contemporary, cannot be well read at all (Vendler 56). The l. of the Authorized Version and of the Gr. and Lat. classics has been influential on Eng. poetry for centuries, and the manifold strategies and effects of allusion (q.v.) must not be overlooked. Virgil's l. in eclogue, georgic, and epic was admired and imitated well past the Ren.

Historical changes in the lang. make the use of good dictionaries mandatory. In Eng., the *OED* (both eds.) is the most generous and its quotations invaluable, but other dictionaries are also needed (e.g. of U.S. Eng., for etymology; see Kenner on Pound's use of Skeat). The elementary philological categories of widening and narrowing, and raising and lowering, in meaning are useful. (Cf. "wanton," where solely modern senses must not be applied to Milton's use, or even as late as Bridges' "Wanton with long delay the gay spring leaping cometh" ("April, 1885"); "gay" is well known.) Hidden semantic and connotative changes must be esp. watched, along with favorite words in a given time (Miles) and key words whose meaning was assumed and so not defined (Becker). The l. of some modern poets pays attention to historical linguistics, while that of others is largely synchronic; readers should test.

Etymologies are stories of origins. The etymologist cares whether they are true or false, but a poet need not (Ruthven); mythologies are for the poet as useful as histories. Philology may include certain assumptions about poetic l. (see Barfield against Max Müller). Etymologies may include histories of war and struggle (for nationalism involves lang. just as class does). Poets may exploit the riches of etymology (see Geoffrey Hill's *Mystery of the Charity of Charles Péguy* on Eng. and Fr. l.). Etymology may function as a "mode of thought" (Curtius) or as a specific "frame for trope" (see Hollander on Hopkins) or both. Invented or implied etymologies can also be useful ("silva" through Dante's well-known "selva" links by sound and sense with "salveo," "salvatio," etc.). Milton plays earlier etymological meaning against later meaning, such play functioning as a trope for the fallen state of lang.: the prelapsarian "savage" hill, for example, is closely connected to its Lat. root "silvaticus" and Eng. "silvan," but with postlapsarian "savage . . . obscur'd" solitude, the later meaning of "savage" as "barbaric" is implied (*PL* 4.172, 9.1085). Eng. is unusually accommodating, combining as it does both Latinate and Germanic words. Other important word-roots should also be noted (cf. the etymological appropriateness of "sherbet" in Eliot's "Journey of the Magi").

L. may be considered along an axis of old to new, with archaism (q.v.) at one end and innovation (neologism being one part) at the other. Archaism may be introduced to enlarge the l. of poetry, sometimes through native terms (Spenser, Hopkins). Or it may be used for certain genres (e.g. literary imitations of the oral ballads) or for specific effects, ironic, allusive, or other. Innovation may remain peculiar to one poet or may enlarge the stock of the poetic lexicon. Neologisms (new-coined words) tend now to be associated with novelty more than freshness, and sometimes with strained effects. The very word indicates they are not common currency. Some periods are conducive to expanding l. in general (the mid 14th c.; the late 16th c.) or to expanding l. in some areas (computer terms, nowadays, though they have yet to be entered in general poetic l.).

Where poets do not invent or resuscitate terms, they draw on vocabulary from different contemporaneous sources (see the *OED* categories). Foreign, local, and dialectal words are noted below. The precision of terms drawn from such areas as theology, philosophy, or the Bible must not be underestimated, for controversy can center on one word. Studies working outward from single words (e.g. Empson; Lewis; Barfield on "ruin") are valuable reminders of historical and conceptual significance in l.

Shakespeare has contributed most to the enlargement of our stock of words; critics regularly note how often he provides the first example of a given word in the *OED*. His many mintings incl.

adaptations from the stock of both Eng. (e.g. "lonely," presumably from Sidney's "loneliness") and other langs. ("monumental," from Lat.), apparent inventions ("bump"), shifts in grammatical function ("control" as a verb rather than a noun), and other strategies. His mastery of l. is seen above all in his boldness and innovation. He possesses easily the largest known vocabulary of any poet (by some calculations, about 21,000 words, as against Milton's 8,000), but it is his extraordinary *use* of so large a word-hoard (as against ordinary recognition of a large reading or speaking vocabulary) that is so remarkable (Jespersen).

Most new words are now generally drawn from scientific or technical sources, though poetry makes comparatively little use of them. In the 18th c., poets could say that "Newton demands the Muse" (see Nicholson), but poets today do not generally say that "Einstein demands the Muse." A. R. Ammons is one of the few modern poets exploiting the possibilities of new scientific l.—e.g. "zygote" (1891, *OED*) rhymed with "goat"; "white dwarf" (1924, *OED* 2d ed.); and "black hole" (1969). Of the large stock of slang and colloquial expressions, many are evanescent or inert, though special uses may be effective. Shakespeare's gift for introducing colloquial l. is a salutary reminder not to reject colloquialisms per se.

Along the axis of old to new, the most interesting question is why and how some l. begins to sound dated. Archaisms and innovations alike are easy to hear. So also is the l. we designate as, say, 18th-c. or Tennysonian or Whitmanian. But what is it that distinguishes the poetic l. of a generation ago, and why do amateur poets tend to use the l. of their poetic grandparents? The aging of words or the passing of their claim on our allegiance is of continuing interest to poets as part of the diachronic aspect of their art.

Different types of poetry require different lexical praxis, though such requirements vary according to time and place. Oral poetry (q.v.) makes use of stock phrases or epithets (q.v.) cast into formulae (see FORMULA). Some of Homer's epithets became renowned, e.g., "poluphloisbos" ("loud-roaring") for the sea (see Amy Clampitt's echo of this). Compound epithets in OE poetry are known by the ON term "kenning" (q.v.) and sometimes take the form of a riddle. Different genres also require different praxis (Fowler 71), a requirement much relaxed today. Epic required a high-style l., as did the sublime (qq.v.; see Monk). Genres of the middle and low style drew from a different register. Satire (q.v.) usually works in the middle style but allows much leeway, esp. in Juvenalian as against Horatian satire. Any l. may become banal—e.g. that of the 16th-c. sonneteers or that of some pastoral (q.v.) writers (cf. Coleridge, *Letters*, 9 Oct. 1794: "The word 'swain' . . . conveys too much of the Cant of Pastoral"). Connotation or association is governed partly by genre, and is all-important for l.

L. also depends partly on place. The largest division in Eng. is between Great Britain and the U.S.A., but poetry from elsewhere (Ireland, Africa, Asia, Australasia, Canada, the Caribbean) also shows important differences. Establishing a distinctive poetic style in a new country with an old lang. presents peculiar problems which novelty in itself will not solve. Within a country, l. will vary locally, and poets can make memorable uses of local terms (Yeats of "perne" in "Sailing to Byzantium," Eliot of "rote" in *The Dry Salvages*; Whitman uses native Amerindian terms). The question of dialect shades into this. Burns and Hardy draw on local and dialectal words. Hopkins' remarkable l. derives from current lang., dialectal and other, as well as older words; a few (e.g. "pitch") have specific Hopkinsian usage (see Milroy). Foreign words, a special case, work along a scale of assimilation, for standard l. includes many words originally considered foreign. Considerable use of foreign l. (apart from novelties like macaronic verse) implies a special contract with the reader, at least in societies unaccustomed to hearing more than one lang. L. may also vary according to class (see above). It is still a matter of dispute how far it varies according to gender.

Interpretive categories are numerous, and readers should be aware of them as such; even taxonomies are interpretive. Beyond the categories already mentioned, l. may be judged according to the degree of "smoothness" (Tennyson as against Browning is a standard example; see Frye 1957), "smoothness" (recollect Dante above) centering on the large and important question of sound in l. (cf. Seamus Heaney on Auden: "the gnomic clunk of Anglo-Saxon phrasing . . . pulled like a harrow against the natural slope of social speech and iambic lyric" [*The Government of the Tongue* (1989) 124]). Or l. may be judged by the degree of difficulty (Browning, Hopkins, Eliot, Stevens), though once-difficult l. can become familiar. Strangeness in l. can contribute to the strangeness sometimes thought necessary for aesthetic effect (Barfield) or for poetry itself (Genette, arguing with Jean Cohen, also compares the *ostranenie* of the Rus. Formalists, and the lang. of a state of dreaming). Some poets are known for difficult or strange l. (e.g. Spenser, the metaphysical poets, Whitman, Browning) but readers should also note consummate skill in quieter effects of l. (e.g. Frost, Larkin, Bishop).

A distinctive praxis in l. is part of what makes a poet familiar, and the l. of an individual oeuvre may be studied in itself (see Fowler 128–29). The discipline of the art of l. is still best apprehended by studying the comments and revisions of good poets.

GENERAL STUDIES: O. Elton, "The Poet's Dictionary," *E&S* 14 (1929); C. Becker, *The Heavenly City of the 18th-C. Philosophers* (1932), esp. 47–63 on key words; O. Jespersen, *Growth and Structure of the Eng. Lang.*, 9th ed. (1938); W. Empson, *The Structure of Complex Words* (1951); D. Davie, *Purity of Diction in Eng. Verse* (1952); Curtius; B. Groom,

The Diction of Poetry from Spenser to Bridges (1955); Frye; C. S. Lewis, *Studies in Words* (1960); W. Nowottny, *The Lang. Poets Use* (1962)—essential reading; J. Miles, *Eras and Modes in Eng. Poetry*, 2d ed. (1964); K. K. Ruthven, "The Poet as Etymologist," *CQ* 11 (1969); O. Barfield, *Poetic Diction*, 3d ed. (1973)—essential reading; F. W. Bateson, *Eng. Poetry and the Eng. Lang.*, 3d ed. (1973); A. Sherbo, *Eng. Poetic Diction from Chaucer to Wordsworth* (1975); N. Frye, "Charms and Riddles," *Spiritus Mundi* (1976); A. Welsh, *Roots of Lyric* (1978); M. Borroff, *Lang. and the Poet* (1979); G. Genette, *Figures of Literary Discourse* (tr. 1982), esp. 75–102; G. Johnston, "Diction in Poetry," *CL* 97 (1983); Fowler—excellent on l. and genre; J. Boase-Beier, *Poetic Compounds* (1987); C. Ricks, *The Force of Poetry* (1987); A. Ferry, *The Art of Naming* (1988); J. Hollander, *Melodious Guile* (1988); H. Vendler, *The Music of What Happens* (1988).

SPECIALIZED STUDIES: S. H. Monk, *The Sublime* (1935); V. L. Rubel, *Poetic Diction in the Eng. Ren.* (1941); M. H. Nicolson, *Newton Demands the Muse* (1946); J. Arthos, *The Lang. of Natural Description in 18th-C. Poetry* (1949); M. M. Mahood, *Shakespeare's Wordplay* (1957); Wimsatt and Brooks, ch. 16; A. Ewert, "Dante's Theory of Diction," *MHRA* 31 (1959); C. K. Stead, *The New Poetic* (1964); G. Tillotson, *Augustan Poetic Diction* (1964); W. J. B. Owen, *Wordsworth as Critic* (1969); H. Kenner, *The Pound Era* (1971), esp. 94–191; J. Milroy, *The Lang. of G. M. Hopkins* (1977); N. Hilton, *Literal Imagination: Blake's Vision of Words* (1983); M. H. Abrams, "Wordsworth and Coleridge on Diction and Figures," *The Correspondent Breeze* (1984); R. W. V. Elliott, *Thomas Hardy's Eng.* (1984); E. Cook, *Poetry, Word-Play, and Word-War in Wallace Stevens* (1988); C. Ricks, *T. S. Eliot and Prejudice* (1988) R. O. A. M. Lyne, *Words and the Poet* (1989)—Virgil; B. M. H. Strang, "Lang.," *Spenser Encyc.*, ed. A. C. Hamilton et al. (1990). E.C.

LIGHT VERSE. The term l. v. is an *omnium gatherum*; although it once referred principally to *vers de société*, its meaning in the 20th c. includes a wide variety of verse: folk poetry, nonsense verse (q.v.), ribald verse, comic poems, and kitsch. If the earlier meaning of the term was too narrow, it has now been stretched so far that it has lost its shape.

While the word "light" has been applied to lit. or music to mean "requiring little mental effort; amusing; entertaining" since the 16th c. (*OED*), the first serious attempt to define the term l. v. came in the 19th c., when Frederick Locker-Lampson compiled the anthology *Lyra elegantiarum* (1867). An earlier anthology, Thomas D'Urfey's *Pills to Purge Melancholy* (1719), is considered the first l. v. collection in Eng. As the title implies, Locker-Lampson was less interested in comic verse than in *vers de société*, verse distinguished by its wit and polish and intended to amuse polite society. In the 20th c., a series of anthologists broadened the term to include poetry that is homely or comic. In *The Oxford Book of L. V.* (1938), W. H. Auden held that l. v. meant any poetry that is "simple, clear, and gay." He included performed poems, nonsense verse, and poetry of ordinary life; l. v. was no longer aristocratic. While Auden was influential, however, not everyone followed his lead. In 1941 Michael Roberts explicitly distinguished between comic verse, and "l. v. or *vers de société*." In 1958, Richard Armour defined l. v. as "poetry written in the spirit of play," then added that l. v. "can safely be limited to what is called *vers de société*: humorous or witty verse that comments critically on contemp. life. It is usually to some degree funny." Such anthologists as G. Grigson and W. Harmon have extended the definition of l. v. even further to include bad poetry, for example, or Cole Porter's lyrics. Matters have been further complicated by Kingsley Amis, who in the *New Oxford Book of Eng. L. V.* (1978) argued that the best l. v. is necessarily conservative; in this he followed a view set forth by D. Macdonald in his 1960 parody anthology.

One result of this widening and blurring of meaning is that the term l. v. is sometimes used pejoratively to mean trivial or unimportant poetry. Because some critics now include the magnificently dreadful verse of such writers as William McGonigal or Julia Moore, the term is also sometimes used to mean any bad verse. Yet no one would dismiss Chaucer's tales, Jonson's epigrams, Pope's *The Rape of the Lock*, Byron's *Don Juan*, Rossetti's *Goblin Market*, or Eliot's cats—all of which find a place in both standard anthols. and in collections of l. v. Thus the term seems, however, to have retained favorable connotations for sophisticated readers who know enough about what is "weighty" to recognize and enjoy what is "light." Some critics fear this audience may be diminishing. Grigson says that nonsense verse "is in danger now because it does demand an accepted idea of the nature of verse in general, a widely shared idea of the ways in which poetry works."

Many varieties of l. v. depend on whether one defines the term principally by reference to technical virtuosity or by reference to content. Among the important types of l. v. as defined by technique are limericks, clerihews, and double dactyls. The limerick is the oldest of these; the clerihew is the most anarchic, with its formalized irregularity; the double dactyl has attracted distinguished poets. Arguably the Japanese form of *senryū* also fits in this category. If one does not insist on the comic nature of l. v., then other polished fixed-term poems would be included: ballads, double ballade, rondeau, sestina, and triolet. Finally, there is poetry distinguished by clever use of certain techniques: verse with interlocking rhymes like Edward Lear's limericks; acrostics like Jonson's "Argument" to *Volpone*; verse with extensive alliteration or assonance (qq.v.), as was occasionally practiced in the Middle Ages—e.g. poems where every word in the line begins with the same letter.

(A modern example of alliterative ingenuity is the poem by Alaric Watts which begins "An Austrian army awfully arrayed / Boldly by battery besieged Belgrade" and proceeds through the alphabet.) There is also visual poetry, pattern poetry, and concrete poetry (qq.v.) such as Lewis Carroll's "Mouse's Tail" in *Alice in Wonderland.* In all such verse the sophistication of the craft—ingenuousness, skill, and polish—is apparent, the technique serving as foil to the content. But the problem with defining l. v. *solely* on technical grounds is that it then seems to include poetry not usually considered light. George Herbert's "Easter Wings" is shaped as craftily as Carroll's "Mouse's Tail," while Gerard Manley Hopkins' "Spring and Fall: To a Young Child" plays with words in much the same punning way as many limericks do. But neither is l. v.

If one turns from technique to content, however, other problems arise. Auden focuses on content in his three categories of l. v.: "poetry written for performance, to be spoken or sung before an audience"; "poetry intended to be read, but having for its subject-matter the everyday social life of its period or the experiences of the poet"; "such nonsense poetry as, through its properties and technique, has a general sense of appeal." His first category includes folk poetry such as riddles, folksongs, or the songs of Thomas Moore or Ira Gershwin. Yet one might object that all poetry is intended to be spoken to an audience, even if the audience is the reader speaking to himself. The lyrics W. S. Gilbert wrote for Arthur Sullivan's music are hardly folksongs, although clearly l. v. Furthermore, the boundary between folksongs and popular songs or between popular songs and light opera is no longer clear, as the careers of Bob Dylan and Stephen Sondheim demonstrate. Auden's category of nonsense verse, however, is uncontroversial, for nonsense is among the oldest of l. v. As Grigson points out, "The moment lit. develops, nonsense lit. must be expected as both a counter-genre and an innocent game."

One of the few points on which most critics agree is that l. v. requires the formal confines of verseform. If one transmogrifies a limerick or double dactyl into prose, one destroys part of its essence. Nonsense verse cast into prose may retain its humor, but it is no longer l. v. The poet Ogden Nash illustrates this point: he regularly stretches metrical forms with his irregular line length, yet his work depends on the reader's recognizing the rules he violates. In "Very Like a Whale," for example, his mock-attack on the lang. of poets is made more effective by the insistent rhymes nestled among the aggressively antimetrical lines. If the complaint were paraphrased in prose, it would be prosaic in every sense of that word, but in verse it amuses, felicitously.

Auden's other category is verse rooted in everyday life. Cl. Gr. offers the verse inscriptions—epigrams and epitaphs—collected in *The Gr. Anthology*; today one might include clever graffiti. Occasional verse would also fit in this category, from Thomas Gray's "On a Favourite Cat Drowned in a Tub of Goldfishes" to William McGonigle's poem on the beaching of the great Tay whale. Clearly the slighter the occasion, the lighter the verse. And because their targets are quotidian, satire, burlesque, and mock-heroic could also be considered l. v. One might of course object that mere homeliness or topicality does not in itself make a work l. v., that not all comic verse is light, and that not all l. v. is comic or *vers de société.* While most readers recognize l. v. as poetry that amuses by reflecting an aspect of their everyday lives in a clever way, probably none would agree on a body of works that meet a univocal set of criteria defining them as l. v.; rather, the works merit inclusion in the class by what Wittgenstein calls "family resemblances." The criteria are diverse and overlap, so that not every poem need meet every criterion.

ANTHOLOGIES: *Wit and Mirth, or Pills to Purge Melancholy*, ed. T. D'Urfey (1719); *Lyra elegantiarum*, ed. F. Locker-Lampson (1867); *Speculum Amantis, Musa Proterva*, both ed. A. H. Bullen (1888, 1889); *A Vers de Société Anthol.*, ed. C. Wells (1900); *Oxford Book of L. V.*, ed. W. H. Auden (1938); *Faber Book of Comic Verse*, ed. M. Roberts (1942); *What Cheer*, ed. D. McCord (1945); *The Worldly Muse*, ed. A. J. M. Smith (1951)—serious l. v.; *Penguin Book of Comic and Curious Verse*, ed. J. M. Cohen (1952); *Silver Treasury of L. V.*, ed. O. Williams (1957); *Parodies: An Anthol.*, ed. D. Macdonald (1960); *New Oxford Book of Eng. L. V.*, ed. K. Amis (1978); *Faber Book of Epigrams & Epitaphs* (1978), *Faber Book of Nonsense Verse* (1980), *Oxford Book of Satirical Verse* (1980), all ed. G. Grigson; *Oxford Book of Am. L. V.*, ed. W. Harmon (1979); *The Tygers of Wrath*, ed. X. J. Kennedy (1981); *Norton Book of L. V.*, ed. R. Baker (1986).

GENERAL: L. Untermeyer, *Play in Poetry* (1938); R. Armour, *Writing L. V.*, 2d ed. (1958); T. Augarde, *Oxford Guide to Word Games* (1984). F.T.

LIMERICK. The most popular form of comic or light verse (q.v.) in Eng., often nonsensical and frequently bawdy: in a famous joke, when Bennett Cerf, then head of Random House, was asked how they chose the winner of their l. contest, he said it was simple: they threw out all that were indecent, and the winner was the one that was left. The verseform of the l. is very exacting: 5 lines of accentual verse rhyming *aabba*, the first, second, and fifth lines having 3 stresses, the third and fourth 2; the rhythm is effectually anapestic or amphibrachic. The first syllables of each line (and unstressed syllables elsewhere in the lines, though less frequently so) may be omitted. The pattern, once caught, is unforgettable (slashes denote stresses; *x*s denote weak syllables or "slacks"):

(x) x / x x / x x /

(x) x / x x / x x /
(x) x / x x /
(x) x / x x /
(x) x / x x / x x /

It may be that the third and fourth lines are actually hemistichs, so that the form is actually 4 lines of 3-3-4-3 stresses, i.e. the form of hymn meter known as *Short Measure*, though the rhythm is not the same. Also possible is that lines 1–2 are hemistichs of a 6-stress long line and lines 3–5 the sections of a 7-stresss long line, the long-line couplet having 6 + 7 stresses, i.e. *Poulter's Measure*, though again, the rhythm differs.

The l. is unique in that it is the only Eng. stanza form used exclusively for light verse. It was popularized by Edward Lear (*A Book of Nonsense* [1846]); however, the etymology of the term "l.," never used by Lear, is unknown. Theories concerning its origin range from the belief that it was an old Fr. form brought to the Irish town of Limerick in 1700 by returning veterans of the Fr. war, to the theory that it originated in the nursery rhymes pub. as *Mother Goose's Melodies* (1791). What is certain is that the l. may be found in a volume entitled *The History of Sixteen Wonderful Old Women* (1821) and in *Anecdotes and Adventures of Fifteen Gentlemen* (1822), possibly by R. S. Sharpe. Lear cites the latter volume as having given him the idea for his ls. Whatever its origin, the l. has a secure place in the history of Eng. verse. In the wake of Lear, such notable authors as Tennyson, Swinburne, Kipling, Stevenson, and W. S. Gilbert attempted the form, and by the beginning of the 20th c. it had become a fashion. The chief tendency in the modern l. has been the use of the final line for surprise or witty reversal, in place of the simply repeated last line of Lear's day.—L. Reed, *The Complete L. Book* (1925); *Oxford Book of Nursery Rhymes*, ed. I. and P. Opie (1951); A. Liede, *Dichtung als Spiel*, v. 2 (1963); W. S. Baring-Gould, *The Lure of the L.* (1972); *The L.: 1700 Examples* (1970, 1974), *The New L.* (1977), both ed. G. Legman; C. Bibby, *The Art of the L.* (1978); G. N. Belknap, "Hist. of the L.," *PBSA* 75 (1981); *Penguin Book of Ls.*, ed. E. O. Parrott (1984).

T.V.F.B.; A.PR.

LINE (Sanskrit *pāda*, Gr. *stichos*, Lat. *versus* "verse"). The concept of the l. is fundamental to the concept of poetry itself, for the l. is the differentia of verse and prose: throughout most of recorded history, poetry has been cast in verse, and verse set in ls. (see VERSE AND PROSE). That is, verse is cast in sentences and ls., prose in sentences and paragraphs. The sense in prose flows continuously, while in verse it is segmented so as to increase information density and perceived structure. It is impossible that there could be verse not set in ls. It is possible there could be poetry not set in ls., if one defines poetry in terms of content or compression of content; and certainly there are hybrid forms such as rhythmical prose and the prose poem (q.v.). But we must assume that the preponderance of the world's poetry has been cast in verse precisely to take advantage of those resources which verse—for which l. is a virtual metonym—has to offer.

Structure. Readers and auditors of poetry perceive the l. as a rhythmical unit and a unit of structure. As a unit of measure, it is linked to its neighbors to form higher-level structures, and as a structure itself, it is built of lower-level units. In metrical verse the l. is usually segmented into elements—hemistichs, measures or metra, or feet—but it is the l. which generates these intralinear units and not vice versa. Ls. are not made simply by defining some unit of measure such as a foot and then stringing units together, because the units are susceptible of differential constraint at differing points in the l. The l., that is, has a shape or contour, and a structure as a whole, over and above the sum of its constituent elements. In various verse systems these elements are bound together in differing ways, such as structural alliteration in OE. It is true that, in the handbooks, meters are usually specified by the type of foot and number of feet per l.—i.e. monometer (a l. of 1 foot), dimeter (2 feet), trimeter (3 feet), tetrameter (4 feet), pentameter (5 feet), hexameter (6 feet), heptameter (7 feet). But these simplistic descriptors do not capture the internal dynamics of a l. form such as the Gr. hexameter.

How do readers and auditors recognize the l. as a rhythmical unit? In isometric verse, meter measures out a constant spacing, either of a certain number of events or a certain span of time (depending on one's theory of meter) which the mind's internal counter tracks in cognition; meter also provides predictable internal structure. The l. can also be bound together by syntactic and rhetorical structures such as parallelism and antithesis which have their own internal logic of completion; these structures may or may not be threaded into meter: in biblical Heb. verse they are, in Whitman they are not.

Line End. But probably the most common strategy is to mark the *end* of the l., since without some sort of signal, we would not know where one l. ended and a second one began. Traditionally the signal has been thought to be a "pause," though not a metrical (l.- internal) one but rather some kind of rhetorical or performative one. (Some have suggested that its duration is something like "half a comma.") But this claim derives not from prosody (verse theory), or from claims made in poetics about the nature or ontological status of poetry as sound (q.v.), but rather from assumptions put in play about performance (q.v.), about the reading aloud of verse. On this account, performers of poetry recognize the l. as a unit (by seeing it as such when read from the page) and mark its end in delivery with a linguistic pause or paralinguistic cue such as elongation of the final

syllable. Auditors hear these cues, which cross or ignore syntax, as boundary markers.

An alternative conception of the signal, which depends rather on assumptions about the status of poetry as a visual text on a page, is "white space to the right of the l." This is perceived in reading, is irrelevant to performance, and *may* by taken not only as a terminal marker but even as a part of structure (see below).

But both these answers specify phenomena *after* the l. ends. A more powerful conception of the signal, not committed to any doctrine about either performance or text, would be, "some kind of marker," not after the final syllable of the l. but *in* the final syllables of the l. In Gr. and Lat. poetry, where the meter is quantitative, auditors could recognize the ends of ls. because they were marked by an alteration in the meter, either an increase in the formality in its closing syllables, a *cadence*, or unexpected shortening at l.-end, *catalexis*. In the hexameter, for example, the l. closes on a spondee—two heavy syllables in succession—rather than a dactyl, a closural pattern which is obligatory. In post-Cl. verse, l.-ends have been most often marked by some distinctive sound echo, chiefly rhyme, though important also are the several other strategies of sound-repetition which approach rhyme (are rhyme-like) or exceed rhyme: homoeoteleuton, assonance, identical rhyme. Short ls. in particular seem to demand the support of rhyme, else they will be mistaken for the cola that are parts of longer ls.; this is a major factor in the devel. of the lyric. Also important are strategies for end-of-l. semantic emphasis: Richard Wilbur, for example, sometimes employs the strategy of putting important words at l.-end, words which may offer an elliptical synopsis or ironic commentary on the argument of the poem. All these strategies, taken together as devices for end-fixing, i.e. *marking* or *weighting* l.-ends so as to highlight or foreground them perceptually, constitute one of the most distinctive categories of metrical universals, conspicuous in a wide range of verse systems.

The phenomenon of extra syllables at l.-end is fairly extensive and is taken account of in several prosodies: l. endings are classified depending on whether the last stress falls on the final syllable of the l. (a "masculine" ending) or the penultimate one (a "feminine" one). The distinction dates from at least Occitan poetry, and may date from ancient Gr. Classifying types of l. endings was also one of the central "metrical tests" which several British and Ger. Shakespeare scholars of the later 19th c. hoped would yield a definitive chronology of the plays. Spedding in 1847 suggested the "pause test," which tabulated frequencies of stopped vs. enjambed ls. Bathurst (1857), Craik, and Hertzberg discussed "weak endings"; Ingram (1874) distinguished types of these as "light" endings (enjambed ls. ending on a pronoun, verb, or relative bearing only secondary stress, plus a slight pause) and "weak" (enjambment on a proclitic, esp. a conjunction or preposition, allowing no pause at all—see Brogan for citations). But statistical evidence which is purely internal is vulnerable to textual criticism and, where not supported by other types of evidence, vulnerable altogether.

Line and Syntax. The syllables of the l. are the arena for deploying meter or rhythm; they are also of course the ground of syntax and sense: these two structures overwrite the same space. Some verse-forms regularly align l. units with sense units; such ls. are known as *end-stopped.* The ("closed") heroic couplet (q.v.), for example, ends the first l. at a major syntactic break and the second at a full stop (sentence end). But few meters do so *every* l.: the demands of narrative or discursive continuation are simply too strong. Systematic contrast or opposition of l. units and sense units (l. end and sentence flow) is the complex phenomenon known as *enjambment* (q.v.). In enjambed verse, syntax pulls the reader through l.-end into the next l., while the prosodic boundary bids pause if not stop. In reading, the mind makes projections, in that pause, based on what has come before, about what word is most likely to appear at the beginning of the next l., expectations which a masterful poet will deliberately thwart, forcing rapid rereading: in Milton such error is the emblem of man's postlapsarian state. Certainly in modern times, at least, it has been thought that one of the chief functions of l. division is to stand in tension with or counterpoint to the divisions of grammar and sense, effecting, in the reader's processing of the text, multiple simultaneous pattern recognition. Most of the time this results in heightened aesthetic pleasure, but it may also work in reverse: in Gwendolyn Brooks' poem "We Real Cool," the severe and reiterated rupture of subjects from predicates across l.-end dislocates sentences, emphasizing subjects (pronouns) severed from action (verbs), generating extreme discomfort. Even in enjambed verse, the word at l.-end, which the French call the *contre-rejet*, receives some sort of momentary foregrounding or emphasis. This is not a matter of the word marking the end but of the end marking the word.

Aural Line and Visual Line. Since the advent of writing in Greece, ca. 750 B.C., and certainly since the advent of printing in the West, ca. 1450 A.D., the l. has had a visual reality in poetry as seen, but long before that time and without interruption throughout print culture it has continued to have an auditory reality in poetry as heard. This gives the l.—poetry itself—a fundamentally duality, in the eye and ear, as seen/heard (Hollander), a duality complicated, however, by the appearance of free verse (q.v.). Arguments about the historical or ontological precedence of the aural over the visual l. therefore miss the point: these forms are, for us, now, both realities; the important question is how they interaffect each other—whether, for example, the break at l.-end demands an auditory

pause in performance. Levertov has argued that there must be linkages of this sort, so that print texts of poems are in fact "accurate scorings" of them as read aloud. Margaret Atwood concurs when she calls the l. "a visual indication of an aural unit."

Line Forms. Up to the advent of free verse, l.-forms and -lengths are mainly determined by genre specification, so that a poet who wishes to write, say, in elegiac distichs (hexameter plus pentameter) knows or learns, by reading her predecessors, what forms have already been tried, for what kinds of subjects, with what kinds of tone, and with what success. In metrical verse, that is, l.-forms were mainly determined by history and convention, by the interplay of tradition and the individual talent; and the prosody which generated them was aurally based. Free verse, by contrast, foregrounded visual space and posited the l. within a two-dimensional matrix where blanks, white spaces, drops, gaps, vectors, and dislocations became possible. This is not to say that free verse abandoned aurality, for some poets, such as the Beats, Olson with his "projective verse," and the proponents of Sound Poetry (q.v.) and "Text-Sound," continued to speak programmatically of the l. as based on the energy of the breath. It is only to say that visual prosody was made central to free verse at least in claim, and if aural prosody was at the same time not dispensed with, no poet seemed to wish or prosodist to be able, at least for the first century of free-verse practice, to give a coherent account of the relations of the one to the other.

To think, however, that the free-verse l. can function solely as visual prosody, without the resources of aural prosody to aid it, is to commit what Perloff calls the "linear fallacy," making some free verse indistinguishable from prose chopped up into ls. The sonic and rhythmic devices of aural prosody offer poetry effects *not available* to prose: and if these are to be discarded, visual prosody must provide others, else the distinction between prose and verse collapses. One might respond that several Eng. poets—Jonson and Pope, to name but two—wrote out drafts of their poems first in prose before versifying them; and verse translators routinely work through intermediary prose versions. But this shows simply that they wished to clarify the argumentative or narrative structure first to get it out of the way, so as to concentrate on *poetic* effects—else why versify at all.

Whether or not visual prosody has been able to generate other devices, and whether or not free verse in fact abandons aural prosody, are disputed questions. It is often held that visual prosody can do much with "placing"—i.e. framing or highlighting words or phrases by isolating them as ls. The l. is thus reduced to the measure, though in a differing sense (if that is possible), since measure traditionally implied regularity. But probably few would hold that the visual l. does not affect the aural l. at all. Levertov says that the free-verse l.-break affects both rhythm and intonation: it is both "a form of punctuation *additional* to the punctuation that forms part of the logic of completed thoughts," which introduces "an a-logical counter-rhythm into the logical rhythm of syntax," and also, more importantly, a device which affects "the *melos* of the poem" by altering pitch patterns, since "a pitch pattern change *does occur* with each variation of lineation" (italics original). If so, then linkage between the two prosodies is preserved. Again, the full demonstration of these matters still remains to be given.

But what, after all, is the status of the visual l.? In the days of the mechanical typewriter, l.-end seemed a right-hand terminus: the next l. could only begin by the long carriage return to the left-hand margin and down. But on the computer, we learned, ls. are stored in a file in one continuous row of characters, with a special character between each l. which is recognized by the computer as a command to return to the left-hand margin *for formatting on screen* (i.e. on the page). This shows that the visual display of the ls. was merely epiphenomenal—a function of the medium of presentation—not inherent to the nature of the thing itself. Is this an analogy for cognition? We know that, in perception, ls. are processed sequentially. The l. is a frame, and frames are arrayed seriatim. In cognition, however, their semantic content is rearranged in short-term memory so as to make possible synchral (aserial) recognition of information. The articulation (jointure) of the frames one to the next is subject to some flexibility (thus extrametrical syllables at the end of one l. and inversions at the beginning of the next are allowed); at the same time, there is good evidence that the ends of the frames are usually more highly constrained than their beginnings, a fact which supports the frames as perceptual units, not mere arbitrary sectionings of the Heraclitean flow. It is these articulated frames across which sentences are stretched to make the distinctive discourse which is verse.

C. A. Langworthy, "Verse-Sentence Patterns in Eng. Poetry," *PQ* 7 (1928); J. S. Diekhoff, "Terminal Pause in Milton's Verse," *SP* 32 (1935); A. Oras, "Echoing Verse Endings in *Paradise Lost*," *So. Atlantic Studies S. E. Leavitt* (1953); C. L. Stevenson, "The Rhythm of Eng. Verse," *JAAC* 28 (1970); D. Crystal, "Intonation and Metrical Theory," *TPS* (1971); C. Ricks, "Wordsworth," *EIC* 21 (1971); B. Stáblein, "Versus," *MGG*, 13.1519–23; H. McCord, "Breaking and Entering," and D. Laferrière, "Free and Non-Free Verse," *Lang&S* 10 (1977); J. Lotman, *The Structure of the Artistic Text* (tr. 1977), ch. 6; D. Levertov, "On the Function of the L.," *ChR* 30 (1979); *Epoch* 29 (1980), 161 ff.—symposium; M. Perloff, "The Linear Fallacy," *GaR* 35 (1981), response in 36 (1982); Brogan; J. C. Stalker, "Reader Expectations and the Poetic L.," *Lang&S* 15 (1982); P. P. Byers, "The Auditory Reality of the Verse L.," *Style* 17 (1983);

R. Bradford, "'Verse Only to the Eye'?: L. Endings in *Paradise Lost*," *EIC* 33 (1983); S. A. Keenan, "Effects of Chunking and L. Length on Reading Efficiency," *VisLang* 18 (1984); *The L. in Postmodern Poetry*, ed. R. Frank and H. Sayre (1988).

T.V.F.B.

LINGUISTICS AND POETICS. We are concerned here with the two senses of the term "poetics" (hereafter p.) which have been developed since about 1960: p. as a theory of the essential property of lit., "literariness"; and p. as the theoretical or systematic study of the texts, and genres of texts, which fall under the category of lit. The intellectual bases for the conjunction of l. and p. were laid in Switzerland and in Russia in the early years of this century by the linguistic theory of Ferdinand de Saussure, reconstructed by two of his Geneva students from their notes of his lectures and pub. in 1916 as *Cours de linguistique générale*; by the aesthetic theory of the Rus. Formalists, articulated most memorably in Victor Shklóvsky's essay "Art as Technique" (1924); and by the narrative analysis of Vladimir Propp, *Morphology of the Folktale* (1928), which is procedurally relevant to poetry as well as to narrative in offering a model for the application of l. to poetry. But the period of greatest activity in l. and p. began in the 1960s, when these works became available in translation—Saussure (Eng. tr. 1959), Shklóvsky and the Formalists (Eng. and Fr. trs. 1965), Propp (Fr. 1957, Eng. 1958). The climate in l. in the 1960s was one of great ambition and confidence, and p. was blessed with the attention of some creative thinkers, notably Roman Jakobson and Roland Barthes; the work of the anthropologist Claude Lévi-Strauss, adapting Propp's model for the structural analysis of myth (q.v.), was also important in showing the way for one kind of linguistic analysis of lit. The 1960s saw the birth of what came to be known as "structuralist p." (see Culler), embodying general theory of lit. and also linguistic theories of both poetry and narrative as genres, with the devel. of linguistic methodologies for analysis of the genres. This period moreover saw the foundation of a more pragmatic, less theoretical discipline of "stylistics" or "linguistic crit." devoted to the linguistic analysis of literary texts.

Citing the Formalists and Jakobson as his authorities, Tzvetan Todorov, the Fr. translator of the Formalists, argued that the business of p. was not the crit. or interp. of individual works but the articulation and codification of the abstract properties which make every literary work possible and which make it "literary." P. then was to be a general science devoted to identifying the defining quality of the literary work. There is an act of idealization in this definition of the goals of p. which is strikingly similar to that proposed at the time by Chomsky in his definition of the goals of l.: the linguist is ultimately concerned with abstract universals of lang., not with individual sentences, which are regarded merely as manifestations of underlying structural principles. The distinction, in l. and in p., derives ultimately from Saussure's separation of *langue* (lang. as system) from *parole* (utterance). The generative analogy caught on with the structuralists, who extrapolated a concept of "literary competence" presumed to be a cognitive property of readers who had "internalized" the universal rules of lit.

What is the substance of literary universals? The most powerful suggestion came from Roman Jakobson ("L. and P.," in Sebeok). A literary work is first and foremost a "verbal message": p. seeks that which makes it literary, given that it is inherently verbal. His proposal, natural enough for one who was essentially a linguist rather than a literary scholar or an aesthetician, is conceived in terms of a specific kind of deployment of the structural resources of lang. The power and facility of Jakobson's synthesis are clear in his theory of metaphor and metonymy (qq.v.), which are seen as poetic figures based on the same structural principles which allow us to understand the nature of two kinds of aphasic disorder (and, so abstract are these principles, much else in "ordinary lang."). These principles, drawn ultimately from Saussure and fundamental to all modern l. and semiotics, are what are generally called "paradigm" and "syntagm" or, by Jakobson, "selection" and "combination": in a memorable dictum, "*the poetic function projects the principle of equivalence from the axis of selection into the axis of combination.*" This superficially cryptic formula is quite simple in its meaning. Language-users, constructing phrases, sentences, and texts, are obliged to perform two operations: to *select* units from sets of alternatives—"cat" rather than "kitty," "rug" rather than "mat," "white" rather than "black," the phoneme /i/ rather than /a/, and so on—and to *combine* their selections sequentially along the syntagmatic chain according to the structural rules of the lang.: "The white cat purred on the rug." An utterance becomes poetic when a language-user selects items which have some paradigmatic relationship (Jakobson's "equivalence," but it could be any systematic relationship, even opposition) and lays them in an extra level of patterning along the same syntagm, the paradigmatic equivalences being salient. In "The white cat sat on the black mat," the subject and object phrases are equivalent syntactically, Determiner + Adjective + Noun, making them parallel; the adjectives are antonyms made even more contrastive by the parallelism; the three words "cat," "sat," and "mat" are phonologically equivalent by virtue of being monosyllables sharing the phoneme /a/. Jakobson's example is "I like Ike." When such patterns draw attention to the verbal texture of the message itself, lang. becomes poetic.

These structural principles form the basis of many traditional prosodic and rhetorical devices, e.g. oxymoron, chiasmus, alliteration, rhyme; Jakob-

son illustrates them copiously in the seminal "L. and P." But this is only part of the point. Jakobson's theory encouraged structural thinking in general and the application to literary forms of notions such as binary opposition, underlying system, syntagmatic ordering, and transformation. The same intellectual conditions that produced Jakobson and Lévi-Strauss's minute structural (but minimally interpretive) analysis of Baudelaire's "Les Chats" also made welcome the story-grammar model of Propp, and thereby grounded, through the combination of Propp and l., the structuralist narratology of the 1960s.

To return to Jakobson's "poetic principle," his proposal hypothesizes the *effect* as well as the mechanics of poetic structure. "The set (*Einstellung*) [this gloss could be construed in the sense of 'orientation,' a term he uses elsewhere in the paper] toward the MESSAGE as such, focus on the message for its own sake, is the POETIC function of lang." The linguistic form of the text is "foregrounded" (Mukařovský), made perceptually salient. The poetic function of lang., "by promoting the palpability of signs, deepens the fundamental dichotomy of signs and objects." The brilliance of this comment lies in its synthesis of Saussurean l. and formalist aesthetics: at a stroke Jakobson has provided a linguistic mechanism for the perceptual difficulty and the distancing of meaning which Shklóvsky—a central figure in Jakobson's first intellectual milieu, pre-Revolutionary Russia—specified as conditions for the aesthetic effect of "defamiliarization": "The technique of art is to make objects 'unfamiliar,' to make forms difficult, to increase the difficulty and length of perception because the process of perception is an aesthetic end in itself and must be prolonged. *Art is a way of experiencing the artfulness of an object; the object is not important*" (1924). Jakobson's l., then, supports the importance of Shklóvsky's aesthetics as a reference-point in modern p.

One of the most controversial aspects of Jakobson's poetic theory is his claim that p. is *part of* l.: "P. deals with problems of verbal structure. . . . Since l. is the global science of verbal structure, p. may be regarded as an integral part of l. . . . Insistence on keeping p. apart from l. is warranted only when the field of l. appears to be illicitly restricted." Setting aside the questions of whether l. could in principle by itself provide an adequate p., and whether Jakobson's p. is itself adequate, we may consider in what sense his p. is part of l. according to his own definitions. Certain basic structural principles, expressed at an extreme level of generality, are repeatedly found manifested in details of linguistic form at the different levels of phonology, syntax, and semantics. Certain configurations—alliteration, for example—are said to be "poetic"; they are found in some texts, such as poems by Baudelaire and Shakespeare, in respect of which the designation "literary" is given and unquestioned, and in certain other texts such as "I like Ike" which would not be called literary—the poetic function is a relative rather than an absolute matter. The same linguistic principles, or variants of them, are found in other linguistic domains, but without the "projection of the principle of equivalence from the axis of selection into the axis of combination." Conceived thus, p. could be said to be a part of l., since its defining concepts and procedures are solely derived from l.; but so could aphasiology, phonology, and the other fields that Jakobson treated in this brilliantly rational yet reductive way.

The analyses of poems given by Jakobson and his co-authors employ linguistic theory literally; when Jakobson identifies something as a feminine rhyme or an indefinite article, the grammatical designation is literal. However, other branches of structuralist p. appeal to linguistic concepts in a less direct, more metaphorical way. For example, Propp's *Morphology of the Folktale* (1928, tr. 1958) describes the general structure of stories on the *analogy* of grammatical structure; and that is explicitly the procedure of the structural narratology of Barthes, A. J. Greimas, and Todorov. The appropriation of linguistic concepts for the description of larger structures in lit., in myth, and in other fields such as architecture, fashion, and cinema—areas for which no adequate metalanguage had previously existed—was the central feature of structuralism in the 1960s. This movement embodies a second relationship of l. and p., a second way in which the appeal to l. has progressed and refined literary theory.

Saussure had maintained that the l. he sketched was only one part of a more general discipline, "sémiologie," devoted to "the life of signs within society." Roland Barthes, half a century later, developed the general framework of "semiology," instanced some of the social constructs and conventions which belonged in this field, and proposed a positioning for both lang. and lit. within this framework. The implication in Saussure is that lang. and l. form just one of the fields within semiology, but what emerges from Barthes is that lang. and l. are privileged, paradigmatic branches of semiology (or what we would now call "semiotics"). Structure in other branches was modeled on linguistic structure, so the procedural assumption followed that linguistic methods and concepts which had been developed for the study of lang. could be extended to other semiotic fields: thus paradigm, syntagm, constituent structure, and transformation could be found in culinary conventions, fashion, etc. Obviously this was a significant methodological gain, allowing the instant foundation of new disciplines with their metalanguage ready-made. The most highly developed of these was literary structuralism.

Lit., on that model, relates to lang. in two distinct ways. First, it *is* lang., literally, in that its medium is words and sentences. And it is also *like* lang., in that it is a second-order semiotic system

patterned on higher units which mirror the structure of the lang. medium. The plot of a narrative text, for instance, is not actually a sentence (though of course it is mediated through sentences), but it can be presented as sentence-like in structure, the actions which advance the story being like verbs, and the characters who perform these actions being like nouns in grammatical functions such as subject and object. The work of Barthes, Greimas, Todorov, and others illustrates the use of l. as a source of structural analogies for the constitutive elements of poetry and of narrative. Although this analogical, basically semiotic procedure is easiest to illustrate with reference to "syntactic" studies of plot, it has been followed fruitfully in theorizing on other dimensions of lit. and on the nature of lit. (Eco; Lotman).

The applications of l. to lit. reviewed so far have been bold and fruitful, and rightly prominent in scholarly debate. It might be said that the main contribution of structural and semiotic p. has been to present certain postulates about l. and p. with such notable force and clarity that metatheoretical debate in the field has been wonderfully facilitated. But such clarity of definition is bought at the price of deliberate limitation of goal, and structuralist p. does not exhaust the range of linguistic contributions to the study of lit. A major restriction is the decision to limit attention to the *formal* aspects of lang.—to syntactic, phonological, and (structural) semantic patterns. Little consideration is given to the pragmatic, functional, or social dimensions of lang. in lit. Lang. is not only formal pattern, it is also *discourse*, mediating social and personal roles and relationships, constructing and reproducing ideology, historically and culturally situated. To cope with such factors—which after all have been among the traditional concerns of literary studies—we need a kind of l. which both the Saussurean and the Chomskyan schools have excluded, a l. of *parole*. This sort of l. has begun to be developed on several fronts in recent years. In fact, the need for such a l. to counter the restrictions of formalist p. had already been argued by Bakhtin/Voloshinov (perhaps one and the same person) in Russia around 1930. Saussurean l. was strongly attacked, and a notion of lang. as the interplay of socially and ideologically accented voices was developed. The Bakhtinian model has proved very fruitful since translations made it available to the West in the early 1970s; the technical l. of discourse has only recently begun to catch up.

A wider and more catholic range of connections between l. and p. is found if we extend the field to work in what is often generally called "stylistics" or "linguistic crit." This work is trans-generic, and many writings on fiction and drama are also relevant to poetry because the discussion often concerns general theory and methodology. In the 1960s there was heated debate about the propriety of applying l. to lit., the linguists making such a proposal meeting the same Anglo-Saxon critical opposition that resisted Fr. structuralism. But in the last 20 years there has nevertheless been published a considerable body of work in linguistic crit. giving practical demonstration of the value of l. to the student of lit. There have been empirical studies of every aspect of lang. in lit., written from the points of view of a wide diversity of linguistic models. Almost always there have been new descriptive insights and new analytic terminology. Some of the more conservative work, which is eclectic in selecting tools from l. and linking them with time-honored rhetorical concepts, has been among the most useful in practical yield. But most linguistic critics have been less eclectic, choosing to focus a particular linguistic model on some specific dimension of poetic structure.

Our understanding of Eng. metrical forms, for example, has been enhanced by a number of technical studies using different phonological theories, e.g. the 4-stress-level suprasegmental phonology of Trager and Smith (Chatman) or the generative phonology of Chomsky and Halle as developed by analogy in generative metrics. At the level of poetic syntax, there have been a number of applications of transformational grammar—e.g. Ohmann's treatment of style in terms of deep and surface structure (in Chatman and Levin, eds., 1967); Freeman's and Austin's studies of transformational preferences as the constitutive elements of artistic design (in Freeman, ed., 1981); and Thorne's generative studies of e e cummings and of Donne (in Freeman, ed., 1970)—initiating debates on linguistic deviation in poetry and the proposition that a poem autonomously creates its own lang. or at least a special dialect. Semantic and pragmatic structures have also been explored, e.g. Levin's comparison of componential semantics and speech-act theory in relation to metaphor—which requires a more thoroughly interdisciplinary approach—or Riffaterre's work on the semiotics of poetry, analysis which focuses intensely on semantic structure (and which is grounded in a sophisticated devel. of Jakobson), or Pratt's approach using the concepts of speech act and of natural narrative.

These are but a few examples of a broad range of studies which were valuable in advancing the theory and procedures of linguistic crit. and of p. Of course it must be conceded that, in a period when l. itself was rapidly changing, it was inevitable that some linguistic critics committed themselves to theoretical models that became outmoded. However, this area of l. and p. has made many permanent contributions by debating issues and testing analyses in the lang. of poetry. The continuing liveliness of linguistic crit. has more recently been attested by completely new orientations, reaction against the predominant formalism of earlier research, and a growing concern with the social contexts and functions of literary discourse. These new interests come at a stage in the

devel. of l. when pragmatics (Sperber and Wilson), discourse analysis (specifically the analysis of belief systems), and functional grammar (Halliday) offer theories and methods for further advances in the conjunction of l. and p.

F. de Saussure, *Cours de linguistique générale*, (1916), 5th ed. (1955), crit. ed., ed. R. Engler (1967–74), tr. R. Harris (1983); V. Shklóvsky, "Art as Technique" (1924), rpt. in *Rus. Formalist Crit.*, ed. and tr. L. Lemon and M. Reis (1965); M. Bakhtin, *Problems of Dostoevsky's P.* (1929, tr. 1973); V. Voloshinov, *Marxism and the Philosophy of Lang.* (1929–30, tr. 1973); G. Trager and H. L. Smith, *An Outline of Eng. Structure* (1951); "Eng. Verse and What It Sounds Like," spec. iss. of *KR* 18 (1956); C. Lévi-Strauss, *Structural Anthropology* (1958, tr. 1968); R. Jakobson, "L. and P.," in Sebeok, rpt. in Jakobson, v. 5; Sebeok; F. L. Utley, "Structural L. and the Literary Critic," *JAAC* 18 (1960); S. R. Levin, *Linguistic Structures in Poetry* (1962), "The Conventions of Poetry," *Literary Style: A Symposium*, ed. S. Chatman (1971), *The Semantics of Metaphor* (1977); R. Jakobson and C. Lévi-Strauss, "Charles Baudelaire's 'Les Chats'" (1962), rpt. in *The Structuralists from Marx to Lévi-Strauss*, ed. R. and F. DeGeorge (1972); R. Barthes, *Elements of Semiology* (1964–67), *S/Z* (1970, tr. 1975); S. Chatman, *A Theory of Meter* (1965); A. J. Greimas, *Sémantique structurale* (1966); J. Mukařovský, "Standard Lang. and Poetic Lang." and R. Ohmann, "Lit. as Sentences" in Chatman and Levin, eds. (1967); N. Chomsky and M. Halle, *The Sound Pattern of Eng.* (1968); G. N. Leech, *A Linguistic Guide to Eng. Poetry* (1969); *L. and Literary Style*, *Essays in Mod. Stylistics*, both ed. D. Freeman (1970, 1981); T. Todorov, *The Poetics of Prose* (1971, tr. 1977); *Literary Style: A Symposium*, and *Approaches to P.*, both ed. S. Chatman (1971, 1973); R. Fowler, *The Langs. of Lit.* (1971), *Lit. as Social Discourse* (1981), *Linguistic Crit.* (1986); S. Chatman and S. R. Levin, eds., *Essays on the Lang. of Lit.* (1967), "L. and Lit.," *Current Trends in L. X*, ed. T. A. Sebeok (1973); P. Kiparsky, "The Role of L. in a Theory of Poetry," *Daedalus* 102 (1973), rpt. in *Lang. as a Human Problem*, ed. E. Haugen and M. Bloomfield (1974); Culler; U. Eco, *A Theory of Semiotics* (1976); A. A. Hill, *Constituent and Pattern in Poetry* (1976); T. Hawkes, *Structuralism and Semiotics* (1977); J. Lotman, *The Structure of the Artistic Text*, tr. R. Vroon (1977); M. Pratt, *Toward a Speech Act Theory of Literary Discourse* (1977); M. Riffaterre, *Semiotics of Poetry* (1978); Brogan, items E709–827—comprehensive list of work in structural and generative metrics to 1981, with numerous other studies in the linguistic analysis of lit. listed *passim*; J. Culler, "Lit. and L.," *Interrelations of Lit.*, ed. J.-P. Barricelli and J. Gibaldi (1982); M. Halliday, *An Intro. to Functional Grammar* (1985); D. Sperber and D. Wilson, *Relevance* (1986); *The L. of Writing*, ed. N. Fabb et al. (1987); D. Birch, *Lang., Lit. and Critical Practice* (1989); *Lang., Discourse and Lit.*, ed. R. Carter and P. Simpson (1989); K. Wales, *A Dictionary of Stylistics* (1989).

R.F.

LOVE POETRY.

I. WESTERN
II. EASTERN
   A. *Arabic and Persian*
   B. *Egyptian*
   C. *Hebrew*
   D. *Indian*
   E. *Chinese*
   F. *Japanese*

I. WESTERN. In evaluating l. p., we must ask first whether the lang. is private and original or formulaic and rhetorical. Is the poet speaking for him- or herself, or is the voice a persona? The poem, if commissioned by friend or patron, may be a projection into another's adventures, or it may be an improvised conflation of real and invented details. A love poem cannot be simplistically read as a literal, journalistic record of an event or relationship; there is always some fictive reshaping of reality for dramatic or psychological ends. A love poem is secondary rather than primary experience; as an imaginative construction, it invites detached contemplation of the spectacle of sex.

We must be particularly cautious when dealing with controversial forms of eroticism like homosexuality. Poems are unreliable historical evidence about any society; they may reflect the consciousness of only one exceptional person. Furthermore, homoerotic images or fantasies in poetry must not be confused with concrete homosexual practice. We may speak of tastes or tendencies in early poets but not of sexual orientation: this is a modern idea.

L. p. is equally informed by artistic trad. and contemp. cultural assumptions. The pagan attitude toward the body and its pleasures was quite different from that of Christianity, which assigns sex to the fallen realm of nature. The richness of Western l. p. may thus arise in part from the dilemma of how to reconcile mind or soul with body. Moreover, the generally higher social status of women in Western as opposed to Eastern culture has given l. p. added complexity or ambivalence: only women of strong personality could have produced the tormented sagas of Catullus or Propertius. We must try to identify a poem's intended audience. In antiquity the love poet was usually addressing a coterie of friends or connoisseurs; since romanticism, however, the poet speaks to him- or herself, with the reader seeming to overhear private thoughts. We must ask about pornographic material in l. p. whether it reflects the freer sensibilities of a different time or whether the poet set out to shock or challenge his contemporaries. Much l. p. is clearly testing the limits of decorous speech, partly to bring sexual desire under the scrutiny and control of imagination. In the great Western theme of the transience

of time, vivid sensuous details illustrate the evanescence of youth and beauty; the poet has a godlike power to defeat time and bestow immortality upon the beloved through art. Romantic impediments give the poem a dramatic frame: the beloved may be indifferent, far away, married to someone else, dead, or of the wrong sex. However, difficulty or disaster in real life is converted into artistic opportunity by the poet, whose work profits from the intensification and exploration of negative emotion.

The history of European l. p. begins with the Gr. lyric poets of the Archaic age (7th–6th cs. B.C). Archilochus, Mimnermus, Sappho, and Alcaeus turn poetry away from the grand epic style toward the quiet personal voice, attentive to mood and emotion. Despite the fragmentary survival of Gr. solo poetry, we see that it contains a new idea of love which Homer shows as foolish or deceptive but never unhappy. Archilochus' account of the anguish of love is deepened by Sappho, whose poetry was honored by male writers and grammarians until the fall of Rome. Sappho and Alcaeus were active on Lesbos, an affluent island off the Aeolian coast of Asia Minor where aristocratic women apparently had more freedom than later in classical Athens. Sappho is primarily a love poet, uninterested in politics or metaphysics. The nature of her love has caused much controversy and many fabrications, some by major scholars. Sappho was married, and she had a daughter, but her poetry suggests that she fell in love with a series of beautiful girls, who moved in and out of her coterie (not a school, club, or cult). There is as yet no evidence, however, that she had physical relations with women. Even the ancients, who had her complete works, were divided about her sexuality.

Sappho shows that l. p. is how Western personality defines itself. The beloved is passionately perceived but also replaceable; he or she may exist primarily as a focus of the poet's consciousness. In "He seems to me a god" (fr. 31), Sappho describes her pain at the sight of a favorite girl sitting and laughing with a man. The lighthearted social scene becomes oppressively internal as the poet sinks into suffering: she cannot speak or see; she is overcome by fever, tremor, pallor. These symptoms of love become conventional and persist more than a thousand years (Lesky). In plain, concise lang., Sappho analyzes her extreme state as if she were both actor and observer; she is candid and emotional yet dignified, austere, almost clinical. This poem, preserved for us by Longinus, is the first great psychological document of Western lit. Sappho's prayer to Aphrodite (fr. 1) converts cult-song into love poem. The goddess, amused at Sappho's desperate appeal for aid, teasingly reminds her of former infatuations and their inevitable end. Love is an endless cycle of pursuit, triumph, and ennui. The poem, seemingly so charming and transparent, is structured by a complex time scheme of past, present, and future, the ever-flowing stream of our emotional life. Sappho also wrote festive wedding songs and the first known description of a romantic moonlit night. She apparently invented the now-commonplace adjective "bittersweet" for the mixed condition of love.

Early Gr. l. p. is based on simple parallelism between human emotion and nature, which has a Mediterranean mildness. Love-sickness, like a storm, is sudden and passing. Imagery is vivid and luminous, as in haiku (q.v.); there is nothing contorted or artificial. Anacreon earned a proverbial reputation for wine, women, and song: his love is not Sappho's spiritual crisis but the passing diversion of a bisexual bon vivant. L. p. was little written in classical Athens, where lyric was absorbed into the tragic choral ode. Plato, who abandoned poetry for philosophy, left epigrams on the beauty of boys. The learned Alexandrian age revived l. p. as an art mode. Theocritus begins the long literary trad. of pastoral (q.v.), where shepherds complain of unrequited love under sunny skies. Most of his *Idylls* contain the voices of rustic characters like homely Polyphemus, courting the scornful nymph Galatea, or Lycidas, a goatherd pining for a youth gone to sea. Aging Theocritus broods about his own love for fickle boys, whose blushes haunt him. In his *Epigrams*, Callimachus takes a lighter attitude toward love, to which he applies sporting metaphors of the hunt. In Medea's agonized passion for Jason in the *Argonautica*, Apollonius Rhodius tries to mesh l. p. with epic. Asklepiades adds new symbols to love trad.: Eros and arrow-darting Cupid. Meleager writes with equal relish of cruel boys and voluptuous women, such as Heliodora. His is a poignant, sensual poetry filled with the color and smell of flowers.

The *Greek Anthology* demonstrates the changes in Gr. l. p. from the Alexandrian through Roman periods. As urban centers grow and speed up, nature metaphors recede. Trashy street life begins, and prostitutes, drag queens, randy tutors, and bathhouse masseuses crowd into view. Love poets become droll, jaded, less lyrical. Women are lusciously described but given no personalities or inner life. For the first time, l. p. incorporates ugliness, squalor, and disgust: Leonidas of Tarentum and Marcus Argentarius write of voracious sluts with special skills, and Antipater of Thessalonika coarsely derides scrawny old lechers. Boy-love is universal: Straton of Sardis, editor of an anthol. of pederastic poems, celebrates the ripening phases of boys' genitals. By the early Byzantine period, however, we feel the impact of Christianity, in more heartfelt sentiment but also in guilt and melancholy.

The Romans inherited a huge body of Gr. l. p. Catullus, the first Lat. writer to adapt elegy (q.v.) for love themes, is obsessed with Lesbia, the glamorous noblewoman Clodia, promiscuously partying with midnight pickups. "I love and I hate": this tortured affair is the most complex contribution

to l. p. since Sappho, whom Catullus admired and imitated. The poet painfully grapples with the ambiguities and ambivalences of being in love with an aggressive, willful Western woman. He also writes tender love poems to a boy, honey-sweet Juventius. There is no Roman l. p. between adult men. Propertius records a long, tangled involvement with capricious Cynthia, a fast-living new woman. There are sensual bed scenes, love-bites, brawls. After Cynthia dies (perhaps poisoned), the angry, humiliated poet sees her ghost over his bed. Tibullus writes of troubled love for two headstrong mistresses, adulterous Delia and greedy Nemesis, and one elusive boy, Marathus. In Virgil's *Eclogue 2*, the shepherd Corydon passionately laments his love-madness for Alexis, a proud, beautiful youth; the poem was traditionally taken as proof of Virgil's own homosexuality. Horace names a half dozen girls whom he playfully lusts for, but only the rosy boy Ligurinus moves him to tears and dreams. In the *Amores*, Ovid boasts of his sexual prowess and offers strategies for adultery. *The Art of Love* tells how to find and keep a lover, including sexual positions, naughty words, and feigned ecstasies. *The Remedies for Love* contains precepts for falling *out* of love. The love-letters of the *Heroides* are rhetorical monologues of famous women (Phaedra, Medea) abandoned by cads. Juvenal shows imperial Rome teeming with effeminates, libertines, and pimps; love or trust is impossible. The Empress prowls the brothels; every good-looking boy is endangered by rich seducers; drunken wives grapple in public stunts. Martial casts himself as a facetious explorer of this lewd world where erections are measured and no girl says no. The *Dionysiaca*, Nonnus' late Gr. epic, assembles fanciful erotic episodes from the life of Dionysus. Also extant are many Gr. and Lat. *priapeia*: obscene comic verses, attached to phallic statues of Priapus in field and garden, which threaten thieves with anal or oral rape.

In medieval romance, love as challenge, danger, or high ideal is central to chivalric quest. From the mid 12th c., woman replaces the feudal lord as center of the militaristic *chansons de geste* (q.v.). Fr. aristocratic taste was refined by the courtly love of the Occitan (Provençal) troubadours (q.v.), who raised woman to spiritual dominance, something new in Western l. p. Amorous intrigue now lures the hero: to consummate his adultery with Guinevere, Chrétien de Troyes' Lancelot bends the bars of her chamber, then bleeds into her bed. The symbolism of golden grail, bleeding lance, and broken sword of Chrétien's *Perceval* is sexual as well as religious. Wolfram von Eschenbach's Ger. Parzival is vowed to purity, but adulterous Anfortas suffers a festering, incurable groin wound. Sexual temptations are specifically set to test a knight's virtue in the Fr. romances *Yder* and *Hunbaut* and the ME *Sir Gawain and the Green Knight*. The adultery of Gottfried von Strassburg's Tristan and Isolde, with their steamy lovemaking, helped define Western romantic love as unhappy or doomed. The Trojan tale of faithful Troilus and treacherous Cressida was invented by Benoît de Sainte-Maure and transmitted to Boccaccio and Chaucer. Heavily influenced by Ovid, *The Romance of the Rose* (Guillaume de Lorris and Jean de Meun) uses dreamlike allegory and sexual symbols of flower, garden, and tower to chart love's assault. The pregnancy of the Rose is a first for European literary heroines. Abelard wrote famous love songs, now lost, to Heloise. Dante's youthful love poems to Beatrice in the *Vita nuova* begin in troubadour style, then modulate toward Christian mysticism. In the *Inferno*'s episode of Paolo and Francesca, seduced into adultery by reading a romance of Lancelot, Dante renounces his early affection for courtly love. Med. Lat. lyrics express homoerotic feeling between teacher and student in monastic communities. There are overtly pederastic poems from the 12th c. and at least one apparently lesbian one, but no known vernacular or pastoral medieval poetry is homosexual. The goliardic *Carmina Burana* contains beautiful lyrics of the northern flowering of spring and love, as well as cheeky verses of carousing and wenching, some startlingly detailed. The Fr. *fabliau*, a ribald verse-tale twice imitated by Chaucer, reacts against courtly love with bedroom pranks, barnyard drubbings, and an earthy stress on woman's hoary genitality. Villon, zestfully atumble with Parisian trollops, will later combine the devil-may-care goliard's pose with the fabliau's slangy comedy.

Ren. epic further expands the romantic elements in chivalric adventure. Boiardo, Ariosto, and Tasso open quest to an armed heroine, a motif adopted by Spenser, whose *Faerie Queene*, emulating Ovid's *Metamorphoses*, copiously catalogues incidents of normal and deviant sex. Petrarch, combining troubadour lyricism with Dante's advanced psychology, creates the modern love poem. His Laura, unlike saintly Beatrice, is a real woman, not a symbol. Petrarch's nature, vibrating to the lover's emotions, will become the romantic pathetic fallacy. His conceits, paradoxes, and images of fire and ice, which spread in sonnet sequences (q.v.) throughout Europe, inspired and burdened Ren. poets, who had to discard the convention of frigid mistress and trembling wooer. Ronsard's sonnets, addressed to Cassandre, Marie, and Hélène, first follow Petrarchan formulas, then achieve a simpler, more musical, debonair style, exquisitely attuned to nature. In the *Amoretti*, Spenser practices the sonnet (introduced to England by Wyatt and Surrey), but his supreme love poem is the *Epithalamion*, celebrating marriage. Like Michelangelo, Shakespeare writes complex l. p. to a beautiful young man and a forceful woman: the fair youth's homoerotic androgyny is reminiscent of Shakespeare's soft, "lovely" Adonis and Marlowe's longhaired, white-fleshed Leander, romanced by Neptune. Richard Barnfield's sonnets and *Affectionate Shepherd* openly offer succu-

lent sexual delights to a boy called Ganymede, a common Ren. allusion. The traditional allegory, based on the Song of Songs, of Christ the bridegroom knocking at the soul's door, creates unmistakable homoeroticism in Donne's Holy Sonnet XIV, George Herbert's "Love (III)," and spiritual stanzas by St. John of the Cross. In ardent poems to his fiancée, later his wife, Donne, with Spenser, demonstrates the new prestige of marriage: before this, no one wrote l. p. to his wife. Furthermore, Donne's erudition implies that his lady, better educated than her medieval precursors, enjoys flattery of her intellect as well as of her beauty. Aretino's sonnets daringly use vulgar street terms for acts of love. Marino's *Adonis* makes baroque opera out of the ritualistic stages of sexual gratification. Waller and Marvell use the *carpe diem* argument to lure shy virgins into surrender; the Cavalier poets adopt a flippant court attitude toward women and pleasure. Carew's "A Rapture" turns Donne's ode to nakedness into a risqué tour of Celia's nether parts. Libertines emerge in the late 17th c.: Rochester, a Restoration wit, writes bluntly of raw couplings with ladies, whores, and boys. Milton's *Lycidas* revives the Cl. style of homoerotic pastoral lament. *Paradise Lost*, following Spenser and Donne, exalts "wedded Love" over the sterile wantonness of "Harlots" and "Court Amours" (4.750–70).

The Age of Reason, valuing self-control and witty detachment, favored satire over l. p. Rousseau's delicate sentiment and pagan nature-worship created the fervent moods of "sensibility" and woman-revering romanticism. Goethe, identifying femaleness with creativity, writes of happy sensual awakening in the *Roman Elegies* and jokes about sodomy with both sexes in the *Venetian Epigrams*, with its autoerotic acrobat Bettina; withheld pornographic verses imitate ancient *priapeia*. Schiller dedicates rhapsodic love poems to Laura, but his hymns to womanhood sentimentally polarize the sexes. Hölderlin addresses Diotima with generalized reverence and reserves his real feeling for Mother Earth. Blake calls for sexual freedom for women and for the end of guilt and shame. Burns composes rural Scottish ballads of bawdy or ill-starred love. Wordsworth's Lucy poems imagine woman reabsorbed into roiling nature. In *Christabel* Coleridge stages a virgin's seduction by a lesbian vampire, nature's emissary. The younger Eng. romantics fuse poetry with free love. In *Epipsychidion* Shelley is ruled by celestial women radiating intellectual light. Keats makes emotion primary; his maidens sensuously feed and sleep or wildly dance dominion over knights and kings. Byron's persona as a "mad, bad" seducer has been revised by modern revelations about his bisexuality. In the "Thyrza" poems, he woos and changes the sex of a favorite Cambridge choirboy; in *Don Juan* his blushing, girlish hero, forced into drag, catches the eye of a tempestuous lesbian sultana. Heine's love ballads are about squires, shepherdboys, hussars, and fishermaidens; later verses record erotic adventures of the famous poet wined and dined by lady admirers.

The Fr. romantics, turning art against nature in the hell of the modern city, make forbidden sex a central theme. Gautier celebrates the lonely, self-complete hermaphrodite. Baudelaire looses brazen whores upon syphilitic male martyrs; sex is torment, cursed by God. Baudelaire's heroic, defiant lesbians are hedonistically modernized by Verlaine and later rehellenized by Louys. In *Femmes* Verlaine uses vigorous street argot to describe the voluptuous sounds and smells of sex with women; in *Hombres* he lauds the brutal virility of young laborers, whom he possesses in their rough workclothes. He and Rimbaud co-wrote an ingenious sonnet about the anus. Mallarmé's leering faun embodies pagan eros; cold, virginal Herodias is woman as castrator. In contrast, Victorian poetry, as typified by the Brownings, exalts tenderness, fidelity, and devotion, the bonds of married love, preserved beyond the grave. Tennyson and the Pre-Raphaelites revive the medieval cult of idealized woman, supporting the Victorian view of woman's spirituality. Tennyson's heroines, like weary Mariana, love in mournful solitude. His *Idylls* retell Arthurian romance. *In Memoriam*, Tennyson's elaborate elegy for Hallam, is homoerotic in feeling. Rossetti's sirens are sultry, smoldering. Swinburne, inspired by Baudelaire, reintroduces sexual frankness into highbrow Eng. lit. His Dolores and Faustine are promiscuous *femmes fatales*, immortal vampires; his Sappho, sadistically caressing Anactoria, boldly proclaims her poetic greatness. Whitman broke taboos in Am. poetry: he names body parts and depicts sex surging through fertile nature; he savors the erotic beauties of both male and female. Though he endorses sexual action and energy, Whitman appears to have been mostly solitary, troubled by homosexual desires, suggested in the "Calamus" section of *Leaves of Grass*. Reflecting the Victorian taste for bereavement, Hardy's early poetry features gloomy provincial tales of love lost: ghosts, graveyards, suicides, tearful partings. Homoerotic Gr. idealism and epicene fin-de-siècle preciosity characterize the poems of Symonds, Carpenter, Hopkins, Wilde, Symons, and Dowson. Renée Vivien, the first poet to advertise her lesbianism, writes only of languid, ethereal beauty.

L. p. of the 20th c. is the most varied and sexually explicit since Cl. lit. T. S. Eliot diagnoses the sexual sterility or passivity of modern man. Yet Neruda writes searing odes to physical passion, boiling with ecstatic elemental imagery. D. H. Lawrence similarly roots the sex impulse in the seasonal cycles of the animal world. Recalling long-ago, one-night pickups of handsome, athletic youths, Cavafy declares sex the creative source of his poetry. For Yeats, woman's haunting beauty is the heart of life's mystery; in "Leda and the Swan," rape is the metaphor for cataclysmic his-

torical change. Rilke contemplates the philosophical dilemma of love, the pressure upon identity, the tension between fate and freedom. Valéry makes lang. erotic: the poet is Narcissus and, in *La Jeune Parque*, the oracle raped by her own inner god. Éluard sees woman erotically metamorphosing through the world, permeating him with her supernatural force. Lorca imagines operatic scenes of heterosexual seduction, rape, or mutilation, and in "Ode to Walt Whitman" denounces urban "pansies" for a visionary homosexuality grounded in living nature. Fascinated but repelled by strippers and whores, Hart Crane records squalid homosexual encounters in subway urinals. Amy Lowell vividly charts the works and days of a settled, sustaining lesbian relationship, while H. D. projects lesbian feeling into Gr. personae, often male. Edna St. Vincent Millay is the first woman poet to claim a man's sexual freedom: her sassy, cynical lyrics of Jazz Age promiscuity with anonymous men are balanced by melancholy love poems to women. Auden blurred the genders in major poems to conceal their homosexual inspiration; his private verse is maliciously bawdy. William Carlos Williams is rare among modern poets in extolling married love and kitchen-centered domestic bliss.

For Dylan Thomas, youth's sexual energies drive upward from moldering, evergreen earth. Theodore Roethke presents woman as unknowable Muse, ruling nature's ghostly breezes and oozy sexual matrix. Delmore Schwartz hails Marilyn Monroe as a new Venus, blessing and redeeming "a nation haunted by Puritanism." The free-living Beat poets, emulating Black hipster talk, broke poetic decorum about sex. Adopting Whitman's chanting form and pansexual theme, Allen Ginsberg playfully celebrates sodomy and master-slave scenarios. In "Marriage," Gregory Corso imagines the whole universe wedding and propagating while he ages, destitute and alone. The Confessional poets weave sex into autobiography. Robert Lowell lies on his marriage bed paralyzed, sedated, unmanned. Anne Sexton aggressively breaks the age-old taboo upon female speech by graphically exploring her own body in adultery and masturbation. Sylvia Plath launched contemp. feminist poetry with her sizzling accounts of modern marriage as hell. With its grisly mix of Nazi fantasy and Freudian family romance, "Daddy," after Yeats' "Leda," may be the love poem of the century. John Berryman's *Sonnets* records a passionate, adulterous affair with a new Laura, her platinum hair lit by the dashboard as they copulate in a car, the modern version of Dido's dark "cave." *Love and Fame* reviews Berryman's career as a "sexual athlete" specializing in quickie encounters. The sexual revolution of the 1960s heightened the new candor. Hippie poetry invoked Buddhist avatars for love's ecstasies. Denise Levertov and Carol Bergé reverse trad. by salaciously detailing the hairy, muscular male body. Diane di Prima finds sharp, fierce imagery for the violent carnality of sex. Charles Bukowski writes of eroticism without illusions in a tough, gritty world of scrappy women, drunks, rooming-houses, and racetracks. Mark Strand mythically sees man helplessly passed from mother to wife to daughter: "I am the toy of women."

The 1960s also freed gay poetry from both underground and coterie. James Merrill, remembering mature love or youthful crisis, makes precise, discreet notations of dramatic place and time. Paul Goodman, Robert Duncan, Frank O'Hara, Thom Gunn, Harold Norse, and Mutsuo Takahashi intricately document the mechanics of homosexual contact for the first time since Imperial Rome: cruising, hustlers, sailors, bodybuilders, bikers, leather bars, bus terminals, toilets, glory holes. Gay male poetry is about energy, adventure, quest, danger; beauty and pleasure amidst secrecy, shame, and pain. Lesbian poetry, in contrast, prefers tender, committed relationships and often burdens itself with moralistic political messages. Adrienne Rich and Judy Grahn describe intimate lesbian sex and express solidarity with victimized women of all social classes; Audre Lorde invokes Af. myths to enlarge female identity. Olga Broumas, linking dreamy sensation to Gr. sun and sea, has produced the most artistically erotic lesbian lyrics. Eleanor Lerman's *Armed Love*, with its intellectual force and hallucinatory sexual ambiguities, remains the leading achievement of modern lesbian poetry, recapitulating the tormented history of Western love from Sappho and Catullus to Baudelaire.

ANTHOLOGIES: *Poetica erotica*, ed. T. R. Smith (1927); *Erotic Poetry: The Lyrics, Ballads, Idylls, and Epics of L., Cl. to Contemp.*, ed. W. Cole (1963); *The Body of L.: An Anthol. of Erotic Verse from Chaucer to Lawrence*, ed. D. Stanford (1965); *Dein Leib ist mein Gedicht: Deutsche erotische Lyrik aus funf Jahrhunderten*, ed. H. Arnold (1970); *Sexual Heretics: Male Homosexuality in Eng. Lit. 1850–1900*, ed. B. Reade (1970); *An Anthol. of Swahili L. P.*, ed. and tr. J. Knappert (1972); *The Gr. Anthol.*, ed. P. Jay (1973); *The Male Muse: A Gay Anthol.*, ed. I. Young (1973); *A Book of L. P.*, ed. J. Stallworthy (1974); *The Woman Troubadours*, ed. M. Bogin (1976); *Eros baroque: Anthologie de la poésie amoureuse baroque 1570–1620*, ed. G. Mathieu-Castellani (1979); *Penguin Book of Homosexual Verse*, ed. S. Coote (1983); *Priapea: Poems for a Phallic God*, tr. W. H. Parker (1988); *Gay and Lesbian Poetry in Our Time*, ed. C. Morse and J. Larkin (1988); *The Song of Eros: Ancient Gr. L. Poems*, tr. B. P. Nystrom (1991); *Games of Venus*, ed. P. Bing (1991).

HISTORY AND CRITICISM: O. Waser, "Eros," Pauly-Wissowa 1.6.484 ff.; R. Reitzenstein, *Zur Sprache der lateinischen Erotik* (1912); H. Brinkmann, *Gesch. der lateinischen Liebesdichtung im Mittelalter* (1925); M. Praz, *The Romantic Agony* (1933); Lewis; A. Day, *The Origins of Lat. Love-Elegy* (1938); J. H. Wilson, *The Court Wits of the Restoration* (1948); D. L. Page,

Sappho and Alcaeus (1955); G. W. Knight, *The Mutual Flame* (1955); D. de Rougemont, *L. in the Western World*, 2d ed. (1956); M. J. Valency, *In Praise of L.: An Intro. to L.-P. of the Ren.* (1958); H. Bloom, *The Visionary Company* (1961); A. Lesky, *A Hist. of Gr. Lit.* (1963); D. Bush, *Mythology and the Ren. Trad. in Eng. Poetry*, 2d ed. (1963); J. B. Broadbent, *Poetic L.* (1964); G. E. Enscoe, *Eros and the Romantics* (1967); Dronke; S. Minta, *L. P. in 16th-C. France* (1977); T. A. Perry, *Erotic Spirituality: The Integrative Trad. from Leone Ebreo to John Donne* (1980); J. H. Hagstrum, *Sex and Sensibility: Ideal and Erotic L. from Milton to Mozart* (1980), *The Romantic Body: L. and Sexuality in Keats, Wordsworth, and Blake* (1985); R. O. A. M. Lyne, *The Lat. L. Poets from Catullus to Horace* (1981); P. J. Kearney, *A Hist. of Erotic Lit.* (1982); A. Richlin, *The Garden of Priapus: Sexuality and Aggression in Roman Humor* (1983); P. Veyne, *Roman Erotic Elegy: L., Poetry, and the West* (1983, tr. 1988); I. Singer, *The Nature of L.*, v. 1, *Plato to Luther*, 2d ed. (1984), v. 2, *Courtly and Romantic* (1984), v. 3, *The Mod. World* (1987); A. J. Smith, *The Metaphysics of L.: Studies in Ren. L. P. from Dante to Milton* (1985); G. Woods, *Articulate Flesh: Male Homoeroticism and Mod. Poetry* (1987); D. O. Frantz, *Festum voluptatis: A Study of Ren. Erotica* (1989); A. R. Jones, *The Currency of Eros: Women's L. Lyric in Europe, 1540–1620* (1990); *The Song of Eros: Ancient Gr. L. Poems*, tr. B. P. Nystrom (1990); C. Paglia, *Sexual Personae: Art and Decadence from Nefertiti to Emily Dickinson* (1990). C.P.

II. EASTERN. A. *Arabic and Persian.* Ar. l. p. in the pre-Islamic period is found chiefly in the extended ode (*qaṣīda*), which typically begins with expressions of nostalgic longing for departed loved ones, or incorporates accounts of amorous conquests which serve the predominant mode of *fakhr* (boasting of personal achievements). In the Islamic period l. p. (generically termed *ghazal* was raised to independent status by the urban poets of the Hijaz (e.g. ʿUmar ibn Abī Rabīʿa [d. 711]) and the ʿUdhri poets of the desert (e.g. Jamīl [d. 710] and the semi-legendary Qays of the Banī ʿAmr, better known as Majnūn, "the madman"), who introduced the motif of the dedicated love-service to an often capricious lady. L. p. flourished under the early Abbasid caliphs (750–850) with Abū Nuwās (d. 813?), who mingled bacchic with erotic motifs and excelled in obscene l. p. (*mujūn*), both hetero- and homosexual, and al-ʿAbbās ibn al-Aḥnaf (d. ca. 808), the celebrated poet of courtly love at the court of Hārūn al-Rashīd. In later periods erotic motifs were generally incorporated into the panegyric *qaṣīda*; but independent l. poems are found in the mystical poetry of Ibn al-Fāriḍ (d. 1235) and Ibn al-ʿArabī (d. 1240), who adapted the lang. of secular bacchic and l. p. to express the mystical longing for union with the divine beloved. In Andalusia and Sicily l. p. retained its independent status, particularly in the Andalusian *muwashshaḥa* and *zajal*.

Persian l. p. first appears both in the exordium of the panegyric *qaṣīda* and in brief, independent lyrics (*ghazal, tarāna*) usually composed to be sung. The *ghazal* was adapted by Sanāʾī (d. 1130–31) for themes of mystical love and was later used extensively for this purpose, notably by Fakhr al-Dīn ʿIrāqī (d. 1289) and Jalāl al-Dīn Rūmī (d. 1273); the *ghazals* of Ḥāfiẓ (d. 1389) incorporate both secular and mystical topics to create deliberate ambiguity. The Persian verse romance (*maṣnavī*), which centers on the love quest, was also adapted to mystical purposes, chiefly by Jāmī (d. 1498).

Both Ar. and Persian l. p. deal primarily with the sufferings and aspirations of the lover; with few exceptions (notably in the *kharjas* of some Andalusian *muwashshaḥāt*) the female voice is absent, and lesbian love does not furnish a poetic theme. Homoeroticism is more prominent in Persian l. p., where it is generally depicted in the context of idealized love, than in Ar., where it is largely confined to *mujūn*, which often deals explicitly with sexual matters; similarly, mystical l. p. is more widespread in Persian than in Ar., and mystical overtones come to permeate secular poetry in later periods. J.S.M.

L. p. has re-emerged in the 20th c. as a privileged theme. Among contemp. Arab poets particularly known for erotic poetry are the Lebanese poets Saʿīd ʿAql (b. 1912) and Nizār Qabbānī (b. 1923). The oeuvres of the Syrian-Lebanese poet Adūnīs (ʿAlī Ahmad Saʿīd, b. 1930) and the Palestinian Tawfīq Ṣāyigh (1923–71) include important love poems.

In Persian poetry there are major erotic themes in the poetry of Ahmad Shamlu (b. 1925) and Mahdi Akhavāni Sālis (1928–90), but Furūgh Farrukhzād (1935–67) is the most innovative and forthright love poet of modern times, with an unprecedented, autobiographical treatment of erotic themes. Her collection *Another Birth* (1964) has been particularly influential. M.B.

Ibn al-ʿArabī, *The Tarjumān al-Ashwāq*, tr. R. A. Nicholson (1911, rpt. 1978); Ibn al-Fāriḍ, *The Mystical Poems*, tr. A. J. Arberry (1956); *The Seven Odes*, tr. A. J. Arberry (1957); Jalāl al-Dīn Rūmī, *Mystical Poems*, tr. A. J. Arberry, 2 v. (1968, 1979); J. T. Monroe, *Hispano-Ar. Poetry* (1974); L. F. Compton, *Andalusian Lyrical Poetry and Old Sp. Love Songs* (1976); A. Karimi-Hakkak, *Anthol. of Mod. Persian Poetry* (1978); A. Schimmel, *As Through a Veil: Mystical Poetry in Islam* (1982); M. C. Hillmann, *A Lonely Woman: Forugh Farrokhzad and Her Poetry* (1987); S. Jayyusi, *Mod. Ar. Poetry: An Anthol.* (1987). J.S.M.; M.B.

B. *Egyptian.* Extant from the Ramesside period (ca. 1305–1080 B.C.) of the New Kingdom are four small collections totaling about 60 poems, noteworthy because they are among the few examples of clearly personal and secular lyric poetry to survive from ancient Egypt. They are also one of the first clear instances of human female speakers (as opposed to deities) in the lit.; though most

likely written by males, about half of the poems are spoken by women young or old. The authors in all cases are unknown. The speakers of both sexes in most cases appear to be young, expressing the full range of love situations between the sexes (never between members of the same sex). There is the catalogue of the beloved's attributes, head to toe; the lovelonging for the absent lover ("I love you through the daytimes, / in the dark, / Through all the long divisions of the night"); the depths of emotion ("Love of you is mixed deep in my vitals, / like water stirred into flour for bread"); the game of love ("I think I'll go home and lie very still, / feigning terminal illness"); the love ecstasy; the broken heart ("Whose turn is it now / making soft eyes up into his face?"). The range of emotion extends from the chasteness, simplicity, and tentativeness of a first love, like that of Romeo and Juliet ("A girl's sleepy feelings wakened by you— / you've made a whole world for me!"), on through the frankly physical attractions of love, directly and sensuously expressed, to at times the openly erotic ("Would your fingers follow the line of my thighs, / learn the curve of my breasts, and the rest?"). In general the love songs are romantic, idyllic, humorous, even satirical, sometimes naive, almost always graceful. Since most of the moves on the chessboard of love are represented, these songs seem broader in scope than the poems in the biblical Song of Solomon (see below), with which they are often compared, but which they antedate by a thousand years.—J. L. Foster, *Love Songs of the New Kingdom* (1974).

J.L.F.

C. *Hebrew.* Only one ancient Heb. love poem is extant, the biblical Song of Songs (also known as the Song of Solomon and the Canticles). In traditional Jewish exegesis, the Song was interpreted as an allegory of God's love for Israel (in Christian exegesis—God's love for the Church). It is, however, a secular, erotic love song of the sort well known from Egyptian l. p., which it resembles and from which Israelite l. p. probably derives. In both places, such songs were probably sung by professional singers as entertainment at private festivities. Traditionally the Song was attributed to King Solomon (mid 10th c. B.C.E.). Scholars are divided on its actual dating, but its lang. seems to belong to the 4th–3d c. B.C.E. Also at issue is the Song's unity, with some commentators regarding it as a collection of short love poems, others viewing it a unified work. The uniform style and the frequent appearance of repetends imply single authorship, though there is no clearly unifying structure. The speakers in the Song are an unmarried couple; since both lovers speak, they are best regarded as personae. The main voice is that of the young woman who speaks to and about her love, praising his beauty in a series of sometimes startling similes and expressing her desire in bold and sensual terms; he responds in kind. The song as a whole is a dialogue in which the words of the lovers often intermesh closely and echo each other. The lovers' relationship is surprisingly egalitarian and their eroticism quite unabashed. No other ancient Heb. l. p. is extant. Ezekiel 33:32, however, refers to a singer of a "song of desire," and Isaiah 5:1–4 seems to be based on a love-song theme.—M. H. Pope, *Song of Songs* (1977); M. Falk, *Love Lyrics from the Bible* (1982); M. V. Fox, *The Song of Songs and the Ancient Egyptian Love Songs* (1985).

M.V.F.

Medieval Heb. liturgical poetry continued the old trad. of treating God and Israel as lovers or spouses. Secular l. p., imitating the Ar. *ghazal* (q.v.), appeared in 10th-c. Spain, particularly by Samuel the Nagid (993–ca. 1055), Moses Ibn Ezra (ca. 1055–ca.1135), and Judah Halevi (ca. 1075–1141). Many poems are apparently homoerotic; but the sex of the lovers is not stressed, for the poems focus more on love than on sexuality, and on the poet's adoration of ideal beauty embodied in the beloved. The erotic prelude to formal panegyric (*nasīb*) was also adopted from Ar. In strophic poems (*muwashshah*), love may be the theme of the whole poem or an intro. to panegyric. In the 11th c. the (usually strophic) epithalamium appears; though intended for the wedding feast, such poems are sometimes unexpectedly erotic. They were composed in the Middle East until the present century.

Liturgical poets in Spain and later in other Mediterranean countries added Ar. love themes to those deriving from the *Song of Songs.* Such poems enjoyed immense popularity, esp. in the Ottoman Empire, where, from the 16th c., mysticism inspired a rich poetic lit. In medieval Castile, Todros ben Judah Abulafia (1247–ca. 1298) continued the trads. of Ar. courtly l. p., but also wrote hedonistic verse as well as spiritual l. p. In Italy, Immanuel of Rome (ca. 1260–1328) introduced the sonnet into Heb.; though influenced by the *dolce stil nuovo,* he also contributed some truly salacious verse. In 17th-c. Italy, the brothers Immanuel (1618–1710) and Jacob (1615–67) Frances introduced baroque taste into Heb. l. p.

Except for Mica Joseph Lebensohn (1828–52), who was influenced by the Ger. romantics, the first period of modern Heb. lit. (*Haskala,* 1781–1881) was nearly devoid of lyrical l. p. The "Love of Zion" movement (1870–1900) generally subordinated love to the Zion-allegory. Only with Ḥayyim Naḥman Bialik (1873–1934) did Heb. poetry free itself from Haskala rhet. and sentimentality. Yet even his l. p. rejects erotic union, the beloved being portrayed as madonna or whore. In the poetry of Nathan Alterman (1910–70), the most outstanding of the modernist poets who rebelled against Bialik's poetics, erotic union is seen only as a transcendental certainty after death, remaining tragically unachievable in the lovers' lifetime. Only with the political and linguistic normalization of the Israel period (1948–) did Heb. l. p. become deromanticized and normalized. Under

the influence of Eng. poetry, the poets of the "Generation of the State," esp. Yehuda Amichai (b. 1924), used a lower diction and dealt with real-life relationships, not only as an object of longing but also as a solution and a refuge for the individual.

S. Burnshaw et al., *The Mod. Heb. Poem Itself* (1965); *Penguin Book of Heb. Verse,* ed. T. Carmi (1981); *Wine, Women, and Death: Med. Heb. Poems on the Good Life* (1986), *The Gazelle: Med. Heb. Poems on God, Israel, and the Soul* (1990), both ed. R. P. Scheindlin. R.P.S.

D. *Indian.* The luxuriant dreamworld of erotic romance that first comes to mind when we think of Indian l. p. is but one aspect of the complex universe of love represented in the vast body of poetry that exists in various Indian langs. such as Sanskrit, Prakrit, Tamil, Hindi, and Bengali.

Indian poets realized the unique power and aesthetic potential of the emotions of passionate sexual love. The erotic mood that emerges from such love was expressed in the antithetical modes of "separation" and "consummation." To experience this mood in the interplay of its two modes was considered the height of aesthetic joy. In Indian poetry, an act of remembering is the focal technique for relating the antithetical modes of love-in-separation and love-in-consummation. This conception of love is dominant throughout the trad.: one finds it in verses scattered through the anthologies, as well as in longer poems and dramas.

The emotions of erotic love are close to the intensity of religious devotion. The poets Kālidāsa (5th c.) and Bhartṛhari (7th c.) address their poems to the god Śiva, the erotically potent divine ascetic. Erotic poems thus become religiously powerful. The lyric poems of medieval devotional lit., which exist in every Indian lang., give expression to a religion that is predominantly emotional. For example, the Bengali devotional cults focus chiefly on episodes in the early romantic life of Kṛṣṇa, as they are described, not in the *Mahābhārata,* but in medieval sectarian epics known as *purāṇas.*

In the devotional poetry addressed to Kṛṣṇa, the emphasis is on complete surrender to his love, accompanied by all the emotions of attachment. What makes Indian devotional poetry unique is the notion that the god may be enticed by the love of the devotee and reciprocate in passionate ways. The idea of incarnation makes it possible to think of the god in human ways and in human relationships.

Although the promise of Kṛṣṇa's embrace exists, what inspires the poets is not the satisfaction of love, but the passionate yearning of a lover separated from his beloved—a suffering of such intensity that it liberates the devotee from worldly concerns. Jayadeva's Sanskrit poem *Gītagovinda* and the devotional lyrics of *bhakti* poetry exemplify how completely religious, erotic, and aesthetic meaning is integrated in this popular trad. B.S.M.

E. *Chinese.* Chinese l. p. in the ancient elite trad., whether in the form of short lyrics expressing agonized passion for women or of longer *fu* celebrating the erotic pursuit of goddesses, was often read as political and moral allegory. But explicit love themes, heightened by sexual images of flowers, seasons, etc., are central to certain types of popular song such as *yuefu.* The personae of popular love songs are almost always female, singing the joys of spring and youth. Such erotic motifs were incorporated, with vividly described love scenes, into the elite Palace Style Poetry of the Six Dynasties (4th–6th c. A.D.). Prince Xiao Gang (503–51) wrote about seductive sleeping beauties and even gave detailed accounts of a catamite engaged in a homosexual act. Later in the Tang Dynasty, Li Shangyin's (ca. 813–58) l. p. fused allusive lang. with emotional suggestiveness, creating ambiguities in sharp contrast to the explicit Palace Style verse. In the Song Dynasty (960–1279), a new form of poetry, *ci,* noted for its romantic themes and association with courtesan culture, became the lyric of emotion *par excellence.* But the *ci* was viewed as a "dying genre" from ca. 1300 to ca. 1600. Late in the Ming Dynasty it was revived by the Yunjian School of *ci* under the impetus of the poet Chen Zilong (1608–47). Chen exchanged many poems with his lover, the woman poet and courtesan Liu Shih (1618–64). Their vivid and dramatic "alternating poems," framed as personal letters, tell the story of a passion felt by two equally talented authors. The idea that Cl. Chinese poetry does not deal with love is a myth. Love remains an important theme for both male and female poets since the 17th c., except that eroticism in poetry was usually interpreted by the Qing (1644–1911) readers as political allegory. The woman poet Wu Zao (fl. 1800) broke taboos in Chinese poetry: she celebrates intimate lesbian love and introduces sexual frankness into *ci* lyric. But in general, gay and lesbian love is not a poetic subject, though it appears frequently in prose fiction and memoirs from the late Ming on.

J. Scott, *Love and Protest* (1972); H. Frankel, *The Flowering Plum and the Palace Lady* (1976); *New Songs from a Jade Terrace,* tr. A. Birrell (1982); *Women Poets of China,* tr. K. Rexroth and L. Chung, rev. ed. (1982); K. S. Chang, *The Late-Ming Poet Ch'en Tzu-lung* (1991). K.S.C.

F. *Japanese.* Early Japanese versions are distinctive in involving violations of the incest taboo. Cl. Japanese poets found erotic potential in wetness, women's tangled hair, Spring, dreams, and color. Since at court men visited women after dark, lovers saw each other indistinctly or not at all until the relation lasted on. Physical description was replaced by poetic or other good taste as erotic attraction. Courtesans and prostitutes first appear with *haikai* (see RENGA), which also features male homosexuality. The female is not a poetic subject, but does appear in prose and erotic pictures. In the performing arts, cross-dressing is important.

Exceptions to these generalizations include the trad. of the passionate woman from ancient to modern times and eccentric individuals like the high cleric Ikkyū Sōjun. In general, however, even today eroticism is tolerated in private rather than public versions.—Nishiyama Sōin, *Sôin Topphyaku Koi no Haikai* (1671); Yosano Akiko, *Midaregami* (1904); R. H. Brower and E. Miner, *Japanese Court Poetry* (1961); E. Miner, *Japanese Linked Poetry* (1979); "Himerareta Bungaku," ed. S. Yamada, spec. iss. of *Koku Bungaku Kaishaku to Kanshō* (1983)—on erotic writing and art. E.M.

LYRIC.

I. ORIGIN AND DEFINITIONS
II. HISTORICAL DEVELOPMENTS
III. CONTEMPORARY DEVELOPMENTS

I. ORIGIN AND DEFINITIONS. L. is one of the three general categories of poetic lit., the others being narrative (or epic) and dramatic. Although the differentiating features between these arbitrary categories are sometimes moot, l. poetry may be said to retain most prominently the elements which evidence its origins in musical expression—singing, chanting, and recitation to musical accompaniment (see SONG). Though drama and epic may also have had their genesis in a spontaneously melodic expression which adapted itself to a ritual need and thus became formalized, music in dramatic and epic poetry was at best secondary to other elements, being mainly a mimetic or mnemonic device. In the case of l., however, the musical element is intrinsic to the work intellectually as well as aesthetically: it becomes the focal point for the poet's perceptions as they are given a verbalized form to convey emotional and rational values. The primary importance of the musical element is indicated in the many generic terms which various cultures have used to designate nonnarrative and nondramatic poetry: the Eng. "l.," derived from the Gr. *lyra*, a musical instrument; the Cl. Gr. *melic*, or *mele* (air, melody); the Ch. *shi* or *ci* (word song).

To speak of the "musical" qualities of l. poetry is not to say that such poetry is written always to be sung. Neither does the appellation of "musical" indicate that l. possesses such attributes as pitch, harmony, syncopation, counterpoint, and other structural characteristics of a tonal, musical line or sequence—though such terms have often been (loosely) applied. To define the quality of lyricism in this way is to limit a l. to the manner of its presentation or to its architectonic aspects. This is largely the approach which Cl. critics have taken in their treatment of l. poetry. On the other hand, equating poetic lyricism with the nonarchitectural or "emotional" qualities of music is even less profitable, because it leads to such question-begging definitions of the l. as "the essence of poetry," "pure poetry," or, most vaguely, "poetry." To declare that "the characteristic of the l. is that it is the product of the pure poetic energy unassociated with other energies, and that l. and poetry are synonymous terms" (Drinkwater) is as extreme a definition of lyricism as the converse claim that a passage is lyrical simply because it possesses "the quality of metrical construction or architecture" (Murray). Both are extreme.

Most of the confusion in the modern critical usage of "l." (i.e. usage after 1550) is due to an overextension of the term to cover a body of poetic writing that has radically altered its nature over the centuries of its devel. The first critical use of the word *mele* by the Greeks was for the purpose of broadly distinguishing between various nonnarrative and nondramatic types of poetry: the melic poem was intended to be sung to musical accompaniment, in contrast to the iambic and elegiac poems, which were chanted. The first general use of "l." to characterize a selection of poetic lit. encompassing several genres did not come until the Alexandrian period, when "l." became a generic term for any poem which was composed to be sung; it was this meaning which "l." largely retained until the Ren. The preoccupation of pre-Ren. critics with the metrics of melic or l. poetry was entirely appropriate to the principle upon which the category was established.

But with the Ren., poets began suiting their work to a visual as well as an auditory medium; even while such critics as Minturno, Scaliger, Sidney, and Puttenham were formulating their discussions of l. poetry, the l. was becoming something quite different from the Cl. melic poem. No longer a performing bard, scop, skald, or troubadour, the poet ceased to "compose" his or her poem for musical presentation but instead "wrote" it for a collection of readers. The l. poem, nominal successor to a well-established poetic method, inherited and employed specific themes, meters, attitudes, images, and myths; but in adapting itself to a new means of presentation, the l. found itself bereft of the very element which had been the foundation of its lyricism—music.

At the time the l. was undergoing this crucial metamorphosis, 15th- and 16th-c. critics chose either to ignore the genre or to treat it by the same quantitative or metrical criteria as the classicists had done. Until the end of the 17th c., therefore, critics failed to distinguish between the true or melodic l., such as the "songs" of Shakespeare, Campion, and Dryden, and the nonmusical, verbal ls. of Donne, Marvell, and Waller. Both the straightforward, plain song-poem and the more abstrusely phrased print-poem were called "l." The neoclassical concern in 18th-c. France and England with the tragic and epic genres was sufficiently overwhelming that the l. receded as a topic of critical discussion; and when the romantic movement came, with its championing of lyrical modes, terminological confusion continued in the equation of "l." with "poetry" by Wordsworth, Goethe, Coleridge, Poe, and other poets and theo-

rists. The 19th-c. devel. of scientific methodology, with consequent insistence on accuracy of terms and precision of generic distinctions, translated itself in lit. crit. into a concern with the intrinsic and characteristic nature of the l. The definitions by Drinkwater and Murray given above represent the overinclusive and overexclusive criteria which resulted from this concern; and critical attempts to re-establish the melodic or musical substance of l. poetry were a third, and equally unsuccessful, method of dealing with the paradoxical nature of a "musical" poetry which was no longer literally "melodic." Such, in greatly simplified lines, is the history of the verbal ambiguity by which post-Ren. critics concealed their basic failure to define both the precise aspects of the l. genre which distinguish it from narrative and dramatic poetry and those which justify the inclusion under one term of all the disparate types of poem commonly called "lyrical."

Critical attempts to define l. poetry by reference to its secondary (i.e. nonmusical) qualities have suffered by being descriptive of various historical groupings of ls. rather than definitive of the category as a whole. Among the best known and most often cited proscriptions regarding the l. are that it must (1) be brief (Poe); (2) "be one, the parts of which mutually support and explain each other, all in their proportion harmonizing with, and supporting the purpose and known influence of metrical arrangement" (Coleridge); (3) be "the spontaneous overflow of powerful feelings" (Wordsworth); (4) be an intensely subjective and personal expression (Hegel); (5) be an "inverted action of mind upon will" (Schopenhauer); or (6) be "the utterance that is overheard" (Mill).

Though the attributes of brevity, metrical coherence, subjectivity, passion, sensuality, and particularity of image are frequently ascribed to the l., there are schools of poetry obviously l. which are not susceptible to such criteria. Milton's *L'Allegro* and *Il Penseroso*, as well as the most famous Eng. elegies, are "brief" in only the most relative sense. Much of the *vers libre* of the present age contradicts the rule of metrical coherence. Imagist ls. are hardly "empassioned" in the ordinary sense of the word. The "lucubrations" of the metaphysicals are something less than sensual in the romantic sense of the term. The problem of subjectivity must always plague the critic of the Elizabethan love l. And, finally, the common artistic admission that the universal can be expressed best, and perhaps solely, through the particular image largely invalidates any distinction between the l. and nonlyric on a metaphoric or thematic basis.

The irreducible denominator of all l. poetry must, therefore, comprise those elements which it shares with the musical forms that produced it. Although l. poetry is not music, it is representative of music in its sound patterns, basing its meter and rhyme on the regular linear measure of the song; or, more remotely, it employs cadence and consonance to approximate the tonal variation of a chant or intonation. Thus the l. retains structural or substantive evidence of its melodic origins, and this factor serves as the categorical principle of poetic lyricism.

In the 20th c., critics, predicating the musical essence of l. as its vital characteristic, have come close to formulating an exact and inclusive definition of l.: "Words build into their poetic meaning by building into sound . . . sound in composition: music" (R. P. Blackmur). "A poet does not compose *in order to* make of lang. delightful and exciting music; he composes a delightful and exciting music in lang. *in order to* make what he has to say peculiarly efficacious in our minds" (Lascelles Abercrombie). Lyrical poetry is "the form wherein the artist presents his image in immediate relation to himself" (James Joyce). "Hence in lyrical poetry what is conveyed is not mere emotion, but the imaginative prehension of emotional states" (Herbert Read). It is "an internal mimesis of sound and imagery" (Northrop Frye). Thus, in modern critical usage it may be said that "l." is a general, categorical, and nominal term, whereas in the pre-Ren. sense it was specific, generic, and descriptive. In its modern meaning, a l. is a type of poetry which is mechanically representational of a musical architecture and which is thematically representational of the poet's sensibility as evidenced in a fusion of conception and image. In its older and more restricted sense, a l. was simply a poem written to be sung; this meaning is preserved in the modern colloquialism of referring to the words of a song as its "ls."

II. HISTORICAL DEVELOPMENTS. However useful definitions of the l. may be, they cannot indicate the great flexibility of technique and range of subjects which have helped this category to comprise the preponderance of poetic lit. There are literally dozens of l. genres, ranging from the ancient *partheneia* to modern *vers libre*, leaving no topic, whether a cicada or a locomotive, untreated. Though it is manifestly impossible to say everything about the historical devel. of the l. in short, certain general facts prove interesting as pieces in an evolving pattern of theories about and treatment of the lyrical mode between various ages, cultures, and individuals. The l. is as old as recorded lit., and its history is that of human experience at its most animated.

The Ancient Middle East and the Western Tradition. It is reasonable to suppose that the first "lyrical" poems came into existence when human beings discovered the pleasure that arises from combining words in a coherent, meaningful sequence with the almost physical process of uttering rhythmical and tonal sounds to convey feelings. Both the instinctive human tendency to hum or intone as an expression of mood and the socialization of this tendency in primitive cultures by the chanting or singing of nonsense syllables in tribal rites are well-documented. At that remote point in

time when the syllables ceased to be nonsense and took on meaning, the first l. was composed, though in what Neanderthal or Cro-Magnon cave this took place, no one will ever know. This impossibility, however, has scarcely impeded speculation about the folk origins of lit., ranging from those of Herder to A. B. Lord and Andrew Welsh. The earliest recorded evidence of l. poetry suggests that such compositions emerged from ritual activity accompanying religious ceremonies and were expressive of mystical experience. Scholars have found evidence to support a theory of the religious derivation of poetry in general and the l. in particular in such lits. as Eskimo, American Indian, and Polynesian.

1. *Sumerian.* The earliest ls. in the Middle East were composed in Sumer (modern Iraq). The Old or Cl. Sumerian Period (2300–2000 B.C.) produced the divine invocation; the New Period (2000–1700 B.C.) added further lyrical riches in the form of proverbs, hymns, lamentations, incantations, and lovesongs. Begun in a much earlier oral trad., the ls. were set down ca. 2500 years ago. Preserved on clay tablets in cuneiform, Sumerian ls. were precursors of the more famous Heb. songpoems. One of the earliest of all lovesongs is the Sumerian "To the Royal Bridegroom" (ca. 2025 B.C.):

> Bridegroom, dear to my heart,
> Goodly is your beauty, honeysweet,
> Lion, dear to my heart,
> Goodly is your beauty, honeysweet.

The poetic devices of these lines—parallelism, inversion, repetition—characterize Sumerian l. poetry and the trads. derived from it.

2. *Assyro-Babylonian* poetry includes such l. types as hymns and prayers: to Ishtar, to Sin the Moon God, to Shamash the Sun God, to Gula the goddess of healing, and other deities. Lyrical elements are also present in the *Gilgamesh* epic. Samples of folk poetry preserved in texts of magic or healing also exhibit a characteristic parallelism. The Hittites, who spoke an Indo-European lang. and lived in modern Turkey, borrowed cuneiform from the Babylonians, their neighbors in Mesopotamia. The earliest of their known ls. from the Old Kingdom (ca. 1700–1600 B.C.) is a short song. The New Kingdom (ca. 1400–1200) produced an extant hymn to Istanu the Sun-God and other ls. embedded in epics.

3. *Egyptian.* The most complete written evidence of early l. activity is Egyptian: the Pyramid Texts (ca. 2600 B.C.) include specimens of the funeral song (elegy—q.v.), song of praise to the king (ode—q.v.), and invocation to the gods (hymn—q.v.); tomb inscriptions from the same period include work songs of shepherds, fishers, and chair bearers. Also among the earliest l. writings of the Old Kingdom are the dialogue, proverb, and lament or complaint (qq.v.). The Middle Kingdom (ca. 1950–1660 B.C.) proliferated these l. types, esp. the hymn, and added victory song. Prophecies, proverbs, and encomia were popular. Works from the New Kingdom (ca. 1570–1070 B.C.) incl. songs of revelry, love songs, and the epitaph. "Harpers' songs" contained elements of the *carpe diem* theme. Although relatively uncomplex, the Egyptian l. contained in nascent form many elements which were to become characteristic of later l. poetry. The lines seem to have some form of free rhythm without meter. Alliteration and parallelism were frequently used, as was paronomasia. Irony and paradox were present in a primitive form; and these first of all lyrics were already treating such subjects as death, piety, love, loneliness, jealousy, martial prowess, and joy. Furthermore, the personal tone of the l., though not ubiquitous, was apparent in such poems as those enclosed in *The Dispute with His Soul of One Who Is Tired of Life.* Remains of other ancient lits. do not show much advancement over the Egyptian.

4. *Hebrew.* The most complete of the ancient bodies of l. poetry is the Heb., which, while owing something to Egyptian and Babylonian sources, nevertheless marked a distinct advance in l. technique. These ls., well known to modern readers because of their religious associations and important for their effect on the patristic lyricists of the Middle Ages, are among the most strikingly beautiful ever written. Though textual evidence indicates that some Heb. l. poetry was written as early as the 10th c. B.C. (notably the *Song of Deborah*), many poems were of much later date; and the earliest Jewish lit. crit. dealing with the l. was as late as the 1st c. A.D. Philo Judaeus (ca. 20 B.C.–A.D. 50) claimed Egyptian origin for some Heb. l. techniques by declaring that Moses was taught "the whole theory of rhythm, harmony, and meter" by the Egyptians. Flavius Josephus (ca. A.D. 37–95), discussing the hymn of Moses in Exodus 15:1–2 said that it was written in hexameters, and the hymns and songs of David in other Cl. meters. Later discussions of Heb. meters were carried on by Origen, Eusebius, and Jerome; Gr. metric nomenclature was overlaid on Heb. prosody not because it accurately fit but because Heb. prosody was poorly understood while Gr. metrics was prestigious: indeed, the ancient Heb. l. meters are only partially understood even today (Gray). It is known, however, that the ls. were accompanied by such instruments as the harp, sackbut, and cymbals; and suggestions of the manner in which hymns, elegies, songs of joy, and victory songs were composed and performed may be found in 1 Samuel 16:23 and 2 Samuel 1:17–27, 6:5, and 15–16.

The ancient Heb. poets were proficient in the use of paronomasia, parallelism, and alliteration (qq.v.), perfecting and using these devices in a variety of ways. Parallelism is obvious in such ls. as Psalm 19 ("The heavens declare the glory of God, and the firmament sheweth his handiwork") and Proverbs 21:17 ("He that loveth pleasure shall be a poor man: he that loveth wine and oil shall not

be rich"); but it is also more subtly used, as in Jeremiah 6:24 ("Anguish hath taken hold of us, and pain, as of a woman in travail"). The use of tropes is highly developed, with simile and metaphor predominating; the apostrophe and hyperbole increase the personal tone of the l. far beyond the Egyptian. Many of the ls. seem intensely subjective, e.g. Psalm 69, but even these poems reflect what Frye has called "the sense of an external and social discipline." Yet the personal tone remains and is essential to the lyricism of such passages as Isaiah 5:1, Psalm 137, and 2 Samuel 1:19.

The Heb. lyricists developed a number of types and subtypes of the l. genre, which are classified by method of performance, source of imagery, and subject matter. These include the psalm (q.v.; from Gr. *psallein*, "to pull upon a stringed instrument"); the pastoral (q.v.), which draws heavily upon the agrarian background of Heb. culture; and the vision or apocalyptic prophecy, which employs the indirection of the trope. Other types incl. the proverb, the epigram (qq.v.), and similar forms of "wisdom" lit.; the descriptive love l.; the triumph; various sorts of threnody; the lament; panegyrics of different kinds; and even a lyrical dialogue (or "drama") in the Book of Job. Some overlapping of types is obvious (the triumph was frequently a panegyric on the hero), but the ambiguity is an historical one and terminological distinctions have yet to be drawn.

5. *Greek* l., like Egyptian and Hebrew, had its origins in religious activity; the first songs were probably composed to suit an occasion of celebration or mourning. Gr. ls. were chanted, sung, or sung and danced; each mode is traceable to some form of religious practice. The dithyramb, for example, may have been composed to commemorate the death of some primitive vegetable god or the birth of Dionysus; in any case, it was originally sung to the accompaniment of the flute playing a melody in the Phrygian mode, which the Greeks considered the most emotional. In time, the dithyramb took on a more particular form, with formalized dance steps corresponding to passages in the text: these rhythmical and thematic patterns may have been the prototype of the fully developed ode, or song of celebration, with its divisions of strophe, antistrophe, and epode as written by Pindar, Sophocles, and others. Similar tracings of the devel. of other lyrical modes in Greece from the Heroic Age to the Homeric to the Periclean have been attempted.

The essential element of the Gr. l. was its meter, which was of two kinds: the stichic, that spoken or recited; and the melic, that suited for singing or singing and dancing. Stichic meters were well demarcated lines of equal length and repetitive rhythm that can be divided into units (feet, metra). Melic meters were composed of phrases of varying length or movement; each phrase, called a colon could be combined into a unit rhythmically complete or rounded, the *periodos*. Some cola are rhythmically repetitious in themselves and may be broken into dimeter or trimeter; but in the melic poems, it was the period or stanza that constituted the l. unit (Dale). Melic meters were obviously subject to wide adaptation, and most of the best known Gr. l. meters are named for the poets who developed and customarily used them: the Sapphic (q.v), Alcaic, Anacreontic, and Pindaric. The earliest Gr. ls. may emerge from the mists of oral tradition, but even in Homer and Hesiod there is an artistic concern with the l. mood and subjects. In the *Iliad*, for instance, there are such embryonic ls. as Helen's laments, Achilles' speech at the death of Patroklos, and the elegiac speeches at Hector's funeral. But it is the hymn which is perhaps the first developed of the definably l. genres, being composed in significant numbers before 700 B.C. The Homeric Hymns from this period indicate the religious nature of the first ls.: they are addressed to Artemis, Dionysus, Heracles, Helios, *et al.*, and the pattern of some became a distinct type of l. hymn (e.g. the "paean," a hymn to Apollo). The hymns employ devices appropriate to the apostrophe but are not very expansive in their tropes, chiefly using the attributive epithet, as in this hymn to Hera: "I sing of golden-throned Hera, whom Rhea bare. Queen of the immortals is she, surpassing all in beauty" (tr. Evelyn-White).

The Homeric epigrams are attributed to this period also, thereby setting up an archetype for the later iambics. The period from 700–500 B.C. saw the rise of elegiac and political verse, written by Solon among others, and the personal lampoon in iambics, by Archilochus, Hipponax, and Simonides of Amorgos. After 660 B.C., melic poetry developed, primarily in two strains: the Aeolian, or personal, ls. written at Lesbos by Sappho and Alcaeus; and the Dorian, or objective, by Alcman, Arion, Stesichorus, and Ibycus. This group of ls. may also be categorized by method of performance as solo or choral, but the dividing line is not sharp. Although the ancient distinctions of melic poetry on the basis of meters may indeed indicate separate categories, the modern definition of the l. would be hard pressed to differentiate between a number of poems by Alcman, Ibycus, and Sappho.

The 5th c. in Greece produced some of the best of l. poets, Simonides, Pindar, and Bacchylides; it was then too that the l. found such magnificent expression in the choral odes of Sophocles, Aeschylus, and Euripides. Melic poetry became national in tone, with the Dorian mode prevailing; there was an abundance of such l. types as the hymn, paean, dithyramb, processional, dance song, triumph, ode, and dirge. Other popular genres were the *partheneia* (songs sung by virgins to flute accompaniment); *nomos* (war song); *kommos* (a dirge sung in Attic drama by an actor and the chorus alternately); *prosodion* (processional song of solemn thanksgiving); *hyporcheme* (dance song); *epinikion* (victory song); *threnos* (dirge);

wedding songs for men; and the *scolion* (banquet song accompanied by the lyre).

Gr. critics were less concerned with l., or melic, poetry than with tragedy and the epic; the few extant comments predicate the musical nature of the genre. Plato's denunciation of all poetry, esp. "representational" tragedy, included the melic, which Plato considered "untrue" or false in its depiction of reality. Stripped of musical coloring and laid bare as ideas, Plato says, the melic poems reveal the ignorance of the poet, which clothes itself in "rhythm, meter, and harmony." (*Republic* 10.4). Aristotle, however (*Poetics* 1–4), remarked the absence of a generic term which might denote such nonepic and nondramatic kinds of poetry as the works in iambic, elegiac, and similar meters, which imitated "by means of lang. alone," as contrasted with the melic poems, which used rhythm, tune, and meter "all employed in combination." This statement indicates the existence of poetry, l. in the modern sense, which was not melic in the Gr. sense; but the Alexandrian use of "l." to indicate such disparate types as the dithyramb, iambics, elegies, and Sapphics, while a broad attempt to repair the deficiency noted by Aristotle, was inexact and only resulted in confusion.

6. *Latin*. Roman critical remarks on the l. would indicate that the term was used in the sense of melic poetry or poetry sung to the lyre. Horace shows a belief that l. poetry is less substantial in content than epic poetry, being the *jocosa lyra*, a view with which Quintilian concurred, though for him the ode might be worthy of more significant themes. To Horace's mind, the "dainty measures" were suited to "the work of celebrating gods and heroes, the champion boxer, the victorious steed, the fond desire of lovers, and the cup that banishes care"; they included the iambic and elegiac distich. These general criteria for form and content were adopted by most Lat. commentators following Horace—Ovid, Petronius, Juvenal, Pliny the Younger—so that, as for l. theory, the Romans were advanced but little beyond the Greeks.

In practice, Roman poets tended to imitate the Alexandrian l. poets, who composed works primarily meant to be read rather than performed. Hadas has pointed out that this practice tended to produce ls. more enigmatic and allusive than earlier "sung" poems had been; and it may be generally observed that Roman l. poets are more notable as formulators of a "personal" or subjective poetry than are the Greeks. While Roman poets modeled their poems on the hymns of Callimachus, the Idyls of Theocritus, the epigrams of Anacreon, and the elegiac laments of Bion and Moschus, they adapted the l. to produce a more subjective or autobiographical utterance. Conventional and minor Roman lyricists were content with the school of "fastidious elegance" which kept them copying the Greeks, and which Catullus mocked, but the Roman genius emphasized particularized experiences—Propertius in his observations, Catullus in his amours, Virgil in his rustic pleasures, Ovid in the sorrows of his exile, Tibullus in his love pangs, Martial and Juvenal in their private asperities. The private insight, the subjective focusing of experience is more keenly apparent in Roman ls. than in other ancient works: e.g. in Ovid's *Tristia* 1.8 ("To their sources shall deep rivers flow"); Martial's *Epigrams* 1.8 ("Thou hast a name that bespeaks the season of the budding year, when Attic bees lay waste the brief-lived spring"); Catullus' *To Hortalus* 65 ("Though I am worn out with constant grief and sorrow calls me away"); and Tibullus, *To Delia* 1.2 ("More wine; let the liquor master these unwonted pains"). Coincidentally, there are many more "occasional" ls. among the Roman poets which celebrate private rather than public festivals, with a greater proportion of such genres as the prothalamium and epithalamium (wedding songs), the *vale* (farewell), the epigram, the satire, and the epistle.

The strict approach to meter which typifies the l. writers of the pre-Augustan and Augustan periods of Roman lit. began to weaken by the middle of the 1st c. A.D., and greater flexibility of form resulted. The rigid preoccupation of the Horatian and Virgilian schools with the exact meters dictated by the Gr.-derived system of quantitative verse was probably an attempt to substitute precision of metrics for the abandoned melodies of the true l. In any case, the ls. of Petronius, unlike the formal measures of the Statian odes, are experimental in form; and during the 2d c., Lat. l. moves toward a nonquantitative or accentual form of verse. The earliest extant examples appear in Christian hymns. In the 3d c., the completely new principle of rhyme can be found in the verses of Commodian; it was then that the principles which were to guide not only Med. Lat. verse but all subsequent modern vernaculars—accent, isosyllabism, and rhyme—were established.

Like the Hebrews and the Greeks, the patristic critics chose to discuss long established forms and practices rather than treating contemp. ones. Patristic l. crit. throws little light on the practice of the times. Eusebius, Jerome, and Origen were concerned with analyzing Heb. l. modes in the categories and terms of Gr. metrics; and the anthologer Isidore of Seville discussed Heb. and Gr. meters in conventional fashion, noting the musical element of l. poetry, but still failed to make any distinctions among the various genres.

Not suprisingly, the first ls. of the Church were hymns patterned on the Heb. Psalter and the Gr. hymns. The earliest were those of St. Hilary of Poitiers, who probably used meter for its mnemonic effectiveness, and who employed the *versus quadratus* or trochaic tetrameter catalectic, subsequently a favorite meter of Prudentius, Fortunatus, and Thomas Aquinas, and the basis for several of the medieval sequences (Beare). St. Ambrose developed the use of iambic dimeters grouped in quatrains (the "Ambrosian" stanza),

which became the basis for all modern stanzas of four 4-beat lines; St. Jerome made Lat. more flexible as the lang. of poetry; and Augustine wrote a didactic psalm against the Donatist heretics which may contain the embryo of all Med. Lat. versification. The Rule of St. Benedict (6th c.) required hymns to be sung at all the canonical hours, an edict which spurred on numerous lyricists. Once more, ls. were composed to be sung or chanted; the indissoluble rapport between the Lat. words and meters and the melodic line must not be forgotten. The sequence or trope depended on the repetition of phrases both verbal and musical; and the involved meters as well as the simple rhymes of the hymns which Abelard wrote for Heloise and her nuns to sing both derive from their avowedly musical nature:

Christiani, plaudite,
(Resurrexit Dominus)
Victo mortis principe
Christus imperat.
Victori occurite,
Qui nos liberat.

The Church hymns of the 12th and 13th cs. are among the most perfect ls. ever produced in the history of liturgical lit.: the *Stabat mater* and *Dies irae* must be included in any list of the world's great ls., and there are numerous other examples of accomplished l. art: sequences, *cantiones*, nativities, and a variety of hymns. The importance of patristic songs cannot be exaggerated in the history of the l.

7. *European Vernaculars.* In Europe, the 12th and 13th cs. saw the growth in popularity of the wandering minstrel: the quasi-ecclesiastical goliards who wrote secular songs in Lat.; the *trouvères* (q.v.) in northern France and troubadours (q.v.) in the south; the Minnesinger in Germany. The l. was widely sung, or sung and danced. The troubadours, composing in the vernacular, produced the *chanso* (song, often of love), *sirventes* (topical songs of satire, eulogy, or personal comment), the *planh* (complaint), *tenso* (debate), *pastorela* and *pastourelle* (pastoral episode), *alba* or *aube* (dawn song), and the *balada* and *dansa* (dance songs). Much study has been devoted to the differences between the *chansons courtois* and *chansons populaire*, the *caroles* and *rondets* for dancing, and the chansons designed for singing only, the *chansons de toile* (work songs), and their subspecies. Medieval ls. survive in abundance, exerting a special charm for the modern reader in their mixture of naïveté and sophistication. They range from slapstick "macaronic" songs (multiple langs.) to the simple understatement of "Foweles in the frith"; from the direct but delightful "Sumer is icumen in" to Chaucer's complex rondeau, "Now welcom somer, with thy sonne softe," and ballade, "Hyd, Absalon, thy gilte tresses clere." Although this period produced such masters of the written "art-lyric" as Chaucer, Bertrand de Born, Chrétien de Troyes, Walther von der Vogelweide, Rutebeuf, Pierre Vidal, and Sordello, l. poetry was still a thing of the people, composed to be sung and enjoyed. The melodic element of the l. genre in the medieval period has not been equaled since the dawn of the Ren.

8. *Renaissance.* After 1400 the l. and music became increasingly disassociated, as evidenced by the rise of such primarily melodic forms as the madrigal, glee, catch, and round, which subordinated the words to the music. Despite the efforts of later writers who were primarily poets and not composers, such as Swinburne, Hopkins, and Yeats, the l. since the Ren. has remained a verbal rather than a musical discipline, and the traces of its melodic origin have become largely vestigial. The influence of the Lat. metrists on the It. and Fr. lyrical theorists of the Ren. may have helped to produce the latter-day emphasis on meter as a substitute for melody. In any event, Ren. lyricists, writing for an aristocratic audience (of readers), adapted their forms to a different medium, and the l. suffered a sea change after the 15th c.

Ren. ls., diffused as they are through several countries and centuries, nevertheless share characteristics which evidence their common origin in the Occitan ls. of the troubadours. In Spain, l. poets fused their Mozarabic and Occitan trads. to perfect some older forms (the *cantiga* and possibly the *cossante*) and to develop some new ones (the *bacarola*, *bailada*, and others). After its inception at the Sicilian court of Frederick II, the *sonetto* (see SONNET) rose to full flower in Dante's *Vita nuova* and Petrarch's sonnets to Laura. Petrarch's *Canzoniere*, the prototype of It. l. poetry, contained sonnets, sestinas, ballatas, and even a few madrigals. In his sonnets, Petrarch struck the passionate thematic chords that were to echo for centuries in the ls. of countless imitators in Italy, France, and England. Despite excellent *canzoni* written by Ariosto and others, it is Petrarch whose name remains synonymous with Ren. It. artverse. More popular vernacular verse—e.g. the *strambotto* and *rispetto*—was also composed in Italy in great quantity.

What Petrarch was to l. poetry in Italy, Ronsard was to the poetry of Ren. France. The leader of the *Pléiade*, a stellar group of poets incl. Joachim du Bellay, Ronsard pub. his version of the sonnets to Laura in *Les Amours* in 1552, and the *Sonnets pour Hélène* in 1578. Ronsard also composed odes, mythological and philosophical *Hymnes*, and elegiacs and pastorals. Scorning the older forms of the rondeau and rondelle, Ronsard explored the whole gamut of lyric images and emotions in his sonnet cycles. Though the earlier sonnets contained frank Petrarchan notes (e.g. "Cent et cent fois penser un penser mesme"), the later works were perfectly Ronsard's own ("Adieu, cruelle, adieu, je te suis ennuyeux").

In England, the publication of Tottel's Miscellany in 1557 marked the beginning of the most lyrical of England's poetic eras. A collection of

songs, sonnets, and other kinds of verse, the Miscellany evidenced the musicality of earlier Eng. poetry and established a form, the miscellany (see ANTHOLOGY), to be imitated repeatedly, from *The Phoenix Nest* (1593) and *Englands Helicon* (1600) to the eventual broadside ballads and Bishop Percy's *Reliques* (1765). Wyatt and Surrey were among the first in England to test the possibilities of the sonnet's thematic and metrical subtleties; and dozens of Eng. sonneteers—Sidney, Daniel, Spenser, Shakespeare *et al.*—pub. lengthy sonnet sequences (q.v.) more or less directly patterned on Petrarch's. Certain old forms of the lyric were redeveloped in England, e.g. the prothalamium and epithalamium; and adulation for Horace and other Lat. l. poets created a widespread vogue for the ode (q.v.). The song (q.v.) also remained popular, both in its melodic and ballad forms, being written by such poets as Sidney, Shakespeare, Campion, and Jonson.

In general, it may be said of Ren. l. poetry that it was a succinct example of the philosophy of humanism. The l.'s preoccupation with the subjective self dovetailed neatly with the humanistic interest in the varied forms of human emotion; and the new geographical concerns of the Ren. supplemented the pastoralism of the traditional l. to produce an imagery that fused scientific and poetic perspectives. The effect of printing on the l. poets was also central; and though the shifting nature of l. poetry was not apparent to those Ren. critics who discussed that "divine" art—Minturno, Scaliger, Tasso, Sebillet, Gascoigne, Sidney—the changes were fundamental for poetic practice as well as theory.

9. *Modern.* Although the past 300 years in the history of the Western l. may be divided into certain chronological "periods" (i.e. Ren., Restoration, Augustan, *Fin de siècle*) or certain distinctive "movements" (metaphysical, neoclassical, romantic, symbolist, modernist), these terms reveal little about the true nature of l. theory and practice. Far more accurate is the designation of all l. poetry after 1600 as "modern." The range of this body of ls., from the most objective or "external" to the most subjective or "internal," may be included in three chief l. types: (a.) the Lyric of Vision or Emblem, (b.) the Lyric of Thought or Idea, and (c.) the Lyric of Emotion or Feeling.

a. The L. of Vision or Emblem, although it has its antecedents in Cl. Gr., Anglo-Saxon, and Ch. poetry, is nevertheless fundamentally the product of the Age of Type. It is this sort of visual poetry (q.v.) that Ezra Pound called "Ideograms" and Apollinaire called "calligrammes" (qq.v.). This is the most externalized kind of l., utilizing the pictorial element of print to represent the object or concept treated in the poem. The visual l. therefore exists in itself, without reference to a private sensibility, whether of poet or reader. The first use of the visual in the modern l. came in the Ren., when poets printed poems in the shape of circles, spires, and pillars (see PATTERN POETRY). Later, George Herbert showed wings, altars, and floor patterns in poems on these subjects; and the prevalence of pictorial ls. among Fr. and Eng. poetasters of the 17th and 18th cs. drew Dryden's scorn in *MacFlecknoe* and Addison's laughter in *The Spectator.* In the later 19th c. the symbolists were influenced by this practice; in the 20th c. the imagists explored it under Ch. influence, and more recent Am. poets, under the influence of dada, incl. Amy Lowell, H. D., William Carlos Williams, e e cummings, May Swenson, and John Hollander.

b. Somewhat more personal but still objective in tone and method is the L. of Thought or Idea, which may be divided into the expository or informative and the didactic or persuasive (even though some critics believe "l." and "didactic" to be contradictory terms). This school of lyricists is classically oriented, believing with Horace that poetry must be *utile* as well as *dulce,* and consequently emphasizing musicality of form to balance prosaic content. Expository L. writers incl. Boileau, Dryden, Cowper, Schiller, Tennyson, and such modern poets as Rainer Maria Rilke in his early descriptive works, Juan Ramón Jiménez, Jorge Guillén, Rafael Alberti, St. John Perse, T. S. Eliot, Robert Lowell, and Elizabeth Bishop. The preoccupation of 19th- and 20th-c. poets with sound and verseform has produced *vers libre* (q.v.), which is as obvious an effort to accompany poetic statement with musical techniques as are the heroic couplets (q.v.) of the neoclassical poetry of statement.

The didactic or persuasive l. includes the allegorical, satiric, exhortatory, and vituperative species. L. allegory (q.v.) is apparent in the animal myths of La Fontaine, Herrick's use of Cupid, Mandeville's bees, Heine's Atta Troll, Arnold's merman, Davidson's dancers in the house of death, and Frost's departmental ants. The satiric l. includes, of course, the l. parody, such as Lewis Carroll's burlesques of Wordsworth and Swinburne's mockery of Tennyson; but it also includes directly satiric verse: Donne's verses on women, Majakovskij's *Bedbug,* Bertolt Brecht's acrid observations on romantic love (see SATIRE). The exhortatory l. is often patriotic or moralistic, as in the Elizabethan panegyrics on England, Burns's call to the Scots, Gabriel D'Annunzio's fervent championing of life and freedom, or Kipling's tributes to Britannia. The vituperative l. aims its darts everywhere: against critics (Pope's *Epistles*), convention (Rimbaud's *Illuminations*), war (Owen and Sassoon), poverty and suffering (Antonio Machado's *Del Camino*), a parent (Sylvia Plath's "Daddy"), or the world in general (Allen Ginsberg's *Howl*). Pound's *Cantos* seem to combine all of the subtypes—allegory, satire, exhortation, and vituperation—uniquely.

c. The most subjective or "internal" strain of modern l. poetry is the L. of Emotion or Feeling. It is this type which has in the modern world become synonymous with "poetry" through the

agency of romantic poetics. The L. of Emotion comprises three major groups: (i.) the sensual l., (ii.) the "imaginative" l. (which intellectualizes emotional states), and (iii.) the mystical l. The mystical l. is antipodal to the l. of emblem: these two varieties of "vision" l., one literal and the other metaphorical, mark the extreme limits of objectivity and subjectivity.

(i.) The sensual l. enjoys an unbroken continuity from the 16th c. to the 20th in the sonnets of Ronsard and the *Pléiade*, the love poetry of the Elizabethans and metaphysicals, the erotica of the Restoration and 18th c., the synaesthetic images of Keats and the romantics, the symbolist glorification of the self and its peculiar sensations, the neurotic sensualism of the Nineties, and the "new" sensualism of the Lost Generation, the existentialists, and the Beat Generation. Ranging in theme from *carpe diem* to *memento mori*, the sensual trad. is sustained in differing forms by Shakespeare, Donne, Collins, Herder, Heine, Baudelaire, Mallarmé, Whitman, D'Annunzio, Millay, and Dylan Thomas. (ii.) The "imaginative" or "intellectualized" L. of Emotion furnishes a host of examples. The Ger. lyricists provide a large number (notably Goethe, Schiller, Rilke, Hauptmann, and George). The "verbalized feelings" of the Eng. romantics and the Fr. symbolists have their modern counterparts in the ls. of Apollinaire and Valéry as well as in the poetry of Garcia Lorca and the It. hermetics Ungaretti and Montale. Many of the ls. of Puškin and Pasternak fall into this category. Writers of British ls. of this type are Auden, MacNeice, Empson, Spender, and Larkin; Americans include Emerson, Dickinson, Frost, Jeffers, MacLeish, Stevens, Moore, Wilbur, and Hecht. (iii.) Finally, the poetry of mysticism is significant in modern l. hist., perhaps as an attempt to find some substitute for the Gr. myths which provoked the Cl. l., or for the Christian mythography which stimulated the med. l. Foremost among the mystical lyricists are Herbert, Vaughan, Traherne, Smart, Blake, Hopkins, Thompson, Baudelaire, Claudel, Yeats, and Rilke.

III. CONTEMPORARY DEVELOPMENTS. The popularity of l. poetry in the 20th c. has increased with its employment in the causes of self-expression, feminism, and racial and social equality. The abundance of l. writing has been accompanied by a plethora of critical analyses and theories. The most influential critical approaches to the l. in the decades after World War II have been those of scholars and academic critics in America and Europe.

During the 1940s and '50s, the "New Critics" in America emphasized the formal integrity of the poem and its self-containment, diminishing the importance of authorial reference and reader response. Their emphasis on paradox and irony, lucubration, and the poem qua poem led to a critical neglect of lyricism. At the same time, in Europe the influence of Ferdinand de Saussure on words as signs and langs. as systems pointed the way to deeper analyses of poetic lang., such as those by Roman Jakobson, and to the deconstructionist approaches of Jacques Derrida.

When deconstruction was imported to America, critics began reexamining the l. mode. Finding unconvincing and inadequate the critical postulation that the speaker of the l. was the poet himself, or even a fictive persona, such critics as Jonathan Culler declared that "the fundamental aspect of l. writing . . . is to produce an apparently phenomenal world through the figure of voice." Paul de Man expanded this to say, "the principle of intelligibility, in l. poetry, depends on the phenomenalization of the poetic voice (which is) the aesthetic presence that determines the hermeneutics of the l." The I-Speaker of the l. thus ceases to be Shelley, Valéry, Li-Bo, or Bashō—or even a mask for or a Joycean image reflective of the poet. The speaker is a device for making the invisible visible. De Man replaces such standard terms as "personification" with "catachresis" in order to explain the function of the frequent trope of address ("apostrophe") in l. poetry. The poet-surrogate is replaced by the figurative voice, a mantic or shamanistic presence that makes the verbal world of the l. a visible world to the mind of the reader.

Thus, from its primordial form, the song as embodiment of emotion, the l. has been expanded and altered through the centuries until it has become one of the chief literary instruments which focus and evaluate the human condition. In flexibility, variety, and polish, it is perhaps the most proficient of the poetic genres. In the immediacy and keenness of its expression, it is certainly the most effective. These qualities have caused the 19th and 20th cs. to look upon the l. as largely their own, but l. poetry has belonged to all ages.

F. B. Gummere, *The Beginnings of Poetry* (1901); J. Erskine, *The Elizabethan L.* (1903); F. E. Schelling, *The Eng. L.* (1913); J. Drinkwater, *The L.* (1915); G. B. Gray, *The Forms of Heb. Poetry* (1915); A. Erman, *The Lit. of the Ancient Egyptians*, tr. A. M. Blackman (1927); P. S. Allen and H. M. Jones, *The Romanesque L.* (1928); G. Murray, *The Cl. Trad. in Poetry* (1930); J. Pfeffer, *Das lyrische Gedicht als aesthetisches Gebilde* (1931); Manitius, 3.III.F; K. Kar, *Thoughts on the Med. L.* (1933); Jeanroy; H. Färber, *Die Lyrik in der Kunsttheorie der Antike* (1936); F. Brittain, *The Med. Lat. and Romance L. to* A.D. *1300* (1937); G. Errante, *Sulla lirica romanza delle origine* (1943); E. K. Chambers, *Early Eng. Ls.* (1947); Le Gentil; C. M. Ing, *Elizabethan Ls.* (1951); J. Wiegand, *Abriss der lyrischen Technik* (1951); G. M. Kirkwood, "A Survey of Pubs. on Gr. L. Poetry, 1932–53," *CW* 47 (1953), foll. by D. E. Gerber for 1952–67 and 1967–75 in *CW* 61 (1968), 70 (1976); G. Benn, *Probleme der Lyrik* (1954); M. Hadas, *Ancilla to Cl. Reading* (1954); A. E. Harvey, "The Classification of Gr. L. Poetry," *ClassQ* n.s. 5 (1955); *The Harley Ls.*, ed. G. L. Brook, 2d ed. (1955); H. Friedrich, *Die Struktur*

der modernen Lyrik (1956)—influential; G. Murphy, *Early Irish Ls.* (1956); J. L. Kinneavy, *A Study of Three Contemp. Theories of L. Poetry* (1956); Beare; Frye; Bowra; R. Dragonetti, *La Technique poétique des trouvères dans la chanson courtoise* (1960); R. Kienast, "Die deutschsprachige Lyrik des Mittelalters," *Deutsche Philologie im Aufriss*, ed. W. Stammler, 2d ed., v. 2 (1960); L. Nelson, *Baroque L. Poetry* (1961); C. S. Lewis, *Eng. Lit. in the 16th C.* (1962); J. M. Cohen, *The Baroque L.* (1963); B. Markwardt, "Lyrik (Theorie)," *Reallexikon* 2.240–52; Jeanroy, *Origines*; *Zur Lyrik-Diskussion*, ed. R. Grimm (1966); F. Goldin, *The Mirror of Narcissus in the Courtly Love L.* (1967); P. Boyde, *Dante's L. Poetry* (1967), *Dante's Style in his L. Poetry* (1971); Y. Winters, "The 16th-C. L. in England: A Critical and Historical Reinterp.," *Elizabethan Poetry*, ed. P. J. Alpers (1967); Dronke; Dale; R. Woolf, *The Eng. Religious L. in the Middle Ages* (1968); *Forms of L.*, ed. R. Brower (1970), esp. P. de Man, "L. and Modernity"; J. Mazzaro, *Transformations in the Ren. Eng. L.* (1970); R. Oliver, *Poems Without Names* (1970); B. Watson, *Ch. Lyricism* (1971); W. Killy, *Elemente der Lyrik* (1972); H. M. Richmond, *Ren. Landscapes* (1973); J. J. Y. Liu, *Major Lyricists of the Northern Sung* (1974); Bec; H. A. Stavan, *La Lyrisme dans la poésie française de 1760 à 1820* (1976); A. Welsh, *Roots of L.* (1978); D. Woolf, *The Concept of the Text* (1978); H. Parry, *The L. Poems of Gr. Tragedy* (1978); B. K. Lewalski, *Protestant Poetics and the 17th-C. Religious L.* (1979); S. Cameron, *L. Time* (1979); *The Interp. of Med. L. Poetry*, ed. W. T. H. Jackson (1980); S. Owen, *The Great Age of Ch. Poetry* (1980); K. S. Chang, *The Evolution of Ch. T'zu Poetry* (1980); Trypanis; Sayce; W. R. Johnson, *The Idea of L.: L. Modes in Ancient and Mod. Poetry* (1982); Fowler; D. A. Campbell, *Gr. L.*, 4 v. (1982–), *The Golden Lyre* (1983); I. H. Levy, *Hitomaro and the Birth of Japanese Lyricism* (1983); W. E. Rogers, *The Three Genres and the Interp. of L.* (1983); H. Spanke, *Studien zur lateinischen und romanischen Lyrik des Mittelalters* (1983); *Gesch. der deutschen Lyrik vom Mittelalter bis zur Gegenwart*, ed. W. Hinderer (1983); *Lyrik des Mittelalters: Probleme und Interpretationen*, ed. H. Bergner et al., 2 v. (1983); A. Williams, *Prophetic Strain: The Greater L. in the 18th C.* (1984); M. H. Abrams, "Structure and Style in the Greater Romantic L.," *The Correspondent Breeze* (1984); *CHCL*, v. 1, chs. 6, 8; J. Herington, *Poetry into Drama* (1985); T. G. Rosenmeyer, "Ancient Literary Genres: A Mirage?" *YCGL* 34 (1985); D. Lindley, *L.* (1985); R. Alter, *The Art of Biblical Poetry* (1985); *L. Poetry: Beyond New Crit.*, ed. C. Hošek and P. Parker (1985); P. S. Diehl, *The Med. European Religious L.* (1985); *The Vitality of the L. Voice*, ed. S. Lin and S. Owen (1986)—Chinese; R. L. Fowler, *The Nature of Early Gr. L.* (1987); Hollier; G. Kaiser, *Gesch. der deutschen Lyrik von Goethe bis Heine: Ein Grundriss in Interpretationen*, 3 v. (1988); D. Lamping, *Das lyrische Gedicht* (1989); A. R. Jones, *The Currency of Eros: Women's Love L. in Europe, 1540–1620* (1990); M. Perloff, *Poetic License: Essays on Modernist and Postmodernist L.* (1990), esp. ch. 1; *Ls. of the Middle Ages*, ed. J. J. Wilhelm (1990); M. Dickie, *L. Contingencies* (1991).

J.W.J.

LYRIC SEQUENCE.

I. RENAISSANCE TO ROMANTIC
II. MODERN BRITISH AND AMERICAN

I. RENAISSANCE TO ROMANTIC. A collocation of lyrics, the l. s. is generally thought to have gained its vernacular identity in the Ren. and after from Francis Petrarch (1304–74), who wrote and arranged his s. called *Rime sparse* or, alternatively, *Canzoniere* over 40 years. Both titles ("Scattered Rhymes" and "Songbook") ironically point away from the rigorous ordering of Petrarch's 366 amatory and devotional lyrics, which manifests several patterns at once (formal, fictional, calendrical) to establish a continuum—or dis-continuum, considering the new-found structural importance of the white space between the lyrics—that greatly exceeds the unities of earlier lyric collections. Petrarch's diverse models included the Book of Psalms; the lyric volumes of Catullus, Propertius, and other Augustan poets; Dante's prosimetric *Vita nuova* (ca. 1292–1300) and his short s. *Rime petrose* (ca. 1296); and the 12th- and 13th-c. *chansonniers* of the Occitan troubadours.

For two centuries after Petrarch's death the possibilities opened by the l. s.—to write lyric *in extenso*, allowing each poetic integer to hold its autonomy as it participates in a larger unity—attracted the efforts of poets in all the countries of Europe; Wilkins lists most of these, an astonishing array. The l. s. effectively became to lyric (q.v.) what tragedy was to drama, or what the novel would be to narrative—not merely a "form" but a complex of generic capacities. Some particularly interesting l. ss. are written at the margins or in the hiatuses of the convention: the 24-sonnet s. (1555) by the Fr. poet Louise Labé dissents from the sexual politics of Petrarchism; the Italians Giovanni Salvatorino and Girolamo Malipiero, believing Petarch's speaker insufficiently Christian, adapt the forms and much of the lang. of the *Rime sparse* to purely devotional experience in their respective works *Thesoro de sacra scrittura* (ca. 1540) and *Il Petrarca spirituale* (1567); and, near the end of the century, after most European cultures had temporarily exhausted the s., a generation of Eng. sonneteers between Thomas Watson's *Hekatompathia* (1582) and Michael Drayton's *Idea* (1600) undertook a frenzy of experimentation, refinement, and superstitious imitation of Petrarch (see SONNET SEQUENCE). While a number of poets essentially non-Petrarchan in topics and ideology are attracted by the l. s. (e.g. George Gascoigne, the first to use the term "s." in Eng. in *Alexander Nevile* [1575], and Ben Jonson in *The Forrest* [1616]), nearly every s. in this period interprets some aspect of Petrarch's rich achievement.

Even such monuments of their vernaculars as Pierre de Ronsard's *Sonnets pour Hélène* (1578, enl. 1584), Luís de Camões' *Rimas* (pub. posthumously 1595, though written as early as the 1540s), and Shakespeare's *Sonnets* (ca. 1593–99, pub. 1609) are, in the first analysis, responses to the *Rime sparse* as a compendium of organizational strategies—formal, characterological, devotional, political, speculative.

The principal trend in the 17th-c. l. s. develops its religious dimensions. Virtuoso events include Jean de la Ceppède's *Théorèmes* (1613, 1622), George Herbert's *The Temple* (pub. 1633), and John Donne's *Holy Sonnets*, incl. "La Corona," a 7-unit crown of sonnets (1633). The first major North Am. l. s., Edward Taylor's *Preparatory Meditations* (written 1682–1725), arrives near the end of this line. The 18th and 19th cs., perhaps because of an attenuation of the formal resources of Petrarchism, perhaps from ineluctable shifts in European poetics and ideologies, produce little of compelling interest in the form of the l. s. itself (though much important poetry within the form). The romantics composed numerous l. ss., but the unities of these works are generally looser than those of earlier specimens; one notices the tendency, as in Novalis and Wordsworth, for example, to announce a l. s. to be in effect a single poem, thus affirming the idea of its unity but waiving the actual tensions generated between strong integers and an equally strong unifying structure.

The most successful 19th-c. European and Am. l. ss. tend to be like Tennyson's *In Memoriam* (1850) in that they do not imitate Petrarch's outward forms and gestures but find contemporary analogues for his inventions—which perhaps inevitably means discarding the sonnet (q.v.) in favor of other constituents. Walt Whitman's *Leaves of Grass* (1855, frequently enl.) is the most visionary of these versions; his influence on the l. s. is virtually that of a New World Petrarch, and has led to an extraordinary abundance of works in the modern period, esp. in the U.S. and in Latin America.

E. H. Wilkins, "A Gen. Survey of Ren. Petrarchism," *Studies in the Life and Works of Petrarch* (1955); G. T. Warkentin, "*Astrophil and Stella* in the Setting of its Trad.," *DAI* 34 (1974): 5211A; C. T. Neely, "The Structure of Ren. Sonnet Ss.," *ELH* 45 (1977)—limited to Eng.; *Arethusa* 13 (1980)—five essays and bibl. on the Augustan volumes; H. Vendler, *The Odes of John Keats* (1983)—the Odes as a l. s.; N. Fraistat, *The Poem and the Book* (1985)—on Eng. romantic collections, ed., *Poems in Their Place* (1986); J. Freccero, "The Fig Tree and the Laurel," *Literary Theory / Ren. Texts*, ed. P. Parker and D. Quint (1986); Hollier; T. P. Roche, Jr., *Petrarch and the Eng. Sonnet Ss.* (1989)—esp. on the religious elements of the early modern s.; R. Greene, *Post-Petrarchism* (1991); *The Ladder of High Designs*, ed. D. Fenoaltea and D. L. Rubin (1991).

R.GR.

II. MODERN BRITISH AND AMERICAN. The modern poetic (or lyric) s. (m. p. s.) is a grouping of mainly lyric poems and passages (or fragments), usually heterogeneous in form, that tend to interact as an organic whole. Although a m. p. s. may well include narrative, dramatic, and ratiocinative elements, its structure is finally lyrical. That is, it is comparable to the characteristic structure of the modern lyric as it has evolved under symbolist, surrealist, and imagist transformations. Its dynamics are primarily emotive and associative, the result of strategic juxtaposition of separate poems and passages without a superimposed logical or fictional continuity. It is, in fact, a modern lyric poem writ large, its several parts providing the same sort of emotive or apperceptive thrusts of affective lang. in relation to the whole as do the shifts of tonal coloration and intensity in a single lyric. The m. p. s. was developed all but unconsciously by poets in search of longer structures that would be both true to inner associative process and responsive to the moral uncertainties and violent disruptions of the age. It has evolved into the dominant modern mode for longer poetic structures and often has an epic or tragic function.

The m. p. s., whose first fully achieved examples in Eng. are Whitman's *Song of Myself* and at least some of Dickinson's fascicles, is the genre in which major poets have excelled: Hardy (*Poems of 1912–13*), Yeats (the Irish Civil War poems, *Words for Music Perhaps, Last Poems and Two Plays*), Pound (*Homage to Sextus Propertius, Hugh Selwyn Mauberley, A Draft of Thirty Cantos, Pisan Cantos*), Eliot (*The Waste Land*, "The Hollow Men," "Ash Wednesday," *Four Quartets*), MacDiarmid (*A Drunk Man Looks at the Thistle*), Williams (*Paterson*), Stevens ("Auroras of Autumn"), Auden (*Look, Stranger!*, "Hora Canonicae"), Lowell (*Life Studies*). Other ss. incl. some of the best work of Hart Crane, Olson, Guthrie, Berryman, Ginsberg, Plath, Sexton, and Kinnell in America, and Bunting, Jones, Kinsella, Hill, Montague, and Hughes in Britain and Ireland.

The formal character of the m. p. s. is unusually protean. On the one hand, it is generally distinguished from the modern "long poem" (q.v.) by its free deployment of different prosodic forms and shifting focal points. (True, Tennyson's *Maud* offers far more shifts of form, tone, and intensity than do Stevens' "Auroras of Autumn," Auden's "Sonnets from China," and Berryman's *Dream Songs*; but its narrative framework contains it within the limits of an extended dramatic monologue). On the other hand, the m. p. s. presents a more psychologically evocative dynamic than a "linked series" does (see section I above). Such a series—whether formally consistent, as in Shakespeare's or Donne's sonnet ss. or as in Tennyson's *In Memoriam*, or whether more varied, as in Herbert's *The Temple*, Blake's *Songs of Innocence and Experience*, or Housman's *A Shropshire Lad*—may exhibit a certain redundancy or overly thematic slant that interferes with true lyrical structure.

And frequently such linked series lack a strong sense of initial (or psychological) pressure—i.e. a directive source of disturbance and response energizing the process of association and realization. In any case, they lack the means to sustain, redirect, or counter that pressure with full responsiveness and control.

The m. p. s. frees us to look at poems of all periods, long and short, in a new way, for its dynamics are those of lyrical structure itself, defined as the overall directive energy of movement—the progression, juxtaposition, and interrelation of all the lyric centers, dynamic shifts, and tonal notes—in a poem or s. The object of lyrical structure is neither to resolve a problem nor to conclude an action but to achieve the keenest, most open realization possible. The best ss. depend for their life on the interrelation of lyric centers—units that present specific qualities and intensities of emotionally and sensuously charged awareness in the *lang.* of a passage, not in the supposed feelings of the author or of any implied "speaker." In the modern period, such centers may be juxtaposed freely. From this perspective, the m. p. s. is a true descendant of the Pindaric ode, Shakespearean dramatic poetry, and Coleridgean organicism in its formal and dynamic openness, yet uniquely modern in its associative and improvisatory aspects.—M. L. Rosenthal and S. M. Gall, *The M. P. S.* (1983); D. A. Sloane, *Aleksandr Blok and the Dynamics of the L. Cycle* (1988). M.L.R.; S.M.G.

M

MADRIGAL. A name given to an It. poetic and musical form in the 14th c. and to a different form in the 16th c. and later. The early m., of which Petrarch's "Nova angeletta" (*Rime* 106) is an example, consisted typically of two or three 3-line stanzas with no set rhyme scheme followed by a couplet or *ritornello*. The lines often comprised eleven syllables, sometimes seven. Musical settings used two or three voices, the lower of which may have been performed on an instrument. The same music was repeated for the stanzas, with different music for the *ritornello*.

In the 16th c., the musical m. used various kinds of poetic forms—the *canzone*, the *ballata*, the sonnet—as well as the m. proper, which had become a very free monostrophic verseform of about a dozen 7- and 11-syllable lines with no fixed order or rhyme scheme, although it usually ended in a couplet. The musical settings varied considerably but usually consisted of three to six voices in a mixture of polyphonic and homophonic textures. Many 16th-c. musical ms., notably those by Luca Marenzio, made a point of illustrating the sense or emotion of the text by means of musical conventions, so each phrase of the text was set to its own music. The subjects tended to be pastoral and amatory. Many of these texts are anonymous, but m. verses by Guarini, Bembo, Alamanni, and Sannazzaro were set, as well as sonnets from Petrarch and stanzas from the narrative poems of Ariosto and Tasso.

This later style of m. influenced the Sp. *villancico* and was adopted by composers in Germany, Poland, Denmark, and the Netherlands. Italianate ms. flourished especially in England. Although ms. were imported as early as the 1530s, the Eng. seem not to have begun composing their own until after the publication in 1588 of *Musica Transalpina*, an anthol. of It. music with words in Eng. translation. The It. models made feminine rhyme a consistent feature of these translations. The earliest Eng. poem called a m. appears in Sir Philip Sidney's *Old Arcadia* (ca. 1577–80), a 15-line poem of mixed six- and ten-syllable lines with masculine rhymes. Some later Eng. writers such as Barnabe Barnes use the term m. loosely to designate lyrics in various forms (see also *Englands Helicon*, 1600). Eng. m. composers set verses in many forms, some of which did resemble the It. m.

By the middle of the 17th c., the musical m. was dead in England, and in Italy had been transformed by Caccini, Monteverdi, and others into "concertato"-style pieces with instrumental continuo which would rapidly evolve into the baroque aria. But amateur singers in England revived some ms. as early as the mid-18th c. They have been performed ever since.

K. Vossler, *Das deutsche M.* (1898); E. H. Fellowes, *The Eng. M. School*, 37 v. (1913–24), rev. as *The Eng. Madrigalists* (1956–76), *Eng. M. Composers*, 2d ed. (1948); A. Einstein, *The It. M.*, 3 v. (1949); A. Obertello, *Madrigali Italiani in Inghilterra* (1949); N. Pirrotta et al., "M.," *MGG*; J. Kerman, *The Elizabethan M.* (1962); *Eng. M. Verse*, ed. E. H. Fellowes, 3d ed. rev. F. W. Sternfeld and D. Greer (1967); B. Pattison, *Music and Poetry of the Eng. Ren.* 2d ed. (1970); J. Roche, *The M.* (1972); Wilkins; P. Ledger, *The Oxford Book of Eng. Ms.* (1978); A. Newcomb, *The M. at Ferrara 1579–1597*, 2 v. (1980); K. von Fischer et al., "M.," *New Grove*; J. Chater, *Luca Marenzio and the It. M.*, 2 v. (1981); J. Haar, *Essays on It. Poetry and Music in the Ren.* (1986); W. Maynard, *Elizabethan Lyric Poetry and its Music* (1986); E. Doughtie, *Eng. Ren. Song* (1986); I. Fenlon and J. Haar, *The It. M. in the Early 16th C.* (1988). E.D.

MASQUE. The m. is primarily a Ren. form brought to perfection in England. It drew upon various kinds of entertainment that had arisen throughout Europe in the Middle Ages, and in England upon the mummings and disguising that had become popular amusements on festive occasions, put on in the streets as well as in halls and courts. The character of such productions took form in Italy when masked and ornately costumed figures presented as mythological or allegorical beings entered a noble hall or royal court, singing and dancing, complimenting their hosts and, in the course of the entertainment, calling upon the spectators to join them in dancing. In bringing the royal and noble personages into the company of beings of a supernal nature, the entertainments were in effect affirming the relation of royalty to divinity.

The ambitiousness of such a purpose was taken to justify very great expenditures for these productions. In Italy Brunelleschi and Leonardo were called upon to design scenery and to devise machines providing spectacular effects, creating heavenly and on occasion infernal vistas. The masques offered at the Eng. court matched these in splendor.

The It. term used for these entertainments, *maschere*, was taken over by the Eng. as *maske* as early as 1513. While spectacle and song dominated, we should regard the dance as the fundamental impetus, so that there is a point to characterizing both Stuart and earlier masques as substantially invitations to a dance.

Ben Jonson is credited with inventing the anti-m., a briefer entertainment ordinarily preceding the main one, and offering effectually the other side of the coin—comic treatment of the unruly forces and elements monarchy subdues, the grotesque persons contrasting with the gorgeous ones of the main m.

Circe, a *ballet de cour* produced in Paris in 1581, helped set the mode for later Eng. masques in providing greater dramatic and thematic unity than had previously been usual. Jonson's conception of the m. as primarily the work of the poet went further in aiming for dramatic unity; in the preface to *Hymenaei* (1606), he speaks of the theme and words of the m. as its soul, the spectacle and mime and dancing its body. Inigo Jones, however, continued to stress the spectacular, and after ceasing collaboration with Jonson in 1631 was free to give full scope to the most elaborate visual effects. The Stuarts supported his and others' work generously—Shirley's *Triumph of Peace* (1634) was as richly produced as anything under Henry VIII. The interest in thematic and dramatic unity, however, continued, and was strengthened by the new custom of concentrating the action on a stage at one end of the hall.

The Elizabethan theater had made some place for maskings; the informal, improvised kind that was apparently very popular is briefly represented in *Romeo and Juliet*. More elaborate forms are present in *Cymbeline, The Winter's Tale*, and *The Tempest*. But the conventions which developed for combining music and dancing with allegory and spectacle excluded the tensions of drama; much of the controversy between Jonson and Inigo Jones concerned the question of how far the presentation of character and meaning were to challenge the choreography. In the Jacobean period the establishment of the proscenium arch led to the devel. of a special form that has been called the "substantive theatre m." With the Restoration, masques were often assimilated in the new operatic works but were also sometimes offered as a special kind of opera, as one may think 18th-c. productions of *Comus* were produced.

Over the past two decades, research on masques has taken new directions, stressing their political content and arguing that they offer representations more emblematic than mimetic. Ignorance of such emphases may lead one to mistake sophistication for quaintness or pragmatic statement of policy for fulsome flattery. Ren. courts did not use masques simply to praise and amuse themselves, but rather to explore in a quasi-mystical way the sources and conditions of power. And while critics have analyzed in some detail how masques are expressions of royal power, they have paid less attention to such expressions by the commons in civic pageantry. To correct this deficiency, some scholars have begun to study how masques are connected to the pageant form, whether in the commercial theater or in civic shows. Work also continues on iconographic understanding of the masques, on actual production techniques, and on the relationship between the elements of lang., dance, and music.

E. K. Chambers, *The Med. Stage* (1903); A. Solerti, *Musica, ballo e drammatica alla corte medicea dal 1600 al 1637* (1905); H. Prunières, *Le Ballet de cour en France avant Benserade et Lully* (1914); *Designs by Inigo Jones for Masques and Plays at Court*, ed. P. Simpson and C. F. Bell (1924); E. Welsford, *The Court M.* (1927); A. Nicoll, *Stuart Masques and the Ren. Stage* (1938); J. Arthos, *On a Mask Presented at Ludlow Castle* (1954); A. Sabol, *Songs and Dances for the Stuart M.* (1959); S. Orgel, *The Jonsonian M.* (1965), *Illusions of Power* (1975); J. G. Demaray, *Milton and the M. Trad.* (1968); A. Fletcher, *The Transcendental M.* (1971); *Twentieth-C. Crit. of Eng. Masques, Pageants and Entertainments*, ed. D. M. Bergeron (1972); S. Orgel and R. C. Strong, *Inigo Jones*, 2 v. (1973); R. C. Strong, *Splendor at Court* (1973), *Art and Power* (1984); D. J. Gordon, *The Ren. Imagination* (1976); G. W. G. Wickham, *Early Eng. Stages, 1300–1660*, 2d ed. (1980); M. C. McGuire, *Milton's Puritan M.* (1983); S. P. Sutherland, *Masques in Jacobean Tragedy* (1983); S. Kogan, *The Hieroglyphic King* (1986); J. Limon, *The M. of Stuart Culture* (1990). J.A.; F.T.

METAPHOR.

I. DEFINITIONS. M. (Gr. "transference") is a trope, or figurative expression, in which a word or phrase is shifted from its normal uses to a context where it evokes new meanings. When the ordinary meaning of a word is at odds with the context, we tend to seek relevant features of the word and the situation that will reveal the intended meaning. If there is a conceptual or material connection between the word and what it denotes—e.g. using cause for effect ("I read Shakespeare," meaning his works) or part for whole ("give me a hand," meaning physical help)—the figure usually has another name (in these examples, metonymy and synecdoche respectively). To understand ms., one must find meanings not predetermined by language, logic, or experience. In the terminology of traditional rhet., these figures are "tropes of a word," appearing in a literal context; in "tropes of a sentence," the entire context is figurative, as in allegory, fable, and (according to some) irony.

Following I. A. Richards, we can call a word or phrase that seems anomalous the "vehicle" of the trope and refer to the underlying idea that it seems to designate as the "tenor." An extended m., as the name implies, is one that the poet develops in some detail. A conceit is an intricate, intellectual, or far-fetched m.; a diminishing m., one of its types, uses a perjorative vehicle with reference to an esteemed tenor. An extreme or exaggerated conceit is a catachresis. Mixed m., traditionally derided because it jumbles disparate vehicles together, has recently found some critical acceptance (see Richards). Dead m. presents fossilized ms. in ordinary usage (e.g. "he *missed* the *point*").

What Quintilian said of tropes remains true today: "This is a subject which has given rise to interminable disputes among the teachers of lit., who have quarreled no less violently with the philosophers than among themselves over the problem of the genera and species into which tropes may be divided, their number and their correct classification." To say that m. is any trope that cannot be classified as metonymy, hyperbole, etc., is to provide only a negative definition. Any attempt to define m. positively—e.g. "the application of a word or phrase to something it does not literally denote, on the basis of a similarity between the objects or ideas involved"—will inevitably apply to other tropes.

Some critics accept this consequence and call all tropes ms. But even this definition begs the question of how to distinguish the figurative from the literal, or at best relegates it to accepted usage. If two species are members of the same genus on the basis of similarity, are they not "like" the genus and therefore like each other? Extension of this argument leads to the conclusion that all figures should be understood literally. On the other hand, every object is unique; "it is only the roughness of the eye that makes any two persons, things, situations seem alike" (Pater). Perfect literalness might be achieved by giving each object a unique name. In relation to that standard, a common noun is a m. because it provides a name that can be applied to different entities on the basis of a likeness between them. Hence "literal" lang. can be considered metaphorical.

Despite these arguments and repeated attempts to create a more satisfactory classification of figures, the principal definitions of the major tropes have remained unchanged since the Cl. period. The most innovative attempts to clarify the status of m. have come from philosophers, linguists, and historians, who have explored m.'s relation to propositional truth and meaning, to the origins of lang. and myth, to world views, scientific models, social attitudes, and ordinary usage. For their purposes, conventional and dead ms. provide adequate examples for analysis. Scholars and literary critics, less concerned with theory than with practice, have usually accepted the imprecise accounts of m. handed down by trad. since the 5th c. B.C. and focused their attention on its effects in particular poems. To say this is not to claim that the uses of m. in poetry are categorically different from those in other domains of lang. use, but simply to call attention to the institutionalized character of poetry. One learns to construe and even create poetic ms. through training at school, and the expectations in place when one starts reading a poem differ from those active in reading a newspaper or conversing. A summary of what critics have usually said about m., with some account of objections to traditional views, will provide a context for a brief discussion of historically important theories of m. and contemp. treatments of the subject.

II. CRITICAL VIEWS. Aristotle's discussion of simile in the *Rhetoric* was until recently the starting-point for most treatments of m. In saying that Achilles "sprang at the foe as a lion," Homer used a simile; had he said "the lion sprang at them," it would have been a m. In one case, according to Aristotle, the comparison is explicit (using "like" or "as"); in the other, the word "lion" is "transferred" to Achilles, but the meaning is the same. Quintilian endorsed Aristotle's view of metaphor as a condensed simile: "in the latter we compare some object to the thing which we wish to describe, whereas in the former the object is actually substituted for the thing." What Max Black calls the "comparison view" of m. is based on the grammatical form "A is B"; m. is seen as a condensed

simile, meaning that A is (like) B. The "substitution view," as Black says, takes "the lion sprang" as the paradigmatic form of m.: rather than predicating a likeness, m. uses a figurative word in place of a literal one. Both of these views are compatible with the reductive conception of m. as "saying one thing and meaning another," thus implying that the poet has gone out of the way to say something other than what was meant (perhaps in the interests of decorating an ordinary thought), and suggesting that in reading poetry, one must dismantle ms. in order to find out what the poem "really means."

A different conception of m. is necessary to sustain Aristotle's claim, endorsed through the centuries, that m. is the most significant feature of poetic style: "that alone cannot be learnt; it is the token of genius. For the right use of m. means an eye for resemblances." New ms. are said to spring from the poet's heightened emotion, keen perception, or intellectual acuity; their functions are aesthetic (making expression more vivid or interesting), pragmatic (conveying meanings concisely), and cognitive (providing words to describe things that have no literal name, or rendering complex abstractions easy to understand through concrete analogies). Emphasis on the value of concreteness and sensory appeal in m. is frequent. Some modern critics treat m. under the general rubric "imagery" (*Bild* has a corresponding importance in Ger. crit., as does *image* in Fr. crit.).

Opposing the comparison and substitution views, 20th-c. critics and philosophers developed more intricate accounts of the verbal and cognitive processes involved in metaphoric usage. Despite their differences, the "interaction" view (Richards and Black), "controversion theory" (Beardsley), and "fusion" view (espoused by New Critics) all hold that m. creates meanings not readily accessible through literal language. Rather than simply substituting one word for another, or comparing two things, m. invokes a transaction between words and things, after which the words, things, and thoughts are not quite the same. M., from these perspectives, is not a decorative figure, but a transformed literalism, meaning precisely what it says. Fusion theorists argue that it unifies the concrete and abstract, the sensual and the conceptual in a concrete universal or symbol. An entire poem, if it is organically unified, can therefore be called a m.

A less audacious explanation of the uniqueness of m. will be discussed in the next section (for a defense of the traditional view, see Fogelin); at present, let us simply note what problems this view solves, and what ones it creates. Treating all tropes as ms., the fusion theory frees crit. from the inconsistent classification systems handed down from antiquity. Synecdoche, as a species—genus or part—whole relation, can be imputed to any comparison whatever (everything being "like" everything else in some generic respect, or part of it, if the level of abstraction is high enough). Metonymy can be an empirically observed association (cause—effect), an entailment (attribute for subject), or a contingent relation (object for possessor). Since no single principle of classification governs these distinctions, one figurative expression can exemplify two or more tropes, as Foquelin observed (*La Rhétorique françoise* [1555]). Fusion theorists eliminate this problem and seek the new meanings released by m., rather than reducing them to uninformative categories.

This freedom from pedantry can, however, entail a loss of precision leading to the neglect, if not the dismissal or misperception, of many tropes. The blurring of traditional distinctions leads to changes in the ways figures are construed. "Pale death," which would now be considered a trite m. involving personification, was in the Ren. a metonymy (death, not personified, is the cause of the effect paleness). The verbal vitality created by tropes other than m. is categorized as "imagery" if it is sensory, and "style" if it is not. Similes, otherwise identical to ms., are automatically accorded a lower status merely because they use the word "like" or "as"—a sign of timidity in the eyes of the fusion theorist, who may see in simile new possibilities for describing the nature of lang. (see Brogan).

Singleminded emphasis on the meaning of m., apart from the semantic and grammatical details of its realization, can lead both modern theorists and traditionalists to questionable interpretive practices. In most textbooks, the only examples of m. provided are in the form "A is B," despite the fact that this is not the most common grammatical form, "the A of B" ("th'expense of spirit") and "the A B" ("the dying year") being more frequent (see Brooke-Rose). When they encounter metaphorical verbs, adjectives, and adverbs, critics who seek new meanings in poetry are tempted to transfer the figuration of these word classes over to the nouns with which they are associated. If a speaker snarls a reply or has a green thought, the critic concludes that the poet has said that the speaker is a dog or wolf, and that the thought is a plant. Johnson's line describing those who gain preferment—"They mount, they shine, evaporate, and fall"—may be clarified through reference to mists and May flies (ephemerae, living for a day), but its power and meaning evaporate if one concludes that the metaphoric verbs are a roundabout way of saying that the rich and famous are like insects.

The alternative assumption would be that the poet wanted to connect figurative attributes to a noun that remains literal; the nominal "A is B" equation was after all available to the poet, along with the other grammatical forms that create identity (apposition, demonstrative reference) and presumably would be used if that were the meaning. The tendency to think that ms. always equate or fuse entities (nouns) reduces the varied effects of m. in poetry to a single register. Poets may intend their figurative renderings of process, attribute, and attitude to evoke a range of relations, from suggestiveness to total fusion. If so,

they are not well served by theorists who translate every figurative velleity into a declaration of equivalence.

The affective and aesthetic functions of m., usually mentioned in traditional accounts, have been emphasized by a few modern critics who oppose the assumption that the purpose of m. is to convey meanings. Forrest-Thomson argues that "the worst disservice crit. can do to poetry is to understand it too soon." Modern poets in particular try to forestall this haste by using ms. that do not lend themselves to assimilation by the discursive elements of the text. Thus they try to preserve poetry from reduction to paraphrastic statement. Shklovsky goes further, asserting that the purpose of new ms. is not to create meaning but to renew perception by "defamiliarizing" the world: unlikely comparisons retard reading and force us to reconceive objects that ordinary words allow us to pass over in haste.

III. HISTORY. Four approaches have dominated all attempts to improve on the account of m. provided by the rhetorical trad. Some writers propose more logical classifications of the tropes. Others undertake semantic analysis of the ways in which features of a word's meaning are activated or repressed in figurative usage. These two modes of analysis blend into each other, but they can be distinguished from treatments of m. that emphasize its existential entailments—its relation to reality and to hist. rather than to logic and lang. The crudity of this fourfold classification must justify the brevity of the following discussion, which touches on only those treatments of m. that, from a contemp. perspective, seem crucial.

For Aristotle, m. has two functions and two structures. In the *Poetics*, its function is to lend dignity to style, by creating an enigma that reveals a likeness, or by giving a name to something that had been nameless ("the ship *plowed* the sea"). But in the *Rhet.*, m. appears as a technique of persuasion, used to make a case appear better or worse than it is in fact. Modern critics would say that "kill," "murder," and "execute" have the same denotation but differ in connotation; for Aristotle, one of the three words would be proper in relation to a particular act, and the other two would be ms. It is from its rhetorical uses that m. acquires its reputation as a dangerous deviation from the truth, being for that reason castigated by Hobbes, Locke, and other Enlightenment philosophers.

The four kinds of m. distinguished by Aristotle in the *Poetics* are of two structural types. One results from substitution (of species for genus, genus for species, or species for species), the two terms having a logical or "natural" relation to each other. The other type has an analogical or equational structure: A is to B as C is to D. Although only two of the four terms need be mentioned (A and D, or B and C), we must infer the other two in order to derive the meaning: "the evening of life" enables us to reconstruct "evening is to day as X is to life." Here we find the bifurcation that will henceforth characterize discussions of tropes: one type is based on accepted conceptual relationships (here, genus—species), and the other type includes all tropes that cannot be so defined. Species—genus and species—species relations are part of common knowledge; to cross from genus to genus, we need four terms that create what might be called a hypothetical likeness, one not given by logic or nature.

The species—genus relation is one of many that make it possible to infer the tenor from the vehicle. Identification of other such relations (e.g. cause—effect, quantitative change) led to the proliferation of names for the tropes in rhet. Once they separated themselves from Aristotle's generic term "m.," it was necessary to define m. in such a way that it would not include the other tropes. Quintilian's solution—to say that m. is a substitution involving any permutation of the terms "animate" and "inanimate"—is not as unreasonable as it first appears. The dividing-line between these two domains, which is a fundamental feature of lang. and culture, cannot easily be crossed by species—genus, part—whole, or subject—adjunct relations. Furthermore, animate—inanimate ms. are strikingly frequent in poetry, as are animate—animate ms. involving humans and nature.

Looking back on Quintilian from the perspectives provided by Vico, the romantic poets, and semioticians, we can see that his untidy classification (which defines m. by reference to its subject matter, the other tropes by reference to "categorical" relationships) reveals something about the role of m. that escapes notice in any purely formal analysis. But when subject matter becomes the primary basis of classification, as it is in 19th- and 20th-c. studies such as those of Brinkmann, Konrad, and Spurgeon, the specificity of the tropes dissolves in the all-embracing category "imagery."

For the Ren., Quintilian's *Institutio oratoria* (pub. 1470) provided an orderly exposition of rhet. in place of the patchwork syntheses inferrable from other Cl. texts; it was Quintilian's renown that made him one target of Ramus's campaign to reform the curriculum. To provide a rational classification of the tropes, Ramus correlated them with Aristotle's ten "categories," the most general ways in which any subject may be described (Howell). Aristotle had himself defined m. by appeal to the category species—genus. Ramus concluded that there were four basic tropes: metonymy, irony ("a change in meaning from opposites to opposites"), m. ("a change in meaning from comparisons to comparisons"), and synecdoche. In Vico's *New Science* (1725 et seq.; tr. 1977), these four become the basis of a history of lang. and civilization. In the age of the gods, metonymy ruled: lightning and thunder were a great effect of unknown origin, and mankind imagined the agent Jove as the cause. The age of heroes was one of synecdoche: men who held themselves to

be sons of Jove embodied his abstract attributes. The age of men is the age of m., in which likenesses are taken from bodies "to signify the operation of abstract minds"; and philosophy gives rise to what we call "literal" meaning.

Ramus is the precursor of modern attempts to reduce tropes to a rationale, and Vico occupies the same position in relation to modern discussions of m.'s importance in the devel. of lang., though their successors are not always aware of this lineage. That lang. was originally metaphorical, mythic, and poetic is a common theme in romanticism—e.g. in Rousseau, Herder, Schelling, and Shelley—but there is little evidence that Vico was their source (the idea can be found in Lucretius, among others). Müller, Werner, and Cassirer exemplify the Ger. thinkers who have developed this theory; Langer and Wheelwright contributed to its popularity in Am. crit. On the basis of recent research on the asymmetry of cerebral functions, Jaynes argues that the visions of Old Testament prophets and the voices of the gods heard by heroes in the *Iliad* came from the brain's right hemisphere; when processed by the left hemisphere, they became the ms. that marked the birth of human consciousness. Nietzsche's contrary thesis—that lang. was originally concrete and literal in reference, and that the abstract vocabulary now considered literal is in fact metaphorical—has recently attracted critical attention (de Man). But as Vico pointed out (par. 409), it makes little sense to speak of lang. as either literal or metaphorical before it incorporates a distinction between the two. Even Gadamer's carefully worded claims about the historical and conceptual primacy of m. cannot escape Vico's objection (Cooper).

The theories of m. proposed by Richards, Black, and Beardsley, which incorporate insights into the workings of lang. and meaning derived from 20th-c. analytic philosophy, provide an alternative to traditional accounts of m. as a substitution, comparison, or fusion of meanings. Sentences, Richards says, are neither created nor interpreted by putting together words with unique meanings. Any ordinary word has several meanings and a number of loosely associated characteristics; often it will be both noun and verb, or noun and adjective. The varied traits or semes of a word's meaning are sometimes sorted into two groups—denotations (characteristics essential to a distinct sense of the word) and connotations—but in practice this distinction is hard to maintain. (In his precise and revealing analysis of this issue, Mac Cormic describes words as "fuzzy sets.") Only when placed in a context does a word take on one or more meanings, at which time some of its traits become salient and others are suppressed. It is often difficult to decide when we have crossed the line between literal and figurative usage. In the series *green dress, green field, green memory, green thought,* the adjective gradually moves from one denotation to another and then takes on a connotative emphasis (that may or may not be listed as a "meaning" in the dictionary) before becoming clearly metaphorical.

Richards looked on m. as a "transaction between contexts," in which tenor and vehicle combine in varied ways to produce meanings. The distinction between literal and figurative is of little importance in his theory. Beardsley makes it more precise by discussing the apparent contradictions that lead us to identify a m. and to seek connotations relevant to the construction of an emergent meaning. He argues persuasively that m. is intensional: we find it in words, not in the objects to which they refer. Black's "interaction theory" contains an important distinction between the "focus" and "frame" of a m. (the figurative expression and the sentence in which it occurs). The focus brings with it not just connotations but a "system of associated commonplaces"—what Eco will later call "encyclopedic knowledge"—that interacts with its frame to produce implications that can be shared by a speech community. To say "Achilles is a lion" can mean that Achilles is courageous only by virtue of the position that the lion occupies in common lore (a hunter, not an herbivore; sociable with its own kind, unlike the tiger, but not a herd animal; monogamous; a lone hunter, unlike the wolf). The lion cannot represent courage until, through a prior mapping of culture on nature, he is the king of beasts. Black emphasizes the "*extensions* of meaning" that novel ms. bring to lang., but his theory has proved most useful in understanding the inherited and dead ms. that structure a society's way of thinking and talking about itself.

IV. CONTEMPORARY THEORIES. Every innovative critical theory of the past two decades has generated a new delineation of m.—either as the "other" of its own conceptual domain, or as the very ground of its new insights. One of the most influential innovations has been Jakobson's opposition of m. to metonymy. In his view, m. results from the substitution of one term for another; it is characteristic of poetry and some literary movements, such as romanticism and symbolism. Metonymy, based on contiguity, appears more frequently in prose and typifies realism. Though they often cross disciplinary boundaries, recent theories of m. can be classified as (a) linguistic or semiotic (based on intralinguistic relationships, or relations between signs of any sort); (b) rhetorical or pragmatic (involving a difference between sentence meaning and speaker meaning); (c) philosophical (emphasizing relations between words and reality, or sense and reference); and (d) extended (treating nonlinguistic relationships in other disciplines).

A. *Linguistics and Semiotics.* The most ambitious recent attempt to identify ling. features of m. appears in *A General Rhet.*, produced by "Group Mu" of the University of Liège. They treat all unexpected suppressions, additions, repetitions, or permutations of ling. elements, from phonemes

to phrases, as figures, the nonfigurative being a hypothetical "degree zero" discourse from which rhet. deviates. M. results from an implicit decomposition of words into their semes (lexical features), some of which will be cancelled, and others added, when one word is substituted for another. The natural route for such substitutions is through species and genus, as Aristotle observed, and Group Mu concludes that a m. consists of two synecdoches, the progression being either species—genus—species (the intermediate term being a class that incl. the first and last terms) or whole—part—whole (here the central term is a class formed where the first and last overlap). S. R. Levin, using a more flexible scheme for the transfer and deletion of semantic features, shows that there are six ways to interpret a m. (his example is "the stone died"). He points out that the grammatical structure of many ms. allows for the transfer of features in two directions: "the brook smiled" can either humanize the brook or add sparkle and liquidity to the idea of smiling. Although he analyzes m. as an intralinguistic phenomenon, Levin recognizes that Aristotle's fourth type, analogy, often depends upon reference to reality—a fact that Group Mu overlooked. Thus m. appears to escape formalization within a system. Umberto Eco's semiotic solution to this problem is to imagine an encyclopedia that describes all the features of reality not included in the semanticist's dictionary. For Riffaterre, mimetic reference is only a feint that the literary text makes before refocusing itself in a network of semiotic commonplaces.

B. *Pragmatics.* Proponents of speech-act theory hold that m. cannot be explained through reference to relationships between words and their ling. contexts. They make a categorical distinction between "word or sentence meaning" and "speaker's utterance meaning." M., in their view, arises from a disparity between the literal meaning of the words used and what is intended by the speaker or writer. Words always retain their invariant "locutionary" definitions, but when used to make ms., the hearer notices that there is something odd about them and infers unstated suggestions or meanings. As evidence for this theory, Ted Cohen says that a speaker's intention may lead us to infer that a literally true sentence is a m. (e.g. "no man is an island"). As evidence against it, L. J. Cohen (in Ortony) points out that speech acts lose their speaker-meaning in indirect discourse ("Tom told George he was sorry" is not an apology), but ms. remain ms. when repeated ("Tom said George was a fire-eater"), indicating that the meaning of a m. is in the words, not in the occasion of their use. Grice's theory of "conversational implicature" provides a set of rules and maxims for normal talk that, when violated, may alert us to the fact that someone is speaking figuratively. His theory, like Searle's, locates m. in a difference between utterance meaning and speaker meaning—the domain of "pragmatics"—and is subject to the same sorts of crit. that speech-act theory has elicited (see Sadock in Ortony; Cooper).

C. *Philosophy.* Searle and Grice are philosophers, but their theories entail empirical claims of relevance to linguists. Davidson's treatment of m. is more strictly philosophical. Meaning, in his view, involves only the relation between lang. and reality. He is willing to accept the pragmatic "distinction between what words mean and what they are used to do," but he denies the existence of metaphoric meaning: "ms. mean what the words, in their most literal interp., mean, and nothing more." To this one might reply that if we did not realize they were patently false, we would not know they were figurative. Recognizing them as such, we may discover truths about the world, but this is not a consequence of some meaning inherent in the words. Nelson Goodman disagrees. In *Langs. of Art* (1979), he conceives of m., like many other activities, as exemplification. Rather than applying a label to a thing, we use the thing as an example of the label, as when the lively appearance of the literal brook is seen as an instance of smiling (above). This reverses the direction of denotation: the example refers to the word, rather than vice versa, and the word may bring with it a whole schema of relationships that will be sorted anew in the metaphorical context. Goodman concludes that m. is "no more independent of truth or falsity than literal use."

D. *Other Extensions.* Postponing discussion of deeper philosophical differences that divide theorists, one cannot help but note that they tend to privilege different moments in the interpretive process. At first glance, or outside time, m. is false (Davidson). Realizing that the creator of a m. means something else, one might create a theory of the difference between sentence and speaker meaning (Searle, Grice). When engaged in deciphering, a reader enacts the interaction theory—discovering new meanings—and the falsity of m. is forgotten. Truth usually results from testing many examples to find one rule; in m., meaning emerges from repeated consideration of a single example, uncovering all its possibilities; and a hypothesis or generalization is the product of the process, not its inception. In accordance with information theory, the low probability of a word or phrase in a particular context implies that it carries a great deal of meaning.

This improbable interaction of conceptual domains has also attracted the attention of theorists in other disciplines, leading them to transport the figure outside its usual literary domain. Like ms., scientific models serve heuristic functions when familiar structures are used to map uncharted phenomena. Mary Hesse and Thomas Kuhn have extended Black's discussion of this subject, which has also attracted the attention of Fr. philosophers of science. Citing the work of Bachelard and Canguilhem, Derrida suggests that the function of m.

in science is not merely heuristic; Boyd (in Ortony) argues that "ms. are *constitutive* of the theories they express, rather than merely exigetical." Lakoff and Johnson show how pervasive ms. are in organizing personal and social experience. For example, words used to describe arguments ("he *attacked* my *position* and he was *right on target*") show that we conceive argument as a form of warfare. A few dozen such equations generate countless ms. in ordinary usage. In such cases the occurrence of a particular m. is less significant than the model ("frame," "schema," "system of commonplaces") that it evokes. When employed to order understanding of the past or to plan for the future, metaphoric analogies take on a narrative dimension (see Schön and Sternberg [in Ortony] for discussion of their influence on social policy). Pepper's *World Hypotheses* (1942), which treats most philosophic systems as elaborations of four "root ms.," proved useful to Hayden White (1973), who argues that a poetic "prefiguration" based on one of the four tropes identified by Ramus and Vico underlies the methods of explanation used by modern historians.

E. *Summary*. The figural use of "m." in modern theory, through which it assimilates not only all other tropes (as in Aristotle) but models, analogies, and narrative methods as well, leads back to the question of whether the literal and figurative can properly be distinguished from one another at all. The simplest and in some ways most logical answers are that all ling. meaning is literal (Davidson) or all figural (Nietzsche). Children do not discriminate between the two in lang. acquisition, and there is little evidence that adult comprehension of literal and metaphorical usage involves different psychological processes (Rummelhart). Rather than attempting to identify rules capable of accounting for literal usage and then explaining figures as transformations or deviations from this set, one can begin from an inclusive set of semantic features from which literal and figurative usage can be derived by imposing further constraints (Weinreich; van Dijk). Alternatively, one can treat literalness "as a limiting case rather than a norm" and develop a pragmatic theory of meaning in which ms. need not be considered different from other usage (Sperber and Wilson).

While contributing to an understanding of its ling. features and conceptual implications, recent theories of m. show that it is not simply one critical problem among others, notable only for the number of disagreements it causes. As that which lies outside the literal, normal, proper, or systematic, m. serves as the topic through which each system defines itself: m. is not simply false, but that which marks the limits of the distinctions between true and false, or between meaningful and deviant. As Derrida says (1972), "each time that a rhet. defines m., not only is *a* philosophy implied, but also a conceptual network in which philosophy *itself* has been constituted." Thus agreement about the status of m. will be deferred until all other philosophical disputes have been resolved. In *The Rule of M.*, the best available survey of the subject, Paul Ricoeur integrates many of the views discussed above in a synoptic theory that preserves while subsuming the oppositions on which they are based. To do so, he is forced to assign literary m. (despite its virtues) a subordinate position in relation to the "speculative discourse" of philosophy, within which m. can reveal the nature of being. Showing how little has changed since Plato described the quarrel between philosophy and rhet., Ricoeur's work has served as a stimulus for a more penetrating analysis of the conceptual operation that gives rise to the distinction between being and textuality, and hence to that between the literal and the figurative (Derrida 1978).

bibliographies: W. A. Shibles, *M.: An Annot. Bibl. and Hist.* (1971); M. Johnson, "Selected Annot. Bibl.," *Philosophical Perspectives on M.* (1981); A. Haverkamp, "Bibliographie," *Theorie der Metapher* (1983)—covers ca. 1870–1981; W. Bohn, "Roman Jakobson's Theory of M. and Metonymy: An Annot. Bibl.," *Style* 18 (1984); J.-P. van Noppen et al., *M.: A Bibl. of Post-1970 Pubs.* (1985).

studies: M. Müller, *Lectures on the Science of Lang.*, 2 ser. (1862, 1865); F. Brinkmann, *Die Metaphern* (1878); V. Shklovsky, "Art as Technique" (1916), tr. in *Rus. Formalist Crit.* (1965); H. Werner, *Die Ursprünge der Metapher* (1919); C. Spurgeon, *Shakespeare's Imagery and What It Tells Us* (1935); I. A. Richards, *The Philosophy of Rhet.* (1936); W. B. Stanford, *Gr. M.* (1936); G. Bachelard, *La Formation de l'esprit scientifique* (1938); H. Konrad, *Étude sur la métaphore* (1939); S. Pepper, *World Hypotheses* (1942); S. Langer, *Philosophy in a New Key* (1942); E. Cassirer, *Lang. and Myth* (1946); M. Foss, *Symbol and M. in Human Experience* (1949); W. Empson, *The Structure of Complex Words* (1951); W. S. Howell, *Logic and Rhet. in England, 1500–1700* (1956); R. Jakobson, "Two Aspects of Lang. and Two Types of Aphasic Disturbances" (1956), rev. in Jakobson, v. 2; C. Brooke-Rose, *A Grammar of M.* (1958); *M. and Symbol*, ed. L. C. Knights and B. Cottle (1960); H.-G. Gadamer, *Wahrheit und Methode* (1960, tr. 1975); T. Kuhn, *The Structure of Scientific Revolutions* (1962); M. C. Beardsley, "The Metaphorical Twist," *PPR* 22 (1962); M. Black, *Models and Ms.* (1962); Jakobson; P. Wheelwright, *M. and Reality* (1962); J. Cohen, *Structure du langage poétique* (1966); M. Hesse, *Models and Analogies in Science* (1966); H. Weinreich, "Explorations in Semantic Theory," *Current Trends in Linguistics*, v. 3 (1966); M. B. Hester, *The Meaning of Poetic M.* (1967); G. Canguilhem, *Études d'histoire et de philosophie des sciences* (1968); N. Goodman, *Langs. of Art* (1968); D. C. Allen, *Image and Meaning: Metaphoric Trads. in Ren. Poetry* (1968); Group Mu; C. Turbayne, *The Myth of M.* (1970); T. Hawkes, *M.* (1972); J. Derrida, "La mythologie blanche," *Marges de la philosophie* (1972, tr. 1982), "The *Retrait* of M.," *Enclitic* 2 (1978); T. A. van Dijk, *Some Aspects of Text Grammars* (1972), "For-

mal Semantics of Metaphorical Discourse," *Poetics* 4 (1975); Lausberg; H. White, *Metahistory* (1973); H. P. Grice, "Logic and Conversation" [1975], *Studies in the Ways of Words* (1989); P. Ricoeur, *La Métaphore vive* (1975, tr. as *The Rule of M.*, 1977); T. Cohen, "Figurative Speech and Figurative Acts," *JP* 71 (1975); J. Jaynes, *The Origin of Consciousness in the Breakdown of the Bicameral Mind* (1976); S. R. Levin, *The Semantics of M.* (1977), *Metaphoric Worlds* (1988); R. Berry, *Shakespearean M.* (1978); D. Davidson, "What Ms. Mean," and P. de Man, "The Epistemology of M.," in "Spec. Iss. on M.," *CritI* 5 (1978); V. Forrest-Thomson, *Poetic Artifice* (1978); M. Riffaterre, *Semiotics of Poetry* (1978); M. Black, "How Ms. Work," and N. Goodman, "M. as Moonlighting," *CritI* 6 (1979); D. Rummelhart, "Some Problems with the Notion of Literal Meanings," and J. Searle, "M.," in *M. and Thought*, ed. A. Ortony (1979); P. de Man, *Allegories of Reading* (1979); *On M.*, ed. S. Sacks (1979); G. Lakoff and M. Johnson, *Ms. We Live By* (1980); J. Culler, "The Turns of M.," *The Pursuit of Signs* (1981); Morier; U. Eco, "The Scandal of M.," *PoT* 4 (1983); G. Bouchard, *Le Procès de la métaphore* (1984); L. Gumpel, *M. Reexamined* (1984); *The Ubiquity of M.*, ed. W. Paprotté and R. Dirven (1985); E. R. Mac Cormac, *A Cognitive Theory of M.* (1985); J. V. Brogan, *Stevens and Simile: A Theory of Lang.* (1986); D. E. Cooper, *M.* (1986); D. Sperber and D. Wilson, *Relevance* (1986); E. Kittay, *M.* (1987); *M., Communication, and Cognition*, ed. M. Danesi, Monograph Series of the Toronto Semiotic Circle 2 (1988); M. C. Haley, *The Semeiosis of Poetic M.* (1988); R. J. Fogelin, *Figuratively Speaking* (1988); C. Hausman, *M. and Art* (1989); G. Lakoff and M. Turner, *More than Cool Reason* (1989).

W.M.

METONYMY (Gr. "change of name," Lat. *denominatio*). A figure in which one word is substituted for another on the basis of some material, causal, or conceptual relation. Quintilian lists the kinds usually distinguished: container for thing contained ("I'll have a glass"); agent for act, product, or object possessed ("reading Wordsworth"); cause for effect; time or place for their characteristics or products ("a bloody decade," "I'll have Burgundy"); associated object for its possessor or user ("the crown" for the king). Other kinds are sometimes identified: parts of the body for states of consciousness associated with them (head and heart for thought and feeling), material for object made of it (ivories for piano keys), and attributes or abstract features for concrete entities. In common usage, one subset of m. is synecdoche.

Because metonymies are common in ordinary usage and new ones are often easy to decipher, they have attracted less critical attention than metaphor (q.v.). When the effect of m. is to create a sense of vividness or particularity—as in Gray's "drowsy tinklings" of sheep (the sound of their bells), or Keats' "beaker full of the warm South"—the figure is often treated as an instance of imagery. But the metonymies "drowsy" and "South" are not concrete images. Auden's lines "the clever *hopes expire* / Of a *low* dishonest *decade*" show that a clear surface meaning can arise from metonymic associations of cause, attribute, and effect that are far from simple. Conversely, m. can create riddles, hard to fathom but clear once understood, as in Dylan Thomas's "Altarwise by owl-light," the title of which suggests facing east in the dark.

Some linguists argue that m. and synecdoche can often be understood as nonfigurative expressions that result from verbal deletions intended to reduce redundancy (Ruwet). "A glass of Burgundy wine" becomes "a glass of Burgundy," or simply "Burgundy" or "a glass" when the sentence or context implies the rest. In "I just read (a novel by) Balzac," the phrase in parentheses can be deleted because the context conveys its sense.

Attempts to produce a definition of m. that would show what generic features its different types have in common have been part of the larger project of deriving a systematic rationale underlying tropes. The meaning assigned to m. in such cases is determined by the categorical features used to define the tropes (e.g. linguistic, logical, semiotic, psychological) and the number of tropes considered fundamental. Ramus, Vico, and their modern followers hold that there are four; Bloom retains six; Jakobson treats two, m. and metaphor. The meaning of m. expands as the number of tropes decreases.

In the fourfold classification, the other three are synecdoche, metaphor, and irony (qq.v.). Ramus (*Arguments in Rhet. Against Quintilian*, 1549) limited m. to "a change in meaning from causes to effects, from subjects to adjuncts, or vice versa." Some argue that these two types of m. can be equated with Aristotle's four causes: efficient (which would explain metonymies of agent for act), material (encompassing metonymies of time, place, matter, and container), formal (m. of abstract for concrete, attribute for subject), and final. Among modern adherents of fourfold classification, Kenneth Burke would limit m. to "reduction" (incorporeal and corporeal); Hayden White defines it as part-whole reduction; Bloom treats it as a change from full to empty. In these three definitions, m. is not reversible: to substitute incorporeal for corporeal, or empty for full, would be another trope (synecdoche, for Burke and Bloom). These critics use the names of the tropes figuratively, applying them to passages or to entire texts.

In practice, it is often difficult to distinguish m. from synecdoche (part for whole, genus for species, or the reverse). Some hold that synecdoche entails a "one—many" substitution, or, in logical terms, a change of extension, whereas m. is a one-for-one replacement involving a change of intension (definition; see Henry). Finding no satisfactory means of distinguishing the two, many

critics use "m." as a generic term for both and contrast it with metaphor on one of several grounds.

Boris Eichenbaum (*Anna Akhmatova*, 1923) held that metaphor operates at a "supra-linguistic" level, that of the idea: a word is pulled from its semantic field to superimpose a second level of meaning on the literal level. M., he said, is a displacement, or lateral semantic shift, that lends words new meanings without leaving the literal plane. Jakobson (1935) extended the dichotomy, suggesting that frequent use of metaphor unites the poet's mythology and being, separated from the world. Poets who prefer m., on the other hand, project their being on an outer reality that their emotion and perception displace from the normal. The shifting, sequential character of m., he said, was more common in prose than in poetry (see Lodge).

Reworking the distinction in 1956, Jakobson described metaphor as a metalinguistic operation (roughly speaking, a process through which an idea or theme is actualized in words). M., he said, was a change that operated on the hierarchy of linguistic units, either affecting their order or substituting part of a word's meaning, or one associated with it, for the word itself. Both tropes can result from substitution of one word for another (on the paradigmatic axis) or from combination (the succession of words on the syntagmatic axis). In metaphor, for example, a single word can be substituted for another, or the two can be successive ("A is B"). A series of metaphors may point toward a single theme, as in Shakespeare's sonnet that successively likens old age to autumn, sunset, and a dying fire. The same possibilities exist in m., but in the latter, there is no metalinguistic idea unifying the chain of metonymies.

Jakobson held that these two tropes could be used to classify mental disorders (e.g. aphasia), literary movements (romanticism and symbolism being based on metaphor, realism on m.), styles in painting and cinematography, operations of the unconscious (Freud's "identification and symbolism" being metaphor, whereas "displacement" and "condensation" are m.), and cultural practices (e.g. the two types of magic identified by J. G. Frazer—one based on similarity, the other on contiguity). Adapting Jakobson's taxonomy to his own psychoanalytic theory, Lacan treated discourse as a continuous m., displaced from the real, in which metaphoric, unconscious signifiers sometimes appear (see Ruegg and Vergote).

Attempts to revise or simplify the m.–metaphor opposition as conceived by Jakobson and Lacan have taken several forms. Henry defines m. as the result of a psychological focus that substitutes the name of one of a word's semantic elements (semes) for the word itself; metaphor in his view is a combination of two metonymies. Le Guern argues that the "contiguity" of Jakobson's m. involves reference to reality, whereas metaphor is the product of a purely linguistic or conceptual operation. De Man pushes this difference further, seeing m. not only as referential but as contingent or accidental, in opposition to the pull toward unification of essences that underlies most uses of metaphor. Bredin agrees that m. refers to the world but sees such "extrinsic relations" as "a kind of ontological cement holding the world together," not as contingencies. Metaphor is also referential in his view, both figures being opposed to the "structural" or intralinguistic relations underlying synecdoche.

In seeking a logic underlying m. and other tropes, theorists are forced to redefine them, reassigning some of the types they have traditionally included to other tropes and discarding yet others. In so doing, they represent tropes as rule-governed transformations of literal usage. The theoretical clarity thus obtained results from treating rhet. and poetics as branches of philosophy or linguistics.

R. Jakobson, "Marginal Notes on the Prose of the Poet Pasternak" (1935) rpt. in *Lang. in Lit.* (tr. 1987), "Two Aspects of Lang. and Two Types of Aphasic Disorders," *Fundamentals of Lang.* (1956), rev. ver. in Jakobson, v. 1; K. Burke, "Four Master Tropes," *A Grammar of Motives* (1945); J. Lacan, *Écrits: A Selection* (1966, tr. 1977); A. Henry, *Métonymie et métaphore* (1971); Lausberg; M. Le Guern, *Sémantique de la métaphore et de la métonymie* (1973); H. White, *Metahistory* (1973); N. Ruwet, "Synecdoches et métonymies," *Poétique* 6 (1975); H. Bloom, *A Map of Misreading* (1975), *Wallace Stevens* (1977); D. Lodge, *The Modes of Mod. Writing* (1977), Pt. 2; P. Ricoeur, *The Rule of Metaphor* (1977); P. de Man, *Allegories of Reading* (1979); M. Ruegg, "Metaphor and M.," *Glyph* 6 (1979); J. Culler, "The Turns of Metaphor," *The Pursuit of Signs* (1981); G. Genette, *Figures of Literary Discourse* (1982), ch. 6; A. Vergote, "From Freud's 'Other Scene' to Lacan's 'Other,'" *Interpreting Lacan*, ed. J. Smith and W. Kerrigan (1983); H. Bredin, "M.," *Poetics Today* 5.1 (1984); W. Bohn, "Jakobson's Theory of Metaphor and M., An. Annot. Bibl.," *Style* 18 (1984); J. Hedley, *Powers in Verse* (1988). W.M.

MOCK EPIC, MOCK HEROIC. Terms used in a broad sense to describe a satiric method in poetry and prose and, more specifically, a distinct subgenre or kind of poetry which seeks a derisive effect by combining formal and elevated lang. with a trivial subject. The m.-h. poem consciously imitates the epic style, follows a Cl. structure and heroic action for deflationary purposes, and employs some of the standard paraphenalia of the epic, e.g. invocations, dedications, celestial interventions, epic similes, canto divisions, and battles. But the neoclassical form is not a subtype of epic nor even a mockery of epic, but rather, in Warren's terms, an "elegantly affectionate homage offered by a writer who finds it irrelevant to his age."

The Homeric *Batrachomyomachia* (Battle of the Frogs and Mice) served as a model for many an 18th-c. battle in m.-e. strain. The 1717 ed. of this

work by Thomas Parnell belongs with many such neoclassical, burlesque battles of pygmies or cranes or rats or hoops or books or sexes. Chaucer had employed the m.-h. style in *The Nun's Priest's Tale,* but Boileau's *Le Lutrin* is commonly viewed as the most influential modern poem magnifying a trivial subject on an ambitious scale. Dryden's *Mac Flecknoe* and Pope's *Rape of the Lock* and *Dunciad* are classic examples of Eng. m.-h. poetry aiming their shafts at literary pretense and social folly. Mock odes, mock elegiacs (Gray's ode "On the Death of a Favourite Cat"), and mock eclogues abound in Eng. poetry, but the m. h. held supremacy among them until it blended with later burlesque and satiric modes.—R. P. Bond, *Eng. Burlesque Poetry, 1700–1750* (1932); A. Warren, *Rage for Order* (1948); K. Schmidt, *Vorstudien zu einer Gesch. des komischen Epos* (1953); G. deF. Lord, *Heroic Mockery* (1977)—on mockery in epic; M. Edwards, "A Meaning for M.-H.," *Poetry and Possibility* (1988); U. Broich, *The 18th-C. M.-H. Poem* (1990); G. G. Colomb, *Designs on Truth: The Poetics of the Augustan M.-E.* (1992). R.P.F.; T.V.F.B.

MODERN LONG POEM. Ezra Pound's 1909 definition of the epic as "the speech of a nation through the mouth of one man" points as well to the central tension which animates the 20th-c. l. p. Typically, the m. l. p. attempts to identify and synthesize the various voices and details of a culture or "tribe" but, unlike the traditional epic, finds itself, "in a society no longer unified by a single, generally accepted code of values . . . justifying its argument by the direct appeal of the author's own experiences and emotions" (Bernstein). Such an appeal brings about both an inevitable foregrounding of the poet as "hero" and an almost ritual acknowledgment of limitation that follows when the desire to speak for a culture stands revealed as a drive toward self-portraiture as well. This tension also generates the technical innovations for which these poems are noted; if no single narrative exists to explain a culture, then it follows that such discontinuous yet accumulative forms as cantos, letters, catalogs, songs, notes, passages, dreams, or journal entries might be useful in tracking a repeatedly engaged, nonguaranteed movement toward an explanatory tale. Individual cantos or letters themselves often become strikingly tentative arrangements of the shifting, resistant materials of a culture.

The most influential example of this "modern" form dates, however, from the 19th-c.: Whitman's "Song of Myself" (1855). Whitman's attempt in that poem to sing the "Me myself" through embracing the many-featured world not only gave his culture a possible articulation ("through me many long dumb voices") but also called attention to itself as a single, still unfinished "call in the midst of the crowd" (Miller). The most striking 20th-c. example of the l. p., Ezra Pound's *Cantos* (1925–72), painfully highlights that tension. In working out an extraordinary number of linguistic rituals through which to gather, juxtapose, and give voice to the "luminous details" of our culture's religious, political, and literary heritage, Pound's poem has become a sourcebook for poets. Yet in their tragic acknowledgment of the limits of one person's ability to gather a "live trad.," the *Cantos* have been a sober warning as well. Shorter but no less ambitious, T. S. Eliot's *The Waste Land* (1922), deliberately juxtaposing the voices and tags of a culture in order to expose the intimate workings of a tormented psyche, has been an equally significant model for poets drawn to the personal implications of the form. William Carlos Williams' *Paterson* (1946–58) offers a clear statement of yet another line of devel. Responding to Pound and Eliot, Williams gathered the half-heard voices of his poem from local New Jersey culture. In so doing, he insisted that the question of the source of one's luminous details must also be asked.

Individual responses to the pull between self-portraiture and cultural synthesis, or between the resistances of local speech and those of a broader trad., vary. Hart Crane's *The Bridge* (1930), David Jones's *The Anathémata* (1952), and James Merrill's *The Changing Light at Sandover* (1983) tend toward synthesis while also bearing traces of the poet's foregrounded hand. The different structural models of these poems—the curve of Brooklyn Bridge suggesting a possible "mystical synthesis of America"; a meditative dream during a celebration of the Mass opening out to the interlocking cycles of history and prehistory; or voices gathered from an ouija board first playfully then insistently dictating a Manichean view of the universe—acknowledge as well individual sources for each gathering. Conversely, l. ps. framed as self-portraits and struggling therefore to chart such concerns as the interplay of multiple voices within a psyche (John Berryman's *Dream Songs* [1969]), or the intricate overlap of self, family, and world (Louis Zukofsky's *"A"* [1927–79]), or the "curious" ways the "shipwreck of the singular" is linked to a larger world (George Oppen's *Of Being Numerous* [1968]), or the rich, if inevitable distortions of the act of portraiture itself (John Ashbery's "Self-Portrait in a Convex Mirror" [1975]) are also important clusters of cultural information.

Other l. ps. highlight the struggle to read and interpret. H. D.'s description of London during World War II, compiled as *Trilogy* (1944–46), leaps from a world in which structures are shattered and remade by nightly bombing into a new sort of reading which itself takes shattered elements of the world's stories and "alchemizes" them into something new. The same range is displayed in Robert Duncan's "Passages" (1968–87), where the struggle to weave himself into what he calls the "grand ensemble" of humanity's made things—poems, debates, religious heresies, and fantasies—is foregrounded. Equally notable l. ps.

have been written that seek to read deeply into one locality: Hugh MacDiarmid's reply to Eliot, *A Drunk Man Looks at the Thistle* (1926), a rapidly moving meditation on the thorny flowering of Scotland; Charles Olson's investigation of the manifold historical implications of Gloucester, Mass. in *The Maximus Poems* (1960–75); or Basil Bunting's autobiographical account in *Briggflatts* (1966) of leaving a landscape and an early love and circling back, years later, to read both with a gathering, resonant intensity.

R. H. Pearce, *The Continuity of Am. Poetry* (1961), ch. 3; L. S. Dembo, *Conceptions of Reality in Mod. Am. Poetry* (1966); H. Kenner, *The Pound Era* (1971); J. E. Miller, *The Am. Quest for a Supreme Fiction* (1979); M. A. Bernstein, *The Tale of the Tribe* (1980); C. Nelson, *Our Last First Poets* (1981); M. L. Rosenthal and S. Gall, *The Mod. Poetic Sequence* (1983); M. Perloff, *The Dance of the Intellect* (1985), ch. 1, *Poetic License* (1990), ch. 6; B. Rajan, *The Form of the Unfinished* (1985); M. Dickie, *On the Modernist L. P.* (1986); T. Gardner, *Discovering Ourselves in Whitman* (1989); J. Walker, *Bardic Ethos and the Am. Epic Poem* (1989); S. Kamboureli, *On the Edge of Genre* (1991); J. Conte, *Unending Design* (1991). T.G.

MONOLOGUE is used in a number of senses in discussing poetry, all of which suggest the idea of a person speaking alone, with or without an audience. Thus prayers and laments are ms., as are many lyric poems. At the same time, m. has a clearly dramatic element, since no speaker ever speaks in complete isolation. All vocalization, indeed all lang., implies an audience, either external or internalized, but m. characteristically emphasizes the subjective and personal element in speech. The term is also often used to refer to sections from longer works that emphasize a single voice representing a unitary point of view. M. is not therefore restricted to a specific genre but rather a point of view, and it may describe any sustained speech by a single person. In ordinary usage the term often suggests lang. used to preclude other speech or to prevent conversation.

In poetry, m. has clear connections to drama. Thus Strindberg's play *The Stronger*, which consists of but one speech by a single character, involves the audience directly in dramatic revelations about the personalities of three people in a love triangle. Similar single-voiced revelations derive from characters as diverse as Sophocles' Oedipus, Marlowe's Tamburlaine, and Beckett's Hamm. "Soliloquy" refers to a form of m. in which an actor speaks alone on the stage. It represents a character's attempt to verbalize his thoughts, consciously or distractedly, as in Othello's self-revelations and Hamlet's meditative speeches. In some cases, characters address the audience directly, as in Falstaff's address on honor. Numerous plays, notably Marlow's *Doctor Faustus*, Goethe's *Faust*, and Byron's *Manfred*, open with an overheard speech by the main character. Thus ms. either heard or overheard produce a complex semantic interplay between character and audience. Some of the most powerful passages in dramatic poetry (q.v.) are cast in m. form.

The technique is ancient, but too closely connected with the ritualistic origins of drama to have a distinguishable origin. Elegies, diatribes, and comic harangues often have strongly monologic elements, while epistles and philosophical poems do not, since they are meant to be read rather than spoken. Significant biblical examples include *The Song of Deborah*, Jeremiah's lament for Jerusalem, and the Psalms. Prophetic utterances were often cast in m. form, focusing attention on the connection between impassioned verse and prophecy. Lengthy speeches in a refined, direct rhet. appear in Gr. epics and odes. Gr. drama includes numerous examples. Some miming included single-voiced speech, although dialogue (q.v.) was the preferred mode. Theocritus' idylls, and the resulting emphasis on first-person speakers, produced admirable examples of m., as did elegiac poems by Propertius. All types of rhet. involving declamation encouraged m. The prosopopoeia, an ancient rhetorical form that impersonated the speech of an actual or imagined person, found its way into poems like Ovid's *Heroides*.

M. tends always to emphasize the personal and potentially ironic element in poetry. Germanic lit. and its OE derivatives, such as *The Wanderer* and *The Wife's Lament*, employ the subjectivity often associated with the form. Devotional and meditative poetry in all trads. tends to canonize m., as in speeches by the Virgin to the Cross or direct addresses to deity, to the point that such orations become fixed subgenres. In the later Middle Ages and the Ren., variations on the classical "complaint" and "address" increase the dramatic element of earlier religious verse. Dunbar, Lindsay, and Skelton continue Chaucer's earlier use of such forms. Imitations of Horace and Ovid develop these trads. even further. M. appears throughout the 16th c. in works by Wyatt, Gascoigne, Raleigh, and Drayton. Elizabethan dramatic verse as in Shakespeare, Marlowe, and Corneille achieves part of its power from the uncompromising directness of m. and its elements of internal dialogue. Milton's *L'Allegro* and *Il Penseroso* point up the tendency for m. to imply or demand a response. The form is weakened by the social tendencies of Restoration and Augustan verse, while romantic poetry de-emphasizes the ironic element of m. by identifying the poet directly with the speaker of the poem. The dramatization of the lyric "I" in poems like Coleridge's *Frost at Midnight* and Wordsworth's "Tintern Abbey" suggests a possible separation between poet and speaker, pointing toward the ironic ms. of the Victorians.

Such emphasis on the subjective posture of the poem's voice, and a new element—impersonation—are characteristic of the best later examples of m. Partly for this reason, the dramatic ms. of

Robert Browning represent the most significant use of the form in postromantic poetry. Browning himself called poems like *My Last Duchess* and *Porphyria's Lover* "dramatic lyrics," emphasizing the blurring of genres implied in the form. Browning's use of m. derives its power from his ability to depict in detail powerful situations in remarkably few lines. His stanzaic ms. draw on the ballad trad., while his blank-verse masterpieces derive directly from dramatic models. The dramatic m. gains additional force from the fact that a silent auditor often constrains or controls the speaker's words, contributing to complex levels of irony within the poem. Browning's ms. have sometimes been regarded as brief closet dramas with a single speaker, while the term "monodrama" has been used to describe poems by Tennyson, such as *Ulysses* and *Tithonus*, which present the varying emotions of a single mind rather than a powerfully conceived dramatic situation. Matthew Arnold, William Morris, the Rossettis, and Swinburne all wrote powerful lyrics "spoken" by dramatic voices, but it is clearly Browning's use of the m. that has had the most significant impact on subsequent poetry.

In the 20th c., Browning's dramatic form of the m. has been adopted most directly by Ezra Pound, who also noted his debt to Propertius. T. S. Eliot, in poems like *The Love Song of J. Alfred Prufrock* and *Gerontion*, creates a m. spoken without a direct dramatic auditor. This form is directly related to the rise of the persona or mask in modern poetry. In recent years one critic has called such poems "mask-lyrics" (Rosmarin). Poems by Pound and Eliot, for example, provide a monologic and dramatic speaker like Tennyson's, one more closely identified with the lyric voice of the poet than with the dramatic voice of an imagined character. Such poems can be distinguished from strictly dramatic ms., which set up a necessarily ironic distance between poet, speaker, and reader. E. A. Robinson, Edgar Lee Masters, and Robert Frost all contributed variations on the form; Robert Lowell, Sylvia Plath, John Ashberry, and Diane Wakowski subsequently employed a monologic mode that reveals a tension between the poet and the speaker of the poem. M. is a characteristic of all poems that strive to deny dialogue or control possible responses to the utterance. While recent speculation has suggested that all lang. contains a dialogical aspect that denies the possibility of pure m. (Bakhtin), poets continue to explore the implications of a single voice speaking alone and advocating a unitary, determinate position.

F. Leo, *Der Monolog im Drama* (1908); E. W. Roessler, *The Soliloquy in Ger. Drama* (1915); I. Hürsel, *Der Monolog im deutsche Drama von Lessing bis Hebbel* (1947); I. B. Sessions, "The Dramatic M.," *PMLA* 62 (1947); H. Schauer and F. W. Wodtke, "Monolog," *Reallexikon*, v. 2; R. Langbaum, *The Poetry of Experience*, 2d ed. (1974); A. D. Culler, "Monodrama and the Dramatic M.," *PMLA* 90 (1975); R. W. Rader, "The Dramatic M. and Related Lyric Forms," *CritI* 3 (1976); A. Sinfield, *Dramatic M.* (1977); J. Blundell, *Menander and the M.* (1980); K. Frieden, *Genius and M.* (1985); L. D. Martin, *Browning's Dramatic Monologues and the Post-Romantic Subject* (1985); P. Parker, "Dante and the Dramatic M.," *SLRev* 2 (1985); A. Rosmarin, *The Power of Genre* (1985), ch. 2; W. Clemen, *Shakespeare's Soliloquies* (1987); J. T. Mayer, *T. S. Eliot's Silent Voices* (1989); E. A. Howe, *Stages of Self: The Dramatic Ms. of Laforgue, Valéry, and Mallarmé* (1990). B.A.N.

MUSIC AND POETRY (Gr. *mousike*, "the art of the Muses"). Our best evidence about primitive song suggests that melodies and rhythms precede words, that the first step toward p. was the fitting of words to pre-existent musical patterns. Primitive cultures did not make the distinctions we now make between m. and p.: the Egyptian "hymn of the seven vowels," for example, appears to have exploited the overtone pitches present in the vowels of any lang. The ancient Gr. linguistic system of pitch accents was strikingly similar to the tetrachordal system of ancient Gr. m. Spoken Gr. moved between two stable pitches, a high pitch (indicated in post-Alexandrian texts by the acute accent) and a low pitch (indicated by the absence of accent); these pitches framed an area from which the sliding pitch indicated by the circumflex accent arose; the grave accent may also indicate such a medial pitch. Gr. m. also moved between two fixed pitches; these tones, a perfect fourth apart, framed a middle area containing two sliding microtonal pitches. The linguistic pitch system operated independently from the rhythmic system we now call quantitative meter; high-pitched syllables did not necessarily correspond with long positions in the meter. But the scraps of ancient m. we possess do show a general correspondence between pitch-accent and melodic shape, and studies of "accentual responsion" in the lyrics of Sappho and Pindar suggest that the poet's choice of words in an antistrophe may have been constrained by an attempt to have those words correspond to the melodic pitch-pattern established in the strophe.

The Greeks used the same word, *moûsike*, to describe dance, m., p., and elementary education. *Moûsike* was essentially a "mnemonic technology," a rhythmic and melodic way of preserving the wisdom of the culture; alphabetic writing, the next advance in mnemonic technology, forced changes. It was adopted as a musical notation soon after its introduction, with letters of the alphabet written above the vowels in a poetic line to indicate pitches. Thanks to the quantitative conventions of Gr. meter, no separate rhythmic notation was necessary. The visual separation of pitches and words in the new notation began to separate the once unified arts; alphabetic writing led to both rhetorical and musical theory, the latter of which, thanks

to Pythagorean mysticism, quickly became concerned with advanced theoretical and mathematical problems virtually divorced from performance.

Roman p., in which the normal word-accents of Lat. words were arbitrarily distorted as those words were wedged into what had once been the rhythms of Gr. m., was another step in the separation. What began as *moûsike*, an organically unified art, had now become not two but four elements: performed m., m. theory, p., and rhetorical theory. Christian thought altered the relative prestige of the four elements: the Church fathers embraced the elaborate mathematical m. theory of the ancient world and allegorized its numbers; they banished instrumental m. from the Church and sought to alter and control vocal m.; on the literary side, by contrast, pagan p. itself had to be saved by allegory, while rhetorical theory was treated with suspicion. The drift in Lat. p. away from quantitative verse toward accentual-syllabic verse, in which the hymns of Ambrose and Augustine are important documents, was a motion away from writing Lat. words to Gr. tunes toward writing Lat. words to Christian tunes whose origin was probably Hebrew.

In the early Middle Ages, liturgical chant became longer, more complex, and more ornate, despite attempts by Charlemagne and Gregory to arrest its devel. When the lengthy melismatic passages sung to the last *a* of the word *alleluia* proved hard to memorize, because church singers had a much less accurate notational system than the now-forgotten letters of the ancient Greeks, monks began writing words for them; the resulting works were called *sequences* or *proses*, though they employ many devices we would call poetic. By fitting new words to a pre-existing melody, such sequence poets as Notker Balbulus (ca. 840–912) again altered poetry, moving it still closer to modern stanzaic form, including rhyme. The troubadours and *trouvères* (qq.v.), composer-poets writing in the vernacular, took over and extended the formal innovations of the sequence, producing increasingly complex stanzaic forms with elaborate rhyme schemes. In their art, poetic form was more complex than musical form, and by the time Dante defined p. as a combination of m. and rhet., "m." had become a somewhat metaphorical term. Not only were the It. poems in forms derived from the troubadours normally written without a specific tune in mind, but poetic form itself had become sufficiently demanding to occupy the attention once devoted to making words fit a pre-existing tune.

Musicians, who were now increasingly called upon to compose settings of pre-existing words, made an important technical advance in the invention of polyphony. They may have gotten the idea of combining two or more melodies from the literary notion of allegory, realizing that the mystical simultaneity of an Old Testament story and its New Testament analogue could become, in m., actual simultaneity. One result, oddly parallel to the dropping away of m. from the troubadour trad., was that texts became less audible in polyphonic vocal m. than in the monodic singing of all previous m. In early polyphony, the *tenor* or lowest part often sustained one vowel for many long notes, while the more rapidly moving upper parts sang as many as 40 short notes on one vowel. Predictably, these upper parts often picked up new texts, including Fr. texts glossing or commenting ironically on the liturgical Lat. text being sustained in the lower part.

Influenced by Christian versions of the ancient Gr. numerological theory, in which the universe was conceived as created and ordered by numbers, medieval poets and composers frequently constructed their pieces by complex, mystical, mathematical formulae. Fr. isorhythmic motets, tricky crab canons in which one line is the other sung backwards, anagram poems concealing the names of mistresses—all elaborate forms whose principles of construction cannot be heard in performance—flourished as representations of the numerical mystery of the universe, or (for adepts in both arts) as secret displays of technical ingenuity. In the service of such causes, musicians treated texts as a tailor treats cloth: they cut them up, stretched them out, redistributed their rhythms in ways that entirely destroyed the original poetic form, obscured the rhyme scheme, and made the content impossible to hear—esp. in motets, where three different texts in two different langs. were sung simultaneously. Guillaume Machaut (1300–77), who was at once the leading composer and the leading poet of his period, wrote such motets, but also simpler monodic songs such as *chansons*, *virelais*, and *lais* in which expression of the text was an artistic concern.

In Ren. p. and m., techniques initially developed as virtuoso modes of construction, such as rhyme in p. and chordal harmony in m., began to acquire expressive values. A new rereading of the ancient poets and rhetoricians, with fresh interest in persuasion, emotion, and the moral force of sounds, was an important factor. Medieval composers had often worked out their m. before pasting in a text, but Ren. composers normally started with a text and worked in various ways at animating or expressing it. Josquin des Prés (ca. 1450–1521), who used dissonant harmonies at painful moments in the text, pointed the way toward the witty rhetorical musical expression of the It. madrigal school, which developed a number of harmonic and melodic "word-painting" conventions for setting words dealing with running, weeping, dying, and so forth. When Cardinal Bembo's edition of 1501 restored Petrarch as a model for lyric p., composers of secular songs were compelled to increase the musical sophistication of their settings, and in searching for musical equivalents of Petrarchan oxymorons—"freezing fires" or "living deaths"—they developed a more expressive use of harmony. Despite this general motion to-

ward expression, however, the highly elaborate methods of construction typical of medieval art survived, as virtuosity or mysticism, in both arts, esp. in England, where the hidden numerical schemes of Spenser's p. and the abstract patterns of John Bull's keyboard fantasias provide extreme examples.

The increased attention to the rhet. and meaning of p. on the part of composers did not satisfy the literary reformers now called the "musical Humanists" (the *Camerata* of Bardi in Florence and the *Académie* of Baïf in Paris). Fired by ancient myths concerning the capacity of m. to arouse various passions, these men concluded that it would do so most effectively by submitting to the rule of the text: they opposed independent musical rhythm, arguing that m. should exactly follow the rhythm of the poem; they opposed the staggered declamation typical of the madrigal, favoring homophonic, chordal singing or monody. Such composers as Monteverdi paid lip service to the aims of this reform program, but did not allow it to deprive their art of the techniques it had developed since the Middle Ages. Operatic recitative is the most familiar legacy of musical humanism, but Monteverdi's operas show as much attention to musical construction as to literary expression. By the later 17th c., opera singers had become more important in the public view than either composers or librettists, and arias designed for vocal display became a central part of operatic practice.

While most Ren. poets possessed some technical understanding of m., thanks to the importance of m. in the traditional school curriculum, poets in later centuries often lacked such knowledge, and their mimetic theories of musical expression proved increasingly inadequate. In 18th-c. vocal m., such composers as Bach continued to employ versions of the mimetic "word-painting" techniques of the madrigal; Pope's witty lines on "sound and sense" in the *Essay on Criticism* are a poetic analogy. But composers, unlike poets, were able to use materials that originated in such local mimesis as building-blocks from which to construct a larger structure. Trained by such rhetorically organized texts as Fux's *Gradus ad Parnassum* (1725), a treatise on counterpoint praised by Bach, they were also learning to combine canonic procedures with an increasingly stable tonal grammar; these devels. liberated instrumental m., which could now embody several kinds of purely musical meaning. The willingness of later 18th-c. concert-goers to attend purely instrumental performances demonstrated once and for all the inadequacy of Ren. theories that had maintained that music's only legitimate function was to animate texts. Mimetic theorists, however, shifted their ground. No longer able to maintain that composers were imitating words, they now insisted that they were imitating or expressing feelings, a doctrine that led to the *Affektenlehre*, a systematic catalogue of musical formulae for expressing passions.

Two fundamentally opposed conceptions of m. were now coexisting uneasily: poets and philosophers continued to insist on the mimetic function of m., now calling it a lang. of the passions, but composers and some theorists, by developing the tight musical syntax we now call the tonal system, had given m. a grammar of its own, a meaning independent of imitation that made possible such larger forms as the "sonata-allegro." The romantic poets, just as ignorant of musical technique as their Augustan predecessors, now embraced m. for the very qualities that had made it unattractive to those older poets, its supposed vagueness, fluidity, and "femininity." They sought in their p. to imitate these myths about m., not the logical, witty m. actually being written by such composers as Haydn. In the cause of a more "musical" p., the romantics loosened Eng. syntax while Haydn and Mozart were tightening and refining musical syntax. But eventually these romantic and literary myths about m. began to affect composers, and in the m. of Berlioz, Liszt, and Wagner, all of whom acknowledge literary influences, a similar loosening of musical syntax takes place. Later 19th-c. composers frequently embraced poetic aims: "program" m., the idea of the "leitmotif," the revived claim that m. could express emotions and tell stories.

Wagner's opponent Eduard Hanslick insisted on the autonomy of m., espousing the revolutionary idea that musical structure itself was the real subject of m. Contemporaneous poetic theories of autonomy were somewhat similar in their drive to separate p. from its subject matter. But while Hanslick rejected all attempts to describe m. as a lang., the poetic autonomists (Poe, Wilde, Pater) claimed to want to make p. more like m. Fr. symbolist p., in its fascination with sound and its attempt to maximize the extent to which words in a poem acquire their meaning from that particular poetic context alone, attempts to realize the program announced in Verlaine's familiar declaration: "De la musique avant toute chose." Still, the waning of the tonal system in 20th-c. m. and Schönberg's success in devising a new system for composition suggested again the limitations of attempts to describe m. in ling. terms. 20th-c. relations between the arts have often followed the old axes of numerical construction: Schönberg was profoundly influenced by the mathematical constructive procedures, in p. and m., of Machaut; in his "expressionist" period, he used poetic line-lengths to determine musical structure; Berg organized his "Lyric Suite" on a sonnet by Mallarmé, but suppressed the text; Auden, in seeking a musical sophistication of technique, invented poetic forms closely related to the serial techniques of modern m. Despite the large differences in the way m. and p. are practiced in the modern world, Pound's cranky insistence that "poets who will not study m. are defective" acknowledges the advantages of a long and fruitful partnership.

BIBLIOGRAPHIES: *A Bibl. on the Relations of Lit. and the Other Arts 1952–1967* (1968); S. P. Scher, "Literatur und Musik: Eine Bibliographie," *Literatur und Musik: Ein Handbuch zur Theorie und Praxis*, ed. S. P. Scher (1982).

STUDIES: G. Reese, *M. in the Middle Ages* (1940), *M. in the Ren.*, 2d ed. (1954); T. S. Eliot, *The M. of P.* (1942); B. Pattison, *M. and P. of the Eng. Ren.* (1948); C. S. Brown, *M. and Lit.: A Comparison of the Arts* (1948), "Musico-Literary Research in the Last Two Decades," *YCGL* 19 (1970), ed., spec. iss. of *CL* 22 (1970); *Source Readings in M. Hist.*, ed. O. Strunk (1950); A. Einstein, *Essays on M.* (1956); J. Hollander, "The M. of P.," *JAAC* 15 (1956), *The Untuning of the Sky: Ideas of M. in Eng. P., 1500–1700* (1961), *Vision and Resonance*, 2d ed. (1985), esp. chs. 1, 2, 4; G. Springer, "Lang. and M.: Parallels and Divergencies," *For Roman Jakobson*, ed. M. Halle (1956); *Sound and P.*, ed. N. Frye (1957); D. Feaver, "The Musical Setting of Euripides' *Orestes*," *AJP* 81 (1960); J. Stevens, *M. and P. in the Early Tudor Court* (1961), *Words and M. in the Middle Ages* (1986); C. M. Bowra, *Primitive Song* (1962); A. Wellek, "The Relationship between M. and P.," *JAAC* 21 (1962); K. G. Just, "Musik und DIchtung," *Deutsche Philologie im Aufriss*, ed. W. Stammler, 2d ed. (1962), 3.699–738; F. W. Sternfeld, *M. in Shakespearean Tragedy* (1963); D. T. Mace, "Musical Humanism, the Doctrine of Rhythmus, and the St. Cecilia Odes of Dryden," *JWCI* 27 (1964); H. Petri, *Literatur und Musik: Form- und Strukturparallein* (1964); G. Reichert, "Literatur und Musik," *Reallexikon* 2.143–63; M. Pazzaglia, *Il verso e l'arte della Canzone nel De vulgari Eloquentia* (1967); S. P. Scher, *Verbal M. in Ger. Lit.* (1968), "Lit. and M.," *Interrelations of Lit.*, ed. J.-P. Barricelli and J. Gibaldi (1982), ed., *M. and Text: Critical Inquiries* (1991); L. Lipking, *The Ordering of the Arts in 18th-C. England* (1970); E. Wahlström, *Accentual Responsion in Gr. Strophic P.* (1970); J. M. Stein, *Poem and M. in the Ger. Lied from Gluck to Hugo Wolf* (1971); P. Johnson, *Form and Transformation in M. and P. of the Eng. Ren.* (1972); H. Van der Werf, *The Chansons of the Troubadours and Trouvères* (1972); D. J. Grout, *A Hist. of Western M.*, 2d ed. (1973); R. Hoppin, *Med. M.* (1978); Michaelides; *Dichtung und Musik*, ed. G. Schnitzler (1979); D. Hillery, *M. and P. in France from Baudelaire to Mallarmé* (1980); J. A. Winn, *Unsuspected Eloquence: A Hist. of the Relations between P. and M.* (1981); E. B. Jorgens, *The Well-Tun'd Word* (1982); W. Mullen, *Choreia: Pindar and Dance* (1982); L. Kramer, *M. and P.: The 19th C. and After* (1984); L. Schleiner, *The Living Lyre in Eng. Verse from Elizabeth through the Restoration* (1984); B. Stimpson, *Paul Valéry and M.* (1984); J. Neubauer, *The Emancipation of M. from Lang.* (1986); D. M. Hertz, *The Tuning of the Word* (1988); G. Comotti, *M. in Gr. and Roman Culture*, tr. R. V. Munson (1989); M. L. Switten, *M. and Lit. in the Middle Ages: An Annot. Bibl.* (1990); *The Jazz Poetry Anthol.*, ed. S. Feinstein and Y. Komunyakaa (1991); C. O. Hartman, *Jazz Text* (1991).

J.A.W.

MYTH.

I. FUNCTIONS AND DIFFUSION
II. MEANING AND SIGNIFICANCE
III. MYTH AND LITERATURE
  A. *Formal*
  B. *Causal*
  C. *Historical*
  D. *Psychological*

I. FUNCTIONS AND DIFFUSION. M. is a narrative or group of narratives which recount the activities of a culture's gods and heroes. These narratives are the product of communal and (often) sacred impulses to sanction and reflect the cultural order existing at the time of their creation. As such, they arouse, at least initially, large-scale beliefs concerning either their veracity or meaningfulness. Prior to—and indeed for long after—the introduction of writing (in Greece, in the 7th c. B.C., for example), the transmission of myths was an important function of oral trad. So far as is known, there has been no culture which has not generated a set of myths uniquely its own.

Originating as it appears to do in the oral stage of human culture, m. constitutes the culture's effort to retain through the exercise of memory its knowledge of itself. Much if not all of oral poetry (q.v.) originates in ritual functions and purposes, and ritual is essentially a physical rendering or equivalent of m. Perhaps the most graphic instance of this is the Babylonian *enuma elish*, a creation m. whose ritual recitation at the New Year was thought to be simultaneously an enactment of the original events of the m. and a celebration of those events designed to restore life to the dead god Marduk. Oral poetry and m. in many cultures often treat the same subjects—the nature of the gods, the origin of the world, man, society, law. For example, the Japanese account in the *Kojiki* of the three Kami emerging from a white cloud, the Australian stories of variously named Supreme Beings who existed in *alcheringa* (dream) time and produced plants and animals while wandering the earth, the ancient Gr. narratives of Cronos and Zeus, and the North Am. Indian tales of Coyote and his animal helpers, all purport to tell how gods came into being, how the world was created, and how ancestors emerged.

Lit. as it is known today, regardless of lang. or national identity, appears to have developed out of or to have grown up in close relation to m. as exemplified by those myths held and venerated by individual ethnic and national groups. Thus, Gr. lit. is shot through with mythic materials—characters, subject matter, plots, and actions. Thus Sophocles, in the most celebrated instance from antiquity, takes up mythic materials and narratives already familiar to his audience about the ill-fated Oedipus to be reworked and given visual embodiment in his dramatic trilogy. In Asia, traditional literary forms such as Japanese *No* drama are rooted in ancestral legends and rites. In Africa,

contemporary artists such as Amos Tuotola deliberately reach back to retrieve and renew the myths of their region.

As a result of cultural diffusion, a culture's mythic materials also incorporate motifs, figures, and incidents from other regions, then are given other forms in their subsequent lives when transmitted through the lits. of later cultures. Mythic figures such as Ulysses, Prometheus, Hercules, and Orpheus have been traced in subsequent manifestations throughout much of Europe, particularly in Ger., Fr., and Eng. lit. Hence, while any national lit. may be closely linked to its cultural myths, it is not exclusively wedded to them. Cultural pluralism and historical stratification rather than geographical or ethnic purity condition both m. and lit., perhaps most particularly in their nascent states. Eng. Ren. lit., for instance, would not be what it is but for its reliance on Cl. myths and the lit. which recorded while also modifying them. Shakespeare's *Venus and Adonis* is an obvious example. Corneille and Racine, like Vittorio Alfieri later, in works such as *Medee, Polyeucte, Andromeque,* and *Phedre* develop their literary versions of Gr. myths just as do Heinrich von Kleist and E. T. A. Hoffmann later. The same is true of non-Cl. instances, such as Cain and Abel, Moses, and Joseph from the Bible, Faust and Tristan from medieval mythography, and Cuchulain, Finn, and Taliesin from Celtic mythology: all these appear in one guise or another in many of the world's lits.

The cultural centrality of myth's chief characters encourage expansion and connection of myths to produce groups of narratives, e.g. the labors of Hercules, the exploits of Zeus, and the adventures of Finn Mac Cool. Myth's inclination to embrace the entire lives of its characters expands in lit. into epics such as *Beowulf* and *Gilgamesh,* as well as into other narrative genres such as the *Bildungsroman.* Another common trait of m. is the manifest impossibility of many of the events and beings described. Fifty-headed monsters, shape-changing deities, talking animals, descents to the underworld, and chariot-drawn flights through the sky all testify to myth's characteristic concern with experiences beyond the normal or natural, a concern extended into the numerous subsequent literary genres which treat of the marvelous and incredible.

II. MEANING AND SIGNIFICANCE. The meaning and significance of m. have occupied thinkers as far back as recorded history. The Gr. Sophists regarded myths as allegorical or symbolic means of conveying truths about nature and the world, as well as human ethics. Euhemerus (3d C. B.C.) saw myths rather as covert accounts of purely naturalistic or historical occurrences and personages. Philo Judaeus and St. Augustine manifest the inclination of Christian and Hebrew theologians to adopt both a literal and an allegorical approach to the myths of the Old Testament while viewing pagan myths as falsehoods. Christian humanists of the Ren. in Europe blurred this by concentrating upon the iconic nature of Christian sacred narratives and upon the poetic and metaphoric character of other myths. A reaction to this accommodation and celebration of m. occurred with the 18th c.: Voltaire, for example, attacked both Christian and pagan m. as superstition founded upon a failure of human reason, but inability to explain why such a failure should occur led him to view m. as a means used by a particular social class (for Voltaire, unscrupulous priests of any religion, but particularly Christianity) for maintaining its power by manipulating the ignorance of the lower classes.

At almost the same time, Giambattista Vico's *Scienza nuova* (1725) provided an approach to m. which blended the allegorical and the euhemeristic. Vico saw society as existing in several emergent stages based upon dominant attitudes, and the figures of m. as constituting class symbols of society. The effect was an attitude toward m. combining the literal and the symbolic. This book became seminal to romantic theorists, particularly Goethe and Herder in Germany. They found m. to be a self-sustaining structure of the human spirit which is a necessary and essential mode of belief and of conceiving reality. In the 20th c., Ernst Cassirer's *Philosophy of Symbolic Forms* continues this outlook while making it conform to neo-Kantian philosophy. In essence, Cassirer argues that m. as product is the result of m. as process. The latter is a mode of mental operation radically different from the familiar norm of subject-object relations, material causality, and rational inference. For him as for a number of modern thinkers, m. is man's original way of looking at the world. Its cornerstone is the general intuition that things, the undifferentiated entirety of the natural world, possess an inherent power that is both magical and extraordinary. This condition of consciousness is the spirit's teleological endeavor to shape and determine the nature of spiritual reality.

III. MYTH AND LITERATURE. Equal in importance to these theories of the nature and significance of m. are the persistent entwinements of m. and the world's lits. Four interrelated strands may be discerned in this relationship: formal, causal, historical, and psychological.

A. *Formal.* The formal relation of m. and lit., as suggested above, encompasses the innumerable shared traits of narrative, character, image, and theme. Prometheus' exploits, for example, are rendered by Apollodorus as well as Aeschylus and Shelley. Without confusing one account with any other, the similarities are obvious. All are accounts of certain events befalling certain individuals whose character traits determine the action. Each is also marked by the presence of images and ideas which lend meaning or significance to the tales. The nailing of Prometheus to Mount Caucasus, the self-blinding of Oedipus, and the wily deceitfulness of the Am. Indian trickster resonate in m. as much as in literary works such as *Prometheus Unbound, Oedipus at Colonus,* or Ted Hughes' recent *Crow.*

Such affinities as these, however, are matters only of similarity and not of identity. From the formal standpoint, the resemblances occur in a less explicit and less sustained fashion in m. than in lit. Actions are more arbitrary, motivation simpler and more enigmatic, and continuity and form marked more by the perfunctory and the ruptured than by design and resolution. The lack of formal identity of traits can be seen in the contrast between, say, the biblical tale of Samson and the developed genre of the short story. The fullest and most resourceful example of this sort of approach is found in the work of Northrop Frye, esp. his *Anatomy of Crit.* (1957), *Fables of Identity* (1963), and *The Great Code* (1982).

B. *Causal.* Both the welter and the persistence of resemblances between m. and lit. raise the obvious question of a causal connection. Since the origins of m. and the generic forms of oral lit. are lost in prehistory, any answer is only provisional if not speculative. Nevertheless, the number, degree, and diversity of similarities between specific myths and individual works of lit. argues a reasonably high probability that at least some, perhaps many, are matters neither of accident or coincidence. One way of viewing this relationship would be to regard m. as logically and temporally prior to lit. This approach sees the formal properties of m. mentioned above as simple, elementary forms which gradually become more complex and diverse as the societies which preserve them develop and transform, and as individual authors come to replace anonymous communal narrators. Such a view, which treats m. as man's earliest expressive mode, is the one most fully developed by Cassirer, Susanne Langer, and Philip Wheelwright. Lit. together with religion, science, history, and philosophy then emerge historically from myth. It is as if a group of undifferentiated iron filings were over time sorted out into various different patterned groupings through the introduction of a number of magnets. The causality implicit in this view is a form of pluralistic teleology.

Another view focuses on their temporal or logical coincidence rather than the temporal priority of one over the other. For scholars such as E. A. Havelock, Milman Parry, and A. B. Lord, the oral stage of human culture is of great importance. This view stresses oral memory as being both different from and greater than that found in the literate stage of cultures. The assistance metrical lang. provides to remembering laws, genealogies, and cultural practices puts it at the very heart of oral culture. Consequently, oral poetry with its conspicuous formulaic, rhythmic, and paratactic traits perpetuates not only the stories but the knowledge of a people. Scholars have drawn evidence of this from such disparate sources as pre-Socratic philosophy, Babylonian law tablets, and Bulgarian folk epics. In serving to reinforce social knowledge, such poetry perpetuates the narratives themselves, particularly when the two are not sharply differentiated. In short, the oral culture possesses in its myths, legends, and folktales what is essentially a lit., except for the absence of a definitive, recorded text. M. and lit., according to this view, are functionally differentiated in that they have different sociological roles stemming from technological transformations. M. is sacred, at least so long as believed by the culture at large, whereas lit. is secular or profane in the original sense. The role of the former is to encourage and render possible actual worship, while the latter provides entertainment in a manner which includes moral reinforcement, social responsibility, and religious piety. The two come together in the concept and act of celebration which unites work and play, activities later separated by literate cultures.

C. *Historical.* A third relationship is one in which m. serves both directly and indirectly as source, influence, and model for lit. The most familiar instance of this, of course, is the extent to which Western European lit. is permeated by Cl., and particularly Gr., myths. In addition to the sheer multitude of such instances, esp. striking is the variety of forms which the relationship is capable of taking. In the hands of Aristophanes, the bawdy yet solemn women's fertility festivals provide the material for his satiric and parodic genius in *The Thesmophoriazusae,* while in the 20th c., mythic heroes like Prometheus and Orpheus focus the moral satire, wit, and irony of iconoclastic writers such as André Gide and Jean Cocteau.

Lit. often draws on m. as a direct source for events and characters, in which case the relationship is one of transcriptive retelling. It also, and more frequently, uses m. to stimulate original conceptions and formulations. Thus the original legend of Faust dealt with a knave, but in Goethe's hands he becomes a figure representative of man's aspirations, while Thomas Mann later made of the same mythic protagonist a much more enigmatic and morally ambiguous figure. The range of the diffusion of m. throughout its own culture and into other cultures is impressive. Equally striking is the lack of a single direction or migratory pattern for this diffusion. Writers are frequently drawn to their own culture's myths, but they are also fascinated with the structural, thematic, and narrative possibilities in myths quite alien to their immediate culture.

D. *Psychological.* The fourth relationship between m. and lit. embraces the affective dynamics of production and response. Linguistically, neither m. nor lit. is an empirically accurate record of historical events. Yet both are taken seriously and held to possess meaning and significance. Belief or something similar to belief seems to be involved in both cases, though not in the way of reportorial accuracy or truth. Even members of a culture holding or revering a particular m. are not always prepared to assert its absolute veracity even when convinced of its centrality as a model or paradigm of past or future events. And persons

outside a particular culture are likely to view its myths as obviously either false or else fictions, tales capable of socio-religious reinforcement, psychological arousal by way of imaginative curiosity, or intrapsychic explorations. As such, myths are substantially of the same psychological order as are works of lit. Both underscore the overwhelmingly central role of fictions in human affairs and the need to define and classify their manifold functions. Each is the product of the affective and constructive powers of the human psyche. Lit. impinges on its reader or audience through its capacity to construct persons, scenes, and even worlds which arouse responses uncircumscribable by rational knowledge or empirical description. The same is true of m. Its roots in religious ritual and the convictions they arouse suggest that one of myth's central functions is to provide narratively and dramatically human contact with that trans-rational but experiential power variously called "mana," "orenda," or the numinous. Though largely secular in function and role, lit. still preserves the emotional core of myth's spiritual power. Both are capable of perpetuating and focusing the significance of those sensations of awe, wonder, and, above all, vitality which are mankind's response to the nature of experience, both external and internal.

F. Cornford, *Thucydides Mythistoricus* (1907); J. Harrison, *Themis* (1912); O. Rank, *The M. of the Birth of the Hero* (1914); B. Malinowski, *M. in Primitive Psychology* (1926), *Sex, Culture, and M.* (1962); *M. and Ritual*, ed. S. Hooke (1933); J. Thomson, *The Art of the Logos* (1935); F. R. R. S. Raglan, *The Hero* (1936); D. Bush, *Mythology and the Romantic Trad. in Eng. Poetry* (1937), *Mythology and the Ren. Trad. in Eng. Poetry*, 2d ed. (1963); E. Cassirer, *Lang. and M.* (1946), *The Philosophy of Symbolic Forms*, 3 v. (tr. 1953–57); H. Frankfort et al., *The Intellectual Adventure of Ancient Man* (1946); J. Campbell, *The Hero with a Thousand Faces* (1948), *The Masks of God*, 4 v. (1959); C. Jung and C. Kerenyi, *Essays on a Science of Mythology* (1949); T. Gaster, *Thespis* (1950); E. Fromm, *The Forgotten Lang.* (1951); E. Neumann, *The Origins of the Hist. of Consciousness* (1954); Wimsatt and Brooks, ch. 31; Frye; S. Langer, *Philosophy in a New Key*, 3d ed. (1957); *M.*, ed. T. Sebeok (1958); M. Eliade, *The Sacred and the Profane* (1959), *M. and Reality* (1963); *M. and Mythmaking*, ed. H. Murray (1960); C. Bowra, *Primitive Song* (1962); J. de Vries, *Heroic Song and Heroic Legend* (1963); *M. and Lit.*, ed. J. B. Vickery (1966); N. Frye, "Lit. and M.," *Relations of Literary Study*, ed. J. Thorpe (1967)—with bibl.; P.-M. Schuhl, "M. in Antiquity," F. L. Utley, "M. in Biblical Times," J. Seznec, "M. in the Middle Ages and the Ren.," B. Feldman, "M. in the 18th and Early 19th Cs.," M. Eliade, "M. in the 19th and 20th Cs.," and F. Hard, "M. in Eng. Lit.: 17th and 18th Cs.," all in *DHI*; H. Slochower, *Mythopoesis* (1970); G. S. Kirk, *M.: Its Meaning and Functions* (1970), *The Nature of Gr. Myths* (1975); P. Maranda, *Mythology* (1972); K. K. Ruthven, *M.* (1976); A. Cook, *M. and Lang.* (1980); J. B. Vickery, "Lit. and M.," *Interrelations of Lit.*, ed. J.-P. Barricelli and J. Gibaldi (1982); H. Adams, *Philosophy of the Literary Symbolic* (1983); *Sacred Narrative*, ed. A. Dundes (1984); *M. in Lit.*, ed. A. Kodjak et al. (1985); H. Blumenberg, *Work on M.* (1985); J. Puhlvel, *Comparative Mythology* (1987); P. Veyne, *Did the Greeks Believe in Their Myths?* (1988).

J.B.V.

N

NARRATIVE POETRY.

I. DEFINITION
II. HISTORY
III. THEORY AND VERSIFICATION

I. DEFINITION. N. is a verbal presentation of a sequence of events or facts (as in *narratio* in rhet. and law) whose disposition in time implies causal connection and point. Traditionally, a distinction has been made between factual or literal and fictive or literary n. (Bacon's "Feigned History"), though in recent years such distinctions have come to seem harder and harder to draw. But it is perhaps still useful, whatever our philosophic doubts, to note that we continue to act as though there are kinds of n., judged so, if not according to generic standards, at least according to standards of intention and context. Thus we do not confuse legal arguments or histories of photography with what are called literary works, fictions, ns. that may in fact exploit legal arguments or histories of photography for their own ends. So we can, at least practically, talk about literary n. as a subtype of n. in general, and, further, talk of subtypes of literary n.: prose fiction and n. p. The grammatical emphasis on poetry in the latter phrase suggests that we must also say that the impact of versification is crucial to the total effect of n. poems.

II. HISTORY. Except in relatively short stretches, n. p., exclusive of epic, does not warrant a history in the sense of evolution or even continuity in change. But unfortunately, epic (incl. the issue of oral trad. versus written composition) cannot always be excluded. It may lie at the roots of medieval romance and conditions the so-called epyllion

in its nonepic permutations (see below). Its dominance is everywhere visible in the ancient world, even in work that has no connection with it. Poets who write in other modes feel called upon to excuse themselves, sometimes ironically (Ovid and Horace), sometimes irritably (Callimachus). Prudentius, 4th-c. Christian, feels compelled to open his *Psychomachia* with an epic invocation to Christ, his muse. And it does not stop there, or centuries later with Milton or Pope or Wordsworth (*The Prelude*), but enters Pound's modernist Homerics that open *The Cantos* and sends a faint Virgilian echo through *The Waste Land* ("A Game of Chess," 1.92-93).

If a history of epic and its impact on western lit. can be written (see, for example, Greene), that is not the same thing as writing a history of n. p. The question here for a historian of other-than-epic modes is where epic ends and something else, still n. p., begins. There is no precise answer, only a sense of local intention in the context of a particular cultural trad. Many n. poems draw heavily for their effects on epic convention (*Batrachomyomachia, The Rape of the Lock*) and probably belong very much to epic history; but many others merely gesture toward epic conventions, out of deference to literary habit or irony, and belong to some other mode of n. p. (Chaucer in *Troilus and Criseyde*). Perhaps the *locus classicus* of the dismissively ironic gesture is found in *Eugene Onegin*, where the narrator waits until the end of the next-to-last canto to invoke the epic muse, and adds, with a Byronic yawn: "Enough! The burden's off my back, / My debt to classicism paid, / Though late, my introduction made" (7.55). But works like Ovid's *Metamorphosis* (and perhaps *The Faerie Queene*) seem to belong both to epic history and to the history of other modes of n. p.

What are some of these other modes? One would surely be the folkloric and tendentious beast-fable in verse, exploited by sophisticated poets—Horace in the fable of the town and country mice (*Satires* 2.6), Chaucer in the "Nun's Priest's Tale," La Fontaine and Gay in their fables, and perhaps, in a violent, near-surreal, and particularly modern form, Ted Hughes in *Crow* (1971).

Another mode is found in poems whose interest is less suspenseful action than highly figured erotic description and lyric pathos—"Hero and Leander," "The Rape of Lucrece," "Endymion"—poems whose lineage can be traced back through Ovid, Catullus, and Virgil to Callimachus and Theocritus, to Alexandrian modes like "n. elegy" (Wheeler 120 ff.) and the elegantly stylized, mythic tale termed by some modern scholars the epyllion, "little epic" (Williams 242 ff.).

Still another mode, exemplified in the cl. Sanskrit n. poem, is based on ancient works, the *Mahābharata* and the *Vālmīki Rāmāyana*, but, like the epyllion, became something quite different from its ancestors. Here the term "epic" intervenes to complicate if not confuse the issues. The term is often applied to *Mahābharata* and *Rāmāyana*, and in some ways fits, but in important ways does not if the model is Homer or Virgil. And the term for the later, literary n. poems, *mahākāvya*, has been translated both as "great poem" and "court epic," suggesting a problem of classification. Here again, the term "epic" can mislead as to poetic intention, though this mode takes for its subject the doings of heroes and gods.

Histories of poetic trads. like the Chinese and Japanese must account for the fact that, as Levy puts it, "lyric and n. modes of expression interpenetrate to a far greater extent than they do in European literary trads." (3). Indeed, the history of Japanese court poetry must explain why n. played so subordinate a role in literary culture. Describing the ethos of this trad., Miner observes that "the characteristic figure of cl. Western lit. is the orator, and of the Japanese the diarist" (9). The poetry that emerges from private meditation on personal feelings in the presence of nature and the divine is likely to be overwhelmingly lyrical.

The concept of interpenetration of modes suggests yet another category of n. poems that consciously exploit a sort of hybridization, namely *verse novels*. In Eng. one thinks first of Elizabeth Barrett Browning's *Aurora Leigh*, the aim of which is, according to Edmond, "to give that attention to everyday life which the novel manages so easily, without relinquishing the manner, power, and concentration of poetry" (131). A far more complex example is Nabokov's *Pale Fire*, set forth as a long poem swathed in the apparatus of commentary, the effect being like nesting n. within n. within n., the poetry absorbed into the larger, prose whole. Interpenetration takes place, too, beginning with the Eng. romantics, between the ballad and the lyric. If romantic ballads often take on the pathos of the personal lyric ("Peter Bell"), the lyric often assimilates the apparent directness, objectivity, and simplicity of the ballad (the "Lucy" poems; Coleridge's "conversation poems"). It would be hard to underestimate the effect of the latter kind of interpenetration on the lyric to this day. It could be argued that a characteristic type of poem by James Wright exemplifies a late version of romantic ballad-like lyric. It is plain that a hist. of the ballad would have to navigate genres other than the n., which suggests the difficulty of doing a history of any literary mode.

To sum up: a history of n. p. has to be plural, and rooted in local intention, not framed as a problem of genre. Where genre (or any typal concept) is involved, history must be concerned with shared intentions, not necessarily with shared forms or terminological correctness.

III. THEORY AND VERSIFICATION. Early Western theorists like Plato and Aristotle approached n. p. as a problem in classification, and specifically, the exact classification of epic. This approach was bound to raise difficulties. Plato noticed that the epic was partly n., partly dramatic. He might just

as easily have noticed that other "purer" modes were likewise and inevitably mixed as artistic exigency required. N. entered tragic drama with the messenger who had a tale to tell, and entered Pindar's epinikia as mythic "evidence" to refute or affirm received opinion. Much of the problem would have disappeared had theory allowed for a view of n. as part of a continuum of poetic discourse in which one or another mode dominated according to poetic intention, others being called into play as needed.

Though we have not wholly profited from the confusions of our forebears, we have recently become considerably more sophisticated and systematic in our approach to the topic of n. Martin has noted that in less than two decades "the theory of n. has displaced the theory of the novel as a topic of central concern in literary study" (15). In the process, our sense of discourse of all kinds has complicated itself immensely; this is reflected in the daunting multiplication (and sometimes duplication) of technical terms meant to sort out and define micro- and macrocosmic elements of discourse in general and literary n. in particular. For example, each aspect of the tripartite division of ancient Western critical trad.—author, text, audience—has been further subdivided into a complex cluster of elements. Thus, in a fairly common model (no one theory or terminology wholly dominates the field), author has been analyzed into actual author, implied author, and n. speaker or persona; these, like a set of Rus. nesting dolls, give definition and point(s) of view to the voice or voices implied by the text, itself analyzed into discourse, story, plot or narration, description, argument, and grammatical, semantic and (deep or surface) structural patterns that create the effects associated with literary n. These effects are actively experienced or received by (ideally competent) readers and analyzed into the mirror image of the authorial cluster.

This model by no means exhausts theoretical interest. The dynamics of composition and of reading (or listening) are important topics of a n. theory, as is time (actual reading time as against time within the n.) and its ordering. And it should come as no surprise to find narratological theory much engaged with the topics of epistemology and method, which call into question theory itself and its actual powers to describe. The long struggle of lit. crit., since the 19th c., to free itself from subjectivity has never been fiercer or crossed so many disciplines. Thus modern theory calls upon anthropology, sociology, psychology, and philosophy to give its methods stable grounds.

Insofar as it is a subclass of literary n., n. p. is subject to analysis in all the categories noted above. When verse is added to the mix, attention is apt to focus on stylistic, grammatical, and semantic patterns, and on the figures and tropes most often associated with poetic discourse. But theory, were it to stop here, would miss what seems to be the real differentia of poetry, measure, which is that which most clearly marks off verse from prose. Before dismissing this as a self-evident proposition, it is well to remind ourselves that keeping time, however simply, makes a difference, even in the banality of Johnson's little parody:

> I put my hat upon my head,
> And walked into the Strand
> And there I met another man
> Whose hat was in his hand.

Here, meter rouses expectation, and expectation conditions the experienced meaning of the verse. Nor is it merely that the best thing about doggerel is that we remember it; some of the strongest comic effects in *Hudibras* are built on such minimal versifying.

But n. poems in general deploy other prosodic and figurative elements to play on, and with, the constant of measure and the expectation it rouses. Measure, in fact, invites such devices as caesura, enjambment, rhyme, assonance, stanza, anaphora, and metonymy; it allows, positively encourages, linguistic behavior much less frequently found in prose. Though it is easy to recall prose that is more "poetic" than some poetry, it is hard to imagine the following locution outside a poem: "The future fearing, while he feels the past" (Crabbe, *The Elder Brother*). In this, by one of the most prosaic of Eng. poets, meter is joined with alliteration, assonance, antithesis, and chiasmus to create the effect. If such symmetries are seldom found in prose, how much less so are those of more richly textured n. poems—this, for example, from Yeats's "The Wanderings of Oisin": "Wrapt in the wave of that music, with weariness more than of earth, / The moil of my centuries filled me" (3.69–70). Measure then, tends to be the differentia that provides the ground for other differentiae that we associate with poetry, and this is as true for n. as for lyric and dramatic poetry.

Useful as this distinction might be, it is yet so general that it can only be the beginning of analysis in the face of the immense variety found in n. p. John Clare's "The Badger" and Kālidāsa's *Kumārasambhava* (The Birth of the War God) are both n. poems, but while the former is close to reporting (however highly charged), the latter, working in the cl. Sanskrit poetic trad., depends heavily on elaborate lyrical description. Clare drives his audience with unrelenting momentum to the brutal end, compelling them to experience the logic of the cruelty he depicts; Kālidāsa slows his audience down so that they will experience the intricate but static beauty of the ideal world, the subject of his poem. What Bruno Snell has said about the Virgil of the *Eclogues* could be fittingly applied to Kālidāsa: "he does not narrate facts or events at all; he is more interested in the unfolding and praising of situations" (*Discovery of the Mind* [1953], 291).

The variety of n. p. is such that it must defeat any attempt to sort it into a manageable taxonomy.

Evidence to suggest the futility of classification: in 1819, Wordsworth published "The Waggoner" and Byron the first cantos of *Don Juan,* while Shelley composed "The Mask of Anarchy" and Keats "Lamia." What besides verse measure and the fact that they are in Eng. brings these poems within the compass of a definition more specific? And more useful? The distance between Crabbe's deliberate plainness and Keats's deliberate brilliance suggests that, if these poems are members of the same family, they are so in Wittgenstein's sense, as A and Z are members of the same family, sharing "family resemblances."

This much conceded, we can examine some of the positive effects that measure can have on the impact of the ns. that unfold within its patterns. Among the simplest, measure can—by jolting expectation—underscore, or create, the tone of surprise, as in this from Byron's "The Corsair": "He thought of her afar, his only bride: / He turn'd and saw—Gulnare, the homicide!" (3.13). The sudden pause, the contrast between what was thought and what is seen: these effects are heightened by the regular meter, which yet contains the disruption in the expected flow of events. More complex effects are found everywhere—e.g. in these two lines from James Dickey's "The Shark's Parlor," where tremendous intensity is built by the tension between a halting measure (a word or clusters of words isolated from each other by abnormally wide spacing and sometimes syntactic disjunction) and the forward thrust of the action:

> The front stairs   the sagging boards
> still coming in   up   taking
> Another step   toward the empty house
> where the rope stood   straining
> (*Buckdancer's Choice* [1965], 41)

Movement can be quickened as well as slowed, however, as in the famous chase in Burns' "Tam O'Shanter."

Measure itself not only controls pace and expectation—and in units as small as a single word—it also provides the ground for other effects from relatively simple anaphora to the most complex of tropes. It is highly unlikely that the following lines from the Radin-Patrick (1937) tr. of Puškin's *Eugene Onegin* would sit comfortably in prose:

> The cuckold with the pompous port,
> Completely satisfied with life,
> Himself, his dinner, and his wife.
> (1.12)

The zeugma here very much depends on strict metrical regularity. But it also depends on rhyme, and the force of the little catalogue that peaks at the last word—all these symmetries heighten the asymmetry of "dinner" and "wife."

This is to say that, grounded in meter, the n. poem has at its disposal virtually all the means available to dramatic and lyric poetry. There are indeed instances where modes blend perfectly—for example, Herbert's "The Pilgrimage," about which one might (momentarily) be tempted to emulate Polonius and find designations like "allegorical n. lyric."

Poems like "The Pilgrimage" lie at the approximate center of a very broad spectrum, ranging from very plain to very elaborate style. On the one side, there are the ns. of the medieval biblical paraphrasts, Chaucer of "Sir Thopas," Butler, Crabbe, Wordsworth on occasion, Scott, Byron, and, in our own time, George Keithley (*The Donner Party* [1972]). On the other, there are the ns. of Ovid, of the author of *Sir Gawain and the Green Knight,* of Shakespeare, Keats, Arnold, and Christina Rossetti (in "Goblin Market"). But even this broad classification breaks down before a poem like *Troilus and Criseyde,* which, in its amplitude, incorporates both plain and elaborate styles and all the genres and modes, lyric, n., dramatic, descriptive—proof, if any were needed, that pure types are not likely to keep their purity over any distance, and that types, depending on need, comfortably assimilate one another.

The best modern theory is cognizant of the complexity lying beyond terms like genre, type, and mode, and is cognizant, too, that the dynamics of composing, performing, reading, hearing, and analyzing forbid the satisfying symmetries of taxonomy. Even categories not long ago taken as given—authorship and audience—have been opened to admit complex functions and qualifications. If modern theorists of n. have mostly concentrated their efforts on prose, it is perhaps because analysis of prosody, as of figures, schemes, and tropes, has customarily been applied to the lyric. If it is understood that n. poems are susceptible of the same kind of analysis, with the difference that prosodic effects feed into the general effect produced by a particular narration, then we can approach individual texts unworried by classificatory anxieties, reasonably confident that we are looking at real similarities and equally real differences. L.N.

A. L. Wheeler, *Catullus and The Trads. of Ancient Poetry* (1934); L. R. Zocca, *Elizabethan N. P.* (1950); Frye; Lord; T. Greene, *The Descent from Heaven* (1963); R. Scholes and R. Kellogg, *The Nature of N.* (1966); *Elizabethan N. Verse,* ed. N. Alexander (1967); F. Kermode, *The Sense of an Ending* (1967), *The Art of Telling* (1983); Kālidāsa, *The Kumārasambhava,* tr. M. R. Kale (1967); *Sanskrit Poetry: From Vidyākara's* Treasury, tr. D. H. H. Ingalls (1968); E. Miner, *Intro. to Japanese Court Poetry* (1968); G. Williams, *Trad. and Originality in Roman Poetry* (1968); *N. P.,* ed. C. L. Sisson (1968); *Jour. of N. Technique* 1– (1970–); M. H. Abrams, *Natural Supernaturalism* (1971); E. Vinaver, *The Rise of Romance* (1971); K. D. Uitti, *Story, Myth, and Celebration in OF N. P., 1050–1200* (1973); E. Miner et al., *To Tell a Story: N. Theory and Practice* (1973); E. S. Rabkin, *N. Suspense* (1974); P. H. Lee, *Songs of the Flying Dragons: A Critical Reading* (1975);

Kālidāsa, *The Transport of Love: The Meghadūta*, tr. L. Nathan (1976); D. Perkins, *Hist. of Mod. Poetry*, v. 1 (1976), ch. 4; S. Chatman, *Story and Discourse* (1978), *Coming to Terms* (1990); *Finding The Center: N. P. of the Zuni Indians*, tr. D. Tedlock (1978); G. Genette, *N. Discourse* (tr. 1980), *N. Discourse Revisited* (tr. 1988); W. J. T. Mitchell, *On N.* (1981); D. A. Russell, *Crit. in Antiquity* (1981); P. Boitani, *Eng. Med. N. in the 13th and 14th Cs.* (1982); M. Perloff, "From Image to Action: The Return of Story in Postmodern Poetry," *ConL* 23 (1982); *The Ballad as N.*, ed. T. Pettitt (1982); J. M. Ganim, *Style and Consciousness in Middle European N.* (1983); *Oxford Book of N. Verse*, ed. I. and P. Opie (1983); R. Alter, "From Line to Story in Biblical Verse," *PoT* 4 (1983); D. Bialostosky, *Making Tales: The Poetics of Wordsworth's N. Experiments* (1984); A. S. Gamal, "N. P. in Cl. Ar. Lit.," *Quest of an Islamic Humanism*, ed. A. H. Green and M. al-Nowaihi (1984); P. Ricoeur, *Time and N.* (1984); C. Christ, *Victorian and Mod. Poetics* (1984); "The Music of What Happens: A Symposium on N. P.," *NER/BLQ* 8 (1985); I. M. Kikawada and A. Quinn, *Before Abraham Was* (1985); B. Connelly, *Arab Folk Epic and Identity* (1986); W. Martin, *Recent Theories of N.* (1986); H. Dubrow, *Captive Victors: Shakespeare's N. Poems and Sonnets* (1987); H. White, *The Content of the Form* (1987); D. Damrosch, *The N. Covenant* (1987)—biblical genres; D. J. Levy, *Chinese N. P.: Late Han through T'ang Dynasties* (1988); R. Edmond, *Affairs of the Hearth* (1988); J. Walker, *Bardic Ethos and the Am. Epic Poem* (1989); *Expansive Poetry*, ed. F. Feirstein (1989); M. J. Toolan, *N.* (1989); H. Fischer, *Romantic Verse N.* (tr. 1991). T.V.F.B.

NEW FORMALISM. A reaction, in late 20th-c. Am. poetry, against free verse (q.v.) and a return to metrical verse and fixed stanza forms.

In the great paradigm shift of the later 19th and early 20th cs. inaugurated by romanticism and carried forward by symbolism into the several movements collectively known as modernism, poets turned away from the canons of traditional prosody (meter, rhyme, stanza) toward alternative forms of verse which were looser metrically if metrical at all, freer rhythmically, or constructed on the basis of visual prosody rather than aural. These new prosodies of "open form" take their origin in Fr. *vers libre* (q.v.) of the 1880s and after and run, via modernist free verse, Beat poetry, concrete poetry, prose poetry, sound poetry, and language poetry, into the 1960s and '70s. In its later stages, such work is collected in anthologies like Hall's 1960 *The New Am. Poetry* (covering 1945–60), Berg and Mezey's 1969 *Naked Poetry*, and Halpern's 1975 *New Am. Poetry Anthol.*

This is not to say that poetry in traditional forms was not being written in America in the 1940s, '50s, '60s, and '70s, however; some poets, like Lowell, Hecht, Merrill, and Berryman, never strayed far from strict form, and others, esp. John Hollander, turned to it increasingly over time. Nor was it a question of liberals versus conservatives—of political Left vs. Right—for both Pound and Eliot were staunch conservatives in politics while radical in poetics; conversely, both Auden and Lowell were leftist in ideology while formalist in prosody. The only issue was whether the form of the poem was to be given by trad. or the moment.

Hence a more scrupulous history of 20th-c. Am. verse would say that modernist free verse was succeeded by several versions of formalist verse (itself following in the wake of New Criticism), which in turn provoked the reaction of the San Francisco Ren. Consequently the New Formalist movement of the '80s is more directly a response to the Sixties. Its impetus may be found in books like Charles Martin's *Room for Error* (1978), Timothy Steele's *Uncertainties and Rest* (1979), and Brad Leithauser's *Hundreds of Fireflies* (1982); it found its own codification in anthologies like Dacey and Jauss's *Strong Measures* (1986) and Lehman's *Ecstatic Occasions, Expedient Forms* (1987); and it evoked new journals like *The Formalist* (1990–).

In retrospect, what the modernists repudiated was not so much a set of worn-out forms as, more deeply, the social attitudes and literary conventions that legitimized those forms and held them in place. Modernist poets sought, for the most part, not to destroy poetic form but to Make It New. Yet even in revolt the more technically scrupulous of the modernists, e.g. Pound, still took inspiration from some of the most elaborate prosodies of *l'ancien régime*, e.g. Occitan and OE. All of them soon discovered, however, that having rejected old forms, they had no choice but to invent new ones in order to go on at all: there is no escape from form, and the reaction to form can only be expressed in another set of forms; for not to have form is not to have any means by which to express the reaction one wants to articulate. No Form, No Speech.

Free verse is thus no more or less artificial than traditional metered verse; the choice of any form is a priori the same: the choice of form. And, in time, with the old attitudes and old moralities manifestly gone, gone altogether, poets discovered they could return to traditional forms once again, forms which a generation of poets who came of age in the Sixties never knew. By 1980, Gioia remarks, "for the first time in the history of modern Eng., most published young poets could not write with minimal competence in traditional meters" since "most of the craft of traditional Eng. versification had been forgotten." The revival of Formalist modes thus became a new initiative amid the general poverty of Am. poetry in the 1980s. In this initiative, old forms were Made New in a very real sense: stripped by time of much of their associative baggage, they became simply opportunities, strategies—ways of proceeding.—

A. Shapiro, "The N. F.," *CritI* 14 (1987); D. Wojahn, "Yes, But: Some Thoughts on the N. F.," *Crazyhorse* 32 (1987); D. Gioia, "Notes on the N. F.," *HudR*

40 (1987); *Expansive Poetry*, ed. F. Feirstein (1989).
T.V.F.B.

NONSENSE VERSE. Some readers consider that any poetry which tells a fantastic story or which describes a fictive world in which the natural laws of the world as we know it do not operate (comparable to the prose example of Lewis Carroll's *Alice in Wonderland*) is n. v. And there are certainly, in the world's poetries, ample numbers of bizarre, fantastic, mythic, or surreal stories in verse which describe some autonomous world which clearly operates according to a set of laws which have their own internal logic. The impossibility trope is common to the rhetorical figure of adynaton and to medieval absurdity-genres such as *coq-à-l'âne*, *frottola*, and *barzelletta*. Any versified account that is nonreferential or unrealistic to a considerable degree could be called n.

These certainly have their interest. It is however naive to believe that n. v. does not "make sense"; much of it does, in its own way. "N.," a modern critic has remarked, "is not no-sense." Rather, we must say, n. v. is verse which does not yield the *same kind* of denotative sense that sentences do in ordinary lang. or prose or even most poetry, where the words chosen are of known lexical meaning (as recorded in dictionaries; see LEXIS) and are arranged in normal syntax. N. v. may in fact yield sense in only vestigial, disconnected, or centrifugal ways, or it may yield sense in unexpected, unpredictable, or hitherto unknown ways. But these are shard-sense or new-sense, not no-sense, which would be the verbal equivalent of a series of random numbers. Users of lang. live in meaning, and will create sense wherever conceivably possible.

Still, the term "n. v." is more properly reserved for verse in which the dislocation is less that of plot or fictive world than of lang. itself. N. v. is most often constituted by unusual words—e.g. neologisms, portmanteau words—or unusual syntax or both. But even a poem which presented wholly unrecognizable morphemes in a wholly unrecognizable syntactic order would evoke at least some threads of sense. The reason is writ large in linguistic competence. All native speakers of a lang. know, intuitively, the phonotactics of their particular lang.: these are the vowel-and-consonant patterns which are permissible in that lang. vs. those which are not, e.g. initial *pl-*, *pr-*, *ps-*, or *st-* in Eng. but not *pf-* or *pg-* and only rarely *skl-* or *sv-*. Even the invention of new words is not carried on in a vacuum; such words automatically suggest vestiges of meaning which their component vowels or consonant patterns carry over from familiar words having similar ones (Bolinger). Further, every lang. develops intricate constellations of synchronic meaning, based on the associations words come to have with each other, quite apart from the diachronic affiliations they have, based on their family histories (Malkiel). So, for example, many Eng. words (though certainly not all) beginning *gl-* are associated with light, e.g. *glisten, gleam, glitter, glow, glower* (but not *gland, glitch*). Also of importance are the rules of morphotactics, namely the rules by which words may be formed in a lang.—e.g. compounding, prefixing, and suffixing. These rules also are finite, specified, and internalized by native speakers.

Of relevance too are the rules of syntax. Whether a lang. is positional or inflectional or some mixture of both, readers have expectations for word order which they maintain even when the words that fill the syntactic slots are unknown. Eng. readers, for example, normally expect the syntactic sequence Subject—Verb—Object, and they expect sentences to be complete and to be drawn from the inventory of only nine patterns available in Eng. Readers are much less disoriented by unfamiliar words in familiar syntax than by the opposite. This explains why most n. v., in which the object is to create amusement or delight not confusion or alienation, relies mainly on novelty in phonotactics and morphology rather than syntax. N. v. draws upon and requires linguistic competence, i.e. the complex of information about sound-combinations, word-formation, and syntactic order which is part of the ordinary processing skill of all native speakers. N. v. simply extends these rules in order to create new semantic structures.

The classic example of n. v. in Eng. is Lewis Carroll's "Jabberwocky," esp. its famous opening:

> 'Twas brillig and the slithy toves
> Did gyre and gimble in the wabe;
> And mimsy were the borogroves
> And the mome raths outgrabe.

The gloss of this poem provided by Humpty Dumpty in ch. 5 of *Through the Looking Glass* is only partly enlightening: many readers would have felt on their own that "slithy" implies both "lithe" and "slimy," or that "gyre" means "to go round and round like a gyroscope," as it does in Yeats. Dumpty reports that "toves" are "something like badgers . . . something like lizards . . . something like corkscrews," and that a "rath" is "a sort of green pig," but these denotations are arbitrary; readers are free to devise their own creatures for these words regardless of whether they take these nouns as having expressive sound patterns (see ONOMATOPOEIA). It is of interest that, of the 23 words in the four lines above, only 11 are novel: the syntax, while paratactic, is perfectly normal Eng.: four main clauses in sequence, the second one, for example, consisting of N + V + V + PrepP.

The affiliations between n. v. and other verse-forms still remain to be mapped out, as do relations on the level of content, given the caveat in paragraph two of this entry. As for the former, n. limericks are a well established form. N. v. may also be related to some types of visual poetry (q.v.). One thinks of the fragmented shapes of e e cum-

mings' poems, for example, but the deep structures of these poems turn out to be traditional—a great many of them are exploded sonnets. As for the latter, the connections are much wider: children's verse covers a range of forms which obviously pertain, and numerous other forms of verse which are highly imaginative or the result of various states of altered consciousness also employ dislocated, reconstructed, or novel lang. The line between n. and the verbal products of dada, surrealism, and cubism, for example, is virtually impossible to draw. The verse generated by computer programs which reassemble lexical items in ordinary syntax to create new poems, or else use the same process to parody famous poems of the past, is also n. v.—of a sort. But is it poetry? We will not ask that here.

N. Verses: An Anthol., ed. L. Reed (1925); E. Cammaerts, *The Poetry of N.* (1926); A. L. Huxley, *Essays New and Old* (1927); *Surrealism*, ed. H. Read (1936); D. K. Roberts, *Nonsensical and Surrealist Verse* (1938); L. E. Arnaud, *Fr. N. Lit. in the Middle Ages* (1942); E. Partridge, *Here, There and Everywhere* (1950); E. Lear, *The Complete N. of Edward Lear*, ed. H. Jackson (1951); *Penguin Book of Comic and Curious Verse*, ed. J. M. Cohen (1952); E. Sewell, *The Field of N.* (1952); G. Orwell, *Shooting an Elephant* (1954); *Anthol. du n.*, ed. R. Benayoun (1957); *A N. Anthol.*, ed. C. Wells (1958); *A Book of N. Songs*, ed. N. Cazden (1961); W. Forster, *Poetry of Significant N.* (1962); D. F. Kirk, *Charles Dodgson, Semeiotician* (1962); A. Liede, *Dichtung als Spiel*, 2 v. (1963); D. Bolinger, *Forms of Eng.* (1965); D. Sonstroem, "Making Earnest of Game: G. M. Hopkins and N. Poetry," *MLQ* 28 (1967); *The N. Book*, ed. D. Emrich (1970); R. Hildebrandt, *N.* (1970); A. Schöne, *Englische N. und Gruselballaden* (1970); D. Petzoldt, *Formen und Funktionen englische N.* (1972); Y. Malkiel, "From Phonosymbolism to Morphosymbolism," *Fourth LACUS Forum*, ed. M. Paradis (1978); S. Stewart, *N.: Aspects of Intertextuality in Folklore and Lit.* (1979); J.-J. Lecercle, *Philosophy Through the Looking-Glass* (1986); G. Deleuze, *The Logic of Sense* (tr. 1989). T.V.F.B.

O

OCTAVE (rarely, octet). A stanza of 8 lines. Os. appear as isolable stanzas, such as particularly the It. *ottava rima* (rhyming *abababcc*) and the Fr. *ballade* (*ababbcbc*), as well as the single o. in Fr. called the *huitain* and the single *ballade* stanza in Eng. called the "*Monk's Tale* stanza" after Chaucer, who probably also derived his 7-line "rhyme royal" (q.v.) from the o. of the *ballade*. The o. is the stanza of the first rank in Occitan poetry, a favorite of the troubadours; and although the *trouvères* of northern France are less exclusive, they still show a decided preference for the o.: roughly a third of all extant OF lyrics are set in one or another form of o. In Sp. the o. of octosyllables rhyming *abbaacca*, less often *ababbccb* or *abbaacac*, is called the *copla de arte menor*, that of 12-syllable lines the *copla de arte mayor*. In It., the Sicilian o., in hendecasyllables rhyming *abababab*, first appears in the 13th c. In ON skaldic poetry, the most important stanzas are *drottkvætt* and *hrynhent*, os. of 6- and 8-syllable lines, respectively, bound tightly by alliteration, stressing, and rhyme.

Os. are also important components of larger stanzas: the first eight lines of the sonnet are also called an o. Shelley uses os. for *The Witch of Atlas*, as does Keats for "Isabella." "I Have finished the First Canto, a long one, of about 180 os.," says Byron in a letter, of *Don Juan*. Gerard Manley Hopkins uses the scheme *ababcbca* for *The Wreck of the Deutschland*. John Berryman uses heterometrical os. in *abcbddba* for *Homage to Mistress Bradstreet* (1953). Louis Zukofsky's *80 Flowers* is a sequence of 81 os.—Schipper; R. Beum, "Yeats's Os.," *TSLL* 3 (1961); R. Moran, "The Os. of E. A. Robinson," *CLQ* 7 (1969). T.V.F.B.

ODE (Gr. *aeidein* "to sing," "to chant"). In modern usage the term for the most formal, ceremonious, and complexly organized form of lyric poetry, usually of considerable length. It is frequently the vehicle for public utterance on state occasions, e.g. a ruler's accession, birthday, or funeral, or the dedication of some imposing public monument. The ode as it has evolved in contemporary lits. generally shows a dual inheritance from classical sources, variously combining the measured, recurrent stanza of the Horation ode, with its attendant balance of tone and sentiment (sometimes amounting to a controlled ambiguity, as in Marvell's "Horation Ode" on Cromwell), and the regular or irregular stanzaic triad of Pindar, with its elevated, vertiginously changeable tone (as in Collins' "Ode on the Poetical Character"), in interesting manifestations as late as Robert Bridges and Paul Claudel. Both forms have frequently been used for poems celebrating public events, but both have just as frequently eschewed such events, sometimes pointedly, in favor of private occasions of crisis or joy. The serious tone of the ode calls for the use of a heightened diction and enrichment by poetic device, but this lays it open, more readily than any other lyric form, to burlesque.

In Gr. lit., the odes of Pindar (522–442 B.C.) were designed for choric song and dance. The

words, the sole surviving element of the integral experience, reflect the demands of the other two arts. A *strophe*, a complex metrical structure whose length and pattern of heterometric lines varies from one ode to another, reflects a dance pattern, which is then repeated exactly in an *antistrophe* (the dancers repeating the steps but in the opposite direction), the pattern being closed by an *epode*, or third section, of differing length and structure. The ode as a whole (surviving examples range from fragments to nearly 300 lines) is built up by exact metrical repetition of the original triadic pattern. These odes, written for performance in a Dionysiac theater or perhaps in the Agora to celebrate athletic victories, frequently appear incoherent in their brilliance of imagery, abrupt shifts in subject matter, and apparent disorder of form within the individual sections. But modern crit. has answered such objections, which date from the time of Pindar himself, by discerning dominating images, emotional relationships between subjects, and complex metrical organization. The tone of the odes is emotional, exalted, and intense; the subject matter, whatever divine myths can be adduced to the occasion.

Apart from Pindar, another pervasive source of the modern ode in Gr. lit. is the cult-hymn, which derived from the Homeric hymns and flourished during the Alexandrian period in the work of Callimachus and others. This sort of poem is notable not for its form but for its structure of argument: an invocation of a deity (later of a personified natural or psychological entity), followed by a narrative genealogy establishing the antiquity and authenticity of the deity, followed by a petition for some special favor, and concluding with a vow of future service. A complete modern instance of this structure is Keats's "Ode to Psyche." Yet another source of the modern ode's structure of prayerful petition is the Psalms and other poems of the Hebrew Bible, which increasingly influenced Eng. poetry by way of Milton, the crit. of John Dennis, the original and translated hymns of Isaac Watts (see HYMN), and Bishop Robert Lowth's *Lectures on the Sacred Poetry of the Hebrews* (1753).

In Lat. lit., the characteristic ode is associated with Horace (65–8 B.C.), who derived his forms not from Pindar but from less elaborate Gr. lyrics, through Alcaeus and Sappho. The Horatian ode is tranquil rather than intense, contemplative rather than brilliant, and intended for the reader in his library rather than for the spectator in the theater. Horace himself wrote commissioned odes, most notably the *Carmen saeculare* for Augustus, all of which more closely approximated the Pindaric form and voice, but his influence on modern poetry is felt more directly in the trad. of what might be called the sustained epigram, esp. in the period between Jonson and Prior. Among the Eng. poets of note, only Mark Akenside habitually wrote odes in the Horatian vein, but in the 17th c. poets as diverse as Herrick, Thomas Randolph, and—most important among them—Marvell with his "Cromwell Ode" wrote urbane Horatians.

The third form of the modern ode, the Anacreontic, is descended from the 16th-c. discovery of a group of some 60 poems, all credited to Anacreon, although the Gr. originals now appear to span a full thousand years. In general the lines are short and, in comparison with the Pindaric ode, the forms simple, the subjects being love or drinking, as in the 18th-c. song "To Anacreon in Heaven," whose tune was appropriated for "The Star-Spangled Banner."

Throughout Europe the history of the ode commences with the rediscovery of the classic forms. The humanistic ode of the 15th and earlier 16th c. shows the adaptation of old meters to new subjects by Fifelfo in both Gr. and Lat., and by Campano, Pontano, and Flaminio in Neo-Latin. The example of the humanistic ode and the publication in 1513 of the Aldine edition of Pindar were the strongest influences upon the vernacular ode in Italy; tentative Pindaric experiments were made by Trissino, Alamanni, and Minturno, but without establishing the ode as a new genre. More successful were the attempts in France by members of the *Pléiade*: after minor trials of the new form by others, Ronsard in 1550 published *The First Four Books of the Odes*, stylistic imitations of Horace, Anacreon, and (in the first book) Pindar. Influenced by Ronsard, both Bernardo Tasso and Gabriele Chiabrera later in the century succeeded in popularizing the form in Italy, where it has been used successfully by, among others, Manzoni, Leopardi (in his *Odicanzone*), Carducci (*Odi barbare*, 1877), and D'Annunzio (*Odi navale*, 1892). In France, the example of Ronsard was widely followed, notably by Boileau in the 17th c. and by Voltaire and others in formal occasional verse in the 18th. The romantic period lent a more personal note to both form and subject matter, notably in the work of Lamartine, Musset, and Victor Hugo. Later, highly personal treatments of the genre may be found in Verlaine's *Odes en son honneur* (1893) and Valéry's *Odes* (1920). In Sp., odes have figured in the work of Pablo Neruda (1904–73), who wrote three volumes of them: *Odas elementales* (1954), *Nuevas odas elementales* (1956), and *Tercer libro de las odas* (1957).

The ode became characteristically Ger. only with the work of G. R. Weckherlin (*Oden und Gesänge*, 1618–19), who, as court poet at Stuttgart, attempted to purify and refashion Ger. letters according to foreign models. In the mid 18th c. Klopstock modified the classical models by use of free rhythms, grand abstract subjects, and a heavy influence from the Lutheran psalms. Later Goethe and Schiller returned to classical models and feeling, as in Schiller's "Ode to Joy," used in the final movement of Beethoven's Ninth Symphony. At the turn of the century, Hölderlin in his complex, mystical, unrhymed odes united classical themes with the characteristic resources of the

Ger. lang. Since Hölderlin, few noteworthy odes have been written in Ger., with the possible exception of those of Rudolph Alexander Schröder (*Deutsche Oden*, 1912).

The few attempts at domesticating the ode in 16th-c. England were largely unsuccessful, although there is probably some influence of the classical ode upon Spenser's "Fowre Hymnes," "Prothalamion" and "Epithalamion." In 1629 appeared the first great imitation of Pindar in Eng., Ben Jonson's "Ode on the Death of Sir H. Morison," with the strophe, antistrophe, and epode of the Cl. model indicated by the Eng. terms "turn," "counter-turn," and "stand." In the same year Milton began the composition of his great ode, "On the Morning of Christ's Nativity," in regular stanzaic form. The genre, however, attained great popularity in Eng. only with the publication of Abraham Cowley's *Pindarique Odes* in 1656, in which he attempted, like Ronsard and Weckherlin before him, to make available to his own lang. the spirit and tone of Pindar rather than to furnish an exact transcription of his manner. Cowley was uncertain whether Pindar's odes were regular, and the matter was not settled until 1706, when the playwright William Congreve published with an ode of his own a "Discourse" showing that they were indeed regular. With the appearance in 1749 of a scholarly tr. of Pindar by Gilbert West, the fashion for Cowleyan Pindarics died away. With Dryden begin the great formal odes of the 18th c.: first the "Ode to the Memory of Mrs. Anne Killigrew," and then, marking the reunion of formal verse and music, the "Song for St. Cecilia's Day" and "Alexander's Feast." For the 18th c. the ode was the perfect means of expressing the sublime. Using personification and other devices of allegory, Gray and Collins in the mid-18th c. marshall emotions ranging from anxiety to terror in the service of their central theme, the "progress of poetry," making the ode a crisis poem that reflects the rivalry of modern lyric with the great poets and genres of the past.

The romantic ode in Eng. lit. is a poem written on the occasion of a vocational or existential crisis in order to reassert the power and range of the poet's voice. It begins with Coleridge's "Dejection: An Ode" (1802) and Wordsworth's pseudo-Pindaric "Ode: Intimations of Immortality" (written 1802–4, pub. 1815). Wordsworth's "Intimations Ode," with its varied line lengths, complex rhyme scheme, and stanzas of varying length and pattern, has been called the greatest Eng. Pindaric ode. Of the other major romantic poets, Shelley wrote the "Ode to the West Wind" and Keats the "Ode on a Grecian Urn," "Ode to a Nightingale," and "To Autumn," arguably the finest three odes in the lang. They are written in regular stanzas derived not from Horace but from Keats's experiments with the sonnet form. Since the romantic period, with the exception of a few brilliant but isolated examples such as Tennyson's "Ode on the Death of the Duke of Wellington," the ode has been neither a popular nor a really successful genre in Eng. Among modern poets, the personal ode in the Horatian manner has been revived with some success, notably by Allen Tate ("Ode to the Confederate Dead") and W. H. Auden ("In Memory of W. B. Yeats," "To Limestone").

Schipper, v. 2, sects. 516–25, and *History* 366 ff.—on the Pindaric; G. Carducci, "Dello svolgimento dell'ode in Italia," *Opere*, v. 16 (1905); E. R. Keppeler, *Die Pindarische Ode in der dt. Poesie des XVII und XVIII Jhs.* (1911); R. Shafer, *The Eng. Ode to 1660* (1918); I. Silver, *The Pindaric Odes of Ronsard* (1937); G. N. Shuster, *The Eng. Ode from Milton to Keats* (1940); G. Highet, *The Cl. Trad.* (1949); N. Maclean, "From Action to Image: Theories of the Lyric in the 18th C.," in Crane; C. Maddison, *Apollo and the Nine: A Hist. of the Ode* (1960); Bowra; K. Viëtor, *Gesch. der deutschen Ode*, 2d ed. (1961); A. W. Pickard-Cambridge, *Dithyramb, Tragedy and Comedy*, 2d ed. (1962); S. Commager, *The Odes of Horace: A Critical Study* (1962); K. Schlüter, *Die englische Ode* (1964); H. D. Goldstein, "*Anglorum Pindarus*: Model and Milieu," *CompLit* 17 (1965); P. Habermann, "Antike Versmasse und Strophen- (Oden-) formen im Deutschen," and J. Wiegand and W. Kohlschmidt, "Ode," *Reallexikon*; J. Heath-Stubbs, *The Ode* (1969); G. Hartman, "Blake and the Progress of Poetry," *Beyond Formalism* (1970); G. Otto, *Ode, Ekloge und Elegie im 18. Jahrhundert* (1973); J. D. Jump, *The Ode* (1974); Wilkins; M. R. Lefkowitz, *The Victory Ode* (1976); P. H. Fry, *The Poet's Calling in the Eng. Ode* (1980); Morier; J. Culler, *The Pursuit of Signs* (1981), ch. 7; K. Crotty, *Song and Action: The Victory Odes of Pindar* (1982); W. Mullen, *Choreia* (1982); H. Vendler, *The Odes of John Keats* (1983); M. H. Abrams, *The Correspondent Breeze* (1984), ch. 4; J. W. Rhodes, *Keats's Major Odes: An Annot. Bibl. of Crit.* (1984); Terras; A. P. Burnett, *The Art of Bacchylides* (1985); D. S. Carne-Ross, *Pindar* (1985); N. Teich, "The Ode in Eng. Lit. Hist.," *PLL* 21 (1985); S. Curran, *Poetic Form and British Romanticism* (1986), ch. 4; W. Fitzgerald, *Agonistic Poetry: The Pindaric Mode in Pindar, Horace, Hölderlin, and the Eng. Ode* (1987); Hollier, 198 ff.; *Selected Odes of Pablo Neruda*, ed. and tr. M. S. Peden (1990); G. Davis, *Polyhymnia* (1991).

S.F.F.; P.H.F.

ONOMATOPOEIA (Ger. *Klangmalerei, Lautsymbolik*) is the traditional term for words which seem to imitate the things they refer to, as in this line from Collins' "Ode to Evening": "Now the air is hushed, somewhere the weak-eyed bat / With short shrill shriek flits by on leathern wing." In the strict sense, o. refers to words which imitate sounds (e.g. *dingdong, roar, swish, murmur, susurrus*), but other qualities such as size, motion, and even color may be suggested; and the term is most often used with wider reference, to denote any word whose sound is felt to have a "natural" or direct relation with its sense. Since their phonetic shape seems

motivated rather than arbitrary, onomatopoeic words exert significant limitations on Saussure's doctrine of the arbitrariness of the sign. Both Jespersen and Sapir showed evidence that Saussure greatly overstated his case.

O. is one of four types of verbal effects which in poetics are usually called "expressive" or "mimetic" but which in linguistics are widely referred to as instances of "phonetic symbolism" or "sound symbolism." All four terms are objectionable: the first begs the question, the second explains little, and the third and fourth are simply confused: expressive and onomatopoeic words are, in the terminology of Charles Sanders Peirce, *icons* rather than *symbols*; it is precisely the point that they are not symbolic in the way ordinary words are. Beyond terminology, however, the four types of effects are closely interrelated: these are (1) o. itself (wordsounds imitate sounds in nature); (2) articulatory gesture or kinesthesia (movements of the vocal or facial muscles or the shape of the mouth is suggestive; see SOUND); (3) synaesthesia, phonesthemes, and other associative phenomena (heard sounds trigger other sensory impressions); and (4) morphosymbolism and iconic syntax (see below).

It has long been fashionable among literary critics to disparage o. as a crude and over-obvious poetic device, but the linguistic evidence for iconicity as an important process at work in langs. worldwide is formidable. There is good evidence that certain iconic effects operate across a wide spectrum of langs. and may be linguistic universals. Jespersen's astonishingly long list of words having the unrounded high front vowel /i/ and all connoting "small, slight, insignificant, or weak" is famous; in Eng., this phoneme is used almost universally by babies if not parents as a suffix for small, familiar, or comforting things (e.g. *mommy, daddy, baby, teddy, beddie, dooggie, kitty, ouchie, munchie*). Woodworth's analysis of deixis in 26 langs. revealed a "systematic relationship between vowel quality and distance," namely that a word having proximal meaning has a vowel of higher pitch than one having distal meaning.

Further, iconic effects in lang. have been shown to be not merely phonological but also morphological and syntactic. O. is therefore simply the most conspicuous instance of a broad range of natural linguistic effects, not a merely "poetic" device. In ordinary speech, sound is motivated by sense: once sense is selected, sound follows. But in iconic speech, sense is motivated by sound: words are chosen for their sound, which itself determines meaning. When Pope says that "the sound must seem an echo to the sense," he means, says Dennis Taylor, not merely that sound follows sense but that "the sense makes us read the sound as confirming the sense." Now meaning proceeds from the very mouth of lang.

It is usually said of o. that the sound of the word imitates a natural phenomenon in the world which the word represents, because the thing is itself a sound, such as *whirr*. But the sound of the word is not often precisely the sound of the thing. As Chatman shows, the connection is not exact: a word cannot enact even a natural sound if that sound is not an available phoneme in the lang. itself, and most onomatopoeic words are only approximations of the natural sounds. Even the words which we tell our children are the sounds the animals make are highly conventionalized in every lang.: Americans teach their children that dogs say "Bow, wow" or "Arf, arf" or "woof," but the Fr. say their dogs say "Ouah, ouah." In Eng., pigs are said to say "oink, oink" (though every adult knows better), while in Flemish they say "gron, gron" (which is closer). The conventionality inherent in every lang. can never be entirely filtered out, and the confusions are legion.

But o. is not solely a relation between words and things, for this formulation leaves out sense. The relation between words and things in the world has not two poles but four. The horizontal, linear relation linking the poles of words and things is crossed by the vertical line linking the poles of sense and sound. Since o. only operates through the agency of words, iconic effects in lines of poetry are created by words which always already bear meaning. "The pattern of sound does not reinforce an already established meaning," says Stein, "so much as it helps shape and modulate that meaning." Dr. Johnson complained that the "same sequence of syllables" can convey very different senses, and both I. A. Richards and John Crowe Ransom constructed "dummies" of onomatopoeic lines to show that alterations of wording without radical alteration of the underlying sound pattern of the line produce wildly different meanings.

The point is that sound-symbolic effects do not operate without words to trigger them, words which first establish the semantic field. Sibilants are often described as sonorous or soft, but they may also be used to convey sinister connotations (the hissing of snakes, slithering through the grass), sadness, and lubricious sexuality. Sounds can never precede meaning: they can only operate on meanings already lexically created. Dr. Johnson's criticism is still trenchant: "it is scarcely to be doubted, that on many occasions we make the musick which we imagine ourselves to hear; that we modulate the poem by our own disposition, and ascribe to the numbers the effects of the sense" (*Rambler*, nos. 94, 92; cf. *Idler* 60; *Life of Pope*).

In general, it may be said that all forms of sound symbolism operate by analogy or association of the physical properties of sound to the physical properties of real objects, as the former are constrained by the conventions specific to each lang., and to meanings. It was Lord Kames who observed in 1762 that "in lang., sound and sense being intimately connected, the properties of the one are readily communicated to the other."

From the wider perspective of sound symbolism

or iconicity, o. is part of a much larger set of associative relations between word(sound)s and meanings. Here one feature of a word, its sound, becomes an analogue, a correlate in another mode, for a feature of some other word or natural process. Now "relationships between sounds will map relationships of sense" (Kenner), a point made by Shelley in the *Defense of Poetry*. This is a much more satisfactory heuristic within which to map out a typology of iconic effects in that it will link seemingly unrelated effects. Such associative processes are no less real than the physical features of objects; once codified as conventions, they become facts of behavior and are taken as fact by speakers of a lang.

The process of association operates in both directions. Once speakers associate a certain sound with a certain meaning or field(s) of meaning, words of similar sound come to be associated with those meanings, and words of similar meaning conversely will come to use similar sounds over time. It is one of the axioms of verse structure that similarity in word-sounds implies some similarity in meanings; all sound-marking devices (e.g. alliteration) and sound patterns in poetry use sound to connect meanings. And similarity is apparently preferentially selected over contrast: Benjamin Lee Whorf pointed out that auditors are likely to notice if words for a certain experience share similar characteristics with it, but usually do not notice a relation of contrast or conflict. Indeed, contrasts will be overlooked: Jakobson pointed out the interesting phenomenon whereby words denoting roughly the same concept often form binary pairs with opposed vowels: tip—*top*, *slit*—*slot*. There is extensive linguistic evidence that certain associations and, more significantly, oppositional structures operate across langs., generating such contrasts as "dark" vs. "bright" vowels. Jakobson attributed such binary oppositional forms to the nature of cognition itself.

Associative processes operate extensively among words themselves, particularly in morphology, though also in syntax. The sound shape of a word is almost never created *ex nihilo*: it is most often formed in relation to some existing word; and after it comes into existence, it is continuously subject to influence from other words in the lang. system not only diachronically but also synchronically. Any entity, no matter how arbitrary when it enters the system, is thereafter subject to continuous accommodation to and influence by all the other entities in the system. Many writers use the term "o." for what in fact amounts to "reverse o." (Bolinger's term): here "not only is the word assimilated to the sound, but the sound is also assimilated to the 'wordness' of the word." Bolinger identified a series of morphosemantic processes based mainly on association wherein the form influences the meaning or the meaning influences the form of words. Constellations of words form over time, all having similar meanings and similar sound: one example is the series of Eng. words beginning with *gl-* which have to do with light; another is the set of words ending in *-ash*. All this evidence strongly suggests that any word "assumes all the affective and associative privileges enjoyed by the most obvious o." (Bolinger). Malkiel has also mapped some of this terrain, arguing that the phenomenon of phonosymbolism is flanked by a parallel and even wider field of "morphosymbolism," namely iconic or mimetic processes at work in word-formation, phrase formation, and syntax. His many examples suggest that lang. encompasses "form symbolism as against sound symbolism," and that there is a steady gradation between the two: phonic and morphic processes of iconicity "work in unison." Consequently, "when we speak of sound-suggestiveness, we speak of the entire lang., not just of a few imitative or self-sufficient forms" (Bolinger). Iconic syntax has only recently been examined (Haiman), but clearly some structures are mimetic, e.g. any sentence which refers to a situation where items appear in sequence and presents its lexical terms in the same sequence.

The subject of iconicity in lang. was a topic of lively interest to the ancients. It is of course central to Plato's *Cratylus*, and it also appears in Aristotle (*Rhet.* 3.9), Demetrius (*On Style*), Dionysius of Halicarnassus (*On Literary Composition*), and Quintilian (*Institutio oratoria* 9.3, 9.4), among others. In Canto 32 of the *Inferno*, Dante seeks the sorts of appropriate words (*s'io avessi le rime aspre e chiocce*) which he classifies in his prose treatise on diction and prosody, *De vulgari eloquentia*. Saussure impressed indelibly upon the mind of the 20th c. the idea that the relation between word and thing is arbitrary. But of course desire cuts so much deeper than fact. As Ransom repeatedly said, poetic lang. aspires to the condition of nature: it "induces the provision of icons among the symbols." Poets continually desire to make lang. appropriate, so that words partake of the nature of things. And the agency is the fact that words *are* things, have physical bodies with extension in space and duration in time, like people, like things.

G. von der Gabelentz, *Die Sprachwissenschaft* (1891); M. Grammont, "Onomatopoées et mots expressifs," *Trentenaire de la société pour l'étude des langues romanes* (1901); A. H. Tolman, "The Symbolic Value of Eng. Sounds," *The Views About* Hamlet (1904); O. Jespersen, "Sound Symbolism," *Lang.* (1922), "Symbolic Value of the Vowel *I*," *Linguistica* (1933); J. R. Firth, *Speech* (1930), "Modes of Meaning," *E&S* 4 (1951); J. C. Ransom, *The World's Body* (1938), "Positive and Near-Positive Aesthetics," *KR* 5 (1943), both rpt. in Ransom; E. Sapir, *Sel. Writings* (1949); D. T. Mace, "The Doctrine of Sound and Sense in Augustan Poetic Theory," *RES* 2 (1951); A. Stein, "Structures of Sound in Milton's Verse," *KR* 15 (1953); H. Wissemann, *Untersuchungen zur Onomatopoiie* (1954); Wellek and Warren, ch. 13; W. T. Moynihan, "The Auditory Correlative," *JAAC* 17 (1958);

P. Delbouille, *Poésie et sonorités*, 2 v. (1961, 1984); Z. Wittoch, "Les Onomatopées forment-elles une système dans la langue?" *AION-SL* 4 (1962); L. P. Wilkinson, *Golden Lat. Artistry* (1963); I. Fónagy, *Die Metaphern in der Phonetik* (1963); G. Bebermeyer, "Lautsymbolik," *Reallexikon* 2.4–8; C. Ricks, "Atomology," *Balcony* 1 (1965); D. Bolinger, *Forms of Eng.* (1965), esp. "The Sign Is Not Arbitrary," "Word Affinities," and "Rime, Assonance, and Morpheme Analysis"; S. Chatman, *A Theory of Meter* (1965); L. B. Murdy, *Sound and Sense in Dylan Thomas's Poetry* (1966); M. B. Emeneau, "Onomatopoetics in the Indian Linguistic Area," *Lang.* 45 (1969); J.-M. Peterfalvi, *Recherches expérimentales sur le symbolisme phonétique* (1970); A. A. Hill, "Sound-Symbolism in Lexicon and Lit.," *Studies in Ling. G. L. Trager* (1972); J. D. Sadler, "O.," *ClassJ* 67 (1972); J. A. Barish, "Yvor Winters and the Antimimetic Prejudice," *NLH* 2 (1970); G. L. Anderson, "Phonetic Symbolism and Phonological Style," *Current Trends in Stylistics* (1972); J. Derbolav. *Platons Sprachphilosophie im Kratylos und in den späteren Schriften* (1972); D. H. Melhem, "Ivan Fónagy and Paul Delbouille: Sonority Structures in Poetic Lang.," *Lang&S* 6 (1973); E. L. Epstein, "The Self-Reflexive Artefact," *Style and Structure in Lit.*, ed. R. Fowler (1975); W. K. Wimsatt, Jr., "In Search of Verbal Mimesis," *Day of the Leopards* (1976); P. L. French, "Toward an Explanation of Phonetic Symbolism," *Word* 28 (1977); Y. Malkiel, "From Phonosymbolism to Morphosymbolism," *Fourth LACUS Forum* (1978), supp. by D. B. Justice, "Iconicity and Association," *RPh* 33 (1980); L. I. Weinstock, "O. and Related Phenomena in Biblical Hebrew," *DAI* 40 (1979): 3268A; R. Jakobson and L. Waugh, "The Spell of Speech Sounds," *The Sound Shape of Lang.* (1979), rpt. in Jakobson, v. 8; D. A. Pharies, "Sound Symbolism in the Romance Langs.," *DAI* 41 (1980): 231A; R. A. Wescott, *Sound and Sense: Essays on Phonosemic Subjects* (1980); M. Borroff, "Sound Symbolism as Drama in the Poetry of Wallace Stevens," *ELH* 48 (1981), "Sound Symbolism as Drama in the Poetry of Robert Frost," *PMLA* 107 (1992); Morier; Brogan, 97–108—survey of studies; R. Lewis, *On Reading Fr. Verse* (1982), ch. 7; N. L. Woodworth, "Sound Symbolism in Proximal and Distal Forms," *Linguistics* 29 (1991). T.V.F.B.

ORAL POETRY, esp. oral traditional poetry (o. t. p.), is poetry composed and transmitted mainly but not exclusively by people who cannot read or write. O. p. is traditional when each generation has received it orally from the preceding one, back to the dawn of poetry, and transmits it to the next as long as the trad. lives. O. t. p. has its own methods of composition and transmission which differentiate it from written lit. We should, however, exclude from this category poetry composed in the manner of written lit. "for oral presentation" (see PERFORMANCE). Such poetry, though delivered orally, does not differ from written poetry in its manner of composition. The origins of o. p. are those of poetry itself and are to be sought in ritual; for the rhythms, sound patterns, and repetitive structures of poetry help to support and give power to the words and actions of ritual.

There are three general divisions by genre: ritual, lyric, and narrative. The non-narrative types of o. t. p. include, under ritual: (a) incantation, (b) lullabies, (c) wedding ritual songs, (d) laments, (e) songs for special festivals, and (f) praise poems, or eulogies. Lyric songs are preeminently love songs.

The two main types of oral traditional narrative poetry are epic and ballad. The epic is stichic, the same metric line being repeated with some variation for the entire song; whereas the ballad is stanzaic. Consequently, the tempo of narration of epic is rapid, but that of the ballad is slower. The epic tends to be longer than the ballad, because it tells its story from beginning to end, often with a fullness of detail which is typical. The ballad concentrates frequently on the most intense or dramatic moment of a story, and some ballads consist entirely of dialogue. Although there are comic ballads, the genre as a whole has an elegiac tone. Some of the great oral traditional epics, such as the *Iliad* or the *Chanson de Roland*, are tragic or have tragic overtones, but on the whole, traditional epic is optimistic; the hero generally triumphs gloriously over the enemy.

The most distinctive characteristic of o. t. p. is its variability or fluidity of text, but in some cultures a certain degree of word-for-word memorization is reported, particularly in the shorter forms. In the longer songs, such as epics, the absence of a fixed text makes word-for-word memorization impossible, there being nothing stable to memorize. In the shorter songs the text tends to become more stable in the practice of a single singer and in a song which is frequently sung. In the case of magic incantations, it is sometimes said that the exact reproduction of a text is necessary to make the magic effective, but the existence of variants seems to indicate that sometimes only certain sound patterns such as alliteration and assonance, or certain types of words or word-combinations, are preserved rather than an entire poem, however short. When traditional singers tell us that a text must be repeated exactly, word-for-word, we know from comparison of performances that they mean essential characteristic for essential characteristic, for their concept of a word is different from that of lettered people.

In those cultures where fluidity of text is attested beyond any doubt, where the absence of a fixed text is well documented, the singer-poet learns in the course of years a special technique of composition by "formula" and "theme." It would be misleading to call it improvisation, which implies creating a story or song on the spur of the moment. It would seem, of course, that there are such oral traditional improvisations of both content and text. They are generally topical in nature. Some scholars maintain, however, that

even these "improvisations" are traditional and hence composed in the formulaic technique.

The formula (q.v.) is "a word or group of words regularly used under given metrical conditions to express a given essential idea" (Parry). The most often used phrases, lines, or even couplets—those which a singer hears most frequently when he is learning—establish the patterns for the poetry, its characteristic syntactic, rhythmic, metric, and acoustic molds and configurations. In time the individual practitioner of the art can form new phrases—create formulas—by analogy with the old as needed. When he has become proficient in thinking in the traditional patterns, incl. the traditional phrases and everything else like them, he is a full-fledged singer of o. t. p. In essence, he has learned to speak—or to sing—the special lang. of that poetry. He composes naturally in the forms of his trad., unconsciously, and often very rapidly, as a native speaker speaks a lang.

O. p., of whatever genre, is paratactic. Its style has been called an "adding" style, because the majority of its lines *could* terminate in a period, insofar as their syntax is concerned; instead, however, another idea is often "added" to what precedes. A comparatively small percentage of necessary run-on lines, in contrast to the number of cases of nonperiodic enjambment, is therefore another distinctive and symptomatic feature of oral traditional style. This does not mean that the traditional singer only adds ideas, however; he can develop them as well, and he can return to themes introduced earlier.

Even as the formulas and their basic patterns make composing of lines possible in performance, so the associative use of parallelism (q.v.) in sound, syntax, and rhythm aids the oral poet in moving from one line to another. A line may suggest what is to follow it. Thus clusters of lines are formed and held together by sound, structure, and association of meaning. Such units are easily remembered. At times the complexity of structural interconnections between verses in oral traditional style is so great that it seems that one could have attained it only with the aid of writing. The truth is that these intricate architectonics of expression were developed first in oral traditional verse, establishing from very archaic times the techniques which man with writing inherited and then believed himself to have "invented" with the stylus, the quill, and the pen. The lang. of o. t. p., formulaic though it be, is in fact a dynamic, organic lang., an organism of man's imaginative life.

Because poems composed in the formulaic style have no fixed text, they are not, and cannot be, memorized, for the text is never the same even in performances by the same singer. There may be changes in the narrative or other context as well. To say that the text remains "essentially the same" or "more or less the same" over time is not to say that the text is fixed.

Although there are many repeated incidents, scenes, and descriptions, these too remain flexible, susceptible to expansion or contraction; a journey may be related briefly or with copious details—the description of armor may occupy one line or fifty and still be termed the same "theme," as such repeated incidents and descriptions are called. A "theme" is not merely a repeated subject, however, but a repeated *passage*. It has a more or less stable core of lines or parts of lines, surrounded by various elements adapting it to its context.

The theme is multiform and has existence only in its multiforms. Habit and frequent use may give its form in the practice of a single singer some degree of stability, but no given form of it is sacrosanct. Themes, like formulas, are useful in any song in which the incident or description may belong. The journey framework may be employed in any number of stories; the assembly of men or of gods is common to many tales in song. In learning a song which the singer hears for the first time, he does not think of the text; he constructs his own. He need remember only the names of persons and places and the sequences of events. In formulas and themes he has the building blocks and techniques for rapid composition.

There are "themes" in ritual and lyric songs as well. Recent research has demonstrated that nonnarrative o. t. p. composed formulaically has passages consisting of a more or less stable core of lines and parts of lines that are used in several songs. For such research to be meaningful, it is necessary to have a large body of texts for comparison, since many variants of songs are possible.

In the o. t. p. of some cultures in India and Africa there is reportedly more memorization than in those discussed above, although it is said that they too are composed formulaically. An analysis of some sample texts has indicated, however, that they also exhibit a more or less stable core with surrounding variations.

It is not surprising that a special manner of composition and transmission would influence in a profound way the poetic structure and the poetics of o. t. p. One of the most obvious ways in which this is manifested is in the repetitions of formulas. Written poetry does not tolerate a high degree of repetition, but nothing is more characteristic of o. t. p. than repetition, because it is endemic in the method of oral formulaic composition itself. Translators of Homer, for example, normally avoid Homer's repetitions of noun-epithet formulas, using several different epithets where Homer used only one. In this they are tacitly acknowledging the difference between the poetics of written poetry and that of o. t. p. Another device in some o. p. is to repeat on occasion the second half of a line at the beginning of the next line. This is not acceptable as a regular phenomenon in written poetry, but it is natural to o. t. p. On the other hand, one must stress that rhetorical devices such as anaphora and epiphora, and figures of speech such as similes, metaphors, and even ring compo-

sition (qq.v.) were used in o. t. p. before the invention of writing and were inherited by written poetry from its predecessor.

The configuration of themes that forms a song in oral formulaic traditional composition is similar to the single theme in its fluidity. Like the themes that make it up, the song, reflecting the desires of the singer at the moment of performance, may be sung long or short, even as tales can be told at length or in brief. It may be ornamented to greater or lesser degree. It too has multiforms, which are usually called variants or versions. The term "multiform" is more accurate than "variant" or "version," terms which imply an "original" that has undergone some kind of change resulting in the text before us.

In o. t. p. one can distinguish three meanings of the word "song," not only in the narrative genres but also in ritual and lyric. The first is that of any performance, for each performance is unique and valid in its own right. The second might be called that of the specific subject matter; for example, the song of the capture of Bagdad by Sultan Selim, which would be designated by a title, "Sultan Selim Captures Bagdad." Combining the first and the second, one can say that there will be as many texts of the specific song as there are performances, whether they are recorded or not. The third meaning of "song" could be called the "generic." The story of the capture of Bagdad (the specific song) falls into the category of a number of stories dealing with the capture of cities, just as the *Odyssey*, for example, falls into the generic category of songs recounting the return of the hero after long absence from home. The texts of this particular generic song are very numerous and reach far back into the depths of human history.

The generic song is of considerable importance in o. t. p. It is not merely a convenient method of classification; it represents the significant core of ideas in a song that survive reinterpretation and specific application to "history," a core held together by tensions from the past that give a meaning to the song not apparent on its surface, no matter how lowly or local any given performance may be. Because of this core, one might say that every song in oral trad. retains the essence of its origins within it, in this way reflecting the origin of the very genre to which it belongs.

In o. t. p. the question of authorship is complicated, yet it is clear that, to use the first of the three meanings of song given above, the performer, the traditional singer, is the "author" of his particular performance. In this case the performer is composer as well. One has, therefore, multiple authors, even as one has multiple texts, of any specific or generic song. From that point of view, there are as many authors as there are performances. But of any given text there is but one author, namely its performer-composer. This is a different concept of multiple authorship from that historically employed in Homeric and other epic criticism since Wolf. Moreover, this concept should not be confused with the theory of "communal" authorship once put forth by scholars of the romantic period.

The date of any text of an oral traditional poem is, consequently, the date of its performance, that being the date of its composition. The date of the specific song would be the date on which some traditional singer for the first time adapted existing themes and configurations to other specific people and events—that is to say, the date of the first performance. This is ordinarily beyond our ken. The date of the generic song is lost in prehistory.

In some cultures there are poems, or songs, usually topical in nature, made up on the spot on demand or composed in live contests of poetry. The poems in this category are ephemeral, created for the moment and usually not transmitted to anyone else. Therefore, since they are improvised for a specific circumstance and do not "enter into tradition," they might be called "oral non-traditional" poems. Because the practice of "improvising" such poems has long existed, however, some scholars feel that the style of these ephemeral creations can also be traditional.

O. p. long played an integral role in the life of human beings and social communities; its practice provided that spiritual activity necessary to man's existence; its bonds with everyday life were manifold and close. Its deeper qualities are becoming clear as they are sifted from the transitional periods in which they were first recorded. The knowledge of how o. t. p. is composed and transmitted has brought with it new modes for its evaluation. And these modes have led back to the symbols and meanings of poetry itself in its origins.

The study of o. p. begun by Milman Parry in the 1930s has engendered many other studies, and debate continues. His theory consisted of first making as exact a description as possible of the process of composition and transmission of oral traditional narrative poetry in order to determine its basic and necessary characteristics, and then applying that knowledge to texts from ancient and medieval times for which there is little or no information about how they were composed. He concentrated on the living practice of oral traditional epics in Yugoslavia, but he also collected a large number of oral traditional lyric songs as well. A description of the collecting and a digest of the contents of the collection can be found in the intro. to v. 1 (1954) of *Serbocroatian Heroic Songs*, ed. by Albert B. Lord. More details on singing and collecting are given in David E. Bynum's Prolegomena to v. 6 (1980) and 14 (1979), ed. by him. The music of a selection of the lyric songs was transcribed by Béla Bartók and pub. with a full study of them by him in 1951.

Parry died in 1935; Lord's book *The Singer of Tales*, published in 1960, gave a description of the Yugoslav practice and applied the principles gained from its study to the Homeric poems, the Anglo-Saxon

Beowulf, the *Chanson de Roland,* and the medieval Gr. epic of *Digenis Akritas.* Later the theory was applied to other poetries as well, incl. the works of Hesiod and the Homeric Hymns in ancient Gr. trad., the MHG *Nibelungenlied,* the med. Sp. *Cantar de mio Cid,* Eng. and Scottish ballads, Pre-Islamic and Cl. Arabic poetry, Chinese traditional lyrics, the quatrains of Latvian *dainas,* and many others.

Subsequent scholarship focused largely on the definitions of o. p. and of the formula, the problem of composition by formula and theme vis-a-vis improvisation, memorization, and the context of performance. Ruth Finnegan has advocated a broad definition of o. p. with emphasis on the literal meaning of "oral" and on performance. Among reports of o. p. in various parts of the world, those of Jeff Opland on praise poetry among the Xhosa in South Africa have contributed to a deepening understanding of the place of that kind of poetry in the general scheme of o. p. Opland has also applied his experience with African o. p. to Anglo-Saxon poems and their possible indebtedness to eulogy. Biblical and Near Eastern studies have been influenced by the approach of scholars in the field of o. t. p. Robert C. Culley and William Whallon were among the pioneers in applying oral-formulaic theory to the Old Testament, and the work of Werner H. Kelber is a valuable contribution to the study of the Gospels from that same point of view. Ching-Hsien Wang broadened the field to include Chinese traditional lyrics. James T. Monroe has written on the oral composition of Pre-Islamic poetry, and Michael J. Zwettler's book, in addition to its main subject, has an excellent intro. on oral-formulaic theory. A thorough bibl. of the scholarship on the theory with full intro. is available in Foley (1985).

Parry's work was concerned primarily with oral traditional epic song. He was also interested in the way of life of the people who practiced that kind of poetry, and he was very much aware of the importance of the circumstances of performance and the role of the audience. These aspects of o. p. have been written about by both anthropologists and others concerned with social studies. A special philosophical branch of studies of orality—rather than oral lit.—has also developed from Parry's writings. It includes the work of Eric A. Havelock, H. Marshall McLuhan, and Walter J. Ong. These scholars have examined the effect of literacy and mass media on the way in which humans view the universe and the world in which they live, as well as how they think. Their writings have some pertinence to the study of o. p.

BIBLIOGRAPHY: J. M. Foley, *Oral Formulaic Theory and Research: An Intro. and Annot. Bibl.* (1985).

JOURNAL: *Oral Tradition* (1986–).

STUDIES: B. Bartók and A. B. Lord, *Serbo- Croatian Folk Songs* (1951); R. H. Webber, "Formulaic Diction in the Sp. Ballad," *Univ. of Calif. Pubs. in Mod. Philology* 34 (1951); C. M. Bowra, *Heroic Poetry* (1952); A. B. Lord, *The Singer of Tales* (1960), "Perspectives on Recent Work on Oral Lit.," *Oral Lit.: Seven Essays,* ed. J. J. Duggan (1975), *Epic Singers and Oral Trad.* (1991); H. M. McLuhan, *The Gutenberg Galaxy* (1962); E. A. Havelock, *Preface to Plato* (1963); G. S. Kirk, *The Songs of Homer* (1962); R. C. Culley, *Oral Formulaic Lang. in the Biblical Psalms* (1967); M. Curschmann, "O. P. in Med. Eng., Fr., and Ger. Lit.: Some Notes on Recent Research," *Speculum* 42 (1967); J. B. Hainsworth, *The Flexibility of the Homeric Formula* (1968); N. K. Chadwick and V. Žirmunskij, *Oral Epics of Central Asia* (1969); A. C. Watts, *The Lyre and the Harp* (1969); W. Whallon, *Formula, Character, and Context: Studies in Homeric, OE, and OT Poetry* (1969); Parry; J. T. Monroe, "Oral Composition in Pre-Islamic Poetry," *JArabL* 3 (1972); J. Duggan, *The* Song of Roland: *Formulaic Style and Poetic Craft* (1973); M. N. Nagler, *Spontaneity and Trad.* (1974); C.-H. Wang, *The Bell and the Drum: "Shih Ching" as Formulaic Poetry in an Oral Trad.* (1974); B. Peabody, *The Winged Word* (1975); *Oral Lit. and the Formula,* ed. B. A. Stolz and R. S. Shannon, III (1976); R. Finnegan, *O. P.: Its Nature, Significance, and Social Context* (1977), ed., *World Treasury of O. P.* (1978); M. J. Zwettler, *The Oral Trad. of Cl. Arabic Poetry* 1978; D. E. Bynum, *The Daemon in the Wood* (1978); G. Nagy, *The Best of the Achaeans* (1979); V. Vikis-Freibergs and I. Freibergs, "Formulaic Analysis of the Computer-Accessible Corpus of Latvian Sun-Songs," *CHum* 12 (1979); J. Opland, *Anglo-Saxon O. P.* (1980), *Xhosa O. P.: Aspects of a Black South-Af. Trad.* (1983); *Oral Trad. Lit.: Festschrift for A. B. Lord,* ed. J. M. Foley (1981); W. J. Ong, *Orality and Literacy* (1982); R. Janko, *Homer, Hesiod and the Hymns* (1982); W. A. Quinn and A. S. Hall, *Jongleur* (1982); W. H. Kelber, *The Oral and the Written Gospel* (1983); V. Vikis-Freibergs, "Creativity and Trad. in Oral Folklore," *Cognitive Processes in the Perception of Art,* ed. R. Crozier and A. J. Chapman (1984); J. F. Nagy, *The Wisdom of the Outlaw* (1985); S. A. Sowayan, *Nabati Poetry: The O. P. of Arabia* (1985); P. Zumthor, *La Lettre et la voix de la litt. médiévale* (1987), *O. P.: An Intro.* (tr. 1990); A. Renoir, *A Key to Old Poems: The Oral-Formulaic Approach to the Interp. of West-Germanic Verse* (1988); J. M. Foley, *The Theory of Oral Composition: Hist. and Methodology* (1988), *Traditional Oral Epic* (1990), ed., *Comparative Research in Oral Trads.* (1988).

A.B.L.

OTTAVA RIMA (or *ottava toscana;* Ger. *stanze;* Rus. *oktava*). In It. prosody, an octave stanza in hendecasyllables rhyming *abababcc.* Its origin is obscure, being variously attributed to development from the stanza of the *canzone* or the *sirventes* or to imitation of the Sicilian *strambotto.* Wilkins suggested that what we today call o. r. was a popular borrowing in the 14th c. on the part of some minstrel or minstrels of the *strambotto* form for long poems (*cantari*) of less than epic length. However, it was in use in the religious verse of late

13th-c. Italy, and it was given definitive artistic form by Boccaccio in his *Filostrato* (1335?) and *Teseida* (1340–42?). Becoming almost immediately the dominant form of It. narrative verse, o. r. was developed in the 15th c. by Politian, Pulci, and Boiardo and reached its apotheosis in the *Orlando Furioso* (1516) of Ariosto, whose genius exploited its potentialities for richness, complexity, and variety of effect. Later in the same century, Torquato Tasso (*Gerusalemme liberata*) showed his mastery of the form. In subsequent centuries o. r. was employed in Italy by Marino (*Adone*), Tassoni (*La secchia rapita*), Alfieri (*Etruria vendicata*), Tommaseo (*Una serva, La Contessa Matilde*), and Marradi (*Sinfonia del bosco*).

In the broader European context, the poets of Ren. Spain and Portugal followed the It. example in adopting the form for narrative purposes. Notable epics in o. r. are Ercilla's *La Araucana* in Sp. and Camões' *Os Lusíadas* in Port. The form was explored by Eng. Ren. poets, e.g. Wyatt (some 15 poems, most monostrophic), Sidney (*Old Arcadia* 35, 54), Spenser, Daniel (*Civil Wars*), Drayton, Greville, Harington (*Orlando Furioso*), and Fairfax (his Tasso). Milton uses it for the coda of *Lycidas*. But it was not until the romantic period that the form found a true Eng. master in Byron, whose tr. of a portion of Pulci's *Morgante Maggiore* (if not J. H. Frere's *The Monks and the Giants*) seems to have made him aware of the stanza's possibilities. He employed the stanza in *Beppo, The Vision of Judgment*, and, with greatest success, in *Don Juan*. Shelley used it after 1820, chiefly for *The Witch of Atlas*; Keats used it for *Isabella*; Yeats used it for several major works (see below). It was popularized in Rus. poetry by Puškin on the model of Byron.

The work of the great masters of the stanza—Ariosto, Byron, and Yeats—suggests that o. r. is most suited to work of a varied nature, blending serious, comic, and satiric attitudes and mingling narrative and discursive modes. Byron, referring to Pulci, calls it "the half-serious rhyme" (*Don Juan* 4.6). Its accumulation of rhyme, reaching a precarious crescendo with the third repetition, prepares the reader for the neat summation, the acute observation, or the epigrammatic twist which comes with the final couplet:

And Julia's voice was lost, except in sighs,
  Until too late for useful conversation;
The tears were gushing from her gentle eyes,
  I wish, indeed, they had not had occasion;
But who, alas! can love, and then be wise?
  Not that remorse did not oppose temptation:
A little still she strove, and much repented,
And whispering "I will ne'er consent"—
  consented.

(*Don Juan* 1.117)

At 8 lines (cf. the Spenserian of 9), the o. r. stanza is long enough to carry the thread of narrative but not so long that it becomes unmanageable, and it allows ampler room for exposition and elaboration than do quatrains. Yeats, the greatest modern master of o. r., uses it for some 15 of his poems, incl. "Sailing to Byzantium" (with near-rhymes), "Among School Children," and "The Circus Animals' Desertion." Significantly, Yeats develops the form precisely at the same time (1910–19) that the dreamy style of his early period is evolving into the more realistic, colloquial style of the great poems of his middle period.

Schipper; P. Habermann, "Stanze," *Reallexikon I*; G. Bünte, *Zur Verskunst der deutschen Stanze* (1928); V. Pernicone, "Storia e svolgimento della metrica," *Problemi ed orientamenti critici di lingua e di letteratura italiana*, ed. A. Momigliano, v. 2 (1948); G. M. Ridenour, *The Style of* Don Juan (1960); A. Limentani, "Storia e struttura dell'o. r.," *Lettere italiane* 13 (1961); A. Roncaglia, "Per la storia dell'o. r.," *CN* 25 (1965); R. Beum, *The Poetic Art of W. B. Yeats* (1969), ch. 10; E. G. Etkind, *Russkie poety-perevodchiki ot Trediakovskogo do Pushkina* (1973), 155–201; Spongano; Wilkins; Elwert, *Italienische*; I. K. Lilly, "Some Structural Invariants in Rus. and Ger. O. R.," *Style* 21 (1987).

A.PR.; C.K.; T.V.F.B.

P

PANEGYRIC (Gr. *panegyrikos*) originally denoted an oration delivered at one of the Gr. festivals; later it came to designate a speech or poem in praise of some person, object, or event. Much oral poetry (q.v.) is p. in nature, consisting of the praises of heroes, armies, victories, and states; and in most cultures particular subgenres developed for specific occasions, such as the Gr. *epinikion*, a victory ode, or the *epithalamium*, a marriage song. P. is closely related to, and may have developed from, the eulogy, a speech or poem in praise of the dead. In Greece, p. was originally a rhetorical type belonging to the epideictic category of oratory. Its rules are given in the rhetorical works of Menander and Hermogenes, and famous examples include the *Panegyricus* of Isocrates, the p. of Pliny the Younger on Trajan, and the 11 other *XII Panegyrici latini* (4th c.). Pindar's odes have sometimes been described as ps. After the 3d c. B.C., when much of rhetorical theory was appropriated for poetics, p. was accepted as a formal poetic type and its rules were given in handbooks of poetry.

Significant Western examples of ps. incl. Apollinaris Sidonius' poems on the Emperors Avitus, Majorian, and Anthemius; Claudian's on the consulships of Honorius, Stilicho, Probinus, and Olybrius; the p. on the death of Celsus by Paulinus of Nola; Aldhelm's *De laudibus virgitate*; and innumerable Christian Lat. poems in praise of Mary, the cross, the martyrs, etc. It remained popular through the Middle Ages both as an independent poetic form and as an important *topos* in longer narrative poems, esp. epic; and like other such forms persisted into the Ren., with perhaps more emphasis on the praise of secular figures and institutions. Scaliger (*Poetices libri septem* [1561]) distinguishes between p., which tends to deal with present men and deeds, and encomium, which deals with those of the past, but in general the two are indistinguishable. The p. underwent a brief revival in 17th-c. encomiastic occasional verse, e.g. Edmund Waller's 1655 "Panegyrick to My Lord Protector."

Pauly-Wissowa 5.2581–83, 18.2340–62; T. Burgess, *Epideictic Lit.* (1902); Curtius; R. Haller, "Lobgedichte," *Reallexikon*; *XII Panegyrici latini*, ed. R. A. B. Mynors (1964); T. Viljamaa, *Studies in Gr. Encomiastic Poetry of the Early Byzantine Period* (1968); J. Stuart, *Izibongo: Zulu Praise-Poems*, ed. T. Cope (1968); A. Georgi, *Das lateinische und deutsche Presigedicht des Mittelalters* (1969); B. K. Lewalski, *Donne's* Anniversaries *and the Poetry of Praise* (1973); J. D. Garrison, *Dryden and the Trad. of P.* (1975); A. C. Hodza, *Shona Praise Poetry*, ed. G. Fortune (1979); *Leaf and Bone: African Praise-Poems*, ed. J. Gleason (1980); R. S. Peterson, *Imitation and Praise in the Poems of Ben Jonson* (1981); S. MacCormack, *Art and Ceremony in Late Antiquity* (1981); A. Hardie, *Statius and the Silvae* (1983); G. W. Most, *The Measures of Praise* (1985); W. Portmann, *Gesch. in der spätantiken Panegyrik* (1988); A. B. Chambers, *Andrew Marvell and Edmund Waller* (1991); L. Kurke, *The Traffic in Praise* (1991).

O.B.H.; T.V.F.B.

PARADOX. A daring statement which unites seemingly contradictory words but which on closer examination proves to have unexpected meaning and truth ("The longest way round is the shortest way home"; "Life is death and death is life"). The structure of p. is similar to the oxymoron, which unites two contradictory concepts into a third ("heavy lightness"), a favorite strategy of Petrarchism. Ps. are esp. suited to an expression of the unspeakable in religion, mysticism, and poetry. First discussed in its formal elements in Stoic philosophy and cl. rhet., the p. became more widely used after Sebastian Frank (*280 Paradoxa from the Holy Scriptures*, 1534) and has always retained an appeal for the Christian mode of expression, as in Luther and Pascal. In the *Concluding Unscientific Postscript* (1846), Kierkegaard considered God's becoming man the greatest p. for human existence.

The most famous literary example of sustained p. is the *Praise of Folly* by Erasmus (1511). In the baroque period, p. became a central figure; it is particularly important in metaphysical poetry, esp. the poetry and prose of Donne, who makes frequent use of p. and paradoxical lang. in the *Paradoxes and Problems* and *Songs and Sonets* (1633, 1635). The p. is manifest in the lit. of the 17th and 18th cs. in its antithetical verbal structure rather than as argument. Diderot in his late dialogue *Le Paradoxe sur le comédien* (1778)—on the art of acting but with far-reaching implications for poetry—holds that an actor should not feel the passion he expresses but should transcend direct imitation and rise to the conception of an intellectual model. Everything in him should become a controlled work of art, and the emotional state should be left to the spectator. In the romantic period, Schlegel (*Fragments*, 1797) called the p. a basic form of human experience and linked it closely with poetry and irony. De Quincey in his *Autobiographical Sketches* (1834–53) argued that the p. is a vital element in poetry reflecting the paradoxical nature of the world which poetry imitates. Nietzsche made p. a key term of human experience and of his own literary expression. In the lit. of the 20th c., p. often fuses with the absurd, which can be interpreted as an intensified expression of the p.

The term p. is widely employed in 20th-c. crit., esp. in the work of the New Criticism. Cleanth Brooks discusses it in *The Well Wrought Urn* (1947, esp. ch. 1) as a form of indirection which is distinctively characteristic of poetic lang. and structure. As his example, Wordsworth ("Composed Upon Westminster Bridge"), illustrates, Brooks does not use p. in the strict antithetical sense but gives it an unusually broad range by showing that good poems are written from insights that enlarge or startlingly modify our commonplace conceptions and understandings, esp. those residing in overly simplistic distinctions; this "disruptive" function of poetic lang. is precisely what Brooks calls paradoxical. Since the degree of paradoxical disruption is an index of poetic meaning, p. and poetry assume a very close affinity with one another. Subsequently, this New Critical emphasis on p. was taken up in deconstruction. Paul de Man argued that the insistence by the New Critics on the unity, harmony, and identity of a poetic work was irreconcilable with their insistence on irony, p., and ambiguity—or rather, that the insistence on unity was the "blindness" of the New Criticism, whereas the insistence on irony and p. was its "insight." Theory thus establishes a close relationship between p. and irony.

Brooks; A. E. Malloch, "Techniques and Function of the Ren. P.," *SP* 53 (1956); W. V. Quine, *The Ways of P. and Other Essays* (1966); R. L. Colie, *Paradoxia Epidemica: The Ren. Trad. of P.* (1966), "Literary P.," *DHI*; De Man; *Le Paradoxe au temps de la Ren.*, ed. M. T. Jones-Davies (1982); J. J. Y. Liu, *Lang.—P.—Poetics: A Chinese Perspective*, ed. R. J. Lynn (1988).

E.H.B.

PARALLELISM (Gr. "side by side"). The repetition of identical or similar syntactic patterns in adjacent phrases, clauses, or sentences; the matching patterns are usually doubled, but more extensive iteration is not rare. The core of a p. is syntactic; when syntactic frames are set in equivalence by p., the elements filling those frames are brought into alignment as well, esp. on the lexical level (thus the term "semantic p."). Meter and rhyme have both been recognized as species of p. on the phonological level. In a formulation of Jakobson, parallel syntax "activates" p. on other linguistic levels. The extreme case of similarity (Jakobsonian *equivalence*) is repetition; alternatively, p. can be considered the most significant subtype of repetition.

There is nothing which restricts p. (or the cognate trope in Cl. rhet., *isocolon*) as a grammatical phenomenon to verse, and it is common in certain forms of elevated prose, notably oratory, prayer, and, in Chinese, letter-writing. Further, it is a complex question whether p. is properly a trope (or master trope) or, again, strictly a matter of only grammar or also rhet. and logic. In Jakobson's later work p. came to hold a place of ever-increasing importance. The variety of grammatical domains over which p. can work is tremendous. A series of examples from Eng. Ren. verse will illustrate some common features of parallelistic usage. The domain may be a phrase: "Light of my life, and life of my desire"; "Oft with true sighes, oft with uncalled teares, / Now with slow words, now with dumbe eloquence" (Sidney, *Astrophil and Stella* 68, 61).

In the first example, two possessive noun phrases are parallel (as are the components, *light/ life, life/ desire*), while in the second, four temporal phrases are matched. Within the second example, the first two phrases involve repetition (*oft=oft*), as do the second two, though the sequence *now . . . now* has a distinct sense, "alternately." Thus the line has an internal reading, and another reading suggested by the p. In each case the short item comes first: *life* (1 syllable), *desire* (2 syllables); *true sighes* (2 syllables), *uncallèd teares* (4 syllables); *slow words* (2 syllables), *dumbe eloquence* (4 syllables).

The unit of p. may be simple clauses: "I may, I must, I can, I will, I do . . . " (*AS* 47). Here the parallels are perfect, *subject + verb*; but greater variety is equally effective: "My mouth doth water, and my breast doth swell, / My tongue doth itch, my thoughts in labour be" (*AS* 37). In more complex structures, the individual entities are themselves likely to be more complex:

> Let Fortune lay on me her worst disgrace,
> Let folke orecharg'd with braine against me crie,
> Let clouds bedimme my face, breake in mine eye,
> Let me no steps but of lost labour trace,
> Let all the earth with scorne recount my case.
>
> (*AS* 64)

The line here is the unit of p.; at the same time, these are lines 3–7 of a sonnet, and as a block they cut across the two opening quatrains.

Complex sentences may show p., e.g. embedded noun clauses and *if/then* structures:

> I saw that teares did in her eyes appeare;
> I saw that sighes her sweetest lips did part.
>
> (*AS* 87)

> If he do love, I burne, I burne in love:
> If he waite well, I never thence would move:
> If he be faire, yet but a dog can be.
>
> (*AS* 59)

In tight p., where the grammatical texture of the parallel entities is identical, it is common to find a reversal of position, as *object + adverb* here: "I curst thee oft, I pitie now thy case" (*AS* 46). This is one variety of the scheme of chiasmus.

These examples are drawn from literate poetry, and the great variety displayed in them is in part due to that circumstance. In most poetries exhibiting p., however, the ways in which the scheme is disposed are relatively few. These poetries tend to be either oral or early-literate; the later typological category is important: the poetry of the Heb. Bible began to be written down two and a half millennia ago, while Finnish folk poetry has only been recorded for a century and a half, but cl. Heb. and Finnish poetry are comparably close to the oral poetic situation and comparably far from the literate setting.

P. is well represented in traditional poetry in Chinese (and its literary offspring, Vietnamese), in Toda (but not in the other Dravidian tongues), in the Semitic langs. (in addition to ancient Semitic texts in Heb., Ugaritic, and Akkadian, there are several types of "parallel prose" in Arabic), in the Uralic langs. (incl. Finnish), in the Austronesian langs. (Rotinese is the best known), and in the Mayan langs., both medieval and modern. Rus. folk poetry is parallelistic, as is some Altaic (Mongol and Turkic) verse.

Biblical Heb. is the best know system of p., first described by Bishop Robert Lowth in 1753. In Heb. verse (and related ancient Near Eastern poetries) p. plays a role in structuring the line. Most scholars believe that the verse of the Bible is distinct from its prose, although parallelistic coloring is found in the prose as well (e.g. the fable in Judges 9). J. L. Kugel has argued that there is in fact no real dividing line between prose and verse, but his remains a minority view. He takes the two halves of a parallel unit as having an

augmentative relationship: the first element is assertive and the second has the force "not only that, but also this." Another minority view holds that there is a strictly recoverable metrical component to the poetry of the Heb. Bible (e.g. Eduard Sievers), but it is more generally assumed that there is an unrecoverable or opaque metrical element in the verse.

More recently, scholars have recognized syntactic regularities as structural features of the verse (O'Connor): "He-led the-east-wind out-of-heaven. / He-guided out-of-his-power the-south-wind" (Psalms 78:26). Lowth's traditional scheme distinguishing three types of biblical parallel verse—"synonymous," "antithetical," and "synthetic"—is of only limited usefulness. The example above would be synonymous, based on the notion that the elements *led* and *guided, east-wind* and *south-wind,* and *heaven* and *power* can be regarded as synonyms. But the idea of synonymy has been exaggerated in Heb. philology; many 20th-c. trs. of the Bible are seriously marred by a reliance on the false principle of "synonymy." As the term "synonymous p." is used, it means that the lines in question refer to (more or less) identical things. This conception can be made to cover a wide range of biblical verse, but there is much left over. Of the rather large latter category, some sets of parallel lines can be taken as "antithetical," either because they mean the opposite or because they use opposed or antithetical pairs:

> In many people is the glory of a king,
> but without people a prince is ruined.
> Slowness to anger is great under-
> standing,
> but a hasty temper is an exaltation of
> folly.
> (Proverbs 14:28–29)

In the first pair of lines the sense of the first line is the opposite of the second, and the range of reference is the same; in the second pair of lines, the opposed pairs *anger/understanding* and *temper/folly,* are lined up, but the reference is to two quite different types of people. This diversity, combined with the fact that antithetical p. occurs largely in so-called wisdom lit. (notably the Book of Proverbs), makes it suspect as an independent category. Even more suspect is "synthetic p.," which was the category meant to catch all remaining examples of p.; and some more narrowly defined subtypes have also been proposed to supplement the typology of biblical p. In fact, the traditional scheme originated by Lowth has outlived its value. Rather, p. needs to be approached as a syntactic phenomenon more or less independent of the semantic features that overpower Lowth's approach.

One particular difficulty posed by Heb. verse is the actual domain of p.: many accounts suggest that the poetry is made up exclusively of parallelistic couplets, but in fact this is true only of wisdom lit.; in other poetic books, esp. those of the prophets, single lines, triplets, and quatrains are common. Thus the practice of some scholars to refer to a single line as a halfline leads to great confusion when it is applied to the bulk of Heb. verse. The line (or colon or stich) can be defined on bases largely distinct from p.

In general, the relationship between the metrical unit and the parallel entity can take on numerous forms. In traditional poetries all (or nearly all) verse discourse may be parallelistic. The basic domain of p. may then be the line (Rotinese, Mayan) or the half-line (Finnish and some other Uralic poetries), or a variable range (Toda). In Chinese verse, largely isosyllabic, with tonal patterning, the rules governing p. vary widely according to genre; *fu* ("parallel prose") is more loosely parallelistic than the central couplets of *lü shi* ("regulated verse"), the most constrained variety of p. known.

In European poetry since the Ren., biblical influence has reinforced the use of p. to the point that few major verse texts are without some p. William Blake and Christopher Smart both use quasi-biblical p. extensively. A decisive break comes with Walt Whitman's choice of biblical structures to supplant the metrical basis of Eng. prosody itself:

> The prairie-grass accepting its own spe-
> cial odor breathing,
> I demand of it the spiritual correspond-
> ing,
> Demand the most copious and close
> companionship of men,
> Demand the blades to rise of words,
> acts, beings,
> Those of the open atmosphere, coarse,
> sunlit, fresh, nutritious,
> Those that go their own gait, erect,
> stepping with freedom and com-
> mand, leading, not following,
> Those with a never quelled audacity,
> those with sweet and lusty flesh
> clear of taint, choice and chary of
> its love-power,
> Those that look carelessly in the faces
> of Presidents and governors, as to
> say, *Who are you?*
> Those of earth-born passion, simple,
> never constrained, never obedient,
> Those of inland America.
> ("Calamus," *Leaves of Grass*)

Nearly as important is Paul Claudel's use of the *verset* (explicitly acknowledged as biblical in origin) as an avenue between the Fr. alexandrine and the prose-poem mode established by Whitman's contemporary, Baudelaire. Because romantic, modernist, and esp. postmodern lit. reflects awareness of a world-wide range of poetry, it was unavoidable that various traditional types of p. would be imitated in European lit.: the many modes of

Chinese imitation, from Pound and Amy Lowell on, are well known; Longfellow's *Hiawatha* (1855) and John Ashbery's "Finnish Rhapsody" (*April Galleons*, 1987) are each narratives in the style of then-current Eng. trs. of the *Kalevala.*

Whatever the particular ways that p. is disposed in verse discourse, schemes of equivalence, often associated with metaphor, juxtaposition, and near repetition, take on distinctive force in a parallelistic context. Gestures of signification and logic are represented and carried out in ways that often elude linearly based discourse. An important key to understanding p. lies in acknowledging that parallel structures can operate in both prose and verse and over spans from phrase to entire poem.

R. Lowth, *De sacra poesi Hebraeorum praelectiones academicae Oxonii habitae* (1753), tr. *Lectures on the Sacred Poetry of the Hebrews* (1787); E. Sievers, *Metrische Studien I–II* (1901–07); Jakobson, esp. v. 3; R. D. Fraser, "Verbal P. in Ballad and Med. Lyric," *DAI* 33 (1973), 6869A; J. J. Fox, "Roman Jakobson and the Comparative Study of P.," *Roman Jakobson: Echoes of His Scholarship*, ed. D. Armstrong and C. H. Van Schooneveld (1977); M. O'Connor, *Heb. Verse Structure* (1980); J. L. Kugel, *The Idea of Biblical Poetry* (1981); A. Berlin, *The Dynamics of Biblical P.* (1985); Terras; D. Pardee, *Ugaritic and Heb. Poetic P.* (1988); M. R. Lichtmann, *The Contemplative Poetry of G. M. Hopkins* (1989). M.O'C.

PARODY.

I. DEFINITION
II. CRITICAL ISSUES
III. HISTORY

I. DEFINITION. P. imitates the distinctive style and thought of a literary text, author, or trad. for comic effect. Some critics distinguish between critical (satiric) and comic p., while others prefer to speak of p. as the dominant comic form and pastiche as the subordinate, more solemn form. In other words, when the imitation of another work is an end in itself, the result is pastiche; when the imitation serves to mock another work, the result is p. An example of pastiche is the player's speech about Pyrrhus in *Hamlet*, a speech that imitates the style and subject of Marlowe's *Dido, Queen of Carthage* (2.1) but is not intended to make the audience smile. A Shakespearean p. is Falstaff's imitation of John Lyly's euphuistic style in the tavern scene in *1 Henry 4*:

> Harry, I do not only marvel where thou spendest thy time, but also how thou art accompanied: for though the camomile, the more it is trodden on the faster it grows, yet youth, the more it is wasted the sooner it wears.

This speech clearly imitates Lyly's passage:

> Though the camomile the more it is trodden and pressed down, the more it spreadeth; yet the violet the oftener it is handled and touched the sooner it withereth and decayeth.
>
> (*Euphues*, 1578)

Unlike Lyly, who is being sententious, Falstaff is being marvelously ironic in his questioning of the errant Prince of Wales. We laugh because the imitation is so accurate and, in Falstaff's mouth, so inappropriate.

One can distinguish between pastiche and p. on the grounds of purpose; distinguishing between p. and burlesque is more difficult. The usual distinction is said to be method: p. strives for congruence in imitation, burlesque for incongruity. Thus a burlesque may imitate a formal style but use it as a vehicle for vulgar or topical content. P., on the other hand, imitates both style and subject, so that the reader's amusement comes from recognizing how closely the p. follows the subject. As Dwight Macdonald sums up the matter, "If burlesque is pouring new wine into old bottles, p. is making a new wine that tastes like the old but has a slightly lethal effect" (559).

II. CRITICAL ISSUES. Some critics have viewed p. as a more important literary form than burlesque, but traditionally p. was regarded as a species or subclass of burlesque. P. at its best deals with sophisticated stylistic techniques, while burlesque is often cheerfully vulgar. One wonders if it is the overly serious critics who prefer the former. P. is attractive not only for considerations of taste, but also because it is more interesting in the challenges it presents, in its nature as a "meta-fiction" (Rose) which raises questions about such theoretical issues as the process of writing, the role of the reader, the role of authority, and the social context of the text (187). Because the success of p. depends not only on the reader's understanding of the text, but also on the recognition of the source-text it is based on and the comical twist or reversal of those cultural values embedded in the source-text, the readerly transaction is complex. And the p. itself of course reinstantiates the source at the same time that it subverts it.

The nature of p. continues to be a matter of discussion. On the one hand, it is seen as a highly reflexive form that celebrates textuality (q.v.). Macdonald praises p. as "an intuitive kind of lit. crit, shorthand for what 'serious' critics must write out at length. It is Method Acting, since a successful parodist must live himself, imaginatively, into his parodee. It is *jiujitsu*, using the impetus of the opponent to defeat him, although 'opponent' and 'defeat' are hardly the words. Most parodies are written out of admiration rather than contempt" (xiii). His positive view of the form is not universal, however, even among those who enjoy p.: Brett warns that "one must never forget the dependent

nature of p. It is a parasitic art and, though it can hold up the eminent to ridicule, without them it could not exist" (25). To some extent the reaction that a p. draws depends on the critic's opinion of the work it imitates. It may also depend on the critic's attitude toward current ideology, for it is undeniable that p. is a subversive form. Freund defines p. as "a literary instrument of ideological crit. P. destroys established ideologies, such as the heroic or fascistic, by searching them for symptomatic, verbally and structurally fixed constructs and tearing these structures down along with the ideologies manifested in them" (13).

So understood, p. is uniformly subversive, even when affectionate. There is also, however, sacred p.: this takes secular themes such as erotic love and transforms them to divine purposes. Both Virgil's incorporation of the two Homeric epics and also much East Asian practice are forms of elevating p., and Milton uses dissimiles and comparisons of "small things to great" to serious ends.

III. HISTORY. P. was originally "a song sung beside," i.e. a comic imitation of a serious poem. Aristotle (*Poetics* 1448a12) attributes its origin to Hegemon of Thasos (5th c. B.C.), who used epic style to represent men not as superior to what they are in ordinary life but as inferior. Athenaeus (15.699a) states that Hegemon was the first to introduce parodies into the theater, but elsewhere quotes Polemo as saying that p. was invented by the iambic poet Hipponax (6th c. B.C.), who had himself been the victim of caricature at the hands of the sculptors and painters; we have a few lines of his mock-epic on a glutton. Much earlier than these examples of p. was the pseudo-Homeric *Margites*, known to Archilochus, which set forth in hexameters with intermingled iambics the story of a fool. We still have the *Battle of the Frogs and Mice*, which parodies Homer.

But the supreme parodist of antiquity was Aristophanes, who may be thought to have reached his highest level in the *Frogs*, where he parodies the styles of Aeschylus and Euripides. But almost every passage of Aristophanes contains a touch of p. In later comedy this element dwindles. Plato imitates the styles of several prose writers with amusing effect; in the *Symposium* (194e–197e) he puts into Agathon's mouth a speech in the manner of Gorgias. Lucian has a good many touches of p. or burlesque: in Dialogue 20, the *Judgment of Paris*, for example, the comic effect is achieved by making the divine characters talk in the lang. of ordinary life.

Roman humor had a strong element of p.; the phlyax pots and the performances which they presumably illustrate must have appealed to the Romans. In Lat. comedy we find occasional burlesque of the tragic manner, as in the prologues to the *Amphitryon* and the *Rudens*, the mad scene in the *Menaechmi*, and Pardalisca's mock-tragic outburst (*Casina* 621 ff.)—passages which, whatever the original may have been, owe their effect to lang. and meter. A more delicate irony is shown in Syrus' mocking reply to the sententious words of Demea (*Adelphoe* 420 ff.). Lucilius parodies such stylistic techniques of the Roman tragedians as Pacuvius' unusual words and awkward compounds. The fourth poem of Catullus is closely parodied in *Catalepton* 10. Persius ridicules by imitation the styles of Pacuvius and other poets. Petronius gives us a long hexameter poem on the Civil War, parts of which may be meant as a caricature of Lucan.

In the later days of the Empire, the Roman mime parodied the rites of the Christian church. During the later Middle Ages, ps. of liturgy, well-known hymns, and even the Bible were popular. Ren. authors, when not embroiled in the polemics of the Reformation, preferred to parody the classics or such "gothic" phenomena as medieval romance and scholasticism; these include Pulci's *Morgante Maggiore* and Cervantes' *Don Quixote* (p. of romance); Giambattista Gelli's *Circe*, Tassoni's *La Secchia Rapita* and Scarron's *Virgile Travesti* (p. of the classics); and Erasmus' *Praise of Folly* and Rabelais' *Gargantua and Pantagruel* (p. of scholasticism). Of these authors, Rabelais is the most universal, the richest, and the most difficult to classify.

P. became institutionalized during the 17th c. The existence of academies and distinct literary movements, particularly in Italy, France, and England, encouraged debates in which p. was used as a weapon of satire. Boccalini's *Ragguagli di Parnasso* (1612) was the origin for a whole genre employing p. as a device for criticizing contemp. authors.

Eng. p. of the late Middle Ages is employed in the cycle plays, where a scene of common life (e.g. the Mak episode in *The Second Shepherd's Play*) provides comic relief. Chaucer's *Sir Thopas* parodies the grandiose style of medieval romance. Shakespeare burlesqued his own romantic love plots with *rustic amours*, and John Marston, in turn, wrote a rough, humorous travesty of *Venus and Adonis*. One of the best-known 17th-c. ps. was the Duke of Buckingham's *The Rehearsal* (1671), which leveled its shafts mainly at Dryden's *The Conquest of Granada* and at the grand manner of the heroic play. In 1701 John Phillips (*The Splendid Shilling*) used the solemn blank verse of Milton to celebrate ludicrous incidents. Later in the century, Sheridan's *The Critic* (1779) revived dramatic p. Exceptions to the general rule that p. rarely outlives the text or trad. parodied, both *The Rehearsal* and *The Critic* have been revived in the 20th c.

The Golden Age of p. in Eng. poetry paralleled the rise of the romantic and Transcendental movements. Canning, Ellis, and John Hookham Frere produced a series of ps. in the *Anti-Jacobin Journal* (1790–1810). Here the Southey-Wordsworth brand of Fr. revolutionary sympathy for knife-grinders and tattered beggars provided good anti-Jacobin sport. Blake's *Vision of Judgment* and Shelley's *Peter Bell* likewise parodied Southey, Wordsworth, and "elemental" poetry. James Hogg in 1816 took off most

of the Eng. romantics, and in 1812 James and Horace Smith published *Rejected Addresses*, a landmark in Eng. p., in which the styles of Scott, Wordsworth, Byron, Coleridge, Samuel Johnson, and others were skillfully but not uproariously parodied. In the later 19th c., names and titles continue to multiply. Tennyson, Browning, Longfellow, Poe, Swinburne, and Whitman become the chief targets for such p. artists as J. K. Stephen, C. S. Calverly, J. C. Squire, Lewis Carroll, Swinburne (who not only produced *The Higher Pantheism in a Nutshell* à la Tennyson, but also parodied himself in *Nephelidia*), and Andrew Lang. Best of all was Max Beerbohm. In America the names of Phoebe Cary, Bret Harte, Mark Twain, Bayard Taylor, H. C. Bunner, and J. K. Bangs were most prominent before 1900. In the 20th c. *The New Yorker* has carried on the trad. established in the last century by *Punch* and *Vanity Fair.* During the 1920s, p. found a highly congenial locus in the temperaments of such talented practitioners as Corey Ford, Louis Untermeyer, Frank Sullivan, Donald Odgen Stewart, Wolcott Gibbs, James Thurber, Robert Benchley, and E. B. White. There has been less p. since then, but it continues in the work of such writers as Kenneth Tynan, Tom Stoppard, Veronica Geng, and Frederick Crews. After 25 centuries, p. seems unlikely to fade as a comic and critical form. But the existence of sacred p. and of Asian counterparts, for example, should be taken as reminders that amusement is not the sole end of p. Seriousness of parodic ends is fundamental to works like Joyce's *Ulysses.* In addition, the mingling of sacred and profane (or erotic) in Indian poetry makes p. almost inseparable from the most serious of nonparodic lit.

COLLECTIONS: *Ps. of the Works of Eng. and Am. Authors*, ed. W. Hamilton, 6 v. (1884–89); *A P. Anthol.*, ed. C. Wells (1904); *A Book of Ps.*, ed. A. Symons (1908); *A Century of P. and Imitation*, ed. W. Jerrold and R. M. Leonard (1913); *Am. Lit. in P.*, ed. R. P. Falk (1955); *Ps.: An Anthol.*, ed. D. MacDonald (1960); *20th-C. P., Am. and British*, ed. B. Lowrey (1960); *The Brand X Anthol. of Poetry*, ed. W. Zaranka (1981); *Faber Book of Ps.*, ed. S. Brett (1984).

HISTORY AND CRITICISM: A. T. Murray, *On P. and Paratragoedia in Aristophanes* (1891); A. S. Martin, *On P.* (1896); C. Stone, *P.* (1915); E. Gosse, "Burlesque," *Sel. Essays* (1928); G. Kitchin, *A Survey of Burlesque and P. in Eng.* (1931); A. H. West, *L'Influence française dans la poésie burlesque en Angleterre entre 1660–1700* (1931); R. P. Bond, *Eng. Burlesque Poetry, 1700–1750* (1932); L. L. Martz, *The Poetry of Meditation*, 2d ed. (1962)—sacred p.; G. Highet, *Anatomy of Satire* (1962); P. Lehmann, *Die lateinische Parodie im Mittelalter*, 2d ed. (1963); J. G. Riewald, "P. as Crit.," *Neophil* 50 (1966); U. Weisstein, "P., Travesty, and Burlesque," *Proc. IVth Congress of the ICLA* (1966); P. Rau, *Paratragodia* (1967); U. Broich, *Studien zum Komischen Epos* (1968); G. D. Kiremidjian, "The Aesthetics of P.," *JAAC* 28 (1969–70); "P., Gr.," and "P., Lat," *Oxford Cl. Dict.*, ed. N. G. L. Hammond and H. H. Scullard, 2d ed. (1970); G. Lee, *Allusion, P., and Imitation* (1971); W. Karrer, *Parodie, Travestie, Pastiche* (1977); A. Liede, "Parodie," *Reallexikon*; M. Rose, *P.//Meta-Fiction* (1979); W. Freund, *Die Literarische Parodie* (1981); J. Hartwig, *Shakespeare's Analogical Scene: P. as Structural Syntax* (1983); L. Hutcheon, *A Theory of P.* (1985); J. A. Dane, *P.: Critical Concepts vs. Literary Practices, Aristophanes to Sterne* (1988); K. Gravdal, *Vilain and Courtois* (1989); E. G. Stanley, "P. in Early Eng. Lit.," *Poetica* 27 (1988).

R.P.F.; F.T.

PASTORAL.

I. HISTORY
II. THEORY

I. HISTORY. The p. is a fictionalized imitation of rural life, usually the life of an imaginary Golden Age, in which the loves of shepherds and shepherdesses play a prominent part; its ends are sometimes sentimental and romantic, but sometimes satirical or political. To insist on a realistic presentation of actual shepherd life would exclude the greater part of the works that are called p. Only when poetry ceases to imitate actual rural life does it become distinctly p. It must be admitted, however, that the term has been and still is used loosely to designate any treatment of rural life, as when Louis Untermeyer speaks of Robert Frost as a "p." poet. Many critics might agree with Edmund Gosse that the "p. is cold, unnatural, artificial, and the humblest reviewer is free to cast a stone at its dishonored grave." But there must be some unique value in a genre that lasted 2,000 years (and has generated the bibliography below).

For all practical purposes the p. begins with Theocritus' *Idylls*, in the 3d c. B.C. Though the canon of Theocritus' work is unsettled, enough of the poems in the collection made by Artemidorus are certainly his to justify the claim that Theocritus is the father of p. poetry. No. 11, for example, in which Polyphemus is depicted as being in love with Galatea and finding solace in song, becomes the prototype of the love lament; no. 1, in which Thyrsis sings of Daphnis' death, sets the pattern and, to no small degree, the matter for the p. elegy (see below); no. 5 and no. 7 introduce the singing match conducted in *amoebaean verses* ("responsive verses"), whereby verses, couplets, or stanzas are spoken alternately by two speakers. The second speaker is expected not only to match the theme introduced by the first but also to improve upon it in some way (see Koster). And no. 7, in the appearance of contemporary poets under feigned names, contains the germ of the allegorical p. Theocritus wrote his ps. while he was at Ptolemy's court in Alexandria, but he remembered the actual herdsmen of his boyhood and the beautiful countryside of Sicily, so he, like the p. poets who followed him, was a city man longing for the coun-

try. But perhaps no other p. poet has ever been able to strike such a happy medium between the real and the ideal.

Virgil's *Eclogues* refine and methodize Theocritus' idylls. Expressing the sentiment inspired by the beauty of external nature in her tranquil moods and the kindred charm inspired by ideal human relationships (love in particular) in verse notable for its exquisite diction and flowing rhythm, they consolidate and popularize the conventions of p. poetry. During the Middle Ages, the p. was chiefly confined to the *pastourelle*, a type of vernacular dialogue first developed in Occitan poetry, and to a few realistic scenes in the religious plays. The vast body of post-medieval p.—that is, p. elegy, p. drama, and p. romance—is a direct outgrowth of Ren. Humanism.

The p. elegy, patterned after such Cl. models as the *Lament for Adonis*, credited to Bion, the *Lament for Bion*, traditionally ascribed to Moschus but most probably by a disciple of Bion, and Theocritus' first idyll, became conventional in the Ren. Its traditional machinery included the invocation, statement of grief, inquiry into the causes of death, sympathy and weeping of nature, procession of mourners, lament, climax, change of mood, and consolation. Marot and Spenser (*Astrophel* [1595], for Sidney) produced important Ren. examples, and numerous other p. poets, including Pope, Ambrose Philips, and Gay tried their hand at the genre. Milton's *Lycidas* and Shelley's *Adonais* conform rather closely to the classical conventions, of which vestiges can be seen even as late as Arnold's *Thyrsis*. In the Eng. trad. it is *Lycidas* which unquestionably holds the first position.

The p. drama was latent in the idylls and eclogues, for the brief dialogue was easily expandable. Even as early as Boccaccio's *Ninfale feisolano* the dramatic intensity of the eclogue was considerably heightened. With the addition of the crossed love plot and secret personal history, the p. drama emerged, and it grew in popularity as the medieval mystery plays lost ground. Poliziano's *Favola di Orfeo* (1472) is perhaps more correctly classified as an opera, but p. elements are prominent. Agostino de' Beccari's *Il Sacrificio* (1554), the first fully-developed p. drama, led to the heyday of the p. drama in Italy during the last quarter of the 16th c. Tasso's *Aminta* (1573), an allegory of the court of Ferrara, is no doubt the greatest of the kind and has exerted the most far-reaching influence on the trad. Second only to it is Guarini's *Il pastor fido* (The Faithful Shepherd, 1580–89), the first important tragicomedy. In France, the most famous drama is Racan's *Les Bergeries* (1625), founded on d'Urfé's *Astrée*. It was followed by countless *bergeries*, which, after the mode of *Astrée*, were so filled with *galant* shepherds and beautiful nymphs that the type wore itself out with its own artificiality. England's first noteworthy p. dramas, Lyly's *Gallathea* and Peele's *Arraignment of Paris*, were both published in 1584, and the most excellent, Fletcher's *Faithful Shepherdess* (imitating Tasso's *Aminta*), in 1610. In general, the Eng. plays differed from their predecessors in that, in the former, the p. setting and elements are merely a backdrop to courtly characters engaged in courtly intrigues. Because of the constant pressure of Eng. empiricism and the austerity of the Puritan taste, the p. drama in England never reached the extravagant artificiality that it attained on the Continent. The last p. drama in England was the belated *Gentle Shepherd* by Allan Ramsay in 1725. Written in Lowland Scots, detailing Scottish scenes, and using "real" shepherds, it was highly praised by the early romantic poets and critics.

The p. romance usually takes the form of a long prose narrative, interspersed with lyrics, built on a complicated plot, and peopled with characters bearing p. names. In antiquity it is represented by Longus' charming story of sexual initiation, *Daphnis and Chloe*. The modern genre, while anticipated by Boccaccio's prosimetric *Ameto* (1342), is usually dated from Sannazaro's *Arcadia* (1504), a remarkable work, written in musical prose and filled with characters who live in innocent voluptuousness. Popular imitations are Montemayor's *Diana* (1559?) in Portugal and Cervantes' *Galatea* (1585) in Spain. In France the indigenous *pastourelle* held back the p. romance; but Rémy Belleau's *Bergerie* (1572) established the type, and in d'Urfé's *Astrée* the baroque p. romance found its most consummate example, as nymphs bedizened in pearls and satin cavort with chivalric shepherds. The most celebrated Eng. p. romance is Sir Philip Sidney's *Arcadia* (the "Old Arcadia," written ca. 1580; rev. 1584, the "New Arcadia," pub. 1590). Its lofty sentiment, sweet rhythm, ornate rhetoric, elaborate description, and high-flown oratory display one aspect of the Italianate style of Elizabethan courtly lit. In spite of the riddle of its plot, in which the strange turns of fortune and love make all the virtuous happy, it is still good reading as a romance of love and adventure. The literary influence of the *Arcadia* was pervasive: Greene and Lodge, for example, imitated it; Shakespeare drew from it for the character of Gloucester in *King Lear*; on the scaffold Charles I recited an adaptation of a Pamela's prayer; in translation the *Arcadia* contributed to the elaborate plots of the Fr. romances; and traces of it may perhaps be seen even in Richardson and Scott. The sustained elaboration of its structure marks another step in the devel. of the novel away from the short story and the picaresque tale. Robert Greene's *Menaphon* (1589), conventional and imitative, adds little to the genre except some delightful lyrics. Thomas Lodge's *Rosalynde* (1590), in the style of Lyly's *Euphues* but diversified with sonnets and eclogues, was dramatized with little alteration by Shakespeare in *As You Like It*.

Early in the 14th c., the p. eclogue was profoundly influenced by the new learning, when Dante, Petrarch, and Boccaccio wrote Lat. ec-

logues after the mode of Virgil. They continued the allegory of their master, extended its political and religious scope, and introduced the personal lament. About the turn of the 15th c., Baptista Spagnuoli Mantuanus exploited the satirical possibilities of the p. by using rustic characters to ridicule the court, the church, and the women of his day. Late in the century, the It. poet Marino (1569–1625) developed a style paralleling gongorism and euphuism. His Fr. p. idyll, *Adone* (1623), filled with affected wordplay and outrageous conceits, represents a baroque aberration of the genre comparable to the contemporaneous *Astrée.* The *Pléiade* transplanted the classical eclogue into France, where Marot and Ronsard and many imitators produced conventional eclogues. In England, the eclogue makes its first appearance in the work of Alexander Barclay (*Egloges* [1515?, pub. 1570]) and Barnabe Googe (*Es., Epitaphs, and Sonnets* [1563], ed. J. M. Kennedy [1989]), but important p. poetry effectually dates from Spenser's *Shepheardes Calender* (1579), which inspired a host of late 16th-c. imitations. Though Spenser follows the conventions of the classical eclogue, he aims at simplicity and naturalness by making use of rustic characters speaking country lang. During the last quarter of the 16th c., England continued to produce much p. poetry in imitation of Spenser. According to modern taste and judgment, those of most merit are Michael Drayton's *Shepherds Garland* (1593) and William Browne's *Britannia's Pastorals* (1613–16). In his *Piscatory Eclogues* (1633), Phineas Fletcher imitated Sannazaro, who may have taken his cue from Theocritus' fisherman's idyll, no. 21.

The swan song of the p. was sung by the Eng. poets of the 18th c. Revived by Pope and Philips, whose rival ps. appeared in Tonson's *Miscellany* in 1709, the p. attracted a surprising amount of interest. Pope, inspired by Virgil's *Eclogues*, produced one of the showpieces of rococo art—a part of "Summer" being so tuneful that Handel set it to music. Philips, under the rising influence of Eng. empiricism, tried to write pastorals that came closer to the realities of Eng. rural life. The followers of neither poet wrote any p. worthy of mention, and the genre soon died of its own inanition. So artificial and effete had it become that Gay's *Shepherd's Week*, in broad burlesque, was sometimes read as a p. in the true Theocritean style. The outstanding examples of the romantic p. are Ger.: Salomon Gessner's *Daphnis* (1754), *Idyllen* (1756), and *Der Tod Abels* (1758). Wordsworth's *Michael*, reflecting the empirical element of Eng. romanticism, well marks the end of serious attempts in the genre.

II. THEORY: Sustained criticism of the p. begins with the essays of the Ren. Humanists, the most important being Vida's *Ars poetica* (1527), Sebillet's *Art poétique françoys* (1548), Scaliger's *Poetices libri septem* (1561), and "E. K."'s epistle and preface in 1579. The interest of these critics in the p. sprang from their desire to enrich the vernacular by imitating the "ancients" in this genre and to exploit its allegorical potential.

But mere imitation of Theocritus and Virgil did not long suffice, as the debate over Guarini's tragicomedy *Il pastor fido* illustrates. In his *Discorso* (1587), Jason Denores attacked this play because, he argued, it is a bastard genre, unauthorized by Aristotle. In *Il verato* (1588), and *Il verato secondo* (1593), Guarini secured the new form against his adversary, thereby widening the scope of the genre. D'Urfé, in *L'Autheur à la bergere astrée* (1610), further extended the bounds of the p. when he turned critic to defend his baroque romance. The extravagances of Marino's *Adone* made him the main target of neoclassical attack.

In France, critical discussion of the p. followed in the course of the *Querelle des anciens et des modernes.* In 1659, René Rapin argued that p. poets should return to the ancient models, and to his *Eclogae sacrae* he prefixed "Dissertatio de carmine pastorali," wherein he declares that he will gather all his theory from "*Theocritus* and *Virgil*, those great and judicious Authors, whose very doing is Authority enough," since "*Pastoral* belongs properly to the Golden Age." The most significant rebuttal to Rapin's theory is Fontenelle's "Discours sur la nature de l'eglogue" (1688). Whereas Rapin looked for his fundamental criterion to the objective authority of the ancients, Fontenelle, like his master Descartes, sought a subjective standard in and expected illumination from "the natural light of Reason." Fontenelle's method is deductive. He starts with a basic assumption, the self-evident clarity of which he thinks no one will question: "all men would be happy, and that too at an easy rate." From this premise he deduces the proposition that p. poetry, if it is to make men happy, must present "a concurrence of the two strongest passions, laziness and love."

The quarrel between the ancients and the moderns was transferred directly to England; Rapin was translated by Thomas Creech in 1684, and Fontenelle was "Englished by Mr. Motteux" in 1695. The clash between the objective authority of the classics and the subjective standards of reason divided the critics into two schools of opinion, which are best denominated as neoclassical and rationalist. The immediate source of the basic ideas of the Eng. Neoclassical critics of the p.—the chief of whom are Walsh, Pope, Gay, Gildon, and Newbery—is Rapin's "Treatise." Pope, in practice as in theory, epitomizes the neoclassical ideal.

The immediate source of the basic ideas of the Eng. rationalist critics—the chief of whom are Addison, Tickell, Purney, and Johnson—is Fontenelle's "Discours." But the Eng. followers of Fontenelle insist that the p. conform to experience as well as to reason. Though Dr. Johnson's *Rambler* essays on the p. observe both Rationalist and empirical premises: his definition of a p. is that it is simply a poem in which "any action or passion is

represented by its effects on a country life."

Romantic p. theory evolved from Rationalist theory. As the critics became more certain of their empirical grounds, they showed more freedom to disregard the form and the content of the traditional p.; to look on nature with heightened emotion; to endow primitive life with benevolence and dignity; and to place a greater value on sentiment and feeling. For example, in *An Essay on the Genius and Writings of Pope* (1756), Joseph Warton, by arguing that Theocritus was primarily a realistic poet and that the Golden Age depicted in his poetry may be equated with 18th-c. rural life, substitutes cultural primitivism for chronological. In "Discours préliminaire" to *Les Saisons*, Jean-François de Saint-Lambert disregards the distinction between the p. and descriptive poetry and speaks with enthusiasm of the beauty of fields, rivers, and woods and of the felicity of rural life as he knew it in his childhood. In *Lectures on Belles Lettres* (1783), Hugh Blair singles out Salomon Gessner's *Idyllen* as the poems in which his "ideas for the improvement of P. Poetry are fully realized." Blair's essay, along with Wordsworth's *Michael* (which exemplifies much of Blair's theory), ends serious consideration of the p. After that poem and Blair's essay, the genre belongs to the academics.

T. Purney, *A Full Enquiry into the True Nature of P.* (1717), rpt. with intro. by E. Wasserman (1948); N. Drake, "On P. Poetry," *Literary Hours* (1798); E. Gosse, "Essay on Eng. P. Poetry," *Works of Spenser* (1882); J. Marsan, *La Pastorale dramatique en France* (1905); E. K. Chambers, *Eng. Ps.* (1906); W. W. Greg, *P. Poetry and P. Drama* (1906); J. Marks, *Eng. P. Drama* (1908); H. A. Rennert, *The Sp. P. Romances* (1912); J. P. W. Crawford, *Sp. P. Drama* (1915); *Torquato Tasso,* Aminta: *A P. Drama,* ed. E. Grillo (1924)—good intro.; M. K. Bragg, *The Formal Eclogue in 18th-C. England* (1926); J. Hubaux, *Les Thèmes bucoliques dans la poésie latine* (1930); W. P. Jones, *The Pastourelle* (1931); W. Empson, *Some Versions of P.* (1935); A. Hulubei, *L'Eglogue en France au XVIe siècle* (1938); T. P. Harrison, *The P. Elegy* (1939); E. F. Wilson, "P. and Epithalamium in Lat. Lit.," *Speculum* 23 (1948); M. I. Gerhardt, *La Pastorale* (1950)—It., Sp., Fr.; J. E. Congleton, *Theories of P. Poetry in England 1684–1798* (1952); *Eng. P. Poetry,* ed. F. Kermode (1952); H. Smith, *Elizabethan Poetry* (1952); Curtius, 183 ff.; E. Merker, "Idylle," *Reallexikon* 1.742–49; J. Duchemin, *La Houlette et la lyre* (1960); W. L. Grant, *Neo-Lat. Lit. and the P.* (1965); Koster—for amoeban verse; T. G. Rosenmeyer, *The Green Cabinet: Theocritus and the European P. Lyric* (1969); R. Cody, *The Landscape of the Mind* (1969)—Tasso and Shakespeare; M. Friedman, *Marvell's P. Art* (1970); P. V. Marinelli, *P.* (1971); H. E. Toliver, *P. Forms and Attitudes* (1971); T. McFarland, *Shakespeare's P. Comedy* (1972); L. Lerner, *The Uses of Nostalgia* (1972); D. Young, *The Heart's Forest* (1972); N. B. Hansen, *That pleasant place* (1973); R. Williams, *The Country and the City* (1973); R. L. Colie, *Shakespeare's Living Art* (1974), chs. 6–7; Wilkins; R. Poggioli, *The Oaten Flute* (1975); *A Book of Eng. P. Verse,* ed. J. Barrell and J. Bull (1975); H. Cooper, *P.: Med. into Ren.* (1977); N. J. Hoffman, *Spenser's Ps.* (1977); *Schäferdichtung,* ed. W. Vosskamp (1977); R. Böschenstein-Schäfer, *Idylle,* 2d ed. (1977)—Ger.; R. Feingold, *Nature and Society* (1978)—18th-c. Eng.; *Survivals of P.,* ed. R. F. Hardin (1979); C. Hunt, Lycidas *and the It. Critics* (1979); P. Alpers, *The Singer of the* Eclogues (1979), "What is P.?" *CritI* 8 (1982); *Seven Versions of Carolingian P.,* ed. R. P. H. Green (1980); R. Mallette, *Spenser, Milton, and Ren. P.* (1981); C. Segal, *Poetry and Myth in Ancient P.* (1981); Fowler; D. M. Halperin, *Before P.: Theocritus and the Ancient Trad. of Bucolic Poetry* (1983); W. J. Kennedy, *Jacopo Sannazaro and the Uses of P.* (1983); *Milton's* Lycidas, ed. C. A. Patrides, 2d ed. (1983); J. Blanchard, *La Pastorale en France aux XIVe et XVe siècles* (1983); A. V. Ettin, *Lit. and the P.* (1984); D. R. Shore, *Spenser and the Poetics of P.* (1985); S. C. Brinkmann, *Die deutschsprachige Pastourelle: 13. bis 16. Jh.* (1985); P. Lindenbaum, *Changing Landscapes: Anti-P. Sentiment in the Eng. Ren.* (1986); L. Metzger, *One Foot in Eden* (1986)—Romantic ps.; A. Patterson, *P. and Ideology* (1987); C. M. Schenck, *Mourning and Panegyric* (1988); J. D. Bernard, *Ceremonies of Innocence* (1989)—Spenser; O. Schur, *Victorian P.* (1989); S. Chaudhuri, *Ren. P. and its Eng. Devels.* (1989); S. Burris, *The Poetry of Resistance* (1990)—Seamus Heaney; K. J. Gutzwiller, *Theocritus' P. Analogies* (1991).

J.E.C.; T.V.F.B.

PATTERN POETRY, known also as "shaped poetry" (Gr. *technopaigneia,* Lat. *carmina figurata*), is premodern verse in which the letters, words, or lines are arrayed visually to form recognizable shapes, usually the shapes of natural objects. While the origins of p. p. in the West are unknown (a Cretan piece dating from ca. 1700 B.C. and some Egyptian pieces dating from 700 B.C. are not certain p. p.), there are six surviving p. poems by Gr. Bucolic poets, shaped as an axe, an egg, wings, two altars, and a syrinx. In late Cl. Lat. there is a panegyric cycle by P. Optatianus Porfyrius (fl. 325 A.D.) praising the Emperor Constantine the Great, whose court poet Optatian was. These poems are for the most part rectilinear or square, with "intexts" woven into or canceled out from the main text (hence their names *carmina quadrata* and *carmina cancellata*). This subgenre, revived at the Merovingian court by Fortunatus, was also popular in the Carolingian Ren. (Boniface, Alcuin, Josephus Scotus) and was the dominant form of the subgenre through the Middle Ages and into the 12th c., when the popularity of p. p. waned. Extant intexts are shaped as a galley with oars (Optatian) and a crucified Christ (Hrabanus Maurus [784–856]; texts in Migne, *PL,* 107.133 ff.); about 70 late Cl. and Med. pieces are known.

A second and larger wave of p. p. began in the 16th c., initially written mainly in Lat. and Gr. by

learned poets in imitation of the pieces in the *Greek Anthology*, but, later, in virtually all the langs. of Europe and in Hebrew as well, in such new shapes as suns, circles, pyramids, and columns, and in dozens of less common and unique shapes (see Puttenham, Bk. 2 ch. 12). By the 17th c., p. p. had become associated, for the most part, with occasional verse and was used to celebrate such occasions as births, marriages, ordinations, and funerals, though in Eng. such major poets as George Herbert (1593–1633) and Robert Herrick (1591–1674) wrote serious p. p.; Herbert's "The Altar" and "Easter wings" are the best-known examples in the lang. While from this period nearly 2000 pieces are known, a reaction against p. p. set in with the spread of neoclassical poetics in the late 17th c. It is difficult for a modern reader to understand the vehemence of the caustic comments heaped on p. p. in the 18th and early 19th cs., e.g. by Addison in *Spectator* no. 58. When p. p. was accepted at all, it was taken as suitable only for comic verse, e.g. "The Tale of the Mouse" in *Alice in Wonderland.*

But late in the 19th c., serious poets turned once more to visual poetry (q.v.), as in Stéphane Mallarmé's "Un coup de dés" and Apollinaire's *Calligrammes* (q.v.); and thousands of modern works have been composed in its several subgenres. However, modern visual poems are usually associated with avant-garde movements and assume some degree of originality, unlike p. p., which is usually mimetic and less abstract in shape than other modern forms of visual poetry such as concrete poetry (q.v.). Further, the shapes of p. p. often have their accrued traditions which the reader in older times would have known but which modern scholars are only now rediscovering (Ernst). Still, some modern poets, such as Dylan Thomas and John Hollander, have successfully experimented with p. p.

Close analogues to p. p. exist in many non-Western poetries, such as *hüi-wen* in Chinese (from the late Han, 2d c. A.D., up to modern times), *ashide-e* in Japanese (esp. in the early Tokugawa, 16th c.), and *citra-kāvyas* in Sanskrit (from the 7th c. A.D. onward) and other langs. of the Indian subcontinent. These latter are particularly interesting since, like Western p. poems, they are composed in traditional shapes and classified by these into *bandhas*, with each *bandha* having its own associations and traditions.

G. Puttenham, *Arte of Eng. Poesie*, ed. G. D. Willcock and A. Walker (1936); *Bucoli graeci*, ed. A. S. F. Gow, 2 v. (1958)—the *Greek Anthology*; Curtius; J. Addison, *Spectator*, ed. D. F. Bond (1965), nos. 58, 63; K. Jhā, *Figurative Poetry in Sanskrit Lit.* (1975); U. Ernst, "Die Entwicklung der optischen Poesie in Antike, Mittelalter und Neuzeit," *GRM* 26 (1976), "Europäische Figurengedichte in Pyramidenform aus dem 16. und 17. Jahrhundert," *Euphorion* 76 (1982); C. Doria, "Visual Writing Forms in Antiquity: The *Versus intexti*," *Visual Lit. Crit.*, ed. R. Kostelanetz (1979); G. Pozzi, *La Parola depinta* (1981); D. W. Seaman, *Concrete Poetry in France* (1981); J. Adler, "*Technopaigneia, carmina figurata*, and *Bilder-Reime*," *YCC* 4 (1982); W. Levitan, "Dancing at the End of the Rope," *TAPA* 115 (1985); J. Hollander, "The Poem in the Eye," in Hollander, and *Types of Shape*, 2d ed. (1991); E. Cook, *Seeing Through Words* (1986); *P. P.: A Symposium*, ed. D. Higgins, Spec. Iss. of *VLang* 20, 1 (1986); D. Higgins, *P. P.: Guide to an Unknown Lit.* (1987).

D.H.

PENTAMETER. In Cl. prosody, this term should denote a meter of five measures or feet, as its name says, but in fact the Gr. p., which is dactylic, does not contain five of any metra: it consists of two hemiepes with an invariable caesura: – ⏑ ⏑ – ⏑ ⏑ – | – ⏑ ⏑ – ⏑ ⏑ – . Contraction of the shorts in the first half of the line is common; the second half runs as shown. P. is the conventional name for the second verse in the couplet form called the elegiac distich, though this is probably a hexameter shortened internally (West calls it "an absurd name for a verse which does not contain five of anything"). The Cl. Gr. and Lat. p. should not be confused with the Eng. "iambic p.," despite the fact that the Ren. prosodists derived that name from Cl. precedent, for the Eng. line had been written in great numbers for two centuries (Chaucer) before it was given any Cl. name, and the internal metrical structures of the two meters are quite distinct—this follows form the deeper and more systematic differences between quantitative and accentual verse-systems. Other terms lacking Cl. connotations which were formerly and are sometimes still used for the staple line of Eng. dramatic and narrative verse include "heroic verse" and "decasyllable" (see also BLANK VERSE; HEROIC COUPLET); which term of these three one chooses depends on what genealogy one assumes for the Eng. line (Cl., native, Romance, mixed) and what featuers of the line one takes as constitutive—feet, stress count, syllable count, or the latter two, or the latter two as creating the first one (feet). Despite the fact that trochaic tetrameters are fairly frequent in Eng., trochaic ps. are extremely rare—Shakespeare's famous line "Never, never, never, never, never" (*King Lear*) notwithstanding; virtually the only sustained example is Browning's "One Word More."—Wilamowitz; Crusius; Halporn et al.; Snell; West.

J.W.H.; T.V.F.B.

PERFORMANCE (Lat. *recitatio*; Ger. *Vortrag, Rezitation*). The recitation of poetry either by its author, a professional performer, or any other reader either alone or before an audience; the term normally implies the latter.

I. THEORY
II. HISTORY

I. THEORY. The p. of poetry entails a performer,

a setting, an audience, and a p. style. The term itself implies focus on the person performing, but in fact nearly all critical discussions (including this one) mainly concern the audience and its responses. Though poets naturally seem the most likely performers, from ancient times to at least the Ren., a class of professional performers or singers has usually been available for p. who have trained in delivery. The setting for p. may be a literary *salon*, a poetry workshop, a ceremonial civic or state occasion, or a quasi-theatrical p. at which a poet, poets, or performers address a wide public. By extension to electronic audio and visual media, ps. can also be disseminated via radio or television broadcasts; phonograph records, audio tapes, or compact disks; and videotapes or movies. A distinction should be made here between ps. which are live and static recordings thereof: the latter are merely fixed copies of but a single p. reduced in form and recoded into some machine lang.

It is also essential to distinguish between the p. of a poem and its composition. These two processes may or may not overlap. In the first case, the poetry presented in p. has already been transcribed as a written text, whether manuscript, scribal copy, or published book. This is the condition of nearly all modern, literary poetry: composition has been completed and the work has passed into textuality. Here p. and composition are separated in temporal sequence.

In the second case, namely oral poetry (q.v.), no distinction exists between composition and p.: the "text" is spontaneously composed during p. by illiterate bards. Such a "text" is unique in every p. and is not normally recorded in any written form or even on tape except perhaps by scholars from Harvard. Successive recitations by even the same bard may draw upon the same story pattern, but the construction of scenes and selection of verbal details is different in every case; the choice of wording and phrasing is both controlled by—and assists—a stock of relatively fixed "formulas" (q.v.). These are at once both narrative and metrical building blocks, serving to construct both metrical lines and a coherent story. Here the written mode is simply absent. Even if one were to transcribe a recitation, the written record would be palpably derivative from only a single p. It should be noted that, in historical terms, the second class of course preceded the first—i.e. orality preceded the invention of writing, print technology, and the spread of literacy (reading). But even in modern literate cultures where written texts are widely published, spontaneous composition has re-emerged as a species of "secondary orality."

The audience is the least understood component of all performative arts: Western poetics has taken virtually no interest in this subject. It is obvious, however, that audiences often bring with them significant sets of expectations about subject, diction, tone, and versification. As Wordsworth remarked, the poet who would write in a new style must create the audience by which it will be appreciated—or perish. Some audiences are trained, but most are not. The exact degree of audience comprehension of oral texts is unknown: some verse traditions, such as OE, apparently helped auditors recognize meter with musical chords, for example. In general, it would seem reasonable to assume that audiences cannot quickly process archaisms or unusual words, complex meters or heterometric stanza forms, or distanced rhymes or elaborate sonal interlace. On the other hand, sound patterns are very much obscured by orthography, particularly in a lang. such as Eng. Sound patterning can certainly be recognized *as* elaborate in p. even when it is not evident *how*, exactly, the sounds are structured. It is a question just how much of poetic form is perceived in oral transmission.

In one respect, however, audiences have an easier time with the recognition of meaning in oral texts. Chatman isolates a central difference between the reading and scansion of poems on the one hand and their p. on the other: in the former two activities, ambiguities of interp. can be preserved and do not have to be settled one way or the other ("disambiguated"). But in p., all ambiguities have to be resolved before or during delivery. Since the nature of p. is linear and temporal, sentences can only be read aloud once and must be given a specific intonational pattern. Hence in p., the performer is forced to choose between alternative intonational patterns and their associated meanings.

P. styles are one of the most interesting subjects in prosody and have direct connections to acting and articulation in the theater. Jakobson has distinguished between "delivery design" and "delivery instance," the former set by verseform, the latter representing the features that are specific to each individual p. But between these lies the realm of expressive style. The two general classes of styles are realistic (naturalistic) and oratorical (declamatory, dramatic, rhapsodic, incantatory). C. S. Lewis once identified two types of performers of metrical verse: "Minstrels" (who recite in a wooden, singsong voice, letting scansion override sense) and "Actors" (who give a flamboyantly expressive recitation, ignoring meter altogether). And early in the 20th c., Robert Bridges argued that verses should be scanned in one way but read aloud another—clear Minstrelsy.

The triumph of naturalistic technique in modern drama has obscured the fact that artificial modes of delivery are well attested in antiquity, as reported by the grammarian Sacerdos (Keil 6.448). The evidence adduced by W. S. Allen (338–46) for "scanning pronunciation" and the demonstration of Ren. pedagogy by Attridge suggest that the practice of reciting verses aloud in an artificial manner has been more the rule than the exception in the West. Nevertheless, for dramatic verse which is metrical, particularly Shakespeare,

actors learn that attention to scansion will elucidate nuances of meaning in lines that a literal or natural delivery style will not manifest (see Hardison). Consequently, great actors learn how to convey both sense and meter together, so that each supports the other.

II. HISTORY. In Oriental poetry, the trad. of poetry presentation is esp. important in Chinese and Japanese poetry and continues in 20th-c. Japan. Occidental poetry readings from the Greeks to the 19th c. have mainly favored invitational ps. in courtly settings. It is likely that ps. of poetry took place at the Alexandrian court of the Ptolemies (ca. 325–30 B.C.) and, at Rome, in the aristocratic residences of C. Cilnius Maecenas (d. 8 B.C.), who encouraged the work of Virgil, Horace, and Propertius. In Petronius' *Satyricon,* Trimalchio first writes, then recites, his own "poetry" to the guests at his banquet.

The fifth of the five great divisions of Cl. rhet., after *inventio* (discovery), *dispositio* (arrangement), *elocutio* (style), and *memoria* (memorization), was *pronuntiatio* or delivery. This was less developed in antiquity than the first four subjects, though Aristotle discusses it, as do Cicero (*De inventione* 1.9) and the *Rhetorica ad Herennium* (3.9), treating, like most subsequent rhetoricians, voice control and gesture. Quintilian devotes a lengthy chapter to the subject (*Institutio oratoria* 11.3). The practice of reciting Lat. verses was encouraged by all the Med. Lat. grammarians and central to Ren. education.

The Occitan troubadours (q.v.) retained professional performers, *jongleurs,* to recite their verses, though the poets of the Minnesang did not; other itinerant minstrels maintained themselves by recitation throughout the Middle Ages. Written poetry was recited at the 13th-c. court of Frederick II, in the Florentine circle of Lorenzo de'Medici (late 15th c.), and in the late 17th-c. *salons* of the Princes de Condé. In the 18th c., however, the patronage system gave way to one of public consumption of published books, and p. accordingly changed from a courtly to a public function. As a young poet of the late 1770s, Goethe read his work at the Weimar Court; on the occasion of a production of *Faust* to commemorate his 80th birthday in 1829, he personally coached the actors in the delivery of their lines. The 18th c. also witnessed the emergence of elocution as an important part of the theory of rhet.

In the 19th c., public recitations by both poets and their admirers became commonplace. The format was generally quasi-theatrical. Edgar Allan Poe in America, Victor Hugo in France, and Alfred, Lord Tennyson in England are examples of major poets noted for the dramatic quality of their readings. Tennyson is the earliest poet for whom we have an extant recording of a poet reading his own works. The work of Robert Browning was recited in meetings of the Browning Society (founded 1881), an organization which produced hundreds of offshoots in the U.S. in the 1880s and 1890s. A *Goethe Gesellschaft* (founded 1885) held readings in places as distant as St. Petersburg and New York. Wagner's opera *Die Meistersinger von Nürnberg* (musical version, 1867), based on the historical figure of Hans Sachs, brought the late medieval Ger. trad. of p. by members of craft guilds (*puys*) to an international audience. Elocution was even further popularized in the 19th c.; the practice of reading aloud from lit. after dinner in Victorian households was widespread, since they were not yet subjected to the brutalities of television and stereo. Elocution led to the emergence in the 20th c. of "oral interp." as a formal activity in Am. university departments of speech.

The p. of poetry is central to symbolist poetics. Mallarmé read his poetry to a select audience on designated Tuesdays at which the poet himself played both host and reader in oracular style. While Mallarmé's poetry was anything but spontaneously written, his ps. both personalized and socialized the work. Stefan George's mode of delivery was consciously influenced by Mallarmé: the audience was restricted to the poet's disciples (*Kreis*), and the occasion was perceived as cultic and sacral. George read from manuscript in a strictly rhapsodic style which disciples were required to follow.

In the 20th c., naturalistic or realistic delivery styles have owned the field. W. B. Yeats was much concerned with having his work sound spontaneous and natural. Though his delivery style was dramatic and incantatory, he deliberately revised some poems so that they would sound like an ordinary man talking. By contrast, T. S. Eliot's ps. were aristocratic in style and tonally flat. The Wagnerian prescription of having the performer seem spontaneous in expression but personally remote had its best 20th-c. exemplar in Dylan Thomas, whose dramatic, incantatory style contrasted sharply with the plain, conversational style of Frost and Auden. Frost's "sentence sounds" are the intonational patterns of colloquial speech, esp. as frozen into idioms—precisely the kind of speech effects that would be likely to come across well to audiences on Frost's frequent reading tours. The many recordings of 20th-c. poets have by now defeated the instinctive belief of many that the poet will be his own best interpreter, or that the poem will open up at last when once we have heard it aloud. Several poets—Pound, Eliot—read in a monotone specifically intended to thwart those expectations.

Politically motivated poetry readings early in the 20th c. served as models for others to come in the second half of the century. Now the poetry p. is the vehicle for political resistance and social activism. In post-revolutionary Russia, Vladimir Majakovskij sang the praises of the October Revolution in lyrics written to be read aloud; his dramatic ps. attracted mass audiences both in Western Europe and the U.S. Avant-garde movements

of the 1920s and '30s such as dada and surrealism generated ps. of poetry staged simultaneously with music, dance, and film, and so adumbrated the intermedia ps. later in the century. Poetry readings of the 1950s and '60s often took the form of multi-media presentations and random artistic "Happenings." Prominent innovators of the poetry p. in the 1950s were the Beat poets, notably Allen Ginsberg, Gregory Corso, and Lawrence Ferlinghetti, all instrumental figures in the movement now known as the San Francisco Renaissance. Orality and p. were foregrounded in the poetics of Charles Olson, who conceived of the poem as a "field of action" and made his unit of measure the "breath group." Olson's "projective verse" found followers in Robert Duncan, Robert Creeley, and Denise Levertov.

Since 1960, New York and San Francisco have been the two major Am. poetry p. centers, with London, Amsterdam, and West Berlin their European counterparts. In New York City, the poetry-reading movement of the 1960s generally associated with the name of Paul Blackburn served as a stimulus for a new vogue of poetry readings in other parts of the country, esp. in Chicago and the West Coast. Further experimentation with elements of recitation, music, song, digitized or synthesized sound, drama, mime, dance, and video, which are mixed, merged, altered, choreographed, or improvised in seriatim, simultaneous, random, or collage order, characterized the phenomena variously called sound poetry, language poetry, intermedia, or sometimes "p. art" of the 1970s and '80s. David Antin called his improvisations "talk poems."

Since the 1950s, then, the p. of poetry in America has undergone a resurgence. Its tone ranges from conversational idioms to street lang. Poetry readings by one poet have become increasingly rare: "open poetry readings" are events to which anyone may bring work to read. Jazz or rock music, electronic audio and visual effects, and spontaneous dramatic presentations often accompany recitation. Consumption of alcohol or other drugs during the p. is not unknown. The ethos in intermedia events such as these is one of experimentation, liberation, and spontaneity. Like all postmodern literary genres, poetry retains a strong interest in p. as a reaction to academic formalism and its fixation on the text. It remains a paradox, however, that the new oral poetry has by and large chosen to disseminate its own works not on cassette or even video tapes but rather in traditional print—book—form.

The heritage of all the various forms of postmodernism in America has been a turning away from the autonomy of the text and the presumption that a text presents one determinate meaning or its author's intended meaning toward the more fluid, less determinate, free play of readerly responses to texts. Hence critical interest has shifted from written documents to ps. as *experiences*. It should not be thought, however, that meaning is therefore removed: rather, it is merely relocated from the more patient and reflective process of reading and coming-to-understand a poem on the page to the more immediate, rapid, sequential process of trying to follow the poem when delivered aurally. Whether the meaning that is thus provided in p. is more or less extensive or fulfilling to auditors as opposed to readers is a judgment that only auditors and readers may make. Nevertheless, many audiences still consider the p. of poetry a communal, nearly sacral event for heightened speech, investing the poet with the transportive powers of the *vates*. And many readers and teachers of poetry continue to believe that poetry achieves its body only when given material form, as sound, in the air, aloud.

JOURNALS: *Lit. in P.* 1– (1980–); *Text and P. Quarterly* 1– (1981–).

STUDIES: Thieme 377–78—lists 19th-c. Fr. works; F. K. Roedemeyer, *Vom künstlerischen Sprechen* (1924); E. Drach, *Die redenden Künste* (1926); T. Taig, *Rhythm and Metre* (1929); R. C. Crosby, "Oral Delivery in the Middle Ages," *Speculum* 11 (1936), "Chaucer and the Custom of Oral Delivery," *Speculum* 13 (1938); W. B. Nichols, *The Speaking of Poetry* (1937); S. F. Bonner, *Roman Declamation* (1950); D. Whitelock, *The Audience of* Beowulf (1951); K. Wais, *Mallarmé*, 2d ed. (1952); E. Salin, *Um Stefan George*, 2d ed. (1954); F. Trojan, *Die Kunst der Rezitation* (1954); S. Chatman, "Linguistics, Poetics, and Interp.," *QJS* 43 (1957), "Linguistic Style, Literary Style, and P.," *Monograph Series Langs. & Ling.* 13 (1962); Y. Winters, "The Audible Reading of Poetry," *The Function of Crit.* (1957); C. S. Lewis, "Metre," *REL* 1 (1960); R. Jakobson, "Linguistics and Poetics," in Sebeok; Lord; S. Levin, "Suprasegmentals and the P. of Poetry," *QJS* 48 (1962); F. Berry, *Poetry and the Physical Voice* (1962); J. M. Stein, "Poetry for the Eye," *Monatschefte* 55 (1963)—against p.; K. T. Loesch, "Literary Ambiguity and Oral P.," *QJS* 51 (1965), "Empirical Studies in Oral Interp.," *Western Speech* 38 (1969); D. Levertov, "Approach to Public Poetry Listenings," *VQR* 41 (1965); D. Norberg, "La Recitation du vers latin," *NBhM* 66 (1965); *The New Rus. Poets, 1953–1968*, ed. G. Reavey (1966); Dale, ch. 13; W. C. Forrest, "The Poem as a Summons to P.," *BJA* 9 (1969); G. Poulet, "Phenomenology of Reading," *NLH* 1 (1969); H. Hein, "P. as an Aesthetic Category," *JAAC* 28 (1970); P. Dickinson, "Spoken Words," *Encounter* 34 (1970); *The East Side Scene*, ed. A. De Loach (1972); S. Massie, *The Living Mirror: Five Young Poets from Leningrad* (1972); *P. in Postmodern Culture*, ed. M. Benamou et al. (1977); M. L. West, "The Singing of Homer," *JHS* 101 (1981); Brogan, sect. IV—bibl.; B. Engler, *Reading and Listening* (1982); K. Quinn, "The Poet and His Audience in the Augustan Age," *Aufstieg und Niedergang der römischen Welt*, ed. W. Haase, v. 2 (1982); B. Rowland, "*Pronuntiatio* and its Effect on Chaucer's

Audience," *SAC* 4 (1982); O. B. Hardison, "Speaking the Speech," *ShQ* 34 (1983); E. R. Kintgen, *The Perception of Poetry* (1983); D. A. Russell, *Gr. Declamation* (1983); W. G. Thalmann, *Conventions of Form and Thought in Early Gr. Epic Poetry* (1984), ch. 4; J. Herington, *Poetry Into Drama* (1985)—Gr.; D. Wojahn, "Appraising the Age of the Poetry Reading," *NER/BLQ* 8 (1985); J.-C. Milner and F. Regnault, *Dire le vers* (1987); B. Bowden, *Chaucer Aloud* (1987); E. Griffiths, *The Printed Voice of Victorian Poetry* (1988); R. Schechner, *P. Theory*, 2d ed. (1988)—theatrical; M. Davidson, *The San Francisco Ren.* (1989); D. Oliver, *Poetry and Narrative in P.* (1989); G. Danek, "Singing Homer," *WHB* 31 (1989); H. M. Sayre, *The Object of P.: the Am. Avant-garde since 1970* (1989), "P.," *Critical Terms for Literary Study*, ed. F. Lentricchia and T. McLaughlin (1990); S. G. Daitz, "On Reading Homer Aloud," *AJP* 112 (1991); D. Cusic, *The Poet as Performer* (1991); *P.*, spec. iss. of *PMLA* 107 (1992).

T.V.F.B.; W.B.F.

PERIPHRASIS. A circumlocution, a roundabout expression that avoids naming something by its most direct term. Since it is constituted through a culturally perceived relationship to a word or phrase that it is *not*, p. has no distinctive form of its own but articulates itself variously through other figures, esp. metaphor. Quintilian (*Institutio oratoria* 8.6.59) subdivides it by function into two types: the euphemistic or "necessary," as in the avoidance of obscenity or other unpleasant matters (Plato's "the fated journey" for "death"—cf. the modern "passing away"); and the decorative, used for stylistic embellishment (Virgil's "Aurora sprinkled the earth with new light" for "day broke"). The descriptive kind includes most periphrases which approximate a two-word definition by combining a specific with a general term ("the finny tribe" to signify fish). Pseudo-Longinus considered it productive of sublimity but, like Quintilian, warned against its excesses, such as preciosity or pleonasm (28–29). Later writers have characterized it as representing a term by its (whole or partial) definition, as in the expression "pressed milk" for "cheese." P. also appears in poetry that tries to translate culture-specific concepts from one lang. to another without neologism.

Though it is unlikely that any movement or era in poetry has succeeded in suppressing p. altogether, some styles favor it more than others. Curtius (275 ff.) associates it, like other rhetorical ornaments, with mannerism and marks stages in its use and abuse. Oral traditions frequently build formulas around periphrases, as in the patronymic "son of Tydeus" for "Diomedes"; these have important metrical functions and are not ornament.

While widely used in biblical and Homeric lit. and by Hesiod, the devel. of p. as an important feature of poetic style begins with Lucretius and Virgil, and through their influence it became a staple device of epic and descriptive poetry throughout the Middle Ages and into the Ren. Classified by medieval rhetoricians as a trope of amplification, p. suited the conception of style which emphasized *copia* and invention. The OE poetic device of variation typically employs multiple periphrastic constructions, as does the kenning (q.v.), the characteristic device of Old Germanic and ON poetry, which in its more elaborate forms illustrates the connection between p. and riddle.

Given new impetus through the work of the Pléiade, p. proliferated in 17th-c. diction, particularly as influenced by the scientific spirit of the age, and even more so in the stock poetic diction of the 18th c., where descriptive poetry often shows periphrastic constructions (Arthos). Since the 18th c., the form has lost much of its prestige in the romantic and modern reaction against rhetorical artifice; more often than not it survives only in inflated uses for humorous effect, as in Dickens. Yet its occasional appearance in the work of modernists such as T. S. Eliot ("white hair of the waves blown back" for "foam") suggests that, insofar as directness of locution is not always the preferable route (direct speech being, most often, shorn of semantic density and allusive richness), p. has an enduring poetic usefulness. See also LEXIS.—

P. Aronstein, "Die periphrastische Form im Englischen," *Anglia* 42 (1918); J. Arthos, *The Lang. of Natural Description in 18th-C. Poetry* (1949); Curtius; D. S. McCoy, *Trad. and Convention: A Study of P. in Eng. Pastoral Poetry from 1557–1715* (1965); Lausberg—fine compendium of Cl. citations; A. Quinn, *Figures of Speech* (1982).

W.W.P.; J.A.

POET (Gr. *poietes*, "maker," from *poiein*, "to make"; Lat. *poeta*).

I. PERSON AND MAKER
II. SEER AND MAKER
III. SOCIAL ROLE

I. PERSON AND MAKER. The first thing to say is that the p. is a maker of poems, and that the maker is not the same as the person. The quality of the maker's productions is not diminished by his or her personal shortcomings or failings in life, and conversely, the successes or social privilege of the person do not improve mediocre poetic productions: all that is *ad hominem*. The person, in the course of coming into and passing through adulthood, forms a self out of life's experiences, but the maker is formed specifically out of a decision to write poetry, to practice ("Without practice," says Blake, "nothing is possible"), to learn technique, and to continue to write. Lipking has written an elegant and absorbing essay on that moment in the lives of several persons when they first determined to become poets. The decision may or may not be entirely conscious—as everywhere else, that does not matter very much. Thus the p. develops a second self, emergent from the first yet supersessive to it; from such augmentation come enhanced opportunities for complex psychic relations. The

maker is able to quarry the person (i.e. memory and sensibility) for experiences, reflections, insights, sensations, ambiguities, ambivalences, feelings, and thoughts—all potential threads to be woven into the emergent fabric of the poem. Most importantly, the maker can take advantage of these complex psychic interactions by creating a speaker for the poem, a persona that is quite distinct from the person.

One may reasonably wonder if it is even possible to say when one is the one self and when the other. The person is a person ever, obviously, but the maker is only a maker in the act of composing verse. W. H. Auden has attested to this belief and the accompanying fear that, when the p. has completed a poem, he or she may never successfully complete another again. Obviously, technique is precisely the resource one would rely on to overcome such a fear and write another poem, except for the fact that technique is always learned about another problem in another poem, so that to repeat it is not to confront a new poem, wholly, but merely to repeat gestures from the past. On the romantic and organicist theory of poetry, to which there have been many 20th c. free-verse adherents, every poem is unique. Still, without the accrual of learned technique, poems and ps. would never improve. The inferior p. is owned by technique; the great p. seeks, masters, alters, and extends technique (see VERSIFICATION).

The p. is only a maker in the act of composition, then, but we must not construe "composition" too narrowly as merely the act of writing lines. The act of making words, and making words make lines, is a process that may well extend over long periods of time and be carried on, according to the vagaries of thought and passion and the exigencies of circumstance, in several episodes, some conscious, some unconscious, many partly both. "Emotion recollected in tranquility" is too elliptical a description of the true nature of poetic genesis, incubation, articulation, and—most importantly—revision. The maker may mine the person (life) simply and directly: here autobiographical material is taken over directly into the poetry, subject only to the shaping that all verbal material undergoes when cast into verseform. But much more likely, in the longstanding view in Western poetics, is that the p. is not limited to her own experiences but will build up, via imagination, a second reality quite other than the external world we know. The experiences of the person may be merely the springboard for extensive creative worldmaking, in exactly the same way that, in Freud's account of dreams, the dream work effects a virtually complete transformation of the latent content of the dream—the raw materials, the stimulus—into the manifest content, the actual dream—which in the case of poetry is the poem itself. Here one may recall Coleridge's distinction between fancy and imagination: it is not the copious generation of images but rather the selection and arrangement of them—along with thought, emotion, memory, and perception—within the frames of meter and stanza that constitutes the art poetical.

The obvious source to consult on making would be the p. But even if one were to ask a p. about the nature of her methods of composition, it is as much a question whether the answer would emerge from the maker or the person, as whether the methods of the one are ever entirely known to the other. Eliot is a prime example of this: despite his many theoretical pronouncements about impersonality, we always emerge from his poetry with the conviction that he is one of the most autobiographical of poets. In general, theorizing by ps. (e.g. Horace, Sidney, Pope, Coleridge, Eliot, Valéry) about poetry is suspect if not compromised, particularly if they build a theory out of their own work. Most ps. are notoriously unreliable as critics of themselves. Some may know what they are doing, in terms of technique and purpose, carry it out, and report on it accurately. Others may "know," in some less conscious sense, what they are doing but not be able to articulate it theoretically or analytically—this is very common. Still others may not even "know" what they are doing, even in the process of doing it, and for a variety of legitimate reasons, most of which follow from Chomsky's distinction between competence and performance. Some ps. may even know what they are doing but choose to be evasive about it to others—again, for a variety of reasons.

One might assume that utterances from a p. are reliable when confined to the subject of the verse-craft. Utterances on all other subjects made by a p. are being made by the person and carry no greater authority than those made by any other citizen. Granted, the p. as person may well have an unusually perceptive or talented or complex personality compared to most; certainly many readers have felt that the lives and personalities of artists are worth knowing. It does not diminish the truth of this to urge, along with it, the corollary that such knowledge alone tells us nothing about the poetry—that, in other words, "biographical crit." is an oxymoron if not a contradiction in terms. What one wants to know is what the artist made, what she had available to work with, and how readers subsequently have responded to it. These are the realms, respectively, of textual and lit. crit., linguistics and stylistics, and reader-response crit.

But makers make poems, and poems are made, of course, out of words. Hence any p. who wishes to perfect her art must be a student of lang.—words, sounds, rhythms—and of forms. Only by thorough knowledge of the materials and their permissible combinations can she master the craft. The p. who aspires to learn technique will study what forms have been shown to have expressive power by the great ps. of the past; the p. who knows that in every age the p. who wishes to be heard must find new means to express a new

sensibility will experiment. Between these two poles lies all the dialectic of tradition and the individual talent that Eliot spoke of. In such study she will learn, in the simplest terms, *what is possible*, for the artist always works in and with a medium, taking advantage of what the medium offers, working around what the medium denies or restricts.

II. SEER AND MAKER. Since the Greeks, there have been two conceptions of the role of the p.: p. as seer, and p. as maker. That is to say, nearly all the metaphors and analogies that have been offered to describe the nature of *poiesis* concern either the poet's eye or the poet's hand. The former concerns mental representation (eidetic creation and alteration) while the latter concerns the physical aspects of text production. These two descriptive modes are inextricably intertwined throughout the history of Western poetics.

In the first conception, poet as seer (Lat. *vates*, Ir. *fáith*, Welsh *gwawd*), the primary emphasis falls on origination, the chief assumption being that ideas, images, phrases, and words come to the p. from some source outside or beyond herself. From antiquity up to the 18th c., this notion was formulated as the doctrine of divine inspiration, first in a kind of frenzy, a "poetic madness" given by the Muses to Gr. poets, then, for Christian poets, in a state of receptiveness to gifts from God. With the advent of romanticism, however, this conception was replaced by the doctrine of imagination (q.v.); now outer is replaced by inner as the point of origination for mental imagery. The p. is now seen not as passive instrument but as creator, and the locus of origin is reduced to the self. All modern psychological investigation derives from the romantic and postromantic focus on inner life. The imagination throws up new images before the poet's mind: the p. is "seer" of forms and modes of reality not apprehensible to normal sight. Perhaps such seeing is the product of some altered state of experience—dream, reverie, or trance, natural or induced—but the older sense of such states as *given*, bestowed by some higher power, is gone. In such accounts the act of making and the role of technique are minimized. This is not to say they are denied; certainly the ps. themselves, at least, knew better. But in extreme formulations of the doctrine of inspiration, the p. is a mere mouthpiece for not only images but wording and even whole poems—as Blake claimed, as Coleridge reported about "Kubla Khan," as Poe would have liked for us to believe, and as Yeats and the surrealists alleged about "automatic writing."

Despite the frequency of references to poetic imagination, however, there are equally frequent appeals to recognize the poet's work as a skilled craft. The oldest metaphors in the world for the poet's work are those of weaving. Even the Greeks conceded the importance of skill at craft. Pindar acknowledges it very early. In the *Iliad* and the *Odyssey*, on every occasion when Homer appeals to the Muses, the request is for matter of a factual nature rather than for form. That is, Homer always asks for inspiration concerning what to say, never how to say it. Apparently divine possession may provide information, for Homer, but the p. must rely on her own accrued technical skill for the making of the song. This union of special knowledge and technical skill is the distinctive mark of the p. in Homer. And the ancient Gr. concept of the p. as maker is given greater authority by Horace, whose insistence on the importance of "the labor of the file" (revision) made the *Ars poetica* a central text throughout the Middle Ages. The "Scottish Chaucerians" writing in Middle Scots refer to themselves as "Makars," a usage echoed in the early Ren. by references to Wyatt and Surrey as the "Courtly Makers." In the 120th c., it finds echo in Ezra Pound's injunction to modernist poets to "Make It New."

Ultimately, of course, there is no simple division between having ideas and shaping them as words, between envisioning and revising: composition does not divide neatly into two distinct activities, separate and sequential. Both proceed simultaneously, interpenetrating each other so extensively that, at any given moment, one probably could not say whether a bit of phrasing or an image has just been invented *ex nihilo*—if such be possible—or was remembered from prior reading of some earlier p. or the newspaper or, indeed, anything. The mind's reception of words and ideas, and their retention in memory and the subconscious, and their alteration both consciously and subconsciously both at inception and later, is so complex a process and so little available to conscious inspection that simplistic dichotomies of origination and alteration seem worse than simplistic. And the role of the word itself, in this process of envisioning, arranging, and casting into verbal form, then verseform, is, while pivotal, nevertheless little understood (see POETRY).

III. SOCIAL ROLE. In the West, the poet's status, prestige, and cultural authority have changed radically in the long transition from preliterate, oral, monarchical societies to literate, written, democratic ones. In tribal societies, the p. naturally moved in the retinue of the chief, king, or ruler because he was invested with the same kind (though perhaps not the same amount) of cultural authority as the king and spiritual authority as the priest. Indeed, the roles of priest and p. become indistinguishable in sacred chant, ritual, and incantatory speech. As the adviser and chronicler of the king, the p. finds his task largely eulogistic (see Opland), praising the present and past deeds of the tribe and its heroes, preserving information (e.g. genealogical, medical) and proverbial lore, offering spiritual advice. Infused with divine sight and speech by the gods, the p. provides words that surpass the ordinary function of lang. to communicate ordinary knowledge. The p. is the carrier of tribal wisdom, the accumulated knowledge and experience of the race; and the

poetry which embodies it, more tightly bound, becomes transcendent speech. The OE *scop* is still carrying on these functions into and past the 8th c. A.D. As tribal and nomadic societies become geographically established, however, both the ruler and the poet become institutionalized. Court poetry and the recording thereof replace bardic improvisation. In the Middle Ages, academic skill at versifying was crowned with a laurel wreath, a practice which was soon transferred to the court: Petrarch was crowned "poet laureate" in 1341, an honorary title preserved in England and eventually America up to the present. In modern times the communal and sacral function of the p. has been repeated in the call, by Mallarmé echoed by Pound, for the p. to "purify the dialect of the tribe."

W. H. Auden reminds us that much of the role of the p. is culturally defined. Some cultures make a formal social distinction between the sacred and the profane; others do not. In those that do, the p. has a public and sacral status as the conveyor of wisdom and knowledge of a very high order. In those that do not, however—which includes all modern Western industrial societies—the p. can present only knowledge that is personal and private, appealing to his or her readers, in essence, to judge for themselves whether or not the knowledge and experience described is not also their own. Now the poet's calling is merely a profession, valued highly by some, viewed with curiosity or skepticism by many.

The poet's cultural and spiritual authority have always had to be balanced against the practical exigencies of making a living; the disparity between the two indeed suggests that the former are more indirect, if not more illusory, than we would like to think. The poet's association with a monarch or a member of the aristocracy for financial support, most often by composing occasional verse on demand, lasted from ancient times until the 18th c., when the patronage system gave way—its end might almost be dated precisely in Dr. Johnson's letter to Lord Chesterfield—to public consumption of printed books. This in turn lasted into the 20th c., when public preference for other artforms (the novel, cinema) forced many ps. to seek their livelihood in "creative writing" positions in universities or on the lecture circuit, as with Frost's very popular and profitable reading tours. In truth, the public for poetry has been small in every age: the more advanced the sensibility, the smaller the circle. But that market, if small, is on the other hand most often the most influential, appreciative of the arts, and possessed of the greatest means. The very notion of a literary *career* oscillates between the development of a reputation, among those competent to judge, for skill at the craft of verse and the utterance of words worth remembering, and, on the other hand, of the cultivation of a readership which will reward the making sufficiently to provide the necessary leisure for creative work in the future. It fell to Wordsworth to articulate the dilemma for the poet seeking to Make It New—that he or she must create the very audience by which the work may adequately be judged.

Further, since the 18th c., as the capacity for empathy, intensity, and imagination increasingly came into conflict with the values of an urban, industrial, polluted, capitalist, and intolerant society, alienation came to be the primary characteristic of the poet's condition. This is made central to the ethos of romanticism—both Blake and Byron saw the p. as an isolated figure, a visionary, prophet, and political radical. Byron, however, like many other 19th-c. Eng. ps., did not have to struggle for existence. But as the 19th c. wore on, alienation became more than merely romantic, witness Baudelaire ("The Albatross") and the symbolists. The p., too sensitive, too verbal, too aestheticized, becomes the *poète maudit*, "the p. defeated by life," brutalized by the vulgar tastes of the bourgeoisie. By the 20th c., with its triumph of media for the masses, alienation is simple fact. Perhaps to some degree it was always fact: ps. who do not consciously define themselves as following in a tradition must of necessity work to escape the burden of the past and the overbearing pressure of the great accomplishments of their predecessors (Bloom). But in modern times this private and psychological resistance-in-order-to-create has been overwritten with a more extensive resistance to the brutalities of modern culture, or what passes for culture.

Finally, we must recognize the deeper, more systemic, hence less visible forms of marginalization of ps., those which apply not individually but categorically, on account of gender, race, nationality, dialect, and class. Modern feminist critics have demonstrated in detail the kinds of disfigurements suffered by women poets who have had to exist in a patriarchal system which controls the means of production and of social recognition.

T. Carlyle, *On Heroes, Hero-Worship, and the Heroic* (1841), lecture 3; O. Behaghel, *Bewusstes und Unbewusstes in dichterische Schaffen* (1906); I. Zangerle, *Die Bestimmung der Dichters* (1949); W. Muschg, "Dichtertypen," *Festschrift F. Strich* (1952); Abrams; Curtius, Excursus 7; M. C. Nahm, *The Artist as Creator* (1956); A. F. Scott, *The Poet's Craft* (1957); P. Crutwell, "Makers and Persons," *HudR* 12 (1960); G. T. Wright, *The P. in the Poem* (1960); R. Jakobson, "Linguistics and Poetics," in Sebeok, rpt. in Jakobson, v. 2; W. H. Auden, "Making, Knowing, and Judging," *The Dyer's Hand* (1962); J. W. Saunders, *The Profession of Eng. Letters* (1964); P. Vicaire, *Recherches sur les mots désignant la poésie et le poète dans l'oeuvre de Platon* (1964); A. Kambylis, *Die Dichterweihe und ihre Symbolik* (1965)—concepts of the poet's calling in antiquity; J. K. Newman, *The Concept of Vates in Augustan Poetry* (1967); R. Schmitt, *Dichtung und Dichtersprache in indogermanischer Zeit* (1967), ed., *Indoger-*

manische Dichtersprache und Namengebung (1973); E. N. Tigerstedt, "The P. as Creator," *CLS* 5 (1968); T.-T. Chow, "The Early Hist. of the Chinese Word *Shih* (Poetry)," *Wen-lin: Studies in the Chinese Humanities* (1968); A. D. Skiadas, "Über das Wesen des Dichters im platonischen *Ion*," *Symbolae Osloenses* 46 (1971); K. K. Ruthven, *Critical Assumptions* (1979), ch. 5; H. Moisl, "Celto-Germanic *Watu-/Wotu-* and Early Germanic Poetry," *N&Q* 225 (1980); J. Opland, *Anglo-Saxon Oral Poetry* (1980); L. Lipking, *The Life of the Poet* (1981); K. Quinn, "The P. and His Audience in the Augustan Age," *Aufstieg und Niedergang der römischen Welt*, ed. W. Haase. v. 2 (1982); R. Helgerson, *Self-Crowned Laureates* (1983); M. Ellmann, *The Poetics of Impersonality* (1987); M. W. Bloomfield and C. W. Dunn, *The Role of the P. in Early Societies* (1989)—Ir., Welsh, Gaelic, ON, and OE; A. Balakian, *The Fiction of the P.* (1992). T.V.F.B.

POETRY (Lat. *poema, poetria,* from Gr. *poiesis,* "making," first attested in Herodotus).

I. MEANS AND ENDS
II. SOUND AND MEANING
III. HEARD AND SEEN
IV. ONTOLOGY

I. MEANS AND ENDS. A poem is an instance of *verbal art,* a text set in verse, bound speech. More generally, a poem conveys heightened forms of perception, experience, meaning, or consciousness in heightened lang., i.e. a heightened mode of discourse. Ends require means: to convey heightened consciousness requires heightened resources. Traditionally these have been taken as the ones offered by pros., i.e. verseform: lineation, meter, sound-patterning, syntactic deployment, and stanza forms. Except for the three or four hybrid forms so far developed in the West—the prose poem, rhythmical prose and rhymeprose, and the prosimetrum—p. has traditionally been distinguished from prose by virtue of being set in verse (see VERSE AND PROSE). What most readers understand as "p." was, up until 1850, set in lines which were metrical, and even the several forms of *vers libre* and free verse (qq.v.) produced since 1850 have been built largely on one or another concept of the line. Lineation is therefore central to the traditional Western conception of p. (see LINE). Prose is cast in sentences; p. is cast in sentences cast into lines. Prose syntax has the shape of meaning, but poetic syntax is stretched across the frame of meter or the poem's visual space, so that it has this shape as well as meaning. Whether the pros. of the poem is primarily aural or visual or mixed (see below), it creates design.

If either of the criteria indicated by the two words italicized in the first sentence of this entry is removed, texts become "poetic" only in looser, more general, and metaphorical senses. (1) *Verbal* but not *artful*: any verbal text or piece of verbal discourse, even if not meant as "art," can be called "poetic" if it seems to exhibit intensified speech—an impassioned plea, a stirring speech, a moving letter. Often these texts partake of the resources offered by traditional rhet., i.e. devices of repetition and figuration. They become more highly figured and patterned than ordinary speech or prose, and so take on the term "poetic" as a metonym, since verse characteristically deploys these features. (2) *Artful* but not *verbal*: any object skillfully made or intended as art though not a verbal text can be called "poetic" in the metaphorical sense—an intense moment in a play or movie, a romantic gesture, a painting, a piece of music. These foreground the act of attention itself, which is the paradigmatic criterion of aesthetic events.

We may identify these two criteria with the *means* and *ends* of p. The means are verbal and the end is the aesthetication of the experience of the object. The means primarily involve design, or increase of order, particularly (1) repetition and patterning of words and phrases (the province of rhet.) and (2) repetition and patterning below the level of the word (the province of pros.; see SOUND). Means of type (1) comprise rhet.; means of type (2) can occur frequently in rhet. but are made systematic in pros. In p., that is, phonological patterning is made "the constitutive device of the sequence" (Jakobson) and so comes to represent a genuine differentia.

The ends, however, are more difficult to define: in the West, the ends of p. have traditionally been seen as instruction and delight—that is, learning (increase of knowledge) and aesthetic pleasure (the nonutilitarian pleasure of such learning or of contemplation of the object, and the appreciation of a thing well made). But some critics have upheld other ends, such as expression or communication or even promotion of didactic content; and some have denied that p. has any specifiable ends at all. Anti-intentionalist critics would say we can never know the end the poet had in mind in creating the poem, and reader-oriented and deconstructive critics would say that even if we could it would make no difference: each reader creates her own meaning in the act of interpretation. If, however, one cannot at least say, on objective grounds, that the end is *not* persuasion, then p. cannot be distinguished from rhet. Some of these ends p. shares with religious, political, oratorical, and ludic (recreational) forms of discourse; and sometimes it may also share some of the first set of means. But not the second.

Aristotle in the *Poetics* identifies four "causes" of art: the formal, material, efficient, and final. In sculpture, these are the shape of, stone used in, sculptor of, and reason for making the sculpture. In p. these are verseform, words (their sound-shapes, visual shapes), the poet, and the purposes for which poets write and readers read. But p. differs from the other arts in that the material cause is not in itself semantically neutral—as are stone in sculpture, sound in music, and color in

painting—but always already semanticized. The webs of meaning which words create naturally when brought into conjunction with each other are compressed and given additional order when subjected to verseform, effects which are also semantic, increasing the semantic density in p. over prose.

II. SOUND AND MEANING. To understand p., therefore, we must understand words and the word. We might bear in mind the remark made by Mallarmé to Degas—which, had it not existed, it would have been necessary to invent—that poems are not made out of ideas, they are made out of words (P. Valéry, *Art of Poetry* [1939]). Critics who treat p. merely as a structure of ideas, imagery, metaphor, or figurative lang. do not, thereby, fundamentally identify anything "poetic" except perhaps by degree, nor do they distinguish verbal art from other verbal texts which are nonart. Critics who take meaning or "theme" (q.v.) as the essence of p. are encouraged, of course, by our automatic response to the referential and semantic character of words, so strong in ordinary lang. use. But they neglect the medium. In all the arts the medium is mere substance, alien and opaque to expression; and in this respect words appear in p. as pure sound—sounds in and of themselves, having aural textures, and sounds patterned, qua pattern. It is for this reason that Wimsatt remarked that "p. approximates the sensuous condition of paint and music not by being less verbal, less characteristic of verbal expression, but actually by being more than usually verbal, by being hyperverbal" ("The Domain of Crit."). Sound must be taken as sound in p., but also as the creator of meaning. Neither can erase the other, since each requires the other in order to exist. These two dimensions of the word are constant interinanimations of each other in verse.

The verbal medium in p. is not, therefore, a fully physical medium, as is sculpture, nor is it pure meaning, as in prose. It is irrefrangibly a bivalent or double medium, which critics separate at their peril. The critic will seek to show how the physical (aural) and semantic dimensions of words are related in both directions (Levý). Few Western critics except some very recent ones have embraced philosophical scepticism so wholly as to claim that the meanings readers derive from the reading of poems are not in some way directly determined by their material elements, words: the presumption of at least some kind of relation between text structures and reader effects is fundamental to all critical discourse whatsoever. The only issues—and they are great ones—are what kinds, and how.

The two dimensions of words in p. (meaning, sound) have produced, historically, two traditions or lines of descent, two kinds of poetries which in their pure forms one might for simplicity call extrinsic and intrinsic, or centrifugal and centripetal. The first emphasizes the referential, propositional, and mimetic aspect of lang., and leads to descriptive and narrative p., esp. epic but also lyric. The Western doctrine of mimesis held that the poem was an imitation (q.v.) of reality; this function works best when the physical bodies of words become purely transparent, the words dissolving into pure meaning. In this mode we see straight through words into interior visualization. While these kinds of poems are obviously referentially false or fictive, they are nonetheless taken as true at a deeper level (see REPRESENTATION AND MIMESIS), leading Richards to his famous but ineffectual attempt at a synthesis, "pseudo-statements." In this line of descent, imitation takes on primary importance, though this concept must not be understood in any naive sense: "representation" would be better, for the ability of lang. to represent extralinguistic events is simply assumed, as "signifier" entails "signified." The prosodic means, the kinds of opportunities afforded by verseform, are in this trad. minimized.

In the second line of descent, reference is minimized, ignored, or denied, and words become wholly of interest in themselves, as pure sound form or visual form or both. This trad. embraces a variety of aural and visual poetic modes never yet mapped out by critics but which include, at the very least, pure p., language p., and sound p. (qq.v.), and perhaps some types of concrete p. and pattern p. (qq.v.) as well. Now lang. focuses on itself, and Jakobson's poetic function is not only foregrounded but becomes the limiting condition of lang. The medium, made (or claimed to be) transparent in the first trad., is now made opaque. The full range of prosodic devices is exploited, sometimes to excess. This p. approaches the condition of music; its medium becomes the purely sensuous medium of the other arts. Meaning is not suppressed (which is impossible) but is made, at best, derivative from sound- and wordplay. P. in this trad. seeks to naturalize its signs, to generate an Adamic lang. where the relation of signifiers to signifieds is not arbitrary but rather motivated; or at least it seeks to produce the *illusion* that they are so.

But these are not mutually exclusive modes of *poiesis*: the great majority of poems written in the history of the world show some characteristics of both, and the great majority of poets have viewed the second set of features as simply the differentia of p. itself, i.e. that set of enabling conditions which validates the distinction between verse and prose at all. Focus on the verbal medium, in other words—on design or pattern in sound—is for these critics the route to the extension and intensification of meaning which p. makes possible.

III. HEARD AND SEEN. Of all the arts, consequently, p. may well seem the most schizophrenic. This was much less apparent before 1450, when the printing press revolutionized modern consciousness. In ancient Greece, and even in the Middle Ages, written texts were relatively rare and often intended simply as transcriptions of oral

performances. But with the spread of literacy and then the ubiquity of print, visual texts become not merely superimposed on but actually correlate to, quasi-independent from, aural ones. By the 16th c., p. had become an artform manifesting itself in two separate and distinct, though mutually permeable, media, (heard) speech and (seen) print. There is the poem in the ear, as Hollander remarks, and there is the poem in the eye, the poem heard and the poem seen, even if only in the mind's ear (mind's eye). The material cause of p., which was originally and for long aural, is now aural-visual mixed.

It is the point of view taken here that these two representational modes are not entirely congruent and that, more centrally, the visual text is not a mere transcript or notation-system for the oral text. Rather, they are two versions of the prior, originary, abstract entity called "the poem." Neither has ontological precedence. Rather, they are, irremediably, two—interrelated, certainly, even deeply and complexly, perhaps even unstably. "The poem" is not to be identified with a recitation (as against a written text) nor, alternatively, with a written text (as opposed to a performance). Much in each recitation is individual and idiosyncratic, and written texts can always be destroyed. "The poem" logically precedes both of these manifestations. As Fowler puts it, the performance of a poem "is not to be viewed as an implementation of the written record . . . but as an independent realization; because the written record is not the poem, but is itself only an implementation of it. The distinction is not between the poem on paper and the reading of it but between the poem (an abstraction) and two ways of realizing it." Poetic texts, that is, "may exist in both types of substantial realization—phonic and graphic." This follows from the nature of lang. itself: "lang. is *form*, not the physical representation of form." Hence "neither the linguist nor the literary critic is interested in mere differences of substance" (7–9). In sum, "the poem" is a twin thing, a bivalent entity, brought together only in acts of consciousness, whether aural performance, silent reading, or heard reading. The audience of oral p. (q.v.) in preliterate cultures heard p. only as aural experience, but the modern reader confronts the poem both visually and aurally, and as both object and experience—i.e. as an ontologically bivalent verbal sign.

IV. ONTOLOGY. Form, then, precedes realizations of form. The realization modes are heard form (oral recitation) and seen form (written text). But these modes are apprehended in radically different ways. The question of *what* a poem is may be more adequately unpacked by asking, rather, *where* a poem is: this is a question about the ontological location or *situs* of the poem. In the chief model of Western poetics, i.e. that given by Abrams (cf. the more complex but not essentially different model of communication mapped by Jakobson), the correlative processes of writing and reading p. are conceived along a transactional continuum which essentially runs from poet through text to reader or audience. And lit. crit. in the 20th c. has followed this line, steadily shifting its focus of attention rightward from biographical crit. (the author) to formalist crit. (the text) to reception theory (the reader). The text stands in the center of the two processes as (what would seem) an object. But the issue, as we have seen in the paragraphs above, is what kind of object, for it is evident that the object has its existence for us only *as experienced*. Many postformalist critics have naively assumed that formalism posited the poem solely as an aesthetic object, with no quarter given to the aesthetic experience evoked by that object. But this is an inaccurate characterization. Formalist crit. certainly emphasized the focusing of critical attention squarely on the aesthetic object (text) so as to avoid the vagaries and distractions of impressionistic, belle-lettristic, and biographical crit. But formalism never held that meaning was literally "in" the text; rather, most of the New Critics held that certain structures and strategies in the text, once apprehended, would generate strategies of interp. which would guarantee some commonality among reader responses. Readerly freedom in interp. was not expressly denied, but on the other hand it was not granted unlimited license, either. Even René Wellek's influential definition, in *Theory of Lit.* (first ed., 1947), of the poem and its "mode of existence" as a structure of norms is in fact not objectivist but rather phenomenological, having been derived from the work of the Polish phenomenologist Roman Ingarden, as Wellek himself later reiterated (*CritI* 4 [1977] 203).

Postformalist, reader-oriented theories posit "the poem" as something created within the interobjective space between the verbal text—itself dumb, mute—and the active consciousness of the reader. Some would hold that this is true for every reader, so that a hundred readers create a hundred unique "poems" by reading the same print text. The text then becomes something like, but perhaps less stipulative than, a musical score, a set of directions for performance. Deconstruction embraced the notion that the text can never control its own meaning since meaning resides in the interstices between words (the "free play of the signifiers"), so that every reader constructs meaning out of the inert materials offered by a text. But this way anarchy lies. In some other readerly theories the infinite proliferation of "poems" from a single text by readers is greatly reduced by the claim that all readers are well socialized members of speech or reading communities, so that they bring to bear, in reading, extensive (if unconscious) sets of conventions (tacitly agreed-upon rules) for interp.

In the 20th c., rapid progress in linguistics and prosody has had the very salutary effect of showing

us just how extraordinarily complex the verse situation really is. To say that lang. structures are overlaid with prosodic ones in verse, and to see how, is to see that we are still very far from a full and satisfying exegesis of what an intricate complex of formal relations a poem is and, concomitantly, what precisely is involved in a reader's cognition and interp. of a poem. "Reading" is still the great undiscovered country of lit. crit., mentioned by many but mapped by few. One reason may be the evident fact that "reading" covers not one situation but three or four: what happens when a reader reads a poem aloud differs radically from what happens when she reads it silently, or listens to someone else read it aloud, or engages the poem in that backward-and-forward movement of reflection and study which Leo Spitzer once described astutely as the "hermeneutic circle." About the detailed nature of these readerly engagements we have at present only minimal knowledge. About the immense effect that patterning has on perception, verbalization, and cognition even psychology knows only a little. Lit. crit. is not psychology, nor even epistemology, but it must go hand in hand with both.

J. W. Mackail, "The Definition of P.," *Lectures on P.* (1911); R. Ingarden, *The Literary Work of Art* (1931, tr. 1973), *The Cognition of the Literary Work of Art* (tr. 1973); J. C. Ransom, "P.: A Note in Ontology," *The World's Body* (1938), "Wanted: An Ontological Critic," *The New Crit.* (1941); M. T. Herrick, *The Fusion of Horatian and Aristotelian Lit. Crit., 1531–1555* (1946), esp. ch. 4; J. J. Donohue, *The Theory of Literary Kinds*, v. 2 (1949); E. C. Pettet, "Shakespeare's Conception of P.," *E&S* 3 (1950); Abrams; Curtius 152–53; R. Wellek, "The Mode of Existence of the Literary Work of Art," in Wellek and Warren; C. L. Stevenson, "On 'What Is a Poem?'" *Phil. Rev.* 66 (1957); S. Hynes, "P., Poetic, Poem," *CE* 19 (1958); R. Jakobson, "Linguistics and Poetics," in Sebeok, rpt. in Jakobson, v. 2; N. A. Greenberg, "The Use of *Poiēma* and *Poiēsis*," *HSCP* 65 (1961); V. M. Hamm, "The Ontology of the Literary Work of Art," *The Critical Matrix*, ed. P. R. Sullivan (1961)—tr. and paraphrase of Ingarden, continued in *CE* 32 (1970); R. Fowler, "Linguistic Theory and the Study of Lit.," *Essays on Style and Lang.* (1966); J. Levý, "The Meanings of Form and the Forms of Meaning," *Poetics—Poetyka—Poetika*, ed. R. Jakobson et al. (1966); E. M. Zemach, "The Ontological Status of Art Objects," *JAAC* 25 (1966–67); J. A. Davison, *From Archilochus to Pindar* (1968); R. Harriott, *P. and Crit. Before Plato* (1969); R. Häussler, "Poiema und Poiesis," *Forschungen zur römischen Literatur*, ed. W. Wimmel (1970); D. M. Miller, "The Location of Verbal Art," *Lang&S* 3 (1970); T. McFarland, "P. and the Poem," *Literary Theory and Structure*, ed. F. Brady et al. (1973); J. Buchler, *The Main of Light* (1974); E. Miner, "The Objective Fallacy and the Real Existence of Lit.," *PTL* 1 (1976); J. Margolis, "The Ontological Peculiarity of Works of Art," *JAAC* 36 (1977); M. P. Battin, "Plato on True and False Poetry," *JAAC* 36 (1977); S. Fish, "How to Recognize a Poem When You See One," *Is There a Text in This Class?* (1980); E. H. Falk, *The Poetics of Roman Ingarden* (1981); A. L. Ford, "A Study of Early Gr. Terms for P.: 'Aoide,' 'Epos,' and 'Poesis,'" *DAI* 42, 5A (1981): 2120; J. J. McGann, "The Text, the Poem, and the Problem of Historical Method," *NLH* 12 (1981); W. J. Verdenius, "The Principles of Gr. Lit. Crit.," *Mnemosyne* 4 (1983); G. B. Walsh, *The Varieties of Enchantment: Early Gr. Views of the Nature of P.* (1984); R. Shusterman, *The Object of Lit. Crit* (1984), ch. 3; Hollander; T. Clark, "Being in Mime," *MLN* 101 (1986).

T.V.F.B.

POLITICS AND POETRY. In 1968 the poet James Merrill was asked about the relationship between poetry and "political realities." "Oh dear," he answered, "these immensely real concerns do not produce *poetry*." *L'art pour l'art* was at that moment more a posture than a tenable position, though 20 years earlier many poets and critics had operated on the assumption that poetry in fact suffers from contact with political subjects. At that time, skepticism about political subjects derived largely from a strong postwar anti-Stalinist impulse among Western literary intellectuals, who had been sympathetic to Soviet Marxism until the Nazi-Soviet pact of 1939. Am. and Western European critics knew that the major controversy among Rus. poets was the role of private rather than public subject matter. On the one hand was Majakovskij, who gladly accepted every practical writing assignment given him by the Soviet bureaucracy; on the other was Anna Akhmatova, disciplined for refusing to write plainly political poems. Her refusal was a gesture of resistance within the Soviet Union; and the devel. of an exalted lyric, antipolitical poetic in the West after the Second World War was one literary aspect of the Cold War. Then too, the political record of the High Modernist poets was discouraging to Western intellectuals: from 1945 to 1958, Ezra Pound was incarcerated in a Washington insane asylum because of charges of treason. All this immediate postwar context has by now been forgotten, and the alleged incompatibility of pol. and p. is now discussed abstractly, as if the difficulty were exclusively theoretical.

The history of poetry, as Tom Paulin has demonstrated, is full of different kinds of successful political poems in every period of literary distinction: medieval peasant's songs, the *Divine Comedy*, Marvell's Horatian Ode, Milton's political sonnets, Dryden's "Absalom and Achitophel," Blake's "London," Andre Chenier's "Iambes VIII," Wordsworth's *Prelude*, Yeats's "Easter 1916," Davie's "Remembering the Thirties," Heaney's "Punishment," and Douglas Dunn's "Green Breeks." The lineaments of heroism, the limits of national or tribal solidarity, the power of persuasion, the forceful imposition of authority, and war—these Homeric themes have long been

traditional to poetry. Political subjects churn at the center of poetry, and toward the periphery the lesser genres—seduction songs, epithalamia, nature poems—now seem, though less obviously, still fascinatingly political. Among traditional subjects, only the coming of spring and the going with old age do not seem inherently political to us now.

What counts as a political subject is a question of audience. Poems like Brecht's "The God of War" seem only vaguely political in peacetime, but in time of war such a poem would be understood to be insistently partisan. Therefore, we may say, all that can be changed by social consensus or external authority is properly called political. The possibility of change itself, certainly not the state, is after all the source of political passion. Political poems concern situations that might be otherwise. Causes and consequences, choices—these are the special concerns of political poets. This sounds rationalistic, yet political writing concerns the possibility of deliberate action rather than unsought revelation. Dryden's "Absalom and Achitophel," for instance, ends not with action but instead with King David's announced resolution to punish the conspirators; although Dryden said he could not bring himself to "show *Absalom* Unfortunate," narration of the events would have been superfluous.

Critics of the 1970s and '80s have shown repeatedly how subjects formerly thought not to be political are in fact importantly so, partly because the range of social change that can be envisaged has expanded. The literary representation of women is an obvious example. The description of landscape is a less obvious one: Raymond Williams has shown that the landscapes of Eng. poetry are political figurations of the prerogatives of wealth and class. Other critics have focused on all that is taken for granted in poetry, for that is where political issues are treated as already somehow settled. As Empson noted, figurative lang. can shut off inquiry by suggesting that some political or historical event is natural, no more in question than an ocean or mountain. Neo-Marxist interpreters set themselves the task of counteracting the alleged effort of poets to render politics invisible. One weakness of this approach is the presupposition that once political intent has been unveiled, a critic's job is done, as though the revelation that poetry is politically motivated were itself a critical achievement. These critics presume, on implicitly ethical grounds, that the values and objectives of their political adversaries cannot be maintained in the light of day. Those with experience in democratic political controversy or struggle know that adversarial views must be not only revealed but argued down one by one, again and again, or else accommodated.

Mythological figures are more obviously political than natural ones, since they bear both a weight of hermeneutical trad. and explicit cultural authority. When Dryden characterized Charles II as King David, the Duke of Monmouth as Absalom, and the Earl of Shaftesbury as Achitophel, the political significance of the poem was set forth explicitly. Invocations of myth, history, or scripture to structure contemp. political events are different from other sorts of figures, however, in that the criterion of aptness is counterbalanced by one of audacity. The choice of a framing myth, esp. in satire, is usually outrageous: Charles II as David? Lenin as Christ, in Hugh MacDiarmid's "First Hymn to Lenin"? Eisenhower as Satan, in Robert Duncan's "The Fire"? But that is where the controversial aspect of these poems is sharpest; what David, Absalom, and Achitophel are made to say is much less tendentious. The pleasure and success of such poems depend on the framing myth seeming audaciously afield yet, finally, acute. Mythological figures also bear the weight of utopian hope for poets, like Blake, Adrienne Rich, or Judy Grahn, who want to imagine some future social order radically different from the actual one that provokes satirists.

The supposed strain between pol. and p. is actually more a strain between pol. and the evaluative criteria of crit. Mod. crit. has only limited access to a didactic view of poetry. Didactic poems are esteemed according to a universality or generality criterion, whereby a poem succeeds insofar as it speaks to the conditions of life in different historical contexts. By this measure satire either derives from an overall moral norm or it counts as a minor genre, because it names names and deeply loves its historic moment. Johnson's "Preface to Shakespeare" blocks the access of modern critics to a contemporaneity criterion that can treasure the minute particulars of a local social milieu. Dryden, Pope, the Pound of Cantos XIV and XV, and, among contemporaries, Turner Cassity express a kind of curiosity and, beyond the satire, fondness for very specific details of their time. Crit. needs a way of esteeming this poetry, not because the details are heterogeneous, but because they are thorny and abundant evidence of a citizen's passionate engagement less in a party than in a particular moment.

Michael McKeon argues that ideological comprehensiveness is the apt criterion for political poetry: political poets succeed by comprehending a wide range of the demands and conflicts of their time, not by rising above their historical moment. This view addresses the apprehensions of New Critics, such as Cleanth Brooks, in the face of partisan leftist poetry of the 1930s: leftist poems employed, in Kenneth Burke's terms, a rhetoric of exclusion. Those circumstances or considerations that did not fit a leftist ideological perspective were ignored, not incorporated into the poem as a sign of faith with the historical moment. The result, Brooks argues, was a sentimental appeal to a reader's political beliefs.

The relationship between particularity and generality is esp. troublesome in political poetry, as it is in political discourse generally. Which experi-

ences typify a class of experiences? How are exceptions to general practice to be recognized? Soviet writers from 1943 to 1953 were routinely censured for attending too closely to the allegedly exceptional negative aspects of social experience. Whose experience stands for that of large social groups or classes? These questions constantly arise in political discussions, and poets answer them. Furthermore, poets claim that metaphors have great explanatory power, that some particular vehicle stands for a larger tenor; poets derive authority by claiming to speak to universal issues, or for some group of people. And, beyond representation, poets use lang. to suggest some not yet realized idea; in this sense, particulars stand for ideals in poetry. Georg Lukacs constantly defended the power of realistic fiction to characterize not just individuals but types, and thereby to invoke ideas that transcend historical actualities. As Peter Demetz has argued, every concept of type, biblical or not, is at base theological, because it presumes a transcendent order whereby particular phenomena are measured. Lukacs imagined that order to be the future that socialism would bring, or the meaning of history. From Engels to the present, Marxist criticism has insisted on the need for poetry to be not merely specific but typical too.

Pindar's songs in praise of Gr. tyrants provide the most obvious archaic example of Western political poetry (the political significance of the *Iliad* is more abstract). Although Robert Lowell wrote praise-poems for Senators Eugene McCarthy and Robert Kennedy during the 1968 Am. Presidential campaign, and Gary Snyder wrote sympathetically about California Governor Jerry Brown in the late 1970s, modern poets have been generally disinclined to employ their art to praise the state in any form. What we now appreciate in Pindar is less his capacity to praise, which Pound dismissed as a big bass drum, than the back of his hand: his clever management of myths to warn the tyrants against the abuses of power. The most admired praise-poem in Eng., Marvell's Horatian Ode, is esteemed for its mix of praise and blame. Mod. poems of sheer blame, such as Robert Bly's "Asian Peace Offers Rejected Without Publication" and Adrienne Rich's "Rape" have been far more widely appreciated. But poems of true praise are much harder for critics to esteem. More importantly, the middle ground between praise and blame has been badly eroded by an oversupply of extremist crit. Thomas Edwards argues that the best political poems express mixed feelings, if outrage then also complicity, if contempt then also sympathy. This view draws support from two considerations: (1) an ambitious poem should do more than gild the monolith of an ideology; and (2) political issues are often more complicated than one or another party suggests.

However, modern poets and critics tend to admire, even more than mixed feelings, frankly oppositional poems. Indeed political poetry, if not poetry in general, is commonly (and too narrowly) understood now to be oppositional by definition. Even a centrist critic like Frank Kermode has argued that "lit. which achieves permanence is likely to be 'transgressive' . . . [the art] of the stranger in conflict with the settled order." "In his heart of hearts," Auden wrote, "the audience [the European poet] desires and expects are those who govern the country." Most contemp. political poems, however offensive they would be to those who do not read them, are consoling to those who do. Am. poets write about the possibility of civic change, but their poems are—quite rightly—no longer of interest to politicians, statesmen, and political administrators, because few Am. poets begin from a belief in political processes or agents as worthy of sustained scrutiny. They rarely take pol. seriously enough in political terms, nor do they present political problems as difficult to solve, or as ethically problematic. Am. poets like Bly, Duncan, and Ginsberg attribute mean motives and low intelligence to their political adversaries, as though virtue or cleverness could make a great difference. When Am. poets now attempt to extend sympathy to politicians, they invoke a psychoanalytic frame, in which no one is really guilty, but the sympathy is clinical and well outside of pol. Critics might credit as oppositional only that political poetry that challenges the political opinions of its audience and condemn those poems that extend the blunt discourse that is routine in political controversy.

Probably well before Milton's prefatory note to *Paradise Lost* readers sensed an analogy between social and prosodic order. Departure from prosodic norms has been loosely likened to political liberty, and discussions of poetic form have thereby been burdened by political polemics. One of the paradoxes of this sort of thinking is that the modernist poets responsible for breaking the force of metrical convention were not champions of political liberty. F. T. Marinetti, Ezra Pound, and T. S. Eliot were staunchly on the political right, though their efforts in free verse (q.v.) made it difficult for younger poets to continue writing in meter. Donald Davie has suggested that there is indeed a deep connection between the literary modernism of Pound and his attraction to authoritarian pol. Pound, Eliot, and other modernists felt unconstrained by the pressure of immediate poetic precedent, by the desire of readers to find continuity from one generation to the next, or by the daily secular experiences of their readers. These poets made a radical break with the conventions of late 19th-c. British and Am. verse. They dealt with historical subject matter very selectively and from a bird's-eye view, skipping over that which did not engage them. And together with Yeats they asserted claims to suprarational revelations. The willfulness that Pound displayed in lit. crit. and admired in the poetry of Dante he also admired in totalitarian pol. of the right.

Under the influence of Fr. and Rus. literary theory, academic critics in England and America have begun to locate the oppositional effort of poetry not in ideas or statements so much as in technical expressions of noncompliance with the referential and discursive features of descriptive, narrative, and expository prose—all of which supposedly underwrite the prevailing capitalist economic and social order. For many critics, a poet's breaking of genre conventions is itself an admirable act of political defiance. Some critics, such as Cora Kaplan, hold that particular rhetorical figures, such as metonymy, have political significance and value for women writers and readers whose "experience [has been] suppressed in public discourse." The Am. "Language Poets" of the 1970s have argued that the undermining of narrative structure and the disruption of syntactic expectations are the most responsible ways in which poets can contribute to large-scale social change: "lang. control = thought control = reality control," Charles Bernstein has said. "Poetry, like war," his collaborator Ron Silliman writes, "is the pursuit of pol. by other means." These poets see their writing as a poetic and expressly Marxist part of a broad intellectual movement in literary theory and philosophy that includes Derrida, Barthes, Jameson, and Rorty. One major objection to this approach to the nexus of pol. and p. is that it appeals only to academics who see it as a demonstration of the practical implications of literary theoretical texts that have achieved currency, as well as one way of reconciling these theories with neo-Marxism. Also, these poets reduce the political significance of poetry from a wide range of statements made by poets throughout history to but one mode of exemplification, based on an analogy of poetic form and political action.

New Historicist critics have pushed well beyond broad analogical arguments by identifying ways in which political motives determine the writing not just of poets or even of imaginative writers, but rather of all sorts of writers within a culture; they discover intentional connections between particular discursive practices and specifiable political positions. Richard Helgerson has shown how the controversy of the 16th and 17th cs. about quantitative verse in Eng. was understood then to be part of a debate about the kind of nation England would become, the kind of civil laws it would establish. John Barrell has argued that conditions of patronage fostered a particular kind of periodic structure in Ren. Eng. praise-poems. In reconstructing the political significance of poetic forms, this devel. in lit. crit. is restoring continuity to a view of poetry that was distinguished until very recently by the work of poet-critics such as Empson and Davie, who insisted always that a poet's formal choices expressed political considerations alive at the time of writing. But by close historical methods, New Historicists have refined connections between poetic form and ideology and located those connections not only in lang. itself but in entire cultural systems.

There is a need for critics to maintain a broad range of interpretive and, especially, evaluative principles for analyzing the relations of pol. to p., so that the political significance of poetry not be confined either to prophecy or to satire. The obligations of a citizen of a state are various, according to circumstances, and a citizen's prerogatives are usually many. The need is for a crit. that does not see the role of political poets as excluding, whether in the name of eloquence, literary trad., or higher truth, any of the duties or options of citizens.

ANTHOLOGIES: *Political Poems and Songs Relating to Eng. Hist.*, ed. T. Wright (1859); *Poems on Affairs of State*, ed. G. deF. Lord et al., 7 v. (1963–75); *Marx and Engels on Lit. and Art*, ed. L. Baxandall and S. Morawski (1973); *Carrying the Darkness: The Poetry of the Vietnam War* (1985), *Unaccustomed Mercy: Soldier Poets of the Vietnam War* (1989), both ed. W. D. Ehrhart; *P. and Pol.*, ed. R. M. Jones (1985); *Faber Book of Political Verse*, ed. T. Paulin (1986); *Pol. and Poetic Value*, ed. R. von Hallberg (1987).

HISTORY AND CRITICISM: W. Empson, *Some Versions of Pastoral* (1935); K. Burke, *Attitudes Toward Hist.* (1937); C. Brooks, *Mod. Poetry and the Trad.* (1939); A. P. d'Entrèves, *Dante as a Political Thinker* (1952); D. V. Erdman, *Blake: Prophet against Empire* (1954); C. V. Wedgwood, *P. and Pol. Under the Stuarts* (1960); B. Snell, *Poetry and Society in Ancient Greece* (1961); H. Swayze, *Political Control of Lit. in the USSR, 1946–1959* (1962); M. Adler, *P. and Pol.* (1965); C. M. Bowra, *P. and Pol. 1900–1960* (1966); P. Demetz, *Marx, Engels, and the Poets*, tr. J. L. Sammons (1967); J. M. Wallace, *Destiny His Choice* (1968); M. Mack, *The Garden and the City: Retirement and Pol. in the Later P. of Pope* (1969); G. Lukacs, *Writer and Critic*, tr. A. D. Kahn (1970); C. Woodring, *Pol. in Eng. Romantic P.* (1970); T. R. Edwards, *Imagination and Power* (1971); K. W. Klein, *The Partisan Voice: A Study of the Political Lyric in France and Germany, 1180–1230* (1971); V. J. Scattergood, *Pol. and P. in the 15th C.* (1971); L. C. Knights, *Public Voices* (1972); S. N. Zwicker, *Dryden's Political Poetry* (1972), *Pol. and Lang. in Dryden's Poetry* (1984); E. J. Brown, *Mayakovsky, A Poet in the Revolution* (1973); D. Davie, *Thomas Hardy and British Poetry* (1973), *Czeslaw Milosz and the Insufficiency of Lyric* (1986); R. Williams, *The Country and the City* (1973); J. F. Mersmann, *Out of the Vietnam Vortex* (1974); N. Reeves, *Heinrich Heine: P. and Pol.* (1974); M. McKeon, *Politics and Poetry in Restoration England* (1975); *Gesch. der politischen Lyrik in Deutschland*, ed. W. Hinderer (1978); W. Mohr and W. Kohlschmidt, "Politische Dichtung," *Reallexikon* 3.157–220; M. Calinescu, "Lit. and Pol.," *Interrelations of Lit.*, ed. J.-P. Barricelli and J. Gibaldi (1982); C. G. Thayer, *Shakespearean Pol.* (1983); J. K. Chandler, *Wordsworth's Second Nature* (1984); D. Norbrook, *P. and Pol. in Eng. Ren.* (1984); T. Olafioye, *Pol. in Af. Poetry* (1984); *P. and Pol. in the Age of Augustus*, ed. T. Woodman and D. West

(1984); R. von Hallberg, *Am. Poetry and Culture, 1945–1980* (1985); C. Bernstein, *Content's Dream* (1986); D. Davie, *Czeslaw Milosz and the Insufficiency of Lyric* (1986); C. Kaplan, *Sea Changes* (1986); P. Breslin, *The Psycho-Political Muse* (1987); P. Godman, *Poets and Emperors: Frankish Pol. and Carolingian P.* (1987); J. Montefiore, *Feminism and Poetry* (1987); R. Silliman, *The New Sentence* (1987); P. S. Stanfield, *Yeats and Pol. in the 1930's* (1988); T. Des Pres, *Praises and Dispraises* (1988); B. Erkkila, *Whitman the Political Poet* (1988); R. Helgerson, "Barbarous Tongues," *The Historical Ren.*, ed. H. Dubrow and R. Strier (1988); *"The Muses Common-Weale": P. and Pol. in the 17th C.*, ed. C. J. Summers and T.-L. Pebworth (1988); A. Patterson, *Pastoral and Ideology* (1988); F. Kermode, *Hist. and Value* (1989). R.V.H.

PROSE POEM. The extreme conventions of 18th-c. Fr. neoclassicism, with its strict rules for the differentiation of "poetry" from "prose," are to be blamed for the controversially hybrid and (aesthetically and even politically) revolutionary genre of the p. p. With its oxymoronic title and its form based on contradiction, the p. p. is suitable to an extraordinary range of perception and expression, from the ambivalent (in content as in form) to the mimetic and the narrative (or even anecdotal). Or rather, they are to be thanked, since the p. p. occasions even now a rapidly increasing interest. Its principal characteristics are those that would insure unity even in brevity and poetic quality even without the line breaks of free verse: high patterning, rhythmic and figural repetition, sustained intensity, and compactness.

In the p. p. a field of vision is represented, sometimes mimetically and often pictorially, only to be, on occasion, cut off abruptly; emotion is contracted under the force of ellipsis, so deepened and made dense; the rhapsodic mode and what Baudelaire called the "prickings of the unconscious" are, in the supreme examples, combined with the metaphoric and the ontological: the p. p. aims at knowing or finding out something not accessible under the more restrictive conventions of verse (Beaujour). It is frequently the manifestation of a willfully self-sufficient form characterized above all by its brevity. It is often spatially interesting (D. Scott). For some critics, it is necessarily intertextual (Riffaterre), for others, politically oriented (Monroe). It is, in any case, not necessarily "poetic" in the traditional sense and can even indulge in an engaging wit.

The p. p. is usually considered to date from Aloysius Bertrand's *Gaspard de la Nuit* (1842), though he was writing p. ps. earlier, and to be marked by heavy traces of Fr. symbolism, and conditioned by the stringency of the Fr. separation of genres. Among its antecedents are the poeticized prose trs. of the Bible, of classical and folk lyrics, and of other foreign verse; the poeticized prose of such romantics as Chateaubriand and the prose passages of Wordsworth's *Lyrical Ballads*; and the intermixtures of verse and prose in Maurice de Guérin's "Le Centaure," Tieck's *Reisegedichte eines Kranken* and Sainte-Beuve's *Alexandrin familier.* Characteristically, it was the romantics who came to the defense of this hybrid: Hugo's plea for the *mélange des genres* in his preface to *Cromwell* is the natural counterpart to Barbey d'Aurevilly's apology for the p. p.

But the most celebrated example of the p. p. is Baudelaire's *Petits Poèmes en prose*, or *Le Spleen de Paris* (begun 1855, pub. 1869), in which he pays tribute to Bertrand for originating the genre. Baudelaire's texts can complicate figuration to the point of "figuring us" as reader (Johnson, in Caws and Riffaterre). His "Thyrse" offers female poetic windings and arabesques around an upright male prose pole as the highly eroticized primary metaphor of mixing, while the *Petits poèmes* themselves are at once anecdotal and intimate, to the point of mixing the self with the subject. Rimbaud's *Illuminations* (1872–76) celebrate with extraordinary intensity the emergence of poems from less intimate matter, a newness dynamic in its deliberate instantaneity, yet the precursor of the aesthetic of suddenness practiced by Hoffmannsthal in his *Philosophie des Metaphorischen*—the speed of the metaphor is an "illumination in which, for just a moment, we catch a glimpse of the universal analogy"—and by imagists such as Pound. Rimbaldian confusion of first- and third-person perspective ("the lyric process of undergoing oneself and the more properly novelistic business of mapping out a behavior"—C. Scott) sets up, together with his notational rapidity, a kind of vibratory instant (Shattuck). Mallarmé's *Divagations* (begun 1864, pub. 1897) with their intricate inwindings of metaphor, Lautreamont's *Chants de Maldoror* (the first canto in 1868, the rest pub. posthumously in 1890) lush with a sort of fruity violence, Gide's *Nourritures terrestres* (Earthy Eats, 1897), and Claudel's *Connaissance de l'est* (The East I Know, 1900), nostalgic and suggestively pictorial, lead to Valéry's *Alphabet* (1912), whose form has been compared to what Valéry later calls, speaking of the dual function of discourse, "the coming and going between two worlds" (Lawler).

Elsewhere, the p. p. flourishes with a different cast: early on, in Switzerland, with Gessner (*Idylls*, 1756); in Germany, with Novalis and Hölderlin, then Stefan George, Rilke, Kafka, Ernst Bloch, and recently, in former East Germany, Helga Novak; in Austria, Hofmannsthal, Altenberg, and Polgar; in Belgium, Verhaeren; in England, De Quincey, Beddoes, Wilde, and the imagists; in Russia, Turgeynev and the Rus. futurists, esp. Xlebnikov; in Italy, the cubo-futurists such as Marinetti; in Spain, Bécquer, Jiménez, and Cernuda; in Latin America, recently, Borges, Neruda, and Paz; in Denmark, J. B. Jacobsen.

Modernist writing as practiced in France after symbolism and postsymbolism increasingly proble-

matized the genre; the so-called cubist poets Max Jacob, Reverdy, and Blaise Cendrars each gave his own slant to the p. p., emphasizing respectively its "situation," its strangely reticent irresolution, and its simultaneous perceptions. The Fr. surrealists Eluard, Breton, and Desnos provide a rich nostalgia and revelatory illumination by means of a startling juxtaposition of images; Gertrude Stein's "Tender Buttons" reaches a height of the lyric and the everyday held in tension, taking its energy from the androgynous. Among recent 20th-c. Fr. poets, René Char, St.-John Perse, and François Ponge, and then Ives Bonnefoy, Dupin, and Deguy prove the sustained vigor of the genre, proved equally in America by such prose poets (after Whitman) as James Wright, Robert Bly, W. S. Merwin, Russell Edson, John Ashbery, and John Hollander, and such language poets (after Gertrude Stein and W. C. Williams) as Charles Bernstein. See now VERS LIBRE; VERSE AND PROSE.

V. Clayton, *The P. P. in Fr. Lit. of the 18th C.* (1936); G. Díaz-Plaja, *El poema en prosa en España* (1956); S. Bernard, *Le Poème en prose de Baudelaire jusqu'á nos jours* (1959); M. Parent, *Saint-John Perse et quelques devanciers* (1960); U. Fülleborn, *Das deutsche Prosagedicht* (1970), *Deutsche Prosagedichte des 20. Jahrhunderts* (1976), *Deutsche Prosagedichte vom 18. Jahrhundert bis zur letzten Jahrhundertwende* (1985); D. Katz, "The Contemp. P. P. in Fr.: An Anthol. with Eng. Trs. and an Essay on the P. P.," *DAI* 31 (1970), 2921A; *The P. P.: An Internat. Anthol.*, ed. M. Benedikt (1976); R. Edson, "The P. P. in America," *Parnassus* 5 (1976); C. Scott, "The P. P. and Free Verse," *Modernism*, ed. M. Bradbury and J. McFarlane (1976); D. Lehman, "The Marriage of Poetry and Prose," *DAI* 39, 8A (1979): 4938; K. Slott, "Poetics of the 19th-C. Fr. P. P.," *DAI* 41, 3A (1980), 1075; J. Holden, "The Prose Lyric," *Ohio Rev.* 24 (1980); D. Keene, *The Mod. Japanese P. P.* (1980); B. Johnson, *The Critical Difference* (1981), ch. 3; S. H. Miller, "The Poetics of the Postmodern Am. P. P.," *DAI* 42 (1981), 2132; *The P. P. in France: Theory and Practice*, ed. M. A. Caws and H. Riffaterre (1983)—13 essays on Fr. and Eng.; R. E. Alexander, "The Am. P. P., 1890–1980," *DAI* 44, 2A (1983), 489; D. Scott, "La structure spatiale du poème en prose," *Poètique* 59 (1984); M. Perloff, *The Dance of the Intellect* (1985); D. Wesling, *The New Poetries* (1985), ch. 6; S. H. Miller, "John Ashbery's P. P.," *Am. Poetry* 3 (1985); M. S. Murphy, "Genre as Subversion: The P. P. in Eng. and Am.," *DAI* 46 (1986): 1932A; J. Monroe, *A Poverty of Objects: The P. P. and the Politics of Genre* (1987); J. Kittay and W. Godzich, *The Emergence of Prose* (1987); R. G. Cohn, *Mallarmé's P. Ps.* (1987); J. Simon, *The P. P. as a Genre in 19th-C. European Lit.* (1987); R. Silliman, *The New Sentence* (1987); S. Fredman, *Poet's Prose*, 2d ed. (1990); M. S. Murphy, *A Trad. of Subversion* (1992); S. Stephens, *The Poetics and Politics of Irony in Baudelaire's* Petits poèmes en prose (1993).

M.A.C.

PROVERB. A traditional saying, pithily or wittily expressed. Proverbial expression is traditionally given to customs, legal and ethical maxims, "blasons populaires," superstitions, weather and medical lore, prophecies, and other categories of conventional wisdom. Ps. are among the oldest poetic expressions in Sanskrit, Hebrew, Germanic, and Scandinavian lits. "Learned" ps. are those long current in lit., as distinct from "popular" trad. The former come into Western European lit. both from the Bible and the Church Fathers and from such Cl. sources as Aristophanes, Theophrastus, Lucian, and Plautus. Erasmus' *Adagia* (1500) was instrumental in spreading Cl. p. lore among the European vernaculars. The first Eng. collection was John Heywood's *Dialogue conteining . . . all the ps. in the Eng. lang.* (1546). But ps. had been commonly used by OE and ME writers, particularly Chaucer. The Elizabethan delight in ps. is evident in John Lyly's *Euphues* (1578–80) and in countless plays—as it is in Shakespeare. The genres of lit. in which ps. frequently occur are the didactic (e.g. Chaucer's *Tale of Melibeus*, Ben Franklin's *Way to Wealth*, and Goethe's "Sprichwörtliches"); the satirical (Pope); works depicting folk characters (*Don Quixote*, J. R. Lowell's *The Biglow Papers*); works reproducing local or national characteristics (E. A. Robinson's "New England"); and literary *tours de force* (Villon's "Ballade des proverbes").

What distinguishes ps. from other figures such as idioms or metaphors, G. B. Milner proposes, is their structure of "four quarters" in a "balanced relationship . . . both in their form and content." This configuration, evident in "Waste/not, // Want/not" or "Qui seme/le vent // Recolte/la tempete," appears in ancient and non-European as well as in modern langs. Milner associates this balanced four-part form with Carl Jung's paradigm of the structure of the mind. Milner's analysis, however, is found inadequate by Alan Dundes (1981), who proposes the p. as "a traditional propositional statement consisting of at least one descriptive element" which consists of "a topic and a comment." Ps. which contain "a single descriptive element are non-oppositional" while those with two or more "may be either oppositional or non-oppositional." Dundes relates proverbial structures to that of riddles; however, "ps. only state problems, in contrast to riddles which solve them." Dundes calls for empirical testing of his hypothesis with ps. from various cultures.

W. Bonser, and T. A. Stephens, *P. Lit.: A Bibl.* (1930); A. Taylor, *The P.* (1931, reissued with Index, 1962); B. J. Whiting, *Chaucer's Use of Ps.* (1934); W. Gottschalk, *Sprichwörter des Romanen*, 3 v. (1935–38); G. Frank, "Ps. in Med. Lit.," *MLN* 58 (1943); S. Singer, *Sprichwörter des Mittelalters*, 3 v. (1944–47); W. G. Smith, *Oxford Dict. of Eng. Ps.*, 2d ed., rev. P. Harvey (1948); M. P. Tilley, *Dict. of the Ps. in England in the 16th and 17th Cs.* (1950); O. E. E. Moll, *Sprichwörterbibliographie* (1958); H. Weinstock, *Die Funktion elisabethanischer Sprich-*

wörter und Pseudosprichwörter bei Shakespeare (1966); D. MacDonald, "Ps., Sententiae, and Exempla in Chaucer's Comic Tales," *Speculum* 41 (1966); F. Seiler, *Deutsche Sprichwortkunde*, 2d ed. (1967); C. G. Smith, *Shakespeare's P. Lore*, 2d ed. (1968), *Spenser's P. Lore* (1970); G. B. Milner, "What is a P.?" *New Society* (6 Feb. 1969), and "De l'armature des locutions proverbiales: Essai de taxonomie semantique," *L'Homme* 9 (1969); A. W. Weidenbrück, *Chaucer's Sprichwortpraxis*, Diss., Bonn (1970); *Oxford Cl. Dict.*, s.v. "Paroemiographers" (1970); F. A. de Caro and W. K. McNeil, *Am. P. Lit.: A Bibl.* (1970); M. I. Kuusi, *Towards an Internat. Type-system of Ps.* (1972); W. R. Herzenstiel, *Erziehungserfahrung im deutschen Sprichwort* (1973); P. Zumthor, "L'épiphonème proverbial," *RSH* 41 (1976); R. W. Dent, *Shakespeare's Proverbial Lang.: An Index* (1981), *Proverbial Lang. in Eng. Drama Exclusive of Shakespeare, 1495–1616: An Index* (1984); A. Dundes, "On the Structure of the P.," *The Wisdom of Many: Essays on the P.*, ed. W. Mieder and A. Dundes (1981); W. Mieder, *Internat. P. Scholarship: An Annot. Bibl.* (1982), *Supplement 1* (1990), *Supplement 2* (1992), comp., *Prentice-Hall Encyc. of World Ps.* (1986); G. Bebermeyer, "Sprichwort," *Reallexikon* 4.132–51; *Proverbium* 1– (1984–); H. and A. Beyer, *Sprichwörterlexikon* (1985); B. J. Whiting, *Mod. Ps. and Proverbial Sayings* (1989); W. Mieder, *Am. Ps.: A Study of Texts and Contexts* (1989).

D.HO.

PSALM. The poetic form of the p., at least as it has come down to us in the Western literary trad., is an invention of the ancient Near East. P.-writing, associated with temple cult, was an important literary activity in Mesopotamia, Egypt, possibly Ugarit (in present-day Syria), and, one may assume, despite the lack of surviving texts, in Canaan. The p.-poets of ancient Israel took over the form from the surrounding cultures, not hesitating to borrow images, phrases, and even whole sequences of lines, but also refashioning the p. to make it an adequate poetic expression of the new monotheistic vision of reality. The collection that constitutes the biblical *Book of Ps.* is an anthology (q.v.)—or more precisely, a conflation of what were originally four smaller anthologies—assembled sometime in the Second Temple period, but that includes poems probably composed over a span of several hundred years, going back to the beginning of the first millennium B.C. There are certain minor shifts in lang. and poetic structure over the centuries, but far more striking are the continuities of style and convention in the whole biblical lit. of ps.

The term for these poems current in Western langs. is derived from Gr. *psalmos*, a song sung to the accompaniment of a plucked instrument. The two Heb. terms, which are sometimes used in conjunction, are *mizmor*, simply indicating "song," without the necessary implication of instrumental accompaniment, and *tehillah*, "praise" (the plural, *tehillim*, is the usual Heb. title of the biblical book). The two predominant subgenres of ps. in the biblical collection are in fact ps. of praise and supplications. Together these make up more than two-thirds of the *Book of Ps.* There are other subgenres, such as wisdom ps., royal ps., and historical ps., but each of these is represented by only a few instances in the traditional collection and none has exerted the post-biblical influence of the supplication and the p. of praise.

Some of the Heb. ps. are clearly marked for liturgical performance on specified occasions or at specified moments in the temple rite. A view promulgated by Mowinckel holds that many ps. are texts for an annual rite of the enthronement of YHWH, but this notion is no more than a conjecture. Other ps. seem intended to be recited by individuals in moments of anguish or exaltation. This double nature of the ps., alternatively collective and personal, has been a source of their relevance both to the institutional and the individual lives of Christians and Jews ever since. At least some of the ps., to judge by the orchestral directions set at their head, were framed for musical performance, and most of them exhibit symmetries of form far beyond the norm of biblical poetry in other genres. Some ps. are cast as alphabetic acrostics. One encounters refrains or refrain-like repetitions, antiphonal voices, and most common of all, "envelope structures," in which the conclusion explicitly echoes images, motifs, or even whole phrases from the beginning.

This fondness for envelope structures is combined with a recurrent concern with the efficacy and power of lang. in general and of poetic lang. in particular. Thus the typical supplication begins with a foregrounding of the act of supplication ("From the depths I called to Thee, O Lord," etc.) and concludes by recapitulating the initial phrases, usually with the implication: "since I have called to Thee, Thou must surely hear me." The typical thanksgiving p. (the most common subcategory of ps. of praise) begins with a declaration of intent to sing out to the Lord, to tell His praises, and concludes by again affirming the act of praise or thanksgiving, now completed in the symmetrical structure of the poem.

It should also be noted that there is a certain fluidity among the psalmodic subgenres. Every supplication looks toward the possibility of becoming a thanksgiving p., and a few are turned into that retrospectively by the confident affirmation at the end. Even within one subgenre, there are striking differences in emphasis: a supplication may stress sin and contrition, the speaker's terror in a moment of acute distress, a reflective meditation on human transience, and much else.

The beautifully choreographed movements and the archetypal simplicity of style of the biblical ps. have made them a recurrent source of inspiration to later poets. For obvious reasons, the *Book of Ps.* has repeatedly influenced Jewish and Christian

liturgical poetry through the ages. With the resurgence of interest in the Bible after the Reformation, adaptations of ps. became widespread. In France, the versions of Clément Marot are particularly noteworthy. The apogee of psalmodic verse in Western langs. was reached in Ren. England, where the Bible in its new vernacular version became central to the culture. A variety of Eng. poets, from Wyatt and Sidney to Herbert and Milton, tried their hand at metrical versions of ps. In the signal instance of Herbert, the poet's original production owes something abiding in its diction, imagery, and sense of form to the model of the biblical ps. Modern Eng. poetry continues to evince a deep interest in ps. Dylan Thomas's work is strongly marked by ps., as is Donald Davie's *To Scorch or Freeze* (1988).

J. Julian, *A Dict. of Hymnology* (1925); *The Psalmists*, ed. D. C. Simpson (1926); I. Baroway, "'The Lyre of David,'" *ELH* 8 (1941); H. Smith, "Eng. Metrical Psalms in the 16th C. and Their Literary Significance," *HLQ* 9 (1946); J. Paterson, *The Praises of Israel: Studies Literary and Religious in the Ps.* (1950); C. S. Lewis, *Reflections on the Ps.* (1958); S. Mowinckel, *The Ps. in Israel's Worship*, 2 v. (tr. 1962); C. Westermann, *The Praise of God in the Ps.* (1965); J. Gotzen et al., "Psalmendichtung," *Reallexikon*; L. Finscher, "P.," *MGG*; H. Gunkel, *The Ps.: A Form-Critical Intro.*, tr. T. M. Horner (1967); R. C. Cully, *Oral-Formulaic Lang. in the Biblical Ps.* (1967); A. L. Strauss, *Bedarkhei hasifrut* (1970), 66–94; C. Freer, *Music for a King* (1972)—Herbert and metrical ps.; R. Lace, "Le Psaume 1—Une Analyse structurale," *Biblica* 57 (1977); E. Werner et al., "P.," and N. Temperley et al., "Ps., Metrical," *New Grove* 15.320–35, 347–82; C. Bloch, *Spelling the Word: George Herbert and the Bible* (1985); P. D. Miller, Jr., *Interpreting the Ps.* (1986); R. Alter, *The Art of Biblical Poetry* (1985), "*Ps.*," *Literary Guide to the Bible*, ed. R. Alter and F. Kermode (1987); E. L. Greenstein, "Ps.," *Encyc. of Religion*, ed. M. Eliade, v. 12 (1987); L. A. Schökel, *A Manual of Heb. Poetics* (1988).
R.A.

PUN. Every art form knows ps. Painting and sculpture have visual ps. (e.g. optical illusions), and music has melodic, rhythmic, and harmonic ps. (e.g. the "deceptive cadence"), but they are most deeply involved with lang. and literary art. Verbal ps. play with sound and meaning: identical or similar sounds bring together two (or more) meanings. The p. "works" when the context gives both meanings significance. In the lang. of poetry we find lexical ps., grammatical ps., and the sophisticated form of p. based on words or phrases which simultaneously belong to disparate levels of diction, situation, or experience. A lexical p. turns on an ambiguous word: Shakespeare's witty Mercutio, mortally wounded, says that by tomorrow he will be "a grave man" (*Romeo and Juliet* 3.1.96), and melancholy Hamlet, asked by Claudius why the "clouds" of mourning still hang on him, replies that on the contrary he is "too much in the sun" (and "son"; *Hamlet* 1.2.67). Although Noah Webster called punning "a low species of wit," it can be a powerful instrument of poetic lang. when it reveals that two things which bear the same name, i.e. the same phonic sequence, also share deeper affinities. John Donne, meditating on the Crucifixion, used Hamlet's p. to embrace profound religious paradox: "There I should see a Sunne, by rising set, / And by that setting endlesse day beget" ("Goodfriday, 1613. Riding Westward").

A grammatical p. turns on some ambiguity in morphology or syntax, e.g. ambiguous parts of speech. Milton describes Eve eating the forbidden fruit with a chilling p. of this type: "Greedily she ingorg'd without restraint, / And knew not eating Death" (*PL* 9.791–92)—i.e., Eve did not know that with the apple she was eating death, and also did not yet know of the ravenous figure of Death soon on its way to devour *her*.

The third form of p. appears at its simplest in those joking tales in which the punchline plays off a familiar idiom, proverb, or slogan ("shaggy dog" stories). A long tale about a bear wearing tennis shoes who is caught stealing a Chinese man's lumber ends with the man shouting "Stop! Boyfoot bear with teak of Chan!" In poetry this form is the basis for what Ezra Pound called *logopoeia*. Early Gr. lyric frequently used epic diction in distinctly nonheroic contexts: e.g. taking Homer's *lusimelēs* ("limb-loosening," "limb-relaxing"), an epithet of sleep in the *Odyssey*, and applying it to desire (Archilochos) or love (Sappho). In OE poetry certain words such as "lord," "joy," and "glory" can have double reference—one pagan and heroic, the other Christian and homiletic: e.g. "There is no man on earth . . . whose *dryhten* [earthly lord] is so gracious that he never worries as to what *dryhten* [heavenly Lord] will bring about for him" (*The Seafarer*). Many lines in modernist poems such as T. S. Eliot's *The Waste Land* and Ezra Pound's *Hugh Selwyn Mauberley* reverberate with this device. Involving both allusion and irony (qq.v.), such ps. bring together different views of human experience for analogy and contrast.

The word *p.* first appeared in Eng. ca. 1650; its origins are not clear. At the same time, neoclassical taste conceived a dislike for ps. in literary lang. which lasted until modernism. Dryden wrote that Shakespeare's comic wit too often degenerated into "clenches," or ps. (*An Essay of Dramatic Poesy* [1668]); Addison attacked ps. as "false wit" (*Spectator* no. 61 [1711]); Dr. Johnson complained about Shakespeare's fatal "quibbles," ps. again ("Preface to Shakespeare" [1765]). Before that, however, the p. had long been treated seriously as a topic in Cl., medieval, and Ren. rhet., where it was known as the figure of speech called *paronomasia* (Gr.) or *adnominatio* (Lat.). The rhetoricians were primarily interested in the figure which *repeated* words of identical or similar sounds. Roman rhetorical treatises such as the anonymous

Rhetorica ad Herennium (4.13–23) and Quintilian's *Institutio oratoria* (9.3) developed elaborate (if inconsistent) categories and nomenclature for this and other figures—distinguishing, for example, whether the sound-repetition brought with it a completely different word, a secondary or extended meaning of the same word, the same word with a different prefix or vowel-length (in Gr. and Lat.), a different word with the same ending (*homoeoteleuton*), the same word in a different grammatical form or function in the sentence, etc. Such figures were considered useful "ornaments" of lang. when kept under control by decorum, and were most valuable, Aristotle suggested (*Rhet.* 3.11), when they supported structural and conceptual meaning, using aural repetition to help define a revealing antithesis or powerful metaphor. Emily Dickinson inherits that trad. when she repeats different grammatical forms of the verb *stop* (*for*), varying at the same time the emphases on its colloquial and literal meanings ("call for" and "cease"): "Because I could not stop for Death— / He kindly stopped for me—" (no. 712). Milton also draws on the rhetorical trad. when he has Adam address Eve as "Sole Eve, Associate sole" (*PL* 9.227): forewarned of an enemy in the Garden, Adam wishes to dissuade Eve from going off alone, first by reminding her she is both the "only" Eve he has and "alone" Eve without him, then by saying his "only" associate is also an associate "soul"—a complex paronomasia which discovers all that is at risk.

Sr. M. Joseph, *Shakespeare's Use of the Arts of Lang.* (1947); Empson; L. Spitzer, "P.," *JEGP* 49 (1950)—etymology; P. F. Baum, "Chaucer's Ps.," *PMLA* 71, 73 (1956, 1958); J. Brown, "Eight Types of P.," *PMLA* 71 (1956); M. M. Mahood, *Shakespeare's Wordplay* (1957); S. B. Greenfield, *The Interp. of OE Poems* (1972), ch. 4; T. W. Ross, *Chaucer's Bawdy* (1972); Lausberg, 322–36; H. A. Ellis, *Shakespeare's Lusty Punning in* Love's Labour's Lost (1973); Murphy; A. Jolles, "Witz," *Einfache Formen*, 5th ed. (1974); W. C. Carroll, *The Great Feast of Lang. in* Love's Labour's Lost (1976); Morier, s.v. "*paronomase*," "*annomination*"; E. Le Comte, *A Dict. of Ps. in Milton's Eng. Poetry* (1981); Jakobson, v. 3; F. Rubinstein, *A Dict. of Shakespeare's Sexual Ps. and Their Significance* (1981); W. Redfern, *Ps.* (1984); M. Foucault, *Death and the Labyrinth* (tr. 1986); E. Cook, *Poetry, Word-play, and Word-war in Wallace Stevens* (1988); *On Ps.*, ed. J. Culler (1988); B. Vickers, *Cl. Rhet. in Eng. Poetry*, 2d ed. (1989), 143–44; E. Partridge, *Shakespeare's Bawdy*, 3d ed. (1990).

A.W.

PUNCTUATION (Lat. *punctus*, "point"). A system of nonalphabetical signs that express meaning through implied pauses, pitch shifts, and other intonational features. For prose, in Western Europe, p. began as a pedagogical and scribal guide to reading aloud. It evolved to mark syntax for silent readers, as urged by Aldo Manuzio in *Orthographiae ratio* (1566) and Ben Jonson in *The Eng. Grammar* (ca. 1617, pub. 1640). For poetry, however, the diachronic shift from oral delivery to silent reading had less impact on comprehension, because in poetry—whether its sound (q.v.) is actual (heard) or virtual (read)—sound helps create sense.

Both pedagogy and performance (q.v.) loom large in the history of p. Speech seems continuous; thus the earliest Gr. inscriptions do not separate words. Points occasionally separate longer phrases, and Aristotle mentions a horizontal line, the *paragraphos*, used to introduce a new topic. At the Museum in Alexandria, ca. 200 B.C., the librarian Aristophanes proposed terms and marks (not corresponding to their modern counterparts) to distinguish a short section (*comma*) from the longer and longest textual sections (*colon*, *periodos*). His terms survived via the Roman grammarians, notably Donatus in the 4th c. A.D., and via early Christian teachers such as Jerome, who devised p. *per cola et commata* ("by phrases") specifically to facilitate reading aloud of his new Vulgate tr. of the Bible. Over the next millennium, various writers developed systems of p., adapting musical notation (neumes) to indicate pitch shifts as well as pauses. Thus authorities long dead could continue to prescribe correct oral recitations of sacred texts.

For secular poetry, early printers (e.g. Caxton) and theorists (e.g. Puttenham) emphasized the practical functions of p.—esp. breath breaks—rather than syntax or prescriptive interp. During the 17th c., syntactic p. became the norm for imaginative as well as expository prose, whereas drama and poetry tended to retain sound-based flexibility. Before John Urry (1721), for instance, no editor preparing Chaucer for print punctuated ordinary line ends. Similarly, readers of Shakespeare's plays as scripts for performance could realize that no amount of enforced syntactic p. can hold to one set meaning Hamlet's "To die, to sleep / No more and by a sleep to say we end." During the 19th c., however, the study of the modern langs. succumbed to the strictures of academic respectability. Teachers and editors began using syntactic p. to prescribe their own imagined performances of secular poetry, just as clerics a millennium earlier had prescribed oral performances of sacred Lat. prose.

The illuminated works of William Blake provide striking examples of authorial p. as a guide to imagined performance, for Blake laboriously engraved each plate with letters, visual art, and other marks, incl. p. In the first stanza of "London," nonsyntactic periods that close two lines help express the narrator's initial, but illusory, sense of control. Each stanza has progressively lighter p., the second spilling with no end mark into the third, until—as ever more appalling images engulf the narrator—p. disappears altogether from the final stanza. The abyss yawns. Despite Blake's efforts, his 20th-c. editor Geoffrey Keynes repunctu-

ated "London" in accord with modern canons of p. and sense, i.e. in accord with prose syntax, so that the poem in his edition (though not in others) presents a detached narrator calmly cataloguing social problems.

Many standard eds. still impose syntactic prose p. on poetry. In the Twickenham Ed. of the Works of Alexander Pope, for example, each editor records words that differ from the copy-text chosen, but each modernizes p. at will, despite Pope's letter telling a printer to "contrive the Capitals & evry [*sic*] thing exactly to correspond" with an earlier printing.

How do editors justify changing authorial p.? In part, they are applying methodologies developed for Cl. and medieval mss., in which any p. that appears is based on a local scribal dialect that may bear little or no relationship to our present standardized system. In one ms. trad., a virgule (slash, solidus) might mark poetic caesura, for example; in another trad. a virgule may mark the shortest of three pauses, and in yet another the longest of six divisions into sense units. The slanted line went nameless until 1837, when Henry Hallam adopted a zoological term to declare that virgules in Chaucer mss. always mark caesura—an assertion which, though dead wrong, was not definitively disproven until 1982.

In mss., other p. marks also vary from one region to another, even within the same decade. Medieval dialects of neumes varied likewise, yet standardized musical notation is not now expected to explicate a set meaning for either readers or performers of a score. But editors of Cl. and medieval poetry began to supply less experienced readers with syntactic p. suitable to explicative prose. Nearly all Chaucer eds. since 1835, for example, have imposed the full range of mod. p. Quotation marks in particular create demonstrable interpretive problems absent from mss. of Chaucer, wherein a given speech need not be assigned definitely to the narrator or to one or another character. By specifying p. absent from or altering p. present in mss., modern editors continue to prescribe their own silent performances of key poetic passages.

In cultures that have not undergone such a pronounced shift from religious to secular textual education, such as Arabic or Chinese, most p. still signals breath breaks and major new topics. Only occasionally does it indicate change in speaker or the many other vocal intonations commonly prescribed in the West by means of syntactic p. applied—or misapplied—to poetry.

P. Simpson, *Shakespearian P.* (1911); W. J. Ong, "Historical Backgrounds of Elizabethan and Jacobean P. Theory," *PMLA* 59 (1944); A. Pope, *The Rape of the Lock and Other Poems*, 3d ed., ed. G. Tillotson (1962), General Note; M. Treip, *Milton's P. and Changing Eng. Usage, 1582–1676* (1970); E. O. Wingo, *Lat. P. in the Cl. Age* (1972); L. D. Reynolds and N. G. Wilson, *Scribes and Scholars*, 2d ed. (1974)—discursive bibl. on current debates re Gr. and Lat. p.; N. Blake, "P. Marks," *Eng. Lang. in Med. Lit.* (1977); G. Killough, "P. and Caesura in Chaucer," *SAC* 4 (1982); B. Bowden, *Chaucer Aloud* (1987), s.v. "p." in index; T. J. Brown, "P.," s.v. "Writing," *New Encyclopedia Britannica: Macropaedia*, 15th ed. (1990); G. Nunberg, *The Linguistics of P.* (1990); J. Lennard, *But I Digress* (1991); M. B. Parkes, *Pause and Effect: The History of P. in the West* (1992).

B.BO.

PURE POETRY (Fr. *poésie pure*, Ger. *absolute Dichtung*). This term refers most specifically to "La Poésie pure," a doctrine derived from Edgar Allan Poe by the Fr. symbolist poets—Baudelaire, Mallarmé, and Valéry—and widely discussed in the late 19th and early 20th c. In this context, "pure" is equivalent to absolute, on the analogy of absolute music, i.e. structuring of sound without ostensible semantic content. The analogy is significant in that both the theory and practice of the symbolists were influenced by the relations of music and poetry (q.v.).

The doctrine was first enunciated in Poe's essay "The Poetic Principle." For Poe, the essential quality of poetry is a kind of lyricism distinguished by intensity and virtually identical with music in its effects. Since the duration of intensity is limited by psychological conditions, Poe concludes that the long poem is a contradiction in terms and that passages which fail to achieve a high level of intensity should not be included in the category of poetry. Poetry is regarded as being entirely an aesthetic phenomenon, differentiated from and independent of the intellect and the moral sense. The products of the latter faculties, ideas and passions, are more properly relegated to the province of prose, and their presence in a poem is positively detrimental to poetic effect.

In their desire for poetry to attain to the condition of music, the symbolists were wholehearted disciples of Poe; in elaborating his theory, however, they were far more aware of the problem of lang. than Poe had been. The relevance of the doctrine of p. p. to critical theory thus rests almost entirely upon its concern with the symbolic or iconic properties of lang. The impetus toward this line of speculation was given in Baudelaire's rephrasing of Poe: the goal of poetry is of the same nature as its principle, and it should have nothing in view but itself. The aim of the symbolists was to confer autonomy upon poetry by subjecting the semantic properties of lang. to the phonetic properties of words, their sounds, kinship, and connotations.

It would not be accurate, however, to ascribe unity of aim to the whole symbolist movement. For Baudelaire, the autonomy of poetic lang. was incomplete in that meaning involved "correspondances" with an ultimate reality. Poe, in his reference to the "supernal beauty" which p. p. was capable of achieving, had hinted at the possibility of a metaphysical or mystical significance in verbal

music. The Abbé Bremond was more explicit in claiming a mystical value for p. p. Mallarmé's conception of p. p. was of a point at which poetry would attain complete linguistic autonomy, the words themselves taking over the initiative and creating the meanings, liberating themselves from the semiotic tyranny of the lang. and the deliberate intentions of the poet. With Mallarmé, subject matter is a function of an intense preoccupation with the medium.

Speculation in this direction reached its limit in Valéry, who eventually found the processes of poetic composition more interesting than the poetry itself. Valéry's contribution to the doctrine of p. p. poetry focused on the most fundamental yet ineluctable aspects of poetic lang., the relation of sound and sense. In his first exposition (prefaced to a volume of verse by Lucien Fabré), Valéry defines p. p. as poetry which is isolated from everything but its essence. Poe's strictures on long poems and on the didactic motive are repeated. The aim of p. p. is to attain from lang. an effect comparable to that produced on the nervous system by music. This essay gave rise to considerable discussion, however, and Valéry was later to deny, without abandoning the doctrine, that he had advocated p. p. in a literal sense. It represented for him a theoretical goal, rarely attainable in view of the nature of lang., wherein sound and meaning form a union as intimate as that of body and soul.

Poe's ideas had little direct influence on Eng.-speaking critics and poets; the idea of p. p. was mainly an importation from France. But the qualities of vagueness and suggestiveness valued by Poe and the symbolists were contrary to the imagist demand for the utmost precision in the rendering of the image. In 1924 George Moore brought out an anthol. entitled *P. P.*, but while Moore had absorbed the views of the symbolists, his own conception returns to the earlier, plastic trad. of the Parnassians. T. S. Eliot regarded p. p. as the most original devel. in the aesthetic of verse made in the last century, characteristically modern in its emphasis upon the medium of verse and its indifference to content, yet also one which decisively terminated with Valéry. Other critics, however, have continued to find the absolutist, sound-intensive, and antireferential mode in modern poems such as Wallace Stevens' "Sea Surface Full of Clouds," not to mention the more radical experiments which followed the modernist era in Beat poetry, language poetry, "text-sound," intermedia, and "sound poetry." If all poetry does indeed aspire to the condition of music, that aspiration will make itself manifest somewhere in the poetry of every age and culture.

H. Bremond, *La Poésie pure* (1926); F. Porché, *Paul Valéry et la poésie pure* (1926); R. P. Warren, "Pure and Impure Poetry," *KR* 5 (1943); T. S. Eliot, "From Poe to Valéry," *RHR* 2 (1949); F. Scarfe, *The Art of Paul Valéry* (1954); Sr. C. de Ste. M. Dion, *The Idea of "P. P." in Eng. Crit. 1900–1945* (1948); H. W. Decker, *P. P., 1925–1930: Theory and Debate in France* (1962); M. Landmann, *Die absolute Dichtung* (1963); B. Böschenstein, *Studien zur Dichtung der Absoluten* (1968); D. J. Mossop, *P. P.: Studies in Fr. Poetic Theory and Practice 1746 to 1945* (1971); Morier; S. Hart, "Poésie pure in Three Sp. Poets," *FMLS* 20 (1984). S.F.; T.V.F.B.

Q

QUATRAIN. A stanza of 4 lines, normally rhymed. The q. is, with its many variations, the most common stanza form in European poetry, and very probably in the world. It is for the establishment, in the 3rd c. A.D., of the q. as the meter of the hymn (q.v.), developed from the Lat. iambic dimeter and subsequently to become the great staple meter of the Middle Ages, that C. W. Jones once called Ambrose, Bishop of Milan, "the man who most affected Western verse." Many hymnals still show the variants in a metrical index. This meter—iambic tetrameter (alone or mixed with trimeter) lines, in foot verse—has as its isomer the 4-stress line in stress verse, which is the basis of the several forms of ballad meter, another staple (in a lower register) in the Middle Ages and still common in songs today. Hence the irony that when Sternhold and Hopkins, in the middle of the 16th c., sought to find a meter familiar to the masses for their vernacular Psalter, they chose the q. form of ballad meter. The q. is the form of the Sp. *copla*, the Rus. *chastushka*, the Malay *pantun*, the Ger. *Schnaderhüpfel*, and the Sanskrit *sloka*; indeed, most of the Vedic stanzas are qs. It was August Meineke who in 1834 showed that even Horace's *Odes* exhibit q. structure.

Most rhyming qs. fall into one of the following categories: (1) *abab*, alternating or "cross-rhyme," or its variant *xbyb* (in which *x* and *y* represent unrhymed lines), a category which includes ballad meter and the elegiac stanza or "heroic q."; (2) *abba*, "envelope rhyme," hence called the envelope stanza, of which Tennyson's *In Memoriam* stanza is a type; (3) *aabb*, in which an effect of internal balance or antithesis is achieved through the use of opposed couplets, as in Shelley's "The Sensitive Plant"; or (4) the monorhymed q., e.g. Gottfried Keller's "Abendlied," or nearly monorhymed q. such as the Omar

Khayyam stanza, rhyming *aaxa*. The first two have been the most popular forms in Western poetry since the 12th c. (Meyer 1.314 ff.). In Fr. verse the *abab* scheme is the standard form for qs.; the *abba* pattern is rare: Lamartine and Musset use it precisely once each, and Hugo, the most prolific versifier in the Fr. lang., only once in every ten poems written in qs. Martinon shows that *aabb* qs. are hardly stanzas at all. Stefan George in "Komm in den totgesagten park" tests all the possible forms of the rhymed q.

Qs. interlinked by rhyme are also to be found, as are those displaying such complications as the alternation of masculine and feminine rhyme and the use of heterometric line lengths. In Rus. poetry the q. is by far the most common stanza form, taking the forms *AbAb* or *aBaB* (capitals denote feminine rhymes). The q. has been used in Western poetry primarily as a unit of composition in longer poems, but is also commonly used for the two component parts of the octave (q.v.) of the sonnet. As a complete poem, the q. lends itself to epigrammatic utterance; Landor and Yeats have shown mastery in the composition of such qs.—P. Martinon, *Les Strophes* (1912). T.V.F.B.

R

REFRAIN (Ger. *Kehrreim*). A line, lines, or part of a line repeated verbatim at intervals throughout a poem, usually at regular intervals, and most often at the end of a stanza—a burden, chorus, or repetend. If the repetition is not verbatim, the phenomenon is sometimes called "incremental repetition."

The r. is a conspicuous feature of oral poetry, particularly verses sung to the accompaniment of dance or during communal labor, esp. when performed by a group and a leader, the latter taking the stanzas, the former the rs. Rs. occur in the Egyptian *Book of the Dead*, in the Heb. Psalms, in early Gr. lyric poetry, in the Lat. epithalamia of Catullus (61, 64), and in the OE *Deor*; they blossom in medieval poetry, first in Med. Lat. religious hymnody and antiphonal responsion, then in the vernaculars. Their first systematic use in OF verse dates from ca. 1147, but they are equally if not more important in the fixed forms of Occitan poetry, as they are also in the medieval ballads. In ME rs. form the *burden* in carols and the tail in tail-rhyme, in Occitan and OF the *envoi* in the *ballade* and *virelai* and the *rentrement* in the *rondeau*, in It. the *ripresa* in the *ballata*, in Sp. the *estribillo* in the *villancico*. In Port. they appear in the *cantiga*. In both Ren. lyrics and in romantic poetry their use still results from close association with song.

A r. may be as short as a single word or as long as a stanza. Though usually recurring as a regular part of a metrical pattern, it may appear irregularly throughout a poem, in regular form or not, or may even be used in free verse. In stanzaic verse it usually occurs at the end of a stanza but may appear at the beginning (the burden in carols) or in the middle (the *stef* in ON skaldic poetry). It may be repeated each time with a slight variation of wording appropriate to its immediate context (Rossetti's "Sister Helen"; Tennyson's "Lady of Shalott") or in such a way that its meaning develops from one recurrence to the next. Poe discusses this type of structure in "The Philosophy of Composition." The r. may be a tag or a nonsense phrase seemingly irrelevant to the rest of the poem or relevant only in spirit ("With a hey, and a ho, and a hey nonny-no" [*As You Like It* 5.3.18]), or it may emphasize or reinforce emotion or meaning by catching up, echoing, and elaborating a crucial image or theme. It gives pleasure in its repetition of sound, and it serves to segment and correlate rhythmical units and so unify the poem. The full comparative study of rs. still remains to be written.

G. Thurau, *Der R. in der französischen Chanson* (1901); F. B. Gummere, *The Beginnings of Poetry* (1908); F. G. Ruhrmann, *Studien zur Gesch. und Charakteristik des Rs. in der engl. Lit.* (1927); P. W. Gainer, *The R. in the Eng. and Scottish Popular Ballads* (1933); N. H. J. van den Boogaard, *Rondeaux et rs. du XIIe au début de XIVe siècle* (1969)—useful list of all extant OF rs. ca. 1148–1332; T. Newcombe, "The R. in Troubadour Lyric Poetry," *Nottingham Mediaeval Studies* 19 (1975); C. F. Williamson, "Wyatt's Use of Repetitions and Rs.," *ELR* 12 (1982); J. Hollander, *The Figure of Echo* (1981), *Melodious Guile* (1988), ch. 7; S. M. Johnson, "The Role of the R. in OF Lyric Poetry," *DAI* 44 (1983): 747A; E. Doss-Quinby, *Les Rs. chez les trouvères du XIIe siècle au début du XIVe* (1984). L. Magnus, *The Track of the Repetend: Syntactic and Lexical Repetition in Mod. Poetry* (1989)—use with caution. T.V.F.B.; L.P.

RENGA. Japanese linked poetry. Although joined poems (*lien-chü*) were composed earlier in China, they did not have the codified nature of *r.*, nor did they develop out of the practice of poetic sequences, as *r.* did. *R.* developed from integrated sequences of *waka* in Japanese royal collections and from shorter (esp. 100-poem) sequences modeled on the collections. *R.* also had ancestry in the capping of one part of a *waka* by one poet

with a second part by another: e.g. two lines of 7 + 7 syllables added to three of 5, 7, and 5. In the 12th c., waka poets composed *r.*, alternating 3- and 2-line stanzas in a nonserious (*mushin*) fashion. Play led to earnest (*ushin*) *r.* At first, impressive stanzas were sought. Later, the greatest *r.* master, Sōgi (1421–1502), emphasized the integrity of sequences along with variety in impressiveness of stanzas and variance in closeness and distance of connection.

A typical *r.* sequence comprised 100 stanzas composed by about three poets at a single sitting (*za*) of about three hours. A given stanza was therefore composed in less than three minutes. Given the complexities of the *r.* code, that meant (as the last practitioner of *r.* put it) that 20 years of practice were necessary before it could be discovered whether one had talent.

A *r.* stanza related semantically only to its predecessor and therefore also to its successor: each stanza was like a link in a chain. The sequences were of varying length, the standard being the *hyakuin* (100 stanzas), with its multiples, *senku* (1000) and *manku* (10,000). Shorter units incl. *yoyoshi* (44), *kasen* (36), *hankasen* (18), and *iisute* (a few stanzas—these last three being used mostly in *haikai*, discussed below).

To give the rapid composition of *r.* coherence, other features of its code were developed. Individual stanzas had as topic one of the seasons or were deemed miscellaneous, and numerous subtopics (e.g. love, Buddhism) were classified. Spring, autumn, and love topics should run at least three stanzas, whereas two sufficed for summer and winter. Some words were governed by suspension: the word for "dream" could not be used more than once in seven stanzas or the word for "insect" more than once in 100 stanzas.

R. (and *haikai*) were written on the fronts and backs of sheets of paper. Each side but the last required a moon stanza, and every sheet a flower stanza. The front side of the first sheet constituted a stately beginning (*jo*) and the back of the last sheet a "fast close" (*kyū*). In between appeared an agitated or broken development (*ha*), a rhythm adapted from court music and bequeathed to *nō* as well as *haikai*. The opening stanza was factual, dealing with the scene and circumstances of the poets as they sat to compose; the remaining stanzas were fictional.

A given sequence would manipulate features of the code. For example, moon and flower stanzas might not appear in the appointed places or might be more numerous than called for. Some sequences were composed by many poets, some by but one and not at a single sitting.

Nonserious *r.* continued to be composed, and from that aberrant practice *haikai* ([no] *r.*) emerged, achieving greatness in the practice of Matsuo Bashō (1644–94), who preferred 36-stanza sequences. Decorum in *haikai* was lower than in *waka* and *r.* It admitted Sinified diction (Japanese versions of Chinese pronunciation of characters rather than the pure Japanese of *waka* and *r.*, somewhat like the Latinate "event" for Eng. "outcome"). *Haikai* introduced other elements, incl. distinct lowerings of tone, quotidian detail, and humor, making it a highly unstable art, difficult to practice well. The emphasis among *haikai* poets on the opening stanzas (*hokku*) led in time to mod. *haiku* (q.v.).

In brief, *r.* (and *haikai*) are distinguished by the authorship of more than one poet composing with great rapidity according to an extremely complex code. The sonnet sequence is sometimes compared to linked poetry, but the comparison is inexact: there is not multiple authorship at a single sitting or anything like the *r.* code, and usually there is plot. The uniqueness of linked poetry has led to debate whether it is narrative or (perhaps most likely) an unusual variety of lyric narrative.

ANTHOLOGIES AND TRANSLATIONS: *Rengashū* [A Coll. of *R.*], ed. T. Ijichi (1960); *R. Haikai Shū* [A Coll. of *R.* and *Haikai*], ed. K. Kaneko et al. (1974); E. Miner, *Japanese Linked Poetry* (1979); E. Miner and H. Odagiri, *The Monkey's Straw Raincoat and Other Poetry of the Bashō School* (1981).

HISTORY AND CRITICISM: Y. Yamada and S. Hoshika, *R. Hōshiki Kōyō* [Main Elements of *R.* Canons] (1936); Y. Yoshida, *R. Gaisetsu* [A General Explanation of *R.*] (1937); R. Kuriyama, *Haikaishi* [A Hist. of *Haikai*] (1963); J. Konishi, *Sōgi* (1971); S. Kidō, *Rengashi Ronkō* [A Crit. Hist. of *R.*], 2 v. (1973); K. Brazell and L. Cook, "The Art of *R.*," *JJS* 2 (1975)—partial tr. of Konishi; E. Miner (1979 above); Miner and Odagiri (above); Miner et al.

E.M.

REPETITION of sound, syllable, word, phrase, line, strophe, metrical pattern, or syntactic structure lies at the core of any definition of poetry. The notion that too much literal r. is tedious, dull, or just plain bad runs counter to the most widely perceived fundamentals of verbal art and its ubiquitous use by poets.

Lat. *versus* derives from the IE etymon meaning "time," "return"; likewise, the Gr. *strophe* (primarily "turning," "revolving") underlies the use of the word "r." in prosody. In the earliest vernacular treatise on It. verse, Dante's *De vulgari eloquentia*, Dante acutely perceives this relationship, as the It. equivalent *volta* ("return") clearly indicates: the recurrence of a formal pattern guides his description of metrical and strophic units: "Cantio est coniugatio stantiarum" (2.9.1). R. is a basic unifying device in all poetry. Various aspects of form all involve some kind of recurrence of equivalent elements, differing only in what linguistic elements are repeated. Recurrence of syntactic elements is called parallelism (q.v.); recurrence at the word level in Cl. rhet. is *ploce* and *polyptoton*; recurrence of stress and quantity is called meter; and recurrence of vocalic and consonantal sounds is, variously, rhyme, alliteration, assonance, or consonance. The r. of signifiers, as above, may be supplemented by

r. that applies to the signified alone, i.e. synonymy or pleonasm. In all cases, r. manifests the projection of the principle of equivalence from the axis of selection to that of combination, resulting in a heightened quotient of self-reference, which Jakobson considers the poetic function.

Primitive religious chants from all cultures show r. developing into cadence and song, with parallelism and r. still constituting, most frequently as anaphora, an important part of the rhet. of liturgy. When a poem is intended for musical rendition, its structure—both formal and thematic—will tend to be repetitive. Therefore, thematic r. is a structural principle in most song lyrics, and a principle of thematic generation. The resulting paratactic structure can take the form of a list, as in much catalogue verse and in the medieval Occitan genres of the *plazer* and *enueg*, which enumerate pleasant or disagreeable things. One of the most common features of primitive song is the r. of the final line, which underscores its closural force:

> From the country of the Yerewas the
> moon rose;
> It came near; it was very cold;
> I sat down, oh, I sat down,
> I sat down, oh, I sat down.

The term *repetend* usually denotes the irregular recurrence of a word, phrase, or line in a poem (unlike a regular refrain), or a partial rather than complete r. Repetends may be found extensively in the ballad (q.v.) and in such modern poems as Meredith's "Love in the Valley," Poe's "Ulalume" and "The Raven," and Eliot's "Love Song of J. Alfred Prufrock"; an example:

> For a breeze of morning moves,
> And the planet of Love is on high,
> Beginning *to faint in the light* that *she loves*
> On a bed of daffodil sky,
> To faint in the light of the sun *she loves*,
> To faint in his light, and to die.
> (Tennyson, *Maud* 22.7)

The systematic r. of formal elements is a force for continuation (and when emended, for closure) of a poem. Patterns of recurrence set up expectations which are strengthened while they are being fulfilled with each successive instance. When a line, phrase, or even a sound is repeated, the experience of the first occurrence is continuously maintained in the present in each subsequent recurrence.

The poetic effects of patterned expressivity which are conditioned by r. were to Baudelaire a principle of the "organization of the life of the mind" (*Salon de 1859*); he also recognized, further, that a dynamics of pattern must include the fracturing of pattern. The constituents of sound which rhythm exploits consist of four elements: intensity, duration, pitch, and phonemic character. In most systems of versification the primary factor in the creation of rhythm is contained in the intensity or duration of sound, or both, so that the verse relies upon patterns of repeated stress or structured line-lengths or both. Complex phonic patterns are generally secondary rhythmic elements which may coexist with primary ones. Meter is a form of systematic r. most often determined by stress or quantity. In Eng. poetry, since meter is based at least partly on stress, the perception of stress-determined meter depends in turn upon the recognition of thematic elements such as syntax. Nonetheless, meter must be regarded as a formal element, since it patterns a physical property of the sounds. The constant presence of meter in a poem separates it from less highly structured sound, maintaining the distinction between poetic and nonpoetic discourse.

The r. of a complete line within a poem may be related to the envelope stanza pattern, or in other ways, such as at the end of a strophe as a refrain. In the OF epics, such as the *Chanson de Roland*, refrains may bind several separate *laisses* into independent larger units. In a multistrophic poem the modification of the refrain may signal closure. A good number of refrains become complex structural elements through minor modification from strophe to strophe. Sometimes thematic alteration collaborates in the sense of an ending, as in this partial quotation of strophe-endings from Wyatt: "But only liff and libertie . . . Lacking my liff and libertie . . . And all for the lack of libertie . . . And loss of liff for libertie . . . Graunt me but liff and libertie . . . My deth, or liff with libertie." The patterns of recurrence have set up expectations strengthened by successive occurrences, and the innovation "deth" bestows integrity upon the reader's experience formally and thematically.

In Heb. poetry, biblical semantic parallelism has been described as two cola that express the same meaning using two different sets of words. Yet this traditional definition glosses over the progression and intensification of thought from one colon to another, which can be expressed by the formula, "A; and what's more, B"—e.g. "Do not fret because of evil men / or be envious of those who do wrong" (Kugel). The second colon in these lines from Psalm 37 develops the thought expressed in the first, so that similarity leads to meditation upon the resulting semantic reinforcement. Equivalence, then, always comports a difference of rank. Even the simple fact of temporal discontinuity between repeating elements leads to a difference in their functions, via the accumulation of significance and recontextualization. Therefore pure r. does not exist.

In OE poetry the essence of the poetic artifice lies in structural alliteration. Usually the line is made up of two hemistichs with two stresses each, and the first three stresses of the four alliterate. Synonymically or antithetically related words are thus emphasized in their relationship, so that r.

stabilizes important themes, reinforces rhythm, clarifies oral-formulaic aspects of the poems, and enhances periphrase by variat

Any analysis of r. needs to be based on the way in which sound r. interacts with syntactic structure as well as at the level of prosody. Sound patterns often run in parallel with syntactic units, so that the poet stresses the form not only of her verse but also of her syntagms. In traditional metrical verse the meter cannot be effective unless the integrity of the line is frequently reinforced by syntactic integrity. In the final couplet of Shakespeare's Sonnet 18 a complex pattern of alliteration, assonance, paronomasia ("lives . . . life") and syntactic parallelism create a series of mutually reinforcing echoes, including "breathe . . . see," "long . . . live . . . life," and "eyes can see . . . life to thee." Syntactic r. also can aid in the creation of a metaphorical equivalence or the perception of a logical sequence or other correspondences between repeated units, as in Frost's "Nothing Gold Can Stay": "So Eden sank to grief, / So dawn goes down to day." Abetted by verbal r., the alignment of the second line with the first forges the implied relationship "As this . . . so that," an implied logical progression.

In free verse (q.v.), distribution patterns of linguistic features are not only less strictly determined than in metrical verse but also less easily quantifiable. The mere fact of a repeated lineation is itself a form of r. that tells the reader to expect rhythm and to pace the reading of the poem so as to realize, as prominently as possible, the rhythmic parallelism of successive lines: "She owns the fine house by the rise of the bank. / She hides, handsome and richly drest aft the blinds of the window" (Walt Whitman).

Syntactic, grammatical, and thematic parallels can function as significant clues to interp. An author's self-echoes and borrowings often provide material for the analysis of other poems by herself and by others. Milton borrows freely from himself and from others including Lat. poets, often extending the borrowing to entire phrases. The r. of a poetic phrase can have an incantatory effect, as in magic spells, or in the opening three lines of T. S. Eliot's "Ash Wednesday," all of them variants of "Because I do not hope to turn again." The next 28 lines contain no fewer than 11 lines clearly related to these and serve as a unifying factor in a poem otherwise very free in structure. In the following passage from Shakespeare, each noun is picked up in a subsequent line and repeated in some form, with grammatical compression in the third line effecting closure: "And let the kettle to the trumpet speak, / The trumpet to the cannoneer without, / The cannons to the heavens, the heaven to earth" (*Hamlet* 5.2.273–75).

In rhymed verse, rhyme determines the spatialized schemes in which rhythmical series are presented and calls attention to the rhythmically important line endings in which meaning has been chiefly concentrated ever since troubadour verse. The r. of similar endings of words or even of identical syllables (*rime riche*) approaches the r. of a whole word, as in identical or equivocal rhyme. Here the accumulation of r. adds meanings and multiplies the possibilities of semantic variation. In the sestina (q.v.), six words repeated identically function to bind together larger units of interstrophic composition transcending the individual strophes. Homonymic equivalence serves to highlight semantic similarity and difference. The use of a single signifier for two signifieds, a form of r., appears in punning, where homonymy brings phonetic equivalence relations into play so as to underscore difference of meaning, though not at all necessarily opposed meaning. At times the clash between two associations when brought into an apparently repetitive association challenges the referential stability of lang., substituting a virtual meaning for the anticipated meaning. In paronomasia, a case of r. with variation, near-r. occurs through a slight differentiation of sound which contracts a slight or extreme difference of sense by altering as little as a phoneme. See now VERSE AND PROSE.

E. S. Le Comte, *Yet Once More* (1953)—Miltonic; S. Rimmon-Kinan, "The Paradoxical Status of R.," *PoT* 1 (1960); R. Abernathy, "Rhymes, Non-Rhymes, and Antirhyme," *To Honor Roman Jakobson* (1967); B. H. Smith, *Poetic Closure* (1968); P. Kiparsky, "The Role of Linguistics in a Theory of Poetry," *Daedalus* 102 (1973); E. G. Kintgen, Jr., "Echoic R. in OE Poetry," *NM* 75 (1974); N. B. Smith, *Figures of R. in the Old Prov. Lyric* (1976); J. Lotman, *The Structure of the Artistic Text* (tr. 1977), ch. 6; M. Shapiro, *Hieroglyph of Time* (1980); M. Cornell, "Varieties of R. in OE Poetry," *Neophil* 65 (1981); R. Jakobson, "Linguistics and Poetics," in Jakobson, v. 3; J. L. Kugel, *The Idea of Biblical Poetry* (1981); J. Hollander, *The Figure of Echo* (1981); Morier—long entry; J. H. Miller, *Fiction and R.* (1982)—prose; M. Frédéric, *La R.: Étude linguistique et rhétorique* (1985)—the fullest mod. study; S. Metzidakis, *R. and Semiotics: Interpreting Prose Poems* (1986); G. Chesters, *Baudelaire and the Poetics of Craft* (1988); L. Magnus, *The Track of the Repetend: Syntactic and Lexical R. in Mod. Poetry* (1989)—use with caution.

M.S.

REPRESENTATION AND MIMESIS.

I. REPRESENTATION
II. MIMESIS
III. CLASSIC AND ROMANTIC
IV. SAUSSURE AND PEIRCE
V. THE VERBAL MEDIUM
VI. TRANSLATION
VII. DESIGN
VIII. THE MASTER TROPE

I. REPRESENTATION. R. is the process by which lang. constructs and conveys meaning. One major component—perhaps the major component—of the process of r. is *reference*: words (texts) refer, or

create pointers to, the external world, other words (other texts), themselves, or to the process of referring; Hutcheon (in Whiteside and Issacharoff) identifies these four types or directions of reference as, respectively, extratextual, intertextual, intratextual, and metatextual. Reference concerns the ability of lang. to describe, capture, express, or convey—the verbs have often been interchangeable—the external world in symbols which the mind can manipulate.

R. lies at the very heart of the nature of lang., i.e. speech as a system of communication for encoding information symbolically and transmitting it from speakers to auditors (for diagram, see Jakobson; see also Abrams). R. is one of the most difficult problems in philosophy; the issues are not merely central to aesthetics: they are fundamental to epistemology and metaphysics. Since we are compelled to talk about the nature of lang. in words themselves, the structures and limitations inherent in words presumably already constrain our ability to talk about lang. at all: there is no vantage point from which we can stand outside lang. so as to critique it. But conversely we do not know how severe these constraints are, and we do know that lang. is capable of both creativity and growth so as to convey new concepts. *A priori*, it would seem reasonable to explore the nature of hammers as a construction tool by hammering, and the nature of paint as an aesthetic medium by painting, hence the nature of lang. by engaging in verbal discourse. This is what we have, and, this side of transcendence, much is still possible.

One would think it might be possible to escape r. altogether; and indeed experimental poetries such as *poésie pure* in the 19th c. and "sound poetry" in the 20th have sought escape from denotation and reference. All art aspires, Mallarmé remarked, to the condition of music. But in fact r. is almost impossible to escape even in nonrealistic or nonreferential art: so Rudolf Arnheim, commenting on abstractionist painting, suggested that "even the simplest patterns point to the meaning of the objects to which they apply," so that ultimately "there is no form without r." Whether abstract or realistic, art largely engages in r.

II. MIMESIS. In general, lang. has two modes for r., one nonmimetic (nonreferential, nonrealistic), the other mimetic. Nonmimetic r. encompasses all abstract art and all r. of imaginary, surreal, and fantastic worlds, worlds other than the quotidian one we inhabit. The second type of r. is m., though the term "m." has been used in two senses, one general, one specific. In the general sense, which is the one canonized by Auerbach, m. amounts to "realistic r.," namely the verbal capturing or conveying of experience in such a way that the mental image or meaning created by the words is judged similar, analogous, or even identical to what we know about the world from sense-data directly. Its structures are symbolic, propositional, and descriptive. Mimetic r. in this sense corresponds to extratextual reference. "Realistic" is of course not much more explanatory, being a very elastic term, but in the general sense of verisimilitude it is clear enough.

M. in the specific sense concerns verbal sequences in which the forms of the words themselves—their sounds, shapes, or sequence—resemble, enact, or reproduce some aspects of that which they refer to. *Resemblance*—i.e. analogy—is the operative term (see section V below). Mimetic lang. is "expressive," "imitative," or iconic. In poetry the medium for mimetic enactment is of course the body of the poem, its corporeal substance, sound (q.v.), and the means are prosodic.

The two senses of m. amount, respectively, to the distinction between mimetic ends and means. Confusion of these is what we may call the Mimetic Fallacy. Critics who infringe the Mimetic Fallacy fail to grasp that artistic means are not to be judged by same criteria as artistic ends. Dickens can describe boring characters without boring us, and Flaubert can depict the shallow intellectual and moral capacities and limited self-knowledge of Emma Bovary without the depiction itself being shallow, naive, or simplistic. On the contrary, it takes great skill to render a shallow character in an incisive and engaging way. Even in realistic art, where the ends are mimetic, the means need not be, though if they are, they compound the effect.

The two functions of lang., referential and mimetic, can be seen in the simple example of how to represent the hue which is a mixture of pink and purple, which we sometimes see through stained-glass windows in the late afternoon. Taking the representational function of lang., by which arbitrary words describe, we could say that the light "has a color blended of pink and purple," or we could say, in a word, "the color was fuchsia." The mimetic function of lang. (in the specific sense) would seek other means: a la Joyce, we might call the color "pinkpurple" (using contiguity to indicate close relation) or "ppuirpnlek" or, better, "PpuIrpNleK" (letting graphic mixture indicate color mixture). Now the graphic signs create an analogy for that which they wish to convey. The analogy is not exact, admittedly—mixing caps and lowercase switches codes—but it works. Note that mimetic forms violate the rules of representational lang.: by calling attention to themselves, they thicken the normal transparency of the sign, forcing us to focus on the signifier itself so as to grasp the meaning. This opacity, which is fundamental to poetry, is what Jakobson termed the "poetic function."

Mimetic processes in lang. are much more extensive than many critics realize, operating not only in phonology but also in morphology and syntax. These processes include not only "sound symbolism" or "phonetic symbolism" (imitative and associative sounds; see SOUND) but also "morphosymbolism" (word-formation and changes in word-shape; see Bolinger) and iconicity in syntax

(Haiman). Though mimetic phenomena in poetry such as onomatopoeia (q.v.) have traditionally been disparaged as trivial and mechanical, extensive linguistic evidence shows that these are not mere "poetical" devices: mimetic processes operate on all levels of lang. and in most if not all langs. worldwide. In a seminal essay, Wimsatt focuses directly on the means available in the verbal medium for miming as opposed to its ordinary means for referring. Distinguishing between the *graphemic* and *phonetic* modes of lang., Wimsatt identifies eight types of verbal iconicity, most of which have analogues in poetry. Though Dr. Johnson disparaged "representative meter" in poetry, and though the range of imitative prosodic effects in Pope's *Essay on Crit.* is limited, m. extends much deeper, and the *desire* for it is embedded in the very fabric of poetic lang. It was Ransom who said that poetic lang. aspires to the condition of nature: it "induces the provision of icons among the symbols." In perfected poetic speech, every word would be motivated and wholly natural, appropriate in form to its meaning. When devices such as sound patterning and repetition are used for artistic purposes, the inert categories of ordinary grammar are energized and raised to a higher level of valency where order *is* naturalized: here words *become* natural experience. M. is thus the desire for the motivated sign (for short history, see Essick, ch. 2).

III. CLASSIC AND ROMANTIC. M. was established as a central concept in Western poetics by Plato and Aristotle. The term "m." derives from Gr. *mimos*, a mime play or actor therein, and seems to have originally meant the mimicking of an animal or person (a mythical hero, a god, or a fabulous creature such as the Minotaur) through facial expression, speech, song, dance, or some amalgam of these. In Eng. it has been customarily translated as "imitation" (q.v.) in the ordinary sense of "copying," though this is very unsatisfactory in several respects (see section VI below).

The Greeks in fact bequeathed to posterity not one conception of poetry but two. The first strain derives from Plato's Theory of Forms; in it the issue of m. as imitation is central. It is the ideality of Forms which made Platonism appealing to Christianity; God is in effect the last Form. Note that the issue of m. is framed explicitly in *visual* terms, as images, shadows, copies. In this theory, which survived until the late 18th c., verbal experience is characterized as secondary and derivative, a *re*-presentation of actual experience, and often only (as in Plato) a poor imitation at that. For Plato, lit. offers only an image of an image, a copy of a copy of the real, so is suspect. By the 18th c., copying "Nature" meant something else (see CONVENTION), but "imitation" was still embraced as sound poetic praxis. In the 19th, Carlyle and Nietzsche's announcements that God Is Dead sealed the fate of this theory forever by removing any possibility of a Being apart from our experience of the world who would validate the accuracy of mimetic rs. In the absence of all such verification, only making remains. All 20th-c. art, therefore, exists in a postmimetic mode; notice that the chief drive in late-industrial technology is to devise machines of ferocious efficiency at *copying*, at creating facsimiles. Endless repetition becomes the postmodern condition.

The second strain derives from the Gr. conception of *poiesis* as making, the creation of objects from formless matter. This strain eventuated in romantic poetics, where form is made over as internal and individual; in organicism, every entity has a form which proceeds from a principle that is self-contained, even as the oak tree springs from the acorn. The advent of romanticism effected a fundamental epistemological shift by reformulating lit. not as an *imitative* but as a *constitutive* art, presenting not external reality but a fuller, partly interior version of reality which includes the feeling subject. M. as imitation was thus supplanted in romantic poetics by the concept of creative imagination (q.v.), the faculty by which the poet envisions and creates realities never before seen and not a part of the external world.

The romantic (expressive) strain has roots in antiquity in Aristotle's argument that the poet surpasses the historian because she is not tied to mere fact but provides a more accurate (if less "factual") account of reality. This is the line followed in the Ren. by Scaliger (*Poetices* 1.1) and Sidney: since the poet as seer can envision a higher order of nature than the one known to ordinary mortals, she becomes not a slave to or copier of reality but in effect a creator, a second god. The poet, says Sidney, "doth grow in effect another nature." Since the poet does not make propositional statements about the external world directly ("nothing affirmeth"), the poet is not a fabricator of falsehoods ("never lieth"), as Plato charged. Now the poet models "truth" at a deeper level, where recreating the sense of life as felt experience is largely not a matter of descriptive detail; nor is verisimilitude required in order to persuade a reader of the "reality" of the fiction. It is not required that rs. be "true" to fact: it is only required that they be *meaningful*, i.e. that they persuade us of their sense of reality. "The truest poetry," says Shakespeare's Touchstone, "is the most feigning." The traditional view in the West, after Aristotle, was that art at its deepest was, if fictive, still "true to life."

IV. SAUSSURE AND PEIRCE. The most extensive theories of verbal r. in modern times have been those of Ferdinand Saussure, the publication of whose lecture notes (like Aristotle's) subsequently created an entire school in poetics, and Charles Sanders Peirce, whose semiotics has been, if more extensive and philosophically coherent, much less influential in Am. lit. crit. (Sheriff). Both theories place the activity of signification—i.e. symbolization, sign-making and sign-interpreting—at the

center of human cognition and interaction with the external world. (For survey and diagrams of the other main theories, see Whiteside and Issacharoff ch. 7, esp. 184–85.)

Saussure followed all Western thought about lang. since Plato in conceiving the nature of linguistic signs to be arbitrary: since the words for the same referent in two different langs. normally differ, there cannot be any innate, natural, or "motivated" relation between the particular sound shape of a word (the signifier) and its referent in the external world (the signified). He departed from such trad., however, in assigning the locus of meaning to the *difference* between such signs and their referents, both between langs. and within a lang. (see SOUND). Since the class of canines is signified by "dog" in Eng. but "chien" in Fr., the meaning "class of canines" must arise not from present phonetic features—which do not coincide at all—but from the differential between them. Since for Saussure the vast majority of words have no motivated form, meaning becomes detached, floating on a sea of contingencies and subjectivities (so Derrida). Meaning arises only in the interplay of signs, hence is polyvalent, unstable, and indeterminate.

Peirce conceives the world as triadic: his ontology identifies Firstness (any thing which is without relation to any other thing), Secondness (any thing which has its being in relation to any one other thing), and Thirdness (any thing which brings into relation a First and a Second). Thirdness is the realm of signs. For Peirce, all thought is carried on in signs, and hence all r., as thought, is mediate cognition. Things in this triadic world may be either objects, meanings, or signs: a sign mediates between an object and its meaning. Meaning exists *a priori*, apart from human cognition; what remains for us is only to discover and know it in greater or lesser degree. Hence meaning is entirely objective, existing on an ideal plane which Peirce makes clear is essentially Hegelian. Meaning itself, however, is a sign, or interpretant; interpretants too (being perceivers' responses to signs) are triadic in classification, being either thought, action, or emotion, of which only thought is "genuine," the other two being of some lesser reality hence "degenerate."

More generally, the relations of sign to object—i.e. the fundamental modes of r.—possible in Peirce are three: resemblance, producing an *icon*; physical contiguity or action, producing an *index*; and thought (cognition), producing a *symbol* or sign. Since all signifying activity is for Peirce mental and even idealist, the first two of these will indicate resemblance or emotive meanings and may express possibilities which might obtain or actualities which do obtain in the world—but not necessary relations which must obtain logically. Such signs are therefore not genuine, though this does not mean they are not important, for in fact their very ambivalences and ambiguities virtually guarantee that they will be used widely and often, and indeed predominate over genuine signs. Genuine signs, incl. lang., exist in the autonomous realm of interpretive thought, which is fundamentally symbolic in character.

The noncongruence of ontological and epistemological assumptions between Peirce and Saussure could hardly be more stark, and hence the ensuing conflicts more inevitable—though often but dimly understood—when structuralism and semiotics came to dominate Am. lit. crit. in the 1960s and 1970s. For Peirce, subjectivity is a minor issue; for Saussure it is fundamental. For Peirce, meaning is always already existent; for Saussure it is local, unstable, and evanescent. It is one of the great ironies of modern crit. that while it was the work of Saussure which exerted the most influence on Am. literary theory of the 1960s and 1970s, Saussure's account of meaning is idiosyncratic and anomalous among the major theories of meaning which have been put forward in the 20th c. (Whiteside and Issacharoff; Sheriff ch. 4). Saussure's account, focusing on the relation signifier–signified, is binary and linear; meaning is endlessly deferred. All the other major philosophical systems are ternary and triangular: all of them—like linguistics, semantics, semiotics, psychology, and cognitive studies—take meaning as a *given*.

Further, Saussure's doctrine of the arbitrariness of the sign is much more limited than is generally recognized. "It has never been contended," says Bolinger, "that complex utterances are arbitrary in the same sense in which *arbitrary* has been applied to morphemes" (234). Even within the word, arbitrariness has limits: the particular form a word takes when coined is heavily determined by the synchronic forces of linguistic evolution, and, once the word exists, it is subject to a further range of diachronic influences from all the other words in the lang.; even "an arbitrary form, once integrated into the system, assumes all the affective and associative privileges enjoyed by the most obvious onomatopoeia" (231). "When we speak of sound-suggestiveness, then, we speak of the entire lang., not just of a few imitative or self-sufficient forms" (234). In general, it would seem reasonable to characterize lang. as a complex mixture of arbitrary forms, mimetic forms, and forms motivated by both the nature of lang. and the nature of cognition, i.e. the mind.

V. THE VERBAL MEDIUM. Lang., the 20th c. realized with a shock, is a modeling system based on a code: everything which is cast into verbal form is *encoded*. Like other systems of expression both natural (gestures) and artificial (ciphers, mathematics), the elements of the system and the rules for combining them are largely arbitrary. Fundamental to all r., therefore—whether mimetic or not—is the nature of the medium. Every r. is a mediation; all rs. are formalizations and symbolizations. It is precisely in the nature of a symbol that it be radically unlike that for which it stands.

All art exploits the imposition of the medium. It is of the very essence of art that the medium be different from that of its objects. In painting, apples and pears are represented in canvas and paint; in sculpture, flesh is represented in stone. But it is the peculiarity of verbal art that the medium is both physical and symbolic—hence double, and inextricably so. Words, besides having such physical characteristics as shape and length, bear meaning as part of their primary substance. In expository prose the physicality of words is suppressed, whereas in poetry it is foregrounded. If the medium were wholly transparent, the expression would shine straight through and there would be no necessity to have art at all: one needs no pane if one wants the wind to blow through. The *medium* is the *mediation*; lacking it, there is no *translation* of the expression into another form, for all art partakes of the condition of translation (see section VI below). Indeed, we must not think expression itself exempt from this requirement, for expression itself takes place in a medium, and is in this respect aesthetic—i.e. trans-substantial. So lang., like art, is a "secondary modeling system"; and though it is true that lang. is art-full, more deeply, one must say that lang. *is* art in its enabling conditions, processes, and achieved effects. From this perspective, m. looks very different. The notion of imitation in the sense of producing a material simulacrum of an original—a copy—recedes rapidly in favor of the notion of achieving analogous results in a completely different mode.

From the point of view of cognitive science and information theory, one would say that a r. is a type of modeling, and that our recognition that an artwork is an imitation or m. of the external world in some way is one species of pattern recognition. But the re-presentation in every r. is also, more directly, a presentation itself, and in art we must say that this is the more important function. It was Diderot who remarked that in poetry "things are said and represented simultaneously." So every r. is re-present, and re-presents as it represents. It does not stand for what preceded or surrounds it, but extends it into a new modality. The relation of the verbal r. to the original sensory one is not one of imitation (copy) but one of *analogy* or *resemblance*, which is very different. An analogy is a perceived congruity in structural relationships between two sets of phenomena. The subject of analogy is one to which Wallace Stevens devotes two important discussions in *The Necessary Angel* (1951), esp. the first of the "Three Academic Pieces" (1947) and "Effects of Analogy" (1948; see SIMILE). Now the difference in modality is foregrounded, and difference is accepted as an essential part of the process.

One of the recognitions of postmimetic crit. is that r. always does violence to that being represented, in that there is always a gap between the two phenomena; this differential between the representational system and that being represented results from the fact that the medium for r. differs fundamentally. The question is not whether the gap exists, but only about its size. Is it even possible to gauge its size? Many would say no; to judge that would require standing apart from the representing in progress. Such a neutral vantage, Heisenberg showed in physics, we will never have: the presence of the subjective perceiver and the act of looking alter the system and so remove the possibility of objective knowledge irrevocably. Poststructuralist crit. made much of this, but in ordinary lang. use, the size of the gap is not the only or even in some cases the major concern.

Furthermore, in other sign-systems it has certainly proven possible to assess and even decrease the gap between the sign-system and the external world. Chief among such systems is of course mathematics, where, in the Euclidean framework, at least, symbolic rs. are taken to provide information about the external world which is susceptible of verification and which allows prediction; empirical data is in turn fed back into the mathematical model so as to improve the predictive power of the system. Certainly other, nonEuclidean representational systems have been developed; these presumably offer predictions about other universes than our own. Such universes are fictive (not ours) yet may still be "true" (internally coherent). It is important to realize, however, that in mathematics the predictive power of any given equation is dependent not on its individual elements, which are themselves abstract and arbitrary, but rather on their relations and on the construction of the representational system as a whole—its axioms, its rules. It is the same with lang.: single words are meaningless in isolation (so Saussure) but depend on the influences of their neighbors (i.e. the lang. as a whole) for their representational power and accuracy, which is considerable (contra Saussure).

VI. TRANSLATION. From a distance, it is curious that Gr. *m.* should have been translated as Eng. "imitation," for the Gr. denotes a gesturing, a figuring-forth, whereas the translation term approaches the process from the other direction. "Imitation" suggests that the purpose is simply to copy an original; the only question then is how closely the imitation (copy) conforms. But this we cannot judge well, even if it were all that mattered, which it is not (see below). The notion of m. was originally drawn from the mime, who produced stylized gestures *without speech* so that the audience would see in them the characteristic features of the actions of some mythological figure or personality type they already knew. But here the task is simply to match memory to sight, and both elements are of the same system: the mime imitates gesture with gesture. Similarly, drama imitates action and speech with action and speech. But in lang., words are of a altogether different stratum, already charged with meaning.

The nature of verbal r. may become clearer if

we consider the case of translation. "Poems may be poorly translated, as they may have been poorly written to begin with," says Willis Barnstone, "but they are not necessarily poorer or better than the original." The quality of the translated version depends rather less on the quality of the original than on "the translator's skill in writing poetry in his own lang." The tr. is not a copy good or bad; it is a *version* enacted in another medium. The nature and quality of the tr. is controlled by the translator's skill, certainly, but greatly too by the resources offered by both the source and the target langs. All tr., that is, is mediation, so that the aim of faithful r. is not to copy the original—which is impossible in any event, every medium (e.g. lang.) being different, and every medium *necessarily* imposing itself at every moment between the knower and the known—but rather to try to do the thing that the original did in its medium as well as that can be done in the new medium, insofar as the doing in the original is understood by the translator.

If the reader grasps the nature of the doing in the tr., then she may perhaps also grasp, *mutatis mutandis*, something of the nature of the doing in the original. But only if she has knowledge of both langs. can she judge aright how well the doing in the target lang. compares to how well the original achieved its own end in the source lang., given what the two langs. respectively allow. But this account, which relates one verbal r. to another, can be extended to all r. in art, and indeed to lang. itself as r. All verbal r. *is* tr. by its very nature. A verbal r. is in effect a translation target lang., except that the source lang. is not verbal; it is sensory. To judge the quality of the verbal r., we should not compare it to the original (sensory) r., for that will tell us very little. What we want, rather, is information about what kinds of sensory and cognitive knowledge are and are not possible, as against what the linguistic and literary systems allow. The point is not that direct comparison between the two modes of r. is impossible, but simply that it is confounded by the radical differences between the representational media, which are very different codes. Schematically:

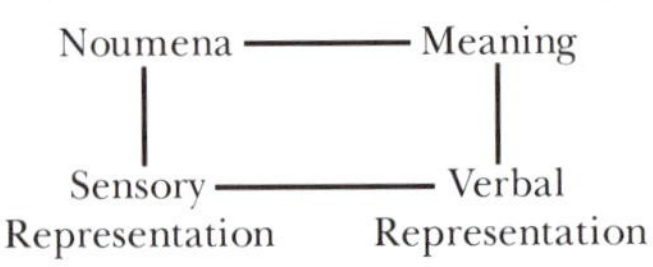

The traditional conception of m. only concerned the SR–VR link. But in the wider field of verbal r., all four relations are important. What we can know about the SR–N link must be derived from psychology, epistemology, physics, and philosophy. We can inspect VR–M directly with discourse about discourse. It is our deepest hope that there is linkage—analogy—between M and N. For some, this is the province of metaphysics; for others, reliable knowledge about the noumenal world comes from religion or transcendent vision. Can words give us any true knowledge about the noumenal world which lies beyond sense-experience? There have been two views put forward in Western philosophy about the relation of lang. (words) to the noumenal world of "things as they are." The transcendentalist view holds that there is a realm beyond words and even sensation which humans can apprehend, if but dimly or briefly, through words, a realm words can gesture toward or give thought entrance upon when words are ordered aright. The chief modern proponent of this view is the Eliot of *Four Quartets*, who speaks of how "words, after speech, reach / Into the silence." It is a view naturally allied with religious faith. The immanentist view, by contrast, is that noumenal knowledge is forever denied us (so Kant) and that lang. and thought are coextensive. As Hazard Adams puts it, "we can only make a world with a lang., indeed in a lang. There is nothing imaginable independent of a medium to imagine in."

Of course, art always seeks to conceal its own artfulness. One of the monuments of modern thinking about r. is René Magritte's famous image of a smoker's pipe over the caption, *Ceci n'est pas une pipe*. Magritte's work addresses precisely our automatic instinct to treat pictorial (and verbal) rs. as if they were the things themselves. If one's child pointed to the picture of the pipe and asked what it is, one would almost never say, "this is a picture of a pipe"; one would instinctively say, "this is a pipe"—thereby subtracting, as a reflex, the distance between the r. of a thing and the thing itself. Of course Magritte's image is not a pipe but rather a r. of a pipe; but when we confront the picture in our ordinary cognitive mode, we attend to the meaning and automatically ignore the fact of the r. itself, i.e. the fact that it is a picture of a thing not a thing. The purpose of Magritte's caption, however, is to force us out of habituated response back into the antecedent recognition of the fact of the r. itself.

VII. DESIGN. Many commentators, esp. of the past two decades, have seemed to think that the verbal r. of experience in lit. does not effect and instantiate any real presence (see Krieger). But it is odd to complain that lit. is of some order of experience secondary to sense experience. We should ask, rather, whether a literary (poetic) artwork represents the same objects, in the same way as, and with the same resulting degree of credibility as sense-experience; if so, how and how much, and if not, why and what instead. Is verbal r. inferior to sensory r.? Plato answers in the affirmative; the romantics answer in the negative. Sense-experience itself is of course a r.; the verbal artefact, however, during the process of reception—reading, hearing, or observing—presents itself directly to cognition, supplanting the external world. It blocks out normal sense-data, replacing it with other data about sensation in a precoded form. "It is hard to write any piece of lit. that corresponds to anything

as such, whatever it be," says Victor Šklovskij, because "art is not the shadow of a thing but the thing itself" (*Theory of Prose* [tr. 1990]). The eye sees the words on the page, in prose, only to ignore them: they are transparent. Poetry, by contrast, is translucent if not opaque: meaning is still conveyed, but now in words whose design arrests our attention.

Fundamental to the nature of all art is that it achieves its effects by increase of order, or design. Regardless of authorial intention, most artworks show themselves to be supercharged with design. It is traditional in Western poetics to think of the faculty or capacity of imagination (q.v.), i.e. image-making, as central to literary r. This is indeed important, esp. for realistic r. and verisimilitude, but it cannot supplant the making (Gr. *poiesis*), the craft, the shaping of the medium which evokes those images and, as it does so, binds their elements more tightly together by form.

In poetry the shaping is of the sensory medium (sound) and carried out largely at the preverbal (phonological) level (meter, rhyme, sound patterning). In its heightened design, texture, and materiality, poetry does not compete with the external world as a species of r.: it offers additional resources and forms of order which are constitutive, not imitative: they enact a heightened form of r. which creates that heightened mode of consciousness which since antiquity has been called "poetic."

Poetry aims to show that relationships exist among the things of the world which are not otherwise obvious; it accomplishes this by binding words together with the subtler cords of sound and meter. These bindings, like the relations they enact, are formal. It was Fenollosa who observed that "the relations between things are more important than the things which they relate." If m. is to occur at all, therefore—given the differences between the representational media—it must address formal rather than substantial features. As Wittgenstein remarks, "what any picture of whatever form must have in common with reality, in order to be able to depict it . . . is logical form, i.e. the form of reality" (*Tractatus* 2.18). Hence the problem of r. is not so much that of the elements of lang. as, again and ever, the problem of form.

Even Aristotle supports this view: m. as "imitation" is only part of the account of art given in the *Poetics*. Aristotle certainly views imitation as a central human activity; nevertheless, he makes it clear that the chief criterion for drama is not spectacle, diction, credible character, or even verisimilitude but the making of a strong plot, i.e. *structure* (q.v.). As for poetry, the account of the means of imitation given in ch. 1 of the *Poetics* is truncated and somewhat garbled, but several modes are identified, and the manner of the imitation is by no means assumed to be direct or literal. The role attributed to imitation in the *Poetics* has distinct limits: only some arts are imitative. Indeed, Aristotle holds that the order of art need not be—and at least once he holds it must not be—the order of reality (else poetry would be inferior to history). On the whole, Aristotle goes to considerable lengths to differentiate the art object from the reality it imitates, and this almost entirely on account of aesthetic *design*. Both imitative and nonimitative arts share this feature, which suggests that, for Aristotle, it is design itself which distinguishes art from life. Even in realistic and referential art, where design is a signifier for the reality outside the artwork, the signifier is a presentational form. In all art, that is, the artwork itself stands as an icon for signification itself.

VIII. THE MASTER TROPE. There has been much interest in the 20th c. in lang. as it manifests itself in psychology, philosophy, prose fiction, and culture. It was Freud who showed that words are often the keys to dream interp. because lang. itself writes the structure of the unconscious. And while the relation of signifier to signified is central in Saussure, much of Lacan's work explores processes based on links between signifier and signifier. Girard has argued that desire itself derives from m. In philosophy, Heidegger conceived poetic lang. as presentational rather than representational, presenting Being itself, a framework built upon by Gadamer. Reference and r. have been major subjects in the study of narrative fiction. In recent years it has been argued by cultural theorists that the study of r. must focus not only on the code but also on the physical and cultural means of production of that code, since these forces undeniably affect both the nature of the code and the uses speakers put it to. All of these inquiries bear directly and deeply on the fundamental issue of how lang. functions, which is to say, on what is possible in verbal r. All of them take lang. as the master trope for human mind and action.

Some modern critics, charmed by the inscrutabilities of Heidegger, blithely dismiss epistemology with a wave of the hand as tiresome Kantianism; such critics reduce m. to mere lang.-use (so G. Bruns in *Renascence* 37 [1985]). But the consequence of such a reduction is to replace truth with use, and too often use ultimately amounts only to power. Valorizations of Heidegger's thought as the only advance upon Kant must be counterbalanced by the recollection that the Ger. thinker was the would-be darling of National Socialism: any theory is properly to be judged, in part, by the ends it is put to. Certainly some postmodern critics have shown that their real interests lie not in the disinterested pursuit of thought for the benefit of humanity but rather in the exercise of power. Even Marxists can be academic capitalists. Against all these fascists stands the lonely ghost of Yvor Winters, whose arguments for the irrefrangibly ethical dimensions of poetry and poetic form (and behind them for civility, reason, and pluralism in critical discourse), while widely ignored, have not diminished in force (Trimpi).

PHILOSOPHICAL TEXTS: F. de Saussure, *Cours de linguistique générale* (1916), 5th ed. (1955), crit.

ed., ed. R. Engler (1967–74), tr. R. Harris (1983); C. K. Ogden and I. A. Richards, *The Meaning of Meaning* (1923); C. S. Peirce, *Coll. Papers*, ed. C. Hartshorne and P. Weiss, 8 v. (1931–58); C. W. Morris, "Foundations of the Theory of Signs" (1938), *Writings on the Gen. Theory of Signs* (1971); W. V. O. Quine, *Word and Object* (1960); L. Wittgenstein, *Tractatus Logico-philosophicus* (tr. 1961); M. Heidegger, *Poetry, Lang., Thought* (tr. 1971); T. W. Adorno, *Aesthetic Theory* (tr. 1984); H.-G. Gadamer, *Truth and Method*, 2d ed. (tr. 1989).

CRITICAL TEXTS: J. W. Draper, "Aristotelian 'M.' in 18th-C. England," *PMLA* 36 (1926); J. Tate, "'Imitation' in Plato's *Republic*," *ClassQ* 22 (1928), "Plato and 'Imitation,'" *ClassQ* 26 (1932); R. McKeon, "Lit. Crit. and the Concept of Imitation in Antiquity," *MP* 34 (1936); Y. Winters, *In Defense of Reason* (1947); W. J. Verdenius, *M.: Plato's Doctrine of Artistic Imitation* (1949); Auerbach—still essential; Abrams; H. Koller, *Die M. in der Antike* (1954); G. F. Else, "'Imitation' in the 5th C.," *ClassP* 53 (1958), *Plato and Aristotle on Poetry* (1986); R. Jakobson, "Linguistics and Poetics" and "Quest for the Essence of Lang.," both rpt. in Jakobson; R. Bernheimer, *The Nature of R.* (1961); B. Hathaway, *The Age of Crit.* (1962); D. L. Bolinger, *Forms of Eng.* (1965); G. Sorban, *M. and Art: Studies in the Origin and Early Devel. of an Aesthetic Vocabulary* (1966); W. Tatarkiewicz, "M.," *DHI*; E. Schwartz, "M. and the Theory of Signs," *CE* 29 (1968); G. Hermeren, *R. and Meaning in the Visual Arts* (1969); L. Golden, "Plato's Concept of *M.*," *BJA* 15 (1975); W. K. Wimsatt, Jr., "In Search of Verbal M.," *Day of the Leopards* (1976); N. Goodman, *Langs. of Art*, 2d ed. (1976); D. Savan, *Intro. to Peirce's Semiotics* (1976); G. Genette, *Mimologiques* (1976), excerpted in Eng. tr. in *PMLA* 104 (1989); H. Felperin, *Shakespearean R.* (1977); M. Riffaterre, *Semiotics of Poetry* (1978); K. K. Ruthven, *Critical Assumptions* (1979), ch. 2; M. Krieger, *Poetic Presence and Illusion* (1979), *Ekphrasis* (1991); J. D. Boyd, *The Function of M. and Its Decline*, 2d ed. (1980); N. Wolterstorff, *Works and Worlds of Art* (1980); *M.: From Mirror to Method*, ed. J. D. Lyons and S. G. Nichols, Jr. (1982)—little on m., but see Krieger; J. Campbell, *Grammatical Man* (1982); A. Nehamas, "Plato on Imitation and Poetry in *Republic* 10," *Plato on Beauty, Wisdom, and the Arts*, ed. J. Moravscik and P. Temko (1982); A. D. Nuttall, *A New M.* (1983); *M. in Contemp. Theory*, ed. M. Spariosu, 2 v. (1984); P. Kivy, *Sound and Semblance* (1984)—r. in music; C.-G. Dubois, "Problems of R. in the 16th C.," *PoT* 5 (1984); J. Weinsheimer, *Imitation* (1984); J. L. Mahoney, *The Whole Internal Universe* (1985); T. R. Martland, "When a Poem Refers," *JAAC* 43 (1985); W. Trimpi, "M. as Appropriate R.," *Renascence* 37 (1985); R. Woodfield, "Words and Pictures," *BJA* 26 (1986); C. Prendergast, *The Order of M.* (1986); T. Clark, "Being in Mime," *MLN* 101 (1986); C. J. Brodsky, *The Imposition of Form* (1988); *CHLC*; J. K. Sheriff, *The Fate of Meaning* (1989); R. N. Essick, *William Blake and the Lang. of Adam* (1989), ch. 2; S. K. Heninger, Jr., *Sidney and Spenser* (1989); K. L. Walton, *M. as Make-believe* (1990); O. Avni, *The Resistance of Reference* (1990); W. J. T. Mitchell, "R.," *Critical Terms for Literary Study*, ed. F. Lentricchia and T. McLaughlin (1990)—unsatisfactory; C. Crittenden, *Unreality* (1991); P. Livingston, *Models of Desire* (1992)—on Girard.

T.V.F.B.

RHETORIC AND POETRY.

I. INTERPRETATION
II. COMPOSITION

The art of oratory or public speaking, rhet. has traditionally had two not altogether separable ends: persuasion, which is audience-directed, and eloquence, which is most often form- and style-directed. Three basic genres have been delineated in oratory: deliberative, forensic, and epideictic, with three concomitant types of orations, speeches given before policy-determining bodies, before courts of law, and before occasional assemblies. Rhet. has been a prominent discipline in Western education since antiquity. Indeed, throughout most of the history of Western civilization, p. was written and read by people for whom rhet. was the major craft of composition. At times the similarities of rhet. and p. have been stressed (p. is the "most prevailing eloquence," remarked Ben Jonson in 1641), at times their difference ("eloquence is written to be heard," John Stuart Mill wrote in 1833, "poetry to be overheard"). A distinction revived by Scaliger in the 16th c. that would limit rhet. to *prose* compositions was overwhelmed by a critical commonplace, also inherited from antiquity, that verse itself is no sure sign of p. To the extent that our own time regards p. as having the ends of rhet.—if not exemplary eloquence then persuasive discourse—the two arts remain all but inextricable.

The relationship between rhet. and p. has always extended both to the composition of p. and to the interpretation (q.v.) of it, even on the most elementary levels. Quintilian's uninnovative but highly influential *Institutio oratoria* (1st c. A.D.) offers the traditional attitude: skill in oratory is founded on "speaking correctly" and "interpreting poets" (1.4.2). The inventive processes of rhet. and p. have been differentiated from time to time, and at least once with revolutionary fervor—"Take Eloquence and wring his neck," Verlaine exclaimed in 1884. These distinctions were usually impelled, like revolutions in interp., by reactions to the intransigence of rhet. and by perceptions of its restrictiveness. Because in our own century the interp. of p. has undergone the more conscious revolution, it will be discussed first in this essay.

I. INTERPRETATION. The rhetorical approach to interp. is, simply, that any discourse should be

understood as if it were a public address. Just as a speech act encompasses such extratextual elements as its speaker's delivery and the audience's response, so rhetorical interpreters have insisted that p. too must be understood as something spoken intentionally, at a certain time, by someone to someone. Discursive arrangement is a gauge of intention, and forms of thought, *logos*, are only one means of securing that intention. There are at least two other means: *ethos*, the audience's perception of the speaker's moral character, and *pathos*, the audience's own emotions. Aristotle (*Rhetoric* 1.2) considered these three to be "modes of proof" because they help to establish the speaker's case. The analytical enterprise of rhet. is uniquely a search for identifiable causes of audience effects, unlike the enterprise of grammar, which is largely a search for the forms of "correctness," or the enterprise of logic (which with grammar and rhet. constituted the Trivium of the ancient liberal-arts curriculum), which is largely a search for the forms of validity. In conducting their search through the three modes of proof, rhetorical interpreters are necessarily historicist and contextual. They conceive of *all* p. as a kind of social act or performance, finding a rhetorical impulse even in that p., such as the symbolist and imagist, which is programmatically non- or even anti-rhetorical (e.g. Gage). They have been attacked in our own time for their prizing of intention and emotion and for their susceptibility to relativist judgment—in the eyes of many, for their failure to view p. *sui generis*.

What p. is, if not rhet., was yet another project of Aristotle, the first critic known to construct a terminology for poetics. Aristotle made *mimesis* (the imitation of human action) the genus of p. and *mythos* (plot) its species. Of rhet., by contrast, persuasion was the genus and audience differentiation the species. Aristotle's efforts to distinguish and arrange the arts more or less horizontally form a sharp contrast to Plato's efforts to synthesize the arts and arrange them hierarchically, with dialectic (a mode of disputation more logical than rhetorical) on top. But Aristotle's division was lost sight of for more than a millennium. It was superseded in the Cl. world by Cicero's elevation of rhet. as an art of *eloquence* (to be traced more completely below) and through the Middle Ages by Horace's *Ars poetica*, which gives p. the ends of rhet. The Horatian position, moreover, reaffirmed the Platonic and Ciceronian views that only knowledge should be the basis of persuasion, and mixed those views with the idea that the poet's powers center in his unique ability to delight. To teach, to delight, to move—the subordinate ends of traditional rhet., subsumed alike by persuasion and eloquence—could be effectively achieved by p. Most medieval manuals of poetry were rhetorics and only the sections on versification made any significant distinction between p. and oratory.

When Aristotle's *Poetics* was rediscovered in the 15th c., it brought with it a formalism that increasingly made the ancient symbiosis of rhet. and p. antagonistic. But initially any felt antagonism was muted by the temper of the Ren., for rhet. had again become dominant in the curriculum, restored to something of its centrality after having been displaced for centuries by logic and dialectic. Ren. poetics at first reaffirmed, then surpassed the didactic, rhetorical, Horatian qualities of the Middle Ages: the poem's utility, its proficiency at teaching or moving—argued Minturno (1559), Scaliger (1561), Sidney (1583)—was achieved through its unique capacity for delighting, esp. through "imitative" means. In these and similar apologetics, p. became a superior rhet., and Virgil or Horace the Ciceronian *perfectus orator*, eloquent by virtue of his largely stylistic ability to make wisdom effective. Rhetorical *imitatio*, the composer's exercise of copying the work of others, became in interpretive theory a readerly role of imitating the model behavior represented in a discourse (the poet, Sidney claimed, might "bestow a Cyrus upon the world to make many Cyruses"), a theoretical position ancient as Plato's *Republic* and sanctioned, if negatively, by the Puritan closing of the theaters in 1642. In this way *imitatio* may have initially blunted perceptions of the precise nature of Aristotle's *mimesis* while ostensibly encompassing it. Gradually, however, a new emphasis on form—a poem's organization, a playwright's use of the "unities"—began to sweep crit. Further stimulating this new emphasis was the revival—with Robortelli's edition of Longinus in 1554—of the concept that the sublimity of p. does not simply persuade but more nearly "transports" its audience (see SUBLIME). This concept also revived interest in an "organic" theory of p., compatible with Aristotelianism and echoed in the modern insistence, extending through Coleridge into the 20th c., that p. must be read as if its form (q.v.) and content were fused. Such an insistence controverts the rhetorical view that form is isolable, interchangeable, and strategic, and content, on the other hand, a manageable body of knowledge, truths, or argument.

Although a certain (mainly Aristotelian) formalism was inaugurated in the poetics of the late Eng. Ren., the movement did not reach its apotheosis until our own century, first with Joel Spingarn in 1910 and Benedetto Croce in 1933, both of whom called for a scrapping of all the older, rhetorically infested terminologies, and then with the New Critics of the 1930s and the later "Neo-Aristotelians," with their insistence that a poem constructs its own autonomous universe cut off from the quotidian requirements of ordinary communication. P. speaks a different lang., Richards theorized in 1929. P. does not communicate, Brooks insisted in 1947. Or if it does, Frye argued in 1957, it does so as a kind of "applied lit." Prophetically, Kenneth Burke offered a "counter statement" to this increasingly dominant formalism as early as

1931, calling for the restoration of a rhetorical perspective in which discursive form could again be seen as strategic and in which content could be seen as a complex fusion of speaker, intention, utterance, and audience.

But the subsequent restoration of rhet. to interp. found three main emphases: the author's relation to the text, the role of the reader, and style. The first distinguished two levels of speaking in the poem, the one on which the narrator of the poem is talking to himself or to another person (see VOICE), and the one on which the poet is speaking to us (Olson, Eliot, Booth, Wright). Increasingly, however, 20th-c. poetics has pursued the second emphasis, focusing on the role of the reader either *of* or *in* the poem—ideal, implied, competent, actual—whose interaction with the text structures it and gives it meaning, or whose presence at least raises questions about the conditions of textuality and communicability (Barthes, Holland, Culler, Iser, Fish, Suleiman and Crossman). Whereas formalists, in their "organic" view of p., insist that p. *means* what it *says*, postformalist critics argue that p. means what it *does*. Nonetheless, these first two emphases involve at best a partial or fragmentary use of rhet. and, often, an antagonism toward its ends. But when the reader is a *listener*, as when p. is performed in an oral culture (Errington, Connelly, Sweeney), the role of rhet. becomes much more extensive—at once more traditional and more Burkean, a general heuristic of communicative strategies—and even reaches beyond Western cultural confines (see ORAL POETRY).

For the stylistic analysis of p., rhet. has traditionally supplied detailed taxonomies of figures, schemes, and tropes ranging from such textural effects as irony to such local effects as alliteration. Catalogues burgeoned particularly among medieval and Ren. rhetoricians, for whom an embellished style (q.v.) was the sum total of eloquence (in Peacham [1593] over 350 figures are described). Four tropes—metaphor, metonymy, synecdoche, irony (qq.v.)—were early conceived as master tropes (Fraunce [1588]) because they generate all figurative uses of lang., an idea reiterated by Burke in the 1940s. Jakobson in 1956 found metaphor and metonymy to be attitudes the mind assumes in coping with degrees of similarity or contiguity between matters, and thus began a movement to view tropes as inherent in intellection. Subsequently, the act of interp. itself came to be seen as tropological (Genette; Rice and Schofer): figures, esp. the master tropes, map mental strategies or processes in the reader's work of unraveling the meaning of a text. The figures and tropes have supplied a taxonomy for anthropology, psychology, linguistics, and history; in modern rhet. they serve as indicators of the inherent plasticity of lang. (Quinn). The plasticity and figurality of lang. have also become concerns of modern deconstructionists (Derrida, de Man) in their obliquely rhetorical examination of the often indeterminate gap between what p. *says* and what it ostensibly *does*.

This brief review may suggest that the ultimate choice is to rhetoricize or not to rhetoricize; to consider p. persuasively audience-directed and stylistically eloquence-directed, or to view it as something other than a conventionally communicative act; to restore all of rhet. or only those fragments available in such modern sciences as linguistics and psychology. The alternatives may be further clarified, and some of the gaps in our survey spanned, by shifting our attention to theories of composition—which by offering attitudes toward the use of lang. also offer an implicit hermeneutic.

II. COMPOSITION. Among Western theories of composition, Aristotle's *Rhetoric* is the oldest. His master stroke in the *Rhetoric*—and one which has been too easily overlooked or too readily absorbed within other theories—is his doctrine that rhetorical practice embodies its own unique mode of thought, observable mainly in the orator's efforts to discover the available means of persuading his audience. This practical reasoning, called "invention" in later theories, deals with probable rather than demonstrable matters: the orator weighs alternatives, substantiates his case, and chooses strategies which he believes will sway. To establish the uniqueness of rhetorical invention, Aristotle advanced the *example* and the *enthymeme* as the counterparts, respectively, of logical induction and syllogism—the point being that the orator composes by giving priority not to form but to audience. Compare the enthymeme with the syllogism: whereas the latter has two premises and a conclusion, with very clear canons of formal completeness and validity (*Only* had we world enough and time, this coyness, Lady, were no crime; but we have *not* world enough and time; therefore, this coyness, Lady, *is* a crime), the enthymeme is a syllogism that either draws its major premise from the audience's beliefs or is so loose or incomplete that it compels the audience silently to supply a condition, premise, or the conclusion (hence, while the opening with the addition of "only" is a syllogistic premise, Marvell's entire poem is actually enthymematic). Accordingly, the audience, its knowledge and emotions, has the priority in rhet. that is held by formal validity in logic, by forms of correctness in grammar, and by form itself in poetry.

In one respect, rhetorical invention became poetic invention by default. Aristotle does not describe the latter, and indeed distinguishes the two largely by implication. His *Poetics* is after all not a handbook of composition but a theory of poetry, of its nature and elements, developed in part by comparison with the drama. One of those elements—thought, the power of an agent to say what can be said or what is fitting to be said (in sum, invention)—Aristotle declines to discuss at length (6.16) because he had already treated it in the *Rhetoric*. Poetic invention, where it does not de-

pend upon plot, would seem to arise from a certain natural plasticity (17), the poet's ability to visualize action and assume attitudes—Aristotle's way of avoiding ascribing poetic invention to either inspiration or poetic madness, the two alternatives Plato saw as the poetic counterparts of rhetorical invention. Nonetheless, the Platonic alternatives have certainly had their advocates through the centuries: the divine *furor* usually associated with Neoplatonism was expressed perfectly by Shakespeare in *A Midsummer Night's Dream* ("The lunatic, the lover, and the poet / Are of imagination all compact") and reached its culmination in the romantic movement of the 19th c. But in the larger historical view, it is rhet., esp. in its developments after Aristotle, which remained the chief discipline whereby writers and speakers learned their craft.

By the time of Cicero, whose Latinity was influential for centuries and whose theories of rhet. were to achieve enormous popularity among Ren. humanists, rhet. had become much more systematized. A unified process of composition implicit in Aristotle became divided into five discrete functions: thought (*inventio*), arrangement (*dispositio*), style (*elocutio*), memory (*memoria*), and delivery (*actio* or *pronuntiato*). Aristotelian rhetorical invention, the search for available means of persuasion, became a pro-and-contra analysis of topics for which forensic oratory was the paradigm. Oratorical arrangement too became more prominent: in forensic oratory, whereas Aristotle had advised only two parts (statement and proof) but allowed four (plus introduction and conclusion), Cicero advised six (exordium, background of the question, statement, proof, refutation, conclusion) and allowed seven (plus a digression). Although Cicero, a poet himself, may have found p. limiting (his persona's famous judgment of p. in *De oratore* 1.70 was exactly reversed by Ben Jonson in *Timber*), nonetheless the two were firmly joined in Cicero's extension of rhet. beyond the end of persuasion, and well beyond the subordinate ends of teaching, pleasing, and moving. Rhet. became the art of *eloquence*, lang. whose artistic force is the formal means whereby its content achieves persuasiveness. As such, rhet. was to cap the statesman's education, and above all be the avenue through which the wisdom of philosophy would be made practical. To accomplish the latter, Cicero rhetoricized philosophy and thereby extended beyond its careful boundaries Aristotle's teachings on rhetorical thought. Rhet., esp. Ciceronian rhet., became a kind of surrogate philosophy which still had great attraction for Ren. humanists fourteen centuries later. In fact, up through the 16th c., Cicero's formalized rhet. and ideal of eloquence were ready tools to fill the practical and apologetic needs of critics and poets—even when his major works were lost.

In the Middle Ages, Cicero's youthful *De inventione* and the pseudo-Ciceronian *Rhetorica ad Herennium* never waned in popularity. Both were only epitomes, offering little more than systematizing. Medieval rhetorics and poetics stressed *dispositio* and *elocutio*, as seen both in St. Augustine's *De doctrina christiana* (426 A.D.) and in Geoffrey of Vinsauf's *Poetria nova* (ca. 1200). The most formalized functions of Ciceronian rhet., functions which directly pertain to the creation of form, seemed to be the critical determinants of eloquence in either art. A concern with rhetorical thought, or any intrusion of *inventio* into systematic philosophy, let alone poetics, was altogether neglected.

But it was precisely that concern with thought which was revived in the Ren. The first published book in Italy was Cicero's masterpiece, *De oratore*, a dialogue in which famous Roman statesmen and lawyers give critical precedence not to arrangement and style, *dispositio* and *elocutio*, but to the strategies of *inventio* in moving others to action. The recovery of Quintilian and the rise to prominence of law as a secular profession gave added impetus to this "new" mode of thought and disputation. Ciceronian legalisms seemed to fire the poets' imaginations as well: *in utramque partem*, the readiness to debate both sides of a question—itself a feature of medieval disputation—becomes a kind of lawyerly embracing of contraries (*controversia*) in the argumentative and ostensibly irresolute fabric of Tudor p. and drama; *qualis sit*, individuating a phenomenon by setting it within a thesis-to-hypothesis (or definite-to-indefinite-question) relationship suffuses Boccaccian fiction and Sidneyan crit.; *ethos* and *ethopoiesis*, the illusion of mind and of behavioral probability, pervade dialogues, mock encomia, and most discussions of courtliness. Schoolroom *imitatio*, including the formal requirements of the forensic oration (esp. the second part, the *narratio* or background of the question), brought fictiveness itself well within rhetorical exercises.

Ultimately, it was Ciceronian *inventio*, including those vestiges within it of Aristotle's distinction between rhetorical and logical modes of thought, which suffered most in the reformations which accompanied the Ren. Rhet. became utterly formalized, far beyond its Ciceronian and even its medieval state. One of the influential books of the early Ren. was *De inventione dialectica* by Rudolphus Agricola (d. 1485). Logic or dialectic, said Agricola, is "to speak in a probable way on any matter"; grammar teaches correctness and clarity, rhet. style. Subsequently the reformers known as Ramists deprived rhet. of *inventio* and *dispositio* (these became solely logical functions) and reduced it to *elocutio* and *actio* (*memoria* was seen as a function of *dispositio*). Though the Ramist reform did not last, rhet. was disintegrated, and it eventually became the subject of such other reformative efforts as Baconian rationalism. Cicero's public mind in search of probabilities was displaced by an isolated, meditative mind totally at odds with traditional *inventio*. Ironically, too, the

reform began to undo Cicero's assertion in *Pro archia poeta* (a document whose discovery by Petrarch in 1333 marked a beginning of the Ren.) that a key difference between p. and rhet. lies in their audiences, p. having a general one, rhet. a specific one. Sidney restated the argument: only p. has the power to draw children from play and old men from the chimney corner. But by the 17th c., rhetorical *inventio* had become unmoored from specific audiences, to the further confusion of rhet. and p.

Moreover, as *inventio* declined in prominence, *elocutio* rose, in fashion at least, not only in the new rhetorics of the 16th c. but in the new poetics, the new literary theories of the time. With the rise of the vernacular over Lat. as the lang. of lit., scholarship, and commerce, rhetorical theories burgeoned with discussions of style, suffused with the restored Ciceronian hierarchy (high, middle, and low or plain styles), further cutting across what few boundaries yet remained between rhet. and p. Although Thomas Wilson, who wrote the first Ciceronian rhet. in Eng. (1553), stayed within rhetorical genres for his examples, other traditional stylists such as Sherry, Peacham, and Fraunce treated *elocutio* by drawing virtually all of their examples from vernacular p. Puttenham's *Arte of English Poesie* (1589) devotes much attention to style and is equally a work on rhetorical *elocutio*, involved as both arts are in what Puttenham regards as the courtly requirements of "dissembling."

Puttenham's book, like many of the Continental poetics of the time (Du Bellay, Ronsard, Peletier), divides theory along the lines of the first three offices of traditional rhet.: *inventio, dispositio, elocutio*. But this rhetoricizing of poetics did little to salvage the rapidly disappearing uniqueness of rhetorical thought, including those poetics that had clear bearing on compositional matters. Geoffrey's advice to medieval poets, to invent by thinking of structure first, was seldom superseded. The "inventive" office, Puttenham taught, was to be performed by the "phantasticall part of man," his imagination, and controlled by choice of genre and by decorum (qq.v.). Audience-anchored doctrines of rhetorical *inventio*—whether the Aristotelian search for the means of persuasion via the probable or the Ciceronian pro-and-contra reasoning through a grid of topics toward eloquence—were to all intents and purposes dead. Nor did either of these doctrines play a significant role in the new literary theories fostered by the recovery of Aristotle's *Poetics*, such as those by the 16th-c. humanists Robortelli and Castelvetro, though two terminologies co-existed. Throughout 17th- and 18th-c. poetics, Aristotelian plot ("fable"), character ("manners"), thought ("sentiments"), and diction continued to exist side-by-side with Ciceronian terminology ("passions," "propriety"). *Inventio* remained the creator's first responsibility, but its considerations of audience centered mainly in decorum. Too, whereas in rhet., *inventio* became the unsystematic action of a solitary mind, in poetics it became largely exculpatory (it was, as Dryden put it in 1667, "the first happiness of the poet's imagination"). In the 18th c., the creative processes began to be scrutinized by the new science of psychology and taught through whatever relicts of ancient rhet. were refashionable. Among those relicts, *elocutio*, or style, retained greater prominence than *inventio*, and for centuries constituted virtually the whole of rhet., only to become the scapegoat of conscious artifice in romantic and postromantic poetics, and ultimately to be revived as an important feature of modern interp.

Two remaining offices of rhet. have received comparatively little attention over the centuries. *Actio*, claimed by Demosthenes as the *sine qua non* of persuasion, did achieve some vogue in the 18th and 19th cs. under the name of "elocution." An effort to scientize delivery, which began with John Bulwer in 1644, occupied the attention of 18th-c. lexicographers and actors (Thomas Sheridan, John Mason) in teaching graceful gesture and correct phonation (now called "pronunciation"). With the teachings of Del Sartre in the 19th c., the movement had an impact, through mannered recitations, on Eng. and Am. education, on p. written to be recited, on styles of acting, and on later "modern" dance. *Memoria*, the storehouse of wisdom as it was known in rhet., and the mother of the Muses, was resistant to much theorizing outside medicine, where it was studied as a faculty of the soul (Yates). Rhyme was early considered not only a figure but a mnemonic device; so was the pithy form of eloquence known as *sententia*. When the two were combined (as in Edgar's speech closing *King Lear*, "The oldest hath borne most; we that are young / Shall never see so much, nor live so long"), a *terminus ad quem* was made memorable. The art of memory also became involved with the creation of fantastic images (the more fantastic, Quintilian advised, the easier to remember) and elaborate "memory theaters" for the rapid recall of complex, even encyclopedic knowledge.

In sum, whether one considers the interp. of p. or its composition, a shared interest in persuasion, eloquence, or even simply form and style has always linked rhet. and p. The fragmentation of rhet. and its dispersal through various disciplines and critical approaches were steady developments in Western culture after the Ren., particularly after the rise of science and of formalist crit. Now the uniqueness of p. is arguably more fully understood than that of rhet. On the other hand, modern efforts to reestablish rhetorical *inventio* (e.g. Perelman) may ultimately serve to reauthenticate rhet. too as *sui generis*. T.O.S.

SOURCES CITED IN TEXT: R. Sherry, *A Treatise of Schemes and Tropes* (1550); T. Wilson, *The Arte of Rhetorique* (1553); S. Minturno, *De poeta* (1559); J. C. Scaliger, *Poetices libri septem* (1561, 1581); P. Sidney, *An Apology for Poetrie* (1583); A. Fraunce, *The Arcadian Rhetorike* (1588); H. Peacham, *The*

Garden of Eloquence, 2d ed. (1593); B. Jonson, *Timber* (1641); J. Bulwer, *Chironomia* (1644); J. Dryden, Preface to *Annus mirabilis* (1667); J. Mason, *An Essay on Elocution* (1748); T. Sheridan, *A Course of Lectures on Elocution* (1762); J. S. Mill, "Thoughts on P. and Its Varieties" [1833], *Dissertations and Discussions* (1859); P. Verlaine, "Art poétique" (1884); J. E. Spingarn, "The New Crit." [1910], *Crit. and America* (1924); I. A. Richards, *Practical Crit.* (1929); K. Burke, *Counter Statement* (1931); B. Croce, *The Defence of P.* (tr. 1933); E. Olson, "Rhet. and the Appreciation of Pope," *MP* 37 (1939); Brooks; T. S. Eliot, *The Three Voices of P.* (1953); R. Jakobson and M. Halle, *Fundamentals of Lang.* (1956); Frye; G. T. Wright, *The Poet in the Poem* (1962); R. Barthes, *S / Z* (1970); N. Holland, *5 Readers Reading* (1975); S. E. Errington, *A Study of Genre* (1975); Culler; Derrida; W. Iser, *The Act of Reading* (1978); S. Fish, *Is There a Text in this Class?* (1980); *The Reader in the Text*, ed. S. R. Suleiman and I. Crossman (1980); J. Gage, *In the Arresting Eye* (1981); W. C. Booth, *The Rhet. of Fiction*, 2d ed. (1983); P. de Man, *The Rhet. of Romanticism* (1984); B. Connelly, *Arab Folk Epic and Identity* (1986); A. Sweeney, *A Full Hearing* (1987).

BIBLIOGRAPHIES: D. Breuer and G. Kopsch, "Rhetorik-Lehrbücher des 16. bis 20. Jahrhunderts," *Rhetorik: Beiträge zu ihrer Gesch. in Deutschland vom 16.–20. Jahrhundert*, ed. H. Schanze (1974); *Historical Rhet.: An Annot. Bibl.*, ed. W. Horner (1980); R. Jamison and J. Dyck, *Rhetorik—Topik—Argumentation: Bibliographie zur Redelehre und Rhetorikforschung im deutschsprachigen Raum 1945 bis 1979/80* (1983); H. F. Plett, *Englische Rhetorik und Poetik, 1479–1660: Eine systematische Bibl.* (1985); J. J. Murphy, *Med. Rhet.: A Select Bibl.*, 2d ed. (1989).

HANDBOOKS AND SURVEYS: R. F. Howes, *Historical Studies of Rhet. and Rhetoricians* (1961); M. H. Nichols, *Rhet. and Crit.* (1962); E. Black, *Rhetorical Crit.* (1965); H. Lausberg, *Elemente der literarischen Rhetorik*, 3d ed. (1967), *Handbuch der literarischen Rhetorik*, 2d ed., 2 v. (1973)—the fullest treatment currently available; L. A. Sonnino, *A Handbook to 16th-C. R.* (1968); P. Dixon, *Rhet.* (1971); W. Jens, "Rhetorik," *Reallexikon* 3.432–56—extensive bibl.; A. J. Quinn, *Figures of Speech* (1982); *Dict. of Literary-Rhetorical Conventions of the Eng. Ren.*, ed. M. Donker and G. M. Muldrow (1982); H. Beristáin, *Diccionario de retórica y poética* (1985); *Methods of Rhetorical Crit.*, ed. B. L. Brock et al., 3d ed. (1989); E. P. J. Corbett, *Cl. Rhet. for the Mod. Student*, 3d ed. (1990); *The Present State of Scholarship in Historical and Contemp. Rhet.*, ed. W. B. Horner, 2d ed. (1990); R. A. Lanham, *A Handlist of Rhetorical Terms*, 2d ed. (1991); *Historisches Wörterbuch der Rhet.*, ed. G. Ueding and W. Jens (forthcoming).

HISTORIES, THEORIES, TEXTS, AND STUDIES: (1) *Classical*: J. C. G. Ernesti, *Lexicon technologiae graecorum rhetoricae*, 8 v. (1795; rpt 1 v., 1962), *Lexicon technologiae latinorum rhetoricae* (1797; rpt 1 v., 1983)—the fullest mod. sources, though rare; *Rhetores graeci*, ed. C. Walz, 9 v. (1832–36); *Rhetores graeci*, ed. L. Spengel, 3 v. (1856); *Rhetores latini minores*, ed. R. Halm (1863); E. M. Cope, *An Intro. to Aristotle's Rhet.* (1867); R. Volkmann, *Die Rhetorik der Griechen und Römer in systematischer Ubersicht*, 2d ed. (1885); C. S. Baldwin, *Ancient Rhet. and Poetic* (1924); W. Rhys Roberts, *Greek Rhet. and Lit. Crit.* (1928); L. S. Hultzén, "Aristotle's *Rhet.* in England to 1600," Diss., Cornell (1932); W. Kroll, "Rhetorik," Pauly-Wissowa, supp., 7.1039–1138; L. Arbusow, *Colores rhetorici* (1948); *Artium Scriptores*, ed. L. Radermacher (1951); M. L. Clarke, *Rhet. at Rome* (1953); D. L. Clark, *Rhet. in Greco-Roman Education* (1957); G. A. Kennedy, *The Art of Persuasion in Greece* (1963)—best Eng. account of Gr. oratory and rhet., *Quintilian* (1969), *The Art of Rhet. in the Roman World, 300 b.c.–a.d. 300* (1972), *Cl. Rhet. and Its Christian and Secular Trad. from Ancient to Mod. Times* (1980), *Gr. Rhet. Under Christian Emperors* (1983); B. Weinberg, "Rhet. After Plato," *DHI*; *Readings in Cl. Rhet.*, ed. T. W. Benson and M. H. Prosser (1969); M. H. McCall, Jr., *Ancient Rhetorical Terms of Similarity and Comparison* (1970); H. Caplan, *Of Eloquence: Studies in Ancient and Med. Rhet.* (1970); A. Scaglione, *The Cl. Theory of Composition from Its Origins to the Present* (1972); *A Synoptic Hist. of Cl. Rhet.*, ed. J. J. Murphy (1972); A. Hellwig, *Untersuchungen zur Theorie der Rhetorik bei Platon und Aristoteles* (1973); G. L. Kustas, *Studies in Byzantine Rhet.* (1973); R. W. Smith, *The Art of Rhet. in Alexandria* (1974); W. Eisenhut, *Einführung in die antike Rhetorik* (1974); J. Martin, *Antike Rhetorik: Technik und Methode* (1974); *Quintilian on the Teaching of Speaking and Writing*, tr. J. J. Murphy (1987); H. L. F. Drijepondt, *Die Antike Theorie der Varietas* (1979); K. Heldmann, *Antike Theorien über Entwicklung und Verfall der Redekunst* (1982); G. Williams, *Figures of Thought in Roman P.* (1980); H. Maguire, *Art and Eloquence in Byzantium* (1981); D. A. Russell, *Crit. in Antiquity* (1981); Norden; M. Fuhrmann, *Die antike Rhetorik* (1984); *CHLC*; *Readings from from Cl. Rhet.*, ed. P. P. Matsen et al. (1990); T. Cole, *The Origins of Rhet. in Ancient Greece* (1991).

(2) *Medieval*: J. M. Manly, "Chaucer and the Rhetoricians," *PBA* 12 (1926); C. S. Baldwin, *Med. Rhet. and Poetic* (1928); A. C. Bartlett, *The Larger Rhetorical Patterns in Anglo-Saxon Poetry* (1935); W. S. Howell, *The Rhet. of Alcuin and Charlemagne* (1941); R. McKeon, "Rhet. in the Middle Ages," *Speculum* 17 (1942); Sr. M. Joseph, *The Trivium*, 3d ed. (1948); H. Kökeritz, "Rhetorical Word-Play in Chaucer," *PMLA* 69 (1954); Curtius, chs. 4, 8; J. J. Murphy, "A New Look at Chaucer and the Rhetoricians," *RES* 15 (1964), *Rhet. in the Middle Ages* (1974), *Med. Rhet.: A Select Bibl.* (above), ed., *Three Med. Rhetorical Arts* (1971); *Readings in Med. Rhet.*, ed. J. M. Miller et al. (1973).

(3) *Renaissance to Modern*: D. L. Clark, *Rhet. and P. in the Eng. Ren.* (1922); I. A. Richards, *Practical Crit.* (1929), *The Philosophy of Rhet.* (1936); M. T. Herrick, *The Poetics of Aristotle in England* (1930),

The Fusion of Horatian and Aristotelian Lit. Crit., 1531–1555 (1946); K. Burke, *Counter Statement* (1931), *A Grammar of Motives* (1944), *A Rhet. of Motives* (1950); C. Winkler, *Elemente der Rede: Gesch. ihrer Theorie 1750–1850* (1931); Patterson; W. Taylor, *A Dict. of the Tudor Figures of Rhet.* (1937); H. D. Rix, *Rhet. in Spenser's Poetry* (1940); W. G. Crane, *Wit and Rhet. in the Ren.* (1941); V. L. Rubel, *Poetic Diction in the Eng. Ren.* (1941)—wider scope than title indicates; Sr. M. Joseph, *Shakespeare's Use of the Arts of Lang.* (1947); R. Tuve, *Elizabethan and Metaphysical Imagery* (1947); Crane; Abrams; W. S. Howell, *Logic and Rhet. in England, 1500–1700* (1956), *18th-C. British Logic and Rhet.* (1971), *Poetics, Rhet., and Logic* (1975); W. J. Ong, *Ramus, Method, and the Decay of Dialogue* (1958); J. B. Broadbent, "Milton's Rhet.," *MP* 56 (1958–59), *Some Graver Subject* (1960); H.-G. Gadamer, *Wahrheit und Methode* (1960); W. C Booth, *The Rhet. of Fiction* (1961); Weinberg; O. B. Hardison, Jr., *The Enduring Monument* (1962); U. Stötzer, *Deutsche Redekunst in 17. und 18. Jh.* (1962); M.-L. Linn, *Studien zur deutschen Rhetorik und Stilistik im 19. Jahrhundert* (1963); C. Neumeister, *Grundsätze der forensischen Rhetorik* (1964); F. A. Yates, *The Art of Memory* (1966); G. Genette, *Figures I, II, III* (1966, 1969, 1972), selections tr. as *Figures of Literary Discourse* (1982); M. L. Colish, *The Mirror of Lang.* (1968); B. Hathaway, *Marvels and Commonplaces* (1968); E. P. J. Corbett, *Rhetorical Analyses of Literary Works* (1969); E. D. Hirsch, Jr., *Validity in Interp.* (1969); C. Perelman and L. Olbrechts-Tyteca, *The New Rhet.* (tr. 1969); W. Barner, *Barock-Rhetorik* (1970); C. Trinkaus, *In Our Image and Likeness*, 2 v. (1970); J. L. Kinneavy, *A Theory of Discourse* (1971); *The Rhet. of Ren. P. from Wyatt to Milton*, ed. T. O. Sloan and R. B. Waddington (1974); W. Trimpi, *The Quality of Fiction* (1974), *Muses of One Mind* (1983); R. A. Lanham, *The Motives of Eloquence* (1976); Group Mu, *Rhétorique de la poésie* (1977); W. J. Kennedy, *Rhetorical Norms in Ren. Lit.* (1978); M. Fumaroli, *L'Âge d'eloquence: Rhétorique et 'res literaria' de la ren. au seuil de l'epoque classique* (1980); P. Valesio, *Novantiqua: Rhetorics as a Contemp. Theory* (1980); J. J. Murphy, *Ren. Rhet.: A Short-Title Catalog* (1981), ed., *Ren. Eloquence* (1983); Group Mu; J. P. Houston, *The Rhet. of P. in the Ren. and 17th C.* (1983); D. Rice and P. Schofer, *Rhetorical Poetics* (1983); T. O. Sloane, *Donne, Milton, and the End of Humanist Rhet.* (1985); A. F. Kinney, *Humanist Poetics* (1986); K. Meerhoff, *Rhétorique et poétique au XVIe siècle en France* (1986); F. Houlette, *19th-C. Rhet.: An Enumerative Bibl.* (1988); D. K. Shuger, *Sacred Rhet.: The Christian Grand Style in the Eng. Ren.* (1988); R. J. Fogelin, *Figuratively Speaking* (1988); B. Vickers, "Rhet. and Poetics," *Cambridge Hist. of Ren. Philosophy*, ed. C. B. Schmitt (1988), *In Defence of Rhet.* (1988), *Cl. Rhet. in Eng. Poetry*, 2d ed. (1989); R. Barilli, *Rhet.*, tr. G. Menozzi (1989); S. Fish, "Rhet.," *Critical Terms for Literary Study*, ed. F. Lentricchia and T. McLaughlin (1990); *The Rhetorical Trad.: Readings from Cl. Times to the Present*, ed. P. Bizzell and B. Herzberg (1990); *The Rhet. of Blair, Campbell, and Whately*, ed. J. L. Golden and E. P. J. Corbett, rev. ed. (1990); *Richards on Rhet.*, ed. A. E. Berthoff (1990); *The Ends of Rhet.*, ed. J. Bender and D. E. Wellbery (1990); N. Johnson, *19th-C. Rhet. in North America* (1991).

NONWESTERN: G. Jenner, *Die poetischen Figuren der Inder von Bhamaha bis Mammata* (1968); E. Gerow, *A Glossary of Indian Figures of Speech* (1971); S. A. Bonebakker, *Materials for the Hist. of Arabic Rhet. from the Hilyat al-muha-dara of Ha-timi* (1975); M.-C. Porcher, *Figures de style en Sanskrite* (1978); K. S. Y. Kao, "Rhet.," in Nienhauser et al.—Chinese.

T.V.F.B.

RHYME ROYAL. A stanza of seven decasyllabic lines rhyming *ababbcc*, first used in Eng. by Chaucer in *Troilus and Criseyde* (hence its alternate name, "*Troilus* stanza"), *The Parlement of Foules*, and four of the *Canterbury Tales*. It is formed on the model of 7-line stanzas in the lyrics of Machaut and Deschamps, or by dropping the fifth line of the *ottava rima* stanza (*abababcc*) as in Boccaccio, or perhaps as a variety of the *chant royal*, a festive form of poetry often composed for the *puy*, a guild festival that flourished in France and England from the late 13th c. well into Chaucer's time. The *puy* traditionally elected a "prince" who presided over mock royal feasts and who crowned the *chauncon reale* as the prize-winning poem of the occasion. The r. r. stanza was also used in ceremonies for the entry of royalty into a city. Later examples survive—John Lydgate's civic show in honor of Henry V's return to London in 1432, and the festive entry of Henry VII into York in 1486. Hence, r. r. apparently had its origin in events that honored royalty both real and imaginary, a form it continued to take both directly and as a literary artifice in the poetry of Chaucer, whose *Parlement of Foules* has been widely regarded as an occasional poem in honor of Richard II.

It was long held that the term r. r. was invented to describe the verse of King James I of Scotland, who used the stanza in his *Kingis Quair* (ca. 1425). However, King James himself never used the term; the earliest use of the term "royal" in association with the stanza appears in John Quixley's tr. (ca. 1400) of Gower's Fr. ballades, which he calls *balades ryale*; and the term *rithme royall* (the terms rhythm and rhyme are synonyms in Med. Lat.) first appears in Gascoigne (*Certayne Notes*, 1575), who uses the term *royal* not with reference to its original use as a form of address but to its gravity of subject, as in Chaucer's *Troilus*.

R. r. in Chaucer is, however, a remarkably flexible form, used as widely and imaginatively as the couplet, in every sort of poetic context. Ample enough for narrative purposes, the stanza is also suited to description, digression, comment, and literary burlesque. It dominated Eng. poetry in the 15th c., being used widely in the poetry of

Lydgate, Hoccleve, Dunbar, Henryson, Hawes, and Barclay, as well as in the drama of Skelton and Bale. As late as the second half of the 16th c., r. r. was the chief Eng. stanza for serious verse, as in Spenser's *Foure Hymnes* and Shakespeare's *Rape of Lucrece.* But Michael Drayton's revision of his r. r. narrative *Mortimeriados* into ottava rima (as *The Barron's Wars*) sometime before 1619 signaled the end of r. r. as a great Eng. measure. In modern times it has been essayed by Wordsworth ("Resolution and Independence"), Morris (*Earthly Paradise*), Auden (*Letter to Lord Byron*), and Masefield; Theodore Roethke adapts the rhyme scheme to *ababccc* for "I Knew a Woman."—Schipper; T. Maynard, *The Connection Between the Ballade, Chaucer's Modification of It, Rime Royal, and the Spenserian Stanza* (1934); P. F. Baum, *Chaucer's Verse* (1961); M. Ito, "Gower and R. R.," *John Gower: The Med. Poet* (1976); M. Stevens, "The Royal Stanza in Early Eng. Lit.," *PMLA* 94 (1979); Brogan.

M.STE.; T.V.F.B.

RIDDLE. An ancient and worldwide form in both oral lit. (the "folk r.") and written lit. (the "literary r."). Because it embodies fundamental forms of metaphor, word play, and paradox, the r. is also important to poetry and poetics generally. A r. takes the form of a question and answer, i.e. a deceptive question and a "right" answer which pierces some central ambiguity in the question. In a "true r.," the question presents a "description" (in one form, something described in terms of something else) and a "block element" (some contradiction or confusion in the description). For example, the r. "What plows and plows, but no furrow remains?" first appears to describe a plow, then blocks that answer ("no furrow remains"), and finally is resolved by another answer (a ship). Such rs. are essentially metaphors with one term concealed, pointing out both similarities and differences between the terms. Some metaphorical rs. leave the block element implicit: "Back of the village sit those who have donned white kerchiefs" (fence posts, each with a cap of snow). Others replace the single comparison with a series of comparisons, which may conflict with one another: "Open like a barn door, / Shut like a bat. / Guess all your lifetime, / You can't guess that" (umbrella).

Another large class of "true rs." is the punning r., based on lexical or grammatical ambiguity: "What turns without moving?" (milk); "Patch upon patch without any stitches" (cabbage). Other rs. are based on apparent anomalies in the laws of nature: "A fire burns in the middle of the sea" (a lamp); "I tremble at each breath of air, / And yet can heaviest burdens bear" (water); "What is full of holes and holds water?" (sponge). Rs. appear both in verse and prose; often they are "framed" by introductory and closing formulas.

The folk r. usually appears less in casual conversation than in more structured social occasions ranging from children's games to various adult rituals (e.g. courtship in tribal South Africa, funeral wakes in the West Indies). In practice the "true r." is associated with other kinds of enigmatic questions: "biblical rs." (which describe a character in the Bible), joking questions such as "conundrums" ("How is a duck like an icicle?" Both grow down.), "wisdom questions" (for which the answer must be known already, as in a catechism), "charades" (which describe a word syllable-by-syllable), and various parody forms such as the "catch-r." (which tricks the answerer into an embarrassing answer).

The literary r. is essentially an imitation of the folk r. by a sophisticated poet, who may develop the possibilities and expand the limits of the basic form. Literary rs. tend toward longer and more elaborate expression (sonnets, e.g., in Ren. Italy). In contrast to folk rs., they may use abstractions as topics (e.g., creation, humility, death, wisdom), exploit the device of prosopopoeia, delight in obscene suggestions, and even give away the answer in the text or title. The literary r. has a long history, appearing in Sanskrit (there are cosmological rs. in the *Rigveda,* parts of which go back to the early first millennium B.C.); in Heb. (a trad. of literary rs. runs from the OT and the Talmud through the Heb. poetry of medieval Spain, which includes the poet Dunash ben Labrat [10th c.], the founder of Sp. Heb. poetry, and the lyric poet Jehuda Halevi [ca. 1085–1140]); in Gr. (esp. in the *Greek Anthology* and in Byzantine lit.); in Arabic (a long trad. of Ar. riddling runs from the 10th c. to the present, the most famous Ar. riddlemaster being Al-Hariri, [ca. 1050–1120], whose *Assemblies* includes several chapters of rs. and other enigmatic questions); in Persian (another long trad., extending from the 10th–11th cs. through the 16th, and perhaps best known from the rs. in the epic *Shahnameh* of Firdūsi [b. 940], in which they are used as a test for the hero); and perhaps in Chinese (there is evidence, though no texts, of literary riddling in China in the 12th–13th cs.). A rich medieval tradition began in Europe with 100 Lat. rs. by "Symphosius" (5th c.). In England, Aldhelm (640–709) wrote 100 rs. in Anglo-Lat. hexameters (*Enigmata*), contributing to the genre the "etymological r.," which uses the text of the r. to explore the meaning of the Lat. name of the answer. The 90 or so OE rs. in the Exeter Book are among the glories of early Eng. poetry—in turn vigorous and energetic, sly and wicked, sharply visual and profoundly paradoxical. The Ren. was another productive period for the literary r., particularly in Italy. Later writers attracted to the form incl. Cervantes, Swift, Schiller, and Heine.

The essence of a r., Aristotle noted, "is to express true facts under impossible conditions," a way of thinking deeply related to metaphor. Like metaphors, rs. teach us something by engendering thought (*Poetics* 22; *Rhetoric* 3.2, 10). They are meant ultimately to reveal rather than conceal.

Children explore the linguistic and cognitive systems of their culture through r.-contests (McDowell), and in traditional societies adults may use r.-sessions in times of crisis as a model for resolving confusion (Abrahams). Metaphorical rs. teach about reading the unknown in terms of the known, and how the one never quite fits the other; punning rs. teach about the back alleys, short cuts, and dead ends in lang., and of the surprises to be found there; rs. pointing out contradictions in the physical world itself teach about the tentative nature of our categories of reality.

The Demaundes Joyous (1511; rpt. 1971)—first Mod. Eng. r. book; J. B. Friedreich, *Gesch. des Räthsels* (1860); K. Ohlert, *Rätsel und Gesellschaftsspiel der alten Griechen*, 2d ed. (1886); W. Schultz, "Rätsel," Pauly-Wissowa; W. Schultz, *Rätsel aus dem hellenischen Kulturkreise* (1909–12); *The Rs. of Aldhelm*, tr. J. H. Pitman (1925); M. De Filippis, *The Literary R. in Italy to the End of the 16th C.*, *The Literary R. in Italy in the 17th C.*, *The Literary R. in Italy in the 18th C.*, U. of Cal. Pubs. in Mod. Philol. 34 (1948), 40 (1953), 83 (1967); A. Taylor, *The Literary R. Before 1600* (1948)—good intro., *Eng. Rs. from Oral Trad.* (1951)—comprehensive; V. Hull and A. Taylor, *A Collection of Ir. Rs.* (1955); Frye; J. F. Adams, "The Anglo-Saxon R. as Lyric Mode," *Crit* 7 (1965); C. T. Scott, *Persian and Ar. Rs.* (1965), "On Defining the R.," *Genre* 2 (1969); D. Bhagwat, *The R. in Indian Life, Lore and Lit.* (1965); A. Hacikyan, *A Linguistic and Literary Analysis of OE Rs.* (1966); D. D. Lucas, *Emily Dickinson and R.* (1969); R. Finnegan, *Oral Lit. in Africa* (1970), ch. 15; *Deutsches Rätselbuch*, ed. V. Schupp (1972); R. D. Abrahams, "The Literary Study of the R.," *TSLL* 14 (1972)—misleading title; R. D. Abrahams and A. Dundes, "Rs.," *Folklore and Folklife*, ed. R. M. Dorson (1972); I. Basgöz and A. Tietze, *Bilmece: A Corpus of Turkish Rs.* (1973); J. Lindow, "Rs., Kennings, and the Complexity of Skaldic Poetry," *SS* 47 (1975); *JAF* 89 (1976)—spec. iss. on rs. and riddling; N. Frye, "Charms and Rs.," *Spiritus Mundi* (1976); K. Wagner, "Rätsel," *Reallexikon*; *The OE Rs. of the Exeter Book*, ed. C. Williamson (1977)—OE texts, *A Feast of Creatures* (1982)—tr. of OE rs.; A. Welsh, *Roots of Lyric* (1978); J. H. McDowell, *Children's Riddling* (1979)—sociolinguistic analysis; *Rs. Ancient and Mod.*, ed. M. Bryant (1983)—useful anthol. esp. of literary rs.; W. J. Pepicello and T. A. Green, *The Lang. of Rs.* (1984)—ling. models of punning rs.; A. R. Rieke, "Donne's Rs.," *JEGP* 83 (1984); N. Howe, "Aldhelm's *Enigmata* and Isidorian Etymology," *ASE* 14 (1985); D. Sadovnikov, *Rs. of the Rus. People* (1986).

A.W.

RING COMPOSITION is a structural principle or rhetorical device in which an element or series of elements are repeated at the beginning and at the end of a poem or narrative unit, thus comprising a "ring" framing a nonannular core. Van Otterlo, the pioneer in this branch of literary study, distinguished between simple framing structures, in which a single repeating element encloses the core material in the pattern *a-x-a*, and annular systems, in which the core material is set off by two or more concentric rings, e.g. *a-b-c-x-c-b-a*. The repeating elements in an annular system are chiasmically ordered. Thus, to employ a geometric metaphor, r. c. provides a mechanism for configuring circles into the linearity of the narrational and receptional processes. The terms "ring structure," "envelope pattern," "framing," and "chiasmus" often refer to the same figure.

R. c. operates on several scales from the microstructural to the macrostructural. Sometimes lexical repetition encapsulates a passage comprising a few lines, a stanza, a fitt, or a short poem; longer episodes or digressions are usually marked off by annular patterns of greater complexity. In their most expanded and intricate forms, ring systems can organize entire epic poems, as Whitman has argued for the *Iliad* and Niles for *Beowulf*. Rings can be made up of repeating material of various sorts—individual words or word roots, themes, images, motifs, or even elaborate narrative sequences. To function successfully as a ring marker, however, the convention of repetition must be sufficiently recognizable within the tradition or the repeating element sufficiently developed within the poem to enable auditors or readers to recognize the recurrence.

R. c. probably arose in response to mnemonic necessity: it provided oral poets with a powerful compositional technique and aural audiences with a means of keeping track of the movement of the story. Structurally, it can function in two ways: (1) to connect the ring-encapsulated passage or episode with the larger story, or (2) to create coherence within such a passage. In either case, it counteracts the paratactic and centrifugal tendencies endemic to oral and oral-derived poetry (Notopoulos). Yet its usefulness can be exaggerated, since the particular pattern that r. c. imposes cannot easily accommodate nonpaired repetitions, linear plot progressions, and other nonsymmetrical movements.

A perspicuous lit. hist. of r. c. cannot yet be written, since its incidence has been assayed in only a few genres and historical periods. The governing scholarly assumption has been that r. c. serves the noetic and stylistic economies of oral-based composition. This assumption has been to a degree validated by the high incidence of the device in the oral or oral-derived poetries (particularly epic and narrative) of the ancient Greeks and Anglo-Saxons. In these two fields, research into r. c. under varying rubrics originated independently during the 1930s and '40s. On the other hand, it had long been known among Celtic scholars that ancient Irish and Welsh poets made extensive use of a ring device called *dúnad(h)*, which requires that a poem begin and end with the same word. In recent decades the device has been documented in other works and traditions arising out of close association with an oral background, such

as the OF *chanson de geste,* the MHG *Nibelungenlied,* 20th-c. Yugoslav oral epic, Scottish balladry, and even the Gospels. Yet its discovery in Pound's *Cantos* and other modern and highly literate works suggests that r. c. may fill more uses over a broader literary spectrum than has been hitherto suspected.

A. Bartlett, *The Larger Rhetorical Patterns in Anglo-Saxon Poetry* (1935); W. van Otterlo, *De Ringcompositie als Opbouwprincipe in de epische Gedichten van Homerus* (1948); J. Notopoulos, "Continuity and Interconnexion in Homeric Oral Composition," *TAPA* 80 (1949); C. Whitman, *Homer and the Heroic Trad.* (1958); P. L. Henry, "A Celtic-Eng. Prosodic Feature," *ZCP* 29 (1962–64); J. Gaisser, "A Structural Analysis of the Digressions in the *Iliad* and the *Odyssey,*" *HSCP* 73 (1969); D. Buchan, *The Ballad and the Folk* (1972); H. W. Tonsfeldt, "Ring Structure in *Beowulf,*" *Neophil* 61 (1977); J. D. Niles, Beowulf (1983); K. Davis, *Fugue and Fresco* (1984); B. Fenik, *Homer and the* Nibelungenlied (1986); A. B. Lord, "The Merging of Two Worlds," *Oral Trad. in Lit.,* ed. J. M. Foley (1986); W. Parks, "Ring Structure and Narrative Embedding in Homer and *Beowulf,*" *NM* 89 (1988). W.W.P.

RONDEAU. Originally the generic term for all Fr. fixed forms (r., rondel, triolet), these being derived from dance-rounds (*rondes* or *rondels*) with singing accompaniment: the refrain was sung by the chorus—the general body of dancers—and the variable section by the leader. The written forbears of the r. are generally thought to be the *rondets* or *rondets de carole* from 13th-c. romances (cf. CAROL). The form by which we know the r. today emerged in the 15th c., and by the beginning of the 16th c. had displaced all competitors. This form, practiced particularly by Clément Marot, consists of 13 lines, octo- or decasyllables, divided into three stanzas of 5, 3, and 5 lines. The whole is constructed on two rhymes only, and the first word (-sound) or words of the first line are used as a *rentrement* (refrain), which occurs as the ninth and fifteenth lines, i.e. at the end of the second and third stanzas, and usually does not rhyme. If we let R. stand for the *rentrement,* the r. has the following scheme: *aabba aabR aabbaR.*

During the course of the 16th c. the r. gradually disappeared. It was restored to fashion at the beginning of the 17th c. by the *précieux* poets, esp. Vincent Voiture, on whose example Théodore de Banville based his 19th-c. revival of the form. Although Musset had experimented with the form earlier in the 19th c., taking some liberties with the rhymes, it was Banville's practice which provided the model for the later 19th-c. explorations of the form. In England, aside from 16th-c. examples (Wyatt in particular), the r., did not really flourish until the end of the 19th c., when under Banville's influence Fr. forms attracted the enthusiasm of poets such as Austin Dobson, Edmund Gosse, W. E. Henley, Ernest Dowson, Thomas Hardy, and Robert Bridges. In Eng. it has, unlike the triolet, often been used as a vehicle for serious verse. In Germany, where it has also been called *Ringelgedicht, Ringelreim,* and *Rundreim,* the r. was used by Weckherlin, Götz, Fischart, and later Hartleben.

The management of the *rentrement* is the key to the r.'s expressive capabilities. Banville says that the *rentrement* is "both more and less than a line, for it plays the major role in the r.'s overall design. It is at once the r.'s subject, its *raison d'être* and its means of expression." Fr. poets, wishing to keep the *rentrement* unrhymed yet fatally drawn to rhyme, found a solution in the punning *rentrement,* which rhymes with itself rather than merely repeating itself. Consequently, in the Fr. r. the *rentrement* tends to remain unassimilated, full of wit, buoyancy, and semantic fireworks. The Eng. poets, on the other hand, sought to integrate the *rentrement* more fully, both by frequently allowing it to rhyme with either the *a* or *b* lines, thus pushing the r. in the direction of that exclusively Eng. form, the roundel, and by exploiting its metrical continuity with the rest of the stanza. The Eng. *rentrement* is also usually longer than the Fr., four syllables rather than one or two. In short, the Eng. r. is altogether graver and more meditative than the Fr., its *rentrement* more clearly a lyric destination, a focus of self-recollection, intimate knowledge, and haunting memory.

As a type of truncated refrain, the *rentrement* probably evolved from copyists' habits of abbreviation, common in the Middle Ages. In Fr. prosody, *rentrements* are usually associated with the r., but whenever the refrains of any poem, in any lang., whether r.-derived or not, are an abbreviated version of the first line either of the poem or of each stanza (e.g. Wyatt, "In *aeternum,*" "Forget not yet," "*Quondam* was I"), then the term *rentrement* can be justifiably applied.

The *r. redoublé,* similar in form, is rare even at the time of Marot, who is known to have composed one in 1526. In the 17th c. a few isolated examples occur in the works of Mme. Deshoulières and Jean de La Fontaine; Banville uses the form in the 19th c. Marot's r. r., 24 lines in six quatrains plus the *rentrement,* may be schematized as follows (R signifying the *rentrement* and capitals and primes denoting whole-line refrains): *ABA'B' babA abaB babA' abaB' babaR.* Each line of stanza 1 is employed in turn as the last line of each of the following four stanzas, which thus serve to develop the content of stanza 1; the final stanza then makes a comment or summation.

T. de Banville, *Petit Traité de poésie française* (1872); J. Gleeson White, *Ballades and Rondeaus* (1887); Kastner; Thieme, 380—lists 10 works 1364–1897; H. L. Cohen, *Lyric Forms from France* (1922); H. Spanke, "Das lateinische R.," *ZFSL* 53 (1929–30); Patterson; M. Françon, "La pratique et la théorie du r. et du rondel chez Théodore de Banville," *MLN* 52 (1937), "Wyatt et le r.," *RQ* 24

(1971); F. Gennrich et al., "R.-Rondo," *MGG* 11.867–84; F. Gennrich, "Deutsche Rondeaux," *BGDSLH* 72 (1950), *Das altfranzösische R. und Virelai im 12. und 13. Jahrhundert* (1963); G. Reaney, "Concerning the Origins of the R., Virelai, and Ballade," *Musica Disciplina* 6 (1952), "The Devel. of the R., Virelai, and Ballade Forms," *Festschrift Karl Fellerer* (1962); M. Rat, "Rondel et r.," *Vie et langage* 14 (1965); N. H. J. van den Boogaard, *Rondeaux et Refrains du XIIe siècle au début du XIVe* (1969)—prints all known rs. ca. 1228–1332, but must be used with caution; F. Deloffre, *Le Vers français* (1969); F. M. Tierney, "An Intro. to the R.," "Origin and Growth of the R. in France," *Inscape* 8 (1970), "The Devel. of the R. in England," "The Causes of the Revival of the R.," *Revue de l'Université d'Ottawa* 41, 43 (1971, 1973); Elwert; C. Scott, "The Revival of the R. in France and England 1860–1920," *RLC* 213 (1980); N. Wilkins, "R.," *New Grove* 16.166–70; Scott; Morier; J. Britnell, "'Clore et rentrer': The Decline of the R.," *FS* 37 (1983).

C.S.; T.V.F.B.

S

SAPPHIC. In early Gr. poetry, an important Aeolic verseform named after Sappho, a Gr. poet from the island of Lesbos of the 7th–6th c. B.C. Prosodically, this form has been of interest to poets throughout most of the history of Western poetry; generically, it has evoked ever-increasing interest in the subjects of gender and love *vis-a-vis* poetry, most recently poetry written by women.

The term "S." refers to both a meter and a stanza form. The S. line, called the "lesser S.," is a hendecasyllable of the pattern – ⏑ – x – ⏑ ⏑ – ⏑ – –. The S. stanza consists of two Lesser S. lines followed by a 16-syllable line which is an extended form of the other two: – ⏑ – x – ⏑ ⏑ – ⏑ – x + – ⏑ ⏑ – –. This latter has been analyzed in several ways; one traditional account sees the last colon as an Adonic colon, – ⏑ ⏑ – –, and treats it as a separate line, giving the S. stanza 4 lines in all. Sappho's contemporary Alcaeus also used the stanza and may have been its inventor. Catullus (84–54 B.C.?) composed two odes in Ss. (Catullus 11 and 51), the second of which is a tr. and adaptation of Sappho frag. 31 (LP); with these poems he probably introduced the S. into Lat. poetry, but it is not certain. It is Horace (65–8 B.C.), however, who provided the S. model for subsequent Roman and European poets; in his *Odes* he uses the form 27 times, second in frequency only to Alcaics. Horace also makes a single use (*Odes* 1.8) of the Greater S. strophe, i.e. an Aristophaneus (– ⏑ ⏑ – ⏑ – –) followed by a Greater S. line of 15 syllables (– ⏑ – – – | ⏑ ⏑ – | – ⏑ ⏑ – ⏑ – –), which can be analyzed as a S. hendecasyllable with an inserted choriamb. His treatment of the S. as a 4-line stanza canonized that form for posterity. Seneca (4 B.C.–A.D. 65) sometimes uses the separate elements in a different order, e.g. by arranging a continuous series of longer lines with an Adonic clausula.

In the Middle Ages, the S. acquired rhyme and was instrumental in the transition from metrical (quantitative) to rhythmical (accentual) meter (Norberg). After the hexameter and the iambic dimeter quatrain (for hymns) it is the most popular verseform of the medieval period: there are 127 examples in *Analecta hymnica*. In the Ren. and after, accentual versions of the S. became three lines of 11 syllables followed by a fourth line of 5, the whole in trochees and dactyls. The fourth line instances the phenomenon of tailing or end-shortening well attested in other stanza forms (see TAIL RHYME).

The revival of Horatian influence on poetics evoked wide interest in the S. stanza among poets and prosodists in Italy, France, Germany, England, and Spain. Leonardo Dati used it for the first time in It. (1441), followed by Galeotto del Carretto (1455–1530), Claudio Tolomei (1492–1555), and others; in the 19th c., experiments were made by Cavallotti and by Carducci (*Odi barbare*). Estéban de Villegas (1589–1669) is the chief practitioner in Spain. In the 18th-c. Ger. revival of interest in quantitative verse, F. G. Klopstock varied an unrhymed stanza with regular positional changes of the trisyllabic foot in the Lesser S. lines; later, August von Platen and others essayed the strict Horatian form. In Eng., Ss. have been written by Sidney (*Old Arcadia*; *Certain Sonnets* 25), Isaac Watts (*Horae lyricae*), Pope ("Ode on Solitude"), Cowper ("Lines under the Influence of Delirium"), Southey ("The Widow"), Tennyson, Swinburne (*Poems and Ballads*), Hardy (the first poem in *Wessex Poems*), Pound ("Apparuit"), and John Fredrick Nims (*Sappho to Valéry*, 2d ed. [1990]). The S. has been the longest lived of the Cl. lyric strophes in the West. But full studies of its history in several vernaculars still remain to be written.

G. Mazzoni, "Per la storia della saffica in Italia," *Atti* dell' Acc. Scienze lett. arti 10 (1894); E. Hjaerne, "Das sapfiska strofen i svensk verskonst," *Sprak och Stil* 13 (1913); Pauly-Wissowa, Supp., 11.1222 ff.; Hardie, pt. 2, ch. 3; Omond; H. Rüdiger, *Gesch. der deutschen Sappho-Übersetzungen* (1934); G. H. Needler, *The Lone Shieling* (1941); D. L. Page, *Sappho and Alcaeus* (1955); Norberg; Bowra; W. Bennett, *Ger. Verse in Cl. Metres* (1963); L. P. Wilkinson, *Golden Lat. Artistry* (1963); Koster;

H. Kenner, "The Muse in Tatters," *Agenda* 6 (1968); N. A. Bonavia-Hunt, *Horace the Minstrel* (1969); R. G. M. Nisbet and M. Hubbard, *A Commentary on Horace*, Odes *Book 1* (1970); R. Paulin, "Six S. Odes 1753–1934," *Seminar* 10 (1974); E. Schäfer, *Deutscher Horaz* (1976); E. Weber, "Prosodie verbale et prosodie musicale: La Strophe saphique au Moyen Age et à la Ren.," *Le Moyen Français* 5 (1979); Halporn et al.; P. Stotz, *Sonderformen der sapphischen Dichtung* (1982); West, 32 ff.; Navarro.

R.A.S.; T.V.F.B.; J.W.H.

SATIRE, as generally defined, is both a mode of discourse or vision that asserts a polemical or critical outlook ("the satiric"), and also a specific literary genre embodying that mode in either prose or verse, esp. formal verse s. From earliest times s. has tended toward didacticism. Despite the aesthetic and often comic or witty pleasure associated with much s., their authors incline toward self-promotion as judges of morals and manners, of behavior and thought. The franchise is theirs, they assume, to pass and execute verbal sentence on both individuals and types. Numerous satirists ridicule or berate shortcomings of their own times within a context whose values—ideally—will outlast the occasions or crises of the moment. Whatever they diagnose as corrupt, they confidently venture to "heal"—in Pope's phrase—albeit severely, "with morals what [their s.] hurts with wit" (*Epistle to Augustus* 262). More subtle than most of his predecessors, Pope nonetheless follows in a long trad. of satirists who denied vindictiveness; they insisted, rather, that they were indignant because of social wrongs, and that they aimed to assure human betterment. Such assertions of regard for the social good, however, are mainly formulaic, and have enabled a large body of s. to escape retribution.

This, however, was certainly not the case—nor was it intended to be the case—in the s. of primitive and ancient cultures. Imprecations, threats of reprisal, and invective (q.v.) were in many primitive cultures believed to be imbued with the "magical power" of "word-slaying" (Elliott). Eskimos composed satiric songs to shame individuals and engaged their enemies in satiric duels as a way of settling differences. In Cl. Greece, Archilochus (7th c. B.C.)—arguably the "first" Gr. satirist—was notorious for the "iambic fury" (Ben Jonson's phrase) of his bitterly personal s. With similar vituperative intent, the magic charms of early Ir. bards supposedly could rhyme both rats and men to death. Further parallels exist in the ritualized abuse with which Gr. Old Comedy began and in the practice of the ancient Arab poets, for whom s. was a lethal weapon to be chanted in actual battle; modern analogues exist in the verbal duels of street slang like "the dozens."

In more sophisticated eras also the satiric voice continues to be heard. The very term *s.* connotes the recognizable structural and topical qualities of a discrete literary genre. But critically charged attributes of s. may also attach themselves to other literary or quasiliterary forms, a usage that occurs in prose and verse (or in mixtures, like the *Satire Ménippée* [1594]). An undercurrent of mockery and parody brightens the chivalric loftiness of Ariosto's 16th-c. *Orlando Furioso*. Following the craft of Theophrastus, 17th-c. authors—e.g. Hall, Overbury, Earle—put forth "character books" as caustic reflections on contemporary manners and persons. Satiric borrowing, similarly, affected the pulpit oratory of a Tillotson or Hoadly, the social condemnation of an Erasmus or Swift, Voltaire or Burke. Both Gay's *Beggar's Opera* and Brecht's 20th-c. adaptation (*Dreigroschenoper*) with gleanings from the 15th-c. Villon show brilliantly how theatrical art and music intermesh for satiric earthiness and imagination. In a related way, the wit of Wilde and Shaw owes much of its effectiveness to an informing satiric intention.

Themes associated with s.—antifeminism, social perversity, duplicity, idiosyncrasy—enrich Aristophanic comedy (4th c. B.C.); comparable themes later served Ren. tragedy as well as neoclassical comedy. For a preeminent example we look to Shakespeare, who even in his tragedies exploited the satiric and comic potential of subjects such as hypocrisy, sycophancy, and feminine wiles. Intolerance and clerical hypocrisy have been set in diverse poetic forms whose tonality ranges from the boisterous and profane to the comic and somber. Toward the end of the 12th c., rowdy "scholar poets," the wandering minstrels of Goliardic verse, outraged the European ecclesiastical establishment. One of the most memorable critics of the medieval Church was Jean de Meun (late 13th c.), whose disenchanted perceptions in his portion of the *Roman de la rose* probably suggested to Chaucer traits for his Friar and Pardoner (as well as for the lustily secular Wife of Bath). Comparably adversarial are the 15th-c. "Scottish Chaucerians" and the 16th-c. John Skelton of *Colin Clout*. Related faults of hypocrisy and greed, powerful incentives for satiric rebuke, continued the anticlerical trad. exemplified in Palingenius, Jonson, and Molière.

Fables and beast epics also contributed to the medieval proliferation of anticlerical s. Often composed in Lat., they drew upon the durable legacy of Aesop. One of the earliest poets in this mode was Nivard of Ghent (fl. 1150), who told many stories about Isengrim the wolf *qua* monk and Reynard the fox as picaresque trickster. The beast fable enjoyed wide popularity, like the anonymous *Roman de Renart* (ca. 1175–1205), in which the cunning fox shares speaking honors with a lion, wolf, bear, cat, and cock, all of whom comment on a roster of human infirmities and abuses. From the perspective of animals, the social, political, and religious peccadilloes of humanity have been depicted by such major poets as Chaucer, La Fontaine, Dryden, and Swift.

When adapted to other literary modes such as prose, the devices of s. make important textural contributions to their host vehicles. Their ultimate significance, however, inheres in a generic process that evolved from the mythic and superstitious archetypes of primitive culture. *Satura* (also *satira*), a nomenclature of Roman times, connoted both license and obligation to expose every kind of human failure, from wanton malice to mindless folly and vanity. Initially, via misprision of folk etymology, *satura* was confused with *satyr*, the rough and lecherous half-man, half-beast of Cl. mythology. This mistaken connection persisted for a long time—into the 18th c.—despite Casaubon's rectification in 1605 (*De satyrica graecorum poesi et romanorum satira*). For the better informed, however, *satura*—by analogy with *lanx satura*—meant an abundant mixture (a hodgepodge of public and private issues that warranted severe crit.). This is essentially the subtext of much modern s. When Quintilian (*Institutio oratoria* 10.1.93) boasted that s. emanated from Roman rather than Gr. genius ("s. is all our own"), he contemplated a genre in hexameters, secondary to be sure, and not nearly so definable as others like elegy, lyric, tragedy, comedy, epic. Still, he validated a distinctive poetic mode, the formal verse s. (*sermones*, "conversations") practiced from the time of Lucilius, Horace, Juvenal, and Persius through the 18th c.

Characteristically, such s. "is a quasi-dramatic poem, 'framed' by an encounter between the Satirist [or the transparent *persona* who speaks for him] and an adversarius" (Elliott 110–11; cf. Randolph 372). The latter may be a friendly antagonist who, recruited for the occasion, becomes an acute interlocutor or courtroom lawyer, *amicus curiae*. Thus mandated, he asks provocative questions that invite from the principal speaker assurances of virtue, righteous indignation, resolution, etc. (cf. Pope's *Imitations of Horace*, in which eloquent exchanges between P. [Pope] and F. [Fortescue] parallel those in Horace (*Satire* 2.1] between Horace and the lawyer Trebatius). Alternatively, a compassionate narrator may initiate a dialogue in which an address to a victim of injustice elicits from the latter an often lengthy, passionate complaint (Juvenal's Umbricius, *Satire* 3; Johnson's Thales, *London*).

Barely concealed, the partisan satirist in dialogues like these is never far from the stage. Because the disguises that he wears are as transparent as they are conventional, readers may connect their own realities with a credible satiric fiction. Prominent among the satirist's postures is him speaking in the voice of an apologist, pretending, for example, that he is by nature mildmannered (like Persius, who describes himself as "half a clown") and therefore satirically ineffectual. The irony of the claim is patent in light of the abrasive, often militant temper of the "apologist." But the deferential fiction succeeds if it arouses a sympathetic response to the satirist who appears as a lonely warrior, an honorable man driven by conscience to excoriate wrongdoings, even at some risk to himself. Such is the intended appeal of Churchill in the mid 18th c.: "Lives there a man, who calmly can stand by, / And see his conscience ripp'd with steady eye?" (*The Apology*). In the words of Juvenal, protesting against his failed society, *difficile est saturam non scribere* (*S.* 1.30). Since Cl. times, apology has helped to shape satiric formulae, as much so in France (Régnier, *Satire* 12; Boileau, *Satire* 9) as in England (Scrope, *Defence of Satire*; Pope, *Epistle to Dr. Arbuthnot*; Johnson, *London*).

Linguistically manipulative, satiric poets exploit the ambivalent and loaded terminology of invective, sarcasm, irony, mockery, hyperbole, and understatement. To control a range of tone and substance that may vary from the obvious to the subtle, many have relied upon the structure of Cl. rhet. codified by Aristotle, Quintilian, and Cicero. Rhet. is a more tightly disciplined mode of discourse than s., but the two have enough similarity of persuasive aim that, with modifications, s. may benefit from the organizing prescriptions of rhet. "The truth," says Cicero (*De oratore* 1.15.70), "is that the poet is a very near kinsman of the orator." Dryden must have taken this premise to heart when he composed *Absalom and Achitophel* (1681), a grand defense of the divine right of kingship. A reader can trace the formal rhetorical progression of the argument with little difficulty—as though Dryden were a Cl. debater—through the sequential stages of *exordium, narratio, confirmatio, reprehensio*, and *peroratio*.

Theoretically, satiric intent is humanistic in spirit if notoriously less generous in tone. Much s., to paraphrase Pope, wishes the world well, and, knightlike, its proponents joust as the self-avowed champions of "truth," "justice," and "reformation." Johnson subsequently extended his own definition of s. ("a poem in which wickedness or folly is censured") in declaring: "All truth is valuable, and satirical crit. may be considered as useful when it rectifies errors and improves judgment: he that rectifies the public taste is a public benefactor" (*Life of Pope*). Even Swift, steeped as he was in Juvenalian *saeva indignatio*, claimed benignity, for "His s. points at no defect, / But what all mortals may correct." (*Verses on the Death of Dr. Swift*).

Not all poets and critics accept the doctrine of *humanitas*, however. For many, the satiric mode destroys, dominated by a "cankered muse" that neither corrects nor reforms, that merely taunts or rejects whatever is gross or absurd. Granted, the fierce energy of Juvenal's s. seems to exude anger, but the underlying incentive is a humane one that should not be mistaken for the pleasure of inflicting pain, which is certainly the impression generated in the 16th c. by Aretino, in his harsh dialogues on tainted It. manners. A similar reaction often accompanies the "hot-blooded rage" of Bishop Hall's *Virgidemia* and Marston's "sharp-

fang'd" *Scourge of Villainie*, and, indeed, the threat that they represented to public complacency seemed real enough to the state censors of the time, who responded by having the poems burned at the stake.

As in some of the miscellaneous Restoration *Poems on Affairs of State*, sadism bursts out of Oldham's *Satires upon the Jesuits* (1681), while personal animus—although clever and laughable—hones the slinging matches between Pope and his "furious Sappho," Lady Mary Wortley Montagu. Claims of constructive intent fail to persuade when the tone is palpably strident or venomous. By the 19th c., in Italy, for example, the social and political structure has changed enough to provoke new causes of satiric complaint. But intense anger flourishes still as in Giuseppe Belli, whose dialect sonnets bitterly decry the deficiencies of all Roman institutions—religious, social, and political.

Much of the popular appeal of s. lies in its adaptability of structure and meter to subject, and in its capacity to evoke laughter. The mock epic, derived from a serious literary vehicle, inflates a trivial theme with a ludicrous pomposity of style and lang. The device, as timehonored as the ancient Gr. *Batrachomyomachia* (*Battle of the Frogs and Mice*), is perpetuated in Pulci's *Morgante Maggiore*, Tassoni's *La secchia rapita*, Boileau's *Le Lutrin*, and numerous others by Marvell, Dryden, and Pope. Related satiric effects are achieved by additional kinds of incongruous imitation: caricature, travesty, burlesque (e.g. diverse poems of Quevedo; Cotton's *Scarronides*; Butler's *Hudibras*; Villiers' *The Rehearsal*); parody (—e.g. Shelley's version of Wordsworth's *Peter Bell*; Byron's of Southey's *A Vision of Judgment*). The staples of satiric inspiration remain abundant. In Byron, we hear echoes of Pope's intolerance of bad writers and taste, and of man's folly generally. Blake and Browning mock the hypocrisy and worldliness of organized religion; Blake wages war esp. on political dunces and oppressors, while Hugo directs political invective against Napoleon III. Thackeray berates speculators and Eng. courts. From opposed ends of the satiric spectrum, e e cummings praises man's resistance to military tyranny while Roy Campbell belittles "inferior races." Yeats and Eliot find in s. a way to voice disillusion with conflicting values of private spirituality or religion and the public disorder of their times.

Meanwhile, Auden reverts to a surprisingly traditional 18th-c. view: "S. is angry and optimistic; it believes that, once people's attention is drawn to some evil, they will mend their ways" (Foreword to *Sense and Inconsequence*, by A. Stewart [1972]). For him as for earlier generations of poets, s. offers a shared basis of self-examination to individuals with "the normal faculty of conscience" (*The Dyer's Hand*, 1962).

COLLECTIONS: *Poems on Affairs of State*, ed. G. deF. Lord et al, 7 v. (1963–75); *Oxford Book of Satirical Verse*, ed. G. Grigson (1980).

HISTORY AND CRITICISM: Dryden, "A Discourse concerning the Original and Progress of S.," (1693); P. Lejay, "Les Origines et la nature de la s. d'Horace," in his ed. of Horace, *Ss.* (1911); G. L. Hendrickson, "Satura tota nostra est," *CP* 22 (1927); J. W. Duff, *Roman S.* (1936); O. J. Campbell, *Comicall Satyre* (1938); V. Cian, *La satira*, 2 v. (1939); D. Worcester, *The Art of S.* (1940); M. C. Randolph, "The Structural Design of the Formal Verse S.," *PQ* 21 (1942); M. Mack, "The Muse of S.," *YR* 41 (1951); I. Jack, *Augustan S.* (1952); J. D. Peter, *Complaint and S. in Early Eng. Lit.* (1956); Frye; J. Sutherland, *Eng. S.* (1958); A. B. Kernan, *The Cankered Muse* (1959), *The Plot of S.* (1965), "S.," *DHI*; R. C. Elliott, *The Power of S.* (1960); G. Highet, *The Anatomy of S.* (1962); H. Schroeder, *Russische Verssatire im 18. Jahrhundert* (1962); W. S. Anderson, *Anger in Juvenal and Seneca* (1964), *Essays on Roman S.* (1982); *S.: A Critical Anthol.*, ed. J. Russell and A. Brown (1967); M. Hodgart, *S.* (1969); L. Lecocq, *La S. en Angleterre de 1588 à 1603* (1969); H. D. Weinbrot, *The Formal Strain* (1969), *Alexander Pope and the Trads. of Formal Verse S.* (1982), ed., *18th-C. S.* (1988); A. Pollard, *S.* (1970); *S.: Mod. Essays in Crit.*, ed. R. Paulson (1971); C. Sanders, *The Scope of S.* (1971); R. Seldon, "Roughness in S. from Horace to Dryden," *MLR* 66 (1971), *Eng. Verse S. 1590–1767* (1978); P. K. Elkin, *Augustan Defence of S.* (1973); M. Coffey, *Roman S.* (1976); J. Brummack, "S.," *Reallexikon* 3.601–14; E. A. and L. D. Bloom, *S.'s Persuasive Voice* (1979); T. Lockwood, *Post-Augustan S.* (1979); M. Seidel, *Satiric Inheritance* (1979); N. Rudd, *The Ss. of Horace*, 2d ed. (1982); *Die englische S.*, ed. W. Weiss (1982); M. T. Hester, *Kinde Pity and Brave Scorne: John Donne's Satyres* (1982); Fowler; V. Carretta, *The Snarling Muse* (1983); F. A. Nussbaum, *The Brink of All We Hate: Eng. Ss. on Women 1660–1750* (1984); *Eng. S. and the Satiric Trad.*, ed. C. Rawson (1984); A. G. Wood, *Literary S. and Theory* (1984); W. Kupersmith, *Roman Satirists in 17th-C. England* (1985); Terras; C. Rawson, *Order from Confusion Sprung* (1985)—pt. 2; J. S. Baumlin, "Generic Contexts of Elizabethan S.," *Ren. Genres*, ed. B. K. Lewalski (1986); L. Guilhamet, *S. and the Transformation of Genre* (1987); H. Javadi, *S. in Persian Lit.* (1988); R. E. Pepin, *Lit. of S. in the 12th C.* (1988).

E.A.B.

SESTET (It. *sestette, sestetto*; Fr. *septain*). (a) The minor division or last 6 lines of a sonnet (q.v.), preceded by an octave (q.v.). Sometimes the octave states a proposition or situation and the s. a conclusion, but no fast rules for content can be formulated. The rhyme scheme of the s. varies: in an It. sonnet it is usually *cdecde*, in an Eng., *efefgg*, but there are others. (b) Any separable 6-line section of a stanza. "S." is not generally used for an isolable 6-line stanza; Eng. has no modern term for the class, so the older "sexain" is still used.

R.O.E.

SESTINA (It. *sestine, sesta rima*). The most complicated of the verseforms initiated by the troubadours, the s. is composed of six stanzas of six lines each, followed by an *envoi* of three lines, all of which are unrhymed, and all decasyllabic (Eng.), hendecasyllabic (It.), or alexandrine (Fr.). The function of rhyme (i.e. sound repetition) in the s. is superseded by a recurrent pattern of end-words, i.e. lexical repetition. The same six end-words occur in each stanza, but in a shifting order which follows a fixed pattern: each successive stanza takes its pattern from a reversed (bottom up) pairing of the lines of the preceding stanza (i.e. last and first, then next-to-last and second, then third-from-last and third). If we let the numbers 1 through 6 stand for the end-words, we may schematize the pattern as follows:

stanza 1 : 123456
stanza 2 : 615243
stanza 3 : 364125
stanza 4 : 532614
stanza 5 : 451362
stanza 6 : 246531
envoy : 531 or 135.

More commonly, the envoy, or *tornada*, is further complicated by the fact that the remaining three end-words, 246, must also occur in the course of its three lines, so that it gathers up all six together.

The invention of the s. is usually attributed to Arnaut Daniel (fl. 1190), and the form was widely cultivated both by his Occitan followers and by poets in Italy (Dante, Petrarch, Gaspara Stampa), Spain, and Portugal (Camões, Ribeiro). A rhymed version (*abcbca* in the first stanza) was introduced into France by Pontus de Tyard (*Erreurs amoureuses*, 1549), a member of the *Pléiade*. Sidney dispenses with rhyme for "Yee Gote-heard Gods," a double s. in the *Old Arcadia* (1590). In Germany, it was the poets of the 17th c. who were attracted to the s. (Opitz, Gryphius, Weckherlin). In France and England the s. enjoyed a revival in the 19th c., thanks to the Comte de Gramont and to Swinburne, both of whom developed rhymed versions, Gramont's all on the same two-rhyme model (*abaabb* in the first stanza), Swinburne's not surprisingly more variable, given that he also composed a double s. of 12 12-line stanzas ("The Complaint of Lisa"). Gramont prefaced his collection with a hist. of the s., in which he describes it as "a reverie in which the same ideas, the same objects, occur to the mind in a succession of different aspects, which nonetheless resemble one another, fluid and changing shape like the clouds in the sky." The s. interested the Fr. and Eng. Parnassians less than the other Romance fixed forms, but has had a certain popularity in the 20th c.; Pound called the s. "a form like a thin sheet of flame, folding and infolding upon itself," and his "S.: Altaforte" and "S. for Isolt," along with Auden's "Paysage moralisé" and *Kairos and Logos* cycle, MacNeice's "To Hedli," and Roy Fuller's "S." are distinguished modern examples.—F. de Gramont, *Sestines, précédés de l'histoire de la sextine* (1872); Kastner; F. Davidson, "The Origin of S.," *MLN* 25 (1910); A. Jeanroy, "La 's. doppia' de Dante et les origines de la sestine," *Romania* 42 (1912); H. L. Cohen, *Lyric Forms from France* (1922); L. Fiedler, "Green Thoughts in a Green Shade," *KR* 18 (1956); J. Riesz, *Die Sestine* (1971)—the fullest study; I. Baldelli, "S.," *Enciclopedia Dantesca*, ed. G. Petrocchi et al., 5 v. (1970–78); M. Shapiro, *Hieroglyph of Time: The Petrarchan S.* (1980); Morier; A. Roncaglia, "L'invenzione della sestina," *Metrica* 2 (1981); Elwert, *Italienische*, sect. 82; J. F. Nims, *A Local Habitation* (1985). A.PR.; C.S.

SIGNIFYING is the dominant satiric form in Afro-Am. verbal expression. It is a rhet. feature which works in implicit and explicit codes, characteristically undergirded by wit, allusion, sarcasm. Even though the s. technique may sometimes be utilized with a complimentary objective, in its basic usage it is doubtful that it can be totally disassociated from sarcasm. The tone of s. may range from harmless teasing to bitter and caustic diatribe. S. was nurtured significantly in Afro-Am. vernacular poetic forms such as toasts, sagas, and blues. There is a toast tradition known specifically as the s. monkey poems which takes satiric shots at all manner of subjects. Among the sagas, "Titanic Shine" is perhaps the most well known. The blues is replete with s. lines of various contour such as "Ain't a baker in town / Can bake a sweet jelly roll like mine" and "Take this hammer and carry it to my captain: / Tell him I'm gone." Although not restricted at all to poetry, the use of the s. tradition in Afro-Am. writing can be traced back to the earliest poets, such as Jupiter Hammon (b. 1711) on up through Paul Laurence Dunbar (1872–1906), carried mainly by Langston Hughes in the 20th c., then taken to its greatest period of exploration with emerging poets of the 1960s like Haki Madhubuti (don l. lee), Sonia Sanchez, Nikki Giovanni, and June Jordan. Gates has used the s. trad. to formulate a mode of crit. for Afro-Am. lit.—C. Mitchell-Kernan, "S.," *Motherwit from the Laughing Barrel: Readings in the Interp. of Afro-Am. Folklore*, ed. A. Dundes (1973); G. Smitherman, *Talkin and Testifyin* (1977); H. L. Gates, Jr., "The Blackness of Blackness: A Critique on the Sign and the S. Monkey," *Figures In Black* (1987), *The S. Monkey* (1988). E.A.P.

SIMILE. A figure of speech most conservatively defined as an explicit comparison using "like" or "as"—e.g. "black, naked women with necks / wound round and round with wire / like the necks of light bulbs" (Elizabeth Bishop). The function of the comparison is to reveal an unexpected likeness between two seemingly disparate things—in this case, the reduction of tribal African women to objects in a certain cultural context, just as light

bulbs are objects in the (ironically similar) Am. context. As this s. illustrates, in the hands of a skilled poet, s. is capable of great power and is not merely "a pastime of very low order," as William Carlos Williams once remarked. Moreover, s. is probably the "oldest readily identifiable poetic artifice in European lit." (Holoka), stretching back through Homer and Mycenaean epic poetry to Sumerian, Sanskrit, and Chinese.

Nevertheless, critics and theorists radically disagree as to what distinguishes s. from factual comparisons on the one hand and metaphor on the other. While some theorists argue that factual comparisons ("My eyes are like yours") and ss. (such as Chaucer's "hir eyen greye as glas") differ only in degree, others argue that they differ in kind (a confusion which Wordsworth successfully exploits when he follows the "wreaths of smoke / Sent up, in silence" with "as might seem / Of vagrant dwellers in the houseless woods"). Similarly, some critics adhere to the traditional view that metaphor is a compressed s., distinguishable from s. only in being an implicit rather than an explicit comparison (Miller), whereas others conclude that not all metaphors and ss. are interchangeable—that metaphor is a "use of lang.," whereas comparison itself is a "psychological process" (Ortony). These questions have entered the domain of linguistics, where at least one theorist has argued in favor of a single "deep structure" of comparison, variously realized as either s. or metaphor, which is capable of distinguishing both ss. from factual comparisons and also metaphors from mere copulative equations such as "My car is a Ford" (Mack). The latter solution supports a growing sense that s. may be marked not only by "like" or "as," but also by many other comparative markers, incl. verbs such as "resemble," "echo," and "seem," connectives such as "as if" and "as though," and phrases such as "the way that" (Darian). From this perspective it seems likely that s. encompasses analogy, rather than being a discrete form of comparison. At the very least, the current exploration into the range of s. suggests that it may be a far more pervasive aspect of both lang. and perception than has previously been thought.

One of the most interesting and salient facts about s. is that in Western culture, at least, there has been a traditional prejudice against s. in favor of metaphor. Wheelwright rightly suggests that this trad. begins with Aristotle, who judges s. inferior for two reasons: since it is longer than metaphor, it is "less pleasing"; and since s. "does not affirm that this *is* that, the mind does not inquire into the matter" (*Rhetoric* 3.4.1406b). Yet even Wheelwright, who wishes to rescue us from the "tyranny of grammarians" who have collapsed s. into metaphor from "syntactical, not semantic considerations," still judges metaphor superior to s. by virtue of its "energy-tension" and compression. As Derrida has noted, "there is no more classical theory of metaphor than treating it as an 'economical' way of avoiding 'extended explanations': and, in the first place, of avoiding s." Although this particular prejudice has a long history, the 20th c. has been esp. rigid in privileging metaphor over s. At least prior to poststructuralist crit., many 20th-c. critics and theorists heralded metaphor as a model for understanding lang., thought, and philosophy. Not surprisingly, Frye regards s. as a "displaced" metaphor, which, for him, corresponds to the displacement of mythic identity into naturalism. In essence, following Coleridge's famous passage in the *Biographia literaria*, 20th-c. critics have tended to associate s. with the lower order of "fancy," and metaphor (rather than Coleridge's own term, "symbol") with the higher order of "imagination." Only recently have a few critics begun to regard s. not just as "literary embellishment" but as "a tool for serious thinking, scientific and otherwise" which "transcribes a paradigm of the creative act itself, whether in poetry or physics" (Darian).

Given the long denigration of s. by critics, it is all the more remarkable that s. has been so widely and so consistently used by poets. The earliest recorded Western lit., Sumerian, uses ss. in virtually every genre (Kramer). Among them are ss. that seem uncannily familiar, such as "as wide as the earth" and "as everlasting as the earth," though the second one had far more power in a culture that believed the earth to be "eternally enduring." Ss. are used throughout the *Rig Veda*—e.g. "In the East the brilliant dawns have stood / Like posts set up at sacrifices" (Cook). Certainly Homer uses ss., though his nearly formulaic ones (such as Thetis' rising out of the sea like a mist) hover somewhere between epithet and metaphor; and in this regard it is esp. provocative that Aristotle allies "proportional metaphor, which contains an epithet" with comparions (McCall). As Greene has pointed out, Virgil's "characteristic trope" is the s. Chaucer frequently turns to s., esp. for humor, as when describing "hende Nicholas" as being "as sweete as is the roote / Of lycorys, or any cetewale"; and Shakespeare achieves great irony by negating conventional ss. in "My mistress' eyes are nothing like the sun."

Furthermore, despite the prevailing emphasis in the 19th c. on the symbol, Shelley also habitually turns to s. In the "Hymn to Intellectual Beauty," the "unseen Power" visits

> with as inconstant wing
> As summer winds that creep from
> flower to flower,—
> Like moonbeams that behind some
> piny mountain shower,
> It visits with inconstant glance
> Each human heart and countenance;
> Like hues and harmonies of evening,—
> Like clouds in starlight widely
> spread,—
> Like memory of music fled— (3-10)

Equally curious—given the early 20th-c. bias in favor of imagism that followed Pound's injunction to "use no unnecessary word" (and which encouraged the excision of ss.)—is Wallace Stevens' increasing use of the s. in key passages of his poetry. He concludes his *Collected Poems*, for example, with "It was like / A new knowledge of reality," an especially reverberating s. that calls attention to the problematic relation, even resemblance, between poetry and what it represents. In contrast, ss. are almost nonexistent in Old Icelandic and OE (Ker). *Beowulf* contains only two ss.-the beam that brightens within the mere just as the sun shines from heaven, and the sword that melts like ice when the frost is loosened (1570a–72b, 1607b–11a)—although both stress the hero's final victory in the mere and the concomitant act of creation in restoring order to society.

The most revered form of s. is the epic s., a lengthy comparison between two highly complex objects, actions, or relations. Homer is credited with inaugurating the epic s., there being no known s. before *The Iliad* of such length or sophistication as the following:

> As is the generation of leaves, so is that
> of humanity.
> The wind scatters the leaves on the
> ground, but the live timber
> burgeons with leaves again in the season of spring returning.
> So one generation of men will grow
> while another dies.
> (6.146-49)

While the epic s. may be used by Homer for contrast or digression, as well as for thematic amplification, subsequent poets such as Virgil, Dante, Ariosto, Spenser, and Milton have refined the device, making it more integral to the structure of the epic. In part this fact reflects the written trad. within which later poets are composing, versus the oral trad. preceding and perhaps surrounding Homer.

Consequently, it is not surprising that later poets would frequently resuscitate specific Homeric ss. (Holoka), as does Virgil when comparing the "whole crowd" to the "forest leaves that flutter down / at the first autumn frost" (*Aeneid* 6.305-10), or as Milton does when describing Lucifer's "Legions" as "lay[ing] intrans't / Thick as Autumnal Leaves that strow the Brooks / In *Vallombrosa*" (*Paradise Lost* 1.301-3), a s. that continues for ten more lines in a highly complex syntactic and imagistic movement which manages to suggest that both the temporality and mortality implicit in falling leaves are themselves a consequence of Lucifer's insurrection. While many critics agree that the epic s. achieves its highest form in Milton, one critic has even argued that in Milton ss. duplicate, at least in small, "God's primal creative act" (Swift). Such an argument challenges the notion of metaphor or even myth (q.v.) as the primal creative act and reflects the growing seriousness with which this ancient (and contemporary) figure of speech is finally being regarded. See now REPRESENTATION AND MIMESIS.

H. Fränkel, *Die homerischen Gleichnisse* (1921); W. P. Ker, *Form and Style in Poetry* (1928); C. M. Bowra, *Trad. and Design in the* Iliad (1930); J. Whaler, "The Miltonic S.," *PMLA* 46 (1931); I. F. Green, "Observations on the Epic Ss. in the *Faerie Queene*," *PQ* 14 (1935); Frye; K. Widmer, "The Iconography of Renunciation: The Miltonic S.," *ELH* 25 (1958); C. S. Lewis, "Dante's Ss.," *NMS* 9 (1965); P. Wheelwright, *The Burning Fountain*, rev. ed. (1968); M. H. McCall, *Ancient Rhet. Theories of S. and Comparison* (1969); S. N. Kramer, "Sumerian Ss.," *JAOS* 89 (1969); S. G. Darian, "Ss. and the Creative Process," *Lang&S* 6 (1973); D. Mack, "Metaphoring as Speech Act," *Poetics* 4 (1975); J. Derrida, "White Mythology," *NLH* 6 (1975); J. P. Holoka, "'Thick as Autumnal Leaves,'" *MiltonQ* 10 (1976); R. H. Lansing, *From Image to Idea: The Ss. in Dante's* Divine Comedy (1977); G. A. Miller, "Images and Models, Ss. and Metaphors," and A. Ortony, "The Role of Similarity in Ss. and Metaphors," *Metaphor and Thought*, ed. Ortony (1979); J. N. Swift, "Ss. of Disguise and the Reader of *Paradise Lost*," *SAQ* 79 (1980); K. O. Murtaugh, *Ariosto and the Cl. S* (1980); A. Cook, *Figural Choice in Poetry and Art* (1985); J. V. Brogan, *Stevens and S.: A Theory of Lang.* (1986); S. A. Nimis, *Narrative Semiotics in the Epic Trad.: The S.* (1987); S. J. Wolfson, "'Comparing Power': Coleridge and S.," *Coleridge's Theory of Imagination Today*, ed. C. Gallant (1989); W. Prunty, *Fallen from the Symboled World* (1989).

J.V.B.

SONG (Lat. *carmen*, Fr. *chanson*, Ger. *Lied*). A term used broadly to refer to verbal utterance that is musically expressive of emotion; hence more narrowly, the combined effect of music and poetry or, by extension, any poem that is suitable for combination with music or is expressive in ways that might be construed as musical; also occasionally used to designated a strictly musical composition without text, deemed "poetic" in its expressivity or featuring markedly "vocal" melodic writing for instruments. From the musical standpoint, "s." has been restricted almost exclusively to musical settings of verse; experiments in setting prose have been very limited. Further, "s." has usually meant compositions for solo voice or a small group of voices (typically one or two voices to a part) rather than a full choir, and for voice(s) alone or in combination with only one or two instruments rather than a full orchestra. Hence, the resulting balance, favoring the audibility of the text and thus appreciation of the nuances of its combination with music, is a defining characteristic of the genre; for literary purposes, these characteristics have also fostered perception of s. as personal

utterance projecting a limited emotional stance experienced by a single persona.

As a literary term, "s." is related to "lyric," originally the single event consisting of a poem sung to the lyre, and eventually used in lit. in divergent senses to refer on the one hand to any poem actually set or intended to be set to music (ditty), and on the other to any poem focusing on the arousal of emotion—the latter taking its derivation from the kind of poem typically sung to the lyre (or to any other musical accompaniment) as s. "Lyric," however, has attained much wider currency than has "s." "Lyric" is the commonly accepted term today for both these meanings, whereas "s.," as a literary term, means an utterance partaking in some way of the condition of music.

For poetry, "s." has been applied in numerous ways corresponding to the nature of the implied relationship with music. The principal categories for treating the setting of poetry in s. are two: formal (including metrical, linear, strophic representation) and semantic (including verbal representation—rendering the meaning of individual words—and expressive—the rendering of a musical simulacrum of the tone or mood of the poem). They need not, of course, exclude each other, and indeed it is frequently difficult if not impossible to separate what may be a metrical rendering from its verbal or expressive function. The categories are useful, however, as some ss. favor one or the other, in turn influencing what are considered "s.-like" elements or effects in poetry.

The association of poetry with music in the ss. of the late Eng. Ren. provides prototypes of almost every meaning of s. In some (as in Thomas Campion, who wrote both words and music for his ss.), the rendering of the formal dimensions of poetry is precise: musical meter is aligned with poetic meter, lines of verse are of uniform length and set to musical phrases of the same length (words are not repeated or extended by musical means), and the strophic repetition of the poem is rendered through repetition of music (as in traditional hymn singing). Poetry that lends itself to settings of this sort is typically predictable in all of these dimensions; hence, such a poem may be designated "s." In the madrigal and in some lute ss., by contrast, such formal properties are likely to be ignored and musical devices instead correlated with individual words to enhance meaning. This might mean repetition of words of special poignancy ("weep, weep") or highlighting of such words through exaggerated duration or unusually high or low pitch; frequently such representation is accomplished through a technique called word-painting, which aligns individual words with musical figures that can be said to depict their meaning (a descending scale for the word "down"; a dissonance for the word "grief"). Inevitably such practices also lead to predictability, in this case in diction. In the poetic miscellanies of the period, "s." and "sonnet" sometimes seem to be used interchangeably and often refer to poems with one or more of these characteristics. At worst they are poems filled with cliche and cloyingly regular in formal properties; at best they achieve a delicate balance between the demands of successful musical rendition and fresh invention.

Ss. featuring more general expressivity of mood or tone in music appear less frequently in this period, although the lutenist-composer John Dowland achieved some remarkable successes in this mode. Perhaps most famous is his "Lachrimae," which existed as an instrumental composition before being provided with its now-famous text, "Flow, my tears." The pervasively doleful mood of the piece is created musically in the accompaniment through its preponderance of descending melodic lines, its minor harmonies, its low register, and the slow, deliberate pace of its phrasing; the poem seems, in effect, to make verbal what the musical rhet. of emotion has already manifested. The role of music, then, in this type of s. is less specifically text-dependent than in other types, and the required balance between music and poetry depends to a greater extent upon the availability of appropriate instrumental resources to combine with the voice.

The *Lied* of 19th-c. Ger. lit. best exemplifies the fully developed expressive setting. In the hands of Franz Schubert, and to a great extent of those who followed him (Schumann, Brahms, Wolf), the role of the accompanying instrument was enhanced to create a highly emotional s. evocative of the overall tone or mood of the poem. Many give credit to the devel. of the modern grand piano for the success of the *Lied*; certainly the notion of expressive setting was not new, as the role of the instrumental accompaniment in Monteverdi's "Combattimento di Tancredi e Clorinda" demonstrates. Such pieces, however, violated the required intimacy of voice and single instrument characteristic of s., and it was not until the devel. of a single instrument with the expressive range of the piano that this mode of s. could flourish. The genre also depended upon—and stimulated—a poetry that provided the appropriate moods, expressed in terms that could be adequately mimicked by music. This is found in the poetry of Ger. romanticism, with its frequent evocation of nature or of ordinary human activity as the locus of emotion. For Schubert, the presence in the poem of a running brook or a woman spinning wool as the background to an emotion-filled reverie provided a means for music to enhance, significantly, what the poem could only suggest. In this context, poetry can be said to be songlike if it presents an intense, sustained, clear emotional stance, called forth by an activity that takes place in time. Typically such poems feature only one such stance or a decided shift from one to another; striking ambiguity or paradox is less songlike insofar as these conditions are less readily imitated in music.

Curiously, poems that depend extensively on

the so-called musicality of words (e.g. Edith Sitwell's "abstract poetry" and the later experiments in sound poetry) are not necessarily songlike, because the sounds of the words draw attention to themselves and thereby detract from the poem's ability to evoke an emotional state. John Hollander's "Philomela" does call forth an emotional state through the association of its sounds with the images they portray, but this work was created for a specific composer and performer and thus may be thought a small libretto rather than a s. in the literary sense.

The most extended use of "s." to refer to a kind of poetry takes the connection well beyond any mechanical representation or concurrence to questions of intent or of the relation to strains of creativity. Thus Jacques Maritain uses "s." to designate the entire genre of lyric poetry, as distinct from narrative or dramatic, referring to "the Poem or the S. as *the poetry of internal music* . . . the immediate expression of creative intuition, the meaning whose intentional content is purely a recess of the subjectivity awakened to itself and things—perceived through an obscure, simple, and totally nonconceptual apperception" (394). Such conceptions of the nature of s. center on the ability of music to tap some source of understanding or sympathy that is not touched by lang. Lawrence Kramer speaks of "the mythical union of a lower reality embodied in lang. and a higher one embodied in music," stating that "through s., usually the s. of a disincarnate voice or of a figure touched by divinity, lang. is represented as broaching the ineffable" (2); it is this sense that is implied in the use of music to evoke the supernatural, whether through strictly instrumental means or through charms, as is common in drama. Music has traditionally been associated with magic and, of course, with religious experience (despite the objections at various times of both Catholic and Puritan), and it has throughout known history been thought of as the lang. of love. The fusion, therefore, of music and poetry in s. has been thought to bring about the most perfect communication possible, combining the ineffable expressivity of music with the rational capabilities of words. And by derivation, poems that are perceived as visionary, conjuring some understanding beyond the normal capacities of words, may be called ss. Spenser's "Epithalamion" and "Prothalamion," Blake's *Ss. of Innocence and Experience*, and Whitman's "S. of Myself" come to mind.

It should be noted that "s." has come to designate certain purely musical compositions too, presumably those, like poems called s., that partake in some measure of the shared experience of music and poetry. Most frequent in this usage are such 19th-c. compositions as Mendelssohn's "Ss. without Words" for piano—short, expressive pieces, typically with a striking, singable melody and the sense that one could describe in words a suitable emotional frame of reference. Their proximity to the *Lied* is probably not coincidence; s., or *Lied*, in that context *means* once again that combination of words and music producing a compressed and intense expression of the rhet. of emotion, and if words are merely implied, the effect is nevertheless present and the composition known as s.

Several specialized types of s., established by use, have similarly given their names to poetic types, esp. elegy, lament, hymn, lay or *lai*, ballad, carol, rondeau, and canzonet.

F. Gennrich, *Grundriss einer Formenlehre des mittelalterlichen Liedes* (1932); T. S. Eliot, "The Music of Poetry" (1942); W. R. Bowden, *The Eng. Dramatic Lyric, 1603–42* (1951); J. Maritain, *Creative Intuition in Art and Poetry* (1953); R. Lebèque, "Ronsard et la musique," *Musique et poésie au XVI siècle* (1954); G. Müller and G. Reichert, "Lied," *Reallexikon* 2.42–62; D. Cooke, *The Lang. of Music* (1959); J. Hollander, *The Untuning of the Sky: Ideas of Music in Eng. Poetry, 1500–1700* (1961); A. Sydow, *Das Lied* (1962); C. M. Bowra, *Primitive S.* (1962); R. H. Thomas, *Poetry and S. in the Ger. Baroque* (1963); *Penguin Book of Lieder*, ed. S. S. Prawer (1964); P. J. Seng, *The Vocal Ss. in the Plays of Shakespeare* (1967); R. Taylor, *The Art of the Minnesinger* (1968); B. H. Bronson, *The Ballad as S.* (1969); D. Ivey, *S.: Anatomy, Imagery, and Styles* (1970)—on musical setting of Eng., Fr., Ger., and It. poetry, 17th to 20th cs.; J. M. Stein, *Poem and Music in the Ger. Lied* (1971); E. Brody and R. A. Fowkes, *The Ger. Lied and Its Poetry* (1971); M. C. Beardsley, "Verse and Music," in Wimsatt; E. Garke, *The Use of Ss. in Elizabethan Prose Fiction* (1972); J. H. Long, *Shakespeare's Use of Music* (1972); H. van der Werf, *The Chansons of the Troubadours and Trouvères* (1972); D. Fischer-Dieskau, *Schubert's Ss.: A Biographical Study* (1977); C. Ericson-Roos, *The Ss. of Robert Burns* (1977); *Med. Eng. Ss.*, ed. E. J. Dobson and F. Ll. Harrison (1979); B. H. Fairchild, *Such Holy S.: Music as Idea, Form, and Image in the Poetry of William Blake* (1980); "S." and "Lied" in *New Grove*; M. Booth, *The Experience of Ss.* (1981); R. C. Friedberg, *Am. Art S. and Am. Poetry*, 2 v. (1981); S. Ratcliffe, *Campion: On S.* (1981); J. A. Winn, *Unsuspected Eloquence: A Hist. of the Relations between Poetry and Music* (1981); W. R. Johnson, *The Idea of Lyric* (1982); E. B. Jorgens, *The Well-Tun'd Word: Musical Interps. of Eng. Poetry, 1597–1651* (1982), ed., *Eng. S., 1600–1675*, 12 v. (1986–89), esp. v. 12, *The Texts of the Ss.*; L. Kramer, *M. and P.: The 19th C. and After* (1984); L. Schleiner, *The Living Lyre in Eng. Verse from Elizabeth through the Restoration* (1984); S. Banfield, *Sensibility and Eng. S.: Critical Studies of the Early 20th C.*, 2 v. (1985); Hollander; M. M. Stoljar, *Poetry and S. in Late 18th-C. Germany* (1985); E. Doughtie, *Eng. Ren. S.* (1986); W. Maynard, *Elizabethan Lyric Poetry and Its Music* (1986); J. Stevens, *Words and Music in the Middle Ages* (1986); D. Seaton, *The Art S.: A Research and Information Guide* (1987)—esp. "Aesthetics, Analysis, Crit."; J. W. Smeed, *Ger. S. and Its*

Poetry, 1740–1900 (1987); D. M. Hertz, *The Tuning of the Word* (1988); E. H. Winkler, *The Function of S. in Contemp. British Drama* (1990); *Lyrics of the Middle Ages*, ed. J. J. Wilhelm (1990). E.B.J.

SONNET (from It. *sonetto*, a little sound or song). A 14-line line poem normally in hendecasyllables (in It.), iambic pentameter (in Eng.), or alexandrines (in Fr.) whose rhyme scheme has, in practice, varied widely despite the traditional assumption that the s. is a fixed form. The three most widely recognized versions of the s., with their traditional rhyme schemes, are the It. or Petrarchan (octave: *abbaabba*; sestet: *cdecde* or *cdcdcd* or a similar combination that avoids the closing couplet), the Spenserian (*abab bcbc cdcd ee*), and the Eng. or Shakespearean (*abab cdcd efef gg*). Weeks (1910) showed in a sample of just under 6000 Eng. ss. that 60% used the *abba abba* pattern for the octaves, and 22% *abab cdcd.*

With respect to the It. pattern (by far the most widely used of the three), it will be observed that a two-part division of thought is invited, and that the octave offers an admirably unified pattern and leads to the *volta* or "turn" of thought in the more varied sestet. The *abbaabba* octave is actually a blend of 3 brace-rhyme quatrains, since the middle four lines, whose sounds overlap the others, reiterate the identical envelope pattern but with the sounds reversed, i.e. *baab.* Normally, too, a definite pause is made in thought development at the end of the eighth line, serving to increase the independent unity of an octave that has already progressed with the greatest economy in rhyme sounds. Certainly it would be difficult to conceive a more artistically compact and phonologically effective pattern. The sestet, on the other hand, with its element of unpredictability, its usually more intense rhyme activity (three rhymes in six lines coming after two in eight) and the structural interdependence of the tercets, implies an acceleration in thought and feeling, a mood more urgent and animated.

The Spenserian and Shakespearean patterns, on the other hand, offer some relief to the greater difficulty of rhyming in Eng. and invite a division of thought into 3 quatrains and a closing or summarizing couplet; and even though such arbitrary divisions are frequently ignored by the poet, the more open rhyme schemes tend to impress the fourfold structure on the reader's ear and to suggest a stepped progression toward the closing couplet. Such matters of relationship between form and content are, however, susceptible of considerable control in the hands of a skilled poet, and the ultimate solution in any given instance may override theoretical considerations in the interests of artistic integrity.

Most deviations from the foregoing patterns have resulted from liberties taken in rhyming, but there have been a few innovations in use of the s., among them the following: *alternating*, where the tercets alternate with the quatrains (Mendès); *caudate*, with "tails" of added lines (Hopkins, Samain, Rilke); *chained* or *linked*, each line beginning with the last word of the previous line; *continuous, iterating*, or *monorhymed* on one or two rhyme sounds throughout (Giacomo da Lentini, Mallarmé, Gosse); *corona*, a series joined together by theme (It.) or rhyme or repeated lines (Sp., Eng.) for panegyric; *curtal*, a s. of ten lines with a halfline tailpiece, divided 6 + 4 1/2 (Hopkins); *dialogue*, a s. distributed between two speakers and usually pastoral in inspiration (Cecco Angiolieri, Dobson); *double*, a s. of 28 lines (Monte Andrea); *enclosed*, in which the tercets are sandwiched between the quatrains (Baudelaire, Rambosson); *interwoven*, with medial as well as end rhyme; *retrograde*, reading the same backward as forward; *reversed* (also called *sonettessa*), in which the sestet precedes the octave (Baudelaire, Verlaine, Huch)—for a reversed Shakespearean s., see Brooke's "Sonnet Reversed"; *rinterzato*, a s. with eight short lines interspersed, making a whole of 22 lines (Guittone d'Arezzo); *terza rima* (q.v.), with the linked-tercets *aba bcb* rhyme scheme; *unrhymed*, where the division into quatrains and tercets is still observed, but the lines are blank (Du Bellay, Becher). In Eng. the most conspicuous variant, the 16-line poems of George Meredith's sequence *Modern Love* (1862), is clearly related to the s. in its themes and *abba cddc effe ghhg* rhyme scheme. In the 20th c., John Hollander has devised a cycle of 169 13-line poems in 13-syllable unrhymed lines (*Powers of Thirteen*, 1983).

Historically, the s. began as some variant of the It. pattern; it is probable that the form resulted either from the addition of a double refrain of six lines (two tercets) to the 2-quatrain Sicilian *strambotto* or from conscious modeling on the form of the *canzone.* In any event (for the origins remain uncertain), the earliest antecedents of the "true" It. s. are credited to Giacomo da Lentini (fl. 1215–33), whose hendecasyllables usually rhymed *abababab cdecde.* Although others of Lentini's contemporaries (the Abbot of Tivoli, Jacopo Mostacci, Piero delle Vigne, Rinaldo d'Aquino) used the form and established the octave-sestet divisions (with quatrain-tercet subdivisions), it remained for Guittone d'Arezzo (1230–94) to invent the *abbaabba* octave, which became traditional through its use by Dante (*Vita nuova; Rime*) and Petrarch (*Canzoniere*). Antonio da Tempo, in his *Summa artis rithmici* (1332), is the first to enunciate theoretical discussion of the s. as a type. The ss. of Dante to Beatrice, and of Petrarch to Laura ("spells which unseal the inmost enchanted fountains of the delight which is the grief of love"—Shelley) normally opened with a strong statement which was then developed; but they were not unmarked by the artificiality of treatment that stemmed from variations on the Platonic love themes, an artificiality that was to be exported with the form in the 15th–16th cs. as the s. made its way to Spain,

Portugal, France, the Netherlands, Poland, and England, Germany, Scandinavia, and Russia, until its use was pan-European and the number of poets not attempting it negligible. Following Petrarch there was in Italy some diminution of dignity in use of the form (as in Serafino dall'Aquila [1466–1500]), but with the work of Michelangelo, Bembo, Castiglione, and Tasso, the s. was reaffirmed as a structure admirably suited to the expression of emotion in lyrical mood, adaptable to a wide range of subject matter (e.g. love, politics, religion), and employed with skill by many writers in the centuries to follow (Alfieri, Foscolo, Carducci, D'Annunzio).

It was the Marquis de Santillana (1398–1458) who introduced the s. form (in hendecasyllables, even) to Spain, although it was not established there until the time of Juan Boscán (1490–1552) and, esp., Garcilaso de la Vega (1503–36) and Lope de Vega (1562–1635) and other dramatists of the *siglo de oro.* Sá de Miranda (1485–1558) and his disciple, Antonio Ferreira, brought the s. to Portugal, where it is better known in the *Rimas* of Camões (1524–80) and, more recently, in the exquisite work of Anthero de Quental (1842–91). Clément Marot (1496–1544) and Mellin de Saint Gelais (1491–1558) introduced it to France, but it was Joachim du Bellay (1522–60) who was most active, writing (in the Petrarchan pattern) the first non-Italian cycle, *L'Olive,* as well as *Les Regrets* and *Les Antiquités de Rome* (tr. by Spenser as *The Ruins of Rome,* an important source for Shakespeare's ss.). Ronsard (1524–85), who experimented with the form in alexandrines, and Philippe Desportes (1546–1606) wrote many ss. and were instrumental in stimulating interest both at home and in England; while Malherbe (1555–1628) put the weight of his authority behind the *abbaabba ccdede* pattern in alexandrines, which became the accepted line length. After a period of decline (general throughout Europe) in the 18th c., Gautier (1811–72), Nerval (1808–55), and Baudelaire (1821–67) revived the form, which soon reached new heights in the work of Verlaine, Mallarmé, Rimbaud, Heredia, and Valéry. Germany received the form relatively late, in the writings of G. R. Weckherlin (1584–1653) and, esp. insofar as creative achievement is concerned, Andreas Gryphius (1616–64). There followed a period of disuse until Gottfried Bürger (1747–94) revived the form and anticipated its use by Schlegel, Eichendorff, Tieck, and other romantic writers. The ss. of August Graf von Platen (1796–1835), *Sonette aus Venedig,* rank among the best in modern times, while in more recent years the mystical sequence, *Sonette an Orpheus* (1923), of Rilke and the writings of R. A. Schröder have brought the Ger. s. to another high point.

In England the s. has had a fruitful history. Wyatt (1503–42) brought the form from Italy but showed an immediate preference (possibly influenced by the work of minor writers while he was abroad) for a closing couplet in the sestet. Wyatt did, however, adhere to the Petrarchan octave; it was Surrey (1517–47) who established the scheme *abab cdcd efef gg,* a pattern more congenial to the comparatively rhyme-poor Eng. lang. in that it filled the 14 lines by 7 rhymes not 5. This pattern was used extensively in the Ren. but by no means exclusively, for there was wide variation in rhyme schemes and line lengths. It was brought to its finest representation by Shakespeare. A rhyme scheme more attractive to Spenser (and in its first 9 lines paralleling his Spenserian stanza [q.v.]) was *abab bcbc cdcd ee,* in effect a compromise between the more rigid It. and the less rigid Eng. patterns. The period also saw many s. sequences (q.v.), beginning with Sidney's *Astrophil and Stella* (1580) and continuing in the sequences of Daniel (*Delia*), Drayton (*Idea*), Spenser (*Amoretti*), and Shakespeare, with a shift to religious themes shortly thereafter in Donne's *Holy Ss.* It remained for Milton to introduce the true It. pattern, to break from sequences to occasional ss., to give greater unity to the form by frequently permitting octave to run into sestet (the "Miltonic" s., but anticipated by the Elizabethans), and a greater richness to the texture by employing his principle of "apt numbers, fit quantity of syllables, and the sense variously drawn out from one verse into another," as in his blank verse. And s.-like structures of 14 lines have even been discerned in the stichic verse of *Paradise Lost,* a practice later echoed by Wordsworth and Hardy (Johnson). Milton's was the strongest influence when, after a century of disuse, the s. was revived in the late 18th c. by Gray, Thomas Warton, Cowper, and Bowles; and reestablished in the early 19th by Wordsworth (also under Milton's influence but easing rhyme demands by use of an *abbaacca* octave in nearly half of his more than 500 ss.) and by Keats, whose frequent use of the Shakespearean pattern did much to reaffirm it as a worthy companion to the generally favored Miltonic-Italian. By this time the scope of s. themes had broadened widely; in Leigh Hunt and Keats it even embraced an unaccustomed humor. S. theory was also developing tentatively during this period (as in Hunt's "Essay on the S."), to eventuate in an unrealistic extreme of purism in T. W. H. Crosland's *The Eng. S.* (1917) before it was more temperately approached by later writers. Since the impetus of the romantic revival, the form has had a continuing and at times distinguished use, as in D. G. Rossetti (*The House of Life*), Christina Rossetti, E. B. Browning (*Ss. from the Portuguese*), and Swinburne. Few writers in the present century (W. H. Auden and Dylan Thomas might be named) have matched the consistent level of production found in the earlier work, although an occasional single s., such as Yeats's "Leda and the Swan," has rare beauty.

The s. did not appear in America until the last quarter of the 18th c., in the work of Colonel David Humphreys, but once introduced, the form

spread rapidly if not distinctively until Longfellow (1807–82), using the It. pattern, lifted it in dignity and lyric tone (esp. in the *Divina Commedia* sequence) to a level easily equal to its counterpart in England. Following him there was wide variety in form and theme, with commendable work from such writers as Lowell, George Henry Boker, and Paul Hamilton Hayne. Of the later writers E. A. Robinson, Edna St. Vincent Millay, Merrill Moore, Allen Tate, and e e cummings (who wrote a considerable number) hold a recognized place, although, space permitting, many others might be named who stand well above what Robinson called

> . . . these little sonnet men
> Who fashion, in a shrewd mechanic
> way,
> Songs without souls, that flicker for a
> day,
> To vanish in irrevocable night.

During the past century, s. themes in both Europe and America have broadened to include almost any subject and mood, even though the main line of development has remained remarkably stable. Structurally, even within the traditional patterns, the s. has reflected the principal influences evident in modern poetry as a whole: the sprung rhythm of Hopkins and free-verse (q.v.) innovations have frequently led to less metronomic movement within the iambic norm; alternatives to exact rhymes have replenished the stock of rhyme-pairs and have sophisticated acoustic relationships; and a more natural idiom has removed much of the artificiality that had long been a burden. This adaptability within a tradition of eight centuries' standing suggests that there will be no diminution of interest in and use of the form in the foreseeable future, and that the inherent difficulties that have kept the numbers of truly fine ss. to an extremely small percentage of those written will deter neither versifier nor genius from testing for her- or himself the challenge of what Rossetti called "a moment's monument,— / Memorial from the Soul's eternity / To one dead deathless hour."

H. Welti, *Gesch. des Sonettes in der deutschen Dichtung* (1884); Schipper, 2.835 ff.; L. Biadene, *Morfologia del sonetto nei secoli XIII e XIV* (1889); M. Jasinski, *Histoire du s. en France* (1903); L. T. Weeks, "The Order of Rimes of the Eng. S.," *MLN* 25 (1910): 176–80, 231—data; Thieme, 381 ff.—lists 17 Fr. works, 1548–1903; F. Villey, "Marot et le premier s. français," *RHL* 20 (1920); R. D. Havens, *The Influence of Milton on Eng. Poetry* (1922)—surveys 18th- and 19th-c. Eng. ss.; W. L. Bullock, "The Genesis of the Eng. S. Form," *PMLA* 38 (1923); L. G. Sterner, *The S. in Am. Lit.* (1930); E. Oliphant, "S. Structure: An Analysis," *PQ* 11 (1932); L. C. John, *The Elizabethan S. Sequences* (1938); L. Zillman, *John Keats and the S. Trad.* (1939); H. Smith, *Elizabethan Poetry* (1952); W. Mönch, *Das Sonett: Gestalt und Gesch.* (1955)—the most comprehensive study to date, with extended bibl.; E. Rivers, "Certain Formal Characteristics of the Primitive Love S.," *Speculum* 33 (1958); E. H. Wilkins, *The Invention of the S. and Other Studies in It. Lit.* (1959); F. T. Prince, "The S. from Wyatt to Shakespeare," *Elizabethan Poetry*, ed. J. R. Brown and B. Harris (1960); J. W. Lever, *The Elizabethan Love S.*, 2d ed. (1965); E. Núñez Mata, *Historia y origen del soneto* (1967); S. Booth, *An Essay on Shakespeare's Ss.* (1969); B. Stirling, *The Shakespeare S. Order: Poems and Groups* (1969); *Das deutsche Sonett: Dichtungen, Gattungspoetik, Dokumente*, ed. J. U. Fechner (1969); J. Levý, "The Devel. of Rhyme-Scheme and of Syntactic Pattern in the Eng. Ren. S.," "On the Relations of Lang. and Stanza Pattern in the Eng. S.," rpt. in his *Paralipomena* (1971)—difficult to obtain but valuable; M. Françon, "L'Introduction du s. en France," *RPh* 26 (1972); J. Fuller, *The S.* (1972); L. M. Johnson, *Wordsworth and the S.* (1973); F. Jost, "The S. in its European Context," *Intro. to Comp. Lit.* (1974); Wilkins; R. L. Colie, *Shakespeare's Living Art* (1974); C. Scott, "The Limits of the S.," *RLC* 50 (1976); F. Kimmich, "Ss. Before Opitz," *GQ* 49 (1976); D. H. Scott, *S. Theory and Practice in 19th-C. France* (1977); H.-J. Schlütter, *Sonett* (1979); S. Hornsby and J. R. Bennett, "The S.: An Annot. Bibl. from 1940 to the Present," *Style* 13 (1979); J. Geninasca, "Forme fixe et forme discursive dans quelques ss. de Baudelaire," *CAIEF* 32 (1980); Brogan, 455 ff.; Morier; L. M. Johnson, *Wordsworth's Metaphysical Verse* (1972)—blank verse ss.; *Penguin Book of Ss*, ed. C. Withers (1979); W. L. Stull, "'Why Are Not Ss. Made of Thee?'" *MP* 80 (1982)—the religious s. trad.; H. S. Donow, *The S. in Eng. and Am.: A Bibl. of Crit.* (1982); Fowler; *Russkij sonet*, ed. B. Romanov, and *Russkij sonet*, ed. V. S. Sovalin (both 1983)—anthols.; A. D. Ferry, *The "Inward" Lang.* (1983); F. Rigolot, "Qu'est-ce qu'un s.?" *RHL* 84 (1984); Elwert, *Italienische*, sect. 83; C. Kleinhenz, *The Early It. S.: The First Century (1220–1321)* (1986); Hollier; P. Oppenheimer, *The Birth of the Mod. Mind* (1989); G. Warkentin, "S., S. Sequence," *The Spenser Encyc.*, ed. A. C. Hamilton et al. (1990); A. L. Martin, *Cervantes and the Burlesque S.* (1991); *Six Masters of the Sp. S.*, ed. and tr. W. Barnstone (1992).

T.V.F.B.; L.J.Z.; C.S.

SONNET SEQUENCE or cycle. A subset of the lyric sequence (q.v.) consisting of a series of sonnets, of any number, that may be organized according to some fictional or intellectual order. The sequence made entirely of sonnets is rarer than readers often suppose, and seldom holds an author's or a culture's attention for long before deliberate variations emerge. The rise of the s. s. in most European langs. coincides with that of Petrarchism: because Petrarch's late 14th-c. *Canzoniere* is made largely but not exclusively of sonnets (317 of 366 poems), many of its imitators and adapters in Fr., Eng., Port., and Sp. saw their roles

as involving the domestication of his sonnet-form; hence the first Petrarchans in the vernaculars (e.g. Wyatt and Surrey in Eng.) are often the first sonneteers in their langs. as well (e.g. Du Bellay and Ronsard in Fr., Sá de Miranda and Camões in Port., Boscán and Garcilaso in Sp.). The most extreme vogue for s. ss. was that of Eng. poets in the later 16th c.: examples include Watson's *Hekatompathia* of 18-line sonnets (1582), Sidney's *Astrophil and Stella* (written early 1580s, pub. 1591), Spenser's *Ruins of Rome* (1591—an adaptation of Du Bellay's *Antiquitez de Rome* [1558] and drawn upon by Shakespeare) and *Amoretti* (with its completing *Epithalamion* [1595]), Constable's *Diana* (1592), Daniel's *Delia* (1592), Drayton's much-revised *Idea* (1593), and Shakespeare's *Sonnets* (written 1590s, pub. 1609). In the 17th c., while poets such as the Spaniard Francisco de Quevedo, Sidney's niece Mary Wroth, and the Mexican nun Sor Juana Inés de la Cruz continue to extend the reach of the amatory and philosophical s. s., the orientation of the s. s. at large (like that of the lyric sequence) turns toward devotional writing. Aside from Quevedo and John Donne, notable religious sonneteers incl. the Ger. Andreas Gryphius, the It. Tommaso Campanella, the Dutchman Constantijn Huygens, and the Frenchman Jean de la Ceppède.

In the early modern period generally, the s. s. is often thought to have a special, almost automatic claim to overall integrity—whether topical (as in Du Bellay's *Les Regrets* [1558]), meditative (as in the "corona" used by Donne and others), or vaguely chronological (as in the common usage of the Eng. word "century" for 100 sonnets). As scholars such as Fumerton and Jones have recently demonstrated, the s. s. can be as much a cultural as a literary construction—a canvas for the working-out of collective interests, a ritual experience, a type of public space—with potential analogues in painting, religion, and architecture, among other disciplines. The job of cultural mediation enacted by the s. s. perhaps indicates why poetic amateurs of note—such as the It. sculptor and painter Michelangelo Buonarroti in the 1530s and 1540s, the Eng. Puritan polemicist Henry Lok in the 1590s, or the Am. philosopher George Santayana in the 1890s—are drawn to this form as a uniquely deprivileged space: it enables them to think through emotional, philosophical, or religious issues in a formally determined, publicly accessible medium. In fact, the first s. s. in Eng.—Ann Lok's *Meditation of a Penitent Sinner* (1560), inspired by the Scottish Puritan John Knox—is the ideologically charged work of a poetic amateur, intervening in contemporary religious debates in the mode of a deeply personal meditation (Roche).

Like the lyric sequence, the s. s. seems to have had few important instances in the 18th c., but became a major romantic and postromantic vehicle. Notable examples incl. Wordworth's several s. ss.; E. B. Browning's *Sonnets from the Portuguese* (1850); Meredith's narrative *Modern Love* (1862), in which the "sonnets" have 16 lines; D. G. Rossetti's *House of Life* (1881); Darío's "sonetos" and "medallones" in *Azul . . .* (1888), a book that impelled Sp.-Am. *modernismo*, which had a recurrent fascination with the sonnet in loosely organized collocations; Pessoa's *35 Sonnets* in Eng. (1918); Rilke's *Sonette an Orpheus* (1923); and cummings' several s. ss. in his early volumes *Tulips and Chimneys* (1923), *&* (1925), and *XLI Poems* (1925). With the 20th-c. modernisms came another hiatus, followed by a renewed sense of the s. s.'s potential for organizing experience, esp. love, though in the later 20th c. it is perhaps impossible for the s. s. to occur without formal irony, cultural critique, or anachronistic pathos. Thus Nicolás Guillén's political volumes are founded on his early experiments as a sonneteer, a role to which he returns for ironic effect (as in "El abuelo" in *West Indies, Ltd.* [1934]); John Berryman's adulterous *Sonnets* (written 1940s, pub. 1968) seek out a self-conscious Petrarchism (esp. no. 15, an adaptation of *Canzoniere* 189). Robert Lowell became all but exclusively a sonneteer in late career: his experiments in recasting the s. s. *Notebook 1967–68* as *Notebook* (1970), *History* (1973), and *For Lizzie and Harriet* (1973) might be considered the climax of his work, culminating in *The Dolphin* (1973) and *Day by Day* (1977). Among more recent adaptations in Eng. are Seamus Heaney's ten "Glanmore Sonnets" (in *Field Work*, 1979) and his 8-sonnet elegy "Clearances" (in *The Haw Lantern*, 1987); Tony Harrison's dissonant rewriting of the formal trad. in *Continuous: 50 Sonnets from the School of Eloquence* (1982); Marilyn Hacker's amatory *Love, Death, and the Changing of the Seasons* (1986), incl. an updated crown of sonnets; and Bill Knott's cultural polemic in *Outremer* (1989).

Elizabethan Sonnet-Cycles, ed. M. F. Crow (1896); L. C. John, *The Eng. S. Ss.* (1938); W. Mönch, *Das Sonett: Gestalt und Gesch.* (1955); A. Gryphius, *Lyrische Gedichte*, ed. H. Palm, v. 3 of *Werke* (1961); *European Metaphysical Poetry*, ed. F. J. Warnke (1961); D. Stone, *Ronsard's Sonnet Cycles* (1966); B. Stirling, *The Shakespeare Sonnet Order* (1968); S. Booth, *An Essay on Shakespeare's Sonnets* (1969); T. Cave, *Devotional Poetry in France 1570–1613* (1969), *The Cornucopian Text* (1979); essays on Ronsard, Scève, and Du Bellay in *YFS* 47 (1972); P. E. Blank, Jr., *Lyric Forms in the S. Ss. of Barnabe Barnes* (1974); J. de la Ceppède, *From the Theorems*, tr. K. Bosley (1983); R. A. Katz, *The Ordered Text* (1985)—Du Bellay's s. ss.; J. Fineman, *Shakespeare's Perjured Eye* (1986); Scherr 235; P. Fumerton, "'Secret' Arts: Elizabethan Miniatures and Sonnets," *Representing the Eng. Ren.*, ed. S. Greenblatt (1988); Hollier; T. P. Roche, Jr., *Petrarch and the Eng. S. Ss.* (1989); G. Warkentin, "S. S.," *The Spenser Encyc.*, ed. A. C. Hamilton et al. (1990); W. C. Johnson, *Spenser's* Amoretti (1990); A. R. Jones, *The Currency of Eros* (1990).

R.GR.

SOUND.

I. THEORETICAL STATUS. Teachers of poetry for centuries have exhorted students to ignore the poem on the page and pay attention, rather, to the words in the lines as sounded aloud. This is still salubrious advice. It does not mean, however, that spoken s. is the ontological locus of "the poem." It only means that—for poems written in traditional, aural pros., at least—students of poetic form must not be distracted by the visual or graphic format, since writing is a very inexact representation of sounded speech.

In Western poetics, there have been three views about the status of s. in poetry. (1) The traditional literary view, and the presently reigning view in linguistics, is that poetry (lang.) is s., and the written or printed text is a derivative phenomenon merely meant to represent s., hence of only secondary interest. (2) The reverse view, namely that the written form is primary, the s. form secondary, is the point of view of deconstruction and, interestingly, most Ren. grammarians, who treated Lat. grammar and pros. as a set of rules applying to a *written* lang. (3) The more radically diremptive (but also synthetic) view of poetry (and lang.) is that the aural and written modes are equivalent but simply differ, both deriving from the ontologically prior nature of lang. itself. The first of these views holds that s. exists only in performance (q.v.) and in time, and that the words on the page are a mere notation or score, as in music. The second holds that lang. has behind it no self-certifying voice and that important aspects of poetry follow from purely visual features of textuality.

The third point of view, which is that taken in the present account, holds that poetry (the Word) is an ontologically bivalent entity. On this account, lang. is a set of structured formal relations which may be manifested in either or both of two physical media, one sonic the other graphic, neither of which has any particular logical or ontological priority despite the fact that the aural mode happened to appear first. Poetry, in short, is a structure or system of s., just as lang. is not the physical ss. of speech but the set of rules which makes the speaking possible. Even s. itself is only a form, not a thing with an independent existence: it is merely an organization of the air. Unlike light, which seems to be an entity unto itself because it can flow through a vacuum, s. would not exist without a carrier medium. S. is realized in a medium, so is composed of that medium, of course, but in itself is wholly form. Modern readers are accustomed first to seeing a poem on a page, then to reading it silently, and only rarely to hearing it read aloud. Prior to the invention of writing and printing and the spread of literacy, however, the situation was quite the reverse: orality was the condition of poetry. This contrariety should lead us to ask the deeper question about the ontological status or site of poetry, i.e. the question of where a poem exists. Few would want to hold that a poem exists on the page, for then someone could burn the page and claim that "the poem" was thereby destroyed. We might then point out that there are other copies of the poem. Suppose our incendiary then burnt all known texts of the poem—is "the poem" then gone? So long as humans exist who remember the text of the poem exactly and can reproduce it, the poem is not lost. If all who remember the poem should die, then presumably the poem indeed ceases to exist. Barring that, most would be uncomfortable saying that poems exist only in our minds, a view which echoes Croce's theory that "real" artworks exist only in their creator's conception, the physical manifestation being but a poor secondhand version. No, we want to insist that artworks, like people, have both a physical and a social existence.

The analogy with music is instructive: presumably few would hold that sheet music is "music." The musical score is only a set of marks on paper which via a set of known conventions are intended as directions for performance. Many would naturally say that music therefore exists only when it is being played and heard. But of course each performance varies in greater or lesser degree; each is music, but not "the music." All performances are however related by the set of invariant features (formal relations) notated on the score which, despite innumerable minute variations, are constant through all accurate performances and thus constitute "the music."

For theoretical purposes, therefore, the term "s." does not refer, literally and most directly, to the s. of the poem actually heard in performance, for in the recitation of verse, speakers produce any number of linguistic and even paralinguistic features which are optional and variable and may well be unique to each performance, and so cannot be a part of that system of invariants common to all performances because they are cued by the text. It is this latter set of features which constitutes the s. system of the poem. We are interested, as Jakobson said, not in the phone but in the phoneme, in ss. as coded into categories by the phonological structure of the lang. and so treated at an abstract level, as a *system* of signs. Poetic s. responds to the lang. system but is further constituted (coded) on a different principle in verse.

A number of prosodic conventions do *not* treat s. as heard: both correct scansion of a meter and correct recognition of a rhyme, for example, depend on knowing which letters in the printed text

of a poem were sounded in the poem's historical context, not which are sounded now. Several metrical rules in fact take account of syllable ss. regardless of whether they exist or not. In Fr. prosody, there is a rule forbidding a masculine–feminine rhyme: feminine rhymes (all words which end in mute -e) must mate with feminine, and masculine (words ending in any other vowel or a consonant) with masculine, regardless of the fact that the mute ending has not been sounded since the 15th c. And in It. prosody, lines are treated in scansion as having 11 syllables even if they have only 9 or 10. "Sound" therefore is as much a conventional and theoretical construct as any other literary concept, varying in definition from lang. to lang. and subject to further constraints by verse systems.

The distinctive feature of poetry is that it is set in lines, i.e. articulated into segments. This segmentation is however of a different nature than the visual format of the poem might suggest; the visual text is determined in part simply by the exigencies of the codex page. As s., however, in the air, "the poem" is pure soundstream, continuous except for being punctuated at various points by pauses. The poem is thus one long string of ss.; the only reason it is not set as one string of characters on the page is that the page is not wide enough. Similarly, a paragraph in a file on a computer disk is actually stored as one continuous string of characters. This chain is wrapped between margins on the screen only for reading convenience. (The right-margin "soft carriage return" characters which are saved in the file but invisible on screen are in effect invisible rhymes.) This analogy suggests that characteristics specific to the display format are not constitutive of the poem but rather derivative. What is constitutive is the segmentation of the soundstream, i.e. its articulation, the fissures breaking and binding together its features.

II. ARTICULATION, ACOUSTICS, AUDITION. Vocal s. is a very different phenomenon depending on whether one asks about how it is produced (its place and manner of production), how it is structured (its inherent characteristics), and how it is received (processed in cognition). These three sets of phenomena are not congruent. At various times, many prosodists have claimed that one or another of these is relevant to the study of poetry. Here it is important that objective acoustic and linguistic facts be distinguished from subjective perceptions by speakers and auditors (though such perceptions themselves also constitute facts).

Articulation. Any modern introductory linguistics or phonetics textbook will reproduce the trapezoid describing the articulation points of the vowels in the mouth, classed from high to low and front to back, and also the standard chart of consonant groups also based on place and manner of articulation, esp. voicing. Distinctive-feature analysis has been applied to poetic s. to see if patterning can be discovered on that basis. One type of onomatopoeia (q.v.) or expressive s., called kinesthetic (see below), is based on the presumption that the mouth and facial gestures involved in s. production contribute to meaning.

Acoustics. In simplified form, the acoustic facts about the nature of s., as we presently understand them, are as follows. S. is a waveform which passes through (organizes) air (or any medium) from a source to a receiver. Its acoustic dimensions or characteristics are: *frequency, amplitude*, and *intensity*; and one must consider also *duration* and *quality*. Its shape can be visualized as the sine waves one sees on an oscilloscope. One wave (from crest to crest) = one cycle; cycles/sec. = *frequency*, measured in Hertz (Hz). The human voice ranges from ca. 100–200 Hz; middle C is 264 Hz. The human ear can hear ss. in the range from ca. 20 to 20,000 Hz. Frequency correlates (complexly) with perceived pitch: the higher the frequency (the more waves per second), the higher the perceived pitch. Conversely, higher frequency of waves per sec. entails shorter wavelength. The *amplitude* of the wave (the height of its peaks and depth of its valleys) correlates (complexly) with perceived loudness. The human ear is not good at discriminating loudness: a s. must increase in intensity by a factor of 10 to be perceived as twice as loud. *Intensity* (i.e. stress) has to do with the overall power of the s. and is measured in decibels (dB). *Duration* ("length," sometimes "quantity") concerns how long a s. is prolonged or held in speech production: some langs. organize this aspect systematically in phonology, and most of these schematize it in verse as quantitative meter. The *quality* ("timbre," "tone-color") of a s. concerns not the nature of the s. itself but of the source making it: the same note played on a flute sounds differently to us than it does from an oboe or guitar or human voice: quality is thus paralinguistic not linguistic, so it is not clear that it is relevant to the analysis of poetry. Verse systems normally select one of the three features—pitch, stress, or duration—and pattern it systematically as meter; other s. patterning is secondary and compounds the degree of order in the text.

Audition. There have been several significant developments in the study of the audition and mental processing (cognition) of s. which may bear on poetry. The most important concerns the hemispheric specialization of functions in the brain, which leads to the brain's differentiation of modes of s., a phenomenon called "dichotic listening." Lateralization of cerebral functions is by now well established. Acoustic signals are received by the brain and processed according to whether they are linguistic or nonlinguistic, the latter including both ambient environmental ss. and music. The analysis and interp. of lang. ss. is (in right-handed people) a left-brain activity, the same hemisphere responsible for cognition, motor activity, and rational thought. At the same time, musical ss. are interpreted in the right brain

(see Jaynes). Complex aural stimuli such as vocal song, where words and music are delivered simultaneously, are processed on double tracks. Poetry, too, which is coded into sonal patterns which are both lexical-semantic as well as prosodic (e.g. meter, rhyme, assonance), is processed by both hemispheres simultaneously, the former ss. being interpreted by one side of the brain as linguistic and the latter ss. by the other as aesthetic. In short, rhythm in lang. is left-brain but pattern recognition is right-brain. As verse is heard, speech pros. is handled and sense extracted—or created, depending on one's epistemology—by the left brain, while verse-art pros. is handled and aesthetic pattern recognized by the right brain. Both sides of the brain are listening to poetry simultaneously but differently. (Tsur gives a different account of this process, postulating a third, "poetic" mode of speech perception.) It is probable that not all elements in the double code are recognized on only one pass, a fact which legitimizes the close prosodic analysis of poetry as training for heightened response in subsequent readings.

A number of researchers in several disciplines have pursued this concept of differential processing of auditory information. Schaeffer, interested in musical timbre, distinguished two modes of hearing: in ordinary listening, ss. are perceived mainly by identifying their source; the s. itself is only secondarily of interest. But in "reduced hearing" we ignore the source and respond to the inherent properties of the s. itself, as a "s. object" (*objet sonore*). In psychoacoustics, Liberman and his colleagues distinguished "speech mode" from "nonspeech mode": in the former, acoustic signals are reduced to phonological categories, while in the latter, they are not, but some acoustic features can cue both, and in general the coding of acoustic signals is very complex.

III. DECONSTRUCTION. But if s. has been a central interest of modern psychoacoustics and cognitive studies, it has been consciously excluded from mainstream critical thinking in America over the past three decades due to the influence of deconstruction and poststructuralist crit.

Deconstruction is grounded in the work of the Swiss linguist Ferdinand de Saussure, who reminded the 20th c. that the form of the words in any given lang. is purely arbitrary: words have no *natural* or motivated connection to things in the external world. The Eng. word for canines is "dog," but in Fr. it is "chien": this *difference* between langs. shows that the form of any word is determined differentially by each lang. Further, *difference* is also operative within a lang.: in Eng., minimal pairs such as *bit* and *bat* refer to two quite different things by virtue of the fact they differ in one phoneme. Saussure concluded that it is the differential itself which creates and constitutes the meaning of two words, not, as here, their identities (in *b* and *t*), and that it is therefore not the *presence* of features in the words of a lang. but the systematic differentials between them that makes meaning possible at all. Readers of Nietzsche will remember his argument that if presence is a reality, absence is too. Derrida later used this reasoning to attack the notion of *logocentrism*, i.e. that lang. is grounded in anything outside itself which would authorize meaning; for him, there is nothing outside of the text. By this he meant to dismantle any metaphysics of presence underpinning all Western thought.

It was a mistake, however, for Saussure to focus on the differential features of lang. to the exclusion of features of presence. Rather, one wants an account in which both presence and absence are held in tension as antinomies, as opposites which require each other in order to exist at all, and in the absence of one of which the other cannot function. Further, one must also recognize that the particular forms of *bit* and *bat* were derived not out of the blue but by historical process, via the regular laws of etymological development and s. change at work in lang., e.g. Grimm's Law. Words, like people, have a historical life: they change their shape over time in response to both internal and cultural forces, and they beget progeny recognizably theirs.

It is indubitable that the word *bat* has no necessary connection to cylindrical wooden objects made by the Louisville Slugger Company for the purpose of hitting balls to left field; any other word would do so long as we agreed by convention to use it. What we should conclude from this, however, is not that there is *no* attachment between our words and the world, nor even that we make the world, i.e. that we live in a self-composed set of fictions, for in fact we know that words do in some sense generate constructions that do at times accurately reflect the external world as humans experience it. At times they do not—but at times they do. This is true for words in exactly the same sense that, in other representational systems, such as mathematics, one can construct equations that lead to predictions that turn out to be correct. Such equations do in fact describe aspects of reality as we know it. It is certainly possible to construct other sets of equations that do not describe this reality but rather some other realities. All well and good. But the more salient fact is simply that some equations *can* be constructed that are indisputably accurate representations of *this* reality.

Unfortunately, however, we cannot say that a sentence in lang. is precisely equivalent to an equation in mathematics in any easy or obvious way, and it is agreed that the shapes of words do not represent things in any directly reliable way. The reliability we sense in lang. is therefore constitutive not of any single word or even string of words but rather at the level of the system. The "fit," insofar as there ever is a fit between words and the world, must be a homology between the two systems taken as wholes, not between any specific elements of the systems. It is idle, on this

account, to fault a net for having holes. "Since lang. is form and not substance," says Culler, "its elements have only contrastive and combinatorial properties" (*Saussure* 49). It was the former of these that Derrida chose to emphasize. But in so doing he ignored the latter, and one might well counter that lang. does not after all function like an anchor or rivet gun to the world, but rather like a net, in which the holes are indeed large and many—but not the means by which it does its work.

Moreover, lang. is not solely a representational system; it is a communication system having a variety of functions such as the expression of emotion or the certification of social process. In ordinary speech, the representational function is often primary, but in poetry, that function is both reduced by the heightened level of phonological patterning and of s. effects unusual to speech and prose, and also augmented by other functions, producing a more comprehensive and synthetic mode of speech. The referential function of lang. is capitalist: it exploits the usefulness of words only to discard them. The poetic function of words however grants them their being; foregrounds their nature, their *quidditas*; and cherishes their quirky uniqueness, their *haecceitas*.

In the course of his attempt to show the impossibility of our ever standing outside of lang. so as to validate any intention or determinate meaning, Derrida denied the primacy of speech over writing in lang. This Derridean emphasis on *l'écriture* (writing) over *lecture* (speech) reversed the traditional and Saussurean identification of s. as primary in lang. For more than a decade after the pub. of the Eng. tr. of *Of Grammatology* in 1976, poststructuralist critics focused on writing and textuality, acting as if s. did not exist, at least in theory. Stewart claims of Derrida that there is "an emphasis on the phoneme more prevalent in his practice than in his theory." Perhaps so. In any case, Derrida, reacting to formalist notions of expressive lang. and intentional meaning, nevertheless went too far in the other direction. If all is lang., then all is certainly text. But phonemic patterning still remains in the structure of the text even when authorizing voice is subtracted, and it is easy to show, as Stewart does (see esp. 103–6), that it is s. alone which is responsible for many of the playful, homophonic textual effects so dear to poststructuralist critics like de Man.

What should be of interest to literary critics is therefore not the question of whether or how words attach to the world but what they are capable of in consort—i.e. what effects they have on each other and, more importantly still, what effects they are capable of when these mutual interanimations are complicated, enriched, compounded, heightened, and made manifest by pattern—in short, when lang. is made over into poetry. The casting of ordinary lang. into poetic form generates meanings and effects *not otherwise possible* absent that form; these are the whole reason for having poetry at all. Hence we should view poetry not as an imitative or mimetic art but as a *constitutive* one.

IV. PATTERN. We are left, then, with the fact—not under erasure—of pattern, of design. S. patterning in poetry covers a wide range of important functions ranging from the aural "tagging" of syllables in semantically important words in the line, to the tagging of thematically important words in the poem, to even more extensive and formalized structures. Alliteration in particular is simply the most conspicuous manifestation of a broad-scale process of semantic underlining. S. patterning often highlights a sequence of key terms central to the thematic progression of the poem. In Shakespeare's sonnet 129, for example, sonal repetition binds and points up the series "purpose, pursuit, possession, proof, proved, proposed" as a metonymic litany of the evils of "lust in action," itself already marked fivefold as "hated . . . bait . . . laid to make the taker" mad. Lynch found that, in some poems, key components of s. patterns important throughout the poem, like the one above, were even brought together in a thematically "summative word."

In Old Germanic and OE poetry, alliteration patterning of halflines is even made part of the meter, vestiges of which lingered for centuries in ME alliterative verse, Ren. artverse, and post-Ren. popular verse and song. Correlate structures appear in several other verse- systems, incl. Welsh *cynghanedd*, where a sequence of consonants in the first hemistich is repeated in the second; and the binary and ternary forms of the Fr. alexandrine, whose metrical structures may be confirmed and enhanced by s. patterning. S. patterns within the line may coincide with (also contradict) metrical stress, link halflines, heighten the symmetry of the line, and augment or complicate other metrical patterning (Chesters).

It is not required that a s. pattern be consciously perceived for it to have effect, nor is it required that the words even make sense, as one can see in both nonsense verse and Gerard Manley Hopkins. Pattern organizes, highlights, and intensifies meaning in all verbal strings. This suggests that it is necessary, even essential, to study pattern, and dangerous to study it in isolation from meaning for very long. Despite some very elaborate taxonomies of s. patterning set forth in the past by scholars such as Masson, no taxonomy which is complex has yet achieved wide acceptance. It seems more profitable to identify only a few relatively simple processes with broad application on a variety of levels of the text. These would presumably include processes such as *sequence* (ABC > ABC), *chiasmus* (ABC > CBA), *alternation* (ABAB), *envelope* (ABBA), and all the secondary mixtures and complications of these simple or primary forms: a pattern expanded, a pattern contracted, a pattern inserted into another pattern, two patterns alternated, two or more patterns interlaced in more complex ways, and patterns with subpatterns

which become motifs to be repeated, varied, and embellished. Masson in 1953 identified two archetypal s. patterns, sequence and chiasmus, and in 1960 suggested that strength or intensity of patterning could be quantified in terms of "bond density." Laferrière in 1974 explored some of the processes listed above.

Study of s. patterning infringes neither the deconstructive critique of authorizing "voice" in lang. nor the formalist "intentional fallacy." Intention there may or may not be, but design there certainly is, and design inevitably has an effect different from no-design. We may not always be able to specify what that effect is, and the effect may not be identical from one reader/auditor to another, but effects there are, and at least some demonstrably result from textual and performative cues. Stability of response to heightened textual organization is not to be wished away. We must attend to both cues and effects if we wish to say anything accurate about how readers read poetry and what happens when they do.

V. EXPRESSIVENESS. The issue of whether ss. bear meaning directly, over and apart from lexical meaning, is disputed. Traditionally, prosodists pointed to onomatopoeia (q.v.), often derided by other critics, but after Peirce, the concept of iconicity has been taken seriously in semiotics. There is good cross-cultural evidence in linguistics that mimetic processes in lang.—phonological, morphological, and even syntactic—are more diverse and extensive than was previously thought. But a number of modern critics have reacted against the older view that ss. could in themselves be "expressive" of meaning, esp. since lexical meaning is never absent to begin with. I. A. Richards constructed a "dummy" of stanza XV of Milton's Nativity Ode to show that ss. really only reinforce meaning, not create it; and John Crowe Ransom parodied Tennyson's "The murmuring of innumerable bees" as "The murdering of innumerable beeves" to show that even when the overall s. pattern is changed only very slightly, the meaning of the words is changed radically and the "mimetic" effect nearly obliterated (see REPRESENTATION AND MIMESIS).

On the whole, one surmises, mimetic effects in poetry most often reinforce lexical sense, so that the strong presumption will remain that when critics assign expressive values to a s. or s. pattern, they are usually responding more immediately to effects that are lexically induced. In short, the s. pattern discovered is made to fit the interp. rather than the reverse. But a more constructive approach is certainly possible: one can view such effects as providing not the same kind of sense as lexical, but rather a different kind on a different level. In Frost's "Desert Places," for example, where the scene is winter and the first two words are "Snow falling," 16 of the 40 syllables in the first stanza contain sibilants. These, as anyone who knows winter knows, are the ss. that snow makes. The poem is talking about snow, certainly, but at the same time it is *making* snow ss. "Mimetic" ss. are not, therefore, representational but presentational: they add to lexical meaning the enactment of that meaning. Poetic s. always focuses attention on itself even as it delivers meaning, and in this respect, it increases the palpability of the sign and so widens the division between signs and objects (Jakobson), making poetic lang. in fact not more representational, in the simplest sense, but less.

But the range of expressiveness of ss. includes not only imitative ss. but also the further range of associative phenomena, wherein the mind generates meanings by making associations between ss. as objects with other objects, or between ss. as one kind of sensation with other kinds of sensory experience or other mental phenomena (memory, desire, images, ideas). Association probably counts for a great deal in all cognition. One may object that associations are transient, creative, and unpredictable, but they are no less real or important for that, and there is evidence to suggest that they are considerably less idiosyncratic than is usually thought. On the breath, s. is a physical entity and has, like every other real entity, dimensionality: size, length, shape. Dante even classes words as "combed out," "shaggy," "glossy," and "rumpled." It is a demonstrable fact that auditors perceive ss. to have aspects of physical shape which they correlate with those of other physical objects. These correlations are associative.

If many expressive ss. are not imitative but associative, then there is no *necessary* connection between the nature of the s. and the terms used to describe it; this follows from the Saussurean doctrine of the arbitrariness of the sign. On the other hand, empirical testing reveals a good deal of common response across cultures to certain types of associations. Some of this stability of association appears to be built into the structure of lang., and may be—since all speakers worldwide have essentially the same articulatory mechanism in the throat—lang.-independent. Other aspects are lang.-specific, however, for langs. of course differ in the size and variety of their phonetic inventories. Yet other associative processes occur not in speakers' cognitive processing but in lang. itself: both Bolinger and Malkiel have shown that words have not only rich and complex diachronic lines of affiliation but equally extensive relationships synchronically, so that at any point in time, some words are exerting a gravitational pull on others, influencing them lexically (semantically), in s. shape, and even orthographically. Malkiel calls this process, aptly, "morphosymbolism."

To describe and to articulate difference between the ss. we hear, we seem all but forced to choose descriptive adjectives—e.g. light, dark, velvety, red—from the other senses. The literary form of this phenomenon is known as synaesthesia (q.v.). Such cross-sensory description is far from being merely metaphorical: there is some evi-

dence that the neural centers in the brain which process sensory stimuli may share information. It is certain that memory combines such information. The operation of this process in poetry is akin to "s. painting" in music, when ss. or their patterns evoke sensory impressions of geography or topography, e.g. thin, steep, high, low, sunken, vast, open, broad, dark, gloomy. Romantic and symbolist poets of the 19th c. in England, France, Germany, and Russia took a keen interest in the associative and expressive values of s. and in synaesthesia. Both A. W. Schlegel and Rimbaud associated vowels with colors. This interest infected newly-emergent psychology, producing a barrage of works in the late 19th and early 20th c. Later scholars such as Delbouille, however, have sharply criticized much of this work as fanciful, arbitrary, and of dubious linguistic accuracy.

One final type of expressive s. is kinesthetic. Here the feel of the s. in the vocal tract or mouth, or the shape of the mouth, or the facial expression produced in making the s. evokes or is associated with an emotion or meaning suggested by the words. Plosives and sibilants, for example, can evoke actual spitting and hissing in relevant contexts, as when Adam turns on Satan in *Paradise Lost*: "Out of my sight, thou Serpent, that name best / Befits thee with him leagu'd, thy self as false / And hateful."

VI. AESTHESIS. The aesthetic function of poetic s. for the auditor or reader evokes the instinctive pleasure of articulating or hearing ss., or of perceiving s. patterns, or of the repetition of s. This is a pleasure of the mouth and ear, and of the right brain, the intuitive side. Soundplay arises naturally from ordinary lang., where it is prominent in children's games, nonsense verse, lullabies, college cheers, advertising jingles, magical charms, chants, and of course song. In prose, it appears in Attic and baroque rhythmical and rhymed prose (see PROSE RHYTHM) and in such movements as euphuism and mannerism. In artverse, however, literary critics have often taken the view that s. patterning which is not closely tied to sense, or which outstrips sense where sense is stale or thin, or which seems cultivated at the expense of sense is the mark of an inferior poet (e.g. Poe, Swinburne, Dylan Thomas). Such critics often associate excessive s. repetition with the soundplay of childhood, with mental derangement (Smart, Hölderlin) or altered states of consciousness, or with literary attempts to imitate either or both (Blake). Nevertheless, in every age there have been movements in poetry toward the condition of music, i.e. pure patterning of s. without regard for meaning (see PURE POETRY), just as there have been opposed movements toward poetries of clear and direct sense—narrative, descriptive, dramatic. The persistence of all such movements testifies to the fact that poetry is both sense and s., and that artistic and critical fashions will oscillate perennially between the two.

VII. MUSICALITY. Discussions of poetic s. as "musical" have a pedigree in the history of crit. which has been productive in inverse relation to its length. Since s. structuring is systematized in music, and since the relations of music to poetry are as old as accompanied song, the use of musical terminology and even sigla to describe poetic s. has seemed natural and appropriate. Critics who have taken an interest in poetic s. have usually followed the ancient practice of construing s. phenomena on the analogy of music, i.e. as some version of "harmony," "mellifluousness," "melodiousness," "euphony," or their opposites, "cacophony" or "dissonance." This has not been a productive approach, however, for the kinds of structuring of s. characteristic of music are not, as Northrop Frye pointed out, those of verse, precisely. Further, even music itself is a very complex system of structuration of s. (see MUSIC AND POETRY). S. in lang. is very different from s. in music, not to mention their mixed forms: recited verse is one thing, sung verse quite another, and verse recited to the accompaniment of music yet another. It was Wimsatt who remarked that "the music of spoken words in itself is meager"; "the art of words is an intellectual art, and the emotions of poetry are simultaneous with conceptions and largely induced through the medium of conceptions." Like much in Wimsatt, this is too severe, but it aims in the right direction.

VIII. ANALYSIS. Discussion of s. is an important topic among the ancients. The arbitrariness of lang. was of course a concern of Plato in the *Cratylus*; other important ancient commentators on s. and poetry incl. Aristotle (*Rhet.* 3.9.9–11); Demetrius (*On Style*), who uses "parhomoeosis" as the generic for all types of s. correspondence and who discusses such topics as imitative effects, cacophony in Homer, and distinctions between "smooth" and "rough" words; and Dionysius of Halicarnassus (*On Literary Composition*, chs. 14–16, 23; tr. Roberts), who discusses aspects of word choice, esp. s., incl. such phenomena as sigmatism and imitative words. Many theories of poetic s. in antiquity attempt some kind of calculus of euphony, some mechanism whereby s. combinations are ranked on a scale. In Cl. rhet., figures were traditionally divided into the schemes and tropes; the former were said to be arrangements or figures of s., the latter of sense. So alliteration, a scheme, is a s.-pattern, while metaphor, a trope, is a relation of idea or thought. This dichotomy, however, which itself varied greatly over time, is too simple: schemes also schematize meaning, and tropes rarely appear in poetry without additionally schematized s. Both Cicero and Quintilian (*Institutio oratoria* 9.3.75–80, 9.4) treat s. under the rubric of prose rhythm (q.v.). Expressive s. is as significant a topic for Dante (*De vulgari eloquentia*) as it is for Baudelaire.

In the 20th c., empirical studies of poetic s. patterning have appealed to some scholars, and in general it is undeniable that extensive inventories

of accurate data on such s. phenomena as rhyme and alliteration simply do not exist at present. But statistical data in isolation is of little use without interp. Empirical and statistical work—i.e. quantitative analysis via computer—yields only narrow results, and must be interpreted with care, for a great deal of the specific data thus generated is trivial—the background noise of the channel. And s. patterns themselves must always be correlated with meaning, a process which must always require the application of critical judgment. It would seem that, for most purposes, inventories of s. figuration treated in isolation from meaning are not very revealing. Certainly the most fruitful modern studies of rhyme, at least (Wimsatt, Jakobson), have seen that phenomenon as irrefrangibly sonal and semantic. Analyses attempting to integrate s. and meaning on several levels of sonal architecture have been few; only recently have some theorists (e.g. Lotman) begun to integrate s. patterning into a larger, unified field theory of poetic structure.

Some empirical researchers have made the mistake of assuming that counting justifies itself, hence that empirical work is theory-free. Not so: no work at all is possible without theory, for counting presupposes deciding what is to be counted, and what is to be counted is thereby already reified: those decisions constitute theory. Data itself is never theory-free. Nor is it of use without consideration of reader response. Frequency counts and phoneme inventories can reveal which s. patterns exist, certainly, but not which ones are perceived, i.e. *salient.* Some s. patterns either go unnoticed or count for little if they are separated by too great a distance. Some "rhymes" in *Paradise Lost,* for example, are so far apart that most readers are not aware they exist. It must always remain to be shown that a statistically significant pattern is perceived or contributes to meaning. Here one would want to say that it is precisely training in the analysis of poetry which heightens readers' abilities to recognize and respond, more fully, to s. patterning in versified texts.

All analysis must be based on accurate diachronic linguistics, as verified by available evidence from historical phonology; the rhymes in dialect poetry, for example. cannot be understood at all without linguistic study of that dialect. But linguistics alone cannot tell us all we need to know about poetic form, for even such seemingly obvious s. patterns as rhyme or alliteration vary in definition from one lang. to another and, within one lang., from one metrical system to another. What constitutes a particular form of poetic s. is conventional and changes from age to age and poetry to poetry.

Each lang. differs in the phonological resources it offers for aesthetic design. The Germanic langs. (incl. Eng.) are dense in consonant clusters and tend to forestressing, hence consonants have less weight there than the same ones would in the Romance langs., where stress is distributed more evenly and vowels more important. A full frequency analysis of the distributional patterns for a large number of langs. would be of value, though it must be remembered that artverse only selects from a narrow register of the lexicon in any lang. Further, within any single poem, a given pattern only has its effect against the ground provided by the local environment of one line and its neighbors, where much is possible. In general, poetic s. effects must be felt, and assessed, against the whole phonology of a lang., the principles of the verse system, the particular verseform presently in play, and the semantic structure of the specific and adjoining lines. That is, the nature and effect of any given structure are constrained in poetry not by fewer systems of conventions but by more.

IX. FULLNESS. Poetry teaches us that verbal experience is fundamentally double. On the most direct level, there is meaning, but underneath there is s. In ordinary speech, we listen only for the meaning: the s. of words is wholly transparent—we listen straight through it. Most of the time, that is, we live, in languaging, on the level of meaning. The ss. of words are ignored by the meaning-making mind when processing s. as lang. Poetic lang. of course bears meanings also. But in poetry, s. has a life of its own, purely on the level of s., structuring meanings, shading and nuancing meanings, and organizing the temporal experience of reading via patterned repetition. These are not epiphenomena, but full components of cognition; lang. without them is not merely less but *other.*

Not to attend to s. in poetry is therefore not to understand poetry at all. But it is important at the same time that we not conceive the concept of "s." too narrowly, and certainly not merely within whatever framework the currently reigning version of linguistics happens to valorize. Note that Ger. distinguishes between *Klang,* the total sonal impression produced by a piece of lang., and *Laut* or *Tön*, each specific s.: the former would approximate Eng. "intonational pattern," i.e. the amalgam of accent, pitch, speed, timbre, and rhythm—s. taken at the full. Mallarmé remarks in "Crise de vers" that "le vers qui de plusieurs vocables refait un mot total, neuf, étranger à la langue et comme incantatoire, achève cet isolement de la parole: niant, d'un trait souverain, le hasard demeuré aux termes malgré artifice de leur retrempe alternée en le sens et la sonorité, et vous cause cette surprise de n'avoir oui jamais tel fragment ordinaire d'élocution, en même temps que la réminiscence de l'objet nommé baigne dans une neuve atmosphère." That clarified space may recall to us what T. S. Eliot once called the "auditory imagination," that fully sentient consciousness which responds to the ss. of poetry in the richest sense, finding them more fully articulate in their joint effect and interanimations, so that the saying is accomplished more fully.

When Wallace Stevens said, in his 1941 lecture

on "The Noble Rider and the S. of Words," that "a poet's words are of things that do not exist without the words," he was saying that words are not secondary to our experience of the world, they are primary. The words in poetry are words not because they express meaning—they do that in prose—but because they are also s., because they take their life in s. It is as s. that they teach us what words are.

See now ALLITERATION; ASSONANCE; ONOMATOPOEIA; REPETITION; STANZA; STRUCTURE; SYNAESTHESIA; VERSE AND PROSE.

BIBLIOGRAPHIES: No full comparative bibl. exists. Thieme 372—full bibl. for Fr. to 1916; Brogan, esp. pp. 53–108—comparative survey of studies to 1981, supp. in *Verseform* (1989).

STUDIES: I. A. Richards, *Practical Crit.* (1929); C. P. Smith, *Pattern and Variation in Poetry* (1932); Patterson; S. Bonneau, *L'Univers poétique d'Alexandre Blok* (1946); A. Spire, *Plaisir poétique et plaisir musculaire* (1949); D. T. Mace, "The Doctrine of S. and Sense in Augustan Poetic Theory," *RES* 2 (1951); D. I. Masson, "Patterns of Vowel and Consonant in a Rilkean Sonnet," *MLR* 46 (1951), "Vowel and Consonant Patterns in Poetry," *JAAC* 12 (1953), "Word and S. in Yeats's 'Byzantium,'" *ELH* 20 (1953), "Free Phonetic Patterns in Shakespeare's Sonnets," *Neophil* 38 (1954), "Wilfred Owen's Free Phonetic Patterns," *JAAC* 13 (1955), "Thematic Analysis of S. in Poetry," *PLPLS-LHS* 9, pt. 4 (1960), "S. Repetition Terms," *Poetics—Poetyka—Poetika*, ed. D. Davie et al. (1961), "Poetic S.-Patterning Reconsidered," *PLPLS-LHS* 16 (1976)—this last a survey of 8 national lits.; A. Oras, "Surrey's Technique of Phonetic Echoes," *JEGP* 50 (1951), "Echoing Verse Endings in *Paradise Lost*," *So. Atlantic Studies S. E. Leavitt* (1953), "Intensified Rhyme Links in *The Faerie Queene*," *JEGP* 54 (1955); S. S. Prawer, *Ger. Lyric Poetry* (1952); J. J. Lynch, "The Tonality of Lyric Poetry," *Word* 9 (1953); A. Stein, "Structures of S. in Milton's Verse," *KR* 15 (1953); H. Kökeritz, *Shakespeare's Pronunciation* (1953); H. W. Belmore, *Rilke's Craftsmanship* (1954); F. Scarfe, *The Art of Paul Valéry* (1954); W. K. Wimsatt, Jr., "One Relation of Rhyme to Reason," *The Verbal Icon* (1954); Wellek and Warren, ch. 13; J. Hollander, "The Music of Poetry," *JAAC* 15 (1956), *The Figure of Echo* (1981); Frye; K. Burke, "On Musicality in Verse," *Philosophy of Literary Form* (1957); S. Chatman, "Linguistics, Poetics, and Interp.," *QJS* 43 (1957); *S. and Poetry*, ed. N. Frye (1957), esp. A. Oras, "Spenser and Milton: Some Parallels and Contrasts in the Handling of S."; D. Hymes, "Phonological Aspects of Style," in Sebeok; N. I. Herescu, *La Poésie latine: Étude des structures phoniques* (1960); P. Delbouille, *Poésie et sonorités*, 2 v. (1961, 1984); I. Fónagy, "Communication in Poetry," *Word* 17 (1961), *Die Metaphern in der Phonetik* (1963), "The Functions of Vocal Style," *Literary Style: A Symposium*, ed. S. Chatman (1971), *La vive voix: Essais de psychophonétique* (1983); L. P. Wilkinson, *Golden Lat. Artistry* (1963); D. Bolinger, *Forms of Eng.* (1965)—morphosemantic effects; J. Levý, "The Meanings of Form and the Forms of Meaning," *Poetics—Poetyka—Poetika*, ed. R. Jakobson et al. (1966); W. B. Stanford, *The S. of Gr.* (1967); A. M. Liberman et al., "Perception of the Speech Code," *Psych Rev.* 74 (1967); P. Schaeffer, *Traité des objets musicaux* (1968); E. J. Dobson, *Eng. Pronunciation, 1500–1700*, 2d ed., 2 v. (1968); B. Hrushovski, [Do Sounds Have Meaning? The Problem of Expressiveness of Sound-Patterns in Poetry], *Ha-Sifrut* 1 (1968), "The Meaning of S. Patterns in Poetry," *PoT* 2 (1980); A. A. Hill, "A Phonological Description of Poetic Ornaments," *Lang&S* 2 (1969); R. W. Bailey, "Statistics and the Ss. of Poetry," *Poetics* 1 (1971); N. Geschwind, "Lang. and the Brain," *Scientific Am.* 226 (1972); P. Ostwald, *The Semiotics of Human S.* (1973); E. D. Polivanov, "The General Phonetic Principle of Any Poetic Technique," *Sel. Works*, ed. A. A. Leontev (1974); D. Laferrière, "Automorphic Structures in the Poem's Grammatical Space," *Semiotica* 10 (1974); F. W. Leakey, *S. and Sense in Fr. Poetry* (1975); J. Jaynes, *The Origin of Consciousness in the Breakdown of the Bicameral Mind* (1976); C. L. van den Berghe, *La Phonostylistique du français* (1976); Derrida; J. Lotman, *The Structure of the Artistic Text* (tr. 1977), esp. 106 ff., 178 ff.; M. Kaimio, *Characterization of S. in Early Gr. Lit.* (1977); Y. Malkiel, "From Phonosymbolism to Morphosymbolism," *Fourth LACUS Forum*, ed. M. Paradis (1978); R. Jakobson and L. Waugh, "The Spell of Speech Ss.," *The S. Shape of Lang.* (1979), rpt. in Jakobson v. 8; V. Erlich, *Rus. Formalism: History–Doctrine*, 3d ed. (1981); R. P. Newton, *Vowel Undersong* (1981); L. O. Bishop, *In Search of Style* (1982); R. Lewis, *On Reading Fr. Verse* (1982), esp. chs. 4, 7; R. Chapman, *The Treatment of Ss. in Lang. and Lit.* (1984); J. C. Ransom, "Positive and Near-Positive Aesthetics," in Ransom; Hollander, esp. chs. 1, 4; B. Scherr, "Instrumentation," in Terras; C. Scott, *The Riches of Rhyme* (1988); G. Chesters, *Baudelaire and the Poetics of Craft* (1988); G. Stewart, *Reading Voices* (1990); R. Tsur, *What Makes S. Patterns Expressive?* (1992).

T.V.F.B.

SOUND POETRY is the performance intermedium in which verbal and sound art are not just mixed, as in a song, but are actually fused. While any poetic text, when read aloud, employs sound elements to reinforce lexical sense, when sound for its own sake becomes the principal expressive medium, sometimes even at the expense of lexical sense, then it becomes meaningful to describe a work as s. p.

Many traditional forms of oral poetry (q.v.) have used incantations, sometimes even nonsense syllables, which function similarly to s. p., as in the "heigh nonnie no" refrains of many Eng. folksongs, though more commonly they function onomatopoetically, as in the healing chants of some Am. Indian tribes. This is also true of early

poetic passages such as the famous "brekk kekk kekk kekk koax koax" chorus in Aristophanes' *The Frogs* (4th c. B.C.). The artistic potential for s. p. was recognized in antiquity; the early Lat. poet Quintus Ennius (239–169 B.C.) wrote the earliest known passage, a tautogram: "O Tite tute, Tati tibi tanta tyranne tulisti." This sort of alliteration run wild reaches its zenith (or nadir) with the 9th-c. Benedictine, Hugobald, who wrote a poem of 146 hexameters for Charles the Bald praising baldness, the *Ecloga de Calvis*, every word of which begins with a C.

The few 19th-c. pieces of s. p. are almost always light verse, such as Robert Southey's "The Cataract of Lodore." Early in the 20th c., futurists such as F. T. Marinetti and F. Cangiullo, dada poets such as Raoul Hausmann, Rus. constructivists such as V. Xlebnikov, and, above all, the independent Ger. artist K. Schwitters (1887–1948) in works such as the *Ursonata* (1925–27) made s. p. a medium for major works.

While phonograph records were made of some early modern s. p. experiments, the expansion of the field came only with the advent of wire and tape recorders in the years following World War II. Until then almost all s. p. had distinguished performances from text, the text serving the function of a musical score for the performer to follow. Works employing this distinction (and many are still being composed) are now known as "text-sound" works (see Kostelanetz).

But in the 1950s a second variety of s. p. became possible, one in which recorded sound, primarily but not exclusively verbal, is manipulated. The final result is not a performable, written notation or text, but a recording, either intended to be played alone or with another performance occurring over it. Such pieces are now known as "audio poems" (Chopin's term). In the 1950s and 1960s many new s. poets emerged, such as H. Chopin (b. 1922), B. Heidsieck (b. 1928) and F. Dufrêne (1930–85) in France, B. Cobbing (b. 1920) in England, F. Mon (b. 1926) in West Germany, G. Rühm (b. 1930) in Austria, and A. Lora-Totino (b. 1928) in Italy, to name only a few. Also, many concrete poets (see CONCRETE POETRY), musical composers, and visual artists have done important work in s. p., and younger poets continue to be attracted to the genre: a complete global listing of serious s. poets would run to hundreds of names. Since the late 1960s, major festivals of s. p. have developed the audience for it in a dozen countries.

Mention should also be made of the interaction of s. p. with the *hörspiel*, which was originally the ordinary Ger. word for "radio play." However, in several European countries, particularly West Germany, and esp. at Westdeutscher Rundfunk in Cologne, several works are commissioned and broadcast each year which are *neues Hörspiel*—sustained audio-acoustical poems written not just by Germans but by artists of other nationalities as well, such as the Am. composer John Cage (b. 1912), whose four *hörspielen* are among several dozen by Americans which have been broadcast and recorded, but for which there is as yet no available public medium in the USA.

STUDIES: E. Jolas, "From 'Jabberwocky' to 'Letterism,'" *Transition* 48 (1948); A. Liede, *Dichtung als Spiel*, 2 v. (1963); "Internat. Electronic Music Catalog," *Electronic Music Rev.* 2–3 (1967)—lists all known audio poems to 1967; *Neues Hörspiel*, ed. K. Schöning (1969); K. Schwitters, *Das literarische Werk*, ed. F. Lach, 6 v. (1973–82); *S. P.: A Catalog*, ed. S. McCaffery and B. Nichol (1978); *Text-Sound Texts* (1980), and Spec. Issue on s. p., *Precisely* 10 (1981), both ed. R. Kostelanetz; H. Chopin, *Poésie Sonore Internationale* (1981)—book with accompanying cassettes (new ed. will have expanded bibl. and discography); J. Cage, *Roaratorio: ein irischer Circus über* Finnegans Wake (1982)—book and cassette; J. Rothenberg, *Technicians of the Sacred*, 2d ed. (1985); D. Higgins, *Pattern Poetry* (1987)—sect. on s. p. before 1900; R. Döhl, *Das Neue Hörspiel* (1988). Periodicals devoted primarily to s. p. were *Ou*, ed. H. Chopin (1963–68), and *Stereo Headphones*, ed. N. Zurbrugg (1972–)—most issues incl. records.

DISCOGRAPHY: L. Greenham, *Internationale Sprachexperimente der 50/60er Jahre* (1970)—shows how concrete poetry converges with s. p.; *Futura: Poesia Sonora*, ed. A. Lora-Totino, Cramps 5206-301 to 307 (1978)—six-record internat. anthol. with book incl. notations and other information; *Text Sound Compositions*, RELP 1049, 1054, 1072–74, 1102–03 (1968–70)—seven records of the Stockholm festivals of s. p. There are also cassette series of s. p., notably the "New Wilderness Audiographics" series (1975–). D.H.

SPENSERIAN STANZA. The stanza invented by Edmund Spenser for his *The Faerie Queene* (1589–96; 6 Books completed), composed of 9 iambic lines, the first 8 being pentameter and the 9th a hexameter or alexandrine, rhyming *ababbcbcc*. The S. s. bears some similarity to It. *ottava rima* and the Occitan-Fr. *ballade*, and also owes some debt to *rhyme royal* (q.v.) as used by Chaucer, but in Spenser's hands it becomes distinctive, and one of the most original metrical innovations in the history of Eng. verse. The stanza is perfectly suited to the nature of Spenser's poem—at once dreamlike and intellectual, by turns vividly narrative and lushly descriptive—for it is short enough to contain sharply etched vignettes of action and yet ample enough to lend itself to digression, description, and comment. The cross rhyming interlocked by the medial *bb* couplet gives the stanza unity, and the final alexandrine lends its greater weight to effects of closure, which Spenser counterbalances however with concatenation, linking otherwise integral stanzas into a longer sequence.

The S. s. fell into disuse in the 17th c., although variants of it occur in the work of Giles and Phineas Fletcher, and, later in the largely forgotten alle-

gorical narratives of the philosopher Henry More. Some 18th-c. poets revived the stanza, chiefly William Shenstone (*The Schoolmistress*, 1742), James Thomson (*The Castle of Indolence*, 1748), James Beattie (*The Minstrel*, 1771–74), and Robert Burns (*The Cotter's Saturday Night*, 1785). But it is the Eng. romantics who made the stanza once again a major vehicle. Wordsworth has 6 poems in the S. s., including "Guilt and Sorrow," but it remained for the younger romantics to produce verse in Spenserians comparable to *The Faerie Queene*. Byron divides the stanza often so as to achieve frequent shifts in tone in *Childe Harold's Pilgrimage* (1812, 1816). Shelley's *Revolt of Islam* (1818) and *Adonais* (1821) show their author to be the greatest master of the form since its creator. Keats in *The Eve of St. Agnes* (1820) revives the rich sensuousness associated with Spenser's whole art. After Keats the S. s. was little used except by Tennyson for "The Lotos-Eaters" and by William Cullen Bryant for "The Ages" (1821), praised by Poe.—Schipper; H. Reschke, *Die Spenserstanze* (1918); T. Maynard, *The Connection Between the Ballade, Chaucer's Modification of It, Rime Royal, and the S. S.* (1934); Empson 33–34; P. Alpers, *The Poetry of* The Faerie Queene (1967), ch. 2; E. Häublein, *The Stanza* (1978), 31-33; Brogan 453–55 for other citations; W. Blissett, "Stanza, S.," and B. M. H. Strang, "Lang., Gen.," *The Spenser Encyc.*, ed. A. C. Hamilton et al. (1990). A.PR.; T.V.F.B.

SPLIT LINES, shared lines; rarely, "amphibious," "broken." In dramatic poetry, a means for accommodating at once both the importance of rapid change of speakers, at times, and the necessity of making verse lines that are metrical wholes, by splitting one line between two or among three or more speakers. Whole-line exchanges between characters do not sufficiently allow for the rapid give-and-take of emotional or energetic scenes. Shakespeare shows steadily increasing use of s. l. over his career—the late plays show incidences of from 15 to almost 20%, peaking in *Antony and Cleopatra* and *Coriolanus*—and, concomitantly, the point of break in the line steadily moves to the right. Shakespeare certainly knew the Cl. device of stichomythia, which is common in Gr. comedy (but not tragedy), but may also have learned the technique of short and s. l. from Virgil, whose use of broken lines and hemistichs was well known in the Ren. In the theater the effect of s. l. is increase of speed and liveliness without yet disrupting the rhythmic flow.—M. G. Tarlinskaja, *Shakespeare's Verse* (1987), ch. 4; G. T. Wright, *Shakespeare's Metrical Art* (1988), ch. 8. T.V.F.B.

STANZA. The notion of end rhyme and the notion of s. are all but reciprocal: the vowel and final consonant of the rhymed words are repeated, and the repetition not merely identifies line ends clearly but also produces a sound structure, a relationship among lines. The rhyming of adjacent lines produces couplet; larger structures are made possible by the intermixture of further rhymes in complex patterns. Ss. may be isometric, i.e. made up of lines of the same length, or heterometric, i.e. made up of lines of differing lengths. The relationship of syntax to line (q.v.) may vary, within limits, as much in s. as in blank verse (q.v.). Syntactic variation, i.e. changing the grammatical structure of lines so that differing syntactic elements are linked by rhyme, is one of the poet's chief means of achieving variety of effect within and between ss.

Once established, the metrical structure of the s., including the rhyme pattern, tends to be repeated exactly in subsequent ss. Indeed, if we are uncertain of the metrical structure of a line in the first s. of a poem written before the 20th c.—as we might well be of the opening line of Donne's "Twicknam Garden," for example, or the third line of Blake's "Ah! Sun-flower"—we can ordinarily resolve the problem by looking at the corresponding line of a later s. Failures of correspondence occur, of course, even in traditional verse; usually they serve some expressive purpose. Most exceptionally, rhyme can be used throughout a poem in emergent and constantly varying (i.e. irregular) patterns, as in *Lycidas* or "Dover Beach," or, more nearly regularly, as in Herbert's "Man."

Much 20th-c. verse is quasi-stanzaic; that is, it looks on the page as though it were composed of groups of lines (approximately) equal in number, the corresponding lines in each group roughly equal in length. There may be a rhyme pattern, strictly or loosely maintained; or the line ends may be marked by lighter sound repetitions. On the other hand, all forms of rhyme may be avoided. In traditional s., a complete lack of rhyme would be all but unthinkable, esp. if the lines in the s. were heterometric, as is often the case: the lines of such a s. may be defined syntactically, but unless we exaggerate the definition in the reading, unrhymed lines of differing lengths tend to blur in form, to shift toward the cadences of prose.

Verse without end rhyme could in theory be divided syntactically into precisely equal groups, equal pulses of two (or any number of) lines of the *same* length. But whether the ear could hear even these equalities without the added signal of rhyme is uncertain; this is the charge laid against Milton by Dr. Johnson. A part of our deepest sense of blank verse is that it is nonstanzaic: in the drama, the brevity or amplitude appropriate to the individual speech determines the speech's length, while in nondramatic blank verse the unit next larger than the line is the verse paragraph (q.v.), its length also determined by something other than metrical requirement.

In isometric ss., the principal source of formal effect lies in the patterning of rhyme and in the relation between rhyme and syntax. Couplet rhyme completes its pattern promptly and concisely; the closed couplet, in which the sense and

the rhyme structures are completed together, has this effect more fully, of course, than does the open couplet. Of slightly larger structures, *terza rima* (q.v.) is progressive, unfolding continuously via concatenation until it is closed off. Alternating rhyme has a somewhat similar effect, though our sense of the symmetries of three and of four is not the same. *Abba,* the envelope s., circles back to close as it began. Because rhyme in the envelope s. returns to its beginnings, there is a sharper hiatus between, say, *abba* and *cddc* than between *abab* and *cdcd*; the s. must be used appropriately. A longer s., like that of Hardy's "According to the Mighty Working," may appear to propose other patterns, then end by describing a figure comparable to the envelope: in the *abcbca* of that s. we have first three unrhymed lines, then what seems an *xaxa* quatrain, then *xabab,* and finally the addition that reveals the true and full rhyme structure by pairing the last line with the first.

In a s. rhymed *abba,* the second and third lines will have only the slightest effect of couplet unless the quatrain is punctuated heavily (period, semicolon, colon) at the end of lines 1 and 3; but even then, the fact that the *b* lines are even- and odd-numbered, not odd- and even-, will skew the effect. In longer ss.—*ababb, ababcc, ababbcc* (rhyme royal [q.v.]), *abababcc* (*ottava rima* [q.v.])—one can see how the patterns relate, and it is easy to imagine what differing effects might be achieved by breaking up the lines syntactically into varying symmetrical or asymmetrical groups.

One of the most fascinating ss. to study, for the (perhaps unexpected) variety it affords, is the Spenserian (q.v.) s., $ababbcbc_5c_6$. Assuming for the moment that the lines are so constructed as to terminate with the ends of phrases, clauses, or sentences, consider how different in effect the following are likely to be: *ab.ab.bc.bc.c*; *abab.bcbc.c*; *ab.abb.cb.cc*; *aba.bb.cbc.c*; *ababb.cbcc*; *aba.bbc.bcc.* One can find examples of all these in *The Faerie Queene.* One can also find extraordinary ss. in which a full stop line-internally strands an important phrase at the end of a sentence, without giving us the sense of completion normally conferred by rhyme (e.g. *FQ* 1.1.11.7). The effect, once we have accustomed ourselves to rhyme, can be strangely disorienting until we are reassured by the resumption of rhyming (cf. *Lycidas* 76).

The Spenserian s. gives us a glimpse of what may be done in even minimally heterometric ss. Characteristically the longer final line brings a sense of amplitude, of fuller utterance, to its conclusion: the "more" of the longer line is a metaphor ready to attach itself to any appropriate signification. S. movement can also be from longer to shorter lines, from more to less: see Henry Vaughan's "The World," or John Crowe Ransom's "Blue Girls." The "less" of the shorter line may also seem a metaphor, its effect that of an arresting brevity, of concision—or perhaps of humbleness, smallness, or fragility.

The 12-syllable line with which the Spenserian ends is susceptible of perfect balance if it breaks 6–6 or 4–4–4, or of what may seem an uneasy imbalance if it breaks 7–5, or 5–7, or 3–4–5. The balance possible in lines containing an even number of 2-syllable groups is also a potential metaphor ready to confer a special sense of order, of harmony, where the lines' denotation encourages this. The "Hymn" s. of Milton's *Nativity Ode* is from this point of view (as from many others) a magnificent invention: rhymed $aa_3b_5cc_3b_5d_4d_6$, it begins with two 3-line pulses, in each of which the last line is ampler than (though proportionate to) the first and second, reaches what seems a preliminary closure at the end of line 6—but proceeds almost unexpectedly to parallel the rising 3–5 progression with a fuller and more perfectly proportioned 4–6, closing, as does the Spenserian, on a long and potentially balanced line. The last line is prefigured by lines 1–2 and 4–5, and the pentameter lines tend to break syntactically 2–3; the last two lines double these proportions, 2–2 and 3–3. Just as lines of the same length produce equalities in verse, or the sense of equality, so proportionality of one line to another is a potentially meaningful characteristic to which the ear, and the mind, accustoms itself in s. It is a characteristic which, once expected, makes failure of proportion, or the breaking of proportion, the more notable. Again and again in the "Hymn" s., simplicities are compounded into triumphant symmetries. Against this, contrast the—equally meaningful—disproportion (or unresolved proportion) between the lengths of the last two lines of "Dover Beach."

The parallel between a repeatable melody in music and a series of line clusters identical in metrical pattern in poetry is clear, and it is tempting to think of s. as originating in song (q.v.). Stanzaic verse can indeed be written specifically to be sung to a pre-existing melody. But words to ancient melodies survived where the music to which they had been sung or chanted did not, and lyric stanzaic verse as a literary form, valued apart from a possible relationship to music, goes back a very long way. The two traditions continue, naturally, to intertwine. Ultimately we may expect a greater complexity of thought and feeling in verse written to be read than in verse meant, rather, to be sung; in the latter there is also relatively less place for intricate patterning of sound (q.v.) than we find in verse which is its own music.

S. is artificial; no sensible person would deny that. Rhyme itself is artificial: its occurrence at measured intervals is the more so. And whatever metrical pattern is chosen for a given s., this will, as in nonstanzaic verse, impose limits on the rhythms available to the individual line. The shorter the line, generally, the stricter the limits. But syntactic structures have their rhythms too, and while these may be modified in many ways, the fitting together of the two kinds of rhythm,

within relatively small spaces, and in lang. that seems unconstrained, requires much skill.

In stanzaic verse, line-end tends to coincide with phrase-, clause-, or sentence-end; rhyme and enjambment (q.v.), closure and the avoidance of it, are at odds, and can be played against one another only temporarily. Ss. themselves are not often enjambed: that is, s.-end and sentence-end normally arrive together. Long ss. would seem to permit greater internal variety of syntactic construction than short ss., and to a degree this is so. But rhythmic constraints on the individual line remain. And it is no more natural to have ideas expressed in repeated units of (say) 92 syllables than of 40, or 28, or 20. Blank verse, the sense it carries "variously drawn out from one verse into another," may therefore seem more "natural" than stanzaic verse, having available—despite constraints—some syntactic freedom. In conversation we do not speak in rhyme; we would not when we speak be chained to the symmetries of s. But in the hands of a skillful poet, the limits ss. impose are transformed into devices and resources, tools for creating meanings, and beauties, not otherwise attainable. E.R.W.

STUDIES AND SURVEYS: *General*: Thieme, 382—lists Fr. studies; E. Häublein, *The S.* (1978)—best modern study, though brief; Scott, ch. 5; Brogan, sect. G—bibl.; L. Turco, *The New Book of Forms* (1986)—very eclectic. *Classical*: Wilamowitz, chs. 15–16; Maas, sects. 61–72; Koster, ch. 14; Dale, ch. 12, and "Stichos and S.," *Collected Papers* (1969); West.

STANZA INDEXES: *French*: P. Martinon, *Les Strophes* (1912); I. Frank, *Répertoire métrique de la poésie des troubadours*, 2 v. (1953, 1957); W. Pfrommer, *Grundzüge der Strophenentwicklung Baudelaire au Apollinaire* (1963); U. Mölk and F. Wolfzettel, *Répertoire métrique de la poésie française des origines à 1350* (1972).

German: F. Schlawe, *Die deutsche Strophenformen 1600-1950* (1972); A. H. Touber, *Deutsche Strophenformen des Mittelalters* (1975); S. Ranawake, *Höfische Strophenkunst* (1976); H. J. Frank, *Handbuch der deutsche Strophenformen* (1980); W. Suppan, "Strophe," *Reallexikon* 4.245–56.

English: Schipper; Schipper, *History*; J. L. Cutler, "A Manual of ME Stanzaic Patterns," 2 v., Diss., Ohio State Univ. (1949); M. C. Honour, "The Metrical Derivations of the Med. Eng. Lyric," 2 v., Diss., Yale Univ. (1949); E. Häublein, *Strophe und Struktur in der Lyrik Sir Phillip Sidneys* (1971), and Häublein (above); B. O'Donnell, *Numerous Verse* (1989).

Spanish, Galician, and Portuguese: G. Tavani, *Repertorio metrico della lirica galego-portoghese* (1967); T. Navarro, *Repertorio de estrofas españolas* (1968). *Italian*: F. P. Memmo, *Dizionario di metrica italiana* (1983); A. Solimena, *Repertorio metrico dello Stil novo* (1980). *Russian*: G. S. Smith, "The Stanza Typology of Rus. Poetry 1735-1816: A General Survey," *Rus. Lit.* 13 (1983); Scherr. T.V.F.B.

STRUCTURE (Ger. *Aufbau*). S. is an important interpretive and methodological concept for critics who are more interested in the internal dynamics of a literary work (the interrelationships which comprise a literary system) than in its relation to external phenomena, its thematic content, or its genetic origins. Emphasis on s. allows the literary work to be conceived as an autonomous object and to be characterized in terms of its s. or internal relations, whence the importance of s. in all formalist approaches to poetry. This view of s. recalls Aristotle's *Poetics*, which could be said to attempt an analysis of poetic ss. For the new theoretical formulation revives that Cl. idea of form which underlies Aristotle's systematic mapping of genres—epic and lyric poetry, tragedy, and comedy. Thus Aristotle's focus on "poetry in itself and of its various kinds" has certain affinities with important 20th-c. conceptions of s. (though the Aristotelian analysis tends to relegate the formal character of poetic s. to the periphery of investigation in favor of such psychological issues as *catharsis*).

Some of the most crucial aspects of the concept of s. may be seen in the divergence between two dominant theoretical models: the generic and the organic. Genre crit. stresses the relationship between the whole (the overall generic code which articulates the structural rules, formal characteristics, or subcategories of the class) and the part (the particular poem which deviates merely in details). This formulation of the relationship of part to whole is radically unlike that which is given in organic theories of poetry, which assume that each individual poem, each individual poetic s., is unique. In the organic model, the s. of the poem has priority over generic rules, and the identification of that s. (i.e. the s. of its logical argument or its image patterns) becomes the privileged object of study.

The distinction between the internal s. of the poem and its extra-literary or contextual references is set forth most clearly in New Criticism, and most particularly informs Cleanth Brooks' influential study *The Well Wrought Urn* (1947), subtitled *Studies in the S. of Poetry*. Brooks is not concerned with a conventional analysis of content; rather, he endeavors to work out a systematic theory of poetic s. In Brooks' conception, what is essential is the inner, *paradoxical* s. of the poem, with its tensions, stresses, and contradictions. As he writes, the structural "principle is not one which involves the arrangement of the various elements into homogeneous groupings, pairing like with like. It unites the like with the unlike." More than anything else, it is these inner tensions and paradoxes uniting "like with unlike" that define poetic s. Far from seeking to establish some homogeneous grouping (common themes, period styles, recurring images), Brooks seeks to isolate the "pattern of resolved stresses" which comprises the completed s. of the poem. Thus he is able to show that the inner coherence of poetic s. is con-

stituted by balanced tensions, harmonized meanings, and dramatized resolutions.

John Crowe Ransom, too, directs the critic's attention away from extra-linguistic references toward the inner form and verbal autonomy of the poetic s. itself. But he differs from Brooks in that the logical s. of the poem—its paraphrasable core—is bound up with its local texture. Nothing illustrates Ransom's distinction between s. and texture so well as his architectural metaphor: poetic s. corresponds to the walls, beams, and supports of a house, texture to the paint, wallpaper, and surface decoration. Whereas Brooks emphasizes s., for Ransom s. gives added importance to texture. This conceptual reversal accentuates the sensuous immediacy and vital concreteness of the poem.

It is worth noting that Ransom's distinction was taken up by Monroe C. Beardsley (*Aesthetics*, 1958), who extended it to the plastic arts, music, and narrative. Nevertheless, it should be observed that the systematic investigation of the internal s. of the poem has been called into question by critics of formalism. New-Historicist critics, for example, take as their purpose to demonstrate that the poem comprises something more than its organized ss. There is a shift from consideration solely of the inner s. of a poem to consideration of its historical context or situation. To put it another way, the inner articulation of poetic s. becomes better understood when the poem is relocated within its sociohistorical context.

The separation of the formal poetic s. from the object of reference is to be found not only in Am. New Criticism, however, but also in much Slavic and Fr. structuralist thought as well—a significant parallel. However, in structuralism the still too organic model is discarded for a new linguistic paradigm. In contrast to the usage of the term in New Criticism, s. in structuralism denotes the domain of the signifier as such (the sounds or letters that are meaningful in a given lang. system). What is important in structuralism is the primacy of the signifier itself. This displacement brackets and suspends the object of reference; it detaches the signifier from what is signified. As a result, the fundamental locus becomes the complex differential relationships of the signifiers within the boundaries of the linguistic system, not the relation of the signifier to the external world. The dominant model is that of the Swiss linguist Ferdinand de Saussure (1857–1913) with his seminal distinction between *langue* (the linguistic possibilities which make up a total lang. system) and *parole* (the local and contingent speech acts which are performed by individuals). On a theoretical level, structuralism as a whole owes very much to this initial assumption. Thus defined, the analytic methods of structuralism always involve a deliberate effort to restore to the object of study the hidden and unarticulated rules of its synchronic functioning (or disfunctioning). As Roland Barthes notes, "s. is therefore actually a *simulacrum* of the object, but a directed, *interested* simulacrum, since the imitated object makes something appear which remained invisible or, if one prefers, unintelligible in the natural object" ("The Structuralist Activity").

Such a construct of a synchronic system allows Roman Jakobson and Claude Lévi-Strauss to delineate a basic poetic s. in Baudelaire's "Les Chats." In particular, the emphasis on synchronic s. makes visible the tensions between two sets of symmetrical/asymmetrical relations. This decoding device foregrounds "a division of the poem into two sestets separated by a distich whose s. contrast[s] vigorously with the rest." The frame is provided by a system of oppositions in dynamic progression which moves via the distich from the first sestet to the second. Jakobson and Lévi-Strauss reconstruct the superimposed formal levels (phonetic, phonological, syntactical, and semantic) of the two sestets. In the first, real cats occupy an important place; in the second, an unexpected reversal opens up an imaginary space beyond the factual and physical world where surreal cats stand out. From a structuralist point of view, these intentional ambiguities combine to produce a new utterance: the sensual and exterior world of the first sestet is maintained at the same time that it is transferred to the intellectual and interior world of the second. Through this example the two critics are able to explore a series of oppositions (fact/myth, constriction/dilation, exteriority/interiority) which are reconciled in the poem in various combinatory forms of linguistic organization. There is no doubt that Lévi-Strauss and Jakobson's reading of "Les Chats" is a paradigmatic example of structuralist method. Yet it should be remarked that other critics have contested their decipherment of the formal features in question. Michael Riffaterre, for example, suggests that the s. of the sonnet is "a sequence of synonymous images, all of them variations on the symbolism of the cat as representative of the contemplative life."

The same application of a linguistic or synchronic model to poetry is to be found in the poetics of Tzvetan Todorov, where the task of the critic is the discovery and description of an overall "s. of significations whose relations can be apprehended." For Todorov, like Jakobson and Lévi-Strauss, the speech act is reconstituted as a signifying system or set of linguistic relations so that the content of a poem becomes lang. itself (rather than the referential object). For example, the *Odyssey* is more appropriately described as a poem about the formal ss. of lang. than as an epic narration of Odysseus' adventures. Seen in this way, all referential aspects of the poem (the speeches, the song of the Sirens, and esp. prophecies) assume the character of a linguistic s.—that is to say, an event of lang. As Todorov puts it: "every nondiscursive event is merely the incarnation of a discourse."

The conception of formal gaps or breaks in Gérard Genette's *Figures of Literary Discourse* provides yet another definition of the nature and function of poetic s. It is characteristic of Genette's theory of figures that his readings emphasize the spaces and gaps inscribed within the poetic text. This distinguishes Genette's microscopic analyses of poetic s. from Todorov's more general theoretical propositions about poetic lang. According to Genette, "poetry finds its place and its function *where lang. falls short*, in precisely those shortcomings that constitute it." For Genette, poetic s. is the particular space that disconnects two or more forms. Yet poetic s. is also a negation of the gap, for it projects a utopia of lang. in which the gap between signifier and signified would be effaced. Poetry, Genette writes, is "gap from the gap, negation, rejection, oblivion, effacement of the gap, of the gap that *makes* lang.; illusion, dream, the necessary and absurd utopia of a lang. without gap, without hiatus—without shortcomings." At the same time, it should be noted that Genette's theory of poetics is unmistakably linked with that of Todorov, in that both critics attempt to define the formal properties of poetic ss.

Saussure's fundamental analysis of the signifier is also explicitly taken up in the recent works of Jacques Derrida and other theorists (most notably the *Tel Quel* group and, in particular, Julia Kristeva). But in this group of writers a new conception emerges which distinguishes them from the contributions of structuralism. This is the deconstruction of those hypostases of s. which fix it conceptually as a centered s. or transcendental signified. Such hypostasized ss.—Derrida sees them as metaphysical substances or mythical plenitudes—derive from a nostalgia for origins, a longing for a metaphysic of absolute presence. Derrida attempts to escape the closure of such a centered s. by emphasizing the free play of the signifier. This indeterminacy (q.v.) of the heterogeneous text is intimately connected with what he has called *différance*, a term which implies both difference and deferral: an interminable temporal movement of signification that cannot be arrested in an absolute presence or closure of "meaning." However, it must be recognized that Derrida's ultimate emphasis on the free play of the signifier is still tied to the way the question of s. is posed within the linguistic premises and traditional logic of Western metaphysics. The decoding of poetic script, the textual decipherment of poetry's specific ss., can be achieved in the first place only in terms of the older modes of lang. and thought which Derrida sets out to deconstruct—that is to say, in terms of what he calls the "always-already-written" (the *trace*). This suggests that his deconstructionist readings and analyses remain somehow structural, in spite of Derrida's effort to avoid the metaphysical closure of centered s. or absolute presence.

Brooks; Crane—important critique of Brooks; W. Emrich, "Die Struktur der moderne Dichtung," *WW* (1952–53); R. S. Crane, *The Langs. of Crit. and the S. of Poetry* (1953); S. Fishman, "Meaning and S. in Poetry," *JAAC* 14 (1956); M. C. Beardsley, *Aesthetics* (1958); *Sens et usage du terme s.*, ed. R. Bastide (1962); R. Wellek, "Concepts of Form and S. in 20th-C. Crit.," *Concepts of Crit.* (1963); H. Meyer, "Über der Begriff Struktur in der Dichtung," *NDH* 10 (1963); J. Lotman, *The S. of the Artistic Text* (1971; tr. 1977); R. Barthes, "The Structuralist Activity," *Critical Essays*, tr. R. Howard (1972); T. Hawkes, *Structuralism and Semiotics* (1977); R. Jakobson and C. Lévi-Stauss, "Charles Baudelaire's 'Les Chats,'" *The Structuralists*, ed. R. and F. DeGeorge (1972); M. Riffaterre, "Describing Poetic Ss.," *Structuralism*, ed. J. Ehrmann (1970); D. Wunderlich, "Terminologie des Strukturbegriffs," *Literaturwiss. und Linguistik*, ed. J. Ihwe, v. 1 (1972); F. Jameson, *The Prison-House of Lang.* (1972)—excellent analysis; F. Martinez-Bonati, "Die logische Struktur der Dichtung," *DVJ* 47 (1973); Culler; T. Todorov, *The Poetics of Prose*, tr. R. Howard (1977); T. D. Young, "Ransom's Critical Theories: S. and Texture," *MissQ* 30 (1977); J. Derrida, "S., Sign and Play," *Writing and Difference*, tr. A. Bass (1978), "Living on: *Border Lines*," *Deconstruction and Crit.* (1979); G. Genette, *Figures of Literary Discourse*, tr. A. Sheridan (1982); R. Williams, "Structural," *Keywords* (1983); J. Kristeva, *Revolution in Poetic Lang.*, tr. M. Waller (1984); J. C. Ransom, "Crit. as Pure Speculation" (1941), in Ransom; J. C. Rowe, "S.," *Critical Terms for Literary Study*, ed. F. Lentricchia and T. McLaughlin (1990).

P.M.; T.V.F.B.

STYLE. How are we to distinguish between what a poem says and the lang. in which it says it? On the one hand, there is no such thing as a "content" utterly separable from words; on the other hand, something can be said about the words which does not refer directly to the content. The relation between the two has been described metaphorically, and looking at these metaphors we discover two kinds. The first suggests (focusing on the creation of the poem) that the relation is mechanical or rhetorical, that s. is something added, more or less at the writer's discretion. On the other hand, if we alter the perspective to focus upon the text, i.e. on s. as the relationships of the lang. in a poem, we find an organic metaphor.

The first kind is common in Ren. and Neoclassical crit., which derives from Cl. rhet. (Aristotle, Cicero): the Eng. term *rhetoric* derives from Gr. *rhetorike*, elliptical for the art of the orator. One part of rhet. was *elocutio*, the selection and placement of words. Longinus in sections 16–43 of *On the Sublime* (1st c. A.D.?) cites the three rhetorical sources of the sublime (q.v.) as figures, diction, and syntax. Puttenham (1589) compares "ornament" to flowers, jewels, and embroidery, but the term had a wider meaning then than now, embracing virtually all the strategies of lang. in a piece of writing. Common too is the comparison of s. to a

garment: "lang. is but the apparel of Poesy," says Sir William Alexander (1634).

The organic view begins to be expressed most insistently in the 19th c. One formulation of the concept is biographical, the other textual. Coleridge states both. Images, he says in the *Biographia literaria*, "become proofs of original genius only as far as they are modified by a predominant passion; or by associated thoughts or images awakened by that passion." Among the ancient critics only Longinus (sect. 17) had expressed such a view. After Coleridge, many critics use or imply the organic metaphor. For Pater, successful s. is, instead of a constructed house, a body "informed." To John Henry Newman, s. is a "thinking out into lang." Middleton Murry considers s. "organic—not the clothes a man wears, but the flesh and bone of his body." In Leo Spitzer's account of the "philological circle," the details of a linguistic structure and its postulated cause or "inner significance" are inextricable.

S., then, has been considered something added, or the parts of a whole, or an extension of mind and character. But this riddle may be more apparent than real, since choice is fundamental to rhet. (persuasion), unity, and authorship. As Spitzer believes, the details of art are not inchoate, chance aggregation, but rather parts of a deliberately related whole. The distinction mechanic / organic may be resolved by treating these as two perspectives on the creative process, as a matter not of opposed formulations but of consideration of the appropriate expressive level or category. The relationship of individual parts and their association in a text to the authorial selection and arrangement of those parts is clarified by the concept of synonymity and by information theory. According to the argument for synonymity, different linguistic forms can produce identical meanings. Differences in expression are then differences in emphasis and not necessarily differences in meaning (Hirsch). According to information theory, natural langs. are inherently redundant. Redundancy makes it possible to convey a message in more than one form, according to the preference or disposition of the individual writer, in contrast to a non-redundent code like telephone messages (Milic, "S."). Thus one can understand designed cohesion (emphasis and correlation) without the mystification of prejudicial metaphors. Choice and text are not inextricably intertwined, but related in the creative process of a writer's choices and a text's coherence.

On the one hand, observation of any text (or discourse: see Kinneavy), but esp. of a closely woven "aesthetic" text, reveals how the meaning of words depends upon the contexts created by the other words in the text. Meaning entails correlation, every word bearing the pressure of all the other words. Hence Riffaterre believes that "the unity of meaning peculiar to poetry" is "the entity of the text," a traditional formalist argument.

On the other hand, study of the strategies and devices of composition reveals the innumerable linguistic techniques by which one can "ornament" thought. Cl. rhetoricians divided such techniques into tropes and schemes, words and syntax. More than 200 figures of speech were recognized by the Tudor rhetoricians. But rhet. was never limited to simile, synecdoche, anaphora, and asyndeton. From its origins in Aristotle and earlier it embraced the process of creation (*inventio*: finding topics) as well as larger expanses of discourse—*dispositio* (organization), logic, point of view, and ethical and emotional appeals. Indeed, textual strategies continue to be created and discovered—for example, graphicology (typographical, visual, and multimedia devices), or a text's deliberate incompleteness, its gaps (Iser).

But s. involves more than the author's production of a text, more than choices and wholes (or lack thereof). S. is also habitual, which is of two kinds: (1) lang. habits shared by a group of people, and (2) lang. habits unique to an individual writer. Group habits can be further divided into three kinds: period, nation, and genre. People share modes of expression over a period of time; nations share certain linguistic and literary habits; and certain specific modes of expression become established by convention (q.v.) as genres (q.v.). Because of the magnitude of the data, national ss. are least understood, but reliable studies are increasing. Period s. is often international in scope, e.g. futurism.

Author s. is also now being studied with increasing thoroughness. We recognize in this orientation the familiar theory that s. reflects the individual author, or "Le s. est l'homme meme" (Buffon). According to Puttenham, writers choose their subjects "according to the metal of their minds." Over the past half-century, stylistics has been accumulating statistical data of extensive scope and detail on authors' s. in full-length studies of individual authors (e.g. Milic, Cluett). Following romanticism, and esp. since Freud, this idea has been extended to unconscious expression. Recent studies have also examined the relationships between authors and their ages, nations, and genres—the interaction of writers and the lang. available to them. To Ohmann, s. is "epistemic"—it reflects the epistemological assumptions of each author. Auerbach argues that s. is a reflection of the tension between the force of an author's individuality and the pressure of social forces, in which society is the more powerful force of the two. Cruttwell believes that only one or two ss. are possible at any given time. Eventually we may have a reliable base of information about the history of s. through the comparison of thorough studies of individual writers in their age. But the problem is complicated by the dynamic reality of writing, which often evolves; hence the lang. of an author must be studied chronologically. Improved methods of quantitative analysis using mathematics and

computers (Abercrombie, Oakman) have made possible the gathering of the enormous data necessary for accurate stylistic description of authors and groups. Such data bring us closer to the truly comparative approach to s. advocated by Spencer and Gregory, Enkvist, and Dressler.

They also greatly clarify attempts to trace the history of s. in terms of register, or patterns derived from class. The issue of the appropriateness of lang. to situation and subject was familiar to Cl. rhetoricians and was systematized in the Ren. doctrine of the three ss., high, middle, and low. According to Richard Sherry in 1550, "there hath bene marked inespecially thre kindes of endigh tynge: The Greate, the smal, the meane." Writers should fit their lang. to their subject, a lofty s. for an eloquent subject, as in tragedy and epic; middle for elegies; low for satire (Puttenham; see STYLE). This is the doctrine of stylistic decorum (q.v.). The grand s. has received particular attention in the past, esp. in Neoclassical crit.

Related to the problem of stylistic levels of decorum is the parallel issue of the kinds of s., particularly plain vs. ornate. Quintilian believed the three ss. were suited to each of the three functions of rhet. and assigned the plain s. to instruction. These categories have always been notoriously imprecise, but linguistic studies are making them clearer. Adolph, for example, compares two translations of the same passages to conclude that Elizabethan writers tend toward unusual collocations and syntax, while Restoration writers tend toward normal. This history is both vexed and benefitted by the recognition of the existence of varieties not yet systematically studied—Ciceronian and antiCiceronian, curt and loose, Puritan antirhetorical, scientific plain, periodic, pointed, utilitarian, and so on, with description being complicated by 20th-c. innovations such as Joyce's sentence in progress (Bennett, *Prose*).

Extraordinarily diverse and complex, s. has been studied intensively during the last 30 years by literary critics, linguists, and sociologists searching for a comprehensive, disciplined understanding. Emergent from these studies is a view of s. as text features in context—as the material of dialogue both within a text (referentiality and reflexivity) and between those dynamic features and readers in their cultural conditions—of nation, period, and genre.

W. Pater, "S.," *Appreciations* (1889); G. L. Hendrickson, "The Origin and Meaning of the Ancient Characters of S.," *AJP* 26 (1905); J. M. Murry, *The Problem of S.* (1925); Sr. M. Joseph, *Shakespeare's Use of the Arts of Lang.* (1947); L. Spitzer, *Linguistics and Lit. Hist.* (1948); Auerbach; Curtius; M. C. Bradbrook, "50 Years of Crit. of Shakespeare's S.," *ShS* 7 (1954); P. Cruttwell, *The Shakespearean Moment* (1954); Wellek and Warren, ch. 14; W. Staton, "The Characters of S. in Elizabethan Prose," *JEGP* 57 (1958); W. Nowottny, *The Lang. Poets Use* (1962); W. Trimpi, *Ben Jonson's Poems* (1962)—on the plain s.; C. Ricks, *Milton's Grand S.* (1963); N. Enkvist, "On Defining S.," *Linguistics and S.*, ed. N. Enkvist et al. (1964); S. Ullmann, *Lang. and S.* (1964); L. T. Milic, *S. and Stylistics* (1967), *A Quantitative Approach to the S. of Jonathan Swift* (1967), "S., Literary," *Encyc. of Communication* (1987); D. L. Peterson, *The Eng. Lyric from Wyatt to Donne* (1967)—on the plain s.; R. Adolph, *The Rise of Mod. Prose Style* (1968); R. A. Sayce, "S. in Lit.," *DHI*; G. Hough, *S. and Stylistics* (1969); A. M. Patterson, *Hermogenes and the Ren.: Seven Ideas of S.* (1970); *Patterns of Literary S.*, ed. J. Strelka (1971); J. Kinneavy, *A Theory of Discourse* (1971); J. R. Bennett, *Prose S.* (1971); N. Goodman, "The Status of S.," *CritI* 1 (1975); E. D. Hirsch, Jr., "Stylistics and Synonymity," *CritI* 1 (1975); R. Cluett, *Prose S. and Critical Reading* (1976); *Current Trends in Textlinguistics*, ed. W. Dressler (1978); W. Iser, *The Act of Reading* (1978); I. Fairley, "Experimental Approaches to Lang. in Lit.," *Style* 13 (1979); R. Oakman, *Computer Methods for Literary Research* (1980); D. A. Russell, "Theories of S.," *Crit. in Antiquity* (1981); Fowler; M. Riffaterre, *Text Production* (1983); L. Urdang and F. Abate, *Literary, Rhetorical, and Linguistic Terms Index* (1983); S. Levinson, *Pragmatics* (1983); H. Seidler, "Stil," *Reallexikon* 4.199–213; J. Abercrombie, *Computer Programs for Literary Analysis* (1984); W. Keach, *Shelley's S.* (1984); *The Concept of S.*, ed. B. Lang, rev. ed. (1987); D. K. Shuger, *Sacred Rhet.: The Christian Grand Style in the Eng. Ren.* (1988). J.R.B.

SUBLIME.

I. CLASSICAL
II. ENLIGHTENMENT TO MODERN

I. CLASSICAL. "S.," a Lat.-derived word meaning literally "(on) high, lofty, elevated," owes its currency as a critical and aesthetic term to the anonymous Gr. treatise *Peri hypsous* (*hypsos*, "height, elevation") once ascribed to the rhetorician Cassius Longinus of the 3d c. A.D. but now generally agreed to belong to the 1st c., perhaps around 50 A.D. Whatever his name and origin, its author was certainly a rhetorician and a teacher of the art, but one of uncommon mold. His essay, with its intimacy of tone (it is addressed to a favorite pupil, a young Roman) and breadth of spirit, stands more or less isolated in its own time, but has had a recurrent fascination for the modern mind since the 17th c.

The idea of sublimity had its roots in the rhetorical distinction, well established before "Longinus," of three levels of style (q.v.), high, middle, and low. His achievement was to draw it out of the technical sphere, where it had to do primarily with style, and to associate it with the general phenomenon of greatness in lit., prose and poetry alike. "Longinus" regards sublimity above all as a thing of the spirit, a spark that leaps from the soul of the writer to the soul of his reader, and only secondarily as a matter of technique and expression.

"Sublimity is the echo of greatness of spirit." Being of the soul, it may pervade a whole work (speech, history, or poem: "Longinus" pays little attention to genre distinctions); or it may flash out at particular moments. "Father Zeus, kill us if thou wilt, but kill us in the light." "God said, 'Let there be light,' and there was light." In such quotations as these, "Longinus" shows among other things his sharp eye for the particular passage and his capacity for empathy with the actual work, qualities which are in fact rare in ancient crit. and which presage the modern spirit.

The distinguishing mark of sublimity for "Longinus" is a certain quality of feeling. But he will not allow it to be identified simply with emotion, for not all emotions are true or noble. Only art can guard against exaggerated or misplaced feeling. Nevertheless, art plays second fiddle to genius in his thinking. He enumerates five sources of the s.: great thoughts, noble feeling, lofty figures, diction, and arrangement. The first two, the crucial ones, are the gift of nature, not art. "Longinus" even prefers the faults of a great spirit, a Homer, a Plato, or a Demosthenes, to the faultless mediocrity that is achieved by following rules.

In later antiquity and the Middle Ages, the treatise remained unknown, or at least exercised little influence. In the Ren., it was first published by Robortelli in 1554, then tr. into Lat. in 1572 and into Eng. in 1652 (by John Hall). But it made no great impression until the late 17th c. Paradoxically enough, it was Boileau, the high priest of Fr. neoclassicism, who launched the *Peri hypsous* on its great modern career and thus helped to prepare the ultimate downfall of Classicism. His tr. (1672) had immense reverberation, esp. in England. The Eng., always restive under the "Fr. rules," instinctively welcomed "Longinus" as an ally. As neoclassicism advanced and subjectivity became increasingly central to Eng. thinking, not only about lit. but about art in general, the s. became a key concept in the rise of romanticism in poetry and the concurrent establishment of aesthetics as a new, separate branch of philosophy.

G.F.E.; T.V.F.B.

II. ENLIGHTENMENT TO MODERN. In the 18th c., the s. represented merely one type of experience that could be described under the general philosophical rubric of sensationism. For a host of writers producing everything from aesthetic treatises to Gothic novels, it was synonymous with irresistible forces that produced overwhelming sensations. In the 18th and 19th cs., the s. came increasingly to be a term of aesthetic approbation, as attested by the interest in both s. landscapes and paintings of s. landscape. In the popular view, the term amounted to a description: it represented primarily a subject matter, the wild and desolate natural scene or the natural force that dwarfed the individual human figure. Its effect was simultaneously to make one conscious of one's own comparative weakness in the face of natural might and to produce a sense of the strength of one's own faculties. As John Baillie put it in his *Essay on the S.* (1747), "Vast Objects occasion vast Sensations, and vast Sensations give the Mind a higher Idea of her own Powers."

Along with the increasing currency of the term in the 18th c., two particularly strong arguments about the place of the s. and s. nature emerged: Edmund Burke's *Philosophical Enquiry into the Origin of Our Ideas of the S. and Beautiful* (1757) and Immanuel Kant's *Critique of Judgement* (1790; commonly referred to as his Third Critique, after the *Critique of Pure Reason* [1781] and *Critique of Practical Reason* [1788]). In the history of what we now call aesthetics, these two works were esp. important for according significance to pleasure in objects that were not, strictly speaking, beautiful. Burke developed the sensationist position into an affectivism that continually connected the s. with the issue of an individual's relationship to society, and Kant made his discussion of the s. a cornerstone of a formalist account of aesthetics.

Burke's *Enquiry* sets out the affectivist position on the s. in an argument that emphasizes the power of experience. In the "Intro. on Taste" added to the 2d ed. (1759), Burke made two claims for the importance of taste. First, his emphasis on the regular operation of the senses makes taste as meaningful and as generalizable as reason: "as the conformation of their organs are nearly, or altogether the same in all men, so the manner of perceiving external objects is in all men the same, or with little difference." Second, his emphasis on the origin of the passions treats taste as a field of determinate knowledge in which the "remembrance of the original natural causes of pleasure" can be distinguished from the acquired tastes that fashion and habit promote. People, he observes, are not likely to be mistaken in their reactions to sensation even though they may often be confused in their reasoning about them.

Burke traced the attractions of the beautiful and the s. to human impulses that are ultimately utilitarian. The beautiful he saw as a manifestation of the human instinct towards sociability, with sociability in turn serving the purpose of the continuation of the species. The s. he treated as a manifestation of the instinct for self-preservation, the response of terror that "anticipates our reasonings, and hurries us on by an irresistible force." The beautiful represents what we love (and love specifically for submitting to us and flattering our sense of our own power); the s. represents all that we fear for being greater and more powerful than we are.

If the notion of sympathy had for writers like Adam Smith suggested how persons might identify with the interests of others, Burke's discussion of the s. and the beautiful emphasizes relationships between individuals and objects far more than intersubjective relationships. Yet Burke nonetheless argues for the social utility of our feelings

of the s. and beautiful. He increasingly aligns the beautiful not merely with the sociable and pleasing but also with a relaxation of the bodily functions that eventually becomes disabling. The s., by contrast, presents difficulties that require "exercise or labour" to be overcome. Although the s. feelings of astonishment or awe may resemble pain, the excitation and exertion that they produce yield a very real pleasure—a consciousness of one's own powers and even a physical exercise that keeps the various organs of sensation in tone. Burke's account may be empiricist in suggesting that objects have regular and predictable operations on the senses, but it ultimately de-emphasizes knowledge of the external world and stresses instead the uses of objects in gratifying and challenging the individual human organism.

Kant, in his *Critical Observations on the Feeling of the Beautiful and S.* (1784), does not depart strikingly from the Burkean position. He identifies the beautiful and the s. as terms under which contrasting kinds of objects of experience might be subsumed; and he sees the enjoyment or displeasure in these objects as having essentially a psychological dimension. In this, his remarks are consistent with the familiar critical view that shifted discussions of pleasure in art and nature from an emphasis on production—what the artist must do to achieve certain effects—to an emphasis on reception—how the response to certain objects raises questions of a viewer's or reader's psychology.

With the Third Critique, however, Kant reoriented aesthetic discussion. Burke and other writers (including the Kant of the *Observations*) had described the s. and the beautiful in terms of both natural and manmade objects; Homeric and Miltonic poetry could serve as examples of sublimity as well as the seemingly infinite expanse of the ocean or a powerful animal in whom "the terrible and s. blaze out together." Kant reduced the metaphorical reach of the term "s." and identified it exclusively with natural objects. The net effect of this reduction was to enable him to argue that the s. is not—or not particularly—important for establishing human inferiority relative to natural might. Rather, the pleasure that one takes in s. nature reveals a pleasure in judging objects that are not the vehicles of a message and not expressions of anyone's intentions. If a poem or a statue cannot exist without the intentional action of its maker, natural sublimity appeals to human viewers in a fashion that stands outside of such communication of intention.

Kant's claim for s. intentionlessness obviously opposes itself to the "argument from design," which reads the book of nature as revealing divine intention. Its primary significance, however, is not so much to argue against belief in divinity as to identify aesthetic judgment as a faculty which is, in interesting ways, unable to ground itself in claims about the prior value of external objects. The s. becomes the primary vehicle for the Kantian argument about the importance of "purposiveness without purpose" in aesthetic objects. While natural beauty might appear to have been formed by design (echoing Addison's sense of the mutually enhancing relationship of art and nature to one another), the s., lacking the form of beautiful nature, bespeaks pleasure in an object that is without bounds not merely in appearing infinite but in having no form. The aesthetic judgment, that is, does not respond to the intrinsic beauty of an object in appreciating natural sublimity, but neither does it merely provide a screen on which individual psychology is projected. Rather, the s. in its intentionlessness and formlessness, makes visible the judgment's role as a form-giving faculty.

Most recently, the s. has gained prominence in deconstruction and rhetorical crit. In the work of Jacques Derrida, for instance, it has figured prominently in his challenge to Kantian formalism. Indeed, for many critics it has come to represent something like an inversion of the Kantian claim about it: namely, the view that the s. represents an "excess" in lang. that keeps it from ever assuming any fixed form or meaning. F.F.

PRIMARY WORKS: *Longinus on the S.*, ed. and tr. W. Rhys Roberts, 2d ed. (1907)—best and fullest ed.; *Anonimo del S.*, ed. and tr. A. Rostagni (1947). Tr.: W. H. Fyfe (1927; Loeb ed.); B. Einarson (1945); G. M. A. Grube (1958); D. A. Russell (1965; text and commentary, 1964).

SECONDARY WORKS: T. R. Henn, *Longinus and Eng. Crit.* (1934); S. T. Monk, *The S.: A Study of Crit. Theories in 18th-C. England* (1935); B. Weinberg, "Trs. and Commentaries on Longinus to 1600, A Bibl.," *MP* 47 (1949–50); F. Wehrli, "Der erhabene und der schlichte Stil in der poetisch-rhetorischen Theorie der Antike," *Phylobolia für P. von der Mühll* (1946); E. Olson, "The Argument of Longinus' *On the S.*," in Crane; Wimsatt and Brooks, esp. chs. 6, 14; W. J. Hipple, *The Beautiful, The S., and the Picturesque* (1957); J. Brody, *Boileau and Longinus* (1958); M. H. Nicolson, *Mountain Gloom and Mountain Glory* (1959), "S. in External Nature," *DHI*; E. Tuveson, *The Imagination as a Means of Grace* (1960); M. Price, "The S. Poem," *Yale Rev.* 58 (1969); Saisselin; A. Litman, *Le S. en France, 1666–1714* (1971); T. E. B. Wood, *The Word "S." and Its Context, 1650–17600* (1972); D. B. Morris, *The Religious S.* (1972); A. O. Wlecke, *Wordsworth and the S.* (1973); W. P. Albrecht, *The S. Pleasures of Tragedy* (1975); J. Derrida, "Economimesis," *Mimesis: Des Articulations* (1975), *La Verité en peinture* (1983), tr. as *The Truth in Painting* (1987); T. Weiskel, *The Romantic S.* (1976); S. A. Ende, *Keats and the S.* (1976); P. H. Fry, *The Reach of Crit.* (1983); P. de Man, "Hegel on the S.," *Displacement*, ed. M. Krupnick (1983); J.-L. Nancy, "L'Offrande s.," *Po&sie* [*Poesie*] 30 (1984); F. Ferguson, "The Nuclear S.," *Diacritics* (1984), *Solitude and the S.* (1992); A. Leighton, *Shelley and the S.* (1984); "The S. and the Beautiful: Reconsiderations," spec. iss. of *NLH* 16 (1985); N. Hertz, *The End of the Line* (1985); E.

Escoubas, "Kant ou la simplicité du s.," *Po&sie* 32 (1985); J.-F. Lyotard, "Le S., à present," *Po&sie* 34 (1985); S. Knapp, *Personification and the S., Milton to Coleridge* (1985); P. Lacoue-Labarthe, "La Verité s.," *Po&sie* 38 (1986); L. W. Marvick, *Mallarmé and the S.* (1986); *The Am. S.*, ed. M. Arensberg (1986); *La Via al S.*, ed. M. Brown et al. (1987); T. M. Kelley, *Wordsworth's Revisionary Aesthetics* (1988); P. Crowther, *The Kantian S.* (1989); P. Boitani, *The Tragic and the S. in Med. Lit.* (1989); *Das Erhabene*, ed. C. Pries (1989); G. L. Stonum, *The Dickinson S.* (1989 1990); S. Guerlac, *The Impersonal S.* (1990); R. Wilson, *Am. S.* (1991); J. F. Diehl, *Women Poets and the Am. S.* (1991); V. A. De Luca, *Words of Eternity: Blake and the Poetics of the S.* (1991).
T.V.F.B.; G.F.E.; F.F.

SYMBOL.

The word "s." derives from the Gr. verb *symballein*, "to put together," and the related noun *symbolon*, "mark," "token," or "sign," referring to the half-coin carried away as a pledge by each of the two parties to an agreement. Hence it means basically a joining or combination and, consequently, something once so joined or combined that stands for or represents, when seen alone, the entire complex. Since almost anything can be seen as standing for something else, the term has, and has engendered, a broad range of applications and interpretations. In the study of lang., for example, words are symbols of what they stand for, but the more common linguistic terminology after Saussure is "signifier" and "signified." A related distinction is that between "sign" and "s.," where the former refers to a relatively specific representation of one thing by another—a red traffic light, for example, means "stop"—while the latter refers to a more polysemous representation of one thing by another—as when the sea, for example, is used to stand for such different feelings as the danger of being overwhelmed (by analogy with drowning), or the excitement and anxiety of making a transition (as in a journey), or the power and fulfillment of strength (as in mighty), and so on.

For the present purpose, however, the meanings and uses of symbolism can be analyzed in terms of two main categories: on the one hand is the study of the s. in such larger contexts as lang. and the interp. of lang. (philology, rhet., linguistics, semantics, semiotics, hermeneutics), of philosophy (metaphysics, epistemology, aesthetics), of social science (sociology, anthropology, psychology—see Bryson's anthols., and Duncan), and of history and religion; on the other hand is the study of the s. in its more specific contexts as an aspect of art, of literary theory and crit., and of lit. (see Wimsatt [1965] and Hayes).

I. IN LARGER CONTEXTS. Historically as well as logically the larger field comes first. Where people once tended habitually to see the physical world in terms of emotional and spiritual values, they now tend to separate their values from the world. It has become one of the clichés of modern crit. that, partly due to the anti-imagistic crusade of the Protestant Reformation, partly to the growth of science and its search for "objective" knowledge, partly to the changes in focus from sacred to secular gradually effected in school curricula, and partly to the mere passage of time, not only have many traditional ss. been rendered meaningless to poets and readers alike, but also the very power of seeing the physical world in terms of values has diminished. Thus symbolism has been called in the 20th c. the "lost" or "forgotten" lang. (Bayley, Fromm).

It may be said, then, that the evolution of symbolism began with the evolution of primitive humanity. It was not until the med. period, however, that ss. and the interp. of ss. became a specific branch of learning. The patristic trad. of biblical exegesis, heavily under the influence of the Platonic and Neoplatonic schools of thought, developed standards and procedures for the doctrinal interp. of Holy Writ according to four levels of meaning—literal, allegorical, moral and tropological, and anagogic. The purpose was twofold: to reconcile the Old Testament with the New, and to reconcile various difficult portions of each with Catholic teaching. Thus, for example, while "The Song of Solomon" is a mildly erotic wedding poem on the literal level, its true meaning on the allegorical level is the "marriage" of Christ and the Church.

This exegetical trad. evolved during the 16th c. into Ren. nature philosophy and, under the influence of the mysticism of the German, Jakob Boehme (1575–1624), and the Swede, Emanuel Swedenborg (1688–1772), into the doctrine of correspondences, which viewed the external world as a system of ss. revealing the spiritual world in material form. By the romantic period, the view that nature is the visual lang. of God or Spirit became established as one of the mainstays of poetry, but two fundamental shifts had occurred: the material and spiritual worlds were seen as merged rather than related simply as representation to thing represented; and, as a result, the meaning of ss. became less fixed and more ambiguous (see Seward, Sewall, Hirst, Wimsatt, Todorov, and Adams).

Out of this romantic trad. has grown a large and influential 20th-c. movement which has tried to reunite what the Reformation and science ostensibly had sundered. Following the lead of such writers as Urban, Cassirer, Langer, and Wheelwright, modern philosophers and critics have developed a set of concepts whereby the lang. of ss. can be regarded as having as much epistemological status as—if not more than—the lang. of fact and reason. The question therefore is: If the latter

refers to the "real" or "objective" world, and is subject to test and verification, what does the former refer to, and does it too have analogous evaluative procedures?

The answers range along a spectrum: at the one extreme are the positivists, who say that the lang. of fact and science is the only true lang., and therefore that all other langs. are nonsense; at the other end are the mystics, who claim that the lang. of fact and science is trivial, and that the only true lang. is that of symbols. Northrop Frye, however, falls off the spectrum altogether: neither world for him, at least as literary critic, exists. He has simply postulated that there is an "order" of lit., that this order has an objective existence in the totality of literary texts, and that it is based upon the fundamental seasonal monomyth of Death and Rebirth.

Closer to the center are two other positions which seek either to balance this subject-object split or to reconcile it. The first is exemplified by I. A. Richards, who accepted the distinction between scientific and poetic lang. but then proceeded to accord to the latter a status and value of its own. Thus he distinguished between "referential" lang., the lang. of fact and science, and "emotive" lang., the lang. of poetry. The status and value of the latter were found in its ability to arouse and organize our emotions, thus giving poetry some sort of psychological and therapeutic if not metaphysical "truth." The New Critics, not entirely satisfied with this distinction, claimed further that poetry gives us another and "higher" kind of truth, a truth of human existence which is more complex and profound than that of mere fact and science.

The reconcilers, exemplified chiefly by Cassirer and Langer and their followers, claim that *all* langs., whether of science or poetry, or of any in between, are various ways in which the human mind constructs reality for itself, and therefore that all our knowledges give *pictures* of reality rather than reality *itself.* For this school, humanity has not "lost" or "forgotten" the lang. of ss.; it has merely come to prefer one kind of symbolic lang. to another.

Such is the process of history, however, that this ambitious theory itself has been turned inside out, and later movements have claimed that, since all our langs. are equally symbolic, they are all equally meaningless—at least insofar as the quest for "objective" truth is concerned. Thus we find the theory of "paradigms" in the philosophy of science (Kuhn, Rorty) which says that scientific hypotheses are merely arbitrary constructs which may appear true in one era but are supplanted by other hypotheses in another era—"truth" being more a matter of cultural convenience, mental set, and consensus than of objective verifiability. And we find the theory of "deconstruction" in ling. and lit. theory and crit. (see Culler), which claims that, since the relation between signifier and signified is arbitrary, lang. itself cannot be made to carry determinate meanings.

Another and somewhat more "rational" approach, represented by Wimsatt (1954), Kermode, and Fingesten, and anticipated by Whitehead, is that ss., since they come between ourselves and reality, can be the agents of distortion and error as well as of knowledge and insight. Thus they urge that the subjective be balanced by the objective.

II. IN POETRY. To the hapless reader confronting the problem of how to recognize, understand, and interpret ss. in poetry, these philosophical disputes may seem not only bewildering but also irrelevant. The fact is, however, that one's practical approach to ss. will be governed in large part by one's theory, for a critic's use of any given term is determined by the assumptions she or he makes about lit., lang., and reality, and by the kind of knowledge sought.

Olson (in Crane), for example, as a neo-Aristotelian, is primarily concerned with literary works in their aspect as artistic wholes of certain kinds, and he therefore regards symbolism as a device which is sometimes used in the service of the work's artistic effect—to aid in the expression of remote ideas, to vivify what otherwise would be faint, to aid in framing the reader's emotional reactions, and the like. Yeats, by contrast, is primarily interested in the suggestive powers of poetry, and so he extends his definition of symbolism to include not only images, metaphors, and myths, but also all the "musical relations" of a poem—rhythm, diction, rhyme, sound. Or again, Wheelwright, Langer, Cassirer, and Urban, as anti-positivists, are concerned to defend poetry as having epistemological status, so they stress symbolism's powers of bodying forth nondiscursive meaning, truth, or vision. Kenneth Burke, as a student of lang. in terms of human motives, deduces the form of a literary work from speculations as to how it functions in relation to the poet's inner life, and so he emphasizes the way in which various elements of that work symbolize an enactment of the poet's psychological tensions.

But the simplest way to begin interpretation is to regard ss., although they may derive from literal or figurative images or both, as a kind of figurative lang. in which what is shown (normally referring to something material) means, by virtue of some sort of resemblance, suggestion, or association, something *more* or something else (normally immaterial).

A. *Identification.* When interpreting ss. in a poem, it is helpful to begin by identifying its imagery and analyzing the source of that imagery in experience, whether from the natural world, the human body, human-made artifacts, and so on. We then proceed to ask whether the imagery in question is literal or figurative. If it is literal, it may belong to any aspect or combination of aspects in the work, whether plot, character, setting, point of view, and the like, and we are to interpret it on this literal level. If we find that such a reading seems in some way incomplete, that we are not fully doing justice to what the work ultimately

seems to be about, then we are to go further and look for the following clues: (1) the connection between image and s. is made explicit, as when the speaker in Arnold's "Dover Beach," after describing the actual seashore scene, goes on to talk about the "Sea of Faith"; (2) the work accumulates additional meanings in context by means of suggestion and association, as in Marvell's "The Garden," where it develops that the speaker is responding to this particular garden with the Garden of Paradise in mind; (3) the literal action portrayed in the poem is embedded within the larger contexts of lit. and myth, as in Tennyson's "Ulysses," where the speaker is presented as an actual person but can only be truly understood within the larger contexts of Homer and Dante.

If the imagery, on the other hand, is not literal, it may be of two sorts: (1) it is presented as if it were literal, but as it develops we see that it is, rather, a dream, a vision, a fantasy, or an imaginary action, and hence must be understood entirely on a symbolic level, as is the case in Yeats's "Sailing to Byzantium," where the speaker talks about crossing the sea and coming to the holy city, which seems literally improbable; (2) there is a literal action and situation, but certain metaphors and similes are also presented in relation to one another and to the literal action so as to produce an additional level of meaning—by way of expanded, recurring, or clustered figures (see Burke and Brower).

Thus, symbolism resembles figures of speech in having a basic doubleness of meaning between what is meant and what is said (tenor and vehicle), but it differs in that what is said is *also* what is meant. The "vehicle" is also a "tenor," and so a s. may be said to be a metaphor in reverse, where the vehicle has been expanded and put in place of the tenor, while the tenor has been left to implication (cf. Bartel). And this applies even to recurring figures within a literal action, because such figures are embedded in a context of more complex relationships within the work as a whole rather than occurring simply as figures *per se.*

Similarly, symbolism resembles allegory (q.v.). Technically, allegory refers to the use of personified abstractions in a literary work. Spenser's *The Faerie Queene* is a standard example: the Redcross Knight represents the Christian soul, Duessa the duplicity of temptation, Una the true Church, and so on. Not only the characters may be allegorical, however; the setting and actions may also follow suit. Thus the work as a whole may be allegorical. The difference between this form and symbolic works is that allegory begins with the tenor, the vehicle being constructed to fit, while s. begins with the vehicle and the tenor is discovered, elicited, or evoked from it. Beginning with Goethe and Coleridge, this distinction was turned into a value judgment, with allegory being condemned as didactic and artificial and s. being praised as natural and organic. This judgment became a commonplace of romantic and modern crit., until a line of defense was established for allegory by more recent critics such as Honig, Fletcher, Hayes, de Man (in Singleton), Brett, Bloomfield, Todorov, and Adams.

B. *Interpretation.* Once we know an image is symbolic, we need to see how it became so and therefore what it means. As a final practical suggestion, we will inquire into the various ways in which links may be established between image and idea to form a s.

(1) The connection, as in metaphor and simile, may be based on resemblance, as mentioned above. A great many natural and universal ss. arise in this way: accomplishing something is like climbing a mountain, making a transition in life is like a journey to a new land, etc. Examples are to be found everywhere in poetry (Bevan, Kimpel, Frye, Douglas, Embler) as well as in everyday usage.

(2) The link may evolve into an associative connection by virtue of repetition, as when a metaphor or simile is repeated so often, either in the work of a single author or in literary trad., that the vehicle can be used alone to summon up the tenor to which it was usually attached, somewhat in the manner of a code. An interesting example is found by comparing Mallarmé's swan imagery in "Le Vierge, le vivace et le bel aujourd'hui" with Yeats's in "Leda and the Swan" and "Among School Children."

(3) The connection may be based on the internal relationships which obtain among the elements of a given work, whereby one thing becomes associated with another by virtue of structural emphasis, arrangement, position, or devel. (which is, of course, true to some degree in all works containing ss.). Examples are the wall as division between the primitive and civilized in Frost, the guitar and the color blue as the aesthetic imagination in Stevens, and the island as complacency and the sea as courage in Auden.

(4) The connection may be based on primitive and magical associations, as when the loss of a man's hair symbolizes the loss of strength (Samson) or the rejection of worldly desires (monastic and ascetic practice), not because of any resemblance between them but rather because a mythic and ritualistic relationship has been established between secondary sex characteristics, virility, and desire. The underlying sterility/fertility symbolism in Eliot's *The Waste Land* is a conspicuous example.

(5) The connection may be derived from a particular historical convention, such as the transmutation of lead to gold as redemption, the lily as chastity and the rose as passion, or the fish as Christ (see Hirst). A noted poetic instance is Yeats's use of the Rosy Cross, derived from Rosicrucianism, to symbolize the joining of flesh and spirit.

(6) The connection may derive from some private system invented by the poet—for example, the phases of the moon as the cycles of history combined with the psychology of individuals in Yeats, or embalmment as an obstacle that cannot

be overcome in the attempt to resurrect the spirit in Dylan Thomas (see Olson 1954).

Critics rightly warn that symbolic associations of imagery should be neither too explicit nor too fixed, for implications of this sort are better felt than explained, and they vary from work to work depending on the individual context (see, for example, Carlson, Mischel, Cary, Wimsatt [1965], and Todorov).

W. B. Yeats, "The Symbolism of Poetry," *Ideas of Good and Evil* (1903); H. Bayley, *The Lost Lang. of Symbolism*, 2 v. (1912); I. A. Richards, *Science and Poetry* (1926); D. A. Mackenzie, *The Migration of Ss.* (1926); A. N. Whitehead, *Symbolism* (1927); H. F. Dunbar, *Symbolism in Med. Thought* (1929); E. Bevan, *Symbolism and Belief* (1938); W. M. Urban, *Lang. and Reality* (1939); K. Burke, *The Philosophy of Literary Form* (1941), *Lang. as Symbolic Action* (1966); S. K. Langer, *Philosophy in a New Key* (1942), *Problems of Art* (1957); C. M. Bowra, *The Heritage of Symbolism* (1943); E. Cassirer, *An Essay on Man* (1944), *Lang. and Myth* (1946), *The Philosophy of Symbolic Forms*, 3 v. (1953–57); E. W. Carlson, "The Range of Symbolism in Poetry," *SAQ* 48 (1949); R. Hertz, *Chance and S.* (1949); M. Foss, *S. and Metaphor in Human Experience* (1949); E. Fromm, *The Forgotten Lang.* (1951); R. A. Brower, *The Fields of Light* (1951); T. Mischel, "The Meanings of 'S.' in Lit.," *ArQ* 8 (1952); E. Olson, "A Dialogue on Symbolism," in Crane, and *The Poetry of Dylan Thomas* (1954); Special Issue on Symbolism, *YFS* 9 (1952–53); Special Issue on Symbolism, *JAAC* 12 (1953); H. D. Duncan, *Lang. and Lit. in Society* (1953), *Ss. and Social Theory* (1969); C. Feidelson, Jr., *Symbolism in Am. Lit.* (1953); B. Kimpel, *The Ss. of Religious Faith* (1954); W. K. Wimsatt, Jr., *The Verbal Icon* (1954), *Hateful Contraries* (1965); P. Wheelwright, *The Burning Fountain* (1954), *Metaphor and Reality* (1962); *Ss. and Values* (1954), *Ss. and Society* (1955), both ed. L. Bryson et al.; F. F. Nesbit, *Lang., Meaning and Reality* (1955); W. Y. Tindall, *The Literary S.* (1955); Wellek and Warren, ch. 15; Frye; F. Kermode, *Romantic Image* (1957); J. Cary, *Art and Reality* (1958); E. Honig, *Dark Conceit* (1959); J. W. Beach, *Obsessive Images* (1960); *Literary Symbolism*, ed. M. Beebe (1960); *Metaphor and S.*, ed. L. C. Knights and B. Cottle (1960); B. Seward, *The Symbolic Rose* (1960); *Symbolism in Religion and Lit.*, ed. R. May (1960); E. Sewell, *The Orphic Voice* (1961), *The Human Metaphor* (1964); H. Musurillo, *S. and Myth in Ancient Poetry* (1961); *Truth, Myth, and S.*, ed. T. J. J. Altizer et al. (1962); T. J. Kuhn, *The Structure of Scientific Revolutions* (1962); R. Ross, *Ss. and Civilization* (1962); *Myth and S.*, ed. B. Slote (1963); A. Fletcher, *Allegory* (1964); D. Hirst, *Hidden Riches* (1964); *Literary Symbolism*, ed. H. Rehder (1965); G. Hough, *An Essay on Crit.* (1966); W. Embler, *Metaphor and Meaning* (1966); N. Goodman, *Langs. of Art* (1968); *Perspectives on Literary Symbolism* (1968), *Lit. Crit. and Myth* (1980), both ed. J. P. Strelka; R. Wellek, "S. and Symbolism in Lit.," *DHI*; C. Hayes, "S. and Allegory," *GR* 44 (1969); *Interp.*, ed. C. S. Singleton (1969); R. L. Brett, *Fancy and Imagination* (1969); M. Douglas, *Natural Ss.* (1970); P. Fingesten, *The Eclipse of Symbolism* (1970); C. Chadwick, *Symbolism* (1971); J. R. Barth, *The Symbolic Imagination* (1977); *S., Myth, and Culture*, ed. D. P. Verene (1979); R. Rorty, *Philosophy and the Mirror of Nature* (1979); *Allegory, Myth, and S.*, ed. M. Bloomfield (1981); Morier; J. D. Culler, *On Deconstruction* (1982); T. Todorov, *Theories of the S., Symbolism and Interp.* (1982); H. Adams, *Philosophy of the Literary Symbolic* (1983); R. Bartel, *Metaphors and Ss.* (1983). N.F.

SYNAESTHESIA. The phenomenon wherein one sense modality is felt, perceived, or described in terms of another, e.g. describing a voice as velvety, warm, heavy, or sweet, or a trumpet-blast as scarlet ("To the bugle," says Emily Dickinson, "every color is red"). Evidence for s. in lit. is ancient and cross-cultural, but critical conceptualization of it in the West dates only from the 18th c., and a specific term for it only appeared in 1891 (*Century Dict.*); in the literary sense it seems to have been first employed by Jules Millet in 1892. S. was popularized by two sonnets, Baudelaire's "Correspondances" (1857) and Rimbaud's "Voyelles" (1871), and by Huysmans' novel *A rebours* (1884), and from these sources became one of the central tenets of symbolism; but the device had been widely employed earlier in Ger. and Eng. romantic poetry, and it also can be found in some of the earliest lit. of the West (in *Iliad* 3.152, the voices of the old Trojans are likened to the "lily-like" voices of cicalas; in *Iliad* 3.222, Odysseus' words fall like winter snowflakes; and in *Odyssey* 12.187, in the "honey-voice" of the Sirens). In Aeschylus' *Persians* (395), "the trumpet set all the shores ablaze with its sound." In the Bible, Hebrews 6.5 and Revelations 1.12 refer to "tasting" the word of God and "seeing" a voice. Dante refers to a place "where the sun is silent" (*Inferno* 1.60). Donne mentions a "loud perfume," Crashaw a "sparkling noyse." Shelley refers to the fragrance of the hyacinth as "music," Heine to words "sweet as moonlight and delicate as the scent of the rose."

S. as the expression of intersense analogues has been exploited in lit. for a variety of effects, particularly increase of textural richness, complication, and unification. It is evident that metaphor, esp. in the tenor and vehicle model, and simile too can approximate the same kinds of suggestion, albeit in looser and more taxonomic forms. Shelley, apparently the first Eng. poet to use s. extensively, uses it particularly in connection with visionary and mystical states of transcendental union ("Alastor," "Epipsychidion," "The Triumph of Life"); here s. suggests not only a greater "refinement and complexity of sensuous experience" but also a "harmony or synthesis of all sensations" and kind of "supersensuous unity" (O'Malley). Cf. Baudelaire's "métamorphose mystique / De tous

mes sens fondus en un" ("Toute Entière").

One important species of s. is *audition colorée*, in which sound (or even silence) is described in terms of colors. Silence is "perfumed" (Rimbaud), "black" (Pindar), "dark" (Macpherson, *Ossian*), "green" (Carducci), "silver" (Wilde), "blue" (D'Annunzio), "chill" (Edith Sitwell), "green water" (Louis Aragon). This phenomenon is common in lit., the most famous example being Rimbaud's sonnet "Voyelles" (Vowels) beginning: "A noir, E blanc, I rouge, U vert, O bleu, voyelles." Such terms as "golden voice," *coloratura soprano*, "chromatic scale," Ger. *Klangfarbe* ("sound-color") show the assimilation of *audition colorée* into both common and scholarly usage. More important still is the "light–dark" opposition in vowels first demonstrated by Wolfgang Köhler in 1910 and subsequently shown to exist in many of the world's langs.: Köhler argued that this opposition is not merely metaphorical but in fact a feature of all the senses resulting from some "central physiological perceptual correlate."

The related term *synaesthesis* appears in the late 19th c. in the course of evolving psychological theories of beauty to mean a wholeness in perception, or anti-atomism in epistemology. I. A. Richards takes this term into his psychological theory of crit. as part of his neurologically derived account of literary value (*Principles*): he too uses it in the sense of "wholeness" to refer to the synergistic nature of sense-experience, wherein wholes, "sensation-complexes," are greater than the sum of their parts.

J. Millet, *Audition colorée* (1892); V. Ségalen, "Les synesthésies et l'école symboliste," *MdF* 42 (1902); I. Babbitt, *The New Laokoon* (1910), ch. 6—attacks s. as decadent; W. Köhler, "Akustische Untersuchungen," *Zeitschrift für Psychologie* 54–72 (1910–15); E. von Erhardt-Siebold, "Synästhesien in der englischen Dichtung des 19. Jahrhunderts," *Englische Studien* 53 (1919–20), "Harmony of the Senses in Eng., Ger., and Fr. Romanticism," *PMLA* 47 (1932); A. Wellek, "Das Doppelempfinden im abendländischen Altertum und Mittelalter," "Zur Gesch. und Kritik des Synästhesie-Forschung," *Archiv für die gesamte Psychologie* 79–80 (1931); W. D. Stanford, *Gr. Metaphor* (1936); S. de Ullmann, "Romanticism and S.," *PMLA* 60 (1945); A. G. Engstrom, "In Defense of S. in Lit.," *PQ* 25 (1946); E. Noulet, *Le premier visage de Rimbaud* (1953); M. Chastaing, "Audition colorée," *Vie et langage* 105, 112 (1960, 1961); G. O'Malley, *Shelley and S.* (1964); R. Étiemble, *Le Sonnet des voyelles* (1968); L. Schrader, *Sinne und Sinnesverknüpfungen* (1969)—s. in It., Sp., and Fr., incl. bibl.; G. Cambon, "S. in the *Divine Comedy*," *DSARDS* 88 (1970); P. Ostwald, *The Semiotics of Human Sound* (1973); L. Vinge, *The Five Senses* (1975); L. E. Marks, *The Unity of the Senses: Interrelatins among the Modalities* (1978); Morier, s.v. "Correspondances"; D. Johnson, "The Role of S. in Jakobson's Theory of Lang.," *IJSLP* 25–26 (1982); N. Ruddick, "S. in Emily Dickinson's Poetry," *PoT* 5 (1984); J. H. Ryalls, "S.," *Semiotica* 58 (1986); J. P. Russo, *I. A. Richards* (1989).

T.V.F.B.; A.G.E.

SYNECDOCHE (Gr. "act of taking together," "understanding one thing with another"). A rhetorical figure in which part is substituted for whole (hired "hands" for hired men), species for genus (live by the "sword" for weapons) or vice versa; or individual for group (the "Roman" won for the Roman army). Lausberg and Group Mu would limit s. to these types, characterizing the figure as a change of quantity, or of a word's semantic features or "semes"—particularizing or generalizing, material or conceptual. But in the trad. of Cl. rhet., s. also includes material for the object made of it ("steel" for sword) and abstract quality for its possessor ("pride" for the person displaying it); the figure of *antonomasia* substitutes a proper name for a common one so as to capture its associations.

In 20th-c. attempts to reduce the varied figures inherited from Cl. rhet. to an intelligible order, one group of critics has followed Ramus and Vico, arguing that there are four basic tropes—metaphor, metonymy, s., and irony. Others, following Jakobson, have claimed that there are only two—metaphor and metonymy. In the latter classification, s. is treated as a subclass of metonymy; in both, there is an evident connection, conceptual or physical, between the figurative word and what it designates, whereas no such connection exists in the case of metaphor.

Todorov and Group Mu reawakened interest in s. with their claim that it is the fundamental figure—based on an increase or decrease of a word's semes (lexical features)—from which metaphor and metonymy are derived. Attempting to draw a clear distinction between metonymy and s., Sato and others suggest that s. be limited to semantic or conceptual relations, with all material connections, such as part–whole, being assigned to metonymy. Meyer (1985) would limit the latter to contextual or accidental connections, s. being a more abstract relation. Others see s. as a figure of integration, metonymy being then one of fragmentation or reduction.

Reacting against such generalizations, Ruwet and Le Guern argue that most examples of s. are not in fact figurative: they are either fully lexicalized (i.e. part of ordinary usage) or can be understood literally. Expressions such as "give me a hand" and "all hands on deck" no longer strike us as figurative. The use of "tree" or "oak" for oak tree, or "weapon" for pistol, need not be considered s., since the generic name can be applied literally to the species. Linguists find that purported instances of s. often result from deletion of a phrase that, if included, would result in redundancy. "A herd of thirty head" does not require "of cattle." Here modern critics reveal why Cl. and Medieval rhetoricians found it difficult to distinguish tropes (nonliteral uses of lang.) from figures

of thought (varied modes of literal expresssion). "Thirty head" can be understood as ellipsis, while "the animal that laughs" (man) and "the gods of blood and salt" (Mars and Neptune) are better classified as periphrasis than as s. plus qualification. Thus s. can be viewed as a stylistic phenomenon, its effect being dependent on whether or not it is expected in its context (Klinkenberg).

Exhaustively studied in linguistics and rhet., s. has attracted less attention in poetics. The former disciplines seldom analyze examples other than nouns, nor do they consider the (hypothetical) genus–species and species–species substitutions that, according to Aristotle, are esp. important in naming that-for-which-no-word-exists. Frost's line "The *shattered* water made a misty din" and his reference to "the *crumpled* water" pushed by a swimming buck are examples. In Pound, the "night sea *churning* shingle" illustrates a precise species-for-species substitution; we lack a generic word for the motion. In some cases, one feels that a s. creates an *ad hoc* genus, as in Williams' description of the falls in *Paterson*: the river "*crashes* from the edge of the gorge / in a *recoil* of spray." If others disagree with the classifications here proposed for these examples, they thereby confirm Klinkenberg's assertion that the identification of s. is often a matter of perception if not of critical rationalization.—K. Burke, "Four Master Tropes," *A Grammar of Motives* (1945); T. Todorov, "Synecdoques," *Communications* 16 (1970); Lausberg; M. Le Guern, *Sémantique de la métaphore et de la métonymie* (1973); H. White, *Metahistory* (1973); N. Ruwet, "Synecdoques et métonymies," *Poétique* 6 (1975); N. Sato, "Synecdoque, un trope suspect," *Revue d'esthétique* 1 (1979); B. Meyer, "Synecdoques du genre?" *Poétique* 57 (1980); Group Mu; J.-M. Klinkenberg, D. Bouverot, B. Meyer, G. Silingardi, J.-P. Schmitz in *Le Français moderne* 51 (1983); B. Meyer, "La synecdoque d'espèce," *Langues et Littératures* 3 (1983), "Sous les pavés, la plage: Autour de la synecdoque du tout," *Poétique* 62 (1985).

W.M.

SYNTAX, POETIC. All human lang. derives its expressive power in part from s., the placement of words in arbitrary but conventional sequences. More than most other lang. users, poets exploit this potential when they write. Other properties of the linguistic medium such as its rhythmic and phonetic form are, of course, equally adaptable and have been more commonly subjected to systematic study by scholars. Nonetheless, poetic s. merits close analysis whether as an isolated feature or in interaction with other linguistic dimensions of poetic form such as lexis, meter, or rhyme.

Increasingly explicit descriptions of the syntactic norms of nonpoetic lang. use have stimulated study of poets' reliance on or disregard of them when composing verse. Consider a single syntactic factor: the word-class membership (the "part of speech") of the words a poet employs. At the simplest level, frequent recourse to words from any one category, even if perfectly acceptable according to the syntactic norms of the lang involved, may still yield a texture that is usually "nominal" or "verbal." An approach to the characterization of syntactic style along these lines has shown that successive literary periods have favored different proportional mixes of nouns, adjectives, and verbs (Miles). Other poets, though, challenge readers' expectations about lexical selection more directly—for example by using words assigned to one syntactic class as if they were members of another (the rhetorical figure known as *anthimeria*). Thus "did" acts as a noun in e e cummings' line "he danced his did," and "grief" becomes a temporal expression in Dylan Thomas's "a grief ago."

Diversity also characterizes poets' deployment of phrasal and clausal structures. On the one hand, repeated use of even the commonest syntactic template may lead to its stylistic dominance in that particular context: the linguistic conciseness so valued in the Augustan heroic couplet, for example, derives in large part from the frequent elision of material from one of two adjacent clauses whenever the reader can infer it from the other (Davie). Thus Dryden, writing "Concurrent Heathens prove the Story True: / The Doctrine, Miracles; which must convince" (*Religio Laici* 147-48), relies on his readers to supply the full form of the second clause, "Miracles prove the Doctrine True."

Other poets, by contrast, employ syntactic structures alien to nonpoetic contexts. At one extreme we find a relatively small inventory of syntactic licenses, e.g. inversion, which poets from many periods have drawn on freely (Dillon). Tennyson's title line "Fair is her cottage in its place" inverts the predicate adjective "fair" with the sentence's subject, "her cottage." Tennyson is fond of such permutations; cf. "Hateful is the dark-blue sky" ("The Lotos-Eaters" 84) and "Calm is the morn without a sound" (*In Memoriam* 11.1). But other instances are common, e.g. in Chaucer ("Short was his gowne" [*Gen. Prol.* 93]), Lovelace ("Thus richer than untempted kings are we" ["The Grasshopper" 37]), Milton ("But peaceful was the night" [Nativity Ode 61]), and Ransom ("Tawny are the leaves turned" ["Antique Harvesters" 1]).

At the other extreme, some poets may delete or permute material, or may so construct whole sentences that readers must struggle to parse them at all. Although usually associated with the style of 20th-c. poets such as cummings (Cureton, Fairley), extreme syntactic complexity may also be found in the works of Shelley and even of Pope, where perhaps we would least expect it (Austin). Pope writes in his "Pastorals," "Now leaves the trees, and flow'rs adorn the ground," ("Spring" 43), eliding the verb from the first clause in violation of the norms of Eng., while Shelley's *Adonais* includes the almost impenetrable sentence:

Thy extreme hope, the loveliest and
 the last,
The bloom, whose petals nipped before they blew
Died on the promise of the fruit, is
 waste.
(51-53)

Naturally, it is important not to confuse mere verbal repetition—even in a poem composed in a fixed form—with genuine syntactic patterning. The refrain line "Do not go gentle into that good night" recurs, as the form demands it should, four times in Thomas's villanelle of the same name. Yet Thomas avoids monotony by denying readers the strict syntactic parallelism they might otherwise expect: lines 1 and 18 constitute self-contained imperative clauses, but lines 6 and 12 are enjambed completions of indicative sentences begun in the preceding lines.

Some poems depend heavily on syntactic patterning as a structural device. Chidiock Tichborne's "Elegy," for example, encapsulates its striking sequence of metaphorical paradoxes in a series of syntactically congruent one-line clauses. Nor need dependence on this particular aspect of syntactic style be confined to individual works; in the Heb. poetry of the Old Testament, parallelism (q.v.) constitutes the basis for an entire poetic trad. Both parallel structures and the subtler, less extensive device of using concentric or mirror-image patterning figure prominently in Augustan syntactic practice. Thus Matthew Prior writes in "To A Lady," "Deeper to wound, she shuns the fight: / She drops her arms, to gain the field." In texts from other literary periods, however, syntactic patterning may be less blatantly, less conventionally displayed. As a result it may achieve a highlighting or foregrounding effect, throwing into relief the passage in which it occurs.

Poets' motives for selecting particular syntactic forms range from the deliberately iconoclastic to the imitative. R. D. Havens has exhaustively documented the way in which Milton influenced the syntactic style of poets for more than two centuries after his death. Both in that instance and, indeed, in the attention paid by Milton himself (among many others) to the techniques of Lat. poets, we must concede the considerable degree to which poets' syntactic choices may be attributable to *tradition*, whether openly acknowledged or silently deferred to.

Syntactic patterns of all kinds, and even the absence of a pattern where one might otherwise be expected, may also function in contextually appropriate ways. Attributing such "mimetic" or "iconic" force to poetic s. has generated considerable controversy since about 1960 (see Brogan for citations). In part, this is due to the fact that its advocates have tended to apply what is an essentially critical method in an unnecessarily mechanistic, pseudo-scientific way (Fish). Nevertheless, lively debate on this question seems likely to continue.

The technically and functionally diverse ways in which poets have turned to use the syntactic raw material of lang. should not be regarded as existing in isolation, however. Enjambment (q.v.) occurs where syntactic form fails to conform with an independently determined metrical line boundary; similar factors determine the strength of a caesura within a line. The rhymes of Hudibrastic verse derive their impact from linking phonetically similar but syntactically contrasted constructions. Semantically, s. may play a crucial role in creating or sustaining poetic ambiguity (Empson). The prominence of poetic s. relative to such other features with which it may interact cannot be predicted out of context. Where a sonnet adheres strictly to the metrical and rhyme conventions of that form, the poem's s. may contribute little to its impact; but when, esp. as in late 19th and 20th c. poetry, such formal constraints weaken, poetic s. may assume an increasingly significant role.

E. A. Abbott, *A Shakespearean Grammar* (1879); W. Franz, *Die Sprache Shakespeares*, 4th ed. (1939); Empson; D. Davie, *Articulate Energy* (1955); J. Miles, *Eras and Modes in Eng. Poetry* (1957); F. Berry, *Poet's Grammar* (1958); Sebeok; C. Ricks, *Milton's Grand Style* (1963); S. R. Levin, "Internal and External Deviation in Poetry," *Word* 21 (1965); *Essays on the Lang. of Lit.*, ed. S. Chatman and S. Levin (1967), sect. 3; A. Scaglione, *The Cl. Theory of Composition from its Origins to the Present* (1972); G. Dillon, "Inversions and Deletions in Eng. Poetry," *Lang&S* 8 (1975); I. Fairley, *E. E. Cummings and Ungrammar* (1975); S. Fish, *Is There A Text In This Class?* (1980); R. Cureton, "The Aesthetic Use of S.," *DAI* 41, 11A (1980), 4698; Brogan, sect. F; R. Jakobson, *Poetry of Grammar and Grammar of Poetry*, v. 3 of Jakobson; G. Roscow, *S. and Style in Chaucer's Poetry* (1981); T. R. Austin, *Lang. Crafted* (1984); V. Bers, *Gr. Poetic S. in the Cl. Age* (1984); B. Mitchell, *OE S.*, 2 v. (1985), *Critical Bibl. of OE S.* (1990); F. C. Robinson, *Beowulf and the Appositive Style* (1985); C. Miller, *Emily Dickinson: A Poet's Grammar* (1987); D. Donoghue, *Style in OE Poetry* (1988); J. P. Houston, *Shakespearean Sentences* (1988).

T.R.A

T

TAIL RHYME (Med. Lat. *versus tripertiti caudati*, Fr. *rime couée*, ME *rime couwee*; Ger. *Schweifreim*; rarely, caudate rhyme). A popular medieval verse-form usually of 6 or 12 lines (or multiples) in which a rhyming couplet is followed by a t. line, the rhyme of which unites the stanza, i.e. *aabccb* or *aabaab*, or *aabccbddbeeb* or *aabaabaabaab*. T. r. appears in Med. Lat. and OF verse, from the one or other or both of which it was transmitted to ME, where it flourished in the 14th c. It is well established by the 12th c., with examples perhaps as early as the 10th. The older view was that t. r. devolved from the medieval sequence, but the generally accepted view now (Guest, Jeanroy, Stengel, Meyer) is that it was created by sectioning of the Med. Lat. long line, most likely the trochaic or iambic tetrameter (the ME equivalent being the fifteener of the *Ormulum*) via internal or leonine rhyme.

In Med. Lat., t. r. is used almost exclusively for religious lyric (not at all for narrative verse), the most common form being of 6 lines rhyming $a_8a_8b_7a_8a_8b_7$ (8 + 7 syllables; *aaabaaab* is also common), the premier example being the *Stabat mater*. Most Eng. prosodists have assumed that the ME form derived from Fr., but the distribution of genres makes this seem unlikely, though it was popular in AN. Moreover, true t. r. must not be confused with the Fr. *douzaine* (*aabaabbbabba* in octosyllables). In the Fr. forms the *b* line is of the same length as the *a*s, whereas in both Med. Lat. and ME the t. line is shorter: this is distinctive of the form. In ME the 6-line form (Saintsbury's "Romance Sixes": $a_4a_4b_3c_4c_4b_3$, counting stresses) is still used for lyric, but the 12-line form became dominant for romance; half of the 70 extant metrical romances use this form, which enjoyed a vogue in East Anglia in the 14th c. (Trounce). The fact that Chaucer parodies t. r. in the *Sir Thopas* also certifies its popularity.

After 1500 t. r. disappears as a common form, though there are many later examples in sixains or mutations, e.g. Drayton's "Nymphidia" and Tennyson's "The Lady of Shalott." Burns revived it and Wordsworth uses it more often than any other stanzaic form save ballad meter (14 poems). The more general point is that many other arrangements of longer and shorter lines exhibit the phenomenon of tailing—conspicuously, the Sapphic (q.v.)—regardless of whether or not the t. line is used as a refrain.—Schipper; Meyer, v. 1; C. Strong, "Hist. and Relations of the T. R. Strophe in Lat., Fr., and Eng.," *PMLA* 22 (1907); A. McI. Trounce, "The Eng. T.-R. Romances," *MÆ* 1–3 (1932–34); U. Dürmüller, *Narrative Possibilities of the T. R. Romances* (1975); A. T. Gaylord, "Chaucer's Dainty 'Dogerel,'" *SAC* 1 (1979).

T.V.F.B.

TANKA (also called *waka* or *uta*). A Japanese form originating in the 7th c. which consists of 31 morae (conventionally construed syllables) in lines of 5, 7, 5, 7, and 7. Hypersyllabic but not hyposyllabic lines are allowed. The 3 "upper lines" (*kami no ku*) and 2 "lower lines" (*shimo no ku*) are distinguished in some poems, and a distinction is sometimes made between a dominant 5–7 or 7–5 rhythm (*go-shichi*, *shichi-gocho*). Although used singly or multiply as envoys (*hanka*, *kaeshiuta*) to "long poems," these "short poems" have been the premier Japanese form for centuries.

Their brevity has raised some question as to how seriously Western readers should take *t.* In practice, *t.* composition is, however, multiple or contextual. From early times they were considered parts of ordered and often integrated collections (*kashū*), units in a series (as of a hundred, *hyakushuuta*, or a less formal series, *rensaku*).

T. and prose narrative contexts have proved extraordinarily congenial. Given the assumption that poetry grows from moving experience and the device of using headnotes in lieu of titles to identify the experience, the prose introductions to poems could also become stories of poems. And given the use of poetry in social intercourse at court, accounts of court life naturally included poems exchanged. Such conditions led to ready combination of *t.* with prose narrative of factual, fictional, and mixed kinds. Among the prose kinds hospitable to *t.* there are tales of poems (*utamonogatari*), of which the *Ise Monogatari* (*Tales of Ise*) is the most famous; poetic diaries (*utanikki*), of which the *Tosa Nikki* (*Tosa Diary* [ca. 935]) is the first; and fictional stories, of which the *Genji Monogatari* (*Tale of Genji*) is the greatest.

After experimentation with various methods of classification, incl. some ultimately of Chinese origin, *t.* were arranged in collections giving pride of place to poems on the seasons and on love. Travel, laments, and complaints were among other hardy topics. Some other topics disappeared, and yet others achieved later popularity, e.g. religious (both Shinto and Buddhist) and miscellaneous, in which no single topic predominated.

In the 14th c., *t.* declines in quality, if not prestige, in favor of linked poetry (*renga*, q.v.). Until modern times, cl. diction and even topics were thought unnecessary, and as a result the necessary refinement led to serious attenuation arrested

only now and again by originality. The modern decline of linked poetry (*haikai* as well as of *renga*) and the popularity of *haiku* (q.v.) were matched by a revival of *t.* invigorated with new subject matter and lang. *T.* societies exist throughout Japan, and the royal household maintains the custom of a New Year's poetry contest on announced topics. The form was revived to greatness by Yosano Akiko (1878–1942); among other modern *t.* poets of note are Masaoka Shiki (1867–1902), Saitō Mokichi, Kitahara Hakushū (1885–1942), Kubota Utsubo (1877–1967), and Maeda Yūgure (1889–1951). Premodern poets of *t.* are too numerous to mention and are usually considered *waka* poets.

Although *t.* has excited intermittent interest in Western writers, for whatever reason, *t.* has never had the influence of *nō* or *haiku.* All the same, it is the definitive literary form in Japanese poetry—which cannot be understood without knowledge of the assumptions, criteria, and achievements of *t.*

Ishikawa Takuboku, *A Handful of Sand*, tr. S. Sakanishi (1934); E. Miner, *The Japanese Trad. in British and Am. Lit.* (1958); R. H. Brower and E. Miner, *Japanese Court Poetry* (1961); Yosano Akiko, *Tangled Hair*, tr. S. Goldstein and S. Shinoda (1971); D. Keene, *World within Walls* (1976), *Dawn to the West* (1984); Nobuyuki Yuasa, *The Zen Poems of Ryōkan* (1981); H. Sato and B. Watson, *From the Country of Eight Islands* (1981): M. Ueda, *Mod. Japanese Poetry and the Nature of Lit.* (1983); A. V. Heinrich, *Fragments of Rainbows* (1983)—poetry of Saitō Mokichi; J. Konishi, *A Hist. of Japanese Lit.*, 5 v. (1984–); S. Kodama, *Am. Poetry and Japanese Culture* (1984); H. C. McCullough, *Kokin Wakashū* (1985); Saitō Mokichi, *Red Lights*, tr. S. Shinoda and S. Goldstein (1989)—selected *t.* sequences.

E.M.

TASTE. When used in an aesthetic context, the term "t." may refer to (1) a person's capacity to perceive and discriminate aesthetic qualities or (2) a person's aesthetic preferences or likings. Both the capacity and the preferences may be treated factually: we have histories that describe national ts. or changes in t. from epoch to epoch; we have psychological studies of the nature of the capacity and sociological studies of the genetic, cultural, and economic forces that determine the preferences of both individuals and groups. However, the term "t." is also frequently used normatively; it then becomes a synonym for "good t." or "correct t." Thus "X has t." may mean "X has the capacity for perceiving and appreciating the truly excellent in art and can be depended upon to make valid aesthetic judgments."

Historically, the term "t." has been most closely associated with aesthetic theories that define their subject as the investigation of such qualities as beauty and sublimity, whether in nature or the fine arts, and of the aesthetic responses that these qualities arouse. The term first became important in European crit. in the late 17th c. Addison in his 1712 *Spectator* papers on t. and on the pleasures of the imagination (nos. 409, 411–21) nicely formulated the main topics discussed by critics belonging to the "School of T." (Spingarn's term). Addison defines t. as "that faculty of the soul which discerns the beauties of an author with pleasure, and the imperfections with dislike." He points out that the term is a metaphor, found in most langs., based on a likeness of "mental t." to the "sensitive t. which gives us a relish of every different flavor that affects the palate." The chief signs of a well-developed state of this faculty are an ability to discriminate differences and to take pleasure in excellencies. Though t. is a natural faculty, it can be improved by reading the authors whose works have stood the test of time, by conversation with persons of refined t., and by a familiarity with the views of the best critics ancient and modern. The critic whom Addison singles out for particular praise is "Longinus." "Longinus" is almost the only critic who has described a class of excellencies that are "more essential to the art" than the excellencies produced by adherence to "mechanical rules which a man of very little t. may discourse upon." The excellencies that Addison is particularly interested in are aesthetic qualities, which in his papers on the pleasures of the imagination he enumerates as novelty, beauty, and grandeur (sublimity).

An important line of 18th-c. critics (chiefly Hutcheson, Hume, Gerard, Burke, Kames, Blair, Reynolds, and Alison) explored in detail this new approach to aesthetic problems. All of these critics were concerned, at least in part, with aesthetic qualities (to those listed by Addison a number of others were added, e.g. the picturesque, the witty, the humorous, the pathetic) and the nature of the faculty (t.) that perceives and enjoys them. Some of the questions concerning t. that these critics tried to answer were the following: Is t. a natural or acquired faculty? What is its relation to genius? Is it an independent faculty, a special internal sense, or is it derivative from man's other faculties? Is it a single faculty or a combination of simpler faculties? What is the relation of t. to reason, emotion, and morality? What is the relation of t. to the rules? To what extent can t. be changed or corrected and by what methods? Is there a standard that determines the correctness of t.? If there is such a standard, how can it be validated? How are divergencies in t. to be explained?

Eng. and Fr. 18th-c. critics, most of whom were empiricists, gave a bewildering variety of answers to these questions. The variety became even more bewildering when Ger. transcendental philosophers and their followers in other countries began to speculate on beauty, sublimity, and other aesthetic qualities. The complex meaning of Kant's explanation of beauty ("purposiveness without a purpose") or Hegel's ("the sensuous appearance of the Idea") can be understood only in the light of each philosopher's transcendental assumptions. The transcendentalists also discovered fac-

ulties in the human mind undreamed of in empirical philosophy, and t. achieved a dignity that it never had before. Coleridge, for example, echoing Kant, defines "t." as "the intermediate faculty which connects the active with the passive powers of our nature, the intellect with the senses; and its appointed function is to elevate the *images* of the latter, while it realizes the *ideas* of the former"; t. is "a sense, and a regulative principle, which may indeed be stifled and latent in some, and be perverted and denaturalized in others, yet is nevertheless universal in a given state of intellectual and moral culture; which is independent of local and temporary circumstances, and dependent only on the degree in which the faculties of the mind are developed" ("On the Principles of Genial Crit.," 1814).

T. remained an important concept for later beauty theorists, particularly for those who, like Poe and Pater, defended an "art for art's sake" position. But toward the end of the 19th c., Tolstoy severely attacked the philosophies of beauty and t. that had dominated 18th- and 19th-c. aesthetics, and in the 20th c., I. A. Richards rejected the "phantom aesthetic state" (*Principles of Lit. Crit.* [1924], ch. 2). Perhaps the most vigorous opposition to the principles of the School of T. has come from certain postmodernist theorists who wish to reestablish a relationship between lit. and other aspects of human experience. For such critics the assumptions that the aesthetic is an autonomous realm and that it should be appreciated and judged by the "disinterested contemplation" of cultivated t. are associated with elitism, aristocratic exclusiveness, aestheticism, a withdrawal from life, and a commitment to an ordained canon of literary works which, when analyzed, can be shown to reflect and reinforce a set of power relations involving social, political, and economic factors no longer consonant with the most enlightened ethical ideals of modern times. In spite of such attacks t. is still widely used to refer to the faculty by which a person perceives aesthetic qualities; but following the work of Frank Sibley and his commentators, t. is no longer confined to the perception of beauty and a few other qualities. The number of qualities considered "aesthetic"—e.g. unified, balanced, grotesque, serene, delicate, elegant, garish, ugly, gaudy, exquisite—has increased tremendously. The enumeration, classification, and analysis of such concepts offer fresh challenges to aestheticians.

Throughout the history of crit. a perennial problem has been the relation of t. (in the sense of preference or liking) to evaluation. It is proverbial that there is no disputing about ts. (*de gustibus non est disputandum*), and theorists have enumerated a great variety of determinants of aesthetic preferences—the temperament and experience of the individual, cultural institutions, and conscious or unconscious assumptions about the nature and value of the arts. The diversity of preferences seems to imply a relativism in t. which calls into question the 18th-c. hope for the discovery of a standard of correct t. based on a consensus of the ages as to which works of lit. should rank as masterpieces, or on the uniformity of a natural faculty present universally in human nature. Lit. has many different values, and different species of lit. have different values. As long as each theorist insists on building a system that prescribes a class of values that lit. "ought" to have and excludes all other possible values, a universally acceptable definition of "good t." cannot be formulated. A commitment to some form of critical pluralism would seem a much more sensible solution; at least it would avoid creating confusion, conflict among dogmatisms, and skepticism in observers.

Critical Essays of the 17th C., ed. J. E. Spingarn, v. 1 (1908), Intro.; A. F. B. Clark, *Boileau and the Fr. Cl. Critics in England, 1660–1830* (1925); F. P. Chambers, *Cycles of T.* (1928), *The Hist. of T.* (1932); E. E. Kellett, *The Whirligig of T.* (1929), *Fashion in Lit.* (1931); E. N. Hooker, "The Discussion of T. from 1750 to 1770 and the New Trend in Lit. Crit.," *PMLA* 49 (1934); L. Venturi, *Hist. of Art Crit.*, tr. C. Marriott (1936); J. Steegmann, *The Rule of T.* (1936); J. Evans, *T. and Temperament* (1939); H. H. Creed, "Coleridge on 'T.,'" *ELH* 13 (1946); G. Boas, *Wingless Pegasus* (1950); F. L. Lucas, *Lit. and Psychology* (1951); H. A. Needham, *T. and Crit. in the 18th C.* (1952); A. Bosker, *Lit. Crit. in the Age of Johnson*, 2d ed. (1953); T. Munro, *Toward Science in Aesthetics* (1956); W. J. Hipple, *The Beautiful, the Sublime and the Picturesque in 18th-C. British Aesthetic Theory* (1957); B. Markwardt, "Geschmack," *Reallexikon*, v. 1—with extended bibl.; B. Jessup, "T. and Judgment in Aesthetic Experience," *JAAC* 19 (1960); F. Sibley, "Aesthetic Concepts," *Philosophy Looks at the Arts*, ed. J. Margolis (1962); R. Saisselin, *T. in 18th-C. France* (1965), and "T.," in Saisselin; G. Tonelli, "T. in the Hist. of Aesthetics from the Ren. to 1770," *DHI*; W. C. Booth, *Critical Understanding* (1979); P. Bourdieu, *Distinction: A Social Critique of the Judgment of T.* (1984); B. H. Smith, *Contingencies of Value* (1988); M. Moriarty, *T. and Ideology in 17th-C. France* (1988). F.GU.

TERZA RIMA (Ger. *Terzine*). It. verseform consisting of interlinked tercets, in which the second line of each tercet rhymes with the first and third lines of the one following, *aba bcb cdc*, etc. The series of tercets formed in this way may be of any length, and is brought to a conclusion by a single final line which rhymes with the second line of the tercet preceding it, *yzy z*. T. r. has a powerful forward momentum, while the concatenated rhymes provide a reassuring structure of continuity, though they may on occasion imprison the poet in a movement of mindless flux (Vigny, "Les Destinées"; Hofmannsthal, "Ballade des äusseren Lebens"). T. r. may equally represent permanence in change. In all its realizations, however, t. r. suggests processes without beginning or end, a *perpetuum mobile*

in which linkage and continuation are seamlessly articulated.

T. r. (in hendecasyllables) was introduced by Dante as an appropriate stanza form for his *Divina Commedia.* The symbolic reference to the Holy Trinity is obvious, and the overtones of tireless quest and of the interconnectedness of things to be found in t. r. were particularly apposite. Most probably Dante developed t. r. from the tercets of the *sirventes,* but whatever the origin of the form, it found immediate popularity with Boccaccio, who used it in his *Amorosa visione,* and with Petrarch (*I Trionfi*). After Dante, t. r. became the preferred meter for allegorical and didactic poems such as Fazio degli Uberti's *Dittamondo* and Federico Frezzi's *Quadriregio.* Some later poems are written in a variety of t. r. called *capitolo ternario.* The implicit difficulty of t. r. discouraged its widespread use after the 14th c., although Monti in the late 18th c. and Foscolo in the early 19th c. wrote noteworthy poems in t. r.

In France, t. r. first appeared in the work of Jean Lemaire de Belges ("Le Temple d'honneur et de vertus," 1503; "La Concorde des deux langages," 1511) and was taken up by poets of the *Pléiade*—Pontus de Tyard, Jodelle, Baïf. The decasyllable almost invariably used by these 16th-c. poets yielded to the alexandrine in the 19th-c. revival of t. r., a revival subscribed to by Parnassians (Gautier, Banville, Leconte de Lisle, Richepin) and symbolists (Verlaine, Ephraim Mikhaël, Pierre Quillard) alike.

The form makes even greater demands on poets who write in a lang. less rich in rhymes than It. or Fr. Chaucer first experimented with t. r. in Eng. for parts of his early "Complaint to his Lady," but its first significant use is by Wyatt, followed by Sidney, Daniel, and Milton. The romantics experimented with it, Byron for "The Prophecy of Dante" and Shelley for "Prince Athanase" and "The Triumph of Life." Shelley's "Ode to the West Wind" is composed of five sections, each rhyming *aba bcb cdc ded ee.* Since the romantics, t. r. has been used, usually with variation of line-length and looser rhymes, by Browning, Hardy, Yeats, Eliot, Auden ("The Sea and the Mirror"), Roy Fuller ("Centaurs"; "To my Brother"), and Archibald MacLeish ("Conquistador"). Other European poets of the 19th and 20th cs. who employed t. r. include the Dutch poets Potgieter and van Eeden and the Germans A. W. Schlegel, Chamisso, Liliencron, Heyse, George, and Hofmannsthal.

H. Schuchardt, *Ritornell und Terzine* (1875); Schipper, *History* 381; L. E. Kastner, "Hist. of the T. R. in France," *ZFSL* 26 (1904); P. Habermann, "Terzine," *Reallexikon I*; J. S. P. Tatlock, "Dante's *T. R.*," *PMLA* 51 (1936); T. Spoerri, "Wie Dantes Vers enstand," *Vox romanica* 2 (1937); L. Binyon, "T. R. in Eng. Poetry," *English* 3 (1940); J. Wain, "T. R.," *RLMC* 1 (1950); R. Bernheim, *Die Terzine in der deutsche Dichtung* (1954); M. Fubini, "La Terzina della *Commedia,*" *DDJ* 43 (1965); P. Boyde, *Dante's Style in his Lyric Poetry* (1971); I. Baldelli, "Terzina," *Enciclopedia Dantesca,* ed. G. Petrocchi et al., v. 5 (1978); Wilkins; J. D. Bone, "On Influence and on Byron and Shelley's Use of T. R. in 1819," *KSMB* 32 (1982); J. Freccero, "The Significance of T. R.," *Dante, Petrarch, Boccaccio,* ed. A. S. Bernardo and A. L. Pellegrini (1983). L.J.Z.; C.S.

TETRAMETER. (Gr. "of four measures"). I. CLASSICAL. In Gr. and Lat., the basic meter for recitation forms is the trochaic t. catalectic, i.e. four trochaic metra (– ᴗ – x , where x = *anceps*) with truncation (catalexis) of the final *anceps.* There is a diaeresis after the second *anceps,* which in comedy is sometimes replaced by a caesura before it or, more rarely, the third *breve.* In Gr. drama this meter is often associated with scenes of excitement. Several prosodic phenomena, such as (1) the occasional responsion of – ᴗ ᴗ ᴗ with a trochaic metron in Aristophanes; (2) the fact that even in the strict versification of Solon, Havet's bridge in his ts. is slightly less stringent than Porson's bridge in his trimeters; and (3) the lower rate of the substitution of ᴗ ᴗ for ᴗ or x as compared to the trimeter suggest that the trochaic t. was less constrained in its access to the rhythms of Gr. speech than some other meters. Ancient trad. (e.g. Aristotle, *Rhetoric* 1407; Marius Victorinus 4.44) considered the trochaic t. a faster meter than the iambic trimeter. At any rate, the greater speed of the t. is probably more than a matter of conventional performance tempo and may reflect a universal feature of falling rhythm in ordinary speech. The trochaic t. catalectic was employed by the archaic iambographers; Aristotle (*Poetics* 1449a2) states that it was used in tragedy before the iambic trimeter, but in extant drama it is much less frequent than the latter.

Besides its use in trochaic, the t. length was also used in antiquity with anapaestic and iambic metra. The anapaestic t. catalectic is used in comic dialogue. It is characterized by metron diaeresis as well as regular median diaeresis, frequent contraction of all but the last pair of *brevia,* and resolution of *longa.* The iambic t. catalectic was used by Hipponax and is fairly frequent in comedy for the entrance and exits of choruses and for contest scenes. Diaeresis after the second metron is preferred but caesura after the third *anceps* is common; resolution and substitution are frequent. The acatalectic iambic t. is used only by Sophocles in the *Ichneutai* and in Ion's satyr-play *Omphale.*

The Lat. adaptation of the Gr. trochaic t. catalectic is the trochaic *septenarius,* used commonly for comic dialogue and favored by Plautus. It stands in the same relation to its Gr. model as does the *senarius,* showing the same regard for linguistic stress. The absence of polysyllabic oxytones and infrequency of proparoxytones in Lat. means that the frequent trochaic closes of paroxytonic words will effect a prevailingly trochaic rhythm. As a popular form it is known as the *versus*

quadratus. Beare summarizes the long-held view that it was this meter, widely used for popular verseforms such as the marching songs of Caesar's legions, and common in late antiquity, which became the basis for much of Med. Lat. versification. It is the meter of the *Pervigilium veneris* and a number of Christian hymns, notably Venantius Fortunatus' *Pange lingua,* and was surpassed only by the iambic dimeter hymn quatrain (itself octosyllabic when regular) and the sequence as the most popular meter of the Middle Ages.—J. Rumpel, "Der trochäische T. bei den griechischen Lyrikern und Dramatikern," *Philologus* 28 (1869); H. J. Kanz, *De tetrametro trochaico* (1913); Beare; Crusius; F. Perusino, *Il tetrametro giambico catalettico nella commedia greca* (1968); A. M. Devine and L. D. Stephens, *Lang. and Metre* (1984); West.

A.M.D.; L.D.S.

II. MODERN. The t. in the prosodies of the modern vernaculars is based on feet rather than metra, hence is but half as long as the Cl. type, typically eight syllables for the iambic and trochaic t., the two commonest forms. Anapestic t. is always experimental, as in Byron's "The Destruction of Sennacherib." The t. strictly speaking shows regularity of metrical patterning (stress-alternation); with freer metrical patterning, it is simply an octosyllable, as in Fr. and Sp. versification, or else accentual verse, as in ballad meter and hymn meter. The Fr. *tétramètre* is however a four-sectioned alexandrine. The t. in both iambic and trochaic forms has retained closer ties with popular verse (it is common for songs, as in Shakespeare) and with orality than the longer line forms, the decasyllable and the dodecasyllable. It is almost always rhymed: in a famous footnote to his 1933 Leslie Stephen lecture, A. E. Housman singles out as one of the mysteries of versification "why, while blank verse can be written in lines of ten or six syllables, a series of octosyllables ceases to be verse if they are not rhymed." In Eng. poetry it appears in couplets through Wordsworth; with Tennyson, it begins to be used in quatrains. Well-known Eng. examples of iambic ts. incl.: one (only) of Shakespeare's sonnets (145), Donne's "The Extasie," Marvell's "To His Coy Mistress," Coleridge's "The Pains of Sleep," Keats' "Eve of St. Mark," Tennyson's *In Memoriam,* Browning's "Porphyria's Lover," and Arnold's "Resignation."

But it is in Rus. prosody that the iambic t. has achieved its greatest realization: Puškin wrote nearly 22,000 ts., over half his entire output, for *The Bronze Horseman* (1833) and *Evgenij Onegin* (1825–31). In Rus. it is a "solid, polished, disciplined thing," says Nabokov, who sees the Eng. form as a "hesitating, loose, capricious" line, always in danger of having its head chopped off, "maimed for life" in the 17th c. by Hudibrastic verse, and ever after emasculated as a serious meter by light verse, mock heroics, didactic verse, and the hymns (54). Statistical profiling of the filling of metrical ictuses with stresses shows that in Rus. there have been three subforms of the t.: in the 18th-c. variant, ictuses 1 and 4 are filled with stresses the most frequently, with 2 heavier than 3; in the mid 19th-c. variant, 2 and 4 are heaviest and nearly equal, with 1 nearly as heavy and 3 very weak; in Belij's 20th-c. variant, 1 and 4 are again heaviest, but 3 is heavier than 2. In Eng., Jonson and Pope write a line in which ictuses 2 and 4 are much stronger than 1 and 3, while in Milton the weights rise steadily through the line (Bailey).

In Eng., trochaic ts. are almost as common as iambic, a fact of interest because trochaic pentameters, for example, are almost nonexistent. One of the more notorious instances of modern trochaic t., Longfellow's *Hiawatha,* is however less wooden metrically than many of its detractors have claimed, and was in any case meant to imitate the meter of the Finnish folk-epic, the *Kalevala.* Catalexis is also remarkably common—as much so in the vernaculars as in Lat.—so that one finds, almost as often as runs of 8s, runs of 7s (Carew, "A Prayer to the Wind"; Shakespeare, *MV* 2.7.65–73) and distinctive mixtures of 8s and 7s in both trochaic (Shakespeare, *Passionate Pilgrim* 21) and iambo-trochaic (Milton, *L'Allegro* and *Il Penseroso*; Shelley, "Lines Written among the Euganean Hills"). The introduction of trochees into iambic ts. is common in Eng., but wholly alien to Rus. prosody.—Schipper; P. Habermann, "T.," *Reallexikon I*; V. Nabokov, *Notes on Prosody* (1964); J. Bailey, *Toward a Statistical Analysis of Eng. Verse* (1975); Morier.

T.V.F.B.

TEXTUALITY, a key concept in poststructuralism, signals a new way of understanding writing, reading, and the relations between them. It stands in opposition to the idea of the "work," its unity, and its humanistic underpinnings, and thus underwrites an attack on the metaphysical presuppositions of the traditional conception of lit. in the West.

The concept of the "work" entails meaning, unity, and the authority of a transcendent source. A work is complete, it exists in space, it is wrought by the creative power of the artist, and its meaning is stable across time and culture. A text, on the other hand, inhabits and is inhabited by lang., without a privileged outside—an origin or source—to guarantee or authorize its meaning. The source of each text is always another text, but there is always another text before that. No text lies outside the endless play of lang., and no text is complete: each exhibits traces or "sediments" of some other text in an endless repetition of originary lack. To humanistic ("logocentric") assertions of a transcendent referent (the transcendental signified) that organizes human experience and renders lang. meaningful, t. opposes the notion that at the origin there is "always already" lang., writing, a trace of some other text. The terms "trace," "supplement," and "writing" indicate an absence in the text, its impossibility of self-presence. Each text is haunted by this absence, which opens it up to an entangled web of relations with every other

text and which permits the articulation of a "subtext." The subtext is not what is "meant" or "expressed," but rather that which tends to "dissimulate or forbid" and which it nonetheless makes evident at certain points of stress or conflict. The subtext functions as a text's unconscious—what it does not know it knows—and indicates a reading against the grain.

The subtext was not always conceived as a strategic dismantling of the text. In the work of the Rus. Formalists, and in the early stages of structuralism, the subtext was one of the visible components of the text, one of the parts that fitted into the whole. The stable linguistics that gave rise to structuralism perceived the subtext as a partially hidden segment of the text, elucidation of which would provide a synchronic view of the whole. Later views of t., which perceived the text diachronically rather than synchronically, in terms of what is missing or absent rather than merely hidden from view, think of the subtext as a destabilizing element in the play of significations. The subtext is not assimilable to the text; it works against and undermines a text's potential meaning.

T. is thus fraught with dissonance. Each text is a locus of conflict which cannot be decided without repression. More recently, t. has become associated with questions of power: not only the power play between text and subtext, but of the competing claims and ideologies which make themselves evident in a text. The major effect of t. is to problematize the question of knowledge—the relation between *what* we know and *how* we know. T. assumes the impossibility of thought without lang., thus effectively subsuming knowledge within lang. itself. Disciplinary knowledge, like the work, also lacks a transcendental signified and is not authorized by any epistemological high ground. Each discipline constitutes itself as a discipline by repressing its linguistic, rhetorical nature, but t. disrupts this movement of repression, highlights it, and focuses on what a field of knowledge tends to "dissimulate" or "forbid." T. assumes the "t." of all disciplines and thus the tropological (rhetorical) nature of all knowledge. Texts read and write one another and translate one another without regard for primacy, secondariness, or disciplinary borderlines. T. transforms the relations between reading and writing and even the very nature of academic institutions: in the world of the work, knowledge is transmitted; in the world of t., knowledge is produced, and that production is always open to question. Barthes' claim that "there is no father-author" and Derrida's statement that "writing is an orphan" (themselves descriptive of the condition of t.) open texts and disciplines to an indeterminacy that infects disciplines with a rhetorical self-consciousness and disrupts the borderlines that made possible their self-definition. In affirming that there is no outside to t. ("il n'y a pas de hors-texte"), t. generates a problematizing of knowledge and the conditions of power which knowledge authorizes.

J. Derrida, *De la grammatologie* (1967, tr. 1976)—with essential preface by G. Spivak, and "Signature, Event, Context," *Marges de la philosophie* (1972); M. Foucault, *The Archaeology of Knowledge and the Discourse on Lang.* (tr. 1972), *Power/Knowledge* (tr. 1980); R. Barthes, *Le plaisir du texte* (1973), "De l'oeuvre au texte," *Revue d'esthétique* (1974); E. Said, "Abecedarium Culturae," *Beginnings* (1975), "The Problem of T.: Two Exemplary Positions," *CritI* 4 (1978); *Textual Strategies*, ed. J. Harari (1979)—excellent intro., bibl.; M. Riffaterre, *La production du texte* (1979, tr. 1983); S. Stewart, "Some Riddles and Proverbs of T.," *Criticism* 21 (1979); *Untying the Text*, ed. R. Young (1981); J. Culler, *On Deconstruction* (1982); *The Question of T.*, ed. W. Spanos and P. Bové (1982); J. MacCannell, "The Temporality of T.: Bakhtin and Derrida," *MLN* 100 (1985); H. Baran, "Subtext," in Terras; *Textual Analysis*, ed. M. A. Caws (1986); *Unnam'd Forms: Blake and T.*, ed. N. Hilton and T. A. Vogler (1986); S. Weber, *Demarcating the Disciplines*, (1986), ed., *Institution and Interp.* (1987); G. Harpham, *The Ascetic Imperative in Culture and Crit.* (1987); C. Norris, *Derrida* (1987); G. Jay, "Paul de Man: Being in Question," *America the Scrivener* (1990).

H.R.E.

TEXTURE. T. signifies the palpable, tangible details inscribed in the poetic text. It refers to the distinguishing elements in a poem which are separate and independent of its structure, the elements that persist when the argument of a poem has been rendered into a prose paraphrase. The term has close affinities with the concept of surface detail in painting and sculpture. Such a conception is designed to solve the difficulties posed by schematic and over-generalized theories of poetics. A poem has t. to the degree to which the phonetic and linguistic characteristics of its surface promote stylistic density. At one level t. involves the familiar poetic techniques of assonance and alliteration; at another level it assumes the form of sensory intensities and tactile associations (e.g. harshness or softness—cf. EUPHONY). It is to these surface qualities that t. corresponds and is made more complex by metrical patterns.

John Crowe Ransom's theory of poetry, in particular, with its stress on the dense t. of meanings in poetry, privileges the notion of t. For Ransom the t. of the poem is specifically related to "a sense of the real density and contingency of the world." By definition, then, t. is intended to correct the exaggerations of "logic" in poetry that cause the colorful local details to disappear into the grayness of systematized abstraction. Thus, in his formulation, poetic t. is characterized by its "sensuous richness," by its "fullness of presentation," by its "immediacy," and by its "concreteness." This is quite distinct from what Ransom saw as Cleanth Brooks' exclusive reliance on poetic structure. In Ransom's crit., the function of the concrete detail is not to authorize the abstract generality of struc-

ture (as, for example, in Brooks' claim that paradoxical structure underpins all poetry from the *Odyssey* to *The Waste Land*). Rather, the detail becomes formally and explicitly disjoined from the structure when the poet chooses her words, metaphors, images, and other devices. Thus Ransom's idea of poetic t., his transformation of it into a specific grounding for a theory of poetics, reverses the relationship between structure and detail in Brooks' model. This reversal serves to reimmerse poetics in the immediate and sensory experience of a contingent reality (something he felt Brooks' paradoxically abstract poetics could not do). Now the filled density of the discrete detail guarantees an experience which exceeds structure as it comes nearer to the physical body of the poem.

However, it would be an exaggeration to say that the new emphasis on t. encourages a blanket repudiation of structure. The poem, Ransom says, always retains a logical structure—what he calls the "substance"—which coincides with its prose paraphrase: "the poem actually continues to contain its ostensible substance which is not fatally diminished from its prose state: that is its logical core or paraphrase." What is significant for him is the close relationship between this logical structure of the poem and its accompanying local t.: both must be present within the poem. This is the ultimate meaning and importance which Ransom gives to t. in his metaphysic of particularity (though he does speak elsewhere of the "almost incessant tendency of poetry to *over-particularization*").

In general usage, textural detail is "irrelevant" to structure, being discovered or selected in the act of composition at a level independent of organization. From such a viewpoint it follows that one important function of detail is precisely to *impede* the argument of the poem. The poet's accommodation of the details to the demands of meter and euphony affords new and quite unexpected insights which then become themselves the most prominent feature of t.; t. thus generates a set of unforeseen and unique meanings out of the reach of structure. On the whole, these specifically poetic meanings are what formalist crit. has made its main object of study.—J. C. Ransom, *The World's Body* (1932), "The Inorganic Muses," *KR* 5 (1943), "Crit. as Pure Speculation," and "Wanted: An Ontological Critic," both in Ransom; Brooks; Wimsatt and Brooks; M. C. Beardsley, *Aesthetics* (1958); J. E. Magner, *John Crowe Ransom: Critical Principles and Preoccupations* (1971); T. D. Young, "Ransom's Critical Theories: Structure and T.," *MissQ* 30 (1977). P.M.

THEME. In common usage "t." refers simply to the subject or topic treated in a discourse or a part of it. Thus to speak of the t. of a poem may be only to give a brief answer to the question, "What is this poem about?" The t. of a poem may be trees, a Grecian urn, liberty, the growth of a poet's mind, the vanity of human wishes. But in literary studies, t. is also used in a number of more specialized senses, esp. as a recurrent element (or particular type of recurrent element) in literary works, and as the doctrinal content of a literary work. The first of these senses is studied under the rubric of Thematics.

Critics who use "t." in the second sense assume that all (good) literary works have a doctrinal content. In fables, apologues, parables, moralities, meditations, allegories, and many lyrics and dramatic monologues, this doctrinal content may receive explicit statement; in mimetic or representational genres it may be present only implicitly. In either case it is claimed that the doctrinal content is the true subject of the poem and is the element that deepens and gives significance to the experience of reading the poem. Further, the "central t." (a poem may reflect a number of related ts.) may be regarded as a formal principle—it governs the author's selection and ordering of all of the other components of the work (the concrete particulars of the poem operate as exemplars or symbols of the thematic content). A critic might point to a poem's t. by using a word or phrase, but for greater fullness and precision she will try to summarize a poem's doctrinal content in a sentence—a general proposition, an expression of attitude or evaluation, or a precept.

This conception of t. has appeared in crit. under a variety of names: "moral," "message," "precept," "thesis," "meaning," "interpretation," "sentence," "idea," "comment," etc. Over the history of crit., one or another of these terms has formed part of the vocabulary of most critics who assign a primary position to the extrinsic values of poetry. Much medieval, Ren., and neoclassical crit. was didactically oriented. Medieval literary theory, for example, conceived of poetry as an adjunct to religion and philosophy. The aim of the poet, like that of a preacher, should be to present persuasively a valid moral precept; his means is the use of attractive parable, allegory, or exemplum. The moral precept is the "t.," "nucleus," "sentence," "fruit," or "grain" of the poem.

Ren. didactic crit., of which Sidney's *Apologie for Poetrie* (1595) is a good example, was similar to the medieval position. The aim of human life, says Sidney, is virtuous action, and poetry is a discipline worthy of man's most serious attention because it is more effective than any other human learning in molding human behavior morally. Neoclassical crit., following Horace, identified pleasure and instruction as the double aims of poetry. This position resulted in a continued stress on the instrumental values of poetry, and the terms "moral" and "t." were used to point to the final cause of a poem and to its principle of unity. Thus Dryden: "The first rule which Bossu prescribes to the writer of an Heroic Poem, and which holds too by the same reason in all Dramatic Poetry, is to make the moral of the work: that is, to lay down to yourself what that precept of morality shall be, which you would insinuate into the people.... 'Tis the

moral that directs the whole action of the play to one centre; and that action or fable is the example built upon the moral, which confirms the truth of it to our experience" ("The Grounds of Crit. in Tragedy" [1679]).

Modern critics identified with the New Criticism regard with suspicion expressions such as the "moral" or the "message" of a poem. But they do talk of poetry as a kind of knowledge, and, when induced to speak of the uses of poetry, they speak of its cognitive and moral values. As a consequence, they have found the term "t." (or "meaning," "significance," "interpretation") indispensable for pointing to the values and principle of unity in a poem. However, they warn that the poem, or at least the good poem, is not a mere rhetorical device for ornamenting a prosaic t. or making it more persuasive. The good poem does not assert its t. Rather, the poem should be regarded as fictive discourse which dramatizes a human situation or moral problem. The purpose of the good poet is to explore the problem or situation in a particularized context. The net result may be simply a detailed diagnosis of the nature and complexities of the problem, though more often the poet comes up with a moral judgment or evaluation which is a possible solution to the problem. Such tentative solutions may also be called "ts."; however, these are only hypotheses which the good poem clarifies, tests, qualifies, and subjects to the fires of irony. In this process of testing, the original t. may be so qualified that no general or paraphrastic statement of it will represent it accurately.

The assumption of traditional doctrinal thematism—that all good poems must have an explicit or implicit moral, religious, or philosophical t. as their unifying principle—has been attacked by a variety of other theorists. Proponents of aestheticism, which by now has had a history of over 200 years, have argued that the function of poetry is not to provide an interpretation of life but to generate an aesthetic experience (thematic elements may of course contribute to the intensity of this experience and may have independent cognitive value as well). Also, critics of the Chicago School vigorously condemned the monism of thematic crit., which, they argued, reduces all the varied forms of lit. to a single type.

Modern nihilism has produced absurdism and a variety of other schools of poetry whose manifestoes and practice show little respect for the assumptions of doctrinal thematism. Perloff analyzes many of the strategies used by these poets to frustrate a reader's attempts to discover a thematic unity in their poems—foregrounding, for example, the qualities of the medium rather than content, the part rather than the whole, the concrete detail rather than the concept, vagueness and uncontrolled suggestiveness rather than explicitness and precision, obscurity and mystification rather than intelligibility, and incoherence (through omissions, inconsistencies, discontinuities) rather than unity and closure. The result is an indeterminate text. Some of its effects are mystery, bewilderment, promise of a meaning which is never fulfilled (unless, paradoxically, meaninglessness is itself considered a meaning). Indeterminacy, undecidability, unreadableness are also key concepts in much poststructuralist crit. But while Perloff studies the poetics of a group of poets who deliberately created indeterminate texts, other theorists have argued that the lessons of modern epistemology, semiotics, and the psychology of the reading process are that all texts are indeterminate. A text may *appear* to be centered—to present a coherent thematic development—but a close reading will always reveal ambiguities, irrelevancies, contradictory implications, and obscurities that deconstruct the surface coherence and send the reader wandering down the paths of a labyrinth from which, she soon discovers, there is no exit. The search for a determinate thematic synthesis of a text is futile; interpretations cannot be verified or falsified; the reader should give up her demand for significance and simply enjoy the linguistic play for which a polysemous text provides the opportunity. If she must have significance, some deconstructionists suggest, she may regard all texts as allegories of the reading process with undecidability or indeterminacy as the explicit or implicit central t. of every text.

R. S. Crane, *The Langs. of Crit. and the Structure of Poetry* (1953); Frye; M. C. Beardsley, *Aesthetics* (1958); M. Krieger, *The Tragic Vision* (1960); S. Sacks, *Fiction and the Shape of Belief* (1964); G. Graff, *Poetic Statement and Critical Dogma* (1970); C. Brooks, *A Shaping Joy* (1971); Culler; R. Levin, *New Readings vs. Old Plays* (1979); M. Perloff, *The Poetics of Indeterminacy: Rimbaud to Cage* (1981); J. Culler, *On Deconstruction* (1982); C. Norris, *The Deconstructive Turn* (1983). F.GU.

TONE. (1) In the general sense, "t." denotes an intangible quality which is metaphorically predicated of a literary work or of some part of it such as its style, and often felt to pervade and "color" the whole, like a mood in a human being, so that the t. becomes its pervading "spirit," "atmosphere," or "aura." To describe the t., critics usually choose adjectives, such as serious, witty, ironic, dignified, sincere, refined, apologetic, playful, vigorous, majestic, quaint, delicate, passionate, leisurely, tranquil, tender, gay, savage, melancholy, grim, playful; more complex descriptors, which attempt to map onto the text larger blocks of a given version of lit. hist., are such attributives as "sentimental," "classical," and "romantic." This usage attributes to the text what other theorists regard as either (a) a projection of the fictive speaker's attitude toward his audience—which is the sense of "t." given it for modern crit. by I. A. Richards—as manifested by markers within the text, or (b) the poet's intention in writing the poem, which may or may not be identical to (a). In either case some degree of

intentionality on some level is presumed and, as a corollary, some conception of the poem as an expressive or affective instrument—i.e. a piece of rhet., meant to move its audience. From the rhetorical point of view, successful management of t., on which the effectiveness of a discourse largely depends, consists primarily in the tactful selection of content and in the adjustment of style to influence a particular audience. Other critics have held that any stylistic feature of a text—word choice, syntax, imagery, metaphors, or other figurative devices—can contribute to t. as expressing the attitude of the speaker. This sense of the term "t." is closely related to that of voice. But one must keep distinct the t. of the poem's speaker and that of the poem itself.

(2) More specifically, "t." means t. of voice, i.e. the inflections given to words by speakers in normal discourse and heard by auditors. It is possible to utter any simple sentence, such as "I never said I loved you," with any one of a number of different intonational contours and thereby mean quite different things (one of which is "But I do love you"). The t. of a speaker's voice thus reveals information about her attitudes, beliefs, feelings, or intent, or, barring that, at least about the real meaning of the utterance. T. may add to, qualify, or even reverse the meaning of what is said. Sarcasm, for example, repudiates the plain lexical sense of the words with a contrary intonational pattern—speakers almost always give preference to intonation over denotation in deciding what an utterance means. These inflections—stress patterns, pitch changes, pauses, and extensions of duration for emphasis—are natural and perspicuous to auditors in oral speech, but written speech (and thereby lit.) has but few orthographic markers for conveying inflections (which in structural linguistics were called "suprasegmentals"), so t. has to be inferred from the context by attentive readers, reasoned out or argued for as a plausible interp. of the statement. It is by virtue of the fact that written speech but poorly captures inflection that opportunity is created for richness of ambiguity and irony.

Some poets have very keen powers of exhibiting the t. of the utterances of their poetic personas and of shifting those ts. rapidly. In Eng. one thinks most immediately of Donne, who in a poem like "The Indifferent" can rapidly shift tonal structures, creating effects of paradox, irony, humor, and satire. Most of the *Songs and Sonets* develop this suppleness of tonal fingering, which is one prominent characteristic of the metaphysical style. Some valuable modern research has been carried out to determine the phonological and syntactic mechanisms of these shifts, which seem to be effected in large part by concluding either a series of parallel members (phrases or clauses) or a main clause with its associated subordinate clauses at points of closure or transition, such as at the volta in the sonnet (Rich).

(3) In a more restricted sense of (2), "t." may mean simply pitch, as when Chinese, for example, is referred to as a t.-lang., the poetry of which has a tonal prosody, a versification based on pitch. This is an extension of the meaning of "t." in music. In short, t. as a characterization of most or all of a discourse derives from t. of voice in the linguistic sense, intonation, which derives from t. as pitch.—I. A. Richards, *Practical Crit.* (1929); J. S. Bastiaenen, *The Moral T. of Jacobean Drama* (1930); I. C. Hungerland, *Poetic Discourse* (1958); L. A. Marre, "Spenser's Control of T.," Diss., Univ. of Notre Dame (1971); M. D. Rich, *The Dynamics of Tonal Shift in the Sonnet* (1975). T.V.F.B.; F.GU.

TRAGEDY.

I. HISTORY OF THE CONCEPT
II. HISTORY OF PERFORMANCE

I. HISTORY OF THE CONCEPT. T. is a particular form of Western drama originating in or around Athens in the second half of the 6th c. B.C. Various authors from Aristotle to René Girard, via Nietzsche and Gilbert Murray, have speculated on its derivation from a "tragico-lyrical" chorus of some kind, i.e. the dithyramb or even some earlier form of ritual violence or sacrifice (the Gr. root of the term for t. refers to goats). Such rites have been understood variously—by Murray as symbolizing the passage of spring and a regenerative cycle; by Nietzsche as marking a rupture between some "Dionysiac" involvement of humans with the natural world and their "Apollonian" rational distancing from it; by Girard as marking some pathway from inhuman and asocial violence toward organized society and culture. The tragic protagonist is a scapegoat, *pharmakos*, whose death or ejection from the social group somehow cleanses, rejuvenates, or indeed creates ordered society. Such speculations are interesting, but they tend to be tautological (since the evidence comes from the ts. themselves) and slightly reactionary, in that their tendency has been to view the rationality ascribed to t. as a kind of fall from the grace of the sacred irrationality that preceded it. Let us simply say, with the preponderance of the critical trad., that ts. act out the failures of human effort, however grand, before some process (call it "fate," "the gods," etc.) of which humans are, by definition, supposed ignorant—although the very existence of ts. itself proposes some kind of knowledge, for they could not otherwise have been composed.

The earliest reasonably assured historical fact is that Thespis, an Athenian, performed some kind of t. at the City (or Greater) Dionysia between 536 and 533 B.C. The subsequent devel. of t. suggests that this performance was probably a kind of duet or dialogue between the poet himself, playing the protagonist, and a chorus. Aristotle informs us that Thespis also invented the tragic mask, at first simply face paint, then cloth, and later perhaps of clay. Arion of Corinth is said to have developed the dithyramb. Cleisthenes, tyrant of Sicyon at war

with Argos, gave to Dionysus the "tragic choruses" that had honored the Argive hero Adrastos. At Phleius, a little later, it is believed that Pratinas composed the first satyr plays. It also seems clear that, from the first, authors of ts. were competing with one another; reports of several contests are extant.

This paucity of information about the origins of t. should constrain speculation. In fact, matters are even worse. Although some information has come to us from papyri and occasional remarks (e.g. by Herodotus), our main informant remains Aristotle, who wrote his *Poetics* a century and a half later, in the mid 4th c. B.C. And that work is not entirely reliable. Although his stated purpose was to analyze t.'s structure and function, comparing it on the one hand to epic and on the other to comedy—the section on comedy has been lost—there is evidence that he distorted his account of t.'s history for political and ideological ends (Jones). Not unnaturally, he also emphasized his own favorites among the large corpus of plays to which he had access; one of his choices conspicuously retains pride of place: *Oedipus Rex*. In his view, this t. exemplified the genre: narrating a serious, complete, and unique human action (howbeit introduced *in medias res*), revealing both its typicality and its limitation, involving people of high estate and issues central to the *polis* as a whole, depicting characters neither wholly good nor entirely bad, and passing through change(s) of fortune, recognition, and catastrophe. The play illustrated precisely, he found, the six essential parts of t.: plot, character, diction (poetic style and order), thought, spectacle, and music (see discussion below).

It is not in the least surprising, however, that Aristotle's 4th c. B.C. *Poetics* should have had political goals. Throughout the 5th c. B.C., Athens (and Greece as a whole) was in a state of political and social uncertainty, constantly fraught with conflict. Both ts. and their authors were centrally involved in these affairs, almost before the reforms of Athenian "democracy" in 510. In 500 the revolt of the Ionian cities of Asia Minor provoked the Persian Wars. The revolt ended in 494 with the fall of Miletos, but the wars dragged on until 449, though the city-states were free of serious threat after 480–79. But then other wars broke out.

Both Aeschylus and Sophocles took part in these events. Aeschylus fought at both Marathon and Salamis; Sophocles was *strategos* with Pericles in the Samian war of 441–39, and apparently again in 428, was frequently employed as an ambassador, and was a member of the ruling council after 413. Clearly, political and military events cannot be irrelevant to the devel. of tragic drama: they provide the context essential to any understanding of t.'s role in Athenian society. For ts. quickly became a profoundly serious forum for dealing with political and religious issues. In 472, for example, Aeschylus' *Persians* concerned Greece's war, as had Phrynichos' *Capture of Miletos* (performed 493–92, just a year after the fall of that city). Again, one of the common interps. of Aeschylus' *Oresteia* (the only extant trilogy) has been that it shows the passage from a society dominated by divine justice to one relying on human justice within a city the authority for whose status nonetheless remains divine. (It is the goddess Athena who finally exonerates Orestes as she simultaneously institutes the new governing order.) No one, we may suppose, doubted the political significance of ts.

There are other obstacles as well to generalizations we may wish to propose, not least of which is the dearth of extant plays. Of the enormous output of 5th- and 4th-c. ts., only the merest handful survive in full—7 each by Aeschylus and Sophocles, and 18 by Euripides. One of the earliest authors, Phrynichos, like Choirilos, was reputed to have composed some 160 ts.; and from the 4th c., Astydamas II was reputed to have composed some 240 plays. Both Aeschylus and Euripides composed between 70–90 plays, Sophocles between 120–30, and they were but three among many. Some 60 authors are known from the Hellenistic period, and records survive of competition late into the 1st c. B.C. For the 5th and 4th cs. alone, this implies several hundred plays at minimum. The surviving corpus thus represents a minuscule proportion of the whole. Worse, perhaps, their survival was due to much later Alexandrian anthologizers, whose purposes were not what we might now consider "literary" but rather philological, grammatical, or rhetorical.

We must thus exercise extreme care in drawing any sweeping or universal conclusions about the nature of Gr. t. The material hazards of transmission, the nondramatic purposes of anthology composition, and the importance of one critic's predilections alone would guarantee that such could not be the case. Yet another barrier to comprehension has been revealed by recent scholarly work which shows that the post-Ren. and Enlightenment emphasis on individual character and psychological "self-understanding" falsified both the plays and the ancient idea of t. in particular, and of personhood and its place in society in general (Belsey; Reiss; Vernant and Vidal-Nacquet). We cannot but conclude, then, that for practical reasons (survival of the corpus), for contextual reasons (e.g. Aristotle's preferences and their consequences for that survival), and for conceptual reasons (our unfamiliarity with Gr. beliefs about humanity, society, and the world), we need to understand Gr. t. not as some generalizable artistic form but as something particular which it requires an effort to comprehend in something like its own terms.

One would say, therefore, that Christopher Leech is quite right to tell us bluntly to beware of overly grand pronouncements about t. It is simply "a concept that we deduce from the contemplation of a heap of ts." (24). Moreover, if this be so, and even had the concept as such existed for

antiquity (which it did not), it would have been different from any notion we may now have, because it would derive from different examples. How much such a (false) abstract concept may affect not only our judgment but even our contemplation of *facts* has been amply demonstrated by George Steiner, who argues for a view of t. that is quite un-Gr. (though using it to assert the subsequent "death" of t.). In his words, "any realistic notion of tragic drama must start from the fact of catastrophe. Ts. end badly" (8). From the early Ren. to the present, many others have maintained the same view. But for the Greeks catastrophe was a technical device that did not have to conclude a t., as we can see in the plays themselves (e.g. *Eumenides, Oedipus at Colonus*, or *Helen*) and can learn by reading Aristotle, who preferred a more fortunate ending. To accept such a concept as defining t. would mean "we must either exclude a very large part of extant Gr. t. or redefine *t.* or *badly*" (Reiss 13). Nor does the caveat apply only to Gr. t. Was Corneille's *Le Cid* a t.? Or his *Horace*? What of Racine's *Bérénice* or Tate's *King Lear*? Certainly the first two *contain* catastrophe, but it would be a nice critic who could define their conclusions as unambiguous in that regard. As for *Bérénice*, does it even contain a catastrophe? (For that matter, does the *Alcestis* of Euripides?) And while we may think Tate's *Lear* simply an absurd botching of a glorious predecessor, neither late 17th- nor 18th-c. audiences thought so.

We ought therefore to be careful before adopting any universal concept of t. or of the tragic. Aristotle's *critical* text did establish a trad., but modern philology, fresh discoveries, greater contextual awareness, all advise us to examine it with care. We should first give their separate due to authors and audiences of the Gr. 5th c. B.C., of the Sp., Fr., and Eng. 16th–17th cs., and of the Ger. and Scandinavian 19th c. We may add, too, those of a 20th c. that has seen a line stretching from Büchner to Hauptmann to Brecht. At the same time, such a list itself implies that despite the differences of history, society, and culture, something has nonetheless enabled people to believe, if not in the identity of t. through time and place, certainly in at least *some* similarity of function or meaning. But we can explore that similarity precisely by emphasizing those very historical and cultural differences. Indeed, only by so doing can we obtain some clear notion of what t. has actually *done*, of what has been the cultural *function* of ts. in the environments in which they have really existed. For it is manifestly the case that a dramatic form called "t." has persistently recurred at moments in the Western trad. since the Greeks.

The customary critical claim has been that the trad. was continuous and essentially homogeneous. Scholars, critics, and dramatists have all asserted that t. derived the rules for its comprehension from Aristotle, and for actual practice and production from the Greeks and Seneca. From Aristotle to Scaliger, from Heinsius to Hume, and from Hegel to Croce, they affirmed that its purpose was that of *katharsis*, understood variously (and often vaguely) as some "purging" of the spectators' emotions of pity and terror, as a kind of "medical" reduction of their force, or as a "religious" emotional purification which made the spectators wiser and more tolerant. (Modern critics tend to see these claims as a response to Plato's argument that far from calming human emotions, mimetic art roused them to greater violence.) Also from Aristotle come the ideas that t. functioned by means of *mimesis* or representation (see REPRESENTATION AND MIMESIS); that its ordering structure was that of a linear "plot" or *mythos*, leading the spectator from a beginning *in medias res* through a middle involving some confusions and at least one "change of fortune" (*peripeteia*) to an end which embodied a "recognition" (*anagnorisis*) of previous ignorance and a new understanding; and that its main protagonist (a person of high estate, and neither wholly "good" nor "bad") underwent this experience because of some "tragic flaw" (*hamartia*).

Part of the difficulty with all this has been subsequent critics' inability to agree on the meaning of even these fundamental terms. *Katharsis* has been the single most disputed term, yet occurs only once in Aristotle's *Poetics*. (Most commentators have agreed, however, that it refers to the spectators, not the actors.) *Mimesis* has posed similar problems, though it is now generally agreed that it does not mean representational copying, but rather some depiction of what is essential to human action, enabling any particular instance to be generalized as in some way typical. *Mythos* denotes the form taken by such depiction.

Hamartia has also proved provocative. From the Ren. until quite recently, *hamartia* was interpreted as something close to *hubris*, a kind of overweening pride, whose exemplary Ren. figures were Marlowe's Tamburlaine and Faustus: indeed, such an interp. works well when applied to the growing individualism depicted in late Ren. and Enlightenment t. The term was traditionally interpreted, then, in a fundamentally ethical sense: i.e. it was a moral failure for which the protagonist was personally responsible. More recently, this has been shown not to work for Gr. drama, with its entirely different view of subject and character, and to be an erroneous interp. of Aristotle (Vernant and Vidal-Nacquet). Closer analysis of the *Poetics* has shown that the term is connected not so much with the sense of a lack of will (though it *can* imply that) as with that of "unwittingness." Indeed, Halliwell has shown that it ranges from willful evil (but is *that* then a "failure"?) to simple lack of knowledge (215–30). We can then perhaps say that *hamartia*, the so-called "tragic flaw," simply named the apparent single cause of *any* failure by the protagonist to act or to know, presented as such to the spectators' understanding. We will soon see the

importance of such audience understanding.

Much disputation may be avoided if we understand that these descriptive terms had their origin in Aristotle's effort to comprehend how t. functioned in his own time and place. The way that Ren. and then Enlightenment critics took them over may tell us a lot about what *they* wanted, but it tells us very little about antiquity. To understand t., therefore, we might cease looking first at Aristotle's terms and begin instead with the historical contexts of the works in question. And here we can see immediately something rather notable about t.'s appearances. Apart from "t." itself, only two things are common to its several appearances in Western societies.

First, t. has occurred at moments of precarious social and political consolidation, which quickly proved to be moments of transition from one form of society to another. In Greece, for example, the heyday of Aeschylus, Sophocles, and Euripides fell precisely between the Persian and the Peloponnesian Wars. The first war signaled a passage from archaic Gr. society toward the consolidation of the city-states, while the latter marked the decay of those conflict-prone states before the consolidation of the Macedonian hegemony and the coming of the "Hellenistic world." Similarly, in the European 16th and 17th cs., the great age of t. was also an age of warfare, incl. the revolts against the Hapsburgs of Spain and Empire, the religious wars in France, the Thirty Years War, the Eng. Civil War, and finally the Frondes. It was a period that marked a transition between the death throes of feudalism and the birth of capitalism. In Germany, the period between the mid 18th and mid 19th cs. (from Lessing, Schiller, and Goethe to Kleist, Hölderlin, and Grillparzer) shows the beginning of a transformation from the feuding principalities of the 18th c. to the Prussian Customs Union of 1830, toward eventual unification and empire under Bismarck.

Second, t. has *in every case* been followed by the consolidation of a political theory of extraordinary power: Plato and Aristotle, Hobbes and Locke, Hegel and Marx. It is as if t. had discovered in its confused environment, and then related, some new conceptual and discursive process enabling certain doubts to be overcome, clearing up the incomprehensions inherent in earlier social and political decay and dissolution, and facilitating the establishment of a new order of rationality. Many recent commentators have observed at length just how much Gr. t. focused upon "lack of security and misplaced certainty in and about lang." (Goldhill; cf. Segal). The same can be said of Ren. t. (Reiss) and has been shown for the Ger. romantics (Benjamin).

Within those limits, what links ts. is their presentation of a protagonist whose powerful wish to achieve some goal seems inevitably to come up against limits against which she or he is powerless. The limits may be self-created, or imposed from without (by people or by some impersonal force), or they may even represent an inability to establish any precise sense of what may be at issue (this is common in the Ren. ts. of Buchanan, for instance). The result is an impasse for the protagonist—defeat, humiliation, often death. Whatever particular interp. we may make of this, in general such drama is itself the sign of the transitional moment after the collapse of a stable order and the re-establishment of another.

II. HISTORY OF PERFORMANCE. To assess either the nature of tragic performance or the playing space in 5th-c. Athens with precision or in detail is difficult. Trad. indicates that performances originally took place in the *agora* (marketplace), which we have no reason to doubt, and all the less so, perhaps, because we know how the ts. of Lope de Vega, Calderón, Tirso de Molina, or Rojas Zorilla were played in the innyards of Golden Age Spain, and how those of Marlowe, Shakespeare, Middleton, Webster, and Beaumont and Fletcher were performed in open-air theaters in England. By the early 5th c., the performances of the Dionysia took place in a wooden structure erected on stone foundations at the foot of the south cliff of the Acropolis. This and other impermanent structures have left almost no trace. The first stone structure dates from the late 4th c. B.C.; the theater of Dionysus visible in our own time is a much later structure on a site that has seen many buildings. Vase paintings may provide some additional evidence, but we need to be wary of drawing conclusions from an artform with its own conventions. The plays themselves, of course, provide most evidence.

The chorus seems to have performed in front of a slightly raised stage, accompanied by a musician (*auletes*) playing on a double pipe. (Later, the stage was raised much higher.) The chorus was composed of citizens drawn from a single *deme*, while the actors themselves were semiprofessional. No women performed. Thespis is thought to have invented the mask; Aeschylus introduced a second actor, Sophocles a third. The former is also credited with being the first to compose a tetralogy on a single theme.

Throughout the 5th c., as explained in Part I, t. had remained deeply caught up in issues involving both the internal political order of an Athens whose form of government and power relations were a constant matter of debate and conflict, and also external relations whose instability reached a climax in the war with Sparta in 431. Final defeat came in 404. But the whole period was one of political and military struggle against which the Athenians had mounted various offensives. Among these t. was one. Contrary to what has all too often been claimed, t. has not been a demonstration of human incapacity in the face of some powerful or incomprehensible event. It has rather been a means of explaining what might otherwise be incomprehensible, a way to show just why a given group or individual failed or was defeated

by some set of circumstances. The Athenians sought to create a new, trustworthy political and epistemological order. To do that, in addition to actual practice, they had to create some ordered conceptual process able to "enclose" whatever might escape such knowledge and such order (since no conceptual or political process can in fact function save by selection and exclusion), and to place it outside their new system. To do that, they had to find a means either of indicating that nothing escaped such order, or else of asserting it could explain and therefore understand such events as did escape it. T. was one of those means, and the audience had so to understand it. It performed order in a theater where semi-professional actors, the representative chorus of the *deme*, and the citizen body gathered together on the slopes of the Acropolis within shouting distance of the Pnyx, the everyday arena of political debate (Else 1965).

As Athens approached its final loss of stability in the late 5th c. B.C. (a process that culminated in Euripides' *Bacchae*, where the order of state finally cedes to a dissolution created by Dionysus himself), t. became increasingly ambiguous. Both the *Bacchae* and *Oedipus at Colonus* were performed posthumously, precisely around the period of the Athenians' final defeat. Both represent the end of the great days of Gr. t. The "classics" started to be performed once again after 385, but in this period of "revival," theater was a matter for traveling professional troupes and was no longer central to the city's political and cultural fabric. Never again in antiquity was it to recover such a role. Perhaps such lack of cultural centrality explains why the 500 years between 5th-c. Athens and Nero's Rome are represented by little more than fragments and titles of ts. and the names of their authors.

T. was "rediscovered" by the European Ren.—by Italy, and then by France, England, the Netherlands, and Spain. Seneca was far and away the single most important influence, esp. as to his bombastic style, but Sophocles and Euripides both had some mild impact. Initially a school exercise in rhetorical composition and performance, a means to improve the vernaculars, and a way to communicate the political and ethical commonplaces of the ancients, the humanist t. of the Ren. rapidly took on an aura of national political commentary. It also helped create vernaculars able to compete in quality of expression with both Gr. and Lat., and provided a means for writers to elaborate an elite literary genre the perfection of which would contribute to establishing equality between the new European cultures and those of antiquity. At the same time, esp. in France, ts. became tools in genuine political battles: Protestant and Catholic confrontations occurred on stage as well as on the battlefield. Bèze's *Abraham sacrifiant* of 1550, for instance, was "answered" by Jodelle's *Cléopâtre captive* of 1552, which praised Henri II and equated him with Octavian. In Spain the ts. of honor written by dramatists such as Lope de Vega and Calderón reflected both the glory of imperial Spain and the ambiguities brought on by growing internal instability and external threats. In England, the hubristic individualist ts. of a Marlowe, and Shakespearian ts. that seem to place their protagonists up against conflicts they cannot resolve, gave way to the darkening tones and violence of Jacobean revenge t., as though in preparation for the political and military struggles to come. In France, humanist t. was replaced by the generation of the elder Corneille: Mairet, Scudéry, Du Ryer, Tristan l'Hermite, and particularly Rotrou. In Corneille, esp., one can follow that era's confrontations between feudal nobility and central monarchical authority, between conspiratorial conflict and State stability. That trad. was continued (albeit in his own way) by Racine, whose ts. may readily be interpreted as a set of "experiments" performed upon different political situations and conditions. At the same time, these plays show how individuals and circumstances threatening to social stability were overcome and removed.

It is probably fair to pick out five or six playwrights from this era who are considered generally, or within their own country, to epitomize the genre. Shakespeare is no doubt premier among them. Writing in blank verse, using subjects drawn from history or legend, he presents characters ranged against obstacles frequently not simply of their own choosing, but even of their own making: *Hamlet* and *Macbeth* introduce supernatural elements, to be sure, but these do not gravely effect the instance of choice. Richard II glories in his eloquent railing, but fails through mistaking his place in a now altered divine order. *King Lear* is possibly the most terrifying of all tragedies, in that Lear himself seems responsible for setting the heavens against him, and brings disaster on innocent and culpable victims alike in a general collapse of the kingdom. In France, the debate was always between the elder Corneille and Racine, between the opulent and Rubenesque Corneille and the clean-cut, ascetic, and Vermeerlike Racine, whose clarity of diction, paucity of display, and tautness of alexandrine, were all quite unique. All three playwrights created the modern psychological tragic figure, as the Schlegels and other Ger. romantic critics claimed Lope de Vega and Calderon did in Spain (though recent work suggests otherwise) and, less familiarly, Joost van den Vondel in Holland, whose 1556 *Lucifer* was a source of Milton's epic and a possible precursor of Goethe's *Faust.* Vondel's vigorous diction, vehement sense of place, feeling for the ancient and modern theater, and above all creation of powerful characters allow the Dutch to call him their "Shakespeare."

T. in 18th-c. Europe tended to be a rather dry attempt to recapture that earlier active t., and tended therefore to be but a pale reflection of the political and conceptual order in whose creation

earlier t. had participated. Such work was represented in Italy, for example, by Alfieri, in France by Crebillon and Voltaire, and in England by Addison (*Cato*) and Lillo (*The London Merchant*). These plays present a kind of moralistic support for what was by then the accepted order of Enlightenment rationality and political organization. Only in Germany from the mid 18th c. on do we find t. achieving the kind of constructive process seen in 5th-c. Athens or in late Ren. Italy, England, Spain, Holland, and France. Lessing was the first to launch this creative t., and the process culminated, brilliantly, in Schiller, Goethe, Kleist, and the (failed) efforts of Hölderlin.

After 1850, Ibsen and Strindberg are doubtless the most likely candidates for consideration as writers of t. Yet even here the characters do not really *compose* a realm of knowledge and action as they did in the great periods of t. Rather do they deal with systematic truths that pre-exist their activities—much as did the protagonists of the 18th c. Whether the same may be asserted of O'Neill and Miller, of Beckett and Pinter, or of Lorca and other modern authors is a question not perhaps to be answered here. One may suggest, however, that in Bertolt Brecht the questions haunting the great ages of t. *do* once again become "constitutive." In his theatrical practice he rediscovers t. as an effort to create a new systematic process in a period marked by the overthrow of cultural and political order. Coming after Büchner, Hauptmann, and Piscator, Brecht shows his protagonists striving to understand a historical movement they themselves create even as they are created by it. History in Brecht becomes fundamentally ambiguous, for the characters themselves have to make it meaningful. History and social action, that is, create one another simultaneously. Once again in Brecht's hands drama becomes creative of a new order and of a new understanding of the social and political practice enabled by such order.

G. Freytag, *Die Technik des Dramas* (1863), 6th ed. tr. as *Technique of the Drama* (1898)—"Freytag's Pyramid"; F. Nietzsche, "The Birth of T." (1872), rpt. *Complete Works*, v. 1 (1924); A. C. Bradley, *Shakespearian T.* (1904); W. B. Yeats, "Tragic Theatre," *Essays* (1924); A. Pickard-Cambridge, *Dithyramb, T. and Comedy* (1927); H. C. Lancaster, *Fr. T. in the Time of Louis XV and Voltaire*, 2 v. (1950); J. V. Cunningham, *Woe or Wonder* (1951); Wimsatt and Brooks, chs. 3, 25; K. Muir, *Shakespeare and the Tragic Pattern* (1958); G. F. Else, *Aristotle's* Poetics: *The Argument* (1959), *The Origin and Early Form of Gr. T.* (1965), *Plato and Aristotle on Poetry* (1986); I. Ribner, *Patterns in Shakespearean T.* (1960), *Jacobean T.* (1962); H. D. F. Kitto, *Gr. T.*, 3d ed. (1961); G. Steiner, *The Death of T.* (1961); E. Olson, *T. and the Theory of Drama* (1961); E. A. Havelock, *Preface to Plato* (1962); J. H. F. Jones, *On Aristotle and Gr. T.* (1962); W. Benjamin, *Ursprung des deutschen Trauerspiels* (1963, tr. J. Osborne as *The Origin of Ger. Tragic Drama*, 1977); *T.: Mod. Essays in Crit.*, ed. L. Michel and R. B. Sewall (1963); M. T. Herrick, *It. T. in the Ren.* (1965); A. Lesky, *Gr. T.* (1965), *Gr. Tragic Poetry*, tr. M. Dillon (1983); R. Williams, *Mod. T.* (1966); B. M. Knox, *The Heroic Temper: Studies in Sophoclean T.* (1966), *Word and Action* (1979); N. Frye, *Fools of Time: Studies in Shakespearean T.* (1967); J. M. R. Margeson, *Origins of Eng. T.* (1967); G. M. Sifakis, *Studies in the Hist. of Hellenistic Drama* (1967); E. G. Ballard, "Tragic, Sense of the," *DHI*; C. Leech, *T.* (1969); J. M. Bremer, *Hamartia* (1969); L. Michel, *The Thing Contained: Theory of the Tragic* (1970); R. Girard, *La Violence et le sacré* (1972); B. Vickers, *Towards Gr. T.* (1973); F. R. Adrastos, *Festival, Comedy and T.: The Gr. Origins of Theatre*, tr. C. Holmes (1975); R. B. Sewall, *The Vision of T.*, 2d ed. (1980); T. J. Reiss, *T. and Truth* (1980); M. C. Bradbrook, *Themes and Conventions in Elizabethan T.*, 2d ed. (1980); Trypanis, chs. 3, 9; *T., Vision and Form*, ed. R. W. Corrigan, 2d ed. (1981); J. Orr, *Tragic Drama and Mod. Society* (1981); M. S. Silk and J. P. Stern, *Nietzsche on T.* (1981); C. Segal, *T. and Civilization* (1981), *Dionysiac Poetics and Euripides' Bacchae* (1982), *Interpreting Gr. T.* (1986); J. P. Vernant and P. Vidal-Nacquet, *T. and Myth in Ancient Gr.*, tr. J. Lloyd (1981), *Gr. T. and Political Theory* (1986); Fowler; O. Mandel, *A Definition of T.* (1982); S. Booth, *King Lear, Macbeth, Indefinition, and T.* (1983); W. B. Stanford, *Gr. T. and the Emotions* (1983); E. Faas, *T. and After: Euripides, Shakespeare, Goethe* (1984); J. Dollimore, *Radical T.* (1984); C. Belsey, *The Subject of T.* (1985); S. L. Cole, *The Absent One: Mourning Ritual, T., and the Performance of Ambivalence* (1985); H. P. Foley, *Ritual Irony: Poetry and Sacrifice in Euripedes* (1985); J. Herington, *Poetry into Drama* (1985); G. Braden, *Ren. T. and the Senecan Trad.* (1985); *CHCL*, v. 1; *Gr. T. and Political Theory*, ed J. P. Euben (1986); S. Goldhill, *Reading Gr. T.* (1986); S. Halliwell, *Aristotle's* Poetics (1986); M. C. Nussbaum, *The Fragility of Goodness* (1986); M. Heath, *The Poetics of Gr. T.* (1987); N. Loraux, *Tragic Ways of Killing a Woman* (1987); A. Poole, *T.: Shakespeare and the Gr. Example* (1987); M. Gellrich, *T. and Theory: The Problem of Conflict since Aristotle* (1988); E. Hall, *Inventing the Barbarian* (1989); M. J. Smethurst, *The Artistry of Aeschylus and Zeami: A Comparative Study of Gr. T. and Nō* (1989); *Nothing to Do with Dionysos: Athenian Drama in Its Social Context*, ed. J. J. Winkler and F. I. Zeitlin (1989); B. Zimmermann, *Gr. T.: An Intro.*, tr. T. Marier (1990); J. P. Euben, *The T. of Political Theory* (1990); T. C. W. Stinton, *Coll. Papers on Gr. T.* (1990); J. Gregory, *Euripides and the Instruction of the Athenians* (1991).

T.J.R.

TRIMETER (Gr. "of three measures"). Aristotle (*Poetics* 1449a24, *Rhetoric* 1408b33) regards the iambic t. as the most speechlike of Gr. meters. It is first employed mixed with dactylic hexameters in the *Margites* ascribed to Homer, though some

scholars regard the first line of the "Nestor's cup" inscription, ca. 750–700 B.C., as a t. Despite ancient trad. (e.g. Pseudo-Plutarch, *De musica* 28), Archilochus (fl. 650 B.C.) is not its inventor, although he was the first to use the word "iamb" and developed the t. as a medium for personal invective (q.v.), a practice in which he was followed by Semonides. The Athenian lawgiver Solon used it for political poetry. The t. is the basic dialogue meter of Gr. tragedy, satyr-play, and comedy. It consists of three iambic metra (x represents *anceps*): x – ᴗ – | x – ᴗ – | x – ᴗ –. The penthemimeral caesura (after the fifth position [second *anceps*]) is much more frequent than the hephthemimeral (after the seventh position [second *breve*]); median diaeresis is permitted occasionally in tragedy, though in Euripides only when accompanied by elision. Resolution of a *longum* is permitted at differing rates in all feet but the last. The final element may be short (*brevis in longo*). Most of these departures from the basic iambic pattern are subject to a complex of finely graded phonological, lexical, and syntactic constraints which form a hierarchy of strictness which decreases from the archaic iambographers through early tragedy and later Euripides to satyr-play and finally comedy. The t. is subject to a number of bridges, the strictness of which follows the same generic and stylistic hierarchy.

The Lat. adaptation of the Gr. t. is the *senarius*, first used by Livius Andronicus, a common dialogue meter of early Lat. drama and frequent in funerary inscriptions. This version of the meter, however, is organized as six feet rather than three metra. The most striking departure from the Gr. t. is the permissibility of spondees in the second and fourth feet. This variation and many differing constraints on word boundaries are motivated by the differing nature of the Lat. stress accent. Iambic-shaped words, even those of the type not regularly subject to iambic shortening, are severely restricted in the interior of the line, since otherwise they would produce conflict of accent with metrical ictus. Spondee-shaped (or – ending) words, however, could be permitted in trochaic segments of the verse, since their stress pattern would preserve the iambic rhythm. In contrast to the *senarius*, the Lat. lyric poets, such as Catullus and Horace, and Seneca in his dramas, follow the pattern of the Gr. t. more closely, excluding spondees in even feet, restricting resolution and substitution, and not admitting iambic shortening.

The prosodies of the modern vernaculars followed the Lat. metrical practice of scanning in feet rather than metra, so that the t. of the Germanic langs. (incl. Eng.) of the later Middle Ages, Ren., and modern period is most often a very short line of three binary feet or six syllables—too short to be capable of sustained effects in narrative or dramatic verse, but very suitable for song. Literary examples incl. a dozen songs by Wyatt ("I will and yet I may not," "Me list no more to sing") and Surrey, Jonson's "Dedication of the King's New Cellar" (with feminine rhymes), Elizabeth Barrett Browning's "The Mourning Mother," and a dozen poems by Shelley, one of them, "To a Skylark," trochaic t. with an alexandrine close. In Fr. prosody, the line now called the *trimètre* is not a t. in this sense: it is an alexandrine of 12 syllables divided into three rhythmical phrases, and made its appearance only with the advent of romanticism.

Schipper; F. Lang, *Platen's T.* (1924); J. Descroix, *Le Trimètre iambique* (1931); G. Rosenthal, *Der T. als deutsche Versmasse* (1934); P. W. Harsh, *Iambic Words and Regard for Accent in Plautus* (1949); Maas, sects. 101–16; D. S. Raven, *Lat. Metre* (1965); C. Questa, *Introduzione alla metrica di Plauto* (1967); W. S. Allen, *Accent and Rhythm* (1973); West; S. L. Schein, *The Iambic T. in Aeschylus and Sophocles* (1979); A. M. Devine and L. D. Stephens, *Lang. and Metre* (1984).

A.M.D.; L.D.S.; T.V.F.B.

TRIOLET. A Fr. fixed form composed of eight lines and using only two rhymes, disposed in the following scheme: *ABaAabAB* (a capital letter indicates a repeated line), e.g. W. E. Henley:

> Easy is the triolet,
> If you really learn to make it!
> Once a neat refrain you get,
> Easy is the triolet.
> As you see!—I pay my debt
> With another rhyme. Deuce take it,
> Easy is the triolet,
> If you really learn to make it!

The challenge of the form lies in managing the intricate repetition so that it seems natural and inevitable, and in achieving in the repetitions variety of meaning or, at least, a shift in emphasis.

The word "t." is not found until 1486, but the form as we know it is much older, the *Urform* in fact of the whole rondeau family. It can be traced back to the 13th c., e.g. in the *Cléomadès* of Adenet-le-Roi, and was subsequently cultivated by such medieval poets as Eustache Deschamps and Jean Froissart. After lapsing from favor, it was revived in the 17th c. by Vincent Voiture and La Fontaine, and again in the latter half of the 19th c. as part of Théodore de Banville's general promotion of the Romance fixed forms. It challenged the skills of poets such as Daudet, Mallarmé, Rimbaud, Robert Bridges, Austin Dobson, Edmund Gosse, W. E. Henley, Andrew Lang, Hardy, and Arthur Symons.

Banville overstates the case when he writes of the satirical capabilities of the t.; it is too playful a form to achieve anything more than benign kinds of ridicule. The frequency of dialogue, particularly in the Eng. t., is also dispersive of any satirical intent. Gosse's summary of the t. is more accurate: "nothing can be more ingeniously mischievous, more playfully sly, than this tiny trill of epigrammatic melody, turning so simply on its own innocent axis." The refrain of the t. may express,

among other things, a cocky superiority, or a lively defense of a *status quo*; it may achieve a certain lyric intensity in the fourth line. The fifth and sixth lines both support the refrain and resist it; they support it structurally by re-establishing some formal stability after the irregularities of the third and fourth lines and thus providing a platform for the final appearance of the refrain; they resist it by allowing a temporary release from its apparent stranglehold, usually accompanied by an expansion of the subject-matter. The Eng. poets create out of the t. a structure of some complexity involving variations both of line-length and meter; the Fr. poets are less adventurous, sticking to octosyllables and often finding it easier to work with sequences of t.-stanzas than with single units. T.-stanzas must find a delicate balance between formal autonomy and formal interdependence.—T. de Banville, *Petit Traité de poésie française* (1872); E. Gosse, "A Plea for Certain Exotic Forms of Verse," *Cornhill Magazine* 36 (1877); J. Gleeson White, *Ballades and Rondeaus* (1887); Kastner; H. L. Cohen, *Lyric Forms from France* (1922); P. Champion, *Histoire poétique du XVe siècle*, 2 v. (1923); L. Spitzer, "T.," *RR* 39 (1948); P. J. Marcotte, "An Intro. to the T.," "More Late Victorian T. Makers," "A Trio of T. Turners," *Inscape* 5-6 (1966–68); C. Scott, "The 19th-C. T.: Fr. and Eng.," *Orbis Litterarum* 35 (1980); Morier. A.PR.; C.S.

TROUBADOUR. Medieval Occitan lyric poet. The term expresses the agent of the verb *trobar* "to find, invent, compose verse"; the etymon of *trobar* may have been hypothetical Med. Lat. *tropare* "to compose a trope" (a liturgical embellishment, Gr. *tropos*) or hypothetical Sp. Arabic *trob* "song" (cl. Arabic *ṭarab* "to sing") or Lat. *turbare* "to disturb, stir up." These proposed etyma correspond to theories of the origin of courtly love in Arabic or in Cl. or Med. Lat.

Extant t. production dates from ca. 1100 to ca. 1300 A.D., beginning with William IX, Duke of Aquitaine and Count of Poitiers (1071–1127), alternately bawdy and courtly, and continuing with Jaufre Rudel (fl. 1125–48), whose distant love tantalizingly blends secular and religious qualities; the biting moralist Marcabru (fl. 1130–49); the love-poet Bernart de Ventadorn (fl. 1147–70?); the witty Peire Vidal (fl. 1183–1204), who travelled as far as Hungary; the political satirist and war poet Bertran de Born (ca. 1150–1215); and the jolly, worldly Monk of Montaudon (fl. 1193–1210). Composition in a difficult style or *trobar clus* is associated with the names of Peire d'Alvernhe (fl. 1149–68), Raimbaut d'Aurenga (d. 1173), and Arnaut Daniel (fl. 1180–95); Giraut de Bornelh (fl. 1162–99), "the master of the ts.," practiced both *trobar clus* and *trobar leu*, or the easy style. In the early 13th c. appear two *trobairitiz* or women troubadours, the moody Castelloza and the more vivacious Comtessa de Dia, who left several songs apiece. Peire Cardenal (fl. 1205–72) followed the satirical trad. of Marcabru and Bertran de Born but was more concerned with religious issues in the period of the Albigensian Crusade; late in the century, Guiraut Riquier (fl. 1254–82) lamented that he was among the last of the ts. In all we know some 450 ts. by name. In the 14th–15th cs., Occitan poetry became an academic prolongation of the earlier trad.; those who wrote it are not called ts., but poets.

Though we have the melodies of only one-tenth of them, it is assumed that virtually all t. poems were set to music. The t. wrote both text and melody, which were performed by the joglar, who served as the messenger for a particular t. by singing his song before its addressee. We have circumstantial information about 101 individual ts. in the prose *vidas* and *razos* which accompany the poems in some mss. This information is considered reliable in objective matters, such as the t.'s place of birth, place of death, and social class, but unreliable in regard to his amorous adventures, which were largely invented by the prose-writers on the basis of what they read in the poems. The outstanding example of such imaginative biography is the *vida* of Jaufre Rudel, which has him perish of love in the arms of the countess of Tripoli.

The ts. exerted influence in both form and content on the Fr. *trouvères* as early as the 12th c., and in the 13th c. on poets writing in Galician-Portuguese and in German. In Italy their influence was felt in the Sicilian school presided over by Frederick II Hohenstaufen, and then by Dante and his friends who created the *dolce stil nuovo*. Through Petrarch, who acknowledged his debt, they affected the development of poetry of the Ren. and beyond.—M. R. Menocal, "The Etymology of Old Prov. *trobar, trobador*: A Return to the 'Third Solution,'" *RPh* 36 (1982). W.D.P.

TROUVÈRE. Medieval lyric poet of Northern France. The term corresponds to Occitan *troubadour* (q.v.), and since the troubadours composed mostly lyric poetry, t. is commonly applied only to lyric poets and not to the authors of OF narrative. We know over 200 ts. by name, and over 400 troubadours. Troubadour lyrics were written ca. 1100–1300; extant t. production began later, ca. 1190–1300, and was much influenced by the troubadours in form and content. The lyric corpus in the two langs. is comparable in size (about 2500 songs); the ts. cultivated esp. the genres of the courtly *chanson* or love song; religious verse; and the *pastourelle*, while avoiding the *sirventes* or satire; genres in the popular style, better preserved in Fr. than in Occitan, incl. the *mal mariée*, the *chanson de toile*, and dance-songs such as the *rondet de carole*, the *ballette*, and the *estampie*. We have the melodies for about three-quarters of t. lyrics, but for only one-tenth of those of the troubadours.

ANTHOLOGIES: *Lirica francese del medio evo*, ed. C. Cremonesi (1955); *Poètes et romanciers du moyen âge*, ed. A. Pauphilet (1958); *Penguin Book of Fr.*

Verse, v. 1: To the 15th C., ed. B. Woledge (1961); *La Poésie lyrique d'oïl*, ed. I.-M. Cluzel and L. Pressouyre (1969); *Lyrics of the Troubadours and Ts.*, ed. F. Goldin (1973); *Lirica cortese d'oïl*, ed. G. Toja, 2d ed. (1976); *Chanter m'estuet*, ed. S. N. Rosenberg (1981); *A Med. Songbook*, ed. F. Collins, Jr. (1982); *Poèmes d'amour des 12e et 13e siècles*, ed. E. Baumgartner (1983); *Mittelalterliche Lyrik Frankreichs, II: Lieder der Ts.*, ed. D. Rieger (1983); *The Med. Lyric*, ed. M. Switten et al., 3 v. (1987–88).

HISTORY AND CRITICISM: J. Frappier, *La Poésie lyrique française aux XIIe et XIIIe siècles* (1954); R. Dragonetti, *La Technique poétique des ts. dans la chanson courtoise* (1960); G. Lavis, *L'Expression de l'affectivité dans la poésie lyrique française du moyen âge* (1972); H. van der Werf, *The Chansons of the Troubadours and Ts.* (1972)—music; P. Zumthor, *Essai de poétique médiévale* (1972); Bec—study with texts; G. Zaganelli, *Aimer, soffrir, joir* (1982).

W.D.P.

U

UT PICTURA POESIS. Few expressions of aesthetic crit. have led to more comment over a period of several centuries than *u. p. p.*, "as is painting so is poetry" (Horace, *Ars poetica* 361). Since Horace mentions the subject thrice (362–65, 1–47, 343–45), we may assume he had some particular interest in it, though investigations of Horatian dicta *vis à vis* contemporaneous Roman painting mostly still remain to be made. Suggestions of the simlitude of poetry and painting were certainly made before Horace, who almost certainly knew—even if he may not have assumed that his audience would recall—the more explicit earlier statement of Simonides of Keos (first attested in the *Auctor ad Herrennium* [4.39] and recorded by Plutarch as a commonplace [*De gloria Atheniensium* 3.347a] more than a century after *Ars poetica*): "poetry is a speaking picture, painting a silent [mute] poetry."

The views of Aristotle—esp. that poetry and painting as arts of imitation should use the same principal element of composition (structure), namely, *plot* in tragedy and *design* (outline) in painting (see *Poetics* 6.19–21)—furnished additional authority for Ren. and later attempts to measure the degree and the nature of the kinship of the arts (the "parallel" of the arts) and to determine the order of precedence among them (the "paragone" of the arts). Moreover, as Lee observes in his analysis of the humanistic doctrine of painting, for which the Horatian dictum served as a kind of final sanction, "writers on art expected one to read [*u. p. p.*] 'as is poetry so is painting.'"

The Horatian simile, however interpreted, asserted the likeness, if not the identity, of painting and poetry; and from so small a kernel came an extensive body of aesthetic speculation and, in particular, an impressive theory of art which prevailed in the 16th, 17th, and most of the 18th c. While a few poets assented to the proposition that painting surpasses poetry in imitating human nature in action as well as in showing a Neoplatonic Ideal Beauty above nature, more of them raided the province of painting for the greater glory of poetry and announced that the pre-eminent painters are the poets. Both Cicero (*Tusc. Disp.* 5.114) and Lucian (who praises Homer as painter [*Eikones* 8]) gave ancient authority for that view, which Petrarch and others reinforced. Among the poets described as master-painters have been Theocritus, Virgil, Tasso, Ariosto, Spenser, Shakespeare, and Milton, not to mention numerous later landscapists in descriptive poetry, the Pre-Raphaelites, and the Parnassians. Painter and critic, Reynolds instanced Michelangelo as the prime witness to "the poetical part of our art" of painting (*Discourse* 15, 1790). Thus a "poetical" or highly imaginative painter could be compared with the "painting" poets.

U. p. p. offered a formula—the success of which "one can hardly deny," Wellek has remarked—for analyzing the relationship of poetry and painting (and other arts). However successful, the Horatian formula proved useful—at least was used—on many occasions as a precept to guide artistic endeavor, as an incitement to aesthetic argument, and as a basic element in several theories of poetry and the arts. Alone and with many accretions, modifications, and transformations, *u. p. p.* inspired a number of meaningful comments about the arts and poetry and even contributed to the theory and praxis of several painters, most notably "learned Poussin." Moreover, like other commonplaces of crit., the Horatian formula stimulated and attracted to itself a variety of views of poetry and painting that are hard to relate to the original statement.

The Horatian simile has of course evoked opposition. Plutarch himself questions its validity (*Mor.* 748A). In *Plastics* (1712), Shaftesbury warned, "Comparisons and parallel[s] . . . between painting and poetry . . . [are] almost ever absurd and at best constrained, lame and defective." The chief counterattack came in *Laokoön* (1766), where Lessing contended that the theories of art associated with *u. p. p.* had been the principal, if not the only, begetter of the confusion of the arts which he deplored in the artistic practice and theory of the time. In this he was anticipated by Da Vinci,

who raises these issues in his *Notebooks* (*Literary Works,* ed. J. P. Richter [1970], 1.48–68, 79–81). Saisselin has shown clearly that the "relations between the sister arts . . . were more complex than a reading of Lessing might lead one to believe." Since then similar charges have been raised by other critics, e.g. Babbitt.

On the other hand, since the late 19th c. the kinship of poetry and painting has appeared in a more favorable light in connection with the arts of the East—particularly in generalizations about the "poetic feeling" of Oriental painting and the pictorial characteristics of Chinese and Japanese poetry and, with the ever-increasing knowledge of Eastern art, in historical and critical studies setting forth the close relationships between Oriental poetry and painting. In China, poets were often painters; and critics, particularly in the 11th and 12th cs., stated the parallelism of poetry and painting in lang. close to that of Simonides and Horace. According to Chou Sun, "Painting and writing are one and the same art." Writing implied calligraphy, which linked painting with poetry. Thus, a poet might "paint poetry" and a painter write "soundless poems." These Eastern views led a number of Occidental poets to follow Japanese rules for poems and Chinese canons of painting in their poems—"images" directly presented to the eye, "free" impressions in a few strokes of syllables and lines, evocations of mood, lyrical epigrams, and abstractionist representations. Still, these poems reflecting the Eastern tendency to regard poetry and painting as "two sides of the same thing" were experimental and specialized works tapping but a few of the resources of the two arts. Moreover, the critical analysis of "the same thing," with its "two sides" remains at least as difficult as the explanation of the Horatian observation, "as is painting so is poetry."

W. G. Howard, "*U. p. p.,*" *PMLA* 24 (1909), ed., *Laokoön: Lessing, Herder, Goethe* (1910); I. Babbitt, *The New Laokoön* (1910); E. Manwaring, *It. Landscape in 18th-C. England* (1925); E. Panofsky, *Idea* (1929, tr. 1968); C. Davies, "*U. p. p.,*" *MLR* 30 (1935); R. W. Lee, "*U. p. p.,*" *Art Bull.* 22 (1940), "*U. p. p.,*" *Dict. of World Lit.,* ed. J. T. Shipley, rev. ed. (1953); C. M. Dawson, *Romano-Campanian Mythological Landscape Painting* (1944); K. Schefold, *Pompejanische Malerei* (1952, Fr. tr. 1972)—Roman poetics vs. pictorial styles; P. W. Lehmann, *Roman Wall Paintings* (1953); Wellek and Warren, ch. 11; H. H. Frankel, "Poetry and Painting: Chinese and Western Views," *CL* 9 (1957); J. R. Spencer, "*Ut pictura rhetorica,*" *JWCI* 20 (1957)—rel. to the rhetorical trad.; Wimsatt and Brooks, ch. 13; J. H. Hagstrum, *The Sister Arts* (1958); R. G. Saisselin, "*U. p. p.*: DuBos to Diderot," *JAAC* 20 (1961); P. H. v. Blanckenhagen, *The Paintings from Boscotrecase* (1962); H. D. Goldstein, "U. p. p.: Reynolds on Imitation and Imagination," *ECS* 1 (1967–68); R. Park, "*U. p. p.*: The 19th-C. Aftermath," *JAAC* 28 (1969); D. T. Mace, "*U. p. p.,*" *Encounters,* ed. J. D. Hunt (1971); H.-C. Buch, *"U. p. p.": Die Beschreibungsliteratur und ihre Kritik von Lessing bis Lukács* (1972); "U. P. P.: A Bibl.," *BB* 29 (1972); W. Trimpi, "The Meaning of Horace's U. p. p.," *JWCI* 36 (1973), "Horace's 'U. p. p.': The Argument for Stylistic Decorum," *Traditio* 34 (1978); J. Graham, "*U. p. p.,*" *DHI*; C. D. Reverand, "*U. p. p.* and Pope's *Satire* II.i," *ECS* 9 (1975–76); W. K. Wimsatt, Jr., "Laokoön: An Oracle Reconsulted," *Day of the Leopards* (1976); E. H. Gombrich, *Art and Illusion,* 5th ed. (1977); E. Gilman, *The Curious Perspective* (1978): R. A. Goodrich, "Plato on Poetry and Painting," *BJA* 22 (1982); J.-M. Croisille, *Poésie et art figuré de Néron aux Flaviens* (1982), and rev. in *JRS* 73 (1983); A. Dolders, "*U. p. p.*: A Sel. Annot. Bibl. of Books and Articles Pub. 1900–1980," *YCGL* 32 (1983); H. Markiewicz, "U. p. p.: A Hist. of the Topos and the Problem," *NLH* 18 (1987).

S.A.L.; T.V.F.B.; W.T.

V

VARIATION is used in three senses in the study of poetry. (1) In OE, the term refers to a technique of poetic composition by which the metrical pattern of half-lines, itself partially formulaic, is deliberately not repeated from the first half-line to the second. Since the number of half-line types is deliberately kept low (only about half a dozen), and since these types may well have been recognizable to or even identified for auditors in performance (e.g. by accompanying harp notes), v. seems to have been a deliberate attempt to avoid monotony in line-construction. (2) Metrical v. is often cited as an explanation for the fact that most actual lines of poetry do not entirely match the pattern of the meter they are said to be written in. (3) More generally, v. is often held to be a desirable characteristic of structure which sustains reader interest. Critics who see literary works as developing, exploring, or asserting "themes" sometimes adapt the analogy of "theme and v." from music, as in the construction of a symphony, where v. is recognized as one of only a few compositional strategies open to any composer. Auditor and reader recognition of a v. as in some respects different from but in others conforming to a prior theme is simply one form of pattern recognition, a fundamental cognitive pro-

cess applicable across media.—C. P. Smith, *Pattern and V. in Poetry* (1932); W. K. Wimsatt, Jr., "When Is V. 'Elegant?'" *The Verbal Icon* (1954); C. S. Brown, "Theme and Vs. as a Literary Form," *YCGL* 27 (1978); S. L. Tarán, *The Art of V. in the Hellenistic Epigram* (1979); F. C. Robinson, Beowulf *and the Appositive Style* (1985); E. R. Sisman, *Haydn and the Cl. V.* (1993). T.V.F.B.

VERS LIBRE. Because of its prosodic relatedness to *vers libres classiques* and *vers libéré*, this term is best reserved for 19th-c. Fr. free verse and those modernist free-verse prosodies that acknowledge a debt to it (e.g. the It. futurists, the Anglo-Am. *vers-libristes*, Pound, Eliot, and the imagists). The directions mapped out by the *vers-libristes* of the late 19th c. have been variously explored and adapted by 20th-c. practitioners such as Apollinaire, Blaise Cendrars, Pierre-Jean Jouve, Pierre Reverdy, Éluard, Robert Desnos, René Char, Yves Bonnefoy, and Michel Deguy.

The emergence of v. l. is specifically datable to 1886, the year in which the review *La Vogue*, edited by Gustave Kahn, published, in rapid succession, Rimbaud's free-verse *Illuminations*, "Marine," and "Mouvement" (possibly written in May, 1873), translations of some of Whitman's *Leaves of Grass* by Jules Laforgue, Kahn's series of poems entitled "Intermède" (to become part of his *Les Palais nomades*, 1887), ten of Laforgue's own free-verse poems (later collected in his *Derniers Vers*, 1890) and further examples by Paul Adam and Jean Moréas. To this list of initiators, Jean Ajalbert, Édouard Dujardin, Albert Mockel, Francis Vielé-Griffin, Émile Verhaeren, Adolphe Retté, Maurice Maeterlinck, Camille Mauclair, and Stuart Merrill added their names in the years immediately following.

One might believe that the relative freedoms of *vers libres classiques* combined with those of *vers libéré* would produce the absolute freedom of v. l., but this is not quite so. V. l. indeed indulges in heterometricity and free-rhyming, and its lines are rhythmically unstable; but it goes further still: it rejects the indispensability of rhyme with its line-demarcative function and instead relates lineation not to number of syllables but to the coincidence of units of meaning and units of rhythm, or to integral impulses of utterance, or else simply to the optimal expressive disposition of its textual raw materials. And indeed, the *vers-libristes* seek to abandon the principle of syllabism itself, by making the number of syllables in a line either irrelevant or indeterminable or both. The undermining of the syllabic system is facilitated by the ambiguous syllabic status of the *e atone* (mute *-e*)—should it be counted when unelided?—and by doubts about the syllabic value of contiguous vowels. Laforgue summarizes the *tabula rasa* of v. l. in a letter to Kahn of July 1886: "I forget to rhyme, I forget about the number of syllables, I forget about stanzaic structure."

Paradoxically, though, syllabic amorphousness produces rhythmic polymorphousness, and polysemy; in other words, a single line of v. l. is potentially several lines, each with its own inherited modalities. In addition, because of its heterometricity, v. l. can maximize rhythmic shifts between lines, creating a verse-texture of multiplied tonalities. Within this paradox lies another fruitful contradiction. One of the original justifications for v. l. was its inimitability, its resistance to abstraction and systemization; thus it could theoretically mold itself to the uniqueness of a personality, a psyche, a mood. Again Kahn: "For a long time I had been seeking to discover in myself a personal rhythm capable of communicating my lyric impulses with the cadence and music which I judged indispensable to them" (Preface to *Premiers Poèmes*, 1897). And yet, v. l. equally proposes a range of rhythmic possibilities which the reader is left to resolve into any one of a number of specific recitations. Given the significance of typographical arrangement in v. l., this contradiction might be reformulated as a polarization of the visual and the oral, of the linguistic and the paralinguistic, of the text as text, demanding to be read on its own terms, and the text as script, a set of incomplete instructions to the reader's voice. One further contradiction might be mentioned: for all v. l.'s ambiguation of syllabic number, with its transference of focus from syllable to accent, from number of syllables to number of measures, many free-verse poems are constructed on a "constante rythmique" (rhythmic constant), an intermittently recurrent measure which can only be defined syllabically.

Two broad currents of development can be distinguished in v. l.: one derives its rhythmic purchase from its varying approximation to, and distance from, recognizably regular lines and often cultivates ironic modes of utterance; the other seeks to undermine the primacy of the line by promoting rhythmic units larger than the line—the *verset* or the stanza—or smaller than the line—the individual measure; this latter strain is often informed by a rhapsodic voice. But in both currents, the line's role as guardian of metrical authority and guarantor of verse as ritual and self-transcendence is removed.

In both currents, too, the stanza finds itself without pedigree, infinitely elastic, insuring no structural continuity. The stanza of v. l. ends not in conformity with some visible structural imperative—though who may say what invisible imperatives operate—but because a movement of utterance comes to an end, and because only by ending can a sequence of lines define its own field of structural and prosodic activity. The stanzas of v. l. are a pursuit of unique kinds of formality constantly renewed, not the repeated confirmation of a certain stanzaic blueprint.

V. l. can claim, with some justification, to have "psychologized" verse-structure, to have made the act of writing apparently simultaneous with

the changing movements of mind: "A poem is not a feeling communicated just as it was conceived before the act of writing. Let us acknowledge the small felicities of rhyme, and the deviations caused by the chances of invention, the whole unforeseen symphony which comes to accompany the subject" (Laforge, *Mélanges posthumes*). By allowing the aleatory and the improvised to inhabit verse, by exploiting the psychological layering produced by variable rhyme-interval and variable margin, by locating verse at the intersection of multiplied coordinates (rhyme, rhymelessness, repetition, the metrical, the nonmetrical), by using linguistic structures to attract and activate paralinguistic features (tempo, pause, tone, accentual variation, emotional coloring), v. l. establishes its affinities with the stream of consciousness of contemporary fiction and proffers a stream of consciousness of poetic reading.

G. Kahn, "Préface," *Premiers Poèmes* (1897); F. Marinetti, *Enquête internationale sur le v. l. et manifeste du futurisme* (1909); Thieme, 386; T. S. Eliot, "Reflections on V. L.," *New Statesman* (1917), rpt. in *To Criticize the Critic* (1965); M. Dondo, *V. L., a Logical Devel. of Fr. Verse* (1922); E. Dujardin, *Les Premiers Poètes du v. l.* (1922); J. Hytier, *Les Techniques modernes du vers français* (1923); Patterson; L.-P. Thomas, *Le Vers moderne: ses moyens d'expression, son esthétique* (1943); H. Morier, *Le Rythme du v. l. symboliste*, 3 v. (1944); P. M. Jones, *The Background of Mod. Fr. Poetry* (1951), Part 2; Z. Czerny, "Le V. l. français et son art structural," *Poetics, Poetyka, Poetika*, ed. D. Davie et al. (1961); *Le Vers français au 20e siècle*, ed. M. Parent (1967); F. Carmody, "La Doctrine du v. l. de Gustave Kahn," *CAIEF* 21 (1969); J. Mazaleyrat, "Problèmes de scansion du v. l.," *Philologische Studien für Joseph M. Piel* (1969); Mazaleyrat; J. Filliolet, "Problématique du v. l.," in *Poétique du vers français*, ed. H. Meschonnic (1974); Elwert; Scott; Morier; D. Grojnowski, "Poétique du v. l.: *Derniers Vers* de Jules Laforgue (1886)," *RHLF* 84 (1984); C. Scott, *A Question of Syllables* (1986), ch.6, *V. l.: The Emergence of Free Verse in France* (1990). C.S.

VERSE AND PROSE.

I. DISTINCTIONS
II. ESSENCE AND FORM
III. INTERANIMATIONS
  A. *Speech Forms*
  B. *Prose Forms*
  C. *Verse Forms*
IV. HISTORY

I. DISTINCTIONS. V. and pr. are two of the three terms central to any discussion of, and distinctions about, the nature and modes of verbal art. The third is "poetry" (q.v.), which is the most difficult—and crucial—concept of the three. Northrop Frye once remarked that establishing a viable distinction between v. and pr. would allow us to write "page two" of the "elementary textbook expounding the fundamental principles" of crit. (*Anatomy* 13). Page one, insofar as it is possible, would answer the question, "What is lit.?" For Frye, the v.–pr. distinction seemed "the most far-reaching of literary facts"; nevertheless, he said in 1957, page two still remained blank. It does not seem so now.

The chief functions of pr. in the modern world are the written representation and communication of information about events, processes, and facts that obtain in the external world. Many readers also implicitly believe that lit. itself, even poetry, makes truth-claims about the world despite the fact that on the surface it is a "fiction": if they believed it didn't, they would find it little worth reading no matter how great its entertainment value. Many poets, e.g. Auden, have assented strongly to this view. Most modern critics, however, would not assent to such a view, or at least not directly: I. A. Richards, for example, held that propositions asserted in poetry are only "pseudo-statements," and most of the New Critics followed Richards in insisting on an absolute distinction between the langs. of science and poetry. Frye himself maintained that lit. "makes no real statements of fact" and is judged not on its truth or falsehood but on its "imaginative consistency."

Apart from judgments about truth-value, however, both common readers and critics recognize a distinction between v. and pr. Most speakers use "poem" as a synonym for "composition in v.": they expect poetry to be cast in verse. Yet the attributive term "poetic" is often applied as well to works not in v., works which readers feel are of greater insight, intensity, or depth of meaning than ordinary writing. And everyone knows that pr., as Eliot once remarked, is written in pr. Such confusing usage raises the logical questions of whether all v. is poetry, or whether all poetry is in v. If the former is true, the latter, its converse, is not, necessarily. The contrapositive, however, "if X is not poetry, then it is not in v.," will be true if the proposition is true. Put another way, the questions are, to begin with, is verseform necessary for "poetry"? And second, is it sufficient?

We must first recognize that the two modes, v. and pr., intersect the concept "poetry" and its opposite, nonpoetry. Crossing these yields four categories, which Eng. usage does not capture at all well. Intensified or heightened lang. in verseform, i.e. "v. poetry," represents what most people automatically think of as "poetry"; quotidian lang. in verseform is "v. nonpoetry," sometimes accepted as poetry but considered doggerel, sometimes denied to be poetry at all. Heightened lang. not in v. is sometimes called "poetry" or, better, "poetic," and if it has rhythmic or sound patterning at all, sometimes "prose poetry" (see below); quotidian lang. not in v. is, for lack of a term, just "pr." To define "poetry" as "a collective term for all poems" (Hynes) simply begs the question. Some readers find the differentia of poetry in heightened lexis and syntax (qq.v.); some find it in versification. Either or both will lead to height-

ened emotion and compressed meaning. If the figuring of lexis and syntax is accomplished via strategies of repetition that are regular enough to be rhythmic, however, the two modes converge toward the middle and merge, producing intermedia.

II. ESSENCE AND FORM. Since antiquity there have been two positions taken on the distinction between v. and pr.; for convenience we may call these the *essentialist* and the *formalist* positions.

Essentialists—"affectivists" might be a better term—do not consider verseform essential to the definition of poetry and view poets as more and sometimes other than versifiers. For centuries, from Quintilian (1st c. A.D.) to romanticism, it was a critical commonplace that Lucan was a rhetorician or historian who wrote in v. and that Plato, Xenophon (*Cyropaedia*), and Heliodorus (*Ethiopian History*) were poets. The major Western proponents of this view incl. Plato, Aristotle, Cicero, Horace, Sidney, Wordsworth, Shelley, Arnold, and Croce.

Aristotle himself argues at the outset of the *Poetics* that metrical form is not a sufficient criterion for "poetry"; for him, the fact that the works of philosophers and historians are in hexameters does not make them poetry. Form does not supersede function in the Aristotelian view. Admittedly, it is difficult to see what Aristotle's conception of poetry is, fully, because the *Poetics* concerns itself mainly with dramatic and secondarily with narrative lit., giving only scant attention to what we would now call lyric; and the very sketchy taxonomy of types of poetry and music given in ch. 1 of the *Poetics* is both confusing and incomplete, either deliberately (Aristotle rightly points out that some forms do not have names) or by virtue of corruption of the text. Aristotle of course grounds his theory on the human instinct or drive for imitation, an assumption which would seem to lead naturally to a referential philosophy of lang. and a mimetic theory of lit. (see REPRESENTATION AND MIMESIS).

Nevertheless, it is clear that Aristotle makes plot structure or fictiveness the crucial criterion of literariness, as Horace does grandiloquence of lang. and sublimity (see SUBLIME). Sidney considers verseform neither a necessary nor a sufficient cause: "It is not ryming and versing that maketh Poesie. One may bee a Poet without versing and a versifier without Poetry"; verseform is "but an ornament and no cause of Poetry" (Smith 1.159). For Shelley, "the popular division into pr. and v. is inadmissible in accurate philosophy." From the essentialist point of view, if the criterion of verseform as the differentia of poetry is abandoned, readers will turn to heightening of diction and figuration of syntax as the criteria. There is, after all, little else.

Formalists consider verseform to be either necessary or sufficient—mainly the former—for the achievement of precisely those effects of heightened intensity, compression, or figured speech which are commonly considered the hallmark of "poetry." They believe that the resources of verseform are not available to pr., or only minimally so. The difference may seem merely a difference of degree, since of course rhythmical structure and sound patterning can be accomplished in pr. But whereas in pr. the constitutive principle is syntax, and through that, sense, in v. the constitutive device of the sequence (so Jakobson) is design itself, design manifested in sound and rhythm and leading to sense and order, i.e. the organization of readerly experience in the processing of the text. Consequently, the difference in degree of formal structure between v. and pr. raises v. onto another plane and creates a difference in kind.

The chief Western formalists incl. Gorgias, Scaliger, Coleridge (*Biographia literaria* ch. 14, perhaps the central text), Hegel, Richards, and Ransom, along with the Rus. Formalists, the New Critics, and Jakobson. Central to their position is the denial of any naive distinction between form and content; they do not consider verseform supererogatory. It is well known that several major Eng. poets, incl. Jonson, Pope, and Whitman, used as a compositional technique the practice of first making a pr. paraphrase of the argument, then casting that into v. But this should not be taken to mean that verseform is merely rearrangement of the words or some superadded wrapper, as Wordsworth seemed to think. Rather, we must see that poets who versify pr. texts, their own or others', or who translate pr. texts into v., are remaking one verbal mode into an altogether different one. For Coleridge, the very act of introducing meter and rhyme into a discourse fundamentally alters the nature of the expression, not merely the form: all relations between words (hence all meanings) are changed by a change in their principle of selection. The eye altering alters all.

Thoughts do not come into being independent of verbal mode, and consequently change of mode entails change of thought. As Masson put it, meaning "is conditioned beforehand by the form of the expression selected." This is the antithesis of Croce's thesis that the aesthetic idea precedes its externalization in a medium. Rather, the physical medium—its limitations, its possibilities, its strategies for formulating concepts, its orders—is an indispensable part of cognition, hence of the creative results of cognition. The New Critical insistence on the irrefrangibility of form and meaning, which is based on Richards and Coleridge and fundamental to formalist method, still seems necessary and valid, though not perhaps sufficient.

At a deeper level, however, it would appear that the formalist and the essentialist perspectives on the problem of poetic form are not two answers to the same question but answers to two different questions about the same issue. The formalist answer concerns itself with the poem as artifact, the essentialist with the poem as experience. "Verse," we may recall, etymologically means "turn," namely the turn at the end of the line (q.v.): v. is

therefore lang. (1) given rhythmic order and (2) set into lines. But this does not mean that all v. is metrical, for meter is but one form of v. prosody among several, and even metrical v. has several subtypes varying in strictness. Consequently it is a mistake to say that what is not metrical is not poetry or even not v. The point is that v., pr., and poetry are not mutually exclusive or even correlate categories: v. and pr. are modes, while poetry, like drama and fiction, is, for lack of a better word, a genre. Any of the three literary genres may be written in either of the modes *or any mixture thereof.* The "modes," however, are not merely forms of writing, but rather forms of structure, since rhythm manifests itself in a linguistic sequence regardless of whether spoken or read. The distinction between v. and pr. is not one between media and essences, precisely, but between structures and effects. Can the formal devices of v. produce effects not obtainable in pr.? The preservation of the distinction demands that the answer be yes.

Lines of v. as manifested on the page are, after all, rhythmic entities before they are graphic entities: if the graphic lines do not show at least some kind of equivalence at the level of sound, they might as well be set as pr., whereas if lines which do show patterning are reset as pr. paragraphs, the meter or rhythmic structure can still be discerned, the line-divisions rediscovered, and the discourse reset as lineated v. This shows that the rhythmic structuring that we associate with v. is inherent in the syntactic strings regardless of presentational mode and would be left intact if print did not exist at all. One of the most interesting and revealing exercises in the study of poetry is to select passages and read them aloud, or else unlineate them and present them as pr., asking auditors or readers to judge whether they are pr. or v.; this was a salon game in the 18th c. Finally, it is worth recalling that much ancient and medieval v. was transcribed without lines, sometimes even without spacing, in order to save costly parchment—written by default, as it were, in pr.

III. INTERANIMATIONS. But making a simple binary distinction between v. and pr. has two shortcomings. First, it may give the misleading impression that what constitutes "v." or "pr." is merely a fact to be discovered rather than a cultural and aesthetic convention which varies from one lang. or v. system to another and, even within one lang., from one age to another. Second, it obscures all the more complex and more interesting mixtures, blends, and intermedia which result from each literary mode influencing the other, not to mention the interesting effects of speech forms. Indeed, all the varieties of spoken and written verbal art, pure and mixed, may be schematized as a constellation of types generated from the three gravitational centers, speech forms, pr. forms, and v. forms.

A. *Speech Forms* invade both v. and pr. in drama. In v. drama, esp. in blank v., speech is so rapid that the audience usually cannot discriminate ends of v. lines and has little sense of overt meter; as Wright suggests, there is only the more general sense of a rhythmical current. This does not make blank v. "v. only to the eye," as Dr. Johnson complained, for there is no evidence that auditors of poetry recognize even stricter verseforms (e.g. sonnets) quickly. It does suggest, however, that the visual form of a poem, its textuality or manifestation in print mode, is an undeniable part of its nature. It also confirms that rhythm itself, which is a phenomenon independent of presentational mode, is a necessary condition of v. In plays where verbal modes are used systematically by a playwright to differentiate characters or the social rank of characters—the paradigm case is Shakespeare's *Midsummer Night's Dream,* where the nobles usually speak blank v., the fairies couplets, and the mechanicals pr.—it is only that subtle but essential "rhythmical current" which enables auditors to distinguish versified speeches from pr. speeches at all. Note, however, that plays cast entirely in v., even heroic couplets, are no more "artificial" than plays written in pr. because all literary artworks naturalize their verbal mode, automatize them, so that auditors or readers take the mode as a given. It is only a question then of what effects v. mode may offer which pr. mode cannot, or vice versa. Failure to grasp this point occasioned much critical confusion in the controversies over blank v. and rhyme in the 17th c.

It would seem, then, that blank v. obtains such power precisely because it strikes a balance between rhythmic current and syntactic sense. That is the secret of its success. In Shakespeare's late plays, where the blank v. achieves more complex and less definable rhythms, the balanced play of phrase against line (Wright) becomes harder to hear, and all clear distinctions between v. and pr. verge on dissolution.

Speech forms invade strict verseforms to produce not a balance or fusion but contrast and tension. Heavy rhyming in short-lined v. with brisk, colloquial, and racy lexis produces the striking comic effects of light v. and satire, as for example in Butler's *Hudibras,* Byron's *Don Juan,* and Ogden Nash, and sometimes in Housman ("Terence, This Is Stupid Stuff"). The weight of expectation in the verseform is countered by the lexical surprise of the rhymes: sense springs open at these appointed places. But the ring of natural speech is quite possible in unrhymed lines as well, of course, e.g. Shakespeare's "I never saw my father in my life" (*Comedy of Errors*).

Pr. too contains representations of speech, such as dialogue and monologue (qq.v.); these are two keys to the success of the novel. Closely allied is what used to be called "stream of consciousness" pr., the staple of Joyce's *Ulysses.* These representations have the clear ring of speech and are fundamentally distinct from the nearly voiceless character of discursive pr., which has altogether different rhythms, lexis, and syntax from ordinary speech.

Few speakers, for example, produce sentences beginning with conjunctive adverbs or absolutes, or sentences with extensive subordination. Pr. is not speech; it is speech more logically and elaborately wrought, an accomplishment made possible by a medium where reflection and rereading are encouraged and where the receiver-reader rather than the speaker controls the pace of the delivery of information.

The other conspicuous manifestation of speech forms in literary art concerns sound patterning. Increase in sound patterning, apart from purely rhythmic effects, appears in speech for a variety of effects. Among the first of these is the mnemonic one, for it is as certain as it is unexplained that sound figuration in short speech forms such as aphorisms, epigrams, and proverbs greatly enhances memorization. Longer forms include (apart from speech disorders) both conscious and unconscious patterning, as in echolalia and glossolalia. Most auditors do not perceive aphorisms or proverbs as pr. and probably would not classify the longer forms so either, though they do recognize them as verbally artful.

In lit., sound patterning is used almost automatically for passages of visionary or prophetic passages, from the Old Testament to Blake, and also for literary imitations (often comic) of the speech of illiterate people (e.g. Mistress Quickly in Shakespeare), drunks (Falstaff), and insane persons (some of Tom o' Bedlam's speeches in *King Lear*). It also appears prominently, of course, in the works of poets who themselves have been thought to have been insane (Christopher Smart, Hölderlin). In pr., the great modern masters are Sterne, Stein, and Joyce. In v., as sound patterning increases, apart from structural sound such as alliteration (q.v.) in OE or rhyme in modern Eng., meaning density is both increased and counterbalanced by pure pattern and the nonsemantic perception thereof. But critics routinely condemn poets such as Swinburne and Dylan Thomas in whose work sound-patterning far outstrips sense, a fact which suggests that matter—meaning, import, "prosaic" sense—is, in mainstream critical judgment, superordinate as a criterion to versecraft. The poets most valorized in our own time, such as Gerard Manley Hopkins, have pressed against convention in both directions.

B. *Prose Forms*. We must remember that modern notions of "pr." are localized and conventional, formed largely by the invention of printing, the shape of the codex page, and (esp.) the devel. of the novel. In the ancient and medieval worlds, the kinds of expository texts now automatically cast into pr. were often cast in v., incl. works on botany, zoology, astronomy, physics, history, genealogy, law, medicine, philosophy, mathematics, rhetoric, and grammar. So were fictional texts. All the literary genres were once versified; now only some are.

Further, pr. itself must not be thought of as the neutral ground or zero degree against which v. deviates; pr. of whatever form, literary or quotidian, is already an artificial and stylized form, heavily influenced by the conditions of writing and by the rhythms of discursive thought. And deviation is not finally the most productive means for distinguishing v. from pr. or for conceptualizing either of them, for both modes are already deviations from speech, and both contain an enormous range of variation within their domains. Even speech itself varies greatly with context. Ordinary conversational speech is mainly fragmentary and discontinuous in character, highly elliptical, often paratactic, sometimes repetitive and other times extensively reliant on tacit conventions of mutual assumption and implication. But there are many contexts in which speech is highly stylized and figured, such as sermons or political oratory, where gifted or trained speakers can compose figured discourse extemporaneously. The strategies for such figuration are the figures codified in Cl. rhetoric. Stylization, therefore, may be applied to speech, to pr., and to v.; it is degree of figuration or stylization that matters for heightening lang., not the presentational mode.

It is for these reasons that Frye and others have held that meter is in fact closer to speech than pr., in being a less complex form of stylization, and have used this claim to explain why pr. does not develop in some cultures, whereas v. has been developed in every known culture, and why, even when pr. does develop, it is "normally a late and sophisticated devel. in the history of a lit."

Pr. rhet. written into v. produces forms such as the neoclassical closed or heroic couplet (q.v.). In the couplets of Pope, the fitting of sentence structure with meter is so finely wrought as to seem all but inevitable, *sprezzatura* executed nearly to perfection. Here metrical structure and rhyme-binding are close and tight, and syntax structural and rhetorical, while lexis (pure word choice) is, by contrast, altogether natural. This *discordia concors* is what led Matthew Arnold to call Pope one of the masters of Eng. pr. That remark happens to apply more accurately to the enjambed or open couplet than the closed; in any event, it clarifies the more important point that quotidian diction and syntax inside any verseform always *pose the threat* of pr. Conversely, elevated diction or convoluted syntax heighten the impression of poetry even in simple or conventionalized verseforms like the sonnet—as in the sonnets of Hopkins.

C. *Verse Forms* also influence or overwrite pr.: in some cultures, such as Chinese and Arabic, rhythmical pr. and even rhymeprose have been extensively cultivated. The chief effect of using rhetorical devices in pr. is simply to impose lexical and syntactic structure. Increased rhetorical patterning leads to parallelism, balance, symmetry, contrast, and "point" in phrases and clauses. If the patterning is extended down to the level of syllables and sound, pr. achieves effects which are rhythmical if not metrical, precisely as in v. Such

effects are conspicuous in Ciceronian pr., with its clausulae, in the medieval *cursus*, in 16th-c. euphuism (incl. elaborate sound-patterning), in 17th-c. mannerist and baroque pr., and in all religious and meditative pr. deriving from the penitential and sermon traditions and so influenced by the parallelistic structure of the Hebrew prosody of the Old Testament, such as Sir Thomas Browne's *Urn Burial* or Jeremy Taylor's *Holy Dying*.

Western poets have also experimented with v. novels; indeed, the line between such v. novels as E. B. Browning's *Aurora Leigh* or Nabokov's *Pale Fire* and narrative poems such as Puškin's *Evgenij Onegin* is, if one ignores page lineation, indeterminate. Subtler manifestations of v. influence also appear: the two paragraphs printed as pr. in Fitzgerald's *This Side of Paradise* which are in fact Spenserian stanzas remind one of the sonnets embedded in *Romeo and Juliet*, and there is blank verse in Dickens' *The Old Curiosity Shop*.

IV. HISTORY. In the Middle Ages, intermingling of modes and genres was more pervasive than at any other time in the history of Western letters and not seriously rivaled again until the late 19th c. The importance of the processes of translation, imitation, adaptation, and paraphrase in rhet. insured that the boundaries between v. and pr. would be fluid. Translations and paraphrases were made sometimes by the same writer, other times at a distance of several centuries. Particularly conspicuous are the metrical saints' lives, the medieval equivalent of the modern popular novel: pr. versions abound throughout the Middle Ages. Several writers explored the *opus geminatum*, a work written in two versions, one v., one pr., typically for two different audiences (learned and lay); examples incl. Aldhelm's *De virginitate* and Bede's life of St. Cuthbert. These have their poetic parallel in works of metrical (quantitative) poetry adapted to rhythmical (accentual) form for the illiterate masses. Writers of metrical texts might also place a pr. paraphrase in a facing column, as Hrabanus Maurus does in the *De laudibus sanctae crucis*. School training in Cl. rhet. included standard exercises in translation and paraphrase, the *copia*, back and forth from Lat. to vernacular and from v. to pr., in both directions and modes: the young Shakespeare endured these exercises at school as did the young Augustine before him. Rhet. and poetic, later distinct, were mutually permeable in the Middle Ages.

In addition to works written in alternative v. and pr. modes, a variety of mixtures and blends were explored, some of which developed into important genres with long lives. Chief among the mixtures is the prosimetrum, a pr. text with lyric insets in a variety of meters, inspired in Med. Lat. by the example of Boethius' *Consolation of Philosophy* and extended to vernacular forms such as the OF *chante-fable*, a performance sung and spoken by two minstrels in alternation. From such medieval exemplars, as well as the model of the *canticum and diverbia* (accompanied song and spoken dialogue) of Lat. drama (Plautus), medieval drama developed the practices of (1) using v. and pr. modes to differentiate characters and (2) intercalating lyric insets or songs in texts whether v. or pr. Both practices were carried into Ren. drama, where even subtler transitions are possible: a character shifting from loftier to more mundane thoughts may shift from one mode to the other; or, under the pressure of shattering emotion, a character may (as more than once in Marlowe) run from v. into pr. in mid-sentence.

It is of interest that pr. was not developed extensively in drama until late: despite the appearance of early works such as Gascoigne's *Supposes* and later comedies, the success of Shakespearean blank v. and the 17th-c. heroic couplet was such that the first Eng. tragedy in pr., George Lillo's *The London Merchant*, did not appear until 1731; and pr. was not the staple mode of drama until the 19th c., a result of the novel. Ren. drama, following medieval precedent, first developed rhymed verseforms (esp. the fourteener), some of them very elaborate, and ornate rhetorical diction. Thereafter it moved (esp. in the hands of Shakespeare) mainly in the direction of natural speech; pr. was used mainly for contrastive functions in versified plays, and became the medium of choice only after dramatists had become more conscious of the page than of the theater.

In addition to mixtures were medieval blends: v. rhythm and rhyme influenced pr. to produce rhymed pr. (Norden) and the rhythmical pr. of ecclesiastical correspondence known as the *cursus*. Both of the two principal outlets for pr. in the Middle Ages, letter-writing (*ars dictaminis*) and preaching (*ars praedicandi*), tended to be rhythmical. Pr. also influenced v.: *prosa* was the standard technical term for the medieval sequence, on account of the wording of its close (see Curtius).

In the modern age, the two chief blended forms have both been movements from v. toward pr. First is "free v." (q.v.), which is the awkward and misleading modern term for a heterogeneous group of nonmetrical but still rhythmical or (at the very least) lineated verbal sequences. Many reactionary critics of the late 19th and early 20th cs. (e.g. Saintsbury) objected to free v. as scarcely v. at all, since it lacked the badge of meter, which would certify strict control of the medium, strict rules making for strong order in a rigidly hierarchical world view. Free v. is obviously literary and minimally v. by the simple criterion of being set in (graphic) lines even if it has no other rhythmical structure, heightened diction, or figured syntax, though many varieties have sought at least some kind of rhythm. That fact, however, is only a fact and not a value judgment; absence of metrical structure may be deemed a pejorative by critics hostile to sweeping changes in the cultural conditions which valorized meter, or an approbative by avant-garde critics who welcome those changes.

The blurring of the line (!) between v. and pr. is simply one literary index of an age which prefers to avoid sharp distinctions, or at least distinctions as traditionally drawn, and to prefer overlapping forms, blended forms, boundary conditions, and all more complex or more fluid composites.

Second is the prose poem (q.v.), developed in France in the late 19th c. and cultivated intensively again in America in the deluge after the Sixties. As with the rhythmical pr. which developed in antiquity, it is difficult to tell whether the modern prose poem developed from the direction of pr. or v. Since it has been cultivated mainly by poets, it would seem the latter. If so, then prose poets refuse v. lineation and the regularity of v. rhythm yet nevertheless write rhythmical or rhetorical figuration back onto pr. in an effort to attain a level of incantatory speech and incandescent consciousness which ratiocinative pr. cannot achieve, because it has not the means.

D. Masson, "Pr. and V.," *Essays Biographical and Critical* (1856); R. L. Stevenson, "On Style in Lit.," *Contemp. Rev.* 47 (1885); D. Winter, "V. and Pr.," *JEGP* 5 (1903–5); J. W. Mackail, "The Definition of Poetry," *Lectures on Poetry* (1911); B. Petermann, *Der Streit um Vers und Prosa in der französischen Literatur des 18. Jhs.* (1913); Thieme, 374–75—lists Fr. works 1548–1912; T. S. Eliot, "Pr. and V.," *The Chapbook* 22 (1921); P. J. Hartog, *On the Relation of Poetry to V.* (1926); W. P. Ker, *Form and Style in Poetry* (1928), ch. 10; A. M. Clark, "Poetry and V.," *Studies in Literary Modes* (1946); M. C. Costello, *Between Fixity and Flux* (1947); Curtius, ch. 8; Frye; S. Hynes, "Poetry, Poetic, Poem," *CE* 19 (1958); N. Frye, *The Well-Tempered Critic* (1963); P. Klopsch, "Prosa und Vers in der mittellateinischen Literatur," *MitJ* 3 (1966); T. McFarland, "Poetry and the Poem," *Literary Theory and Structure*, ed. F. Brady et al. (1973); I. A. Richards, *V. versus Pr.* (1978); Norden; P. Habermann and K. Kanzog, "Vers, Verslehre, Vers und Prosa," *Reallexikon*; M. Perloff, *The Futurist Moment* (1986), ch. 5; W. Godzich and J. Kittay, *The Emergence of Pr.* (1987); T. Steele, *Missing Measures* (1990), ch. 3. T.V.F.B.

VERSE EPISTLE (Gr. *epistole*, Lat. *epistula*). A poem addressed to a friend, lover, or patron, written in a familiar style and in hexameters (Cl.) or their modern equivalents. Two types of v. es. exist: the one on moral and philosophical subjects, which stems from Horace's *Es.*, and the other on romantic and sentimental subjects, which stems from Ovid's *Heroides*. Though the v. e. may be found as early as 146 B.C. (L. Mummius Achaicus' letters from Corinth and some of the satires of Lucullus), Horace perfected the form, employing common diction, personal details, and a plain style to lend familiarity to his philosophical subjects. His letters to the Lucius Calpurnius Piso and his sons (ca. 10 B.C.) on the art of poetry, known since Quintillian as the *Ars poetica*, became a standard genre of the Middle Ages and after. Ovid used the same style for his *Tristia* and *Ex Ponto* but developed the sentimental e. in his *Heroides*, which are fictional letters from the legendary women of antiquity—e.g. Helen, Medea, Dido—to their lovers (tr. D. Hine, 1991). Throughout the Middle Ages, the latter seems to have been the more popular type, for it had an influence on the poets of courtly love and subsequently inspired Samuel Daniel to introduce the form into Eng., e.g. his *Letter from Octavia to Marcus Antonius*. Such also was the source for Donne's large body of memorable v. es. ("Sir, more than Kisses, letters mingle souls") and Pope's *Eloisa to Abelard*.

But it was the Horatian e. which had the greater effect on Ren. and modern poetry. Petrarch, the first humanist to know Horace, wrote his influential *Epistulae metricae* in Lat. Subsequently, Ariosto's *Satires* in terza rima employed the form in vernacular It. In all these epistles Christian sentiment made itself felt. In Spain, Garcilaso's *Epístola a Boscán* (1543) in blank verse and the *Epístola moral a Fabio* in terza rima introduced and perfected the form. Fr. writers esp. cultivated it for its "graceful precision and dignified familiarity"; Boileau's 12 es. in couplets (1668–95) are considered the finest examples. Ben Jonson began the Eng. use of the Horatian form (*Forest*, 1616) and was followed by others, e.g. Vaughan, Dryden, and Congreve. But the finest examples in Eng. are Pope's *Moral Essays* and the *Epistle to Dr. Arbuthnot* in heroic couplets. The romantics did not value the v. e., though Shelley, Keats, and Landor on occasion wrote them. Examples in the 20th c. incl. W. H. Auden's *New Year Letter* and Louis MacNeice's *Letters from Iceland*.

H. Peter, *Der Brief in der römische Lit.* (1901); J. Vianey, *Les Epîtres de Marot* (1935); W. Grenzmann, "Briefgedicht," *Reallexikon*; J. A. Levine, "The Status of the V. E. Before Pope," *SP* 59 (1962); W. Trimpi, *Ben Jonson's Poems* (1962); J. Norton-Smith, "Chaucer's Epistolary Style," *Essays on Style and Lang.*, ed. R. Fowler (1966); *John Donne: The Satires, Epigrams and V. Letters*, ed. W. Milgate (1967); N. C. de Nagy, *Michael Drayton's* England's Heroical Es. (1968); R. S. Matteson, "Eng. V. Es., 1660–1758," *DAI* 28 (1968): 5023A; D. J. Palmer, "The V. E.," *Metaphysical Poetry*, ed. M. Bradbury and D. Palmer (1970); C. Levine, "The V. E. in Sp. Poetry of the Golden Age," *DAI* 35 (1974): 3690A; M. Motsch, *Die poetische Epistel* (1974); A. B. Cameron, "Donne's Deliberative V. Es.," *ELR* 6 (1976); C. C. Koppel, "Of Poets and Poesy: The Eng. V. E., 1595–1640," *DAI* 39 (1978): 2292A; M. R. Sperberg-McQueen, "Martin Opitz and the Trad. of the Ren. Poetic E.," *Daphnis* 11 (1982); J. E. Brown, "The V. Es. of A. S. Pushkin," *DAI* 45 (1984): 201A; C. Guillén, "Notes toward the Study of the Ren. Letter," *Ren. Genres*, ed. B. K. Lewalski (1986); M. Camargo, *The ME Verse Love E.* (1991); W. C. Dowling, *The Epistolary Moment* (1991)—on the 18th c. R.A.H.; T.V.F.B.

VERSE PARAGRAPH. If a p. is defined as one or more sentences unified by a dominant mood or thought, then poetry, like prose, can be seen as moving forward in units which could be called ps. Many lyrics might be described as single v. ps., the sonnet as one or, if the *volta* be sufficiently marked, two. Further, because for centuries stanza was syntactically (as well as metrically) defined, the sense in elaborate stanzaic forms like the Spenserian and *ottava rima* (qq.v.) tends to assume p. form. However, most traditional stanzas are isometrical or isomorphic—i.e. identical in number of lines, meter, and rhyme scheme—and variety of effect is difficult to achieve in such poems. The result, esp. in long works by inferior poets, can be a numbing monotony of effect.

A distinctive characteristic of the p. in prose is of course that it does not have a settled length, that each individual p. may take the form most appropriate to the thought requiring expression. In poetry such freedom can best be achieved in narrative and descriptive poetry, where the ps. are often indicated by indentation of or spacing between lines. They are most prominent, however, in blank verse. (Rhymed v. ps. do occur—e.g. the irregular *canzoni* of Milton's *Lycidas* or the indented sections of varying numbers of couplets within the subdivisions of Pope's *Essay on Man.*) But it is in nondramatic blank verse that the v. p. as we customarily think of it reaches its fullest devel. Lacking the somewhat arbitrary organization provided by an established rhyme scheme, blank verse must provide units supporting the organization of idea such that the narration, description, or exposition unfolds in a series of stages felt as justly proportioned. In this sense the v. p. is a syntactic period, frequently a complex or periodic sentence, deployed in enjambed stichic verse so that the beginnings and ends of the syntactic frames conspicuously do not coincide with those of the metrical frames (the lines), with the result that meter and syntax are in counterpoint or tension.

The v. p. is a common feature of Old Germanic heroic poetry and is an important element in Eng. poetry as early as *Beowulf*, where sentences often begin at the caesura. But by general consent the greatest master of the v. p. is John Milton. Many of the most characteristic effects of *Paradise Lost*—its majesty, its epic sweep, its rich counterpoint of line and sentence rhythms—are produced or enhanced by Milton's v. ps. To sustain his ps., Milton employed enjambment (q.v.), interruption, inversion, and suspension, the device of the periodic sentence whereby the completion of the thought is delayed until the end of the period. The average sentence in Milton covers 17 lines, but often may cover 25 to 30. So powerful was Milton's influence on later poets that his voice, his distinctive rhythms, even his vocabulary and syntactic strategies can be recognized in much Eng. metrical verse of the 18th and 19th cs. (Havens). Whitman too, and the subsequent free verse (q.v.) for which he provided one model, makes much use of the v. p. (e.g. the first 22 lines of "Out of the Cradle Endlessly Rocking"). Thus far, however, no one distinctive free-verse p. has been devised—since free verse lacks the background of the constant meter against which the v. p. can be perceived to play, it is doubtful one could be—and when the v. p. is spoken of, one is still likely to think automatically of Milton.—G. Hübner, *Die stilistische Spannung in Milton's* P.L. (1913); R. D. Havens, *The Influence of Milton on Eng. Poetry* (1922); J. Whaler, *Counterpoint and Symbol* (1956); E. Weismiller, "Blank Verse," *A Milton Encyc.*, ed. W. B. Hunter, Jr., et al. (1978); W. H. Beale, "Rhet. in the OE V. P.," *NM* 80 (1979); J. Hollander, "'Sense Variously Drawn Out,'" in Hollander. A.PR.; E.R.W.; T.V.F.B.

VILLANELLE (from It. *villanella*, a rustic song or dance, *villano*, a peasant). Introduced into France in the 16th c. (Grévin, Du Bellay, Desportes), the v. first had as its only distinguishing features a pastoral subject and use of a refrain; in other respects it was without rule, although a sequence of four 8-line stanzas with a refrain of one or two lines repeated at the end of each stanza was a popular option. The form only became standardized in the 17th c., when prosodists such as Richelet based their definition on "J'ay perdu ma tourterelle" by Jean Passerat (1534–1602), a poem in tercets on only two rhymes, in which the first and third line of the first tercet are repeated alternately as the third line of the following tercets, and appear together at the end of the final stanza, thus creating a quatrain. Passerat's v. is of 19 lines and can be schematized thus: A_1bA_2 abA_1 abA_2 abA_1 abA_2 abA_1A_2 (A_1 and A_2 denote different [rhyming] refrain lines). Obviously the v. is essentially stanzaic in nature, and this is how Fr. poets have treated it, extending it and contracting it at will; in his presentation of the form, Théodore de Banville quotes a v. of Philoxène Boyer "La Marquise Aurore," which has 25 lines, while "L'Ornière," one of Maurice Rollinat's large output of vs., has as many as 85 lines; Leconte de Lisle, on the other hand, uses only 13 lines in his "V." and 18 in "Dans l'air léger," which omits the final A_2 and takes other liberties with the final stanza's rhyme scheme.

While the Fr. poets who revived the v. in the later 19th c. treated it as a stanza type, their Eng. counterparts, however, invested it with the status of a fixed form. Although Austin Dobson tried to present the v. to the Eng. as he found it in Banville, declaring "there is no restriction as to the number of stanzas," his compatriots stuck rigidly to the 19-line Passerat model popularized by Joseph Boulmier (*Vs.*, 1878). Enthusiasts for the form were legion (Edmund Gosse, Dobson, Andrew Lang, W. E. Henley, Ernest Dowson, Hardy, Wilde, E. A. Robinson). While the v. continued to attract pastoral subjects (Dowson, Wilde), it also became a vehicle for *vers de société* and, in a small way, part of the attempt to find an equivalent for the Horatian

ode in the Romance fixed forms (e.g. Dobson, *Tu ne quaesieris*, Odes 1, 11). Lang remarked: "There is a foreign grace and a little technical difficulty overcome in the Eng. *ballade* and *v.*, as in the Horatian sapphics and alcaics."

Surprisingly, perhaps, the fortunes of the v. have prospered in the 20th c. Following Leconte de Lisle, Auden ("If I Could Tell You," "Miranda's Song"), Dylan Thomas ("Do not go Gentle into that Good Night") and Roy Fuller ("The Fifties," "Magic") among others (Empson, Roethke, Plath) have explored the v.'s capacity to deal with serious, even metaphysical subjects while adhering to the strict 19-line model. A more recent tendency, deriving in part perhaps from Pound's free-verse "V.: The Psychological Hour," has sought to introduce greater flexibility into the traditional form by exploiting enjambment, metrical variation, and half-rhymes (James Merrill, Richard Hugo). There have even been efforts at a prose v., 19 sentences matching the pattern in repetition but not rhyme.

Banville describes the v. as "a plait of gold and silver threads into which is woven a third, rose-colored thread." The *A* refrains certainly have a metallic, unyielding character. Of Dowson's vs. Pound writes: "the refrains are an emotional fact which the intellect, in the various gyrations of the poem, tries in vain and in vain to escape." Banville's rose-colored thread, on the other hand, is to be found in the *b* lines, which attempt to withstand the conspiracy of the refrains and assert change and mortality, and for that reason have a peculiar poignancy and vulnerability.—T. de Banville, *Petit Traité de poésie française* (1872); E. Gosse, "A Plea for Certain Exotic Forms of Verse," *Cornhill Magazine* 36 (1877); A. Dobson, "A Note on Some Foreign Forms of Verse," *Latter Day Lyrics*, ed. W. Davenport Adams (1878); J. Gleeson White, *Ballades and Rondeaus* (1887); Schipper; Kastner; H. L. Cohen, *Lyric Forms from France* (1922); Scott; Morier; R. E. McFarland, "Victorian Vs.," *VP* 20 (1982), "The Revival of the V.," *RR* 73 (1982), "The Contemporary V." *MPS* 11 (1982); M. Pfister, "Die V. in der englischen moderne," *Anglia* 219 (1982). C.S.

VIRELAI (also called *chanson baladée* and *vireli*). Originally a variant of the common dance song with refrain, of which the rondeau (q.v.) is the most prominent type, this medieval Fr. lyric form developed in the 13th c. and at first may have been performed by one or more leading voices and a chorus. It begins with a refrain, followed by a stanza of four lines, of which the first two have a musical line (repeated) different from that of the refrain. The last two lines of the stanza return to the music of the refrain. The opening refrain, words and music, is then sung again. The v. usually continues with two more stanzas presented in this same way. A v. with only one stanza would be a *bergerette*. In Italy the 13th-c. *laude*, and in Spain the *cantigas*, follow the same form. The syllables *vireli* and *virelai* were probably nonsense refrains, originally, which later came to designate the type.

The large number of variations and optional elements both in the *lai* and in the v. (as practiced by Guillaume de Machaut and Jean Froissart, Christine de Pisan, and Eustache Deschamps) has produced much uncertainty among 20th-c. Fr. prosodists about how both forms should be defined, so that one must approach any modern definition with great caution. Most recent commentators follow Théodore de Banville (1872), who, relying on the authority of the 17th-c. prosodist le Père Mourgues (*Traité de la poésie française*, (1685), tried to settle matters once and for all by defining the *lai* as a poem in which each stanza is a combination of 3-line groups, two longer lines followed by a shorter one, with the longer lines sharing one rhyme-sound and the shorter lines another (*aabaabaab ccdccdccd*, etc.). Then, calling upon a false etymology of v.—from *virer* (to turn) and *lai*—he defined the v. as a *lai* in which the rhyme-sounds are "turned" from stanza to stanza; that is, the rhyme of the shorter lines becomes the rhyme of the longer lines in the following stanza (*aabaabaab, bbcbbcbbc*, etc.). Calling the v. thus defined the *v. ancien*, Banville goes on to describe the *v. nouveau*, which bears no relation to the *v. ancien* and is, if anything, more akin to the villanelle (q.v.). The *v. nouveau* opens with a refrain, whose two lines then recur separately and alternately as the refrains of the stanzas following, reappearing together again only at the end of the final stanza, but with their order reversed. The stanzas of the *v. nouveau* may be of any length and employ any rhyme scheme, but the poem is limited to two rhyme-sounds only. Here again, Banville merely follows le Père Mourgues, whose "Le Rimeur rebuté" is used as an illustration. John Payne's "Spring Sadness" (*v. ancien*) and Austin Dobson's "July" (*v. nouveau*) are the only evidence that these two forms have excited any interest.

T. de Banville, *Petit Traité de poésie française* (1872); E. Gosse, "A Plea for Certain Exotic Forms of Verse," *Cornhill Magazine* 36 (1877); J. Gleeson White, *Ballades and Rondeaus* (1887); Kastner; H. L. Cohen, *Lyric Forms from France* (1922); Le Gentil; P. Le Gentil, *Le V. et le villancico* (1954); M. Françon, "On the Nature of the V.," *Symposium* 9 (1955); G. Reaney, "The Devel. of the Rondeau, V., and Ballade," *Festschrift Karl Fellerer* (1962); F. Gennrich, *Das altfranzösische Rondeau und V. im 12. und 13. Jahrhundert* (1963); F. Gennrich and G. Reaney, "V.," *MGG* 13.1802–11; N. Wilkins, "V.," *New Grove*; Morier. U.T.H.; C.S.

VISUAL ARTS AND POETRY. The parallels that have traditionally been drawn between the v. a. (painting in particular) and p. are based on Western conceptions of mimetic representation (see REPRESENTATION AND MIMESIS; IMITATION). Both painting and poetry have long been thought to be

imitative arts, devoted to the depiction of human life and the external world. It is with this mimetic function clearly in view that Aristotle groups the various arts together in the opening of the *Poetics*; in a similar vein, Horace suggests in the *Ars poetica* (361) that "as a painting, so also a poem" (see UT PICTURA POESIS). Perhaps the most explicit comparison, however, is a remark attributed by Plutarch to Simonides of Ceos, who is said to have called poetry a "speaking picture" and painting a "silent [mute] poem." This intriguing epigram—both parallel and distinction—emphasizes the representational force of both arts while also calling attention to their essential dissimilarities. It remained for Lessing to point out, in *Laokoön*, that poetry is fundamentally a temporal art, whereas the primary attribute of painting is spatial. Lessing and a host of subsequent theorists have warned of the dire effects of disregarding these crucial distinctions, but writers and painters have always been fascinated by the relations that serve to join words and images.

The most tangible relation between the v. a. and p. occurs when words and images are combined or when words themselves also constitute a visual image (see VISUAL POETRY). Pattern poetry (q.v.) such as George Herbert's "Easter Wings" or John Hollander's *Types of Shape* might be thought of as doubly representational, with visual shape and poetic argument simultaneously reinforcing each other. Concrete poetry (q.v.), moreover, normally privileges the "visibility" above the "readability" of the verbal image. In both shaped and concrete poetry, however, the text has become more fully iconic (in the semiotic sense) because it substantially replicates the object it represents.

Despite Suzanne Langer's observation that "There are no happy marriages in art—only successful rape," poets, painters, and engravers have managed to combine their work in a variety of interesting ways. The rarest occurrence—and also one of the most complex—is composite art, a fusion of writing and visual images from the hand of the same artist. William Blake, generally agreed to be the greatest poet-painter, created a startlingly wide range of compositions that vary from engraved poems with little illustrative accompaniment to ambitious visual images with minimal verbal reinforcement. Even when word and image are closely intertwined (as in some of the more successful *Songs of Innocence and Songs of Experience*), the precise function of this composite form is not necessarily consistent. The illustration may reinforce the meaning of the poem, or it may complicate or undermine it (as in many states of "The Tyger"); it is also possible that the poem will make little sense without the visual images that penetrate and frame it. In their interdependence of word and image, Blake's designs represent a provocative reworking of the traditional emblem, in which motto, verse, and engraving all serve to illuminate the same idea.

Blake illustrated the texts of other poets as well as his own, and invariably his contribution must be seen as a powerful reinterpretation of Dante or Milton or Gray or Young. The visual illustration of poetry is normally not this ambitious, of course; even when the poet and engraver agree to appear in print together, the result—as in the early illustrations of Pope's *Rape of the Lock*—can be flat and insipid. Mutual illumination, as in Richard Bentley's witty and urbane designs for six poems by Thomas Gray, is much harder to find, but when it does occur it reinforces our sense of a shared taste, sensibility, or ideological conviction that can find expression in either artistic form.

Poets who paint and painters who write do not necessarily combine word and image in a single text; it is more likely, in fact, that they will produce separate works, incl. works that are difficult to compare or reconcile with each other. This is certainly the case with e e cummings, whose large body of work on canvas seems unusually conventional when compared with his experiments in verse. But even where these relations are tenuous, the entangled careers of these "double agents" are often worth examining. It is useful to know, for example, that Pope studied with Charles Jervas, that Turner and Michelangelo were poets, that D. H. Lawrence—so often considered within the trad. of the novel alone—was, like Dante Gabriel Rossetti, an accomplished painter *and* poet.

Moreover, the v. a. serve as a topic for p., even when the author is not a poet-painter nor even closely connected with the world of art (as Ariosto and Aretino were with Titian, for example). Poetry can take the form of a treatise on art, as in Du Fresnoy's *De arte graphica*, translated by Dryden. A poet such as Marvell may structure his satire in the guise of "Last Instructions to a Painter"—an esp. popular genre within the 17th c.—or as a tour through a portrait gallery (as in Pope's "Epistle to a Lady" or Browning's "My Last Duchess"). Just as prevalent as satires are encomiastic poems in which the writer attempts to judge and praise the skills of the painter. One of the shrewdest is Lovelace's poem on Lely's double portrait of Charles I and his son, the Duke of York, in which he is able to gauge movement toward more naturalistic representation in portraits of the mid 17th c. In similar (if less discriminating) ways Waller would address himself to Van Dyck, Dryden to Kneller and Anne Killigrew, Pope to Jervas, Gray to Bentley, and Wordsworth to Reynolds. Baudelaire, himself a distinguished critic of art, devoted sections of *Les Phares* to eight painters and sculptors, and Browning—in "Andrea del Sarto" and "Fra Lippo Lippi"—provided his own imaginative and provocative reassessment of early Ren. painting.

Poems or parts of poems that describe specific paintings are traditionally said to partake of *ekphrasis* (q.v.). Ekphrastic poems (literally) speak to, for, or about a work of art; they are verbal representations of visual representations. The

paintings (or other objects) they describe may be actual or fictive. Sometimes these ekphrastic fragments provide the only depiction we have of lost objects, thus voicing a pre-emptive verbal presence in place of an absent visual source. More often they are figurative encounters between words and image: encounters, as Mitchell has argued, that are not necessarily innocent of masculine, verbal assaults on a silent (or muted) object of desire or terror (Keats's still unravished bride of quietness; Shelley's Medusa). The relationship between the verbal and visual representation may or may not be clear. William Carlos Williams's "Portrait of a Lady" appears to be inspired by, or indebted to, Fragonard's *The Swing*. "Your thighs are appletrees / whose blossoms touch the sky," it begins, but the speaker's answer to "Which sky?" is, problematically, "The sky / where Watteau hung a lady's / slipper." Even when the connection is clear, the image that serves as the nominal starting point for the poem may quickly disappear as the poet pursues his or her own line of argument. In "Walker Skating" (the title alone suggests the paradoxes to follow), Brian Morris contemplates a handsome but chilly portrait sometimes attributed to Raeburn:

> A grave liver if ever I saw one:
> His speech would be more stately than
> his suit.
> With folded arms and peremptory hat
> The Reverend Robert Walker walks on
> water.
> Ineffable superiority.

Morris focuses on how external representation (dress, paint) reveals the qualities that lurk within; he attempts to open up the painting, to envision the "sharp teeth, sea-snakes, / And the cold eye of the treacherous eel" that lie beneath the scored ice, baleful reminders of "How frail a foundation, laird of the kirk, / Is laid for your assurance. . . ."

Ekphrastic poetry is necessarily limited to a description (or verbal bodying forth) of a specific visual object; as Krieger has noted, "the poem must convert the transparency of its verbal medium into the physical solidity of the medium of the spatial arts." In so doing, ekphrastic poetry must have recourse to descriptive vividness and particularity, the corporeality of words, and the patterning of verbal artifices, all of which are essential to the much broader category of literary pictorialism. According to Hagstrum, in order to be called pictorial a description or image "must be, in its essentials, capable of translation into painting or some other visual art." It need not resemble a particular painting or school of painting, but "its leading details and their manner and order of presentation must be imaginable as a painting or sculpture." Visual detail constitutes the pictorial, but these details must also be ordered in a picturable way. The pictorial, moreover, necessarily "involves the reduction of motion to stasis or something suggesting such a reduction." It also implies some "limitation of meaning": meaning must seem to arise from the *visibilia* that are present in the poetical context.

Pictorialism therefore places stringent demands upon its readers; it requires us to develop the ability to *see* what the poet describes, to develop (in Pope's famous phrase) "quick poetic eyes." Pictorial poetry sometimes evokes the paintings (or *kinds* of paintings) that lie behind the verbal icon, as in the Claudes and Salvatore Rosas that are evoked in the shadowy texts of Collins' odes (Swinburne called Collins the "perfect painter of still life or starlit vision"). Or the pictorial poet may focus on the discrete image, as Joseph Warton did in an unpublished passage in his Winchester College gathering book:

> The Solemn Silence of the Pyramids.
> The Dark
> gloomy Scenes in Mines. The Fall of
> the Nile.
> Distant Noises. Indian Brachmins wan-
> dering by
> their Rivers. Medea's nightly Spells.
> Meteors
> in the Night. Griping of a Serpent or a
> Crocodile.
> A Lamp in a lone Tow'r . . . . The Fall
> of the River
> Niagara. Oedipus and his Daughters in
> the *Storm* a fine
> Subject for a Picture. Woman with
> Child meeting a
> devouring serpent in a Desart.

The relationship between these pictorial scenes and poetry is clear from Warton's titles: the images were suitable both as "Subjects for a Picture" and as "Similes" to be used in poems ("Loathsom as the twining of a Serpent round one's Body"). As Warton argued in his essay on Pope, "The use, the force, and the excellence of lang. certainly consists in raising *clear, complete*, and *circumstantial* images, and in turning *readers* into *spectators*."

Implicit in the practice of pictorialism is the belief that poetry and painting may both address themselves to the same issues (allegorical subjects, for example), depict the same scenes (often landscapes), or elicit similar emotional or aesthetic responses in the reader-turned-spectator or spectator-turned-reader (this is particularly noticeable in 17th- and 18th-c. experiments to depict "the passions" in poetry, music, painting, and drama). It is important for students of poetry to recognize, moreover, that these common elements—which are so often crucial to the rhetorical energy and vividness of verbal texts—have also played an instrumental role in legitimizing painting's claim to be a liberal art (Lee). If pictorialism serves as an index of the ways in which poetry has been influenced by or aspires to be translated into

some aspect of the v. a., we must also recall how frequently poetry has directly inspired work in painting or the graphic arts, and not merely in book illustration alone. (The Shakespeare and Milton galleries of the late 18th c. are perhaps the best-known examples of how painters and engravers have attempted to recreate—sometimes slavishly, but often not—the imaginative world of their articulate precursors.) Recent analyses of paintings, moreover, have increasingly turned to semiotics and "visual poetcs" as they attempt to interpret the role of time and narrative in the visual arts (*Style*; Steiner).

It is precisely on this common ground, however—similar subject matter or aesthetic response—that more speculative attempts to link the v. a. and p. have so often faltered. Even attempts (such as Mario Praz's) to draw parallels between poetry and painting based on their contemporaneity have encountered serious opposition. Wellek and Warren caution that "the arts did not evolve with the same speed at the same time," and—like Steiner—they seriously question the usefulness of translating Wölfflin's famous characterizations of "closed" and "open" forms in Ren. and baroque painting into the vocabulary of literary analysis. More recent theory has therefore attempted to confront both similarity and difference in interartistic comparisons in a much more rigorous way. One influential theory, first proposed by Frank, draws attention to "spatial form" (or at least the thwarted temporality and consecutiveness) of modern lit., incl. the poetry of Eliot and Pound. Frank argues that, in their disjunctiveness and breach of narrative, modernist poems mirror the instability, the loss of control over meaning and purpose that we experience in an increasingly scientific and technological world.

This thesis has been radically broadened by Mitchell, who argues that "far from being a unique phenomenon of some modern lit., and far from being restricted to the features which Frank identifies in those works (simultaneity and discontinuity), spatial form is a crucial aspect of the experience and interp. of lit. in all ages and cultures." Mitchell's task is to convince us that temporality and spatiality are not necessarily at odds with each other, that we shall in fact encounter interpretive difficulties whenever we attempt to analyze the function of one without considering the role of the other. In *Iconology*, esp., he attempts to expose the ideological assumptions on which interartistic difference is based. He proposes that "there is no *essential* difference between poetry and painting, no difference, that is, that is given for all time by the inherent natures of the media, the objects they represent, or the laws of the human mind." Each culture will simply sort out the distinctive qualities of its mimetic symbols according to the various differences that are in effect at the time. The *paragone* "or debate of poetry and painting is never just a contest between two kinds of signs, but a struggle between body and soul, world and mind, nature and culture," a critique that he applies to the aesthetic treatises of Lessing and Burke.

A quite different approach lies in the analysis of artistic structures. Wellek and Warren conceded at least this to the historical comparison of the arts, that it might be profitable to examine "how the norms of art are tied to specific social classes and thus subject to uniform changes, or how aesthetic values change with social revolutions." But they held out more hope for what they called "the most central approach to a comparison of the arts," which would be based on "an analysis of the actual objects of art, and thus of their structural relationships." When Steiner analyzes an ekphrastic poem by Williams within the context of the Brueghel painting on which it is based, for example, she eschews history in order to study the "ever-changing set of correlations by painters and writers, who are free to stress different elements of the structures of their art in order to achieve" a correspondence. An interartistic parallel, she argues, "is not dictated by the pre-existent structures of the arts involved; instead, it is an exploration of how these two structures can be aligned." Semiotics, she claims, having defined the terms in which we examine different sign systems, has made the analogy between lit. and painting a more interesting area of investigation.

A more historical approach has been offered by Abel, who has deftly shown how the 18th-c. concern with thematic subject matter gave way to quite different values during the romantic period: "As the study of iconography is appropriate to the relationship between two arts in which content is primary, and the emphasis on different signs is appropriate to arts in which the nature of the sign is considered to be primary, the study of relationships among the signs themselves is appropriate to arts in which the power of imagination (q.v.) is held the primary feature." The similarity between Baudelaire's poems and Delacroix's paintings does not derive from their subjects or from the actual pattern of their signs but "from their com mon emphasis on establishing interrelationships achieved in the different ways dictated by their different signs." Two works may be alike as wholes but not in their individual features; despite their stylistic differences, "the functions of these different methods are the same in both the arts: to balance form and movement in an interrelated whole."

A final, more empirical approach—combining reader-response crit. and the theory of spatial form with a common-sense appraisal of what has worked best in earlier scholarship—has been proposed by Gilman, who isolates three areas of discourse that, when combined, he believes will form a workable comparative method. These are "the study of cultural concerns that influence expression in the various artistic media; the construction of likely analogies (with proper regard for the limitations of the analogical relationship); and the testing

of the fruitfulness of such analogies in the experience of the witness" (the reader and viewer). If the analogy is worth making, he argues, "the sister arts should strike sparks whose light is only visible from an interdisciplinary point of view."

J. Dryden, *A Parallel of P. and Painting* (1695); G. Lessing, *Laokoön* (1766); I. Babbitt, *The New Laokoön* (1910); K. Borinski, *Die Antike in Poetik und Kunsttheorie*, 2 v. (1914–24); C. B. Tinker, *Painter and Poet* (1938); T. M. Greene, *The Arts and the Art of Crit.* (1940); R. W. Lee, *"Ut Pictura Poesis"* (1940); J. Frank, "Spatial Form in Mod. Lit." (1945), rpt. in *The Widening Gyre* (1963); E. Souriau, *La Correspondance des arts* (1947); W. Stevens, *The Necessary Angel* (1951); H. A. Hatzfeld, *Lit. through Art* (1952); Curtius; S. Langer, *Feeling and Form* (1953), "Deceptive Analogies: Specious and Real Relationships Among the Arts," *Problems of Art* (1957); Wellek and Warren, ch. 11; Wimsatt and Brooks, ch. 13; J. H. Hagstrum, *The Sister Arts* (1958); G. Bebermeyer, "Literatur und bildende Kunst," *Reallexikon* 2.82–103; I. Jack, *Keats and the Mirror of Art* (1967); M. Krieger, "The Ekphrastic Principle," *The Play and Place of Crit.* (1967); T. Munro, *The Arts and their Interrelations* (1967); *A Bibl. on the Relations of Lit. and the Other Arts 1952–1967* (1968); L. Lipking, *The Ordering of the Arts in 18th-C. England* (1970), "Quick Poetic Eyes: Another Look at Literary Pictorialism," *Articulate Images*, ed. R. Wendorf (1983)—substantial bibl.; M. R. Pointon, *Milton and Eng. Art* (1970); M. Praz, *Mnemosyne* (1970); I. Tayler, *Blake's Illustration to the Poems of Gray* (1971); D. Rosand, "*Ut Pictor Poeta*: Meaning in Titian's *Poesie*," *NLH* 3 (1972); J. D. Hunt, *The Figure in the Landscape* (1976); M. R. Brownell, *Alexander Pope and the Arts of Georgian England* (1978); *The Relationship of Painting and Lit.: A Guide to Information Sources*, ed. E. L Huddleston and D. A. Noverr (1978); W. J. T. Mitchell, *Blake's Composite Art* (1978), *Iconology* (1986); E. B. Gilman, *The Curious Perspective* (1978); E. Abel, "Redefining the Sister Arts," and W. J. T. Mitchell, "Spatial Form in Lit.: Toward a General Theory," *The Lang. of Images* (1980); R. Halsband, *"The Rape of the Lock" and its Illustrations, 1714–1896* (1980); *Ezra Pound and the V. A.*, ed. H. Zinnes (1980); M. A. Caws, *The Eye in the Text* (1981); L. Gent, *Picture and Poetry 1560–1620* (1981); W. Steiner, *The Colors of Rhet.* (1982), *Pictures of Romance* (1988); J. M. Croisille, *Poésie et art figuré de Néron aux Flaviens* (1982); R. A. Goodrich, "Plato on P. and Painting," *BJA* 22 (1982); W. Marling, *W. C. Williams and the Painters 1909–1923* (1982); U. Weisstein, "Lit. and the V. A.," *Interrelations of Lit.*, ed. J.-P. Barricelli and J. Gibaldi (1982)—with bibl.; W. Trimpi, *Muses of One Mind* (1983); H. M. Sayre, *The Visual Text of W. C. Williams* (1983); W. S. Heckscher, *Art and Lit.: Studies in Relationship* (1985); E. B. Loizeaux, *Yeats and the V. A.* (1986); C. Pace, "'Delineated lives': Themes and Variations in 17th-c. Poems about Portraits," *Word & Image* 2 (1986); M. Roston, *Ren. Perspectives in Lit. and the V. A.* (1987), *Changing Perspectives in Lit. and the V. A.* (1990); D. Scott, *Picturalist Poetics* (1987); M. A. Cohen, *Poet and Painter* (1987); *Style* 22, 2 (1988); *Poets on Painters*, ed. J. D. McClatchy (1988); *Teaching Lit. and Other Arts*, ed. J.-P. Barricelli et al. (1990); C. Hulse, *The Rule of Art* (1990); T. J. Hines, *Collaborative Form* (1991).

R.W.

VISUAL POETRY.

I. FUNCTIONS
II. DEVELOPMENT
III. FREE VERSE
IV. VIDEO POETRY

V. p. is poetry composed for the eye as well as, or more than, for the ear. All printed poetry is v. p. in a broad sense, in that when we read the poem the v. form affects how we read it and so contributes to our experience of its sound, movement, and meaning. In traditional verse, however, the written text serves mainly a notational role, and its v. aspects are subordinate to the oral form they represent. In v. p. in the strict sense, on the other hand, the v. form of the text becomes an object for apprehension in its own terms. As Mooij points out, "written poetry allows for devices of foregrounding not available to oral poetry." Among the devices for creating v. form that written lang. furnishes the poet are lineation, line-length, line-grouping, indentation, intra- and interlinear white space, punctuation (q.v.), capitalization, and size and style of type.

In general, the v. form of a poem may be figurative or nonfigurative; if figurative, it may be mimetic or abstract. The v. form of most poems is nonfigurative: such poems are isometrical or heterometrical, hence consist of regular or irregular blocks of long or short lines. Open arrangements of lines on the page space are usually also nonfigurative. This is not to say that such poems are not v. p., however; their v. form may still be important. In the case of short poems, the shape of the whole poem is apprehended immediately as open or dense, balanced or imbalanced, even or uneven, simple or intricate. It is worth remembering that the overwhelming majority of lyric poems are meant to fit on a codex page, hence to meet the reader's eye as a simultaneously apprehensible whole. In stanzaic poems, the regular partitioning of the text may convey a sense of order and control and generate an expectation of regular closure. Further, the individual stanzas themselves are apprehensible v. units. Stanzas in symmetrical shapes may suggest stability or stillness, while asymmetrical shapes may suggest instability or movement in a direction. Stanzas of complex shape may convey a sense of elaborate artifice. For the reader steeped in poetry, the v. forms of stanzas also carry resonances and echoes from trad.: they recall antecedent poems written in stanzas of similar shape. The basic shape of the Sapphic (q.v.) stanza, for example, is recognizable even in extreme variations.

Printing stanzas with lines of different lengths or rhymes indented different amounts enhances the sense of their order and pattern, as in John Donne's *Songs and Sonets*.

In Cl. and Ren. pattern poetry (q.v.) we find examples of figurative v. form that is mimetic: the printed text takes the shape of objects, such as altars or wings; there are also 20th-c. examples of mimetic v. form, among them Apollinaire's calligrammes (q.v.) and some concrete poetry (q.v.). Poems in the shape of geometric figures such as circles and lozenges, another kind of pattern poetry, realize the possibility of figurative v. form that is abstract: in the Ren., 15 such forms are enumerated by Puttenham. Less rigidly geometric forms are not uncommon in conventional poetry (Ranta).

I. FUNCTIONS. The viability of v. p. as a literary mode depends directly on the functions that can be served by v. form. These fall into two classes: (a) those which reinforce the sense of the poem's unity and autonomy, and (b) those which tend to be disintegrative and intertextual. In group (a) we can enumerate six functions: (1) to lend prominence to phonological, syntactic, or rhetorical structures in the text (this would include scoring for performance and the use of white space to express emotion, invite contemplation, or signal closure); (2) to indicate juxtapositions of images and ideas; (3) to signal shifts in topic, tone, or perspective; (4) to render iconically the subject of the poem or an object referred to in it (incl. the use of white space as an icon of space, whiteness, distance, void, or duration); (5) to present the reader with an abstract shape of energy; and (6) to help foreground the text as an aesthetic object. In group (b) we can discriminate a further six functions: (1) to signal a general or particular relation to poetic trad.; (2) to allude to various other genres of printed texts; (3) to engage and sustain reader attention by creating interest and texture; (4) to cross-cut other textual structures, producing counterpoint between two or more structures occupying the same words; (5) to heighten the reader's awareness of the reading process; and (6) to draw attention to particular features of the text and, more generally, to defamiliarize aspects of lang., writing, and textuality. The v. form of a poem can of course realize several different functions, even inconsistent ones, at once.

II. DEVELOPMENT. Historically, "all poetry is originally oral, and the earliest inscriptions of it were clearly ways of preserving material after the trad. of recitation had changed or been lost" (Hollander). Subsequently, "the development has been from . . . v. organization of phonological data . . . to a v. organization that carries meaning without reference to the phonological" (Cummings and Simmons). Finally, "once the inscribed text was firmly established as a standard . . . end-product of literary art and typical object of literary appreciation, it was only natural that the literary artist would exploit the rich aesthetic possibilities offered by the inscribed medium" (Shusterman). V. effects have been exploited at least from ca. 300 B.C. in various modes of v. p. and in mixed-media works.

Perhaps the best known of the modes of v. p. is pattern poetry; *versus intexti*, first composed in the 4th c. and reaching their fullest devel. in the work of Hrabanus Maurus (9th c), constitute a lesser known subgenre of pattern poetry. Such poems were composed on a grid, 35 squares by 35, each square containing a letter, with type size and, later, color and outlining used to distinguish v. images from the background of the rest of the text. In the Ren. flourished the mixed-media genre of the emblem (q.v.), which typically comprised a short motto, a picture, and an explanatory, moralizing poem. An ancient v. genre, the acrostic, subverts the convention of reading from left to right and from top to bottom. Inscriptions, originally cut in stone with no regard for the appearance of the text, acquired beautiful lettering in Roman monumental art, which was reproduced and imitated in the Ren. In the 16th and 17th cs., esp. in northern Italy, they flourished briefly as a literary genre in printed books (Sparrow). The form, used mainly for religious and political eulogy, was really lineated prose—prose composed and printed in centered lines of uneven lengths, with the line-divisions supporting the sense.

The 20th c. has seen an abundance of highly v. works. These were heralded, just before the turn of the century, by Stéphane Mallarmé's late work "Un coup de dés jamais n'abolira le hasard," a v. composition in two-page spreads employing various type sizes and abundant white space. Early in the 20th c. appeared the calligrammes of Apollinaire, in which lettering (often handwriting) of different sizes typically sketches the shape of an object (e.g. a smoking cigar). The typographical experiments of futurism and dada, the typewriter compositions of the Am. poet e e cummings, and concrete poetry all use v. form to overcome the transparency of the medium, to make written lang. palpable and the poem thing-like.

III. FREE VERSE. V. form plays a more important role in the prosody of free verse (q.v.) than in that of metrical verse. One recognized function in free verse is scoring the text for performance. Charles Olson in his 1950 essay "Projective Verse" claimed that there should be a direct relationship between the amount of white space and the length of pause; different marks of punctuation could also be used to signal different lengths of pause. But regardless of whether it signals pause, intra- or interlinear white space can fulfill such mimetic, expressive, and rhetorical functions as (1) iconically rendering space, distance, length of time, void, silence, or whiteness; (2) signaling emotion too great for words; and (3) inviting the reader to take time for contemplation of the preceding text, signaling closure. Many free-verse poets exploit these possibilities through arrangement of text in the page space, as does Denise Levertov in this passage

from "The Five-Day Rain":

> Sequence broken, tension
> of sunlight broken.
> So light a rain
>
> fine shreds
> pending above the rigid leaves.

In other free-verse poems, esp. many of the experiments of e e cummings, white space and unconventional typography have the effect of defamiliarizing split or isolated textual elements. Lineation is often used to juxtapose images, as in Ezra Pound's famous imagist poem "In a Station of the Metro." In short-line free verse with normative strong enjambment (q.v.), lineation sets up a counterpoint to the syntactic structure of the text, i.e. to the phrasing with which it would normally be spoken. Much modernist free verse tends to eschew line-initial capitals, which otherwise give v. prominence to the line as a unit. In long-line free verse, when capitals are used only at the beginnings of sentences, lineation tends to be submerged in syntax.

Free-verse poets, notably W. C. Williams, sometimes arrange their lines in "sight-stanzas." Whereas traditional stanzas can be described in terms of meter and rhyme scheme, such couplets, tercets, and quatrains are perceptible as stanzas only by virtue of their having equal numbers of lines and creating iterated v. patterns. In verse arranged in sight-stanzas, tight syntactic and semantic connections typically extend across stanza boundaries, while major syntactic and semantic boundaries occur with stanzas. The separation of v. from other semantic aspects of the text's form liberates a previously subservient (pattern-marking) element from previously privileged (pattern-making) elements, while the v. order may compensate for considerable looseness of syntax or argument.

IV. VIDEO POETRY. The computer has introduced a new medium of text production and display, one which opens new possibilities for v. p. Besides facilitating the creation of spatial form, the integration of graphic elements with text, the automated generation of text, and the use of color, the computer, by putting the pace of appearance and disappearance of segments of text under control of the poet, allows temporal rhythms to be realized visually. In such electronic texts, temporality, hitherto a primary aspect of the prosody of oral poetry, becomes central to v. p.

G. Puttenham, *The Arte of Eng. Poesie* (1589), rpt. in Smith; J. Sparrow, *Visible Words* (1969); R. Massin, *La Lettre et l'image*, 2d ed. (1973); *Speaking Pictures*, ed. M. Klonsky (1975)—anthol.; J. J. A. Mooij, "On the 'Foregrounding' of Graphic Elements in Poetry," *Comparative Poetics*, ed. D. W. Fokkema et al. (1976); J. Ranta, "Geometry, Vision, and Poetic Form," *CE* 39 (1978); *V. Lit. Crit.*, ed. R. Kostelanetz (1979); R. Kostelanetz, *The Old Poetries and the New* (1981); Morier, s.v. "Blanchissement," "Vide"; R. Shusterman, "Aesthetic Blindness to Textual Visuality," *JAAC* 41 (1982); H. M. Sayre, *The V. Text of W. C. Williams* (1983); M. Cummings and R. Simmons, "Graphology," *The Lang. of Lit* (1983); C. Taylor, *A Poetics of Seeing* (1985); S. Cushman, *W. C. Williams and the Meanings of Measure* (1985), ch. 2; Hollander; W. Bohn, *The Aesthetics of V. P., 1914–1928* (1986); R. Cureton, "V. Form in e e cummings' *No Thanks*," *Word & Image* 2 (1986); *The Line in Postmodern Poetry*, ed. R. Frank and H. Sayre (1988). E.B.

VOICE. To stress v. in discussions of poetry may be simply a reminder of the large extent to which poetry depends on sound. The qualities of vocal sounds enter directly into the aesthetic experience of performance, of poetry readings, but no less do those sounds resonate in the "inner ear" of a fully attentive silent reading. T. S. Eliot felt that one may hear at least three voices of poetry: that of the poet in silent meditation, that of the poet addressing an audience, and that of a dramatic character or persona created by the poet. Implicit in Eliot's division is the notion that behind these various vs. lies one original v.—or what Aristotle called *ethos*—that expresses the poet's intentions and organizes the various personae.

Within romantic theories of poetry, v. plays a significant role as the embodiment of the author's expression. Whether the poet speaks in Wordsworth's "lang. of men" or through Blake's prophecies, v. is the vehicle through which private vision is translated to the world. This romantic spirit has been revived by more recent poets for whom the oral trad. represents a positive model of poetry's unmediated access to an audience. At the same time, contemp. theories of "projective verse" have stressed the role of v. and breath in the construction of the poetic line.

The expressive view of v. has been qualified by much modern theory, beginning with New Criticism and its prohibitions against biographical crit. and the "intentional fallacy." More recent caveats have been offered by reader-response and reception theories, which view v. less as the product of a speaking subject than as the site of multiple narrative positions. In his influential *Rhetoric of Fiction* (1961), Wayne C. Booth uses the phrase "implied author" to describe the constructed or fictive nature of intentional acts. Authorial v. in this view is an idealized projection of the historical author, a figure to whom the reader must give tacit approval if verisimilitude is to be maintained. In an "Afterword" to the 2d ed. (1983), Booth distinguishes five senses of "author," and therefore five senses of authorial "v."

A more complex discussion of literary v. has been offered by structuralist analyses of narrative. For Tzvetan Todorov and Gérard Genette, most treatments of v. have been conducted under the rather vague concept of "point of view," which tends to equate narrator with author, i.e. the per-

spective through which the story is being told with the one who tells it. Genette uses the term v. to describe relations established between narrator and audience, relations that define "the way in which the narrating itself is implicated in the narrative." By using the term v., Genette foregrounds the role that grammar plays in creating person. When a character uses the passive rather than the active v. or when the verb tense changes from present to past, a new level of narration is introduced for which traditional notions of point of view are inadequate.

Structuralist narratology deals with v. as a construction within a specifically aesthetic frame. For Marxist philologists and linguists such as V. N. Volosinov and Mikhail Bakhtin, v. and speech are social in origin and exist as exchanges between specific historical individuals. Whereas structuralist thinkers such as Ferdinand de Saussure had mapped the terrain of lang. as a neutral system of phonemic differences (*la langue*), Volosinov (*Marxism and the Philosophy of Lang.* [1973]) argued that utterances are choices made in response to specific social and ideological conditions. Speech does not occur within a closed system of signifiers, as Saussure maintained, but within a constantly shifting ideological landscape. Bakhtin extended (some feel invented) Volosinov's historical definition of lang. and applied it to the novel (*The Dialogic Imagination* [tr. 1981]).

Implicit in both Marxist and structuralist analyses of literary texts is the idea that lang. constitutes subjects rather than serving as the instrument through which an original subject speaks. The philosophical ramifications of this fact are worked out within the poststructural thought of Jacques Derrida. Like Volosinov and Bakhtin, Derrida builds his critique on Saussure's linguistics, but instead of faulting an ahistorical definition of lang., he argues against Saussure's metaphysical privileging of speech over writing. In *Of Grammatology* (1973) and other essays, Derrida studies the way that Western thought since Plato has taken for granted the ontological priority of speech and presence, even (as in Heidegger) when that very presence is being called into question. Derrida attacks the assumption that v. or speech is directly linked to some metaphysical essence or quality that precedes its textual inscription. Any system of thought based on such a presupposition recapitulates the entire history of metaphysical thought and thus fails to recognize the radical difference that structures all discourse. Since difference is visible in the printed word but inaudible in the spoken, writing (*écriture*) becomes Derrida's preferred model for lang. as a system of signs in which presence is bracketed.

Although neither Marxist, structuralist, nor poststructuralist views relate directly to the specific use of v. in poetry, these forms of crit. theorize the intentional nature of literary lang. insofar as it expresses what a specific author "wants" to say. It remains a paradoxical fact of postwar literary life that the revival of romantic theories of poetry with their strong emphasis on orality and presence has coincided with a theoretical critique of those very assumptions.

T. S. Eliot, "The Three Vs. of Poetry," *On Poetry and Poets* (1957); C. Olson, "Projective Verse," *Selected Writings* (1960); W. J. Ong, *The Barbarian Within* (1962); F. Berry, *Poetry and the Physical V.* (1962); A. D. Ferry, *Milton's Epic V.* (1963); R. Delasanta, *The Epic V.* (1967); J. Derrida, *Of Grammatology* (tr. 1976); T. Todorov, *The Poetics of Prose* (1977); G. Genette, *Narrative Discourse* (1980); M. M. Bakhtin, *The Dialogic Imagination* (tr. 1981); *Lyric Poetry: Beyond New Crit.*, ed. C. Hošek and P. Parker (1985); P. Zumthor, *La Poésie et la voix dans la civilisation médiévale* (1987); R. O. A. M. Lyne, *Further Voices in Virgil's* Aeneid (1987); E. Griffiths, *The Printed V. of Victorian Poetry* (1989); D. Appelbaum, *V.* (1990).

F.GU.; M.D.

WIT. In Plato's *Republic* and in Aristotle's *Poetics* the word *euphuia* occurs in senses ranging from "shapeliness" to "cleverness." In Cicero and Quintilian the Lat. equivalent is *ingenium* in senses that would seem to generate the whole historical range of the meanings of w. in Eng. Its equivalents in other langs. also undergo historical semantic change: *esprit* (or *bel esprit*) in Fr., *agudeza* and *gracia* in Sp., *ingegno, acutezza*, and *argutezza* in It., *Witz* and *Schärfe* in Ger., and *um* and *ostroumie* in Rus. All these terms have different histories, yet in the 17th and 18th cs. a fairly close correspondence of meaning develops among them. By this time, perhaps under the influence of Quintilian (10.1.130), w. (*ingenium*) is contrasted with *judgment* (*iudicium*): w. (or fantasy and its congeners) must be controlled by the discipline of judgment to produce proper works of art. The later, specialized senses of w. as risible sparkle or mere levity have parallels in other langs., and have no great artistic interest other than our own currently felt meaning.

In its heyday as a critical term, w. referred to the inventive or imaginative faculty and, in particular, to the ability to see similarity in disparates (cf. Aristotle, *Poetics* 1459a). Indeed, w. could be prized for its ability to discover brilliant, paradoxi-

cal, or far-fetched images and figures, esp. metaphor, metonymy, irony, paradox, pun, or antithesis, notably as propounded by Baltasar Gracián in *Agudeza y arte de ingenio* (1642; expanded 1648) and Emmanuele Tesauro in *Il cannocchiale aristotelico* (1654). Their contemporary Thomas Hobbes in his *Leviathan* (1651) asserted in a general psychological vein that "*Naturall W.* consisteth in two things: *Celerity of Imagining* (that is, swift succession of one thought to another) and *steddy direction* to some approved end." This use of w. shows historical continuity from early times when both ingenuity and judgment could be encompassed in the same term, as also in the use of *esprit* by Descartes and Pascal. For imaginative lit., however, Eng. *w.* and its parallel terms in other tongues retained into the late 18th c. the specialized meaning of ingeniousness: it is this meaning which is the most fruitful to trace.

A prime text in Eng. is Pope's "Essay on Crit." (1711): "*True W.* is *Nature* to Advantage drest, / What oft was *Thought*, but ne'er so well *Exprest.*" It sums up in its context the central sense of w. to be found in poet-critics from Dryden to Johnson and indicates a rejection of the "false w." or mere cleverness of the previous as well as the current age. Yet Pope, as others elsewhere, granted license to "Great Wits" who could "*snatch a grace* beyond the Reach of Art." Empson counts 46 occurrences of the word "w." in the "Essay on Crit." and sorts out at least six different meanings. Such polysemy is not unusual, nor is it distracting so long as we recognize the use of *w.* in a technical aesthetic sense to mean the imaginative or striking figure, the flash of verbal intuition, the marmoreal phrase, the pointed dictum. While Pope constitutes a norm or middle point, however, the history of the literary sense of w. in Britain begins with Shakespeare and the metaphysical poets (cf. esp. T. Carew's "Elegie upon the Death of . . . Dr. John Donne" [1640] and A. Cowley's "Of W." [1656]), continues through Dryden (e.g. "Preface to *Annus mirabilis*, 1666), Addison's *Spectator* papers on w. (e.g. 58–61), and Dr. Johnson to the romantics, who transformed its meaning.

In Fr. the word *esprit* is polyvalent in many of the same ways; indeed Boileau's *Art poétique* (1674) was clearly a prime model for Pope. *Esprit* was a more unstable or modish word than *w.* (a symptomatic text is the anonymous *Entretiens galants* [1681]), yet it survived in its focused meaning at least through Voltaire. In Italy *acutezza* or *argutezza* and *arguzia* and in Spain *agudeza* were generally treated as the rhetorical ornament enhancing the thought (*concetto, concepto*). In belated Rus. lit. the most famous use of the parallel term *um* occurs in Griboedov's comedy *Gore ot uma* (Woe from W., 1824). The key words in It., Sp., and Rus. have survived in contemp. speech, though bereft of their literary specificity, while Eng. "w." entered quite a new realm of meaning parallel to that other historically complex word "humor" (see Lewis).

The early Ger. romantics, in their profound originality, gave new life to traditional terms such as *Witz*. Jean Paul Richter (*Vorschule der Ästhetik*, 1804) and Friedrich Schlegel (in his three sets of *Fragmente* or *Aphorismen*, 1789–1802) used *Witz* not merely in its rhetorical sense but, more significantly, to denote a whole world view of creativity in such a way as to vie even with *Phantasie* and *Ironie*. Schiller's *Spieltrieb* ("play drive") and Solgar's *Ironie* are also important in the romantic overthrow of neoclassical orthodoxy (Wellek, v. 2).

A whole new constellation of literary terminology was forming in the earlier 19th c. of which we are the heirs. *Imagination* (q.v.) came to take on the sense of discovery and invention formerly embraced by *w.*, leading to a reduction in the sense of the latter (and its cognates in other langs.) to mere humor, a tendency which became widespread. In the course of the 20th c., notions related to the older meaning of w. have come to the fore, e.g. Freud's psychoanalytic concept of *Witz*, T. S. Eliot's revaluation of "metaphysical" w., C. Brooks' emphasis on irony and paradox as the principal devices of literary complexity and structure, and a persistent strain of parody (q.v.) as a means to what might be called intertextual w., as in Joyce's *Ulysses* and Mann's *Doktor Faustus*. Thus the meaning of w., though it may not have come full circle, has in the 20th c. regained some critical force and, through its literarily serious connection with irony and parody, begun to approach again its old kinship with imagination.

S. Freud, *Jokes and their Relation to the Unconscious* (1905; tr. 1960); *Critical Essays of the 17th C.*, ed. J. E. Spingarn, v. 1 (1908)—see Intro.; M. A. Grant, *The Ancient Rhetorical Theories of the Laughable* (1924); W. G. Crane, *W. and Rhet. in the Ren.* (1937); C. Brooks, *Mod. Poetry and the Trad.* (1939): S. H. Monk, "A Grace Beyond the Reach of Art," *JHI* 5 (1944); E. B. O. Borgerhoff, *The Freedom of Fr. Classicism* (1950); T. S. Eliot, "The Metaphysical Poets" and "Andrew Marvell" in *Essays*; W. Empson, *The Structure of Complex Words* (1951); J. E. Spingarn, *A Hist. of Lit. Crit. in the Ren.*, 2d ed. (1954); Wellek, v. 1–2; Wimsatt and Brooks, ch. 12; C. S. Lewis, *Studies in Words* (1960)—comprehensive sketch of the semantic hist.; A. Stein, "On Elizabethan W.," *SEL* 1 (1961); G. Williamson, *The Proper W. of Poetry* (1961); S. L. Bethell, "The Nature of Metaphysical W.," *Discussions of John Donne*, ed. F. Kermode (1962); T. N. Marsh, "Elizabethan W. in Metaphor and Conceit," *EM* 13 (1962); J. A. Mazzeo, *Ren. and 17th-C. Studies* (1964); *The Idea of Comedy*, ed. W. K. Wimsatt, Jr. (1969); *Historisches Wörterbuch der Philosophie*, ed. J. Ritter and K. Gründer (1971), s.v. "Ingenium"; M. C. Wanamaker, *Discordia Concors* (1975); E. B. Gilman, *The Curious Perspective* (1978); A. J. Smith, *Metaphysical W.* (1992); J. Sitter, *Arguments of Augustan W.* (1992).

W.V.O'C.; L.NE.

DEER PARK PUBLIC LIBRARY
3 2244 11122 4255

808.103 The new Princeton
NEW handbook of poetic
terms.

$18.95 PB

DATE			

DEER PARK PUBLIC LIBRARY
44 LAKE AVENUE
DEER PARK, NY 11729

BAKER & TAYLOR

RECEIVED FEB 27 1998